Lecture Notes in Computer Science 5706

Commenced Publication in 1973
Founding and Former Series Editors:
Gerhard Goos, Juris Hartmanis, and Jan van Leeuwen

Carol Peters Thomas Deselaers
Nicola Ferro Julio Gonzalo
Gareth J.F. Jones Mikko Kurimo
Thomas Mandl Anselmo Peñas
Vivien Petras (Eds.)

Evaluating Systems for Multilingual and Multimodal Information Access

9th Workshop of the Cross-Language Evaluation Forum, CLEF 2008
Aarhus, Denmark, September 17-19, 2008
Revised Selected Papers

Springer

Volume Editors

Carol Peters
ISTI, CNR, Pisa, Italy; carol.peters@isti.cnr.it

Thomas Deselaers
RWTH Aachen University, Germany; deselaers@informatik.rwth-aachen.de

Nicola Ferro
University of Padua, Italy; nicola.ferro@unipd.it

Julio Gonzalo
Anselmo Peñas
LSI-UNED, Madrid, Spain; {julio,anselmo}@lsi.uned.es

Gareth J.F. Jones
Dublin City University, Ireland; gareth.jones@computing.dcu.ie

Mikko Kurimo
Helsinki University of Technology, Finland; mikko.kurimo@tkk.fi

Thomas Mandl
University of Hildesheim, Germany; mandl@uni-hildesheim.de

Vivien Petras
Humboldt University Berlin, Germany; vivien.petras@ibi.hu-berlin.de

Managing Editor
Danilo Giampiccolo, CELCT, Trento, Italy; giampiccolo@celct.it

Library of Congress Control Number: 2009934437

CR Subject Classification (1998): I.2.7, H.2.8, I.7, H.4, H.5, H.5.2, I.1.3

LNCS Sublibrary: SL 3 – Information Systems and Application, incl. Internet/Web and HCI

ISSN 0302-9743
ISBN-10 3-642-04446-8 Springer Berlin Heidelberg New York
ISBN-13 978-3-642-04446-5 Springer Berlin Heidelberg New York

springer.com

Printed in Germany

Typesetting: Camera-ready by author, data conversion by Scientific Publishing Services, Chennai, India
Printed on acid-free paper SPIN: 12753536 06/3180 5 4 3 2 1 0

Preface

The ninth campaign of the Cross-Language Evaluation Forum (CLEF) for European languages was held from January to September 2008. There were seven main evaluation tracks in CLEF 2008 plus two pilot tasks. The aim, as usual, was to test the performance of a wide range of multilingual information access (MLIA) systems or system components. This year, 100 groups, mainly but not only from academia, participated in the campaign. Most of the groups were from Europe but there was also a good contingent from North America and Asia plus a few participants from South America and Africa. Full details regarding the design of the tracks, the methodologies used for evaluation, and the results obtained by the participants can be found in the different sections of these proceedings.

The results of the CLEF 2008 campaign were presented at a two-and-a-half day workshop held in Aarhus, Denmark, September 17–19, and attended by 150 researchers and system developers. The annual workshop, held in conjunction with the European Conference on Digital Libraries, plays an important role by providing the opportunity for all the groups that have participated in the evaluation campaign to get together comparing approaches and exchanging ideas.

The schedule of the workshop was divided between plenary track overviews, and parallel, poster and breakout sessions presenting this year's experiments and discussing ideas for the future. There were several invited talks. Noriko Kando, National Institute of Informatics Tokyo, reported on the activities of NTCIR-7 (NTCIR is an evaluation initiative focussed on testing IR systems for Asian languages), while John Tait of the Information Retrieval Facility, Vienna, presented a proposal for an Intellectual Property track which would focus on cross-language retrieval of legal patents in CLEF 2009. In the final session, Donna Harman, US National Institute of Standards and Technology, presented her impressions of the main trends emerging from the 2008 workshop and campaign, and Martin Braschler of Zurich University of Applied Sciences gave a talk describing a survey he had made on the search functionality of enterprise websites. The presentations given at the CLEF workshop can be found on the CLEF website at www.clef-campaign.org.

The workshop was preceded by two related events. On September 16, the ImageCLEF group, with the sponsorship of the Quaero program (www.quaero.org), organized a one-day workshop on Multimedia Information Retrieval Evaluation. The workshop included presentations of the activities of both Quaero and Theseus, two international projects working on the development of next-generation Internet search engines. The Morpho Challenge 2008 meeting on "Unsupervised Morpheme Analysis" was held on the morning of September 17. Morpho Challenge 2008 was part of the EU Network of Excellence PASCAL Programme and was run in collaboration with CLEF.

The CLEF 2008 and 2009 campaigns were organized as activities of TrebleCLEF, a Coordination Action of the Seventh Framework Programme. TrebleCLEF is building on and extending the results achieved by CLEF. The objective is to support the development and consolidation of expertise in the multidisciplinary research area of multilingual information access and to promote a dissemination action in the relevant application communities. TrebleCLEF is also attempting to promote more user-and usage-focused investigations within CLEF.

At the time of writing the organization of CLEF 2009 is well underway. In line with the TrebleCLEF philosophy, the campaign this year includes three new tracks focused on analyzing user behavior in a multilingual context (LogCLEF), on studying the requireements of multilingual patent search (CLEF-IP), and on improving our understanding of MLIA systems and their behavior with respect to languages (GridCLEF).

These post-campaign proceedings represent extended and revised versions of the initial working notes distributed at the workshop. All papers were subjected to a reviewing procedure. The final volume was prepared with the assistance of the Center for the Evaluation of Language and Communication Technologies (CELCT), Trento, Italy, under the coordination of Danilo Giampiccolo. The support of CELCT is gratefully acknowledged. We should also like to thank all our reviewers for their careful refereeing.

May 2009

Carol Peters
Thomas Deselaers
Nicola Ferro
Julio Gonzalo
Gareth J. F. Jones
Mikko Kurimo
Thomas Mandl
Anselmo Peñas
Vivien Petras

Reviewers

The Editors express their gratitude to the colleagues listed below for their assistance in reviewing the papers in this volume:

- Eneko Agirre, University of the Basque Country, Spain
- Abolfazl AleAhmad, University of Tehran, Iran
- Hadi Amiri, University of Tehran, Iran
- Ebru Arisoy, Bogazici University, Turkey
- Stefan Baerisch, GESIS Leibniz-Institut for Social Sciences, Bonn, Germany
- Delphine Bernhard, Darmstadt University of Technology, Germany
- Johan Bos, University of Rome "La Sapienza", Italy
- Burcu Can, University of York, UK
- Nuno Cardoso, University of Lisbon, Portugal
- Paula Carvalho, Linguateca and University of Lisbon, Portugal
- Leda Casanova, CELCT, Italy
- Tolga Ciloglu, Middle East Technical University, Turkey
- Paul D. Clough, University of Sheffield, UK
- Luis F. Costa, SINTEF ICT, Portugal
- Thomas M. Deserno, RWTH Aachen University, Germany
- Giorgio Di Nunzio, University of Padua, Italy
- Corina Forascu, Institute for Research in Artificial Intelligence, Romania
- Miguel Garcia-Cumbreras, University of Jaen, Spain
- Fredric C. Gey, University of California at Berkeley, USA
- Ingo Glöckner, FernUniversität in Hagen, Germany
- Harald Hammarström, Chalmers University, Sweden
- Allan Hanbury, Technical University of Vienna, Austria
- Donna Harman, National Institute of Standards and Technology, USA
- Sven Hartrumpf, FernUniversität in Hagen, Germany
- Jesús Herrera, Universidad Complutense de Madrid, Spain
- William Hersh, Oregon Health and Science University, Portland, USA
- Jayashree Kalpathy-Cramer, Oregon Health and Science University, USA
- Chunyu Kit, Hong Kong City University, China
- Dietrich Klakow, University of Saarland, Germany
- Jana Kludas, University of Geneva, Switzerland
- Zornitsa Kozareva, USC Information Sciences Institute, USA
- Martha Larson, Delft University of Technology, The Netherlands
- Ray Larson, University of California at Berkeley, USA
- Johannes Leveling, FernUniversität in Hagen, Germany
- Patricio Martínez, University of Alicante, Spain
- Paul McNamee, Johns Hopkins University, USA

- Henning Müller, University of Applied Sciences Western Switzerland, Sierre and University of Geneva, Switzerland
- Diego Molla, Macquarie University, Australia
- Manuel Montes, INAOE, Mexico
- Günter Neumann, German Research Centre for Artificial Intelligence, Germany
- Eamonn Newman, Dublin City University, Ireland
- Petya Osenova, Bulgarian Academy of Sciences, Bulgaria
- Simon Overell, Imperial College London, UK
- Alvaro Rodrigo, UNED, Madrid, Spain
- Paolo Rosso, Polytechnic University of Valencia, Spain
- Andrew Salway, Dublin City University, Ireland
- Mark Sanderson, University of Sheffield, UK
- Diana Santos, Linguateca and SINTEF ICT, Norway
- Murat Saraclar, Bogazici University, Turkey
- Jacques Savoy, University of Neuchâtel, Switzerland
- Gianmaria Silvello, University of Padua, Italy
- Theodora Tsikrika, CWI, Amsterdam, The Netherlands
- Jordi Turmo, Polytechnic of Catalonia, Spain
- Christa Womser-Hacker, University of Hildesheim, Germany
- Fabio Massimo Zanzotto, Unversity of Rome "Tor Vergata", Italy

CLEF 2008 Coordination

CLEF is coordinated by the Istituto di Scienza e Tecnologie dell'Informazione, Consiglio Nazionale delle Ricerche, Pisa. The following institutions contributed to the organization of the different tracks of the CLEF 2008 campaign:

- Adaptive Informatics Research Centre, Helsinki University of Technology, Finland
- Athena Research Center, Athens, Greece
- Business Information Systems, Univ. of Applied Sciences Western Switzerland, Sierre, Switzerland
- Centre for the Evaluation of Human Language and Multimodal Communication Technologies (CELCT), Trento, Italy
- Centruum vor Wiskunde en Informatica, Amsterdam, The Netherlands
- Computer Science Department, University of the Basque Country, Spain
- Computer Vision and Multimedia Lab, University of Geneva, Switzerland
- Database Research Group, University of Tehran, Iran
- Department of Computer Science, Aachen University of Technology, Germany
- Department of Computer Science and Information Systems, University of Limerick, Ireland
- Department of Information Engineering, University of Padua, Italy
- Department of Information Science, University of Hildesheim, Germany
- Department of Information Studies, University of Sheffield, UK
- Department of Medical Informatics and Clinical Epidemiology, Oregon Health and Science University, USA
- Department of Medical Informatics, Aachen University of Technology, Germany
- Department of Medical Informatics, University Hospitals and University of Geneva, Switzerland
- Evaluations and Language Resources Distribution Agency Sarl, Paris, France
- German Research Centre for Artificial Intelligence, Saarbrücken, Germany
- GESIS Leibniz-Institut for the Social Sciences, Bonn, Germany
- Information Science, University of Groningen, The Netherlands
- Institute of Computer Aided Automation, Vienna University of Technology, Austria
- Intelligent Systems Lab Amsterdam, University of Amsterdam, The Netherlands
- Laboratoire d'Informatique pour la Mécanique et les Sciences de l'Ingénieur (LIMSI), Orsay, France

- Lenguajes y Sistemas Informáticos, Universidad Nacional de Educación a Distancia, Madrid, Spain
- Linguateca, Sintef, Oslo, Norway
- Linguateca, CISUC, Department of Information Engineering, University of Coimbra, Portugal
- Linguateca, XLDB, Department of Information Engineering, University of Lisbon, Portugal
- Linguistic Modelling Laboratory, Bulgarian Academy of Sciences, Bulgaria
- Microsoft Research Asia
- National Institute of Standards and Technology, Gaithersburg, USA
- Research Computing Center of Moscow State University, Russia
- Romanian Institute for Computer Science, Romania
- School of Computing, Dublin City University, Ireland
- School of Computer Science and Mathematics, Victoria University, Australia
- TALP Research Center, Universitat Politécnica de Catalunya, Barcelona, Spain
- UC Data Archive and School of Information Management and Systems, UC Berkeley, USA

CLEF 2008 Steering Committee

- Hans Uszkoreit, German Research Center for Artificial Intelligence, Germany
- Felisa Verdejo, LSI-UNED, Madrid, Spain
- José Luis Vicedo, University of Alicante, Spain
- Ellen Voorhees, National Institute of Standards and Technology, USA
- Christa Womser-Hacker, University of Hildesheim, Germany

Table of Contents

Part I: Multilingual Textual Document Retrieval (Ad Hoc)

TEL@CLEF

Persian@CLEF

Robust-WSD

Ad Hoc Mixed: TEL and Persian

Part II: Mono- and Cross-Language Scientific Data Retrieval (Domain-Specific)

Part III: Interactive Cross-Language Retrieval (iCLEF)

Part IV: Multiple Language Question Answering (QA@CLEF)

Mono and Bilingual QA

Answer Validation Exercise (AVE)

Question Answering on Script Transcription (QAST)

Part V: Cross-Language Retrieval in Image Collections (ImageCLEF)

ImageCLEFphoto

ImageCLEFmed

ImageCLEFWiki

Part VI: Multilingual Web Track (WebCLEF)

Part VII: Cross-Language Geographical Retrieval (GeoCLEF)

Part VIII: Cross-Language Video Retrieval (VideoCLEF)

Part IX: Multilingual Information Filtering (INFILE@CLEF)

Part X: Morpho Challenge at CLEF 2008

What Happened in CLEF 2008

Carol Peters

Istituto di Scienza e Tecnologie dell'Informazione (ISTI-CNR), Pisa, Italy
carol.peters@isti.cnr.it

Abstract. The organization of the CLEF 2008 evaluation campaign is described and details are provided concerning the tracks, test collections, evaluation infrastructure, and participation. The main results are commented and future evolutions in the organization of CLEF are discussed.

1 Introduction

The objective of the Cross Language Evaluation Forum is to promote research in the field of multilingual system development. This is done through the organisation of annual evaluation campaigns in which a series of tracks designed to test different aspects of mono- and cross-language information retrieval (IR) are offered. The intention is to encourage experimentation with all kinds of multilingual information access – from the development of systems for monolingual retrieval operating on many languages to the implementation of complete multilingual multimedia search services. This has been achieved by offering an increasingly complex and varied set of evaluation tasks over the years. The aim is not only to meet but also to anticipate the emerging needs of the R&D community and to encourage the development of next generation multilingual IR systems.

This volume contains a series of papers describing the research activities and experiments that were conducted in the CLEF 2008 campaign. The main features of this campaign are briefly outlined below in order to provide the necessary background to these papers. In the final sections, we comment on the main results obtained and discuss our ideas for the future of CLEF.

2 Tracks and Tasks in CLEF 2008

CLEF 2008 offered seven tracks designed to evaluate the performance of systems for:

- multilingual textual document retrieval (Ad Hoc)
- mono- and cross-language information retrieval on structured scientific data (Domain-Specific)
- interactive cross-language retrieval (iCLEF)
- multiple language question answering (QA@CLEF)
- cross-language retrieval in image collections (ImageCLEF)
- multilingual retrieval of web documents (WebCLEF)
- cross-language geographical information retrieval (GeoCLEF)

C. Peters et al. (Eds.): CLEF 2008, LNCS 5706, pp. 1–14, 2009.

Two new tracks were offered as pilot tasks:

- cross-language video retrieval (VideoCLEF)
- multilingual information filtering (INFILE@CLEF)

In addition, Morpho Challenge 2008 was organized in collaboration with CLEF as part of the EU Network of Excellence Pascal Challenge Program[1].

Here below we give a brief overview of the various activities.

Multilingual Textual Document Retrieval (Ad Hoc). The aim of this track is to promote the development of monolingual and cross-language textual document retrieval systems. From 2000 - 2007 the track used exclusively collections of European newspaper and news agency documents. This year the focus of the track was considerably widened: we introduced very different document collections, a non-European target language, and an information retrieval (IR) task designed to attract participation from groups interested in natural language processing (NLP). The track was thus structured in three distinct streams. The first task offered monolingual and cross-language search on library catalog records and was organized in collaboration with The European Library (TEL)[2]. The second task resembled the ad hoc retrieval tasks of previous years but this time the target collection was a Persian newspaper corpora. The third task was the robust activity which this year used word sense disambiguated (WSD) data. The track was coordinated jointly by ISTI-CNR and Padua University, Italy; Hildesheim University, Germany; and the University of the Basque Country, Spain, with the collaboration of the Database Research Group, University of Tehran, Iran.

Cross-Language Scientific Data Retrieval (Domain-Specific). This track studies how the structure of data (i.e. metadata, controlled vocabularies) can be exploited to improve search in a collection. In 2008, mono- and cross-language domain-specific retrieval was studied in the domain of social sciences using structured data (e.g. bibliographic data, keywords, and abstracts) from scientific reference databases. The target collections provided were: GIRT-4 for German/English, Cambridge Sociological Abstracts for English, and the ISISS corpus provided by the Institute of Scientific Information for Social Sciences of the Russian Academy of Science. A multilingual controlled vocabulary (German, English, Russian) suitable for use with GIRT-4 and ISISS together with a bi-directional mapping between this vocabulary and that used for indexing the Sociological Abstracts was provided. It was decided to terminate this task in 2008 as we felt that it had fulfilled its purpose in providing us with the opportunity to compare differences between free-text search over languages with structured document retrieval. In fact, a main finding has been that search on metadata-based documents (just title, abstracts, thesaurus descriptors) can achieve similar results as for full-text archives (ca. 50% in precision as highest result). The track was coordinated by GESIS-IZ Social Science Information Centre, Bonn, Germany.

---

[1] See http://www.cis.hut.fi/morphochallenge2008/

[2] See http://www.theeuropeanlibrary.org/

CLEF 2000	▪ **mono-, bi- & multilingual text doc retrieval (Ad Hoc)** ▪ **mono- and cross-language information on structured scientific data (Domain-Specific)**
CLEF 2001 New	▪ **interactive cross-language retrieval (iCLEF)**
CLEF 2002 New	▪ **cross-language spoken document retrieval (CL-SR)**
CLEF 2003 New	▪ **multiple language question answering (QA@CLEF)** ▪ **cross-language retrieval in image collections (ImageCLEF)**
CLEF 2005 New	▪ **multilingual retrieval of Web documents (WebCLEF)** ▪ **cross-language geographical retrieval (GeoCLEF)**
CLEF 2008 New	▪ **cross-language video retrieval (VideoCLEF)** ▪ **multilingual information filtering (INFILE@CLEF)**

Fig. 1. Evolution of CLEF Tracks

Interactive Cross-Language Retrieval (iCLEF). In iCLEF, cross-language search capabilities are studied from a user-inclusive perspective. A central research question is how best to assist users when searching information written in unknown languages, rather than how best an algorithm can find information written in languages different from the query language. Since 2006, iCLEF has moved from news collections (a standard for text retrieval experiments) in order to explore user behaviour in a collection where the cross-language search necessity arises more naturally for average users. The choice fell on Flickr[3], a large-scale, online image database based on a large social network of WWW users, with the potential for offering both challenging and realistic multilingual search tasks for interactive experiments. The search interface provided by the iCLEF organizers was a basic cross-language retrieval system for the Flickr image database presented as an online game: the user is given an image, and must find it again without any a priori knowledge of the language(s) in which the image is annotated. The game was publicized on the CLEF mailing list and prizes were offered for the best results in order to encourage participation.

The main novelty of the iCLEF 2008 experiments was the shared analysis of a search log from a single search interface provided by the organizers (i.e. the focus was on log analysis, rather than on system design). The 2008 experiments resulted in a truly reusable data set (the first time in iCLEF!), with 5,000 complete search sessions recorded and 5,000 post-search and post-experience questionnaires. The track was coordinated by UNED, Madrid, Spain; Sheffield University, UK; Swedish Institute of Computer Science, Sweden.

[3] See http://www.flickr.com/

Multilingual Question Answering (QA@CLEF). This track has been offering monolingual and cross-language question answering tasks since 2003. QA@CLEF 2008 proposed both main and pilot tasks. The main scenario was event-targeted QA on a heterogeneous document collection (news articles and Wikipedia). A large number of questions were topic-related, i.e. clusters of related questions possibly containing anaphoric references. Besides the usual news collections, articles from Wikipedia were also considered as sources of answers. Many monolingual and cross-language sub-tasks were offered: Basque, Bulgarian, Dutch, English, French, German, Italian, Portuguese, Romanian and Spanish were proposed as both query and target languages; not all were used in the end.

After 6 years, a lot of resources and know-how have been accumulated. However, the tasks offered have proved to be difficult for the systems which have not shown a very good overall performance, even those that have participated year by year. In addition, a result of offering so many language possibilities has been that there have always been very few systems participating in the same task, with the same languages. This has meant that comparative analysis is extremely problematic. Consequently, the QA organisers have decided to redefine the task for CLEF 2009 to permit the evaluation and comparison of systems even when they are working in different languages. The new setting will also take as reference a real user scenario, in a new document collection in which multilinguality is more natural.

The additional exercises in 2008 were the following:

- The Answer Validation Exercise (AVE) in its third edition was aimed at evaluating answer validation systems based on recognizing textual entailment.
- QAST was focused on Question Answering over Speech Transcriptions of seminars. In this 2nd year pilot task, answers to factual and definitional questions in English were extracted from spontaneous speech transcriptions related to separate scenarios in English, French and Spanish.
- QA-WSD provided questions and collections with already disambiguated Word Senses in order to study their contribution to QA performance.

The track was organized by a number of institutions (one for each target language) and jointly coordinated by CELCT, Trento, Italy and UNED, Madrid, Spain.

Cross-Language Retrieval in Image Collections (ImageCLEF). This track evaluated retrieval of images from multilingual collections; both text and visual retrieval techniques were exploitable. Five challenging tasks were offered in 2008:

- A photo retrieval task: a good image search engine ensures that duplicate or near duplicate documents retrieved in response to a query are hidden from the user. Ideally the top results of a ranked list will contain diverse items representing different sub-topics within the results. This task focused on the study of successful clustering to provide diversity in the top-ranked results. The target collection contained images with captions in English and German; queries were in English.
- A medical image retrieval task: this is a domain-specific retrieval task in a domain where many ontologies exist; the target collection was a subset of the Goldminer collection containing images from English articles published in Radiology and

Radiographics with captions and html links to the full text articles. Queries were provided in English, French and German.

- A visual concept deception task: the objective was to identify language-independent visual concepts that would help in solving the photo retrieval task. A training database was released with approximately 1,800 images classified according to a concept hierarchy. This data was used to train concept detection/annotation techniques. Participants were required to determine the presence/absence of the concepts for each of the 1,000 images in the test database.
- An automatic medical image annotation task: image annotation or classification can be important when searching for images from a database of radiographs. The aim of the task was to find out how well current language-independent techniques can identify image modality, body orientation, body region, and biological system on the basis of the visual information provided by the images.
- A Wikipedia image retrieval task: this was an ad hoc image search task where the information structure can be exploited for retrieval. The aim was to investigate retrieval approaches in the context of a larger scale and heterogeneous collection of images (similar to those encountered on the Web) that are searched for by users with diverse information needs.

The University and University Hospitals of Geneva, Switzerland; RWTH Aachen, Germany; Oregon Health and Science University, USA; Victoria University, Australia; Sheffield University, UK; Vienna University of Technology, Austria; CWI, The Netherlands, collaborated in the track organization.

Multilingual Web Retrieval (WebCLEF). In the past three years this track has focused on evaluating systems providing multi- and cross-lingual access to web data. WebCLEF 2008 repeated the track setup of the 2007 edition offering an information synthesis task, where, for a given topic, participating systems were asked to extract important snippets from web pages (fetched from the live web and provided by the task organizers). The systems had to focus on extracting, summarizing, filtering and presenting information relevant to the topic, rather than on large scale web search and retrieval per se. The focus was on refining the assessment procedure and evaluation measures. This task had lots of similarities with (topic-oriented) multi-document summarization and with answering complex questions. An important difference is that at WebCLEF 2008, topics could come with extensive descriptions and with many thousands of documents from which important facts have to be mined. In addition, WebCLEF worked with web documents, that may be very noisy and redundant. The track was coordinated by the University of Amsterdam, The Netherlands.

Although the Internet would seem to be the obvious application scenario for a CLIR system, WebCLEF had a rather disappointing participation in 2008. For this reason, we decided to drop this track – at least for 2009.

Cross-Language Geographical Retrieval (GeoCLEF). The purpose of GeoCLEF is to test and evaluate cross-language geographic information retrieval for topics with a geographic specification. How best to transform into a machine readable format the imprecise description of a geographic area found in many user queries is still an open research problem. As in previous years, GeoCLEF 2008 examined geographic search of a text corpus. Some topics simulated the situation of a user who poses a query when looking at a map on the screen. For these topics, the system received the content part and a rectangular shape which defines the geographic context. In GeoCLEF 2006 and 2007, it was found that keyword based systems often do well on the task and the best systems worked without any specific geographic resource. In 2008 the best monolingual systems used specific geo-reasoning; there was much named-entity recognition (often using Wikipedia) and NER topic parsing. Geographic ontologies were also used (such as GeoNames and World Gazeteer), in particular for query expansion. However, as in previous years, in the cross-language tasks, the best systems used no specific geo components; standard approaches like BM25 and blind relevance feedback worked well. A new pilot task on Wikipedia, GikiP, was also offered. The track was organized by Hildesheim University; Germany; Linguateca, Norway and Portugal; Sheffield University, UK.

Cross-Language Video Retrieval (VideoCLEF). VideoCLEF used a video corpus containing episodes of a dual language television program in Dutch and English. Three tasks were offered: (1) Automatic assignment of subject tags (i.e., classification), (2) Automatic translation of metadata for visualization, and (3) Automatic selection of semantically representative keyframes. The dual language programming of Dutch TV offered a unique scientific opportunity, presenting the challenge of how to exploit speech features from both languages. Participants were supplied with archival metadata including title and description, shot boundaries, mshot-level keyframes and automatic speech recognition (ASR) transcripts in both Dutch and English. The video content was chosen to reflect the cultural heritage domain and the subject labels used in the automatic classification task were selected to be representative of cultural heritage themes. The track was coordinated by the University of Amsterdam; data was provided by The Netherlands Institute of Sound and Vision; University of Twente, provided the speech transcripts; Dublin City University, Ireland, provided the shot segmentations and the key frames.

Multilingual Information Filtering (INFILE@CLEF). INFILE (INformation, FILtering & Evaluation) was a cross-language adaptive filtering evaluation track sponsored by the French National Research Agency. INFILE offered monolingual and cross-language tasks, using a corpus of 100,000 Agence France Press (AFP) comparable newswire stories for Arabic, English and French. Evaluation was performed by an automatic interrogation of test systems with a simulated user feedback. A curve of the evolution of efficiency was computed along with more classical measures tested in TREC. The track was coordinated by the Evaluation and Language resources Distribution Agency (ELDA), France.

Unsupervised Morpheme Analysis (MorphoChallenge). The objective of MorphoChallenge is to design a statistical machine learning algorithm that discovers which morphemes (smallest individually meaningful units of language) form words. The scientific goals are: (i) to understand the phenomena underlying word construction in natural languages; (ii) to discover approaches suitable for a wide range of languages: (iii) to advance machine learning methodology. The aim of MorphoChallenge 2008 was similar to that of MorphoChallenge 2007, where the goal was to find the morpheme analysis of the word forms in the data. Two tasks were offered. CLEF data for English, Finnish and German was used in the second task in which information retrieval experiments were performed where the words in the documents and queries were replaced by their proposed morpheme representations. The search was then based on morphemes instead of words. The activity was coordinated by Helsinki University of Technology, Finland.

Details on the technical infrastructure and the organisation of all these tracks can be found in the track overview reports in this volume, collocated at the beginning of the relevant sections.

3 Test Collections

The CLEF test collections are made up of documents, topics and relevance assessments. The topics are created to simulate particular information needs from which the systems derive the queries to search the document collections. System performance is evaluated by judging the results retrieved in response to a topic with respect to their relevance, and computing the relevant measures, depending on the methodology adopted by the track.

A number of different document collections were used in CLEF 2008 to build the test collections:

- CLEF multilingual corpus of more than 3 million news documents in 14 European languages. This corpus is divided into two comparable collections: 1994-1995 - Dutch, English, Finnish, French, German, Italian, Portuguese, Russian, Spanish, Swedish; 2000-2002 - Basque, Bulgarian, Czech, English, Hungarian. The Basque data was new this year. Parts of this collections were used in the Ad Hoc, QuestionAnswering, GeoCLEF and MorphoChallenge tracks.
- Data from The European Library /TEL): approximately 3 million library catalog records from the national libraries of Austria, Britain and France, used in the Ad Hoc track.
- Hamshahri Persian newspaper corpus; nearly 170,000 documents used in the Ad Hoc track;
- The GIRT-4 social science database in English and German (over 300,000 documents), the Russian ISISS collection for sociology and economics (approx. 150,000 docs), Cambridge Sociological Abstracts in English (20,000 docs). These collections were used in the domain-specific track.
- Online Flickr database, used in the iCLEF track
- The ImageCLEF track used collections for both general photographic and medical image retrieval:
    - » IAPR TC-12 photo database of 20,000 still natural images (plus 20,000 corresponding thumbnails) with captions in English, and German;
    - » ARRS Goldminer database – nearly 200,000 images published in 249 selected peer-reviewed radiology journals
    - » IRMA collection in English and German of 12,000 classified images for automatic medical image annotation
    - » INEX Wikipedia image collection, approximately 150,000 images associated with unstructured and noisy textual annotations in English

- Videos in Dutch and English of documentary television programs, approximately 30 hours, used in the VideoCLEF track.
- Agence France Press (AFP) comparable newswire stories in Arabic, French and English for the INFILE track

4 Technical Infrastructure

The DIRECT system developed by the University of Padua managed the technical infrastructure for several of the CLEF 2008 tracks: Ad Hoc, Domain-Specific, GeoCLEF. DIRECT (Distributed Information Retrieval Evaluation Campaign Tool[4]) is a digital library system designed to manage the scientific data and information resources produced during an evaluation campaign. A preliminary version of DIRECT was introduced into CLEF in 2005 and subsequently tested and developed in the CLEF 2006 and 2007 campaigns. In CLEF 2008 DIRECT provided procedures to handle:

- the track set-up, harvesting of documents, management of the registration of participants to tracks;
- the submission of experiments, collection of metadata about experiments, and their validation;
- the creation of document pools and the management of relevance assessment;
- the provision of common statistical analysis tools for both organizers and participants in order to allow the comparison of the experiments;
- the provision of common tools for summarizing, producing reports and graphs on the measured performances and conducted analyses.

DIRECT was used by over 130 participating groups from 20 countries, who submitted 490 experiments. Within the DIRECT framework, 80 assessors created over 200 topics in seven different languages and assessed about 250,000 documents, including documents in languages like Russian, which uses the Cyrillic alphabet, and Persian, which is written from right to left.

5 Participation

A total of 100 groups submitted runs in CLEF 2008, a big increase on the 81 groups of CLEF 2007: 69 from Europe, 12 from North America; 15 from Asia, 3 from South America and 1 from Africa. The breakdown of participation of groups per track is as follows: Ad Hoc 26; Domain-Specific 6; iCLEF 6; QAatCLEF 29; ImageCLEF 42; WebCLEF 3; GeoCLEF 11; VideoCLEF 5; INFILE 1; Morpho Challenge 6. The increase in participation was almost entirely due to a rise in interest from Europe – participation from the other continents remained more or less stable and we had our first ever group from an African country: Uganda.

A list of groups and indications of the tracks in which they participated can be found in the CLEF2008 Working Notes on the CLEF website. Figure 1 shows the variation in participation over the years and Figure 2 shows the shift in focus as new tracks are added.

---

[4] http//direct.dei.unipd.it/

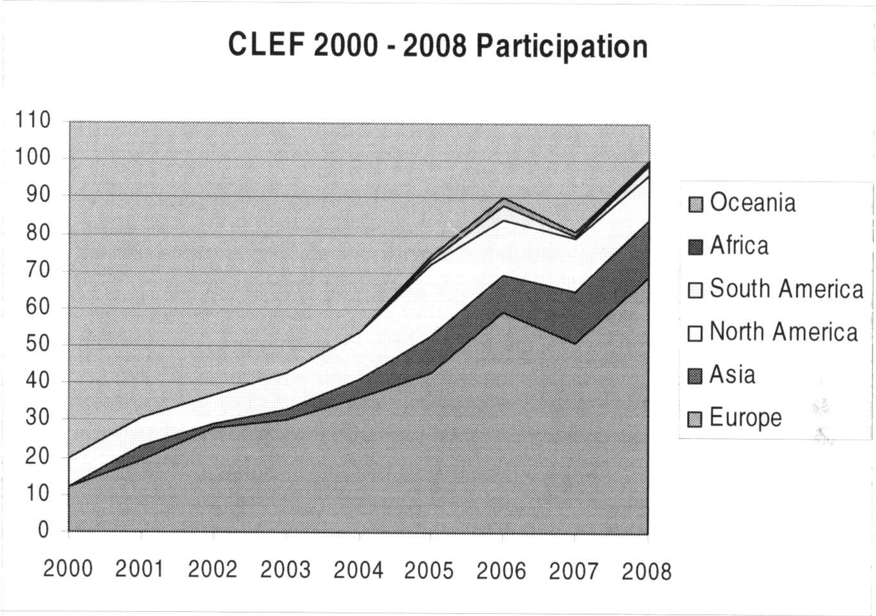

Fig. 2. CLEF 2000 – 2008: Variation in Participation

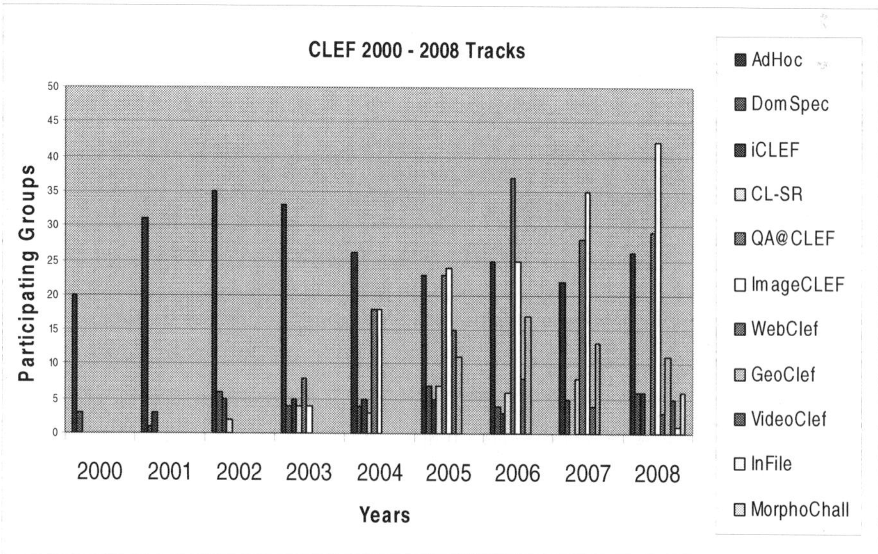

Fig. 3. CLEF 2000 – 2008: Participation per Track in Tracks

6 Results

CLEF has been running for almost ten years now with the main goal of sustaining the growth of excellence in multilingual language processing and information access across language boundaries. We can sum up the main results in the following points:

- Investigation of core issues in MultiLingual Information Access (MLIA) which enable effective transfer over language boundaries, including the development of multiple language processing tools (e.g. stemmers, word decompounders, part-of-speech taggers); creation of linguistic resources (e.g. multilingual dictionaries and corpora); implementation of appropriate cross-language retrieval models and algorithms for different tasks and languages;
- Creation of important reusable test collections and resources in diverse media for a large number of European languages, representative of the major European language typologies;
- Significant and quantifiable improvements in the performance of MLIA systems;

However, although CLEF has done much to promote the development of multilingual IR systems, the focus has been on building and testing research prototypes rather than developing fully operational systems. The challenge that we are now attempting to tackle is how to best transfer these research results to the market place. How we are now trying to face this challenge is described in the following section.

7 CLEF and TrebleCLEF

CLEF is organized mainly through the voluntary efforts of many different institutions and research groups. However, the central coordination has always received some support from the EU IST programme under the unit for Digital Libraries and Technology Enhanced Learning, mainly within the framework of the DELOS Network of Excellence. CLEF 2008 has been organized under the auspices of TrebleCLEF, a Coordination Action of the Seventh Framework Programme, Theme ICT 1-4-1[5].

For many years, CLEF has thus been a forum where researchers can perform experiments, discuss results and exchange ideas; most of the results have been published but the extensive CLEF-related literature is mainly intended for the academic community. Contacts with interested application communities have been notably lacking. In fact, evaluation campaigns have their limitations. They tend to focus on aspects of system performance that can be measured easily in an objective setting (e.g. precision and recall) and to ignore others that are equally important for overall system development. Thus, while in CLEF, much attention has been paid to improving performance in terms of the ranking of results through the refining of query expansion procedures, term weighting schemes, algorithms for the merging of results, equally important criteria of speed, stability, usability have been mainly ignored.

At the beginning of 2008 we launched a new activity which aims at building on and extending the results already achieved by CLEF. This activity, called TrebleCLEF aims at stimulating the development of operational MLIA systems rather than research

[5] See www.trebleclef.eu

prototypes. TrebleCLEF is promoting research, development, implementation and industrial take-up of multilingual, multimodal information access functionality in the following ways:

- by continuing to support the annual CLEF system evaluation campaigns with tracks and tasks designed to stimulate R&D to meet the requirements of the user and application communities
- by constituting a scientific forum for the MLIA community of researchers enabling them to meet and discuss results, emerging trends, new directions.
- by acting as a virtual centre of competence providing a central reference point for anyone interested in studying or implementing MLIA functionality and encouraging the dissemination of information:

The first major results of this activity will be seen in 2009 with the publication of three Best Practices studies:

- Best Practices in Language Resources for Multilingual Information Access
- Best Practices in System and User-oriented Multilingual Information Access
- Best Practices for Test Collection Creation, Evaluation Methodologies and Language Processing Technologies

We are also organizing a Summer School on Multilingual Information Access in June 2009 and a MLIA Technology Transfer Day at the end of the year. With TrebleCLEF we hope to bridge the gap between research activities promoted in CLEF and the application of the results in a real-world context.

8 The Future of CLEF

Since CLEF began the associated technologies, services and users of multilingual IR systems have been in continual evolution, with many new factors and trends influencing the field. For example, the growth of the Internet has been exponential with respect to the number of users and languages used regularly for global information dissemination. The expectations and habits of users are constantly changing, together with the ways in which they interact with content and services, often creating new and original ways of exploiting them. Language barriers are no longer seen as inviolable and there is a growing dissatisfaction with the technologies currently available to overcome them.

This constantly evolving scenario poses challenges to the research community which must react to these new trends and emerging needs. CLEF initially assumed a user model reflecting simple information seeking behaviour: the retrieval of a list of relevant items in response to a single query that could then be used for further consultation in various languages and media types. This simple scenario of user interaction has allowed researchers to focus their attention on studying core technical issues for CLIR systems and associated components.

If we are to continue advancing the state-of-the-art in multilingual information access technologies, we now need to rethink and update this user model. We have to study and evaluate multilingual issues from a communicative perspective rather than a purely retrieval one. We need to examine the interactions between four main entities: users, their tasks, languages, and content in order to understand how these factors impact on the design and development of MLIA systems. It is not sufficient to successfully cross the language boundary, results must be retrieved in a form that is interpretable and reusable. Future cross-language system evaluation campaigns must activate new forms of experimental evaluation - laboratory and interactive – in order to foster the development of MLIA systems more adherent to the new user needs. We need a deeper understanding of the interaction between multicultural and information proactive users, multilingual content, language-dependent tasks, and the enabling technologies consisting of MLIA systems and their components.

At the same time, benchmarking efforts must prove their usefulness for industrial take-up; evaluation initiatives risk being seen as irrelevant for system developers if the data they investigate are not of realistic scale and if the use cases and scenarios tested do not appear valid.

Future editions of CLEF should thus introduce a new series of evaluation cycles which move beyond the current set-up, impacting on:

- Methodology definition: evolution of the current evaluation paradigm, developing new models and metrics to describe the needs and behavior of the new multicultural and multi-tasking users;
- System building: driving the development of MLIA systems and assessing their conformity with respect to the newly identified user needs, tasks, and models;
- Results assessment: measuring all aspects of system & component performance including response times, usability, and user satisfaction
- Community building: promoting the creation of a multidisciplinary community of researchers which goes beyond the existing CLEF community by building bridges to other relevant research domains such as the MT, information science and user studies sectors, and to application communities, such as the enterprise search, legal, patent, educational, cultural heritage and infotainment areas;
- Validation of technology: providing a reasonably comprehensive typology of use cases and usage scenarios for multilingual search, validated through user studies, to enable reuse of appropriate resources and to enable common evaluation schemes;
- Technology transfer: guaranteeing that the results obtained are demonstrated as useful for industrial deployment.

Achieving this goal will require further synergy between various research communities including machine translation, information retrieval, question answering, information extraction, and representatives from end user groups.

Acknowledgements

It would be impossible to run the CLEF evaluation initiative and organize the annual workshops without considerable assistance from many groups. CLEF is organized on a distributed basis, with different research groups being responsible for the running of the various tracks. My gratitude goes to all those who have been involved in the coordination of the 2008 campaigns. A list of the main institutions involved is given in the following pages. Here below, let me thank just some of the people responsible for the coordination of the different tracks. My apologies to all those I have not managed to mention:

- Abolfazl AleAhmad, Hadi Amiri, Eneko Agirre, Giorgio Di Nunzio, Nicola Ferro, Thomas Mandl, Nicolas Moreau, Alessandro Nardi and Vivien Petras for the Ad Hoc Track
- Vivien Petras and Stefan Baerisch for the Domain-Specific track
- Paul Clough, Julio Gonzalo and Jussi Karlgren for iCLEF
- Danilo Giampiccolo Pamela Forner, Dan Cristea, Corina Forascu, Nicolas Moreau, Petya Osenova, Anselmo Peñas, Iñaki Alegria, Bogdan Sacaleanu, Prokopis Prokopidis, Paulo Rocha and Richard Sutcliffe for QA@CLEF
- Allan Hanbury, Paul Clough, Thomas Arni, Mark Sanderson, Henning Müller, Thomas Deselaers , Thomas Deserno, Michael Grubinger, Jayashree Kalpathy–Cramer, and William Hersh for ImageCLEF
- Valentin Jijkoun and Maarten de Rijke for Web-CLEF
- Thomas Mandl, Fredric Gey, Ray Larson, Mark Sanderson, Diana Santos, Paula Carvalho for GeoCLEF
- Martha Larson and Gareth Jones for VideoCLEF
- Djamel Mostefa for INFILE
- Marco Duissin, Giorgio Di Nunzio and Nicola Ferro for developing and managing the DIRECT infrastructure.

I also thank all those colleagues who have helped us by preparing topic sets in different languages and the members of the CLEF Steering Committee who have assisted me with their advice and suggestions throughout this campaign.

Furthermore, I gratefully acknowledge the support of all the data providers and copyright holders, and in particular:

- The Los Angeles Times, for the American-English newspaper collection.
- SMG Newspapers (The Herald) for the British-English newspaper collection.
- Le Monde S.A. and ELDA: Evaluations and Language resources Distribution Agency, for the French newspaper collection.
- Frankfurter Rundschau, Druck und Verlagshaus Frankfurt am Main; Der Spiegel, Spiegel Verlag, Hamburg, for the German newspaper collections.
- Hypersystems Srl, Torino and La Stampa, for the Italian newspaper data.
- Agencia EFE S.A. for the Spanish news agency data.
- NRC Handelsblad, Algemeen Dagblad and PCM Landelijke dagbladen/Het Parool for the Dutch newspaper data.
- Aamulehti Oyj and Sanoma Osakeyhtiö for the Finnish newspaper data.
- Russika-Izvestia for the Russian newspaper data.
- Hamshahri newspaper and DBRG, Univ. Tehran, for the Persian newspaper data.
- Público, Portugal, and Linguateca for the Portuguese (PT) newspaper collection.
- Folha, Brazil, and Linguateca for the Portuguese (BR) newspaper collection.
- Tidningarnas Telegrambyrå (TT) SE-105 12 Stockholm, Sweden for the Swedish newspaper data.

- Schweizerische Depeschenagentur, Switzerland, for the French, German & Italian Swiss news agency data.
- Ringier Kiadoi Rt. (Ringier Publishing Inc.).and the Research Institute for Linguistics, Hungarian Acad. Sci. for the Hungarian newspaper documents.
- Sega AD, Sofia; Standart Nyuz AD, Sofia, Novinar OD, Sofia and the BulTreeBank Project, Linguistic Modelling Laboratory, IPP, Bulgarian Acad. Sci, for the Bulgarian newspaper documents
- Mafra a.s. and Lidové Noviny a.s. for the Czech newspaper data
- AFP Agence France Presse for the English, French and Arabic newswire data used in the INFILE track
- The British Library, Bibliothèque Nationale de France and the Austrian National Library for the library catalog records forming part of The European Library (TEL)
- InformationsZentrum Sozialwissen-schaften, Bonn, for the GIRT social science database.
- SocioNet system for the Russian Social Science Corpora
- Institute of Scientific Information for Social Sciences of the Russian Academy of Science (ISISS RAS) for the ISISS database
- University and University Hospitals, Geneva, Switzerland and Oregon Health and Science University for the ImageCLEFmed Radiological Medical Database.
- Aachen University of Technology (RWTH), Germany, for the IRMA annotated medical images.
- The Radiology Dept. of the University Hospitals of Geneva for the Casimage database and the PEIR (Pathology Education Image Resource) for the images and the HEAL (Health Education Assets Library) for the Annotation of the Peir dataset.
- Mallinkrodt Institue of Radiology for permission to use their nuclear medicine teaching file.
- University of Basel's Pathopic project for their Pathology teaching file.
- Michael Grubinger, administrator of the IAPR Image Benchmark, Clement Leung who initiated and supervised the IAPR Image Benchmark Project, and André Kiwitz, the Managing Director of Viventura for granting access to the image database and the raw image annotations of the tour guides.
- USC Shoah Foundation Institute, and IBM (English) and The Johns Hopkins University Center for Language and Speech Processing (Czech) for the speech transcriptions.
- The Institute of Sound and Vision, The Netherlands, for the English/Dutch videos, the University of Twente for the speech transcriptions, and Dublin City University for the shot segmentation.

CLEF 2008: Ad Hoc Track Overview

Eneko Agirre[1], Giorgio Maria Di Nunzio[2], Nicola Ferro[2], Thomas Mandl[3], and Carol Peters[4]

[1] Computer Science Department, University of the Basque Country, Spain
e.agirre@ehu.es
[2] Department of Information Engineering, University of Padua, Italy
{dinunzio,ferro}@dei.unipd.it
[3] Information Science, University of Hildesheim, Germany
mandl@uni-hildesheim.de
[4] ISTI-CNR, Area di Ricerca, Pisa, Italy
carol.peters@isti.cnr.it

Abstract. We describe the objectives and organization of the CLEF 2008 Ad Hoc track and discuss the main characteristics of the tasks offered to test monolingual and cross-language textual document retrieval systems. The track was changed considerably this year with the introduction of tasks with new document collections consisting of (i) library catalog records derived from The European Library, and (ii) and non-European language data, plus a task offering the chance to test retrieval with word sense disambiguated data. The track was thus structured in three distinct streams denominated: TEL@CLEF, Persian@CLEF and Robust WSD. The results obtained for each task are presented and statistical analyses are given.

1 Introduction

The Ad Hoc retrieval track is generally considered to be the core track in the *Cross-Language Evaluation Forum (CLEF)*. It is the one track that has been offered each year, from 2000 through 2008, and will be offered again in 2009. The aim of this track is to promote the development of monolingual and cross-language textual document retrieval systems. From 2000 - 2007, the track used exclusively collections of European newspaper and news agency documents[1] and worked hard at offering increasingly complex and diverse tasks, adding new languages each year. The results have been considerable; it is probably true to say that this track has done much to foster the creation of a strong European research community in the cross-language text retrieval area. It has provided the resources, the test collections and also the forum for discussion and comparison of ideas and approaches. Groups submitting experiments over several years have shown flexibility in advancing to more complex tasks, from monolingual to bilingual and multilingual experiments. Much work has been done

[1] Over the years, this track has built up test collections for monolingual and cross-language system evaluation in 14 European languages (see the Introduction to this volume for more details).

C. Peters et al. (Eds.): CLEF 2008, LNCS 5706, pp. 15–37, 2009.

on fine-tuning for individual languages while other efforts have concentrated on developing language-independent strategies. In fact, one of the papers in this section reports some interesting post-workshop experiments on previous CLEF Ad Hoc test collections in 13 languages, comparing the performance of different indexing approaches: word, stems, morphemes, n-gram stems and character n-grams [27].

This year the focus of the track was considerably widened: we introduced very different document collections, a non-European target language, and an information retrieval (IR) task designed to attract participation from groups interested in natural language processing (NLP). The track was thus structured in three distinct streams:

- TEL@CLEF
- Persian@CLEF
- Robust WSD

The first task was an application-oriented task, offering monolingual and cross-language search on library catalog records and was organized in collaboration with The European Library (TEL)[2]. The second task resembled the Ad Hoc retrieval tasks of previous years but this time the target collection was a Persian newspaper corpus. The third task was the robust activity which this year used word sense disambiguated (WSD) data, and involved English documents and monolingual and cross-language search in Spanish.

In this paper we first present the track setup, the evaluation methodology and the participation in the different tasks (Section 2). We then describe the main features of each task and show the results (Sections 3 - 5). The final section provides a brief summing up. For information on the various approaches and resources used by the groups participating in this track and the issues they focused on, we refer the reader to the rest of the papers in the Ad Hoc section of these Proceedings.

2 Track Setup

The Ad Hoc track in CLEF adopts a corpus-based, automatic scoring method for the assessment of system performance, based on ideas first introduced in the Cranfield experiments in the late 1960s [10]. The **tasks** offered are studied in order to effectively measure textual document retrieval under specific conditions. The **test collections** are made up of **documents**, **topics** and **relevance assessments**. The topics consist of a set of statements simulating information needs from which the systems derive the queries to search the document collections. Evaluation of system performance is then done by judging the documents retrieved in response to a topic with respect to their relevance, and computing the recall and precision measures. The distinguishing feature of the CLEF Ad Hoc track is that it applies this evaluation paradigm in a multilingual setting.

[2] See `http://www.theeuropeanlibrary.org/`

This means that the criteria normally adopted to create a test collection, consisting of suitable documents, sample queries and relevance assessments, have been adapted to satisfy the particular requirements of the multilingual context. All language dependent tasks such as topic creation and relevance judgment are performed in a distributed setting by native speakers. Rules are established and a tight central coordination is maintained in order to ensure consistency and coherency of topic and relevance judgment sets over the different collections, languages and tracks.

2.1 Test Collections

The three streams of the Ad Hoc track created very distinct test collections this year. The details are given in this section.

The Documents. Each of the three Ad Hoc tasks used a different set of documents.

The TEL task used three collections derived from:

- the British Library (BL); 1,000,100 documents, 1.2 GB;
- the Bibliothéque Nationale de France (BNF); 1,000,100 documents, 1.3 GB;
- the Austrian National Library (ONB); 869,353 documents, 1.3 GB.

We refer to the three collections (BL, BNF, ONB) as English, French and German because in each case this is the main language of the collection. However, each collection is to some extent multilingual and contains documents (catalog records) in many additional languages.

The TEL data is very different from the newspaper articles and news agency dispatches previously used in the CLEF Ad Hoc track. The data tends to be very sparse. Many records contain only title, author and subject information; other records provide more detail. The title and (if existing) an abstract or description may be in a different language to that understood as the language of the collection. The subject information is normally in the main language of the collection. About 66% of the documents in the English and German collection have subject headings, only 37% in the French collection. Dewey Classification (DDC) is not available in the French collection; negligible (approx. 0.3%) in the German collection; but occurs in about half of the English documents (456,408 docs to be exact). Whereas in the traditional Ad Hoc task the user searches directly for a document containing information of interest, here the user tries to identify which publications are of potential interest according to the information provided by the catalog card.

The Persian task used the Hamshahri corpus of 1996-2002 newspapers as the target collection. This corpus was made available to CLEF by the Data Base Research Group (DBRG) of the University of Tehran. Hamshahri is one of the most popular daily newspapers in Iran. The Hamshahri corpus is a Persian test collection that consists of 345 MB of news texts for the years 1996 to 2002 (corpus size with tags is 564 MB). This corpus contains more than 160,000 news articles about a variety of subjects and includes nearly 417000 different words. Hamshahri articles vary between 1KB and 140KB in size[3].

The robust task used existing CLEF news collections but with word sense disambiguation (WSD) information added. The word sense disambiguation data was automatically added by systems from two leading research laboratories, UBC [2] and NUS [9]. Both systems returned word senses from the English WordNet, version 1.6.

The document collections were offered both with and without WSD, and included the following[4]:

- LA Times 94 (with word sense disambiguated data); ca 113,000 documents, 425 MB without WSD, 1,448 MB (UBC) or 2,151 MB (NUS) with WSD;
- Glasgow Herald 95 (with word sense disambiguated data); ca 56,500 documents, 154 MB without WSD, 626 MB (UBC) or 904 MB (NUS) with WSD.

The Topics. Topics in the CLEF Ad Hoc track are structured statements representing information needs. Each topic typically consists of three parts: a brief "title" statement; a one-sentence "description"; a more complex "narrative" specifying the relevance assessment criteria. Topics are prepared in xml format and identified by means of a Digital Object Identifier (DOI)[5] of the experiment [30] which allows us to reference and cite them.

For the TEL task, a common set of 50 topics was prepared in each of the 3 main collection languages (English, French and German) plus Dutch and Spanish in response to demand. Only the Title and Description fields were released to the participants. The narrative was employed to provide information for the assessors on how the topics should be judged. The topic sets were prepared on the basis of the contents of the collections.

In Ad Hoc, when a task uses data collections in more than one language, we consider it important to be able to use versions of the same core topic set to query all collections. This makes it easier to compare results over different collections and also facilitates the preparation of extra topic sets in additional languages. However, it is never easy to find topics that are effective for several different collections and the topic preparation stage requires considerable discussion between the coordinators for each language in order to identify suitable common candidates. The sparseness of the data made this particularly difficult for the TEL task and tended to lead to the formulation of topics that were quite broad in scope so that at least some relevant documents could be found in each collection. A result of this strategy is that there tends to be a considerable lack of evenness of distribution of relevant documents over the collections. For each topic, the results expected from the separate collections can vary considerably, e.g. a topic of particular interest to Britain, such as the example given in Figure 1, can be

[3] For more information, see `http://ece.ut.ac.ir/dbrg/hamshahri/`

[4] A sample document and dtd are available at `http://ixa2.si.ehu.es/clirwsd/`

[5] `http://www.doi.org/`

```
<?xml version="1.0" encoding="UTF-8" standalone="no"?>
<topic>
    <identifier>10.2452/451-AH</identifier>

    <title lang="en">Roman Military in Britain</title>
    <title lang="de">Römisches Militär in Britannien</title>
    <title lang="es">El ejército romano en Britania</title>
    <title lang="fr">L'armée romaine en Grande-Bretagne</title>
    <title lang="nl">Romeinse Leger in Groot-Brittannie</title>

    <description lang="en">Find books or publications on the Roman invasion or military
        occupation of Britain.</description>
    <description lang="de">Finden Sie Bücher oder Publikationen über die römische
        Invasion oder das Militär in Britannien.</description>
    <description lang="es">Encuentre libros o publicaciones sobre la invasión romana
        o la ocupación militar romana en Britania.</description>
    <description lang="fr">Trouver des livres ou des publications sur l'invasion et
        l'occupation de la Grande-Bretagne par les Romains.</description>
    <description lang="nl">Vind boeken of publicaties over de Romeinse invasie of
        bezetting van Groot-Brittannie.</description>
</topic>
```

Fig. 1. Example of TEL topic in all five languages: topic 10.2452/451-AH

```
<?xml version="1.0" encoding="UTF-8" standalone="no"?>
<topic>
    <identifier>10.2452/599-AH</identifier>
    <title lang="en">2nd of Khordad election</title>
    <title lang="fa">انتخابات دوم خرداد</title>

    <description lang="en">Find documents that include information about the 2nd of Khordad
        presidential elections.</description>
    <description lang="fa">سندهايي را پيدا كن كه شامل اطلاعاتي در مورد انتخابات دوم خرداد ماه سال 76
        هستند</description>

    <narrative lang="en">Any information about candidates and their sayings, Khatami's unexpected
        winning in the 2nd of Khordad 1376 presidential election is relevant.</narrative>
    <narrative lang="fa">سندهاي مربوط شامل اطلاعاتي در مورد نامزدها و گفته هاي آنها، پيروزي
        غيرمنتظره خاتمي در انتخابات رياست جمهوري در دوم خرداد ماه سال 76 است</narrative>
</topic>
```

Fig. 2. Example of Persian topic: topic 10.2452/599-AH

expected to find far more relevant documents in the BL collection than in BNF or ONB.

For the Persian task, 50 topics were created in Persian by the Data Base Research group of the University of Tehran, and then translated into English. The rule in CLEF when creating topics in additional languages is not to produce literal translations but to attempt to render them as naturally as possible. This was a particularly difficult task when going from Persian to English as cultural differences had to be catered for.

For example, Iran commonly uses a different calendar from Europe and reference was often made in the Persian topics to events that are well known to Iranian society but not often discussed in English. This is shown in the example of Figure 2, where the rather awkward English rendering evidences the uncertainty of the translator.

The WSD robust task used existing CLEF topics in English and Spanish as follows:

- CLEF 2001; Topics 41-90; LA Times 94
- CLEF 2002; Topics 91-140; LA Times 94
- CLEF 2003; Topics 141-200; LA Times 94, Glasgow Herald 95

```
<top>
    <num>10.2452/141-WSD-AH</num>

    <EN-title>
        <TERM ID="10.2452/141-WSD-AH-1" LEMA="letter" POS="NNP">
            <WF>Letter</WF>
            <SYNSET SCORE="0" CODE="05115901-n"/>
            <SYNSET SCORE="0" CODE="05362432-n"/>
            <SYNSET SCORE="0" CODE="05029514-n"/>
            <SYNSET SCORE="1" CODE="04968965-n"/>
        </TERM>

        <TERM ID="10.2452/141-WSD-AH-2" LEMA="bomb" POS="NNP">
            <WF>Bomb</WF>
            <SYNSET SCORE="0.888888888888889" CODE="02310834-n"/>
            <SYNSET SCORE="0" CODE="05484679-n"/>
            <SYNSET SCORE="0.111111111111111" CODE="02311368-n"/>
        </TERM>

        <TERM ID="10.2452/141-WSD-AH-3" LEMA="for" POS="IN">
            <WF>for</WF>
        </TERM>

        ...

    </EN-title>

    <EN-desc>
        <TERM ID="10.2452/141-WSD-AH-5" LEMA="find" POS="VBP">
            <WF>Find</WF>
            <SYNSET SCORE="0" CODE="00658116-v"/>

            ...

        </TERM>

        ...

    </EN-desc>

    <EN-narr>
        ...
    </EN-narr>
</top>
```


Fig. 3. Example of Robust WSD topic: topic `10.2452/141-WSD-AH`

- CLEF 2004; Topics 201-250; Glasgow Herald 95
- CLEF 2005; Topics 251-300; LA Times 94, Glasgow Herald 95
- CLEF 2006; Topics 301-350; LA Times 94, Glasgow Herald 95

Topics from years 2001, 2002 and 2004 were used as training topics (relevance assessments were offered to participants), and topics from years 2003, 2005 and 2006 were used for the test.

All topics were offered both with and without WSD. Topics in English were disambiguated by both UBC [2] and NUS [9] systems, yielding word senses from WordNet version 1.6. A large-scale disambiguation system for Spanish was not available, so we used the first-sense heuristic, yielding senses from the Spanish wordnet, which is tightly aligned to the English WordNet version 1.6 (i.e., they share synset numbers or sense codes). An excerpt from a topic is shown in Figure 3, where each term in the topic is followed by its senses with their respective scores as assigned by the automatic WSD system[6].

Relevance Assessment. The number of documents in large test collections such as CLEF makes it impractical to judge every document for relevance. Instead

[6] Full sample and dtd are available at `http://ixa2.si.ehu.es/clirwsd/`

approximate recall values are calculated using pooling techniques. The results submitted by the groups participating in the Ad Hoc tasks are used to form a pool of documents for each topic and language by collecting the highly ranked documents from selected runs according to a set of predefined criteria. Traditionally, the top 100 ranked documents from each of the runs selected are included in the pool; in such a case we say that the pool is of depth 100. This pool is then used for subsequent relevance judgments. After calculating the effectiveness measures, the results are analyzed and run statistics produced and distributed.

The stability of pools constructed in this way and their reliability for post-campaign experiments is discussed in [7] with respect to the CLEF 2003 pools. New pools were formed in CLEF 2008 for the runs submitted for the TEL and the Persian mono- and bilingual tasks. Instead, the robust tasks used the original pools and relevance assessments from previous CLEF campaigns.

The main criteria used when constructing the pools were:

- favour diversity among approaches adopted by participants, according to the descriptions of the experiments provided by the participants;
- choose at least one experiment for each participant in each task, from among the experiments with highest priority as indicated by the participant;
- add mandatory title+description experiments, even though they do not have high priority;
- add manual experiments, when provided;
- for bilingual tasks, ensure that each source topic language is represented.

One important limitation when forming the pools is the number of documents to be assessed. Last year, for collections of newspaper documents, we estimated that assessors could judge from 60 to 100 documents per hour, providing binary judgments: relevant / not relevant. Our estimate this year for the TEL catalog records was higher as these records are much shorter than the average newspaper article (100 to 120 documents per hour). In both cases, it can be seen what a time-consuming and resource expensive task human relevance assessment is. This limitation impacts strongly on the application of the criteria above - and implies that we are obliged to be flexible in the number of documents judged per selected run for individual pools.

Thus, in CLEF 2008, we used a depth of the top 60 ranked documents from selected runs in order to build pools of more-or-less equivalent size (approx. 25,000 documents) for the TEL English, French, and German and the Persian task[7]. Our CLEF2008 Working Notes paper reports summary information on the 2008 Ad Hoc pools used to calculate the results for the main monolingual and bilingual experiments. For each pool, we show the number of topics, the number of runs submitted, the number of runs included in the pool, the number of documents in the pool (relevant and non-relevant), and the number of assessors.

[7] Tests made on NTCIR pools in previous years have suggested that a depth of 60 is normally adequate to create stable pools, as long as a sufficient number of runs from different systems have been included.

In addition the distribution of relevant documents across the topics is compared for the different Ad Hoc pools [4].

For the TEL documents, we judged for relevance only those documents that are written totally or partially in English, French and German (and Spanish for searches on the English collection), e.g. a catalog record written entirely in Hungarian was counted as not relevant as it was of no use to our hypothetical user; however, a catalog record with perhaps the title and a brief description in Hungarian, but with subject descriptors in French, German or English was judged for relevance as it could be potentially useful. Our assessors had no additional knowledge of the documents referred to by the catalog records (or surrogates) contained in the collection. They judged for relevance on the information contained in the records made available to the systems. This was a non trivial task due to the lack of information present in the documents. During the relevance assessment activity there was much consultation between the assessors for the three TEL collections in order to ensure that the same assessment criteria were adopted by everyone.

The relevance judgments for the Persian results were done by the DBRG group in Tehran. Again, assessment was performed on a binary basis and the standard CLEF assessment rules were applied, e.g. if in doubt with respect to the relevance of a given document, assessors are requested to ask themselves whether the document in question would be useful in any way if they had to write a report on the given topic.

As has already been stated, the robust WSD task used existing relevance assessments from previous years. The relevance assessments regarding the training topics were provided to participants before competition time.

This year, we tried a slight improvement with respect to the traditional pooling strategy adopted so far in CLEF. During the topic creation phase, the assessors express their opinion about the relevance of the documents they inspect with respect to the topic. Although this opinion may change during the various discussions between assessors in this phase, we consider these indications as potentially useful in helping to strengthen the pools of documents that will be judged for relevance. These documents are thus added to the pools. However, the assessors are not informed of which documents they had previously judged in order not to bias them in any way.

Similarly to last year, in his paper, Stephen Tomlinson, has reported some sampling experiments aimed at estimating the judging coverage for the CLEF 2008 TEL and Persian test collections. He finds that this tends to be lower than the estimates he produced for the CLEF 2007 collections. With respect to the TEL collections, the implication is that at best 55% of the relevant documents are included in the pools - however, most of the unjudged relevant documents are for the 10 or more queries that have the most known answers [33]. According to studies on earlier TREC collections which gave similar results, in any case this "level of completeness" should be acceptable. For Persian the coverage is much lower - around 25%; this could be a result of the fact that all the Persian topics tend to be relatively broad. This year's Persian collection is thus considered to be less stable than usual.

2.2 Result Calculation

Evaluation campaigns such as TREC and CLEF are based on the belief that the effectiveness of *Information Retrieval Systems (IRSs)* can be objectively evaluated by an analysis of a representative set of sample search results. For this, effectiveness measures are calculated based on the results submitted by the participants and the relevance assessments. Popular measures usually adopted for exercises of this type are Recall and Precision. Details on how they are calculated for CLEF are given in [8]. For the robust task, we used additional measures, see Section 5.

The individual results for all official Ad Hoc experiments in CLEF 2008 are given in the Appendices of the CLEF 2008 Working Notes [14],[15], [16].

2.3 Participants and Experiments

As shown in Table 1, a total of 24 groups from 14 different countries submitted official results for one or more of the Ad Hoc tasks - a slight increase on the

Table 1. CLEF 2008 Ad Hoc participants

Participant	Institution	Country
chemnitz	Chemnitz University of Technology	Germany
cheshire	U.C.Berkeley	United States
geneva	University of Geneva	Switzerland
imag	Inst. for Infocomm Research	France
inaoe	INAOE	Mexico
inesc	INESC ID	Portugal
isi	Indian Statistical Institute	India
ixa	Univ. Basque Country	Spain
jhu-apl	Johns Hopkins University Applied Physics Lab	United States
karlsruhe	University of Karlsruhe	Germany
know-center	Knowledge Relationship Discovery	Austria
opentext	Open Text Corporation	Canada
tehran-IRDB	IR-DB Research Group	Iran
tehran-NLP	NLP-Software Engineering Grad. Lab	Iran
tehran-NLPDB	NLP-DB Research Group	Iran
tehran-NLPDB2	NLP-DB Group	Iran
tehran-SEC	School of Electrical Computing-1	Iran
twente	Univ. of Twente	Netherlands
ucm	Universidad Complutense de Madrid	Spain
ufrgs	Univ. Fed. do Rio Grande do Sul	Brazil
uniba	Universita' di Bari	Italy
unine	U.Neuchatel-Informatics	Switzerland
xerox	Xerox Reseearch - Data Mining	France
xerox-sas	Xerox SAS	Italy

Table 2. Breakdown of experiments into tracks and topic languages

(a) Number of experiments per track, participant.

Track	**# Part.**	**# Runs**
TEL Mono English	13	37
TEL Mono French	9	29
TEL Mono German	10	30
TEL Biling. English	8	24
TEL Biling. French	5	16
TEL Biling. German	6	17
Mono Persian	8	53
Biling. Persian	3	13
Robust Mono English Test	8	20
Robust Mono English Training	1	2
Robust Biling. English Test	4	8
Robust Mono English Test WSD	7	25
Robust Mono English Training WSD	1	5
Robust Biling. English Test WSD	4	10
Total		**289**

(b) List of experiments by topic language.

Topic Lang.	**# Runs**
English	120
Farsi	51
German	44
French	44
Spanish	26
Dutch	3
Portuguese	1
Total	**289**

22 participants of last year[8]. A total of 289 runs were submitted with an increase of about 22% on the 235 runs of 2007. The average number of submitted runs per participant also increased: from 10.6 runs/participant of 2007 to 12.0 runs/participant of this year.

Participants were required to submit at least one title+description ("TD") run per task in order to increase comparability between experiments. The large majority of runs (215 out of 289, 74.40%) used this combination of topic fields, 27 (9.34%) used all fields[9], 47 (16.26%) used the title field only. The majority of experiments were conducted using automatic query construction (273 out of 289, 94.47%) and only in a small fraction of the experiments (16 out 289, 5.53%) were queries been manually constructed from topics. A breakdown into the separate tasks is shown in Table 2(a).

Seven different topic languages were used in the Ad Hoc experiments. As always, the most popular language for queries was English, with Farsi second. The number of runs per topic language is shown in Table 2(b).

3 TEL@CLEF

The objective of this activity was to search and retrieve relevant items from collections of library catalog cards. The underlying aim was to identify the most

[8] Two additional Spanish groups presented results after the deadline for the robust tasks; their results were thus not reported in the official list but their papers are included in this volume [26], [28].

[9] The narrative field was only offered for the Persian and Robust tasks.

effective retrieval technologies for searching this type of very sparse data. When we designed the task, the question the user was presumed to be asking was "Is the publication described by the bibliographic record relevant to my information need?"

3.1 Tasks

Two subtasks were offered: Monolingual and Bilingual. By monolingual we mean that the query is in the same language as the expected language of the collection. By bilingual we mean that the query is in a different language to the main language of the collection. For example, in an EN $\rightarrow$ FR run, relevant documents (bibliographic records) could be any document in the BNF collection (referred to as the French collection) in whatever language they are written. The same is true for a monolingual FR $\rightarrow$ FR run - relevant documents from the BNF collection could actually also be in English or German, not just French.

In CLEF 2008, the activity we simulated was that of users who have a working knowledge of English, French and German (plus wrt the English collection also Spanish) and who want to discover the existence of relevant documents that can be useful for them in one of our three target collections. One of our suppositions was that, knowing that these collections are to some extent multilingual, some systems may attempt to use specific tools to discover this. For example, a system trying the cross-language English to French task on the BNF target collection but knowing that documents retrieved in English and German will also be judged for relevance might choose to employ an English-German as well as the probable English-French dictionary. Groups attempting anything of this type were asked to declare such runs with a ++ indication.

3.2 Participants

13 groups submitted 153 runs for the TEL task: all groups submitted monolingual runs (96 runs out of 153); 8 groups also submitted bilingual runs (57 runs out of 153). Table 2(a) provides a breakdown of the number of participants and submitted runs by task.

3.3 Results

Monolingual Results. Table 3 shows the top five groups for each target collection, ordered by mean average precision. The table reports: the short name of the participating group; the mean average precision achieved by the experiment; the DOI of the experiment; and the performance difference between the first and the last participant.

Bilingual Results. Table 4 shows the top five groups for each target collection, ordered by mean average precision. The table reports: the short name of the participating group; the mean average precision achieved by the experiment; the DOI of the experiment; and the performance difference between the first and the last participant.

Table 3. Best entries for the monolingual TEL tasks

Track	Rank	Participant	Experiment DOI	MAP
English	**1st**	unine	10.2415/AH-TEL-MONO-EN-CLEF2008.UNINE.UNINEEN3	37.53%
	2nd	inesc	10.2415/AH-TEL-MONO-EN-CLEF2008.INESC.RUN3	36.23%
	3rd	chemnitz	10.2415/AH-TEL-MONO-EN-CLEF2008.CHEMNITZ.CUT_SIMPLE	35.61%
	4th	jhu-apl	10.2415/AH-TEL-MONO-EN-CLEF2008.JHU-APL.JHUMOEN4RF	35.31%
	5th	cheshire	10.2415/AH-TEL-MONO-EN-CLEF2008.CHESHIRE.BKAHTELMENTDT2F	34.66%
	Difference			8.28%
French	**1st**	unine	10.2415/AH-TEL-MONO-FR-CLEF2008.UNINE.UNINEFR3	33.27%
	2nd	xerox	10.2415/AH-TEL-MONO-FR-CLEF2008.XEROX.J1	30.88%
	3rd	jhu-apl	10.2415/AH-TEL-MONO-FR-CLEF2008.JHU-APL.JHUMOFR4	29.50%
	4th	opentext	10.2415/AH-TEL-MONO-FR-CLEF2008.OPENTEXT.OTFR08TD	25.23%
	5th	chesire	10.2415/AH-TEL-MONO-FR-CLEF2008.CHESHIRE.BKAHTELMFRTDT2FB	24.37%
	Difference			36.52%
German	**1st**	opentext	10.2415/AH-TEL-MONO-DE-CLEF2008.OPENTEXT.OTDE08TDE	35.71%
	2nd	jhu-apl	10.2415/AH-TEL-MONO-DE-CLEF2008.JHU-APL.JHUMODE4	33.77%
	3rd	unine	10.2415/AH-TEL-MONO-DE-CLEF2008.UNINE.UNINEDE1	30.12%
	4th	xerox	10.2415/AH-TEL-MONO-DE-CLEF2008.XEROX.T1	27.36%
	5th	inesc	10.2415/AH-TEL-MONO-DE-CLEF2008.INESC.RUN3	22.97%
	Difference			55.46%

Table 4. Best entries for the bilingual TEL tasks

Track	Rank	Participant	Experiment DOI	MAP
English	**1st**	chemnitz	10.2415/AH-TEL-BILI-X2EN-CLEF2008.CHEMNITZ.CUT_SIMPLE_DE2EN	34.15%
	2nd	chesire	10.2415/AH-TEL-BILI-X2EN-CLEF2008.CHESHIRE.BKAHTELBFRENTDT2FB	28.24%
	3rd	ufrgs	10.2415/AH-TEL-BILI-X2EN-CLEF2008.UFRGS.UFRGS_BI_SP_EN2	23.15%
	4th	twente	10.2415/AH-TEL-BILI-X2EN-CLEF2008.TWENTE.FCW	22.78%
	5th	jhu-apl	10.2415/AH-TEL-BILI-X2EN-CLEF2008.JHU-APL.JHUBIDEEN5	21.11%
	Difference			61.77%
French	**1st**	chesire	10.2415/AH-TEL-BILI-X2FR-CLEF2008.CHESHIRE.BKAHTELBDEFRTDT2FB	18.84%
	2nd	chemnitz	10.2415/AH-TEL-BILI-X2FR-CLEF2008.CHEMNITZ.CUT_SIMPLE_EN2FR	17.54%
	3rd	jhu-apl	10.2415/AH-TEL-BILI-X2FR-CLEF2008.JHU-APL.JHUBINLFR5	17.46%
	4th	xerox	10.2415/AH-TEL-BILI-X2FR-CLEF2008.XEROX.GER_FRE_J	11.62%
	5th	xerox-sas	10.2415/AH-TEL-BILI-X2FR-CLEF2008.XEROX-SAS.CACAOENGFREPLAIN	6.78%
	Difference			177.87%
German	**1st**	jhu-apl	10.2415/AH-TEL-BILI-X2DE-CLEF2008.JHU-APL.JHUBIENDE5	18.98%
	2nd	chemnitz	10.2415/AH-TEL-BILI-X2DE-CLEF2008.CHEMNITZ.CUT_MERGED_SIMPLE_EN2DE	18.51%
	3rd	chesire	10.2415/AH-TEL-BILI-X2DE-CLEF2008.CHESHIRE.BKAHTELBENDETDT2FB	15.56%
	4th	xerox	10.2415/AH-TEL-BILI-X2DE-CLEF2008.XEROX.FRE_GER_J	12.05%
	5th	karlsruhe	10.2415/AH-TEL-BILI-X2DE-CLEF2008.KARLSRUHE.AIFB_ONB_EN	6.67%
	Difference			184.55%

For bilingual retrieval evaluation, a common method is to compare results against monolingual baselines. We have the following results for CLEF 2008:

- X → EN: 90.99% of best monolingual English IR system;
- X → FR: 56.63% of best monolingual French IR system;
- X → DE: 53.15% of best monolingual German IR system.

While the best result for English, obtained with German topics, is very good and can be considered as state-of-the-art for a cross-language system running on well-tested languages with reliable processing tools and resources such as English and German, the results for the other two target collections are fairly disappointing.

3.4 Approaches and Discussion

In the TEL experiments, all the traditional approaches to monolingual and cross-language retrieval were attempted by the different groups. Retrieval algorithms included language models, vector-space and probabilistic approaches, and translation resources ranged from bilingual dictionaries, parallel and comparable corpora, to on-line MT systems. Groups often used a combination of more than one resource.

One of the most interesting and new features of the TEL task was the multilinguality of the collections. Only about half of each collection was in the national language (English, French or German), with virtually all other languages represented by one or more entries in one or another of the collections. However, only a few groups took this into specific consideration trying to devise ways to address this aspect and, somewhat disappointingly, their efforts do not appear to have been particularly rewarded by improved performance.

This is shown by the group from the Technical University of Chemnitz, who had overall the best results in the bilingual tasks (1st for XtoEN; 2nd for XtoFR and DE) although they did not do so well in the monolingual tasks. In their official submissions for the campaign, this group attempted to tackle the multilinguality of the collections in several ways. First, they tried to identify the language of each record in the collections using a language detector. Unfortunately, due to an error, they were unable to use the indices created in this way[10]. Second, in both their monolingual and cross-language experiments they implemented a retrieval algorithm which translated the query into the top 10 (in terms of occurrence) languages and merged these multilingual terms into a single query. They ran experiments weighting the query in different ways on the basis of estimated distribution of language content in the collections. In the monolingual experiments, rather disappointingly, the results showed that their purely monolingual baseline always out performed experiments with query translations and language weights. This finding was confirmed with the bilingual experiments where again the better results were achieved with the baseline configurations. They attributed their good overall results for bilingual to the superiority of the Google online translation service. These experiments are described in their Working Notes submission [23]. In their paper in this volume, they describe a series of post workshop experiments for both mono- and cross-language tasks. Disappointingly, they found that their experiments on generating multilingual queries actually resulted in poorer retrieval effectiveness in all cases [22].

[10] This meant that they had to recreate their indices and perform all official experiments at the very last moment; this may have impacted on their results.

Another group that attempted to tackle the multilinguality of the target collections was Xerox. In their official runs, this group built a single index containing all languages (according to the expected languages which they identified as just English, French and German although, as stated, the collections actually contain documents in other languages as well). This, of course, meant that the queries also had to be issued in all three languages. They built a multilingual probabilistic dictionary and for each target collection gave more weight to the official language of the collection [11]. Although their results for both monolingual and bilingual experiments for the French and German collections were always within the top five; they were not quite so successful with the English collection. In their post-campaign experiments described in this volume, they propose an approach to handling target collections in multiple languages. However, and similarly to the work by the group from Chemnitz, their experiments showed that exploiting information in languages different from the official language of the collection gave no advantage[12].

Most groups actually ignored the multilinguality of the single collections in their experiments. Good examples of this are three veteran CLEF groups, UniNE which had, overall the best monolingual results, JHU which appeared in the top five for all bilingual tasks, and Berkeley which figured in the top five for all experiments except for monolingual German. UniNe appeared to focus on testing different IR models and combination approaches whereas the major interest of JHU was on the most efficient methods for indexing. Berkeley tested a version of the Logistic Regression (LR) algorithm that has been used very successfully in cross-language IR by Berkeley researchers for a number of years together with blind relevance feedback [18],[27], [24].

As has been mentioned, the TEL data is structured data; participants were told that they could use all fields. Some groups attempted to exploit this by weighting the contents of different fields differently. See, for example [25]. The combination used in the experiments of this group is based on repeating the title field three times, the subject field twice and keeping the other document fields unchanged.

To sum up, it appears that the majority of groups took this task as a traditional Ad Hoc retrieval task and applied traditional methods. However, it is far too early to confirm whether this is really the best approach to retrieval on library catalog cards. This task is being repeated in CLEF 2009 and we hope that the results will provide more evidence as to which are the most effective approaches when handling catalog data of this type.

4 Persian@CLEF

This activity was coordinated in collaboration with the Data Base Research Group (DBRG) of Tehran University. It was the first time that CLEF offered a non-European language target collection. Persian is an Indo-European language spoken in Iran, Afghanistan and Tajikistan. It is also known as Farsi.

We chose Persian as our first non-European target language for a number of reasons: its challenging script (a modified version of the Arabic alphabet with elision of short vowels) which is written from right to left; its morphology (extensive use of suffixes and compounding); its political and cultural importance. However, the main influencing factor was the generous offer from DBRG to provide an important newspaper corpus (Hamshahri) as the target collection and to be responsible for the coordination of the activity. This collaboration has proved very fruitful and intellectually stimulating and is being continued in 2009.

4.1 Tasks

The activity was organised as a typical Ad Hoc text retrieval task on newspaper collections. Two tasks were offered: monolingual retrieval; cross-language retrieval: English queries to Persian target. For each topic, participants had to find relevant documents in the collection and submit the results in a ranked list.

4.2 Participants

Eight groups submitted 66 runs for the Persian task: all eight submitted monolingual runs (53 runs out of 66); 3 groups also submitted bilingual runs (13 runs out of 66). Five of the groups were formed of Persian native speakers, mostly from the University of Tehran; they were all first time CLEF participants. The other three groups were CLEF veterans with much experience in the CLEF Ad Hoc track. Table 2(a) provides a breakdown of the number of participants and submitted runs by task.

4.3 Results

Table 5 shows the top five groups for each target collection, ordered by mean average precision. The table reports: the short name of the participating group; the mean average precision achieved by the experiment; the DOI of the experiment; and the performance difference between the first and the last participant.

Table 5. Best entries for the Persian tasks

Track	Rank	Participant	Experiment DOI	MAP
Monolingual	1st	unine	10.2415/AH-PERSIAN-MONO-FA-CLEF2008.UNINE.UNINEPE2	48.98%
	2nd	jhu-apl	10.2415/AH-PERSIAN-MONO-FA-CLEF2008.JHU-APL.JHUFASK41R400	45.19%
	3rd	opentext	10.2415/AH-PERSIAN-MONO-FA-CLEF2008.OPENTEXT.OTFA08T	42.08%
	4th	tehran-nlpdb2	10.2415/AH-PERSIAN-MONO-FA-CLEF2008.TEHRAN-NLPDB2.UTNLPDB3INEXPC2	28.83%
	5th	tehran-nlpdb	10.2415/AH-PERSIAN-MONO-FA-CLEF2008.TEHRAN-NLPDB.UTNLPDB1MT	28.14%
	Difference			74.05%
Bilingual	1st	jhu-apl	10.2415/AH-PERSIAN-BILI-X2FA-CLEF2008.JHU-APL.JHUENFASK41R400	45.19%
	2nd	tehran-nlpdb	10.2415/AH-PERSIAN-BILI-X2FA-CLEF2008.TEHRAN-NLPDB.UTNLPDB1BT4G	14.45%
	3rd	tehran-sec	10.2415/AH-PERSIAN-BILI-X2FA-CLEF2008.TEHRAN-SEC.CLDTDR	12.88%
	4th	–	–	–
	5th	–	–	–
	Difference			250.85%

As stated above, a common method for bilingual retrieval evaluation is to compare results against monolingual baselines. We have the following results for CLEF 2008:

– X → FA: 92.26% of best monolingual Farsi IR system.

This appears to be in line with state-of-the-art performance for cross-language systems.

4.4 Approaches

As was to be expected a common theme in a number of the papers was the most effective way to handle the Persian morphology. The group with the best results in the monolingual task tested three approaches; no stemming, a light stemmer developed in-house, and a 4-gram indexing approach. Their best performance was achieved using their light stemmer which has been made freely available on their website. However, they commented that the loss in performance with the no stemming approach was not very great. This group also tested three probabilistic models: Okapi, DFR and statistical language model (LM). The best results were obtained with the latter two [18]. The participant with the second best results compared several different forms of textual normalization: character n-grams, n-gram stems, ordinary words, words automatically segmented into morphemes, and a novel form of n-gram indexing based on n-grams with character skips. He found that that character 4-grams performed the best [27]. This participant also performed some interesting post-workshop experiments on previous CLEF Ad Hoc test collections in 13 languages comparing the results. The findings of [18] were confirmed by [34] in his Working Notes paper. This participant also tested runs with no stemming, with the UniNE stemmer and with n-grams. Similarly, he reported that stemming had relatively little impact.

Somewhat surprisingly, most of the papers from Iran-based groups do not provide much information on morphological analysis or stemming in their papers. One mentions the application of a light Porter-like stemmer but reported that the algorithm adopted was too simple and results did not improve [5]. Only one of these groups provides some detailed discussion of the impact of stemming. This group used a simple stemmer (PERSTEM[11]) and reported that in most cases stemming did improve performance but noted that this was in contrast with experiments conducted by other groups at the University of Tehran on the same collection. They suggest that further experiments with different types of stemmers and stemming techniques are required in order to clarify the role of stemming in Persian text processing [21]. Two of the Persian groups also decided to annotate the corpus with part-of-speech tags in order to evaluate the impact of such information on the performance of the retrieval algorithms [20],[21]. The results reported do not appear to show any great boost in performance.

Other experiments by the groups from Iran included an investigation into the effect of fusion of different retrieval technique. Two approaches were tested:

[11] http://sourceforge.net/projects/perstem

combining the results of nine distinct retrieval methods; combining the results of the same method but with different types of tokens. The second strategy applied a vector space model and ran it with three different types of tokens namely 4-grams, stemmed single terms and unstemmed single terms. This approach gave better results [1].

For the cross-language task, the English topics were translated into Persian. As remarked above, the task of the translators was not easy as it was both a cross-language and also a cross-cultural task. The best result - again by a CLEF veteran participant - obtained 92% of the top monolingual performance. This is well in line with state-of-the-art performance for good cross-language retrieval systems. This group used an online machine translation system applied to the queries[12] [27].

The other two submissions for the cross-language task were from Iran-based groups. We have received a report from just one of them [5]. This group applied both query and document translation. For query translation they used a method based on the estimation of translation probabilities. In the document translation part they used the Shiraz machine translation system to translate the documents into English. They then created a Hybrid CLIR system by score-based merging of the two retrieval system results. The best performance was obtained with the hybrid system, confirming the reports of other researchers in previous CLEF campaigns, and elsewhere.

5 Robust – WSD Experiments

The robust task ran for the third time at CLEF 2008. It is an Ad Hoc retrieval task based on data of previous CLEF campaigns. The robust task emphasizes the difficult topics by a non-linear integration of the results of individual topics into one result for a system, using the geometric mean of the average precision for all topics (GMAP) as an additional evaluation measure [32,35]. Given the difficulty of the task, training data including topics and relevance assessments was provided for the participants to tune their systems to the collection.

This year the robust task also incorporated word sense disambiguation information provided by the organizers to the participants. The task follows the 2007 joint SemEval-CLEF task [3], and has the aim of exploring the contribution of word sense disambiguation to monolingual and cross-language information retrieval. Note that a similar exercise was also run in the question answering track at CLEF 2008. The goal of the task is to test whether WSD can be used beneficially for retrieval systems, and thus participants were required to submit at least one baseline run without WSD and one run using the WSD annotations. Participants could also submit four further baseline runs without WSD and four runs using WSD.

The experiment involved both monolingual (topics and documents in English) and bilingual experiments (topics in Spanish and documents in English). In

[12] http://www.parstranslator.net/eng/translate.htm

addition to the original documents and topics, the organizers of the task provided both documents and topics which had been automatically tagged with word senses from WordNet version 1.6 using two state-of-the-art word sense disambiguation systems, UBC [2] and NUS [9]. These systems provided weighted word sense tags for each of the nouns, verbs, adjectives and adverbs that they could disambiguate.

In addition, the participants could use publicly available data from the English and Spanish wordnets in order to test different expansion strategies. Note that given the tight alignment of the Spanish and English wordnets, the wordnets could also be used to translate directly from one sense to another, and perform expansion to terms in another language.

5.1 Participants

Eight groups submitted 63 runs for the Robust tasks: all groups submitted monolingual runs (45 runs out of 63); 4 groups also submitted bilingual runs (18 runs out of 63). Moreover, 7 groups participated in the WSD tasks, submitting 40 out of 63 runs, 30 monolingual and 10 bilingual. Table 2(a) provides a breakdown of the number of participants and submitted runs by task. Two further groups were late, so they are not included in the official results but they do have papers in this volume [26], [28].

5.2 Results

Monolingual Results. Table 6 shows the best results for this task. The performance difference between the best and the last (up to 5) placed group is given (in terms of average precision).

Bilingual Results. Table 7 shows the best results for this task. The performance difference between the best and the last (up to 5) placed group is given (in terms of average precision). All the experiments were from English to French.

Table 6. Best entries for the robust monolingual task

Track	Rank	Participant	Experiment DOI	MAP	GMAP
English	1st	unine	10.2415/AH-ROBUST-MONO-EN-TEST-CLEF2008.UNINE.UNINEROBUST4	45.14%	21.17%
	2nd	geneva	10.2415/AH-ROBUST-MONO-EN-TEST-CLEF2008.GENEVA.ISILEMTDN	39.17%	16.53%
	3rd	ucm	10.2415/AH-ROBUST-MONO-EN-TEST-CLEF2008.UCM.BM25_BO1	38.34%	15.28%
	4th	ixa	10.2415/AH-ROBUST-MONO-EN-TEST-CLEF2008.IXA.EN2ENNOWSDPSREL	38.10%	15.72%
	5th	ufrgs	10.2415/AH-ROBUST-MONO-EN-TEST-CLEF2008.UFRGS.UFRGS_R_MONO2_TEST	33.94%	13.96%
	Difference			33.03%	51.64%
English WSD	1st	unine	10.2415/AH-ROBUST-WSD-MONO-EN-TEST-CLEF2008.UNINE.UNINEROBUST6	44.98%	21.54%
	2nd	ucm	10.2415/AH-ROBUST-WSD-MONO-EN-TEST-CLEF2008.UCM.BM25_BO1_CLAUSES_09	39.57%	16.17%
	3rd	ixa	10.2415/AH-ROBUST-WSD-MONO-EN-TEST-CLEF2008.IXA.EN2ENUBCDOCSPSREL	38.99%	15.52%
	4th	geneva	10.2415/AH-ROBUST-WSD-MONO-EN-TEST-CLEF2008.GENEVA.ISINUSLWTDN	38.13%	16.25%
	5th	ufrgs	10.2415/AH-ROBUST-WSD-MONO-EN-TEST-CLEF2008.UFRGS.UFRGS_R_MONO_WSD5_TEST	34.64%	14.17%
	Difference			29.84%	52.01%

Table 7. Best entries for the robust bilingual task

Track	Rank	Participant	Experiment DOI	MAP	GMAP
English	1st	ufrgs	10.2415/AH-ROBUST-BILI-X2EN-TEST-CLEF2008.UFRGS.UFRGS_R_BI3_TEST	36.38%	13.00%
	2nd	geneva	10.2415/AH-ROBUST-BILI-X2EN-TEST-CLEF2008.GENEVA.ISIESENTD	30.36%	10.96%
	3rd	ixa	10.2415/AH-ROBUST-BILI-X2EN-TEST-CLEF2008.IXA.ES2ENNOWSDPSREL	19.57%	1.62%
	4th	uniba	10.2415/AH-ROBUST-BILI-X2EN-TEST-CLEF2008.UNIBA.CROSS1TDNUS2F	2.56%	0.04%
	5th	–	–	–	–
	Difference			1.321.09%	32.400.00%
English WSD	1st	ixa	10.2415/AH-ROBUST-WSD-BILI-X2EN-TEST-CLEF2008.IXA.ES2EN1STTOPSUBCDOCSPSREL	23.56%	1.71%
	2nd	ufrgs	10.2415/AH-ROBUST-WSD-BILI-X2EN-TEST-CLEF2008.UFRGS.UFRGS_R_BI_WSD1_TEST	21.77%	5.14%
	3rd	geneva	10.2415/AH-ROBUST-WSD-BILI-X2EN-TEST-CLEF2008.GENEVA.ISIESPWSDTDN	9.70%	0.37%
	4th	geneva	10.2415/AH-ROBUST-WSD-BILI-X2EN-TEST-CLEF2008.UNIBA.CROSSWSD12NUS2F	7.23%	0.16%
	5th	–	–	–	–
	Difference			225.86%	3.112.50%

Evaluating the bilingual retrieval evaluation, we have the following results for CLEF 2008:

– X $\rightarrow$ EN: 80.59% of best monolingual English IR system (MAP);
– X $\rightarrow$ EN WSD: 52.38% of best monolingual English IR system (MAP).

5.3 Analysis

In this section we focus on the comparison between WSD and non-WSD runs. Overall, the best GMAP result in the monolingual system was for a run using WSD, but the best MAP was obtained for a non-WSD run. Several other participants were able to obtain their best MAP and GMAP scores using WSD information. In the bilingual experiments, the best results in MAP and GMAP were for non-WSD runs, but several participants were able to profit from the WSD annotations.

In the monolingual experiments, cf. Table 6, the best results overall in both MAP and GMAP were for unine. Their WSD runs scored very similar to the non-WSD runs, with a slight decrease of MAP (0.16 percentage points) and a slight increase of GMAP (0.27 percentage points) [17]. The second best MAP scoring team attained MAP and GMAP improvements using WSD (from 38.34 MAP – 15.28 GMAP in their best non-WSD run to 39.57 MAP – 16.18 GMAP in their best WSD run) [31]. The third best scoring team in MAP achieved lower scores on both MAP and GMAP using WSD information [19]. The fourth best team obtained better MAP results using WSD information (from 38.10 to 38.99 MAP), but lower GMAP (from 15.72 to 15.52) [29]. Regarding the rest of participants, while ufrgs and uniba obtained improvements, know-center did not, and inaoe only submitted non-WSD runs. Two additional groups (IRn and sinai) sent their results late. Both groups had their best scores for non-WSD systems. You will find more details in the relevant papers in this volume.

In the bilingual experiments, cf. Table 7, the best results overall in both MAP and GMAP were for a system which did not use WSD annotations (36.39, compared to 21.77 MAP for their best run using WSD) [13]. The second scoring team also failed to profit from WSD annotations (30.36 compared to 9.70 MAP) [19].

The other two participating groups did obtain improvements, with ixa attaining 23.56 MAP with WSD (compared to 19.57 without) [29] and uniba attaining (7.23 MAP) [6].

All in all, the exercise showed that some teams did improve results using WSD annotations (up to approx. 1 MAP point in monolingual and approx. 4 MAP points in bilingual), providing the best GMAP results for the monolingual exercise, but the best results for the bilingual were for systems which did not use WSD (with a gap of approx. 13 MAP points). In any case, further case-by-case analysis of the actual systems and runs will be needed in order to get more insight about the contribution of WSD.

6 Conclusions

The Ad Hoc task in CLEF 2008 was almost completely renovated with new collections and new tasks. It focused on three different issues:

- real scenario: document retrieval from multilingual and sparse catalogue records to meet actual user needs (TEL@CLEF)
- linguistic resources: "exotic languages" to favour the creation of new experimental collections and the growth of regional IR communities (Persian@CLEF)
- advanced language processing: assessing whether word sense disambiguation can improve system performances (Robust WSD)

For all three tasks, we were very happy with the number of participants. However, overall, the results have been fairly inconclusive.

From the results of the TEL task, it would appear that there is no need for systems to apply any dedicated processing to handle the specificity of these collections (very sparse, essentially multilingual data) and that traditional IR and CLIR approaches can perform well with no extra boosting. However, we feel that it is too early to make such assumptions; many more experiments are needed.

The Persian task continued in the tradition of the CLEF Ad Hoc retrieval tasks on newspaper collections. The first results seem to confirm that the traditional IR/CLIR approaches port well to "new" languages - where by "new" we intend languages which have not been subjected to a lot of testing and experimental IR studies previously.

The robust exercise had, for the first time, the additional goal of measuring to what extent IR systems could profit from automatic word sense disambiguation information. The conclusions are mixed: while some top scoring groups did manage to improve the results using WSD information by approx. 1 MAP percentage point (approx. 4 MAP percentage points in the cross-language exercise) and the best monolingual GMAP score was for a WSD run (0.27 percentage points), the best scores for the rest came from systems which did not use WSD information. Given the relatively short time that the participants had to try effective ways of using the word sense information we think that these results are fairly positive.

However, in our opinion, a further evaluation exercise is needed for participants to further develop their systems.

All three tasks are being run again in CLEF 2009 both in order to provide participants with another chance to test their systems after refinement and tuning on the basis of the CLEF 2008 experiments and also to be able to create useful and consolidated test collections.

Acknowledgements

The TEL task was studied in order to provide useful input to The European Library (TEL); we express our gratitude in particular to Jill Cousins, Programme Director, and Sjoerd Siebinga, Technical Developer of TEL. Vivien Petras, GESIS Leibniz-Institut for the Social Sciences, Bonn, Germany, and Nicolas Moreau, Evaluation and Language Resources Distribution Agency, France, were responsible for the creation of the topics and the supervision of the relevance assessment work for the ONB and BNF data respectively. We thank them for their valuable assistance.

We should also like to acknowledge the enormous contribution to the coordination of the Persian task made by the Data Base Research group of the University of Tehran and in particular to Abolfazl AleAhmad and Hadi Amiri. They were responsible for the preparation of the set of topics for the Hamshahri collection in Farsi and English and for the subsequent relevance assessments.

The robust task was partially funded by the Ministry of Education (project KNOW TIN2006-15049) and the European Commission (project KYOTO ICT-2007-211423). We would like to acknowledge the invaluable suggestions and practical assistance of Arantxa Otegi, German Rigau and Piek Vossen. In particular, Arantxa was responsible for the XML coding of the word sense disambiguation results. We also want to thank Oier Lopez de Lacalle, who runs the UBC WSD system, and Yee Seng Chan, Hwee Tou Ng and Zhi Zhong, who run the NUS WSD system. Their generous contribution was invaluable to run this exercise.

References

1. Aghazade, Z., Dehghani, N., Farzinvash, L., Rahimi, R., AleAhmad, A., Amiri, H., Oroumchian, F.: Fusion of Retrieval Models at CLEF 2008 Ad-Hoc Persian Track. In: Peters, C., et al. (eds.) CLEF 2008. LNCS, vol. 5706, pp. 97–104. Springer, Heidelberg (2009)
2. Agirre, E., Lopez de Lacalle, O.: UBC-ALM: Combining k-NN with SVD for WSD. In: Proceedings of the 4th International Workshop on Semantic Evaluations (SemEval 2007), Prague, Czech Republic, pp. 341–345 (2007)
3. Agirre, E., Magnini, B., Lopez de Lacalle, O., Otegi, A., Rigau, G., Vossen, P.: SemEval-2007 Task 01: Evaluating WSD on Cross-Language Information Retrieval. In: Peters, C., Jijkoun, V., Mandl, T., Müller, H., Oard, D.W., Peñas, A., Petras, V., Santos, D. (eds.) CLEF 2007. LNCS, vol. 5152, pp. 908–917. Springer, Heidelberg (2008)
4. Agirre, E., Di Nunzio, G.M., Ferro, N., Peters, C., Mandl, T.: CLEF 2008: Ad Hoc Track Overview. In: Borri, F., Nardi, A., Peters, C. (eds.) Working Notes for the CLEF 2009 Workshop (2009), http://www.clef-campaign.org/

5. AleAhmad, A., Kamalloo, E., Zareh, A., Rahgozar, M., Oroumchian, F.: Cross Language Experiments at Persian@CLEF 2008. In: Peters, C., et al. (eds.) CLEF 2008. LNCS, vol. 5706, pp. 105–112. Springer, Heidelberg (2009)
6. Caputo, A., Basile, P., Semeraro, G.: SENSE: SEmantic N-levels Search Engine at CLEF 2008 Ad Hoc Robust-WSD Track. In: Peters, C., et al. (eds.) CLEF 2008. LNCS, vol. 5706, pp. 126–133. Springer, Heidelberg (2009)
7. Braschler, M.: CLEF 2003 - Overview of results. In: Peters, C., Gonzalo, J., Braschler, M., Kluck, M. (eds.) CLEF 2003. LNCS, vol. 3237, pp. 44–63. Springer, Heidelberg (2004)
8. Braschler, M., Peters, C.: CLEF 2003 Methodology and Metrics. In: Peters, C., Gonzalo, J., Braschler, M., Kluck, M. (eds.) CLEF 2003. LNCS, vol. 3237, pp. 7–20. Springer, Heidelberg (2004)
9. Chan, Y.S., Ng, H.T., Zhong, Z.: NUS-PT: Exploiting Parallel Texts for Word Sense Disambiguation in the English All-Words Tasks. In: Proceedings of the 4th International Workshop on Semantic Evaluations (SemEval 2007), Prague, Czech Republic, pp. 253–256 (2007)
10. Cleverdon, C.: The Cranfield Tests on Index Language Devices. In: Sparck Jones, K., Willett, P. (eds.) Readings in Information Retrieval, pp. 47–59. Morgan Kaufmann Publisher, Inc., San Francisco (1997)
11. Clinchant, S., Renders, J.-M.: XRCE's Participation in CLEF 2008 Ad-Hoc Track. In: Borri, F., Nardi, A., Peters, C. (eds.) Working Notes for the CLEF 2008 Workshop, http://www.clef-campaign.org/
12. Clinchant, S., Renders, J.-M.: Multi-language Models and Meta-dictionary Adaptation for Accessing Multilingual Digital Libraries. In: Peters, C., et al. (eds.) CLEF 2008. LNCS, vol. 5706, pp. 83–88. Springer, Heidelberg (2009)
13. Costa Acosta, O., Geraldo, A.P., Orengo, V.M., Villavicencio, A.: UFRGS@CLEF 2008: Indexing Multiword Expressions for Information Retrieval. In: Borri, F., Nardi, A., Peters, C. (eds.) Working Notes for the CLEF 2008 Workshop (2008), http://www.clef-campaign.org/
14. Di Nunzio, G.M., Ferro, N.: Appendix A: Results of the TEL@CLEF Task. In: Borri, F., Nardi, A., Peters, C. (eds.) Working Notes for the CLEF 2008 Workshop (2008), http://www.clef-campaign.org/
15. Di Nunzio, G.M., Ferro, N.: Appendix B: Results of the Persian Task. In: Borri, F., Nardi, A., Peters, C. (eds.) Working Notes for the CLEF 2008 Workshop (2008), http://www.clef-campaign.org/
16. Di Nunzio, G.M., Ferro, N.: Appendix C: Results of the Robust Task. In: Borri, F., Nardi, A., Peters, C. (eds.) Working Notes for the CLEF 2008 Workshop (2008), http://www.clef-campaign.org/
17. Dolamic, L., Fautsch, C., Savoy, J.: UniNE at CLEF 2008: TEL, Persian and Robust IR. In: Borri, F., Nardi, A., Peters, C. (eds.) Working Notes for the CLEF 2008 Workshop (2008), http://www.clef-campaign.org/
18. Dolamic, L., Fautsch, C., Savoy, J.: UniNE at CLEF 2008: TEL and Persian IR. In: Peters, C., et al. (eds.) CLEF 2008. LNCS, vol. 5706, pp. 178–185. Springer, Heidelberg (2009)
19. Guyot, J., Falquet, G., Radhouani, S., Benzineb, K.: Analysis of Word Sense Disambiguation-Based Information Retrieval. In: Peters, C., et al. (eds.) CLEF 2008. LNCS, vol. 5706, pp. 146–154. Springer, Heidelberg (2009)
20. Jadidinejad, A.H., Mohtarami, M., Amiri, H.: Investigation on Application of Local Cluster Analysis and Part of Speech Tagging on Persian Text. In: Borri, F., Nardi, A., Peters, C. (eds.) Working Notes for the CLEF 2008 Workshop (2008), http://www.clef-campaign.org/

21. Karimpour, R., Ghorbani, A., Pishdad, A., Mohtarami, M., AleAhmad, A., Amiri, H., Oroumchian, F.: Improving Persian Information Retrieval Systems Using Stemming and Part of Speech Tagging. In: Peters, C., et al. (eds.) CLEF 2008. LNCS, vol. 5706, pp. 89–96. Springer, Heidelberg (2009)
22. Kuersten, J., Wilhelm, T., Eibl, M.: CLEF 2008 Ad-Hoc Track: Comparing and Combining Different IR Approaches. In: Peters, C., et al. (eds.) CLEF 2008. LNCS, vol. 5706, pp. 75–82. Springer, Heidelberg (2009)
23. Kuersten, J., Wilhelm, T., Eibl, M.: CLEF 2008 Ad-Hoc Track: On-line Processing Experiments with Xtrieval. In: Borri, F., Nardi, A., Peters, C. (eds.) Working Notes for the CLEF 2008 Workshop (2008), `http://www.clef-campaign.org/`
24. Larson, R.: Logistic Regression for Metadata: Cheshire takes on Adhoc-TEL. In: Peters, C., et al. (eds.) CLEF 2008. LNCS, vol. 5706, pp. 38–41. Springer, Heidelberg (2009)
25. Machado, J., Martins, B., Borbinha, J.: Experiments on a Multinomial Language Model versus Lucene's off-the-shelf Ranking Scheme and Rochio Query Expansion (TEL@CLEF Monolingual Task). In: Peters, C., et al. (eds.) CLEF 2008. LNCS, vol. 5706, pp. 50–57. Springer, Heidelberg (2009)
26. Martínez-Santiago, F., Perea-Ortega, J.M., García-Cumbreras, M.A.: Evaluating Word Sense Disambiguation Tools for an Information Retrieval task. In: Peters, C., et al. (eds.) CLEF 2008. LNCS, vol. 5706, pp. 113–117. Springer, Heidelberg (2009)
27. McNamee, P.: JHU Ad Hoc Experiments at CLEF 2008. In: Peters, C., et al. (eds.) CLEF 2008. LNCS, vol. 5706, pp. 170–177. Springer, Heidelberg (2009)
28. Navarro, S., Llopis, F., Muñoz, R.: IRn in the CLEF Robust WSD Task 2008. In: Peters, C., et al. (eds.) CLEF 2008. LNCS, vol. 5706, pp. 134–137. Springer, Heidelberg (2009)
29. Otegi, A., Agirre, E., Rigau, G.: IXA at CLEF 2008 Robust-WSD Task: using Word Sense Disambiguation for (Cross Lingual) Information Retrieval. In: Peters, C., et al. (eds.) CLEF 2008. LNCS, vol. 5706, pp. 118–125. Springer, Heidelberg (2009)
30. Paskin, N. (ed.): The DOI Handbook – Edition 4.4.1. International DOI Foundation (IDF) (2006), `http://dx.doi.org/10.1000/186`
31. Pérez-Agüera, J.R., Zaragoza, H.: Query Clauses and Term Independence. In: Peters, C., et al. (eds.) CLEF 2008. LNCS, vol. 5706, pp. 138–145. Springer, Heidelberg (2009)
32. Robertson, S.: On GMAP: and Other Transformations. In: Yu, P.S., Tsotras, V., Fox, E.A., Liu, C.B. (eds.) Proc. 15th International Conference on Information and Knowledge Management (CIKM 2006), pp. 78–83. ACM Press, New York (2006)
33. Tomlinson, S.: Sampling Precision to Depth 10000 at CLEF 2008. In: Peters, C., et al. (eds.) CLEF 2008. LNCS, vol. 5706, pp. 163–169. Springer, Heidelberg (2009)
34. Tomlinson, S.: German, French, English and Persian Retrieval Experiments at CLEF 2008. In: Borri, F., Nardi, A., Peters, C. (eds.) Working Notes for the CLEF 2008 Workshop (2008), `http://www.clef-campaign.org/`
35. Voorhees, E.M.: The TREC Robust Retrieval Track. SIGIR Forum 39, 11–20 (2005)

Logistic Regression for Metadata: Cheshire Takes on Adhoc-TEL

Ray R. Larson

School of Information
University of California, Berkeley, USA
ray@ischool.berkeley.edu

Abstract. In this paper we will briefly describe the approaches taken by the Berkeley Cheshire Group for the Adhoc-TEL 2008 tasks (Mono and Bilingual retrieval). Since the Adhoc-TEL task is new for this year, we took the approach of using methods that have performed fairly well in other tasks. In particular, the approach this year used probabilistic text retrieval based on logistic regression and incorporating blind relevance feedback for all of the runs. All translation for bilingual tasks was performed using the LEC Power Translator PC-based MT system. This approach seems to be a fit good for the limited TEL records, since the overall results show Cheshire runs in the top five submitted runs for all languages and tasks except for Monolingual German.

1 Introduction

The CLEF Adhoc-TEL collections are different from most of the data used for testing in the various CLEF tasks. The three sub-collections – British Library (BL), Biblioteque Nationale de France (BNF), and the Austrian National Library (ONB) – each represent about 1 million bibliographic records from The European Library union catalog (TEL). The records, we can assume, were originally in some version of MARC (Machine Readable Cataloging) before they were converted to a much more simplified bibliographic record based on the Dublin Core metadata schema. Each of the subcollections use somewhat differing encoding of the (assumed) original MARC data, not always including all of the fields that might be useful in retrieval.

Although each the collections were considered to be "mainly" in a particular language (English for BL, French for BNF, and German for ONB), according to the language codes of the records, only about half of each collection was in that main language, with virtually all other languages represented by one or more entries in one or another of the collections. German, French, English, and Spanish records were available in all of collections. Although this overlap of languages presents an interesting multilingual search (and evaluation) problem, it was not addressed in our experiments this year.

This paper concentrates on the retrieval algorithms and evaluation results for Berkeley's official submissions for the Adhoc-TEL 2008 track. All of the runs were automatic without manual intervention in the queries (or translations). We

C. Peters et al. (Eds.): CLEF 2008, LNCS 5706, pp. 38–41, 2009.

submitted six Monolingual runs (two German, two English, and two French) and nine Bilingual runs (each of the three main languages to both of the other main languages (German, English and French). In addition we submitted three runs from Spanish translations of the topics to the three main languages.

This paper first describes the retrieval algorithms used for our submissions, followed by a discussion of the processing used for the runs. We then examine the results obtained for our official runs, and finally present conclusions and future directions for Adhoc-TEL participation.

2 Retrieval Approaches for Adhoc-TEL

The basic form and variables of the *Logistic Regression* (LR) algorithm used for all of our submissions was originally developed by Cooper, et al. [2], along with its original adaptation for Blind relevance feedback developed by Chen [1] with further adaptation for the Cheshire II IR system[3]. The full formal definition of the "TREC2" algorithm for Logistic Regression-based search and Blind relevance feedback is available in the CLEF working notes version of this paper[4]. The required page limits for this paper do not permit a full description here.

The Cheshire II system uses the XML structure of the documents to extract selected portions for indexing and retrieval. Any combination of tags can be used to define the index contents. For our submitted runs for the TEL Adhoc tasks we only used a single index, that contains most of the content-bearing parts of records (titles, notes, subjects, etc.), for all of our submitted runs.

For all indexing we used language-specific stoplists to exclude function words and very common words from the indexing and searching. The German language runs *did not* use decompounding in the indexing and querying processes to generate simple word forms from compounds. The Snowball stemmer was used by Cheshire for language-specific stemming. In our runs the language-specific stoplists and stemming were limited to the *main language* of the collection. Even though each collection included multiple languages, these were treated as if they were in the main language for the collection.

2.1 Search Processing

Searching the Adhoc-TEL collection using the Cheshire II system involved using TCL scripts to parse the topics and submit the title and description or the title alone from the topics. For monolingual search tasks we used the topics in the appropriate language (English, German, and French), for bilingual tasks the topics were translated from the source language to the target language using the LEC Power Translator PC-based machine translation system.

The scripts for each run submitted the topic elements as they appeared in the topic to the system for TREC2 logistic regression searching with blind feedback. When both the "title" and "description" topic elements were used, they were combined into a single probabilistic query. Table 1 shows which elements were used in the "Type" column, T for title only and TD for title and description.

3 Results for Submitted Runs

The summary results (as Mean Average Precision) for the submitted bilingual and monolingual runs for English German and French are shown in Table 1, the Recall-Precision curves for these runs are also shown in Figure 1 (left side for monolingual and right side for bilingual). In Figure 1 the names for the individual runs represent the language codes, which can easily be compared with full names and descriptions in Table 1 (since each language combination has only a single run).

Table 1 indicates runs that had the highest overall MAP for the task by asterisks next to the run name.

Obviously the "weak man" in our current implementation remains monolingual German. This may be due to decompounding issues, but the higher results

Table 1. Submitted Adhoc-TEL Runs

Run Name	Description	Type	MAP
M-DE-TD-T2FB	Monolingual German	TD auto	0.1742
M-DE-T-T2FB	Monolingual German	T auto	0.1980 *
M-EN-TD-T2FB	Monolingual English	TD auto	0.3466 *
M-EN-T-T2FB	Monolingual English	T auto	0.2773
M-FR-TD-T2FB	Monolingual French	TD auto	0.2438 *
M-FR-T-T2FB	Monolingual French	T auto	0.1931
B-ENDE-TD-T2FB	Bilingual English⇒German	TD auto	0.1556 *
B-ESDE-TD-T2FB	Bilingual Spanish⇒German	TD auto	0.1165
B-FRDE-TD-T2FB	Bilingual French⇒German	TD auto	0.1291
B-DEEN-TD-T2FB	Bilingual German⇒English	TD auto	0.1847
B-ESEN-TD-T2FB	Bilingual Spanish⇒English	TD auto	0.2694
B-FREN-TD-T2FB	Bilingual French⇒English	TD auto	0.2825 *
B-DEFR-TD-T2FB	Bilingual German⇒French	TD auto	0.1885 *
B-ENFR-TD-T2FB	Bilingual English⇒French	TD auto	0.1749
B-ESFR-TD-T2FB	Bilingual Spanish⇒French	TD auto	0.1767

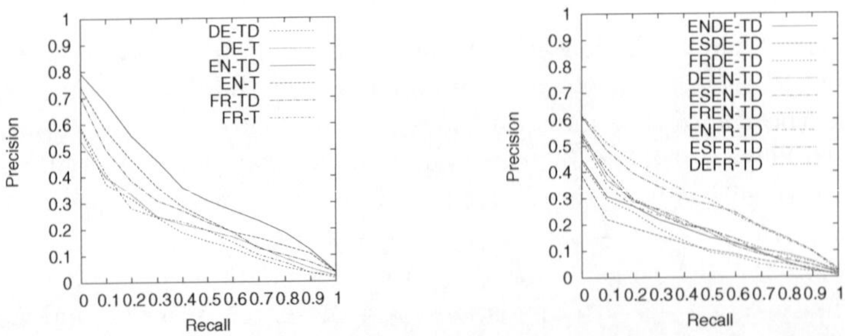

Fig. 1. Berkeley Monolingual Runs (left) and Bilingual Runs (right)

for title-only monolingual seem anomalous, since for each other language, the combination of title and description performed better than title alone.

In spite of this relatively poor performance in monolingual German, we had the rather surprising results that for bilingual English to German our submitted run B-ENDE-TD-T2FB was ranked third overall among the bilingual "to German" runs submitted, and our German to French bilingual run B-DEFR-TD-T2FB was ranked first in the bilingual "to French" task well ahead of our English to French run. This would seem to indicate that the our translation system works quite well with the Adhoc-TEL topics.

4 Additional Analysis and Conclusions

We conducted a small experiment to test whether the fusion approaches used for our GeoCLEF entries (see our GeoCLEF paper in this volume) would improve the performance of our Adhoc-TEL results for Monolingual English. We ran a number of tests using the fusion of Logistic Regression and OKAPI ranking algorithms, and compared them to the submitted results. Unlike our GeoCLEF results, where the same fusion method was used, there was no improvement in MIP for any of the "pivot values" tested. Our best results remain our submitted results for the Monolingual task using Logistic regression with Blind feedback alone.

In looking at the overall results for the various Adhoc-TEL tasks, it would appear that the basic logistic regression with blind relevance feedback approach, coupled with the LEC translation system is a fairly good combination. Since Adhoc-TEL is a new task, we took a fairly conservative approach using methods that have worked well in the past.

In our experiments for other tracks (GeoCLEF for example) we reintroduced fusion approached for retrieval that performed quite well and could be easily applied to this task as well. For future work we intend to test these approaches as well as some other approaches that would incorporate external supplementary topical indexing for the books (primarily) represented by Adhoc-TEL records.

References

1. Chen, A.: Cross-Language Retrieval Experiments at CLEF 2002. LNCS, vol. 2785, pp. 28–48. Springer, Heidelberg (2003)
2. Cooper, W.S., Gey, F.C., Dabney, D.P.: Probabilistic retrieval based on staged logistic regression. In: 15th Annual International ACM SIGIR Conference on Research and Development in Information Retrieval, Copenhagen, Denmark, pp. 198–210. ACM, New York (1992)
3. Larson, R.R.: Probabilistic retrieval, component fusion and blind feedback for XML retrieval. In: Fuhr, N., Lalmas, M., Malik, S., Kazai, G. (eds.) INEX 2005. LNCS, vol. 3977, pp. 225–239. Springer, Heidelberg (2006)
4. Larson, R.R.: Logistic regression for metadata: Cheshire takes on Adhoc-TEL: CLEF working notes (2008), `http://www.clef-campaign.org/2008/working_notes/larson_Adhoc_TEL.pdf`

Query Expansion via Library Classification System

Alessio Bosca and Luca Dini

CELI s.r.l. – Torino - C. Moncalieri, 21
{alessio.bosca,dini}@celi.it

Abstract. Managing the development and delivery of multilingual electronic library services is one of the major current challenges for making digital content in Europe more accessible, usable and exploitable. Digital libraries and OPAC-based traditional libraries are the most important source of reliable information used daily by scholars, researchers, knowledge workers and citizens to carry on their working (and leisure) activities. Facilitating access to multilingual document collections therefore is an important way of supporting the dissemination of knowledge and cultural content. CACAO offers an innovative approach for accessing, understanding and navigating multilingual textual content in digital libraries and OPACs, enabling European users to better exploit the available European electronic content. This paper describes the participation of the CACAO project consortium in the TEL@CLEF 2008 task and proposes a novel approach for exploiting library classification systems as a mean to drive query expansion.

1 Introduction

For more than 10 years there has been an increasing amount of digitized cultural heritage which is in principle freely available worldwide. Especially the digitizing of literature has been boosted in the recent years accompanied by new activities of major commercial enterprises like Google, Microsoft and Yahoo or European projects such as Europeana (see [9]). However most of the digitizing efforts have been undertaken by a few European countries, especially in the richer northern and western parts of Europe. Therefore the majority of the content is today only available within monolingual retrieval systems depending on the primary language of the producers. Multilingual retrieval options will improve the availability of cultural goods for a majority of the European citizens and provide an equal access throughout Europe to these digital resources.

CACAO (Cross-language Access to Catalogues And On-line libraries) is an EU project funded under the eContentplus program that proposes an innovative approach for accessing, understanding and navigating multilingual textual content in digital libraries and OPACs, enabling European users to better exploit the available European electronic content. By coupling sound Natural Language Processing techniques with available information retrieval systems the project aims at the delivery of a non-intrusive infrastructure to be integrated with current OPACs and digital libraries. The result of such integration will be the possibility for the user to type in queries in his/her own language and retrieve volumes and documents in any available language.

C. Peters et al. (Eds.): CLEF 2008, LNCS 5706, pp. 42–49, 2009.

This paper describes the participation of the CACAO consortium in the TEL task of the CLEF 2008 campaign. In addition to an overview of CACAO project, the main scientific contribution consists in a novel approach for query enrichment and expansion by leveraging the native library classification system.

In our participation in the TEL@CLEF task we registered in both the monolingual and the bilingual retrieval tasks as these tasks provided the perfect opportunity to test the baseline version of the CACAO cross language information retrieval system (CLIR) and obtain feedbacks for its enhancement, although the project itself was at an early stage of development. The obtained results show that the proposed solution yields encouraging outcomes although suffers from lack of maturity.

This paper presents the approach proposed by CACAO project for providing cross-lingual access to digital catalogues and it is organized as follows: an overview of the CACAO architecture is presented in Section 2. Section 3 introduces the novel approach proposed for query expansion. Section 4 presents the TEL@CLEF participation and results; Section 5 concludes the paper and proposes some future work.

2 CACAO System

The architecture of the CACAO system is an integration of several subsystems coordinated by a central manager that triggers scheduled activities (i.e. data harvesting or processing) and reacting to external stimuli represented by end users queries. The *Harvesting* subsystem is in charge of collecting data from digital libraries, abstracting from the multiplicity of standards and protocols, and storing them in a repository. The *Corpus Analysis* subsystem performs specific analysis and transformation on the data collected from libraries and infers new information that is then used to support query processing and resource retrieval (e.g. query expansion, terms disambiguation). The *CLIR* subsystem is in charge of analyzing the monolingual user query in input and transforming and enriching it by means of translations and expansions. *Web Services* subsystem represents external modules providing specific services (e.g. linguistic analysis, translations).

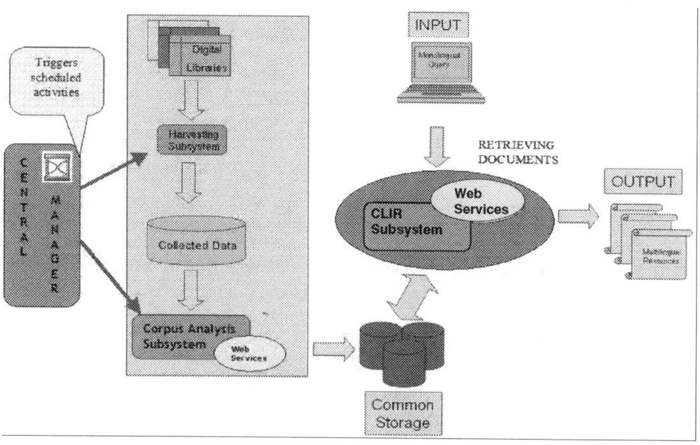

Fig. 1. CACAO architecture

2.1 CACAO CLIR System

With respect to CLEF campaign participation, the subsystem directly involved in the competition is the CLIR engine therefore a short description of its modules is here provided. The CLIR capabilities and results emerge from the interactions of a set of internal and external modules. The internal components consist of "linguistic agnostic" modules that contain the core logic of the system and focus on the management of the information directly harvested from libraries and the data inferred from it (i.e. search indexes, corpus-based semantic vector, association maps of terms with library categories). The external components instead provide CACAO architecture with linguistic analysis capabilities and resources (i.e. lemmatization and named entities recognition, bilingual translations via dictionaries, thesauri) and consist of Web Services. This architectural solution allows for a flexible integration of new resources in order to extend the system support to new languages.

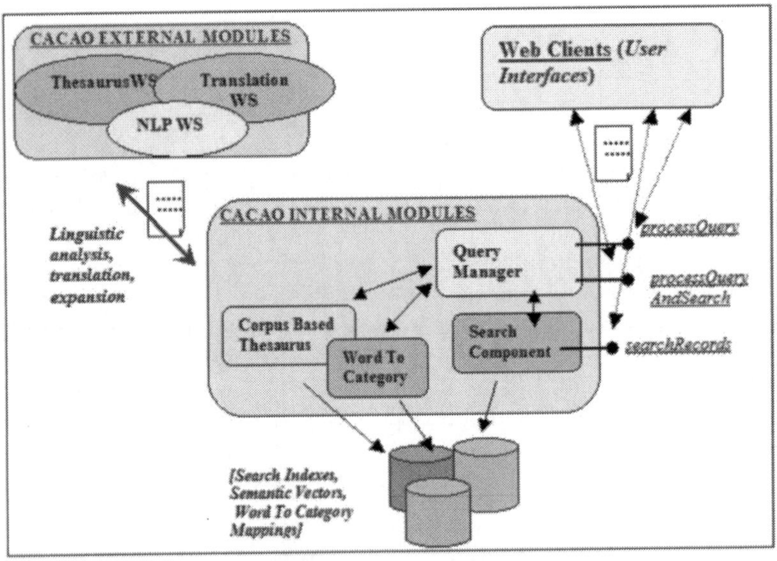

Fig. 2. CLIR system overview

The internal modules comprised in the CLIR system consist of the *Query Manager*, the *Search Component*, the *WordToCategory Component* and the *Corpus-Thesaurus Component*. The external modules provide the CLIR system with the capability of dealing with specific languages and consist of a *Natural Language Processing WS*, a *Translation WS* and a *Thesaurus WS*.

2.1.1 Internal Modules

The *Query Manager* is the top level module providing the entry point for the CLIR services; it receives as input the monolingual request from users and returns an XML document listing the retrieved documents as well as details on the translation and expansion of the search terms. It implements the query management strategies and coordinates the interactions of the other modules.

The *Search Component* module performs the task of retrieving the documents from the collection of records harvested from libraries. This component uses as a search base the index built off-line by the Corpus Analysis subsystem exploiting the Lucene open source engine (see [4]).

The *Corpus-Thesaurus Component* provides the CLIR system with information on the semantic distance between words computed by the *Corpus Analysis* subsystem exploiting the Random Indexing approach (see [2]). This module is used for disambiguation and expansion activities.

Word2Category Component provides a mapping facility between library categories and bag of words and is used as a mean to expand queries. An extended description of this component and the approach it adopts for query expansion is provided in section 3.

2.1.2 External Modules

NLP Web Service is exploited by the *Query Manager* system in order to enhance the query terms with linguistic information, by reducing each word to one syntactically disambiguated stem as well as to identify the so-called named entities (i.e. person names and geographical names).

The goal of *Translation WS* consists in translating the query terms expressed by users in their own language into the different target languages covered by the system.

The *Translation WS* provides data such as synonyms, hypernyms or hyponyms for a given term and it is used for query expansion purposes.

2.1.3 CLIR Query Processing Strategy

The input query is analyzed through the *NLP WS* and lemmatized; in the lemmatization process named entities are also identified, since they should be treated differently with respect to translations and expansions.

The query terms enriched by linguistic information are translated by means of the Translation WS, thus obtaining a list of translation candidates. The candidates are then disambiguated using the corpus based semantic vectors (computed by the *Corpus Analysis* subsystem on the harvested metadata) and according the following approach:

As a first step the system automatically groups the keywords in sets of semantically related terms by means of their semantic distance. This process allows the system to group together all the keywords bearing a common meaning; then the translation candidates of each keyword group are analyzed in order to prune away all the elements with a low similarity to the center of the translation group, computed as the sum of the vector representation of terms (a variation of the algorithm proposed by [5]).

Different expansion strategies can then be applied on the original terms or on the translated ones; CACAO CLIR allows for query expansion mechanisms based on external Thesauri or exploiting the corpus based semantic vectors with the *Corpus-Thesaurus Component* (by adding the N nearest neighbors of each keyword group) or by means of *WordToCategory* module. In the next section a detailed description of the strategy expansion based on the *WordToCategory* component is provided.

3 WordToCategory Query Expansion Strategy

Word2Category is intended as a software resource supporting the activity of query enrichment by exploiting the natural clustering that the digital records have in libraries. The *Corpus Analysis* subsystem in fact, while processing the harvested data, computes a mapping resource that associates words with one or more “librarian” classifications by observing the word distribution across the different classification categories. The approach adopted involves collecting terms from the actual titles of documents that librarians have classified under a given category and selecting from them all the ones that satisfy certain linguistic and statistical relevance requirements; from the titles of documents associated with a given classification category only the words identified by the *NLP WS* module as nouns, adjective or verbs and with a term frequency greater than a lower bound threshold are retained.

The query expansion strategy adopted by *Word2Category* module builds on this mapping resource and aims at identifying the classification categories that are relevant for the user input query and at exploiting such information as an additional search parameter in order to enhance the document retrieval process. The module uses the mapping resource in order to find all the classification categories that are related to any query term and employs the intersection of such categoriy sets as an additional parameter for searching documents.

The Dewey Decimal Classification (DDC) is the most adopted Classification System (CS) in the world (see [6]) and has been chosen as the reference system by CACAO. In the DDC all knowledge is organized into ten main classes. Each main class is subdivided into ten divisions, and each division has ten sections. A notation deeper in the hierarchy identifies a more specific topic. For example 300 identifies the class for "Social Sciences", 330 the division "Economics" and 332 the section "Financial Economics". Further specifications can be added after a point.

As an example, processing with *Word2Category* the query *“Roman Military in Britain”* and exploiting the mapping resource computed from the “British Library” corpus of the TEL track yields in output the following DDC categories:

Table 1. DDC Categories retrieved

<table>
<tr><td>300 - Social sciences</td><td>370 - Education</td><td>909 - World history</td></tr>
<tr><td>306 - Culture & institutions</td><td>320 - Political science</td><td rowspan="2">930 - History of ancient world (to ca. 499)</td></tr>
<tr><td>301 - Sociology & anthropology</td><td>900 - History</td></tr>
</table>

4 TEL@CLEF 2008 Experiments

In order to import the TEL@CLEF collections metadata within the CACAO system and process the proposed topics as standard input queries using the native CLIR system a few slight adaptations were necessary; thus a specific harvesting module and a component devoted to preprocess CLEF topics were integrated in the previously described architecture.

GoNetwork s.r.l. (one of the CACAO partners) deployed a specific harvester module for importing the XML corpus documents. The data thus imported from CLEF collections is stored in an internal repository and then processed by the *Corpus Analysis* subsystem with the standard CACAO procedures; the relevant textual information of each record (in this context the *dc:subject*, and *dc:title* fields of TEL metadata) has been lemmatized using the XIP incremental parser from XEROX (see [1]) and all the data has been then indexed using the Lucene open source engine (see [4]). By means of lexical semantics technologies a corpus based word space model has been created for each of the TEL@CLEF collections. Following the approach described in the previous section the *CorpusAnalysis* subsystem computes a resource that relates words to classification categories.

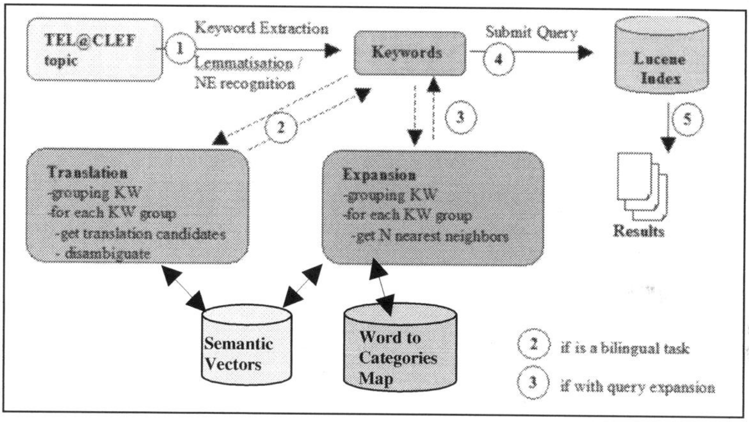

Fig. 3. CLEF Topic Processing

The TEL@CLEF topics are expressed with 2 fields; the first one contains a list of few keywords (*title*) while the second one consist of a sentence better detailing user information needs (*description*). Since the approach adopted by CACAO system for dealing with user queries is based on free keywords, the *description* field of TEL topics has to be pre-processed in order to extract a set of relevant keywords from the sentence, while the *title* field already fits the model. For this purpose a simple keyword extractor module has been exploited for each of the main languages present in the corpus (English, French and German).

Each description sentence has been analyzed in order to extract two different kinds of information, one representing the content type of the items to be retrieved (as novels, poetry or photo collections) and the other conveying additional detail on user interests.

Keywords obtained in this preprocessing phase are translated for the bilingual subtasks and in the experiments involving query expansion the terms are enriched (either in the original or in the target language) by means of *Corpus-Thesaurus* and *Word2Category* components.

4.1 Experimental Results

For every target TEL collection we submitted 2 runs for each monolingual and bilingual subtask, one involving query expansion by means of *Corpus-Thesaurus* and one not (see [8]). After the conclusion of the CLEF campaign, the CACAO consortium obtained from the conference organizers the software evaluation tool in order to perform additional experiments using the *Word2Category* approach for expansion and evaluating them.

The experiments involving the proposed approach have been performed only on the "British Library" collection since it is the only one with a significant presence of Dewey classifications in the metadata. According to the "CLEF 2008: Ad Hoc Track Overview" (see [7]) Dewey Classification (DDC) is not available in the French collection, negligible (~0.3%) in the German collection, but occurs in about half of the English documents (456,408 docs to be exact).

The results of these additional experiments have been included in the following table, along with previous results of CLEF. The table reports the Mean Average Precision (MAP), the R-Precision (the precision after R results, where R is the number of relevant document for the query) and the precision at 5% and at 20% of the retrieved results. Runs ID with _base suffix do not involve query expansion; the ones with prefix _SV perform the expansion exploiting the *Corpus-Thesaurus* component and the one with prefix _w2c exploiting the *Word2Category* approach.

Table 2. Target Collection: British Library

run ID	**MAP**	**R-precision**	**Precision**		**# Relevant docs retrieved**	**Input language**
			@5	**@20**		
CLEF_base	17,27 %	0.212	0.416	0.275	1625 / 2533	En
CLEF_SV	13,3%	0.17	0.336	0.227	1441 / 2533	En
New_base	22.3%	0.253	0.532	0.344	1561 / 2533	En
New_w2c	22.2%	0.255	0.532	0.345	1583 / 2533	En
New_SV	21.3%	0.24	0.5	0.324	1564 / 2533	En

5 Conclusion and Future Work

The strategy for query expansion presented in this paper proposes to exploit the natural clustering that the digital records have in libraries by observing the word distribution across the different classification categories and identifying the classification categories that are relevant for the user input query; in the proposed approach such information is used as an additional search parameter in order to enhance the document retrieval process.

The experimental results show that the introduction of *Word2Category* approach slightly increases the overall number of documents retrieved although it does not enhance the performances in term of precision. In future work we intend to investigate further the approach presented experimenting with the expansion of the retrieved categories by navigating the taxonomy of classifications.

Acknowledgments

This work has been supported and founded by CACAO EU project (ECP 2006 DILI 510035).

References

1. At-Mokhtar, S., Chanod, J.-P., Roux, C.: Robustness beyond shallowness: in-cremental dependency parsing. NLE Journal (2002)
2. Sahlgren, M.: An Introduction to Random Indexing. In: Proceedings of the Methods and Applications of Semantic Indexing Workshop at the 7th International Conference on Terminology and Knowledge Engineering, TKE 2005, Copenhagen, Denmark, August 16 (2005)
3. Ergane. An online multilingual dictionary, `http://download.travlang.com/Ergane/`
4. Lucene. The Lucene search engine, `http://jakarta.apache.org/lucene/`
5. Curtoni, P., Dini, L.: Celi participation at CLEF 2006: Cross language delegated search. In: CLEF 2006 Working notes (2006)
6. Dewey Decimal System, `http://www.oclc.org/dewey/`
7. Agirre, E., Di Nunzio, G.M., Ferro, N., Mandl, T., Peters, C.: CLEF 2008: Ad Hoc Track Overview. In: Working Notes for the CLEF 2008 Workshop (2008)
8. Bosca, A., Dini, L.: CACAO PROJECT AT THE TEL@CLEF 2008 TASK. In: Working Notes for the CLEF 2008 Workshop (2008)
9. Europeana project, `http://dev.europeana.eu/about.php`

Experiments on a Multinomial Language Model versus Lucene's Off-the-Shelf Ranking Scheme and Rocchio Query Expansion (TEL@CLEF Monolingual Task)

Jorge Machado, Bruno Martins, and José Borbinha

Departmento de Engenharia Informática, Technical University of Lisbon, Portugal
{jorge.r.machado,bruno.martins,jose.borbinha}@ist.utl.pt

Abstract. We describe our participation in the TEL@CLEF task of the CLEF 2008 ad-hoc track, where we measured the retrieval performance of the IR service that is currently under development as part of the DIGMAP project. DIGMAP's IR service is mostly based on Lucene, together with extensions for using query expansion and multinomial language modelling. In our runs, we experimented combinations of query expansion, Lucene's off-the-shelf ranking scheme and the ranking scheme based on multinomial language modelling. Results show that query expansion and multinomial language modelling both result in increased performance.

Keywords: Language Model, Vector Space Model, Lucene, Rocchio QE, Stemming.

1 Introduction

One task of the ad-hoc track at the 2008 edition of the Cross Language Evaluation Forum (CLEF) addresses the problem of searching and retrieving relevant items from collections of bibliographic records from The European Library (TEL@CLEF). Three target collections were provided, each corresponding to a monolingual retrieval task where we participated:

- TEL Catalogue records in English. Copyright British Library (BL)
- TEL Catalogue records in French. Copyright Bibliothèque Nationale de France (BnF)
- TEL Catalogue records in German. Copyright Austrian National Library (ONB)

The evaluation task aimed at investigating the best approaches for retrieval from library catalogues, where the information is frequently very sparse and often stored in unexpected languages.

This paper describes the participation of the Technical University of Lisbon at the TEL@CLEF task. Our experiments aimed at measuring the retrieval performance of the IR service that is currently under development as part of DIGMAP[1], an EU-funded project which addresses the development of services for virtual digital libraries of

[1] http://www.dgmap.eu

C. Peters et al. (Eds.): CLEF 2008, LNCS 5706, pp. 50–57, 2009.

materials related to historical cartography [7]. DIGMAP collects bibliographic metadata from European national libraries and other relevant third-party providers (e.g. collections with descriptions available through OAI-PMH), aiming to provide advanced searching and browsing mechanisms that combine thematic, geographic and temporal aspects. In case of success, the ultimate goal of the project is to become fully integrated into The European Library.

The DIGMAP text retrieval service is mostly based on Lucene, together with extensions for using query expansion and multinomial language modeling. A previous version of the system was described in the MSc thesis of Machado [4] and we are now in the process of developing extensions for geo-temporal information retrieval. In CLEF, we experimented combinations of query expansion, Lucene's off-the-shelf ranking scheme and the ranking scheme based on multinomial language modeling.

2 The Experimental Environment

The underlying IR system used in our submissions is based on Lucene[2], together with a multinomial language modeling extension developed at the University of Amsterdam and a query expansion extension developed by Neil Rubens. The following subsections detail these components.

2.1 Lucene's Off-the-Shelf Retrieval Model

We started with Lucene's off-the-shelf retrieval model. For a collection D, document d and query q, the ranking score is given by the formula bellow:

$$ranking(q,d) = \sum_{t \in q} \frac{tf_{t,q} \cdot idf_t}{norm_q} \cdot \frac{tf_{t,d} \cdot idf_t}{norm_d} \cdot coord_{q,d} \cdot weight_t \quad (1)$$

where:

$$\begin{aligned} tf_{t,X} &= \sqrt{termFrequency(t,X)}, \\ idf_t &= 1 + \log \frac{|D|}{documentFrequency(t,D)}, \\ norm_q &= \sqrt{\sum_{t \in q} tf_{t,q} \cdot idf_t^2}, \\ norm_d &= \sqrt{|d|}, \\ coord_{q,d} &= \frac{|q \cap d|}{|q|} \end{aligned} \quad (2)$$

Lucene has been extensively used in previous editions of the CLEF, NTCIR and TREC joint evaluation experiments.

[2] http://lucene.apache.org

2.2 Lucene Extension Based on Multinomial Language Modeling

We experimented with a Lucene extension that implements a retrieval scheme based on estimating a language model (LM) for each document, using the formula described by Hiemstra [2]. This extension was developed at the Informatics Institute of the University of Amsterdam[3]. For any given query, it ranks the documents with respect to the likelihood that the document's LM generated the query:

$$ranking(d,q) = P(d \mid q) \propto P(d) \cdot \prod_{t \in q} P(t \mid d) \tag{3}$$

In the formula, d is a document and t is a term in query q. The probabilities are reduced to rank-equivalent logs of probabilities. To account for data sparseness, the likelihood P(t|d) is interpolated using Jelinek-Mercer smoothing:

$$P(d \mid q) = P(d) \cdot \prod_{t \in q} ((1-\lambda) \cdot P(t \mid D) + \lambda \cdot P(t \mid d)) \tag{4}$$

In the formula, *D* is the collection and λ is a smoothing parameter (in our experiments set to the default value of 0.15). The model needs to estimate three probabilities: the prior probability of the document, *P(d)*; the probability of observing a term in a document, *P(t|d)* and the probability of observing the term in the collection, *P(t|D)*. Assuming the query terms to be independent, and using a linear interpolation of a document model and a collection model to estimate the probability of a query term, the probabilities can be estimated using maximum likelihood estimates:

$$\begin{aligned} P(t \mid d) &= \frac{termFrequency(t,d)}{\mid d \mid} \\ P(t \mid D) &= \frac{documentFrequency(t,D)}{\sum_{t' \in D} documentFrequency(t',D)} \\ P(d) &= \frac{\mid d \mid}{\sum_{d' \in D} \mid d' \mid} \end{aligned} \tag{5}$$

This language modeling approach has been used in past experiments within the CLEF, NTCIR and TREC joint evaluation campaigns – see for example Ahn et. Al [6].

2.3 Rocchio Query Expansion

The fact that there are frequently occurring spelling variations and synonyms for any query term degrades the performance of standard techniques for ad-hoc retrieval. To overcome this problem, we experimented with the method for pseudo feedback query

---

[3] http://ilps.science.uva.nl/Resources/

expansion proposed by Rocchio [3]. The Lucene extension from the LucQE project[4] implements this approach. On test data from the 2004 TREC Robust Retrieval Track, LucQE achieved a MAP score of 0.2433 using Rocchio query expansion.

Assuming that the top D documents returned for an original query qi are relevant, a better query qi+1 can be given by the terms resulting from the formula bellow:

$$q_{i+1} = \alpha \cdot q_i + \frac{\beta}{|D|} \cdot \sum_{d_r \in D} termWeight(d_r) \quad (6)$$

In the formula, α and β are tuning parameters. In our experiments, they were set to the default values of 1.0 and 0.75. The system was allowed to add up to 200 terms extracted from the 10 highest ranked documents (i.e. the |D| parameter) from the original query q_i. The query expansion method was tuned through experiments with the ad-hoc collections and relevance judgments from previous CLEF editions

2.4 Processing the Topics and the Document Collections

Before the actual indexing, the document collections (i.e. the bibliographic records) were passed through the following pre-processing operations:

- **Field Weighting.** The bibliographic records composing the collections from the TEL@CLEF experiment contain structured information in the form of document fields such as *title* or *subject*. We use the scheme proposed by Robertson et. al [5] to weight the different document field according to their importance. Instead of changing the ranking formulas in order to introduce boosting factors, we generate virtual documents in which the content of some specific fields is repeated. The combination used in our experiments is based on repeating the *title* field three times, the *subject* field twice and keeping the other document fields unchanged.
- **Normalization.** The structured documents were converted to unstructured documents for the process of indexing, removing the XML tags and putting the element's contents in separate sentences.

Topic processing was fully automatic and the queries submitted to the IR engine were generated using all parts of the topics (i.e. title, description and narrative). The generation of the actual queries from the query topics was based on the following sequence of processing operations:

- **Parsing and Normalisation.** All characters were reduced to the lowercase unaccented equivalents (i.e. "Ö" reduced to "o" and "É" to "e" etc.) in order to maximise matching.
- **Stop Word Removal.** Stopword lists were used to remove terms that carry little meaning and would otherwise introduce noise. The considered stop words came from the minimized lists distributed with Lucene, containing words such as articles, pronouns, prepositions, conjunctions or interjections. For English, French and German, these lists contained 120, 155 and 231 terms, respectively.

---

[4] http://lucene-qe.sourceforge.net/

- **Retrieval.**The resulting queries were submitted to the IR system, which had been used to index the document collections. In some of the submitted runs, variations of the Porter [1] stemming algorithm specific to the language of the collection were used on both the queries and the documents. The stemming algorithms came from the Snowball package[5].

Lucene internally normalizes documents and queries to lower case, also removing stop-words. However, explicitly introducing these operations when processing the topics, has the advantage of facilitating the development of more advanced topic processing (e.g. adding query expansion methods).

3 The Experimental Story

We submitted 12 official runs to the CLEF evaluation process, a total of 4 runs for each of the languages/collections under consideration in the monolingual task. The conditions under test for each of the submitted runs are as follows:

1. Baseline run using the off-the-shelf retrieval model from Lucene.
2. Lucene with the language modeling extension.
3. Lucene with the language modeling extension and language-specific stemming algorithms.
4. Lucene's off-the-shelf retrieval model with the extension for doing Rocchio query expansion.

We also discuss here the results of some unofficial runs that resulted from experiments that we performed with our retrieval engine. The test conditions for these unofficial runs are:

5. Lucene with the language modeling extension and Rocchio query expansion.
6. Lucene with the language modeling extension, Rocchio query expansion and stemming.
7. Lucene's off-the-shelf retrieval model with Rocchio query expansion and stemming.

4 Results

Table 1 shows the obtained results for the official runs that make up our TEL@CLEF experiments. The results show that, in terms of the mean average precision (MAP), run 3 consistently outperforms our other submissions. The language modeling approach, complemented with the use of stemming, indeed seems beneficial to the retrieval task at study, significantly improving over the baseline run (e.g., a t-test over the MAP results for runs 1 and 3 returns p-values of 0.0041, 0.26 and 0.0001 respectively for the English, French and German collections). Run 4 (i.e., query expansion)

[5] http://snowball.tartarus.org/

Table 1. Results for the official runs submitted to TEL@CLEF

	English				French				German			
	RUN 1	RUN 2	RUN 3	RUN 4	RUN 1	RUN 2	RUN 3	RUN 4	RUN 1	RUN 2	RUN 3	RUN 4
num_q	50	50	50	50	50	50	50	50	50	50	50	50
num_ret	50000	50000	50000	50000	50000	50000	50000	50000	48368	48368	49138	50000
num_rel	2533	2533	2533	2533	1339	1339	1339	1339	1637	1637	1637	1637
num_rel_ret	1858	1884	2056	2060	830	791	1028	891	736	752	943	921
map	0.2976	0.2969	**0.3623**	0.3048	0.2174	0.2020	**0.2341**	0.2190	0.1218	0.1404	**0.2298**	0.1605
gm_ap	0.2015	0.2008	0.2418	0.1939	0.0746	0.0648	0.0941	0.0553	0.0427	0.0534	0.0964	0.0475
R-prec	0.3106	0.3118	**0.3649**	0.3130	0.2463	0.2297	**0.2547**	0.2406	0.1446	0.1606	**0.2432**	0.1838
bpref	0.3126	0.3095	0.3619	0.3415	0.2215	0.2068	0.2315	0.2427	0.1203	0.1374	0.2346	0.1759
recip_rank	0.8263	0.8271	0.8318	0.7936	0.6143	0.5984	0.6309	0.5768	0.5069	0.5950	0.7007	0.5382
ircl_prn.0.00	0.8474	0.8580	0.8580	0.8259	0.6386	0.6224	0.6564	0.6120	0.5368	0.6431	0.7292	0.5764
ircl_prn.0.10	0.6917	0.6470	0.6912	0.6305	0.4804	0.4428	0.4730	0.4800	0.3512	0.3918	0.5392	0.3349
ircl_prn.0.20	0.4997	0.4979	0.5527	0.4829	0.3680	0.3450	0.3520	0.3636	0.2411	0.2562	0.4381	0.2658
ircl_prn.0.30	0.3753	0.3858	0.4537	0.3976	0.3035	0.3010	0.3057	0.2974	0.1505	0.1687	0.3102	0.2268
ircl_prn.0.40	0.3160	0.3166	0.3824	0.3127	0.2236	0.2134	0.2644	0.2318	0.1109	0.1348	0.2417	0.1880
ircl_prn.0.50	0.2654	0.2775	0.3439	0.2611	0.1812	0.1774	0.2265	0.1962	0.0749	0.0861	0.1839	0.1613
ircl_prn.0.60	0.1935	0.2093	0.2870	0.2245	0.1453	0.1331	0.1857	0.1553	0.0581	0.0741	0.1583	0.1251
ircl_prn.0.70	0.1351	0.1448	0.2464	0.1803	0.1089	0.0896	0.1285	0.1107	0.0408	0.0571	0.0879	0.0723
ircl_prn.0.80	0.1106	0.1170	0.1937	0.1362	0.0713	0.0634	0.0978	0.0825	0.0336	0.0354	0.0690	0.0483
ircl_prn.0.90	0.0668	0.0752	0.1153	0.0806	0.0403	0.0456	0.0734	0.0463	0.0154	0.0223	0.0236	0.0227
ircl_prn.1.00	0.0149	0.0177	0.0345	0.0320	0.0099	0.0130	0.0391	0.0124	0.0044	0.0072	0.0041	0.0034
P@5	0.6000	0.5720	0.6160	0.5920	0.3720	0.3560	0.3640	0.3680	0.3040	0.3640	0.4800	0.2960
P@10	0.4840	0.4920	0.5160	0.5020	0.2900	0.2800	0.3020	0.3160	0.2440	0.2680	0.4040	0.2560
P@15	0.4347	0.4293	0.4667	0.4373	0.2520	0.2427	0.2680	0.2600	0.2213	0.2373	0.3547	0.2453
P@20	0.4000	0.3930	0.4250	0.3910	0.2360	0.2270	0.2430	0.2270	0.2020	0.2110	0.3150	0.2260
P@30	0.3500	0.3373	0.3800	0.3333	0.2067	0.2020	0.2147	0.1853	0.1793	0.1847	0.2540	0.1973
P@100	0.2072	0.2124	0.2442	0.2048	0.1102	0.1036	0.1230	0.1064	0.0850	0.0892	0.1204	0.1096
P@200	0.1308	0.1330	0.1559	0.1396	0.0638	0.0626	0.0780	0.0664	0.0496	0.0518	0.0729	0.0686
P@500	0.0663	0.0681	0.0758	0.0728	0.0304	0.0292	0.0374	0.0322	0.0242	0.0246	0.0344	0.0333
P@1000	0.0372	0.0377	0.0411	0.0412	0.0166	0.0158	0.0206	0.0178	0.0147	0.0150	0.0189	0.0184

also consistently outperformed the baseline run with the off-the-shelf Lucene retrieval scheme, although run 2 (i.e. language modeling without stemming) failed to improve over the baseline. Statistical significance tests returned a low confidence for the results when comparing run 1 against run 2 or run 3.

The charts at Figure 1 show precision-recall curves for the official runs, separating the results according to the language (i.e. English, French and German submissions, from left to right).

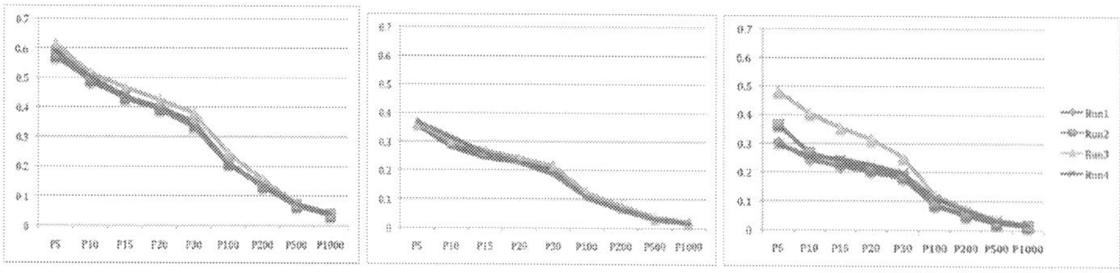

Fig. 1. Precision vs. Recall curves for the official runs submitted to TEL@CLEF

Table 2 shows the results obtained from the unofficial runs that were described in the previous section. The values show that, in terms of MAP, naively combining the language modeling approach with query expansion results in a poor retrieval performance. Results also show that complementing run 4 (Lucene's standard retrieval model, plus Rocchio query expansion) with stemming can be beneficial, particularly in the case of the English collection.

Table 2. Results for the unofficial runs using the TEL@CLEF collections

	English			French			German		
	RUN 5	RUN 6	RUN 7	RUN 5	RUN 6	RUN 7	RUN 5	RUN 6	RUN 7
num_q	50	50	50	50	50	50	50	50	50
num_ret	50000	50000	50000	50000	50000	50000	50000	50000	50000
num_rel	2533	2533	2533	1339	1339	1339	1637	1637	1637
num_rel_ret	1343	1448	2124	583	598	976	734	842	1065
map	**0.1776**	**0.2301**	**0.3527**	**0.1175**	**0.1404**	**0.2258**	**0.1035**	**0.1591**	**0.2437**
gm_ap	0.0942	0.1270	0.2499	0.0263	0.0347	0.0858	0.0255	0.0440	0.0844
R-prec	**0.2241**	**0.2733**	**0.3576**	**0.1519**	**0.1893**	**0.2358**	**0.1395**	**0.1996**	**0.2616**
bpref	0.2923	0.3328	0.3768	0.1768	0.2038	0.2496	0.1862	0.2616	0.2685
recip_rank	0.7107	0.7434	0.8324	0.4470	0.5429	0.5721	0.4676	0.6222	0.6221
ircl_prn.0.00	0.7591	0.7882	0.8717	0.5070	0.5866	0.6158	0.5101	0.6560	0.6751
ircl_prn.0.10	0.4847	0.5838	0.7180	0.3435	0.3859	0.4937	0.3414	0.4677	0.5178
ircl_prn.0.20	0.3039	0.4375	0.5532	0.2449	0.2810	0.3488	0.1773	0.3297	0.4183
ircl_prn.0.30	0.2362	0.3227	0.4487	0.1418	0.1698	0.2882	0.1176	0.2196	0.3491
ircl_prn.0.40	0.1815	0.2441	0.3711	0.1041	0.1274	0.2477	0.0895	0.1521	0.2991
ircl_prn.0.50	0.1363	0.1829	0.3155	0.0873	0.1073	0.2137	0.0720	0.1013	0.2381
ircl_prn.0.60	0.0779	0.1163	0.2596	0.0519	0.0632	0.1681	0.0454	0.0570	0.1901
ircl_prn.0.70	0.0438	0.0735	0.2092	0.0191	0.0326	0.1161	0.0140	0.0198	0.1109
ircl_prn.0.80	0.0220	0.0361	0.1616	0.0063	0.0241	0.0873	0.0053	0.0076	0.0666
ircl_prn.0.90	0.0110	0.0114	0.1048	0.0033	0.0058	0.0498	0.0007	0.0014	0.0265
ircl_prn.1.00	0.0004	0.0017	0.0503	0.0006	0.0014	0.0229	0.0007	0.0014	0.0063
P@5	0.5160	0.5920	0.6600	0.3080	0.3640	0.3880	0.3360	0.4640	0.4640
P@10	0.4220	0.4860	0.5460	0.2420	0.2520	0.3280	0.2520	0.3580	0.4060
P@15	0.3547	0.4107	0.4720	0.1853	0.2067	0.2733	0.2027	0.2787	0.3427
P@20	0.3120	0.3620	0.4360	0.1480	0.1740	0.2490	0.1730	0.2400	0.3110
P@30	0.3500	0.3027	0.3760	0.2067	0.1360	0.2147	0.1793	0.1940	0.2540
P@100	0.2072	0.1572	0.2350	0.1102	0.0648	0.1230	0.0850	0.0904	0.1204
P@200	0.1308	0.1006	0.1548	0.0638	0.0389	0.0780	0.0496	0.0559	0.0729
P@500	0.0663	0.0498	0.0760	0.0304	0.0292	0.0374	0.0242	0.0246	0.0344
P@1000	0.0372	0.0290	0.0425	0.0166	0.0158	0.0206	0.0147	0.0150	0.0189

5 Conclusions

The obtained results support the support the hypotheses that using Rocchio query expansion and a ranking scheme based on language modeling can be beneficial to the CLEF ad-hoc task. Our official runs only made use of relatively simple techniques, but we're now in the process of implementing additional features into our retrieval engine. These include geographic information retrieval extensions with basis on Local Lucene[6] and advanced query expansion methods using bibliographic information.

References

1. Porter, M.F.: An algorithm for suffix stripping. In: Sparck Jones, K., Willett, P. (eds.) Readings in Information Retrieval, pp. 313–316. Morgan Kaufmann, San Francisco (1997) (1980)
2. Hiemstra, D.: Using Language Models for Information Retrieval: Ph.D. Thesis, Centre for Telematics and Information Technology, University of Twente (2001)
3. Rocchio, J.J.: Relevance Feedback in Information Retrieval. In: The SMART Retrieval System. Experiments in Automatic Document Processing, pp. 313–323. Prentice Hall, Englewood Cliffs (1971)

[6] http://sourceforge.net/projects/locallucene

4. Machado, J.: Mitra: A Metadata Aware Web Search Engine for Digital Libraries: M.Sc. Thesis, Departamento de Engenharia Informática, Technical University of Lisbon (2008)
5. Robertson, S., Zaragoza, H., Taylor, M.: Simple BM25 extension to multiple weighted fields. In: Proceedings of the Thirteenth ACM international Conference on information and Knowledge Management, CIKM 2004, Washington, D.C., USA, pp. 42–49. ACM, New York (2004)
6. Ahn, D.D., Azzopardi, L., Balog, K., Fissaha, A.S., Jijkoun, V., Kamps, J., Müller, K., de Rijke, M., Sang, E.T.K.: The University of Amsterdam at TREC 2005: Working Notes for the 2005 Text Retrieval Conference (2005)
7. Pedrosa, G., Luzio, J., Manguinhas, H., Martins, B.: DIGMAP: A service for searching and browsing old maps. In: Proceedings of the 8th ACM/IEEE-CS Joint Conference on Digital Libraries, JCDL 2008, Pittsburgh PA, PA, USA, p. 431. ACM, New York (2008)

WikiTranslate: Query Translation for Cross-Lingual Information Retrieval Using Only Wikipedia

Dong Nguyen, Arnold Overwijk, Claudia Hauff, Dolf R.B. Trieschnigg, Djoerd Hiemstra, and Franciska de Jong

University of Twente,
The Netherlands
{dong.p.ng,arnold.overwijk}@gmail.com,
{c.hauff,trieschn,f.m.g.dejong}@ewi.utwente.nl
hiemstra@cs.utwente.nl

Abstract. This paper presents WikiTranslate, a system which performs query translation for cross-lingual information retrieval (CLIR) using only Wikipedia to obtain translations. Queries are mapped to Wikipedia concepts and the corresponding translations of these concepts in the target language are used to create the final query. WikiTranslate is evaluated by searching with topics formulated in Dutch, French and Spanish in an English data collection. The system achieved a performance of 67% compared to the monolingual baseline.

Keywords: Cross-lingual information retrieval, query translation, word sense disambiguation, Wikipedia, comparable corpus.

1 Introduction

This paper introduces WikiTranslate; a system that performs query translation using only Wikipedia as a translation resource. Most Wikipedia articles contain *cross-lingual links*: links to articles about the same concept in a different language. These cross-lingual links can be followed to obtain translations. The aim of this research is to explore the possibilities of Wikipedia for query translation in CLIR.

The main research question of this paper is: *Is Wikipedia a viable alternative to current translation resources in cross-lingual information retrieval?*

We treat Wikipedia articles as representations of concepts (i.e. units of knowledge). WikiTranslate maps the query to Wikipedia concepts. Through the cross-lingual links translations of the concepts in another language are retrieved. This raises the following sub questions: *How can queries be mapped to Wikipedia concepts?* and *How to create a query given the Wikipedia concepts?*

Our method uses the unique structure of Wikipedia, enabling us to investigate new possibilities to perform query translation. Wikipedia has the following advantages compared to the existing resources used to perform query translation (e.g. bilingual dictionaries, parallel corpora etc.):

- Better coverage of named entities and domain specific terms [1], which might make it suitable to handle translations of proper names.
- Continuous contributions of a large community keep the information up-to-date.

C. Peters et al. (Eds.): CLEF 2008, LNCS 5706, pp. 58–65, 2009.

- Wikipedia articles provide more context in comparison with sources like online dictionaries. This can be used to perform word sense disambiguation [2].
- Presence of redirect pages; pages that represent alternative names of concepts (e.g. synonyms, abbreviations and spelling variants [1]) and that consist of a link that directs to the main article it represents. They may be used for query expansion.

However, the coverage of common words in Wikipedia is smaller than translation dictionaries and some terms have many senses, some very specific and uncommon, making word sense disambiguation more difficult. For example in Wikipedia the term *house* has senses like a novel, song, operating system or a game.

The overview of this paper is as follows. First an overview of Wikipedia and related work in the field of CLIR is given. Then WikiTranslate is introduced and the experimental setup is described. Results are then presented and discussed.

2 Related work

Kraaij et al. [3] make an important observation about CLIR. The final query delivered to the system does not have to be a single translation. Including synonyms and related words can in fact improve performance. One approach to accomplish this is with query expansion or using parallel corpora (e.g. [4,5]). In the first step of Sheridan et al. [5], the best matching documents in the source language are retrieved. Next, frequently occurring words in comparable documents in the target language are selected to compose the final query. Lavrenko et al. [4] follows the same approach except that their method creates a relevance model in the target language.

Wikipedia is an online, multilingual encyclopedia to which everyone can contribute. Its characteristics make it suitable as a semantic lexical resource [1]. Wikipedia has been used for automatic word sense disambiguation [6] and for translation. Su et al. [7] use it to translate out of vocabulary words and Schönhofen et al. [8] use it to translate queries. The notion that it can be treated as a comparable corpus is new and has not been researched much yet except by Potthast et al[9]. Wikipedia can be seen as a comparable corpus since articles are represented in different languages and connected through cross-lingual links.

3 Proposed Approach

The approach used by WikiTranslate consists of two important steps: mapping the query in source language to Wikipedia concepts and creating the final query in the target language using these found concepts.

The first step maps the query to Wikipedia concepts. First, the most relevant concepts to the query are extracted after a search with the whole query (step 1a). Next, a search on every term of the query is performed (step 1b) using the internal links from the concepts retrieved with step 1a (called LINKS) or using the text and title of the Wikipedia articles (called CONTENTS).

The second step creates the translation. First, we add articles that redirect to the found Wikipedia concepts to include synonyms and spelling variants (step 2a). Furthermore articles retrieved with step 1a are given more weight (step 2b). Finally, the final query is created using the found concepts (step 2c).

This approach differs from traditional approaches, since we make use of the text and internal links, which are not available for approaches based on dictionaries and parallel corpora. This approach also differs from other approaches using Wikipedia. Su et al. [7] and Schönhofen et al. [8] have only used Wikipedia to enhance their translations. An advantage of our approach is that it allows extraction of phrases from the topics, since the titles of Wikipedia articles are often phrases. Furthermore by adding the top documents from step 1a, the most relevant concepts to the whole query are added. Also related concepts can be added, creating a kind of query expansion effect.

4 Experimental Setup

Lucene is used as the underlying retrieval system to retrieve Wikipedia articles. From each article the title, text and cross-lingual links are extracted. The first paragraph of an article is extracted as well, which is called *description*. Because long articles tend to score lower, instead of searching on the whole text, the search scope can be limited to the first paragraph, since the first paragraph usually contains a summary of the article. If the article is a redirect page, the title of the referred page is also stored. Wikipedia articles that represent images, help pages, templates, portal pages and pages about the use of Wikipedia are excluded. To enhance comparability, the same preprocessing method is used for all languages. We choose stemming, although there is no uniform best way of preprocessing for all languages [10]. Stemming is best for Dutch and Spanish, but 4-gramming is more suitable for English and French [10]. We use Snowball stemmers to perform stemming [11]. Words are removed with the lists from the Snowball algorithm [11].

To illustrate the steps of the proposed approach we translate the following topic (C230 from the Ad hoc task of CLEF 2004) from Dutch to English:

```
<title> Atlantis-Mir Koppeling </title>
<desc> Vind documenten over de eerste space shuttle aankoppeling tussen
de Amerikaanse shuttle Atlantis en het Mir ruimte station. </desc>
```

(English: *Atlantis-Mir Docking, Find documents reporting the first space shuttle docking between the US shuttle Atlantis and the Mir space station*).

4.1 Step 1: Mapping the Query to Wikipedia Concepts

This step is based on [4] and [5] as we also retrieve the best matching documents in the source language and use them to create a new query.

First the original query is put in Lucene, retrieving the most relevant Wikipedia concepts. The concepts can be retrieved by searching on the title, text, description or a combination of these fields. The top documents will be considered as relevant and will be used for translations. With this method word sense disambiguation is performed automatically [8]. We set a minimum score and maximum number of documents to be included to determine which top documents will be included.

Our example finds the concepts "*space shuttle atlantis*" and "*mir (ruimtestation)*" with the following stemmed query:

```
(title:atlantis text:atlantis) (title:mir text:mir) (title:koppel
text:koppel) (title:eerst text:eerst)..(title:station text:station)
```

We also search for every term of the query separately, because with the previous step some terms may not be found. For example, the query *history of literature* yields mostly articles about literature (missing the term *history*). To avoid this problem, every term in the query is searched separately to find Wikipedia concepts. This step is quite similar to the mapping of a query to dictionary entries, but Wikipedia offers new ways of mapping them. Two different methods are used to map concepts to an individual term.

The first method, which we will call LINKS, uses the internal links of relevant concepts found in step 1. The expectation is that these terms are related to the top relevant documents of the first search. Therefore the internal links from the top documents of the first search are extracted. The search on every term is first only performed on these links. If no concepts are found, or the found concepts are hardly relevant (i.e. have a low score), then the search is performed on the whole Wikipedia corpus. It is also possible to go deeper: including the internal links of the internal links from the top documents etc.

The second method (called CONTENTS) searches with the whole query, but gives the searched term more weight. An exact match has precedence over this step. For the term *tussen* (English: *between*) from our example, the following query is used:

```
((+title:tuss)^1.6) (descr:atlantis) .. (descr:ruimt) (descr:station)
```

The following concepts are recognized for our example topic: *America, Atlantis (disambiguation), Coupling, Mir, Mir (disambiguation), Russian Federal Space Agency, Shuttle, Space Shuttle Atlantis, Space Shuttle program,* and *Station.*

4.2 Step 2: Creating the Translated Query

The translation can be expanded by adding the redirect pages referring to the found concepts (adding synonyms etc.). For the concept "*space shuttle atlantis*" the following translations are added: "*atlantis (space shuttle), ov-104, ss atlantis* etc".

The expectation is that the concepts retrieved by step 1a returns the most relevant concepts. Therefore these concepts are given a higher weight than the other concepts For every found concept the translation can be obtained through the cross-lingual links. From every translation, terms like *disambiguation*, *category*, etc. and non-word characters are removed. Translations like *w#y* and *w(y)* are split into *w* and *y*.

There are different possibilities to put the translations together. We can include every found translation as a phrase query (e.g. "x y"), as an OR query of its terms (e.g. x y), or both (e.g. "x y" x y). The final translation of our example topic looks as follows (without step 2a):

```
"station"^1.0 station^1.0 "russian federal space agency"^1.0 ....
space^3.0 shuttle^3.0 atlantis^3.0.. "mir"^3.0 mir^3.0
```

Note that concepts from step 1a (*space shuttle atlantis* and *mir (ruimtestation)*) are given a higher weight (3.0). Other concepts have a standard weight (1.0).

5 Evaluation

WikiTranslate is evaluated on retrieval of English documents using translated Dutch, French and Spanish queries. The system is first evaluated with the data of CLEF 2006, 2005 and 2004. The best performing system is also evaluated with data of CLEF 2008. Note that a different data collection is used in the evaluation of 2008.

Experiments have been carried out using only the title of the topic (T), or using the title and description (T+D) of a topic. Tests are performed with the following systems: No word sense disambiguation (NO_WSD), word sense disambiguation using links (LINKS), word sense disambiguation through text (CONTENT) and word sense disambiguation through text and weighted query terms (CONTENT_W). The basic underlying system uses parameters that are determined experimentally. Furthermore, no query expansion is applied and every translation is added as a phrase query and as an OR-query of its terms. A stop list is used to filter particularly query words (e.g. "*find*", "*documents*", "*describe*", "*discuss*" etc.) from the description.

To compare the results of the different systems, the results are averaged per system and task over every tested language. Table 1 shows the results. For each run the MAP of the bilingual system is compared with the MAP of the monolingual system (%Mono).

Table 1. Summary of runs 2004, 2005 and 2006

Task	ID	% Mono.
T	NO_WSD	72.71%
T	LINKS	71.88%
T	CONTENT	74.89%
T	CONTENT_W	72.70%

Task	ID	% Mono.
T+D	NO_WSD	68.98%
T+D	LINKS	71.44%
T+D	CONTENT	73.18%
T+D	CONTENT_W	74.98%

CONTENT_W with T + D performs best. Averaging the runs with these settings over the years 2004, 2005 and 2006 shows us that Spanish had an average performance of 71.89% and French had an average of 76.78%.

When creating the final query, different options with using phrase queries and OR queries are possible. Including translations only as a phrase query results in a average MAP decrease of 0.0990. Random tests are used to determine the effect of different steps. Including redirects showed an average MAP decrease of 0.118. Filtering non-related words lead to an MAP increase of 0.0926.

The system CONTENT_W (using T+D) has been submitted to the CLEF ad hoc task 2008. The results can be found in table 2.

Table 2. Results run 2008

Language	MAP
English (monolingual)	0.3407
French	0.2278 (66.86%)
Spanish	0.2181 (64.02%)
Dutch	0.2038 (59.82%)

The French run (which contains 50 topics), which had the best performance, is analyzed in more detail. 12 translations performed better than the original topics and 38 performed worse. An overview can be found in figure 1.

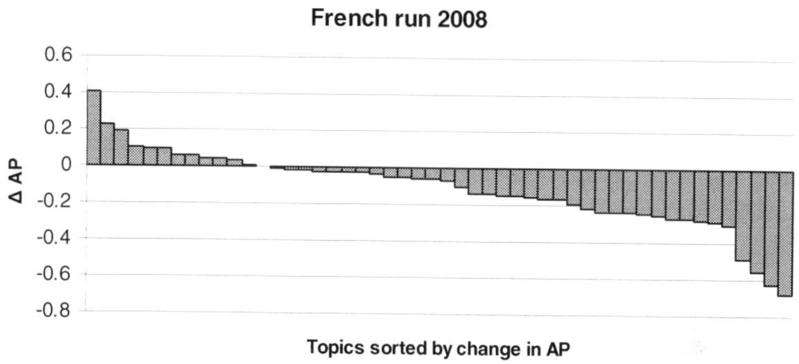

Fig. 1. A comparison of original (French) and translated (English) topics

When analyzing the queries we see that sometimes new, but relevant terms are added with the new translations. For example the translation for topic 477 contains the term "*investment*" which wasn't included in the original English topic about web advertising. Furthermore the translations *internet* and *advertising* were given a higher weight. The translation had an increase of AP from 0.1954 (from 0.0345 to 0.2299).

However the translations of some queries are totally wrong. One of the worst performing is topic 457. The translation showed an AP decrease of 0.2340 (from 0.2626 to 0.0286). When looking at the translation of this topic, we see that the system had difficulties translating the term *fictives* (English: *fictional*). It mapped the concepts "*Planets in science fiction*" and "*Fictional brands*" to this term.

6 Discussion

It is difficult to make a solid comparison with the performances of other systems. First of all since the approach of WikiTranslate is different than other approaches, it is reasonable to have a lower performance than state of the art systems that use well researched methods. We also used a standard information retrieval system (Lucene) and have not paid further attention to this. At the ad hoc task of CLEF 2004, 2005 and 2006 French, Spanish and Dutch are not chosen as a source language, which makes it even harder to compare. However, since the system achieves performances around 70 and 75% of the monolingual baseline, which are manually created queries, these results are very reasonable. The performance of the system with the dataset of 2008 is significantly lower. This might be due to use of a different data collection [12].

Table 1 shows that word sense disambiguation doesn't improve when we only use the title, but it improves if we also use the description. For the task T + D the performance depends on the right stop words lists. Without filtering these words the performance decreases. This can be explained because WikiTranslate retrieves concepts related to these terms, but not related to the query.

Including every found translation only as a phrase query significantly decreases the performance of the system. Query expansion using spelling variants and synonyms also decreases the performance of the system. Because every concept is expanded, wrongly recognized concepts are also expanded, including a lot of non related translations. Furthermore when we manually look at the redirects, some redirects are very global or not very related to the concept.

WikiTranslate performs particularly well with translating proper nouns. Translations that are missed are most of the times adjectives and common words. However, these terms are sometimes crucial (e.g. *longest*). Sometimes translations were missed, because the system wasn't able to find the corresponding concepts due to shortcomings of the used stemmers.

The analysis of one single run showed that some topics performed even better than the original ones. This indicates that this method is very promising.

7 Conclusion and Future Work

In this paper the system WikiTranslate is introduced that performs query translation using only Wikipedia as translation source. WikiTranslate maps queries to Wikipedia concepts and creates the final query through the obtained cross-lingual links. The best approach uses the text and titles of the articles.

We have demonstrated that it is possible to achieve reasonable results using only Wikipedia. We believe that it can be valuable alternative to current translation resources and that the unique structure of Wikipedia can be very useful in CLIR. The use of Wikipedia might also be suitable for Interactive CLIR, where user feedback is used to translate, since Wikipedia concepts are very understandable for people.

Wikipedia allows translating phrases and proper nouns especially well. In addition it is very scalable since the most up to date version of Wikipedia can be used. The coverage of Wikipedia for the languages Dutch, French and Spanish seems to be enough to get reasonable results. The major drawback of Wikipedia is the bad coverage of common words. To cope with missed translations other resources like EuroWordNet [13] or a bilingual dictionary might be incorporated.

We believe that with further research a higher performance can be achieved. The method to map concepts can be refined by also using pages like disambiguation pages, and by filtering concepts which are not very related to the other retrieved concepts (used by [8]). Also the query weighting method can be refined.

It would be also interesting to explore other methods of query expansion using Wikipedia. Internal links that occur often at the retrieved concepts or internal links in the first paragraph of the retrieved concepts could be added. However since query expansion can cause query drift, it might be better to give the added concepts a lower weight. Furthermore we should only expand very relevant and related concepts.

Acknowledgments. This paper is based on research partly funded by IST project MESH (http://www.mesh-ip.eu) and by bsik program MultimediaN (http://www.multimedian.nl).

References

1. Zesch, T., Gurevych, I., Mühlhäuser, M.: Analyzing and Accessing Wikipedia as a Lexical Semantic Resource. In: Data Structures for Linguistic Resources and Applications, pp. 197–205 (2007)
2. Sanderson, M.: Word sense disambiguation and information retrieval. In: Proceedings of the 17th annual international ACM SIGIR conference on Research and development in information retrieval, pp. 142–151. Springer-Verlag New York, Inc., Dublin (1994)
3. Kraaij, W., Nie, J.-Y., Simard, M.: Embedding web-based statistical translation models in cross-language information retrieval. Comput. Linguist. 29, 381–419 (2003)
4. Lavrenko, V., Choquette, M., Croft, W.B.: Cross-lingual relevance models. In: Proceedings of the 25th annual international ACM SIGIR conference on Research and development in information retrieval, Tampere, Finland, pp. 175–182. ACM, New York (2002)
5. Sheridan, P., Ballerini, J.P.: Experiments in multilingual information retrieval using the SPIDER system. In: Proceedings of the 19th annual international ACM SIGIR conference on Research and development in information retrieval, Zurich, Switzerland, pp. 58–65. ACM, New York (1996)
6. Mihalcea, R.: Using Wikipedia for Automatic Word Sense Disambiguation. In: The North American Chapter of the Association for Computational Linguistics (NAACL 2007), Rochester (2007)
7. Su, C.-Y., Lin, T.-C., Shih-Hung, W.: Using Wikipedia to Translate OOV Term on MLIR. In: The 6th NTCIR Workshop, Tokyo (2007)
8. Schönhofen, P., Benczúr, A., Bíró, I., Csalogány, K.: Performing Cross-Language Retrieval with Wikipedia. In: CLEF 2007, Budapest (2007)
9. Potthast, M., Stein, B., Anderka, M.: A Wikipedia-based Multilingual Retrieval Model. In: Macdonald, C., Ounis, I., Plachouras, V., Ruthven, I., White, R.W. (eds.) ECIR 2008. LNCS, vol. 4956, pp. 522–530. Springer, Heidelberg (2008)
10. Hollink, V., Kamps, J., Monz, C., de Rijke, M.: Monolingual Document Retrieval for European Languages. Inf. Retr. 7, 33–52 (2004)
11. Stemming algorithms for use in information retrieval, `http://www.snowball.tartarus.org`
12. Agirre, E., Di Nunzio, G.M., Ferro, N., Mandl, T., Peters, C.: CLEF 2008: Ad Hoc Track Overview. In: Working Notes for the CLEF 2008 Workshop (2008)
13. Vossen, P.: EuroWordNet: a multilingual database for information retrieval. In: Proceedings of the DELOS workshop on Cross-language Information, Zurich, Switzerland (1997)

UFRGS@CLEF2008: Using Association Rules for Cross-Language Information Retrieval

André Pinto Geraldo and Viviane P. Moreira

Instituto de Informática - UFRGS
Porto Alegre - RS - Brazil
{apgeraldo,viviane}@inf.ufrgs.br

Abstract. For UFRGS's participation on the TEL task at CLEF2008, our aim was to assess the validity of using algorithms for mining association rules to find mappings between concepts on a Cross-Language Information Retrieval scenario. Our approach requires a sample of parallel documents to serve as the basis for the generation of the association rules. The results of the experiments show that the performance of our approach is not statistically different from the monolingual baseline in terms of mean average precision. This is an indication that association rules can be effectively used to map concepts between languages. We have also tested a modification to BM25 that aims at increasing the weight of rare terms. The results show that this modified version achieved better performance. The improvements were considered to be statistically significant in terms of MAP on our monolingual runs.

Keywords: association rules, experimentation, performance measurement.

1 Introduction

This paper reports on monolingual and bilingual ad-hoc information retrieval experiments that we have performed for the TEL task at CLEF2008. Our aim was to use algorithms for mining association rules to map concepts between languages, on a Cross-Language Information Retrieval (CLIR) scenario. These algorithms are widely used for data mining purposes. A common example is market-basket data, i.e. the items that a customer buys at one transaction. For such data, an association rule would state, for example, that "90% of customers that purchase bread also purchase milk".

The motivation is that such algorithms are computationally cheaper than other co-occurrence-based techniques such as Latent Semantic Indexing [4]. Our goal was to use automatic methods that did not employ resources such as dictionaries, thesauri or machine translation.

The remainder of this paper is organised as follows: Section 2 proposes an approach for using algorithms for mining association rules for CLIR; Section 3 presents some modifications we implemented on the Okapi BM25 formula to improve retrieval results; Section 5 discusses the experiments and results; and Section 6 presents the conclusions.

C. Peters et al. (Eds.): CLEF 2008, LNCS 5706, pp. 66–74, 2009.

2 Association Rules for CLIR

An association rule (AR) is an implication of the form $X \Rightarrow Y$, where $X = \{x_1, x_2, ..., x_n\}$, and $Y = \{y_1, y_2, ..., y_m\}$ are sets of items. The problem of mining ARs in market-basket data was firstly investigated by Agrawal [2]. In the rule "90% of customers that purchase bread also purchase milk", the antecedent is bread and the consequent is milk. The number 90% is the confidence factor (*conf*) of the rule. The confidence of the rule can be interpreted as the probability that the items in the consequent will be purchased given that the items in the antecedent are purchased. An AR also has a support level associated to it. The support (*sup*) of a rule refers to how frequently the sets of items $X \cup Y$ occur in the database. Eq. 1 shows how support and confidence of an AR are calculated.

$$conf(X \Rightarrow Y) = \frac{n(X \cup Y)}{n(X)} \qquad sup(X \Rightarrow Y) = \frac{n(X \cup Y)}{N} \tag{1}$$

Where n is the number of transactions and N is the total number of transactions in the database.

The problem of mining ARs is to generate all rules that have support and confidence greater than predefined thresholds. We have used the Apriori Algorithm [3] to extract the ARs. The algorithm calculates the support of the individual items and then proceeds by combining the individual items two-by-two, three-by-three and so on. If the support of the itemset is lower than the threshold *minsup*, this itemset is discarded. More formally, let I be an itemset, for each subset $v \subseteq I$ the algorithm will generate a rule of the form $v \Rightarrow (I - v)$ if $sup(I)/sup(v)$ is greater than *minsup*.

Our proposal is to map the problem of finding ARs between items in a market-basket scenario to the problem of finding cross-linguistic equivalents between a pair of languages on a parallel corpus. This approach is based on co-occurrences and works under the assumption that cross-linguistic equivalents would have a significant number of co-occurrences over a parallel corpus. In our approach, the transaction database is replaced by a text collection; the items that the customer buys correspond to the terms in the text; and the shopping transactions are represented by documents.

The proposed approach to use algorithms for mining ARs for CLIR can be divided into five phases depicted in Figure 1. Next we explain each phase.

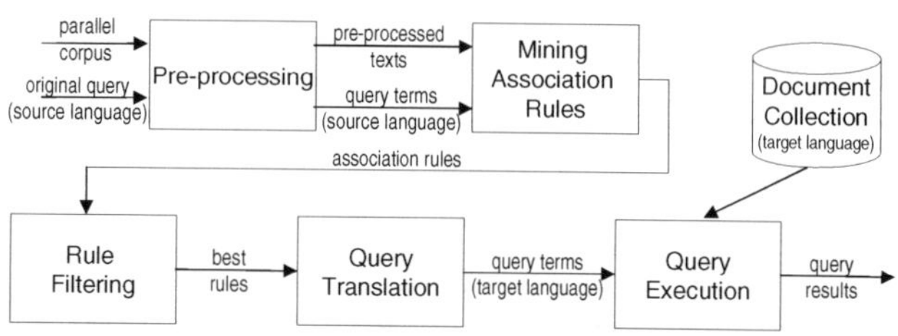

Fig. 1. Proposed approach for using association rules for CLIR

(i) Pre-processing. The inputs for this phase are a collection of parallel documents and the original query in the source language. During this phase the original text and its cross-language equivalent are initially treated separately. We remove stop-words, apply stemming, break the documents into sentences, and tag all terms in one of the languages with a prefix (e.g. all English words are tagged with an "E="). The aim is to avoid generating rules between words in the same language. The last step is to merge each sentence with its translation. The output of this phase is a set of pre-processed parallel sentences. During this phase, an inverted index containing all stems in the document collection and the list of sentences in which they appear is also built. The index will be used in the next phase to enable selection of the sentences over which the Apriori algorithm is run. The pre-processing phase is shown in Figure 2.

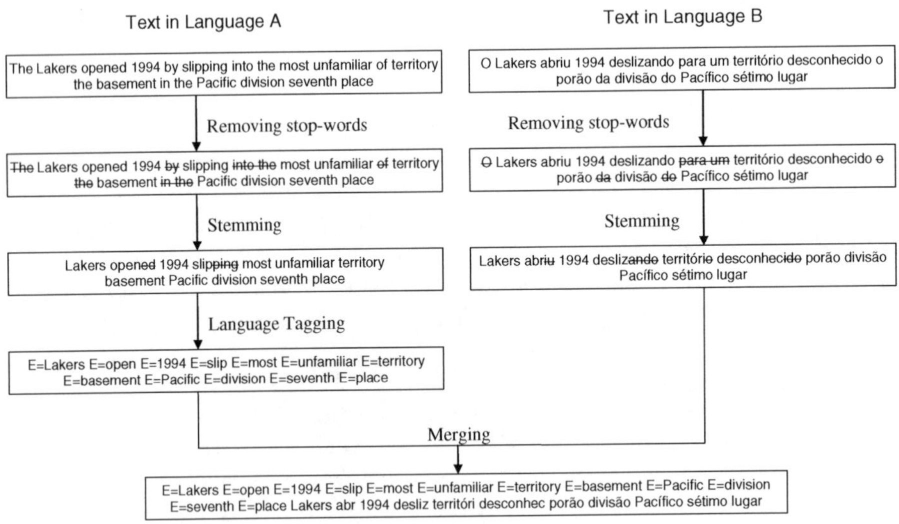

Fig. 2. Steps in the pre-processing phase

(ii) Mining ARs. This step consists in generating ARs for the terms in the query. We run the Apriori algorithm over the pre-processed parallel sentences. In order to speed up rule generation, only sentences that contain the query terms are considered. As a result, the support for all rules will be 100%, which means that we can no longer use this metric as an indication of rule usefulness. The output of this phase is a set of ARs for each query term.

(iii) Rule Filtering. The aim of this step is to keep the rules that most likely map a term in the source language to its translation in the target language. The series of heuristics listed below was developed by observing empirical data. They are applied on the ARs generated for each query term. Table 1 shows the application of these heuristics.

Table 1. Example of filtering association rules for the term "civil". The numbers between brackets are the confidence of the AR.

civil ⇒ E=war (26.1)	Discarded - Low confidence (c)
civil ⇒ E=civilian (29.6)	**Selected** - Complement to 100 (d)
civil ⇒ guerr (25.6)	Discarded - Antecedent and consequent in the same language (a)
civil ⇒ E=civil (70.5)	**Selected** - AR with highest confidence (b)

(a) Discard rules in which the antecedent and the consequent are in the same language. Since we are trying to map terms between languages, these rules are not of interest.

(b) Select the AR with the highest confidence, which will be called M. This rule is more likely to be the correct mapping.

(c) Select the ARs that have confidence of at least 80% of M.

(d) Select ARs with confidence equal to $(100 - M \pm 0.1)$, as it was observed that words in a language that are normally translated into two (or more) words in another language tend to have complementary confidences.

(iv) Query Translation. Each term in the original query is replaced by all possible translations that remain after the filtering process. The output of this step is the query in the target language.

(v) Query Execution. The last step if to execute the queries in a search engine. At this stage, the CLIR problem has been reduced to a traditional monolingual query processing. The output is a list of retrieved documents.

It is worth pointing out that the collection used as a basis for the mining of ARs need not be the same used for document retrieval. It is possible to extract the ARs from a bilingual corpus and to use a different test collection for document retrieval.

Our approach mines the ARs on demand, according to a lazy strategy as proposed by Veloso et al. [10]. Thus, we only generate rules for the terms in the query, and as we only consider the sentences in which the query terms appear for rule generation, the number of rules is significantly reduced. On the other hand, this strategy delays query processing. To speed up this process, we could build a cache of ARs, eliminating the need to mine for all the rules at query time.

3 Modifying BM25 to Emphasise Rare Terms

Okapi BM25 [8] is a ranking function used by search engines to rank documents according to their similarity to a given query. This is a very popular ranking function and it is implemented in many IR systems. In order to improve our IR results, we have implemented modifications to the original BM25 formula, shown in Eq. 2.

$$BM25(D,Q) = \sum_{i=1}^{n} log\left(\frac{N - n(q_i) + 0,5}{n(q_i) + 0,5}\right) * \frac{f(q_i,D) * (k_1 + 1)}{f(q_i,D) + k * \left(1 - b + b * \frac{|AL|}{tf_{dt}}\right)} \quad (2)$$

where: N is the number of documents in the collection

$n(q_i)$ is the number of documents indexed by term q_i

$f(q_i, D)$ is the frequency of term q_i on document D
AL is the number of terms in document D
k_1 and b are parameters, usually chosen as 2.0 and 0.75, respectively.

Our modification on BM25 aims at promoting rare terms, i.e. terms that occur in few documents. The modification is divided into two steps. The first step is to reduce the weight of common terms in the collection and it is accomplished by adding a new multiplier to the original function. The weights of the multipliers were defined by observing query results on the LA Times collection and are shown in Eq. 3 . We call them "Intermediate Scores" or scoreI.

$$scoreI(D,Q) = (0.00005p_i^4 - 0.019p_i^3 + 0.0211p_i^2 - 0.0926p_i + 1,1697) * BM25(D,Q) \quad (3)$$

where: $pi = n(q_i)/N$ is number of documents indexed by the term

The improvement in terms of query results obtained by Eq. 3 is only marginal. It will only achieve significant results when stop-words are not removed or in collections with very few documents. As a consequence, a second phase is applied.

The second step aims at promoting rare terms more emphatically. Let m be the average number of occurrences of the terms in the collection. Using m, the number of occurrences of each term $n(q_i)$, and the intermediate scoreI, the modified version of BM25, called BM25+, is shown in Eq 4. It is important to notice that the seven conditionals in Eq. 4 are not mutually exclusive. For example, a term appearing in just one of 10,000 documents would receive all increments in the BM25+ function.

$$BM25+(D,Q) = \begin{cases} scoreI(D,Q) + 0.05 * min\left(4, \frac{n(q_i)}{m}\right) & if n(q_i) < m \\ scoreI(D,Q) + 0.1 & if n(q_i) < 1000 \\ scoreI(D,Q) + 0.2 & if n(q_i) < 500 \\ scoreI(D,Q) + 0.3 & if n(q_i) < 100 \\ scoreI(D,Q) + 0.5 & if n(q_i) < 50 \\ scoreI(D,Q) + 0.8 & if n(q_i) < 20 \\ scoreI(D,Q) + 1.5 & if n(q_i) < 6 \end{cases} \quad (4)$$

4 Experiments

This section describes our experiments submitted to the CLEF-2008 campaign. Section 4.1 details the resources used, and Section 4.2 presents the results.

4.1 Description of Runs and Resources

We worked on the English TEL collection, which contains catalogue data from the British Library. The details of the test collection are described in [1]. Our aim was to test the feasibility of our proposed approach for using ARs to map concepts between languages. Our bilingual experiments use Spanish queries to retrieve documents in English.

The procedure is the same as described in Section 2. Since our approach needs a sample of parallel documents and the TEL collection does not have parallel documents, we had to translate a sample of the original documents using Google

Translator. The sample size was 25% of the collection (250,025 documents) and it taken by picking one in every four documents in sequence.

We removed stop-words according to the lists available from Snowball . The Porter Stemmer [7] was used on the English texts and the Spanish version of the Porter Stemmer (Snowball [9]) was used on the Spanish documents. The IR system we used was Zettair [11], which is a compact and fast search engine developed by RMIT University (Australia) distributed under a BSD-style license. Zettair implements a series of IR metrics for comparing queries and documents. We used Okapi BM25 as some preliminary tests we performed on other data collections showed it achieved the best results.

The time taken to run all queries was approximately 12 seconds including the mining of the ARs, rule filtering, query translation and processing by the search engine. The tests were performed on a Pentium 4 2.8GHz with 512 Mb of RAM running Windows XP.

Since our goal was to test our approach on a cross-linguistic setting, our monolingual runs serve only as a baseline. Four official runs were submitted:

- UFRGS_BI_SP_EN - uses our proposed method for ARs
- UFRGS_BI_SP_EN2 - uses our proposed method for ARs and BM25+
- UFRGS_MONO_EN1 - baseline monolingual run
- UFRGS_MONO_EN2 - monolingual run using BM25+

In order to compare our approach to query translation using a machine translation system, we carried out two unofficial runs using Google Translator [5] to translate the queries from Spanish into English. One of the runs uses our proposed modification on BM25 (run tagged GoogleTrans_BM25+), and the other one uses the standard formula (run tagged GoogleTrans).

4.2 Results

Our results are summarised in Figure 3. Comparing the official monolingual and bilingual runs, we notice that the bilingual executions achieve up to 86% of the corresponding monolingual performance in terms of Mean Average Precision (MAP). A T-test showed that the difference in performance between monolingual and bilingual runs is not statistically significant if measured by MAP. Compared to other participants, our bilingual version was ranked in third place. These results indicate that our approach for mapping concepts between languages using ARs is adequate. We observed a high correlation (0.8) between the performance of the monolingual and the bilingual runs in terms of MAP. This indicates that topics that score high on the monolingual run also tend to score high on the cross-language run.

When comparing performance in terms of Pr@10, however, our bilingual runs are statistically worse than their monolingual counterparts. This fact can be observed in Figure 3, as the superiority of the monolingual runs is more evident at low recall levels. From recall 0.5 onwards, all runs have very similar results.

Comparing the results obtained by the original BM25 formula and BM25+, we can see that our modification achieves better results both for ARs and Google-Translator. Improvements were noticed in terms of MAP and PR@10. However,

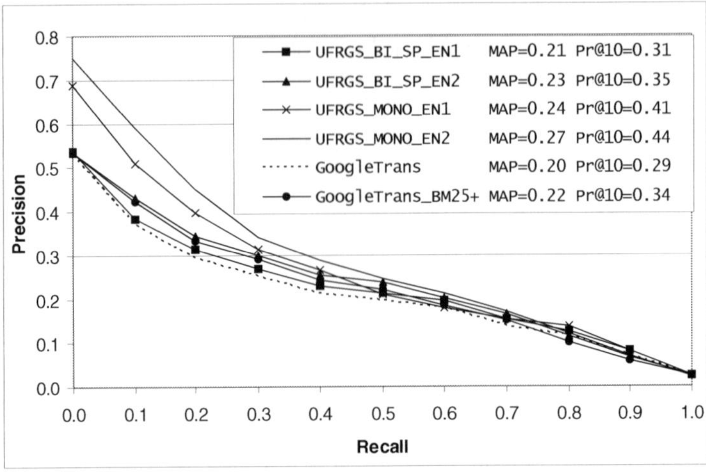

Fig. 3. Recall-precision curves for the submitted runs

this difference was only considered statistically significant for the monolingual run in terms of MAP. In all five topics with person names (458, 471, 478, 486, 500) the performance of BM25+ was worse than the performance of the original implementation. The topic in which BM25+ yielded the worst performance was topic 500 "Paul Gauguin y Tahití". This loss can be attributed to not finding some rare terms including proper names. As a result, other terms that could resolve the issue had their weight diminished.

Our results are equivalent to the results of translating the queries via Google Translator. Machine translation systems apply much more complex Natural Language Processing techniques. Thus, this equivalence in the results favours our simpler approach.

Comparing our results to the mean of other participating groups we noticed that for topics with names of locations our results tend to be better (16 out of 20 for monolingual runs and 15 out of 20 for bilingual runs). We believe this is due to the weighting scheme benefiting rare terms. This was noticed specially for topic 480. This comparison also showed that our results tend to be better for topics with higher number of relevant documents within the collection.

Comparing our official bilingual runs to the average of all participants we observed that our results were better than average for "easy" topics. By easy we mean topics with a high average MAP. Analysing the five highest scoring topics on average of all bilingual Spanish-English runs our result is 40% better. The cases in which our performance was worse usually had a mistranslated term. For example, in topic 460 the expression "Películas de terror" was translated to "Terror movies" when the correct translation should be "Horror movies". In this specific case our performance was about 44% worse than the average.

Manually analysing a sample of words translates by our approach, we found that a correct translation was produced 90% of times. By correct, we mean that the translated term was found in a bilingual dictionary amongst the translations for the original term. We attribute our good results in the TEL task to finding correct translations. The query expansion effect brings significant gains, however it happens in few topics.

5 Conclusions

This paper reported on monolingual and bilingual ad-hoc information retrieval experiments that we have performed for the TEL task. Our aim was to validate our proposal of using algorithms for mining association rules for CLIR. The results of the experiments show that our bilingual runs achieve 86% of the performance of the monolingual runs. More importantly, is that the difference in MAP is not statistically significant, which shows our approach is feasible. Since we used automatic translation to generate a sample of parallel documents and it is widely known that these algorithms are far from perfect, it is possible that our results would be better if had translation was used. This fact still needs further investigation.

We have also tested a modification we proposed over Okapi BM25 to increase the weight of rare terms. The results show that the modified version, which we called BM25+, achieves better results.

The experiments reported here provided encouraging results. However, there are still a number of open issues that will be explored as future work; they include: assessing the impact of the size of the sample used for translation in the results; comparing results obtained using an automatic translator to generate a parallel collection against the results obtained using a higher quality (hand-translated) parallel corpus.

Acknowledgements

The authors would like to thank Marcos Gonçalves for his valuable advice. This work was partially supported by CNPq Universal 484585/2007-0. André Geraldo is funded by a studentship from CNPq.

References

1. Aguirre, E., et al.: CLEF 2008: Ad Hoc Track Overview. In: Peters, C., et al. (eds.) CLEF 2008. LNCS, vol. 5706, pp. 15–37. Springer, Heidelberg (2009)
2. Agrawal, R., Imielinski, T., Swami, A.: Mining Association Rules between Sets of Items in Large Databases. In: Proc. of the ACM SIGMOD Conference on Management of Data, Washington, D.C (1993)
3. Agrawal, R., Srikant, R.: Fast Algorithms for Mining Association Rules. In: Proceedings of the 20th VLDB Conference, Santiago, Chile, pp. 487–499 (1994)
4. Deerwester, S., Dumais, S., Furnas, G., Landauer, T., Harshman, R.: Indexing by Latent Semantic Analysis. Journal of the American Society for Information Science 41(6), 1–13 (1990)

5. Google Translator, http://www.google.com/translate_t (accessed on: February 8, 2009)
6. Hipp, J., Güntzer, U.: Is pushing constraints deeply into the mining algorithms really what we want?: an alternative approach for association rule mining. ACM SIGKDD Explorations Newsletter 4(1), 50–55 (2002)
7. Porter, M.F.: An Algorithm for Suffix Stripping. Program 14(3), 130–137 (1980)
8. Robertson, S., Walker, S.: Okapi at TREC-3. In: Proceedings of the Third Text REtrieval Conference (TREC). Gaithesburg, Maryland (1994)
9. Snowball. Spanish Stemmer, http://snowball.tartarus.org/algorithms/spanish/stemmer.html (retrieved August 08, 2008)
10. Veloso, A., Meira Jr., W., Gonçalves, M.A., Zaki, M.: Multi-label Lazy Associative Classification. In: Kok, J.N., Koronacki, J., Lopez de Mantaras, R., Matwin, S., Mladenič, D., Skowron, A. (eds.) PKDD 2007. LNCS (LNAI), vol. 4702, pp. 605–612. Springer, Heidelberg (2007)
11. Zettair, www.seg.rmit.edu.au/zettair/ (retrieved 11/06/07, 2007)

CLEF 2008 Ad-Hoc Track: Comparing and Combining Different IR Approaches

Jens Kürsten, Thomas Wilhelm, and Maximilian Eibl

Chemnitz University of Technology
Faculty of Computer Science, Chair Computer Science and Media
Straße der Nationen 62
09107 Chemnitz, Germany
{jens.kuersten,thomas.wilhelm,eibl}@cs.tu-chemnitz.de

Abstract. This article describes post workshop experiments that were conducted after our first participation at the *TEL@CLEF task*. We used the *Xtrieval* framework [5], [4] for the preparation and execution of the experiments. We ran 69 experiments in the setting of the CLEF 2008 task, whereof 39 were monolingual and 30 were cross-lingual. We investigated the capabilities of the current version of Xtrieval, which could use the two retrieval cores Lucene and Lemur from now on. Our main goal was to compare and combine the results from those retrieval engines. The translation of the topics for the cross-lingual experiments was realized with a plug-in to access the Google AJAX language API. The performance of our monolingual experiments was better than the best experiments we submitted during the evaluation campaign. Our cross-lingual experiments performed very well for all target collections and achieved between 87% and 100% of the monolingual retrieval effectiveness. The combination of the results from the Lucene and the Lemur retrieval core showed very consistent performance.

Keywords: Evaluation, Experimentation, Data Fusion, Cross-Language Information Retrieval.

1 Introduction and Outline

The *Xtrieval* framework [5],[4] was used to prepare and execute this years retrieval experiments for the *TEL@CLEF task*. The core retrieval functionality is provided by Apache Lucene[1] and by the Lemur Toolkit[2]. For the *Ad-Hoc track* three different multilingual corpora with content mainly in German, English and French were provided by *The European Library* (TEL)[3]. Each collection consists of approximately one million library records. These library records only contain very sparse information and have descriptions in multiple languages [1].

[1] http://lucene.apache.org
[2] http://www.lemurproject.org
[3] http://www.theeuropeanlibrary.org

C. Peters et al. (Eds.): CLEF 2008, LNCS 5706, pp. 75–82, 2009.

The remainder of the paper is organized as follows. Section 2 describes the general system setup and section 3 provides all individual configurations as well as the results of our monolingual experiments. Cross-lingual experiments are presented in section 4. In sections 5 and 6 we summarize the results and sum up our observations.

2 Experimental Setup

We conducted monolingual experiments on each of the collections and also submitted experiments for the bilingual subtasks. For the translation of the topics the Google AJAX language API[4] was accessed through a JSON[5] programming interface. Our official experiments for the *TEL@CLEF task* are described in [3] and are not listed in this article. Since we made major changes in the Xtrieval framework we report numerous post workshop experiments here. Due to the fact that the current version of the Xtrieval framework does not yet support relevance feedback when using the Lemur Toolkit retrieval core, we did not apply pseudo-relevance feedback in any experiment shown in this article.

2.1 Experiment Data Fusion

Since the combination of several tokenization approaches was very successful in our experiments of the past years [5], we present numerous combination experiments in this work. The main goal was to investigate whether the combination of runs with different retrieval cores (Lemur and Lucene) could provide consistent improvement in terms of retrieval effectiveness or not. We decided to use our implementation of the Z-Score [7] data fusion operator. It has shown quite consistent results in comparison to other fusion operators [4], [5], [7]. Due to the fact that we did not want to go beyond the scope of this article we did not take into account different data fusion operators. We consider the combination of different tokenization and retrieval approaches as especially important for the planned grid experiments task in next years evaluation campaign.

3 Monolingual Experiments

We ran 39 experiments in total, 13 for each of the target collections. A standard processing chain for indexing and retrieval was applied for all experiments. For all experiments a language-specific stopword list was applied[6]. All terms were stemmed with algorithms appropriate for the language. We used different stemmers for each language: Porter[7] and Krovetz [2] for English, Snowball[7] and a n-gram variant decompounding stemmer[8] for German and the Snowball[7] implementation as well as a stemmer for French that is described in [6].

[4] http://code.google.com/apis/ajaxlanguage/documentation

[5] http://json.org

[6] http://members.unine.ch/jacques.savoy/clef/index.html

[7] http://snowball.tartarus.org

[8] http://www-user.tu-chemnitz.de/~wags/cv/clr.pdf

3.1 Lemur Toolkit Retrieval Core

For each language we compared two different stemming approaches and applied four retrieval algorithms, namely *Okapi*, *TF-IDF (VSM)*, *KL* and *Inquery*. The configuration and retrieval effectiveness of the experiments are shown in table 1. We only listed the two best performing experiments to keep the table well arranged. We compared our experiments to the best result submitted for the evaluation campaign (first row of each block in table 1).

Table 1. Monolingual Configurations and Results for Lemur Retrieval Core

id	*lang*	*stemmer*	*model*	*map*
top_ah08_en	EN	-	-	0.3753
cut_postws_en1	EN	porter	okapi	0.3452 (-08.02%)
cut_postws_en2	EN	krovetz	inquery	0.3334 (-11.16%)
cut_postws_en_merged1	EN	porter/krovetz	okapi/inquery	0.3135 (-16.47%)
top_ah08_fr	FR	-	-	0.3327
cut_postws_fr1	FR	porter	okapi	0.2339 (-29.69%)
cut_postws_fr2	FR	savoy	okapi	0.2348 (-29.43%)
cut_postws_fr_merged1	FR	porter/savoy	okapi	0.1951 (-41.36%)
top_ah08_de	DE	-	-	0.3571
cut_postws_de1	DE	porter	tfidf	0.2312 (-35.26%)
cut_postws_de2	DE	pgdec	okapi	0.3062 (-14.26%)
cut_postws_de_merged1	DE	porter/pgdec	tfidf/okapi	0.1213 (-66.03%)

The results of the monolingual experiments show that the different stemming techniques only had a small affect on retrieval effectiveness for English and French, whilst the n-gram decompounder clearly outperformed the standard porter stemmer for German. It can also be seen that the combination of the best runs decreased retrieval effectiveness in every case.

3.2 Lucene Retrieval Core

We also investigated whether the different stemming approaches affected the retrieval effectiveness when using Lucene as retrieval core. Additionally we tried using the language information from the documents to constrain the results (LC) to the expected languages German, English, French and Spanish. Again we compared our experiments to the best result submitted for the evaluation campaign (first row of each block).

The results in table 2 show that using several stemming techniques did not significantly affect the retrieval effectiveness on the English and French collection. But for the German collection the n-gram decompound stemmer (pgdec) clearly outperformed the run with the porter stemmer. Contrary to the merged experiments with the Lemur core the combined experiments when using the Lucene core improved retrieval effectiveness for all three collections. But the last merged run on each collection shows a significant decrease in retrieval effectiveness when

Table 2. Monolingual Configurations and Results for Lucene Retrieval Core

id	*lang*	*stemmer*	*LC*	*map*
top_ah08_en	EN	-	-	0.3753
cut_postws_en3	EN	porter	no	0.3758 (+0.13%)
cut_postws_en4	EN	krovetz	no	0.3731 (-0.58%)
cut_postws_en_merged2	EN	porter/krovetz	no	0.3880 (+03.38%)
cut_postws_en_merged3	EN	porter/krovetz	yes	0.3559 (-05.17%)
top_ah08_fr	FR	-	-	0.3327
cut_postws_fr3	FR	porter	no	0.2511 (-24.53%)
cut_postws_fr4	FR	savoy	no	0.2472 (-25.70%)
cut_postws_fr_merged2	FR	porter/savoy	no	0.2637 (-20.74%)
cut_postws_fr_merged3	FR	porter/savoy	yes	0.2276 (-31.59%)
top_ah08_de	DE	-	-	0.3571
cut_postws_de3	DE	porter	no	0.2327 (-34.84%)
cut_postws_de4	DE	pgdec	no	0.2783 (-22.07%)
cut_postws_de_merged2	DE	porter/pgdec	no	0.3079 (-13.78%)
cut_postws_de_merged3	DE	porter/pgdec	yes	0.1561 (-56.29%)

the results were constrained to contain the language tags for English, German, French and Spanish.

3.3 Combined Experiments

In this section we present the results of merged experiments when using both the Lucene and the Lemur retrieval cores. In table 3 we compare our experiments to the best result from section 2.1 and 2.2.

The merged experiments with the Lemur and Lucene retrieval cores show that the retrieval effectiveness could be improved on the English and French collection although the gain was not significant. On the German collection there was one experiment that did significantly outperform the best run from a single retrieval core.

Table 3. Monolingual Configurations and Results for Combined Retrieval Cores

id	*lang*	*stemmer*	*retrieval core*	*map*
cut_postws_en_merged2	EN	porter/krovetz	lucene	0.3880
cut_postws_en_merged_e1_e3	EN	porter	lucene/lemur	0.3800 (-02.06%)
cut_postws_en_merged_e2_e4	EN	krovetz	lucene/lemur	0.3837 (-01.11%)
cut_postws_en_merged_e1_e4	EN	porter/krovetz	lucene/lemur	0.3908 (+0.72%)
cut_postws_fr_merged2	FR	porter/savoy	lucene	0.2637
cut_postws_fr_merged_e1_e3	FR	porter	lucene/lemur	0.2634 (-0.11%)
cut_postws_fr_merged_e2_e4	FR	savoy	lucene/lemur	0.2561 (-02.88%)
cut_postws_fr_merged_e1_e4	FR	porter/savoy	lucene/lemur	0.2669 (+01.21%)
cut_postws_de_merged2	DE	porter/pgdec	lucene	0.3079
cut_postws_de_merged_e1_e3	DE	porter	lucene/lemur	0.2582 (-16.14%)
cut_postws_de_merged_e2_e4	DE	pgdec	lucene/lemur	0.3209 (+04.22%)
cut_postws_de_merged_e2_e3	DE	porter/pgdec	lucene/lemur	0.3318 (+07.76%)

4 Cross-Lingual Experiments

We ran 30 experiments in total, 10 for each of the target collections. We applied the same text processing chain as stated in section 2 for all our cross-lingual experiments. To evaluate the influences of the multilingual contents of the corpora we used the best result for each collection from section 2 as reference. In the following subsections we compare two different cross-lingual retrieval approaches: (a) simple translation of the query to the main target collection language, i.e. generating a query in one language and (b) translating the query to all languages in the set of English, French, German and Spanish, i.e. generating a multilingual query in four languages.

4.1 Lemur Toolkit Retrieval Core

In table 4 we listed our cross-lingual experiments using the Lemur retrieval core. Those experiments were compared to the best (in terms of *MAP*) monolingual run from section 2 by applying the same system configurations and alternating the translation pairs and strategies.

Table 4. Cross-lingual Configurations and Results for Lemur Retrieval Core

id	*lang*	*stemmer*	*translation*	*map*
cut_postws_en1	EN	porter	none	0.3452
cut_postws_de2en_1	DE→EN	porter	(a)	0.3205 (-07.16%)
cut_postws_fr2en_1	FR→EN	porter	(a)	0.3571 (+03.45%)
cut_postws_de2en_2	DE→EN	porter	(b)	0.2017 (-41.57%)
cut_postws_fr2en_2	FR→EN	porter	(b)	0.2123 (-38.50%)
cut_postws_fr2	FR	savoy	none	0.2348
cut_postws_de2fr_1	DE→FR	savoy	(a)	0.2328 (-0.85%)
cut_postws_en2fr_1	EN→FR	savoy	(a)	0.2335 (-0.55%)
cut_postws_de2fr_2	DE→FR	savoy	(b)	0.1561 (-33.52%)
cut_postws_en2fr_2	EN→FR	savoy	(b)	0.1663 (-29.17%)
cut_postws_de2	DE	pgdec	none	0.3062
cut_postws_en2de_1	EN→DE	pgdec	(a)	0.2523 (-17.60%)
cut_postws_fr2de_1	FR→DE	pgdec	(a)	0.2922 (-04.57%)
cut_postws_en2de_2	EN→DE	pgdec	(b)	0.1503 (-50.91%)
cut_postws_fr2de_2	FR→DE	pgdec	(b)	0.1601 (-47.71%)

The performance of the cross-lingual experiments was very strong on all target collections and even improved on the best monolingual run in one case. In general better retrieval performance was achieved when the query was only translated to the target language of the corresponding collection (i.e. strategy (a)).

4.2 Lucene Retrieval Core

In table 5 our cross-lingual experiments using the Lucene retrieval core are listed. We compare those experiments to the best (in terms of *MAP*) monolingual run

Table 5. Cross-lingual Configurations and Results for Lucene Retrieval Core

id	*lang*	*stemmer*	*translation*	*map*
cut_postws_en_merged2	EN	porter/krovetz	none	0.3880
cut_postws_de2en_3	DE→EN	porter/krovetz	(a)	0.3675 (-05.28%)
cut_postws_fr2en_3	FR→EN	porter/krovetz	(a)	0.3895 (+0.39%)
cut_postws_de2en_4	DE→EN	porter/krovetz	(b)	0.1816 (-53.20%)
cut_postws_fr2en_4	FR→EN	porter/krovetz	(b)	0.1607 (-58.58%)
cut_postws_fr_merged2	FR	porter/savoy	none	0.2637
cut_postws_de2fr_3	DE→FR	porter/savoy	(a)	0.2323 (-11.91%)
cut_postws_en2fr_3	EN→FR	porter/savoy	(a)	0.2284 (-13.39%)
cut_postws_de2fr_4	DE→FR	porter/savoy	(b)	0.1680 (-36.29%)
cut_postws_en2fr_4	EN→FR	porter/savoy	(b)	0.1863 (-29.35%)
cut_postws_de_merged2	DE	porter/pgdec	none	0.3079
cut_postws_en2de_3	EN→DE	porter/pgdec	(a)	0.2806 (-08.87%)
cut_postws_fr2de_3	FR→DE	porter/pgdec	(a)	0.2673 (-13.19%)
cut_postws_en2de_4	EN→DE	porter/pgdec	(b)	0.1352 (-56.09%)
cut_postws_fr2de_4	FR→DE	porter/pgdec	(b)	0.1375 (-55.34%)

from section 2 by applying the same system configurations and alternating the translation pairs and strategies.

Similar to the experiments using the Lemur retrieval core the retrieval effectiveness of the cross-lingual experiments is very good on the English and German target collection. The runs on the French collection had a small decrease of about 13% in terms of *MAP*. Again the experiments using the translation strategy (a) clearly outperformed the other runs.

4.3 Combined Experiments

In this section we present the results of merged experiments when using both the Lucene and the Lemur retrieval cores in the cross-lingual setting. In table 6 we compare our experiments to the best monolingual result (in terms of *MAP* from section 2.3.

Table 6. Cross-lingual Configurations and Results for Combined Retrieval Cores

id	*lang*	*translation*	*retrieval core*	*map*
cut_postws_en_merged_e1_e4	EN	none	lucene/lemur	0.3908
cut_postws_de2en_merged_e1_e4	DE→EN	(a)	lucene/lemur	0.3643 (-06.78%)
cut_postws_fr2en_merged_e1_e4	FR→EN	(a)	lucene/lemur	0.3919 (+0.26%)
cut_postws_fr_merged_e1_e4	FR	none	lucene/lemur	0.2669
cut_postws_de2fr_merged_e1_e4	DE→FR	(a)	lucene/lemur	0.2515 (-05.77%)
cut_postws_en2fr_merged_e1_e4	EN→FR	(a)	lucene/lemur	0.2451 (-08.17%)
cut_postws_de_merged_e2_e3	DE	none	lucene/lemur	0.3318
cut_postws_en2de_merged_e2_e3	EN→DE	(a)	lucene/lemur	0.2913 (-12.21%)
cut_postws_fr2de_merged_e2_e3	FR→DE	(a)	lucene/lemur	0.2643 (-20.34%)

In table 6 one can see that the cross-lingual experiments when combining both retrieval cores achieve a good overall retrieval effectiveness. On the English target collection the results are the same as for the best monolingual run. For the French and German collection the decrease in effectiveness is about 6% and 12% respectly. So we can draw the conclusion that the cross-language retrieval performance depends on the language and its distribution in the collection.

5 Result Analysis - Summary

The following list provides a summary of the analysis of our post workshop retrieval experiments in the *TEL@CLEF* setting at CLEF 2008:

- *Monolingual:* Our post workshop experiments achieved good performance on all target collections. Nevertheless there is room for improvement especially on the French and German target collections. The performance of all experiments were improved in comparison to our officially submitted experiments.
- *Cross-lingual:* We showed that cross-lingual experiments almost reach the retrieval effectiveness of monolingual runs. Our experiments on generating multilingual queries deteriorated retrieval effectiveness in all investigated configurations.
- *Combination:* The experiments with merging results from the two retrieval cores Lemur and Lucene showed consistent retrieval effectiveness. Unfortunately the combination of the results did not improve retrieval effectiveness in all cases.

6 Conclusion and Future Work

This year, we participated in the *TEL@CLEF task* for the first time. An important observation in all our experiments for this years CLEF campaign was that the translation service provided by Google seems to be extremely superior to any other approach or system. This should motivate the cross-language community to investigate and improve their current approaches.

In our future work we will complete the integration of the Lemur Toolkit as retrieval core, i. e. we will implemented an appropriate pseudo-relevance feedback approach. Additionally we will also try to integrate the Terrier Toolkit. For next year we also plan to run experiments using a language detector during indexing.

Acknowledgments

We would like to thank Jaques Savoy and his co-workers for providing numerous resources for language processing. Also, we would like to thank Giorgio M. di Nunzio and Nicola Ferro for developing and operating the DIRECT system[9]. Finally we are also very grateful to our colleagues Robert Knauf and Marc Ritter for helping us with the final preparation of this article.

[9] http://direct.dei.unipd.it

This work was partially accomplished in conjunction with the project *sachs-Media*, which is funded by the *Entrepreneurial Regions*[10] program of the German Federal Ministry of Education and Research.

References

1. Agirre, E., Di Nunzio, G.M., Ferro, N., Mandl, T., Peters, C.: CLEF 2008: Ad Hoc Track Overview. In: Working Notes for the CLEF 2008 Workshop, September 17-19, Aarhus, Denmark (October 2008)
2. Krovetz, R.: Viewing Morphology as an Inference Process. In: SIGIR 1993: Proceedings of the 16th annual international ACM SIGIR conference on Research and development in information retrieval, pp. 191–202. ACM, New York (1993)
3. Kürsten, J., Wilhelm, T., Eibl, M.: CLEF 2008 Ad-Hoc Track: On-line Processing Experiments with Xtrieval. In: Working Notes for the CLEF 2008 Workshop, September 17-19, Aarhus, Denmark (October 2008)
4. Kürsten, J., Wilhelm, T., Eibl, M.: Extensible Retrieval and Evaluation Framework: Xtrieval. In: LWA 2008: Lernen - Wissen - Adaption, Workshop Proceedings - FGIR, Würzburg, October 2008, pp. 107–110 (2008)
5. Kürsten, J., Wilhelm, T., Eibl, M.: The Xtrieval Framework at CLEF 2007: Domain-Specific Track. In: Peters, C., Jijkoun, V., Mandl, T., Müller, H., Oard, D.W., Peñas, A., Petras, V., Santos, D. (eds.) CLEF 2007. LNCS, vol. 5152, pp. 174–181. Springer, Heidelberg (2008)
6. Savoy, J.: A stemming procedure and stopword list for general French corpora. Journal of the American Society for Information Science, 944–982 (1999)
7. Savoy, J.: Data Fusion for Effective European Monolingual Information Retrieval. In: Peters, C., Clough, P., Gonzalo, J., Jones, G.J.F., Kluck, M., Magnini, B. (eds.) CLEF 2004. LNCS, vol. 3491, pp. 233–244. Springer, Heidelberg (2005)

[10] The Innovation Initiative for the New German Federal States

Multi-language Models and Meta-dictionary Adaptation for Accessing Multilingual Digital Libraries

Stephane Clinchant and Jean-Michel Renders

Xerox Research Centre Europe
FirstName.LastName@xrce.xerox.com

Abstract. Accessing digital libraries raises the important issue of how to deal with the multilinguality of the documents. Inside a target collection, documents can be written in very different languages and the record associated to a particular document often contains field descriptors in different languages. This paper proposes a principled way to solve this issue, by proposing a multi-language model approach to information retrieval, as well as an extension of the dictionary adaptation mechanism to cover multiple languages (including the source language). In experiments related to the TEL task of the CLEF2008 Ad-hoc track, runs based on the assumption of a purely bilingual approach, translating the query only in the official language of the collection, appeared to result in performance (mean average precision) larger or equal to the ones of the other participants. But, contrarily to our initial intuition, in the case of the TEL task, the experiments showed that exploiting information in languages different from the official language of the collection turns out to offer no advantage.

1 Introduction

Accessing multilingual digital libraries raises important challenges in cross-lingual information retrieval. The main one is related to the heterogeneous and partial nature of the records that are used as surrogates of the real documents to be retrieved: relying on a title and some description fields that do not necessarily follow the same guidelines is not an easy task to extrapolate the whole content of the document. A second challenge lies in the multilinguality of the documents. Inside a target collection, documents can be written in very different languages and the record associated to a particular document often contains field descriptors (titles, subject, ...) in different languages. This paper deals with this second challenge, by proposing a principled way to tackle multilinguality.

Basically, we extend the classical language modeling approach to information retrieval, by allowing more than one language to be represented in the document model. In other words, a document in the target collection will be defined by a probability distribution over the words of a meta-vocabulary, obtained by the union of the vocabularies over different languages (in this paper: English, French

C. Peters et al. (Eds.): CLEF 2008, LNCS 5706, pp. 83–88, 2009.

and German). This means that there is a unique index per collection – and not as many indexes as languages in the collection.

Now, for a query written in some source language, we have to build a multi-language model of this query in order to compare it with the multi-language model of the target documents. This will be done by using a meta-dictionary, namely a probabilistic translation matrix that gives for each word in the source language, all the potential equivalent words in the meta-vocabulary (including the source language itself). However, as the target collection has numerous biases (in the domains covered and in the proportion of words in different languages), there are a lot of spurious or inadequate translations. We already proposed in [3] a way to automatically filter out (or re-weight) these inadequate translations by a dictionary adaptation method. It is rather easy to extend this dictionary adaptation method, which is originally designed for a pure bilingual framework, to a true multi-lingual case, including the source language itself.

The next section explains briefly our approach (multi-language model and meta-dictionary); the reader interested to more details should refer to the working notes [4]. We then shortly mention the pre-processing we adopted for the *CLEF 2008/TEL* Task. After that, we present the results that we obtained; it should be noted that these results do not correspond to our official runs, as those runs were impacted by bugs that precluded us to index the whole collection; these bugs are now fixed, so that we are able to present here the correct evaluation figures. We conclude the paper by an analysis of the results.

2 Dealing with Multilingual Documents

The framework of our retrieval experiments is the Language Model approach to Information Retrieval [5]. Digital library collections, are clearly multilingual: for the case of the *TEL* collection (which will serve as our illustrative example in the rest of the paper), a document can be described by French words in a field and in German in an other field. Following the language modelling approach, we decide not to split a document into parts according to the language: a document is a sequence of tokens, which may be of any language; accordingly, a single language model is associated to the document, which is a probability distribution over the words (actually lemma's) of three concatenated vocabularies (English, French and German). In the following, this concatenation of vocabularies will be called the "meta-language". Thus, the feature space of different languages is aggregated into a single description space. This way, we do not build different indexes for a collection (according to the identified languages) but a single index is built containing all the languages.

However, building a single index to cope with multilinguality is just halfway to the solution, as the query is in general expressed only in one language. Indeed, since collections are multilingual, a query word need to be translated into the "meta-language", including its original language. This is done by building probabilistic meta-dictionaries (from a single source language to the meta-language).

These probabilistic dictionaries are built as a combination of monolingual resources (thesauri) and bilingual lexicons extracted from parallel corpora and completed by approximate string matching equivalences for lemmas not covered by the parallel corpus. An important issue is how to weight the different translation probabilities when we merge the monolingual thesaurus and the pair-wise bilingual dictionaries. We have chosen to merge them linearly. We believe that those linear weights should depend on the target collection and on the given task.

Once the meta-dictionary is built from these standard monolingual and bilingual resources, we propose to adapt it for a specific (query, target collection) pair, following the method we presented previously [3]. This amounts to filter out irrelevant, spurious meta-translations, as well as increasing the probabilities of more coherent word translations or synonyms.

3 Pre-processing and Global Approach

Our approach is evaluated in the framework of the TEL Task of CLEF 2008 - Ad-hoc Track. For a complete description of the task and the collections, see the overview paper [2]. We have participated to all 'monolingual' and 'bilingual' tasks. None of the tasks were truly monolingual or bilingual, which motivated our method to cope with multilinguality. For the 3 main languages (English, German, French), we used our home-made lemmatiser and word-segmenter (decompounder) for German. From the fields available for a document record, we only kept the title as well as the subject fields. Classical stopword removal was performed. As monolingual resources, we used the Open Office thesauri[1]. As multilingual resources, we used a probabilistic dictionary, called ELRAC, that is a combination of a very standard one (ELRA) and a lexicon automatically extracted from the parallel JRC-AC (Acquis Communautaire) Corpus. Finally, we carried out our experiments relying on the Lemur Toolkit [1] and on the dictionary adaptation algorithm [3].

4 Analysis of Multilinguality

The main question of the experiments presented is to assess whether it is worthwhile to take into account multilinguality. Indeed, we could simply ignore the fact that the target collection is multilingual and make the assumption that the official language of the collection is the only one to be taken into account. Alternatively, but in a much more complex way, we could try to identify all the languages present in the target collection, and index separately the collection parts corresponding to the different languages; after this, the query must be translated in all the identified languages of the collection, matched against the different sub-parts of the collection, and results coming from the different taget languages must be merged; the last step is often very difficult to solve, as relevance scores for different languages are not directly comparable. Our approach is somewhat in-between, as it takes multilinguality into account, but in an unique

[1] Available on *http://wiki.services.openoffice.org/wiki/Dictionaries*

Table 1. Performance of runs without translation (MAP)

Query Language / Target Collection	MAP
EN to BNF	8.77
DE to BNF	0.3
FR to BNF	29.3
EN to BL	33.42
FR to BL	1.53
DE to BL	1.72
EN to ONB	4.55
FR to ONB	0.44
DE to ONB	25.3

framework, without necessity to do late fusion on relevance scores coming from different languages.

We start with a first series of experiments[2] in order to evaluate the performance of **untranslated** queries. Recall that the documents are multilingual and thus our indexing. The goal is just to see if there is any hope to go further than the pure bilingual case, by exploiting languages that are not the official language of the collections. Table 1 shows those results evaluated in Mean Average Precision. Those results show that, without any translation, it is very hard to get relevant documents when the source query language is different from the main target language of the collection, except for English. So, if we take the BNF collection as a particular example, the question amounts to know if it is useful and/or possible to get 8.77% as an extra performance with respect to the pure bilingual (English - French), that do not exploit the English part of the BNF Collection.

The next table (Table 2) shows the results for pure bilingual runs: a query is completely translated to the target collection official language. The second column of the table shows the source and target languages we used for the runs. Results are given without (W/0 adapt) and after (W/adapt) dictionary adaptation and, finally, with pseudo-relevance feedback after adaptation (Fb).

Note that these pure bilingual runs show performance that superior or equal to the ones of the other participants (see [2]).

Finally, Table 3 shows the results of our complete multi-lingual approach, using a meta-dictionary which is biased towards the language of the target collection. For example, the meta-dictionary for the BL collection will give more weights to English words than French or German words. We chose the mixture weights as 0.8 for the target collection and 0.1 for the two remaining languages. Using this meta-dictionary, translated queries become multilingual. The first column of the table shows the source languages and the target collections we

[2] Recall that the results reported here are not the ones obtained from our official runs; they are also different from those presented in the Working Notes [4], where we simply applied a "patch" to correct our bugs. Here, bugs reported in [4] are completely fixed and the results can be compared with results of the other participants.

Table 2. Dictionary Adaptation Experimental Results in Mean Average Precision for Bilingual Runs

Translation	Initial Dictionary	W/O adapt	W/ adapt	P-values	Fb
EN to BNF	English To French	23.86	26.63	0.02	27.52
DE to BNF	German To French	22.50	25.01	0.10	25.80
FR to BL	French To English	26.0	29.87	0.0008	31.7
DE to BL	German To English	24.50	27.42	0.002	29.44
EN to ONB	English To German	21.58	23.85	0.05	25.40
FR to ONB	French To German	23.79	25.52	0.07	26.34

Table 3. Dictionary Adaptation Experimental Results in Mean Average Precision for Multilingual Runs

Translation	W/O adapt	W/ adapt	P-values
EN to BNF	22.6	26.9	0.001
DE to BNF	19.5	20.6	0.27
FR to BL	25.6	28.47	0.015
DE to BL	23.46	25.19	0.06
EN to ONB	20.5	21.8	0.18
FR to ONB	20.9	22.3	0.05

used for the runs. Results are given without (W/0 adapt) and after (W/adapt) adaptation.

Multilingual runs turn out not to be better than the bilingual runs : that is the main conclusion of this table and this set of experiments. In the next section, we propose some explanations for this observation.

5 Conclusion

The leitmotiv of our work was to deal with multilinguality. Our goal was to get a single retrieval model and to have a single index for all the languages of one specific collection. The purpose of our work was to go beyond a pure bilingual approach which omits parts of the collection that are not in the official language, while still being much less complex than separate indexing and late fusion.

Runs based on the assumption of a purely bilingual approach, translating the query only in the official language of the collection, appeared to result in performance (mean average precision) larger or equal to the ones of the other participants. These runs are based on a dictionary adaptation mechanism, followed by a pseudo-relevance feedback step.

However, in the case of the TEL task, the experiments showed that the best results are pure monolingual or pure bilingual: exploiting information in a language different from the official language of the collection turns out to offer no advantage in this case. We can see two potential reasons to this phenomenon. It is possible that the way queries were constructed and synchronised over collections is responsible for this effect (queries were chosen in such a way that there is a significant amount of relevant documents in each collection). Another reason is that other fields, such as the subject and the description fields, are expressed in the official language of the collection and are able to capture adequately the content of the document, so that leveraging other languages is not useful.

Acknowledgements

This work was partly supported by the IST Programme of the European Community, under the SMART project, FP6-IST-2005-033917.

References

1. http://www.lemurproject.org/
2. Agirre, E., Nunzio, G.D., Ferro, N., Mandl, T., Peters, C.: Clef 2008 ad-hoc track overview. In: Working Notes of CLEF 2008. Avalaible On-line on the CLEF Web Site (2008)
3. Clinchant, S., Renders, J.-M.: Xrce's participation to clef 2007 - domain specific track. In: Working Notes of CLEF 2007. Avalaible On-line on the CLEF Web Site (2007)
4. Clinchant, S., Renders, J.-M.: Xrce's participation to clef 2008 ad-hoc track. In: Working Notes of CLEF 2008. Avalaible On-line on the CLEF Web Site (2008)
5. Zhai, C., Lafferty, J.: A study of smoothing methods for language models applied to ad hoc to information retrieval. In: Proceedings of SIGIR 2001, pp. 334–342. ACM, New York (2001)

Improving Persian Information Retrieval Systems Using Stemming and Part of Speech Tagging

Reza Karimpour[1], Amineh Ghorbani[1], Azadeh Pishdad[1], Mitra Mohtarami[1], Abolfazl AleAhmad[1], Hadi Amiri[1], and Farhad Oroumchian [2]

[1] Electerical and Computer Engineering Faculty, University of Tehran
[2] University of Wollongong in Dubai
{r.karimpour,a.ghorbani,a.pishdad,m.mohtarami,a.aleahmad, h.amiri}@ece.ut.ac.ir, farhadoroumchian@uowdubai.ac.ae

Abstract. With the emergence of vast resources of information, it is necessary to develop methods that retrieve the most relevant information according to needs. These retrieval methods may benefit from natural language constructs to boost their results by achieving higher precision and recall rates. In this study, we have used part of speech properties of terms as extra source of information about document and query terms and have evaluated the impact of such data on the performance of the Persian retrieval algorithms. Furthermore the effect of stemming has been experimented as a complement to this research. Our findings indicate that part of speech tags may have small influence on effectiveness of the retrieved results. However, when this information is combined with stemming it improves the accuracy of the outcomes considerably.

Keywords: Natural language, Persian information retrieval, Part of speech.

1 Introduction

Exploiting meta-information of the terms in the retrieval process may result in precision and recall improvements. Part of speech information clarifies the role of each term in queries and documents. It may also help in assigning different priorities to different query terms. In addition stemming can collapse many surface words in languages such as Arabic and Persian into a single representation and improve the recall of the system.

The general objective of the present study is to further investigate the potential benefits of incorporating part of speech information into both query and document processing and to observe the consequences of such incorporation in Persian information retrieval. Another objective is to investigate the interaction of stemming and part of speech tagging in such environment.

Improving the performance of retrieval engines has been a major concern for years leading to development of many efficient and effective algorithms and systems [1, 2, 3]. However, the retrieval effectiveness of some European languages such as English have been studied in more depth than Middle Eastern languages such as Persian (Farsi). In addition document retrieval has been an interesting topic for those working in natural language processing (NLP) [4, 5] but not much work has been done on the use of these techniques for Persian document retrieval.

C. Peters et al. (Eds.): CLEF 2008, LNCS 5706, pp. 89–96, 2009.

In recent years, there has been some interest on Persian information retrieval but none of those approaches have used part of speech tagging, although POS has been applied successfully to information retrieval in other language [6]. On the other hand studies in Persian POS tagging have reported accuracy rates of up to 95% using statistical methods such as TnT or with post-processing with MLE taggers [7, 8, 9, 10]. Therefore it seems reasonable to use these taggers in the development of a new generation of retrieval engines for Persian language.

In this research we utilize POS tagging methods to preferentially match the specified types of terms in documents and queries. We also try to control the impact of certain types of words that seems not to have a major contribution to the overall results.

2 Part of Speech Tagging

Part of speech tagging selects the most likely sequence of syntactic categories for the words in a sentence. It determines the tags that best represent the grammatical characteristics of the words, such as part of speech, grammatical number, gender, person, etc. This task is not trivial since many words are ambiguous. Most of the retrieval models ignore the role of the content words in the sentences and treat them uniformly. Although a lot could be realized from the role a word plays in a sentence and its surrounding words. Besides this, the role of each word depends on what the user means by the words in the query [11].

In different languages and tagging systems, the number of tags varies from a dozen to several hundred depending on the specificity of the information provided by the tag. For example a tag-set may just categorize nouns as singular and plural while another tag-set may provide more detail such as name of location or person. Obviously, not all of these tags have the same impact on the retrieval of the relevant documents [12]. Therefore the computation of a proper tag-set with the right size and granularity for a particular collection of a language is an issue worthy of studying.

In this study, we take advantage of the Bijankhan [13] corpus which is a manually tagged Persian text collection. In its original form it includes 550 different tags. This collection has been processed and prepared for machine learning applications. The new collection has over 2 million words and only 40 POS tags [7].

It has been reported that in some applications of IR, nouns are more important than the other tokens [14, 15]. However, sometimes even stop words can be useful [14]. The importance of various POS tags is very subjective. For example in some areas such as biology or advertisement that emphasize the differences among things and their characteristic, adjectives are more important. While in other applications such as music which are mostly adverb-rich, the role of adverbs become more important [11]. Some studies also have investigated the role of verbs in document analysis [16].

After analyzing the impact of these 40 different tags, eventually we find out that nouns, verbs, adjectives and adverbs are the most important POS Tags in Persian retrieval. In the result this section we will show the impact of using these tags on the performance.

In this study the TnT POS tagger[1] is used to determine the part of speech of Persian words. TnT is a very efficient statistical part-of-speech tagger that is trainable on different languages and virtually any tag-set. TnT requires a pre-tagged document collection for training phase. The system incorporates several methods of smoothing as well as handling unknown words. Employing the tagger to either a new language or new tag-set is a simple process [17].

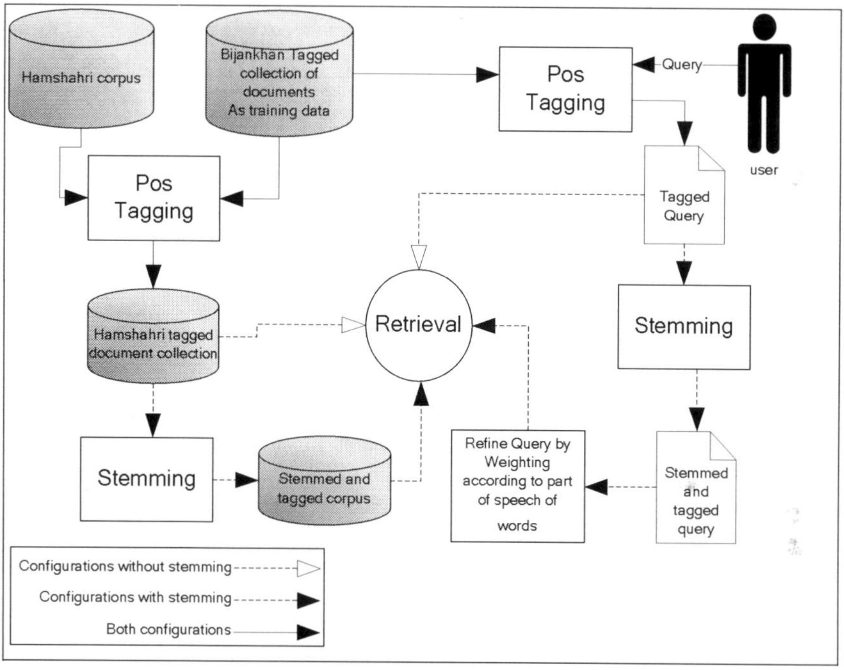

Fig. 1. The Framework of our approach

3 Methodology and Implementation

This study and experiments have been conducted as part of the Persian track at CLEF 2008 [18]. As a retrieval engine, we have utilized the Indri retrieval system [3] which is provided as part of the Lemur project[2]. TNT POS tagger was trained on Bijankhan POS collection with 40 tags. Subsequently the Hamshahri corpus [19] and its CLEF queries were tagged using this tagger (see Figure 1).

After experimenting with different tagging schemas, the corpus and the queries were stemmed in order to evaluate the effect of stemming and its interaction with POS tagging in retrieval context. Stemming was performed by employing simple

[1] TnT can be found at http://www.coli.uni-saarland.de/~thorsten/tnt/

[2] The Lemur Project. 2001-2008. University of Massachusetts and Carnegie Mellon University. [www.lemurproject.org]

grammatical rules using PERSTEM Persian stemmer [20]. Consequently we prepared 4 different variations of the Hamshahri document collection which included, normal (neither stemmed nor tagged), stemmed, tagged (terms tagged with related parts of speech), and both stemmed and tagged.

We conducted two types of experiments. In one set of experiments all terms were treated equally. That is, there were no preferences among the term types except for their statistical weight calculated by the Indri system. In the second set of experiments, we defined preferences among the term types based on their POS tags. For example in one experiment, nouns could have received a weight of 3 while verbs might have received a weight of 1, which means that the nouns were given three times more importance than the verbs. Experiments also differed based on what sections of the queries were used. Some experiments used only the title section of the queries and some others used both the title and the description sections of the CLEF queries. Table 1 lists the configurations used in our experiments.

Table 1. Different configurations

Config.	*Corpus*	*Query*
1	Normal	Title (Neither stemmed nor tagged)
2	Tagged	Title with equal weighting for all POS tags
3	Tagged	Title plus description with equal weighting for all POS tags
4	Stemmed	Stemmed title without POS tagging
5	Stemmed	Stemmed Title plus description
6	Stemmed (stop words removed)	Stemmed Title plus description (stop words removed)
7	Stemmed and tagged	Stemmed title with equal weighting for all POS tags
8	Tagged	Title with various weighting schemes for different POS tags

4 Results

Before discussing the results, it should be noted that since the Hamshahri collection has tagged automatically as described above we do not have any measurement of the accuracy of the tagging yet, however basic observations and sampling has shown reasonable accuracy.

Table 2 summarizes the outcomes of our experiments. The result of the base line system without employing tagging or stemming has an average precision of 27% and R-precision at 36%. By matching the tagged corpus with the tagged title of the queries the average precision climbs to 35% and the R-Precision increases by 1%. This is an interesting result since no part of speech preferences has been implemented in this run. This search is based on matching similar terms with similar roles in documents and queries. When the description field of the queries is added to the model, the performance of the system experiences a minor setback with the average precision at

29% which is still higher than the normal corpus and the R-Precision declins to 34% which again is a little higher than that of the normal retrieval performance. Generally we observed that adding descriptions in all configurations would degrade the performance of the system. The reason for this reduction is the negative effect of the extra terms in the query description that misleads the retrieval. We concluded that these misleading terms add more ambiguity than those POS tags can clarify.

Table 2. Main Results

Config.	*Average precision*	*R-Precision*
Normal corpus	0.2716	0.3627
Tagged (title)	0.3505	0.3784
Tagged (title + description)	0.2989	0.3497
Stemmed title	0.3625	0.4102
Stemmed (title + description)	0.1723	0.2157
Stemmed(title + description + stop words)	0.1672	0.2106
Stemmed and tagged (title)	0.3944	0.4151
Different weightings (average)	0.2263	0.2655

The results we obtained indicate that Persian retrieval benefits from stemming. Stemming the documents and queries alone returned one of the best results of our experiments with the average precision at 36% and R-Precision at 41%. This is in contrast with experiments conducted before by other groups in University of Tehran on the same corpus. However, when the title and description were used as query, the performance fell sharply. This configuration had one of the worst performances, even lower than the base line system. The reason of this poor outcome again was the extra text in the description which seemed to be too general and ambiguous. In this case stemming made the situation worse because it collapsed many surface words into a single representation and added to the ambiguity. In general, the effect of the stemming in Persian retrieval is still a research question. More experiments need to be performed with different types of stemmers as well as further scrutinizing the stemming techniques and their effect on Persian text retrieval. At the moment our conclusion is that the aggressive stemming is not useful and the simple stemming is sufficient.

Stop word removal is normally a very powerful tool in improving the precision. However, when stop word removal was applied to stemming of the title and the description of the queries, it did not improve the precision.

The best result of our experiments was achieved by stemming the tagged corpus and the title of the queries. This configuration produced an average precision of 39% which was the best. The R-precision in this case stays at 41%. In other words, combining simple stemming and part of speech tagging improves the average precision but does not change the R-precision. This shows that the stemming is more powerful than the part of speech tagging when it comes to precision.

Table 3. Weighting schemes

Noun	*Verb*	*Adjective*	*Adverb*	*Average Precision*	*R-Precision*
3	2	1	1	0.2635	0.3097
3	0	3	0	0.2597	0.2888
0	2	0	0	0.1108	0.1256
0	0	1	0	0.1198	0.1186
0	0	0	1	0.0977	0.1111
20 less used tags omitted, others equal weight				0.2745	0.3097

We explored the idea of POS tag preferences and their effect on precision. In these experiments, a weight of zero to three was given to each POS tag which then was multiplied with the actual weight of the term itself. So, we could emphasis or de-emphasis the contribution of the terms with certain part of speech tags. We explored many different combinations of preferences for different tags but in general we did not find any meaningful improvement in these experiments. Yet, on the contrary we found that many combinations have strong negative effect on precision. Table 3 depicts the results of some of these experiments. Assigning a weight of zero to a tag is the same as omitting the terms with that tag from the corpus and the queries. For example, (Noun=3, Verb=2, Adjective=1, Adverb=1) means the terms that are noun have been weighted three times the terms that are adjective or adverb. Similarly, the terms that are verbs have been weighted twice those of adjectives or adverbs. We also carried out experiments on the contribution of each tag to the overall performance of the retrieval. In some experiments as much as 20 least significant tags were omitted from the queries but it negatively affected the precision and recall. In general the average precision for all the tag weighting schemes was 0.22 and the average R-Precision was 0.26. The best run achieved a precision of 0.26 with R-precision of 0.31 which is much lower than one can achieve by simple stemming. The reason for such behavior can be explained by the importance of different tags in the Persian language. Despite our original study that led us to the omission of the 20 least important tags, they actually played a role in the retrieval. Thus omitting them or down playing their contribution declines the performance of the system.

5 Conclusion and Future Work

This study attempted to measure the effectiveness of part of speech tags and stemming on Persian information retrieval. Different configurations were tested and the results demonstrated that retrieving documents by matching the terms and their part of speech in documents with the terms and their part of speech in queries improves the performance. However, it was evident that while some parts of speech are more important than others, eliminating the least important ones or reducing their overall impact on the query processing degrades the performance of the system. The best results were achieved by giving equal importance to all POS tags.

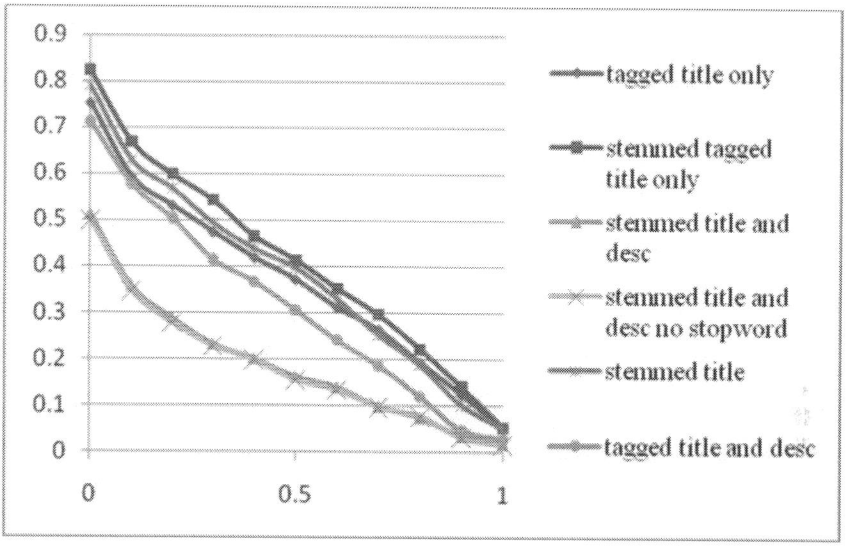

Fig. 2. R-Precision of the different configurations

The effect of stemming was also studied and it became clear that simple stemming in these experiments greatly improves precision. This study also observed that combining simple stemming and POS matching yields the best performance.

A future study would be utilizing retrieval models and systems other than Indri in order to make sure that the obtained results are not system dependent. However, given our previous experiences with different retrieval models on Persian language, we do not consider this as a major issue.

References

1. Witten, I., Moffat, A., Bell, T.: Managing Gigabytes: Compressing and Indexing Documents and Images. IEEE Transactions on Information Theory 41(6) (1995)
2. Singhal, A., Buckley, C., Mitra, M.: Pivoted document length normalization. In: Proc. 19th ACM SIGIR, pp. 21–29. ACM, New York (1996)
3. Strohman, T., Metzler, D., Turtle, H., Croft, W.: Indri: A Language-Model Based Search Engine for Complex Queries. Technical Report IR-407, CIIR, UMass Amherst (2005)
4. Liddy, E.D.: Automatic Document Retrieval. Encyclopedia of Language and Linguistics. Elsevier Press, Amsterdam (2005)
5. Lewis, D., Jones, K.: Natural Language Processing for Information Retrieval. Communications of the ACM 39(1), 92–101 (1996)
6. Amiri, H., AleAhmad, A., Oroumchian, F., Lucas, C., Rahgozar, M.: Using OWA Fuzzy Operator to Merge Retrieval System Results. In: Computational Approaches to Arabic Script-based Languages (2007)
7. Amiri, H., Hojjat, H., Oroumchian, F.: Investigation on a Feasible Corpus for Persian POS Tagging. In: Proc. 12th International CSI Computer Conference, CSICC (2007)

8. Raja, F., Amiri, H., Tasharofi, S., Sarmadi, M., Hojjat, H., Oroumchian, F.: Evaluation of Part of Speech Tagging on Persian Text. In: The Second Workshop on Computational Approaches to Arabic Script-Based Languages, Stanford University, U.S.A (2007)
9. Mohtarami, M., Amiri, H., Oroumchian, F.: Using Heuristic Rules to Improve Persian Part of speech Tagging Accuracy. In: Proc. 6th International Conference on Informatics and Systems, INFOS 2006 (2006)
10. Oroumchian, F., Tasharofi, S., Amiri, H., Hojjat, H., Raja, F.: Creating a Feasible Corpus for Persian POS Tagging. Technical Report, No. TR3/06, University of Wollongong, Dubai Campus (2006)
11. Shah, C., Bombay, I.I.T., Mumbai, P., Maharashtra, I., Bhattacharyya, P.: A Study for Evaluating the Importance of Various Parts of Speech (POS) for Information Retrieval (IR). In: Proc. International Conference on Universal Knowledge and Languages, ICUKL (2002)
12. Carlberger, J., Kann, V.: Implementing an Efficient Part-Of-Speech Tagger. Software Practice and Experience 29(9), 815–832 (1999)
13. BijanKhan, M.: The Role of the Corpus in Writing a Grammar: An Introduction to a Software. Iranian Journal of Linguistics 19(2) (2004)
14. Turney, P., Littman, M.: Unsupervised Learning of Semantic Orientation from a Hundred-Billion-Word Corpus. National Research Council of Canada (2002)
15. Paik, W., Liddy, E., Yu, E., McKenna, M.: Interpretation of Proper Nouns for Information Retrieval. In: Proc. Workshop on Human Language Technology, pp. 309–313. Association for Computational Linguistics Morristown, NJ (1993)
16. Klavans, J.L., Kan, M.Y.: The Role of Verbs in Document Analysis. In: Proc. Coling-ACL, vol. 36, pp. 680–686. Association for Computational Linguistics (1998)
17. Brants, T.: TnT–a Statistical Part-of-Speech Tagger. In: Proc. 6th Conference on Applied Natural Language Processing (ANLP 2000), Seattle, WA, pp. 224–231 (2000)
18. Agirre, E., Nunzio, G.M.D., Ferro, N., Mandl, T., Peters, C.: CLEF 2008: Ad Hoc Track Overview. In: Peters, C., et al. (eds.) CLEF 2008. LNCS, vol. 5706, pp. 15–37. Springer, Heidelberg (2009)
19. Aleahmad, A., Hakimian, P., Mahdikhani, F., Oroumchian, F.: N-gram and Local Context Analysis for Persian Text Retrieval. In: Proc. IEEE International Symposium on Signal Processing and its Applications, Sharjah, UAE, pp. 1–4 (2007)
20. Dehdari, J., Lonsdale, D.: A Link Grammar Parser for Persian. Aspects of Iranian Linguistics, vol. 1. Cambridge Scholars Press, Cambridge (2008)

Fusion of Retrieval Models at CLEF 2008 Ad Hoc Persian Track

Zahra Aghazade[1], Nazanin Dehghani[1], Leili Farzinvash[1], Razieh Rahimi[1], Abolfazl AleAhmad[1], Hadi Amiri[1], and Farhad Oroumchian[2]

[1] University of Tehran, School of Electrical and Computer Engineering, North Karegar Street, Tehran, Iran
{z.aghazadeh,n.dehghany,l.farzinvash, r.rahimi,a.aleahmad,h.amiri}@ece.ut.ac.ir
[2] University of Wollongong in Dubai, PO Box 20183, Dubai, UAE
FarhadOroumchian@uowdubai.ac.ae
http://ece.ut.ac.ir/dbrg

Abstract. Metasearch engines submit the user query to several underlying search engines and then merge their retrieved results to generate a single list that is more effective to the users information needs. According to the idea behind metasearch engines, it seems that merging the results retrieved from different retrieval models will improve the search coverage and precision. In this study, we have investigated the effect of fusion of different retrieval techniques on the performance of Persian retrieval. We use an extension of Ordered Weighted Average (OWA) operator called IOWA and a weighting schema, NOWA for merging the results. Our experimental results show that merging by OWA operators produces better MAP.

Keywords: Information Retrieval, Information Fusion, Persian Text Retrieval.

1 Introduction

With the rapid growth of the volume of the data, improving the effectiveness of information retrieval systems is essential. In this study, we try to use the idea behind metasearch engines in order to improve the results of Persian information retrieval. We consider each retrieval model as a decision maker and then fuse their decisions with an OWA operator in order to increase the effectiveness. This work has been done as our first participation in the CLEF evaluation campaign [1]. For the *ad hoc* Persian track we submitted eleven experiments (runs). Our main goal was to study the effect of fusion operators and whether fusing retrieval models can bring additional performance improvements. The collection that is used in this study is a standard test collection of Persian text which is called Hamshahri and was made available to CLEF by University of Tehran [2], [3].

In Section 2, we present a brief description of the retrieval methods that have been used in our experiments. Previous experiments have demonstrated that

C. Peters et al. (Eds.): CLEF 2008, LNCS 5706, pp. 97–104, 2009.

these methods have good performance on Persian retrieval.In Section 3, OWA operator and its extensions that are used for merging the results are described. One key point in the OWA operator is to determine its associated weights. In this study, we use a weighting model which is based on Normal distribution and an IOWA extension. There are two approaches to fuse the retrieved lists: (1) Combine the results of distinct retrieval methods, (2) Combine the results of the same method but with different types of tokens. Runs that submitted to CLEF 2008 use the first approach and results show that using this approach does not lend itself to a significant improvement. It seems although the retrieval methods are different but their performances and result sets are similar. In another word, those retrieval methods provide the same vision of the data. After CLEF results were published, we tried the second approach and we were able to improve the effectiveness up to 5.67% and reached the 45.22% MAP on the test set. Section 4 describes the experiments and their results.

2 Retrieval Methods

In this work, for the purpose of fusion, we needed different retrieval methods. After studying different retrieval toolkits, finally we choose *Terrier* [4]. Different methods have been implemented in *Terrier* toolkit. Among these methods, we selected nine of them. The weighting models and a brief description of them

Table 1. A description of retrieval methods

Weighting Model	Description
BB2	Bose-Einstein model for randomness, the ratio of two Bernoulli's processes for first normalization, and Normalization 2 for term frequency normalization
BM25	The BM25 probabilistic model
DFR_BM25	This DFR model, if expanded in Taylor's series, provides the BM25 formula, when the parameter c is set to 1.
IFB2	Inverse Term Frequency model for randomness, the ratio of two Bernoulli's processes for first normalization, and Normalization 2 for term frequency normalization
In_expB2	Inverse expected document frequency model for randomness, the ratio of two Bernoulli's processes for first normalization, and Normalization 2 for term frequency normalization
In_expC2	Inverse expected document frequency model for randomness, the ratio of two Bernoulli's processes for first normalization, and Normalization 2 for term frequency normalization with natural logarithm
InL2	Inverse document frequency model for randomness, Laplace succession for first normalization, and Normalization 2 for term frequency normalization
PL2	Poisson estimation for randomness, Laplace succession for first normalization, and Normalization 2 for term frequency normalization
TF_IDF	The tf*idf weighting function, where tf is given by Robertson's tf and idf is given by the standard Sparck Jones' idf

Table 2. Comparison between different weighting models

Weighting Model	MAP	R-Precision
BB2	0.3854	0.4167
BM25	0.3562	0.4009
DFR_BM25	0.3562	0.4347
IFB2	0.4017	0.4328
In_expB2	0.3997	0.4329
In_expC2	0.4190	0.4461
InL2	0.3832	0.4200
PL2	0.4314	0.4548
TF_IDF	0.3574	0.4017

(from [5]) are illustrated in Table 1. Table 2 depicts the result obtained from running the above nine methods described in Table 1 on the training set of queries.

3 OWA Fuzzy Operator

This section describes the Order Weighted Average (OWA) operator, normal distribution-based weighting and IOWA extension.

3.1 OWA Definition

An OWA operator of dimension n is a mapping $OWA : R^n \rightarrow R$, that has an associated n vector $w = (w_1, w_2, ..., w_n)^T$ such that $w_j \in [0, 1]$ and $\sum_{j=1}^{n} w_j = 1$. Furthermore,

$$OWA(a_1, a_2, ..., a_n) = \sum_{j=1}^{n} b_j w_j \tag{1}$$

where b_j is the jth largest element of the collection of the aggregated objects $a_1, a_2, ..., a_n$ [6].

3.2 IOWA

An IOWA operator is defined as follows:

$$IOWA(< u_1, a_1 >, < u_2, a_2 >, ..., < u_n, a_n >) = \sum_{j=1}^{n} w_j b_j \tag{2}$$

where $w = (w_1, w_2, ..., w_n)^T$ is a weighting vector, such that $\sum_{j=1}^{n} w_j = 1$, $0 \leq w_j \leq 1$ and b_j is the a_i value of the OWA pair $< u_i, a_i >$ having the jth largest u_i, and u_i in $< u_i, a_i >$ is referred to as the order inducing variable and a_i as the argument variable. It is assumed that a_i is an exact numerical value while u_i can be drawn from any ordinal set Ω [7]. The weighting vector which is used in our experiment will be defined in Section 4.

3.3 NOWA

Suppose that we want to fuse n preference values provided by n different individuals. Some individuals may assign unduly high or unduly low preference values to their preferred or repugnant objects. In such a case, we shall assign very low weights to these false or biased opinions, that is to say, the closer a preference value (argument) is to the mid one(s), the more the weight it will receive; conversely, the further a preference value is from the mid one(s), the less the weight it will have.

Let $w = (w_1, w_2, ..., w_n)^T$ be the weight vector of the OWA operator; then we define the following [8]:

$$w_i = \frac{1}{\sqrt{2\Pi}\sigma_n} e^{-[(i-\mu_n)^2/2{\sigma_n}^2]} \tag{3}$$

where μ_n is the mean of the collection of $1, 2, ..., n$, σ_n, $(\sigma_n > 0)$ is the standard deviation of the collection of $1, 2, ..., n$. μ_n and σ_n are obtained by the following formulas, respectively:

$$\mu_n = \frac{1}{n}\frac{n(n+1)}{2} \tag{4}$$

$$\sigma_n = \sqrt{\frac{1}{n}\sum_{i=1}^{n}(i-\mu_n)^2} \tag{5}$$

Consider that $\sum_{j=1}^{n} w_j = 1$ and $0 \leq w_j \leq 1$ then we have:

$$w_i = \frac{\frac{1}{\sqrt{2\Pi}\sigma_n} e^{-[(i-\mu_n)^2/2{\sigma_n}^2]}}{\sum_{j=1}^{n}\frac{1}{\sqrt{2\Pi}\sigma_n} e^{-[(j-\mu_n)^2/2{\sigma_n}^2]}} = \frac{e^{-[(i-\mu_n)^2/2{\sigma_n}^2]}}{\sum_{j=1}^{n} e^{-[(j-\mu_n)^2/2{\sigma_n}^2]}} \tag{6}$$

4 Experiment

For the experiments, CLEF has obtained the standard Persian test collection which is called Hamshahri. Hamshahri collection is the largest test collection of Persian text. This collection is prepared and distributed by University of Tehran. The third version of Hamshahri collection is 660 MB in size and contains more than 160,000 distinct textual news articles in Persian [9]. There were 50 training queries with their relevance judgments and 50 test queries prepared for the Persian *ad hoc* track. For the CLEF, we choose nine methods of document retrieval described above and fuse the top hundred retrieved results from each of them. The evaluation of the single IR models are depicted in Table 2.

We use OWA operator based on normal distribution weighting for merging the lists. In this problem, we have nine decision makers, so the weighting vector is as the following:

$$n = 9, \mu_9 = 5, \sigma_9 = \sqrt{\frac{20}{3}}, ornes(w) = 0.5, disp(w) = 2.1195, \tag{7}$$

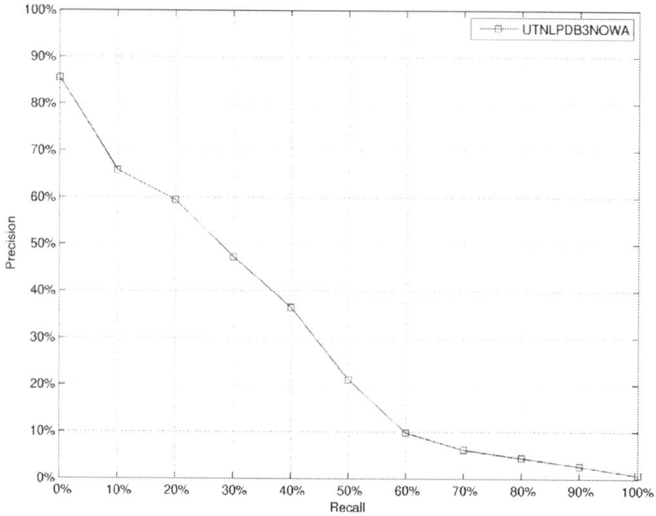

Fig. 1. The result of running NOWA published by CLEF 2008

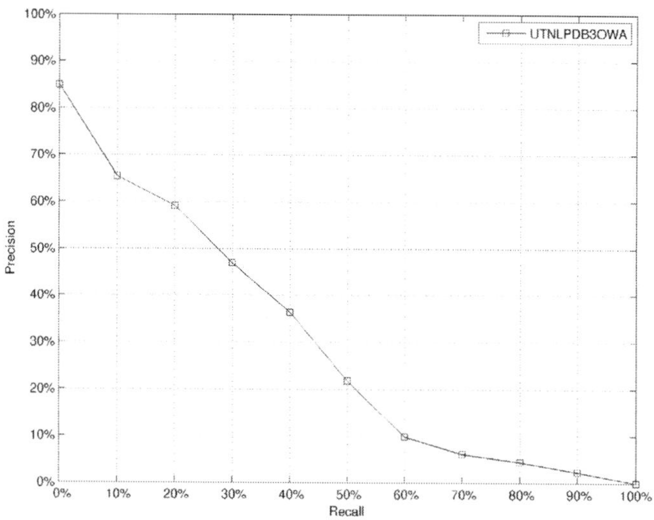

Fig. 2. The result of running IOWA published by CLEF 2008

$$w = (0.0506, 0.0855, 0.1243, 0.1557, 0.1678, 0.1557, 0.1243, 0.0855, 0.0506)^T \quad (8)$$

The precision-recall diagram obtained after submitting the OWA run to CLEF is illustrated in figure 1. IOWA extension was also tested. We used 50 training queries in order to calculate the weighting vector for this method. We ran the nine selected retrieval methods on the collection. The following weighting vector is obtained by using the average precision of each method as its weight. These percisions are obtained from Table 2:

0.4167/3.8409, 0.4009/3.8409, 0.4347/3.8409, 0.4328/3.8409, 0.4329/3.8409, 0.4461/3.8409, 0.42/3.8409, 0.4548/3.8409, 0.402/3.8409 (3.8409 is the sum of the obtained average precisions)

Figure 2 illustrates the precision-recall diagram of IOWA run with the above weighting vector.

5 Analyzing the Results and More Experiments

We submitted top hundred retrieved documents for our runs to CLEF, while CLEF evaluates the results by top thousand documents which decreased average precision about 10% in average. Therefore, in future we intend to calculate our Precision-Recall charts and other measurements based on the retrieved documents. The results published by CLEF for our fusion runs show that using fusion techniques on these methods does not yield to improved results over the individual methods. By analyzing the lists obtained from the retrieval methods, we observed that these result lists for these nine different methods have high overlap among them. On the other hand, fusion methods work well when there are significant differences between decision makers. Therefore, we have concluded that although the methods are different they are not significantly different from each other and basically they provide the same view of the collection.

After the CLEF results were published, we decided to investigate the second approach for fusion and looked the effect of different tokens in retrieval. For this purpose we chose a vector space model and ran it on the training set three times with three different types of tokens namely 4-grams, stemmed single terms and unstemmed single terms. To obtaining best results, we ran PL2 method of *Terrier* toolkit on 4-gram terms, Indri of Lemur toolkit [10] on stemmed terms and TF_IDF of *Terrier* toolkit on unstemmed terms. Then we applied the above OWA methods and as shown in Table 3, we obtained 9.97% improvements over individual runs.

After that, we continued this approach and did more experiments with the CLEF test set. On the test set, this approach lead only to 5.67% improvements on the average precision over individual runs using NOWA method and 5.6% using IOWA method. Table 4 demonstrate the obtained results.

Table 3. Comparison between different weighting models on the training set

Retrieval Method	MAP	R-Precision	Dif
TF_IDF with unstemmed single terms	0.4163	0.4073	
PL2 with 4-gram terms	0.4100	0.3990	
Indri with stemmed terms	0.4100	0.4183	
IOWA	0.5160	0.4928	+9.97%
NOWA	0.5030	0.4839	+8.67%

Table 4. Comparison between different weighting models on the test set

Retrieval Method	MAP	R-Precision	Dif
TF_IDF with unstemmed single terms	0.3847	0.4122	
PL2 with 4-gram terms	0.3669	0.3939	
Indri with stemmed terms	0.3955	0.4149	
IOWA	0.4515	0.4708	+5.6%
NOWA	0.4522	0.4736	+5.67%

6 Conclusion

Our motivation for participation in the *ad hoc* Persian track of CLEF was to investigate the influence of fusion techniques on the effectiveness of Persian retrieval methods. First we used nine retrieval methods and then fused the results by NOWA and IOWA. The obtained results showed that although there were some improvements on the overall performance but it was not significant. In the second stage, we changed our approach to use different types of tokens with the same method. To reach this goal, we focused on working with different types of terms instead of different methods. Results indicates that fusion produces better results under such circumstances although this improvement was under 10% on the training set and 6% on the test set.

In future, we will continue investigating the effects of different token types and retrieval engines on Persian retrieval and will try to fine tune an engine based on fusion.

References

1. Agirre, E., Di Nunzio, G.M., Ferro, N., Mandl, T., Peters, C.: Multilingual Textual Document Retrieval (Ad Hoc). In: Evaluating Systems for Multilingual and Multimodal Information Access 9th Workshop of the Cross-Language Evaluation Forum, CLEF 2008, Aarhus, Denmark (2008)
2. Oroumchian, F., Darrudi, E., Taghiyareh, F., Angoshtari, N.: Experiments with Persian Text Compression for Web. In: 13th international World Wide Web conference on Alternate track papers & posters, pp. 478–479. ACM, New York (2004)
3. Hamshahri Corpus, `http://ece.ut.ac.ir/dbrg/hamshahri`
4. Terrier Information Retrieval Platform, `http://ir.dcs.gla.ac.uk/terrier`
5. Amati, G., van Rijsbergen, C.J.: Probabilistic models of information retrieval based on measuring the divergence from randomness. ACM Transactions on Information Systems 20(4), 357–389 (2002)
6. Yager, R.R.: On ordered weighted averaging aggregation operators in multicriteria decision making. IEEE Transactions on Systems, Man and Cybernetics 18, 183–190 (1988)
7. Yager, R.R., Filev, D.P.: Induced ordered weighted averaging operators. IEEE Transactions on Systems, Man and Cybernetics–Part B 29, 141–150 (1999)
8. Min, D., Xu-rui, Z., Yun-xiang, C.: A Note on OWA Operator Based on the Normal Distribution. In: International Conference on Management Science and Engineering, pp. 537–542 (2007)

9. Amiri, H., AleAhmad, A., Oroumchian, F., Lucas, C., Rahgozar, M.: Using OWA Fuzzy Operator to Merge Retrieval System Results. In: The Second Workshop on Computational Approaches to Arabic Script-based Languages, Stanford University, USA (2007)
10. INDRI - Language modeling meets inference networks, `http://www.lemurproject.org/indri`

Cross Language Experiments at Persian@CLEF 2008

Abolfazl AleAhmad[1], Ehsan Kamalloo[1], Arash Zareh[1], Masoud Rahgozar[1], and Farhad Oroumchian[2]

[1] School of Electrical and Computer Engineering, University of Tehran
[2] Faculty of Computer Science and Engineering, University of Wollongong in Dubai

Abstract. In this study we will discuss our cross language text retrieval experiments of Persian ad hoc track at CLEF 2008. Two teams from University of Tehran were involved in cross language text retrieval part of the track using two different CLIR approaches that are query translation and document translation. For query translation we use a method named Combinatorial Translation Probability (CTP) calculation for estimation of translation probabilities. In the document translation part, we use the Shiraz machine translation system for translation of documents into English. Then we create a Hybrid CLIR system by score-based merging of the two retrieval system results. In addition, we investigated N-grams and a light stemmer in our monolingual experiments.

Keywords: Persian English cross language, Farsi bilingual text retrieval.

1 Introduction

The Persian language is categorized as a branch of Indo-European languages and is the official language of Iran, Afghanistan and Tajikistan and is also spoken in some other countries in the Middle East. This language has some characteristics that necessitate usage of different information retrieval algorithms. Morphological analysis of the language is relatively hard because of its grammatical rules [1]. For example the word "خبر" is an Arabic word that is used in Persian. This word has two plural forms in Persian "اخبار" and "خبرها", the first plural form obeys Arabic grammatical rules and the second plural form is obtained by use of Persian rules.

After creation of 50 new bilingual topics and standardization of Hamshahri collection according to CLEF standards, we could investigate CLIR on Persian. Persian@CLEF 2008 is our first attempt to evaluate cross language information retrieval on the language. Our aim is to investigate two main approaches of cross language text retrieval on Persian that are query translation and document translation.

We used the Hamshahri collection [2, 3] for evaluation of our retrieval methods. Documents of this collection are actually news articles of Hamshahri newspaper from year 1996 to 2002. The collection contains 160,000+ documents from variety of subjects. The documents size varies from short news (under 1 KB) to rather long articles (e.g. 140 KB) with the average of 1.8 KB. Also we used Apache Lucene [4] and Lemur toolkit [5] for indexing and retrieval on the collection.

The remaining parts of this paper are organized as follows: section 2 introduces our monolingual experiments, section 3 discusses our query translation method and its

C. Peters et al. (Eds.): CLEF 2008, LNCS 5706, pp. 105–112, 2009.

results, section 4 contains document translation experimental results and finally we will conclude our paper in section 5.

2 Experiments on Monolingual Persian Text Retrieval

There exist some morphological analyzers for Persian [6, 7] but their performance is much less than morphological analyzers of other languages like English. So, in our monolingual experiments we tried to investigate some alternative methods like n-grams. Also, we used a stop word list in monolingual part of our experiments to improve retrieval results. In order to create the stop word list, we manually checked most frequent words of the collection and extracted actual stop words. Then we added some other words from the Bijankhan Persian corpus [8, 9] that were marked with tags like proposition and conjunction. The final stop word list contains 796 items.

In the monolingual experiments, we submitted top 100 retrieved documents of six monolingual runs that are summarized in table 1 and their description is as follows:

- *Run #1*: This run uses the Lucene retrieval engine with vector space retrieval model using a light stemmer.
- *Run #2*: This run is the same as the previous run but the light stemmer is not used.
- *Run #3*: This run uses the Lemur retrieval engine with Language Modeling and 3-grams of the queries and documents are used.
- *Run #4*: This run is the same as the previous run but it uses 4-grams
- *Run #5*: This run is the same as the previous run but it uses 5-grams
- *Run #6*: This run is the same as the previous run but N-grams are not used

Table 1. Persian monolingual retrieval systems

Run#	Run Name	tot-ret	rel-ret	MAP	Retrieval Model	Retrieval System
1	SECMLSR	5161	1967	26.89	Vector Space	Lucene
2	SECMLUSR	5161	1991	27.08	Vector Space	Lucene
3	UTNLPDB1M3G	5161	1901	26.07	Language Modeling	Lemur
4	UTNLPDB1M4G	5161	1950	26.70	Language Modeling	Lemur
5	UTNLPDB1M5G	5161	1983	27.13	Language Modeling	Lemur
6	UTNLPDB1MT	5161	2035	28.14	Language Modeling	Lemur

In all of these runs we used title part of the 50 Persian topics that was made available at CLEF 2008. In the first run, we used a light Persian stemmer that works like the Porter algorithm but it could not improve our results because of the simple algorithm of the stemmer. As an example, consider the word "فیلم" that was a term in topic no 559. This word is a noun that means 'film' in English but our light stemmer considers the final 'م' letter of the word as a suffix and converts it to 'فیل' that means 'elephant' in English.

Also, it worth mentioning that we do not cross word boundaries for building N-grams. For example 4-gram of the word “ویمبلدون” is “ ویمب +یمبل + مبلد + بلدو +لدون” by use of our method.

3 CLIR by Query Translation

This section illustrates our query translation experiments at Persian ad hoc track of CLEF 2008. As the users query is expressed in English and the collection's documents are written in Persian, we used an English-Persian dictionary with 50,000+ entries for translation of the query terms. In addition, we inserted some proper nouns into the dictionary. The query translation process is accomplished as follows.

Let M be the number of query terms, then we define user's query as $Q = \{q_i\}$ $(i=1,\ldots,M)$, then we looked up each q_i in the dictionary and after finding translations of q_i we split the translations into its constituent tokens. Then we eliminate those tokens that are included in our Persian stop word list.

If we define T as the translation function that returns Persian translations set of a given English term q_i as described above, then we have $|T(q_1)|\times|T(q_2)|\times\cdots\times|T(q_M)|$ different possible translations for the query Q and as one can expect $|T(q_i)|>1$ for most of query terms. So, we need a retrieval model which enables us to take translation probabilities into consideration. For this purpose we use the Probabilistic Structured Query (PSQ) method [10] and for calculation of PSQ weights, in section 3.1 we propose our method for translation probability estimation. Then our query translation CLIR experimental results are presented in section 3.2.

3.1 Combinatorial Translation Probability

Translation probability is generally estimated from parallel corpus statistics. But as no parallel corpus is available for Persian, in this section we introduce a method which estimates English to Persian translation probabilities by use of the Persian collection itself. As most user queries contain more than two terms (e.g. in the Hamshahri collection all queries has two or more terms), the main idea is to use co-occurrence probability of terms in the collection for translation probability calculation of adjacent query terms.

Consider M as the number of user's query terms then we define the users query as $Q = \{q_i\}$ $(i=1,\ldots,M)$. For translation of Q, we look up Q members in an English-Persian dictionary to find their Persian equivalents. Considering T as the translation function, then we define set of translations of Q members as $E=\{T(q_1),T(q_2),\ldots,T(q_M)\}$, then the probability that two adjacent query terms q_i and q_{i+1} are translated into $E[i,x]$ and $E[i+1,y]$ respectively, is calculated from the following equation:

$$P(q_i \to E[i,x] \wedge q_{i+1} \to E[i+1,y]) = \frac{|D_{E[i,x]} \cap D_{E[i+1,y]}|}{c + Min(|D_{E[i,x]}|,|D_{E[i+1,y]}|)} \tag{2}$$

$$(x = 1..|T(q_i)|,\ y = 1..|T(q_{i+1})|)$$

Where $D_{E[i,x]}$ is a subset of collection's documents that contains the term *E[i,x]* and the constant *c* is a small value to prevent the denominator to become zero. In the next step we create translation probability matrix W_k for each pair of adjacent query terms:

$$W_k = \{w_{m,n}\} \ (m = 1..|T(q_k)|, n = 1..|T(q_{k+1})|)$$

Where $w_{m,n}$ is calculated using equation (2). Then Combinatorial Translation Probability (CTP) is a $|T(q_1)| \times |T(q_M)|$ matrix that is calculated by multiplication of all of the W_k matrices:

$$CPT(Q) = W_1 \times \ldots \times W_k \ (k = 1 \ldots M - 1)$$

In other words, CTP matrix contains probability of translation of *Q* members into their different possible translations in Persian. Given the *CTP(Q)* matrix, the algorithm in table 2 returns the TDimes matrix which contains dimensions of $E = \{T(q_1), T(q_2), \ldots, T(q_M)\}$ matrix that correspond to top *n* most probable translations of the query $Q = \{q_i\}$ $(i=1,\ldots,M)$.

Table 2. Calculation of the TDimes matrix

1. Let *TopRows[n]* be the row number of *n* largest members of CTP
2. Let *TopColumns[n]* be the column number of *n* largest members of CTP
3. For i ← [1,…,n]
   3.1. Let R = TopRows [i]
   3.2. Let C = TopColumns [i]
   3.3. TDimes[i,M] = C
   3.4. For j ← [M-1,…,1]
      If (j=1)
         Let TDimes[i,j]= R
      else
         Let TDimes [i,j]= the column number of the largest element of Rth row of W_{i-1}
4. Output the *TDimes* matrix

Having TDimes matrix, we are able to extract different translation of the users query from $E = \{T(q_1), T(q_2), \ldots, T(q_M)\}$ and their weight from CTP. For example if we consider an English query that has three terms then the most probable Persian translation of the query terms would be *E[1,TDimes [1,1]]*, *E[2,TDimes [1,2]]* and *E[3,TDimes [1,3]]* respectively and the translated query's weight would be *CTP[TopColumns[1],TopRows[1]]* .

3.2 Query Translation Experimental Results

We translated the queries through term lookup in an English-Persian dictionary as described before and using methods of PSQ [10] and section 3.1. All of our query translation experiments were ran using title of the English version of the 50 topics except run #8 in which we used title + description of the topics. In this part of our experiments we had eight runs that are summarized in table 3 and their description is as follows:

- *Run #1*: In this run we concatenate all meanings of each of the query terms to formulate a Persian query.
- *Run #2*: The same as previous run but uses top 5 Persian meanings of each of the query words for query translation.
- *Run #3*: The same as previous run but uses the first Persian meaning of each of the query words for query translation.
- *Run #4*: Uses all Persian meanings of query terms for query translation for calculating CTP. Then we used the PSQ method with top 10 most probable Persian translations of the query.
- *Run #5*: In this run we first look up top 5 meanings of query terms in the dictionary and then we convert them into 4-grams for calculating CTP. Then we use PSQ method with top 10 most probable Persian translations of the query to run 4-gram based retrieval.
- *Run #6*: The same as previous run but we use 5-grams instead of 4-grams.
- *Run #7*: This run is the same as run #3 but in this run we use the Lucene vector space retrieval model.
- *Run #8*: This run is the same as run #7 but in this run we use title + description. We eliminate common words such as 'find', 'information', from the topics description.

We used the Lemur toolkit [5] for implementation of our algorithm for run #1 to run #5. The default retrieval model of the lemur's retrieval engine (Indri) is language modeling. The Indri retrieval engine supports structured queries and we could easily implement the PSQ method using CPT for translation probability estimation. Also, run #7 and run #8 are implemented by use of the Lucene retrieval engine.

Table 3. English-Persian query translation experiments

Run#	Run Name	tot-rel	rel-ret	MAP	Dif	Retrieval Model	Retrieval System
1	UTNLPDB1BA	5161	758	6.73	baseline	LM	Lemur
2	UTNLPDB1BT5	5161	974	10.19	+ 3.46	LM	Lemur
3	UTNLPDB1BT1	5161	930	12.4	+ 5.67	LM	Lemur
4	UTNLPDB1BA10	5161	1150	14.07	+ 7.34	LM	Lemur
5	UTNLPDB1BT4G	5161	1196	14.46	+ 7.73	LM	Lemur
6	UTNLPDB1BT5G	5161	1166	14.43	+ 7.70	LM	Lemur
7	CLQTR	5161	677	8.93	+ 2.20	Vector Space	Lucene
8	CLQTDR	5161	592	6.01	- 0.72	Vector Space	Lucene

Also Figure 1 depicts the precision-recall graph of the eight runs for top 100 retrieved documents that are calculated by use of the Trec_Eval tool. According to the 'comparison of median average precision' figure that was released at Persian@CLEF 2008, this method could over perform monolingual retrieval results for some topics like topic no 570. This is because of the implicit query expansion effect of this method. The topic's title is 'Iran dam construction' and after its translation into Persian, the CTP method adds the word 'آب' to the query that means water in English.

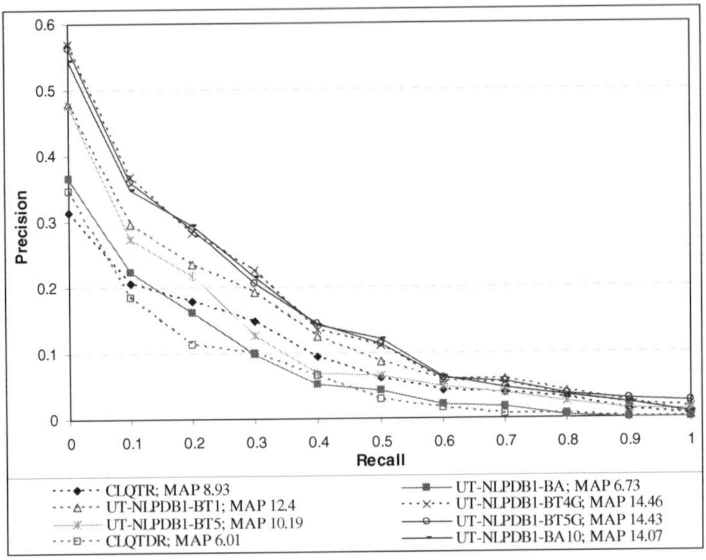

Fig. 1. Precision-Recall of the six query translation runs

4 CLIR by Document Translation

TIn order to translate the Hamshahri collection's documents from Persian into English, we used the Shiraz machine translation system that is prepared at the New Mexico State University [12]. The Shiraz machine translation system is an open source project that is written with the C language [13]. This system uses a bilingual Persian to English dictionary consisting of approximately 50,000 terms, a complete morphological analyzer and a syntactic parser. The machine translation system is mainly targeted at translating news material.

Document translation is not a popular approach because this approach of CLIR is not computationally efficient. This fact was also apparent in our experiments. We ran the Shiraz machine translation on a PC with 2G of RAM and an Intel 3.2G CPU and it took more than 12 days to translate nearly 80 percent of the collection. Finally we could translate 134165 out of 166774 documents of the collection and we skipped translation of long documents to save time. In our document translation experiments we had one run, named CLDTDR, by use of document translation that is described below:

- *Run #9*: In this run we use the English version of the 50 topics of Persian@CLEF 2008. Then we retrieved translated documents of the collection using the Lucene vector space retrieval engine. This run utilizes title + description part of the topics.

Furthermore, we tried a hybrid CLIR method by score-based merging of the results of query translation and document translation methods. For this purpose we used merge results of the CLDTDR and UTNLPDB1BT4G runs. The two runs used different retrieval engines and hence their retrieval scores were not in the same scale. To address this problem we used the following equation to bring the scores of the two retrieval lists into the same scale:

$$Score_i = \frac{x_i - Min(L_{i,q})}{Max(L_{i,q}) - Min(L_{i,q})}$$

In which x_i and $Score_i$ are the old and the normalized scores, $Min(L_{i,q})$ and $Max(L_{i,q})$ are the minimum and maximum scores in the ith retrieved list for the query q (i=1,2 for the two runs). This normalization normalizes the scores into the range [0, 1]. Then for obtaining the merged results we chose top 100 documents with highest weight from the two lists.

Table 4 and Figure 2 show performance of our query translation, document translation and hybrid CLIR systems and compare them with one of our monolingual systems as a baseline.

Table 4. Comparison of CLIR retireval experiments

Run Name	tot-el	rel-et	MAP	CLIR/Mono	Retrieval Model	Retrieval Sys.
SECMLUSR	5161	1967	27.08	baseline	Vector Space	Lucene
UTNLPDB1BT4G	5161	1196	14.46	53 %	LM	Lemur
CLDTDR	5161	1234	12.88	48 %	Vector Space	Lucene
Hybrid CLIR	5161	1478	16.19	60 %	LM + Vector Space	Lemur + Lucene

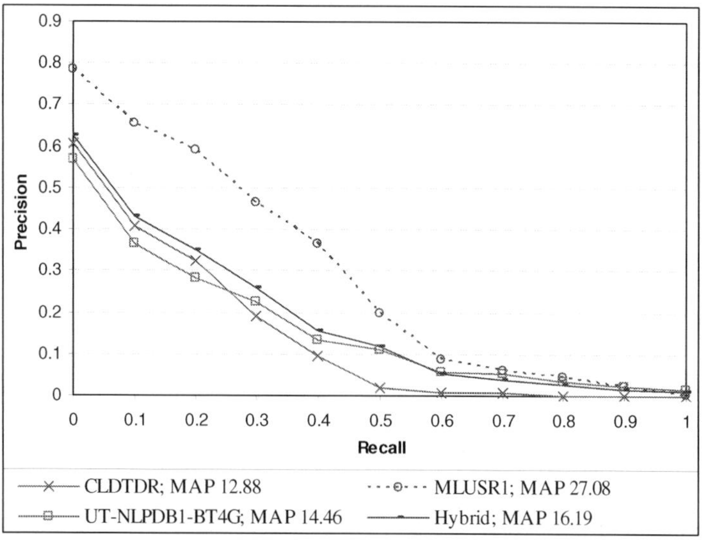

Fig. 2. Precision-Recall of CLIR experiments

5 Discussion and Future Works

In Persian ad hoc track of ninth CLEF campaign, in addition to some monolingual retrieval systems, we evaluated a number of cross language information retrieval systems. In monolingual part of our experiments we evaluated N-grams and a light stemmer on the Persian language and in cross language part we evaluated query translation and document translation approaches of English-Persian cross language information retrieval. We used combinatorial translation probability method for query translation that uses statistics of the target language for estimating translation probabilities. Result of our hybrid cross language information retrieval experiments also suggests usefulness of combining document translation and query translation.

Acknowledgements

We would like to thank CLEF 2008 organizers for their supports in development of the Hamshahri collection.

References

1. Taghva, K., Coombs, J., Pareda, R., Nartker, T.: Language Model-Based Retrieval for Persian Documents. In: International Conference on Information Technology: Coding and Computing, ITCC 2004 (2004)
2. Aleahmad, A., Amiri, H., Rahgozar, M., Oroumchian, F.: Hamshahri: a standard persian text collection. Journal of Knowledge-based systems (2008) (submitted)
3. Peters, C., Deselaers, T., Ferro, N., Gonzalo, J., Jones, J.F., Kurimo, M., Mandl, T., Peñas, A., Petras, V.: Evaluating Systems for Multilingual and Multimodal Information Access. In: 9th Workshop of the Cross-Language Evaluation Forum, Aarhus, Denmark (2008)
4. Apache Lucene project, `http://lucene.apache.org/` (cited September 1, 2008)
5. Lemur Toolkit, `http://www.lemurproject.org/` (cited September 1, 2008)
6. Tashakori, M., Meybodi, M.R., Oroumchian, F.: Bon: The Persian Stemmer. In: Shafazand, H., Tjoa, A.M. (eds.) EurAsia-ICT 2002. LNCS, vol. 2510, pp. 487–494. Springer, Heidelberg (2002)
7. Mokhtaripour, A., Jahanpour, S.: Introduction to a new Farsi stemmer. In: 15th ACM international conference on Information and knowledge management, Arlington, Virginia, USA, pp. 826–827 (2006)
8. Amiri, H., Hojjat, H., Oroumchian, F.: Investigation on a Feasible Corpus for Persian POS Tagging. In: 12th international CSI computer conference, Tehran, Iran (2007)
9. Bijankhan Corpus, `http://ece.ut.ac.ir/dbrg/bijankhan/` (cited September 1, 2008)
10. Pirkola, A.: The effects of query structure and dictionary setups in dictionary-based cross-language information retrieval. In: 21st Annual International ACM SIGIR Conference on Research and Development in Information Retrieval, pp. 55–63. ACM Press, New York (1998)
11. Darwish, K., Oard, D.W.: Probabilistic structured query methods. In: 26th International ACM SIGIR Conference on Research and Development in Information Retrieval, pp. 338–344. ACM Press, New York (2003)
12. Amtrup, J.W., Mansouri Rad, H., Megerdoomian, K., Zajac, R.: Persian-English Machine Translation: An Overview of the Shiraz Project. In: Memoranda in Computer and Cognitive Science, New Mexico State University (2000)
13. Shiraz Project, `http://crl.nmsu.edu/Research/Projects/shiraz` (cited September 1, 2008)

Evaluating Word Sense Disambiguation Tools for Information Retrieval Task

Fernando Martínez-Santiago, José M. Perea-Ortega, and Miguel A. García-Cumbreras

SINAI Research Group*, Computer Science Department, University of Jaén, Spain
{dofer,jmperea,magc}@ujaen.es

Abstract. The main interest of this paper is the characterization of queries where WSD is a useful tool. That is, which issues must be fulfilled by a query in order to apply an state-of-art WSD tool? In addition, we have evaluated several approaches in order to apply WSD. We have used several types of indices. Thus, we have generated 13 indices and we have carried out 39 different experiments, obtaining that some indices based on WSD tools even outperforms slightly the non disambiguated baseline case. After the interpretation of our experiments, we think that only queries with terms very polysemous and very high IDF value are improved by using WSD.

1 Introduction

Nowadays, the information unit managed by most IR models is the word. A theoretical good idea is the elaboration of IR systems based on concepts better than words or the lemmas of those words. We define a concept as a lexicographic-independent representation of an idea or object. Given a language, it does not care the vocabulary available in order to represent such a concept. Thus, a concept-based IR system translates words into concepts. State-of-art WSD tools obtain about 60% of precision/recall [1] [2] for "*fine-grained all words*" task[1]. Is this enough to improve an IR system? For which topics an improvement can be achieved and which topics performance deteriorates? Which features might be good predictors for improvement by WSD? After the interpretation of our experiments, we think that only queries with terms very polysemous and very high IDF value are improved by using WSD.

2 Experimental Framework

In the experiments carried out in this paper we have used two disambiguated collections provided by the NUS [1] and UBC [2] teams, and the default collection

* http://sinai.ujaen.es

[1] Fine-grained all words is the name of a usual WSD task. In this paper, we have used WSD in a very similar way.

C. Peters et al. (Eds.): CLEF 2008, LNCS 5706, pp. 113–117, 2009.

without WSD data provided by the Cross Language Evaluation Forum (CLEF) in its *Robust WSD* task [3].

For each disambiguated collection we have generated four different indices:

- **A1 index type.** This index stores the concatenation of each stem and its *synset code* which has the highest score. By example, *televis04745188n.*
- **B1 index type.** This index stores only the *synset code* which has the highest score for each disambiguated token. Thus, only the *synset code* is stored. By example, *04745188n.*
- **A2 index type.** It is the same as A1 index type but adding the two *stem+synset* which have the highest score.
- **B2 index type.** It is the same as B1 index type but adding the two *synset codes* which have the highest score.

In addition to these 8 indexes (4 for UBC team and 4 for NUS team), we have generated four common indices (common-A1, common-B1, common-A2 and common-B2), merging a token from each disambiguated collection. Therefore, we have generated a total of 12 different indices for the experiments with WSD data.

For the default collection without WSD data we have preprocessed it as usual (stemming and stopwords removed). Since disambiguated collections take into account WordNet multi-words as index tokens, we have marked such multi-word in the document collection. This approach performed worst but needs to be reported for better comparison.

3 Experiments and Results

We report only the most relevant results obtained by using both disambiguated collections and the systems developed by the NUS and UBC teams.

The selected set of experiments is depicted in Table 1. As we expect, applying WSD in "*a blind way*" to improve IR does not work as well as we would like. We do not think that results get better making use of other WSD tool, since the collections were disambiguated using a state-of-art disambiguation software. On the other hand, the *synsets*-based indices improve the indices based on *term+synset.* We conclude that taking into account the synonyms, it leads to substantial improvement. Finally, the mixed approach merging NUS and UBC synsets, outperforms very slightly the experiments based on the NUS WSD system. We think that both systems are similar but NUS WSD system outperforms the UBC one, so the addition of the UBC system does not lead to any improvement at all.

3.1 When Should We Apply WSD?

State-of-art WSD systems must not be applied in the same way as other usual IR techniques such as pseudo-relevance feedback (PRF) or stemming, for example. Our research interest revolves around the question: Which queries benefit from WSD and how can we recognize these queries?

Table 1. The most outstanding results (TD only)

Experiment	WSD system	Index unit	AvgP
NUS-indexA1-TD	NUS team	stem+sysnset (type A1)	0.368
NUS-indexB1-TD	NUS team	synset only (type B1)	**0.376**
NUS-indexA2-TD	NUS team	two first stem+synsets (type A2)	0.321
NUS-indexB2-TD	NUS team	two first synsets (type B2)	0.315
UBC-indexA1-TD	UBC team	stem+sysnset (type A1)	0.315
UBC-indexB1-TD	UBC team	synset only (type B1)	0.323
UBC-indexA2-TD	UBC team	two first stem+synsets (type A2)	0.282
UBC-indexB2-TD	UBC team	two first synsets (type B2)	0.279
MIXED-A1-TD	NUS+UBC team	stem+sysnset (type A1)	0.366
MIXED-B1-TD	NUS+UBC team	synset only (type B1)	**0.381**
MIXED-A1-TD	NUS+UBC team	two first stem+synsets (type A2)	0.314
MIDED-B1-TD	NUS+UBC team	two first synsets (type B2)	0.326
Baseline case	none	stem of the word	**0.374**

In order to carry out a more detailed analysis of the results, we compared the *baseline* and "*NUS-indexB1*" (disambiguation by using NUS WSD system) cases. *NUS-indexB1* obtained better average precision than the baseline case in 58 queries. It means an improvement of 36.2% of queries by using disambiguated queries. If we count only the queries improved more than 10%, a remarkable 28.6% (46 queries) is obtained. Thus, we aim to recognize a common set of properties in order to define these sets of queries for applying WSD properly for the IR task. The first hypothesis we investigated was: "*very polysemous queries will be improved by WSD*". If we take into account the original 160 queries, the average number of senses per word is 2.39 (*stopwords* have been removed). If we take into account the 58 queries improved by using WSD, the average number of senses per word is 2.37. Finally, the average number of senses per word is 2.43 for not improved queries by using WSD (102 queries). These results are disappointing. A more detailed analysis reveals that non-empty words such as "*find*" or "*information*" are very common. In addition, these words are polysemous and they have a very poor semantic weight.

Table 2 shows some queries where the difference between the baseline case and disambiguated index is noteworthy. The next step is the evaluation at the term level. In order to get an idea of the situation, we analyze some words. Results are depicted in Table 3. This is a preliminary work, but there some interesting issues:

- There are words with very low IDF and very polysemous. For instance, "*give*" is not a very interesting word for usual IR systems. Anyway, if the IR system uses an index based on *synsets*, then the IDF of each word increases because of polysemy: obviously, in an index based on *synsets*, every sense of each word will obtain an IDF higher than the corresponding word in an index based on stems or lemmas.

Table 2. Some queries where the difference between the baseline case and disambiguated index is noteworthy

Query id	Query text (Title+Description)	AvgP (baseline)	AvgP (*NUS-indexB1*)	Avg. of word senses
10.2452/180-AH	Bankruptcy of Barings.	0.025	0.765	3.71
10.2452/151-AH	Wonders of Ancient World.	0.061	0.571	3.54
10.2452/190-AH	Child Labor in Asia.	0.887	0.123	2.07
10.2452/252-AH	Pension Schemes in Europe.	0.444	0.15	4

Table 3. Some terms and data about these terms

Term	Query id	IDF	*synset* (NUS)	NUS Confidence	Correct sense?
bankruptcy	10.2452/180-AH	4.76	0386165-n	0.52	Yes
ancient	10.2452/151-AH	4.42	01665065-a	0.43	Yes
world	10.2452/151-AH	1.64	06753779-n	0.13	No
child	10.2452/190-AH	2.94	07153837-n	0.79	No
give	10.2452/252-AH	1.597	01529684-v	0.37	No

- On the other hand, there are words like "*bankruptcy*" or "*ancient*" in which the IDF is high, so the IDF of the corresponding disambiguated *synset* will be high, too. If the WSD software has a high confidence in order to assign the correct sense, then it might be a good candidate for disambiguation.

In order to obtain reliable conclusions we need a very elaborate list of words and a lot of information about how the word is being disambiguated in the query and in the document collection, and how the correct/erroneous disambiguation of the word affects to the final score of the document. Anyway, we would go so far as to say a first approximation: words with low IDF and a high number of senses must not be disambiguated. On the other hand, words with high IDF and high disambiguation confidence must be disambiguated. The selective application of WSD might be beneficial for IR.

4 Conclusions and Future Work

State-of-art WSD is not an useful tool for every query, for every term of every query, but we think that some queries could be improved by using WSD. In this paper we investigate queries where WSD gets better results. We find that there are situations where WSD must be used, but these scenarios are very specific. Since some queries are improved by WSD and some queries not at all, if we want to apply WSD in a good way we have to manage two indices per collection: the disambiguated one and the stem-based one. In addition, IR system will have to carry out an additional analysis of the user query in order to take a decision about which of both indices seem more suitable for each user query. Even more, we think that, given a user query, some words should be disambiguated and others do not. This requires a more detailed study based on the confidence of the WSD system, the semantic weight of the term, and other factors that we are investigating nowadays.

Finally, we have reported a set of experiments: we have created indices based on the best sense per term, two first senses per term, term+sense, NUS-best sense+UBC+best sense and the only experiment that outperforms the base line is the one based on NUS best sense index.

Acknowledgments

This work has been supported by a grant from the Spanish Government, project TIMOM (TIN2006-15265-C06-03), and the RFC/PP2006/Id_514 granted by the University of Jaén.

References

1. Cai, J.F., Lee, W.S., Teh, Y.W.: Nus-ml:improving word sense disambiguation using topic features. In: Proceedings of the Fourth International Workshop on Semantic Evaluations (SemEval 2007), Prague, Czech Republic, pp. 249–252 (2007)
2. Agirre, E., de Lacalle, O.L.: Ubc-alm: Combining k-nn with svd for wsd. In: Proceedings of the Fourth International Workshop on Semantic Evaluations (SemEval 2007), Prague, Czech Republic, pp. 342–345 (2007)
3. Agirre, E., Nunzio, G.M.D., Ferro, N., Mandl, T., Peters, C.: CLEF 2008: Ad Hoc Track Overview. In: Peters, C., et al. (eds.) CLEF 2008. LNCS, vol. 5706, pp. 15–37. Springer, Heidelberg (2008)

IXA at CLEF 2008 Robust-WSD Task: Using Word Sense Disambiguation for (Cross Lingual) Information Retrieval

Eneko Agirre, Arantxa Otegi, and German Rigau

IXA NLP Group - University of Basque Country, Donostia, Basque Country
arantza.otegi@ehu.es

Abstract. This paper describes experiments for the CLEF 2008 Robust-WSD task, both for the monolingual (English) and the bilingual (Spanish to English) subtasks. We tried several query and document expansion and translation strategies, with and without the use of the word sense disambiguation results provided by the organizers. All expansions and translations were done using the English and Spanish wordnets as provided by the organizers and no other resource was used. We used Indri as the search engine, which we tuned in the training part. Our main goal was to improve (Cross Lingual) Information Retrieval results using WSD information, and we attained improvements in both mono and bilingual subtasks, with statistically significant differences on the second. Our best systems ranked 4th overall and 3rd overall in the monolingual and bilingual subtasks, respectively.

1 Introduction

Our experiments intended to test whether word sense disambiguation (WSD) information can be beneficial for Cross Lingual Information Retrieval (CLIR). We carried out different expansion and translation strategies of both the topics and documents with and without word sense information. For this purpose, we used thef open source Indri search engine, which is based on the inference network framework and supports structured queries [7].

The remainder of this paper is organized as follows. Section 2 describes the experiments carried out, Section 3 presents the results obtained, Section 4 describes some related work and, finally, Section 5 draws the conclusions and mentions future work.

2 Experiments

In short, our main experimentation strategy consisted on trying several expansion and translation strategies, all of which used the synonyms in the English and Spanish wordnets made available by the organizers as the sole resources

C. Peters et al. (Eds.): CLEF 2008, LNCS 5706, pp. 118–125, 2009.

(i.e., we did not use any other external resource), with and without word sense information. Our runs have consisted of different combinations of expanded (translated) topics and documents. The steps of our retrieval system are the following. We first expand and translate the documents and topics. In a second step we index the original, expanded and translated document collections. Then we test different query expansion and translation strategies, and finally we search for the queries in the indexes in various combinations. All steps are described sequentially.

2.1 Expansion and Translation Strategies

WSD data provided to the participants was based on WordNet version 1.6. Each word sense has a WordNet synset assigned with a score. Using those synset codes and the English and Spanish wordnets, we expanded both the documents and the topics. In this way, we generated different topic and document collections using different approaches of expansion and translation, as follows:

- Full expansion of English topics and documents: expansion to all synonyms of all senses.
- Best expansion of English topics and documents: expansion to the synonyms of the sense with highest WSD score for each word, using either UBC or NUS disambiguation data (as provided by organizers).
- Full translation of English documents: translation from English to Spanish of all senses.
- Best translation of English documents: translation from English to Spanish of the sense with highest WSD score for each word, using either UBC or NUS disambiguation data.
- Translation of Spanish topics: translation from Spanish to English of the first sense for each word, taking the English variants from the WordNet.

In the subsequent steps, we used different combinations of these expanded and translated collections.

2.2 Indexing

Once the collections had been pre-processed, they were indexed using Indri. While indexing, the Indri implementation of the Krovetz stemming algorithm was applied to document terms. We created several indexes: one with the original collection words, and one with each collection created after applying different expansion (and translation) strategies, as explained in Section 2.1. No stopword list was used, but only nouns, adjectives, verbs and numbers were indexed.

2.3 Query Construction

We constructed queries using the title and description topic fields. Based on the training topics, we excluded some words and phrases from the queries, such as *find, describing, discussing, document, report* for English and *encontrar, describir, documentos, noticias, ejemplos* for Spanish. After excluding those words and taking only nouns, adjectives, verbs and numbers, we constructed several queries for each topic as follows:

1. Original words.
2. Both original words and expansions for the best sense of each word.
3. Both original words and all expansions for each word.
4. Translated words, using translations for the best sense of each word. If a word had no translation, the original word was included in the query.

The first three cases are for the monolingual runs, and the last one for the bilingual run which translated the query. Table 1 shows some examples of each case for the sample topic.

In the first case, we constructed a simple query combining the original words using the Indri operator `#combine` (see *case 1* in Table 1). Note that multiword expressions (as present in WordNet), such as *alternative medicine*, are added to the query joined with the `#1` operator (ordered window).

For the rest of cases, we have used some other operators available in the structural Indri Query Language. For *case 2*, where we include original words as

Table 1. Query examples using the title and description fields of a topic. Check Section 2.3 for further explanations.

English topic	*<EN-title>Alternative Medicine</EN-title>* *<EN-desc>Find documents discussing any kind of alternative or natural medical treatment including specific therapies such as acupuncture, homeopathy, chiropractics, or others</EN-desc>*
Spanish topic	*<ES-title>Medicina Alternativa</ES-title>* *<ES-desc>Encontrar documentos que traten sobre algún tipo de tratamiento medico alternativo o naturista, incluyendo terapias concretas como la acupuntura, la homeopatía, la quiropráctica, u otras</ES-desc>*
case 1	`#combine(#1(alternative medicine) kind alternative natural medical treatment including specific therapies acupuncture homeopathy chiropractics others)`
case 2	`#weight(0.6 #combine(#1(alternative medicine) kind alternative natural medical treatment including specific therapies acupuncture homeopathy chiropractics others) 0.4 #combine(#syn(#1(complementary medicine) #1(alternative medicine)) #syn(variety form sort) #syn(option choice) #syn(include) #syn(therapy) #syn(stylostixis) #syn(homoeopathy) #syn(chiropractic)))`
case 3	`#weight(0.6 #combine(#1(alternative medicine) kind alternative natural medical treatment including specific therapies acupuncture homeopathy chiropractics others) 0.4 #combine(#wsyn(1 #1(complementary medicine) 1 #1(alternative medicine)) #wsyn(1 form 1 variety 1 sort) #wsyn(1 option 1 choice) #wsyn(0 nonsynthetic 0 uncontrived 0 misbegot 0 unaffected 0 spurious 0 bastardly 0 lifelike 0 bastard 0 wild 0 rude 0 spontaneous 0 misbegotten 0 unstudied 0 raw) #wsyn(0 aesculapian) #wsyn(0 discussion 0 discourse 0.414874001229255 handling) #wsyn(0 admit 0 #1(let in) 1 include) #wsyn(1 therapy) #wsyn(1 stylostixis) #wsyn(1 homoeopathy) #wsyn(1 chiropractic)))`
case 4	`#combine(#syn(#1(alternative medicine) #1(complementary medicine)) type treatment #syn(medicate medicine) #syn(alternate alternative) #syn(naturistic nudist) include concrete #syn(acupuncture stylostixis) #syn(homeopathy homoeopathy) quiropráctica )`

well as synonyms (obtained after expansion) in the query, we constructed two subqueries, one with original words, and another one with the expanded words. Both subqueries are combined into a single query using the `#weight` operator, where original words are weighted with 0.6, and synonyms with 0.4. We did not fine-tune this weights. We used the synonym operator (`#syn`) to join the expanded words of each sense, as they are meant to be synonyms.

In the case of full expansion (*case 3*), instead of `#syn`, we used `#wsyn` (weighted synonym). This operator allows to give different weights to synonyms, which we took from the score returned by the disambiguation system, that is, each synonym was weighted according to the WSD weight of the corresponding sense of the target word.

For *case 4*, we constructed the query using the first sense of each word of the Spanish topics in order to get their translated English words. In the Spanish topic of the example, as *quiropractica* had not any sense assigned, we could not get its translation and therefore, we included the original Spanish word in the query (see *case 4* in Table 1).

2.4 Retrieval

We carried out several retrieval experiments combining different kinds of indexes with different kinds of queries. We used the training data to perform extensive experimentation, and chose the ones with best MAP results in order to produce the test topic runs. The submitted runs are described in Section 3.

In some of the experiments we applied pseudo-relevance feedback (PRF) with the following default parameters: fbDocs:10, fbTerms:50, fbMu:0 and fbOrigWeight: 0.5. Unfortunately, we did not have time to tune those parameters for the official deadline.

3 Results

Table 2 summarizes the results of our submitted runs. We present them here, as follows:

- monolingual without WSD:
  En2EnNowsd; original terms in topics; original terms in documents.
  En2EnNowsdPsrel; same as `En2EnNowsd`, but with PRF.
- monolingual with WSD:
  En2EnNusDocsPsrel; original terms in topics; both original and expanded terms in documents, using best sense according to NUS word sense disambiguation; PRF.
  En2EnUbcDocsPsrel; original terms in topics; both original and expanded terms in documents, using best sense according to UBC word sense disambiguation; PRF.
  En2EnFullStructTopNusDocsPsrel; both original and fully expanded terms in topics; both original and expanded terms in documents, using best sense according to NUS word sense disambiguation; PRF.

- bilingual without WSD:
  Es2EnNowsd; original terms in topics (in Spanish); translated terms in documents (from English to Spanish).
  Es2EnNowsdPsrel; same as `Es2EnNowsd`, but with PRF.
- bilingual with WSD:
  Es2EnNusDocsPsrel; original terms in topics (in Spanish); translated terms in documents, using the best sense according to NUS word sense disambiguation; PRF.
  Es2EnUbcDocsPsrel; original terms in topics (in Spanish); translated terms in documents, using the best sense according to UBC word sense disambiguation; PRF.
  Es2En1stTopsNusDocsPsrel; translated terms in topics (from Spanish to English) for first sense in Spanish; both original and expanded terms of the best sense according to NUS disambiguation data; PRF.
  Es2En1stTopsUbcDocsPsrel; translated terms in topics (from Spanish to English) for first sense in Spanish; both original and expanded terms of the best sense according to UBC disambiguation data; PRF.

The results show that the use of WSD data has been effective. With respect to monolingual retrieval, `En2EnUbcDocsPsrel` obtains the best results from our runs, although the difference with respect to `En2WnNowsdPsrel` is not statistically significant[1]. Regarding the bilingual results, `Es2En1stTopsUbcDocsPsrel` is the best, and the difference with respect to `Es2EnNowsdPsrel` is statistically significant. These results confirm the results that we obtained on the training data. Although not shown here, those results showed that the use of WSD led to significantly better results with respect to using all senses (full expansion).

Table 2. Results for submitted runs

		runId	map	gmap
monolingual	no WSD	En2EnNowsd	0.3534	0.1488
		En2EnNowsdPsrel	**0.3810**	0.1572
	with WSD	En2EnNusDocsPsrel	0.3862	0.1541
		En2EnUbcDocsPsrel	**0.3899**	0.1552
		En2EnFullStructTopsNusDocsPsrel	0.3890	0.1532
bilingual	no WSD	Es2EnNowsd	0.1835	0.0164
		Es2EnNowsdPsrel	**0.1957**	0.0162
	with WSD	Es2EnNusDocsPsrel	0.2138	0.0205
		Es2EnUbcDocsPsrel	0.2100	0.0212
		Es2En1stTopsNusDocsPsrel	0.2350	0.0176
		Es2En1stTopsUbcDocsPsrel	**0.2356**	0.0172

[1] We used paired Randomization Tests over MAPs with α=0.05.

Although it was not our main goal, our systems ranked high in the exercise, making the 7th best in the monolingual no-WSD subtask, 9th in monolingual using WSD, 5th best in the bilingual no-WSD subtask, and 1st in bilingual using WSD. Overall, our best runs ranked 4th overall and 3rd overall in the monolingual and bilingual subtasks, respectively.

After analyzing the experiments and the results, we have found that the approach of expanding the documents works better than expanding the topics. The extensive experimentation that we performed on the use of structured queries did not yield better results than just expanding the documents.

In our experiments we did not make any effort to deal with hard topics, and we only paid attention to improvements in Mean Average Precision (MAP) metric. In fact, we applied the settings which proved best in training data according to MAP, and we did not pay attention to the Geometric Mean Average Precision (GMAP) values.

4 Related Work

Several teams have managed to successfully use word sense data. Stokoe et al. [6] developed a system that performed sense-based information retrieval which, when used in a large scale IR experiment, demonstrated improved precision over the standard term-based vector space model. They noted that with a word-sense disambiguation accuracy of only 62.1% the experiments showed an absolute increase of 1.73% and a relative increase over TF*IDF of 45.9%. The authors thing that their results support Gonzalo et al. [1] less conservative claim that a breakeven point of 50-60% would be adequate for improved IR performance.

Liu et al. [3] used WordNet to disambiguate word senses of query terms. They employed high-precision disambiguation of query terms for selective query expansion. Whenever the sense of a query term was determined, its synonyms, hyponyms, words from its definition and its compound words were considered for possible additions to the query. Experimental results showed that their approach yielded between 23% and 31% improvements over the best-known results on the TREC 9, 10 and 12 collections for short (title only) queries, without using Web data. In subsequent work [4], they showed that word sense disambiguation together with other components of their retrieval system yielded a result which was 13.7% above than produced by the same system but without disambiguation.

Kim et al. [2] assigned coarse-grained word senses defined in WordNet to query terms and document terms by an unsupervised algorithm which used co-occurrence information constructed automatically. Promising results were obtained when combined with pseudo relevance feedback and state-of-the-art retrieval functions such as BM25.

Finally, Pérez-Agüera and Zaragoza [5] devise a novel way to use word sense disambiguation data. They make explicit some of the term dependence information using a form of structured query, and use a ranking function capable of taking the structure information into account. They combined the use of query expansion techniques and semantic disambiguation to construct the structured queries, yielding queries that are both semantically rich and focused on the query. They report improved results on the same dataset reported here.

Compared to previous work, our own is less sophisticated, but we provide indications that word sense disambiguation on the documents, accompanied by expansion, produces better results than a similar strategy on the queries. All in all, our approach is complementary to other work, and suggests that experimentation on the document side can offer further improvements.

5 Conclusions and Future Work

We have reported our experiments for the Robust-WSD Track at CLEF. All our runs ended up in good ranking, taking into account that these have been our first experiments in the field of information retrieval. This is remarkable, as we did not use any external resources, except the WSD information and Spanish and English wordnets provided by the organizers. Note also that we did not do any proper parameter tuning (e.g. in the relevance feedback step) on the training part.

Our main goal was to get better (CL)IR results using WSD and we achieved it, obtaining remarkable gains in bilingual IR, and smaller gains in monolingual IR. We discovered that using WSD information for document expansion is a good strategy, in contrast to most previous IR work, which has focused on WSD of topics.

For the future, we plan to improve the bilingual results, mainly incorporating external resources like bilingual dictionaries. Our main goal will be to pursue more sophisticated methods for expansion and indexing of documents using WSD information, beyond the simple combinations tried in this paper.

Acknowledgements

This work has been supported by KNOW (TIN2006-15049-C03-01) and KYOTO (ICT-2007-211423). Arantxa Otegi's work is funded by a PhD grant from the Basque Government.

References

1. Gonzalo, J., Verdejo, F., Chugur, I., Cigarran, J.: Indexing With WordNet Synsets Can Improve Text Retrieval. In: Proceedings of the 36th Annual Meeting of the Association for Computational Linguistics and 17th International Conference on Computational Linguistics (1998)
2. Kim, S., Seo, H., Rim, H.: Information retrieval using word senses: Root sense tagging approach. In: Proceedings of SIGIR 2004 (2004)
3. Liu, S., Liu, F., Yu, C., Meng, W.: An effective approach to document retrieval via utilizing WordNet and recognizing phrases. In: Proceedings SIGIR 2004 (2004)
4. Liu, S., Yu, C., Meng, W.: Word Sense Disambiguation in Queries. In: Proceedings of ACM Conference on Information and Knowledge Management, CIKM 2005 (2005)

5. Pérez-Agüera, J.R., Zaragoza, H.: UCM-Y!R at CLEF 2008 Robust and WSD tasks. In: Peters, C., et al. (eds.) CLEF 2008. LNCS, vol. 5706. Springer, Heidelberg (2009)
6. Stokoe, C., Oakes, M.P., Tait, J.: Word Sense Disambiguation in Information Retrieval Revisited. In: Proceedings of SIGIR 2003 (2003)
7. Strohman, T., Metzler, D., Turtle, H., Croft, W.B.: Indri: A language model-based search engine for complex queries. In: Proceedings of the International Conference on Intelligence Analysis (2005)

SENSE: SEmantic N-levels Search Engine at CLEF2008 Ad Hoc Robust-WSD Track

Annalina Caputo, Pierpaolo Basile, and Giovanni Semeraro

Department of Computer Science
University of Bari
70126 Bari, Italy
{acaputo,basilepp,semeraro}@di.uniba.it

Abstract. This paper presents the results of the experiments conducted at the University of Bari for the Ad Hoc Robust-WSD track of the Cross-Language Evaluation Forum (CLEF) 2008. The evaluation was performed using SENSE (SEmantic N-levels Search Engine), a semantic search engine that tries to overcome the limitations of the ranked keyword approach by introducing *semantic levels*, which integrate (and not simply replace) the lexical level represented by keywords.

We show how SENSE is able to manage documents indexed at two separate levels, keyword and word meaning, in an attempt of improving the retrieval performance.

Two types of experiments have been performed by exploiting both only one indexing level and all indexing levels at the same time. The experiments performed combining keywords and word meanings, extracted from the WordNet lexical database, show the promise of the idea and point out the value of our institution.

In particular the results confirm our hypothesis: The combination of two indexing levels outperforms a single level. Indeed, an improvement of 35% in precision has been obtained by adopting the N-levels model with respect to the results obtained by exploiting the indexing level based only on keywords.

1 Introduction

Information Retrieval (IR) systems are generally concerned with the selection of documents, from a fixed collection, which satisfy a user's one-off information need (query). The traditional search strategy performed by IR systems is ranked keyword search: For a given query, a list of documents, ordered by *relevance*, is returned. Relevance computation is primarily driven by a string-matching operation: If any query word is found in a document belonging to the collection, a match is made and the document is considered as relevant.

Ranked keyword search has been quite successful in the past, in spite of its obvious limits basically due to polysemy, the presence of multiple meanings for one word, and synonymy, different words having the same meaning. The result is that, due to synonymy, relevant documents can be missed if they do not contain

C. Peters et al. (Eds.): CLEF 2008, LNCS 5706, pp. 126–133, 2009.

the exact query keywords, while due to polysemy wrong documents could be deemed as relevant. These problems call for alternative methods that work not only at the lexical level of the documents, but also at the *meaning* level.

Therefore, in our interpretation semantic information could be captured from a text by looking at *word meanings*, as they are described in a reference dictionary (e.g. WordNet [6]). We propose an IR system which manages documents indexed at multiple separate levels: keywords and senses (word meanings). The system is able to combine keyword search with semantic information provided by the word meaning level.

The rest of the paper is structured as follows: The N-levels model used in SENSE is described in Section 2, while Section 3 presents an overview of the word meaning level. The details of the system setup for the CLEF competition are provided in Section 4. Finally, the experiments are described in Section 5. Conclusions and future work close the paper.

2 N-levels Model

The main idea underlying the definition of an open framework to model different semantic aspects (or levels) pertaining document content is that there are several ways to describe the semantics of a document. Each semantic facet needs specific techniques and ad-hoc similarity functions. To address this problem we propose a framework in which a different IR model is defined for each level in the document representation. Each level corresponds to a *logical view* that aims at describing one of the possible semantic spaces in which documents can be represented. The adoption of different levels is intended to guarantee acceptable system performance even when not all semantics representations are available for a document.

We suppose that a keyword level is always present and, when also other levels are available, these ones are used to offer enhanced retrieval capabilities. Furthermore, our framework allows to associate each level with the appropriate representation and similarity measure. The following semantic levels are currently available in the framework:

Keyword level - the entry level in which the document is represented by the words occurring in the text.

Word meaning level - this level is represented through *synsets* obtained by WordNet, a semantic lexicon for the English language. A synset is a set of synonym words (with the same meaning). Word Sense Disambiguation algorithms are adopted to assign synsets to words.

Analogously, N different levels of representation are needed for representing queries. The N query levels are not necessarily extracted simultaneously from the original keyword query issued by the user: A query level can be obtained when needed. We also extended the notion of relevance $R(q, d)$, which computes the *degree of similarity* between each document d in the collection and the user query q. The relevance must be evaluated at each level by defining a proper *local similarity function* that computes document relevance according to the weights defined by the corresponding local scoring function. Since the ultimate goal is to obtain a *single* list of documents ranked in decreasing order of relevance, a *global ranking function* is needed to merge all the result lists that come from each level. This function is independent of both the number of levels and the specific local scoring and similarity functions because it takes as input N ranked lists of documents and produces a unique merged list of the most relevant documents.

The aggregation of lists in a single one requires two steps: The first one produces the N normalized lists and the second one merges the N lists in a single one. The two steps are thoroughly described in [2]. In CLEF we adopt Z-Score normalization and CombSUM [4,5] respectively as score normalization and rank aggregation function.

3 Word Meaning Level

In SENSE, features at the word meaning level are *synsets* obtained from WordNet, a semantic lexicon for the English language. In order to assign synsets to words, we adopted a Word Sense Disambiguation (WSD) strategy. In the case of CLEF, SENSE used the synsets provided by the organizers of the Ad Hoc Robust-WSD track. The documents provided by the organizers contain a list of the possible synsets for each word with a score representing the confidence with which each sense can be associated with that word [1]. We use this factor to weigh the synsets in the meaning index structure.

The idea behind the adoption of WSD is that each document is represented, at the meaning level, by the senses conveyed by the words in its content, together with their respective occurrences. Documents are represented by using a synset-based vector space. Consequently, the vocabulary at this level is the set of distinct synsets recognized by the WSD procedure in the collection, while the weight of each synset for a document is computed according to the local scoring function defined in the next section.

3.1 Synset Scoring Function

Given a document d_i and its synset representation computed by the WSD procedure, $X = [s_1, s_2, \ldots, s_k]$, the basic idea is to compute a weight for each $s_j \in X$.

The weight, called SFIDF (synset frequency, inverse document frequency), is computed according to a strategy resembling the tf-idf score for words:

$$\text{SFIDF}(s_j, d_i) = \underbrace{\text{TF}(s_j, d_i)}_{\text{synset frequency}} \cdot \underbrace{log\frac{|C|}{n_j}}_{\text{IDF}} \tag{1}$$

where $|C|$ is the total number of documents in the collection and n_j is the number of documents containing the synset s_j. $\text{TF}(s_j, d_i)$ computes the frequency of s_j in the document d_i.

Finally, the synset confidence factor (α) is used to weigh the SFIDF value. Thus, the final local score for a synset s_j in d_i is:

$$\text{SFIDF}(s_j, d_i) \cdot (1 + \alpha) \tag{2}$$

3.2 Synset Similarity Function

The local similarity functions for both the meaning and the keyword levels are computed using a modified version of the LUCENE default document score. For the meaning level, both query and document vectors contain synsets instead of keywords. Given a query q and a document d_i, the synset similarity is computed as:

$$synsim(q, d_i) = C(q, d_i) \cdot \sum_{s_j \in q} (\text{SFIDF}(s_j, d_i)(1 + \alpha) \cdot N(d_i)) \tag{3}$$

where:

- $C(q, d_i)$ is the number of query terms in d_i;
- $\text{SFIDF}(s_j, d_i)$ and α are computed as described in the previous section;
- $N(d_i)$ is the document length normalization factor.

4 System Setup

We adopted the SENSE framework to build our IR system for CLEF evaluation. We used two different levels: keyword level using word stems and word meaning level using WordNet synsets. All the SENSE components involved in the experiments are implemented in JAVA using the last available version of Lucene API (2.3.2). Experiments were run on an Intel Core 2 Quad processor at 2.4 GHz, operating in 32 bit mode, running Linux (UBUNTU 7.10), with 2 GB of main memory.

In according to CLEF guidelines we performed two different tracks of experiments: Ad Hoc Robust-WSD Mono-language and Cross-language. Each track required two different evaluations: with and without synsets. We exploited several combinations of levels and query expansion methods, especially for the meaning level. All query expansion methods are automatic and do not require manual operations. Moreover, we used different boosting factors for each topic field and gave more importance to the terms in the fields TITLE and DESCRIPTION. More details on the track are reported in the track overview paper [1].

In particular for the Ad-Hoc Mono-language track we performed the following runs:

1. **MONO1TDnus2f:** the query is built using word stems in the fields TITLE and DESCRIPTION of the topics. All query terms are joined adopting the OR boolean operator. The terms in the TITLE field are boosted using a factor 2.

2. **MONO11nus2f:** similar to the previous run but in this case we add the NARRATIVE field and we adopt different term boosting values: 4 for TITLE, 2 for DESCRIPTION and 1 for NARRATIVE. These boost factors are used for all the following runs.
3. **MONO12nus2f:** for this instance we adopt the Lucene Phrase Query in addition to the query expansion described in MONO11nus2f. This kind of queries are able to exploit term proximity in the computation of relevance score. We build proximity query using the terms contained into the TITLE and DESCRIPTION fields. In detail: for TITLE we build a proximity query using all the terms into the field, while for DESCRIPTION we build a proximity query for each sentence.
4. **MONO13nus2f:** as the previous run but we adopt a different strategy to build Phrase Query. We exploit PoS-tag in order to build proximity queries. We produce a proximity query for each sequence of PoS-tags that matches the following patterns: *adjective-noun-verb*, *verb-adjective-noun*, *verb-noun*, *noun-verb* and *adjective-noun*. For example, into the sentence: *'The wrapping artist Christo took two weeks...'* we build a proximity query using the following terms: *"artist Christo took"*.
5. **MONO14nus2f:** this experiment adopts a combination of all the previous methods.
6. **MONOwsd1nus2f:** the query is built by expanding the synsets in the TITLE and DESCRIPTION fields of the topics. This run exploits the hypernyms and hyponyms. In particular, we include only the direct hyponyms and the hypernyms that have a path length less or equal to two. For synsets we adopt a different boost factor taking into account both the field and synsets distance.
7. **MONOwsd11nus2f:** in this instance each word is expanded using the whole set of synsets in WordNet and we compute a boosting factor using the ZIPF distribution that approximates properly the natural distribution of meanings. The ZIPF formula is:

$$f(k; N; s) = \frac{1/k^s}{\sum_{n=1}^{N} 1/n^s} \quad (4)$$

   where:
   - N is the number of synsets;
   - k is the synset rank. The synsets in WordNet are ranked according to the their frequency in a reference corpus;
   - s is the value of the exponent characterizing the distribution: After tuning experiments we set s equal to 2.
8. **MONOwsd12nus2f:** in this experiment we exploit the N-level architecture of SENSE. For the keyword level we adopt the query expansion described in MONO14nus2f and for the word meaning level the MONOwsd1nus2f.
9. **MONOwsd13nus2f:** as the previous run but, for the word meaning level we adopt the method described in MONOwsd11nus2f.

For the Ad-Hoc Cross-language track we performed the following runs:

1. **CROSS1TDnus2f:** the query is built using word stems in the TITLE and DESCRIPTION fields of the topics. In the Cross-language track the topics are in Spanish, thus a translation of terms in English is required. The SENSE system was not specifically developed for the Cross-language retrieval task hence in this instance we adopted a very trivial method in order to translate the query in English. We exploited WordNet dictionary to translate a word. In detail, we query Spanish WordNet using the Spanish word w_s and retrieve the whole set of synsets S related to the word w_s; then we use the set S to query English WordNet and retrieve, for each synset in S, the set of the English synonyms W_e. Finally, we build the query using the words in W_e. The boost factors have the same values used in the Mono-language track.
2. **CROSS1nus2f:** as described in the previous run adding the NARRATIVE field.
3. **CROSSwsd1nus2f:** in this case we adopt the same method presented in MONOwsd1nus2f but we use directly the synsets in Spanish Topic. It is important to notice that terms in a Spanish query are disambiguated using the first sense in Spanish WordNet.
4. **CROSSwsd11nus2f:** in this instance we exploit the N-levels architecture. For the keyword level we adopt the method described in CROSS1nus2f and for word meaning level the method proposed in MONOwsd1nus2f.
5. **CROSSwsd12nus2f:** this run differs from the CROSSwsd11nus2f for the use of a different Spanish-English translation method. We use directly the Spanish WordNet synset instead of the Spanish word. We query English WordNet using the synsets into the topic and retrieve, for each synset, the set of synonymous English words.

For all the runs we removed the stop words from both the index and the topics. In particular, we built a different stop words list for topics in order to remove non-informative words such as *find*, *reports*, *describe* which occur with high frequency in topics and are poorly discriminating.

5 Experimental Session

The experiments were carried out on the CLEF Ad Hoc WSD-Robust dataset derived from the English CLEF data, which comprises corpora from "Los Angeles Times" and "Glasgow Herald", amounting to 166, 726 documents and 160 topics in English and Spanish. The relevance judgments were taken from CLEF.

The goal of our evaluation is to prove that the combination of two levels outperforms a single level. In particular, the combination of keyword and meaning levels turns out to be more effective than the keyword level alone.

To measure retrieval performance, we adopted Mean-Average-Precision (MAP) calculated by the CLEF organizers using the DIRECT system on the basis of 1,000 retrieved items per request. Table 1 shows the results for each run with an overview on the exploited features.

Table 1. Results of the performed experiments

Run	MONO	CROSS	N-levels	WSD	*MAP*
MONO1TDnus2f	X	-	-	-	0.168
MONO11nus2f	X	-	-	-	*0.192*
MONO12nus2f	X	-	-	-	0.145
MONO13nus2f	X	-	-	-	0.154
MONO14nus2f	X	-	-	-	0.068
MONOwsd1nus2f	X	-	-	X	0.180
MONOwsd11nus2f	X	-	-	X	0.186
MONOwsd12nus2f	X	-	X	X	**0.220**
MONOwsd13nus2f	X	-	X	X	**0.227**
CROSS1TDnus2f	X	X	-	-	0.025
CROSS1nus2f	X	X	-	-	0.015
CROSSwsd1nus2f	X	X	-	X	0.071
CROSSwsd11nus2f	X	X	X	X	0.060
CROSSwsd12nus2f	X	X	X	X	**0.072**

The results confirm our hypothesis: The combination of two levels outperforms a single level. In particular, the combination of keyword and meaning levels (MONOwsd12nus2f and MONOwsd13nus2f) is more effective than the single keyword level (MONO1TDnus2f and MONO11nus2f). If we consider MONO1-TDnus2f as baseline, we obtain an improvement of 35% in precision using the N-levels model (MONOwsd13nus2f).

It is interesting to notice that just the use of the word meaning level alone is able to outperform the keyword level. This result has a motivation: We chose to index all the synsets for each word (not only the synset with the highest confidence factor). This intuition makes the retrieval process easier.

Regarding the Cross-language track, our system achieves a low precision. This was an expected result because it is not designed specifically for this kind of task. Moreover, the method adopted for topic translation is based only on the use of WordNet as dictionary. In particular, performance of the Cross-language without WSD (experiments: CROSS1TDnus2f and CROSS1nus2f) are not satisfying because the system exploits only keywords and the translation process introduces a lot of wrong terms into the query, producing a noise effect. Conversely, the word meaning level is able to help the retrieval process, as shown in CROSSwsd1nus2f, where we used only the word meaning level (without keywords). In the second attempt (CROSSwsd11nus2f) we combined the keyword level with the word meaning level obtaining worse results due to the keyword translation method (as in CROSS1TDnus2f). Finally, we tried to translate the Spanish words using directly the synsets obtaining a good result with respect to the previous one.

We noticed that our system has a low precision compared to the other CLEF competition participants. This is due to the standard relevance function implemented in Lucene and this result was expected. In particular, Lucene performance decreases when the number of terms in the query grows. In fact, the experiment MONO14nus2f produces large queries and results point out that the system achieves a low precision in this experiment with respect to the others that rely exclusively on keywords. This problem also affects the Cross-language experiments because we translate a Spanish word using all the possible English translations (CROSS1TDnus2f) producing a query with a lot of terms. Details concerning this well known behavior of Lucene can be found in [3]. Nonetheless, the goal of our evaluation was to prove the effectiveness of the N-levels model and the experiments confirm our hypothesis.

6 Conclusion and Future Work

We have described and tested SENSE, a semantic N-levels IR system which manages documents indexed at multiple separate levels: keywords and meanings. The system is able to combine keyword search with semantic information provided by the other indexing levels.

The distinctive feature of the system is that, differently from the previous approaches, an adaptation of the vector space model is proposed to integrate, rather than simply replace, the lexical space with semantic spaces. We provided a detailed description of the SENSE model, by defining a local scoring function, a local similarity function for synsets and a global ranking function in order to merge rankings produced by different levels.

We performed an intensive evaluation using the CLEF Ad Hoc Robust-WSD dataset. This dataset supplies both words and synsets for each document and it is the ideal framework to evaluate the N-levels architecture. The experiments show that the N-levels model is effective when the word meaning level is involved.

As future research we plan to improve the performance of the system. We can achieve this goal mainly by improving the relevance function implemented in Lucene. Furthermore, we intend to investigate different IR models, such as language modeling.

References

1. Agirre, E., Di Nunzio, G.M., Ferro, N., Mandl, T., Peters, C.: Clef 2008: Ad hoc track overview. In: Abstracts of the CLEF 2008 Workshop (2008)
2. Basile, P., Caputo, A., Gentile, A.L., Degemmis, M., Lops, P., Semeraro, G.: Enhancing semantic search using n-levels document representation. In: Bloehdorn, S., Grobelnik, M., Mika, P., Tran, D.T. (eds.) SemSearch. CEUR Workshop Proceedings, vol. 334, pp. 29–43. CEUR-WS.org (2008)
3. Cohen, D., Amitay, E., Carmel, D.: Lucene and juru at trec 2007: 1-million queries track. In: Proceedings of the 16th Text REtrieval Conference (TREC 2007) (November 2007)
4. Fox, E.A., Shaw, J.A.: Combination of multiple searches. In: TREC, pp. 243–252 (1993)
5. Lee, J.-H.: Analyses of multiple evidence combination. In: SIGIR, pp. 267–276. ACM, New York (1997)
6. Miller, G.A.: Wordnet: a lexical database for english. Commun. ACM 38(11), 39–41 (1995)

IR-n in the CLEF Robust WSD Task 2008

Sergio Navarro, Fernando Llopis, and Rafael Muñoz

Natural Language Processing and Information Systems Group,
University of Alicante, Spain
{snavarro,llopis,rafael}@dlsi.ua.es
http://gplsi.dlsi.ua.es

Abstract. In our approach to the Robust WSD task we have used a passage based system jointly with a WordNet and WSD based term expansion for the documents and queries. Furthermore, we have experimented with two well known relevance feedback methods - LCA and PRF -, in order to figure out which is more suitable to take profit of the WSD query expansion based on Wordnet. Our best run has obtained a 4th - 0.4008 MAP -. A major finding is that LCA fits better than PRF to this task due to it is able to take advantage of the expanded documents and queries.

1 Introduction

The aim of the CLEF Robust WSD task [1] is exploring the contribution of Word Sense Disambiguation (WSD) to monolingual and multilingual Information Retrieval, in order to find successful methods to take profit of WSD information which helps the systems to increase their levels of robustness.

This paper is structured as follows: Firstly, it presents the main characteristics of the IR-n system focusing on the documents and query expansion module and the relevance feedback strategies, then it moves on to explain the experiments we have made to evaluate the system and the results of our participation. Finally it describes the conclusions and future works.

2 The IR-n System

The IR-n passage-based system differs from other systems of the same category with regard to the method proposed for defining the passage. IR-n defines the passages by a number of consecutive sentences in a document. Passage systems can consider the proximity of words with each other, that appear in a document in order to evaluate their relevance [2].

2.1 Expansion Based on WordNet (WN) Using WSD

The system expands all term non tagged as 'NNP' within the queries and the collection documents. To carry out the expansion, it first selects the most likely

C. Peters et al. (Eds.): CLEF 2008, LNCS 5706, pp. 134–137, 2009.

WN synset returned by the WSD system - in the event of a tie it selects all the synsets with the maximum probability -. And afterwards, it generates the term expansion using all synonyms belonging to the selected synset/s.

In the phase of selecting the synset of a term, optionally IR-n can use two WSD systems in order to limit the synset selection only to those synsets which have been ranked as the most likely by one of the two WSD, and that at least has been ranked at second place by the other WSD system.

Finally, IR-n uses a parameter which lets us configure the weight assigned for the terms added to the query.

2.2 Relevance Feedback

We are comparing in this CLEF edition Pseudo Relevance Feedback (PRF)[3] with Local Context Analysis (LCA) [4] strategy. In the selection of terms, PRF gives more importance to those terms which have a higher frequency in the top relevant documents than in the whole collection. An alternative query expansion method relies on the Local Context Analysis (LCA), based on the hypothesis that a common term from the top-ranked relevant documents will tend to co-occur with all query terms within the top-ranked documents. That is an attempt to avoid including terms from top-ranked, non-relevant documents in the expansion. Furthermore, in the case of polysemous words, this method will help to retrieve documents more related to the sense of the query, since it is logical to think that the user will use words from the domain associated with this sense to complete the query. Indeed we think that in our participation it could be better to use a method based on the terms of the query as LCA, since that the expanded terms based on WN used in the query and in all the documents could boost performance of this relevance feedback strategy, improving its ability for skipping non relevant documents.

3 Training

The training phase aims to establish the optimum values for the parameters of the system for the collection. Below we describe some of them:

- **Relevance Feedback (relFB):** Indicating which relevance feedback uses the system - PRF or LCA.
- **WSD system used for the expansion of the Collection (WSDCOL):** Indicate which WSD system has been used or if none has been used for the documents expansion.
- **WSD system used for the expansion of the Query (WSDQuery):** Indicate which WSD system has been used or if none has been used for the query expansion.
- **Weight for the WN based Expanded Terms (wWN):** Is the weight used for the expanded terms using WN within the query.

For the experiments we have worked with DFR as the weighting schema and a passage size of four sentences based on a previous training phase of the baseline configuration.

The worst results - under the baseline results - has been obtained for those runs which only used expansion for the query - not for the collection -. The best run without relevance feedback has been the one which used NUS WSD system for expanding the query and the collection. Furthermore, we have seen that the method of mixing the two WSD did not improve the run which only have used NUS WSD system.

Finally, in respect the relevance feedback training, as we forecasted, the best MAP results were obtained using LCA. Indeed, the major improvement respect PRF occurs with the run which uses NUS WSD system.

4 Results in 2008 Robust WSD Task

For our participation, we have sent two runs without WSD: the baseline, and the best run using LCA. And four with WSD: the best run without relevance feedback and the three best runs using relevance feedback.

The best runs submitted by the participants with and without WSD have obtained a MAP value of 0.4499 and 0.4515 respectively. Table 1 shows our results and rank position in the Monolingual classification in MAP terms.

Table 1. Results in 2008 Robust WSD Task

runName	relFb	WSD COL	WSD Query	wWN	rk MAP	MAP	GMAP	recall
TestIRnSinColLCA	LCA	no	no	-	3	0.4008	0.1514	0.8851
TestIRnSinCol	no	no	no	-	10	0.3661	0.1473	0.8851
TestIRnUBC_0.2_LCA	LCA	UBC	UBC	0.2	13	0.3748	0.1361	0.8768
TestIRnNUSSoloCol_LCA	LCA	NUS	no	-	14	0.3726	0.1384	0.8722
TestIRnNUS_0.2_LCA	LCA	NUS	NUS	0.2	15	0.3720	0.1389	0.8761
TestIRnNUS_0.2	no	NUS	NUS	0.2	16	0.3664	0.1471	0.8669

On the one hand, opposite to what happens in training phase, all the runs which have used WSD have obtained results for all the measures under the results of the run which have used LCA without WSD. On the other hand these results show us that LCA always improves the results respect the same configuration without its use.

4.1 Analyzing the Results

Reviewing the queries and their WSD expansions, we saw that for all the queries there are terms expansions that decrease the precision of the retrieval. The reason is that the system is expanding terms which are not on the focus of of the need behind the query - in example - 'Find' = 'breakthrough discovery' -. However, there are queries for which the WSD strategy overcome this problem with a suitable expansion of the term with more meaning in the query - in example 'Earthquakes' = 'earthquake quake seism temblor' or 'flood' = 'deluge inundation' -. Also, we have found wrong disambiguations of the most meaningful terms in a query, that deter the precision of the query expansion - in example 'EU' = 'atomic number 63' -.

5 Conclusion and Future Work

The major finding of the experiments performed and the posterior analysis carried out is that we have identified the two main causes for the contradictory results obtained in this task. On one hand the lack of strategy for selecting the terms of the query to be expanded. And on the other hand the mistakes found in the disambiguation of some terms.

We believe that in future works could be interesting to develop a good term selection strategy for the WSD query expansion. It would allow us to have a more confident system to measure whether WSD system are useful for information retrieval.

Finally, it is important to mention that LCA has showed that it is able to improve the results more than PRF for those WSD expansion configurations that better results obtain.

Acknowledgment

This research has been partially funded by the Spanish Government within the framework of the TEXT-MESS (TIN-2006-15265-C06-01) project and by European Union (EU) within the framework of the QALL-ME project (FP6-IST-033860).

References

1. Agirre, E., Di Nunzio, G.M., Ferro, N., Mandl, T., Peters, C.: CLEF 2008: Ad Hoc track overview. In: Peters, C., et al. (eds.) CLEF 2008. LNCS, vol. 5706, pp. 15–37. Springer, Heidelberg (2009)
2. Llopis, F.: IR-n: Un Sistema de Recuperacin de Informacin Basado en Pasajes. PhD thesis, University of Alicante (2003)
3. Robertson, S.E., Sparck Jones, K.: Relevance weighting of search terms. Journal of the American Society for Information Science 27(3), 129–146 (1976)
4. Xu, J., Croft, W.B.: Improving the effectiveness of information retrieval with local context analysis. ACM Trans. Inf. Syst. 18(1), 79–112 (2000)

Query Clauses and Term Independence

José R. Pérez-Agüera[1] and Hugo Zaragoza[2]

[1] Dpto. de Ingeniería del Software e Inteligencia Artificial, UCM
[2] Yahoo! Research Barcelona
jose.aguera@fdi.ucm.es,hugoz@yahoo-inc.com

Abstract. Much current research in IR, Web-Search and Semantic-Web technologies aims at enriching the user query to gain a richer, more semantic understanding of the information need. Almost in all cases this query enrichment step is approached independently of the ranking function; however, this may be far from optimal. In this paper we discuss the problem of term dependency in the context of query expansion and show its dangers in a number of empirical evaluations. Furthermore we propose a simple method (query clauses) that can be applied to several standard ranking functions to exploit a simple type of term dependency.

1 Introduction

Central to modern IR ranking functions is the term independence assumption. This assumption takes on many forms, but loosely it implies that the effect of each query term on document relevance can be evaluated independently of the other query terms. This has the effect of rendering all queries flat, devoid of structure.

In some cases there is an explicit query structure that needs to be taken into account, as in the case XML document retrieval; these cases are studied in the domain of Structured IR. However, there are many cases in which queries have some known semantic structure, such as degree of synonymy between terms, term cooccurrence or correlation information with respect to the query or to specific query terms, etc. This is typical, for example, when the query has been constructed by a query-expansion method, when stemming or normalizing terms, when taking into account multi-terms or phrases, etc. Surprisingly, almost all ranking functions (and experiments) ignore this structural information: after expansion, selection and re-weighting of terms, a flat query (a set of weighted terms) is given to the ranking function which assumes terms are independent and scores documents accordingly.

Taking into account the semantic dependency of terms in queries is important for two reasons. The first reason is obvious: we wish to improve ranking performance and so the ranking function should exploit all the information available, including structural semantic information. But there is an even more compelling reason to take into account semantic dependencies: if we don't, we will hurt the baseline performance as we add more and more information to the query! This, in our opinion, is the current state-of-affairs: countless models and experiments

C. Peters et al. (Eds.): CLEF 2008, LNCS 5706, pp. 138–145, 2009.

are developed to enrich queries, but ranking performance remains the same or decreases; we believe that part of the problem is the use of inadequate ranking functions.

Another important point, in our opinion, is that the query semantic structure cannot be reduced to a set of global term correlations, it needs to be taken into account at query time, since two queries can enforece different relationships between the same terms (for example, two terms may be considered synonyms with respect to one query but not to another).

2 Ranking Independent Terms

One of the reasons of the high performance of modern ad-hoc retrieval systems is their use of document term frequency. It is well known that i) probability of relevance of a document increases as the term frequency of a query term increases, and ii) this increase is non-linear. For this reason most modern ranking functions use an increasing saturating function to weight document terms that are in the query. An example of this is the term saturating function used in BM25:

$$w(d,t) := \frac{tf(d,t)}{tf(d,t) + K_1} \tag{1}$$

where $tf(d,t)$ is the term frequency of term t in document d, and K1 is a constant. Similar nonlinear term frequency functions are found in most IR ranking models such modern variants of the vector space model, the language model, divergence from randomness models etc. On the other hand, all these ranking functions assume that the relevance information of different query terms is independent and therefore the relevance information gained by seeing query terms can be computed separately and added linearly (or log-linearly), for example, in BM25:

$$score(d) := \sum_{t \in q} w(d,t) \cdot idf(t) \tag{2}$$

This independence assumption is usually reasonable for short queries (i.e. *Italian restaurant in Cambridge*), since users use each term to represent a different aspect of the query. However, such assumption breaks down for queries that are sufficiently complex to contain terms with sufficiently close meaning. Consider for example the query *Italian restaurant cafeteria bistro Barcelona.* Having seen the term restaurant twice in a document, which term is more informative: Cambridge or cafeteria ? Loosely speaking, if a group of terms carries the same meaning, the amount of relevance information gained by their presence should diminish as we see other terms in this group, very much like in equation 1 does for term frequency, and unlike 2.

This situation arise very often in modern IR tasks and systems, in particular in the following areas:

- morphological expansion (e.g. stemming, spelling, abbreviations, capitalization),
- extracting multi-terms from the query,

- query term expansion (e.g. user feedback, co-occurrence based expansion),
- lexical semantic expansion (e.g. using WordNet),
- using taxonomies and ontologies to improve search,
- user modeling, personalization,
- query disambiguation (where terms are added to clarify the correct semantic context),
- document classification (where the query is a set of documents),
- structured queries (such as TREC structured topics).

3 Ranking Independent Clauses

We propose to consider two levels of representation: terms and term clauses. Clauses are sets of weighted terms that are thought to represent a particular aspect of the query. The weights represent their relative importance within the clause (in particular, the strength of the dependence with relevance). Thus, a query can thought of as a bag of bags of (weighted) terms:

$$c := \left\{ (t_0, w_0), (t_1, w_1), ..., \left(t_{|c|}, w_{|c|}\right) \right\}$$
$$q := \left\{ c_1, c_2, ..., c_{|q|} \right\}$$

Boolean retrieval models and the INQUERY retrieval model have used query representations even more general than this. Here we restrict ourselves to this representation with two levels to give clear semantics to each level: term and clause. We are going to consider terms within a clause as if they were greatly dependent with respect to relevance; in fact we will consider them as if they were virtually the same term. Second, we consider terms across clauses as being independent with respect relevance as usual.

We first give an intuitive description of the method proposed to rank documents against queries with clauses, using a term-document matrix transformation. In practice the transformation is not necessary, counts can be calculated directly on a standard inverted index; we show this for several standard retrieval models later.

Lets imagine that we are given a query with clauses, such:

$$q := \{ \{ (A, 1.0), (B, 0.7) \}, \{ (C, 1.0) \} \} \quad (3)$$

We could in principle replace in the collection all the occurrences of terms within a clause by a virtual term representing the clause. Alternatively, we could add the term frequencies of all terms within a clause obtaining a set of clause frequencies. If the terms are weighted in the clause, we can take the weights into account when adding term frequencies. For a collection with only four terms A-D, and the previous query, we could construct a modifed representation of documents in terms of the 2 query clauses as follows:

Doc	A	B	C	D
d_1	2	1	0	1
d_2	0	1	1	1
d_3	1	2	0	2

Documento	Clausula 1	Clausula 2
$d_1\|_q$	2.7	0
$d_2\|_q$	0.7	1
$d_3\|_q$	2.4	0

Another way to say this is that we construct a linear projection from the space of terms to the space of clauses. Formally, if we represent a collection by the matrix $D = (tf(d_i, t_j))_{i,j}$ where $tf(d_i, t_j)$ is the term frequency of the jth term in the ith document (i.e. documents are row vectors of term frequencies), and we represent the query as a matrix of weights $C = (c_{ij})$ where c_{ij} is the weight of jth term in ith clause (i.e. a clauses are row vectors of relative term weights).

For the previous query we would have:

$$C := \begin{bmatrix} 1 & 0.7 & 0 & 0 \\ 0 & 0 & 1 & 0 \\ 0 & 0 & 0 & 0 \end{bmatrix}$$

Then we can obtain the transformed representation as the linear projection:

$$d|_C := d \times C^{\mathsf{T}}$$

In practice one can carry out this transformation very efficiently since the term-document matrix is very sparse and the projection can be carried out efficiently on an inverted file. Therefore, one can compute standard term statistics on the fly after a user issues a query. The statistics needed will depend on the ranking model but the most common ones are detailed here. The length of the document is not modified by the projection, nor the average document length. The clause term frequencies *ctf* and clause collection frequencies *ccf* can be computed as:

$$ctf(d,c) := \sum_{(t,w)\in c} w \cdot tf(d,t)$$

$$ccf(d,c) := \sum_{(t,w)\in c} w \cdot \sum_{d} tf(d,t)$$

$$p_{\mathsf{ML}}(c|d) := \frac{ctf(d,c)}{ctf(d,c) + \sum_{t\notin c} tf(d,t)}$$

$$p_{\mathsf{ML}}(c|Col) := \frac{ccf(d,c)}{ccf(d,c) + \sum_{d,t\notin c} tf(d,t)}$$

The most problematic statistic is the inverse clause frequency *icf*, since this is not clearly defined in the weighted case. One possible choice is the number of postings containing at least one term in the clause ; we refer to this as icfOR (c), and we note that it can be computed directly from the size of the clause result set (documents with non-zero ctf). However, this number may be unfairly large for clauses with lowly weighted common terms. Furthermore, in some settings this number may not even be available (for example if we only score the query term AND set or if we drop from the computation documents unlikely to be highly scored). Another possibility is to use the expected idf for a clause term in a document[1]:

$$icf_{\mathsf{E}}(d,c) = \frac{1}{\sum_{t' \in c} w_{t'} \cdot tf(d,t')} \sum_{(t,w) \in c} w \cdot tf(d,t) \cdot idf(t) \quad (4)$$

This has several nice properties, for example terms added with very small weights have very small effect on the clause idf , and terms not occurring in the document have no effect at all. However, it has the dissadvantage that it needs to be computed for every clause in every document, at query time, unlike idf which can be pre-computed for each term.

With these statistics at hand we can compute the relevance score of a document with respect to a query with clauses for a number of retrieval systems; we display several in table 1.

Table 1. Implementing query clauses in several standard ranking models

MODEL	FUNCTION
BM25	$\frac{ctf}{ctf+K} \cdot icf$
VSM	$\frac{ctf \cdot icf}{\lvert\lvert d_{\lvert q} \rvert\rvert}$
DFR (PL2)	$\frac{1}{ctf+1} \left(ctf \cdot \log_2 \frac{ctf}{\lambda} + (\lambda - ctf) \cdot \log_2 e + 0.5 \cdot \log_2 (2\pi \cdot ctf) \right)$
LM (KL)	$p_{\mathsf{smoothed}}(c \mid q) \log \left(p_{\mathsf{smoothed}}(c \mid d) \right)$

4 Evaluation

We carry out a series of experiments to demonstrate the dangers of the term independence assumption for queries with strongly correlated terms, and to test the proposed query-clauses ranking idea. Evaluation is carried out on CLEF Ad-hoc WSD Collection (LA Times 94 and Glasgow Herald 95, both with Word Sense Disambiguation (WSD) data. All runs employ the standard and the query clause version of BM25 (2 and Table 1 respectively).

4.1 Query Expansion Experiments

Our approach is to apply state of the art query expansion methods. In our experience state of the art query expansion methods are superior than semantic expansion methods based on WordNet or corpus statistics; their main advantage is that they lead to expansions that are truly query dependant; semantic information tends to be too vague and it is hard to use without knowing the context

[1] In our empirical evaluation we found this to be a better than using the min or the max clause idf, and better than using the idf of the highest weighted term.

in which a word is used. However, the idea behind our method is that query expansion and semantic information may be used complementary. In particular, semantic information may be useful to decide the semantic query structure (query clauses in our case).

In this experiment we proceed as follows. First we assign to each original query term a different query clause (we assume query terms to be independent in the traditional way). We assign the weight of 1 to these terms. Then we do standard query expansion algorithms and select the usual number of expansion terms (40 in our case) for the query. We use the DFR Bo1 method for this, although similar results can be obtained with the other methods. We then compute a semantic similarity between each original query word and each expansion word. If this similarity is above a threshold α, we include the expanded term in the clause of the original term; we assign to this term a weight equal its expansion weight. All the expanded terms remaining are grouped into an extra query clause. This way, the number of clauses is always equal to $|q| + 1$.

As an example, let's say that the original query was a, b and the terms c, d, e were found to be good expansion terms, with weights wc , wd , we respectively. After computing the 6 semantic distances between original and expanded terms, we would check which were above a threshold α. Say for example that only d was sufficiently similar to b, all other similarities being bellow α. Then we would end up with the query:

$$\{\ \{(a,1)\},\ \{(b,1),\ (d,\ wd\ )\},\ \{(c,\ wc\ ),\ (e,wd)\}\} \tag{5}$$

Semantic similarities are computed based on WordNet. There exists an extensive literature on measures of semantic similarity. We have used the WordNet Similarity package, which contains many semantic measures. In particular we used (after some experimentation) the wup measure which is based on the LCS (Lexical Conceptual Structure) depth of the term pair in WordNet. The threshold α is a free parameter and we have experimented with different thresholds. In order to map the terms to WordNet, we used the WSD information in the corpus.

We can see that the proposed method improves results over the baseline and over query expansion, for all relevance measures including GMAP. This is very encouraging because it is one of the few results to our knowledge that show that

Table 2. Results for clause queries using different similarity thresholds in WordNet. α is the similarity threshold.

	MAP	GMAP	R-PREC	P@5	P@10
BM25 (baseline)	.3614	.1553	.3524	.4325	.3663
BM25 + Bo1	.3835	.1528	.3615	.4613	.3844
BM25 + Bo1 + Clausulas 0.0	.3937	.1620	.3735	.4600	.3869
BM25 + Bo1 + Clausulas 0.3	.3935	.1613	.3726	.4563	.3869
BM25 + Bo1 + Clausulas 0.6	.3926	.1606	.3737	.4600	.3906
BM25 + Bo1 + Clausulas 0.9	.3957	.1618	.3772	.4625	.3975

WordNet information and WSD can be used to improve ad-hoc retrieval in an open domain.

Note that increasing values of α lead to increasing results[2] ; however $\alpha = 1$ lead to poor results during the testing phase. This is somewhat surprising, reinforces the idea that one must be very conservative in query expansion in order to be robust. This requires further investigation.

In our opinion a bottleneck to further improve performance is in the creation of high quality structures. WordNet Similarity methods tend to produce noisy clauses, often putting in correspondence terms that are not related in the context of the query.

5 Discussion and Related Work

In this paper we try to show the importance of term dependence issues, how they show up unexpectedly in simple experiments and how they can have a strong adverse effect in performance. Furthermore we propose a method to represent and take into account a simple form of dependence between terms.

In most of query expansion literature terms are selected (globally from the entire corpus or locally from the top retrieved documents), weighted with respect to their potential relevance and then passed on to a standard retrieval system, which is considered a black box. Here we are concerned only with this black box and not with the expansion process; for this reason we will not review the query expansion literature (an up to date overview can be found in [2]). Some work on user and pseudo-feedback has tackled the issue of term re-weighting, from early Rochio algorithms to more modern probabilistic approaches of relevance feedback. While these works discuss the ranking function, to our knowledge they all assume query term Independence and concentrate on the re-weighting formula. Again, we are not concerned here on the re-weighting of terms (this is left unspecified in our work), and therefore we do not review this literature further (see for example[1]).

A few papers have dealt with the issue of term *correlation* and its effect on retrieval. In [6] the problem of correlation is discussed in depth. They remark that *term correlation* is only an abstract concept and can be understood in a number of ways. They measure term correlation in terms of *term co-occurrence.* Furthermore they propose to represent documents not in the space of terms but in the space of *minterms* which are sets of highly correlated terms. This has the effect of *decorrelating* the terms in the query with respect to hypothetical *concepts* (formally defined as minterms). Instead of computing all term correlations, [4] proposes to mine association rules to compute the most significant term correlations and the rotates the term vectors to reflect the extracted correlations; this yields a more selective term de-correlation. [3] also proposes mining association rules to find term sets of correlated terms. However, the ranking function adjustment proposed is based on the same idea of this paper: collapsing term

[2] Note that query inclusion requires that similarity is greater than α, so even for $\alpha = 0$ many terms are not assigned to clauses.

frequencies within a clause. In fact, if we disregard relative weights, we use the VSM model, and we construct query clauses using association rules in [3], the ranking function here is exactly the same as in [3]. However our work differs from the previously cited papers in that it is not tied to an extraction method or a ranking model, it does not specify the form of the term correlations and furthermore it asumes that term correlations will be *query-dependant.*

Also related to our work are ranking function for structured retrieval (i.e. XML retrieval), because here the issue of term independence arises when aggregating scores from several parts of a document that contain the same terms. Again most papers consider agragation methods that asume independence, but [5] have shown that it is important to take dependance into account and propose an simple agregation method that takes dependance across sections into account. In fact we apply here the same idea to take into account dependence across terms, extending the idea to sets of terms and to other ranking models.

References

1. Carpineto, C., de Mori, R., Romano, G., Bigi, B.: An information-theoretic approach to automatic query expansion. ACM Trans. Inf. Syst. 19(1), 1–27 (2001)
2. Chang, Y., Ounis, I., Kim, M.: Query reformulation using automatically generated query concepts from a document space. Inf. Process. Manage. 42(2), 453–468 (2006)
3. Pôssas, B., Ziviani, N., Wagner Meira, J., Ribeiro-Neto, B.: Set-based vector model: An efficient approach for correlation-based ranking. ACM Trans. Inf. Syst. 23(4), 397–429 (2005)
4. Silva, I.R., Souza, J.N., Santos, K.S.: Dependence among terms in vector space model. In: IDEAS 2004: Proceedings of the International Database Engineering and Applications Symposium (IDEAS 2004), Washington, DC, USA, 2004, pp. 97–102. IEEE Computer Society, Los Alamitos (2004)
5. Taylor, M., Zaragoza, H., Craswell, N., Robertson, S., Burges, C.: Optimisation methods for ranking functions with multiple parameters. In: CIKM 2006: Proceedings of the 15th ACM international conference on Information and knowledge management, pp. 585–593. ACM Press, New York (2006)
6. Wong, S.K.M., Ziarko, W., Raghavan, V.V., Wong, P.C.N.: On modeling of information retrieval concepts in vector space. ACM Trans. Database Syst. 12(2), 299–321 (1987)

Analysis of Word Sense Disambiguation-Based Information Retrieval

Jacques Guyot, Gilles Falquet, Saïd Radhouani, and Karim Benzineb

Centre Universitaire dInformatique, University of Geneva
Route de Drize 7, 1227 Carouge, Switzerland

Abstract. Several studies have tried to improve retrieval performances based on automatic Word Sense Disambiguation techniques. So far, most attempts have failed. We try, through this paper, to give a deep analysis of the reasons behind these failures. During our participation at the Robust WSD task at CLEF 2008, we performed experiments on monolingual (English) and bilingual (Spanish to English) collections. Our official results and a deep analysis are described below, along with our conclusions and perspectives.

1 Introduction

Our aim through this paper is not to propose sophisticated strategies to improve retrieval performances using a word sense disambiguation (WSD) algorithm. Rather we mainly want to explore whether WSD (plus the semantic information in WordNet) can be useful in Information Retrieval (IR) and Cross Lingual Information Retrieval (CLIR). Therefore, we carried out a set of experiences in monolingual and bilingual tasks. Then we deeply analyzed the obtained results and formulated some conclusions and perspectives. In the rest of this paper, we first present the steps of the collection processing (Section 2). Then we describe our indexing and searching strategies (Section 3). The obtained results of our experiments are detailed in section 4. Before concluding (Section 6), we discuss the obtained results and provide some perspectives in section 5.

2 Collection processing

The corpus is a news collection, containing 166000 English documents and 160 topics. All topics are available in English and Spanish. Each topic contains three fields: a title (T), a description (D), and a narrative (N). The corpus was disambiguated using two leading WSD systems: the University of the Basque Country (UBC) [1] and the National University of Singapore (NUS) [5], resulting in two different sets. English documents and queries were processed using the English WordNet, while the Spanish topics were annotated using the Spanish WordNet. The disambiguation process consists of annotating documents and queries by adding sense information to all content words (figure 1). Thus, each occurrence of a word is replaced by an XML element containing the word identifier (TERM

C. Peters et al. (Eds.): CLEF 2008, LNCS 5706, pp. 146–154, 2009.

```
<DOC>
    <DOCNO>GH950102-000000</DOCNO>
    <DOCID>GH950102-000000</DOCID>

    <HEADLINE>
        <TERM ID="GH950102-000000-1" LEMA="alien" POS="JJ">
            <WF>Alien</WF>
            <SYNSET SCORE="0.6" CODE="01295935-a"/>
            <SYNSET SCORE="0.4" CODE="00984080-a"/>
        </TERM>

        <TERM ID="GH950102-000000-2" LEMA="treatment" POS="NN">
            <WF>treatment</WF>
            <SYNSET SCORE="0.827904118008605" CODE="00735486-n"/>
            <SYNSET SCORE="0" CODE="03857483-n"/>
            <SYNSET SCORE="0.172095881991395" CODE="00430183-n"/>
            <SYNSET SCORE="0" CODE="05340429-n"/>
        </TERM>
```

Fig. 1. Example of WordNet-based document annotation

ID), an extracted lemma (LEMA), a part-of-speech (POS) tag (noun, verb, adjective, etc.), the original word form (WF), and a list of senses together with their respective scores. The senses are represented by WordNet *synset* codes.

3 Indexing and Searching Strategies

For several reasons, we chose to index the corpus using our IDX-VLI indexer [6]. Indeed, IDX-VLI can gather a wealth of information (positions, etc.), it has built-in operators, and it is remarkably fast. Still, we only used the basic version of that indexer *i.e.*. We did not use any relevance feedback mechanism, context description, or any other sophisticated tool of that sort. We thus avoided interfering with the direct results of the experiment, and we facilitated the result analysis. Documents and queries content were represented using the Okapi BM25 weighting scheme (with default parameters).

3.1 Documents Processing

We developed and tested the following document processing strategies that we applied to each <TERM> element within each document annotation:

- NAT: Keep only the word form of each element (*i.e.*, rebuild the original text);
- LEM: Keep only the lemma;
- POS: Keep the lemma and the part-of-speech tag;
- WSD: Keep only the synset that has the best score[1];

[1] This amounts to considering that the disambiguation algorithm is "perfect". Alternatively, we could have added all synsets that have a score greater than a given threshold.

– WSDL: Keep the lemma and the best corresponding synset (with the higher score).

During the indexing process, these strategies were applied to all terms including numbers, except for stop-words. Given the poor performance of the POS strategy, we quickly gave up this option.

3.2 Topics Processing

The same strategies were applied to the topics, with an extended stop-word list including words such as *report*, *find*, etc. For each topic, we derived three queries:

– T: Includes only the title field;
– TD: Includes the title and the description fields:
– TDN: Includes the title, the description, and the narrative fields.

4 Experimental Results

4.1 Lemma-Based Strategy

In order to come up with a reasonably good baseline, we tested several approaches to build a Boolean pre-filter from a given topic. We didn't want to have a low baseline: when the baseline is low, the probability achieve a better result using WSD becomes high. This happened to Basile et *al.* and Otegi et *al.* when they used WSD for the bilingual robust WSD task at CLEF 2008 [2][8], and for Stokoe et *al.* when they applied their WSD system on a large-scale TREC data collection [20].

The obtained results of the baseline are described in table 1 where columns contain respectively the run's name, the run's description, and the corresponding result in terms of mean average precision (MAP). These tests were conducted on 150 training topics. The best results were given using the OR filtering.

4.2 WSD-Based Strategy

In addition to the filtering strategies used for the baseline runs, we performed two more runs based on the hyperonym relationship extracted from WordNet. The results obtained on the training corpus are described in table 2.

The obtained results on the training corpus showed that the strategy based on the OR-filtering gives the best result. Therefore, we decided to use it for the official runs described in the following section.

Table 1. Baseline results in terms of MAP

Run name	Run description	MAP
OR	The logical OR of the words (or lemmas)	**0.255**
AND	The logical AND of the words	0.158
NEAR	The logical OR of all pairs (t_i NEAR t_j), where t_i and t_j are two query terms	0.152

Table 2. WSD-based runs results in terms of MAP

Run name	Run description	MAP
OR	The logical OR of the best synset corresponding to a topic word	**0.224**
AND	The logical AND of the best synset corresponding to a topic word	0.151
NEAR	The logical OR of all pairs (s_i NEAR s_j), where s_i and s_j are the best synsets corresponding to two topic words t_i and t_j	0.125
HYPER	The logical AND of each (s_i OR h_i), where s_i is the best synset corresponding to a topic word t_i, and h_i is the direct hypernym of s_i in WordNet	0.143
ORHYPER	The logical OR of each (s_i OR h_i), where s_i is the best synset corresponding to a topic term t_i, and h_i is the direct hypernym of s_i in WordNet	0.1843

Table 3. Official results in terms of MAP for the monolingual task

Used topic field	Lemma-based strategy	WSD-based strategy (NUS corpus)
T	0.3064	0.2120
TD	0.3664	0.2934
TDN	**0.3917**	0.3269

4.3 Official Results

We carried out several runs in the monolingual and the bilingual task. For the purposes of this paper, however, we present only the most significant ones.

Table 3 contains the official results in terms of MAP for the monolingual task. The first column contains the topic fields used during the corresponding run; the second column contains the results of the lemma-based strategy; and the third column contains the results of the WSD-based strategy (using the NUS disambiguation algorithm). The results clearly demonstrate that the use of WSD techniques does not improve the retrieval performances compared to a lemma-based approach. The best result was obtained using all the topic fields with lemma as indexing unit (0.3917).

The results also demonstrated that the retrieval performances obtained using the NUS disambiguation algorithm are higher than those obtained using the UBC disambiguation algorithm.

The best result obtained using WSD occurred when we combined a WSD-based indexing with a lemma-based indexing (0.3814). However, it is lower than the result obtained using lemma only.

For the bilingual task, the baseline consisted of translating topics from English to Spanish using Google translator. The obtained results using only the title of the topic gave a MAP of 0.3036. The use of WSD significantly decreases the retrieval performances (0.0846 of MAP using the NUS algorithm).

5 Findings and Discussion

From our experiences, both on the training and the test corpora, we note the following facts:

- Using D and N topics fields increases the MAP in all cases (with and without WSD). This is most probably due to the ranking method that benefits from the additional terms provided by D and N topics fields.
- On the test run with the UBC system, using only synsets (WSD) decreases the MAP: -4.6% using the T field, and -3.1% using TDN. On the training topics, combining lemmas and synsets (WSD + Lemma) slightly improves the MAP (+0.6%). This is the only case where disambiguation brings an improvement.
- Using different disambiguation algorithms for queries and documents noticeably decreases the results. This should not happen if the algorithms were perfect. It demonstrates that disambiguation acts as a kind of "encoding" process on words, and obviously the best results are obtained when the same "encoding", producing the same mistakes, is applied to both queries and documents. Thus, at this stage, the disambiguation algorithms are not interoperable.

We carefully analyzed around 50 queries to better understand what happened with the disambiguation process. For instance, the query whose title is "*El Nio and the weather*" was disambiguated, using the NUS algorithm, as follows:

- "El" was interpreted as the abbreviation "el." of "elevation";
- "Niño" was interpreted as the abbreviation "Ni" of "nickel", probably because the parser failed on the non-ASCII character "ñ";
- "Weather" was correctly interpreted as the "weather" concept.

Although the disambiguation was incorrect, WSD was as good as LEM because the "encoding" was the same in the documents and in the queries. In addition, WSD was also as good as LEM because there were a few or no documents dealing about nickel that could have produced noise. More generally, when the WSD results were better than the LEM ones, it was not due to semantic processing but to contingencies. For instance, the query title "*Teenage Suicides*" had a better score with WSD because "teenage" was not recognized. Thus, the query became *suicides*, which is narrower than *teenage OR suicide*, and avoided retrieving a large amount of irrelevant documents about teenagers.

The poor performance on Spanish queries is due to: *i*) the above-mentioned lack of interoperability between the different WSD algorithms and *ii*) the low quality of the Spanish WSD itself. This can be illustrated in the following examples:

Topic 41: "Pesticide in baby food" is translated into "Pesticidas en alimentos para bebes", and then converted into the FOOD and DRINK (verb) concepts, because "bebes" is a conjugated form of "beber", which is the Spanish verb for drink.

Topic 43: "El Niño and the weather" is translated into "El Niño y el tiempo", and then converted into the CHILD and TIME concepts, because "Niño" is the Spanish noun for child, and "tiempo" is an ambiguous word meaning both time and weather.

Given those difficulties, outstanding results could not be expected. Looking back on the results, it can be noted that 1793 documents were retrieved out of the 2052 relevant ones (*i.e.*, almost 90% of them). The core issue is to sort out documents so as to reject those whose content does not match users' expectations. A closer look at our results on the training corpus showed that we achieved a solid performance on some topics. This does not mean that our search engine "interpreted" correctly said topics. Rather, it is simply due to the fact that the corpus included only good matches for those topics. Therefore, it was almost impossible to find wrong answers. For instance, on topic 50, which deals with "the Revolt in Chiapas", we retrieved 106 documents out of 107 possible relevant ones, with a MAP of 87%. This is due to the fact that in the corpus, the Chiapas are only known for their revolt (in fact if we Google the word "Chiapas", a good proportion of the results are currently about the Chiapas rebellion). On the other hand, on topic 59, which deals with "Computer Viruses", our search engine retrieved 1 out of 1 possible relevant document, with a MAP of 0.03. This low result is because the 300 documents retrieved before the one we were looking for were indeed about viruses and computers, but did not mention any virus *name* or *damage* as was requested. Therefore term disambiguation does not help search engines to interpret what kinds of documents are expected. A topic, such as the one above, requires the text to be read and correctly interpreted in order to decide whether it is actually a correct match. After a deep analysis, we concluded that the retrieval performance of WSD-based system depends at least on three factors:

1. The quality of the used semantic resource, and in particular its coverage compared with the vocabulary of corpus. This problem can be avoided if we combine WSD-based indexing with keywords-based indexing. So far, the few works that have been successful are those who proceeded using this method [13][18].
2. The quality (accuracy) of the used disambiguation algorithm: As mentioned by several studies [10][19], the main difficulty to improve retrieval performances is due to the inefficacity of disambiguation algorithm, especially when queries are short (one or two words)[21]. Indeed, it is judicious to think that by using a perfect algorithm (with 100% accuracy), retrieval performances will be at least equal to those obtained by keywords-based approach. We postulate that when a query is large enough (more than two words), the probability that a document containing the query terms in a different context or meaning from the intended definitions one is very low. For instance, it is unlikely that a document containing *mouse*, *cheese*, and *cat* is in fact dealing about a computer mouse. This probably renders WSD useless in many situations. Such a query is similar in nature to the narrative-based tests. On the other hand, the WSD approach could be more applicable when

queries include only one or two words (which is the most frequent case in standard searches). So far, the studies regarding this problem have shown that: *i*) ambiguity does not have a strong impact on retrieval performances, especially when queries are quite long (the matching between a query and a document performs already an implicit disambiguation); *ii*) when a disambiguation algorithm is used, it must be very accurate (more than 90%), and *iii*) retrieval performances can be outperformed when indexing is based both on WSD and keywords.

3. The method used to "interpret" the semantic content of documents and queries: in existing approaches, once concepts are extracted, documents and queries are considered as bags of concepts. Therefore, semantic relationships that may exist between the concepts they contain are not exploited. Consequently, documents dealing with a subject close to that of the query could not be found with these approaches. WSD is a very partial semantic analysis that is insufficient to really interpret queries' content. For instance, consider the query "*Computer Viruses*" whose narrative is "Relevant documents should mention the name of the computer virus, and possibly the damage it does." To find relevant documents, a system must recognize phrases that contain virus names ("the XX virus", "the virus named XX", "the virus known as XX", etc.). It should also recognize phrases describing damages ("XX erases the hard disk", "XX causes system crashes", but not "XX propagates through mail messages"). These tasks are very difficult to perform and they are far beyond the scope of WSD. Query expansion (QE) is a possible solution to this problem because they make it possible to extend content representation of the query in order to increase the chance of matching documents [3][14][22]. That said, QE must be controlled in order to carefully choose the concepts to be added to the original query, otherwise the results can be disappointing [3][15]. In [11] and [12], the authors obtained positive results by expanding queries using WSD, but the effect of the use of WSD and QE are not quantified in isolation. In fact, even though the main objective of their study was to evaluate the performance of WSD in IR, they should have examined the accuracy of their disambiguation method in isolation, so that they could quantify its effect when used in their IR experiments. A more comprehensive study was carried out in [9], in which the authors added additional sense information to both documents and queries using WordNet. Their large-scale experiments on a TREC collection produced promising results, clearly demonstrating the positive effect of WSD on retrieval performances. From our personal experiences, a possible solution to these problems is to use domain knowledge not only for WSD, but also for indexing and searching [17]. We notably showed how the use of semantic relationships could provide a precise representation of documents and query content. Relationships can therefore be used during the information retrieval process in order to allow the system to find a relevant document to a given query, even if it does not share any term with that query [16].

6 Conclusion

Our aim through this paper was to explore whether WSD can be useful in IR and CLIR. Our results confirmed that WSD does not allow for any retrieval performance improvement. It is obvious that these failures are primarily due to the weakness of WSD techniques, but also they depend on many other factors, such as the quality of the semantic resource used by WSD algorithm and the method used to "interpret" the semantic documents and queries content. We think that WSD-based indexing is a promising approach for language-independent indexing and retrieval systems. Although an efficient WSD is essential to create good conceptual indexes, we demonstrated in [7] that ambiguous indexes (with several concepts for some terms) are often sufficient to reach a good multilingual retrieval performance, for the reasons mentioned above. We also revealed that non-trivial queries, like those treated in our study, require adding domain knowledge during indexing and querying process. As shown in our previous work, this can be reached using expressive documents and queries languages, respectively during documents and queries content representation [16].

Acknowledgment

This work was supported by the Swiss National Science Foundation.

References

1. Agirre, E., Lopez de Lacalle, O.: UBC-ALM: Decombining k-NN with SVD for WSD. In: Proc. of the 4th International Workshop on Semantic Evaluations (SemEval 2007), Prague, Czech Republic, pp. 341–345 (2007)
2. Basile, P., Caputo, A., Semeraro, G.: UNIBA-SENSE at CLEF 2008: SEmantic N-levels Search Engine. In: Working notes of 9th Workshop of the Cross-Language Evaluation Forum, CLEF 2008, Aarhus, Denmark, September 17-19 (2008)
3. Baziz, M.: Indexation conceptuel le guidée par ontologie pour la recherche d'information. Thèse de doctorat, Universitè Paul Sabatier, Toulouse, France (Décember 2005)
4. Baziz, M., Aussenac-Gilles, N., Boughanem, M.: Désambiguisation et Expansion de Requêtes dans un SRI, Etude de l'apport des liens sémantiques. Revue des Sciences et Technologies de l'information (RSTI) série ISI 8, 113–136 (2003)
5. Chan, Y.S., Hwee, T., Zhong, Z.: NUS-PT: Exploiting Parallel Texts for Word Sense Disambiguation in the English All-Words Tasks. In: Proc. of the 4th International Workshop on Semantic Evaluations (SemEval 2007), Prague, Czech Republic, pp. 253–256 (2007)
6. Guyot, J., Falquet, G., Benzineb, K.: Construire un moteur d'indexation. Revue Technique et science informatique (TSI), Hermes, Paris (2006)
7. Guyot, J., Radhouani, S., Falquet, G.: Conceptual Indexing for Multilingual Information Retrieval. In: Peters, C., Gey, F.C., Gonzalo, J., Müller, H., Jones, G.J.F., Kluck, M., Magnini, B., de Rijke, M., Giampiccolo, D. (eds.) CLEF 2005. LNCS, vol. 4022, pp. 102–112. Springer, Heidelberg (2006)

8. Otegi, A., Agirre, E., Rigau, G.: IXA at CLEF 2008 Robust-WSD Task: using Word Sense Disambiguation for (Cross Lingual) Information Retrieval. In: Working notes of the 9th Workshop of the Cross-Language Evaluation Forum, CLEF 2008, Aarhus, Denmark, September 17-19 (2008)
9. Kim, S.-B., Seo, H.-C., Rim, H.-C.: Information Retrieval Using Word Senses: Root Sense Tagging Approach. In: Proc. of the 27th annual international ACM SIGIR Conference, pp. 258–265. ACM Press, New York (2004)
10. Krovetz, R., Bruce Croft, W.: Lexical ambiguity and information retrieval. ACM Transactions on Information Systems 10(2), 115–141 (1992)
11. Liu, S., Liu, F., Yu, C., Meng, W.: An effective approach to document retrieval via utilizing WordNet and recognizing phrases. In: Proc. of the 27th ACM SIGIR Conference, pp. 266–272. ACM Press, New York (2004)
12. Liu, S., Yu, C., Meng, W.: Word Sense Disambiguation in Queries. In: Proc. of the 14th ACM CIKM Conference, pp. 525–532 (2005)
13. Mihalcea, R., Moldovan, D.: Semantic indexing using wordnet senses. In: Proc. of the ACL-2000 workshop on Recent advances in natural language processing and information retrieval, Morristown, NJ, USA, pp. 35–45. Association for Computational Linguistics (2000)
14. Mihalcea, R., Moldovan, D.: An iterative approach to word sense disambiguation. In: Proc. of the Thirteenth International Florida Artificial Intelligence Research Society Conference, pp. 219–223. AAAI Press, Menlo Park (2000)
15. Qiu, Y., Frei, H.-P.: Concept based query expansion. In: Korfhage, R., Rasmussen, E.M., Willett, P. (eds.) SIGIR, pp. 160–169. ACM, New York (1993)
16. Radhouani, S., Falquet, G., Chevallet, J.-P.: Description Logic to Model a Domain Specific Information Retrieval System. In: Bhowmick, S.S., Küng, J., Wagner, R. (eds.) DEXA 2008. LNCS, vol. 5181, pp. 142–149. Springer, Heidelberg (2008)
17. Radhouani, S., Falquet, G.: Using External Knowledge to Solve Multi-Dimensional Queries. In: Proc. 13th Intl Conf. on Concurrent Engineering Research and Applications (CE 2006), Antibes. IOS Press, Amsterdam (2006)
18. Schütze, H., Pedersen, J.O.: Information Retrieval Based on Word Senses. In: Fourth Annual Symposium on Document Analysis and Information Retrieval (1995)
19. Sanderson, M.: Word Sense Disambiguation and Information Retrieval. In: Proc. of the 17th ACM SIGIR Conference, pp. 142–150 (1994)
20. Stokoe, C., Oakes, M.P., Tait, J.: Word sense disambiguation in information retrieval revisited. In: Proc. of the 26th ACM SIGIR Conference, pp. 159–166. ACM Press, New York (2003)
21. Voorhees, E.M.: Using wordnet to disambiguate word senses for text retrieval. In: Korfhage, R., Rasmussen, E.M., Willett, P. (eds.) SIGIR, pp. 171–180. ACM, New York (1993)
22. Voorhees, E.M.: Query expansion using lexical-semantic relations. In: Proc. of the 17th annual international ACM SIGIR conference on Research and development in information retrieval, pp. 61–69. Springer-Verlag New York, Inc. (1994)

Crosslanguage Retrieval Based on Wikipedia Statistics

Andreas Juffinger, Roman Kern, and Michael Granitzer

Know-Center, Graz
{ajuffinger,rkern,mgranitzer}@know-center.at

Abstract. In this paper we present the methodology, implementations and evaluation results of the crosslanguage retrieval system we have developed for the Robust WSD Task at CLEF 2008. Our system is based on query preprocessing for translation and homogenisation of queries. The presented preprocessing of queries includes two stages: Firstly, a query translation step based on term statistics of cooccuring articles in Wikipedia. Secondly, different disjunct query composition techniques to search in the CLEF corpus. We apply the same preprocessing steps for the monolingual as well as the crosslingual task and thereby acting fair and in a similar way across these tasks. The evaluation revealed that the similar processing comes at nearly no costs for monolingual retrieval but enables us to do crosslanguage retrieval and also a feasible comparison of our system performance on these two tasks.

1 Introduction

The goal of the task was to test whether WSD can be used beneficially for retrieval systems [2]. The organisers believe that polysemy is among the reasons for information retrieval systems to fail. The focus in our contribution to this task, especially within this paper, lies on the following three points:

- How competitive is Apache Lucene[1], a state-of-the-art open source search engine which is used in many business applications, against scientific state-of-the-art?
- How can we reconstruct a query from a retrieval result, what does this cost for the monolingual task, and can this method be used to reconstruct the query in a different language?
- What is the impact of the provided WSD[1,3] on this system, can we identify a statistical significant change in the performance?

In order to be able to compare different translation and disambiguation strategies we propose an approach to crosslanguage retrieval based on Lucene, where each query is preprocessed in the same way independent of the query and target language (Fig. 1(a)). Within our retrieval system we exploit cooccurrences on

[1] http://lucene.apache.org

C. Peters et al. (Eds.): CLEF 2008, LNCS 5706, pp. 155–162, 2009.

corpus level to archive the cross language retrieval functionality. For our experiments in this task we used the English and Spanish Wikipedia[2]. Thereby the mapping between the articles from one language to the other language comes from the author defined cross-language links between the articles of different languages. Every query was then processed as shown in Fig 1(b): Firstly, we queried the Wikipedia index in the query language with terms from different sections of the provided queries. Secondly, we exploit the appropriate English articles from the search result and extract significant English query terms. Note that this query reconstruction step is mandatory for cross-language but is optional for monolingual problems. Thirdly, we used these query terms to query one of the CLEF indexes, either the plain index or the WSD index.

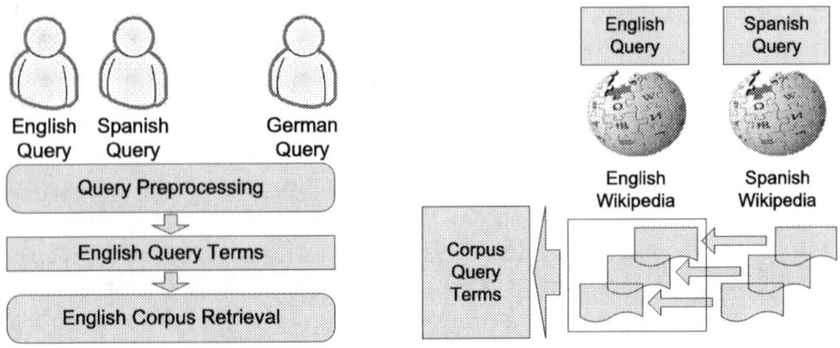

(a) Crosslanguage Retrieval (b) Cooccurrence Exploitation

Fig. 1. Retrieval Methodology

The remaining contribution is structured as follows: Section 2 provides an overview of our system in terms of index structures and methodology used. Section 3 details the proposed query processing technique. Results are outlined in Section 4 and Section 5 concludes this contribution.

2 System Architecture

Our system is based on a number of different indexes as shown in Figure 2. The Multilingual Wikipedia Index is used at the query preprocessing layer and the Plain and WSD Index are the indexes of the CLEF newspaper corpus data. The retrieval system was implemented in Java, based on the Apache Lucene text search engine library. This search engine library provides a high performance text retrieval engine for arbitrary, configurable indexes.

2.1 Multilingual Wikipedia Index

As discussed in the introduction we perform query preprocessing on every incoming query independent of the query language. The index we have built for query

[2] `http://www.wikipedia.org`

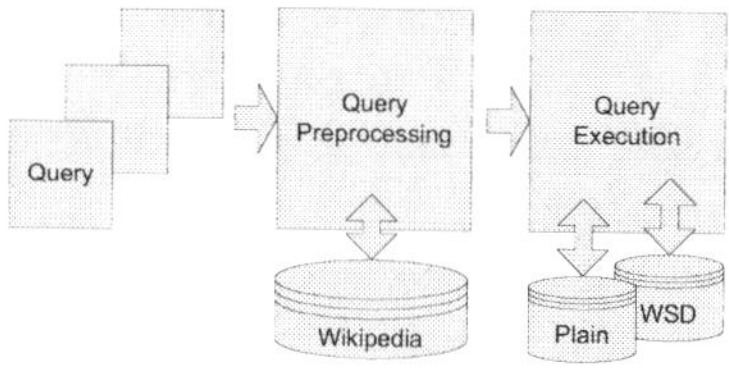

Fig. 2. System Architecture

preprocessing is called Multilingual Wikipedia Index. This index is created using English, Spanish, and possibly other Wikipedia data. Each Wikipedia article in every language is added to the Multilingual Wikipedia Index.

If one article links to another article in a different language we add an internal reference between these two articles. This approach allows to search in one language, and by exploiting the internal references, we are able to retrieve the appropriate articles in another language. To build this index the publicly available XML dumps are parsed with Bliki[3]. The parsed content is then indexed without further preprocessing, stemming or stop-word removal.

2.2 Corpus Retrieval Indexes

In this section we describe the different indexes we have created to retrieve the news articles from the CLEF corpus. Each of the following CLEF indexes contains all documents from the Los Angeles Times (1994) and Glasgow Herald (1995) dataset. Only the content of the documents was processed, title or other metadata has been ignored.

Plain Document Index. For the plain text variant, the data has been processed by the default Lucene indexing chain. The newspaper plain text has been tokenized by whitespaces and then transformed to lower case. For this work we have indexed the content terms in the original word form. Furthermore, we indexed the lemmatised form to evaluate and compare the impact of the word form to our approach.

WSD Document Index. For the word sense disambiguated variant, we used the available WSD information to compute the synonyms for the document terms. To maximize the impact of the WSD information we decided to only take the WordNet [6] Sense with highest WSD value from the data. All synonym terms found were indexed at the same position within the document as the original term to prevail phrase queries.

Technically speaking, the Lucene index is a document term matrix. Each document is thereby represented as a vector of terms. Lucene further allows to put more than one term on every term vector position. All terms on the same position are then transparently interchangeable. Lucene processes phrase queries as follows: Firstly, all documents are searched with a boolean "and" query for all

[3] `http://matheclipse.org/doc/bliki/index.html`

terms in the phrase. Secondly, Lucene retrieves the "distance" between the terms within the term vector of all matching documents. Thirdly, if the "distance" between the query terms equals one the phrase matches. That is the reason why terms at the same position are completly interchangeable in phrase queries. For example the term *baby* and the synonym *infant* indexed on the same position makes it possible, that the phrase query *baby food* would retrieve all documents, where either *baby food* or *infant food* occurs as phrase.

3 CLEF Query Processing

First, each query is processed and interpunctation characters are removed. Next, phrase queries are identified by either underscore characters between the terms or quotation marks surrounding the phrase. For phrase queries, the word order is maintained throughout the whole process. Next the query is tokenized and then stop words are removed from the query. Note that the system uses language specific stop word lists.

3.1 Query Translation

In the first step of query translation the extracted terms are translated to a set of terms in the search corpus target language (English). For each original query term we search in the Multilingual Wikipedia Index. We then collect the ids and the scores from the top 50 search results. Using these ids the appropriate English version of the Wikipedia articles are then retrieved by exploiting the earlier mentioned references between the articles of different languages. In the next step, all English terms from these articles are extracted and we calculate a weight for each term by multiplying the score of the article with the inverse document frequency of the term. In the last step we use the top 5 terms for each separate query term to build the final query.

A major advantage of our approach is that it relies only on term distribution statistics to "translate" terms into English query terms. No additional knowledge base, like dictionaries, taxonomies, and ontologies are used.

3.2 Query Construction

The collected and translated query terms, developed by the whole query preprocessing pipeline as shown in Fig. 3 are used to search for CLEF articles. The final query is thereby a hierarchical disjunction query. For the first level, the top 5 translated terms per original query term are used to formulate a standard boolean disjunction query. In the next level these queries are combined in two different ways: Boolean Disjunction[4] and Disjunction Max[5].

[4] org.apache.lucene.search.BooleanQuery
[5] org.apache.lucene.search.DisjunctionMaxQuery

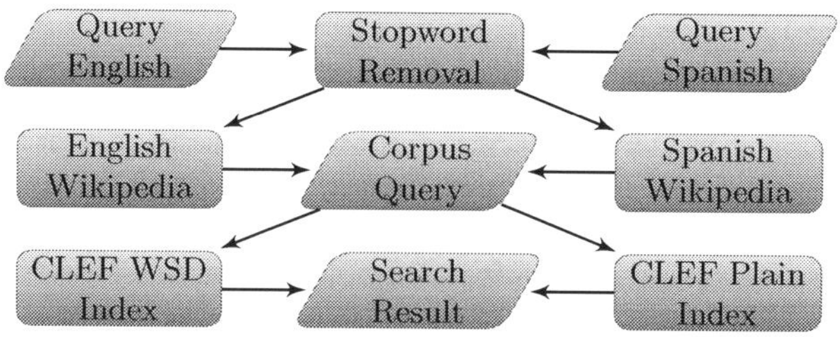

Fig. 3. Overview of the query processing

- *Boolean Disjunction:* This combination calculates the combined document score as a sum of the distinct scores for each single query term and normalizes this sum by the number of query terms. The boolean disjunction therefore calculates the mean of the scores.

$$score = \frac{1}{N} \sum s_i \tag{1}$$

- *Disjunction Max:* This combination calculates the score as the sum of the maximum score for a document for any subquery, plus a tie breaking increment for any additional matching subqueries. So the disjunction max preferes documents with high individual score. A tie breaking factor $t = 0$ leads to a total score whereby only the maximum scoring sub query contributes to the final score. In our case, this is where we assign importance to documents containing multiple or all query terms. That is why we have set the tie breaking factor to the number of subqueries to ensure that each combination counts more than a single retrieval result.

$$score = max_i(s_i) + t * \sum s_i - t * max_i(s_i) \tag{2}$$

Depending on the task, this hierarchical disjunction query is used to search the index with or without WSD information.

4 Experiments and Results

For this work, we have identified and solved drawbacks of our applied method for the Robust WSD Task at CLEF2008, as outlined in [4].

Motivated by the findings of other groups in the challenge and authors [5] we experimented with different retrieval features. We evaluated impact of title, description and narrative for the retrieval results. In our work we evaluated all three possible paths of the query pipeline shown in Fig. 4. From the original query we evaluated the use of different combinations of (T) title, (D) description, and (N) narrative. We further evaluated all of these pipes by using the provided lemma information instead of the wordform.

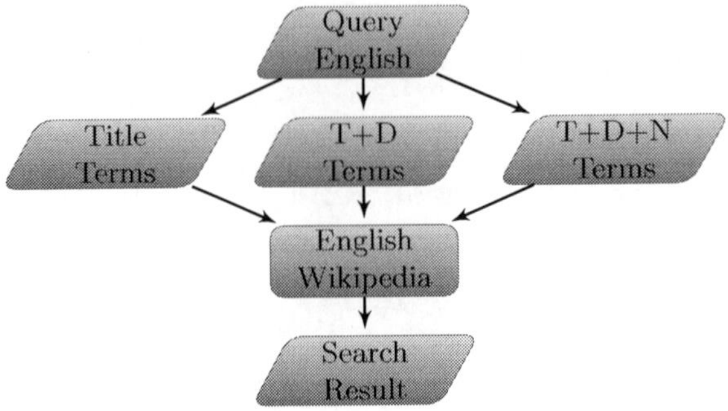

Fig. 4. Query Pipelines Wordform/Lemas

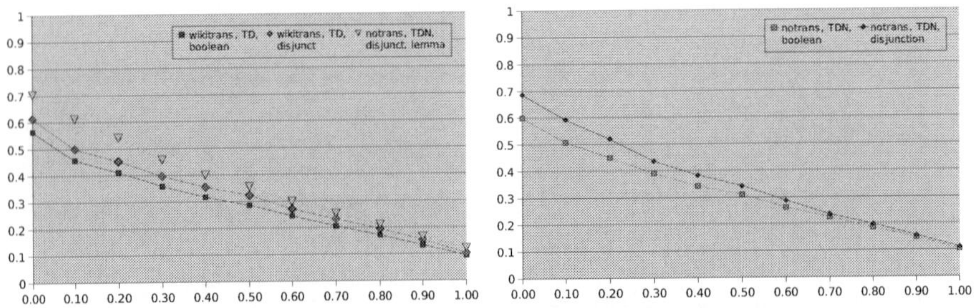

(a) Overall Improvement (b) Disjunction vs. Boolean Queries

Fig. 5. Impact of Parsing and Query Combination

Our original system failed by a high number of queries and by looking closer at the failed queries we found a number of weird English terms in the Wikipedia index for each article. Although such metatags should be ignored by the search engine, due to their low TFIDF[8] weight, we decided to parse the Wikipedia content to get rid of these formatting and style terms. The impact of this alteration was significant and we were able to improve our results by 3.5% for MAP(27.72 vs 32.25) as well as a consistant improvement in the precision at rank curve. This is shown in Figure 5(a); the squares curve denotes the original version and the diamonds curve reflects the increased performance. The triangles curve reflects the performance of our best retrieval system including all improvements we achieved.

A drawback of the original system was that the retrieval scoring did not count multiple matches in a hierachical boolean query. We therefore experimented with different scoring algorithms and query term combination methods. The experiments revealed that we can improve the performance by using *DisjunctionMax* queries for the monolingual task (see Fig. 5(b)) and the crosslanguage task (see Fig. 7(b)).

In combination with the use of lemmas instead of the normal wordform we were able to further improve the MAP by 1.3%. As shown in Fig. 5(a), triangle

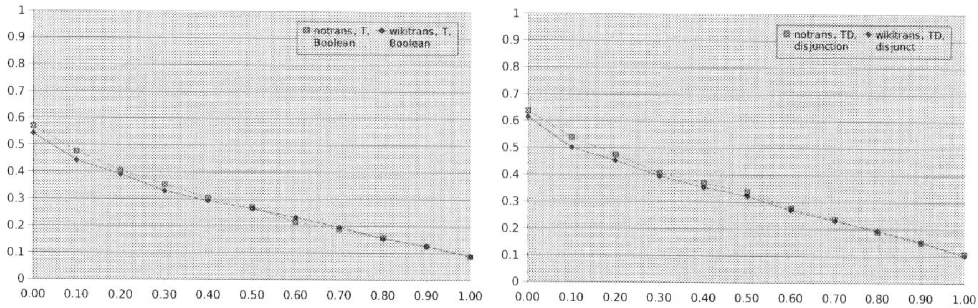

(a) Boolean Lucene Queries (b) Disjunction Lucene Queries

Fig. 6. Translation Impact

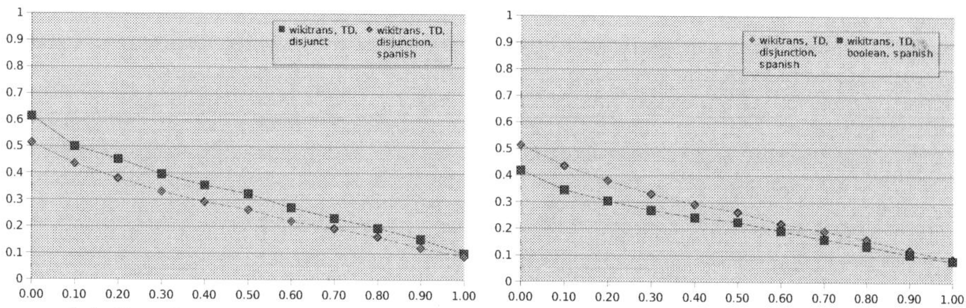

(a) Crosslanguage Performance (b) Disjunct vs. Boolean (ES)

Fig. 7. Crosslanguage Retrieval

curve, we are able to outperform all earlier results with this approach. Based on our implementation we were not able to successfully apply WSD information in these experiments. Although the results are slightly better for a number of queries, we were not able to show a statistically significant improvement.

4.1 Evaluation of the Crosslanguage Methodology

The central point in our methodology is that the system acts similar to all languages, not preferring the corpus language, but allows cross language retrieval. To evaluate the cost of this approach we measured the system performance on English queries with and without Wikipedia translation. Our main intention was thereby that we are able to show that the fairness is not too expensive for the monolingual task but furthermore enables us to do cross language retrieval. As shown in Fig. 6 we were able to reveal that the query translation through Wikipedia has no significant impact on the monolingual task. As shown in Fig. 6(a) and 6(b) this holds for boolean queries as well as for disjunction max queries.

The performance of this system for the crosslingual task is shown in Fig. 7. As one can see the performance for crosslingual retrieval is worse than the performance for the monolingual task. Due to the higher degree of complexity such a result is clear. With a MAP value of about 26% we would have been competitive in the crosslanguage challenge. In comparance with our best monolingual result this MAP is about 6% worse. In comparance with other groups, we are now able to work on eather task and improvements in one task will automatically improve the performance in the other task.

5 Conclusion

Our retrieval system for fair crosslanguage retrieval based on Apache Lucene has proofed to be competitive with other scientific state-of-the-art retrieval techniques with sophisticated weighting schemes [7]. Further we were able to show that our query reconstruction methodology is not a major constraint for the monolingual task, but makes us competitive in the cross language task. In all of our experiments we were not able to show a significant improvement for the retrieval task when using word sense disambiguation information.

Acknowledgement

The Know-Center is funded within the Austrian COMET Program - Competence Centers for Excellent Technologies - under the auspices of the Austrian Ministry of Transport, Innovation and Technology, the Austrian Ministry of Economics and Labor and by the State of Styria. COMET is managed by the Austrian Research Promotion Agency FFG.

References

1. Agiree, E., de Lacall, O.L.: UBC-ALM: Combining k-NN with SVD for WSD. In: Proc. of the 4th Int. Workshop on Semantic Evaluations, pp. 341–345 (2007)
2. Agirre, E., Giorgio, M., Di Nunzio, Ferro, N., Mandl, T., Peters, C.: Clef 2008: Ad hoc track overview (2008)
3. Chang, Y., Ng, H.T., Zhong, Z.: NUS-PT: Exploiting parallel texts for word sense disambiguation in the english all-words tasks. In: Proc. of the 4th Int. Workshop on Semantic Evaluations (2007)
4. Juffinger, A., Kern, R., Granitzer, M.: Exploiting cooccurrence on corpus and document level for fair crosslanguage retrieval. In: Working Notes for the CLEF 2008 Workshop, Aarhus, Denmark, September 17-19 (2008)
5. Anderka, M., Potthast, M., Stein, B.: A wikipedia-based multilingual retrieval model. In: Macdonald, C., Ounis, I., Plachouras, V., Ruthven, I., White, R.W. (eds.) ECIR 2008. LNCS, vol. 4956, pp. 522–530. Springer, Heidelberg (2008)
6. Miller, G.: Wordnet: A lexical database for english. Comm. ACM (1995)
7. Robertson, S., Zaragoza, H., Taylor, M.: Simple bm25 extension to multiple weighted fields. In: Proc. of the 13th ACM international conference on Information and knowledge management (2004)
8. Salton, G., Buckley, C.: Term-weighting approaches in automatic text retrieval. Information Processing and Management (1988)

Sampling Precision to Depth 10000 at CLEF 2008

Stephen Tomlinson

Open Text Corporation
Ottawa, Ontario, Canada
stomlins@opentext.com
http://www.opentext.com/

Abstract. We conducted an experiment to test the completeness of the relevance judgments for the monolingual German, French, English and Persian (Farsi) information retrieval tasks of the Ad Hoc Track of the Cross-Language Evaluation Forum (CLEF) 2008. In the ad hoc retrieval tasks, the system was given 50 natural language queries, and the goal was to find all of the relevant documents (with high precision) in a particular document set. For each language, we submitted a sample of the first 10000 retrieved items to investigate the frequency of relevant items at deeper ranks than the official judging depth (of 60). The results suggest that, on average, the percentage of relevant items assessed was less than 55% for German, French and English and less than 25% for Persian.

1 Introduction

Open Text eDOCS SearchServer™ is a toolkit for developing enterprise search and retrieval applications. The eDOCS SearchServer kernel is also embedded in various components of the Open Text eDOCS Suite[1].

The eDOCS SearchServer kernel works in Unicode internally [4] and supports most of the world's major character sets and languages. The major conferences in text retrieval experimentation (CLEF [1], NTCIR [5] and TREC [7]) have provided judged test collections for objective experimentation with the SearchServer kernel in more than a dozen languages.

This paper describes an experiment conducted with the eDOCS SearchServer kernel (experimental post-6.0 builds) for testing the completeness of the relevance judgments for the monolingual German, French, English and Persian information retrieval tasks of the Ad Hoc Track of the Cross-Language Evaluation Forum (CLEF) 2008.

[1] Open Text eDOCS SearchServer and Open Text eDOCS Suite are trademarks or registered trademarks of Open Text Corporation in the United States of America, Canada, the European Union and/or other countries. This list of trademarks is not exhaustive. Other trademarks, registered trademarks, product names, company names, brands and service names mentioned herein are property of Open Text Corporation or other respective owners.

C. Peters et al. (Eds.): CLEF 2008, LNCS 5706, pp. 163–169, 2009.

2 Methodology

2.1 Data

The CLEF 2008 Ad Hoc Track document sets consisted of XML-tagged records or documents in 4 different languages: German, French, English and Persian (also known as Farsi). For German, French and English, the records were library catalog cards (bibliographic records describing publications archived by The European Library (TEL)). For Persian, the documents were newspaper articles (Hamshahri corpus of 1996-2002). Table 1 gives the collection sizes.

Table 1. Sizes of CLEF 2008 Ad Hoc Track Test Collections

Code	Language	Text Size (uncompressed)	Documents	|Topics|	Rel/Topic
DE	German	1,306,492,248 bytes	869,353	50	33 (lo 2, hi 84)
EN	English	1,208,383,351 bytes	1,000,100	50	51 (lo 7, hi 190)
FA	Persian	628,471,252 bytes	166,774	50	103 (lo 7, hi 255)
FR	French	1,362,122,091 bytes	1,000,100	50	27 (lo 3, hi 224)

The CLEF organizers created 50 natural language "topics" (numbered 451-500 for German, French and English and 551-600 for Persian) and translated them into many languages. Sometimes topics are discarded for some languages because of a lack of relevant documents (though that did not happen this year). Table 1 gives the final number of topics for each language and their average number of relevant documents (along with the lowest and highest number of relevant documents of any topic). For more information on the CLEF test collections, please see the track overview paper [2].

2.2 Base Run

For German, French and English, our base run used a vector of the words in the Title and Description topic fields. These words were stemmed with the lexicon-based inflectional stemming component of SearchServer, which includes decompounding for German.

For Persian, our base run used a vector of the words in the Title, Description and Narrative topic fields. We used a stemmer that was ported from Savoy's [6]. (Also our stopword list for Persian was derived from Savoy's [6].)

For each language, the base run retrieved the top-10,000 ranked documents for each topic (using the ranking approach described in [8]).

2.3 Sample Run

For each language, we created a sample run whose first 100 rows contained the following rows of the base run for the language in the following order:

```
1, 2, 3, 4, 5, 6, 7, 8, 9, 10,
20, 30, 40, 50, 60, 70, 80, 90, 100,
200, 300, 400, 500, 600, 700, 800, 900, 1000,
2000, 3000, 4000, 5000, 6000, 7000, 8000, 9000, 10000,
15, 25, ..., 95,
150, 250, ..., 950,
1500, 2500, ..., 9500,
125, 175, ..., 975,
1250, 1750, ..., 9750.
```

The remainder of the sample run was padded with the top-ranked remaining rows from the base run until 1000 rows had been retrieved (i.e. rows 11, 12, 13, 14, 16, ..., 962 of the base run).

This ordering (e.g. the placement of the sample from depth 10000 before the sample from depth 15) was chosen because of uncertainty of how deep the judging would be (e.g. last year the judging depth was 60 for some languages and 80 for others, and these judging depths are not announced in advance). As long as the top-37 were judged, we would have sampling to depth 10000 (because in the above list, you can count that after 37 samples that depth 10000 is reached). The extra sample points, if judged, would just improve the accuracy (because they are just additional sample points from the top 10000, not deeper sample points).

Our sample run for each language was submitted to the CLEF organizers for assessing in June 2008.

3 Results

When we received the relevance judgments and analyzed them in August 2008, we checked the judging depth of our sample runs. We found that the top-60 rows were judged for each topic for each language.

Tables 2, 3, 4 and 5 show the results of the sampling for each language. The columns are as follows:

- "Depth Range": The range of depths being sampled. The 11 depth ranges cover from 1 to 10000.
- "Samples": The depths of the sample points from the depth range. The samples are always uniformly spaced. They always end at the last point of the depth range. The total number of sample points (over the 11 rows of the table) adds to 60 for all 4 languages.
- "# Rel": The number of each type of item retrieved from the sample points over the 50 topics. The item type codes are R (relevant), N (non-relevant) and U (unjudged, of which there are always 0). An X is used when a sample point was not submitted because fewer than 10000 rows were retrieved for the topic (this just happened for one German topic). The sum of the item type counts is always 50 times the number of sample points for the depth range (because there are 50 topics for each language).

Table 2. Marginal Precision of German Base-TD Run at Various Depths

Depth Range	Samples	# Rel	Precision	Wgt	EstRel/Topic
1-5	1, 2, ..., 5	146R, 104N, 0U	0.584	1	2.9
6-10	6, 7, ..., 10	92R, 158N, 0U	0.368	1	1.8
11-50	15, 20, ..., 50	85R, 315N, 0U	0.212	5	8.5
51-100	55, 60, ..., 100	36R, 464N, 0U	0.072	5	3.6
101-200	150, 200	4R, 96N, 0U	0.040	50	4.0
201-500	250, 300, ..., 500	6R, 294N, 0U	0.020	50	6.0
501-900	550, 600, ..., 900	2R, 398N, 0U	0.005	50	2.0
901-1000	950, 1000	1R, 99N, 0U	0.010	50	1.0
1001-3000	1500, 2000, ..., 3000	1R, 199N, 0U	0.005	500	10.0
3001-6000	3500, 4000, ..., 6000	0R, 300N, 0U	0.000	500	0.0
6001-10000	7000, 8000, ..., 10000	1R, 196N, 3X	0.005	1000	20.0

Table 3. Marginal Precision of French Base-TD Run at Various Depths

Depth Range	Samples	# Rel	Precision	Wgt	EstRel/Topic
1-5	1, 2, ..., 5	91R, 159N, 0U	0.364	1	1.8
6-10	6, 7, ..., 10	63R, 187N, 0U	0.252	1	1.3
11-50	15, 20, ..., 50	51R, 349N, 0U	0.128	5	5.1
51-100	55, 60, ..., 100	50R, 450N, 0U	0.100	5	5.0
101-200	150, 200	2R, 98N, 0U	0.020	50	2.0
201-500	250, 300, ..., 500	9R, 291N, 0U	0.030	50	9.0
501-900	550, 600, ..., 900	6R, 394N, 0U	0.015	50	6.0
901-1000	950, 1000	1R, 99N, 0U	0.010	50	1.0
1001-3000	1500, 2000, ..., 3000	1R, 199N, 0U	0.005	500	10.0
3001-6000	3500, 4000, ..., 6000	1R, 299N, 0U	0.003	500	10.0
6001-10000	7000, 8000, ..., 10000	0R, 200N, 0U	0.000	1000	0.0

- "Precision": Estimated precision of the depth range (R/(R+N+U+X)).
- "Wgt": The weight of each sample point. The weight is equal to the difference in ranks between sample points, i.e. each sample point can be thought of as representing this number of rows, which is itself plus the preceding unsampled rows.
- "EstRel/Topic": Estimated number of relevant items retrieved per topic for this depth range. This is the Precision multiplied by the size of the depth range. Or equivalently, it is (R * Wgt) / 50.

Because each sample point is at the deep end of the range of rows it represents, the sampling should tend to underestimate precision for each depth range (assuming that precision tends to fall with depth, which appears to be the case for all 4 languages).

Table 6 shows the sums of the estimated number of relevant items per topic over all depth ranges in its first row (i.e. it is the sum of the EstRel/Topic entries

Table 4. Marginal Precision of English Base-TD Run at Various Depths

Depth Range	Samples	# Rel	Precision	Wgt	EstRel/Topic
1-5	1, 2, ..., 5	137R, 113N, 0U	0.548	1	2.7
6-10	6, 7, ..., 10	93R, 157N, 0U	0.372	1	1.9
11-50	15, 20, ..., 50	96R, 304N, 0U	0.240	5	9.6
51-100	55, 60, ..., 100	75R, 425N, 0U	0.150	5	7.5
101-200	150, 200	8R, 92N, 0U	0.080	50	8.0
201-500	250, 300, ..., 500	17R, 283N, 0U	0.057	50	17.0
501-900	550, 600, ..., 900	7R, 393N, 0U	0.018	50	7.0
901-1000	950, 1000	2R, 98N, 0U	0.020	50	2.0
1001-3000	1500, 2000, ..., 3000	2R, 198N, 0U	0.010	500	20.0
3001-6000	3500, 4000, ..., 6000	2R, 298N, 0U	0.007	500	20.0
6001-10000	7000, 8000, ..., 10000	0R, 200N, 0U	0.000	1000	0.0

Table 5. Marginal Precision of Persian Base-TDN Run at Various Depths

Depth Range	Samples	# Rel	Precision	Wgt	EstRel/Topic
1-5	1, 2, ..., 5	145R, 105N, 0U	0.580	1	2.9
6-10	6, 7, ..., 10	135R, 115N, 0U	0.540	1	2.7
11-50	15, 20, ..., 50	136R, 264N, 0U	0.340	5	13.6
51-100	55, 60, ..., 100	145R, 355N, 0U	0.290	5	14.5
101-200	150, 200	22R, 78N, 0U	0.220	50	22.0
201-500	250, 300, ..., 500	61R, 239N, 0U	0.203	50	61.0
501-900	550, 600, ..., 900	49R, 351N, 0U	0.123	50	49.0
901-1000	950, 1000	7R, 93N, 0U	0.070	50	7.0
1001-3000	1500, 2000, ..., 3000	11R, 189N, 0U	0.055	500	110.0
3001-6000	3500, 4000, ..., 6000	9R, 291N, 0U	0.030	500	90.0
6001-10000	7000, 8000, ..., 10000	2R, 198N, 0U	0.010	1000	40.0

Table 6. Estimated Percentage of Relevant Items that are Judged, Per Topic

	DE	FR	EN	FA
Estimated Rel@10000	59.9	51.2	95.7	412.7
Official Rel/Topic	32.7	26.8	50.7	103.2
Percentage Judged	55%	52%	53%	25%

in the last column of the corresponding table from Tables 2-5). The official number of relevant items per topic for each language is listed in the second row. The final row of the table just divides the official number of relevant items by the estimated number in the first 10000 retrieved (e.g. for German, 32.7/59.9=55%). This number should tend to be an overestimate of the percentage of all relevant items that are judged (on average per topic) because there may be relevant items that were not matched by the query in the first 10000 rows.

3.1 Remarks

These estimates of judging coverage for the CLEF 2008 collections (55% for German, 52% for French, 53% for English, 25% for Persian) tend to be lower than the estimates we produced for the CLEF 2007 collections last year [10] (55% for Czech, 69% for Bulgarian, 83% for Hungarian) or the estimates we produced for the NTCIR-6 collections (58% for Chinese, 78% for Japanese, 100% for Korean) [11]. The German, French and English estimates are higher than the estimates we produced for the TREC 2006 Legal and Terabyte collections using a similar approach (18% for TREC Legal and 36% for TREC Terabyte) [9], while the Persian estimate is in the same ballpark as the estimates for the (much larger) TREC 2006 collections.

The incompleteness results for German, French and English are similar to what [12] found for depth-100 pooling on the old TREC collections of approximately 500,000 documents. [12] reported that "it is likely that at best 50%-70% of the relevant documents have been found; most of these unjudged relevant documents are for the 10 or so queries that already have the most known answers."

Fortunately, [12] also found for such test collections that "overall they do indeed lead to reliable results." [3] also considers the "levels of completeness" in some older TREC collections to be "quite acceptable" even though additional judging found additional relevant documents. And we can confirm that we have gained a lot of insights from the CLEF test collections over the years, particularly when conducting topic analyses such as described in [8].

For Persian, the topics appear to have been relatively broad (more relevant documents per topic on average) which led to the judging coverage being relatively shallow (less than 25% on average based on the sampling experiment). It may be particularly advisable to conduct a "system omission" study on this collection (like the one described in [12]) which may indicate whether or not the collection is likely to give reliable results for systems that did not contribute to the pooling.

3.2 Error Analysis

We should note that our sampling was very coarse at the deeper ranks, e.g. for German, 1 relevant item out of 200 samples in the 6001-10000 range led to an estimate of 20 relevant items per topic in this range. If the sampling had turned up 0 or 2 relevant items, a minor difference, the estimate would have been 0 or 40 relevant items per topic in this range, leading to a substantially different sum (39.9 or 79.9 instead of 59.9). We leave the computation of confidence intervals for our estimates, along with analysis of the variance across topics, as future work.

4 Conclusions

We conducted an experiment to test the completeness of the relevance judgments for the monolingual German, French, English and Persian information retrieval tasks of the Ad Hoc Track of the Cross-Language Evaluation Forum (CLEF) 2008. For each language, we submitted a sample of the first 10000 retrieved items to investigate the frequency of relevant items at deeper ranks than the official judging depth (of 60). Based on the results, we estimated that the percentage of relevant items assessed was less than 55% for German, 52% for French, 53% for English and 25% for Persian. For German, French and English, these levels of completeness are in line with the estimates that have been made for some past test collections which are still considered useful and fair for comparing retrieval methods. For Persian, the completeless levels are lower than usual. For any test collection, it is prudent to conduct a "system omission" study (like the one described in [12]) which may indicate whether or not the collection is likely to give reliable results for systems that did not contribute to the pooling. Such a study would be particularly advisable for the Persian collection.

References

1. Cross-Language Evaluation Forum web site, `http://www.clef-campaign.org/`
2. Agirre, E., Di Nunzio, G.M., Ferro, N., Mandl, T., Peters, C.: CLEF 2008: Ad Hoc Track Overview. In: Peters, C., et al. (eds.) CLEF 2008. LNCS, vol. 5706, pp. 15–37. Springer, Heidelberg (2009)
3. Harman, D.K.: The TREC Test Collections. In: TREC: Experiment and Evaluation in Information Retrieval (2005)
4. Hodgson, A.: Converting the Fulcrum Search Engine to Unicode. In: Sixteenth International Unicode Conference (2000)
5. NTCIR (NII-NACSIS Test Collection for IR Systems), `http://research.nii.ac.jp/~ntcadm/index-en.html`
6. Savoy, J.: CLEF and Multilingual information retrieval resource page, `http://www.unine.ch/info/clef/`
7. Text REtrieval Conference (TREC), `http://trec.nist.gov/`
8. Tomlinson, S.: Bulgarian and Hungarian Experiments with Hummingbird SearchServer™ at CLEF 2005. In: Peters, C., Gey, F.C., Gonzalo, J., Müller, H., Jones, G.J.F., Kluck, M., Magnini, B., de Rijke, M., Giampiccolo, D. (eds.) CLEF 2005. LNCS, vol. 4022, pp. 194–203. Springer, Heidelberg (2006)
9. Tomlinson, S.: Experiments with the Negotiated Boolean Queries of the TREC 2006 Legal Discovery Track. In: Proceedings of TREC 2006 (2006)
10. Tomlinson, S.: Sampling Precision to Depth 10000 at CLEF 2007. In: Peters, C., Jijkoun, V., Mandl, T., Müller, H., Oard, D.W., Peñas, A., Petras, V., Santos, D. (eds.) CLEF 2007. LNCS, vol. 5152, pp. 57–63. Springer, Heidelberg (2008)
11. Tomlinson, S.: Sampling Precision to Depth 9000: Evaluation Experiments at NTCIR-6. In: Proceedings of NTCIR-6 (2007)
12. Zobel, J.: How Reliable are the Results of Large-Scale Information Retrieval Experiments? In: SIGIR 1998, pp. 307–314 (1998)

JHU Ad Hoc Experiments at CLEF 2008

Paul McNamee

JHU Human Language Technology Center of Excellence
paul.mcnamee@jhuapl.edu

Abstract. For CLEF 2008 JHU conducted monolingual and bilingual experiments in the ad hoc TEL and Persian tasks. Additionally we performed several post hoc experiments using previous CLEF ad hoc tests sets in 13 languages.

In all three tasks we explored alternative methods of tokenizing documents including plain words, stemmed words, automatically induced segments, a single selected n-gram from each word, and all n-grams from words (*i.e.,* traditional character n-grams). Character n-grams demonstrated consistent gains over ordinary words in each of these three diverse sets of experiments. Using mean average precision, relative gains of of 50-200% on the TEL task, 5% on the Persian task, and 18% averaged over 13 languages from past CLEF evaluations, were observed.

1 Introduction

As a tokenization scheme character n-grams possess many advantages. They work in every language, require no training, and are more effective than plain words. It also appears that n-grams are beneficial for normalizing morphological variation, particularly in languages where words have many related surface forms.

Using test sets in the 13 languages used in the ad hoc tracks at previous CLEF evaluations, we compared n-grams to several tokenization alternatives, including a rule-based stemmer (*Snowball*), an unsupervised morphological segmenter (*Morfessor*), and a synthetic form of stemming based on selecting a single character n-gram from each word. Character n-grams of length $n = 5$ were the most effective technique, performing 18% better than unnormalized words, averaged across the set of languages.

Accordingly n-grams were used in official submissions to the CLEF 2008 ad hoc tasks. The JHU HAIRCUT retrieval system was used with a statistical language model similarity metric with a smoothing constant of 0.5. The similarity calculation combines document term frequencies and corpus frequencies (for smoothing) using linear interpolation with a smoothing constant of 0.5 [8]. For retrieval of Farsi text, we explored a variant of n-gram indexing, *skipgrams*, which are n-gram sequences that omit some letters. Farsi has root and template morphology and it was thought that skipgrams might prove effective.

In Section 2 we describe our experiments for the TEL subtask. In Section 3 we analyze our training experiments and official results for the Persian subtask. In Section 4 our experiments on past CLEF collections are described.

C. Peters et al. (Eds.): CLEF 2008, LNCS 5706, pp. 170–177, 2009.

2 TEL Task

The TEL task involved focused on searching electronic card catalog records in English, French, and German using data from the British Library, the Bibliotheque Nationale de France, and the Österreichische Nationalbibliothek (Austrian National Library). Our approach to TEL was to treat the collection as unstructured documents. Fields that did not appear to contain good indexable content were removed, including: publisher, rights, format, description, indentifier, contributor, type, language, coverage, issued, available, extent, spatial, and created. Text from the following fields was retained: ispartof, edition, alternative, tableofcontents, abstract, bibliographiccitation, subject, title, abstract, date, creator, source, and relation. All SGML tags were removed.

Some of these choices were harmful. For example, queries that specified a particular language or document type (e.g., maps) would have benefitted from some of the deleted metadata. The aim of removing these fields was to increase the coherence of each document's indexable terms.

2.1 Indexing Schemes

The tokenization methods explored were:

- **words:** space-delimited tokens.
- **snow:** output of the Snowball stemmer.
- **morf:** the set of morphemes for each word identified by the *Morfessor* algorithm. Morfessor is available online at http://www.cis.hut.fi/projects/morpho/. A model was trained using the document collection's lexicon with digit-containing tokens omitted. The default parameters for the Morfessor algorithm were used [1].
- **lcn4/5:** least common n-gram stem (*i.e.,* rarest word-internal character n-gram) of length $n = 4$ or $n = 5$ [3].
- **4-grams:** overlapping, word-spanning character 4-grams produced from the stream of words encountered in the document or query.
- **5-grams:** length $n = 5$ n-grams created in the same fashion as the character 4-grams.

Common to each tokenization method was conversion to lower case letters, removal of punctuation, and truncation of long numbers to 6 digits.

2.2 Monolingual Results

Our official submissions were based on 4-grams, both with and without relevance feedback, 5-grams (no RF), and stemmed words. Table 1 lists mean average precision (MAP) for these runs and for several unsubmitted runs. In the official run names xx indicates one of de (German), en (English), or fr (French).

While performance did not vary dramatically in English, except for the unnormalized word run which performed the worst, 4-grams were dominant with the French and German collections. Large gains were observed with 4-grams compared to plain words – more than a 50% relative gain in French and over 200% in German.

Table 1. Monolingual Results

	English	French	German	Run designation
words	0.2719	0.2019	0.1073	not submitted
snow	0.3480	0.2290	0.1757	aplmoxxs
morf	0.3171	0.2332	0.1989	not submitted
lcn4	0.3086	0.2223	0.1565	not submitted
lcn5	0.2993	0.2270	0.1810	not submitted
4-grams	0.3382	**0.2950**	**0.3377**	aplmoxx4
5-grams	0.3190	0.2800	0.3102	aplmoxx5
4-grams + RF	**0.3531**	0.2861	0.3176	aplmoxx4rf

Table 2. Official Bilingual Runs

	Target Language		
Source	English	French	German
Dutch	0.2024	0.1746	x
English	x	0.1669	0.1899
French	0.2087	x	0.1829
German	0.2111	0.1608	x
Spanish	0.1856	x	x

2.3 Bilingual Results

We considered the following bilingual pairs: Dutch/French/German/Spanish to English; Dutch/English/German to French; and, English/French to German. For each language pair the source side query was tokenized using only character 5-grams and those n-grams were 'translated' to the target language using a large aligned parallel corpus (content from the Official Journal of the European Journal). The methodology in query term translation was like that in [5]; however, here no pre-translation query expansion was performed. In Table 2 results are presented using mean average precision to compare performance.

Source language did not make a large difference in performance across the three collections. Bilingual performance was approximately 60% of the highest performing monolingual run, which is a bit lower than we have customarily observed in bilingual retrieval against news corpora at CLEF.

3 Persian Language Task

We made submissions for both the monolingual and bilingual subtasks. The bilingual submissions were based on online machine translation software[1] applied to the queries, so only one set of indexes was required. In addition to the methods in Section 2.1 we used skipgrams, 4- or 5-grams with and without one internal skip (denoted by *sk41 & sk51*).

[1] http://www.parstranslator.net/eng/translate.htm

3.1 Skipgrams

Pirkola et al. [7] have proposed n-grams with skips[2] to match terminology for cross-language information retrieval in languages sharing a common alphabet. For example, the English word *calcitonin* can be matched to its Finnish translation *kalsitoniini*, supported in part by matches like l⋆t and n⋆n. Järvelin et al. [2] formalized the notion of skipgrams and investigated methods of comparing lexical terms; however, they focused on the case where a single skip is formed by deleting contiguous letters. This makes sense when only bigrams are considered – then the only place to skip characters is between the first and last letters of the (skip) bigram.

But with longer n-grams there are multiple places where skips can occur, and character skipgram methods can be generalized even further by including the possibility of multiple non-adjacent skips within a single word (though no such experiments are reported here). In these experiments skipgrams are considered as an alternative method for tokenization that might support matches across morphologically related words. When a letter is skipped we replace that letter in the n-gram subsequence with a special symbol (*i.e.*, a dot character (•). This is done in an attempt to avoid unintended conflations with n-gram strings produced by unrelated words. Skipgram tokenization of length four for the word *cream* would include the regular n-grams *crea* and *ream* in addition to *c•eam*, *cr•am*, and *cre•m*.

3.2 Training Data

Each method of tokenization was compared on the 50 training topics. In Table 3 runs without relevance feedback are presented along with runs that made use of automated feedback using various numbers of expansion terms. Based on the training data 5-grams and skipgrams (sk41) were the most effective approaches, although when no relevance feedback was used plain words had the highest score.

Table 3. Training results for Persian (MAP)

	No RF	50	100	200	400	800
4-grams	0.3883	0.4199	0.4231	0.4172	-	-
5-grams	0.3810	0.4225	0.4305	0.4280	-	-
words	0.4091	0.4175	0.3999	0.3905	-	-
morfessor	0.3784	0.3951	0.3801	0.3637	-	-
lcn4	0.3914	0.3975	0.3840	0.3730	-	-
lcn5	0.3978	0.3960	0.3779	0.3723	-	-
sk41	0.3886	0.4000	0.4156	0.4332	0.4372	0.4290
sk51	0.3613	0.3607	0.3817	0.4012	0.4216	0.4280

[2] They use the term s-grams.

3.3 Monolingual and Bilingual Results

In Table 4 mean average precision is reported for eight tokenization methods using the test topics. The n-grams methods are the highest performing approach and the skipgrams perform slightly worse than traditional character n-grams. The highest performing run was character 4-grams using 200 expansion terms which got a MAP score of 0.4564; however the results on the training topics suggested 5-grams would outperform and we based our submissions on them instead. N-grams need more query expansion terms than words to maximize performance, and skipgrams, being even more conflationary require more than regular 4- or 5-grams.

Table 4. Monolingual runs

	No RF	50	100	200	400
words	0.3617	0.4332	0.4299	0.4211	-
morf	0.3559	0.4250	0.4223	0.4156	-
lcn4	0.3629	0.4252	0.4256	0.4180	-
lcn5	0.3506	0.4225	0.4188	0.4085	-
4-grams	0.3986	0.4383	0.4530	0.4564	-
5-grams	0.3821	0.4288	0.4493	0.4558	-
sk41	0.3906	0.3732	0.4053	0.4384	0.4519
sk51	0.3512	0.3238	0.3595	0.4008	0.4250

The results for our official monolingual and bilingual runs are given in Table 5. Tokenization method did not appear to drastically affect the outcome monolingually; however, words and the Morfessor-based runs did markedly worse on the bilingual task compared to the n-gram based methods.

Table 5. Official runs

	Task	Index	RF Terms	MAP
jhufa5r100	mono	5-grams	100	0.4493
jhufask41r400	mono	sk41	400	**0.4519**
jhufawr50	mono	words	50	0.4332
jhufamr50	mono	morf	50	0.4250
jhuenfa5r100	bi	5-grams	100	0.1660
jhuenfask41r400	bi	sk41	400	**0.1892**
jhuenfawr50	bi	words	50	0.0946
jhuenfamr50	bi	morf	50	0.1112

4 Analysis from Past CLEF Collections

We compare plain words, stems, induced morphemes, n-gram stems, and character n-grams using test sets from the CLEF ad hoc tasks between 2002 and 2007. Table 6 gives MAP for each method in 13 languages.

Table 6. Comparison of 7 Tokenization Alternatives (Mean Average Precision)

Language	Data	Queries	Words	Snow	Morf	LCN4	LCN5	4-gram	5-gram
Bulgarian	06-07	100	0.2195		0.2786	0.2937	0.2547	**0.3163**	0.2916
Czech	07	50	0.2270		0.3215	0.2567	0.2477	**0.3294**	0.3223
Dutch	02-03	106	0.4162	0.4273	0.4274	0.4021	0.4073	0.4378	**0.4443**
English	02-03	96	0.4829	**0.5008**	0.4265	0.4759	0.4861	0.4411	0.4612
Finnish	02-03	75	0.3191	0.4173	0.3846	0.3970	0.3900	0.4827	**0.4960**
French	02-03	102	0.4267	**0.4558**	0.4231	0.4392	0.4355	0.4442	0.4399
German	02-03	106	0.3489	0.3842	0.4122	0.3613	0.3656	0.4281	**0.4321**
Hungarian	06-07	98	0.1979		0.2932	0.2784	0.2704	**0.3549**	0.3438
Italy	02-03	100	0.3950	**0.4350**	0.3770	0.4127	0.4054	0.3925	0.4220
Portuguese	05-06	100	0.3232		0.3403	0.3442	0.3381	0.3316	**0.3515**
Russian	03-04	62	0.2671		0.3307	0.2875	0.3053	**0.3406**	0.3330
Spanish	02-03	107	0.4265	**0.4671**	0.4230	0.4260	0.4323	0.4465	0.4376
Swedish	02-03	102	0.3387	0.3756	0.3738	0.3638	0.3467	0.4236	**0.4271**
Average			0.3375		0.3698	0.3645	0.3604	0.3955	**0.3979**
Average (8 Snowball langs)			0.3504	0.3848	0.3608	0.3642	0.3632	0.3885	**0.3956**

4.1 Unnormalized Words

Not attempting to control for morphological processes can have harmful effects. In Bulgarian, Czech, Finnish, and Hungarian, more than a 30% loss is observed compared to the use of 4-grams as indexing terms.

4.2 Snowball Stemming

Snowball does not support Bulgarian, Czech, or Russian and due to character encoding issues with the software we were not able to use it for Portuguese and Hungarian. Stemming, when available, is quite effective, and just slightly below the top-ranked approach of character n-grams.

4.3 Morfessor Segments

As it may be difficult to find a rule-based stemmer for every language, a language-independent approach can be quite attractive. The Morfessor algorithm only requires a lexicon for a language to learn to identify morpheme boundaries, even for previously unseen words. Such automatically detected segments can be an effective form of tokenization [6]. Examples of the algorithm's output are presented in Table 7, along with results for Snowball and character 5-grams.

Compared to plain words the induced morphemes produced by Morfessor led to gains in 9 of 13 languages; 8 of these were significant improvements with $p < 0.05$ (Wilcoxon test). The languages where words outperformed segments were English, French, Italian, and Spanish – each is relatively low in morphological complexity. The differences in French and Spanish were less than 0.004 in absolute terms. Segments achieved more than a 20% relative improvement in Bulgarian, Finnish, and Russian, and over 40% in Czech and Hungarian.

Table 7. Word Normalization Examples

Word	Snowball	Morfessor	5-grams
authored	author	author+ed	_auth, autho, uthor, thore, hored, ored_
authorized	author	author+ized	_auth, autho, uthor, thori, horiz, orize, rized, ized_
authorship	authorship	author+ship	_auth, autho, uthor, thors, horsh, orshi, rship, ship_
afoot	afoot	a+foot	_afoo, afoot, foot_
footballs	footbal	football+s	_foot, footb, ootba, otbal, tbaall, balls, alls_
footloose	footloos	foot+loose	_foot, footl, ootlo, otloo, tloos, loose, oose_
footprint	footprint	foot+print	_foot, footp, ootpr, otpri, tprin, print, rint_
feet	feet	feet	_feet, feet_
juggle	juggl	juggle	_jugg, juggl, uggle, ggle_
juggled	juggl	juggle+d	_jugg, juggl, uggle, ggled, gled_
jugglers	juggler	juggle+r+s	_jugg, juggl, uggle, ggler, glers, lers_

4.4 Least Common N-Gram Stems

Another language-neutral approach to stemming is to select for each word, its least common n-gram. This requires advance knowledge of n-gram frequencies, but this is easily obtainable by constructing a regular n-gram index, or even by scanning a corpus and counting. Lengths of $n = 4$ and $n = 5$ appear about equally effective with a slight advantage for *lcn4*, but this is influenced primarily by the languages with greater morphological complexity, which see larger changes. An 8% relative improvement in mean average precision over words is obtained. As can be seen from Table 1, in languages where rule-based stemming is available its use is preferable. N-gram stemming achieves comparable performance with Morfessor segments..

4.5 Overlapping Character N-Grams

N-grams achieve morphological regularization indirectly due to the fact that subsequences that touch on word roots will match. For example, "juggling" and "juggler" will share the 5-grams _jugg and juggl. While n-gram's redundancy enables useful matches, other matches are less valuable, for example, every word ending in 'tion' will share 5-gram tion_ with all of the others. In practice these morphological false alarms are almost completely discounted because term weighting de-emphasizes them. In fact, such affixes can be so common, that ignoring them entirely by treating them as "stop n-grams" is a reasonable thing to do.

Character n-grams are the most effective technique studied here, giving a relative improvement of 18%. Consistent with earlier work [4] lengths of $n = 4$ and $n = 5$ are equally effective averaged across the 13 languages; however there are some noticeable differences in particular languages. The data is suggestive of a trend that the most morphologically variable languages (*i.e.*, Bulgarian, Czech, Hungarian, and Russian) gain more from 4-grams than 5-grams, while 5-grams have a slight advantage in medium complexity languages.

Snowball stems are roughly as effective as n-grams, on average, but only available in certain languages (*i.e.*, 8 of 13 in this study). The other "alternative" stemming approaches, segments and least common n-grams, appear to gain about half of the benefit that full n-gram indexing sees compared to unnormalized word forms.

5 Conclusions

We examined a variety of methods for lexical normalization, finding that the most effective technique was character n-gram indexing. N-grams achieved consistent gains in mean average precision over unlemmatized words. Relative gains of of 50-200% on the TEL task, 5% on the Persian task, and 18% averaged over thirteen languages from past CLEF evaluations, were observed. In languages such as Czech, Bulgarian, Finnish, and Hungarian gains of over 40% were observed. While rule-based stemming can be quite effective, such tools are not available in every language and even when present, require additional work to integrate with an IR system. When language-neutral methods are able to achieve the same, or better performance, their use should be seriously considered.

References

1. Creutz, M., Lagus, K.: Unsupervised Morpheme Segmentation and Morphology Induction from Text Corpora Using Morfessor 1.0. Helsinki University of Technology Technical Report A81 (2005)
2. Järvelin, A., Järvelin, A., Järvelin, K.: S-grams: Defining Generalized N-grams for Information Retrieval. Information Processing and Management 43(4), 1005–1019 (2007)
3. Mayfield, J., McNamee, P.: Single n-gram stemming. In: Proceedings of ACM SIGIR 2003, pp. 415–416 (2003)
4. McNamee, P., Mayfield, J.: Character N-Gram Tokenization for European Language Text Retrieval. Information Retrieval 7(1-2), 73–97 (2004)
5. McNamee, P., Mayfield, J.: Translating pieces of words. In: Proceedings of the 28th Annual International ACM SIGIR Conference on Research and Development in Information Retrieval, Salvador, Brazil, pp. 643–644 (2005)
6. McNamee, P., Nicholas, C., Mayfield, J.: Don't Have a Stemmer?: Be Un+concern+ed. In: Proceedings of ACM SIGIR 2008, pp. 813–814 (2008)
7. Pirkola, A., Keskustalo, H., Leppänen, E., Känsälä, A., Järvelin, K.: Targeted s-gram matching: a novel n-gram matching technique for cross- and mono-lingual word form variants. Information Research 7(2) (2002)
8. Ponte, J.M., Croft, W.B.: A Language Modeling Approach to Information Retrieval. In: Proceedings of ACM SIGIR 1998, pp. 275–281 (1998)

UniNE at CLEF 2008: TEL, and Persian IR

Ljiljana Dolamic, Claire Fautsch, and Jacques Savoy

Computer Science Department, University of Neuchatel,
Rue Emile Argand 11, 2009 Neuchatel, Switzerland
{Ljiljana.Dolamic,Claire.Fautsch,Jacques.Savoy}@unine.ch

Abstract. In our participation in this evaluation campaign, our first objective was to analyze retrieval effectiveness when using The European Library (TEL) corpora composed of very short descriptions (library catalog records) and also to evaluate the retrieval effectiveness of several IR models. As a second objective we wanted to design and evaluate a stopword list and a light stemming strategy for the Persian (Farsi), a member of the Indo-European family of languages and whose morphology is more complex than of the English language.

1 Introduction

During the last few years, the IR group at University of Neuchatel has focused on designing, implementing and evaluating IR systems for various natural languages, including European [1] and popular Asian languages (namely, Chinese, Japanese, and Korean). The main objective of our work is still to promote effective monolingual IR in many different natural languages.

The rest of this paper is organized as follows: Section 2 describes the main characteristics of the TEL corpus used in the CLEF-2008 ad hoc track. Section 3 outlines the main aspects of the various IR models used with TEL collections as well as an evaluation of our official runs and certain related experiments. Section 4 presents the principal features of the Persian (Farsi) language along with the stopword list and stemming strategy developed for this language, and describes our official runs for this task.

2 Overview of TEL Corpus

In a certain sense this first ad hoc task takes us back to our research roots, due to the need to look for relevant items among the card catalog on the collection located at The European Library (TEL) (see www.TheEuropeanLibrary.org). This collection includes three sub-collections, one in the English language (from British Library), the second in German (Austrian National Library) and the third in French (Bibliothèque nationale de France). The real challenge in our work is to retrieve pertinent records through relying on very short catalog descriptions on the information items involved. In many of these record items

C. Peters et al. (Eds.): CLEF 2008, LNCS 5706, pp. 178–185, 2009.

the only information contained is the title (under the tag <title>) and author, plus manually assigned subject headings (tag <subject>). Other records may however contain a short description of the object (tags <description> and <alternative>). Each record may of course contain other fields not used during the indexing process such as language, document identification, author, publisher, location, issue, date, etc. For more information, see [2].

The average size of each topic description is relatively short (between 10 and 16 terms), and is similar for all three languages (perhaps a bit longer for the French corpus). The descriptors are subdivided into title (T), descriptive (D) and narrative (N) logical sections, and from them we automatically removed certain phrases such as "Relevant document report ..." or "Relevante Dokumente berichten ...", etc. All our runs were fully automatic.

The available topics cover various subjects (e.g., Topic #500: "Gauguin and Tahiti," Topic #468: "Modern Japanese Culture," Topic #471: "Watchmaking," or Topic #477: "Web Advertising", etc.). While topic descriptions do not generally contain many proper names (creators and their works), we found two topics containing personal names ("Henry VIII" and "Gauguin"), and 23 with geographical names (e.g., "Europe," "Eastern," "Bordeaux" or "Greek"). Expressions referring to the Untied States of America are not standardized and may for example take the form "USA," "North America," or "America." Also, time periods are infrequently used (in 7 topics only), with many including expressions that are fairly broad (e.g., "Modern," or "Roman"), while others are more precise ("World War I").

3 IR Models and Evaluation

An essential element in our indexing strategy was the use a stopword list to denote very frequent word forms having no important impact matching topic and document representatives (e.g., "the," "in," "or," "has," etc.). In our experiments the stopword list contained 589 English, 484 French and 578 German terms, and diacritics were replaced by their corresponding non-accented equivalent. Another element was the use of light stemmers developed for the French and German languages, wherein inflectional suffixes attached only to nouns and adjectives were removed. This resulted in more effective retrieval than do more aggressive stemmers that also remove derivational suffixes [3]. These stemmers and stopword lists are freely available at the Web site `www.unine.ch/info/clef`. For the English language we tried both a light stemmer (the S-stemmer proposed by Harman [4] to remove only the plural form '-s') and a more aggressive version [5] based on a list of around 60 suffixes.

In the German language compound words are widely used and present some specific challenges. For example the compound noun "Forschungsprojekt" can be divided into "Forschung" + 's' + "Projekt" (research + project), and the augment (i.e. the letter 's' in our example) is not always present (e.g., "Bankangestelltenlohn" combines "Bank" + "Angestellten" + "Lohn" (salary)). Given the fairly wide use of compound constructions in German and their many different forms, an effective IR system must include an automatic decompounding procedure. The automatic one used in our experiments [1] leaves both the compound form and its composite parts in both the topic and document representatives.

In an effort to obtain high MAP values we considered adopting different weighting schemes for the terms found in documents or queries. This would thus allow us to account for term occurrence frequency (denoted tf), inverse document frequency (denoted *idf*) as well as the document length. In the following experiments we considered the classical $tf \cdot idf$ formulation (with the cosine normalization), as well as probabilistic models such as the Okapi (or BM25) and variants derived from the DFR (*Divergence from Randomness*) family of models. Finally we also implemented a statistical language model (LM) known as a non-parametric probabilistic model (Okapi and DFR are considered as parametric models). For specific details on these IR models, see [6].

To measure retrieval performance we used the mean average precision (MAP) obtained from 50 queries. The best performance obtained under a given condition is shown in bold type in the following tables. We then applied the bootstrap methodology in order to statistically determine whether or not a given search strategy would be better than the performance depicted in bold. Thus, in the tables in this paper we added an asterisk to indicate any statistically significant differences resulting from the use of a two-sided non-parametric bootstrap test ($\alpha = 5\%$).

Table 1 shows the MAP obtained by various probabilistic models for the English collection, using two different query formulations (T or TD) and two stemmers. The last two columns show the MAP obtained when applying our light stemmer to the French corpus. An analysis of this data shows that the best performing IR model was usually the DFR-$I(n_e)B2$ or DFR-PB2 formulation (English corpus, T queries). For the English corpus with the Porter stemmer and TD query formulation, the LM model performed slightly better (0.3701 vs. 0.3643, a statistically non-significant difference).

Table 1. MAP of Various IR Models and Query Formulations (English & French TEL Corpus)

	Mean Average Precision					
	English	English	English	English	French	French
Query	T	TD	T	TD	T	TD
Stemmer	S-stem.	S-stem.	Porter	Porter		
Okapi	0.2795*	0.3171*	0.3004*	0.3329*	0.2659*	0.2998*
DFR-PB2	**0.3076**	0.3540	**0.3263**	0.3646	0.2734	0.3103*
DFR-GL2	0.2935*	0.3300*	0.3125*	0.3478*	0.2734	0.3117*
DFR-$I(n_e)B2$	0.3075	**0.3541**	0.3258	0.3643	**0.2825**	**0.3291**
LM	0.3029	0.3527	0.3180	**0.3701**	0.2747	0.3201
tf idf	0.1420*	0.1783*	0.1600*	0.1871*	0.1555*	0.1821*
% over T		+14.6%		+12.4%		+14.7%
% over S-stem.			+6.2%	+4.2%		

Table 2. MAP of Various IR Models and Query Formulations (German TEL Corpus)

	Mean Average Precision			
	German	German	German	German
Query	T	TD	T	TD
Decompounding?	no	no	yes	yes
Okapi	0.1462*	0.1872*	0.2188*	0.2522*
DFR-PB2	**0.1635**	**0.2097**	0.2193	0.2555
DFR-GL2	0.1462*	0.1878*	0.2309	0.2615*
DFR-$I(n_e)B2$	0.1606	0.2071	0.2248	0.2615
LM	0.1529	0.1972*	**0.2361**	**0.2697**
tf idf	0.1105*	0.1382*	0.1312*	0.1598*
% over T		+28.5%		+15.1%
% over no decomp.			+46.8%	+31.5%

The second last line shows the percentage variations derived from comparing results with the short (T) query formulation, and the last line the performance difference obtained using the S-stemmer. As indicated, increasing query size improves the MAP (around +12.4% to +14.7%). Statistically, when using the MAP obtained by T query formulation as baseline, the TD query format always improves retrieval performance significantly.

According to the MAP, the best indexing seemed to be the stemming technique using Porter's approach. In this case, the MAP with TD query formulation and Porter's stemmer increased by about 4.2% compared to the S-stemmer. Applying our statistical test when comparing the S-stemmer with Porter's approach, only three cases had statistically significant performance differences (underlined in Table 1).

Table 2 shows the MAP obtained with the probabilistic models and with two query formulations (T or TD) to the German collection, and comparing performances with and without our automatic decompounding approach. The best IR models seemed to be the DFR-PB2 (without decompounding) or the LM with our decompounding scheme. By adding terms to the topic descriptions, we could improve MAP (between 15.1% to 28.5%), although the performance differences were never statistically significant. Comparing the average performances shows that applying an automatic decompounding approach improved retrieval effectiveness, on average by 46.8% for short query formulations compared to +31.5% for TD queries) (see last line of Table 2). When analyzing the performance of various models, the differences were usually statistically significant (MAP underlined in Table 2).

An analysis showed that pseudo-relevance feedback (or blind-query expansion) seemed to be a useful technique for enhancing retrieval effectiveness. In this study, we adopted Rocchio's approach [7] (denoted "Roc." in the following tables with $\alpha = 0.75$, $\beta = 0.75$), whereby the system was allowed to add m terms extracted from the k best ranked documents from the original query. From previous experiments we learned that this type of blind query expansion strategy does not always work well. More particularly, we believe that including terms

Table 3. Description and MAP of Our Best Official TEL Runs

Language	Index	Query	Model	Query expansion	MAP	MAP
English	Porter	TD	Okapi		0.3329	Z-score
UniNEen3	S-stem	TD	$I(n_e)B2$		0.3541	**0.3754**
	Porter	TD	LM	Roc. 5 docs/10 terms	0.3913	
French	stem	TD	Okapi		0.2998	Z-score
UniNEfr3	stem	TD	$I(n_e)B2$		0.3291	**0.3327**
	stem	TD	LM	Roc. 5 docs/10 terms	0.3150	
German	decomp.	TD	Okapi	idf 5 docs/10 terms	0.2302	Z-score
UniNEde1	word	TD	GL2	Roc. 5 docs/20 terms	0.2356	**0.3013**
	decomp.	TD	$I(n_e)B2$	Roc. 5 docs/50 terms	0.2757	

occurring frequently in the corpus (because they also appear in the top-ranked documents) may introduce additional noise, and thus be ineffective in discriminating between relevant and non-relevant items. We thus decided to also apply our idf-based query expansion model [8] (denoted "idf" in following tables).

It is usually assumed that combining result lists computed by different search models (data fusion) could improve retrieval effectiveness [9]. Thus in this study we combined three probabilistic models representing both the parametric (Okapi and DFR) and non-parametric (language model or LM) approaches. To produce a combination such as this we evaluated various fusion operators and thus we suggest the "Z-score" approach which applies a normalization procedure to each result list before combining the different document scores (see details in [1]).

4 IR with Persian (Farsi) Language

The Persian (or Farsi) language is a member of the Indo-European family and has relatively few morphological variations. This year we used a corpus comprising Hamshahri newspapers from 1996 to 2002 (611 MB). It contains exactly 166,774 documents covering various subjects (politics, literature, art and economics, etc.) and comprises 448,100 different words. Article size varies between 1 KB and 140 KB and include on average about 202 tokens (127 when counting the number of distinct word types). The corpus is coded in UTF-8 and its alphabet has 28 Arabic letters plus an additional 4 letters used in Persian ("گ" "چ" "ژ" "پ").

We began by building a Persian stopword list containing 884 terms. Unlike most others lists, it contains the collection's most frequently occurring words (determinants, prepositions, conjunctions, pronouns or certain auxiliary verb forms), plus a large number of suffixes already separated from word stems in the collection. Note that that the Persian language does not include definite (the) or indefinite (a, an) articles (indefinite articles are indicated by a suffix ("ی" or simply by "one").

As a stemming strategy we used either a morphological analysis [10] or our simple, fast and light stemmer. It removes only nouns and adjective inflections (number and case only, since Persian does distinguish gender). The general pattern is the following: <possessive> <plural> <other-suffix> <stem>.) In our light stemmer we usually remove possessive, plural and certain suffixes marked as others. The following examples from our light stemmer illustrate certain aspects of the Persian morphology. From the plural form "درختان" ("trees"), we can obtain "درخت" ("tree"). The plural is usually denoted by either "ها" (inanimate) or by "ان" or "ها" (animate nouns). The plural forms for words borrowed from Arabic usually apply the language's own plural formation rule, and in Persian there are certain irregular formations similar to "mouse/mice". For the possessive form, "دستم" ("my hand"), our stemmer returns "دست" ("hand"). For the form "ایرانیان" ("Iranians") we remove both the plural and the derivational suffixes to obtain "ایران" ("Iran"). In this corpus we saw certain circumstances where suffixes might be written together or separated from the word (e.g., "ها"). Adjectives are usually indeclinable whether used attributively or as a predicate. When used as substantives, adjectives take the normal plural endings, while comparative and superlative forms use the endings "تر" and "ترین" .

Unlike the Latin, German or Hungarian languages, Persian uses few case markers (other than the accusative case and certain specific genitive cases). The genitive case may also be expressed by coupling two nouns using the particle known as ezafe (e.g., "پسرمرد" "man's son"). As usually done in English, other relations are expressed using prepositions (e.g., in, with, etc.).

Table 4 shows the MAP obtained by various probabilistic models using the Persian collection, and two different query formulations (T or TD), two stemmers and two indexing strategies (word or 4-gram). Since in documents (and queries) inflectional suffixes are usually clearly delimited (presence of a small space), applying our light stemmer or ignoring stemming does not lead to significantly different retrieval performances. Adding more terms in query formulations improves the MAP (between 4.8% to 14.6%) and the performance differences are usually statistically significant. The use of words as indexing units tends to

Table 4. MAP of Various IR Models and Query Formulations (Persian Corpus)

	Mean Average Precision					
Query	T	TD	T	TD	T	TD
Stemmer	none	none	light	light	4-gram	4-gram
Okapi	0.4065*	0.4266	0.4092*	0.4292*	0.3965*	0.4087*
DFR-PL2	0.4078*	0.4274	0.4120	0.4335	0.3815*	0.4005*
DFR-$I(n_e)C2$	**0.4203**	**0.4351**	**0.4204**	**0.4376**	**0.4127**	**0.4235**
LM	0.3621*	0.3839*	0.3607*	0.3854*	0.3248*	0.3518*
tf idf	0.2727*	0.2824*	0.2717*	0.2838*	0.2608*	0.2700*
% over T		+4.8%		+5.2%		+14.6%
% over "none"			+0.4%	+0.8%	-5.1%	-5.3%

Table 5. Description and MAP of Our Official Persian Monolingual Runs

Language	Index	Query	Model	Query expansion	MAP	MAP
UniNEpe1	word	T	PL2		0.4078	Z-score
	4-gram	T	LM	idf 10 docs/20 terms	0.3783	0.4675
	word	T	Okapi	Roc. 10 docs/20 terms	0.4376	
UniNEpe2	4-gram	TD	$I(n_e)C2$		0.4235	Z-score
	word	TD	PL2		0.4274	**0.4898**
	stem	TD	PL2	idf 10 docs/20 terms	0.4513	
	word	TD	PL2	Roc. 10 docs/20 terms	0.4311	
UniNEpe3	4-gram	TD	Okapi	Roc. 5 docs/100 terms	0.4335	Z-score
	word	TD	LM	idf 10 docs/70 terms	0.4141	0.4814
	word	TD	PL2		0.4274	
UniNEpe4	4-gram	TDN	LM	idf 10 docs/100 terms	0.3738	Z-score
	word	TDN	LM	Roc. 10 docs/20 terms	0.4415	**0.4807**
	word	TDN	PL2		0.4425	

produce better MAP and in certain cases, as underlined in Table 4, performance differences are fairly significant.

Table 5 shows the exact specifications of our four official monolingual runs for IR evaluation task for Persian, based mainly on the three probabilistic models (Okapi, DFR and statistical language model (LM)). The strategy we followed consisted of combining different indexing units (words, stems, and 4-grams), based on various probabilistic IR models (Okapi or DFR) and using three different blind-query expansion techniques (Rocchio, idf-based or none). As for the TEL runs (see Table 3) we suggest combining these probabilistic models using the "Z-score" approach (see details in [1]). Of course other methods can be applied to combine these ranked lists as for example the round-robin (RR), taking the sum of the different document scores (SUM) or sum these scores after normalizing (NormMax) (e.g., divided them by the max). If we consider our first official run (UniNEpe1), the MAP achieved with the "RR" approach is 0.4376, the "SUM" method produces a MAP of 0.4413, the "NormMax" 0.4639 and the "Z-score" 0.4675. The performance differences with the "Z-score" are always significant, at least for this run.

5 Conclusion

In this ninth CLEF campaign we evaluated various probabilistic IR models using two different test collections. The first was composed of short bibliographic notices extracted from the TEL corpora (written in the English, German and French) and the second containing newspapers articles written in Persian. For the latter we also suggested a stopword list and a light stemming strategy.

The results of our various experiments demonstrate that the $I(n_e)B2$ or PB2 models (or $I(n_e)C2$ for the Persian language) derived from the Divergence from Randomness (DFR) paradigm and the LM model seem to provide the best overall retrieval performances (see Tables 1, 2, and 4). The Okapi model used in our experiments usually results in retrieval performances inferior to those obtained with the DFR or LM approaches. A data fusion strategy may however enhance the retrieval performance for the French and German (see Tables 3) or Persian languages (see Table 5), but not for the English corpus.

For the Persian language (Table 4), our light stemmer tends to produce better MAP than does the 4-gram indexing scheme (relative difference of 5.5%). For an approach ignoring a stemming stage the performance difference is however is rather small. Finally Persian new words can be formed using compound construction (e.g., handgun), yet retrieval effectiveness obtained by applying automatic decompounding procedures remains unknown.

Acknowledgments. This research was supported in part by the Swiss National Science Foundation under Grant #200021-113273.

References

1. Savoy, J.: Combining Multiple Strategies for Effective Monolingual and Cross-Lingual Retrieval. IR Journal 7, 121–148 (2004)
2. Agirre, E., Di Nunzio, G.M., Ferro, N., Mandl, T., Peters, C.: CLEF 2008: Ad Hoc Track Overview. In: Peters, C., et al. (eds.) CLEF 2008. LNCS, vol. 5706, pp. 15–37. Springer, Heidelberg (2009)
3. Savoy, J.: Light Stemming Approaches for the French, Portuguese, German and Hungarian Languages. In: Proceedings ACM-SAC, pp. 1031–1035. ACM Press, New York (2006)
4. Harman, D.K.: How Effective is Suffixing? Journal of the American Society for Information Science 42, 7–15 (1991)
5. Porter, M.F.: An algorithm for Suffix Stripping. Program 14, 130–137 (1980)
6. Dolamic, L., Savoy, J.: Stemming Approaches for East European Languages. In: Peters, C., Jijkoun, V., Mandl, T., Müller, H., Oard, D.W., Peñas, A., Petras, V., Santos, D. (eds.) CLEF 2007. LNCS, vol. 5152, pp. 37–44. Springer, Heidelberg (2008)
7. Buckley, C., Singhal, A., Mitra, M., Salton, G.: New Retrieval Approaches Using SMART. In: Proceedings TREC-4, pp. 25–48. Gaithersburg (1996)
8. Abdou, S., Savoy, J.: Searching in Medline: Stemming, Query Expansion, and Manual Indexing Evaluation. Information Processing & Management 44, 781–789 (2008)
9. Vogt, C.C., Cottrell, G.W.: Fusion via a Linear Combination of Scores. IR Journal 1, 151–173 (1999)
10. Miangah, T.M.: Automatic Lemmatization of Persian Words. Journal of Quantitative Linguistics 13, 1–15 (2006)

The Domain-Specific Track at CLEF 2008

Vivien Petras[1] and Stefan Baerisch[2]

[1] Berlin School of Library and Information Science, Humboldt University, Dorotheenstr. 26, 10117 Berlin, Germany
[2] GESIS Leibniz-Institute for the Social Sciences, Lennéstr. 30, 53113 Bonn, Germany
vivien.petras@ibi.hu-berlin.de, stefan.baerisch@gesis.org

Abstract. The domain-specific track evaluates retrieval models for structured scientific bibliographic collections in English, German and Russian. Documents contain textual elements (title, abstracts) as well as subject keywords from controlled vocabularies, which can be used in query expansion and bilingual translation. Mappings between the different controlled vocabularies are provided. In 2008, new Russian language resources were provided, among them Russian-English and Russian-German terminology lists as well as a mapping table between the Russian and German controlled vocabularies. Six participants experimented with different retrieval systems and query expansion schemes. Compared to previous years, the queries were more discriminating, so that fewer relevant documents were found per query. The year 2008 marked the last year of the domain-specific track, a special issue of important experiments and results is planned.

Categories and Subject Descriptors

H.3 [Information Storage and Retrieval]: H.3.1 Content Analysis and Indexing; H.3.3 Information Search and Retrieval.

General Terms

Measurement, Performance, Experimentation.

Keywords: Information Retrieval, Evaluation, Controlled Vocabularies.

1 Introduction

The domain-specific track was running continuously since the inception of CLEF in 2000 (Kluck & Gey, 2001; Kluck, 2004). The collections, topics and assessments efforts were provided by GESIS in Bonn, Germany (formerly Social Science Information Centre) in cooperation with its partners INION (Russia), Cambridge Scientific Abstracts (USA) and the University of Padova (Italy) as the developers and operators of the DIRECT system.

The track focused on mono- and cross-language information retrieval in structured social science bibliographic data collections. The focus was the leveraging of controlled vocabularies and other structured metadata entities to improve information retrieval and translation.

C. Peters et al. (Eds.): CLEF 2008, LNCS 5706, pp. 186–198, 2009.

The participants were provided with four collections for retrieval (1 German, 2 English, and 1 Russian) as well as a number of supplemental mapping and terminology tables for the controlled vocabularies. Each year, 25 new topics were created in German and then translated into English and Russian.

2 The Domain-Specific Task

The domain-specific track included three subtasks:

- *Monolingual retrieval* against the German GIRT collection, the English GIRT and CSA Sociological Abstract collections, or the Russian INION ISISS collection;
- *Bilingual retrieval* from any of the source languages to any of the target languages;
- *Multilingual retrieval* from any source language to all collections / languages.

2.1 The Test Collections

The GIRT databases (version 4) contain extracts from the German Social Science Information Centre's SOLIS (Social Science Literature) and SOFIS (Social Science Research Projects) databases from 1990-2000. The INION ISISS corpus covers social sciences and economics in Russian. The second English collection is an extract from CSA's Sociological abstracts.

German
The German GIRT collection (the social science **G**erman **I**ndexing and **R**etrieval **T**estdatabase) contains 151,319 documents covering the years 1990-2000 using the German version of the Thesaurus for the Social Sciences (GIRT-description, 2007). Almost all documents contain an abstract (145,941).

English
The English GIRT collection is a pseudo-parallel corpus to the German GIRT collection, providing translated versions of the German documents. It also contains 151,319 documents using the English version of the Thesaurus for the Social Sciences but only 17% (26,058) documents contain an abstract.

The Sociological Abstracts database from Cambridge Scientific Abstracts (CSA) holds 20,000 documents, 94% of which contain an abstract. The documents were taken from the SA database covering the years 1994, 1995, and 1996. Additional to title and abstract, each document contains subject-describing keywords from the CSA Thesaurus of Sociological Indexing Terms and classification codes from the Sociological Abstracts classification.

Russian
For the retrieval of Russian collections, the INION corpus ISISS with bibliographic data from the social sciences and economics with 145,802 documents was used. ISISS documents contain authors, titles, abstracts (for 27% of the test collection or 39,404 documents) and keywords from the Inion Thesaurus.

2.2 Controlled Vocabularies

The GIRT collections have descriptors from the GESIS Thesaurus for the Social Sciences in German and English depending on the collection language. The CSA Sociological Abstracts documents contain descriptors from the CSA Thesaurus of Sociological Indexing Terms and the Russian ISISS documents are provided with Russian INION Thesaurus terms. GIRT documents also contain classification codes from the GESIS classification and CSA SA documents from the Sociological Abstracts classification. Table 1 shows the distribution of subject-describing terms per document in each collection.

Table 1. Distribution of subject-describing terms per collection

Collection	*GIRT-4 (German or English)*	*CSA Sociological Abstracts*	*INION ISISS*
Thesaurus descriptors / document	10	6.4	3.9
Classification codes / document	2	1.3	n/a

Vocabulary mappings
Vocabulary mappings are one-directional, intellectually created term transformations between two controlled vocabularies. They can be used to switch from the subject metadata terms of one knowledge system to another, enabling retrieval systems to treat the subject descriptions of two or more different collections as one and the same.

For the English and German collections, mappings between the GESIS Thesaurus for the Social Sciences and the English CSA Thesaurus of Sociological Indexing Terms were provided. The mapping from the English Thesaurus for the Social Sciences to the English CSA Thesaurus of Sociological Indexing Terms was supplied for monolingual retrieval. Additionally, there was also a translation table with the German and English terms from the GESIS Thesaurus for the Social Sciences.

Three new Russian resources were developed in 2008: two translation tables as well as a mapping.

One translation table contains translation between the German and Russian terms from the GESIS Thesaurus for the Social Sciences), which can also be used in conjunction with the German-English translation table. The second translation table lists Russian and English translations (11694 term pairs) for the INION ISISS descriptor list. Finally, mappings from the Russian INION ISISS descriptor list to the GESUS Thesaurus Sozialwissenschaften were made available.

An example of a mapping from the English Thesaurus for the Social Sciences to the English CSA Thesaurus of Sociological Indexing Terms is given below:

<mapping>
<original-term> *counseling for the aged* </original-term>
<mapped-term> *Counseling + Elderly*</mapped-term>
</mapping>

This example shows that a mapping can overcome differences in technical language, the synonym problem and the treatment of singular and plural in different controlled vocabularies.

2.3 Topic Preparation

For topic preparation, colleagues from the GESIS Social Science Information Centre suggested 2-5 topics related to specialized subject areas and potentially relevant in the years 1990-2000 (the coverage of our test collections). Specialized subject areas are based on the 28 subject categories utilized for the GESIS bibliographic service sofid, which bi-annually publishes updates on new entries in the SOLIS and SOFIS databases (from which the GIRT collections were generated). Topics range from general sociology, family research, women and gender studies, international relations, research on Eastern Europe to social psychology and environmental research. An overview of the service including the 28 topics can be found at the following URL: http://www.gesis.org/en/information/soFid/index.htm.

The suggestions were then checked for their breadth, variance from previous years and coverage in the test collections and edited for style and format. In 2008, topics 201-225 for the domain-specific collections were created. Figure 1 shows topic 207 as an example.

<top>
<num>207</num>
<EN-title>*Economic growth and environmental destruction*</EN-title>
<EN-desc>*Find documents on the topic of the connection between economic growth and environmental destruction.*</EN-desc>
<EN-narr>*Relevant documents address the connection between economic growth and environmental destruction, particularly the question of whether continued economic growth generally leads to environmental destruction or if the concept of qualitative growth can prevent this.*</EN-narr>
</top>

Fig. 1. Example topic in English

All topics were initially created in German and then translated into English and Russian. The method works well for German and English, because the German and English collections are virtually equivalent. However, Russian topic preparation is somewhat more difficult as the collection is different in scope, contains shorter documents and a large and non-controlled vocabulary. Consequently, not all Russian topic translations retrieve relevant documents in the database. Table 2 lists all 25 topic titles.

Table 2. English topic titles for the domain-specific track 2008

201 Health risks at work	213 Migrant organizations
202 Political culture and European integration	214 Violence in old age
203 Democratic transformation in Eastern Europe	215 Tobacco advertising
204 Child and youth welfare in the Russian Federation	216 Islamist parallel societies in Western Europe
205 Minority policy in the Baltic states	217 Poverty and social exclusion
206 Environmental justice	218 Generational differences on the Internet
207 Economic growth and environmental destruction	219 (Intellectually) Gifted
208 Leisure time mobility	220 Healthcare for prostitutes
209 Doping and sports	221 Violence in schools
210 Establishment of new businesses after the reunification	222 Commuting and labor mobility
211 Shrinking cities	223 Media in the preschool age
212 Labor market and migration	224 Employment service
	225 Chronic illnesses

3 Overview of the 2008 Domain-Specific Track

Details of the individual runs and methods tested can be found in appendix C of the working notes and in the corresponding articles by the participating groups.

3.1 Participants

Six of the nine registered groups (listed in table 3) submitted runs and descriptions of their experiments (Fautsch, Dolamic & Savoy, 2008; Gobeill & Ruch, 2008; Kürsten, Wilhelm & Eibl, 2008; Larson, 2008; Meij & de Rijke, 2008; Müller & Gurevych, 2008).

Table 3. Domain-specific track 2008 - participants

Abbreviation	*Group Institution*	*Country*
Amsterdam	University of Amsterdam	The Netherlands
Chemnitz	Chemnitz University of Technology	Germany
Cheshire	School of Information, UC Berkeley	USA
Darmstadt	TU Darmstadt	Germany
Hug	University Hospitals Geneva	Switzerland
UniNE	Computer Science Department, University of Neuchatel	Switzerland

3.2 Submitted Runs

The total number of submitted runs decreased slightly compared to 2007, although one more group submitted runs. Table 4 shows the number of runs (numbers from 2007 in brackets).

Table 4. Submitted runs per task in the domain-specific track 2008

Task	Runs
Monolingual	
- against German	10 (13)
- against English	12 (15)
- against Russian	9 (11)
Bilingual	
- against German	12 (14)
- against English	9 (15)
- against Russian	8 (9)
Multilingual	9 (9)

English was the most popular language for monolingual retrieval as well as a starting language for bilingual retrieval. All groups participated in the monolingual English task, and four groups took part in the German and Russian monolingual tasks respectively. Three groups experimented with bilingual against German or English, whereas only 2 groups tackled the bilingual against Russian and multilingual tasks respectively.

3.3 Relevance Assessments

All relevance assessments were processed using the DIRECT system (Distributed Information Retrieval Evaluation Campaign Tool) provided by Giorgio M. Di Nunzio and Nicola Ferro from the Information Management Systems (IMS) Research Group at the University of Padova, Italy.

Documents were pooled using the top 100 ranked documents from each submission. Table 5 shows pool sizes and the number of assessed documents per topic for the three different languages.

Table 5. Pool sizes in the domain-specific track 2008

	Pool size	Documents assessed per topic
German	14793	592
English	14835	593
Russian	13930	557

Because of a late submission, the runs by the Hug group were not included in the pooling process but were analyzed with the existing pools. One assessor was assigned for each language to avoid as many interpersonal assessment differences as possible.

Both the feedback from the assessors as well as the precision numbers show that this year's topics were somewhat more difficult or more discriminating. The average number of relevant topics per task and language (table 6) also corroborate this impression. The average number of relevant documents decreased for all three languages with Russian seeing the largest drop. As in previous years, however, the German and English averages were similar.

Table 6. Relevant documents per language pool

	German	English	Russian
2008	15%	14%	2%
2007	22%	25%	10%
2006	39%	26%	n/a
2005	20%	21%	9% (RSSC)

Figures 2-4 show the number of relevant documents per individual topics for the three languages.

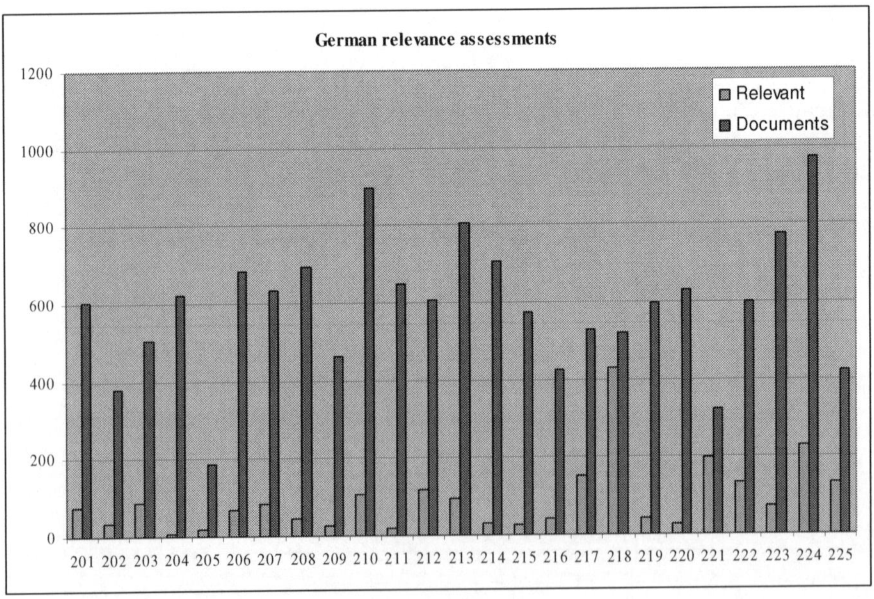

Fig. 2. German assessments per topic 2008

For German, six topics stood out as having more than 20% relevant documents in their pool: 217, 218, 221, 222, 224 and 225.

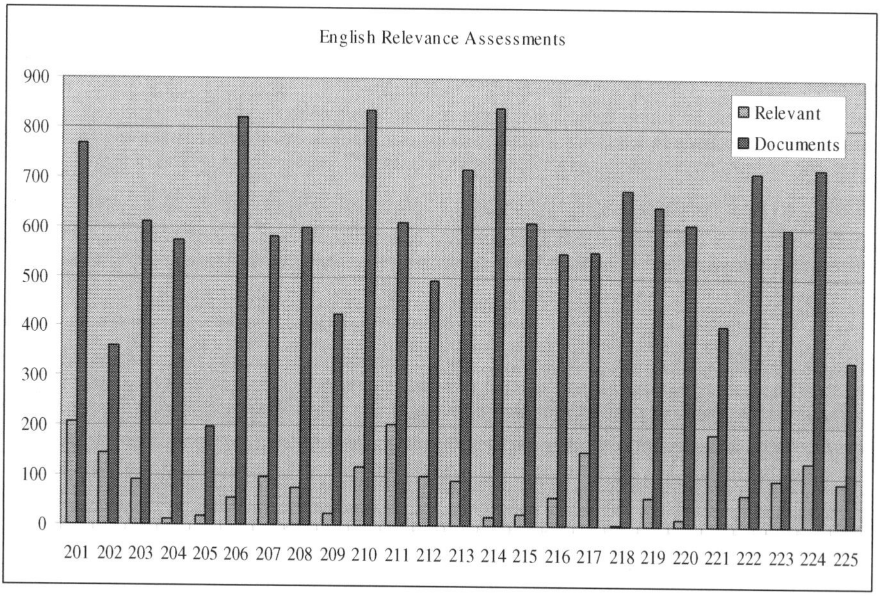

Fig. 3. English assessments per topic 2008

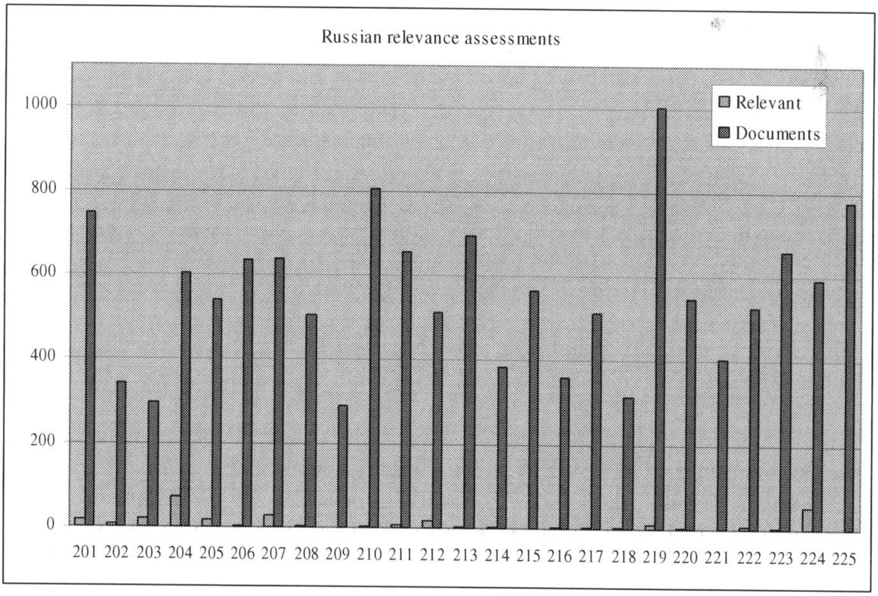

Fig. 4. Russian assessments per topic 2008

For English, seven topics retrieved more than 20% relevant documents (201, 202, 211, 212, 217, 221, 225). Three of these topics (217, 221, 225) overlap with the German results, surprisingly, topic 218, which retrieved the greatest number of relevant documents in German, retrieved the least (percentage-wise) in English. This might be due to different interpretations and assessments of the content of the topic (Generational differences on the Internet).

For the more difficult Russian collection, the highest percentage of relevant documents retrieved was found for topic 204 (12%), followed by 224 (9%) and 203 (7%). The pool for topic 224 (Employment service) contains also more than 20% relevant documents in the German collection and more than 17% in the English collection. One topic (209) did not retrieve any relevant documents in the Russian collection.

A closer look at the correlation between the number of relevant documents per topics and precision and recall might reveal more insight. One interesting question is whether the topics with the most relevant documents available are also the "easiest" for retrieval systems to find in terms of precision and recall measures.

3.4 Results

In the Appendix of the CLEF 2008 Working Notes, varied evaluation measures for each run per task and recall-precision graphs for the top-performing runs for each task can be looked up.

4 Domain-Specific Experiments

The 2008 track saw the use of a broad range of retrieval models, language processing, translation, and query expansion approaches. Statistical language models, probabilistic and vector-space models were employed with translation approaches that leverage thesaurus mappings as well as machine translation systems or web-based translation services. Two of the six participants employed concept models based on semantic relatedness both for translation and query expansion.

4.1 Retrieval Models

The participants utilized a number of different retrieval models. Statistical language models were used as well as different implementations of the probabilistic model and vector-space schemes. The structure of the collection documents, the topics and the controlled vocabularies and the associated mappings were used to different degrees.

The Chemnitz group (Kürsten, Wilhelm & Eibl, 2008) used their Apache Lucene-based Xtrieval framework for the experiments and utilized the Z-score Operator (Savoy, 2005) to combine the results of runs with different language processing and translation approaches.

Darmstadt (Müller & Gurevych, 2008) applied a statistical model implemented in Lucene in addition to two semantic models, SR-Text and SR-Word. The semantic models utilized both Wikipedia and Wiktionary as sources for terms to form concepts that facilitate the use of semantic relatedness in the retrieval process. The CombSUM method by Fox and Shaw (Fox & Shaw, 1994) was used for the merging of results from the multiple retrieval models.

The Geneva group (Gobeill & Ruch, 2008) used their EasyIR system, which supports both regular expression searches and retrieval based on the vector space model.

Berkeley (Larson, 2008) implemented a probabilistic logistic regression model with the Cheshire II system that was also employed for the Ad-hoc and GeoCLEF tracks.

UniNE (Fautsch, Dolamic & Savoy, 2008) employed and evaluated multiple retrieval models. A tf-idf based statistical model was compared with two probabilistic models, the BM25 scheme and four implementations of the Divergence from Randomness model. Additionally, an approach based on a statistical language model was utilized.

The Amsterdam (Meij & de Rijke, 2008) group used a language model approach to map between query terms, controlled vocabulary concepts and document terms. Parsimonization was used to increase the probability weights of specific terms compared to more general terms in the corpus.

4.2 Language Processing

A number of different combinations of stemming, lemmatization and decompounding techniques were utilized by the participants, often in combination with stopword lists.

Chemnitz used combinations of the Porter and the Krovetz stemmers for English and the Snowball stemmer and an N-Gram based decompounding approach for German. The group used a stemmer developed by UniNE for Russian.

The UniNE group used stopword lists of between 430 and 603 words for the three different corpora languages. Stemming for English was done using the SMART stemmer. 52 stemming rules that removed inflections due to gender, number and case were defined for Russian. German words were treated with a lightweight stemmer and decompounding algorithm developed by the group.

Darmstadt used the probabilistic part-of-speech tagging system TreeTagger (Schmid, 1994) for lemmatization. Decompounding was employed for German words. For retrieval, both a compound word and its elements were used in combination.

Geneva used an implementation of a Porter stemmer.

Berkeley employed a stopword list for common words in all languages, but did not use decompounding for German.

Amsterdam did not do any preprocessing on the document collections.

4.3 Translation

Different approaches to translation and the treatment of different languages were used by the groups. Besides the use of machine translations software, the language mappings of the provided controlled vocabularies were used in addition to the use of concepts models from external sources (Wikipedia) for cross-language retrieval.

Darmstadt used the Systran machine translation system and utilized cross-language links in Wikipedia in order to map between concept vectors for different languages in the SR-Text system.

Berkeley used the commercial LEC Power translator with good results.

Chemnitz made use of the Google AJAX language API. In addition to pure translation, a combination of automatic translation and language mappings as provided by the bilingual translation tables was employed.

Geneva did not use translation, but employed the bilingual thesaurus for query expansion as described below.

Amsterdam used a combined approach that leveraged concept models for both translation and query expansion.

4.4 Query Expansion

All participants used query expansion. The techniques employed included the expansion by terms from the top-k documents as well the utilization of concept models, idf-based approaches and the use of Google and the Wikipedia.

Chemnitz used a blind feedback approach that was combined for some runs with query expansion based on thesaurus terms. It was found that such use of the controlled vocabulary did not benefit the retrieval effectiveness.

The UniNE group tested four different blind feedback approaches. The classic Rocchio blind feedback method is compared to two variants of an approach that extends a query with terms selected based on their pseudo document frequency, which are considered for inclusion in the query if they are within 10 words of the search term in the document. Finally, Google and Wikipedia were used for query expansion where the terms included in text snippets were used for query expansion.

Geneva used the bilingual thesaurus for query expansion. The descriptors in the top 10 documents for a German query were collected and transferred into English using the bilingual thesaurus. The resulting terms were used for query expansion.

Amsterdam used a blind relevance feedback approach based on concept models of the thesauri provided for the track that used the concepts defined in the thesauri as a pivot language.

Berkeley used a probabilistic blind feedback approach based on the work by Robertson and Sparck Jones (Robertson, 1976).

Darmstadt implemented a query expansion method based on concept models derived from Wikipedia and Wiktionary.

5 Conclusion

The year 2008 marked the last year of the domain-specific track. Between 2000 and 2008, nine domain-specific tracks were held. The collections changed intermittently but the GIRT English and German collections have remained stable since 2003. The Russian collections were changed during the years. In total, 225 topics with relevant judgements are prepared for the domain-specific collection. This provides a large and well-prepared testing ground for further experimentation.

The results and group papers show that query expansion with blind feedback mechanisms using document, controlled vocabulary terms or external resources is still a major experimentation area for domain-specific retrieval. Interesting distributed retrieval scenarios with different databases can be simulated using the four different collections and six different technical vocabularies provided by the domain-specific test collections.

A special issue commemorating important findings and results of the domain-specific track is planned and more result analysis of all runs will provide further insights in retrieval and evaluation optimization procedures.

With the ad-hoc TEL track, a new track using bibliographic data (catalog records) and different controlled vocabularies is used. Hopefully, the experiences from the domain-specific track can support these new developments.

Acknowledgements

We thank all participants of the domain-specific track for their hard work over the years. We look forward to collecting your input for the special issue.

We would also like to thank Cambridge Scientific Abstracts for providing the documents for the Sociological Abstracts test collection and INION for providing the documents for the ISISS collection.

We greatly acknowledge the support of Natalia Loukachevitch and her colleagues from the Research Computing Center of M.V. Lomonosov Moscow State University in translating the topics into Russian.

Very special thanks also to Giorgio Di Nunzio and Nicola Ferro from the Information Management Systems (IMS) Research Group at the University of Padova for providing the DIRECT system and all their help in the assessments process and for providing the graphs and numbers for the results analysis.

Claudia Henning did the German assessments. Jeof Spiro translated and assessed the English topics. Oksana Schäfer provided the Russian assessments.

References

GIRT Description, GIRT - Mono- and Cross-language Domain-Specific Information Retrieval, GIRT4 (2007), http://www.gesis.org/en/research/information_technology/girt4.htm

Fautsch, C., Dolamic, L., Savoy, J.: UniNE at Domain-Specific IR - CLEF 2008: Scientific Data Retrieval: Various Query Expansion Approaches. In: Peters, C., et al. (eds.) CLEF 2008. LNCS, vol. 5706, pp. 199–202. Springer, Heidelberg (2009)

Fox, E., Shaw, J.: Combination of Multiple Searches. In: Proceedings of the 2nd Text REtrieval Conference (Trec-2), pp. 243–252 (1994)

Gobeill, J., Ruch, P.: First Participation of University and Hospitals of Geneva to Domain-Specific Track in CLEF 2008. In: Peters, C., et al. (eds.) CLEF 2008. LNCS, vol. 5706. Springer, Heidelberg (2009)

Kluck, M., Gey, F.C.: The Domain-Specific Task of CLEF - Specific Evaluation Strategies in Cross-Language Information Retrieval. In: Peters, C. (ed.) CLEF 2000. LNCS, vol. 2069, pp. 48–56. Springer, Heidelberg (2001)

Kluck, M.: The GIRT Data in the Evaluation of CLIR Systems – from 1997 until 2003. In: Peters, C., Gonzalo, J., Braschler, M., Kluck, M. (eds.) CLEF 2003. LNCS, vol. 3237, pp. 376–390. Springer, Heidelberg (2004)

Kürsten, J., Wilhelm, T., Eibl, M.: The Xtrieval Framework at CLEF 2008: Domain-Specific Track. In: Peters, C., et al. (eds.) CLEF 2008. LNCS, vol. 5706, pp. 215–218. Springer, Heidelberg (2008)

Larson, R.R.: Back to Basics - Again - for Domain Specific Retrieval. In: Peters, C., et al. (eds.) CLEF 2008. LNCS, vol. 5706, pp. 203–206. Springer, Heidelberg (2008)

Meij, E., de Rijke, M.: The University of Amsterdam at the CLEF 2008 Domain Specific Track: Parsimonious Relevance and Concept Models. In: Peters, C., et al. (eds.) CLEF 2008. LNCS, vol. 5706. Springer, Heidelberg (2008)

Müller, C., Gurevych, I.: Using Wikipedia and Wiktionary in Domain-Specific Information Retrieval. In: Peters, C., et al. (eds.) CLEF 2008. LNCS, vol. 5706, pp. 219–226. Springer, Heidelberg (2008)

Savoy, J.: Data Fusion for Effective European Monolingual Information Retrieval. In: Peters, C., Clough, P., Gonzalo, J., Jones, G.J.F., Kluck, M., Magnini, B. (eds.) CLEF 2004. LNCS, vol. 3491, pp. 233–244. Springer, Heidelberg (2005)

Schmid, H.: Probabilistic part-of-speech tagging using decision trees. In: Proceedings of International Conference on New Methods in Language Processing, vol. 12 (1994)

Robertson, S., Jones, S., et al.: Relevance Weighting of Search Terms. Journal of the American Society for Information Science 27(3), 129–146 (1976)

UniNE at Domain-Specific IR - CLEF 2008

Claire Fautsch, Ljiljana Dolamic, and Jacques Savoy

Computer Science Department, University of Neuchatel,
Rue Emile Argand 11, 2009 Neuchatel, Switzerland
{Claire.Fautsch,Ljiljana.Dolamic,Jacques.Savoy}@unine.ch

Abstract. Our first objective in participating in this domain-specific evaluation campaign is to propose and evaluate various indexing and search strategies for the German, English and Russian languages, and thus obtain retrieval effectiveness superior to that of language-independent approaches (n-gram). To do so we evaluated the GIRT-4 test-collection using the Okapi model, various IR models based on the Divergence from Randomness (DFR) paradigm, the statistical language model (LM) together with the classical $tf \cdot idf$ vector-processing scheme.

1 Introduction

Domain-specific retrieval is an interesting task due to its ability to access bibliographic notices (usually comprising title and abstract records) extracted from one German social science source, two English sources, and one Russian corpus. These records also contain manually assigned keywords taken from a controlled vocabulary and applied by subject experts with a good knowledge of the discipline to which the indexed articles belong. These descriptors should be helpful in improving document surrogates and consequently extracting more pertinent information, while also discarding any irrelevant aspects. Access to the underlying thesaurus would allow improved retrieval performance (for details about this test-collection see [1]).

2 Indexing and Searching Strategies

For the English, German and Russian languages we used the same stopword lists and stemmers we had selected for our previous CLEF participation [2]. For English these were the SMART stemmer and stopword list (containing 571 items). For German we applied our light stemmer and a stopword list (603 words) available at `http://www.unine.ch/info/clef/`, along with our decompounding algorithm [3]. For Russian we applied a stopword list containing 430 words and our light stemming procedure (applying 53 rules to remove final suffixes representing gender, number and the six Russian grammatical cases).

In order to obtain higher MAP values, we considered certain probabilistic models such as the Okapi (or BM25), and as a second probabilistic approach we implemented variants of the DFR (*Divergence from Randomness*) paradigm.

C. Peters et al. (Eds.): CLEF 2008, LNCS 5706, pp. 199–202, 2009.

We also examined an approach based on a statistical language model (LM), also known as a non-parametric probabilistic model (for a precise definition of these IR models see [4]). For comparison purposes, we also added the classical $tf \cdot idf$ model (with cosine normalization).

Table 1 shows the mean average precision (MAP) obtained using the Russian collection, combined with the short (T) or medium query formulations (TD), along with two different indexing strategies (word-based using a light stemmer (inflectional only) and n-gram scheme). This table shows that when using word-based indexing, the DFR-$I(n_e)B2$ or LM models tend to perform the best (values indicated in bold). With 4-gram indexing the LM model always performs best, while the short query formulation (T) tends to provide better retrieval performance than the medium (TD) topic formulation, even though the performance differences were never statistically significant. A comparison of the word-based and 4-gram indexing systems shows that the relative difference varies from +4.7% to +10% and favors the 4-gram approach, but no statistically significant performance differences were detected.

Table 2 lists the MAP obtained from four probabilistic models and the classical $tf \cdot idf$ vector-space model for the German or English collection and three different query formulations (title-only or T, TD, and TDN). The bottom line

Table 1. Monolingual Evaluation of the Russian Corpus (24 queries)

	Mean Average Precision			
Query	T	TD	T	TD
Indexing	word+light	word+light	4-gram	4-gram
Okapi	0.1477	0.1406	0.1562	0.1500
DFR-GL2	0.1578	0.1388	0.1685	0.1635
DFR-$I(n_e)B2$	0.1531	**0.1529**	0.1460	0.1414*
LM	**0.1592**	0.1393	**0.1759**	**0.1739**
tf idf	0.1091	0.1134	0.1144*	0.1179*
% over T		-7.5%		-2.8%
% over word			+4.7%	+10.0%

Table 2. Monolingual Evaluation of German and English Corpora (25 queries)

	Mean Average Precision				
Language	German	German	German	English	English
Query	T	TD	TDN	T	TD
Indexing	word	word	word	word	word
Okapi	0.3815	0.4069	0.4164	0.2592	0.3039*
DFR-GL2	0.3793	0.4000	0.4031*	0.2578	0.2910*
DFR-$I(n_e)B2$	**0.3940**	**0.4179**	0.4202	**0.2684**	**0.3215**
LM	0.3791	0.4130	**0.4321**	0.2365*	0.2883*
tf idf	0.2212*	0.2391*	0.2467*	0.1715*	0.1959*
% over TD	-6.4%		+1.2%	-15.5%	

shows the percent change when compared to the medium (TD) query formulation. For both the German and the English corpora, the DFR-$I(n_e)B2$ model tends to produce the best retrieval performances. For the German corpus however the performance differences between the other probabilistic models tend not to be statistically significant. Compared to the classical $tf \cdot idf$ vector-space model, the performance differences are always statistically significant (and denoted by a "*" in the tables). For the English collection and TD queries, the performance differences with the best IR model are always significant.

A careful analysis of some queries shows when and why our search strategy failed to rank pertinent articles at the top of the returned list. For the Russian corpus, using different terms or word phrases to express the same concept seemed to be the main source of search failure. This was the case for example with Topic #210 ("Establishment of new business after the reunification" or "Создание новых предприятий после воссоединения") in which the term "воссоединения" is too general and does not specify German reunification. Another source of error was that our light stemmer could not conflate different related surface forms to the same stem. For Topic #213 ("Migrant organization" or "Мигрантские организации") relevant items tend to use the term "миграция" (migration) or "мигранты" (migrants), but do not conflate to the same stem, as was also true with the related noun "иммигрантов" (immigrants). Moreover relevant items may also use the term "самоорганизация" (self-organization) that does not match the second set of search terms (for Russian texts we do not apply a decompounding procedure).

3 Official Results and Conclusion

Table 3 shows our best official runs during the monolingual GIRT task, where a data fusion operator "Z-Score" was applied (see [3]) for each run. For all runs, we automatically expanded the queries using the blind relevance feedback method developed by Rocchio [5] (denoted "Roc"), our IDFQE approach [6] (denoted "idf"), or the first two text snippets returned by Google.

Table 3. Description and MAP Results for Our Best Official Monolingual Runs

Language	Index	Query	Model	Query expansion	MAP	MAP
German	dec.	TD	$I(n_e)B2$	Roc. 10 docs/200 terms	0.3992	Z-score
UniNEDSde1	dec.	TD	LM	Google	0.4265	**0.4537**
	dec.	TD	PB2	idf 10 docs/150 terms	0.4226	
English	stem	TD	$I(n_e)B2$	Roc. 10 docs/100 terms	0.3140	Z-score
UniNEDSen1	stem	TD	$I(n_e)B2$		0.3562	**0.3770**
	stem	TD	LM	Roc. 5 docs/150 terms	0.3677	
Russian	4-gram	TD	$I(n_e)B2$	Roc. 3 docs/150 terms	0.1129	Z-score
UniNEru4	stem	TD	$I(n_e)B2$	Roc. 5 docs/70 terms	0.1652	**0.1890**
	stem	TD	$I(n_e)B2$	idf 3 docs/70 terms	0.1739	

In conclusion, in this domain-specific evaluation campaign we evaluated various probabilistic models using the German, English and Russian languages. For the German and Russian languages we applied our light stemming approach and stopword list. The resulting MAP (see Tables 1 and 2) show that the DFR-$I(n_e)B2$ or the LM model usually provided the best retrieval effectiveness. The performance differences between Okapi and the various DFR models were usually rather small and statistically non-significant

This year we suggest two new query expansion techniques. The first is denoted "idf-window" and based on co-occurrence of relatively rare terms in a closed context (within 10 terms from the occurrence of a search term in a retrieved document). As a second new approach to expand the query, we add the first two text snippets found by Google. Compared to the performance before query expansion (e.g., with the German corpus, TD queries and the LM model the MAP is 0.4130), the Rocchio approach combined with the idf-based blind query expansion does not improve retrieval performance, yet the "idf-window" variant results in better retrieval performance (+3.3%, from 0.4130 to 0.4265). When using the first two text snippets returned by Google we are also able to slightly enhance the MAP (from 0.4177 to 0.42266, or +3.9%, German collection, PB2 model, TD queries, values given in Table 2 and 3).

Acknowledgments. This research was supported in part by the Swiss National Science Foundation under Grant #200021-113273.

References

1. Petras, V., Baerisch, S.: The Domain-Specific Track at CLEF 2008. In: Peters, C., et al. (eds.) CLEF 2008. LNCS, vol. 5706, pp. 187–200. Springer, Heidelberg (2009)
2. Fautsch, C., Dolamic, L., Abdou, S., Savoy, J.: Domain-Specific IR for German, English and Russian Languages. In: Peters, C., Jijkoun, V., Mandl, T., Müller, H., Oard, D.W., Peñas, A., Petras, V., Santos, D. (eds.) CLEF 2007. LNCS, vol. 5152, pp. 196–199. Springer, Heidelberg (2008)
3. Savoy, J.: Combining Multiple Strategies for Effective Monolingual and Cross-Lingual Retrieval. IR Journal 7, 121–148 (2004)
4. Dolamic, L., Savoy, J.: Stemming Approaches for East European Languages. In: Peters, C., Jijkoun, V., Mandl, T., Müller, H., Oard, D.W., Peñas, A., Petras, V., Santos, D. (eds.) CLEF 2007. LNCS, vol. 5152, pp. 37–44. Springer, Heidelberg (2008)
5. Buckley, C., Singhal, A., Mitra, M., Salton, G.: New Retrieval Approaches Using SMART. In: Proceedings TREC-4, pp. 25–48. Gaithersburg (1996)
6. Abdou, S., Savoy, J.: Searching in Medline: Stemming, Query Expansion, and Manual Indexing Evaluation. Information Processing & Management 44, 781–789 (2008)

Back to Basics – Again – for Domain-Specific Retrieval

Ray R. Larson

School of Information
University of California, Berkeley, USA
ray@ischool.berkeley.edu

Abstract. In this paper we will describe Berkeley's approach to the Domain-Specific (DS) track for CLEF 2008. Last year we used *Entry Vocabulary Indexes* and Thesaurus expansion approaches for DS, but found in later testing that some simple text retrieval approaches had better results than these more complex query expansion approaches. This year we decided to revisit our basic text retrieval approaches and see how they would stack up against the various expansion approaches used by other groups. The results are now in and the answer is clear, they perform pretty badly compared to other groups' approaches.

All of the runs submitted were performed using the Cheshire II system. This year the Berkeley/Cheshire group submitted a total of twenty-four runs, including two for each subtask of the DS track. These include six Monolingual runs for English, German, and Russian, twelve Bilingual runs (four X2EN, four X2DE, and four X2RU), and six Multilingual runs (two EN, two DE, and two RU). The overall results include Cheshire runs in the top five participants for each task, but usually as the lowest of the five (and often fewer) groups.

1 Introduction

This paper discusses the retrieval methods and evaluation results for Berkeley's participation in the CLEF 2008 Domain-Specific track. In 2007 we focused on query expansion using Entry Vocabulary Indexes(EVIs)[1,4], and thesaurus lookup of topic terms. Once the relevance judgements for 2007 were released we discovered that these rather complex method actually did not perform as well as basic text retrieval on the topics without additional query expansion. So, this year for the Domain-Specific track we have returned to using a basic text retrieval approach using Probabilistic retrieval based on Logistic Regression with the inclusion of blind feedback, as used in 2006[2].

All of the submitted runs for this year's Domain-Specific track used the Cheshire II system for indexing and retrieval.

2 Retrieval Approaches for Domain-Specific Retrieval

For all of our official submitted runs this year we used the "TREC2" Logistic regression algorithm along with blind feedback. The Algorithms are formally described in our CLEF working notes paper[3].

C. Peters et al. (Eds.): CLEF 2008, LNCS 5706, pp. 203–206, 2009.

Although the Cheshire II system uses the XML structure of documents and extracts selected portions of the record for indexing and retrieval, for the submitted runs this year we used only a single one of these indexes that contains the entire content of the document.

For all indexing we used language-specific stoplists to exclude function words and very common words from the indexing and searching. The German language runs, however, did *not* use decompounding in the indexing and querying processes to generate simple word forms from compounds.

2.1 Search Processing

Searching the Domain-Specific collection used Cheshire II scripts to parse the topics and submit the title and description elements from the topics to the index containing all terms from the documents. For the monolingual search tasks we used the topics in the appropriate language (English, German, or Russian), and for bilingual tasks the topics were translated from the source language to the target language using the LEC Power Translator PC-based program. Overall we have found that this translation program seems to generate good translations

Table 1. Submitted Domain-Specific Runs

Run Name	Description	Exp.	MAP
BRK-MO-DE-TD	Monolingual German	TD auto	0.3155
BRK-MO-DE-TDN	Monolingual German	TDN auto	0.3111
BRK-MO-EN-TD	Monolingual English	TD auto	0.3200
BRK-MO-EN-TDN	Monolingual English	TDN auto	0.3095
BRK-MO-RU-TD	Monolingual Russian	TD auto	0.1306
BRK-MO-RU-TDN	Monolingual Russian	TDN auto	0.1260
BRK-BI-ENDE-TD	Bilingual English⇒German	TD auto	0.1982
BRK-BI-ENDE-TDN	Bilingual English⇒German	TDN auto	0.1726
BRK-BI-RUDE-TD	Bilingual Russian⇒German	TD auto	0.1188
BRK-BI-RUDE-TDN	Bilingual Russian⇒German	TDN auto	0.1087
BRK-BI-DEEN-TD	Bilingual German⇒English	TD auto	0.1668
BRK-BI-DEEN-TDN	Bilingual German⇒English	TDN auto	0.1454
BRK-BI-RUEN-TD	Bilingual Russian⇒English	TD auto	0.1765
BRK-BI-RUEN-TDN	Bilingual Russian⇒ English	TDN auto	0.1748
BRK-BI-DERU-TD	Bilingual German⇒Russian	TD auto	0.0515
BRK-BI-DERU-TDN	Bilingual German⇒Russian	TDN auto	0.0550
BRK-BI-ENRU-TD	Bilingual English⇒Russian	TD auto	0.0857
BRK-BI-ENRU-TDN	Bilingual English⇒Russian	TDN auto	0.0662
BRK-MU-DE-TD	Multilingual German	TD auto	0.0984
BRK-MU-DE-TDN	Multilingual German	TDN auto	0.0984
BRK-MU-EN-TD	Multilingual English	TD auto	0.1057
BRK-MU-EN-TDN	Multilingual English	TDN auto	0.1034
BRK-MU-RU-TD	Multilingual Russian	TD auto	0.0662
BRK-MU-RU-TDN	Multilingual Russian	TDN auto	0.0701

between any of the languages needed for this track, but we still intend to do some further testing to compare to previous approaches (which used web-based translation tools like Babelfish and PROMT). We suspect that, as always, different tools provide a more accurate representation of different topics for some languages, but the LEC Power Translator seemed to do pretty good (and often better) translations for all of the needed languages.

All searches were submitted using the TREC2 Algorithm with blind feedback described above. This year we did no expansion of topics or use of the thesaurus or the classification clusters created last year. The differences in the runs for a given language or language pair (for bilingual) in Table 1 are primarily whether the topic title and description only (TD) or title, description and narrative (TDN).

3 Results for Submitted Runs

The summary results (as Mean Average Precision) for all of our submitted runs for English, German and Russian are shown in Table 1, the Recall-Precision curves for these runs are not included in this paper but may be found in our CLEF working notes paper[3].

We have observed that for the vast majority of our runs using the narrative tends to degrade instead of improve performance. (We observed the same in other tracks as well.)

It is worth noting that the approaches used in our submitted runs provided the best results when testing with 2007 data and topics when compared to our official 2007 runs. In fact we may have over-simplified for this track. Although at least one Cheshire run appeared in the top five runs of the overall summary results available on the DIRECT system, none of them were top-ranked and for many tasks there appeared to be fewer than five participants.

4 Additional Analysis and Conclusions

Given that the re-introduction of fusion approaches in our GeoCLEF entry led to very good results, we decided to try some experiments applying the same fusion approaches for this task. We conducted analyses to try different "pivot values" for the fusion approach (described in our GeoCLEF paper in this volume).

Table 2 shows the results of these experiments, along with the officially submitted Monolingual English runs for reference. As Table 2 shows, the use of the fusion approach provides a small improvement in MAP for pivot values from 0.01 to 0.15, with the peak value at 0.07. This suggests that further experimentation with fusion approaches for this task are warranted. In particular it would be interesting to combine the LR with blind feedback along with one of the search term recommenders used in Domain-Specific CLEF 2007 using a fusion approach.

Table 2. Fusion Experiments for English Monolingual

Run Name	Description	Exp.	MAP
BRK-MO-EN-TD	Monolingual English	TD auto	0.3200
BRK-MO-EN-TDN	Monolingual English	TDN auto	0.3095
EN_POST01	LR + OKAPI Pivot 0.01	TD	0.3207
EN_POST05	LR + OKAPI Pivot 0.05	TD	0.3215
EN_POST07	LR + OKAPI Pivot 0.07	TD	0.3218
EN_POST10	LR + OKAPI Pivot 0.10	TD	0.3215
EN_POST15	LR + OKAPI Pivot 0.15	TD	0.3202
EN_POST20	LR + OKAPI Pivot 0.20	TD	0.3178
EN_POST30	LR + OKAPI Pivot 0.30	TD	0.3103
EN_POST40	LR + OKAPI Pivot 0.40	TD	0.2999
EN_POST50	LR + OKAPI Pivot 0.50	TD	0.2888
EN_POST99	LR + OKAPI Pivot 0.99	TD	0.2290

However, none of the results reported here would change the ranking of the Cheshire system when compared to other participants. We need to find effective methods of using the topical metadata included with the domain-specific collections to enhance performance.

References

1. Gey, F., Buckland, M., Chen, A., Larson, R.: Entry vocabulary - a technology to enhance digital search. In: Proceedings of HLT 2001, First International Conference on Human Language Technology, San Diego, March 2001, pp. 91–95 (2001)
2. Larson, R.R.: Domain specific retrieval: Back to basics. In: Peters, C., Clough, P., Gey, F.C., Karlgren, J., Magnini, B., Oard, D.W., de Rijke, M., Stempfhuber, M. (eds.) CLEF 2006. LNCS, vol. 4730, pp. 174–177. Springer, Heidelberg (2007)
3. Larson, R.R.: Back to basics - again - for domain specific retrieval: CLEF working notes (2008), `http://www.clef-campaign.org/2008/working_notes/Berkeley_Domain_Specific_08.pdf`
4. Petras, V., Gey, F., Larson, R.: Domain-specific CLIR of english, german and russian using fusion and subject metadata for query expansion. In: Peters, C., Gey, F.C., Gonzalo, J., Müller, H., Jones, G.J.F., Kluck, M., Magnini, B., de Rijke, M., Giampiccolo, D. (eds.) CLEF 2005. LNCS, vol. 4022, pp. 226–237. Springer, Heidelberg (2006)

Concept Models for Domain-Specific Search

Edgar Meij and Maarten de Rijke

ISLA, University of Amsterdam
Science Park 107, 1098 XG Amsterdam, The Netherlands
{emeij,mdr}@science.uva.nl

Abstract. We describe our participation in the 2008 CLEF Domain-specific track. We evaluate blind relevance feedback models and concept models on the CLEF domain-specific test collection. Applying relevance modeling techniques is found to have a positive effect on the 2008 topic set, in terms of mean average precision and precision@10. Applying concept models for blind relevance feedback, results in even bigger improvements over a query-likelihood baseline, in terms of mean average precision and early precision.

Keywords: Language modeling, Blind relevance feedback, Concept models.

1 Introduction

Our approach to retrieving documents that are annotated with thesaurus terms is to model the language use associated with concepts from a thesaurus or ontology. To this end we use the document annotations as a "bridge" between vocabulary terms and the concepts in the knowledge source at hand. We model the language use associated with concepts using a generative language modeling framework, which provides theoretically sound estimation methods and builds upon a solid statistical background.

Our concept models may be used to determine semantic relatedness or to generate navigational suggestions, either in the form of concepts or vocabulary terms. These can then be used as suggestions for the user or for blind relevance feedback [8,9,14]. In order to apply blind relevance feedback using our models, we perform a double translation. First, we estimate the most likely concepts given a query and then we use the most distinguishing terms from these concepts to formulate a new query. To find the most distinguishing terms given a concept, we apply a technique based on expectation-maximization (EM) [4] to re-estimate probabilities of one model with respect to another. Events that are well-predicted by the latter model will lose probability mass, which in turn will be given to the remaining events. Recently, we have successfully applied this technique to the estimation of relevance models on a variety of tasks and collections [9,10].

We address two research questions: (i) What are the effects of estimating and applying relevance models to the collection used at the CLEF 2008 Domain-specific track [7]? And (ii) what are the results of applying our concept models for blind relevance feedback? We find that applying relevance models helps for the CLEF 2008 Domain-specific test collection in terms of both mean average precision and early precision, although not

C. Peters et al. (Eds.): CLEF 2008, LNCS 5706, pp. 207–214, 2009.

significantly. Our concept models are able to significantly outperform a baseline query-likelihood run, both in terms of mean average precision and early precision. Moreover, we even improve over relevance models in terms of MAP.

The remainder of this paper is organized as follows. In Section 2 we introduce our retrieval framework. In Section 3 we introduce the details of our models. In Section 4 we describe our experimental setup, parameter settings, and document preprocessing steps. In Section 5 we discuss our results and we end with a concluding section.

2 Language Modeling

In the area of information retrieval, language modeling-based methods have been around for about a decade now [5,12,16]. Such methods are centered around the assumption that a query as issued by a user is a sample generated from an underlying term distribution—the information need. The documents in the collection are modeled in a similar fashion and are usually considered to be a mixture of a document-specific model and a more general background model. At retrieval time, each document is ranked according to the likelihood of having generated the query (query-likelihood).

Lafferty and Zhai [6] propose to generalize the query likelihood model to the KL-divergence scoring method, in which the query is modeled separately. Scoring documents then comes down to measuring the divergence between a query model $P(t|\theta_Q)$ and each document model $P(t|\theta_D)$, in which the divergence is negated for ranking purposes. The query model can be defined using the empirical maximum-likelihood estimate (MLE) on the original query as follows:

$$P(t|\tilde{\theta}_Q) = P(t|Q) = n(t;Q) \cdot |Q|^{-1}, \tag{1}$$

where $n(t;Q)$ is the number of occurrences of term t in query Q and $|Q|$ the length of the query. Under this definition, KL-divergence produces the same document ranking as the query likelihood model [16]. More formally, the score for each document given a query using the KL-divergence retrieval model is:

$$\begin{aligned} \text{Score}(Q,D) &= -\text{KL}(\theta_Q||\theta_D) \\ &= -\sum_{t\in\mathcal{V}} P(t|\theta_Q)\log P(t|\theta_D) + \sum_{t\in\mathcal{V}} P(t|\theta_Q)\log P(t|\theta_Q), \end{aligned} \tag{2}$$

where $\mathcal{V}$ denotes the vocabulary. The expression $\sum_{t\in\mathcal{V}} P(t|\theta_Q)\log P(t|\theta_Q)$—i.e., the entropy of the query—is constant per query and can be ignored for ranking purposes.

2.1 Document Modeling

Each document model $P(t|\theta_D)$ is estimated as the MLE of each term in the document $P(t|D)$, linearly interpolated with a background language model $P(t)$, which in turn is calculated as the likelihood of observing t in a sufficiently large corpus, such as the entire document collection:

$$P(t|\theta_D) = \lambda_D P(t|D) + (1-\lambda_D)P(t). \tag{3}$$

This may be interpreted as a way of accounting for the fact that the (pseudo-)relevant documents contain terms related to the information need as well as terms from a more general model. We smooth using Bayesian smoothing with a Dirichlet prior and set $\lambda_D = \frac{\mu}{|D|+\mu}$ and $(1-\lambda_D) = \frac{|D|}{|D|+\mu}$, where μ is the Dirichlet prior that controls the influence of smoothing [3,18].

2.2 Query Modeling

Relevance feedback can be applied to better capture a user's information need [1,7,15]. In a language modeling context, this can be performed by re-estimating the query model, i.e., $P(t|\theta_Q)$ in Eq. 2 [12,17]. For blind relevance feedback one considers terms in a set of (pseudo-)relevant documents and selects the most informative ones. These terms may then be reweighed and used to estimate a query model.

Relevance modeling is one specific technique for estimating a query model given a set of (pseudo-)relevant documents $\mathcal{D}_Q$. The query and documents are both taken to be samples of an underlying generative model—the relevance model. There are several ways to estimate the parameters of this model given the observed data, each following a different independence assumption [7]. We use method 2, which is formulated as:

$$P(t|\hat{\theta}_Q) \propto P(t) \prod_{q_i \in Q} \sum_{D_i \in \mathcal{D}_Q} P(q_i|\theta_{D_i}) P(\theta_{D_i}|t), \tag{4}$$

where $q_1, \ldots, q_k$ are the query terms, D a document, and t a term. Bayes' rule is used to estimate the term $P(\theta_D|t)$:

$$P(\theta_D|t) = P(t|\theta_D) P(\theta_D) \cdot P(t)^{-1}, \tag{5}$$

where we assume the document prior $P(\theta_D)$ to be uniform. The initial query is interpolated with the expanded part [2,13,17], thus reweighing the initial query terms and providing smoothing for the relatively sparse initial sample $P(t|\tilde{\theta}_Q)$:

$$P(t|\theta_Q) = \lambda_Q P(t|\tilde{\theta}_Q) + (1-\lambda_Q) P(t|\hat{\theta}_Q) \tag{6}$$

3 Concept Models

In order to leverage the explicit knowledge encapsulated in the GIRT/CSASA thesauri used in the CLEF Domain-specific track, we perform blind relevance feedback using the concepts defined therein. To incorporate concepts in the retrieval process, we propose to leverage the conceptual knowledge in the estimation of a query model, which is obtained from a double translation. In this translation, concepts are used as a pivot language; the initial query is translated to concepts and back to expanded query terms:

$$P(t|\hat{\theta}_Q) = \sum_{c \in C} P(t|c) P(c|Q). \tag{7}$$

We assume that the probability of selecting a term is no longer dependent on the query once we have selected a concept given that query. Two components need to be estimated here: $P(t|c)$, to which we refer as a *generative concept model*, and $P(c|Q)$, to which we will refer as *conceptual query model*. These will be detailed in the following sections.

Table 1. Top 6 stemmed terms for the document model belonging to document CSASA-1-EN-9706464 (entitled "American indian ethnic renewal: red power and the resurgence of identity and culture.") from the CLEF Domain Specific collection.

$P(t\|D)$ estimated using MLE	$P(t\|D)$ estimated using Eq. 13
0.061 the	0.54 indian
0.054 of	0.46 ethnic
0.045 indian	
0.038 ethnic	
0.028 in	
0.028 american	

3.1 Conceptual Query Modeling

The conceptual query model $P(c|Q)$ is a distribution over concepts specific to the query. In some settings, concepts are provided with a query or as part of a query. If this is not the case, however, we may leverage the document annotations to approximate this step. We formulate the estimation of concepts relevant to a query by determining which concepts are most likely given the query. To estimate this probability, we consider the top-ranked documents returned by an initial retrieval run, denoted $\mathcal{D}_Q$, and look at the annotations associated with these documents. So, in order to determine the probability of a concept given a query, we look for concepts with the highest posterior probability:

$$P(c|Q) = \Sigma_{D \in \mathcal{D}_Q} P(c|D)P(D|Q). \tag{8}$$

Here, $P(D|Q)$ is determined by applying Bayes' rule on the initial retrieval scores, similar to Eq. 5. We assume that the probability of observing a concept is independent of the query, once we have selected a document given the query; the estimation of this term is addressed below (viz. Eq. 15). As an example, Table 1 shows the top six terms from a (term) document model, before and after parsimonization; clearly, the parsimonious document model is much more specific.

3.2 Generative Concept Models

As to the first component in Eq. 7—the concept model $P(t|c)$—we associate each GIRT/CSASA thesaurus concept with a language model. We determine the level of association between a term t and a concept c by looking at the way annotators have labeled the documents and determine the probability of observing t given c: $P(t|c) = P(t,c) \cdot P(c)^{-1}$. The concepts used to annotate documents may have different characteristics from other parts of a document, such as title and content. The annotations are selected by trained indexers from a concept language while the actual content consists of free text. Since the terms that make up the document are "generated" using a different process than the concepts, we assume that t and c are independent and identical samples given a document D in which they occur. So, the probability of observing both t and c is

$$P(t,c) = \Sigma_D P(D)P(c,t|D) = \Sigma_{D \in \mathcal{D}_C} P(D)P(t|D)P(c|D), \tag{9}$$

where $\mathcal{D}_C$ denotes the set of documents annotated with concept c. When we assume each document in this set to have a uniform prior probability of being selected, we obtain

$$P(t|c) = \frac{P(t,c)}{P(c)} \propto \frac{1}{P(c)} \sum_{D \in \mathcal{D}_C} P(t|D)P(c|D). \tag{10}$$

Hence, it remains to define three terms: $P(c)$, $P(t|D)$, and $P(c|D)$. The term $P(c)^{-1}$ functions as a penalty for frequently occurring and thus relatively non-informative concepts. We estimate this term using standard MLE on the document collection:

$$P(c) = \frac{\sum_D n(c;D)}{\sum_{c'} \sum_{D'} n(c';D')}. \tag{11}$$

Next we turn to $P(x|D)$, where $x \in \{t,c\}$. The size of these models (in terms of the number of words or concepts that receive a non-zero probability) may be large, e.g., in the case of a large document collection or of frequently occurring concepts. Not all observed *events* (i.e., terms or concepts) are equally informative. We have assumed that each document is a mixture of document-specific and more general terms (Eq. 3); we generalize this to also include concepts. We update each document model by reducing the probability mass of non-specific events by iteratively adjusting the individual probabilities in each document, based on a comparison with a large reference corpus (the collection). Formally, we maximize the posterior probability of D after observing x:

$$P(D|x) = \frac{\lambda_C P(x|D)}{(1-\lambda_C)P(x) + \lambda_C P(x|D)}. \tag{12}$$

Note that λ_C may be set differently from λ_D (Eq. 3) and differently for either terms or concepts. In this paper, we fix $\lambda_C = 0.15$ [9]. We then apply the following EM algorithm until the estimates no longer change significantly:

$$\text{E-step:} \qquad e_x = P(D|x) \tag{13}$$

$$\text{M-step:} \qquad P_C(x|D) = \frac{n(x;D)e_x}{\sum_{x'} n(x';D)e_{x'}}.$$

After the EM algorithm converges, we remove those events with a probability lower than a threshold δ. Thus, the resulting document model for terms, $P(t|\hat{\theta}_D)$, to be used in Eq. 10 is given by:

$$P(t|\hat{\theta}_D) = \begin{cases} Z_{D_t} \cdot P_C(t|D) & \text{if } t \in D \text{ and } P_C(t|D) > \delta_t \\ 0 & \text{otherwise,} \end{cases} \tag{14}$$

where Z_{D_t} is a document-specific normalization factor: $Z_{D_t} = 1/\sum_t P_C(t|D)$. Table 1 gives an example of the effects of applying this algorithm to a document from the current document collection. Similarly, the resulting document model for concepts, $P(c|\hat{\theta}_D)$, to be used for $P(c|D)$ in Eq. 10, is given by:

$$P(c|\hat{\theta}_D) = \begin{cases} Z_{D_c} \cdot P_C(c|D) & \text{if } c \in D \text{ and } P_C(c|D) > \delta_c \\ 0 & \text{otherwise,} \end{cases} \tag{15}$$

where Z_{D_c} is a document-specific normalization factor: $Z_{D_c} = 1/\sum_c P_C(c|D)$. We fix $\delta_t = \delta_c = 0.01$.

Table 2. Statistics of the CLEF 2008 Domain-specific test collection

Documents					Topics			Relevant Documents			
Total	Avg. length	Std. dev. length	Avg. concepts	Std. dev. concepts	Total	Avg. length	Std. dev. length	Total	Avg.	Min.	Max.
171319	198.3	42.3	10.1	4.2	25	3	1.7	2133	85	4	206

4 Experimental Setup

Other than replacing HTML entities we did not apply any preprocessing to the document collection. To estimate our concept models, we used the `CONTROLLED-TERM-EN` field in the documents. Given the models introduced in the previous sections, we need to estimate a number of parameters, viz. λ_Q (Eq. 6), $|\mathcal{D}_Q|$ (Eq. 4), $|\mathcal{V}_Q|$ (Eq. 4), and $|C|$ (Eq. 7). We choose to optimize the parameter values by determining the mean average precision for each set of parameters and show the results of the best performing settings. For λ_Q we sweep in the interval [0,1] with increments of 0.1. The other parameters are investigated in the range [1,10] with increments of 1. We determine the MAP scores on the same topics that we present results for, similar to [11,18]. While computationally expensive (exponential in the number of parameters), this approach provides us with an upper bound on the performance one might achieve using the described models.

As our baseline, we employ a run based on the KL-divergence retrieval method and set $\lambda_Q = 1$ (viz. Section 2, Eq. 6). As to μ (Eq. 3), we set this parameter to the average document length. All the results that we report on use this baseline as their initially retrieved document set. Since our concept language models also rely on pseudo-relevance feedback, we use the method introduced by [7] (Eq. 4) as another baseline.

5 Results and Discussion

Table 3 lists the results of our runs. We see that our conceptual language model (CM) has a significant positive effect on the number of relevant documents retrieved. Compared with QL and RM, CM loses in very early precision (P5), but not significantly. It already makes up for this later in the top 10 (P10) and even more so further down the ranking. The differences in P5, P10 and MAP between the three runs are not significant; given the relatively small number of topics (25), it is hard to achieve statistically significant differences.

Table 3. Results of the query likelihood (QL), relevance (RM) and conceptual language model (CM). Percentages indicate relative difference with QL. Significance is tested using a Wilcoxon sign rank test; * indicates a statistically significant difference against QL ($p < 0.05$).

	QL	RM		CM	
Relevant retrieved	1468	1473	+0.3%	**1602**	+9.1%*
P5	0.5280	**0.5680**	+7.6%	0.4880	-7.6%
P10	0.4680	0.4800	+2.6%	**0.4840**	+3.4%
MAP	0.2819	0.2856	+1.3%	**0.2991**	+6.1%

Next we turn to the precision-recall plot for our three runs, QL, RM and CM; see Figure 1. As can be expected, given the numbers in Table 3, at very low recall levels RM and QL both outperform CM; at high recall levels (between 0.5 and 0.9) CM outperforms QL and RM, that perform at very comparable levels.

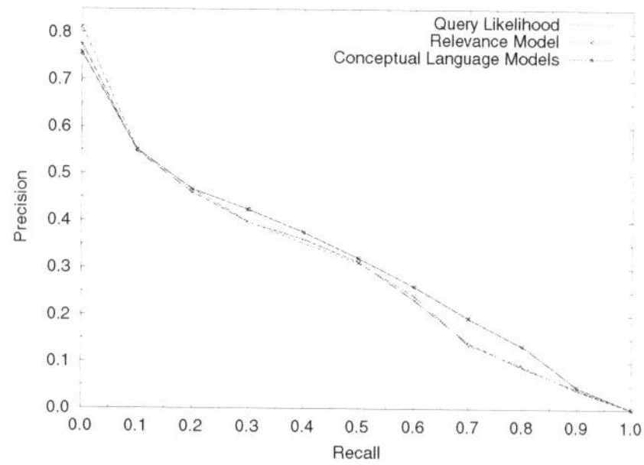

Fig. 1. Precision recall graph

Finally, we turn to a topic level comparison of CM and the baseline run QL; see Figure 2. First, in terms of MAP, CM outperforms QL on 14 out 25 topics, while QL beats CM on 8; there is a large gain for one topic (211: *Shrinking cities*). In terms of P5, CM outperforms QL on only 4 topics, while QL beats CM on 7; here, topics 223 (*Media in the preschool age*) and 210 (*Establishment of new businesses after the reunification*) are especially hard for CM (-0.40 and -0.70, respectively). In terms of P10, CM beats QL on 11 topics, but loses on 8: topics 223 and 210 are still amongst the topics on which CM loses, but the losses are not as dramatic as they were for P5 (-0.20 and -0.40, respectively).

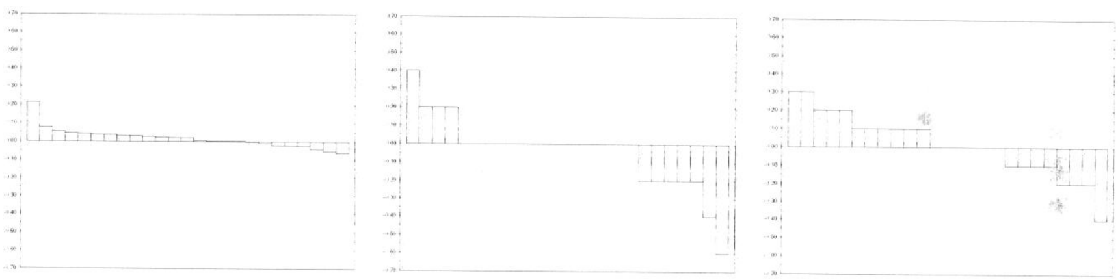

(a) MAP (b) P5 (c) P10

Fig. 2. Per-topic breakdown of the improvement of CM over the QL baseline on various evaluation measures. A positive value indicates an improvement over the baseline.

6 Conclusion

We described our participation in the 2008 edition of the CLEF Domain Specific track. Specifically, we examined blind relevance feedback models and concept models. Applying relevance modeling techniques was found to have a positive effect on the current topics, in terms of mean average precision and precision@10. When applying concept models for blind relevance feedback, we observed an even bigger as well as significant improvement over the query-likelihood baseline, also in terms of mean average precision and early precision. The most noticable effect of our concept models was on recall; in future work, on larger topic sets, we aim to analyze these effects further.

Acknowledgements

This work was carried out in the context of the Virtual Laboratory for e-Science project, which is supported by a BSIK grant from the Dutch Ministry of Education, Culture and

Science and is part of the ICT innovation program of the Ministry of Economic Affairs. The work was also supported by the DuOMAn project carried out within the STEVIN programme under project number STE-09-12, and by the Netherlands Organisation for Scientific Research (NWO) under project numbers 017.001.190, 640.001.501, 640.-002.501, 612.066.512, 612.061.814, 612.061.815, 640.004.802.

References

1. Anick, P.: Using terminological feedback for web search refinement: a log-based study. In: SIGIR 2003 (2003)
2. Balog, K., Weerkamp, W., de Rijke, M.: A few examples go a long way: constructing query models from elaborate query formulations. In: SIGIR 2008 (2008)
3. Chen, S.F., Goodman, J.: An empirical study of smoothing techniques for language modeling. In: ACL 1996 (1996)
4. Dempster, A.P., Laird, N.M., Rubin, D.B.: Maximum likelihood from incomplete data via the EM algorithm. J. Royal Statistical Society. Series B 39(1), 1–38 (1977)
5. Hiemstra, D.: A linguistically motivated probabilistic model of information retrieval. In: Nikolaou, C., Stephanidis, C. (eds.) ECDL 1998. LNCS, vol. 1513, p. 569. Springer, Heidelberg (1998)
6. Lafferty, J., Zhai, C.: Document language models, query models, and risk minimization for information retrieval. In: SIGIR 2001 (2001)
7. Lavrenko, V., Croft, B.W.: Relevance based language models. In: SIGIR 2001 (2001)
8. Meij, E., de Rijke, M.: Thesaurus-based feedback to support mixed search and browsing environments. In: Kovács, L., Fuhr, N., Meghini, C. (eds.) ECDL 2007. LNCS, vol. 4675, pp. 247–258. Springer, Heidelberg (2007)
9. Meij, E., Trieschnigg, D., de Rijke, M., Kraaij, W.: Parsimonious concept modeling. In: SIGIR 2008 (2008)
10. Meij, E., Weerkamp, W., Balog, K., de Rijke, M.: Parsimonious relevance models. In: SIGIR 2008 (2008)
11. Mitra, M., Singhal, A., Buckley, C.: Improving automatic query expansion. In: SIGIR 1998 (1998)
12. Ponte, J.M., Croft, W.B.: A language modeling approach to information retrieval. In: SIGIR 1998 (1998)
13. Rocchio, J.: Relevance feedback in information retrieval. In: The SMART Retrieval System: Experiments in Automatic Document Processing. Prentice Hall, Englewood Cliffs (1971)
14. Trieschnigg, D., Meij, E., de Rijke, M., Kraaij, W.: Measuring concept relatedness using language models. In: SIGIR 2008 (2008)
15. Xu, J., Croft, W.B.: Query expansion using local and global document analysis. In: SIGIR 1996 (1996)
16. Zhai, C.: Risk Minimization and Language Modeling in Text Retrieval. PhD thesis, Carnegie Mellon University (2002)
17. Zhai, C., Lafferty, J.: Model-based feedback in the language modeling approach to information retrieval. In: CIKM 2001 (2001)
18. Zhai, C., Lafferty, J.: A study of smoothing methods for language models applied to information retrieval. ACM Transactions on Information Systems 22(2), 179–214 (2004)

The Xtrieval Framework at CLEF 2008: Domain-Specific Track

Jens Kürsten, Thomas Wilhelm, and Maximilian Eibl

Chemnitz University of Technology
Faculty of Computer Science, Chair Computer Science and Media
Straße der Nationen 62
09107 Chemnitz, Germany
{jens.kuersten,thomas.wilhelm,eibl}@cs.tu-chemnitz.de

Abstract. This article describes our participation at the *Domain-Specific track*. We used the *Xtrieval* framework for the preparation and execution of the experiments. The translation of the topics for the cross-lingual experiments was realized with a plug-in to access the Google AJAX language API. This year, we submitted 20 experiments in total. In all our experiments we applied a standard top-k pseudo-relevance feedback algorithm. We used merged monolingual runs as baseline for comparison to all our cross-lingual experiments. Translating the topics for the bilingual experiments decreased the retrieval effectiveness only between 8 and 15 percent.

Keywords: Evaluation, Cross-Language Retrieval, Domain-Specific Retrieval.

1 Introduction and Outline

The *Xtrieval* framework [3],[2] was used to prepare and execute this year's *Domain-Specific* text retrieval experiments. The core retrieval functionality is provided by Apache Lucene[1]. For the *Domain-Specific track* three different corpora with sociological content in German, English and Russian were employed [4]. For the translation of the topics the Google AJAX language API[2] was accessed through a JSON[3] programming interface. We also used the provided bilingual thesauri to investigate their impact on bilingual retrieval effectiveness.

The remainder of the paper is organized as follows. Section 2 describes the general setup of our system. The individual configurations and the results of our submitted experiments are presented in sections 3 to 5 and in section 6 we summarize our observations.

[1] http://lucene.apache.org
[2] http://code.google.com/apis/ajaxlanguage/documentation
[3] http://json.org

C. Peters et al. (Eds.): CLEF 2008, LNCS 5706, pp. 215–218, 2009.

2 Experimental Setup

We combined several stemming methods for each language in the retrieval stage. Thereby the input streams were tokenized differently, but documents and queries were processed with the same stemming algorithm for each run. We realized a late fusion stage to combine the experiments with our implementation of the *Z-Score* operator [5]. We compared standard retrieval experiments to query expansion based on the provided domain-specific thesauri to investigate their impact in terms of retrieval effectiveness. A standard top-k pseudo-relevance feedback algorithm was used to improve retrieval effectiveness.

3 Monolingual Experiments

We submitted 5 monolingual experiments in total, 2 for the English and the German subtasks and 1 for the Russian subtask. For all experiments a language-specific stopword list was applied[4]. We used different stemmers for each language: Porter[5] and Krovetz [1] for English, Snowball[5] and a n-gram variant decompounding stemmer[6] for German as well as an Java implementation of a stemmer for Russian[4]. For two experiments the provided thesauri were used for query expansion (tqe) by adding each corresponding term from the thesauri for each of the terms of the original query. In table 1, the retrieval effectiveness of our experiments is presented in terms of mean average precision (map).

Table 1. Experimental Results for the monolingual subtask

id	*lang*	*tqe*	*map*
cut_merged	DE	no	0.4367
cut_merged_thes	DE	yes	0.4359
cut_merged	EN	no	0.3891
cut_merged_thes	EN	yes	0.3869
cut_merged	RU	no	0.0955

Our experiments on the German and English collections had very good overall retrieval effectiveness. In contrast to that our experiment on the Russian collection performed very bad. It is also obvious that the thesaurus based query expansion did not improve the retrieval effectiveness, but at least it did not significantly decrease MAP.

4 Bilingual Experiments

We submitted 12 experiments in total for the bilingual subtask and compared the translation from different source languages and the performance of pure topic

[4] http://members.unine.ch/jacques.savoy/clef/index.html
[5] http://snowball.tartarus.org
[6] http://www-user.tu-chemnitz.de/~wags/cv/clr.pdf

Table 2. Experimental Results for the bilingual subtask

id	*lang*	*tqe*	*map*
cut_merged	DE	no	0.4367
cut_merged_en2de	EN→DE	no	0.3702 (-15.23%)
cut_merged_en2de_thes	EN→DE	yes	0.3554 (-18.62%)
cut_merged	EN	no	0.3891
cut_merged_ru2en	RU→EN	no	0.3385 (-13.00%)
cut_merged_ru2en_thes	RU→EN	yes	0.3276 (-15.81%)
cut_merged	RU	no	0.0955
cut_merged_en2ru	EN→RU	no	0.0882 (-07.64%)
cut_merged_en2ru_thes	EN→RU	yes	0.0597 (-37.49%)

translation (PTT) to combined translation (CT). For CT we used the PTT and tried to improve the translation with the help of the bilingual thesauri, i.e. for every term occurring in the bilingual thesauri we added its provided translation to the topic. In table 2 we compare our bilingual experiments with respect to the performance of the corresponding monolingual experiment.

Probably due to the quality of Google's translation service and the strong performance of our monolingual runs the retrieval effectiveness of our bilingual experiments was very good as well. The translation supported by the provided thesauri did not improve the retrieval effectiveness.

5 Multilingual Experiments

For the participation at the multilingual subtask 3 experiments were submitted. Topics in all given languages were used, with one language as source for one experiment. All target collections were queried for each multilingual experiment. The results of the evaluation are shown in table 3.

The retrieval performance of our multilingual experiments was very good, especially in comparison to the experimental results of the years before [3]. We assume this to be due to Google's translation service on the one hand but also to the deployed result list fusion algorithm [5]. It is obvious that the performance is almost equal for the experiments, where we used the German and English topics, while translating the Russian topics performed worst.

Table 3. Experimental Results for the multilingual subtask

id	*lang*	*map*
cut_merged_de2x	DE→X	0.2816
cut_merged_en2x	EN→X	0.2751
cut_merged_ru2x	RU→X	0.2357

6 Result Analysis - Summary

The following list provides a summary of the analysis of our experiments:

- *Monolingual:* The performance of our monolingual experiments was very good for the German and English collections and worse for the Russian collection. Interestingly, the retrieval effectiveness could not be improved by utilizing the provided domain-specific thesauri for query expansion.
- *Bilingual:* Probably due to the used translation service our bilingual experiments performed very well. Astonishingly, we could not improve the retrieval performance by using the provided bilingual thesauri.
- *Multilingual:* Again, mainly due to the quality of the translation and the result list combination capabilities of the *Xtrieval* framework we achieved very impressive results in terms of retrieval effectiveness. The best results were obtained by translating the English and German topics.

Acknowledgments

This work was partially accomplished in conjunction with the project *sachsMedia*, which is funded by the *Entrepreneurial Regions*[7] program of the German Federal Ministry of Education and Research.

References

1. Krovetz, R.: Viewing Morphology as an Inference Process. In: SIGIR 1993: Proceedings of the 16th annual international ACM SIGIR conference on Research and development in information retrieval, pp. 191–202. ACM, New York (1993)
2. Kürsten, J., Wilhelm, T., Eibl, M.: Extensible Retrieval and Evaluation Framework: Xtrieval. In: LWA 2008: Lernen - Wissen - Adaption Workshop Proceedings, Würzburg (October 2008)
3. Kürsten, J., Wilhelm, T., Eibl, M.: The XTRIEVAL Framework at CLEF 2007: Domain-Specific Track. In: Peters, C., Jijkoun, V., Mandl, T., Müller, H., Oard, D.W., Peñas, A., Petras, V., Santos, D. (eds.) CLEF 2007. LNCS, vol. 5152, pp. 174–181. Springer, Heidelberg (2008)
4. Petras, V., Baerisch, S.: The Domain-Specific Track at CLEF 2008. In: Working Notes for the CLEF 2008 Workshop, September 17-19, Aarhus, Denmark (October 2008)
5. Savoy, J.: Data Fusion for Effective European Monolingual Information Retrieval. In: Peters, C., Clough, P., Gonzalo, J., Jones, G.J.F., Kluck, M., Magnini, B. (eds.) CLEF 2004. LNCS, vol. 3491, pp. 233–244. Springer, Heidelberg (2005)

[7] The Innovation Initiative for the New German Federal States.

Using Wikipedia and Wiktionary in Domain-Specific Information Retrieval

Christof Müller and Iryna Gurevych

Ubiquitous Knowledge Processing Lab, Computer Science Department,
Technische Universität Darmstadt,
Hochschulstr. 10, D-64289 Darmstadt, Germany
{mueller,gurevych}@tk.informatik.tu-darmstadt.de
http://www.ukp.tu-darmstadt.de

Abstract. The main objective of our experiments in the domain-specific track at CLEF 2008 is utilizing semantic knowledge from collaborative knowledge bases such as Wikipedia and Wiktionary to improve the effectiveness of information retrieval. While Wikipedia has already been used in IR, the application of Wiktionary in this task is new. We evaluate two retrieval models, i.e. SR-Text and SR-Word, based on semantic relatedness by comparing their performance to a statistical model as implemented by Lucene. We refer to Wikipedia article titles and Wiktionary word entries as concepts and map query and document terms to concept vectors which are then used to compute the document relevance. In the bilingual task, we translate the English topics into the document language, i.e. German, by using machine translation. For SR-Text, we alternatively perform the translation process by using cross-language links in Wikipedia, whereby the terms are directly mapped to concept vectors in the target language. The evaluation shows that the latter approach especially improves the retrieval performance in cases where the machine translation system incorrectly translates query terms.

Keywords: Information Retrieval, Semantic Relatedness, Collaborative Knowledge Bases, Cross-Language Information Retrieval.

1 Introduction

Statistical models are most frequently used in domain-specific information retrieval (**IR**). One of the disadvantages of these models is their lack of flexibility concerning synonymy, i.e. expressing a concept with different terms. There exist several approaches of tackling the problem of synonymy divided into local and global methods.

Local methods like relevance and pseudo-relevance feedback try to refine the representation of the user's information need by using either manual or automatic feedback about already returned documents. However, these methods require that the relevant documents show a significant term overlap, and that the term overlap between relevant and irrelevant documents is small. Also they are not able to close the gap between the vocabulary used in queries and in documents,

C. Peters et al. (Eds.): CLEF 2008, LNCS 5706, pp. 219–226, 2009.

i.e. query terms which do not explicitly occur in the document collection cannot be expanded with related terms.

Global methods expand the query with related terms using either automatically built thesauri based on the document collection or external linguistic knowledge bases like WordNet [1]. Using thesauri which are based on the document collection also suffers from the inability to close the vocabulary gap, if query terms do not occur in the document collection. The use of linguistic knowledge bases for query expansion has shown inconclusive results so far. Voorhees [2] could improve retrieval performance only in some cases even for manually selected expansion terms, while Mandala *et al.* [3] improved the performance on several test collections by combining a linguistic knowledge base with different types of thesauri built from the underlying text collections. The general problem of query expansion is that in fact it is able to improve recall in certain situations, but at the same time precision degrades as also irrelevant terms are added to the query.

Another knowledge-based approach to tackle the problem of synonymy is to use retrieval models which are based on semantic relatedness (**SR**) between query and document terms computed by using linguistic knowledge bases. Although first results of employing SR in IR were inconclusive [4], there have also been several promising results, e.g., [5,6]. One of the main problems with using linguistic knowledge bases for semantically enhanced IR is the low coverage, especially of domain-specific vocabulary.

A new form of resources, so called collaborative knowledge bases [7] have the potential to overcome these limitations. Enabled by Web 2.0 technologies which simplify the editing and annotation process of web content, collaborative knowledge bases are constructed by volunteers on the web and have reached a size which makes them promising for improving IR performance. The most widely used and probably largest collaborative knowledge base is Wikipedia[1] which contains encyclopedic knowledge in a broad range of domains.

For our experiments in the domain-specific track at CLEF 2008 [8], we employ Wikipedia and for the first time Wiktionary[2] as knowledge bases for SR-based IR models. We compare their performance to a statistical model and also combine all three models by adding their respective relevance scores for each document. We perform the experiments for the languages English, German, and Russian. For bilingual IR experiments using English topics on a German document collection, we use (i) machine translation methods for statistical and semantic IR models, and (ii) cross-language links in Wikipedia for one of the semantic IR models.

2 Information Retrieval Models

Besides applying standard preprocessing steps like tokenization and stopword removal, we use the TreeTagger [9] for lemmatization. For the German test data, we also split compounds into their constituents [10], and we use both constituents

[1] `http://www.wikipedia.org`

[2] `http://www.wiktionary.org`

and compounds in the retrieval process. As baseline IR model we use Lucene[3] which is based on the vector space model. We also use Lucene for combining it with the semantic models.

2.1 Semantic Models

In our experiments, we adapt a method proposed by Gabrilovich and Markovitch [11] where article titles in Wikipedia are referred to as concepts and the article texts as textual representation of these concepts. The concept vector of a term consists of its tf value in the respective Wikipedia articles. In order to map a document or a query to its concept vector, we first build the concept vectors for all its terms. We then sum up the concept vectors after normalizing each vector and scaling it with the respective term's tf and idf values. Given the concept vector of a query and a document, we use the cosine of the angle between the two vectors as relevance score. We refer to this model as **SR-Text**.

Additionally, we employ a retrieval model proposed in [12], which we refer to as **SR-Word**. We extended the model by also taking into account the idf value of document terms and the tf value of query and document terms. This model is represented by the following equation:

$$r_{SR}(d,q) = \frac{\sum_{i=1}^{n_d} \sum_{j=1}^{n_q} tf(t_{d,i},d) \cdot idf(t_{d,i}) \cdot tf(t_{q,j},q) \cdot idf(t_{q,j}) \cdot s(t_{d,i},t_{q,j})}{(1+n_{nsm}) \cdot (1+n_{nr})} \tag{1}$$

where n_d is the number of unique terms in the document, n_q the number of unique terms in the query, $t_{d,i}$ the i-th unique document term, $t_{q,j}$ the j-th unique query term, $s(t_{d,i},t_{q,j})$ the SR score for the respective document and query term (using the cosine of the respective terms' concept vectors as score analog to SR-Text), n_{nsm} the number of unique query terms not literally found in the document, and n_{nr} the number of unique query terms which do not contribute a SR score above a predefined threshold. For SR-Text and SR-Word, we compute tf and idf as follows:

$$tf(t) = 1 + \log f(t) \tag{2}$$

where $f(t)$ is the frequency of term t in the corresponding document or query, and

$$idf(t) = \log \frac{n_{docs}}{df(t)} \tag{3}$$

where n_{docs} is the number of documents in the collection and $df(t)$ is the number of documents in the collection containing term t.

Besides Wikipedia we use Wiktionary as a knowledge base for the IR models. Thereby, we refer to each word entry in Wiktionary as a distinct concept, and use the entry's information as the textual representation of a concept analogous to the text of Wikipedia articles (for details see [13]). In order to improve retrieval effectiveness, we combine the concept space of Wikipedia and Wiktionary, so

[3] `http://lucene.apache.org`

that the concept vector of one term consists of concepts from both knowledge bases. For Wikipedia we remove concepts where the respective Wikipedia articles have less than 100 words or fewer than 5 in- or outlinks. For both Wikipedia and Wiktionary, we remove concepts from a term's concept vector if their normalized values are below the predefined threshold of *0.01*. The pruning methods are applied with the goal of noise reduction and better performance. For accessing the collaborative knowledge bases we use freely available Java-based APIs described in [7].

2.2 Combination of Models

As the statistical and semantic models use different types of information represented in queries, documents and possibly external knowledge, we hypothesize that a combination of the models might increase the retrieval effectiveness. We therefore combine their relevance scores computed separately into one relevance score for each document per query. For computing the combined relevance score, we use the *CombSUM* method which was introduced by Fox and Shaw [14] where the combined relevance score is set to the sum of the individual relevance scores. Before combining the scores, they are normalized using the formula:

$$r_{norm} = \frac{r_{orig} - r_{min}}{r_{max} - r_{min}} \tag{4}$$

where r_{orig} is the original relevance score, r_{min} is the minimal and r_{max} is the maximal occurring score for the query.

3 Evaluation

We experiment with several query types by using different combinations of the topic fields. In our training runs using topics from the past CLEF workshops, we found that the retrieval effectiveness improved when query terms are weighted depending on the field in which they occur. We therefore use the following weights for query terms in all experiments: 1 for *title* (**T**), 0.8 for *description* (**D**), and 0.6 for *narrative* (**N**).

We set the threshold for SR values in SR-Word to the following values as they showed the best performance in the training runs: 0.25 for English, 0.11 for German, and 0.23 for Russian.

Besides the officially submitted runs, we performed several experiments where the concept vectors used in SR-Text were normalized again after removing some concepts that had values below the predefined threshold of 0.01. We found that the performance increased slightly for most experiments. We therefore report the results of these new experiments together with some other additional runs.

3.1 Monolingual Retrieval

Table 1 shows the mean average precision (**MAP**) of each model and the combination of all models over query length and language for the monolingual experiments. For English and German, we used the combination of Wikipedia and

Table 1. The MAP values of the monolingual runs. The highest value of the separate models is in bold for each query type.

	English			German			Russian		
	T	TD	TDN	T	TD	TDN	T	TD	TDN
Lucene	**0.2514**	**0.2983**	**0.2987**	0.3405	0.3318	**0.3536**	0.1194	**0.1254**	**0.1286**
SR-Text	0.2020	0.2220	0.2521	0.2761	0.3204	0.3302	**0.1277**	0.1096	0.0745
SR-Word	0.2351	0.2595	0.2526	**0.3605**	**0.3548**	0.3248	0.1211	0.1058	0.0930
Combination	0.2735	0.3104	0.3211	0.3719	0.3820	0.3922	0.1387	0.1383	0.1330

Wiktionary as knowledge base, for Russian we used only Wikipedia as the API for Wiktionary does not allow to parse the Russian Wiktionary edition.

For English, Lucene outperforms the semantic models for all query types. For German this is only the case for the longest query type TDN. Using query types T and TD, the SR-Word model outperforms Lucene. Except for the query type T where SR-Text performs best, the semantic models are outperformed on the Russian data set by Lucene. However, when Lucene is combined with the semantic models by using the CombSUM method, MAP increases for all languages and query types and outperforms the separate models. Compared to Lucene, the highest and statistically significant[4] increase of MAP is 9% for English and 15% for German. For Russian we receive the best result when Lucene using query type TDN is combined with SR-Word using query type T. This results in a MAP of 0.1491 which is an increase of 16% as compared to Lucene. However, the difference is not statistically significant.

The performance of Lucene almost consistently increases for longer query types on all three languages. For SR-Text we also observe a trend to perform better for longer queries except for Russian. For SR-Word the trend is opposite for German and Russian.

For English and German, we also performed experiments using either Wikipedia or Wiktionary separately as knowledge base. The results show that for German the combination of Wikipedia and Wiktionary slightly improves the performance in most cases. For English using only Wikipedia often performs better than using the combination of both knowledge bases. Using Wiktionary separately always performed worse than using Wikipedia or the combination of both.

3.2 Bilingual Retrieval

In the bilingual retrieval, we use English topics with the German document collection. The English topics are translated into German using machine translation[5] (**MT**). For the SR-Text model, we additionally explore a different method using the cross-language links (**CLL**) between language specific editions of Wikipedia. A cross-language link points from an article in one language to the same article in a different language, e.g. an English article might point to its German

[4] We used a paired t-test to determine the statistical significance.

[5] `http://babelfish.yahoo.com/` which is based on the Systran Translator.

Table 2. The MAP values of the bilingual runs. The highest value of the separate models and the combinations is in bold for each query type.

	T	TD	TDN
Lucene	0.1490	0.1638	**0.1746**
SR-Text-MT	0.1173	0.1519	0.1547
SR-Text-CLL	0.1193	0.1288	0.1225
SR-Word	**0.1806**	**0.1760**	0.1688
Lucene + SR-Text-MT	0.1476	0.1783	0.1925
Lucene + SR-Text-CLL	0.1963	**0.2139**	**0.2205**
Lucene + SR-Text-MT + SR-Word	0.1687	0.1891	0.1976
Lucene + SR-Text-CLL + SR-Word	**0.2003**	0.2117	0.2162
Lucene + SR-Text-MT + SR-Text-CLL + SR-Word	0.1944	0.2089	0.2128

counterpart. By using these links, we map a concept vector whose concepts are represented by articles in the English Wikipedia into a concept vector whose concepts are represented by articles in the German Wikipedia. Thus, by transforming the concept vector of an English query using cross-language links, the similarity between the English query and a German document is computed by the SR-Text model without actually translating the query.[6] As Wiktionary also has cross-language links and furthermore many of the word entries contain translations of the term into other languages, it is possible to apply the CLL method to both Wikipedia and Wiktionary. However, we only report the results for using CLLs in Wikipedia.

The results of the bilingual runs are shown in Table 2. Generally, the MAP values in our bilingual runs are much lower compared to the monolingual German runs as both methods, MT and CLL, add noise to the retrieval process. For the query types T and TD, SR-Word is the best performing model. For the query type TDN, Lucene performs slightly better than SR-Word. At first sight, SR-Text using MT seems to yield better results than SR-Text using the CLL method. When combined with the Lucene model, SR-Text-CLL outperforms SR-Text-MT. When we use the respective best performing query type for each model, the combination of Lucene with query type TDN, SR-Text-CLL with query type TD and SR-Word with query type T results in a MAP of 0.2350 which is the best performance of our bilingual runs. Compared to using Lucene alone, this is a significant increase of 35%. This run is not shown in Table 2.

Analyzing the results of individual queries, we found that the CLL method is especially beneficial in cases of substantial translation errors for the query terms. In topic no. 209 where the English title field contains the terms *Doping and sports* the correct German translation of *Doping* would be the same term *Doping*. Instead, it is incorrectly translated by the machine translation system to *Lackieren* which has the meaning of *painting* or *varnishing*. As the Lucene

[6] As we do not actually translate the query terms, we have no information about the document frequency of a query term to compute its idf value. Therefore, we use the term's document frequency in Wikipedia for computing its idf value.

model relies on the translation with the MT system, the combination with SR-Text using the CLL method especially improves the retrieval in these cases. The lower performance of SR-Text-CLL compared to SR-Text-MT when not combined with Lucene might result from missing cross-language links between articles in the German and English Wikipedia. Not even half of the articles in the German Wikipedia link to the respective articles in the English Wikipedia.

4 Conclusions

In our experiments, we have explored the integration of semantic knowledge from collaborative knowledge bases into IR. For the first time, we have employed Wiktionary in combination with Wikipedia for this task. We have evaluated two IR models (SR-Text and SR-Word) based on semantic relatedness by comparing their performance to a statistical model as implemented by Lucene. In these semantic models, the articles in Wikipedia and the word entries in Wiktionary are employed as textual representations of concepts. The SR-Text model computes the similarity of a query and document by summing up the concept vectors of the query and document terms respectively and then computing the cosine of the angle between the query's and the document's concept vector. The SR-Word model combines individual similarities of each query and document term pair that are above a predefined threshold and then applies a set of heuristics to compute the final relevance score.

In the monolingual task, the combination of Lucene and the semantic models increases the MAP by 9% for English, 15% for German, and 16% for Russian as compared to Lucene. In the bilingual task, we translated the English topics into the document language, i.e. German, by using machine translation. For SR-Text, we additionally explored a different method using the cross-language links between different language editions of Wikipedia. This approach especially improved the retrieval performance in cases where the machine translation system incorrectly translated terms. When Lucene was combined with SR-Text-CLL and SR-Word, the MAP increased by 35%. In our future work, we will additionally use the cross-language links in Wiktionary to further improve the IR effectiveness. We also plan to integrate the cross-language links into the SR-Word model.

Acknowledgement

This work was supported by the Volkswagen Foundation as part of the Lichtenberg-Professorship Program under grant No. I/82806 and by the German Research Foundation under grant No. GU 798/1-3.

References

1. Fellbaum, C. (ed.): WordNet: An Electronic Lexical Database. MIT Press, Cambridge, MA (1998)
2. Voorhees, E.M.: Query expansion using lexical-semantic relations. In: SIGIR 1994: Proceedings of the 17th annual international ACM SIGIR conference on Research and development in information retrieval, pp. 61–69. Springer-Verlag New York, Inc., New York (1994)

3. Mandala, R., Tokunaga, T., Tanaka, H.: The Use of WordNet in Information Retrieval. In: Harabagiu, S. (ed.) Proceedings of the COLING-ACL workshop on Usage of WordNet in Natural Language Processing, pp. 31–37. Association for Computational Linguistics, Somerset (1998)
4. Smeaton, A.: Using NLP or NLP Resources for Information Retrieval Tasks. In: Strzalkowski, T. (ed.) Natural Language Information Retrieval, pp. 99–111. Kluwer Academic Publishers, Dordrecht (1999)
5. Lytinen, S., Tomuro, N., Repede, T.: The use of WordNet sense tagging in FAQFinder. In: Proceedings of the AAAI 2000 workshop on AI and Web Search, Austin, TX (2000)
6. Müller, C., Gurevych, I., Mühlhäuser, M.: Integrating Semantic Knowledge into Text Similarity and Information Retrieval. In: Proceedings of the First IEEE International Conference on Semantic Computing (ICSC), Irvine, CA, USA, pp. 257–264 (2007)
7. Zesch, T., Müller, C., Gurevych, I.: Extracting Lexical Semantic Knowledge from Wikipedia and Wiktionary. In: Proceedings of the Conference on Language Resources and Evaluation, LREC (2008)
8. Petras, V., Baerisch, S.: The Domain-Specific Track at CLEF 2008. In: Peters, C., et al. (eds.) CLEF 2008. LNCS, vol. 5706, pp. 186–198. Springer, Heidelberg (2009)
9. Schmid, H.: Probabilistic part-of-speech tagging using decision trees. In: Proceedings of Conference on New Methods in Language Processing (1994)
10. Langer, S.: Zur Morphologie und Semantik von Nominalkomposita. In: Tagungsband der Konferenz zur Verarbeitung natürlicher Sprache, KONVENS, pp. 83–97 (1998)
11. Gabrilovich, E., Markovitch, S.: Computing Semantic Relatedness using Wikipedia-based Explicit Semantic Analysis. In: Proceedings of The Twentieth International Joint Conference for Artificial Intelligence, Hyderabad, India, pp. 1606–1611 (2007)
12. Müller, C., Gurevych, I.: Exploring the Potential of Semantic Relatedness in Information Retrieval. In: Schaaf, M., Althoff, K.D. (eds.) LWA 2006 Lernen - Wissensentdeckung - Adaptivität, 9.-11.10.2006 in Hildesheim. Hildesheimer Informatikberichte, pp. 126–131. Universität Hildesheim, Hildesheim (2006)
13. Zesch, T., Müller, C., Gurevych, I.: Using Wiktionary for Computing Semantic Relatedness. In: Proceedings of the Twenty-Third AAAI Conference on Artificial Intelligence, AAAI 2008, Chicago, Illinois, USA, pp. 861–867 (2008)
14. Fox, E., Shaw, J.: Combination of multiple searches. In: Proceedings of the 2nd Text REtrieval Conference (TREC-2), pp. 243–252 (1994)

Overview of iCLEF 2008: Search Log Analysis for Multilingual Image Retrieval

Julio Gonzalo[1], Paul Clough[2], and Jussi Karlgren[3]

[1] UNED, Madrid, Spain
[2] University of Sheffield, Sheffield, UK
[3] SICS, Kista, Sweden

Abstract. This paper summarises activities from the iCLEF 2008 task. In an attempt to encourage greater participation in user-orientated experiments, a new task was organised based on users participating in an interactive cross-language image search experiment. Organizers provided a default multilingual search system which accessed images from Flickr, with the whole iCLEF experiment run as an online game. Interaction by users with the system was recorded in log files which were shared with participants for further analyses, and provide a future resource for studying various effects on user-orientated cross-language search. In total six groups participated in iCLEF, providing a combined effort in generating results for a shared experiment on user-orientated cross-language retrieval.

1 Introduction

iCLEF is the interactive track of CLEF (Cross-Language Evaluation Forum), an annual evaluation exercise for Multilingual Information Access systems. In iCLEF, Cross-Language search capabilities are studied from a user-inclusive perspective. A central research question is how best to assist users when searching information written in unknown languages, rather than how best an algorithm can find information written in languages different from the query language.

Since 2006, iCLEF has moved away from news collections (a standard for text retrieval experiments) in order to explore user behaviour in scenarios where the necessity for cross-language search arises more naturally for the average user. We chose Flickr, a large-scale, web-based image database based on a large social network of WWW users sharing over two billion images, with the potential for offering both challenging and realistic multilingual search tasks for interactive experiments.

Over the last years, iCLEF participants have typically designed one or more cross-language search interfaces for tasks such as document retrieval, question answering or text-based image retrieval. Experiments were hypothesis-driven, and interfaces were studied and compared using controlled user populations under laboratory conditions. This experimental setting has provided valuable research insights into the problem, but has a major limitation: user populations are

C. Peters et al. (Eds.): CLEF 2008, LNCS 5706, pp. 227–235, 2009.

necessarily small in size, and the cost of training users, scheduling and monitoring search sessions is very high. In addition, the target notion of relevance does not cover all aspects that make an interactive search session successful; other factors include user satisfaction with the results and usability of the interface.

The main novelty of the iCLEF 2008 shared experience has been to focus on the shared analysis of a large search log from a single search interface provided by the iCLEF organizers. The focus is, therefore, on search log analysis rather than on system design. The idea is to study the behaviour of users in an (almost) naturalistic search scenario, having a much larger data set than in previous iCLEF campaigns. The search interface provided by iCLEF organizers is a basic cross-language retrieval system to access images in Flickr, presented as an online game: the user is given an image, and she must find it again without any a-priori knowledge of the language(s) in which the image is annotated. Game-like features are intended to engage casual users and therefore increase the chances of achieving a large, representative search log. More information can be found in [8].

The structure of the rest of the paper is as follows: Section 2 describes the task guidelines; Section 3 describes the features of the search log distributed to participants. In Section 4 we summarize the participation in the track and give some conclusions about the experience.

2 Task Guidelines

2.1 Search Task Definition

First of all, the decision to use Flickr as the target collection is based on (i) the inherent multilingual nature of the database, provided by tagging and commenting features utilised by a worldwide network of users, (ii) although it is in constant evolution, which may affect reproducibility of results, the Flickr search API allows the specification of timeframes (e.g. search in images uploaded between 2004 and 2007), which permits defining a more stable dataset for experiments; and (iii) the Flickr search API provides a stable service which supports full boolean queries, something which is essential to perform cross-language searches without direct access to the index.

For 2008, our primary goal was harvesting a large search log of users performing multilingual searches on the Flickr database. Rather than recruiting users (which inevitably leads to small populations), we wanted to publicize the task and attract as many users as possible from all around the world, and engage them with search. To reach this goal, we needed to observe some restrictions:

- The search task should be clear and simple, requiring no a-priori training or reading for the casual user.
- The search task should be engaging and addictive. Making it an online game - with a rank of users - helps achieve that, with the rank providing a clear indication of success.

- It should have an adaptive level of difficulty to prevent novice users from being discouraged, and to prevent advanced users from being unchallenged.
- The task should be naturally multilingual.

We decided to adopt a known-item retrieval search task: the user is given a raw (unnanotated) image and the goal is to find the image again in the Flickr database, using a multilingual search interface provided by iCLEF organizers. The user does not know in advance in which languages the image is annotated; therefore searching in multiple languages is essential to get optimal results.

The task is organized as an online game: the more images found, the higher a user is ranked. In case of ties, the ranking will also depend on precision (number of images found / number of images attempted). At any time the user can see the "Hall of Fame" with a rank of all registered users.

Depending on the image, the source and target languages, this can be a very challenging task. To have an adaptive level of difficulty, we implemented a hints mechanism. At any time whilst searching, the user is allowed to quit the search (skip to next image) or ask for a hint. The first hint is always the target language (and therefore the search becomes mono or bilingual as opposed to multilingual). The rest of the hints are keywords used to annotate the image. Each image found scores 25 points, but for every hint requested, there is a penalty of 5 points.

Initially a five minute time limit per image was considered, but initial testing indicated that such a limitation was not natural and changed users' search behaviour. Therefore we decided to remove time restrictions from the task definition.

2.2 Search Interface

We designed the so-called *Flickling* interface to provide a basic cross-language search front-end to Flickr. Flickling is described in detail in [1]; here we will summarize its basic functionalities:

- User registration, which records the user's native language and language skills in each of the six European languages considered (EN, ES, IT, DE, NL, FR).
- Localization of the interface in all six languages.[1]
- Two search modes: mono and multilingual. The latter takes the query in one language and returns search results in up to six languages, by launching a full boolean query to the Flickr search API.
- Cross-language search is performed via term-to-term translations between six languages using free dictionaries (taken from: `http://xdxf.revdanica.com/down`).
- A term-to-term automatic translation facility which selects the best target translations according to (i) string similarity between the source and target words; (ii) presence of the candidate translation in the suggested terms offered by Flickr for the whole query; and (iii) user translation preferences.

[1] Thanks go to the CLEF groups at the U. of Amsterdam, U. of Hildesheim, ELDA and CNR for providing native translations of the interface texts.

- A query translation assistant that allows users to pick/remove translations, and add their own translations (which go into a personal dictionary). We did not provide back-translations to support this process, in order to study correlations between target language abilities (active, passive, none) and selection of translations.
- A query refinement assistant that allows users to refine or modify their query with terms suggested by Flickr and terms extracted from the image rank. When the term is in a foreign language, the assistant tries to display translations into the user's preferred language to facilitate feedback.
- Control of the game-like features of the task: user registration and user profiles, groups, ordering of images, recording of session logs and access to the hall of fame.
- Post-search questionnaires (launched after each image is found or failed) and final questionnaires (launched after the user has searched fifteen images, not necessarily at the end of the experience).

Figure 1 shows a snapshot of the search interface. Note that we did not intend to provide the best possible cross-language assistance to search the Flickr collection. As we wanted to focus on user behaviour - rather than on hypothesis testing for a particular interactive facility - our intention was to provide a standard, baseline interface that is not dependent on a particular approach to cross-language search assistance.

2.3 Participation in the Track

Participants in iCLEF2008 can essentially do two tasks: (1) analyse log files based on all participating users (which is the default option) and, (2) perform their own interactive experiments with the interface provided by the organizers. CLEF individuals will register in the interface as part of a team, so that a ranking of teams is produced in addition to a ranking of individuals.

Generation of search logs. Participants can mine data from the search session logs, for example looking for differences in search behaviour according to language skills, or correlations between search success and search strategies.

Interactive experiments. Participants can recruit their own users and conduct their own experiments with the interface. For instance, they could recruit a set of users with passive language abilities and another with active abilities in certain languages and, besides studying the search logs, they could perform observational studies on how they search, conduct interviews, etc. iCLEF organizers provided assistance with defining appropriate user groups and image lists, for example, within the common search interface. Besides these two options, and given the community spirit of iCLEF, we were open to groups having their own plans (e.g. testing their own interface designs) as long as they did not change the overall shared search task (known-item search on Flickr).

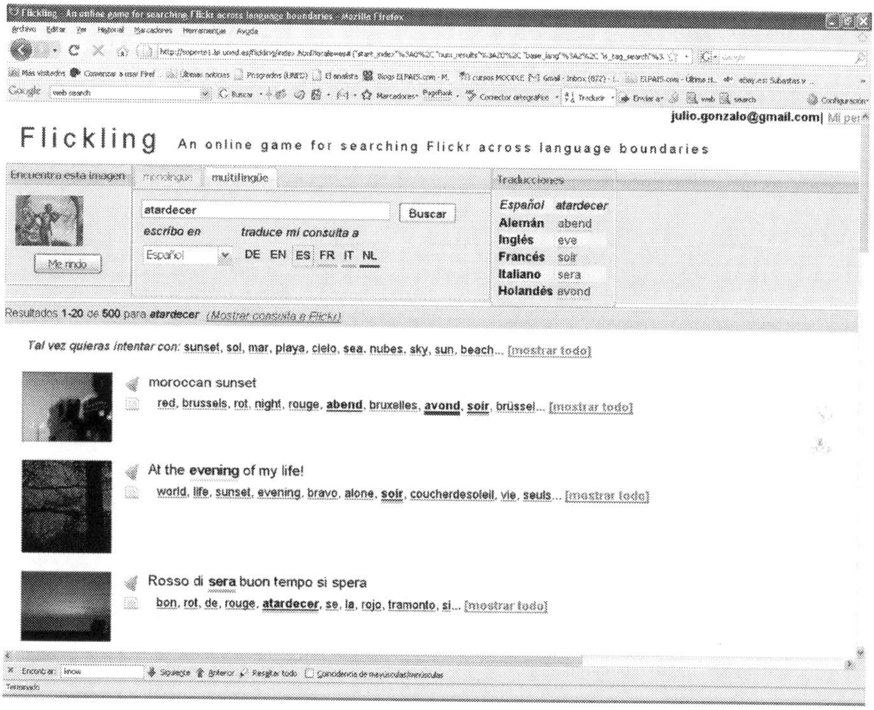

Fig. 1. The Flickling search interface used to harvest search logs

3 Dataset: Flickling Search Logs

Search logs were harvested from the Flickling search interface between the beginning of May and the 15th of June 2008 (see [1] for details on the content and syntax of the logs). In order to entice a large set of users, the "CLEF Flickr Challenge" was publicized in Information Access forums (e.g. the SIG-IR and CLEF lists), Flickr blogs and general photographic blogs. We made a special effort to engage the CLEF community in the experience, with the goal of getting researchers closer to the CLIR problem from a user's perspective. To achieve this goal, CLEF organizers agreed to award two prizes consisting of free registration for the workshop: one for the best individual searcher and one for the best scoring CLEF group.

Dissemination was successful: during the log harvesting period, the interface was visited by useres from 40 different countries from Europe, the Americas, Asia and Oceania (see Figure 2). More than 300 people registered (around 230 were active searchers) and 104 performed searches for at least 10 different images. Out of them, 18 users attempted all 103 images considered for the task. Apart from general users, the group affiliation revealed at least three user profiles: researchers in Information Retrieval, linguistics students (most from the University of Padova) and photography fans (many entering from a Spanish blog specialized in photography, `dzoom.org.es`).

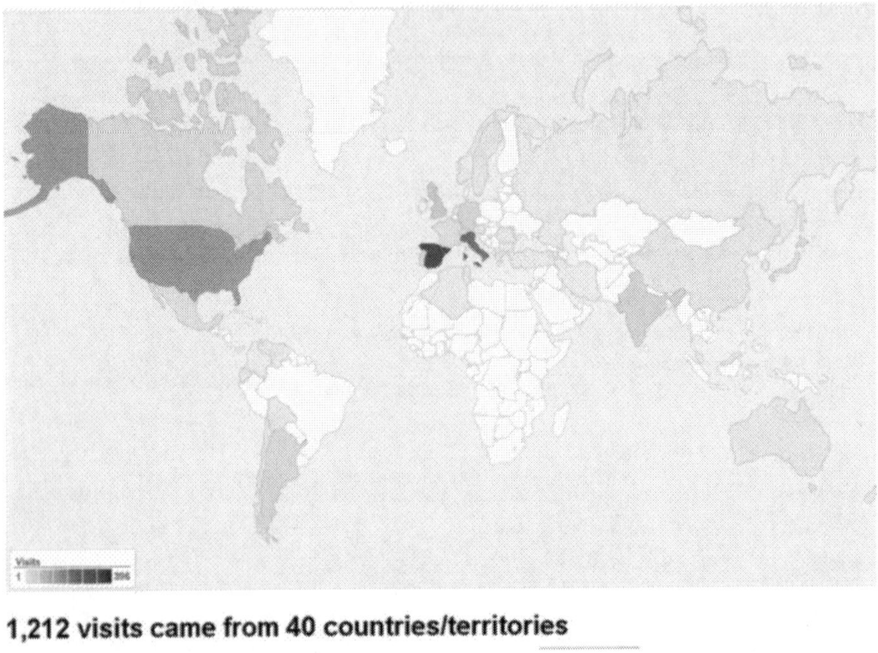

Fig. 2. Geographic distribution of accesses in the search logs

Profiles of user's language skills were very diverse, with a wide range of native and second language abilities. There was a total of 5101 complete search sessions (i.e. a user starts searching for an image and either finds the image or gives up), out of which the image was annotated in an active language (for the user) in 2809 cases, in an unknown language in 1566 cases, and in a passive language (when the user can partially read but cannot write) in 726 cases. Note that, even when the image is annotated in an active language for the user, this is not known by the user a-priori, and therefore the search behaviour is equally multilingual.

On average each search session included around four queries launched in the monolingual search mode, and four queries in the multilingual search mode. Overall, it was possible to collect a large controlled multilingual search log, which includes both search behaviour (interactions with the system) and users' subjective impressions of the system (via questionnaires). This offers a rich source of information for helping to understand multilingual search characteristics from a user's perspective. A reusable data source has been produced for the first time since iCLEF first began.

4 Participation and Findings

Six groups submitted results for this year's interactive track: Universidad Nacional de Educación a Distancia (UNED), the Swedish Institute of Computer

Science (SICS), Manchester Metropolitan University (MMU), the University of Padua (UNIPD), University of Westminster, and the Indian Institute of Information Technology Hyderabad (IIIT-H). Studies ranged from exploring the effects of searcher background on results, studying how much attention searchers pay to language phenomena when searching images, how the effect of constraining the session might influence results, and examining logs to find evidence of user confidence in the search process.

UNED examined the effects of searcher competence in the target language and system learning effects, studying the logs and examining user responses to the questionnaires given to users at the completion of each completed or aborted task [5]. Analyses showed that when users had competence in the target language, their success at searching was higher; with passive knowledge user interaction showed similar success to those with active competence, but requiring more interactions with the system. Finally, users with no competence in the target language found less images and with a higher cognitive effort.

SICS studied the logs to find evidence of different levels of user confidence and competence in the behaviour exhibited and recorded in them [4]. The main conclusion is that to study these effects, the task design must be formulated to better capture and distinguish the difference between user decisions to terminate or continue a search.

MMU studied how users considered language and cross-linguistic issues during a session and how they switched between the cross-lingual and mono-lingual interfaces. This was done through think-aloud protocols, observation, and interviews of users engaged in search tasks [3]. Their main finding is that their users did not make significant use of the cross-lingual functionalities of the system, nor did they think about language aspects when searching for an image. This again speaks to the necessity of careful design for a task which will better capture the complexity of a cross-lingual search task.

UNIPD also recruited users to be observed on-site, and constrained the task (in its first cycle) to require users to make a rapid decision of whether an image was relevant or not [2]. One of the conclusions pertinent to future cycles of the task is that the users are likely to be satisfied with a *similar* image, not necessarily needing the exact item designated correct by the game design. Designing future tasks might be well served in attempting to capture this usage-oriented aspect of user satisfaction.

The submission from the University of Westminster (UK) explored user's interaction with the facility provided by Flickling to add user-specific translation terms [6]. By exploring the user's perceived language skills and usage of the personal dictionary feature, experiments demonstrated that even with modest language skills, users were interacting with and using the dictionary-edit feature. Results point towards further study of collaborative translation in the global web space.

Finally, the group from IIIT-H studied the effects of language skills on user's search behaviour [7]. Results showed that user's typically started with a monolingual interface (the majority of users having Spanish as their mother tongue) but soon moved to the cross-language interface, making use of facilities such as search hints when searching in languages other than their mother tongue. Overall, users mainly searched in their native language in which they felt more confident than searching (far less) in their passive languages. An interesting result was that, on average, users found more images successfully using the monolingual interface.

5 Conclusions

This paper has described a radical approach to studying user-orientated aspects of cross-language image search: iCLEF2008 has attempted to run a large-scale interactive experiment as an online game to generate log files for further study. A default multilingual information access system developed by the organizers was provided to participants to lower the cost of entry and generate search logs recording user's interaction with the system and qualitative feedback about the search tasks and system (through online questionnaires). Although this initial attempt at encouraging greater participation in user-orientated evaluation resulted in submissions from 6 groups (the largest number of groups submitting to iCLEF in recent years), however a much larger number of users did make use of the system during the period of data collection showing potential for further experiments in 2009. The results of the experiments will be used to inform more usage-oriented tasks for future cycles; the methodology has proven to be lightweight and should be helpful for future participants; the logs will be a sustainable and reusable resource for future user-orientated studies of cross-language search behaviour.

Acknowledgements

This work has been partially supported by the Regional Government of Madrid under the MAVIR Research Network (S-0505/TIC-0267) and the Spanish Government under project Text-Mess (TIN2006-15265-C06-02).

References

1. Peinado, V., Artiles, J., Gonzalo, J., Barker, E., López-Ostenero, F.: FlickLing: a multilingual search interface for Flickr. In: CLEF 2008 Workshop Notes, Aarhus, Denmark, September 17-19 (2008)
2. Di Nunzio, G.M.: "Interactive" Undergraduate Students: UNIPD at iCLEF 2008. In: Peters, C., et al. (eds.) CLEF 2008. LNCS, vol. 5706. Springer, Heidelberg (2009)
3. Vassilakaki, E., Johnson, F., Hartley, R.J., Randall, D.: A Study of Users' Image Seeking Behaviour in Flickling. In: Peters, C., et al. (eds.) CLEF 2008. LNCS, vol. 5706, pp. 251–259. Springer, Heidelberg (2009)
4. Karlgren, J.: SICS at iCLEF 2008: User confidence and satisfaction inferred from iCLEF logs. In: Peters, C., et al. (eds.) CLEF 2008. LNCS, vol. 5706, pp. 260–261. Springer, Heidelberg (2009)

5. Peinado, V., Gonzalo, J., Artiles, J., López-Ostenero, F.: UNED at iCLEF 2008: Analysis of a large log of multilingual image searches in Flickr. In: Peters, C., et al. (eds.) CLEF 2008. LNCS, vol. 5706, pp. 236–242. Springer, Heidelberg (2009)
6. Tanase, D.I., Kapetanios, E.: Evaluating the impact of personal dictionaries for cross-language information retrieval of socially annotated images. In: CLEF 2008 Workshop Notes, Aarhus, Denmark, September 17-19 (2008)
7. Vundavalli, S.: Mining the behaviour of users in a multilingual information access task. In: CLEF 2008 Workshop Notes, Aarhus, Denmark, September 17-19 (2008)
8. Clough, P., Gonzalo, J., Kargren, J., Barker, E., Artiles, J., Peinado, V.: Large-Scale Interactive Evaluation of Multilingual Information Access Systems - the iCLEF Flickr Challenge. In: Proceedings of the Workshop on novel methodologies for evaluation in information retrieval, 30th European Conference on Information Retrieval, Glasgow, 30th March-3rd April (2008)

Log Analysis of Multilingual Image Searches in Flickr

Víctor Peinado, Julio Gonzalo, Javier Artiles, and Fernando López-Ostenero

NLP & IR Group at UNED, ETSI Informática UNED
c/ Juan del Rosal, 16. E-28040 Madrid, Spain
{victor,julio}@lsi.uned.es, javart@gmail.com, flopez@lsi.uned.es
http://nlp.uned.es

Abstract. In this paper, we summarize our analysis over the logs of multilingual image searches in Flickr provided to iCLEF 2008 participants. We have studied: a) correlations between the language skills of searchers in the target language and other session parameters, such as success (was the image found?), number of query refinements, etc.; b) usage of specific cross-language search facilities; and c) users perceptions on the task (questionnaire analysis).

We have studied 4,302 complete search sessions (searcher/target image pairs) from the logs provided by the organization. Our analysis shows that when users have active competence in the target language, their success rate is 18% higher than if they do not know the language at all. If the user has passive competence of the language (i.e. can partially understand texts but cannot make queries), the success rate equals those with active competence, but at the expense of executing more interactions with the system.

Finally, the usage of specific cross-language facilities (such as refining translations offered by the system) is low, but significantly higher than standard relevance feedback facilities, and is perceived as useful by searchers.

1 Introduction

In this paper, we summarize our analysis over the logs of multilingual image searches in Flickr provided to iCLEF 2008 participants [1].

In this search log, every session consists of a searcher (a registered user with a profile that includes her native language and her proficiency in English, Spanish, Italian, German, Dutch and French) and a target image (from the Flickr image database, annotated in one or more of those six languages). When the session starts, the user does not know in which language(s) the image is annotated. The interface provides facilities to perform queries simultaneously in up to six languages (via dictionary translation of query terms), to provide controlled relevance feedback (clicking on suggested terms and tags from the images found) and to refine the translations provided by the system (changing the selection of the system or adding new translations to a personal dictionary). The task is, therefore, a multilingual known-item retrieval task. If the user gives up, she can

C. Peters et al. (Eds.): CLEF 2008, LNCS 5706, pp. 236–242, 2009.

ask for hints; the first hint is always the target language (which turns the task into bilingual or monolingual search, depending on the user's language skills). The rest of the hints are keywords used to annotate the image, which is aimed at preventing users from being discouraged with difficult images.

The log consists of more than 5,000 search sessions by more than 200 users with a wide range of skills in the interface languages, coming from four continents. The size of this corpus permits studying the behavior of users in a multilingual search scenario at a scale that had not been possible before.

The UNED team has focused on studying: a) correlations between the language skills of searchers in the target language and other session parameters, such as success (was the image found?), number of query refinements, etc.; b) usage of specific cross-language search facilities; and c) users perceptions on the task (questionnaire analysis).

The structure of the rest of the paper is as follows: Section 2 describes the process performed to regularize the logs and characterize each user's search sessions. In Section 3 we search for correlations between language skills of searchers and other parameters of the search sessions. In Section 4 we report on other aspects of our study, focusing on the usage of cross-lingual refinement facilities and users' perceptions on the task. Finally, in Section 5 we draw some general conclusions.

2 Log Processing and Characterization of the Search Sessions

We have processed the logs provided by the iCLEF organization in order to identify and characterize search sessions. A search session starts when the user is given a target image and finishes when the user either finds the image or gives up and stops searching. In the meantime, the user may log out and log in (even several times) and, essentially, interact with the interface: launch queries, explore the rank of results, ask for hints, read descriptions associated to images, manipulate the translations suggested by the system and therefore improve her personal dictionary, etc.

Once search sessions are identified and open sessions are filtered out (those that were active when the log was produced or those that died because of user inactivity for more than 24 hours), we retained 5,101 search sessions. However, in the following analyses, we are focusing only on the fifteen first search sessions performed for those users who have searched for at least fifteen images. Thus, we are considering a stable population of 76 users and 4,302 search sessions.

We have processed the logs to provide a rich characterization of each session. The essential features are the user's profile (in particular her language skills), the use of the different interface facilities (including translation features), the session number (when was the image searched in the search history of the user), etc. We have also distinguished between the behavior before and after asking for the first hint, which is the language in which the image is annotated, because it represents the frontier between fully multilingual search (the image can be annotated in any of six languages) and bilingual or monolingual search.

See [3] for a comprehensive list of the features that we have extracted.

3 Analysis Considering Language Skills

In our first analysis we have divided search sessions in three groups, according to users' profile with respect to the annotation language of the target image: "active" is the group of sessions where the image was annotated in a language in which the user can read and write. Sessions in "passive" are those where the target language was partially understandable by the user, but the user could not make queries in that language (think, for instance, of French for most Spanish or Italian speakers). Finally "unknown" stands for images annotated in languages completely unfamiliar for the user. In our pool of sessions we found 2,768 for active, 622 for passive and 912 for unknown. These figures are large enough to reach quantitatively meaningful conclusions.

Table 1 shows average values for success rate and number of hints requested for each of these three groups. The most notable result is the degree of success (was the image found?) for each of the groups: active and passive speakers successfully found the image 84% and 83% of the times. Users with no competence in the annotation language obtained 69%, performing 18% worse. It is somehow surprising that users which only have a passive knowledge of the target language perform as well as those with active knowledge, because the first group must necessarily use the translation capabilities of the system to express their query. The unknown group performs only 18% worse, which reveals a consistent difference but not a large gap. Note that the translation capabilities of the interface were not optimal: they used only freely available general-purpose dictionaries with some coverage gaps, and they were not tailored to the domain (the Flickr database).

Note that, as users could ask for hints, it can be the case that "passive" cases reach the same success as "active" ones because they simply ask for much more hints. This is not the case: the average number of hints hardly varies between the three groups, ranging from a minimum of 2.14 hints per session to a maximum of 2.42.

Table 2 shows what we have called the cognitive effort of our users, i.e., interactions with the interface such as the average number of typed queries, the number of times that the user explored the ranking beyond the first page of results (containing 20 items) and the use of the relevance feedback (consisting of related terms provided by Flickr and the tags associated to the ranking images, see Section 4).

In general, it seems that there is a clear ordering between active, passive and unknown sessions: active sessions need less interactions, passive more, and

Table 1. User's behavior according to language skills: average success rate and hints requested

competence	success rate	# hints requested
active	84%	2.14
passive	83%	2.22
unknown	69%	2.42

Table 2. Cognitive effort according to language skills: typed queries, ranking exploration and use of relevance feedback

competence	# typed queries		ranking exploration		relevance feedback	
	mono	multi	mono	multi	mono	multi
active	3.79	3.43	2.32	2.39	0.04	0.03
passive	4.02	3.68	3.02	2.76	0.05	0.02
unknown	3.51	4.15	2.34	3.33	0.07	0.09

unknown even more, specially in a multilingual environment. For instance, the average number of queries posed in the multilingual search mode is 3.43 for active sessions, 3.68 for passive sessions, and 4.15 for unknown sessions. Therefore, passive sessions achieve similar success than active sessions, but with a higher effort. Unknown sessions have even higher effort, but still with a 18% loss in effectiveness. As far as the use of relevance feedback is concerned, the general tendency continues: unknown users tend to perform more interactions.

In some features this tendency is broken, as in ranking exploration: passive sessions tend to explore the rank further than unknown sessions, perhaps because the textual information in the images can be more easily used to do relevance feedback.

Notice that we have not included search time in these tables. Although the logs provide time stamps, we have discarded them because there is no way of knowing when the user was actively engaged in the task or performing some other task while the session remained open. Therefore, time is less reliable as an activity indicator than the number of interactions with the system.

4 Usage of Specific Cross-Lingual Refinement Facilities

FlickLing[2] search interface provides some functionalities which take advantage of some of the Flickr's services[1]. Flickr services suggest new terms related to a given query and FlickLing allows to use these terms to launch a new query or to refine a previous one.

This functionality was used by a small percentage of users, as shown in the last two columns of Table 2. This is in accordance with the common place that relevance feedback facilities are rarely used in search engines (at least in non-specialized search scenarios), even if they can provide more search effectiveness.

But it is interesting to note that if we compare these standard relevance feedback mechanisms with the usage of the personal dictionary, we can point out a positive indication of their usefulness at certain stages of the search process. Refining the translations provided by the system and adding new preferred translation to the personal dictionary (referred in Table 3 as dictionary manipulations) range from 0.10 to 0.20. These figures may seem quite low in absolute terms, but they double the use of relevance feedback in all users' profiles.

[1] See `http://www.flickr.com/services/api` for further information about Flickr API.

Table 3. Usage of personal dictionary: modifications of the personal dictionary and query terms affected by these modifications

competence	dictionary manipulations	query terms modified
active	0.10	0.06
passive	0.11	0.06
unknown	0.20	0.13

4.1 User Perceptions on the Task

Although the primary source of information are the activity logs of the users, the logs also contains the answers to two types of questionnaires: one is presented after each session (in two forms: one if the search failed and another one if the search succeeded), and another one is presented only once, when the user has performed fifteen search sessions (and therefore has a rather complete overall impression of the task). In this paper, we focused on the former ones.

The overall questionnaires collect different aspects of the task. Let's comment some of the most interesting results about the challeging aspect of the task itself, the usefulness of the interface facilities and the strategies used by users to find the correct translations for their queries:

Which, In Your Opinion, Are The Most Challenging Aspects Of The Task? Notably, when we restrict this question to experienced users, which has searched at least for fifteen images, over 85% of the users agree or strongly agree that "Selecting/finding appropriate translations for the terms in my query" is the most challenging aspect of the task.

Which Interface Facilities Did You Find Most Useful? Cross-language facilities —automatic translation of query terms and possibility of improving the translations chosen by the system— are much more valued (more than 70% of support) than standard feedback facilities —namely, the assistant to select new terms from the tags associated to the results and the additional query terms suggested by Flickr—, as shown in Figure 1. This seems to be in accordance with the proportional usage of these two kinds of facilities, although we must remark that the actual usage of those facilities is lower than what would be expected from the questionnaire.

Which Interface Facilities Did You Miss? Three facilities have an agreement rate (agree or strongly agree) above 70%: "a system able to select the translations for my query better", "The classification of search results in different tabs according to the image caption language", and "the possibility to search according to the visual features of the image". Other choices have slightly lower agreement rates: "an advanced search mode giving more control on how Flickr is queried", "bilingual dictionaries with a better coverage", and "more support to decide what the possible translations mean and therefore which ones are more appropriate". The least valued option (yet with an agreement rate above 50%) is

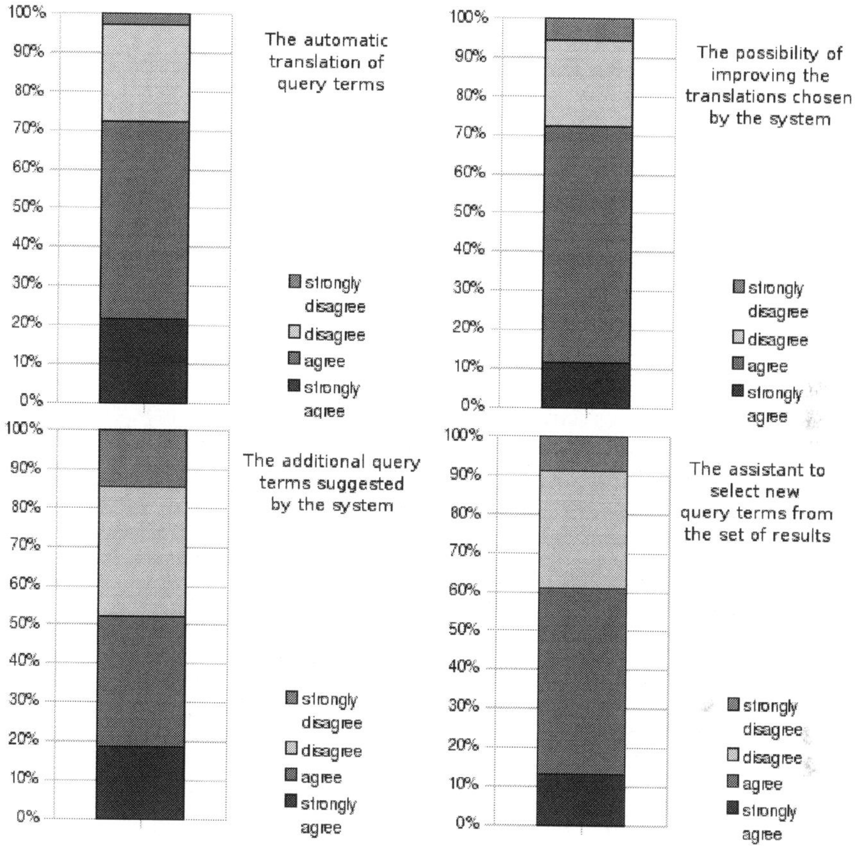

Fig. 1. Which interface facilities did you find most useful?

"detection and translation of multi-word expressions", perhaps due to the nature of the task and the annotations (tags are frequently single words).

It is difficult to extract conclusions from the answers to this question, apart from the fact that users seem to appreciate all features that can seemingly improve the search experience, even if interactive features are not frequently used in practice.

How Did You Select/Find The Best Translations For Your Query Terms? By far the most popular answer is "using my knowledge of target languages whenever possible", which was frequently used by 60% of the users and sometimes by another 30%. In contrast, less than 10% frequently "did not pay attention to the translations. I just trusted the system". This is in sharp contrast with the average behavior of users, which rarely modify the translations chosen by the system, and deserves further investigation. Finally, "using additional dictionaries and other online sources" is used frequently by less than 20% of the users, and "sometimes" by another 20%.

5 Conclusions

The search logs under study in the iCLEF 2008 task provide a more solid base to extract conclusions about the behavior of users in multilingual search scenarios than most previous experiments, which were mostly performed under laboratory conditions and therefore more restricted in size.

At UNED we have analyzed 4,302 complete search sessions (searcher/target image pairs) in the logs provided by the organization. Our analysis shows that when users have active competence in the target language, their success rate is 18% higher than if they do not know the language at all. If the user has passive competence of the language (i.e. can partially understand texts but cannot make queries), the success rate equals those with active competence, but at the expense of executing more interactions with the system.

In general terms, users with no active competence or no competence at all in the annotation language of the image need to perform more interactions with the systems, which means more cognitive effort.

Finally, the perception of experience users about cross-language retrieval interactive facilities is very positive, in spite of the fact that they are not frequently used. This is an indication that advanced search features —in this case, manipulation of translations offered by the system— might not be used frequently, but when they are used they become critical for the success of the task. A consequence is that query translation assistance should be hidden in the default settings of a cross-language search interface, but should be possible to invoke it for certain advanced users or specific search situations.

Acknowledgements

This work has been partially supported by the Regional Government of Madrid under the MAVIR Research Network (S-0505/TIC-0267) and the Spanish Government under project Text-Mess (TIN2006-15265-C06-02).

References

1. Gonzalo, J., Clough, P., Karlgren, J.: Overview of iCLEF 2008: search log analysis for Multilingual Image Retrieval. In: Peters, C., et al. (eds.) CLEF 2008. LNCS, vol. 5706, pp. 227–235. Springer, Heidelberg (2009)
2. Peinado, V., Artiles, J., Gonzalo, J., Barker, E., López-Ostenero, F.: FlickLing: a multilingual search interface for Flickr. In: Working Notes for the CLEF 2008 Workshop, Aarhus, Denmark, September 17-19 (2008)
3. Peinado, V., Gonzalo, J., Artiles, J., López-Ostenero, F.: UNED at iCLEF 2008: Analysis of a large log of multilingual image searches in Flickr. In: Working Notes for the CLEF 2008 Workshop, Aarhus, Denmark, September 17-19 (2008)

Cross-Lingual Image Retrieval Interactions Based on a Game Competition

Giorgio Maria Di Nunzio

Department of Information Engineering – University of Padua
Via Gradenigo, 6/a – 35131 Padova – Italy
dinunzio@dei.unipd.it

Abstract. This is the first year of participation of the University of Padua in the interactive CLEF track. A group of students of Linguistics at the Faculty of Humanities were asked to participate in the experiment. The interaction of the user with a cross-lingual system, the solutions they find for a given task, and the tools that a system should provide in order to assist the user in the task are studied by means of questionnaire analysis together with some log analysis. Interesting insights and results emerged and can be summarized with the following points: the hardest obstacle in finding the given image are the size of the set of images retrieved, the difficulty in describing the image, and finding suitable keywords in one or more languages.

1 Introduction

The CLEF interactive track (iCLEF)[1] has been conducted since 2001 in the context of the Cross Language Evaluation Forum (CLEF)[2] with the aim of studying the interaction with a multilingual information retrieval system from a user point of view. Since 2005, the iCLEF has shifted the focus from the search for textual documents to the search of images [1]. This year, the iCLEF 2008 [2] is focused on the known-item image retrieval based on the Flickr[3] database of images using the Flickling [3] the search interface. The search interface provided by iCLEF organizers is a basic cross-language retrieval system for the Flickr image database, presented as an online game: the user is given an image, and he must find it again without any a priori knowledge of the language (one or more) in which the image is annotated.

The University of Padua (UNIPD) participated in this track for the first time. The aim of this study is to analyze and observe a group of students with peculiar linguistic skills and compare this group to the rest of the participants to understand whether the knowledge of one or more languages is a significant feature during a cross-lingual task and how this difference impacts on the use of the search tools. Suggestions and preferences of the users during the search task are

[1] http://nlp.uned.es/iCLEF/
[2] http://www.clef-campaign.org/
[3] http://www.flickr.com/

C. Peters et al. (Eds.): CLEF 2008, LNCS 5706, pp. 243–250, 2009.

gathered directly by means of questionnaires or indirectly by the study of action logs [4]. In order to have a large number of users, students of Linguistics of the Faculty of Humanities were asked to participate in the game. Participation was not mandatory; nevertheless, an incentive was given in order to convince students to play: extra points in their marks as a reward of their effort. Availability of these students was important for the aim of this study since these are users who use different languages every day.

The paper is organized as follows: Section 2 describes the group of students which have been asked to participate in the experiment. In Section 3 the analysis of the questionnaires gathered on-line and off-line is presented. Section 4 shows the analysis carried out on the logs of the Flickling system. Final remarks and comments are presented in Section 5.

2 Students and Language Skills

The users involved are students from the Faculty of Humanities of UNIPD, of the courses "Linguistics and Modern Cultures" and "Languages for Cultural Mediation". During the first year of their career, students have to attend a basic course on Computers and Computer Science, and in the context of this class they were asked to participate in this game. They were free to participate and interact as long as they wished; however, an incentive (extra points for their final marks) was offered in order to persuade them to use some more of their spare time. Given this particular situation the students were asked not to cheat and follow this simple rule: for the first game, they had to register under the group of "University of Padua - Linguistics"; if they wanted to play again and improve their score, they had to register under the group of "University of Padua 2 - Linguistics". Therefore, results of this second group are highly biased by the fact that these students had already played and knew many of the keywords already used to find the pictures. This second group will not be considered in the analysis.

The number of students of this two courses was around 250, students who regularly attended the lessons were around 120. At the end the number of students who participated in the experiment was 60 which was surprisingly high. Consider also that the students are not familiar with search engines, and only two of them knew Flickr before the game started.

The profile of these students are important for this study because of their language skills. They are probably good in evaluating the translations and the suggestions given by the system. We can roughly divide the students in the following overlapping groups: the main mother tongue language is Italian; the majority of students study English and/or Spanish; German, French and Portuguese are usually the second, or third, language chosen for their studies; a minority of students study eastern country languages, such as Russian, Greek, or Slavic languages. It is also important to underline that there were some foreign students among those who participated in the game.

3 Questionnaire Analysis

During the Flickling game, there are questionnaires that users have to fill-in. Questionnaires are shown: (i) at the end of the search of each image. There are two types of questionnaire: the found image questionnaire (when the image is found), and the give up questionnaire (when the user decides to skip the image because it was not possible to find it); (ii) after a certain number of images: the overall questionnaire, which asks general questions about the whole cross-lingual task, the interface, and possible improvements.

3.1 A Questionnaire for Each Image

There are two types of questionnaires which are shown at the end of the search for an image: the image found questionnaire, and the give up questionnaire. There are six questions for the first questionnaire, and five questions for the second one. The analysis of these two questionnaire aims to provide insights about the differences between the group of UNIPD and the other users.

For the found image questionnaire, in the logs there are 1,607 records for UNIPD and 1,993 records for the others. The main question of this questionnaire was "What problems did you encounter while searching for this image?". The distribution of the answers is similar for both groups, however there are some interesting points to highlight. In general, when an image is found in most of the cases the task resulted easy (55% for UNIPD and 36% for the others). However, there is a significant amount of users who thought that the task was hard because of the size of the image set (10% of the answers of UNIPD confirm this, while it is 20% for the others). There is also the obstacle due to the difficulty of describing the image for many of the users. This problem is also connected to the knowledge of the language in which the image is described and the problems related to the translation of the query. Not surprisingly, this last point is less important for the students of UNIPD (who study languages) compared to common users.

For the give up questionnaire, there are 479 records for UNIPD and 516 records for the others. The main general question was "Why are you giving up on this image?". Two main problems are encountered by both groups: the size of the set of the images retrieved (42% fir UNIPD, while 30% for the others); the difficulty of finding the suitable keywords for the image (38% for UNIPD and 52% for the others).

3.2 Overall Questionnaire

The overall questionnaire, which was presented to the users only after completing 15 searches, consists of 27 single-choice answers plus 2 open questions. The analysis presented here compares the answers of the UNIPD group the answer of the rest of the participants who filled-in the questionnaires. For UNIPD, the questionnaires were gathered both during the on-line game and off-line during the final exams, one month after the end of the game. In particular: 27 students filled in the on-line questionnaire; 17 students filled in the questionnaire on paper during the exams.

Table 1. Questions with significant statistical difference between the two groups

Question 3D	strongly agree	agree	disagree	strongly disagree
UNIPD	17	24	2	1
Others	9	17	8	2
Question 6C	strongly agree	agree	disagree	strongly disagree
UNIPD	6	26	11	1
Others	14	15	5	2
Question 8E	strongly agree	agree	disagree	strongly disagree
UNIPD	7	22	15	0
Others	18	13	4	1

It has to be noted that the students who filled in the quetionnaire on paper are not all the same students of the on-line questionnaire. It was not possible to know exactly who did what (some students even forgot if they had done this questionnaire), but we can roughly say that the overlap between the two groups is less than 10 people (which means that less than 10 people filled in both questionnaires). The number of questionnaires filled in by other users is 36.

In the following, we analyze the questions that presented a statistical significant difference between the two groups, UNIPD and the rest of the participants, using a t-test significance level at 5%. Results are summarized in Table1. A first result is that the search task was considered interesting for UNIPD students while less interesting for the others (Question 3D). A second important point is that the most challenging aspects of the task is handling multiple target languages at the same time (Question 6D). Finally, the possibility to search according to visual features of the images (search images that look like this, search only B/W images, search only for dark images, etc.) is more important for a general user compared to the students of UNIPD (Question 8E).

4 Log Analysis

The logs made available for studying the actions of each user were released as a text file. Each row of the log file contains either an action of the user or an action of the logging system. The log goes from April 24th 2008 until June 16th 2008 for a total of 1,483,806 recorded actions. For the purpose of the analyisis and for a more convenient management, this file was loaded into a table of a PostgreSQL[4] database. The records were also cleaned, in the sense that some of the actions recorded were not useful for the analysis. A detailed list of actions performed on the log can be found in [5].

The log file contains also the scores that each user earns when an image is found. In fact, during the game, users can earn points if they find the image, the amount of points earned depends on how many hints he asks. If the user finds the image without using any hint, 25 points are earned otherwise the user

[4] `http://www.postgresql.org/`

Table 2. Total score for the best participating group with more than ten participants, the average score per user, the number of participants per group

Participant	Points	Average	Participants
Other Users	4,910	51	95
University of Padua - Linguistics	20,465	341	60
dZoom	7,370	184	40
UNED LSI	2,120	176	12

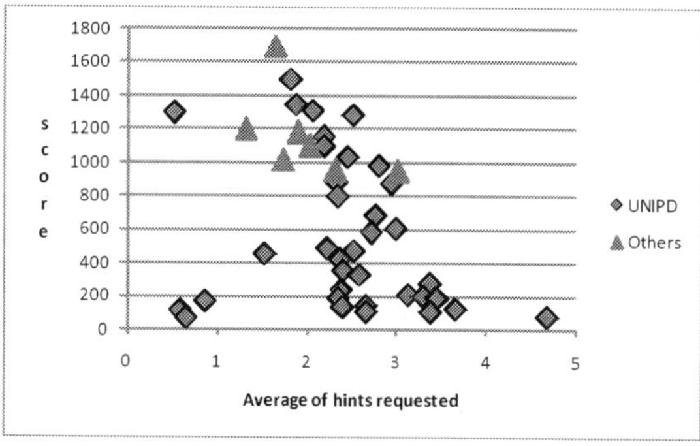

Fig. 1. Scores with respect to number of images viewed and found

can give up the search and go to the next image. Instead of giving up, the user can also ask for hints for the search, each hint costs 5 points (with 1 hint the score goes down to 20, with 2 hints to 15, and so on). At the end of each search, a questionnaire, presented in Section 3, is shown to the user to ask him how easy/hard it was to find (or not find) the image.

In Table 2 the list of the best participants is shown, ordered by the number of users per group (last column), with the respective total score and the average score per user. The UNIPD group had the highest total score, and one of the highest average scores per user. In the following sections, the scores of the UNIPD users are studied in order to understand whether there are different in the strategies among users, how many hints have been requested, how many times a cross-lingual search has been performed and so on.

Found or Skipped? There is in general a positive correlation between the scores and the number of images (the more images, the higher score), however there are differences which can be underlined, for example the scores versus the number of images found are more scattered and some differences among top scorers can be appreciated. This plot also tells that when a comparable number of images are found among different users, the fact that there may be differences in scores is that hints are used more frequently from one user than another.

Hints and Clues. In Figure 2 the highest scores of UNIPD and the other best participants are shown with respect to the average number of hints asked per

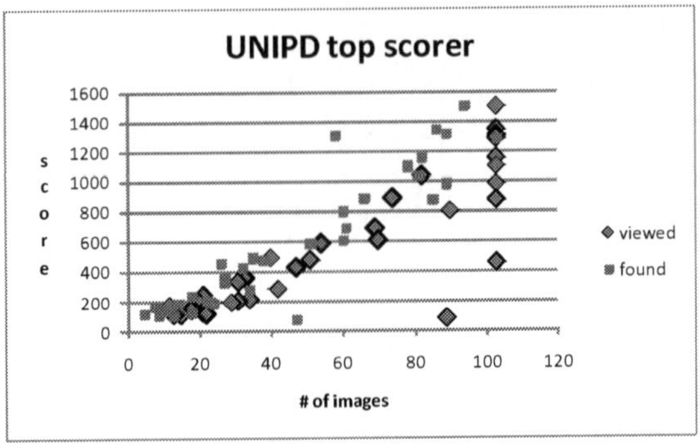

Fig. 2. Scores with respect of the number of hints requested on average

Table 3. UNIPD average mono- and cross-lingual searches of the top five scorers

userid	mono search	cross search
user_50	5.43	3.69
user_51	0.07	16.44
user_57	5.29	4.12
user_01	12.44	1.56
user_07	9.37	0.84

image. This plot shows that the best participants, in terms of scores, used on average about 2 hints per image. It is important to understand what the first hint is: when you ask for the first hint, the system tells you in what language the image you are searching for is described. This means that, on average users needed to know in advance the language of the description of the image before finding it.

Monolingual or Multilingual? In Table 3 and Figure 3 the average of monolingual and cross-lingual searches are shown for the top scorers. It is not easy to find regular patterns in the behavior of the users. On average, the top scorers use from 5 to 6 monolingual searches per image, and from 6 to 7 cross-lingual searches per image; however, "average" users are not common. In fact, it is more frequent the situation where a user prefers either to search in one language or to do a cross-lingual search. Performances, in terms of scores, seem not to be affected by the strategy chosen.

5 Comments and Final Remarks

From the analysis of the questionnaires and the system logs, interesting insights and results emerged and can be summarized as follows.

The hardest obstacle in finding an image was probably the size of the set of images retrieved. In both cases, image found or image skipped, a large number

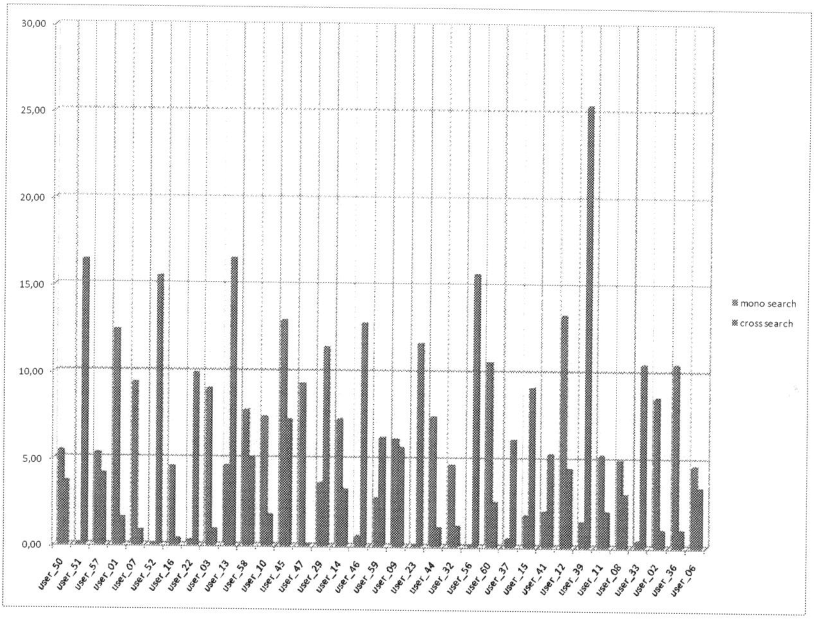

Fig. 3. Average numbers of monolingual and cross-lingual searches per image. Histogram based on the data of Table 3.

of users claimed that it was hard to find the image because there were too many images retrieved. However, from the direct interaction with the students and from some comments written in the questionnaires there were many cases in which the set of retrieved images contained the same "object" of the picture but not the exact picture. In real cases, you probably want to look for some image, not one in particular. The extra effort, which in our opinion is not realistic, that iCLEF participants has to do should be taken into consideration when doing the analysis of the data.

Another hard point was the difficulty in describing the image. Finding suitable keywords is indeed a hard task. It is also likely to have inappropriate tags for the image to find. A solution to this problem could be adding the possibility to search according to visual features of the images. However, the answers in the questionnaires were not so positive about this tool.

Users in general may find difficult to describe the image because the language in which is described is not known. As one could expect, this problem is less evident for UNIPD students. There is also the need of bilingual dictionaries with a better coverage, and a system able to give good suggestions for translating the keywords.

We also saw that there is not a strategy that outperforms the others. Using more monolingual searches than multilingual, a mix of the two, or prefer multilingual searches does not influence the final score. It would be interesting to study how users reformulate queries and whether the reformulation changes from one strategy to another. This was not part of the analysis and is currently future work.

One final comment is about the time for each search. Unfortunately, the calculation of the time was not accurate enough to do this type of analysis, During the observation of the students of UNIPD, the feeling is that users spend much more time in the search compared to a similar realistic situation. We tried to simulate a "real user scenario" with this idea in mind: a user does not spend more than two or three minutes per image and can ask at most one hint, and the user should not be influenced by the final score. This user, the author of the paper himself, is actually user_01 shown in all the previous tables and plots. This strategy easily brings to a low precision, many images are skipped, but in a real scenario the same user would have been satisfied by the search because usually a similar image (to the given image) is found. The time spent for each image is very low (probably the lowest compared to the other users), but in this case there is a bias to take into account when looking at the scores: the expertise in using search engines.

Acknowledgements

The work reported has been partially supported by the TrebleCLEF Coordination Action, as part of the Seventh Framework Programme of the European Commission, Theme ICT-1-4-1 Digital libraries and technology-enhanced learning (Contract 215231).

References

1. Artiles, J., Barker, E., Clough, P., Gonzalo, J., Karlgren, J., Peinado, V.: Large-scale interactive evaluation of multilingual information access systems - the iCLEF flickr challenge. In: Workshop on Novel Methodologies for Evaluation in Information Retrieval (30th European Conference for Information Retrieval (ECIR 2008), Glasgow (2008)
2. Gonzalo, J., Clough, P., Karlgren, J.: Overview of iCLEF 2008: Search Log Analysis for Multilingual Image Retrieval. In: Peters, C., et al. (eds.) CLEF 2008. LNCS, vol. 5706, pp. 227–235. Springer, Heidelberg (2009)
3. Peinado, V., Artiles, J., Gonzalo, J., Barker, E., López-Ostenero, F.: Flickling: a multilingual search interface for flickr. In: Borri, F., Peters, C. (eds.) Cross Language Evaluation Forum (CLEF 2008) Workshop Notes, Aarhus (September 2008)
4. Karlgren, J.: SICS at iCLEF 2008: User confidence and satisfaction inferred from iCLEF logs. In: Working Notes for the CLEF 2008 Workshop, Aarhus (September 2008), `http://www.clef-campaign.org/2008/working_notes/karlgren-paperCLEF2008.pdf`
5. Di Nunzio, G.M.: "Interactive" Undergraduate Students: UNIPD at iCLEF 2008. In: Working Notes for the CLEF 2008 Workshop, Aarhus (September 2008), `http://www.clef-campaign.org/2008/working_notes/CLEF2008WN-Contents.html`

A Study of Users' Image Seeking Behaviour in FlickLing

Evgenia Vassilakaki, Frances Johnson, Richard J. Hartley, and David Randall

Dept. Information & Communications, MMU
{evgenia.vassilakaki}@student.mmu.ac.uk,
{f.johnson,r.j.hartley,d.randall}@mmu.ac.uk
http://www.hlss.mmu.ac.uk/infocomms/

Abstract. This study aims to explore users' image seeking behaviour when searching for a known, non-annotated image in the FlickLing search interface provided by iCLEF2008 track. The main focus of our study was threefold: a) to identify the reasons that determined users' choice of a specific interface mode, b) to examine whether users were thinking about languages when searching for images and to what extent and c) to examine if used, how helpful the translations proved to be for finding the images. This study used questionnaires, retrospective thinking aloud, observation and interviews to meet its research questions.

Keywords: Multilingual Information Retrieval, User Behaviour, User Image Seeking Behaviour, Flickr, FlickLing, iCLEF.

1 Introduction

Cross Language Evaluation Forum (CLEF) is an annual evaluation campaign that aims to promote the development of monolingual and multilingual information retrieval systems for European Languages. The 2008 iCLEF track focuses both on acquiring a large set of search session logs for the participants to mine and on allowing participants to perform their own interactive experiments with the FlickLing interface provided and adopting the task predefined by the Organizers [1]. The aim of this study is to explore users' image seeking behaviour when searching and retrieving known, non-annotated images across languages in FlickLing. In particular, the research questions that will be addressed in this paper are: a) identify the reasons that determined our users' choice of a specific interface (monolingual/multilingual), b) examine if and/or to what extent users were thinking about languages when searching and retrieving images and c) examine if and/or to what extent users were paying attention to translations when searching and retrieving images.

The remainder of this paper is structured as follows: details concerning the different methods that we used in assembling the data and the way that the study was carried out are given in section 2. We provide an analysis of our findings and a discussion of them in sections 3 and 4 respectively. Finally, we conclude summarizing the different image seeking behaviours that our users exhibited while using FlickLing in section 5.

C. Peters et al. (Eds.): CLEF 2008, LNCS 5706, pp. 251–259, 2009.

2 Method

In this section, further details about the test object, the users, the task, the methods employed are presented.

Test Object. Our main reason of interest in participating in the iCLEF2008 Flickr challenge was to investigate the behaviour of users when asked to search and retrieve a known, non-annotated image across languages.

Users. The study was carried out with a sample of 10 users, three male and seven female, ranging in age from 20 to 40. In particular, seven were research postgraduate students, one taught postgraduate student, one lecturer and one member of MMU administrative staff. In addition, four of the users were English native speakers, two Greek, one German, one Spanish, one Arabic and one Luganda. Moreover, one of the users was monolingual, four stated knowledge of a language other than their native and five were multilingual.

The users were also asked to state their level of comprehension for the languages used in FlickLing but also any other additional language. In particular, from the six non English native speakers, two stated an Excellent knowledge of English, three Very Good and one Basic. In regards of German, three of the nine non-German native speakers stated a Basic knowledge of German. Four out of ten users stated knowledge of French, three of whom Basic and one Good. Concerning Italian two out of ten users stated knowledge of Italian language, Basic and Good respectively. Two out of ten stated a Basic knowledge of Dutch and finally, three out of nine non Spanish native speakers stated a Basic knowledge of Spanish.

All ten users have searched in the past for an image on the web. In particular, four stated that they "rarely" have, three "sometimes", two "very often" and one "often". In addition, nine out of ten stated that they have searched for an image on the web in a language other than their native and only one had not. In addition to the users' previous knowledge and experience with Flickr, nine out of ten users answered this question. Three of whom stated that "Yes" they have used Flickr in the past.

Task. Our users were asked to search for the first three (3) given images after login using all the features of the FlickLing interface. The users did not know in advance in which of the six languages (English, German, Dutch, Spanish, French, Italian) the image was described enforcing them to use both monolingual and multilingual modes to find the given image. The images presented to users were not controlled but given randomly from a set of 100 stored in the FlickLing database.

Retrospective Thinking Aloud. Retrospective thinking aloud is a widely used method for usability testing of software and interfaces. Its basic principle is to ask from potential users to complete a certain task with the testing object in question and to describe their thoughts and actions afterwards on the basis of a video recording their task performance [2]. This method focuses on peoples' cognitive processes after having completed a specific task. It is a method that enables the users and not the experts to point out the problems concerning the test object in the usability test. In this context, we used a premiere screen recorder in order to capture the users' search sessions in individual videos and a digital recorder for the retrospective thinking aloud.

Observation. The observation method was used to form specific questions regarding preselected research areas of the test object in an attempt to shed light on specific behaviours of the users on specific occasions. A form was created to assist the work of the facilitator at focusing on specific areas of interest and at the same time reflecting on users' behaviour.

Interviews. The last part of the study consisted of small scale individual interviews with every user after the completion of the retrospective thinking aloud. The interviews lasted no more than 10 minutes for every user. The questions asked varied according to user's answers to the questionnaire, search session, retrospective thinking aloud and the notes gathered throughout the study. The main goal of these questions was to clarify specific actions of the user's image seeking behaviour during the search session and expressions that the user used to describe what he/she was doing.

Experimental Procedure. The study was carried out in 10 individual sessions, at the same lab and each lasted from one to two hours approximately. During each session, users were given written instructions about the experimental procedure and the task itself. After that, users were asked to fill in the questionnaire on background knowledge, login and start completing the task (search for the first three images) while screen recording software was taping the computer screen. Having done that, users were asked to watch the recorded session and describe what they were doing and thinking in retrospect. Finally, a semi-structured interview lasting no more than 10 minutes was carried out with each user.

3 Findings

The analysis of data gathered focused on and will be presented according to the study's three research questions (RQ).

1st RQ: Identify the reasons that determined each time users' choice of a specific interface (monolingual/ multilingual). Out of the ten users, two used the monolingual interface and the rest switched between interfaces. The reasons our ten users gave for their behaviour in the thinking aloud process and interviews are stated below.

Only Monolingual Interface: Two out of ten users did not use at all the multilingual interface, even though they were given images to search in a language unknown to them. The first user, an English native speaker with basic knowledge in French when informed by the system that the image was annotated in French, stated: *"because I am not good in French...I realized that I am never going to find it...So, I decided to give up"*. When asked why the subject didn't use the multilingual interface, the subject answered: *"Because I did not trust my abilities with other languages, to be able to put the decent search words in...Because I did not know the keywords to search in other languages"*. As a final remark, the subject added: *"I was not confident with the languages"*.

The second user, a Luganda native speaker with no knowledge of French, stated: *"I went for the hint and it said that the image is described in French... Well, I thought, I do not speak French, I can't understand that"*. When asked why the subject did not use the multilingual interface, answered: *"If I knew how to use another language, then I could use the multilingual and access the same image in another language"*. When asked how the subject was planning to cope with the problem of searching a French annotated image on monolingual interface by using English keywords, the subject stated: *"I thought that the image was not available and all images should be described in English [as well]. So, I thought that it was inaccessible...that I could not get it"*. These two users were not feeling confident of their language skills and they were not used to searching for images in languages other than English. These users would use the multilingual interface only if they could speak the language in which they were searching and that one would be other than English.

Switching between Monolingual & Multilingual Interface: The remaining eight users switched between monolingual and multilingual interfaces in order to complete the given task. A variety of reasons to justify these actions were reported by the users during the retrospective thinking aloud process and interviews. In particular, users identified the following reasons why: *"In order to increase or decrease the number of results, depending on the results that I had on the beginning of my search"*, *"because I assumed that it would give me the highest possible number of relevant results in relation to my query"*, *"I am trying to find the right combination of keywords"*, *"if you know where the picture was from, or if you know the place then you can like recognize the language in which you can type in"*, *"I was looking to isolate words and translate them"*, *"Simple because I wasn't getting any of the results that I wanted"*, *"I tried to increase my chances of getting the image...I am widening my possibilities"*, *"I am just trying out the system"*, *"So, it was not there* [monolingual English], *I guess it was in other language"* and lastly: *"For me the problem was more kind of how to find where the image was from"*.

Also, hints played a significant part in users' choice of an interface. As stated by the users: *"I switched to monolingual because the hint told me that the image was described in English"* and *"I went to ask for a hint on language just in case because that seemed to save me lots of time"*.

Two users, both English native speakers, stayed on multilingual interface though after taking the hint, they both knew that their image was described in English. When asked why they haven't switched to monolingual, they stated consecutively: *"Because I did not think that would make any difference, because I was assuming that it is in English as well"* and *"Well, because I was there. I did not realize that...I thought, to be honest, I thought, it's not going to make that much difference really"*.

There were also some cases that although users were seemingly using a specific interface (monolingual or multilingual) they stated during retrospective thinking aloud process and confirmed afterward with the interviews that: *"I was not paying attention to the fact that it was multilingual. Maybe, I forgot about that and left it as it was"* and *"I was so focused on trying to see how to describe the image that I was not paying attention to the interface"*.

2nd RQ: to explore if and/or to what extent languages were forming the image seeking behaviour of our users. Two users out of ten used only the monolingual interface searching in English, although they knew that the images may not be described in English. From their reasons stated for this decision it appears that knowledge of and/or confidence in a language other than their native language is a determining factor. When asked if the subject was thinking about languages while searching, the Luganda native speaker answered: *"No. I did not...because when I am searching for images on the Internet, I normally get them in English because I imagine that...I guess it's a little bit of arrogance, I speak English and I imagine that images... That if you put them in Internet, they should have English tags"*.

The other eight users who were switching between monolingual and multilingual interfaces, can be divided in two groups: a) those who were thinking about languages and b) those for whom languages were not a variable when performing the given task. In particular, four of them stated that: *"Now, I made the relationship of country, Florida...I write them* [keywords] *in English"*, *"I had the feeling that the building which I recognized, was described in German"*, *"It was not within my results, so, I guessed that it is in other language"*, *"Because by looking at the tortoise had written on it...it was written in English. So, I assumed that it would be in English...And I was also thinking at this time, I wonder if it is English or not...because the child got a little blue and red hat and I was thinking, maybe the child is French... Yes, I changed into multilingual because I think that maybe it is French, with the outside possibilities that it might be Italian"* and lastly *"Well, that's probable a bit Anglo-centrism. You know, well, it is a picture in England"*.

On the other hand, the remaining four users when asked if they were thinking about languages during the task, they said that: *"To be honest, I was not thinking about languages...I did not consider it a variable that influences my results"*, *"I did not bother about languages...I did not really think about them...I did not focus on languages while performing my searches. Maybe, because I am not used to, is not widely used or maybe I am not using languages when retrieving information on the web"*, *"I was not taking languages under consideration when searching for the images"* and *"For me it was not a question of language...In my mind language was a very small factor in there* [FlickLing]. *It did not really play any important role"*.

3rd RQ: to examine the use of translations and the influence of translations on the users' information seeking behaviour. We are obliged at this point to exclude the two users who used only the monolingual interface and the four users who used the multilingual interface but with no thought to the translations. The remaining four users tried both the monolingual and multilingual interfaces driven by the need to identify the language of the image and the appropriate keywords to retrieve the given images. In particular the four users, when asked if they were paying attention to the translations, stated: *"Yes, but it did not translate anything"*, *"Yes, I did use them"*, *"Yes, at this point I am trying to figure out how this translation thing works"* and *"Yes, I was paying attention to the translations"*.

In addition, when users asked if they could judge the translations that were given to them, users answered: *"Overall, I had the feeling that the translations of the system were not that good...I switched to monolingual because the translations were not doing anything"*, *"I would trust the system to give me the right translations...I would have to for languages unknown to me"*, *"Because I went for the languages that I had a vague idea about and it did not tell me something that I did not really know"* and last *"I was not satisfied with the German translations because I can understand German...it's not the right word in German for a man. So, it should have been something else...in Dutch I don't know what the translation is, so, I had to accept it, whatever it is... Yes, I was satisfied* [Dutch translations] *because the computer knows the Dutch language better than I do... maybe that's not the best translation, so, I just had to accept it. There was anything that I could do about it really"*.

Finally, when the users were asked if the translations were helpful in terms of actually contributing to the retrieval of the image, users stated: *"Ok, I have got the translations but they are not doing anything to me...at the end, I totally disregard the translations"*, *"I think that I stop searching for translations, when I stop having much confidence that it was bringing me the right translations..."*, *"The words that I was trying to isolate like particular words like London, the different translations there were not coming up...and what it was saying, like gigante in Italian for giant, it told something that I already knew. So it was not isolating the words in the way that I wanted it to. It was just telling me the adjectives were, which I did not really need"* and finally *"At the end I was not paying attention to the translations, I was purely interested in finding the image as quickly as possible because once more I did not think the translations would necessarily help me"*.

4 Discussion

The evaluation of CLIR effectiveness often does not involve the end user largely because initial hypotheses often exclude their experiences. It follows that experimental success is not success in the users eyes. Much of what we set out to do was to assess the difference between experimental assumptions and a user perspective rather than provide a test of "success". On the one hand, it may be assumed that since the translation is automated the user has no role to play or possibly that the user has no interest in the translation, providing the system is effective. On the other hand, the non trivial challenges posed in the effort in designing realistic task scenarios, recruiting participants, analyzing large amounts of data to obtain user assessments or to observe search behaviour can be prohibitive. However, we take the view of Petrelli [3] that effective system design must be in accordance with the end users' needs and to best assist users involved in cross-language information retrieval we need to understand their behaviours and the search problems they face. Petrelli's study of users involved in CLIR presented a number of interesting findings. In particular, the users preferred the interface which hid the translation and that language knowledge and sight of the translation affected search behaviour.

Our study based on retrospective thinking aloud revealed a complex picture of the influence (or not) of language skills and confidence therein and of perceptions of the role of the multilingual interface, language and translations in image retrieval. Most revealing and of potential interest to future study of users of CLIR is the finding that less than half of our users appeared to consider identification of the language to be essential in retrieving the image. The majority either lacked confidence in using different languages or were so focused on finding the given images and completing the task that were not thinking at all about languages. Indicative of this was the comment *"...completing the task successfully. What was success for me? That you find the image. In any way I possible could. I was not focusing on translations...I thought my task is to find that image and I will do whatever I could to find it"*. Only four users attempted to identify the language of the images from its context (or from the Hint feature) and use it to their benefit. Of those for which languages played a significant role in the process of identifying keywords to search for the images, the translations were judged to be poor as either the translations were not coming up, were not corresponding to the users' keywords or were judged to be resulting in the retrieval of irrelevant results. As a consequence, users were losing interest and trust in translations resulting in no usage of them or not paying attention to them. Some were treating the multilingual interface as a translator, trying to isolate specific words, translate them on the multilingual interface and use the translations to retrieve the images on the monolingual interface. Another user stated that: *"I think I just saw it as a translation tool and not as an integrated translation thing that already was retrieving images. I did not really use it in this way because in my mind, it was only translating my keywords"*.

One of the initial aims of this study was to look in greater detail at how working with the translations affected search behaviour with regards to the actual search terms entered by users. Unfortunately, this study could not reach a conclusion because only four out of ten users used the translations and this was in a way not anticipated. Our study did reveal many reasons for non use of the multilingual interface, ranging from a lack of confidence in languages to a lack of trust in the system translations to a disregard for the need to search in other languages. However a further factor which may have influenced little use of translations is the interface design in presenting to the user how the multilingual interface worked or how the translations could be used to benefit the search. The feedback we obtained from the users suggested a variety of reasons for using the multilingual interface other than to address a recognized need to search for the image in another language. This may suggest that the purpose of the multilingual interface was not clear to the users as does the observation that it was used as a translator tool to run both the search terms and the translations in the search box.

On the whole, it would appear that users were so focused on completing the task, "obsessed" (as a user stated) of finding the images that even from the beginning of the task, they were not thinking really which interface they are going to use and for what reason. Even users who were concerned about languages, at the end of the task, also admitted that they were not paying much attention to the interfaces because they thought that it was not making any difference.

5 Conclusion

This study aimed at investigating the users' image seeking behaviour when retrieving a known, non-annotated image in Flickling. In particular, we identified the reasons why two of our users were choosing to search only on the monolingual interface and the eight switching between interfaces. We demonstrated that only four users were thinking about languages when trying to retrieve the given images while the rest of our users were more preoccupied with finding the images and completing "successfully" the task. Consequently, we showed that only these four users were paying attention to translations provided by the system. These stated that translations were not helpful or they were not making much difference in finding the given images since the results were irrelevant to what they were looking for.

This small study has also shown that if we are to ask whether a CLIR system should display query translations or not, then the answer is no. Our users were either not interested in the translations or found them to be poor. However taking the findings to such conclusion would be foolhardy given the complexity of the activity highlighted in the users' comments that they were so engaged in finding the image that language or translations played little or no part. Rather than reaching firm conclusions, this small study has suggested the need for more research into users' search behaviour with translations (and in image retrieval) if we are to design CLIR systems which will not place additional or unnecessary cognitive demands on the user and will support effective search behaviour and performance.

References

1. Gonzalo, J., Clough, P., Karlgren, J.: Overview of iCLEF 2008: Search Log Analysis for Multilingual Image Retrieval. In: Peters, C., et al. (eds.) CLEF 2008. LNCS, vol. 5706, pp. 227–235. Springer, Heidelberg (2009)

2. Haak, M., de Jong, M., Schellens, P.J.: Retrospective vs. Concurrent Think-aloud Protocols: Testing the Usability of an Online Library Catalogue. Behaviour & Information Technology 22, 339–351 (2003)
3. Petrelli, D., Beaulieu, M., Sanderson, M., Demetriou, G., Herring, P., Hansen, P.: Observing Users, Designing Clarity: A Case Study on the User-centered Design of a Cross-language Information Retrieval System. JASIST 55, 923–934 (2004)

SICS at iCLEF 2008: User Confidence and Satisfaction Tentatively Inferred from iCLEF Logs

Jussi Karlgren

The Swedish Institute of Computer Science

Abstract. This paper gives a brief description of some initial experiments performed at SICS using the interactive image search query logs provided for participants in the interactive track of CLEF. The SICS experiments attempt to establish whether user confidence and trust in results can be related to logged behaviour.

SICS has participated in this year's iCLEF cycle mainly with an eye on future experimental settings to work on measurement of searcher trust and confidence in the search process and its results, in keeping with previous experimental studies performed at SICS [2,1, e.g.]. SICS has used the Flickling interface [3] and the logs delivered by it to study how searcher actions can be interpreted as exponents of user confidence.

Variables under consideration for this purpose can be *indirect*, such as length of interaction, time spent on query formulation, and other measures which require non-trivial interpretation during the analysis phase. Other variables can be more *direct*, in that they more clearly indicate competence or confidence on the part of the searcher, such as observed edits and additions made to the user dictionaries by the searcher or the number of times a query is reformulated. A confident searcher can be assumed to be more likely to enter edits into the user dictionary and not to reformulate queries to the same extent. In Table 1 such measures are tabulated, with a distinction between sequences of actions that ultimately provide a successfully identified target image and sequences which terminate by the searcher giving up requesting another target. We find that while the differences given are statistically significant by Mann Whitney's U, they are less sizeable than might have been expected, and in some cases do not conform to expectation as to their direction: one would expect unsuccessful sequences use more hints than successful sequences, e.g. Better measures of user confidence in their actions should be measurable somehow, but these logs give little purchase for this type of analysis.

For next year's cycle we plan to investigate whether user actions can be logged to include their sense of *satisfying* a search — by indicating that a found image might be *good enough* even if not identical to the target image.

C. Peters et al. (Eds.): CLEF 2008, LNCS 5706, pp. 260–261, 2009.

Table 1. Indications of user confidence

	Number of sessions with adds to dictionary	Ratio of reformulated queries	Number of hints taken
Successful search sequences	0.087	0.77	2.3
Unsuccessful search sequences	0.090	0.74	1.7

References

1. Karlgren, J.: Changing the subject; one way of measuring trust in information. In: Workshop on Novel Methodologies for Evaluation in Information Retrieval, Glasgow, Scotland (2008)
2. Karlgren, J., Olsson, F.: Trusting the results in crosslingual keyword-based image retrieval. In: Peters, C., Clough, P., Gey, F.C., Karlgren, J., Magnini, B., Oard, D.W., de Rijke, M., Stempfhuber, M. (eds.) CLEF 2006. LNCS, vol. 4730, pp. 217–222. Springer, Heidelberg (2007)
3. Peinado, V., Artiles, J., Gonzalo, J., Barker, E., López-Ostenero, F.: Flickling: a multilingual search interface for flickr. In: Peters, C., et al. (eds.) CLEF 2008. LNCS, vol. 5706, pp. 236–242. Springer, Heidelberg (2009)

Overview of the Clef 2008 Multilingual Question Answering Track

Pamela Forner[1], Anselmo Peñas[2], Eneko Agirre[3], Iñaki Alegria[4], Corina Forăscu[5], Nicolas Moreau[6], Petya Osenova[7], Prokopis Prokopidis[8], Paulo Rocha[9], Bogdan Sacaleanu[10], Richard Sutcliffe[11], and Erik Tjong Kim Sang[12]

[1] CELCT, Trento, Italy
forner@celct.it
[2] Departamento de Lenguajes y Sistemas Informáticos, UNED, Madrid, Spain
anselmo@lsi.uned.es
[3] Computer Science Department, University of Basque Country, Spain
e.agirre@ehu.es
[4] University of Basque Country, Spain
i.alegria@ehu.es
[5] UAIC and RACAI, Romania
corinfor@info.uaic.ro
[6] ELDA/ELRA, Paris, France
moreau@elda.org
[7] BTB, Bulgaria,
petya@bultreebank.org
[8] ILSP Greece, Athena Research Center
prokopis@ilsp.gr
[9] Linguateca, DEI UC, Portugal
Paulo.Rocha@di.uminho.pt
[10] DFKI, Germany
bogdan@dfki.de
[11] DLTG, University of Limerick, Ireland
richard.sutcliffe@ul.ie
[12] University of Groningen
e.f.tjong.kim.sang@rug.nl

Abstract. The QA campaign at CLEF 2008 [1], was mainly the same as that proposed last year. The results and the analyses reported by last year's participants suggested that the changes introduced in the previous campaign had led to a drop in systems' performance. So for this year's competition it has been decided to practically replicate last year's exercise. Following last year's experience some QA pairs were grouped in clusters. Every cluster was characterized by a topic (not given to participants). The questions from a cluster contained co-references between one of them and the others. Moreover, as last year, the systems were given the possibility to search for answers in Wikipedia as document corpus beside the usual newswire collection. In addition to the main task, three additional exercises were offered, namely the Answer Validation Exercise (AVE), the Question Answering on Speech Transcriptions (QAST), which continued last year's successful pilots, together with the new Word Sense Disambiguation for Question Answering (QA-WSD). As general remark, it must be

C. Peters et al. (Eds.): CLEF 2008, LNCS 5706, pp. 262–295, 2009.

said that the main task still proved to be very challenging for participating systems. As a kind of shallow comparison with last year's results the best overall accuracy dropped significantly from 42% to 19% in the multi-lingual subtasks, but increased a little in the monolingual sub-tasks, going from 54% to 63%.

1 Introduction

QA@CLEF 2008 was carried out according to the spirit of the campaign, consolidated in previous years. Beside the classical main task, three additional exercises were proposed:

- the *main* task: several monolingual and cross-language sub-tasks, were offered: Bulgarian, English, French, German, Italian, Portuguese, Romanian, Greek, Basque and Spanish were proposed as both query and target languages.
- the *Answer Validation Exercise* (AVE) [2]: in its third round was aimed at evaluating answer validation systems based on textual entailment recognition. In this task, systems were required to emulate human assessment of QA responses and decide whether an *Answer* to a *Question* is correct or not according to a given *Text*. Results were evaluated against the QA human assessments.
- the *Question Answering on Speech Transcripts* (QAST) [3]: which continued last year's successful pilot task, aimed at providing a framework in which QA systems could be evaluated when the answers to factual and definition questions must be extracted from spontaneous speech transcriptions.
- the *Word Sense Disambiguation for Question Answering* (QA- WSD) [4], a pilot task which provided the questions and collections with already disambiguated Word Senses in order to study their contribution to QA performance.

As far as the main task is concerned, following last year experience, the exercise consisted of topic-related questions, i.e. clusters of questions which were related to the same topic and contained co-references between one question and the others. The requirement for questions related to a topic necessarily implies that the questions refer to common concepts and entities within the domain in question. This is accomplished either by co-reference or by anaphoric reference to the topic, implicitly or explicitly expressed in the first question or in its answer.

Moreover, besides the usual news collections provided by ELRA/ELDA, articles from Wikipedia were considered as an answer source. Some questions could have answers only in one collection, i.e. either only in the news corpus or in Wikipedia.

As a general remark, this year we had the same number of participants as in 2007 campaign, but the number of submissions went up. Due to the complexity of the innovations introduced in 2007 - the introduction of topic-related sets of questions and anaphora, list questions, Wikipedia corpus - the questions tended to get a lot more difficult and the performance of systems dropped dramatically, so, people were disinclined to continue the following year (i.e. 2008), inverting the positive trend in participation registered in the previous campaigns.

As reflected in the results, the task proved to be even more difficult than expected. Results improved in the monolingual subtasks but are still very low in the cross-lingual subtasks.

This paper describes the preparation process and presents the results of the QA track at CLEF 2008. In section 2, the tasks of the track are described in detail. The results are reported in section 3. In section 4, some final analysis about this campaign is given.

2 Task Description

As far as the main task is concerned, the consolidated procedure was followed, capitalizing on the experience of the task proposed in 2007.

The exercise consisted of topic-related questions, i.e. clusters of questions which were related to the same topic and contained co-references between one question and the others. Neither the question types (F, D, L) nor the topics were given to the participants.

The systems were fed with a set of 200 questions -which could concern facts or events (F-actoid questions), definitions of people, things or organisations (D-efinition questions), or lists of people, objects or data (L-ist questions)- and were asked to return up to three exact answers per question, where *exact* meant that neither more nor less than the information required was given.

The answer needed to be supported by the docid of the document in which the exact answer was found, and by portion(s) of text, which provided enough context to support the correctness of the exact answer. Supporting texts could be taken from different sections of the relevant documents, and could sum up to a maximum of 700 bytes. There were no particular restrictions on the length of an answer-string, but unnecessary pieces of information were penalized, since the answer was marked as *ineXact*. As in previous years, the exact answer could be exactly copied and pasted from the document, even if it was grammatically incorrect (e.g.: inflectional case did not match the one required by the question). Anyway, systems were also allowed to use natural language generation in order to correct morpho-syntactical inconsistencies (e.g., in German, changing *dem Presidenten* into *der President* if the question implies that the answer is in nominative case), and to introduce grammatical and lexical changes (e.g., QUESTION: *What nationality is X*? TEXT: *X is from the Netherlands* → EXACT ANSWER: Dutch).

The subtasks were both:

- monolingual, where the language of the question (Source language) and the language of the texts collection (Target language) were the same;
- cross-lingual, where the questions were formulated in a language different from that of the texts collection.

Two new languages have been added, i.e. Basque and Greek both as source and target languages. In total eleven source languages were considered, namely, Basque, Bulgarian, Dutch, English, French, German, Greek, Italian, Portuguese, Romanian and Spanish. All these languages were also considered as target languages.

Table 1. Tasks activated in 2008 (coloured cells)

SOURCE LANGUAGES (questions) \ TARGET LANGUAGES (corpus and answers)	BG	DE	EL	EN	ES	EU	FR	IT	NL	PT	RO
BG	■										
DE		■			■						
EL			■								
EN	■	■	■		■	■	■	■	■	■	■
ES		■		■	■	■	■	■		■	
EU				■	■	■		■			
FR				■	■		■			■	
IT				■	■		■	■			
NL				■	■				■		
PT				■	■		■	■		■	
RO				■							■

As shown in Table 1, 43 tasks were proposed:

- 10 Monolingual -i.e. Bulgarian (BG), German (DE), Greek (EL), Spanish (ES), Basque (EU), French (FR), Italian (IT), Dutch (NL), Portuguese (PT) and Romanian (RO);
- 33 Cross-lingual (as customary in recent campaigns, in order to prepare the cross-language subtasks, for which at least one participant had registered, some target language question sets were translated into the combined source languages).

Anyway, as Table **2.** Tasks chosen by at least 1 participant in QA@CLEF campaigns2 shows, not all the proposed tasks were then carried out by the participants.

As long-established, the monolingual English (EN) task was not available as it seems to have been already thoroughly investigated in TREC campaigns [11]. English was still both source and target language in the cross-language tasks.

Table 2. Tasks chosen by at least 1 participant in QA@CLEF campaigns

	MONOLINGUAL	CROSS-LINGUAL
CLEF-2004	6	13
CLEF-2005	8	15
CLEF-2006	7	17
CLEF-2007	7	11
CLEF-2008	**8**	**12**

2.1 Questions Grouped by Topic

The procedure followed to prepare the test set was the same as that used in the 2007 campaign. First of all, each organizing group, responsible for a target language, freely chose a number of topics. For each topic, one to four questions were generated. Topics could be not only named entities or events, but also other categories such as objects, natural phenomena, etc. (e.g. George W. Bush; Olympic Games; notebooks; hurricanes; etc.). The set of ordered questions were related to the topic as follows:

- the topic was named either in the first question or in the first answer
- the following questions could contain co-references to the topic expressed in the first question/answer pair.

Topics were not given in the test set, but could be inferred from the first question/answer pair. For example, if the topic was *George W. Bush*, the cluster of questions related to it could have been:

Q1: *Who is George W. Bush?*; Q2: *When was he born?*; Q3: *Who is his wife?*

The requirement for questions related to a same topic necessarily implies that the questions refer to common concepts and entities within the domain. The most common form is pronominal anaphoric reference to the topic declared in the first question, e.g.:

Q4: *What is a polygraph?*; Q5: *When was **it** invented?*

However, other forms of co-reference occurred in the questions. Here is an example:

Q6: *Who wrote the song "Dancing Queen"?*; Q7: *How many people were in **the group**?*

Here *the group* refers to an entity expressed not in the question but only in the answer. However the QA system does not know this and has to infer it, a task which can be very complex, especially if the topic is not provided in the test set.

2.2 Document Collections

Beside the data collections composed of news articles provided by ELRA/ELDA (see Table 3), also Wikipedia was considered.

Table 3. Document collections used in QA@CLEF 2008

TARGET LANG.	COLLECTION	PERIOD	SIZE
[BG] Bulgarian	Sega	2002	120 MB (33,356 docs)
	Standart	2002	93 MB (35,839 docs)
	Novinar	2002	
[DE] German	Frankfurter Rundschau	1994	320 MB (139,715 docs)
	Der Spiegel	1994/1995	63 MB (13,979 docs)
	German SDA	1994	144 MB (71,677 docs)
	German SDA	1995	141 MB (69,438 docs)
[EL] Greek	The Southeast European Times	2002	
[EN] English	Los Angeles Times	1994	425 MB (113,005 docs)
	Glasgow Herald	1995	154 MB (56,472 docs)
[ES] Spanish	EFE	1994	509 MB (215,738 docs)
	EFE	1995	577 MB (238,307 docs)
[EU] Basque	Egunkaria	2001/2003	216 MB (119,982 docs)
[FR] French	Le Monde	1994	157 MB (44,013 docs)
	Le Monde	1995	156 MB (47,646 docs)
	French SDA	1994	86 MB (43,178 docs)
	French SDA	1995	88 MB (42,615 docs)
[IT] Italian	La Stampa	1994	193 MB (58,051 docs)
	Italian SDA	1994	85 MB (50,527 docs)
	Italian SDA	1995	85 MB (50,527 docs)
[NL] Dutch	NRC Handelsblad	1994/1995	299 MB (84,121 docs)
	Algemeen Dagblad	1994/1995	241 MB (106,483 docs)
[PT] Portuguese	Público	1994	164 MB (51,751 docs)
	Público	1995	176 MB (55,070 docs)
	Folha de São Paulo	1994	108 MB (51,875 docs)
	Folha de São Paulo	1995	116 MB (52,038 docs)

The Wikipedia pages in the target languages, as found in the version of November 2006, could be used. Romanian had Wikipedia[1] as the only document collection, because there was no newswire Romanian corpus. The "snapshots" of Wikipedia were made available for download both in XML and HTML versions. The answers to the questions had to be taken from actual entries or articles of Wikipedia pages. Other types of data such as images, discussions, categories, templates, revision histories, as well as any files with user information and meta-information pages, had to be excluded.

One of the major reasons for using Wikipedia was to make a first step towards web formatted corpora where to search for answers. In fact, as nowadays so large information sources are available on the web, this may be considered a desirable next level in the evolution of QA systems. An important advantage of Wikipedia is that it is freely

[1] http://static.wikipedia.org/downloads/November_2006/ro/

available for all languages so far considered. Anyway the variation in size of Wikipedia, depending on the language, is still problematic.

2.3 Types of Questions

As far as the question types are concerned, as in previous campaigns, the three following categories were considered:

1. *Factoid questions*, fact-based questions, asking for the name of a person, a location, the extent of something, the day on which something happened, etc. We consider the following 8 answer types for factoids:

- PERSON, e.g.: Q8: Who was called the "Iron-Chancellor"? A8: Otto von Bismarck.
- TIME, e.g.: Q9: What year was Martin Luther King murdered? A9: 1968.
- LOCATION, e.g.: Q10: Which town was Wolfgang Amadeus Mozart born in? A10: Salzburg.
- ORGANIZATION, e.g.: Q11: What party does Tony Blair belong to?: A11: Labour Party.
- MEASURE, e.g.: Q12: How high is Kanchenjunga? A12: 8598m.
- COUNT, e.g.: Q13: How many people died during the Terror of PoPot? A13: 1 million.
- OBJECT, e.g.: Q14: What does magma consist of? A14: Molten rock.
- OTHER, i.e. everything that does not fit into the other categories above, e.g.: Q15: Which treaty was signed in 1979? A15: Israel-Egyptian peace treaty.

2. *Definition questions*, such as "What/Who is X?", are divided into the following subtypes:

- PERSON, i.e., questions asking for the role/job/important information about someone, e.g.: Q16*: Who is Robert Altmann?* A16*: Film maker*
- ORGANIZATION, i.e., questions asking for the mission/full name/important information about an organization, e.g.: Q17: *What is the Knesset?* A17: *Parliament of Israel.*
- OBJECT, i.e., questions asking for the description/function of objects, e.g.: Q18: *What is Atlantis?* A18: *Space Shuttle.*
- OTHER, i.e., question asking for the description of natural phenomena, technologies, legal procedures etc., e.g.: Q19: *What is Eurovision?* A19: *Song contest.*

3. *Closed list questions*: i.e., questions that require one answer containing a determined number of items, e.g.: Q20: Name all the airports in London, England. A20: Gatwick, Stansted, Heathrow, Luton and City.

As only one answer was allowed, all the items had to be present in sequence in the document and copied, one next to the other, in the answer slot.

Besides, all types of questions could contain a temporal restriction, i.e. a temporal specification that provided important information for the retrieval of the correct answer, for example:

Q21: *Who was the Chancellor of Germany from 1974 to 1982?*
A21: *Helmut Schmidt.*
Q22: *Which book was published by George Orwell in 1945?*
A22: *Animal Farm.*
Q23: *Which organization did Shimon Perez chair after Isaac Rabin's death?*
A23: *Labour Party Central Committee.*

Some questions could have no answer in the document collection, and in that case the exact answer was "NIL" and the answer and support docid fields were left empty. A question was assumed to have no right answer when neither human assessors nor participating systems could find one.

The distribution of the questions among these categories is described in Table 4. Each question set was then translated into English, which worked as inter-language during the translation of the datasets into the other tongues for the activated cross-lingual subtasks.

Table 4. Test set breakdown according to question type, number of participants and number of runs

	F	D	L	T	NIL	# Participants	# Runs
BG	159	24	17	28	9	1	1
DE	160	30	10	9	13	3	12
EL	163	29	8	31	0	0	0
EN	160	30	10	12	0	4	5
ES	161	19	20	42	10	4	10
EU	145	39	16	23	17	1	4
FR	135	30	35	66	10	1	3
IT	157	31	12	13	10	0	0
NL	151	39	10	13	10	1	4
PT	162	28	10	16	11	6	9
RO	162	28	10	47	11	2	4

2.4 Formats

As the format is concerned, also this year both input and output files were formatted as an XML file. For example, the first four questions in the EN-FR test set, i.e. English questions that hit a French document collection - were represented as follows:

```
<input>
  <q target_lang="FR" source_lang="EN" q_id="0001"
     q_group_id="1600">Which is the largest bird in Africa?</q>
  <q target_lang="FR" source_lang="EN" q_id="0002" q_group_id="1600">How
     many species of ostriches are there?</q>
  <q target_lang="FR" source_lang="EN" q_id="0003" q_group_id="1601">Who
     served as a UNICEF goodwill ambassador between 1988 and 1992?</q>
  <q target_lang="FR" source_lang="EN" q_id="0004"
     q_group_id="1601">What languages did she speak?</q>
...
</input>
```

An example of system output which answered the above questions was the following:


```
<?xml version="1.0" encoding="UTF-8"?>
<!DOCTYPE output SYSTEM "QA-CLEF-OUTPUT.dtd">
<output>
<a q_id="0001" q_group_id="1600" run_id="syna081enfr" score="0.000">
<answer>version</answer>
<docid>Afrique des Grands Lacs</docid>
<support>
<s_id>Afrique des Grands Lacs</s_id>
<s_string>Comprendre la crise de l'Afrique des grands lacs - dossier RFI
     (version archivée par Internet Archive).</s_string>
</support>
</a>
<a q_id="0002" q_group_id="1600" run_id="syna081enfr" score="0.000">
<answer>500 000</answer>
<docid>ATS.940202.0138</docid>
<support>
<s_id>ATS.940202.0138</s_id>
<s_string>Avec une superficie de seulement 51 000 km2, le Costa Rica
     abrite quelque 500 000 espèces végétales et animales. Il compte
     plus d'espèces d'oiseaux et d'arbres qu'il n'y en a sur l'ensemble
     du territoire des Etats-Unis. </s_string>
</support>
</a>
<a q_id="0003" q_group_id="1601" run_id="syna081enfr" score="0.000">
<answer>NIL</answer>
<docid/>
<support>
<s_id/>
<s_string/>
</support>
</a>
<a q_id="0004" q_group_id="1601" run_id="syna081enfr" score="0.000">
<answer>NIL</answer>
<docid/>
<support>
<s_id/>
<s_string/>
</support>
</a>
...
</output>
```


2.5 Evaluation Measures and Assessment

As far the evaluation process is concerned, no changes were made with respect to the previous campaigns. Human judges assessed the exact answer (i.e. the shortest string of words which is supposed to provide the exact amount of information to answer the question) as:

- R (Right) if correct;
- W (Wrong) if incorrect;
- X (ineXact) if contained less or more information than that required by the query;
- U (Unsupported) if either the *docid* was missing or wrong, or the supporting snippet did not contain the exact answer.

Most assessor-groups managed to guarantee a second judgement of all the runs.

As regards the evaluation measures, the main one was accuracy, defined as the proportion of questions that received a correct answer in first place. In addition most assessor groups computed Confident Weighted Score (CWS) [16] and the Mean Reciprocal Rank (MRR) over up to three assessed answers per question.

In the next section the particularities in evaluation are reported for each target language.

3 Results

As far as accuracy is concerned, scores were generally far lower than usual, as Figure 1 shows. Although comparison between different languages and years is not possible, in Figure 1 we can observe some trends which characterized this year's competition: best accuracy in the monolingual task increased with respect to last year, going up again to the values recorded in 2006. But systems - even those that participated in all previous campaigns - did not achieve a brilliant overall performance. Apparently systems could not manage suitably the new challenges, although they improved their performances when tackling issues already treated in previous campaigns.

More in detail, best accuracy in the monolingual task scored 63.5% almost ten points up with respect to last year, meanwhile the overall performance of the systems was quite low, as average accuracy was 23,63, practically the same as last year. On the contrary, the performances in the cross-language tasks recorded a drastic drop: best accuracy reached only 19% compared to 42% in the previous year, which means more than 20 points lower. Average accuracy was more or less the same as in 2007 – 13% compared to 11%.

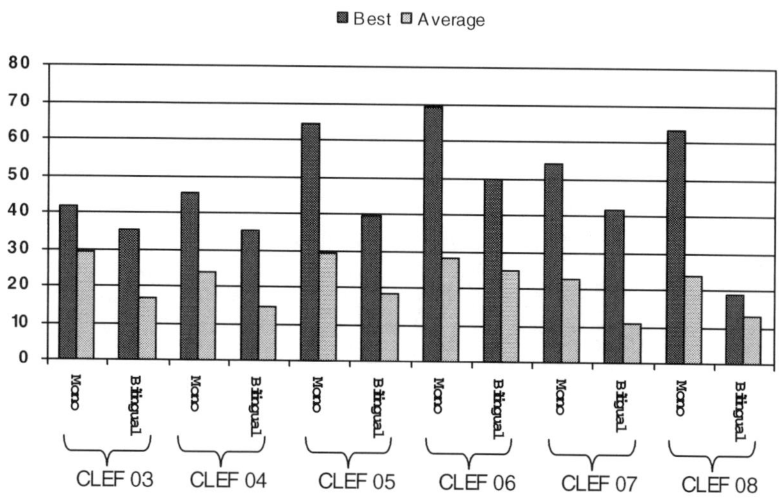

Fig. 1. Best and average scores in QA@CLEF campaigns

3.1 Participation

The number of participants has remained almost the same as in 2007 (see Table 5). As noticed, this is probably the consequence of the new challenges introduced last year in the exercise.

Also the geographical distribution remained almost unchanged, even though there was no participation from Australia and Asia. No runs were submitted neither for Italian or Greek tasks.

The number of submitted runs, increased from a total of 37 registered last year to 51 (see Table 6). The breakdown of participants and runs, according to language, is shown in Table 4 (Section 2.3). As in previous campaigns, more participants chose the monolingual tasks, which once again demonstrated to be more approachable.

Table 5. Number of participants in QA@CLEF

	America	Europe	Asia	Australia	TOTAL
CLEF 2003	3	5	0	0	8
CLEF 2004	1	17	0	0	18
CLEF 2005	1	22	1	0	24
CLEF 2006	4	24	2	0	30
CLEF 2007	3	16	1	1	21
CLEF 2008	**1**	**20**	**0**	**0**	**21**

Table 6. Number of submitted runs

	Submitted runs	Monolingual	Cross-lingual
CLEF 2003	17	6	11
CLEF 2004	48	20	28
CLEF 2005	67	43	24
CLEF 2006	77	42	35
CLEF 2007	37	23	14
CLEF 2008	**51**	**31**	**20**

In the following subsections a more detailed analysis of the results in each language follows, giving specific information on the performances of the participating systems in the single sub-tasks and on the different types of questions, providing the relevant statistics and comments.

3.2 Basque as Target

In the first year working with Basque as target only one research group submitted runs for evaluation in the track having Basque as target language (Ixa group from the University of the Basque Country). They sent four runs: one monolingual, one English-Basque and two Spanish-Basque.

The Basque question set consisted of 145 factoid questions, 39 definition questions and 16 list questions. 39 questions contained a temporal restriction, and 10 had no answer in the Gold Standard. 40 answers were retrieved from Wikipedia, the remains from the news collections. Half of the questions were linked to a topic, so the second (and sometimes the 3rd) question was more difficult to answer.

The news collection was the Egunkaria newspaper during 2000, 2001 and 2002 years and the information from Wikipedia was the exportation corresponding to the 2006 year.

Table 7 shows the evaluation results for the four submitted runs (one monolingual and three cross-lingual). The table shows the number of Right, Wrong, ineXact and Unsupported answers, as well as the percentage of correctly answered Factoids, Temporally restricted questions, Definition and List questions.

The monolingual run (ixag081eueu.xml) achieved accuracy of 13%, lower than expected. It is necessary to underline that Basque is a highly flexional language, making the matching of terms and entities more complex. The system achieved better accuracy in factoids questions (15.9%) and no correct answers were retrieved for list questions. It is necessary to remark that 57 answers were NIL but only four of them were correct. This is one of the issues participants can improve.

Table 7. Evaluation results for the four submitted runs

Run	R #	W #	X #	U #	%F [145]	%T [23]	%D [39]	L% [16]	NIL		CWS	Overall accuracy
									#	% [*]		
ixag08 1eueu	26	163	11	0	15.9	8.7	7.7	0	4	7.0	0.023	13
ixag08 1eneu	11	182	7	0	5.5	4.3	7.7	0	6	6.2	0.004	5.5
ixag08 1eseu	11	182	7	0	6.9	4.3	2.6	0	4	4.8	0.004	5.5
ixag08 2eseu	7	185	8	0	4.8	4.3	0	0	3	3.5	0.003	3.5

Looking to the cross-lingual runs, the loss of accuracy respect to the monolingual system is a bit more than 50% for the two best runs. This percentage is quite similar with runs for other target languages in 2007. The overall accuracy is the same for both (English and Spanish to Basque) but only they agree in five correct answers (each system gives other six correct answers). The second system for Spanish-Basque get poorer results and only is slightly better in inexact answers. These runs get also a lot of NIL answers.

3.3 Bulgarian as Target

This year, contrary to the expectations, only one run by one group (BTB) was performed for Bulgarian. As the table above shows, the result is far from satisfying. Again, the definitions were detected better in comparison to other question types. Also, the difference between the detection of factoids and of temporally restricted questions is negligible.

Table 8. Results for the submitted run for Bulgarian

Run	R	W	X	U	% F	% T	% D	% L	NIL		CWS	MRR	Overall accuracy
	#	#	#	#	[*]	[*]	[*]	[*]	#	% [*]			
btb1	20	173	7	0	8.80	7.14	25.00	0.00	-	0.00	0.01	-	10 %

3.4 Dutch as Target

The questions for the Dutch subtask of CLEF-QA 2008 were written by four native speakers. They selected random articles from either Wikipedia or the news collection and composed questions based on the topics of the articles.

The quartet produced a total of 222 question-answer pairs from which they selected a set of 200 that satisfied the type distribution requirements of the task organizers. An overview of the question types and answer types can be found in Table 9.

This year, only one team took part in the question answering task with Dutch as target language: the University of Groningen. The team submitted two monolingual runs and two cross-lingual runs (English to Dutch). All runs were assessed twice by a single assessor. This resulted in a total of eight conflicts (1%). These were corrected. The results of the assessment can be found in Table 10.

Table 9. Properties of the 200 Dutch questions (134 topics) in the test set

Question types	
Definition	39
Factoid	151
List	
Answer source	
News	20
None (NIL answer)	5
Wikipedia	175
Definition answer types	
Location	3
Object	6
Organization	8
Other	12
Person	10

Factoid answer types	
Count	20
Location	18
Measure	20
Object	19
Organization	18
Other	17
Person	19
Time	20
List answer types	
Location	6
Other	1
Person	2
Time	1

Temporal restriction	
No	187
Yes	13
Question per topic	
1 question	100
2 questions	15
3 questions	6
4 questions	13
Topic types	
Location	15
Object	23
Organization	14
Other	50
Person	32

Table 10. Assessment results for the four submitted runs for Dutch

Run	R #	W #	X #	U #	%F [151]	%T [13]	%D [39]	L% [10]	NIL #	NIL % [*]	CWS	Overall accuracy
gron081nlnl	50	138	11	1	24.5	15.4	33.3	0.0	19	5.3	0.342	25.0
gron082nlnl	51	136	10	3	24.5	15.4	35.9	0.0	15	6.7	0.331	25.5
gron081ennl	27	157	10	6	13.2	7.7	17.9	0.0	30	3.3	0.235	13.5
gron082ennl	27	157	10	6	13.2	7.7	17.9	0.0	30	3.3	0.235	13.5

The two cross-lingual runs gron081ennl andron082ennl produced exactly the same answers.

The best monolingual run (gron082nlnl) achieved exactly the same score as the best run of 2007 (25.5%). The same is true for the best monolingual run (13.5%). The fact that the two scores are in the same range as last year is no big surprise since the task has not changed considerably this year and all scores have been achieved by the same system.

Like in 2007, the system performed better for definition questions than for other question types. The definition questions could be divided in two subtypes: those that asked for a definition (26) and those that contained a definition and asked for the name of the defined object (12). The monolingual runs performed similarly for both subtypes but the cross-lingual runs did not contain a correct answer to any question of the second subtype.

None of the runs obtained any points for the list questions. The answers contained some parts that were correct but none of them were completely correct. We were unable to award points for partially correct answers in the current assessment scheme.

All the runs were produced by the same system and the differences between the runs are small. The cross-lingual runs contained seven correct answers that were not present in any of the monolingual runs (for questions 20, 25, 120, 131, 142, 150 and 200). Eight questions were only answered correctly in a single monolingual run (1, 28, 54, 72, 83, 143, 193 and 199). Thirty-five questions were answered correctly in two runs, three in three runs and seventeen in all four runs. 137 questions failed to receive any correct answer.

3.5 English as Target

The task this year was exactly the same as in 2007 and moreover the three collections were the same: Glasgow Herald, LA Times and Wikipedia. However, given the considerable interest in the Wikipedia which has been shown by Question Answering groups generally, it was decided to increase the number of questions drawn from it to 75% overall, with just 25% coming from the two newspaper collections. This means that 40 of the 160 Factoids came from the newspapers, together with seven of the 30 Definitions and two of the ten Lists. These questions were divided equally between the Glasgow Herald and LA Times. All the remainder questions were drawn from the Wikipedia.

Table 11. Evaluation results for the English submitted runs

Run	R	W	X	U	% F	% T	% D	% L	NIL		CWS	K1	Overall accuracy
	#	#	#	#	[160]	[12]	[30]	[10]	#	%[0]			
dcun081deen	16	168	7	9	5.00	8.33	26.67	0.00	0	0.00	0.00516	0.10	8.00
dcun082deen	1	195	3	1	0.63	0.00	0.00	0.00	0	0.00	0.00013	0.03	0.50
dfki081deen	28	164	5	3	6.25	8.33	60.00	0.00	0	0.00	0.01760	N/A	14.00
ilkm081nlen	7	182	2	9	4.38	0.00	0.00	0.00	0	0.00	0.00175	N/A	3.50
wlvs081roen	38	155	2	5	11.25	0.00	66.67	0.00	0	0.00	0.05436	0.13	19.00

Considerable care was taken in the selection of the questions. The distribution by answer type was controlled exactly as in previous years. As requested by the organisers there were exactly twenty each of Factoid target type PERSON, TIME, LOCATION, MEASURE, COUNT, ORGANIZATION, OBJECT and OTHER. Similarly for Definitions there were eight PERSON, seven ORGANIZATION, seven OBJECT and eight OTHER. For Lists there were four OTHER, two each of PERSON and ORGANIZATION, and one each of LOCATION and OBJECT.

In addition to the above distribution, we also controlled the distribution of topics for the question groups, something which was made practicable by the use of the Wikipedia. Questions were drawn from a number of predefined subject fields: countries towns, roads and bridges, shops, politicians and politics, sports and sports people, foods and vegetables, cars, classical music including instruments, popular music, literature poetry and drama, philosophy, films, architecture, languages, science, consumer goods, and finally organisations. Questions were distributed among these topics. The maximum in any topic was twenty (sports) and the minimum was two (shops). For the majority there were between four and six question groups. For each such topic, one or more questions were set depending on what information the texts contained. As a change from last year, the organisers asked us to include 100 singleton topics. This effectively meant that half the questions in the overall set of 200 were simple "one-off" queries as were set in CLEF prior to 2007 and for the earlier TREC campaigns.

Questions were entered via a web interface developed by the organisers last year. However, this year they improved it considerably, for example allowing modifications to be made to existing entries. This was a great help and a commendable effort on their part.

Five cross-lingual runs with English as target were submitted this year, as compared with eight in 2007 and thirteen in 2006. Four groups participated in three languages, Dutch, German and Romanian. Each group worked with only one source language, and only DCUN submitted two runs. The rest submitted only one run.

All answers were double-judged. Where the assessors differed, the case was reviewed and a decision taken. There were 63 judgement differences in total. Three of the runs contained multiple answers to individual questions in certain cases, and these were all assessed, as per the requirement of the organisers. If we assume that the number of judgements was in fact 200 questions * five runs, i.e. 1,000, we can compute a lower bound for the agreement level. This gives a figure of (1,000-63)/1,000, i.e. 93.7%. The equivalent figure for 2007 (called Agreement Level 2 in the Working Notes for last year) was 97.6%. Given that we have computed a lower bound this year (and not therefore the exact figure) this seems acceptable.

Of the five runs with English as target, *wlvs081roen* was the best with an accuracy of 19.00% overall. They also did very will on the definitions, scoring 66.67%. The only source language for which there was more than one run was German, for which there were three submissions from two groups: *dfki081* scored the best with 14.00% and this was followed by *dcun081deen* with 8.00% and *dcun082deen* with 0.50%. DFKI also did very well on definitions with an accuracy of 60.00. Interestingly, none of the systems answered any of the list questions correctly. Only *dcun082deen* answered one list question inexactly.

If we compare the results this year with those of last year when the task was very similar, performance has improved here. The best score in 2007 was *wolv071roen* with 14% (the best score) which has now improved to 19%. Similarly, *dfki071deen* scored 7% in 2007 but increased this to 14% this year in *dfki081deen*. An attempt was made to set easier questions this year, which might have affected performance. In addition, many more questions came from the Wikipedia in 2008 with only a minority being drawn from the newspaper corpora.

3.6 QA-WSD Subtask for English as Target

The QA-WSD task brought semantic and retrieval evaluation together. The participants were offered the same queries and document collections as for the main QA exercise, but with the addition of word sense tags as provided by two automatic word sense disambiguation (WSD) systems. Contrary to the main QA task, Wikipedia articles were not included, and thus systems need to reply to the questions that have an answer in the news document collection. In the QA-WSD track only English monolingual and Spanish to English bilingual tasks are offered, i.e. English is the only target language, and queries are available on both English and Spanish. The queries were the same as for the main QA exercise, and the participation followed the same process, except for the use of the sense-annotated data.

The goal of the task was to test whether WSD can be used beneficially for Question Answering, and is closely related to the Robust-WSD subtask of the ad-hoc track in CLEF 2008. Participants were required to send two runs for each of the monolingual/bilingual tasks where they participate: one which does not use sense annotations and another one which does use sense annotations. Whenever possible, the only difference between the two runs should be solely the use or not of the sense information. Participants which send a single run would be discarded from the evaluation.

The WSD data is based on WordNet version 1.6 and was supplemented with freely available data from the English and Spanish WordNets in order to test different expansion strategies. Two leading WSD experts run their systems [17][18], and provided those WSD results for the participants to use. The task website [4] provides additional information on data formats and resources.

Table 12. Results of the EN2EN QA-WSD runs on the 49 queries which had replies in the news collections

Run	R #	W #	X #	U #	%F [40]	%T [5]	%D [7]	L% [2]	NIL		CWS	Overall accuracy
									0	% [0]		
nlel08 1enen	8	41	0	0	17.5	0	14.2	0	0	0	0.03	16.32
nlel08 2enen	7	42	0	0	15.0	0	0	0	0	0	0.02	14.29

Table 13. Results of the EN2EN QA-WSD runs on all 200 queries, just for the sake of comparison

Run	R #	W #	X #	U #	%F [160]	%T [5]	%D [7]	L% [10]	NIL		CWS	Overall accuracy
									0	% [0]		
nlel08 1enen	10	188	0	2	5.6	0	3.3	0	0	0	0.00	5.00
nlel08 2enen	8	189	0	3	4.4	0	3.3	0	0	0	0.00	4.00

From the 200 questions provided to participants, only 49 queries had a correct answer in the news collection, the rest having their reply in Wikipedia. The table below provides the results for the participant on those 49 questions.

The first run does not use WSD, while the second uses the sense tags returned by the NUS WSD system. The WSD tags where used in the passage retrieval module. The use of WSD does not provide any improvement, and causes one more error. For the sake of completeness we also include below the results on all 200 queries. Surprisingly the participant managed to find two (one in the WSD run) correct answers for the Wikipedia questions in the news collection.

3.7 French as Target

This year only one group took part in the evaluation tasks using French as a target language: the French group *Synapse Développement.* Last year's second participant, the *Language Computer Corporation* (LCC, USA) didn't send any submission this time.

Synapse submitted three runs in total: one monolingual run and two bilingual runs (English-to-French and Portuguese-to-French).

As last year, three types of questions were proposed: factual, definition and closed list questions. Participants could return one exact answer per question and up to two runs. Some questions (10%) had no answer in the document collection, and in this case the exact answer is "NIL".

The French test set consists of 200 questions where 135 were factoids (F), 30 definitions (D), and 35 closed list questions (L).

Among these 200 questions, 66 were temporally restricted questions (T) and 12 were NIL questions (i.e. a "NIL" answer was expected, meaning that there is no valid answer for this question in the document collection).

Table 14. Results of the monolingual and bilingual French runs

Run	Assessed Answers (#)	R #	W #	X #	U #	%F [135]	%T [66]	%D [30]	L% [35]	NIL Answers #	NIL Answers % [12]	CWS	Overall accuracy
syn08frfr	200	131	77	9	1	54.8	51.5	86.7	37.1	20	50.0	0.30937	56.5
syn08enfr	200	36	157	6	1	15.6	15.1	50.0	0.0	60	8.3	0.02646	18.0
syn08ptfr	200	33	163	4	0	14.1	13.6	43.3	2.9	67	11.9	0.02387	16.5

Table 14 shows the final results of the assessment of the 3 runs submitted by Synapse. For each run, the following statistics are provided:

- The number of correct (R), wrong (W), inexact (X) and unsupported answers (U),
- The accuracy calculated within each of the categories of questions: F, D, T and L questions,
- The number of NIL answers and the proportion of correct ones (i.e. corresponding to a NIL questions),
- The Confidence Weighted Score (CWS) measure.
- The accuracy calculated over all answers.

Figure 2 shows the best scores for systems using French as target in the last five CLEF QA campaigns.

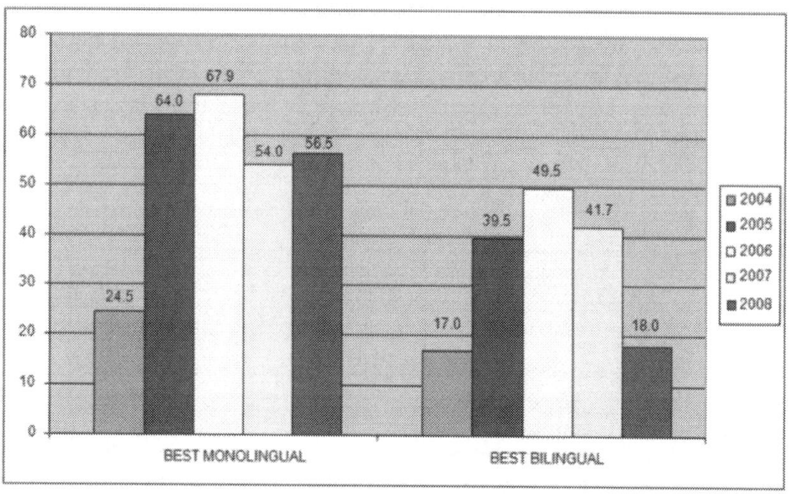

Fig. 2. Best scores for systems using French as target in CLEF QA campaigns

For the monolingual task, the Synapse system returned 113 correct answers (accuracy of 56.5%), comparable to last year (accuracy of 54.0%). The bilingual runs performance is quite low, with an accuracy of 18.0% for EN-to-FR and 16.5% for PT-to-FR.

The level of performance strongly depends on the type of questions. The monolingual runs score very high on the definition questions (86.7%). The lowest performance is obtained with closed list questions (37.1%). It is even more obvious when looking at the bilingual runs. If the systems performed pretty well on the definition questions (50.0% and 43.3% for EN-to-FR and PT-to-FR respectively), they could not cope with the closed list questions. The PT-to-FR system could only give one close list correct answer. The EN-to-FR system could not even answer to any of these questions. The bilingual runs did not reach high accuracy with factoid and temporally restricted questions (50.0% and 43.3% for EN-to-FR and PT-to-FR respectively). This year, the complexity of the task, in particular regarding closed list questions, seems to have been hard to cope with for the bilingual systems.

The complexity of the task is also reflected by the number of NIL answers. The monolingual system returned 20 NIL answers (to be compared with the 12 expected). The bilingual systems returned 60 (EN-to-FR) and 67 (EN-to-FR) NIL answers, i.e. at least 5 times more as expected.

It is also interesting to look at the results when categorizing questions by the size of the topic they belong to. This year, topics could contain from 1 single question to 4 questions. The CLEF 2008 set consists of:

- 52 single question topics,
- 33 topics with 2 questions (66 questions in total),
- 18 topics with 3 questions (54 questions in total),
- 7 topics with 4 questions (28 questions in total).

Table 15, Table 16 and Table 17 give the results of each run according to the size of the topics.

The monolingual system (Table 15) is not sensitive to the size of the topic question set. On the opposite, the performances of the bilingual systems (Table 16 and Table 17) decrease by a half, when comparing the 1- and 2-question sets to the 3- and 4-question sets. A possible explanation is that the bilingual systems perform poorly with questions containing anaphoric references (which are more likely to occur in the 3- and 4-question sets).

Table 15. Results per topic size (FR-to-FR)

Run	Size of topic	Assessed Answers #	Overall accuracy (%)
syn08frfr	1	52	55.8
syn08frfr	2	66	50.0
syn08frfr	3	24	66.7
syn08frfr	4	28	53.6

Table 16. Results per topic size (EN-to-FR)

Run	Size of topic	Assessed Answers #	Overall accuracy (%)
syn08enfr	1	52	21.2
syn08enfr	2	66	22.7
syn08enfr	3	24	13.0
syn08enfr	4	28	10.7

Table 17. Results per topic size (PT-to-FR)

Run	Size of topic	Assessed Answers #	Overall accuracy (%)
syn08ptfr	1	52	25.0
syn08ptfr	2	66	18.2
syn08ptfr	3	24	9.3
syn08ptfr	4	28	10.7

This year, the number and complexity of closed list questions was clearly higher than the previous year. In the same way, there were more temporally restricted questions, more topics (comprising from 2 to 4 questions) and more anaphoric references. It seems that this higher level of difficulty particularly impacted the bilingual tasks. In spite of this, the monolingual Synapse system performed slightly better than last year.

3.8 German as Target

Three research groups submitted runs for evaluation in the track having German as target language: The German Research Center for Artificial Intelligence (DFKI), the Fern Universität Hagen (FUHA) and the Universität Koblenz-Landau (LOGA). All groups provided system runs for the monolingual scenario, DFKI and FUHA submitted runs for the cross-language English-German scenario and FUHA had also runs for the Spanish-German scenario.

Compared to the previous editions, this year monolingual runs registered an increase in accuracy while bilingual runs showed a slight decrease (Figure 3).

The number of topics covered by the test set questions was of 120 distributed as it follows: 74 topics consisting of 1 question, 24 topics of 2 related questions, 10 topics of 3 related questions, and 12 topics of 4 related questions. The distribution of the topics over the document collections (CLEF vs. Wikipedia) is presented in Table 18.

According to Table 19 the most frequent topic types were OTHER (32), OBJECT (29) and ORGANIZATION (24), with first two types more present for the Wikipedia collection of documents (WIKI).

The details of systems' results can be seen in Table 21.

As regards the source of the answers, 97 questions from 57 topics asked for information out of the CLEF document collection and the rest of 103 from 63 topics for information from Wikipedia. Table 20 shows a breakdown of the test set questions by the expected answer type (EAType) for each collection of data.

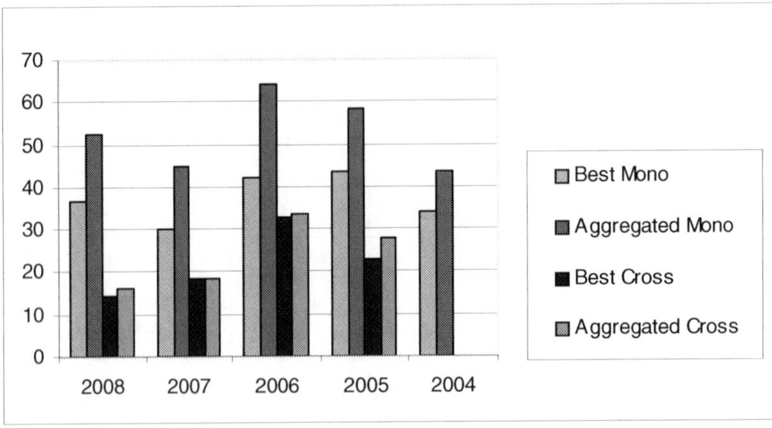

Fig. 3. Results evolution in German as target

Table 18. Topic distribution over data collections

Topic Size	# Topics / CLEF	# Topics / WIKI	# Topics
1	39	35	74
2	10	14	24
3	5	5	10
4	3	9	12
Total	**57**	**63**	**120**

Table 19. Topic type breakdown over data collections

Topic Type	CLEF					WIKI				
	Topic Size				Total	Topic Size				Total
	1	2	3	4		1	2	3	4	
PERSON	5	2	1	1	9	0	1	0	2	3
OBJECT	7	1	0	0	8	16	3	0	2	21
ORGANIZATION	9	1	2	1	13	7	2	1	1	11
LOCATION	8	2	2	1	13	1	3	2	2	8
EVENT	0	0	0	0	0	0	2	0	0	2
OTHER	9	4	0	1	14	11	3	2	2	18
					57					63

Table 20. Question EAType breakdown over data collections

EAType	CLEF	WIKI	Total
PERSON	15	15	30
LOCATION	13	12	25
TIME	13	8	21
COUNT	13	7	20
OBJECT	7	18	25
MEASURE	12	8	20
ORGANIZATION	15	13	28
OTHER	9	22	31
Total	97	103	200

Table 21. System Performance – Details

Run	R	W	X	U	% F	% T	% D	% L	NIL		CWS	MRR	Overall accuracy
	#	#	#	#	[160]	[9]	[30]	[10]	#	% [10]			
dfki081dede$_M$	73	119	2	6	30.62	44.44	80	0	0	0	0.16	0	36.5
dfki082dede$_M$	74	120	2	4	31.25	33.33	80	0	0	0	0.16	0	37
fuha081dede$_M$	45	141	8	6	24.37	44.44	20	0	1	4.76	0.05	0.29	22.5
fuha082dede$_M$	46	139	11	4	25.62	33.33	16.66	0	21	4.76	0.048	0.29	23
loga081dede$_M$	29	159	11	1	13.75	0	20	10	55	5.45	0.031	0.19	14.5
loga082dede$_M$	27	163	9	1	13.12	0	16.66	10	48	4.16	0.029	0.17	13.5
dfki081ende$_C$	29	164	2	5	10	0	43.33	0	0	0	0.038	0	14.5
fuha081ende$_C$	28	163	6	3	15	11.11	13.33	0	81	7.4	0.023	0.24	14
fuha082ende$_C$	28	160	6	6	15	11.11	13.33	0	81	7.4	0.019	0.22	14
fuha081esde$_C$	19	169	9	2	9.43	0	13.33	0	9	0	0.015	0.15	9.54
fuha082esde$_C$	17	173	5	5	8.12	0	13.33	0	61	3.27	0.007	0.13	8.5

3.9 Portuguese as Target

The Portuguese track had six different participants: beside the veteran groups of Priberam, Linguateca, Universidade de Évora, INESC and FEUP, we had a new participants this year, Universidade Aberta. No bilingual task occurred this year.

In this fourth year of Portuguese participation, Priberam repeated the top place of its previous years, with University of Évora behind. Again we added the classification the classification X-, meaning incomplete, keeping the classification X+ for answers with extra text or other kinds of inexactness. In Table 22 we present the overall results (all tables in these notes refer exclusively to the first answer by each system).

Table 22. Results of the runs with Portuguese as target: all 200 questions (first answers only)

Run Name		W (#)	X+ (#)	X- (#)	U (#)	Overall Accuracy (%)	NIL Accuracy		
							#	Precision (%)	Recall (%)
diue081	93	94	8	1	2	46.5%	21	9.5	20
esfi081	47	134	5	7	5	23.5%	20	20.0	20
esfi082	39	137	7	9	6	19.5%	20	15.0	10
feup081	29	165	2	2	2	14.5%	142	8.5	90
feup082	25	169	3	1	2	12.5%	149	8.1	90
idsa081	65	119	8		8	32.5%	12	16.7	20
ines081	40	150	2	1	5	20.0%	123	9.7	90
ines082	40	150	2	1	5	20.0%	123	9.7	90
prib081	127	55	9	3	4	63.5%	8	12.5	10

To provide a more direct comparison with pre-2006 results, in Table 23 we present the results both for first question of each topic (which we believe is more readily comparable to such results) and for the linked questions.

On the whole, compared to last year, Priberam and Senso (UE) improved their results, which were already the best. INESC system and Esfinge (Linguateca) also showed some improvement, at a lower level Raposa (FEUP) showed similar results. The system of Universidade Aberta appeared with good results compared to some veteran systems. We leave it to the participants to comment on whether it might have been caused by harder questions or changes (or lack thereof) in the systems.

Table 23. Results of the runs with Portuguese as target: answers to linked and unlinked questions

Run Name	First questions (# 151)						Linked questions (# 49)	
	R (#)	W (#)	X+ (#)	X- (#)	U (#)	Accuracy (%)	R (#)	Accuracy (%)
diue081	82	59	6	3	1	54.3	11	22.4
esfi081	42	92	5	7	5	27.3	7	14.3
esfi082	33	97	6	9	6	21.9	8	16.3
feup081	29	116	2	2	2	19.2	3	6.1
feup082	25	120	3	1	2	16.6	3	6.1
idsa081	54	85	6		6	35.8	11	22.4
ines081	35	106	2	3	5	23.2	8	16.3
ines082	35	106	2	3	5	23.2	8	16.3
prib081	105	32	9	4	1	69.5	22	44.9

Table 24. Results of the assessment of the monolingual Portuguese runs: definitions

Run	loc	obj	org	oth	per	TOT	%
	1	6	6	8	6	27	
diue081		5	6	8	5	24	89%
esfi081		1	2	4	2	9	33%
esfi082				1	1	2	7%
feup081		1	1	1	1	4	15%
feup082		1	1	1	1	4	15%
idsa081	1	5	1	5	5	17	63%
ines081	1	5	1	7	3	17	63%
ines082	1	5	1	7	3	17	63%
prib081		5	5	6	2	18	67%
combination	1	6	6	8	6	27	100%

Unlike last year , the results over linked questions are significatively different (and below) from those over not-linked. Question 180 was wrongly redacted, referring to Aida's opera *Verdi* instead of the other way around, which also affected two linked questions. Therefore, we accepted both NIL answers to those questions, as well as correct ones.

Table 24 shows the results for each answer type of definition questions, while Table 25 shows the results for each answer type of factoid questions (including list questions). As it can be seen, four out of six systems perform clearly better when it comes to definitions than to factoids. Particularly Senso has a high accuracy regarding definitions.

Table 25. Results of the assessment of the Portuguese runs: factoids, including lists

Run	cou	loc	mea	obj	org	oth	per	tim	TOT	%
	17	38	16	2	10	33	33	24	173	
diue081	6	17	8	1	5	13	8	11	69	35%
esfi081	8	8	2		2	2	14	4	40	20%
esfi082	8	8	2		2	2	13	4	39	20%
feup081	5	4	4		1	2	8	4	28	14%
feup082	5	3	4		1	2	6	3	24	12%
idsa081	9	9	9			6	8	7	48	24%
ines081	4	9	2			1	4	6	26	13%
ines082	4	9	2			1	4	6	26	13%
prib081	11	21	13	1	7	18	22	16	109	55%
combination	16	31	15	1	7	23	27	21	141	82%

We included in both Table 24 and Table 25 a virtual run, called combination, in which one question is considered correct if at least one participating system found a valid answer. The objective of this combination run is to show the potential achievement when combining the capacities of all the participants. The combination run can be considered, somehow, state-of-the-art in monolingual Portuguese question answering. All definition questions were answered by at least one system.

Table 26. Average size of answers (values in number of words)

Run name	Non-NIL Answers (#)	Average answer size	Average answer size (R only)	Average snippet size	Average snippet size (R only)
diue081	179	2.8	3.6	25.9	26.1
esfi081	180	2.6	3.0	78.4	62.5
esfi082	180	1.8	1.7	78.2	62.4
feup081	58	1.8	3.4	64.2	51.6
feup081	51	1.8	3.7	63.3	51.4
idsa081	188	5.0	10.0	28.6	34.4
ines081	77	3.0	7.4	79.6	36.6
ines082	77	3.0	7.4	79.6	36.6
prib081	192	3.2	3.4	27.6	25.1

The system with best results, Priberam, answered correctly 64.8% the questions with at least one correct answer. In all, 130 questions were answered by more than one system.

In Table 26, we present some values concerning answer and snippet size.

Temporally restricted questions. Table 27 presents the results of the 17 temporally restricted questions. As in previous years, the effectiveness of the systems to answer those questions is visibly lower than for non-TRQ questions.

Table 27. Accuracy of temporally restricted questions

Run name	Correct answers (#)	T.R.Q correctness (%)	Non-T.R.Q correctness (%)	Total correctness (%)
diue081	4	23.5	48..6	46.5
esfi081	3	17.6	24.0	23.5
esfi082	3	17.6	19.7	19.5
feup081	1	5.9	15.3	14.5
feup082	1	5.9	13.1	12.5
Idsa081	2	11.8	34.4	32.5
ines081	1	5.9	21.3	20.0
ines082	1	5.9	21.3	20.0
prib081	8	47.1	65.0	63.5

List questions. ten questions were defined as list questions all closed list factoids with two to five each[2]. The results haven't improved with UE getting two correct answers. Priberam three and all other system zero. There were however seven cases of incomplete answers (i.e.. answering some elements of the list only) although only two of them with than one element of the answer.

Answer source. Table 28 presents the distribution of questions by source during their selection. The distribution of sources used by the different runs and their correctness.

[2] There were some open list questions as well, but they were classified and evaluated as ordinary factoids.

Table 28. Answers by source and their correctness

Run	News		Wikipedia		NIL	
	#	% correct	#	% correct	#	% correct
Selection	34	-	144	-	10	-
diue081	35	40%	144	53%	21	10%
esfi081	85	21%	95	28%	20	10%
esfi082	81	17%	99	24%	20	5%
feup081	10	40%	48	33%	142	6%
feup082	9	44%	42	29%	149	6%
idsa081	50	28%	138	36%	12	17%
ines081	31	23%	46	52%	123	7%
ines082	31	23%	46	52%	123	7%
prib081	46	63%	146	66%	8	13%

3.10 Romanian as Target

In the third year of Romanian participation in QA@CLEF, and the second one with Romanian addressed as a target language, the question generation was based on the collection of Wikipedia Romanian pages frozen in November 2006[3]- the same corpus as in the previous edition[4].

Table 29. Question & Answer types distribution in Romanian (in brackets the number of temporally restricted questions)

Q type /expected A type	PER-SON	TIME	LOC.	ORG.	MEAS URE	COU NT	OB-JECT	OTH ER	TOTAL
FACTOID	20 (9)	23 (5)	26 (4)	20 (10)	17 (3)	22 (5)	18 (4)	16 (4)	**162 (44)**
DEF.	8		1	6 (2)			6	7	**28 (2)**
LIST	3		1 (1)	1			2 (1)	3	**10 (2)**
NIL									**8**

The questions were generated starting from the corpus and based on the Guidelines for Question Generation, the Guidelines for Participants [5] and the final decisions taken after email discussions between the organizers. The 200 questions are distributed according to Table 29, where for each type of question and expected answer we indicate also the temporally restricted questions out of the total number of questions. Without counting the NIL questions, 100% of the questions has the answer in Wikipedia collection.

As the Guidelines for Question Generation did not change since the previous edition, there were no major difficulties in creating the Romanian gold standard for the

[3] `http://static.wikipedia.org/downloads/November_2006/ro/`

[4] At http://static.wikipedia.org/downloads/ the frozen versions of Wikipedia exist for April 2007 and June 2008, for all languages involved in QA@CLEF.

2008 QA@CLEF. The working version of the GS was uploaded on the question generation interface developed at CELCT (Italy), by filling all the required fields.

For the topic-related questions (clusters of up to four questions, related to one same topic) we kept about the same number as in the previous edition: in 2007 we had 122 topics and now there are 119 topics. The percentage of topic-linked questions is illustrated in Table 30, showing that 127 questions were grouped under 46 topics, hence 63.5% out of the total 200 questions were linked in topics with more than one question.

Table 30. Topic-related questions

# of questions / Topic type	PERSON	LOC.	ORG.	EVENT	OBJECT	OTHER	**Total topics**	**Total questions**
4 Qs	5	1	1			5	**12**	**48**
3 Qs	5	1		1	1	3	**11**	**33**
2 Qs	5	3	4		2	9	**23**	**46**
1 Q	13	6	19		17	18	**73**	**73**
TOTAL	**28**	**11**	**24**	**1**	**20**	**35**	**119**	**200**

In fact the questions contain not 127, but only 51 anaphoric elements of various types, so that 25.5% of the questions are linked through coreferential relations. The personal, possessive or demonstrative pronouns were used in most of the cases to create anaphoric relations. The antecedents are mainly the focus of the previous question, or the previous answer. Few such questions require inference in order to be correctly answered. For example in order to correctly answer the F-Time question *When was the first Esperanto dictionary for Romanian published?* and then the L-Other *Name all the grammatical cases of this artificial language.*, one needs to correctly link the anaphor "artificial language" to its antecedent which is "Esperanto" and not "Romanian" (also a language but not artificial); this is possible by establishing, based on a text snippet, that Esperanto is an artificial language.

The 8 NIL questions, even though they seem somehow unnatural, were created by including questions about facts impossible from a human perception; for example the question *In which year did Paul Kline publish his work about the natural phenomena called hail?* has no answer in any of the articles about the psychologist. Another type of NIL questions are those based on inference – the question *How many bicameral Parliaments are there in Cuba?* is a NIL question because in all wiki articles one can find that Cuba has a unicameral parliament. Another type of NIL questions (with answer in English, but not in Romanian) we have created cannot be good items neither in a cross-lingual evaluation where the answers are to be find in any language, nor in an evaluation based on an open text collection such as the web. The question *What is a micron?* has no answer in the Romanian wiki articles from 2006, but it can have an answer in other Romanian webpages, and, moreover, in the English wiki articles it has more than a correct answer depending on the domain where the term is used (in the metric system or in vacuum engineering).

For the LIST type we created only questions whose answers are to be found in one same text section. The 2007 evaluation for Romanian showed that "open list" questions (with answers in various sections of an article or even in various articles) are difficult to handle, therefore we made the LIST questions easier.

Since especially at the evaluation time we realized some of the initial questions were badly classified according to their category (F, D or L with their subtypes, as well as with respect to the temporal restrictions), after the final official evaluation we provided the gold standard of questions with all necessary corrections. The final version, available since October 2008 on the CLEF website[5], has no major impact on the official judgements: modifying the standard does not change the types of R, X, U or W answers submitted by the participants.

Systems' analysis and evaluation. Like in the 2007 edition, this year two Romanian groups took part in the monolingual task with Romanian as a target language: the Faculty of Computer Science from the Al. I. Cuza University of Iasi (UAIC), and the Research Institute for Artificial Intelligence from the Romanian Academy (ICIA), Bucharest. Each group submitted two runs, the four systems having an average of 2.4 answers per question for ICIA, and 1.92 for UAIC. The 2008 general results are presented in Tables 31 below.

The statistics includes a system, named *combined*, obtained through the combination of the 4 participating RO-RO systems. Because at the evaluation time we observed that there are correct answers not only in the first position, but also on the second or the third, the *combined* system considers that an answer is R if there exists at least one R answer among all the answers returned by the four systems. If there is no R answer, the same strategy is applied to X, U and finally W answers. This "ideal" system permits to calculate the percentage of the questions (and their type), answered by at least one of the four systems in any of the maximum 3 answers returned for a question.

All three systems crashed on the LIST questions. The best results were obtained by ICIA for DEFINITION questions, whereas UAIC performed best with the FACTOID questions. The *combined* system suggests that a joint system, developed by both groups, would improve substantially the general results for Romanian.

Using in a first stage the web interface for assessing the QA runs, developed at UNED in Spain, the assessment took into consideration one question with all its answers at the time, assuring that the same evaluation criteria are uniformly applied to all answers. The judgment of the answers was based on the same Guidelines as in 2007, therefore we kept the same criteria as in 2007, in order to assure consistency inside the Romanian language, which gives also the possibility to evaluate the systems in their evolution from one year to another. For example, one could easily see that the UAIC systems had most of the answers for the DEFINITION questions evaluated as ineXact, because the answers were judged as being "longer than the minimum amount of information required" and hence "unnecessary pieces of information were penalized". Since all the 2007 and 2008 answers were evaluated this way, we considered it is more important to have uniformly applied rules inside one language than to change the evaluation in order to be consistent across languages. On

---

[5] http://celct.isti.cnr.it/ClefQA/index.php?page=showAllGoldStandard.php

the other hand the ICIA answers judged as ineXact are due to answers that are too long, snippets shortened as such as they do not contain the answer, or because the answer and the snippet has no connections.

Tables 31. Results in the monolingual task, Romanian as target language

Run	R	W		U	F	T	D	L	NIL		CWS	MRR	Overall accuracy
	#	#		#	[162]	[47]	[28]	[10]	#	% [8]			
icia081roro	10	179	1	0	4.938	8.511	7.143	0.0	15	6.667	0.00812	0.08217	5.0
icia082roro	21	168	1	0	6.173	8.511	39.286	0.0	15	6.667	0.02191	0.14319	10.5
uaic081roro	41	128	7	3	24.691	25.532	3.571	0.0	65	7.692	0.03679	0.34324	20.5
uaic082roro	45	125	6	4	26.543	27.660	3.571	10.0	64	9.375	0.04892	0.36799	22.5

Run	FACTOID QUESTIONS					LIST QUESTIONS					DEFINITION QUESTION				
	R	W	X	U	ACC	R	W	X	U	ACC	R	W	X	U	ACC
Combined	*72*	*75*	*12*	*3*	*44.444*	*1*	*9*	*0*	*0*	*10.000*	*14*	*5*	*10*	*0*	*50.000*
icia081roro	8	144	10	0	4.938	0	10	0	0	0.000	2	25	1	0	7.143
icia082roro	10	143	9	0	6.173	0	10	0	0	0.000	11	15	2	0	39.286
uaic081roro	40	113	6	3	24.691	0	9	1	0	0.000	1	6	21	0	3.571
uaic082roro	43	110	5	4	26.543	1	9	0	0	10.000	1	6	21	0	3.571

3.11 Spanish as Target

The participation at the Spanish as Target subtask has decreased from 5 groups in 2007 to 4 groups this year. 6 runs were monolingual and 3 runs were crosslingual. Table 32 shows the summary of systems results with the number of Right (R), Wrong (W), Inexact (X) and Unsupported (U) answers. The table shows also the accuracy (in percentage) of factoids (F), factoids with temporal restriction (T), definitions (D) and list questions (L). Best values are marked in bold face.

Table 32 shows the big overall difference (around relative 50%) between the first system (Priberam) and the rest. However, as in last three editions, INAOE is the best system answering definitions (up to 95% of accuracy this year). We wonder why the rest of participants don't implement their technology.

This year, up to three answers were assessed per question and thus, MRR values are given.

Table 32. Results for Spanish as target

Run	R #	W #	X #	U #	% F [124]	% T [36]	% D [20]	% L [20]	NIL #	F [10]	MRR	Overall accuracy
prib081eses	**86**	105	5	4	**41,13**	**41,67**	75	**20**	3	0,17	**0,4483**	**42,5**
inao082eses	44	152	3	1	19,35	8,33	80	5	4	0,10	0,2342	22
inao081eses	42	156	1	1	15,32	8,33	**95**	5	3	0,13	0,2375	21
qaua082eses	39	156	4	1	22,58	13,89	30	-	6	0,15	0,2217	19,5
mira081eses	32	156	3	9	12,90	2,78	75	-	3	0,21	0,1766	16
mira082eses	29	159	3	9	11,29	2,78	70	-	3	0,23	0,1591	14,50
qaua081enes	25	173	-	2	11,29	16,67	20	5	6	0,19	0,1450	12,50
qaua082enes	18	176	3	3	9,68	8,33	15	-	8	0,15	0,1108	9
mira081fres	10	185	2	3	5,65	-	15	-	3	0,12	0,0533	5

Table 33 shows that the first question of the topic group is answered much more easily than the rest of the questions of the group. The first one is, somehow, self-contained, while the rest contain references to the previous questions and answers.

Table 33. Results for self-contained and linked questions, compared with overall accuracy

Run	% Accuracy over Self-contained questions [139]	% Accuracy over Linked questions [61]	% Overall Accuracy [200]
prib081eses	53,24	18,03	42,50
inao082eses	25,18	13,11	22,00
inao081eses	25,18	9,84	21,00
qaua082eses	22,30	13,11	19,50
mira081eses	21,58	3,28	16,00
mira082eses	21,58	3,28	14,50
qaua081enes	17,27	-	12,50
qaua082enes	12,23	1,64	9,00
mira081fres	6,47	1,64	5,00

Table 34 shows the harmonic mean (F) of precision and recall for NIL questions. The values are very low at this respect.

Table 34. Results for Spanish as target for NIL questions

	F-measure (Self-contained@1)	F-measure (@1)	Precision (@1)	Recall (@1)
prib081eses	0,26	0,17	0.12	0.30
inao082eses	0,14	0.10	0.06	0.40
inao081eses	0,19	0.13	0.08	0.30
qaua082eses	0,27	0.15	0.09	0.60
mira081eses	0,27	0.21	0.17	0.30
mira082eses	0,29	0.23	0.19	0.30
qaua081enes	0,26	0.19	0.11	0.80
qaua082enes	0,20	0.15	0.09	0.60
mira081fres	0,15	0.12	0.07	0.30

The correlation coefficient *r* between the self-score and the correctness of the answers (shown in Table 35) has been similar to the obtained last year in general terms. The table shows also the performance in the Answer Extraction step. Since a supporting snippet is requested in order to assess the correctness of the answer, we have evaluated the systems capability to extract the answer when the snippet contains it. The first column of Table 35 shows the percentage of cases in which the correct answer was finally extracted from the snippet once the snippet was the right one. This information is very useful to diagnose if the lack of performance is due to the passage retrieval or to the answer extraction process. In general, all systems have improved their performance in Answer Extraction compared with previous editions.

Observe that the best system achieves the best *r* score and has the best answer extraction module.

Table 35. Answer extraction and correlation coefficient (r) for Spanish as target

Run	%Answer Extraction	r
prib081eses	90,53	0,4006
mira082eses	80,56	0,0771
inao082eses	80,00	0,1593
mira081eses	80,00	0,0713
qaua082eses	73,58	0,2466
inao081eses	67,74	0,1625
qaua081enes	75,76	0,0944
qaua082enes	58,06	0,0061

mira081fres	55,56	0,0552

With respect to the source of the answers, Table 36 shows that in this second year of using Wikipedia, this collection is now the main source of correct answers for most of the systems (with the exception of U. of Alicante).

Table 36. Results for questions with answer in Wikipedia and EFE

Run	% Of correct answers found in EFE	% Of Correct Answers found in Wikipedia	% Of Correct answers found NIL
prib081eses	36,97	60,50	2,52
inao082eses	24,14	68,97	6,90
inao081eses	25	70	5
qaua082eses	48,53	42,65	8,82
mira081eses	23,26	69,77	6,98
mira082eses	21,62	70,27	8,11
qaua081enes	52,27	29,55	18,18
qaua082enes	48,57	34,29	17,14
mira081fres	33,33	41,67	25

4 Conclusions

This year we proposed the same evaluation setting as in 2007 campaign. In fact, last year the task was changed considerably and this affected the general level of results and also the level of participation in the QA task. This year participation increased slightly but the task proved to be still very difficult. This decrease in participation can be explained by the discouragement of some participants. Some have complained that the task is each year harder (e.g. this year, there were more closed list questions and anaphoric references than last year) that can result in a decrease in the systems performances.

Moreover, the overall decrease in accuracy was probably due to linked questions. This fact confirms that topic resolution is a weak point for QA systems, and a not well defined task in the case of bilingual exercises.

Wikipedia increased its presence as a source of questions and answers. Following last year's conclusions Wikipedia seemed to be a good source for finding answers to simple factoid questions and definitions.

Very few runs obtained any points for the closed list questions. Some answers contained some parts of the expected list that were correct but very few were completely correct. We were unable to award points for partially correct answers to closed list questions in the current assessment scheme.

Only 5 out of 11 target languages had more than one different participating group. Thus from the evaluation methodology perspective, a comparison between systems working under similar circumstances cannot be accomplished and this impedes one of the major goals of campaigns such the QA@CLEF, i.e. the systems comparison which could determine an improvement in approaching QA problematic issues.

In conclusion, it is clear that a redefinition of the task should be thought in the next campaign. This new definition of the task should permit the evaluation and comparison of systems even working in different languages. The new setting should also take as reference a real user scenario, perhaps in new documents collection.

Acknowledgements. A special thank to Danilo Giampiccolo (CELCT, Trento, Italy), who has given his precious advise and valuable support at many levels for the preparation and realization of the QA track at CLEF 2008, and to Giovanni Moretti (CELCT, Trento, Italy), who has designed and maintained the QA website and the online interfaces for data management.

Anselmo Peñas has been partially supported by the Spanish Ministry of Science and Technology within the Text-Mess-INES project (TIN2006-15265-C06-02).

Paulo Rocha was supported by the Linguateca project, jointly funded by the Portuguese Government and the European Union (FEDER and FSE), under contract ref. POSC/339/1.3/C/NAC

References

1. QA@CLEF Website, `http://clef-qa.itc.it/`
2. AVE Website, `http://nlp.uned.es/QA/ave/`
3. QAST Website, `http://www.lsi.upc.edu/~qast/`
4. QA-WSD Website, `http://ixa2.si.ehu.es/qawsd/`
5. QA@CLEF 2007. Guidelines (2007), `http://clef-qa.itc.it/2007/download/QACLEF07_Guidelines-for-Participants.pdf`
6. Hartrumpf, S., Glöckner, I., Leveling, J.: University of Hagen at QA@CLEF 2007: Coreference Resolution for Questions and Answer Merging. In: Peters, C., et al. (eds.) CLEF 2008. LNCS, vol. 5706, pp. 421–428. Springer, Heidelberg (2009)
7. Herrera, J., Peñas, A., Verdejo, F.: Question Answering Pilot Task at CLEF 2004. In: Peters, C., Clough, P., Gonzalo, J., Jones, G.J.F., Kluck, M., Magnini, B. (eds.) CLEF 2004. LNCS, vol. 3491, pp. 581–590. Springer, Heidelberg (2005)
8. Landis, J.R., Koch, G.G.: The measurements of observer agreement for categorical data. Biometrics 33, 159–174 (1997)
9. Magnini, B., Giampiccolo, D., Forner, P., Ayache, C., Jijkoun, V., Osenova, P., Peñas, A., Rocha, P., Sacaleanu, B., Sutcliffe, R.: Overview of the CLEF 2006 Multilingual Question Answering Track. In: Peters, C., Clough, P., Gey, F.C., Karlgren, J., Magnini, B., Oard, D.W., de Rijke, M., Stempfhuber, M. (eds.) CLEF 2006. LNCS, vol. 4730, pp. 223–256. Springer, Heidelberg (2007)
10. Vallin, A., Magnini, B., Giampiccolo, D., Aunimo, L., Ayache, C., Osenova, P., Peñas, A., de Rijke, M., Sacaleanu, B., Santos, D., Sutcliffe, R.: Overview of the CLEF 2005 Multilingual Question Answering Track. In: Peters, C., Gey, F.C., Gonzalo, J., Müller, H., Jones, G.J.F., Kluck, M., Magnini, B., de Rijke, M., Giampiccolo, D. (eds.) CLEF 2005. LNCS, vol. 4022, pp. 307–331. Springer, Heidelberg (2006)

11. Voorhees, E.: Overview of the TREC 2002 Question Answering Track. In: NIST Special Publication 500-251: The Eleventh Text REtrieval Conference (TREC 2002). National Institute of Standards and Technology, USA (2002)
12. Agirre, E., Lopez de Lacalle, O.: UBC-ALM: Combining k-NN with SVD for WSD. In: Proceedings of the 4th International Workshop on Semantic Evaluations (SemEval 2007), Prague, Czech Republic, pp. 341–345 (2007)
13. Chan, Y.S., Ng, H.T., Zhong, Z.: NUS-PT: Exploiting Parallel Texts for Word Sense Disambiguation in the English All-Words Tasks. In: Proceedings of the 4th International Workshop on Semantic Evaluations (SemEval 2007), Prague, Czech Republic, pp. 253–256 (2007)

Overview of the Answer Validation Exercise 2008

Álvaro Rodrigo , Anselmo Peñas, and Felisa Verdejo

Dpto. Lenguajes y Sistemas Informáticos, UNED
{alvarory,anselmo,felisa}@lsi.uned.es

Abstract. The Answer Validation Exercise at the Cross Language Evaluation Forum (CLEF) is aimed at developing systems able to decide whether the answer of a Question Answering (QA) system is correct or not. We present here the exercise description, the changes in the evaluation with respect to the last edition and the results of this third edition (AVE 2008). Last year's changes allowed us to measure the possible gain in performance obtained by using AV systems as the selection method of QA systems. Then, in this edition we wanted to reward AV systems able to detect also if all the candidate answers to a question are incorrect. 9 groups have participated with 24 runs in 5 different languages, and compared with the QA systems, the results show an evidence of the potential gain that more sophisticated AV modules might introduce in the task of QA.

1 Introduction

The first Answer Validation Exercise (AVE 2006) [12] was activated two years ago in order to promote the development and evaluation of subsystems aimed at validating the correctness of the answers given by Question Answering (QA) systems. In some sense, systems must emulate human assessment of QA responses and decide whether an answer is correct or not according to a given supporting text. This automatic Answer Validation is expected to be useful for improving QA systems performance [5]. However, the evaluation methodology in AVE 2006 did not permit to quantify this improvement and thus, the exercise was modified in AVE 2007 [14], where the problem of Automatic Hypothesis Generation was also introduced.

In AVE 2007, participant systems had to emulate QA systems selecting one answer per question from a set of candidate ones. These candidate answers were the ones given by QA systems participating at the QA main track at CLEF. This allowed us to study the use of Answer Validation (AV) systems as the answer selection method used by a multi-stream QA system. Nevertheless, it was not acknowledged the ability of an AV system detecting if all the candidate answers to a question were incorrect. Systems with this ability could ask for new answers to the QA systems, opening the possibility of obtaining a correct answer to the question in a second chance. Besides, NIL answers could be detected.

Since in this edition we were interested in studying this ability, we have introduced new measures in the evaluation framework. The purpose of these new measures is to account the contribution that a detection of questions without correct answers could have in the results of a multi-stream QA system that uses AV for the selection of the final answer. These measures work under the assumption that the AV system could ask

C. Peters et al. (Eds.): CLEF 2008, LNCS 5706, pp. 296–313, 2009.

for more answers in a second chance when no correct answer to a question has been detected.

2 Exercise Description

Participant systems receive a set of pairs {*Answer, Supporting Text*} grouped by *Question* (see Figure 1 for an example). They must consider the *Question* and validate each of these {*Answer, Supporting Text*} pairs. The number of answers to be validated per question depends on the number of participant systems at the QA main track.

Participant systems must return one of the following values for each answer according to the response format (see Figure 2):

- **VALIDATED** indicates that the answer is correct and supported by the given supporting text. There is no restriction in the number of VALIDATED answers returned per question (from zero to all).
- **SELECTED** indicates that the answer is VALIDATED and it is the one chosen as the output to the current question by a hypothetical QA system. The SELECTED answers are evaluated against the QA systems of the Main Track. No more than one answer per question can be marked as SELECTED. At least one of the VALIDATED answers of a question group must be marked as SELECTED.

```
<q id="0001" lang="EN">
        <q_str>What was the nationality of Jacques
        Offenbach?</q_str>
        <a id="0001_1" value="">
                <a_str>Germany</a_str>
                <t_str doc="Offenbach">Offenbach Offenbach Offenbach
                can refer to: The city Offenbach in Hesse,
                Germany.</t_str>
        </a>
        <a id="0001_2" value="">
                <a_str>France</a_str>
                <t_str doc="Jacques Offenbach">His son received the
                name "Jakob Offenbach" at birth, though he changed it
                to Jacques when he settled in France.</t_str>
        </a>
        <a id="0001_3" value="">
                <a_str>Thousand Oaks</a_str>
                <t_str doc="LA111794-0288">Ventura College's
                production of George Bernard Shaw's "Arms and the
                Man" and  Moorpark College's version of the Jacques
                Offenbach operetta "La Vie  Parisienne" are the
                costume shows; in Thousand Oaks, Cal Lutheran
                University is  mounting the contemporary drama "Minor
                Demons."</t_str>
        </a>
        ...
</q>
```

Fig. 1. Excerpt of the English test collection in AVE 2008

```
|q_id a_id [SELECTED|VALIDATED|REJECTED] confidence_score|
```

Fig. 2. Response format in AVE 2008

- **REJECTED** indicates that the answer is incorrect or there is not enough evidence of its correctness. There is no restriction in the number of REJECTED answers per question (from zero to all).

This configuration permitted us to compare the responses of the AV systems with those of the QA systems, and to obtain some evidences about the gain in performance that sophisticated AV modules might give to QA systems (see below).

3 Collections

Like in the past edition of QA@CLEF [3], questions in the QA 2008 track were grouped by topic. In this organization by topics, the first question of each topic is self contained in the sense that there is no need of information outside the question to answer it. However, the rest of the topic questions can refer to implicit information linked to the previous questions and answers of the topic group (anaphora, co-reference, etc.). Therefore, for the AVE 2008 test collections we only made use of the self-contained questions (the first one of each topic group) and their respective answers given by the participant systems in QA.

A goal of the exercise is to promote the development of AV systems that perform an analysis beyond the use of redundancies in answers. Since the fact of grouping all the answers to the same question could lead to provide extra information based on counting answer redundancies, if an answer is contained in another answer, we remove the shorter one. Furthermore, NIL and empty answers were discarded for building the AVE collections. This processing led to a reduction in the number of answers initially given by QA systems (see Tables 1 and 2): from 38.36% in the English development collection to 78.57% in the Bulgarian test collection.

For the assessments, we reused the QA judgments because they were done considering the supporting snippets in a similar way to the AV systems must do. The relation between QA assessments and AVE judgments was the following:

- Answers judged as *Right* in QA have a value equal to *VALIDATED* in AVE.
- Answers judged as *Wrong* or *Unsupported* in QA have a value equal to *REJECTED* in AVE.
- Answers judged as *Inexact* in QA have a value equal to *UNKNOWN* in AVE and are ignored for evaluation purposes.
- Answers not evaluated at the QA main track (if any) are also tagged as *UNKNOWN* in AVE and they are also ignored in the evaluation.

3.1 Development Collections

Development collections were obtained from the QA@CLEF 2006 [9] and 2007 [3] main track questions and answers. Table 1 shows the number of questions and answers

Table 1. Number of questions and answers in the AVE 2008 development collections

	German	English	Spanish	French	Italian	Dutch	Portuguese	Romanian
Questions	295	267	369	318	292	276	348	82
Answers(final)	768	1316	2368	1674	576	724	1163	103
% over available answers	35.1	61.64	55.72	51.54	51.61	45.53	36.34	42.21
VALIDATED	202	151	392	348	102	131	301	45
REJECTED	566	1165	1976	1326	474	593	862	58

Table 2. Number of questions and answers in the AVE 2008 test collections

	German	English	Spanish	French	Bulgarian	Dutch	Portuguese	Romanian	Basque
Questions	119	160	136	108	27	128	149	119	104
Answers(final)	1027	1055	1528	199	27	228	1014	497	541
% over available answers	39.61	57.37	49.98	60.30	21.43	42.54	43.63	48.58	55.09
VALIDATED	111	79	153	52	12	44	208	52	39
REJECTED	854	940	1354	126	9	177	747	406	483
UNKNOWN	62	36	21	21	6	7	59	39	19

for each language together with the percentage that these answers represent over the number of answers initially available, and the number of answers with VALIDATED and REJECTED values.

These collections were available for participants after their registration at CLEF at http://nlp.uned.es/clef-qa/ave/

3.2 Test Collections

Test collections were obtained from the runs sent to QA@CLEF 2008 main track [2]. In this edition, there were runs in 9 languages: German, English, Spanish, French, Bulgarian, Dutch, Portuguese, Romanian and Basque. Thus, a test collection in AVE was generated for each of these languages.

Table 2 shows the number of questions and the number of answers to be VALIDATED (or REJECTED) in the test collections together with the percentage that these answers represent over the answers initially available. The number of UNKNOWN answers in each collection is also given.

4 Evaluation

We used two groups of measures in order to evaluate the performance of AV systems: one group for evaluating the ability of systems detecting correct answers, and another group for evaluating AV systems selecting answers from different streams. Before describing the proposed groups of measures, we discuss some of the decisions taken into account in the evaluation performed at AVE.

4.1 Preliminary Discussion

At the time of thinking how to perform the evaluation of AVE we thought in several options. On one hand, since the AVE task can be seen as a classification of answers in correct or incorrect ones, we thought of an evaluation based on accuracy like the one performed in Machine Learning. According to this evaluation, the detection of incorrect and correct answers is rewarded in the same proportion. However, as it was argued in [13], the unbalanced nature of the collections moved us to use an approach based on the evaluation of correct answers.

In evaluations with unbalanced collections it is usual to follow an approach based on *precision* and *recall*:

- *Precision*: the proportion of answers validated by the system that are actually correct (see formula (1)).
- *Recall*: the proportion of correct answers detected by the system (see formula (2)).

With these measures, two different approaches can be taken:

1. Calculate the values over the whole pool of answers. That is, to calculate a global *precision* and a global *recall*.
2. Calculate the values of *precision* and *recall* per question group, and then, to calculate the average *precision* and the average *recall* over all the question groups.

Each of these approaches have its advantages and disadvantages and the decision of which one to choose depends on the objectives of the evaluation. If we want to evaluate the performance of systems validating answers per question, that is, to evaluate the detection of correct answers to a question, then we should follow option 2. However, this second approach has the following problems:

- Sometimes, in some question groups there is not any correct answer. In this cases, it makes no sense to talk about *the proportion of correct answers detected* since there are no correct answers to be detected. This means that we cannot talk of *recall* in these question groups and as a consequence, it makes no sense to calculate the average *recall* (since the *recall* value does not exist in some question groups).
- In the output produced by a system, the number of answers given as validated in a question group is variable. Thus, in some cases the system can validate zero answers in a question groups (that means that the system rejects all the answers of the question group). Therefore, in these question groups it is not reasonable to calculate *the proportion of answers validated by the system*, which is the *precision*, because no answer has been validated. Again, it does not make sense to calculate an average value under these conditions.

Given the drawbacks explained above, we find that the second approach was more problematic to apply and less informative for our purposes than the first one. For this reason, we opted for a global evaluation over the whole pool of answers as we described in section 4.2.

4.2 Evaluating the Correct Validation

The objective of the first group of measures is to evaluate the ability of AV systems validating answers from a pool of available ones. Thus, the measures of this group are useful for the evaluation of an AV system used for ranking or filtering answers.

As we argued above, instead of using an overall accuracy, the first group of measures is composed by *precision* (1), *recall* (2) and *F-measure* (3) (harmonic mean) over answers that must be VALIDATED (in this first group of measures when a participant system returns SELECTED to an answer, the answer is considered as VALIDATED).

Results can be compared between systems but always taking as reference the following baselines:

- A system that accepts all answers (returns VALIDATED or SELECTED in 100% of cases)
- A system that accepts 50% of the answers (randomly)

$$precision = \frac{|VALIDATED\ correctly|}{|VALIDATED|} \tag{1}$$

$$recall = \frac{|VALIDATED\ correctly|}{|CORRECT|} \tag{2}$$

$$F = \frac{2 * recall * precision}{recall + precision} \tag{3}$$

4.3 Evaluating the Correct Selection

Since our aim was to obtain evidences about the usefulness of using AV for selecting answers from different streams, the second group of measures was created with this purpose in AVE 2007. Thus, this group of measures aims at comparing QA systems performance with the potential gain that AV systems could add to them.

Since answers were grouped by questions and AV systems were requested to SELECT one or none of them, for each question there are two possible situations:

- There is only an answer selected.
- All the answers have been rejected.

Thus, the resulting behavior is comparable to a QA system: for each question there is no more than one SELECTED answer. The first of these measures, which was already used in AVE 2007, is *qa_accuracy* (4): the proportion of questions for which a correct answer has been selected. This is a measure comparable to the accuracy used in the QA Main Track and therefore, we can compare AV systems taking as reference the QA systems performance over the questions involved in AVE test collections.

This measure has an upper bound given by the proportion of questions that have at least one correct answer (in its corresponding group). This upper bound corresponds to a perfect selection of the correct answers given by all the QA systems at the main track. The normalization of *qa_accuracy* with this upper bound is given by *%_best_combination* (5), where the percentage of the perfect selection is calculated.

Besides the upper bound, results of *qa_accuracy* can be compared with the following baseline system: a system that validates 100% of the answers and selects randomly one of them. Thus, this baseline can be seen as the average proportion of correct answers per question group. We called this baseline *random_qa_accuracy* (6). Moreover, another baseline can be also taken into account: since a good AV system should be able to outperform the best QA system, we can consider the best QA system of each language as a baseline.

$$qa\_accuracy = \frac{|answers\ SELECTED\ correctly|}{|questions|} \tag{4}$$

$$\%\_best\_combination = \frac{|answers\ SELECTED\ correctly|}{|questions\ with\ correct\ answers|} * 100 \tag{5}$$

$$random\_qa\_accuracy = \frac{1}{|questions|} \sum_{q \epsilon questions} \frac{|correct\ answers\ of\ (q)|}{answers\ of(q)} \tag{6}$$

The problem of *qa_accuracy* is that it only acknowledges the ability of a system for selecting correct answers and not the ability of detecting that all the answers to a question are incorrect, which is an ability we wanted to acknowledge in this edition. The justification of why to acknowledge this ability arises from the fact that a possible gain in performance could be obtained in these questions. In this situation, the AV system could ask to the QA systems for other answers to these questions, opening the possibility of obtaining correct answers to these questions.

With the purpose of evaluating this behavior, we proposed in AVE 2008 the use of *qa_rej_accuracy* (7), which acknowledges systems capable of detecting correctly questions without correct answers. That is, questions were all the given answers are incorrect.

Thus, with this measure and *qa_accuracy* we can propose *qa_accuracy_max* (8). This measure represents a range with a lower bound expressed by *qa_accuracy* and an upper bound that adds to *qa_accuracy* the accuracy that would be obtained answering correctly all the questions accounted by *qa_rej_accuracy*. Besides, *qa_accuracy_max* can be seen as a measure corresponding to the usual idea of *accuracy* used in classification. That is, the measure accounts both the number of questions were a correct answer has been found (which can be seen in classification as the number of correct examples detected) and the number of questions were all the answers has been correctly rejected (this is similar to the detection of negative examples in classification). However, the unbalanced nature of the collections forces us to consider *qa_accuracy_max* just as an upper bound of an AVE system performance.

With the objective of giving an evaluation measure no so sensible to the unbalanced nature of the AVE collections and able to reward both the correct selections and the correct rejections (questions where the AVE system detects that all the answers are incorrect), we proposed *estimated_qa_performance* (9) in this edition of AVE. This measure can be seen as an estimation of the performance achieved by a system with an upper bound of *qa_accuracy_max*. Thus, *estimated_qa_performance* rewards the accuracy of AV systems detecting questions without correct answers in the proportion they select correct ones (the value of *qa_accuracy*).

$$qa\_rej\_accuracy = \frac{|questions\ REJECTED\ correctly|}{|questions|} \quad (7)$$

$$qa\_accuracy\_max = qa\_accuracy + qa\_rej\_accuracy \quad (8)$$

$$estimated\_qa\_performance = qa\_accuracy + qa\_rej\_accuracy * qa\_accuracy \quad (9)$$

5 Results

Nine groups (the same number that in the last edition) have participated in five different languages (German, English, Spanish, French and Romanian) with 24 runs. Table 3 shows the participant groups and the number of runs they submitted per language. Again, English and Spanish were the most popular with 8 and 6 runs respectively.

Tables 5-9 in the appendix show the results of *precision*, *recall* and *F-measure* over correct answers for all participant systems in each language. Results cannot be compared between languages since the number of answers to be validated and the proportion of correct ones are different for each language (due to the real submission of QA systems). However, they can be compared in each language with two baselines values that are given: the results of a system that always accepts all answers (validates 100% of the answers), and the results of a hypothetical system that validates the 50% of answers.

One of our goals is to obtain some evidences about the potential improvement that AV systems could provide to QA systems. Tables 10-14 in the appendix show the rankings of systems (merging QA and AV systems) according to *estimated_qa_performance* calculated only over the subset of questions considered in AVE 2008. The tables contain also the information about the results of QA and AVE systems using the measures *qa_accuracy*, *%_best_combination*, *qa_rej_accuracy* and *qa_accuracy_max*. The values of *qa_accuracy* and *estimated_qa_performance* are the same in QA systems. Again, results cannot be compared between different languages, but they can be compared with the random baselines and with the results of the best QA system in each language.

Table 3. Participants and runs per language in AVE 2008

	German	**English**	**Spanish**	**French**	**Romanian**	**Total**
Fernuniversität in Hagen (FUH)	2					2
LIMSI				2		2
U. Iasi		2			2	4
DFKI	1	1				2
INAOE			2			2
U. Alicante		1	2			3
UNC		2				2
U. Jaén		2	2	2		2
LINA				1		1
Total	3	8	6	5	2	24

Table 4. Information about the techniques used by the AVE participants.

	U. Iasi	INAOE	FUH	DFKI	U. Jaén	U. Alicante	LIMSI	LINA	UNC
Generates hypotheses	X					X			
WordNet	X		X		X	X			X
Chunking	X				X		X	X	
n-grams, longest common subsequences				X	X	X			X
Phrase transformations	X						X		
NER	X	X	X	X		X	X	X	
Num. expressions	X	X	X	X		X	X	X	
Temp. expressions	X	X	X				X	X	
Coreference resolution									
Dependency analysis	X			X			X		
Syntactic similarity									
Functions (sub, obj, etc)	X			X			X		
Syntactic transformations	X						X		
Word-sense disambiguation			X						
Semantic parsing	X		X						
Semantic role labeling			X						
First order logic representation	X		X						
Theorem prover			X						
Semantic similarity									

The graphic interpretations of these tables are shown in Figures 3-7 in the appendix. In these graphics the value of *qa_accuracy_max* is 1 in the perfect selection baseline. This corresponds to a perfect selection of a correct answer (if any) per question and the detection of all the questions with no correct answers (*qa_rej_accuracy*). However, the value of *estimated_qa_performance* in this baseline is not 1 because it is assumed that the questions detected in *qa_rej_accuracy* will be answered with a precision value equal to the *qa_accuracy* of the perfect selection baseline. This *qa_accuracy* represents the accuracy of the best combination of the QA systems involved, which is not perfect.

In three languages (German, English and Romanian) there has been at least one AV system performing better than the best QA system (the AVE system *ltqa* beat the QA system *dfki082dede* in German as it can be seen in Table 10; the AVE systems *ltqa*, *ofe* and *uaic_2* beat the QA system *wlvs081roen* in English according to Table 13 and the AVE system *uaic_2* defeat the QA system *UAIC082roro* in Romanian according to Table 14). In the languages where the best value of *qa_accuracy* was not obtained by an AV system, the best QA system outperforms in more than a 100% the following QA systems (the QA system *prib081eses* outperforms the QA system *inaoe081eses* in a 116% in Spanish according to Table 11 whereas the QA system *syna081frfr* outperforms in a 147% the QA system *syna081ptfr* in French according to Table 12). If we see an AV system as a multi-stream selector of candidate answers, then AV systems follow a behavior similar to an ensemble of classifiers. An ensemble of classifiers is likely to be more accurate than an individual classifier except in the case of an element of the assemble outperforms in a high percent the rest of the classifiers [1].

5.1 Analysis of Measures

Regarding the use of the new measure *estimated_qa_performance*, the rankings are very similar to the ones obtained ranking by *qa_accuracy*. In fact, there have been only two changes, which are located in the English ranking (see Table 13 in the appendix). Firstly, the system `uaic_2` obtains a better performance than `ofe` according to *qa_accuracy* (0.24 against 0.19). However, according to *estimated_qa_performance*, `ofe` is better than `uaic_2` (0.27 against 0.24). This means that `uaic_2` is better selecting correct answers. Nevertheless, if we consider the possible gain in performance that might be obtained detecting that all the answers to a question are incorrect and asking for new ones to the QA systems, then `ofe` is better. Therefore, the system `ofe` may help to obtain better results in QA than the system `uaic_2`. Besides, it can be seen how the ranking according to *estimated_qa_performance* is more similar to the one given by *F-measure*, which in some way, also considers the precision of a system detecting incorrect answers. The second change in the rankings involves the QA system `dfki081deen`, which has a better performance than the AVE system `jota_2` according to *qa_accuracy*. However, according to *estimated_qa_performance*, the two systems have the same performance. Again, this indicates that AV systems detecting incorrect answers could lead to a better performance in QA.

Then, it seams that *estimated_qa_performance* is a better measure for AV systems than *qa_accuracy* because it takes into account the ability of a system rejecting incorrect answers. Thus, a better estimation of the performance obtained by using AV systems in QA is being given. Furthermore, the rankings are more similar to the ones obtained by using *F-measure*.

5.2 Analysis of the Techniques Used

The participation in AVE 2008 has showed evidences of the growing interest in using AV in QA participant systems at CLEF, since 6 of the 9 groups participants in AVE have also participated in the QA main track [4,7,18,19,11,10]. In fact, two of these participants have used their AV systems as a component in their QA@CLEF systems [4,18], obtaining an improvement in the performance [6,18].

Regarding the techniques used, all the participants have used textual entailment (TE) in their systems except two groups [8,10]. Instead of using TE, these two participants used QA systems for performing the task. Then, their QA systems looked for answers to the questions involved in AVE. Finally, the answers in the AVE test collections were compared with the ones given by their QA systems in order to take the final decision of validation or rejection.

On the other hand, while in the past edition the half of the participants reported the use of automatic hypothesis generation, in this edition only three participants have used it [7,11,19]. They had questions patterns that were instantiated with the corresponding answers in order to build each hypothesis.

Table 4 shows the techniques used by AVE participant systems. Following the tendency showed in the past edition, all the systems have reported the use of lexical processing. Moreover, this year there are more groups using syntactic processing, mainly chunking or dependency analysis. In fact, the system with the best result in each language, except in Spanish, performed some kind of syntactic processing mainly by means of dependency parsing. However, the use of semantic analysis has decreased while the use of WordNet has been increased (50% of participants used it).

Furthermore, there has been a high increase in the use of Named Entities (NE), with 7 of 9 groups considering them. In particular, the participants who generated hypothesis gave also a high importante to the NEs [7,11,19]. They used the restriction that all the NEs in a hypothesis had to be present in the corresponding supporting snippet in order to validate the answer. Therefore, it seems that the recognition of NEs is being used as an important source of information to be taken into account in AV [17].

Some systems have also worked in other QA focused features like the expected type of answer [19,10,18,7]. Thus, participant systems that took this consideration into account detected whether the expected type of answer matched with the type of the given answer. While for some participants it was a feature to be used in combination with other ones [18], for other participants it was taken as a constraint necessary for validating an answer [19].

Regarding the final validation decision, most of participants have used ML following the tendency of the last edition. Besides, ML was used by the participants with the best score in each language. While lexical similarity was the most common used feature, syntactic similarity was included by the half of participants and semantic features were taken into account by very few participants. Furthermore, a participant included also the use of non-overlap features, which showed to be more discriminative than the traditional overlap ones [18].

Only one participant reported the use of a theorem prover this year [4]. Since this participant was interested in having a real time answer validator, his system only checked whether the snippet contained a correct answer to the given question. Thus, some errors were produced when the answer to be validated was incorrect, despite the fact that the snippet contains a correct one. However, he achieved a quick answer validator based on logic.

On the other hand, Support Vector Machines (SVM) and decision trees were the most used classifiers. Nevertheless, there are not evidences about the best performance of one or another of these classifiers.

After a comparison between the tools and the results obtained, it seems that to use more tools or to perform a more complex processing does not guarantee a better performance. For example, according to Table 4, the system *U. Iasi* used more tools than the system *DFKI* (*U. Iasi* used a chunker and a semantic parser while *DFKI* did not). However, according to Tables 8 and 13, the results in English of *U. Iasi* are worse than the ones of *DFKI*. On the other hand, the system *FUH* used also more tools than the system *DFKI* (*FUH* used word-sense disambiguation, semantic parsing and semantic role labeling while *DFKI* did not). Again, the results of *DFKI* are better (see results for German in Tables 5 and 10) despite the fact it used less tools than *FUH*.

Finally, the selection decision was carried out taking the VALIDATED answer with the highest score when more than one VALIDATED answer was found.

6 Conclusions

In AVE 2008 there has been the same number of participants of last year (9) in 5 different languages. However, 8 more runs have been sent, showing a growing interest in the task.

Results show that AV systems could improve the performance of current QA systems. This improvement comes when AV systems are used for selecting the final answer from a set of candidate ones. In fact, according to the results, except in the languages where the best QA system outperforms the others QA systems in more than a 100%, there was an AV system with better performance than QA systems.

In this edition new measures have been introduced in order to obtain a more informative estimation of the potential of AV systems in QA performance. These new measures reward the ability of some systems detecting if all the candidate answers to a question are incorrect. These measures have shown to be very useful when two systems have a similar performance according to *qa_accuracy*. In this situation, the new measure *estimated_qa_performance* has indicated that AV systems with a better precision detecting incorrect answers would be more useful in QA because more answers could be asked to QA systems when all the candidate answers to a question are incorrect. Then, a correct answer might be found in this second chance.

The most used technique continues being lexical processing while the use of syntactic analysis has grown. Nevertheless, very few systems have performed semantic analysis. Besides, a high percent of participants have combined different features using ML. Finally, the best systems performed both lexical and syntactic analysis, as well as they consider NE.

Acknowledgments

This work has been partially supported by the European project Treble CLEF (ICT-1-4-1 215231), the Spanish Ministry of Science and Innovation with the project Text-Mess-INES (TIN2006-15265-C06-02), the Education Council of the Regional Government of Madrid and the European Social Fund. We are grateful to all the people involved in the organization of the QA track (specially to the coordinators at CELCT, Danilo Giampiccolo and Pamela Forner). Special thanks also to Ingo Glöckner, who helped with his comments toward the final editing of the paper.

References

1. Dietterich, T.G.: Machine learning research: Four current directions. AI Magazine 18(4), 97–136 (1997)
2. Forner, P., Peñas, A., Alegria, I., Forascu, C., Moreau, N., Osenova, P., Prokopidis, P., Rocha, P., Sacaleanu, B., Sutcliffe, R., Sang, E.T.K.: Overview of the CLEF 2008 Multilingual Question Answering Track. In: Working Notes for the CLEF 2008 Workshop, Aarhus, Denmark, September 17-19 (2008)
3. Giampiccolo, D., Forner, P., Herrera, J., Peñas, A., Ayache, C., Forascu, C., Jijkoun, V., Osenova, P., Rocha, P., Sacaleanu, B., Sutcliffe, R.F.E.: Overview of the CLEF 2007 Multilingual Question Answering Track. In: Peters, et al [16], pp. 200–236

4. Glöckner, I.: RAVE: A Fast Logic-Based Answer Validator. In: Peters, C., et al. (eds.) CLEF 2008. LNCS, vol. 5706, pp. 468–471. Springer, Heidelberg (2009)
5. Harabagiu, S., Hickl, A.: Methods for Using Textual Entailment in Open-Domain Question Answering. In: Proceedings of the 21st International Conference on Computational Linguistics and 44th Annual Meeting of the ACL, Sydney, pp. 905–912 (2006)
6. Hartrumpf, S., Glöckner, I., Leveling, J.: University of Hagen at QA@CLEF 2008: Efficient Question Answering with Question Decomposition and Multiple Answer Streams. In: Working Notes for the CLEF 2008 Workshop, Aarhus, Denmark, September 17-19 (2008)
7. Iftene, A., Balahur-Dobrescu, A.: Answer Validation on English and Romanian Languages. In: Peters, C., et al. (eds.) CLEF 2008. LNCS, vol. 5706, pp. 448–451. Springer, Heidelberg (2009)
8. Jacquin, C., Monceaux, L., Desmontils, E.: The Answer Validation System ProdicosAV dedicated to French. In: Peters, C., et al. (eds.) CLEF 2008. LNCS, vol. 5706, pp. 452–459. Springer, Heidelberg (2009)
9. Magnini, B., Giampiccolo, D., Forner, P., Ayache, C., Jijkoun, V., Osenova, P., Peñas, A., Rocha, P., Sacaleanu, B., Sutcliffe, R.F.E.: Overview of the CLEF 2006 Multilingual Question Answering Track. In: Peters, et al [15], pp. 223–256
10. Moriceau, V., Tannier, X., Grappy, A., Grau, B.: Justification of Answers by Verification of Dependency Relations - The French AVE Task. In: Working Notes for the CLEF 2008 Workshop, Aarhus, Denmark, September 17-19 (2008)
11. Ferrández, Ó., Muñoz, R., Palomar, M.: Studying the Influence of Semantic Constraints in AVE. In: Peters, C., et al. (eds.) CLEF 2008. LNCS, vol. 5706, pp. 460–467. Springer, Heidelberg (2009)
12. Peñas, A., Rodrigo, Á., Sama, V., Verdejo, F.: Overview of the Answer Validation Exercise 2006. In: Peters, et al [15], pp. 257–264
13. Peñas, A., Rodrigo, Á., Sama, V., Verdejo, F.: Testing the Reasoning for Question Answering Validation. Journal of Logic and Computation 18(3), 459–474 (2008)
14. Peñas, A., Rodrigo, Á., Verdejo, F.: Overview of the Answer Validation Exercise 2007. In: Peters, et al [16], pp. 237–248
15. Peters, C., Clough, P., Gey, F.C., Karlgren, J., Magnini, B., Oard, D.W., de Rijke, M., Stempfhuber, M. (eds.): CLEF 2006. LNCS, vol. 4730. Springer, Heidelberg (2007)
16. Peters, C., Jijkoun, V., Mandl, T., Müller, H., Oard, D.W., Peñas, A., Petras, V., Santos, D. (eds.): CLEF 2007. LNCS, vol. 5152. Springer, Heidelberg (2008)
17. Rodrigo, Á., Peñas, A., Herrera, J., Verdejo, F.: The Effect of Entity Recognition on Answer Validation. In: Peters, et al [15], pp. 483–489
18. Téllez-Valero, A., Juárez-González, A., Gómez, M.M., Villaseñor-Pineda, L.: Using Non-Overlap Features for Supervised Answer Validation. In: Peters, C., et al. (eds.) CLEF 2008. LNCS, vol. 5706, pp. 476–479. Springer, Heidelberg (2009)
19. Wang, R., Neumann, G.: Information Synthesis for Answer Validation. In: Peters, C., et al. (eds.) CLEF 2008. LNCS, vol. 5706, pp. 472–475. Springer, Heidelberg (2009)

Appendix

Tables 5-9 show the values of *precision*, *recall* and *F-measure* over correct answers of AVE participant systems in different languages.

Tables 10-14 show the rankings of systems (merging QA and AV systems) according to *estimated_qa_performance*. The tables contain also the information about the results of QA and AVE systems using the measures *qa_accuracy*, *%_best_combination*, *qa_rej_accuracy* and *qa_accuracy_max*. Figures 3-7 show the graphic interpretations of Tables 10-14.

Table 5. Precision, recall and F measure over correct answers for German

Group	System Id	F	Precision	Recall
DFKI	ltqa	0.61	0.54	0.71
FUH	glockner_1	0.39	0.33	0.49
FUH	glockner_2	0.29	0.25	0.34
100% VALIDATED		0.21	0,12	1
50% VALIDATED		0.19	0.12	0.5

Table 6. Precision, recall and F measure over correct answers for Spanish

Group	System Id	F	Precision	Recall
UA	ofe_2	0.44	0.32	0.67
INAOE	tellez_2	0.39	0.30	0.59
UA	ofe_1	0.38	0.26	0.76
INAOE	tellez_1	0.23	0,13	0.86
100% VALIDATED		0.18	0.10	1
50% VALIDATED		0.17	0.10	0.5
UJA	magc_1(timbl)	0.06	0.15	0.04
UJA	magc_2(bbr)	0.05	0.22	0.03

Table 7. Precision, recall and F measure over correct answers for French

Group	System Id	F	Precision	Recall
LIMSI	bgrau_1	0.61	0.75	0.52
LIMSI	bgrau_2	0.57	0.88	0.42
LINA	monceaux	0.51	0.56	0.46
100% VALIDATED		0.45	0.29	1
50% VALIDATED		0.37	0.29	0.5
UJA	magc_1(timbl)	0.08	0.15	0.06
UJA	magc_2(bbr)	0.08	0.13	0.06

Table 8. Precision, recall and F measure over correct answers for English

Group	System Id	F	Precision	Recall
DFKI	ltqa	0.64	0.54	0.78
UA	ofe	0.49	0.35	0.86
UNC	jota_2	0.21	0.13	0.56
Iasi	uaic_2	0.19	0.11	0.85
UNC	jota_1	0.17	0.09	0.94
Iasi	uaic_1	0.17	0.09	0.76
100% VALIDATED		0.14	0.08	1
50% VALIDATED		0.13	0.08	0.5
UJA	magc_2(bbr)	0.02	0.17	0.01
UJA	magc_1(timbl)	0	0	0

Table 9. Precision, recall and F measure over correct answers for Romanian

Group	System Id	F	Precision	Recall
Iasi	uaic_2	0.23	0.13	0.92
Iasi	uaic_1	0.22	0.12	0.92
100% VALIDATED		0.20	0.11	1
50% VALIDATED		0.19	0.11	0.50

Table 10. Comparing AV systems performance with QA systems in German

System	System type	estimated_qa_performance	qa_accuracy (% best combination)	qa_rej_accuracy	qa_accuracy_max
Perfect selection		0.77	0.52 (100%)	0.48	1
ltqa	AV	0.52	0.43 (82.26%)	0.21	0.64
dfki082dede	QA	0.38	0.38 (72.58%)	0	0.38
dfki081dede	QA	0.37	0.37 (70.97%)	0	0.37
glockner_1	AV	0.32	0.32 (61.29%)	0	0.32
fuha082dede	QA	0.24	0.24 (45.16%)	0	0.24
glockner_2	AV	0.23	0.23 (43.55%)	0	0.23
fuha081dede	QA	0.22	0.22 (41.94%)	0	0.22
loga081dede	QA	0.17	0.17 (32.26%)	0	0.17
fuha082ende	QA	0.16	0.16 (30.65%)	0	0.16
fuha081ende	QA	0.16	0.16 (30.65%)	0	0.16
loga082dede	QA	0.15	0.15 (29.03%)	0	0.15
dfki081ende	QA	0.14	0.14 (27.42%)	0	0.14
fuha081esde	QA	0.12	0.12 (22.58%)	0	0.12
Random		0.11	0.11 (21.13%)	0	0.11
fuha082esde	QA	0.10	0.10 (19.35%)	0	0.10

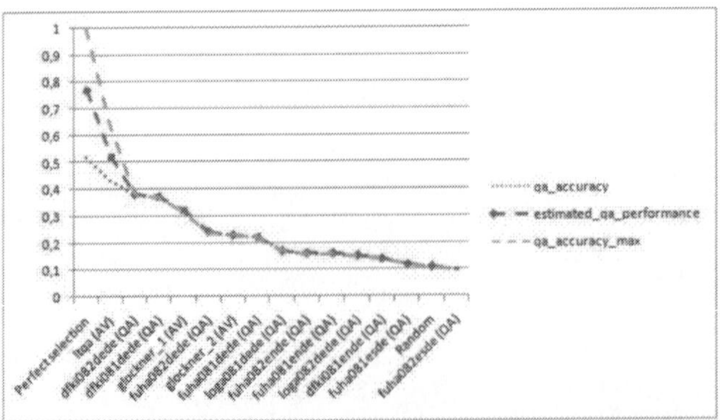

Fig. 3. Graphic comparing AV systems performance with QA systems in German

Table 11. Comparing AV systems performance with QA systems in Spanish

System	System type	estimated_ qa_performance	qa_accuracy (% best combination)	qa_rej_ accuracy	qa_ accuracy_max
Perfect selection		0.85	0.62 (100%)	0.38	1
prib081eses	QA	0.54	0.54 (88.10%)	0	0.54
ofe_1	AV	0.37	0.32 (52.38%)	0.14	0.46
tellez_1	AV	0.34	0.32 (52.38%)	0.06	0.38
ofe_2	AV	0.33	0.27 (44.05%)	0.21	0.48
tellez_2	AV	0.33	0.27 (44.05%)	0.22	0.49
inao081eses	QA	0.25	0.25 (40.48%)	0	0.25
inao082eses	QA	0.25	0.25 (40.48%)	0	0.25
qaua082eses	QA	0.22	0.22 (35.71%)	0	0.22
mira081eses	QA	0.21	0.21 (33.33%)	0	0.21
mira082eses	QA	0.18	0.18 (29.76%)	0	0.18
qaua081enes	QA	0.18	0.18 (28.57%)	0	0.18
qaua082enes	QA	0.13	0.13 (21.43%)	0	0.13
qaua081eses	QA	0.12	0.12 (19.05%)	0	0.12
Random		0.11	0.11 (17.12%)	0	0.11
mira081fres	QA	0.06	0.06 (9.52%)	0	0.06
magc_1(timbl)	AV	0.06	0.04 (7.14%)	0.32	0.36
magc_2(bbr)	AV	0.03	0.02 (3.57%)	0.35	0.37

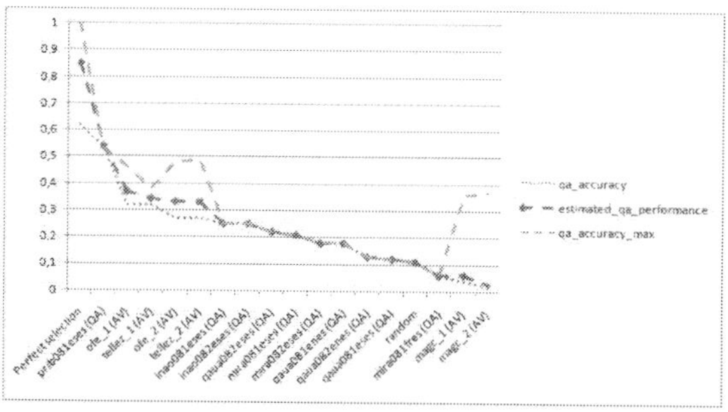

Fig. 4. Graphic comparing AV systems performance with QA systems in Spanish

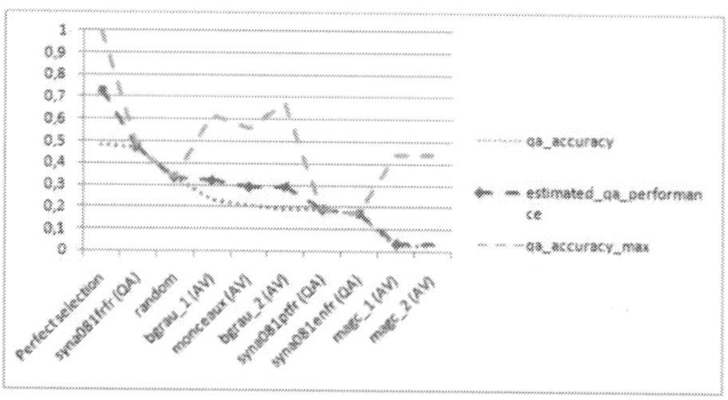

Fig. 5. Graphic comparing AV systems performance with QA systems in French

Table 12. Comparing AV systems performance with QA systems in French

System	System type	estimated_ qa_performance	qa_accuracy (% best combination)	qa_rej_ accuracy	qa_ accuracy_max
Perfect selection		0.73	0.48 (100%)	0.52	1
syna081frfr	QA	0.47	0.47 (98.08%)	0	0.47
Random		0.33	0.33 (68.80%)	0	0.33
bgrau_1	AV	0.32	0.23 (48.08%)	0.39	0.62
monceaux	AV	0.29	0.21 (44.23%)	0.35	0.56
bgrau_2	AV	0.29	0.19 (40.38%)	0.48	0.67
syna081ptfr	QA	0.19	0.19 (40.38%)	0	0.19
syna081enfr	QA	0.17	0.17 (34.62%)	0	0.17
magc_1(timbl)	AV	0.04	0.03 (5.77%)	0.41	0.44
magc_2(bbr)	AV	0.04	0.03 (5.77%)	0.41	0.44

Table 13. Comparing AV systems performance with QA systems in English

System	System type	estimated_ qa_performance	qa_accuracy (% best combination)	qa_rej_ accuracy	qa_ accuracy_max
Perfect selection		0.56	0.34 (100%)	0.66	1
ltqa	AV	0.34	0.24 (70.37%)	0.44	0.68
ofe	AV	0.27	0.19 (57.41%)	0.4	0.59
uaic_2	AV	0.24	0.24 (70.37%)	0.01	0.25
wlvs081roen	QA	0.21	0.21 (62.96%)	0	0.21
uaic_1	AV	0.19	0.19 (57.41%)	0	0.19
jota_2	AV	0.17	0.16 (46.30%)	0.1	0.26
dfki081deen	QA	0.17	0.17 (50%)	0	0.17
jota_1	AV	0.16	0.16 (46.30%)	0	0.16
dcun081deen	QA	0.10	0.10 (29.63%)	0	0.10
Random		0.09	0.09 (25.25%)	0	0.09
nlel081enen	QA	0.06	0.06 (18.52%)	0	0.06
nlel082enen	QA	0.05	0.05 (14.81%)	0	0.05
ilkm081nlen	QA	0.04	0.04 (12.96%)	0	0.04
magc_2(bbr)	AV	0.01	0.01 (1.85%)	0.64	0.65
dcun082deen	QA	0.01	0.01 (1.85%)	0	0.01
magc_1(timbl)	AV	0	0 (0%)	0.63	0.63

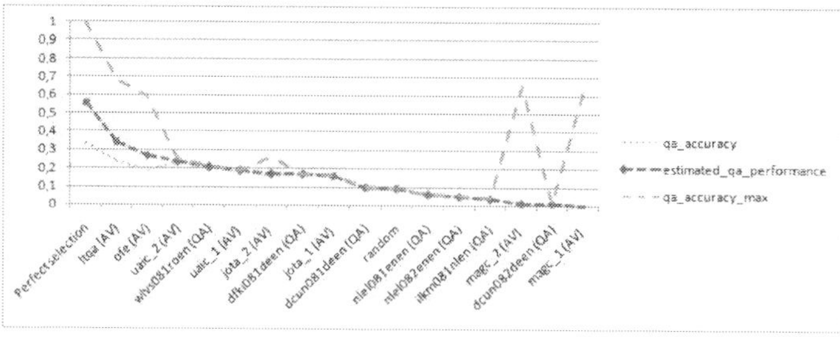

Fig. 6. Graphic comparing AV systems performance with QA systems in English

Table 14. Comparing AV systems performance with QA systems in Romanian

System	System type	estimated_ qa_performance	qa_accuracy (% best combination)	qa_rej_ accuracy	qa_ accuracy_max
Perfect selection		0.65	0.41 (100%)	0.59	1
uaic_2	AV	0.25	0.24 (57.14%)	0.05	0.29
UAIC082roro	QA	0.22	0.22 (53.06%)	0	0.22
UAIC081roro	QA	0.19	0.19 (46.94%)	0	0.19
uaic_1	AV	0.17	0.17 (40.82%)	0	0.17
icia082roro	QA	0.17	0.17 (40.82%)	0	0.17
Random		0.10	0.10 (24.66%)	0	0.10
icia081roro	QA	0.08	0.08 (18.37%)	0	0.08

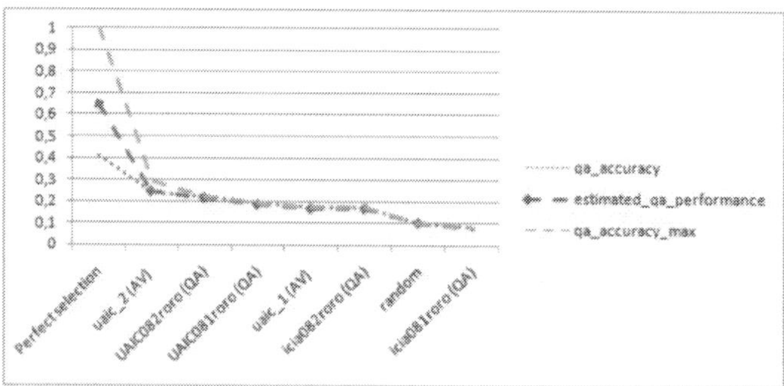

Fig. 7. Graphic comparing AV systems performance with QA systems in Romanian

Overview of QAST 2008

Jordi Turmo[1], Pere R. Comas[1], Sophie Rosset[2], Lori Lamel[2], Nicolas Moreau[3], and Djamel Mostefa[3]

[1] TALP Research Center, Technical University of Catalonia (UPC)
turmo@lsi.upc.edu, pcomas@lsi.upc.edu
[2] LIMSI, Paris, France
rosset@limsi.fr, lamel@limsi.fr
[3] ELDA/ELRA, Paris, France
moreau@elda.org, mostefa@elda.org

Abstract. This paper describes the experience of QAST 2008, the second time a pilot track of CLEF has been held aiming to evaluate the task of Question Answering in Speech Transcripts. Five sites submitted results for at least one of the five scenarios (lectures in English, meetings in English, broadcast news in French and European Parliament debates in English and Spanish). In order to assess the impact of potential errors of automatic speech recognition, for each task contrastive conditions are with manual and automatically produced transcripts. The QAST 2008 evaluation framework is described, along with descriptions of the five scenarios and their associated data, the system submissions for this pilot track and the official evaluation results.

Keywords: Question answering, Spontaneous speech transcripts.

1 Introduction

Question Answering (QA) technology aims at providing relevant answers to natural language questions. Most Question Answering research has focused on mining document collections containing written texts to answer written questions [3,6]. Documents can be either open domain (newspapers, newswire, Wikipedia...) or restricted domain (biomedical papers...) but share, in general, a decent writing quality, at least grammar-wise. In addition to written sources, a lot (and growing amount) of potentially interesting information appears in spoken documents, such as broadcast news, speeches, seminars, meetings or telephone conversations. The QAST track aims at investigating the problem of question answering in such audio documents.

Current text-based QA systems tend to use technologies that require texts to have been written in accordance with standard norms for written grammar. The syntax of speech is quite different than that of written language, with more local but less constrained relations between phrases, and punctuation, which gives boundary cues in written language, is typically absent. Speech also contains disfluencies, repetitions, restarts and corrections. Moreover, any practical

C. Peters et al. (Eds.): CLEF 2008, LNCS 5706, pp. 314–324, 2009.

application of search in speech requires the transcriptions to be produced automatically, and the Automatic Speech Recognizers (ASR) introduce a number of errors. Therefore current techniques for text-based QA need substantial adaptation in order to access the information contained in audio documents. Preliminary research on QA in speech transcriptions was addressed in QAST 2007, a pilot evaluation track at CLEF 2007 in which systems attempted to provide answers to written factual questions by mining speech transcripts of seminars and meetings [5].

This paper provides an overview of the second QAST pilot evaluation. Section 2 describes the principles of this evaluation track. Sections 3 present the evaluation framework and section 4 the systems that participated. Section 5 reports and discusses the achieved results, followed by some conclusions in Section 6.

2 The QAST 2008 Task

The objective of this pilot track is to develop a framework in which QA systems can be evaluated when the answers have to be found in speech transcripts, these transcripts being either produced manually or automatically. There are five main objectives to this evaluation:

- Motivating and driving the design of novel and robust QA architectures for speech transcripts;
- Measuring the loss due to the inaccuracies in state-of-the-art ASR technology;
- Measuring this loss at different ASR performance levels given by the ASR word error rate;
- Comparing the performance of QA systems on different kinds of speech data (prepared speech such as broadcast news (BN) or parliamentary hearings vs. spontaneous in meeting for instance);
- Motivating the development of monolingual QA systems for languages other than English.

In the 2008 evaluation, as in the 2007 pilot evaluation, an answer is structured as a simple [answer string, document id] pair where the answer string contains nothing more than the full and exact answer, and the document id is the unique identifier of the document supporting the answer. In 2008, for the tasks on automatic speech transcripts, the answer string consisted of the <start-time> and the <end-time> giving the position of the answer in the signal. Figure 1 illustrates this point comparing the expected answer to the question *What is the Vlaams Blok?* in a manual transcript (the text *criminal organisation*) and in an automatic transcript (the time segment *1019.228 1019.858*). A system can provide up to 5 ranked answers per question.

A total of ten tasks were defined for this second edition of QAST covering five main task scenarios and three languages: lectures in English about *speech and language processing* (T1), meetings in English about *design of television remote controls* (T2), French broadcast news (T3) and European Parliament debates in English (T4) and Spanish (T5). The complete set of tasks are:

Question: *What is the Vlaams Blok?*

Manual transcript: *the Belgian Supreme Court has upheld a previous ruling that declares the Vlaams Blok a criminal organization and effectively bans it .*
Answer: *criminal organisation*

Extracted portion of an **automatic transcript (CTM file format):**
(...)
20041115_1705_1735_EN_SAT 1 1018.408 0.440 Vlaams 0.9779
20041115_1705_1735_EN_SAT 1 1018.848 0.300 Blok 0.8305
20041115_1705_1735_EN_SAT 1 1019.168 0.060 a 0.4176
20041115_1705_1735_EN_SAT 1 **1019.228** 0.470 criminal 0.9131
20041115_1705_1735_EN_SAT 1 **1019.858** 0.840 organisation 0.5847
20041115_1705_1735_EN_SAT 1 1020.938 0.100 and 0.9747
(...)
Answer: 1019.228 1019.858

Fig. 1. Example query *What is the Vlaams Blok?* and response from manual (top) and automatic (bottom) transcripts

- T1a: QA in manual transcriptions of lectures in English.
- T1b: QA in automatic transcriptions of lectures in English.
- T2a: QA in manual transcriptions of meetings in English.
- T2b: QA in automatic transcriptions of meetings in English.
- T3a: QA in manual transcriptions of broadcast news for French.
- T3b: QA in automatic transcriptions of broadcast news for French.
- T4a: QA in manual transcriptions of European Parliament Plenary sessions in English.
- T4b: QA in automatic transcriptions of European Parliament Plenary sessions in English.
- T5a: QA in manual transcriptions of European Parliament Plenary sessions in Spanish.
- T5b: QA in automatic transcriptions of European Parliament Plenary sessions in Spanish.

3 Evaluation Protocol

3.1 Data Collections

The data for this second edition of QAST is derived from five different resources, covering spontaneous speech, semi-spontaneous speech and prepared speech: The first two are the same as were used in QAST 2007 [5].

- The **CHIL corpus**[1] (as used for QAST 2007): The corpus contains about 25 hours of speech, mostly spoken by non native speakers of English, with an estimated ASR Word Error Rate (WER) of 20%.

[1] http://chil.server.de

- The **AMI corpus**[2] (as used for QAST 2007): This corpus contains about 100 hours of speech, with an ASR WER of about 38%.
- French broadcast news: The test portion of the **ESTER corpus** [1] contains 10 hours of broadcast news in French, recorded from different sources (France Inter, Radio France International, Radio Classique, France Culture, Radio Television du Maroc). There are 3 different automatic speech recognition outputs with different error rates (WER = 11.0%, 23.9% and 35.4%). The manual transcriptions were produced by ELDA.
- Spanish parliament: The **TC-STAR05 EPPS Spanish corpus** [4] is comprised of three hours of recordings from the European Parliament in Spanish. The data was used to evaluate recognition systems developed in the TC-STAR project. There are 3 different automatic speech recognition outputs with different word error rates (11.5%, 12.7% and 13.7%). The manual transcriptions were done by ELDA.
- English parliament: The **TC-STAR05 EPPS English corpus** [4] contains 3 hours of recordings from the European Parliament in English. The data was used to evaluated speech recognizers in the TC-STAR project. There are 3 different automatic speech recognition outputs with different word error rates (10.6%, 14% and 24.1%) . The manual transcriptions were done by ELDA.

The spoken data cover a broader range of types, both in terms of content and in speaking style. The Broadcast News and European Parliament date are less spontaneous than the lecture and meeting speech as they are typically prepared in advance and are closer in structure to written texts. While meetings and lectures are representative of *spontaneous speech*, Broadcast News and European Parliament sessions are usually referred to as *prepared speech*. Although they typically have few interruptions and turn-taking problems when compared to meeting data, many of the characteristics of spoken language are still present (hesitations, breath noises, speech errors, false starts, mispronunciations and corrections). One of the reasons for including the additional types of data was to be closer to the textual data used to assess written QA, and to benefit from the availability of multiple speech recognizers that have been developed for these languages and tasks in the context of European or national projects [2,1,4].

Questions and answer types. For each of the five scenarios, two sets of questions have been provided to the participants, the first for development purposes and the second for the evaluation.

- Development set (11 March 2008) :
  - Lectures: 10 seminars and 50 questions.
  - Meetings: 50 meetings and 50 questions.
  - French broadcast news: 6 shows and 50 questions.
  - English EPPS: 2 sessions and 50 questions.
  - Spanish EPPS: 2 sessions and 50 questions.

[2] http://www.amiproject.org

– Evaluation set (15 June 2008):
  • Lectures: 15 seminars and 100 questions.
  • Meetings: 120 meetings and 100 questions.
  • French broadcast news: 12 shows and 100 questions.
  • English EPPS: 4 sessions and 100 questions.
  • Spanish EPPS: 4 sessions and 100 questions.

Two types of questions were considered this year: factual questions and definitional ones. For each corpus (CHIL, AMI, ESTER, EPPS EN, EPPS ES) roughly 70% of the questions are factual, 20% are definitional, and 10% are NIL (i.e., questions having no answer in the document collection).

The question sets are formatted as plain text files, with one question per line (see the QAST 2008 Guidelines[3]). The factual questions similar to those used in the 2007 evaluation. The expected answer to these questions is a Named Entity (person, location, organization, language, system, method, measure, time, color, shape and material). The definition questions are questions such as *What is the Vlaams Blok?* and the answer can be anything. In this example, the answer would be *a criminal organization.* The definition questions are subdivided into the following types:

– **Person:** question about someone
  Q: *Who is George Bush?*
  R: *The President of the United States of America.*
– **Organisation:** question about an organisation
  Q: *What is Cortes?*
  R: *Parliament of Spain.*
– **Object:** question about any kind of objects
  Q: *What is F-15?*
  R: *combat aircraft.*
– **Other:** questions about technology, natural phenomena, etc.
  Q: *What is the name of the system created by AT&T?*
  R: *The How can I help you system.*

3.2 Human Judgment

As in QAST 2007, the answer files submitted by participants have been manually judged by native speaking assessors, who considered the correctness and exactness of the returned answers. They also checked that the document labeled with the returned docid supports the given answer. One assessor evaluated the results, and another assessor manually checked each judgment of the first one. Any doubts about an answer was solved through various discussions. The assessors used the QASTLE[4] evaluation tool developed in Perl (at ELDA) to evaluate the responses. A simple window-based interface permits easy, simultaneous access to the question, the answer and the document associated with the answer.

[3] http://www.lsi.upc.edu/~qast: News
[4] http://www.elda.org/qastle/

For T1b, T2b, T3b, T4b and T5b (QA on automatic transcripts) the manual transcriptions were aligned to the automatic ASR outputs to find associate times with the answers in the automatic transcripts. The alignments between the automatic and the manual transcription were done using time information. Unfortunately, for some documents time information were not available and only word alignments were used.

After each judgment the submission files were modified, adding a new element in the first column: the answer's evaluation (or judgment). The four possible judgments (also used at TREC[6]) correspond to a number ranging between 0 and 3:

- 0 correct: the answer-string consists of the relevant information (exact answer), and the answer is supported by the returned document.
- 1 incorrect: the answer-string does not contain a correct answer.
- 2 inexact: the answer-string contains a correct answer and the docid supports it, but the string has bits of the answer missing or contains additional texts (longer than it should be).
- 3 unsupported: the answer-string contains a correct answer, but is not supported by the docid.

3.3 Measures

The two following metrics (also used in CLEF) were used in the QAST evaluation:

1. Mean Reciprocal Rank (MRR): This measures how well the right answer is ranked in the list of 5 possible answers..
2. Accuracy: The fraction of correct answers ranked in the first position in the list of 5 possible answers.

4 Submitted Runs

A total of five groups from four different countries submitted results for one or more of the proposed QAST 2008 tasks. Due to various reasons (technical, financial, etc.), three other groups registered but were not be able to submit any results.

The five participating groups were:

- CUT, Chemnitz University of Technology, Germany;
- INAOE, Instituto Nacional de Astrofica, Optica y Electrica, Mexico;
- LIMSI, Laboratoire d'Informatique et de Mécanique des Sciences de l'Ingénieur, France;
- UA, Universidad de Alicante, Spain;
- UPC, Universitat Politècnica de Catalunya, Spain.

All groups participated to task T4 (English EPPS). Only LIMSI participated to task T3 (French broadcast news). Table 1 shows the number of submitted runs

Table 1. Submitted runs per participant and task. T1 (English lectures), T2 (English meetings), T3 (French BN), T4 (English EPPS), T5 (Spanish EPPS).

Participant	T1a	T1b	T2A	T2b	T3a	T3b	T4a	T4b	T5a	T5b
CUT	2	-	-	-	-	-	2	-	-	-
INAOE	-	-	-	-	-	-	1	2	-	-
LIMSI	1	1	1	1	2	3	1	3	2	3
UA	-	-	-	-	-	-	1	3	-	-
UPC	1	2	1	2	-	-	1	6	1	6
Total	4	3	2	3	2	3	6	14	3	9

Table 2. Characteristics of the systems that participated in QAST 2008

System	Enrichment	Question classification	Doc./Passage Retrieval	Factual Answer Extraction	Def. Answer Extraction	NERC
cut1	words, NEs and POS	hand-crafted rules	pass. ranking based on RSV	hand-crafted rules with fallback str. in 1st pass.	hand-crafted fallback strategy	Stanford NER, rules with classification
cut2				same in top-3 pass.		
inaoe1	words and NEs	hand-crafted rules	Lemur	candidate selection based on NEs	-	regular expressions
inaoe2	same plus phonetics					
limsi1	words, lemmas, morphologic derivations, synonyms and extended NEs	hand-crafted rules	ranking based on search descriptors	ranking based on distance and redundancy	specific index for known acronyms	hand-crafted rules with stochastic POS
limsi2				tree-rewriting based distance		
ua1	words, NEs POS and n-grams	hand-crafted rules	ranking based on n-grams	ranking based on keyword distance and mutual information	-	hand-crafted rules
upc1	words, NEs lemmas and POS	perceptrons	ranking based on iterative query relaxation	ranking based on keyword distance and density	-	hand-crafted rules, gazeetters and perceptrons
upc2	same plus phonetics		addition of approximated phonetic matching			

per participant and task. Each participant could submit up to 32 submissions (2 runs per task and transcription). The number of submissions ranged from 2 to 20. The characteristics of the systems used in the submissions are summarized in Table 2. A total of 49 submissions were evaluated with the distribution across tasks shown in the bottom row of Table 2.

5 Results

The results for the ten QAST 2008 tasks are presented in Tables 3 to 12, according to factual questions, definitional questions, and all questions.

Table 3. Results for task T1a, English lectures, manual transcripts (78 factual questions and 22 definitional ones)

System	Factual			Definitional			All	
	#Correct	MRR	Acc	#Correct	MRR	Acc	MRR	Acc
cut1	14	0.18	17.9	2	0.09	9.1	0.16	16.0
cut2	16	0.19	16.7	8	0.26	18.2	0.20	17.0
limsi1	48	0.53	47.4	4	0.18	18.2	0.45	41.0
upc1	39	0.44	38.5	4	0.18	18.2	0.38	34.0

Table 4. Results for task T1b, English lectures, ASR transcripts (78 factual questions and 22 definitional ones)

System	Factual			Definitional			All	
ASR 20%	#Correct	MRR	Acc	#Correct	MRR	Acc	MRR	Acc
limsi1	33	0.34	30.8	3	0.14	13.6	0.30	27.0
upc1	35	0.39	34.6	4	0.18	18.2	0.34	31.0
upc2	35	0.37	33.3	4	0.18	18.2	0.33	30.0

Table 5. Results for task T2a, English meetings, manual transcripts (74 factual questions and 26 definitional ones)

System	Factual			Definitional			All	
	#Correct	MRR	Acc	#Correct	MRR	Acc	MRR	Acc
limsi1	44	0.47	37.8	7	0.22	19.2	0.40	33.0
upc1	29	0.35	31.1	3	0.12	11.5	0.29	26.0

Table 6. Results for task T2b, English meetings, ASR transcripts (74 factual questions and 26 definitional ones)

System	Factual			Definitional			All	
ASR 38%	#Correct	MRR	Acc	#Correct	MRR	Acc	MRR	Acc
limsi1	23	0.21	16.2	6	0.18	15.4	0.20	16.0
upc1	19	0.20	17.6	5	0.19	19.2	0.20	18.0
upc2	16	0.16	10.8	6	0.23	23.1	0.18	14.0

Table 7. Results for task T3a, French BN, manual transcripts (75 factual questions and 25 definitional ones)

System	Factual			Definitional			All	
	#Correct	MRR	Acc	#Correct	MRR	Acc	MRR	Acc
limsi1	45	0.50	45.3	13	0.47	44.0	0.49	45.0
limsi2	45	0.47	41.3	13	0.46	44.0	0.47	42.0

Table 8. Results for task T3b, French BN, ASR transcripts (75 factual questions and 25 definitional ones)

ASR	System	Factual			Definitional			All	
		#Correct	MRR	Acc	#Correct	MRR	Acc	MRR	Acc
a 11.0%	limsi1	42	0.49	44.0	9	0.33	32.0	0.45	41.0
b 23.9%	limsi1	29	0.28	22.7	10	0.34	32.0	0.30	25.0
c 35.4%	limsi1	24	0.24	20.0	7	0.26	24.0	0.24	21.0

Table 9. Results for task T4a, English EPPS, manual transcripts (75 factual questions and 25 definitional ones)

System	Factual			Definitional			All	
	#Correct	MRR	Acc	#Correct	MRR	Acc	MRR	Acc
cut1	12	0.16	16.0	9	0.36	36.0	0.21	21.0
cut2	12	0.16	16.0	11	0.39	36.0	0.22	21.0
inaoe1	41	0.43	37.3	6	0.21	20.0	0.38	33.0
limsi1	44	0.43	33.3	12	0.39	32.0	0.42	33.0
ua1	32	0.30	21.3	4	0.16	16.0	0.27	20.0
upc1	38	0.44	40.0	4	0.16	16.0	0.37	34.0

Table 10. Results for task T4b English EPPS, ASR transcripts (75 factual questions and 25 definitional ones)

ASR	System	Factual			Definitional			All	
		#Correct	MRR	Acc	#Correct	MRR	Acc	MRR	Acc
a 10.6%	inaoe1	32	0.37	33.3	5	0.20	20.0	0.33	30.0
	inaoe2	34	0.38	32.0	5	0.20	20.0	0.33	29.0
	limsi1	24	0.23	18.7	9	0.31	28.0	0.25	21.0
	ua1	12	0.09	4.0	4	0.16	16.0	0.10	7.0
	upc1	18	0.22	20.0	4	0.17	16.7	0.21	19.0
	upc2	16	0.16	13.3	4	0.17	16.7	0.16	14.1
b 14.0%	limsi1	22	0.21	16.0	9	0.33	32.0	0.24	20.0
	ua1	12	0.11	8.0	4	0.16	16.0	0.12	10.0
	upc1	15	0.18	16.0	4	0.16	16.0	0.17	16.0
	upc2	14	0.16	13.3	4	0.16	16.0	0.16	14.0
c 24.1%	limsi1	21	0.21	16.0	8	0.30	28.0	0.23	19.0
	ua1	9	0.10	8.0	5	0.20	20.0	0.12	11.0
	upc1	11	0.11	9.3	5	0.20	20.0	0.14	12.0
	upc2	11	0.11	8.0	4	0.16	16.0	0.12	10.0

For manual transcriptions, the accuracy ranges from 45% (LIMSI1 on task T3a) down to 7% (UPC1 on task T5a). For automatic transcriptions, the accuracy goes from 41% (LIMSI1 on task T3b and ASR a) to 2% (UPC1 on task T5b and ASR c). Generally speaking, a loss in accuracy is observed when dealing

Table 11. Results for task T5a, Spanish EPPS, manual transcripts (75 factual questions and 25 definitional ones)

System	Factual			Definitional			All	
	#Correct	MRR	Acc	#Correct	MRR	Acc	MRR	Acc
limsi1	29	0.32	29.3	13	0.44	36.0	0.35	31.0
limsi2	29	0.32	29.3	13	0.42	32.0	0.35	30.0
upc1	9	0.11	9.3	3	0.05	0.0	0.09	7.0

Table 12. Results for task T5b, Spanish EPPS, ASR transcripts (75 factual questions and 25 definitional ones)

ASR	System	Factual			Definitional			All	
		#Correct	MRR	Acc	#Correct	MRR	Acc	MRR	Acc
	limsi1	20	0.25	24.0	8	0.28	24.0	0.26	24.0
a	upc1	5	0.05	4.0	0	0.00	00.0	0.04	3.0
11.5%	upc2	5	0.06	5.3	2	0.08	8.0	0.07	6.0
	limsi1	18	0.20	17.3	9	0.28	24.0	0.22	19.0
b	upc1	5	0.06	5.3	0	0.00	00.0	0.05	4.0
12.7%	upc2	5	0.06	5.3	2	0.08	8.0	0.07	6.0
	limsi1	20	0.24	22.7	8	0.27	24.0	0.25	23.0
c	upc1	2	0.03	2.7	0	0.00	00.0	0.02	2.0
13.7%	upc2	3	0.03	2.7	1	0.04	4.0	0.04	3.0

with automatic transcriptions. Comparing the best accuracy results on manual transcription and automatic transcriptions, the loss of accuracy goes from 15% for task T2 to 4% for tasks T3 and T4 tasks. This difference is larger for tasks where the ASR word error rate is higher.

Another observation concerns the loss of accuracy when dealing with different word error rates. Generally speaking higher WER results in lower accuracy (e.g. from 30% for T4b_A to 20% for T4b_B). Strangely enough this is not completely true for the T5b task where results for ASR_C (13.7% WER) are 4% higher than for ASR_B (12.7% WER). The WER being rather close, it is probable that ASR_C errors had a smaller impact on the named entities present in the questions.

6 Conclusions

In this paper, the QAST 2008 evaluation has been described. Five groups participated in this track with a total of 49 submitted runs, across ten tasks that included dealing with different types of speech (spontaneous or prepared), different languages (English, Spanish and French) and different word error rates for automatic transcriptions (from 10.5% to 35.4%). For the tasks where the word error rate was low enough (around 10%) the loss in accuracy compared to manual transcriptions was under 5%, suggesting that QA in such documents is potentially feasible. However, even where ASR performance is reasonably good, there remain outstanding challenges in dealing with spoken language and the earlier mentioned differences from written language. The results from the QAST evaluation indicate that if a QA system which performs well on manual transcriptions it also performs reasonably well on high quality automatic transcriptions. The performance on spoken language have not yet reached the level of those in the main QA track.

Acknowledgments

This work has been jointly funded by the Spanish Ministry of Science (TEXTMESS project - TIN2006-15265-C06) and OSEO under the Quaero program.

References

1. Galliano, S., Geoffrois, E., Gravier, G., Bonastre, J.F., Mostefa, D., Choukri, K.: Corpus description of the ESTER Evaluation Campaign for the Rich Transcription of French Broadcast News. In: Proceedings of LREC 2006, Genoa, pp. 315–320 (2006)
2. Gravier, G., Bonastre, J.F., Galliano, S., Geoffrois, E., McTait, K., Choukri, K.: The ESTER evaluation campaign of Rich Transcription of French Broadcast News. In: Proceedings of LREC 2004, Lisbon, pp. 885–888 (2004)
3. Peters, C., Clough, P., Gey, F.C., Karlgren, J., Magnini, B., Oard, D.W., de Rijke, M., Stempfhuber, M. (eds.): CLEF 2006. LNCS, vol. 4730. Springer, Heidelberg (2007)
4. TC-Star (2004-2008), http://www.tc-star.org
5. Turmo, J., Comas, P.R., Ayache, C., Mostefa, D., Rosset, S., Lamel, L.: Overview of QAST 2007. In: Peters, C., Jijkoun, V., Mandl, T., Müller, H., Oard, D.W., Peñas, A., Petras, V., Santos, D. (eds.) CLEF 2007. LNCS, vol. 5152, pp. 249–256. Springer, Heidelberg (2008)
6. Voorhees, E.M., Buckland, L.L. (eds.): The Fifteenth Text Retrieval Conference Proceedings, TREC 2006 (2006)

Assessing the Impact of Thesaurus-Based Expansion Techniques in QA-Centric IR

Luís Sarmento, Jorge Teixeira, and Eugénio Oliveira

Faculdade de Engenharia da Universidade do Porto
Laboratorio de Inteligência Artificial e Ciências de Computadores
Rua Dr. Roberto Frias, s/n 4200-465 Porto, Portugal
las@fe.up.pt, jft@fe.up.pt, eco@fe.up.pt

Abstract. We study the impact of using thesaurus-based query expansion methods at the Information Retrieval (IR) stage of a Question Answering (QA) system. We focus on expanding queries for questions regarding *actions* and *events*, where verbs have a central role. Two different thesaurus are used: the OpenOffice thesaurus and an automatically generated verb thesaurus. The performance of thesaurus-based methods is compared against what is obtained by (i) executing *no expansion* and (ii) applying a simple query generalization method. Results show that thesaurus-based approaches help improving *recall* at retrieval, while keeping satisfactory *precision*. However, we confirm that positive impact for the final QA performance is mostly achieved due to increase in *recall*, which can also be obtained by using simpler methods. Nevertheless, because of its better relative precision thesaurus-based expansion is effective in selectively reducing the number of *irrelevant* text passages retrieved, thus reducing computational load in the answer extraction stage.

1 Introduction

One of the most obvious limitations of many automatic question answering (QA) systems is their relatively low recall: for many questions, QA systems are unable to produce any answer at all. Some of the reasons have to do with insuccess at the Information Retrieval stage, i.e. with the inability to find text passages from which candidate answers can be extracted. Thus, solving recall problems at the IR stage of a QA system allows to globally improve its performance for *all types* of questions. However, QA-centric IR has a set of requirements that make it different from generic IR. First, in generic IR the retrieval unit is the *document*, while in QA-centric IR it is usually a smaller text passage, such as a paragraph or a sentence. Second, in QA-centric IR, fine-tuned ranking is not as crucial and in general IR, because further filtering is performed down the QA pipeline. As mentioned in [1], in standard pipeline QA architectures improving *recall* in IR stage is often more important than improving *precision*: subsequent processing stages may filter out uninteresting text passages obtained, but they will never be able to extract the right answer candidates if the passage that contains the answer is not retrieved.

C. Peters et al. (Eds.): CLEF 2008, LNCS 5706, pp. 325–332, 2009.

In this paper, we wish to extend work reported in [2] on evaluating the impact of thesaurus-based query expansion techniques at the IR stage of a QA system. Our goal is improving the QA performance for factoid questions concerning *actions* or *events*, such as "Who killed J.F.K?" or "When did Brazil last won the World Cup?". These type of questions involve an explicit references to actions through specific verbs (e.g. "to kill", "to win"), which have key roles in retrieving relevant text passages. One expects to increase the chances of finding correct answers if semantically equivalent verbs are used in the retrieval of text passages.

2 Related Work

In [3] eight different *passage-retrieval* algorithms are evaluated in the scope of QA-centric IR. Each passage-retrieval algorithm was run over the top 200 documents retrieved by the same document retriever, operating over the TREC-10 collection. Three document retrievers were experimented (Lucene, PRISE and TREC's "oracle"). Results show that passage retrieval algorithms employing density-based measures for scoring query terms perform better in finding answer bearing passages, but interactions between the passage retrieval and the document retrieval systems may greatly affect the results. Other works trying to evaluate specific techniques for improving recall have also not allowed to draw simple conclusions regarding the impact of such techniques on the overall performance of QA systems. In [4] component evaluation of a QA system showed that *turning off* the stemming component *improved* slightly the overall results. Such slight improvement was observed for about half the types of factoid questions, except for date question ("When... ?") where performance dropped significantly when stemming was turned off. In [5], the author reports that stemming improves the precision in the retrieval of documents containing correct answers, but improvements depend on the type and on the specifity of the question (i.e. number of documents containing the answer). In [1] the authors conclude that indexing stemmed word forms actually leads to inferior document retrieval recall, when compared to baseline (no stemming nor expansion). On the other hand retrieval-time morphological-based query expansion tends to increase document retrieval recall at the cost of bringing more irrelevant documents and placing relevant documents in lower ranks.

The work in [6] shows an example of how Cyc is used for query expansion in the MySentient QA system. MySentient uses Cyc to expand terms to its synonyms (including acronym expansion), to its specializations or generalizations, to possible instances or classes (e.g. "MasterCard" is an *instance-of* "credit card"), and to concepts related by meronomy/holonomy (*is-part-of* or *is-composed-by*). The authors claim that such expansion procedures improve system performance, although no performance figures are given. In [7], Wordnet is used to expand terms found in the question by all terms contained in their synsets. A Boolean search expression is made by combining all expanded terms in a logical OR. The authors observe that such a direct approach may bring problems when synonyms are also highly polysemous words. For example "high" can be a possible synonym of "high school" but since it is much more frequent (and polysemous) it will make the original "high school" term relatively less significant in the search expression. To account for this problem, document ranking is made by pondering the original terms twice as much as the synonyms. However, problematic situations arise when the original word is itself polysemous, leading to totally inappropriate expansions. An approach that tries to solve some of the problem generated by ambiguity is presented in [8]. The proposed technique uses a combination of Blind Relevance Feedback (BRF) and Word-Sense Disambiguation named Sense-based Blind Relevance Feedback (S-BRF). In a first step, sets of paragraphs are retrieved using several combinations of the original terms found in questions. In a second step, the retrieved paragraphs are subjected to linguistic analysis (POS-tagging, multi-word recognition, named-entity recognition) and to word-sense disambiguation over WordNet senses. For each of the original question terms, the *most frequent sense* found on the retrieved paragraphs is chosen. Query expansion is then made by expanding only the previously found sense, using WordNet. S-BRF leads to an increase of 7% in the precision of retrieval of answer-bearing documents, in relation to results obtained using "standard" morphological query expansion. Notably, as reported in [5], pure BRF-based solutions seem to perform quite badly in QA-based retrieval.

When resources like Wordnet of Cyc are not available, systems may follow alternative approaches supported by statistical techniques. In [9] two query expansion methods based on statistical machine translation models are proposed, although focusing on a different yet related problem: *answer retrieval*. In the first method, a "translation model" from question-words to answers-words was learned using a large corpus of question-answer pairs. Using such translation model, each question word can be expanded to a set of words that are expected to occur in the answer. In a second method, an English-Chinese parallel corpus was used to learn English paraphrases. Query expansion was made by adding to the query the n-best paraphrases of the original terms.

3 Question-Answering Framework: RAPOSA

The Question Answering system that we will use to evaluate the impact of query expansion, RAPOSA, follows a classical pipeline architecture, composed of five main modules: the Question Parser, the Query Generator, the Passage Retriever, the Answer Extractor, and the Answer Selector. Since RAPOSA has been extensively described elsewhere ([10] and [11]), we will focus only on the Query Generator module. The *Query Generator* may generate queries according to two different strategies: using *pseudo-stemming* and using *thesaurus-based expansion*. Generation using *pseudo-stemming* involves a simple lexical process: for terms *not* identified as named-entities, the last 2-4 characters are stripped and substituted by wild-cards. Query generation using thesaurus-based expansion relies on a pre-existing verb thesaurus. For factoid questions that explicitly refer to *actions* or *events*, expansion is made by first taking the source verb and finding its lemma, v_s, and then using the verb thesaurus to find up to n verbs related to v_s: v_{r1}, v_{r2} ... v_{rn}. Finally, pseudo-stemming is applied to terms in the query, including source verb v_s and related verbs v_{r1}, v_{r2} ... v_{rn}, in order to match most possible verb inflections for each expansion found.

4 Thesauri for Expansion

We have two thesaurus available for supporting verb expansion: the OpenOffice thesaurus for Portuguese and an automatically generated verb thesaurus. The OpenOffice thesaurus[1] contains 4002 synsets for adjectives, nouns and names. We took the verb synsets and indexed each verb in it to produce (verb → list of all equivalent verbs) mappings for all verbs. We obtained 2783 such mappings.

The automatically generated verb thesaurus was built following a simplified approach to that described in [12]. The basic principle is that "similar" words should have "similar" distributional properties under a given context. For the case of verbs in Portuguese, one can intuitively see that much of the information capable of describing the semantic properties of a verb can be found in the two following words. Within this context we can observe many of the more relevant verb-object relations, as well as typical adverbial constructions. We used n-gram information compiled from a large web-corpus of about 1000 million words to obtain a distributional description of verbs in Portuguese ([13]). N-gram information in this collection is not POS-tagged, but because verb forms in Portuguese are inflected, they can frequently be unambiguously distinguished using a dictionary. We used a dictionary to filter out ambiguous verb forms so that only the 3-grams (w_1, w_2, w_3) matching the following selection pattern were chosen: (w_1 = [unambiguous verb form] & w_2 = * & w_3 = *). Verb forms (at w_1) were lemmatized in order to obtain tuples of the form (verb lemma, w_2 w_3, frequency), and feature information from the various forms was merged. There are 173,607,555 distinct 3-grams available in n-gram database, and 14,238,180 (8.2%) matched the selection pattern, corresponding to 4,958 verbs. Verb v_i is described using a feature vector $[v_i]$ containing the pre-compiled information about co-occurring words. Vector features were weighted by Mutual Information and vectors were then compared using the cosine-metric, to obtain the list of nearest-neighbours. Verbs corresponding to such nearest-neighbours of $[v_i]$ are considered the "verb equivalents" of v_i. The automatically generated thesaurus can be visualized via: `http://pattie.fe.up.pt/cgi-bin/word_map.pl`.

5 Experimental Setup

We took the CLEF 2007 and CLEF 2008 test sets, both having 200 question of several types (factoid, definition and enumeration), and we chose a subset of action/event-related factoid questions. To ensure that all test questions could potentially be answered we selected only those that we knew that our system

[1] Available from http://openthesaurus.caixamagica.pt/. Version used is dated from 2006-08-17.

Table 1. The two sub-sets of action/event-related factoid questions used for testing

test set	DATE	ORG	ORG/PER	PER	GPE	QNT	$\sum$
CLEF-2007	9	5	4	5	3	1	27
CLEF-2008	12	1	3	9	2	0	27

could parse and extract candidate answers. We chose 27 questions from each of the test sets whose expected answer type could be any of the following (see Table 1): date or time expression (DATE), an organization (ORG), a geo-political entity (GPE), a person (PER) or a quantity (QNT). For these types of questions, our system relies on the *simplest* answer extraction strategy that we have available. The question is parsed and the expected type of answer is identified. Answer candidates are those entities extracted from the retrieved text passages whose type is compatible with the expected answer type. The final answer chosen is the *most frequent compatible candidate* found. We configured our system to answer test questions using 4 different options for query generation / expansion:

1. **Run** R_{ps}: queries are generated by *pseudo-stemming*. Up to 150 text passages can be retrieved. This will be our baseline method.
2. **Run** R_{oo}: query expansion is made using the *OpenOffice thesaurus*. The verb is expanded to at most 14 related verbs options, and a maximum of 10 snippets are retrieved per verb. At most 150 text passages are retrieved.
3. **Run** R_{st}: similar to R_{oo}, but using *statistical thesaurus* for expansion.
4. **Run** R_0: in this run, we *remove* the verb from the query. Only the *argument* of the question (e.g. "J.F.K." in "Who killed J.F.K?") is used in the query, thus providing maximum retrieval recall, although possible decreasing precision in retrieval. Up to 150 text passages will be considered.

Answers were searched in the Wikipedia-derived collection provided by the CLEF organization and were manually evaluated. We checked non-nil answers to see if they were *correct*, *incorrect* or *inexact* (i.e. only partially correct). *Unsupported answers* were considered *incorrect*. When the system was not able to produce any answer (i.e. produce the NIL answer) we checked whether the answer was present in the retrieved text passages but it was not extracted. In those cases, we can assume that the problem is related with the answer extraction.

6 Results and Analysis

Table 2 presents statistics about expansion and retrieval. The second column presents the average number of branches provided by the expansion mechanism. Obviously, both R_{ps} and R_0 only generate one query, so branching is 1. On the other hand, R_{st} generates the highest number of query branches (10.9). The third column presents the number of questions for which *no passages* were retrieved (out of 54 question). The last column indicates the average number of passages retrieved when at least one passage was retrieved. Again, R_0 allows retrieving

Table 2. Retrieval Statistics

Run	Avg. Branching	No Passages (in 54)	Avg. # Passages
R_{ps}	1	37	2.6
R_{oo}	3.1	35	3.4
R_{st}	10.9	30	7.7
R_0	1	19	24.6

Table 3. Results obtained for the four query generation / expansion configurations

Run	Correct	Incorrect	Inex.	NIL (No Ext.)	$\sum$	Inex. + No Ext.
R_{ps}	4	10	1	39 (2)	54	3
R_{oo}	3	11	2	38 (2)	54	4
R_{st}	3	9	4	38 (4)	54	8
R_0	10	16	3	25 (5)	54	8

more passages. R_{st} allows retrieving more passages than R_{oo} although the increase is proportionally lower than the corresponding increase in the branching factor. This suggests that some of the verbs provided by the statistic thesaurus might not be correct (or correlated with the argument of the question).

Table 3 presents overall QA results for the 54 questions in the test set. The first three columns report the results in case of non-nil answer (correct, incorrect and inexact). The forth column present the number of NIL answers, and explicitly shows number of cases where the answer *was present* in the text passages but the system was unable to extract it. The last column presents the number of cases where the found answer was inexact or was not extracted, emphasizing the cases where the retrieved text passages contained the correct answer but the extraction stage failed (partially or completely).

Run R_0 clearly outperforms all others both in the number of correct answers, and in the number of non-NIL answers. R_0 also produces more incorrect answers but the relative increase in the number of correct answers is much higher. If we only consider correct answers, none of the runs that use thesaurus expansion methods, R_{oo} and R_{st}, beats the baseline run, R_{ps}, that uses pseudo-stemming. The number of incorrect and NIL answers also does not change significantly between runs R_{ps}, R_{oo} and R_{st}. The only significant difference is the aggregate number of *inexact answers* plus *not extracted* answers, where the figure for run R_{st} is higher that for runs R_{ps}, R_{oo}. This suggest that R_{st} was able to find the appropriate text passages in several cases but the extraction stage was unable to identify the correct answer.

Generally, results confirm that retrieving and analyzing more passages helps finding more corrects answers (higher recall). We verified that, similar to [5], this seems to be specially the case when the number of passages referring to the argument of the question is *very low* (e.g. 1-3). In those situations, query expansion (by any method) helps finding the few decisive passages. R_0 clearly outperforms all others, but at the cost of processing many more text passages (even when limiting retrieved snippets to 150). Results of R_{ps}, R_{oo} and R_{st} do not vary significantly in terms of *correct* and *incorrect* answers. However, if we consider the aggregate number of *inexact* and *not extracted* answers (last column in Table 3) we see that R_{st} could potentially outperform both R_{ps} and R_{oo} if the extraction procedure was made more efficient. This can also be confirmed by the fact that R_{st} was able to retrieve passages for 5 questions more than R_{oo} and 7 more than R_{ps} (see Table 2). Unfortunately, the same can not be said for R_{oo} in relation to R_{ps}: the difference in performance is not significant. Apparently, thesaurus-based expansion is effective only when branching is relatively high, which is the case of R_{st}. Since thesaurus-based expansion uses closely related verbs, the number of passages retrieved and processed in subsequent stages of the QA pipeline does not grow in an uncontrolled fashion, as it does in run R_0 when arguments are very frequent entities.

7 Conclusion and Future Work

Results obtained with this set of questions and the Wikipedia collection do not clearly demonstrate that applying thesaurus-based expansion in QA-centric IR is advantageous for the *overall* QA performance in comparison with not performing any query expansion at all. In fact, simpler "query expansion" methods may lead to better *overall* QA performance. Nevertheless, results suggest that thesaurus-based expansion improves recall at the IR-stage of the QA pipeline, while still keeping *reasonable* levels of precision. Whether such improvement is propagated to the *overall* QA performance depends on whether subsequent answer extraction procedures are successful or not. In any case, the main advantage of using thesaurus-based expansion is allowing to improve retrieval recall while still limiting the number of absolutely irrelevant text passages, thus helping to reduce computational load further down the QA chain.

Future work will necessarily focus on problems related to the extraction of answer candidates. Additionally, we wish to improve our statistic thesaurus by using more linguistic information, namely POS tagging, to find equivalence relations not only between simple verbs, but also between simple verbs and compound verbs. From the point of view of query expansion itself we wish to experiment the impact of expanding the initial verb to larger sets, such as for example by also expanding each of the verbs obtained after expanding the initial verb, and using the much larger resulting set in the queries.

Acknowledgments

This work was partially supported by grant SFRH/BD/ 23590/2005 from Fundação para a Ciência e Tecnologia (Portugal), co-financed by POSI.

References

1. Bilotti, M.W., Katz, B., Lin, J.: What works better for question answering: Stemming or morphological query expansion? In: Proceedings of the Information Retrieval for Question Answering (IR4QA) Workshop. SIGIR 2004, Sheffield, England (July 2004)
2. Sarmento, L., Teixeira, J., Oliveira, E.: Experiments with query expansion in the raposa (fox) question answering system. In: Borri, F., Nardi, A., Peters, C. (eds.) Working Notes for the CLEF 2008 Workshop, Aarhus, Denmark, September 17-19 (2008)
3. Tellex, S., Katz, B., Lin, J., Fern, A., Marton, G.: Quantitative evaluation of passage retrieval algorithms for question answering. In: Proceedings of the 26th Annual International ACM SIGIR Conference on Research and Development in Information Retrieval SIGIR, pp. 41–47. ACM Press, New York (2003)
4. Costa, L., Sarmento, L.: Component evaluation in a question answering system. In: Proceedings of the 5th International Conference on Language Resources and Evaluation (LREC 2006), Genoa, Italy (May 2006)
5. Monz, C.: Document retrieval in the context of question answering. In: Sebastiani, F. (ed.) ECIR 2003. LNCS, vol. 2633, pp. 571–579. Springer, Heidelberg (2003)
6. Curtis, J., Matthews, G., Baxter, D.: On the effective use of cyc in a question answering system. In: IJCAI Workshop on Knowledge and Reasoning for Answering Questions (KRAQ 2005), Edinburgh, Scotland (2005)
7. Hovy, E., Gerber, L., Hermjakob, U., Junk, M., Lin, C.Y.: Question answering in webclopedia. In: Proceedings of the 9th Text REtrieval Conference, Gaithersburg, MD, USA, November 2000, pp. 655–664 (2000)
8. Negri, M.: Sense-based blind relevance feedback for question answering. In: SIGIR 2004 Workshop on Information Retrieval For Question Answering (IR4QA), Sheffield, UK (July 2004)
9. Riezler, S., Vasserman, A., Tsochantaridis, I., Mittal, V.O., Liu, Y.: Statistical machine translation for query expansion in answer retrieval. In: Proceedings of the 45th Annual Meeting of the Association for Computational Linguistics, Prague, Czech Republic, June 23-30 (2007)
10. Sarmento, L.: A first step to address biography generation as an iterative QA task. In: Peters, C., Clough, P., Gey, F.C., Karlgren, J., Magnini, B., Oard, D.W., de Rijke, M., Stempfhuber, M. (eds.) CLEF 2006. LNCS, vol. 4730, pp. 473–482. Springer, Heidelberg (2007)
11. Sarmento, L., Oliveira, E.: Making RAPOSA (FOX) smarter. In: Nardi, A., Peters, C. (eds.) Working Notes of the Cross-Language Evaluation Forum (CLEF) Workshop 2007, Budapest, Hungary (September 2007)
12. Lin, D.: Automatic Retrieval and Clustering of Similar Words. In: Proceedings of COLING-ACL 1998, Montreal, vol. 2, pp. 768–773 (1998)
13. Sarmento, L.: BACO - A large database of text and co-occurrences. In: Calzolari, N., Choukri, K., Gangemi, A., Maegaard, B., Mariani, J., Odjik, J., Tapias, D. (eds.) Proceedings of the 5th International Conference on Language Resources and Evaluation (LREC 2006), Genoa, Italy, May 22-28, pp. 1787–1790 (2006)

Using AliQAn in Monolingual QA@CLEF 2008*

Sandra Roger, Katia Vila, Antonio Ferrández, María Pardiño,
José Manuel Gómez, Marcel Puchol-Blasco, and Jesús Peral

University of Alicante, Department of Software and Computing Systems
San Vicente del Raspeig Road, 03690 Alicante, Spain
{sroger,kvila,antonio,maria,jmgomez,marcel,jperal}@dlsi.ua.es
http://www.dlsi.ua.es/

Abstract. This paper describes the participation of the system AliQAn in the CLEF 2008 Spanish monolingual QA task. This time, the main goals of the current version of AliQAn were to deal with topic-related questions and to decrease the number of inexact answers. We have also explored the use of the Wikipedia corpora, which have posed some new challenges for the QA task.

1 Introduction

AliQAn, an open-domain QA system, has already been described in detail in [4,1,5]. Briefly, it is based on complex pattern matching using natural language processing tools. This year, we have tested a method to decrease the number of inexact answers and we have adapted our system to work on Wikipedia which poses some new challenges. Finally, a new method has been implemented to treat the topic-related questions.

2 Description of the System

2.1 Dealing with Topic-Related Questions

The underlying idea behind our treatment of context-dependent questions is described next. It complements the dependent questions by adding the noun phrases of the first question of each cluster and the noun phrases of the answer for this question. By reasons of simplicity and to avoid introducing noise, we only considered the co-reference between the first question and other of the same cluster. The algorithm that was used had the following steps: (1) answering the first question of one cluster without special treatment; (2) extracting the set

* This paper has been partially supported by the Spanish government, project TIN-2006-15265-C06-01 and project GV06/028, and by the framework of the project QALL-ME, which is a 6th Framework Research Programme of the European Union (EU), contract number: FP6-IST-033860, by the Spanish Government through the research grant AP2007-03072, by the Generalitat Valenciana through the research grant BFPI06/182 and by the University of Comahue under the project 04/E062.

C. Peters et al. (Eds.): CLEF 2008, LNCS 5706, pp. 333–336, 2009.

of noun phrases from that question and its answer; (3) adding this set of noun phrases to all dependent questions; and (4) handling and extracting the answers from these expanded questions.

For instance, if we consider the question 008 (¿Dónde vivía la tribu de los Mojave? (Where did the Mohave tribe live?)), the system returns the answer "Arizona"(step 1). Step 2 produces the noun phrases "la tribu de los Mojave" (Mohave tribe) and "Arizona". Then (step 3), it obtains the noun phrases that correspond to the question 009 (¿Quiénes eran sus enemigos?" (Who were their enemies?)): "sus enemigos" (their enemies). Finally, in step 4, the previous extended set with all the noun phrases is used to find the answer to question 009.

For the final answers for topic-related questions, we obtain the following criteria: (i) if the answer to the extended question (following the steps previously described) was *nil*, then *nil* was returned as final answer; and (ii), in the opposite case, we have ranked the answers (with or without extension) in a decreasing order, thus the first three answers of the ranking are returned.

2.2 Avoiding Inexact Answers

This year, the algorithm to reduce the number of inexact answers has been modified only for the questions whose expected answer type is group, person, first name, place, country or city. In these cases, we try to extract the right answer from the too long answers returned by the AliQAn system.

Before explaining the general algorithm that we have used to handle these answers, we will define some variables: Set A = the answer from AliQAn. If A has no complements then $\Omega = \{$ proper noun or multiword proper nouns $\in A\}$. On the other hand, if A has complement then $\Omega = \{$ proper noun or multiword proper nouns $\in$ head of $A\} \cup \{$ proper noun or multiword proper nouns $\in$ complement of $A\}$. For example, consider the noun phrase "la Sociedad Española de Vexilología" (the Spanish Society of Vexillology); in this case, the cardinality of Ω (i.e. $|\Omega|$) is 2 and its elements are: "Spanish Society" and "Vexillology".

The algorithm for finding the new answer (A') begins by evaluating the set Ω. If $|\Omega| > 1$ then the elements of Ω are ranked according to their weights. The weight is increased or decreased in accordance with different criteria and whether it belongs to a specific dictionary or not. Criteria and dictionary are defined according to the expected answer type. After this, the algorithm selects the element with higher score and it returns the head of the noun phrase corresponding to this element as A'. On the other hand, if $|\Omega| = 1$, then it only returns, as A', the corresponding head of the noun phrase. In the previous example, we suppose that the weight of "Spanish Society" is N_1 and the weight of "Vexillology" is N_2. If $N_1 > N_2$, then the algorithm returns "the Spanish Society" else it returns "Vexillology".

2.3 Exploring Wikipedia

Compared to traditional CLEF corpora (based on articles from newspapers), Wikipedia is a very large document collection and has not enough redundancy of information. In contrast, the articles from newspapers have a fair amount of redundancy because they are usually published, with pretty much relevance, on different days, by different people and using different expressions. Wikipedia collections use hyperlinks to avoid information repetition (i.e. data which is sensitive to be repeated is replaced by links to the original source).

An Information Retrieval (IR) system needs to be more precise in order to filter the fair amount of irrelevant information due to the size of the Wikipedia collections. At the same time, an IR system needs to have high coverage to deal with the low redundancy of these corpora. In addition, Wikipedia, unlike newspaper collections, is highly structured. This structure gives a lot of information about the article topic in the form of tables, references and links. Hence, an IR system needs to consider this structure to take advantage of this information.

Bearing these considerations in mind, we aim to adapt the IR-n system [3] in order to be able to use very large document collections, and to face up to the above-commented new Wikipedia challenges.

In addition, we would like to point out that several problems derived from the codification of the non-latin characters in Wikipedia were solved from the viewpoint of our QA system. The source of these problems is that the Wikipedia collections were coded in UTF-8, while our QA system uses ISO 8859-1 encoding to perform the morpho-syntactic labelling of documents via MACO and SUPAR (more details of these NLP tools in [4]).

The proposed solution for our QA system consists of controlling the correspondences between the two encodings for non-latin characters. Even though it is a simple solution, good results are obtained. Nevertheless, as future work we wish to adapt our system and its related tools to deal with the UTF-8 encoding.

3 Results and Conclusions

Table 1 shows the results for 2008.

This year, the main contributions are:

- Our strategy to deal with topic-related questions is simple, but it obtains good results. This algorithm obtains an accuracy over linked questions of 13.11%. This result is the second-best accuracy, together with inao082eses, with respect to all the Spanish monolingual runs (Table 33 of [2]).
- AliQAn system had a high percentage of inexact answers in previous years. This kind of answers has been improved in this participation: 24 in the year 2005 [4] and 15 in the year 2006 [1] to 4 this year [5], which all correspond to list questions (It is important to say that list questions are not supported by our system). On the other hand, if we consider the questions whose expected answer type that have been treated (group, person, first name, place, country

Table 1. General results obtained in the QA@CLEF 2008

Right	Inexact	Unsupported	Wrong	Overall Accuracy
39	4	1	156	19.50%

and city), the improvement was 10% compared with our baseline (without any treatment).

- Using Wikipedia with our IR & QA systems. On the one hand, our IR system has been adapted to make possible the use of Wikipedia with very large document collections. On the other hand, several problems derived from the codification of the non-latin characters in Wikipedia have been resolved in order to properly use it together with our QA system.
- Wikipedia is more structured than the corpus EFE. Our system is not adapted to work on structured corpora such as Wikipedia, yet. For this reason and for the problem stated above, the main source of correct answers was the corpus EFE (Table 35 of [2]).

Finally, all questions given in this track, except the questions of type list, have been treated by our system and only one has been unsupported. Our paper includes only one run for the Spanish monolingual QA task and it has achieved an overall accuracy of 19.50%. We also would like to point out that this is the first time we deal with Wikipedia and topic-related questions for our participation in the QA@CLEF task. However, we have been able to extract some of the strengths and weaknesses of our system, which we will take into account for future improvements. Furthermore, our future work is focused on the multilingual task, the adaptation of the NLP tools related to our system to directly work with the UTF-8 encoding and the incorporation of knowledge to the phases that can be useful to increase the performance of our system.

References

1. Ferrández, S., López-Moreno, P., Roger, S., Ferrández, A., Peral, J., Alvarado, X., Noguera, E., Llopis, F.: Monolingual and Cross-Lingual QA Using AliQAn and BRILI Systems for CLEF 2006. In: Peters, C., Clough, P., Gey, F.C., Karlgren, J., Magnini, B., Oard, D.W., de Rijke, M., Stempfhuber, M. (eds.) CLEF 2006. LNCS, vol. 4730, pp. 450–453. Springer, Heidelberg (2007)
2. Forner, P., Peñas, A., Alegria, I., Forăscu, C., Moreau, N., Osenova, P., Prokopidis, P., Rocha, P., Sacaleanu, B., Sutcliffe, R., Tjong Kim Sang, E.: Overview of the CLEF 2008 Multilingual Question Answering Track. In: Working Notes of Cross Language Evaluation Forum, CLEF (2008)
3. Llopis, F., Vicedo, J.L., Ferrández, A.: Passage Selection to Improve Question Answering. In: Proceedings of the COLING 2002 Workshop on Multilingual Summarization and Question Answering, Taipei, Taiwan, pp. 1–6 (2002)
4. Roger, S., Ferrández, S., Ferrández, A., Peral, J., Llopis, F., Aguilar, A., Tomás, D.: AliQAn, Spanish QA System at CLEF 2005. In: Peters, C., Gey, F.C., Gonzalo, J., Müller, H., Jones, G.J.F., Kluck, M., Magnini, B., de Rijke, M., Giampiccolo, D. (eds.) CLEF 2005. LNCS, vol. 4022, pp. 457–466. Springer, Heidelberg (2006)
5. Roger, S., Vila, K., Ferrández, A., Pardiño, M., Gómez, J.M., Puchol-Blasco, M., Peral, J.: AliQAn, Spanish QA System at CLEF 2008. In: Working Notes of Cross Language Evaluation Forum, CLEF (2008)

Priberam's Question Answering System in QA@CLEF 2008

Carlos Amaral, Adán Cassan, Helena Figueira, André Martins, Afonso Mendes, Pedro Mendes, José Pina, and Cláudia Pinto

Priberam,
Alameda D. Afonso Henriques, 41 – 2.º Esq., 1000–123 Lisboa, Portugal
{cma,ach,hgf,atm,amm,prm,jfp,cp}@priberam.pt
http://www.priberam.pt

Abstract. This paper describes the changes implemented in Priberam's question answering (QA) system, followed by the discussion of the results obtained in Portuguese and Spanish monolingual runs at QA@CLEF 2008. We enhanced the syntactic analysis of the question and improved the indexing process by using question categories at the sentence retrieval level. The fine-tuning of the syntactic analysis allowed the system to more precisely match the pivots of the question with their counterparts in the answer. As a result, in QA@CLEF 2008, Priberam's system achieved a considerable overall accuracy increase in the Portuguese run.

Keywords: Question answering, Query Expansion.

1 Introduction

The performance of Priberam's system in last year's QA@CLEF reflected internal and external changes. Internally, the most relevant change was the introduction of syntactic question processing [1]. Externally, the CLEF organisation introduced topic-related questions and added Wikipedia as a target document collection [2]. The result was a slight increase of the overall accuracy in the Spanish (ES) run and a significant decrease of the overall accuracy in the Portuguese (PT) run.

The main goal of our participation in QA@CLEF 2008 was to stabilize the system in order to surpass the results it obtained in previous QA@CLEF participations [3,4]. To enhance its performance, we improved the indexing/retrieval process by using question categories (QC) at sentence retrieval level and ontology domains of the expected answer in document retrieval. The fine-tuning of the syntactic analysis, by using the phrases' core nodes as objects (see section 2.2), allowed the system to more precisely match the pivots[1] of the question with their counterparts in the answer, taking into account their syntactic functions. As a result, in QA@CLEF 2008, Priberam's system achieved a considerable overall accuracy increase in the Portuguese run.

[1] As presented in [3], *pivots* are the key elements of the question, and they can be words, expressions, NEs, phrases, numbers, dates, abbreviations, etc.

C. Peters et al. (Eds.): CLEF 2008, LNCS 5706, pp. 337–344, 2009.

The paper is organised as follows: section 2 describes the major adjustments made to the system, such as the work done in improving the syntactic processing of the question and the adaptations to deal with topic-related questions; section 3 analyses and discusses the results of both monolingual runs; section 4 presents the conclusions and future work.

2 Adaptations and Improvements of the System

Briefly, Priberam's QA open-domain system [3,5] relies on a set of linguistic resources (a wide coverage lexicon, a thesaurus and a multilingual ontology) and software tools (which can be used to write and test grammars, to build contextual rules for performing morphological disambiguation or named entity (NE) recognition, to build patterns for question categorization/answer extraction, etc.). The system is based on a five-step architecture: the indexing process, the question analysis, the document retrieval, the sentence retrieval, and the answer extraction. When a question is submitted and matches a given question pattern (QP), a category is assigned to it and a set of question answering patterns (QAPs) becomes active. Then, documents containing sentences with categories in common with the question (earlier determined during indexation via answer patterns (APs)) are analysed; the active QAPs are then applied to each sentence in order to extract the possible answers. Since the overall architecture remains unchanged, this year we focused on (i) the improvement of the indexing/retrieval process, (ii) the refinement of the question syntactic analysis, (iii) the fine-tuning of named entity recognition, and (iv) the treatment of topic-related questions.

2.1 Improvements of the Indexing/Retrieval Process

This year we kept the approach used and described on previous CLEF campaigns [3], but the system was submitted to a lot of fine-tuning and optimization. Some of the enhancements allowed us to go further on what we indexed and queried for without major performance penalties. The most important changes were indexing of QCs at sentence level instead of at document level, the complete indexation of ontology domains at document level and the use of different ratings for document titles and document body (both for Wikipedia and newspaper articles). In [3] we described the work done in two different steps, document retrieval and sentence retrieval. Much of the work done on the second step is now also done on the first step because many of the problems the system experienced in the retrieval process were due to the loss of documents in document retrieval. The following summarizes the most important changes implemented:

1. It is now possible to embed in the QAPs rules for querying the ontology of the target answer (see section 2.2);
2. A document indexed with the QC on the same sentence as the pivots has now a much higher rating;

3. Documents where the pivots (especially NEs) appear in the title have priority over the other documents;
4. Documents that are more recent have higher priority (this is relevant for news corpora);
5. It is now possible to write rules to tag some pivots with higher/lower priority or discard them for retrieval.

2.2 Refinement of the Question Syntactic Analysis

We maintain the approach presented last year, where the syntactic structure of the question was captured by using FLiP's[2] linguistic technology [1]. The main difference is that now we detect the core nodes of the syntactic phrases and use them as the question's objects (its main constituents).

Each syntactic phrase may have one or more core nodes, that may coincide with the head phrase or not, and that are assigned to different object types accordingly to their relevance in extracting the expected answer. Object assignment is done after parsing, using the syntactic information that was treated in that stage. Typically, object assignment establishes a hierarchy of objects: it places the core nodes of subjects at the top, followed by those of the verb's complements, the head of the verb phrase and the adjuncts. It also gives priority to NEs: for example, PT question 33 "Que político é conhecido como Iznogoud?" [Which politician is known as Iznogoud?] retains "Iznogoud" as the object, "é conhecido" as the verbal object and "político" as the restraining object.

This strategy can help solving a few simple instances of syntactic ambiguity, such as those derived from prepositional phrase (PP) attachment, in case of overgeneration or parsing errors [6], since the core nodes remain the same. For instance, in PT question 62 "Qual a largura do Canal da Mancha no seu ponto mais estreito?" [What is the width of the English Channel in its narrowest point?], which has three contracted prepositions ("do", "da" and "no"), the parser could wrongly build the PP "do Canal da Mancha no seu ponto mais estreito" [of the English Channel in its narrowest point]. If the parser could not find its core nodes, the whole PP would be used as the object, thus introducing noise in the document retrieval stage. By establishing core nodes, one can assign the detected NE "Canal da Mancha" as the object and "no seu ponto mais estreito" as the modifying object.

We added a specific object, the interrogative object, which works as a placeholder for the expected answer. We use it along with the QC to narrow the search for target sentences and extract the answer. The use of its ontological domains led to a considerable increase in the accuracy of the retrieval process. For instance, in PT question 1 "Que tipo de animal é o Cocas?" [What kind of animal is Kermit?], the system looks for documents containing words and

[2] FLiP is Priberam's proofing tools package for Portuguese; it includes a grammar checker and style checker, a spell checker, a thesaurus and a hyphenator that enable different proofing levels—word, sentence, paragraph and text—of European and Brazilian Portuguese. An online version is available at `http://www.flip.pt`

expressions belonging to the same ontology level of "animal", the question's interrogative object. Thus, sentences that do not contain the word "animal", but contain words like "sapo" [toad] or "rã" [frog], are retrieved.

2.3 Fine-Tuning of Named Entity Recognition (NER)

The NER engine Priberam has been using in its QA system participated this year in HAREM, an evaluation contest for Portuguese NER.[3] This participation led to an external evaluation of the engine and, consequently had a positive impact on the precision of the answer extraction, namely in the more specific QCs. Besides the NEs already detected (e.g. people, places and organisations), we had to build new rules to recognize NEs that denote written and not written works, things (objects, substances), events, abstractions and numeric values (currencies, quantities, classifications). Rules that recognise time expressions were also improved, because Priberam's NER engine was a participant in the time track of the second HAREM as well.

This, as mentioned above, was particularly important for some QCs such as <WRITTEN WORK>, <NOT WRITTEN WORK>, <STAR>, or <CLASSIFICATION>. For QCs such as <DENOMINATION>, <FUNCTION> or <LOCATION>, the semantic values of NEs were already being used in the indexing process and answer extraction, allowing the system to perform more accurately in these categories. With the addition of the new semantic tags and the creation of new rules that classify NEs using those tags, we were able to narrow the number of candidate answers in the more specific QCs. Thus, for a question such as topic-related PT question 162 "Diga um desses filmes." [Name one of those films.], whose topic is *Jean Vigo*, candidate answers that contained NEs classified as not written works were given a higher score.

Not only does this fine-tuning of the NER improve the answer extraction process, it also improves the syntactic parsing by restricting, for example, the number of PPs, hence preventing overgeneration, which will in turn create a more precise parser (see section 2.2).

The performance of Priberam's NER engine led to its commercial exploration: it is now being used for search refining in the sites of two major Portuguese news media, *TSF* radio station[4] and *Jornal de Notícias* newspaper[5].

2.4 Dealing with Topic-Related Questions

As mentioned in [1], the procedure for dealing with topic-related questions could perform poorly because of the excess of pivots. Moreover, since we just merged the question pivots, we loosed the question syntactical analysis. Like last year, we only analyse the first question from the set and the current question, which means that we do not keep track of the changes to the topic. This had an impact on the Spanish questions but not on the Portuguese ones. In our opinion, topic-related questions are not very interesting for a commercial system at this stage

[3] `http://www.linguateca.pt/HAREM/`

[4] `http://www.tsf.pt`

[5] `http://www.jn.pt`

Table 1. Examples of question analysis of topic-related questions

Question	QC	Objects	Answer
PT 11: Qual é a montanha mais alta do México? [Which is the highest mountain in Mexico?]	<MOUNTAIN>	• México • mais alta	Citlaltépetl
PT 12: E do Japão? [And in Japan?]	NIL	• Japão • NIL	
PT 12 final question analysis:	<MOUNTAIN> (inherited from PT 11)	• Japão (since it is expressed) • mais alta (inherited from PT 11)	(the system does not import the answer to PT 11 as an object because it has the same QC)
PT 81: Quem foi o último rei de Portugal? [Who was the last king of Portugal?]	<FUNCTION>	• último rei de Portugal	D. Manuel II
PT 82: Em que período foi ele rei? [In which period was he a king?]	<CHRONOLOGY>	• ele rei	
PT 82 final question analysis:	<CHRONOLOGY>	• D. Manuel II rei de Portugal	

of QA systems. We developed the module for CLEF purposes only. The strategy we applied this year to topic-related questions was the following (see Table 1 for examples):

1. analyse the first question;
2. save the answer;
3. analyse the current question;
4. handle explicit anaphors (those where the pronoun is expressed);
5. use the last expressed QC;
6. use the argument analysis of the question which expresses the QC;
7. import the missing arguments from the first question to the current question;
8. if the QC changes, also import the answer.

This procedure still has flaws and systematically failed in questions like PT questions 37 "E um não-metal." [And a nonmetal.], 65 "E do pão?" [And of bread?] and 144 "E a segunda" [And the second one?], where the arguments of the first question were not replaced but added. In the Spanish run, topic-related questions suffered with this new schema, since question syntactical analysis is still quite poor when compared to Portuguese.

3 Results

Table 2 presents the results of Portuguese and Spanish monolingual runs submitted by Priberam to the main task of QA@CLEF 2008, according to three question categories, *factoid* (FACT), *definition* (DEF) and *list* (LIST), with the judgments used for evaluation (R=Right, W=Wrong, X=Inexact, U=Unsupported).

Table 2. Results by category of question, including detailed results of topic and non-topic-related questions

		R		W		X		U		Total		Accuracy (%)	
		PT	ES	PT	ES	PT	ES	PT	ES	PT	ES	PT	ES
Non-topic related	FACT	83	55	24	45	4	0	1	2	112	102	**74.1**	**53.9**
	DEF	18	15	4	3	6	0	0	0	28	18	**64.3**	**83.3**
	LIST	3	5	3	8	3	4	0	1	9	18	**33.3**	**27.8**
	Total	104	75	31	56	13	4	1	3	149	138	**69.8**	**54.3**
Topic related	FACT	23	11	23	46	1	1	3	1	50	59	**46.0**	**18.6**
	DEF	0	0	0	1	0	0	0	0	0	1	-	**0.0**
	LIST	0	0	1	2	0	0	0	0	1	2	**0.0**	**0.0**
	Total	23	11	24	49	1	1	3	1	51	62	**45.1**	**17.7**
General (all)	FACT	106	66	47	91	5	1	4	3	162	161	**65.4**	**41.0**
	DEF	18	15	4	4	6	0	0	0	28	19	**64.3**	**78.9**
	LIST	3	5	4	10	3	0	0	1	10	20	**30.0**	**25.0**
	Total	127	86	55	105	14	4	4	4	200	200	**63.5**	**43.0**

Regarding the Portuguese run, the improvement of more than 20% in the accuracy of general factoid questions considerably contributed to the increase of the overall accuracy, which surpassed that of last year (50%). Besides that, an analysis of PT question clusters shows that there was an increase in the number of clusters (37 clusters, in a total of 88 questions, 51 of which topic-related) but that the system was able to extract the correct answers 45% of the times, which means a boost of nearly 30%, when comparing to last year's results.

Despite these general positive results, Table 2 also shows a decrease of accuracy in DEF and LIST questions. The reasons for failures are assembled in Table 3. In the Portuguese run, the main source of error was the extraction of candidate answers, followed by the choice of the final answer. The main reason for errors in extraction of candidate answers is the coverage of QAPs, which are handwritten and therefore limited.

Table 3. Reasons for W, X and U answers

Stage ↓ **Question →**	W+X+U		Failure (%)	
	PT	ES	PT	ES
Document retrieval	4	1	4.1	0.9
Extraction of candidate answers	33	75	46.6	66.4
Choice of the final answer	20	17	27.4	15.0
NIL validation	8	9	11.0	8.0
Topic	4	7	5.5	6.2
Other	4	4	5.5	3.5
Total	73	113	100.0	100.0

With regard to the Spanish run, Table 2 shows that results within non-topic-related questions are quite similar to those of last year, while topic-related questions had a decrease in its accuracy of almost 20%. At this point, it deserves to be said that the number of both question clusters and topic-related questions doubled in 2008 for the ES test set: from 20 clusters and 30 topic-related questions, it passed to 48 clusters and 62 topic-related questions. This fact had, consequently, a strong impact on the Spanish results, both on the falling of topic-related questions accuracy by itself and, mainly, on the global results. Another remarkable fact about the Spanish set is a significant increase of the number of LIST questions compared to last year's set or to the Portuguese set.

In Table 3 we classified as Other all the unsupported answers in the ES run. All of them are certainly correct answers, but at least three of them do not explicitly contain all the needed supporting information in the snippet, although this information does appear in the document. Those errors could be seen as a limitation of the system in the way of presenting the information, and not in the way it processes those questions. One example is ES question 10 "A qué edad murió Wallace Rowling?" [At which age did Wallace Rowling die?]. The QA system correctly answered "67 años" from the snippet "- Sir Wallace Rowling, ex primer ministro de Nueva Zelanda, 67 años.". Although the actual snippet does not support the answer, the document where it comes from is a list of deceased people in 1995 from EFE, but that is not shown in the snippet. Finally, there is an interesting case of extraction problem with the answer to the ES question 75 "Cómo se pronuncia eso?" [How is it pronounced?], whose topic is *TeX*. The correct answer is displayed in Wikipedia between square brackets, and it happens to be ignored by the QA system because of that.

From the analysis of the results, we conclude that the retrieval stage and the question analysis stage are performing very well, that QAPs need to broaden their coverage and that the work done for Portuguese this year must be ported to the Spanish rules.

4 Conclusions and Future Work

Priberam's aim for QA@CLEF 2008 was to consolidate the system and improve its performance. Even though this year there was no real time exercise, from our tests we verified that we doubled the speed of the system and improved the capacity to answer multiple questions simultaneously by enhancing the parallelism of the algorithms. These improvements were crucial for the implementation of the search engine in the sites of *TSF* and *Jornal de Notícias*. The retrieval stage is performing very well and the changes in the syntactical/semantic analysis now cover all the QCs.

During last year, we have been working on anaphora resolution and we had a first prototype of the system a few days before CLEF. As we had no feedback on how the system was behaving, we decided not to submit the results with anaphora resolution. After CLEF, we managed to run the tests and found out that the results were almost the same. This is an interesting result and we are convinced that this is due to the CLEF set of questions being extracted directly from what is written in the documents. Anaphora resolution is important only when dealing with Wikipedia articles between the body of the text and the title (this was already implemented last year).

The work done this year in the Portuguese module must also be done in the Spanish module, specifically on the syntactical/semantic analysis of the questions and NER. As we mentioned in section 2.1, we plan to tag each word/phrase with the QC in the indexing stage. We hope that without a big penalty on index size we can achieve better accuracy and speed. Future work will also include, since we now have a big corpus of questions/answers/false answers, working on algorithms to automatically learn new question patterns from corpora.

Acknowledgements. Priberam would like to thank Synapse Développement, as well as the CLEF organisation and Linguateca.

References

1. Amaral, C., Cassán, A., Figueira, H., Martins, A., Mendes, A., Mendes, P., Pinto, C., Vidal, D.: Priberam's Question Answering System in QA@CLEF 2007. In: Peters, C., Jijkoun, V., Mandl, T., Müller, H., Oard, D.W., Peñas, A., Petras, V., Santos, D. (eds.) CLEF 2007. LNCS, vol. 5152, pp. 364–371. Springer, Heidelberg (2008)
2. Giampiccolo, D., Forner, P., Peñas, A., Ayache, C., Cristea, D., Jijkoun, V., Osenova, P., Rocha, P., Sacaleanu, B., Stutcliffe, R.: Overview of the CLEF 2007 Multilingual Question Answering Track. In: Peters, C., Jijkoun, V., Mandl, T., Müller, H., Oard, D.W., Peñas, A., Petras, V., Santos, D. (eds.) CLEF 2007. LNCS, vol. 5152, pp. 200–236. Springer, Heidelberg (2008)
3. Amaral, C., Figueira, H., Martins, A., Mendes, A., Mendes, P., Pinto, C.: Priberam's Question Answering System for Portuguese. In: Peters, C., Gey, F.C., Gonzalo, J., Müller, H., Jones, G.J.F., Kluck, M., Magnini, B., de Rijke, M., Giampiccolo, D. (eds.) CLEF 2005. LNCS, vol. 4022, pp. 410–419. Springer, Heidelberg (2006)
4. Cassán, A., Figueira, H., Martins, A., Mendes, A., Mendes, P., Pinto, C., Vidal, D.: Priberam's Question Answering System in a Cross-language Environment. In: Peters, C., Clough, P., Gey, F.C., Karlgren, J., Magnini, B., Oard, D.W., de Rijke, M., Stempfhuber, M. (eds.) CLEF 2006. LNCS, vol. 4730, pp. 300–309. Springer, Heidelberg (2007)
5. Amaral, C., Laurent, D., Martins, A., Mendes, A., Pinto, C.: Design and Implementation of a Semantic Search Engine for Portuguese. In: Proceedings of 4th International Conference on Language Resources and Evaluation (LREC 2004), vol. 1, pp. 247–250 (2004)
6. Zhao, S., Lin, D.: A Nearest-Neighbor Method for Resolving PP-Attachment Ambiguity. In: Su, K.-Y., Tsujii, J., Lee, J.-H., Kwong, O.Y. (eds.) IJCNLP 2004. LNCS (LNAI), vol. 3248, pp. 545–554. Springer, Heidelberg (2005)

IdSay: Question Answering for Portuguese

Gracinda Carvalho[1,2,3], David Martins de Matos[2,4], and Vitor Rocio[1,3]

[1] Universidade Aberta, Rua da Escola Politécnica, 147 1269-001 Lisboa, Portugal
gracindac@univ-ab.pt,vjr@univ-ab.pt
[2] L2F/INESC-ID Lisboa, Rua Alves Redol 9 1000-029 Lisboa, Portugal
david.matos@inesc-id.pt
[3] CITI FCT/UNL
[4] Instituto Superior Técnico/UTL

Abstract. IdSay is an open domain Question Answering (QA) system for Portuguese. Its current version can be considered a baseline version, using mainly techniques from the area of Information Retrieval (IR). The only external information it uses besides the text collections is lexical information for Portuguese. It was submitted to the monolingual Portuguese task of the QA track of the Cross-Language Evaluation Forum 2008 (QA@CLEF) for the first time, and it answered correctly to 65 of the 200 questions in the first answer, and to 85 answers considering the three answers that could be returned per question. Generally, the types of questions that are answered better by IdSay system are measure factoids, count factoids and definitions, but there is still work to be done in these areas, as well as in the treatment of time. List questions, location and people/organization factoids are the types of question with more room for improvement.

1 Introduction

The objective of a QA system is to provide an answer, in a short and precise way, to a question in natural language. Answers are produced by searching a knowledge base that usually consists of natural language text. The usefulness of this type of system is to find the exact information in large volumes of text data.

IdSay (I'd Say or I dare Say) is an open domain QA system for Portuguese that was developed from scratch, with the objective of optimizing computational space and time, so that response could be fast. It was submitted to the monolingual Portuguese task of the QA track of the Cross-Language Evaluation Forum 2008 (QA@CLEF) for the first time. IdSay results placed it in third place among the other five systems that had participated in previous campaigns. Details of the task, and comparative results can be found in the overview of the QA track [1].

In Sect. 2 we describe IdSay briefly. In Sect. 3 we analyse the results obtained in QA@CLEF 2008, and in Sect. 4 we end with conclusions and future work.

C. Peters et al. (Eds.): CLEF 2008, LNCS 5706, pp. 345–352, 2009.

2 The IdSay System

Developing a QA system combines the task of treating large quantities of unstructured data (text), and the need to have a good understanding of the text to produce exact and short answers. Therefore it is natural that the areas of IR and natural language processing (NLP) are the foundations of these systems.

This is the approach we intend to follow in building IdSay system. We started by developing the core version of the system, which is based on information retrieval techniques. We chose this option for two main reasons: Firstly because we want to have a baseline to compare and draw conclusions of the effectiveness of the further NLP enhancements we plan to implement. Secondly because we intend to have an efficient retrieval base that can work as independently of the language as possible to reuse with different languages in the future.

The present version of IdSay is as close as possible to simple keyword search. The only external information that we use besides the text collections is lexical information for Portuguese [2]. In the rest of this section we briefly describe IdSay system, starting by the information indexing in Sect. 2.1, followed by an overview of the architecture of the system in Sect. 2.2.

2.1 Information Indexing

IdSay system is based on indexing techniques that were developed from scratch using C++. The IR engine was built with cross-language usage in mind, so we tried to develop it modularly, with the language-specific information clearly separated from generic components. For this purpose we analyse the input text data in successive levels, building an index file for each layer.

Level 1 Document Level. The documents are kept as close to the original text as possible, apart from the compression techniques used. It includes also tokenization and the minimal pre-processing to allow efficient retrieval, namely separation of words with spaces and lowercase conversion.

Level 2 Lemmatization or Stemming. According to the results of our previous work [3], in which lemmatization and stemming were compared, we opted for doing only lemmatization[1]. We intend however, in future versions of the system, to try different stemming techniques and lemmatization using a different lexicon. We do not remove stop words from the texts. This level corresponds to making equivalence classes based on related words at a linguistic level, and therefore it is one of the levels that is more language-specific.

Level 3 Entities. At level 3, which we call the entity level, we find all sequences of words that co-occur often in the text collections, and if their number of occurrences is higher than a given threshold (100 seems to be a reasonable

[1] Both options are available; when we say we use lemmatization, we are talking about the system setup for QA@CLEF.

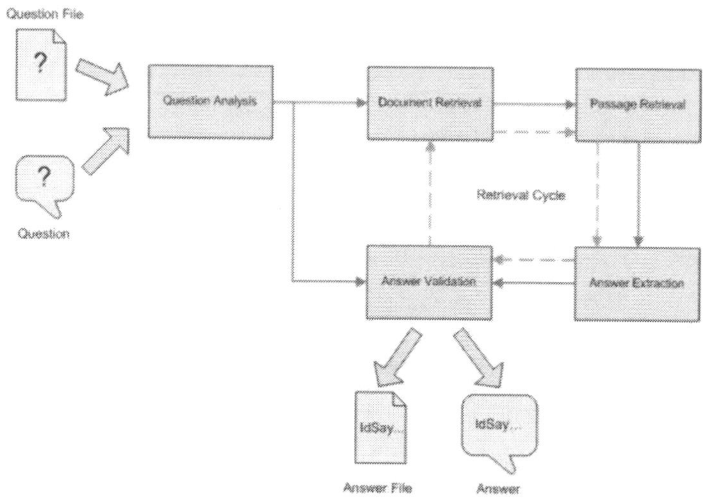

Fig. 1. IdSay system architecture

value), we consider them an entity whether it corresponds to a meaningful entity, like the name of an organization, or to a common string of words. For the time being, we rely on our ranking mechanism to eliminate the second kind of entities from answers, but we may do some further work in this area in the future.

2.2 System Overview

IdSay accepts either a question written by the user (manual interface), or a set of questions in an XML file (automatic interface). Each question is analysed in the question analysis module to determine the question type and other variables to be used in the answer extraction and validation modules. The question analysis also determines a search string with the information of which words and entities to use in the document retrieval module to produce a list of documents that match both. This list of documents is then processed by the passage retrieval module, responsible for the search of passages from the documents that contain the search string, and with length (number of words) up to a given limit (60). The passages are then sent to the answer extraction module, where short segments of text (candidate answers) are produced that are then passed on to the answer validation module. This module validates answers and returns the most relevant ones. If in one of the steps no data is produced, the search string is revised and the loop starts again (retrieval cycle).The global architecture of IdSay is presented in Fig. 1.

The index files for the text collection[2] occupy 1.15 GB of disk space, and took about 4 hours to build. The load time is around 1 minute, and the time

[2] The text collection occupies around 9 GB of disk space, in over 600,000 files. More details on the collection can be found in [1].

to process 200 questions is less than 1 minute. These values correspond to tests using a machine with an AMD Athlon 64 processor (2.21 GHz), with 4GB of RAM, running Windows XP.

3 QA@CLEF 2008 Results

In the present section we analyse the results obtained by IdSay. First we look into the evaluation metrics that describe the overall performance of the system, and proceed with a more detailed question based analysis.

3.1 Evaluation Metrics

The main evaluation metric used in QA@CLEF 2008 is accuracy over the first answer, which is the average of first answers that where judged to be correct. We also calculated the accuracy over all answers because it is also a common measure used for QA systems. Another metric used is MRR (Mean Reciprocal Rank) which is the mean of the reciprocal of the rank of the first answer that is correct for each question, as defined in [4]. Table 1. summarizes the results of IdSay system.

Table 1. IdSay results overview

Accuracy over the first answer	Accuracy over all answers	MRR
32.500%	42.500%	0.37083

3.2 Detailed Analysis of Results

IdSay has different approaches according to different criteria, for instance, specific procedures regarding question category and type. In the present section we analyse our results, covering different characteristics of the questions.

Results by Question Category. Three question categories are considered in QA@CLEF, namely *F* (factoids), *D* (definitions) and *L* (closed list questions).The results obtained by IdSay are summarized in Table 2.

Table 2. Results by category

Question Category	Total Questions	Right	Wrong	ineXact	Unsupported	Accuracy
F	162	47	100	7	8	29.012%
D	28	18	10	0	0	64.286%
L	10	0	9	1	0	0%

The results show a stronger ability for the system to answer definition questions than factoids, which was expected due to the valuable aid of having an encyclopaedic data collection. The low value obtained for list questions is not a surprise because we did not have the time to treat this category of questions, so these are treated as factoids.

Definition Questions. This type of question generally occurs in the form: "O que é X?" [What is X?] or "Quem é X?" [Who is X?], in which we consider X the reference entity. IdSay starts by searching for the reference entity in Wikipedia, looking for a page for this concept. If such a page is found, the beginning of the page is returned as the answer.

The majority of definition questions were of the type "O que ser X?" [What to be X?][3]. IdSay answered correctly to half of them based on Wikipedia pages. If Wikipedia does not provide a definition, we follow the default procedure of searching the data collection in search for occurrences of the reference entity. An example of a correct definition found via the default procedure is (Question#66 O que é o jagertee?) [What is jagertee?], for which the answer was found within the data collection, in a sentence "o jagertee é chá com adição de rum" [jagertee is tee with addition of rum].

There were 7 definition questions of the type "Quem ser X?" [Who to be X?], of which IdSay answered 5 correctly based on Wikipedia pages. The two questions not answered correctly were (Question#23 Quem é FHC?) [Who is FHC?] and (Question#41 Quem é Narcís Serra?) [Who is Narcís Serra?]. The first corresponds to a Wikipedia page that is not found because the keyword FHC is not the name of the page for former Brazilian President Fernando Henrique Cardoso (but rather a redirect). In the second case, there is no Wikipedia page for Narcís Serra, and although in this case two news articles are found with the information, the answers were wrong due to extraction problems.

Factoids Results by Question Type. We consider the following types of questions: *P* - person/organization, *D* - date/time, *L* - location, *C* - count, *M* - measure, *O* - Other. We will start by analysing the results for the types for which we developed special procedures because they involved numeric values: *C*, *M* and *D*. We consider the assessment of the question to be the best answer, using the following priority: R, U, X and W[4].

Table 3 presents the results of IdSay for the type of factoids count, measure and date. We procced with an analysis of these results.

Table 3. Results by question type

Question Type	# Questions	Right	Wrong	Unsupported	ineXact
Count	19	13 (68.4%)	5 (26.3%)	1 (5.3%)	0 (0%)
Measure	12	9 (75.0%)	2 (16.7%)	1 (8.3%)	0 (0%)
Date	24	11 (45.8%)	12 (50.0%)	0 (0%)	1 (4.2%)

Factoids Count. These questions usually start by "Quantos/as X " [How many X]. X usually represents what we are tying to count. The general form of the

[3] We use the lemmatized form of the verb to cover the several tenses occuring in the questions.

[4] For example, if a question has three answers judged W, X and U we consider the U answer.

answer is usually a number followed by X. There were 20 count questions, with very diverse instances of X, namely esposas, faixas, províncias, repúblicas, actos, atletas, estados, filhos, filmes, gêneros, habitantes, jogadores, ossos, refugiados, votos [wives, stripes, provinces, republics, acts, athletes, states, sons, movies, gender, inhabitants, players, bones, refugees, votes].

An example of a correct answer is (Question#70 Quantas províncias tem a Ucrânia?) [How many provinces does Ukraine have?]. In the question, the reference entity Ukraine was identified and the identification of the unit allowed the correct answer to be found: 24 provinces. The case of (Question#10 Quantas províncias tem a Catalunha?) [How many provinces does Catalonia have?] is similar, with 51 documents retrieved that produced the answer "4 provinces" supported by more than one passage. However the answer was considered unsupported, due to the choice of the shortest passage. As an example of a question that produced wrong answers, we can look at (Question#18 Quantos ossos têm a face?[sic]) [How many bones do the face have?]. Although the question is incorrectly formulated (agreement is violated because the verb should be singular), the lemmatization took care of that and produced the search string "bone to have face". However, the answers produced were incorrect (number of bones of parts of the face, as the nose, returned) because the correct answer occurred in a phrase using the construction "é constituída por" [consists of] instead of the verb "ter" [to have].

Factoids Measure. This type of question is similar to the previous one, and generally occurs if the form of "Qual/ais .. o/a X de " [What the X of] in which X is a measure, which can have several units. The answer is generally a numerical value in the correct units for the measure. There were several cases of measures in the question set: altura, área, dotação, envergadura, largura, temperatura, comprimento [height, area, money value, bulkiness, width, temperature, length]. IdSay supports several systems of measures and the corresponding units implemented in the manner of authority lists as described in [5]. It allows the search of the answers of the correct type.

An example of a correct answer is (Question#142 Qual é a área da Groenlândia?) [What is the area of Greenland?], for which only the value of the area "2 170 600 km 2" is returned and in the same passage there are other numbers, that would also be returned if we did not check the area units. The incorrect answers were given for questions that supposedly should produce NIL answers.

Factoids Date. The most common form of occurrence for this type of question is in questions starting by "Quando" [When], though there are also 4 questions staring by "Em que ano" [In which year]. IdSay has a specific treatment of dates, starting with the pre-processing of the texts, and also in the extraction of the answer. However this treatment is not fully developed, for instance the temporal restrictions are not taken into account. Therefore, the results achieved for this type are worse than for the preceding two types. The low accuracy for temporally restricted questions, 18.750%, can also be interpreted in light of this limitation.

An example of a correct answer is (Question#86 Quando é que ele tomou posse?) [When was he empowered?], which is also an example of a question that belongs to a cluster with first question (Question#85 Quantos votos teve o Lula nas eleições presidenciais de 2002?) [How many votes had Lula in the presidential election of 2002?]. Although Question#85 was not successfully answered, the reference to Lula (Brazilian President Luiz Inácio Lula da Silva) is correctly resolved in Question#86 (reference resolution based on the question, not the answer). As for the 12 wrong answers there are several aspects that contribute to that, there are questions about periods that were not treated by the system, and there is a need to treat date information from Wikipedia in a more practical way, e.g. the listed items in such pages are not terminated, so events tend to be mixed up in the resulting text.

Factoids Person. This type of question generally appears in a form starting by "Quem" [Who], but that is not always the case. The results for this type had an overall accuracy of 34%, which is in line with the general performance of the system. Examples of correct answers were (Question#92 Quem fundou a escola estóica?) [Who founded the stoic school?] (Question#143 Quem foi a primeira mulher no espaço?) [Who was the first woman in the space?] for which the system gives the correct answers (Zenão de Cítio and Valentina Tershkova, respectively) but they are accompanied by wrong second and third answers, that have different information related to the subject. We must therefore find a way to filter entities of type person. As stated in Sect. 2, IdSay keeps two separate indexes for words and for entities (two words or more). In the case of these two questions, the number of documents retrieved searching only for words were 11 for Question#92 and 1991 for Question#143. After combining the search for entities the number of documents decreased to 2 and 75, respectively. The case of Question#143 clearly shows an example of the utility in combining the search by single word with the search for entities.

NIL Accuracy. About the NIL accuracy, the reported value of 16.667% (2 right answers out of 12) for IdSay indicates the need of improvement in our mechanism to determine how well a passage supports the answers, to minimize the negative effect of the retrieval cycle in relaxing constraints. However comparatively to the other systems IdSay has the highest performance in NIL accuracy.

4 Conclusions and Future Improvements

We found the results of our first participation at QA@CLEF very encouraging. The fact that these results were obtained with particularly challenging rules (even for veteran participants) seems to reinforce the validity of our approach.

The analysis of our participation at QA@CLEF shows that the retrieval component works reasonably well, but the answer extraction mechanism is less efficient and is generally responsible for the wrong answers produced by the system. We expect that the introduction of NLP techniques will help in this regard. Another area that we identified that can benefit from these techniques is the answer validation module. In this module the ranking of answers by frequency means that we produce the answer that appears most frequently in the passages extracted from the data collection. This means that an answer may be supported by several passages, but we can only give one as support. In this participation we chose the shortest one, but in several cases this option led for the support to be considered unsatisfactory by the assessors. We must therefore introduce an analysis of the passages to determine how strongly they support the answer.

Regarding the setup of the system, we find lemmatization a good choice as a whole, since it provides an efficient search, with just one case of a definition being wrong on its account.

As for short term improvements, these include attributing a confidence score to each answer, treating temporally restricted questions and the improvement of co-references between questions. The scoring mechanism of the answers is already partially implemented, since several supports for an answer are already considered, with different weights attributed to different kinds of occurrences.

As for future enhancements, besides of the introduction of NLP methods, we intend to accommodate semantic relations between concepts by adding further levels of indexing. As an example, we would like to introduce equivalences at a conceptual level, for instance by means of a thesaurus. Another future direction we intend to follow is introducing other languages besides Portuguese.

References

[1] Forner, P., et al.: Overview of the CLEF 2008 Multilingual Question Answering Track. In: Peters, C., et al. (eds.) CLEF 2008. LNCS, vol. 5706, pp. 262–295. Springer, Heidelberg (2009)
[2] Alves, M.A.: Engenharia do Léxico Computacional: princípios, tecnologia e o caso das palavras compostas. Mestrado em Engenharia Informática Faculdade de Ciências e Tecnologia da Universidade Nova de Lisboa (2002)
[3] Carvalho, G., Martins de Matos, D., Rocio, V.: Document retrieval for question answering: a quantitative evaluation of text preprocessing. In: Proceedings of the ACM first Ph.D. workshop in CIKM (ACM), pp. 125–130 (2007)
[4] Magnini, B., et al.: The Multiple Language Question Answering Track at CLEF 2003. In: Peters, C., Gonzalo, J., Braschler, M., Kluck, M. (eds.) CLEF 2003. LNCS, vol. 3237, pp. 471–486. Springer, Heidelberg (2004)
[5] Prager, J.: Open-Domain Question-Answering. Foundations and Trends® in Information Retrieval (Now Publishers) 1(2), 91–231 (2006)

Dublin City University at QA@CLEF 2008

Sisay Fissaha Adafre[1] and Josef van Genabith[2]

[1] National Center for Language Technology, School of Computing, DCU
sadafre,josef@computing.dcu.ie
[2] IBM CAS Dublin

Abstract. We describe our participation in Multilingual Question Answering at CLEF 2008 using German and English as our source and target languages, respectively. The system was built using UIMA (Unstructured Information Management Architechture) as underlying framework.

1 Introduction

This is our first participation in the Multilingual Question Answering Track of CLEF. We took part in the bilingual CLEFQA task (German-English) where German is the source language and English the target language. We used the Bable Fish[1] online translation system to translate the German questions into English. The system is targeted at *Factoid* and *Definition* questions.

QA systems generally consist of online methods that generate answers to questions automatically by directly analysing the text corpus. Systems also make use of external resources in the form of Gazetteers or precompiled tables which are obtained through offline mining of large text corpora or the web. Although it has been shown that outputs of offline mining methods can be used to improve QA results, our focus in designing the current system is on testing our online methods which are based on information extraction methods. Our system does not make use of precompiled tables or Gazetteers but uses Web snippets to rerank candidate answers extracted from the document collections. WordNet is also used as a lexical resource in the system. Typical QA systems employ various Natural Language Processing (NLP) and Machine Learning (ML) tools, a set of heuristics and different lexical resources. Seamless integration of the various components is one of the major challenges of QA system development. In order to facilitate our development process, we used the Unstructured Information Management Architecture (UIMA) as our underlying framework [7].

In the remainder of this paper, we will describe our system. Section 2 provides a description of the system that deals with factoid questions. Section 2.6 summarises the treatment of Definition questions. Section 3 presents the result of the experimental evaluation. Finally, Section 4 presents some concluding remarks.

---

[1] http://babelfish.yahoo.com/

C. Peters et al. (Eds.): CLEF 2008, LNCS 5706, pp. 353–360, 2009.

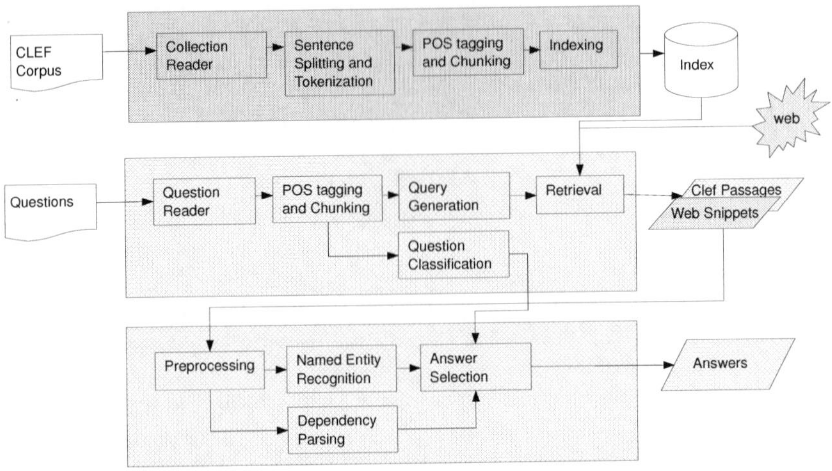

Fig. 1. System Architecture

2 System Description

Our question answering system consists of the following core components: Question Analysis, Passage Retrieval, Sentence Analysis and Answer Selection [8]. Each of these components employs various tools, and a set of heuristic rules. In order to ease the development process, we used UIMA. UIMA facilitates integration of typical natural language processing tasks, such as tokenization, sentence detection, named-entity recognition, POS-tagging and chunking, parsing, etc, for building complex applications such as question answering [7].

The main inputs of the system are the document collection and the questions. Corresponding to each of these inputs, we have collection and question processing components. As shown in Figure 1, the collection processing component carries out preprocessing and indexing of the document collection. The preprocessing step involves splitting documents into sentences, and POS tagging and chunking. Indexing takes care of merging a set of sentences into passages using the Lucene retrieval engine [2]. Question processing consists of question classification and query generation. The question processing also involves POS-tagging and chunking, and parsing of the input questions. The question classifier and query generation component take the linguistically annotated questions and generate question classes, and queries, respectively. Finally, the answer selection component, which matches questions to answers, consists of passage retrieval, sentence analysis, and answer reranking components. We used TreeTagger [5] for POS-tagging and Chunking, and a treebank-based Lexical Functional Grammar(LFG) parser for dependecy parsing [1]. In the next section, we provide descriptions of the components assuming *Factoid* questions.

2.1 Question Analysis

Question analysis consists of a question classification and a query generation component. We employed both machine-learning and rule-based methods for defining question classes. For the machine learning approach, we trained a classifier using the data set provided by Li and Roth [9]. We used the MinorThird implementation of Conditional Random Fields as our classifier [10]. The classification taxonomy consists of two layers in which the top level consists of 6 major classes (ABBREVIATION, ENTITY, DESCRIPTION, HUMAN, LOCATION and NUMERIC VALUE), and the bottom level consist of 50 classes. The resulting labels are used to define expected answer types for the questions.

The rule based approach uses syntactic clues to identify terms that can be used to refine the expected answer types - *focus* terms. Typical patterns include noun phrase chunks following a *wh* word in the question.

Query generation identifies terms that will be used for retrieving passages. This step uses patterns defined on the output of the POS-tagger and Chunker. The query terms consist of noun phrase and verb phrase chuncks, where the focus terms and stop-words are removed.

According to the CLEF guidelines, the questions are organised around a set of topics where each *topic* is associated with a group of questions. However, the *topic* is not explicitly given and must be inferred from the first question or its answer. Therefore, we devised a rule based method for detecting the *topic* of the first question. The rules make use of both surface patterns and syntactic clues to identify the *topic* of the question. The *topic* of the first question is also appended to the query set of each of the remaining questions.

2.2 Passage Retrieval

We have three kinds of sources: the CLEF Document Collection, a Wikipedia Corpus, and the Web. We treat each of these sources differently.

We split the CLEF documents into passages where each passage is composed of 10 consecutive sentences in the document. The sentences in the passages are POS tagged and chunked. The original sentences are indexed and the tagged sentences are stored in a separate field in the index - *CollIndex*. Similarly, the articles of the Wikipedia corpus are also indexed separately - *WikiIndex*. However, the Wikipedia articles are indexed as a whole and are not split into passages.

The query terms generated by the Question Analysis Module is submitted to both *CollIndex* and *WikiIndex*. The top 100 passages from *CollIndex* and the top 10 articles from *WikiIndex* are retained. Both the passages from the CLEF corpus and Wikipedia articles are split into sentences. Each sentence will be assigned the retrieval status value of the corresponding passage or article as an initial score. The sentences are passed to the sentence analysis module.

The system also retrieves web snippets using the same queries. We used Yahoo! APIs for retrieving web snippets. We split the snippets into sentences, and retained only those that contained one or more of the query terms.

The lists we obtain from the three sources contain scores that are computed quite differently. In order to minimise the effects of the variation in scoring methods, we normalise scores in each list as follows:

$$R = \text{score}_{\text{norm}} = \frac{\text{score}_{\text{MAX}} - \text{score}}{\text{score}_{\text{MAX}} - \text{score}_{\text{MIN}}} \tag{1}$$

2.3 Sentence Analysis

We run a named entity recogniser on the sentences retrieved. We trained a CRF-based named entity recogniser on the CoNLL Corpus [6]. It recognises the following four major classes: PERSON, ORGANISATION, LOCATION, MISCELLANEOUS. Since there is a mismatch between the classes generated by the question classifier and the named entity recognizer, we devised a mapping between the outputs of the two systems. Furthermore, we added additional classes that could reliably be identified using simple pattern matching (such as dates) to the major classes. We have identified the following classes: *Date*, *Person*, *Location*, *Organization*, *Numeric*, *Count* and *Description*. These classes are further qualified by using the bottom levels of the question classification taxonomy, or terms extracted using the rule-based component. For example, *Location* is further qualified by *Country*, *City*, or *State*. We have one open class which contains all named entities that do not map to any of the above classes. The sentences are also parsed using a dependency parser. The result is used to extract dependency triples that are used to measure the similarity between Questions and Sentences containing the candidate answer as will be explained in Section 2.5.

2.4 Candidate Answer Extraction

We consider two cases when searching for candidate answers to the questions. The first case relates to when the question class can be mapped to one of the predefined classes. In this case, the named entities whose types match one of the broad question classes are taken as candidate answers. For the non-matching case, all noun phrases are extracted from the sentence and are considered candidates. The list will be filtered and reranked as described in Section 2.5.

2.5 Answer Reranking

We reranked candidate answers based on different sources of evidence, such as syntactic similarity of the sentence with the question, proximity of query terms to the candidate answers, similarity of the semantic type of the candidate answer to the answer type, and centrality of the sentence with respect to a corpus of web snippets retrieved using the query terms extracted from the question. We provide details of each these filtering mechanisms.

Syntactic Similarity. Syntax-based evidence has been used to rerank candidate answers in a number of QA Systems [3,12,13]. For examples, syntactic structures have been used for generating syntactic patterns or for measuring the similarity between the questions and the sentence containing the candidate answer. In the current system, syntactic similarity is measured in terms of the number of shared dependency relations between the sentence and the question. For this, we parsed both the questions and the sentences using a Lexical Functional Grammar(LFG)-based parser [1] developed at Dublin City University. The system takes the output of a syntactic parser (Charniak parser [11]) and generates an F-Structure, a labeled bilexical dependency graph. The output can also be provided in the form of a set of dependency triples. We count how many dependency pairs are shared between the questions and answers, normalise the resulting value and add it to the overall score of the named entities extracted from the sentence.

Term Proximity. This method is based on the assumption that answers are likely to be found in a close proximity to the query terms in a sentence. In other words, if more query terms appear in the vicinity of the candidate answer, the candidate answer is likely to be the true answer and hence receives more weight. We implemented this intuition as follows. For each candidate answer in a sentence, we take a window of 10 terms centred in the candidate answer. We then count how many of the query terms appear in the window. The final score is obtained by dividing the resulting count by the total number of query terms.

Type Filtering. As mentioned in Section 2.3, our type classification is limited and assigns a significant part of the named entity classes to miscellaneous. On the other hand, the types derived from questions are either specific instances of the major classes we identified or may not be covered by the major classes. In order to fill the gap, we devised the following methods for computing semantic similarity between the expected answer type and the candidate answer.

Wikipedia Category. Wikipedia contains a large set of user defined categories which are assigned to its entries. Our method is implemented as a binary feature function. First, we check if the candidate answer has an entry in Wikipedia. If the candidate answer does not match an entry in Wikipedia, we assign a score of 0. If there is a Wikipedia entry corresponding to the answer string, we retrieve the categories associated with the Wikipedia article. We then check if the answer type terms are contained in the category lists. If the answer type terms do not occur in the category list, we assign the candidate term a zero score; otherwise we assign a score of 1 to the candidate answer.

WordNet Hierarchy. We take both the expected answer type and the candidate answer, and check if they have entries in WordNet. We then check if the expected answer type and the candidate answer stand in the *Hypernym* relation with respect to the WordNet hierarchy. We assign a score which is the inverse of the distance between the two concepts in the hierarchy. For example, if the expected answer type is a direct hypernym of the candidate answer, the candidate answer recieves a score of 1.0, else it will be less than 1.0.

WordNet vs Wikipedia Category. This is an extension of *Wikipedia Category* method above. We take the expected answer type, and generate its *WordNet* *hyponyms* sets (5 levels down the hierarchy). We take the Wikipedia categories of the candidate answer. Finally, we compute the fraction of shared entries between these two sets as a measure of semantic similarity.

Web based evidence. We used the Web in two ways: to find answers, and to rerank candidate answers from the *CLEF Collection* and *Wikipedia Collection.*

Web Answers. The Web snippets pass through the same processing pipelines as the snippets (sentences) from the *CLEF Collection* and *Wikipedia Collection.* Since answers must come from the later two collections, the Web answers must be mapped onto the collection (*Answer Projection*). For each candidate answer in the lists obtained from the *CLEF Collection* and *Wikipedia Collection*, we check if there is a matching Web answer in the ranked list of Web answers. If there is, we add the score of the Web answer to the current score of the candidate answer. This assigns more weight to those candidates that are found in the Web answer list.

Web based reranking. This type of evidence for sentence importance is based on the assumption that there is a high degree of redundancy among the top web snippets returned for a given query. In order to take advantage of this fact, we create a *reference corpus* consisting of the top 50 ranking Web snippets. This corpus will be used to estimate relevance of the sentences to the questions. This is estimated as a graph-based similarity score between the target sentence with the web corpus as described in [14].

Combining Scores. The overall scores for reranking candidate sentences are computed as a linear combination of the *Retrieval scores*, *Syntactic similarity*, *Term Proximity*, *Type Filtering* and *Web-based evidence.* Each of these scores have been normalized between 0 and 1 using the formula given in 1. The baseline system simply sums these scores without taking into account the relative importance of the different evidences.

2.6 Definition Questions

The *definition* questions expect short snippets or sentences that provide a concise definition or description of the topic as an answer, unlike *factoids* for which the expected answers are largely named entities. As a result, we adopted a different strategy for definition questions. The system takes the *topic* generated by the question analysis module, and submits it as a query to the retrieval module. The returned passages and Wikipedia articles are split into sentences. Sentences that do not contain the topic are removed from the list. We assign more weight to sentences with copula verb constructions with the *topic* as a subject, e.g. TOPIC is .... Finally, the sentences are reranked using evidence obtained from the web as described in Section 2.5.

Table 1. Results

	Factoid		Definition		Overall	
	Complete	Partial	Complete	Partial	Complete	Partial
Right	8	1	8	0	16	1
Wrong	142	157	16	9	168	195
ineXact	1	1	6	1	7	3
Unsupported	9	1	0	0	9	1

3 Experimental Evaluation

We submitted two runs for our CLEF participation. The first run is the output of the complete system - *Complete.* The second run is the output of the system without the web-based reranking component - *Partial.*

Overall the system returned only 16 *exact* answers, and 25 correct answers counting *unsupported* answers. The web reranking component contributed significantly. The result without the web reranking component is disappointing. This is attributed to a number of problems. Error analysis shows different sources of errors such as Translation, Questions classification, Topic detection, and Named entity recognition. The main problem was lack of proper testing due to time constraints. The system is also limited in scope. As mentioned in Section 1, the system relies primarily on online methods, focusing on a restricted class of named entities. Since it is an evolving system, we believe that its coverage will improve by adding more semantic categories.

Effects of Translation. Although most of the German questions have been accurately translated into English, there are still some translation errors which affected our results. The errors range from incorrect word order, missing constituents to non-translated terms, and have affected both the question classification and query generation components. For example, the following translation error resulted in an incorrect question classification - Description or Definition. E.g. DE: *Wie großist Jerod Ward?* - EN: *Is Jerod Ward how large?.*

Another common error is incorrect word order, e.g. DE: *Wann schrieb Mathieu Orfila sein "Trait des poisons"?* - EN: *When Mathieu Orfila wrote its "Trait poisons"?.* The query generation component identifies *Mathieu Orfila* as the *Focus* of the sentence, as in *Country* in the question form *Which country ....* This error propagated to subsequent analysis steps resulting in few candidate answers.

The last example contains a term that is not translated, i.e. DE: *Geomorphologie* - EN: *Geomorphologie.* Since it is the only term in the query generated by the question analsys component, it returned very few snippets.

4 Conclusion

We presented our QA system adapted for the German-English bilingual CLEFQA task. The system is developed using UIMA as our underlying framework.

The exercise showed that UIMA facilitates building QA system in a short period of time. We also observed that translation quality affects our result since the system is originally designed assuming grammatically correct inputs.

The system is largely based on Information Extraction methods, with various filtering and reranking steps to pin point the correct answers. It is limited in a number of aspects since it is in its early stages of development. Our future plan is to extend the types of question that can be handled, and improve the methods for those already implemented. Specifically, we would like to extend our dependency triple based scoring method to include the full LFG-based parse output. Finally, computation of the overall score is based on simple linear combination of the individual scores ignoring their relative weights. In the future, we will use a ML based approach for computing the overall score using the individual evidences as features.

References

1. Cahill, A., Burke, O.R., Riezler, S., Genabith, J., Way, A.: Wide-Coverage Statistical Parsing Using Automatic Dependency Structure Annotation. Computational Linguistics 34(1), 81–124 (2008)
2. Lucene. The Lucene search engine, `http://lucene.apache.org/`
3. Molla, D., Gardiner, M.: Answerfinder - question answering by combining lexical. In: Australasian Language Technology Workshop (ALTW), Sydney (2004)
4. Monz, C.: From Document Retrieval to Question Answering. PhD thesis, University of Amsterdam (2003)
5. Schmid, H.: Probabilistic part-of-speech tagging using decision trees. In: International Conference on New Methods in Language Processing, UK (1994)
6. Tjong Kim Sang, E.F.: Introduction to the conll-2002 shared task: Language-independent named entity recognition. In: COLING 2002: Proceedings of the 6th conference on Natural language learning (2002)
7. UIMA. Unstructured information management architecture, `http://domino.research.ibm.com/comm/research_projects.nsf/pages/uima.index.html`
8. Kupiec, J.: Murax: A robust linguistic approach for question-answering using an on-line encyclopedia. In: Proceedings of 16th Annual International ACM/SIGIR Conference (1993)
9. Li, X., Roth, D.: Learning question classifiers, 2002. In: Proc. COLING 2002 (2002)
10. Cohen, W.: Methods for identifying names and ontological relations in text using heuristics for inducing regularities from data, `http://minorthird.sourceforge.net`
11. Charniak, E.: A maximum-entropy-inspired parser. In: Proceedings of the first conference on North American chapter of the Association for Computational Linguistics, pp. 132–139. Morgan Kaufmann Publishers Inc., San Francisco (2000)
12. Katz, B., Lin, J.: Selectively using relations to improve precision in question answering. In: Proceedings of EACL 2003 Workshop on Natural Language Processing for Question Answering (April 2003)
13. Lamjiri, A.K., Kosseim, L., Radakrishnan, T.: Comparing the Contribution of Syntactic and Semantic Features in Closed versus Open Domain Question Answering. In: Proc. ICSC 2007, Irvine, California, USA, pp. 679–685 (2007)
14. Adafre, S.F., de Rijke, M.: Estimating importance features for fact mining (with a case study in biography mining). In: RIAO (2007)

Using Answer Retrieval Patterns to Answer Portuguese Questions

Luís Fernando Costa

Linguateca, Oslo Node, SINTEF ICT
Pb 124 Blindern, 0314 Oslo, Norway
luis.costa@sintef.no

Abstract. Esfinge is a general domain Portuguese question answering system which has been participating at QA@CLEF since 2004. It uses the information available in the "official" document collections used in QA@CLEF (newspaper text and Wikipedia) and information from the Web as an additional resource when searching for answers. Where it regards the use of external tools, Esfinge uses a syntactic analyzer, a morphological analyzer and a named entity recognizer. This year an alternative approach to retrieve answers was tested: whereas in previous years, search patterns were used to retrieve relevant documents, this year a new type of search patterns was also used to extract the answers themselves. We also evaluated the second and third best answers returned by Esfinge. This evaluation showed that when Esfinge answers correctly a question, it does so usually with its first answer. Furthermore, the experiments revealed that the answer retrieval patterns created for this participation improve the results, but only for definition questions.

Keywords: Question Answering, Portuguese, Answer Extraction.

1 Introduction

The proposed task in this year at QA@CLEF [1] was quite similar to the previous year. The main novelty was that systems could return up to three answers for each question. Besides taking advantage of that possibility, our participation focused in using an alternative approach to retrieve answers. In previous years, search patterns were used to retrieve relevant documents to particular questions. This year we also used a new type of search patterns to extract the answers themselves.

The following sections describe in detail the system architecture used this year, how the answer retrieval patterns were created and the results obtained in the official runs. There is a final section where the results are discussed and where directions for future work are indicated.

2 Esfinge at CLEF 2008

Esfinge has been participating at CLEF since 2004. These participations are described in detail in [2, 3, 4, 5]. Figure 1 gives an overview of the system used this year:

C. Peters et al. (Eds.): CLEF 2008, LNCS 5706, pp. 361–368, 2009.

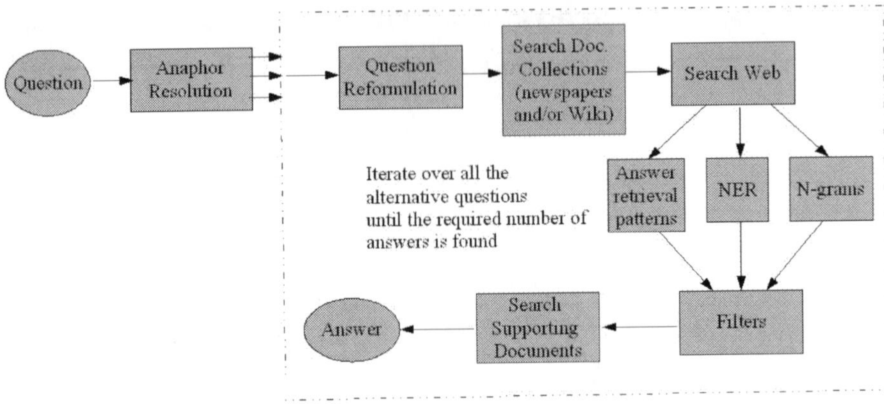

Fig. 1. Modules used in Esfinge

For a question, there can be up to three main iterations involved: the first over an eventual set of anaphor resolved question candidates, then an iteration over the different types of patterns used to retrieve relevant text passages and finally an iteration over the three techniques used to obtain candidate answers.

The *Anaphor Resolution* module is the first module in Esfinge. It adds to the original questions, a list of alternative questions where the anaphors are tentatively resolved. This module uses the analysis of the PALAVRAS parser [6] to identify the anaphoric element in a question and a list of candidates to replace it in the context of the other questions in the same topic. This module is described in much more detail in [5].

Then, for each of the alternative questions generated by the *Anaphor Resolution* module and until Esfinge finds the requested number of answers:

The *Question Reformulation* module transforms the question into patterns that will be later used to retrieve text passages which are relevant to the question. This is done using two different approaches: a purely string matching technique and an alternative approach which uses the analysis of PALAVRAS.

Esfinge starts with the string matching technique. This technique uses patterns which have an associated score giving an indication about how likely the pattern will help to retrieve relevant text passages.

For example the question *Quem foi Baden Powell de Aquino?* (Who was Baden Powell de Aquino?) matches with the patterns (simplified here for illustration purposes):

Quem ([^\s?]) ([^?]*)\??/"$2 $1"/10*
Quem ([^?])\??/$1/1*

Which in turn generate the following (Text pattern/Score) pairs:

"Baden Powell de Aquino foi"/10
foi Baden Powell de Aquino/1

2.1 Searching and Supporting Answers

Text patterns are then searched in the document collections (newspapers and Portuguese Wikipedia) using the *Search Document Collections* module in order to find text passages that are relevant to the question. These patterns are also searched in the Web using Yahoo's search API[1].

Subsequently, Esfinge analyzes all the retrieved relevant texts to obtain candidate answers. Three techniques are used for that purpose: extraction of answers using the answer retrieval patterns created for this year's participation, using named entity recognition (NER) and using an n-grams harvesting module. These techniques are used in sequence until Esfinge finds the required number of answers.

Esfinge begins by extracting answers using answer retrieval patterns. These patterns associate questions with their respective answers. They are different from the patterns used to retrieve relevant documents because they include the position where the answers should appear.

The following pattern is a simplified example of the patterns used by Esfinge. The patterns that are actually used are a bit more general in the sense that they can for instance cater for some alternative verb tenses (*é, são , eram*) and alternative articles (*a, o, as, os, um, uma*):

O que é a __X__? → __X__ (__ANSWER__)

This means that the answer to a question like *O que é a __X__?* (What is the X?) can be retrieved inside parenthesis following the string *__X__*.

First, Esfinge checks which patterns match with the question (left hand side of the rules). The patterns on the right hand side of the rules are then searched in the relevant documents to the question with the purpose of finding candidate answers.

The candidate answers are then ranked according to their frequency, length and the score of the passage from where they were retrieved using the formula: *Candidate answer score* $= \sum (F * S * L)$, through the passages retrieved in the previous modules where F = Candidate answer frequency, S = Score of the passage and L = Candidate answer length[2].

At this stage Esfinge has a list of candidate answers $\{A_1, A_2 \ldots A_n\}$. These candidate answers are then tested using the following filters:

- A filter that excludes answers contained in the question. For instance, the answer *partido* (party) is not a good answer to the question *A que partido pertence Zapatero?* (To which party does Zapatero belong?).
- A filter that excludes answers contained in a list of undesired answers (very frequent words that usually can not answer questions). This list includes words like *parte* (part), *antigo* (old), *pessoas* (people), *mais* (more) and was created based on experiments performed with the system. At present, it contains 96 entries.

An answer that passes all the filters proceeds to the next module which checks whether there are documents in the collections which support the answer.

---

[1] http://developer.yahoo.com/search/web/V1/webSearch.html

[2] This parameter is only used for the answers obtained through the n-grams module. For the other answers this is set to 1.

If Esfinge does not find the requested number of answers using the answer retrieval patterns, it tries to get more answers using the NER system SIEMÊS [7]. This system is used for the questions which imply specific types of answers like *Place*, *People*, *Quantity*, *Date*, *Height*, *Duration*, *Area*, *Organization* or *Distance*. Esfinge uses pattern matching to check whether it is possible to infer the type of answer for a given question. For example, questions starting with *Onde* (Where) imply an answer of type *Place*, questions starting with *Quando* (When) imply an answer of type *Date* and questions starting with *Quantos* (How Many) imply an answer of type *Quantity*. For these questions, Esfinge uses SIEMÊS to tag the relevant text passages in order to count the number of occurrences of named entities belonging to the expected categories. The identified named entities are then ranked, filtrated and checked for the existence of documents which can support them in a similar manner as described previously for the answers obtained using answer retrieval patterns.

In case the previous efforts still do not yield the necessary number of answers, Esfinge uses its last answer retrieval technique: n-grams harvesting. The n-grams obtained through the *N-grams* module are also ranked, filtrated and checked for the existence of documents which can support them in a similar manner as described previously for the other two techniques. The answers obtained through n-gram harvesting, however, are submitted to an additional filter that uses the morphological analyzer jspell [8] to check the PoS of the words contained in the answer. Jspell returns a list of tags for each of the words and Esfinge rejects all answers in which the first and last word are not tagged as one of the following categories: adjectives, common nouns, numbers and proper nouns.

2.2 Alternative Techniques Used to Find Relevant Texts

If at this stage, Esfinge did not retrieve the required number of answers, the next step will be to select a new set of relevant texts (this time using patterns based on the analysis of the question by PALAVRAS [6]). These patterns are created using the main verb of the question, its arguments and adjuncts and entities from previous questions belonging to the same topic. From this stage, Esfinge repeats the steps described in sub-section 2.1.

In case the last step did not yield the required number of answers either, a last attempt is tried which consists in selecting relevant texts using patterns without verbs based on the analysis of the question by PALAVRAS.

3 Creating the List of Answer Retrieval Patterns

The hypothesis we wanted to test this year was whether it would be possible to extract useful answer retrieval patterns from the solutions available from the previous editions of CLEF. Unfortunately, only the solutions for 2007 questions were available for Portuguese (answers and text passages where they occur).

The following is an example of the solution for the question *Quem é o Lampadinha?* (Who is Lampadinha?) :

<pergunta ano="2007" id_org="X" new_id_org="070101" categoria="D" tipo="PERSON" restrição="NO" ling_orig="PT" tarefa_pt="0101" tópico="082">
<texto>Quem é o ***Lampadinha****?</texto>*
<resposta n="1" docid="Professor Pardal 6afa">
um pequeno andróide com uma lâmpada no lugar da cabeça*</resposta>*
<extracto n="1" resposta_n="1">"Pardal é ajudado frequentemente por ***Lampadinha*** *(criado por Barks em 1953),* ***um pequeno andróide com uma lâmpada no lugar da cabeça****, que é considerado sua maior invenção (ao lado do ""chapéu pensador"", um dispositivo em forma de telhado com chaminé habitado por corvos, que o ajuda a ter idéias)."</extracto>*
</pergunta>

From this solution one can derive the following pattern:

*Quem é o __X__ ? → __X__ *, __ANSWER__,*

This pattern means that the answer for a question of the form *Quem é o X ?* (Who is X?) can be retrieved following a comma which appears after an instance of *X*. The asterisk (*) stands for 0 or more characters after the sub-string *__X__* and immediately before a comma.

For the participation at CLEF 2008, 24 answer retrieval patterns were derived. The process used to obtain these patterns was semi-automatic: they were derived automatically from the solution file, but then adjusted manually, not only in order to correct or complete them, but also to generalize them.

4 Results

Since the main goal of this participation was to evaluate the impact of the answer retrieval patterns described in the previous section, two official runs were submitted: *esfi081PTPT* which uses these patterns and *esfi082PTPT* which does not use them.

Table 1 shows the results of the official runs, considering all the questions (Total), only for factoid (F) or definition (D) questions. Table 1 includes results where only the first answers were evaluated, where the first and second answers were evaluated and where all the three answers returned were evaluated. As an illustration, take a question for which only the third answer was correct: this question was only accounted in the columns labeled *First 3 answers*.

Table 1. Results of the official runs

Runs	Right Answers (First Answer)			Right Answers (First 2 Answers)			Right Answers (First 3 Answers)		
	F	D	Total	F	D	Total	F	D	Total
esfi081ptpt	39	10	49	44	14	58	46	16	62
esfi082ptpt	39	2	41	43	3	46	46	6	52

Table 2 summarizes the performance of the three techniques used to obtain answers in the best run (we studied only the first answer for each question). As previously mentioned these techniques are used in sequence until the required number of answers is obtained. The first technique which is tested (answer retrieval patterns) was used in a small number of questions, but with a good precision (46% of right answers for the answers obtained using this technique). NER, the second technique, catered for most of Esfinge's right answers, but with lower precision (32%). The last technique, n-grams harvesting, had the lowest precision (6%) and contributed with less right answers to the overall result.

Table 2. Origin of the answers

Answer origin	Answers	Right Answers
Answer Retrieval Patterns	24	11
NER	93	30
N-gram	63	4
NIL	20	4
Total	200	49

An additional study was performed in order to find how Esfinge supported its right answers (newspaper text or Wikipedia) in the best run. The results, summarized in Table 3, reveal that Esfinge found support in Wikipedia more often than in newspaper texts both for factoid and definition questions.

Table 3. Distribution of the answer support

Type of Question	Answer Supported with Newspaper Text	Answer Supported with Wikipedia	Right Answers
Right Definition Answers	4	6	10
Right Factoid Answers	14	21	35
Right NIL Answers			4
Right Answers	18	27	49

Table 4. Causes for wrong answers in the best run

Problem	Wrong Answers in 2008	Wrong Answers in 2007
Co-reference Resolution	20	25
Wrong or Incomplete Search Patterns	5	63
Document Retrieval Failure	44	33
Named Entity Recognition	13	3
Answer Scoring Algorithm	45	24
Mistake in the Supported Answer Filter	13	7
Others	11	10
Total	151	165

Table 4 provides the results of the detailed error analysis performed for the best runs of Esfinge in 2008 and 2007[3].

5 Discussion of the results and further work

We consider our participation at QA@CLEF this year fruitful and rewarding. In our opinion it was wise that the organization proposed a similar task as last year's since a good number of challenges remain to be achieved.

The main novelty in this years's task which consisted in allowing the return of up to three answers for each question, allowed us to investigate how good the second and third best answers returned by Esfinge are. The conclusion regarding this matter is that when Esfinge answers correctly a question, it does so usually with its first answer. For instance the best run had 49 right answers, but even when considering all the answers returned (3 for each question) the number of right answers amounted only to 62 right answers.

Regarding the study on the performance of the three techniques used to obtain answers, our findings were that the first technique which is tested (answer retrieval patterns) had a good precision (46%) in the small number of questions were it was used; NER, the second technique, catered for most of Esfinge's right answers, at a lower precision (32%); the last technique, N-gram harvesting, had the lowest precision (6%) and contributed with less right answers to the overall result. However, this last technique is only used when it is not possible to find answers with the other two. This means that it is probably used with a considerable number of the hardest questions.

The error analysis shows that, comparing with 2007 experiments, errors occur more often in modules which appear later in the system's workflow. Whereas in 2007, most errors were caused by wrong or incomplete search patterns, this year they were mainly caused by document retrieval failure or the answer scoring algorithm (Table 4).

Nonetheless, the most relevant result obtained in this year's participation was that the answer retrieval patterns clearly improved the results for definition questions (the first answer is correct for 34% of the definition questions and there was a correct answer in one of the three returned answers for 55% of questions of this type), but the same does not applied for the factoid questions. These patterns were used in a small number of questions, but the precision of the answers was quite good (46%). This good precision confirmed our intuition that the best order to search for answers would be the one used in our system: first using answer retrieval patterns, then NER and finally n-grams harvesting.

We believe that there is still improvement potential where it regards the use of answer retrieval patterns. Therefore, we would like to deepen our research on how to create these patterns with a more automated approach (as stated we used a semi-automatic process taking as input last year's solutions). Additionally, there is also interest in investigating how the results can improve when more answer retrieval patterns are used.

---

[3] The 2007 analysis was not performed on an official run (those results were compromised by severe bugs), but on the best run out of the repetitions executed after the bugs were corrected.

Acknowledgments

This work was done in the scope of the Linguateca project, jointly funded by the Portuguese Government and the European Union (FEDER and FSE) under contract ref. POSC/339/1.3/C/NAC.

References

1. Forner, P., Peñas, A., Alegria, I., Forăscu, C., Moreau, N., Osenova, P., Prokopidis, P., Rocha, P., Sacaleanu, B., Sutcliffe, R., Sang, E.T.K.: Overview of the CLEF 2008 Multilingual Question Answering Track. In: Peters, C., et al. (eds.) CLEF 2008. LNCS, vol. 5706, pp. 262–295. Springer, Heidelberg (2009)
2. Costa, L.: First Evaluation of Esfinge - a Question Answering System for Portuguese. In: Peters, C., Clough, P., Gonzalo, J., Jones, G.J.F., Kluck, M., Magnini, B. (eds.) CLEF 2004. LNCS, vol. 3491, pp. 522–533. Springer, Heidelberg (2005)
3. Costa, L.: 20th Century Esfinge (Sphinx) solving the riddles at CLEF 2005. In: Peters, C., Gey, F.C., Gonzalo, J., Müller, H., Jones, G.J.F., Kluck, M., Magnini, B., de Rijke, M., Giampiccolo, D. (eds.) CLEF 2005. LNCS, vol. 4022, pp. 467–476. Springer, Heidelberg (2006)
4. Costa, L.: Question answering beyond CLEF document collections. In: Peters, C., Clough, P., Gey, F.C., Karlgren, J., Magnini, B., Oard, D.W., de Rijke, M., Stempfhuber, M. (eds.) CLEF 2006. LNCS, vol. 4730, pp. 405–414. Springer, Heidelberg (2007)
5. Cabral, L.M., Costa, L.F., Santos, D.: What happened to Esfinge in 2007? In: Peters, C., Jijkoun, V., Mandl, T., Müller, H., Oard, D.W., Peñas, A., Petras, V., Santos, D. (eds.) CLEF 2007. LNCS, vol. 5152, pp. 261–268. Springer, Heidelberg (2008)
6. Bick, E.: The Parsing System "Palavras": Automatic Grammatical Analysis of Portuguese in a Constraint Grammar Framework. Aarhus University Press, Aarhus (2000)
7. Sarmento, L.: SIEMÊS - A Named Entity Recognizer for Portuguese Relying on Similarity Rules. In: Vieira, R., Quaresma, P., Nunes, M.d.G.V., Mamede, N.J., Oliveira, C., Dias, M.C. (eds.) PROPOR 2006. LNCS (LNAI), vol. 3960, pp. 90–99. Springer, Heidelberg (2006)
8. Simões, A.M., Almeida, J.J.: Jspell.pm - um módulo de análise morfológica para uso em Processamento de Linguagem Natural. In: Gonçalves, A., Correia, C.N. (eds.) Actas do XVII Encontro Nacional da Associação Portuguesa de Linguística (APL 2001) (Lisboa, 2-4 de Outubro de 2001), pp. 485–495. APL, Lisboa (2001)

Ihardetsi: A Basque Question Answering System at QA@CLEF 2008

Olatz Ansa, Xabier Arregi, Arantxa Otegi, and Ander Soraluze

IXA Group, University of the Basque Country
olatz.ansa@ehu.es

Abstract. This paper describes *Ihardetsi*, a question answering system for Basque. We present the results of our first participation in the QA@CLEF 2008 evaluation task. We participated in three subtasks using Basque, English and Spanish as source languages, and Basque as target language. We approached the Spanish-Basque and English-Basque cross-lingual tasks with a machine translation system that first processes a question in the source language (i.e. Spanish, English), then translates it into the target language (i.e. Basque) and, finally, sends the obtained Basque question as input to the monolingual module.

1 Introduction

In the QA@CLEF 2008 edition, the Basque language was incorporated for the first time both as source language and as target language. In this context a new monolingual task, Basque-Basque, and two cross-lingual tasks, English-Basque and Spanish-Basque, were organised.

The main goal of our first participation in QA@CLEF for Basque was to evaluate our basic system by comparing it with the state of the art of non-English question answering systems. Besides, the analysis of the results could reveal a number of future system improvements. We took part in the Basque-Basque, Spanish-Basque and English-Basque tasks. Our system, Ihardetsi, is a Basque monolingual system, and we use two machine translation systems [1] for the cross-lingual tasks in order to translate the questions into Basque: one for Spahish-Basque and other for English-Basque.

This paper is structured as follows: the next section presents the corpus processing, section 3 describes the system architecture, section 4 introduces the results and a preliminary analysis of the kind of errors that the system made and conclusions and directions of future work follow in section 5.

2 Corpus Processing

The QA@CLEF 2008 campaign establishes two different document collections for each language: a newswire collection and the Wikipedia. In the case of Basque, a dump of the *Wikipedia 2006* and the *Euskaldunon Egunkaria* newspaper collection (from 2000 until 2002) were provided.

C. Peters et al. (Eds.): CLEF 2008, LNCS 5706, pp. 369–376, 2009.

The document collection has been lemmatized with Morfeus[2] before being indexed. Due to the fact that Basque is an agglutinative language, a given lemma makes many different word forms, depending on the case (genitive, locative, etc.) or the number (singular, plural, indefinite) for nouns and adjectives, and the person (1st, 2nd, etc.) and the tense (present, past, etc.) for verbs. For example, the lemma *lan* ("work") forms the inflections *lana* ("the work"), *lanak* ("works" or "the works"), *lanari* ("to the work"), etc. This means that looking only for the exact word given or the word plus an "s" for the plural is not enough for Basque. And the use of wildcards, which some search engines allow, is not an adequate solution, as these can return not only inflections of the word, but also derivatives, unrelated words, etc. For example, looking for *lan** would also return all the forms of the words *lanabes* ("tool"), *lanbide* ("job"), *lanbro* ("fog"), and many more.

Before analysing the Wikipedia, it needed to be cleaned, e.g. by getting rid of HTML tags. So, we created an XML parser that extracts page title, paragraphs, and lists and then creates a simple XML document, which is very similar to the XML of the newspaper collection.

The entire document collection was lemmatized, tagged with part-of-speech and named entities. The named entity recogniser and classifier (NERC) for Basque, Eihera[3], captures entities such as PERSON, ORGANISATION and LOCATION. The numerical and temporal expressions are captured by the lemmatizer and tagger.

Finally, the document collection was indexed by lemma; the Swish-e[1] search engine was used and the retrieval unit was the passage (for this experiment a passage is a paragraph).

3 System Overview

An XML configuration file controls the processing of the system. The configuration file is a declarative document where all the features involved in a run are described. The set of features is divided into two categories:

1. General requirements. The configruation file includes specifications such as the corpus to be used, the location of the list of questions to be answered, and the metrics and conditions for the evaluation.
2. Descriptors of the QA process itself. This subset of features represents the characteristics of the answering process. Mainly, it determines which modules operate during the answering process, describes them and specifies the parameters of each module. In that way, the process is controlled by means of the configuration file, and different processing options, techniques, and resources can be easily activated, deactivated or adapted.

The principles of versatility and adaptability have guided the development of the system. It is based on web services, integrated using the SOAP communication protocol. Some tools previously developed in the IXA Group are used as

[1] http://swish-e.org/

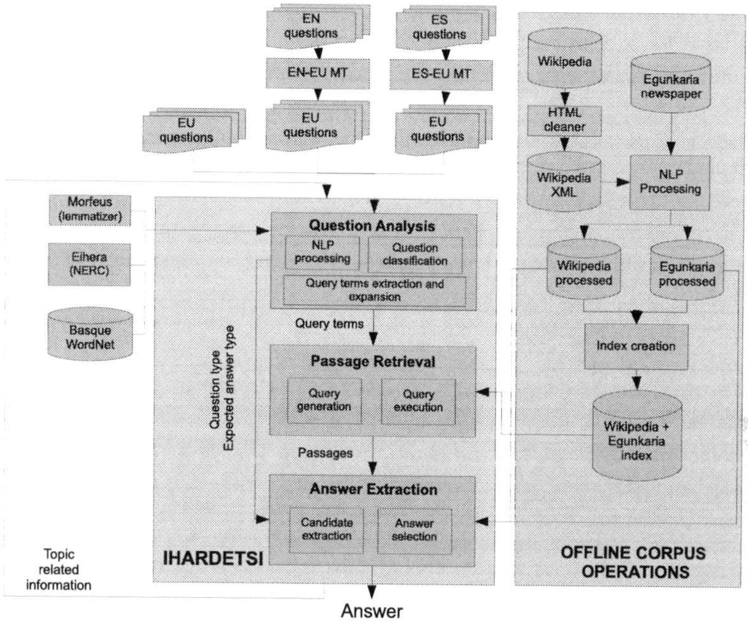

Fig. 1. The system general architecture

autonomous web services. This distributed model allows to parameterise the linguistic tools and to adjust the behaviour of the system during the development and testing phases.

The communication between the web services is done using XML documents. This model has been adopted by some other systems (e.g. [4], [5]). The current monolingual version has three main modules: question analysis, passage retrieval and answer extraction. A complementary module, the question translation, has been added for cross-lingual versions. Fig. 1 shows the architecture of the system.

3.1 Question Translation

A machine translation engine named *Matxin*[2] [6] has been used for question translations. This rule-based engine has been developed for translating from Spanish to Basque. Due to the different structures of the languages the quality of the translations is not sufficient for *dissemination* ("publishable quality"), but it can be used for *assimilation* ("understanding quality"). It has been developed for a general domain and tested with texts from newspapers, but not with questions. A shallow test was carried out on factoid questions from previous editions of CLEF and we considered that the results were good enough for using it in this task. In any case a wider evaluation was required.

[2] A free version of the MT engine is in a public repository (*http://matxin.sourceforge.net/*)

In the English to Basque translation we have used an early version of the English to Basque engine based on the same technology. The quality was poor and in a similar shallow test with factoid questions we detected that the translations of some types of question were wrong, specially when the question marker was composed of two words that appeared as non-contiguous (i.e. WHERE *is he* FROM?). To face this problem a heuristic was applied after the translation process in order to repair bad translations of question markers. The heuristic was implemented using a small number of conditional rules, which work on the original and on the translated sentences.

3.2 Question Analysis

The main goal of this module is to analyse the question and to generate the information needed for the next tasks. Concretely, search terms are extracted for the passage retrieval module, and both the question type (factoid, list or definition) and the expected answer type along with some lexical information are passed to the answer extraction module. To achieve this goal, our question analyser performs the following steps:

Linguistic Processing. The question analysis uses a set of general purpose tools like the lemmatizer and tagger *Morfeus* [2], and the NERC *Eihera* [3].

Question Classification. In order to identify the question type, the question focus and the expected answer type, a set of rules has been defined after the examination of a Basque question set.

The question focus is the word or the word sequence that defines or disambiguates the question. For example, in the question *Which river is in the south of this country?*, the focus is *river* and in question *What is the North Pole?*, the focus is *North Pole*.

The next step is to identify the expected answer type. Our system's answer type taxonomy distinguishes the following classes: PERSON, ORGANISATION, DESCRIPTION, LOCATION, QUANTITY, TIME, ENTITY and OTHER. The assignment of a class to the analysed question is performed using the interrogative word, some heuristics of syntactic nature and the type of the question focus. The question focus is mapped with the semantic file characteristic of the BasqueWN[7], and in this way we obtain the expected answer type (PERSON, ORGANISATION, LOCATION, QUANTITY, TEMPORAL).

Query Terms Extraction and Expansion. All nouns, verbs, adjectives and abbreviations of the question constitute the set of search terms. They are lemmatized and arranged by their *inverse document frequency* (IDF) value in the corpora in descending order.

Optionally, the search terms can be expanded by using synonymy, hyponymy and hypernymy information. To do this, the system uses a service which consults the lexical-semantic database BasqueWN.

3.3 Passage Retrieval

The retrieval unit is a passage and not the entire document. The corpus is indexed by lemma using the swish-e search engine. The corpus is batch-processed (see section 2): all words are lemmatized, and complex lexical units and entities are marked.

This module produces a set of queries taking as input a) the search terms selected by the question analysis module, b) the search terms selected by the question analysis module for the first question of a topic (if the question is not the first) and c) the first three answers of the first question of a topic (if the question is not the first). For each input (a,b,c), a set of queries are created using relaxation techniques [8], and then they are combined to generate the set of final queries. Finally, they are executed until one of the queries retrieves at least one passage, and a confidence score is associated to each selected passage.

3.4 Answer Extraction

Two tasks are performed in sequence: candidate extraction and answer selection. The candidate extraction consists of extracting all the candidate answers from the highest scoring passages. The answer selection consists of choosing the best three answers.

The process of candidate extraction is carried out on the set of passages obtained in the previous step. First, all candidate answers are detected from each retrieved passage and a set of windows are defined around them. The selected window for each candidate answer is the smallest one which contains all the query terms. Then, the candidate answer score is computed like this:

$$score_{CA} = \frac{\sum_{i=1}^{n} w_i}{n} \tag{1}$$

where n is the window size and w_i is the i word weight. w_i is 1 for search terms, 0.8 for the synonyms of the search terms, 0.5 for hyponyms and hypernyms, and 0.3 for other question terms.

Then, the candidate answers extraction process addresses each question type in a different manner, as follows:

- **Question type is Factoid:** the answer selection depends on named entities in most of the cases except when the expected answer type is *Other*. In such a case, all the entities and nouns near the question focus are selected.
- **Question type is Definition:** a set of rules have been defined to extract definitions from retrieved text passages.
- **Question type is List:** we followed a heuristic looking for lists of candidate answers in the same passage.

Once a list of scored candidates has been extracted, it is necessary to group those that are identical and recalculate their scores. In the process of answer selection, we try to map as identical those answers that refer to the same entity.

The formula used to compute the final score of each answer is as follows:

$$final\_score_A = \frac{\sum_{i=1}^{p} score_{CA}}{N} \tag{2}$$

where p is the number of identical answers and N is the number of candidate answers. Finally, the answers are ordered by their final score, and the three best are chosen.

4 Results

This section describes the results obtained in our CLEF 2008 participation. We submitted four runs: one Basque monolingual run, one English-Basque cross-lingual run, and two Spanish-Basque cross-lingual runs. The methodology we employed targeted precision at the cost of recall, therefore we always choose NIL answers for those questions in which we could not locate a candidate answer in the retrieved passages.

4.1 Monolingual System

As expected, the best results were obtained for the monolingual task. Table 1 illustrates the results achieved by our system in the monolingual run.

It is clear that the best results were achieved for factoid questions. This is due to the fact that we focused on this type of questions in the development of the system. There were 145 factoid questions and 50 had a correct or inexact answer in the proposed three answers, 22 had a NIL answer (incorrect) and 73 had an incorrect answer. Analysing these 73 questions we detected that for 17 of them, a correct passage was detected, but the system did not extract the correct answer.

The system answered NIL for 57 questions but only 4 of them were correct. Analysing the reasons we can group them into 5 groups:

- The expected answer type detection failed: 6 questions
- No passage was retrieved: 14 questions
- The passage had the answer but the system could not extract it: 13 questions
- Retrieved passages had not the answer: 16 questions
- Some other reason: 4 questions

Table 1. Results obatined in the Basque monolingual run at QA@CLEF 2008

	1st ANSWER					2nd ANSWER		3rd ANSWER	
	R	W	I	U	ACC.	R	I	R	I
OVERALL	26	163	11	0	13.0%	9	4	5	0
FACTOID	23	113	9	0	15.9%	9	2	5	0
DEFINITION	3	36	0	0	7.7%	0	2	0	0
LIST	0	14	2	0	0.0%	0	0	0	0
TEMPORALLY RESTRICTED	2	19	2	0	8.7%	0	1	1	0

Table 2. Results obtained in cross-lingual runs at QA@CLEF 2008

	EN-EU					ES-EU					ES-EU with synonymy				
	R	W	X	U	ACC.	R	W	X	U	ACC.	R	W	X	U	ACC.
OVERALL	11	182	7	0	5.5%	11	182	7	0	5.5%	7	185	8	0	4.5%
FACTOID	8	130	7	0	5.5%	10	129	6	0	6.9%	7	130	8	0	4.8%
DEFINITION	3	36	0	0	7.7%	1	37	1	0	2.6%	0	39	0	0	0.0%
LIST	0	16	0	0	0.0%	0	16	0	0	0.0%	0	16	0	0	0.0%
TEMPORALLY RESTRICTED	1	22	0	0	4.3%	2	21	0	0	8.7%	1	22	0	0	4.4%

It is remarkable that no other system took part in the Basque as target task, so the obtained results could not be directly compared with another Basque system. Nonetheless, it is interesting to contrast our results with some other languages. For that purpose, we choose QA@CLEF 2007 [9] results as a reference because that was the first time that topic-related questions and the Wikipedia corpus were included. Although our results are far from the best ones, with overall accuracy of 54%, we realized that almost 40% of all the runs got worse results than those of our system.

4.2 Cross-Lingual Systems

Three cross-lingual runs, two for Spanish-Basque and one for English-Basque, have been performed. The aim of the second run for Spanish-Basque was to test whether the semantic expansion of the question (see section 3.2) could compensate the lack of precision in the translation process. The results of the three runs are shown in Table 2.

The main conclusions we want to remark are:

- The results are quite poor. The loss of precision compared to the results of the monolingual system is more than 50%.
- Very similar results are obtained for the basic Spanish-Basque and for the English-Spanish runs (in both there are 11 right answers, 7 right answers in 2nd or 3rd place and 7 inexact in the first place). Due to the better quality of the Spanish-Basque translator we expected to improve the results for this pair of languages, but all runs had similar accuracy. In spite of that the right answers do not correspond always to the same questions; only five of the eleven right answers were common.
- The semantic expansion in the second run for Spanish-Basque did not achieve better results. A slightly smaller precision was observed, because some right answers were lost. However, new right or inexact answers appear but not in the first place. Moreover, against our expectatives no more passages were recovered by using semantic expansion.

5 Conclusions and Future Work

In this paper we describe our participation in the QA@CLEF campaign with a monolingual Basque-Basque and two cross-lingual English-Basque and Spanish-Basque systems. Thanks to this track we have had the opportunity to test our systems. Although the results might look not so good, our general conclusion is positive considering that it was our first participation. In order to compare our results with other systems, the particularities of the Basque language must be taken into account. Moreover, we have been able to identify some of the strengths and weaknesses of each module of the system, and that will be considered for future improvements.

Acknowledgements

We would like to thank Gorka Labaka and Mikel Lersundi who collaborated in this research. Arantxa Otegi's work is funded by a PhD grant from the Basque Government. Part of this work was funded by the Basque Government (Department of Language Policy and Department of Industry —Anhitz project, Etortek IE06-185—) and the Education Ministry (KNOW, TIN2006-15049-C03-01).

References

1. Alegria, I., Arregi, X., Artola, X., Díaz de Ilarraza, A., Labaka, G., Lersundi, M., Mayor, A., Sarasola, K.: Strategies for sustainable MT for Basque: incremental design, reusability, standardization and open-source. In: IJCNLP 2008 Workshop on NLP for Less Privileged Languages, pp. 59–64 (2008)
2. Ezeiza, N., Aduriz, I., Alegria, I., Arriola, J., Urizar, R.: Combining Stochastic and Rule-Based Methods for Disambiguation in Agglutinative Languages. In: COLING-ACL, pp. 380–384 (1998)
3. Alegria, I., Arregi, O., Balza, I., Ezeiza, N., Fernandez, I., Urizar, R.: Design and Development of a Named Entity Recognizer for an Agglutinative Language. In: IJCNLP (2004)
4. Tomás, D., Vicedo, J., Saiz, M., Izquierdo, R.: Building an XML framework for Question Answering. In: CLEF (2005)
5. Hiyakumoto, L.: Planning in the JAVELIN QA System. In: CMU-CS-04-132 (2004)
6. Alegria, I., Díaz de Ilarraza, A., Labaka, G., Lersundi, M., Mayor, A., Sarasola, K.: Transfer-based MT from Spanish into Basque: reusability, standardization and open source. In: Gelbukh, A. (ed.) CICLing 2007. LNCS, vol. 4394, pp. 374–384. Springer, Heidelberg (2007)
7. Atserias, J., Villarejo, L., Rigau, G., Agirre, E., Carroll, J., Magnini, B., Vossen, P.: The MEANING Multilingual Central Repository. In: GWC (2004)
8. Bilotti, M.: Query Expansion Techniques for Question Answering. Master's thesis, Massachusetts Institute of Technology (2004)
9. Giampiccolo, D., Peñas, A., Ayache, C., Cristea, D., Forner, P., Jijkoun, V., Osenova, P., Rocha, P., Sacaleanu, B., Sutcliffe, R.: Overview of the CLEF 2007 multilingual question answering track. In: Peters, C., Jijkoun, V., Mandl, T., Müller, H., Oard, D.W., Peñas, A., Petras, V., Santos, D. (eds.) CLEF 2007. LNCS, vol. 5152, pp. 200–236. Springer, Heidelberg (2008)

Question Interpretation in QA@L^{2}F

Luísa Coheur, Ana Mendes, João Guimarães,
Nuno J. Mamede, and Ricardo Ribeiro

L^{2}F/INESC-ID Lisboa
Rua Alves Redol, 9, 1000-029 Lisboa, Portugal
qa-clef@l2f.inesc-id.pt

Abstract. The Question Interpretation module of QA@L^{2}F, the question-answering system from L^{2}F/INESC-ID, is thoroughly described in this paper, as well as the frame formalism[1] it employs. Moreover, the anaphora resolution process introduced this year, based on frames manipulation, is detailed.

The overall results QA@L^{2}F achieved at the CLEF competition and a brief overview on the system's evolution throughout the 2 years of joint evaluation are presented. The results of an evaluation to the QI module alone are also detailed here.

1 Introduction

QA@L^{2}F is the question-answering (QA) system from L^{2}F/INESC-ID, that participated in 2007 and 2008 in the monolingual QA task of CLEF [2]. To answer questions, the system follows three main steps:

- Corpus Pre-Processing: as in many QA systems, like Senso [3], information sources are partly processed in order to extract potentially relevant information, like named entities and relations between concepts. The latter information represents possible answers to questions and is stored in a database;
- Question Interpretation (QI): the question is analysed and transformed into a frame, which is mapped into an SQL query or used to search relevant snippets;
- Answer Extraction: each question type is mapped into a single strategy. As a result, depending on the question type, different strategies are used to find the answer. However, if no answer is found, the system proceeds and tries to find an answer using alternative strategies. Details on these strategies can be found in [2].

This paper focus on the QI module of QA@L^{2}F, and is organized as follows: section 2 presents related work; section 3 describes the QI module, including the anaphora resolution process; section 4 shows, discusses and compares evaluation results; finally, section 5 concludes and points to future work.

[1] As in [1], we call 'frame' to a set of slot-value pairs; we call 'frame element' to each slot-value pair.

C. Peters et al. (Eds.): CLEF 2008, LNCS 5706, pp. 377–384, 2009.

2 Related Work

Work in questions' processing can be split into two main tasks: question classification and QI. Question classification aims at mapping different question types into proper semantic categories. For instance, [4] proposes a method to classify *what-type* questions based on head noun tagging. QI goal is the conversion of natural language questions into structured information, more suitable for computer processing. Typically, these structures are logical forms or frames, but they can also be questions in natural language that the computer already understands. Clearly, question classification has an important role in QI. Although some systems implement hybrid approaches, involved techniques can be classified as: a) basic QI; b) statistical QI; c) linguistically-motivated QI.

Basic QI includes keyword detection, pattern matching and the use of simple algorithms capable of associating new input to already understood utterances. Although not focused on QA, a classical example that perfectly illustrates a system based on pattern matching, is the well-known ELIZA [5], invented in the early 1960's, aiming at emulating a psychologist.

In what concerns statistical QI, there are several techniques being explored, coming some of them from the Machine Learning framework. The main problem of using statistical techniques in QI is the small size of the potential training data. An example of a work that applies statistical methods to little training data is the one presented in [1], where four different techniques are applied to a training set (not only questions) constituted by 477 sentence/frame pairs. Results from this evaluation ranged from a 0.75 F-score to a 0.83 F-score. It should be noticed, however, that these results derived from the fact that the domain was limited and it was possible to replace each entity of the domain by its correspondent class name. In an open QA system is not obvious that these techniques would obtain similar results.

Finally, linguistically-motivated QI use some level of linguistic information. Some systems implementing this paradigm base their performance on a syntax/semantics interface, where each syntactic rule is associated with a semantic rule and logical forms are generated in a bottom-up, compositional process. Variations of this approach are followed by several systems. Two of the most referenced books in Natural Language Processing, that is [6] and [7], depict this approach. Also, last year, QA@L^2F [2] followed these lines, although a slightly different (but also common) linguistically-motivated technique was used: a semantic module was operating over a dependency structure, obtained after a cascaded syntactic/semantic analysis.

Due to a strong dependency between the semantic and the syntactic analysis, that brought many problems to the semantic analysis, this year QA@L^2F follows a different strategy that combines a) and c) approaches. On the one hand, question classification is based on a sophisticated pattern matching, that uses morpho-syntactic patterns. On the other hand, the module profits from a named entity recognizer based on a deep linguistic analysis of the question, in order to identify relevant entities (people, titles, locations, dates, etc.). This information is merged in order to create a frame. This process is detailed in the next section.

3 Question Interpretation

The QI module of QA@L^{2}F module involves the following steps:

- morphological analysis, performed by Palavroso [8] and MARv [9];
- creation of intermediary frames (pre-frame), representing relevant information extracted from the question. This step is performed by RuDriCo (an improved version of PAsMo [10]), which is a rule-based tool that recognizes multi-word term and collapses them into single tokens; it can also split tokens. RuDricCo's rules are patterns that match against labeled text, being RuDriCo the tool responsible for the sophisticated pattern matching that uses morpho-syntactic patterns, as mentioned before;
- named entity recognition, performed by XIP [11], a tool that returns the input organized into chunks, connected by dependency relations, and also identifies and classifies the named entities in the input. As mentioned before, XIP bases the named entity recognition in a deep linguistic analysis;
- frames creation, in which information from the pre-frame is merged with the information returned by the named entity recognizer.

Figure 1 depicts the entire question interpretation process used in QA@L^{2}F.

The following example shows a RuDriCo rule, which is able to capture questions such as *Quando nasceu Thomas Mann* (*When was Thomas Mann born?*), and responsible for a pre-frame creation.

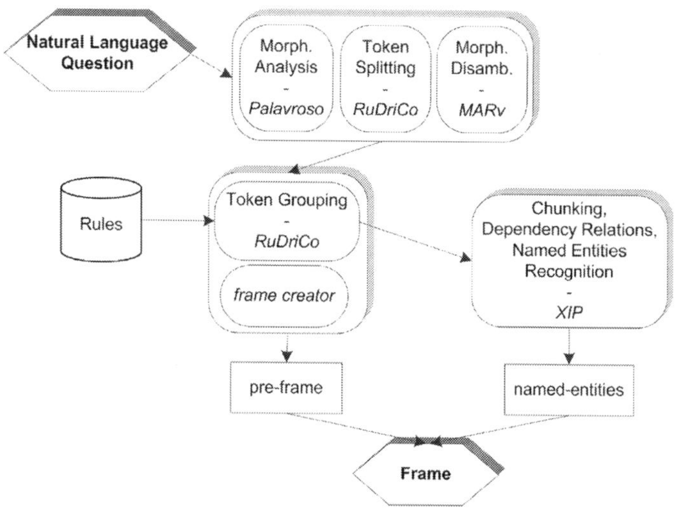

Fig. 1. Question interpretation in QA@L^{2}F

```
S1 ['quando','CAT'/C1]
S2 ['ser','CAT'/C2]? 'que' [L3,'CAT'/C3]?
S4 [L4,'CAT'/'nascer']
S6 [L6,'CAT'/'noun']
S7 [L7,'HMM'/'true']*
S10 ['?','CAT'/C10] -->
   S1 ['onde', 'CAT'/C1, 'type'/'onde_verb']
   S4 [L2,'CAT'/'verb']
   S6@+S7@* [L6@+L7@*, 'type'/'target'].
```

The left side (before the arrow) of this rule matches the question; the right side outputs the frame elements that constitute the pre-frame. XIP outputs the named entities. Both results are merged into a final frame. In the next section, the frames formalism used in QA@L^{2}F is described.

3.1 Frames

Each frame in QA@L^{2}F consists of the following elements:

- the question type: a string that identifies the script to be called;
- the question target: a string that represents the question main entity;
- named entities: a set of strings representing the entities identified by the named entity recognizer;
- auxiliaries: a set of strings constituted by auxiliary (and optional) elements from the question, like the target-type, main verbs, adjectives and adverbs.

Considering the previous question *Quando nasceu Thomas Mann?*, its corresponding frame is:

Frame
when/script-wiki-target.pl
target="thomas mann"
entities people="thomas mann"
auxiliaries verb="nasceu"

The question type is identified by the script `when/script-wiki-target.pl`, the question target is `thomas mann` which is also identified by the named entity recognizer as *people*; the auxiliaries' set is constituted solely by the verb `nasceu`.

The obtained script is then called and uses the other frame elements either to build the SQL query or to obtain the snippets that may contain the answer. Details about this step can be found in [2].

3.2 Anaphora Resolution

The capability of handling anaphora plays an important role along the entire pipeline of QA systems. It can have much impact in the performance of its compounding modules: on one hand, the benefits of analyzing, identifying and solving anaphoric references during the corpus pre-processing stage are shown in the studies driven by [12]; on the other hand, during the question interpretation step, anaphora resolution is also applied in order to deal with follow-up questions, as pointed by [13].

Like other systems that already deal with this linguistic phenomenon (see, for instance, [14] and [3]), QA@L^{2}F integrates a module for pronominal anaphora resolution for follow-up questions. In addition, ellipsis, being a special case of anaphora, is also addressed in this module.

Anaphora resolution is based on insertions/replacements over the frame elements of anaphoric questions. The frame associated with the reference question provides the frame elements to be used in these insertions/replacements. For instance, in pronominal anaphora, the target pronoun is replaced by the target of the reference question. In order to illustrate this procedure, consider the next group of questions[2]:

1. *Onde nasceu a Florbela Espanca?* (*Where was Florbela Espanca born?*)
2. *Quando?* (*When?*)
3. *Onde morreu ela?* (*Where did she die?*)

The following frame was generated by the reference question:

Reference

where/script-wiki-target.pl
target="florbela espanca"
entities people="florbela espanca"
auxiliaries verb="nasceu"

The manipulated frames for each of the follow-up questions are shown next[3]:

Elliptic question

when/script-wiki-target.pl
target="florbela espanca"
entities people="florbela espanca"
auxiliares verb="nasceu"

Pronominal question

where/script-wiki-target.pl
target="florbela espanca"
entities people="florbela espanca"
auxiliares verb="morreu"

This module bases its actions on the assumption that only the information introduced by the reference question can be used in anaphora resolution. However, this is not always the case, since follow-up questions can also provide information to further questions. Developments on this module are, thus, still required.

[2] We will call "reference question" to the first question, "elliptic question" to the second question and "pronominal question" to the last one.

[3] Frame elements that do not result from insertions/replacements from the frame associated with the reference question are displayed in bold italic.

4 Evaluation

QA@L^{2}F was evaluated at CLEF, using Portuguese as source and target languages. This section presents the QI step evaluation as well as the system final results.

In what concerns the QI step, frames were generated and then manually evaluated in terms of its correctness according to the expected frame. The results are the following:

```
Total: 200 questions

Right: 113
Wrong:  87
     Total fail:       14
     Partially wrong: 73
           Wrong script:       27
           Wrong target:       50
           Wrong entities:      7
              Wrong auxiliaries: 50
```

As it can be seen, from the 200 questions that constituted the test set, the QI module succeeded in creating the correct frame in 56,5% of the cases. In 14 of the 87 wrong frames, the module completely failed to create the frame. The other items represent which of the frame components were wrongly identified.

Considering only anaphoric questions, 13 of the 52 follow-up questions where mapped into the correct frame, resulting in an accuracy of 25%. It should be noticed that in these 13 frames, 4 were incorrect due to errors occurred in the generation of the reference frame.

Table 1 shows the final results. The system had better overall results this year: 20% of correct answers, against 14% last year. However, the number of wrong answers continues high (150), although it decreased from 166 since 2007.

Table 2 shows the detailed results for each question type. Just like what happened at the competition in 2007, the system obtained this year the best results in definition questions. Also, the accuracy in factoid questions improved: 22 factoid questions were answered correctly (corresponding to 13.580% of precision), against only 8 (5.03%) in the last year. Moreover, the system answered correctly to one list question: last year no correct answers were given to any question of this type.

It is worth to mention that we did not profit from the fact that the system could return 3 answers. In fact, the distribution of the 230 answers given by our

Table 1. QA@L^{2}F results at CLEF 2008

Right	Wrong	ineXact	Unsupported	Accuracy over the FIRST answer (%)
40	150	5	5	40/200 = 20%

Table 2. QA@L^{2}F results for each question type

Question Type	Total	Right	Wrong	ineXact	Unsupported	Accuracy (%)
Factoids	162	22	132	3	5	22/162 = 13.580%
Lists	10	1	8	1	0	1/10 = 10.0%
Definition	28	17	10	1	0	17/28 = 60.714%
Temporally Restricted	16	1	14	0	1	1/16 = 6.250%

system was the following: 184 single answers, 2 double answers and 14 triple answers. Moreover, several answers were extracted from Wikipedia's tables and, although the page from where they were extracted was correctly identified, they were considered unsupported.

5 Conclusions and Future Work

We presented the QI module of QA@L^{2}F, which uses a linguistically-motivated pattern matching system to obtain part of a frame and that profits from a named entity recognizer to build the whole frame. Moreover, anaphora is solved by manipulating frames, according to the type of the involved questions. Results about these processes were also presented, as well as the results obtained by QA@L^{2}F in QA@CLEF08.

In the near future we intend to improve the QI module, not only by expanding its rules, but also by exploring other techniques. Also, we need to evaluate the impact of each one of the different types of the errors in the system capability of obtaining a correct answer.

Although the entire system needs strong improvements, there are many small things to be done in QA@L^{2}F that can make it achieve better results. In the following we detach some of these improvements:

- Validate the answer type: 10 out of the 150 wrong answers do not have the expected type from the question. Being so, and already possessing a tool that is able to say that something is a *PERSON* or a *LOCATION* (for instance), it is not difficult to validate an answer type, as this is already retrieved from the question. This will certainly give better results, when articulated with redundancy, than only using redundancy by itself.
- Return 3 answers: as said before, only 230 answers (of which 14 triple) where returned this year;
- Improve the anaphora solver: as mentioned before, the system only solves anaphoras based on the frame constructed for the first question of a group of related questions and an anaphora can be related with any question from that group.

References

1. Bhagat, R., Leuski, A., Hovy, E.: Shallow semantic parsing despite little training data. In: Proc. ACL/SIGPARSE *9th* Int. Workshop on Parsing Technologies (2005)
2. Mendes, A., Coheur, L., Mamede, N.J., Ribeiro, R., Batista, F., de Matos, D.M.: QA@L2F, first steps at QA@CLEF. In: Proc. Cross-Language Evaluation Forum 2007. In: Peters, C., Jijkoun, V., Mandl, T., Müller, H., Oard, D.W., Peñas, A., Petras, V., Santos, D. (eds.) CLEF 2007. LNCS, vol. 5152, pp. 356–363. Springer, Heidelberg (2008)
3. Amaral, C., Cassan, A., Figueira, H., Martins, A., Mendes, A., Mendes, P., Pinto, C., Vidal, D.: Priberam's Question Answering System in QA@CLEF 2007. In: Peters, C., Jijkoun, V., Mandl, T., Müller, H., Oard, D.W., Peñas, A., Petras, V., Santos, D. (eds.) CLEF 2007. LNCS, vol. 5152, pp. 364–371. Springer, Heidelberg (2008)
4. Li, F., Zhang, X., Yuan, J., Zhu, X.: Classifying what-type questions by head noun tagging. In: Proc. *22nd* Int. Conference on Computational Linguistics, Coling (2008)
5. Weizenbaum, J.: ELIZA—a computer program for the study of natural language communication between man and machine. Communications of the Association for Computing Machinery 9(1), 36–45 (1965)
6. Allen, J.: Natural language understanding, 2nd edn. Benjamin-Cummings Publishing Co., Inc. (1995)
7. Jurafsky, D., Martin, J.H.: Speech and Language Processing, 2nd edn. Prentice-Hall, Inc., Englewood Cliffs (2006)
8. Medeiros, J.C.: Análise Morfológica e Correcção Ortográfica do Português. Master's thesis, Instituto Superior Técnico, Univ. Técnica Lisboa, Portugal (1995)
9. Ribeiro, R., Mamede, N.J., Trancoso, I.: Using Morphossyntactic Information in TTS Systems: comparing strategies for European Portuguese. In: Mamede, N.J., Baptista, J., Trancoso, I., Nunes, M.d.G.V. (eds.) PROPOR 2003. LNCS, vol. 2721, pp. 143–150. Springer, Heidelberg (2003)
10. Pardal, J.P., Mamede, N.J.: Terms Spotting with Linguistics and Statistics. In: Proc. Int. Workshop "Taller de Herramientas y Recursos Linguísticos para el Espanõl y el Portugués", IX Iberoamerican Conference on Artificial Intelligence, IBERAMIA (2004)
11. Aït-Mokhtar, S., Chanod, J.P., Roux, C.: A Multi-Input Dependency Parser. In: Proc. *7th* Int. Workshop on Parsing Technologies (2001)
12. Vicedo, J.L., Ferrández, A.: Importance of pronominal anaphora resolution in question answering systems. In: ACL 2000: Proc. of the *38th* Annual Meeting of the ACL (2000)
13. Negri, M., Kouylekov, M.O.: Who Are We Talking About? Tracking the Referent in a Question Answering Series. In: Branco, A. (ed.) DAARC 2007. LNCS (LNAI), vol. 4410, pp. 167–178. Springer, Heidelberg (2007)
14. Bouma, G., Kloosterman, G., Mur, J., van Noord, G., van der Plas, L., Tiedemann, J.: Question Answering with Joost at CLEF 2007. In: Peters, C., Jijkoun, V., Mandl, T., Müller, H., Oard, D.W., Peñas, A., Petras, V., Santos, D. (eds.) CLEF 2007. LNCS, vol. 5152, pp. 257–260. Springer, Heidelberg (2008)

UAIC Participation at QA@CLEF2008

Adrian Iftene[1], Diana Trandabăţ[1,2], Ionuţ Pistol[1], Alex-Mihai Moruz[1,2], Maria Husarciuc[1,2], and Dan Cristea[1,2]

[1] UAIC: Faculty of Computer Science, "Alexandru Ioan Cuza" University, Romania
[2] Institute for Computer Science, Romanian Academy Iasi Branch
{adiftene,dtrandabat,ipistol,amoruz,mhusarciuc,
dcristea}@info.uaic.ro

Abstract. 2008 marked UAIC's[1] third consecutive participation at the QA@CLEF competition, with continually improving results. The most significant change to our system with regards to the previous edition is the partial transition to a real-time QA system, as a consequence of the simplification or elimination of the main time-consuming tasks such as linguistic pre-processing. A brief description of our system and an analysis of the errors introduced by each module are given in this paper.

1 Introduction

Question Answering systems, especially the real-time ones, seem to come more and more frequently to the attention of researchers. The need for more advanced web searches and the emergence of the Semantic Web seems to be the main motivations for most groups concerned with QA research.

The team working at the "Al. I. Cuza" University of Iasi, Romania developed its first QA system for QA@CLEF 2006 competition [1], with a result of 9.47 % accuracy. In the 2007 competition, the CLEF organizers introduced the Romanian Wikipedia as a Romanian language corpus, thus it became possible for us to take part in the RO-RO QA track. We scored better than the first year (12 %) [2], [3], but the most significant improvement was the streamlining of the full QA system serving as the base of what would become this year's participation.

The 2008's Romanian corpus was the same as that of the QA@CLEF2007; by building upon previous experience, we also tried to create a real-time QA engine by eliminating of POS[2] tagging and NE[3] identification.

The second important improvement concerns the information retrieval part. In the case of definition type questions, queries were built in a specific way and the score of the snippets extracted from documents with the same title as the entity that must be defined were boosted. Also, the corpus was indexed at both paragraph and document level, and both types of returned snippets were kept; if the search for the answer in paragraph snippets is unsuccessful, the answer is searched in documents snippets.

---

[1] "Al. I. Cuza" University.
[2] Part-Of-Speech.
[3] Named Entities.

C. Peters et al. (Eds.): CLEF 2008, LNCS 5706, pp. 385–392, 2009.

The last main improvement was in the answer extraction module, where we tried to build very specific patterns in order to identify the final answer. For example, the MEASURE type was divided into three subtypes SURFACE, LENGTH, and OTHER_MEASURE. In this way, by creating more specialised patterns was improved the quality of the extraction module. Also, in order to extract for definitions questions, a specialised grammar was used.

The general system architecture is described in Section 2, while Section 3 is concerned with error analysis.

2 Architecture of the QA System

The system architecture is similar to that of our previous systems (see Figure 1). However, we eliminate POS tagging from the pre-processing modules, we use a Romanian grammar in order to identify the definition type answers, and NE recognition was only done on relevant snippets in order to reduce running time.

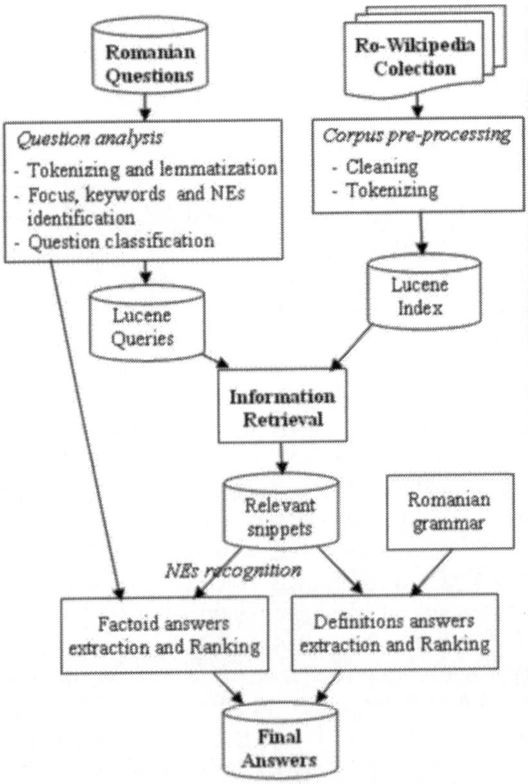

Fig. 1. UAIC system used in QA@CLEF2008

2.1 Corpus Pre-processing

The Wikipedia set of Romanian documents serving as the 2008 Romanian CLEF corpus includes 180.500 html files, with a total size of 1.9 GB. The documents include topic related information, as well as forum discussions, images and user profiles. The first step prior to indexing the documents was a filtering of irrelevant information, in order to improve querying the set of documents by reducing the overall size. This was accomplished by:

- Removing documents containing images, user profiles and forum discussions. The filtering was performed automatically using patterns for the name of documents.
- Removing all the html markups. The only annotated information preserved in the indexed documents is the page title and paragraph information.

These two steps reduced to corpora to 63712 documents totaling 116 MB of text. This reduction significantly made the indexing and query search time to go down.

2.2 Question Analysis

This step is mainly concerned with the identification of the semantic type of the answer (expected answer type). In addition, it also provides the question focus, the question type and a set of relevant keywords. The Question Analyzer performs the following steps (improvements over our previous versions are given in detail):

i) NP-chunking and **Named Entity extraction;**

ii) Question focus identification: in some cases, the first noun in a question is not the focus, as it concerns real world knowledge rather than the answer. The motivation for this comes from the fact that usually, in questions such as "*În ce oraş s-a născut Vladimir Ilici Lenin*?" (En: *In what city was Vladimir Ilici Lenin born*?) the answer is not "*oraşul Simbirsk*" (En: *the city of Simbirsk*); the correct answer is simply "*Simbirsk*", which is known to be a city. The selection of the first noun as focus actually prevents the finding of the answer in this case, as the snippet containing the correct answer does not even contain the word "*oraş*" (En: *city*).

iii) Answer type identification: the answer type was divided into more specific classes: the PERSON answer type into PERSON, MALE and FEMALE; the MEASURE answer type into SURFACE, LENGTH, and MEASURE; LOCATION into CITY, COUNTRY, REGION, RIVER, OCEAN and LOCATION. The nature of these changes comes from the fact that the tool used for NE recognition (GATE[4]) uses the same specific sub-types, and this change improves answer extraction.

iv) Question type inferring

v) Keyword generation: As keywords the focus, as well as verbs, nouns and NEs are considered. For all these words the form from the question but also the lemma form are considered.

vi) Anaphora resolution: For grouped questions we solve entity references in the following way: we insert the answer of the previous question to the list of keywords and we add to the current list of keywords all NEs from the previous questions in the same group.

[4] GATE: http://gate.ac.uk/

2.3 Index Creation and Information Retrieval

The purpose of this module is to retrieve the relevant snippets of text for every question. For this task we used the Lucene[5], an indexing and search tools. Below we have given a brief description of the module:

i) Query creation
Queries are formed using the sequences of keywords, some Lucene operators such as "+" (*mandatory*) and "^" (*boost factor*) and the "title" field (*to select documents after title*), in the case of definition questions. For the question "*Câte zile avea aprilie înainte de 700 î.Hr.?*" (En: *How many days had April before 700 B.C.?*) the query is:

```
+(zile^2 zi) aprilie^3 700^3 î.Hr.^3
```

For the definition question "*Cine este Ares?*" (En: *Who is Ares?*) the query is:

```
(title:Ares) Ares
```

ii) Index creation
The index of the document collection is based upon the document tokens determined in the pre-processing phase. Actually, two indexes have been created, one at paragraph level and one at document level. The paragraph index is more precise in terms of relevant text and is preferred when extracting snippets. If the answer is not found in the paragraph index, the query is applied to the document index instead.

iii) Relevant snippet extraction
Using the queries and the indexes, we extract a ranked list of snippets for every question.

2.4 Answer Extraction

The extraction process depends on the expected answer type: the answer extraction module identifies the named entities in every snippet provided by Lucene and matches them to the answer type. When the answer type is not a named entity, syntactic patterns based on question focus are used instead.

For identifying named entities we use GATE; for the MEASURE and DATE types we start from the collection of patterns developed in the 2007 competition, to which we add more patterns in order to split them into subtypes. Thus, we split the MEASURE pattern that was a number followed by a measure unit into three patterns: LENGTH, SURFACE and OTHER_MEASURE. For example, for the LENGTH pattern we consider numbers followed by either: *kilometru*, *metru* and *centimetru* (En: *kilometre*, *metre*, and *centimetre*) in singular, plural and short form (*km*, *m* and *cm*). The same split operation was done for the DATE type where we consider YEAR and FULL_DATE.

For the 2008 competition we have included a special module in order to extract answers for DEFINITION questions. This module is based on a Romanian grammar [4] built for the LT4eL project[6]. Definitions have been separated into six types in order to reduce the search space and the complexity of grammar rules:

[5] Lucene: http://lucene.apache.org/
[6] LT4eL: http://www.let.uu.nl/lt4el/

- "**is_def**" – Definitions containing the verb "*este*" (En: *is*):
- "**verb_def**" – Definitions containing specific verbs, different from "*este*". These verbs are "*indica*" (En: *denote*), "*arăta*" (En: *show*), "*preciza*" (En: *state*), "*reprezenta*" (En: *represent*), "*defini*" (En: *define*), "*specifica*" (En: *specify*), "*consta*" (En: *consist*), "*fixa*" (En: *name*), "*permite*" (En: *permit*).
- "**punct_def**" – Definitions determined on the basis of punctuation signs like "-", "()", "," etc.
- "**layout_def**" – Definitions that can be deduced from the html layout (eg. they can be included in tables).
- "**pron_def**" – Anaphoric definitions, when the defining term is expressed in a previos sentence and is only referred in the definition.
- "**other_def**" – Other definitions, which cannot be included in any of the previous categories (eg. "i.e.").

3 Results

The UAIC's system best run in the CLEF@QA2008 competition obtained the results presented in Table 1:

Table 1. Results of UAIC's best run

	Result evaluation	
R	RIGHT	62[7]
U	UNSUPPORTED	4
W	WRONG	125
X	INEXACT	9
	TOTAL	**200**

Each answer was evaluated as being RIGHT, UNSUPPORTED (no supporting snippet provided), WRONG or INEXACT (incomplete answer). The precision of our system was 31 %, with 19 % better than the precision obtained last year. The main improvement was obtained in the case of definition questions, where we found the correct answer for 18 questions out of 28, as compared to 0 correct answers in 2007. A detailed analysis of the performance of the system for each question type is presented in Table 2.

When performing the analysis of the errors introduced by the system, we found that most errors were introduced by the answer extraction module. However, the number of errors introduced by this module significantly decreased as compared to the last year's system. Table 3 shows the number of errors introduced by each module.

[7] In the official evaluation, the Romanian evaluator decided to maintain an unitary evaluation with previous CLEF edition, and thus considered the right answer for the definition questions of the form *Who/What is X?* as being just *Y*, and not *X is Y*, as our system returned, thus our answer to such definition questions has been considered inexact. For the other languages, both answers *X is Y* and *Y* were considered correct. Therefore, we consider that most of the definition answers judged as inexact are in fact correct, especially since removing or providing an *X is* to the *X is Y* answers a trivial task, not depending on the QA system real performance.

Table 2. Performance of the UAIC system per question type

	R	W	X	U	Total
Factoid	43	110	5	4	162
List	1	9	0	0	10
Definition	18	6	4	0	28
TOTAL	**62**	**125**	**9**	**4**	**200**

Table 3. Number of errors introduced per module

Module	Submodule	No. of Errors	% of errors
Question analysis	Question/answer type error	6	4.8
	Query formulation error	7	5.6
	Anaphoric related questions	5	4
Indexing	Answer in two paragraphs	9	7.2
Information retrieval	Snippet extraction errors	40	32
Answer extraction	Answer extraction errors	31	24.8
	Inferences needed for answer extraction	3	2.4
	Answer ordering error	13	10.4
	Gold error, answer correctly given	5	4
	Incorrect nil questions	2	1.6
	Incorrect list questions	4	3.2

3.1 Question Analysis Errors

The errors introduced in the question analysis phase are mainly due to:

- incorrect answer type identification,
- incorrect question type identification,
- improper generation of the query,
- incorrect anaphora resolution in group questions.

For the two first classes of errors, most misidentified answer type occurred for the *Object* type (15 misclassifications), since our question analysis module focuses mainly on named entities in order to determine the answer type, while questions with *Object* answer type usually don't contain any NEs. Another misclassification was noticed between the *Count* and *Measure* type (4 cases). However, even if a total number of 38 wrong answer types were misclassified, this didn't stop the system from answering the question correctly. Only 6 questions haven't been correctly answered due to question/answer type misclassification.

The errors occurred due to the query generation are linked to the fact that considering very general noun phrases as mandatory for the query, lots of irrelevant documents are generated. Another error source was the wrong identification of the antecedent in case of the anaphoric related questions, where the answer of a previous question from the anaphoric group was needed in order to complete the query for the current question. Thus, introducing the wrong answer as mandatory keyword in the query returns inconsistent documents. Several anaphorically related questions, however, can be answered correctly if the list of keywords contains only named entities from the previous question and not the previous answer.

3.2 Indexing and Information Retrieval

An important problem of the indexing was that, since we focused on the paragraph indexing, we couldn't answer questions that needed two paragraphs in order to justify or complete the answer. For example, for the question "*How many people can get in the Black Church in Brasov*?" the answer is found by combining the two anaphoric related paragraphs "*Biserica Neagră este cel mai mare edificiu de cult în stil gotic din sud-estul Europei...*" (En: *The Black Church is the biggest religious edifice in gothic style in the south-east of Europe…*) and "*În această biserică încap circa 5.000 de persoane.*" (En: *Around 5.000 persons can fit in this church.*).

An important number of errors were also introduced due to incorrect snippet extraction from the Wikipedia corpus. In order to improve this, the Wikipedia collection needs to be further cleaned, and empty categories, image names or links need to be removed from the indexable corpus. Another possible improvement is by refining the query according to the answer type. We started to work on this kind of refinement this year, when we included in the query for the definition questions the defined term as possible title of the Wikipedia article, especially if it was a named entity.

3.3 Answer Extraction

Although the answer extraction module performed better than in the previous edition, there were still problems, mainly due to the fact that the metrics used in order to rank the possible answers found in the returned snippets are still improvable.

Another problem for the answer extraction module was that it contained no inference submodule, while several questions could not be answered without inference. For instance, for the question "*Which goddess, sister of Ares, is the daughter of Metis*?", the answer is found in the paragraphs "*Ares era în mitologia greacă zeul războiului... era fiul lui Zeus şi al Herei.*" (En: *Ares was, in the Greek mythology, the God of War… He was the son of Zeus and Hera.*) and "*Atena era fiica lui Zeus şi a lui Metis.*"(En: *Athens was the daughter of Zeus and Métis*). The system needs to know that, since *Ares was the son of Zeus*, and *Athens the daughter of Zeus*, *Ares and Athens are brothers*.

For other 6 questions (2 NIL and 4 list questions), the system couldn't find the right answer. However, for 5 other questions, the system provided a better result than the humans that annotated the gold answers. Thus, one of the questions for instance is "*How high the Black Church is*?", and the gold answer is "*38 meters wide*". In another case, the response given by our system is more specific than the one considered by the gold evaluation, for the question "*When did Bulgaria regain its complete independence?*", when the gold answer was "*1908*" and we answered "*October 1908*", since we found the answer in another snippet.

4 Conclusions

This paper presents the Romanian Question Answering system which took part in the QA@CLEF 2008 competition. The evaluation shows an overall accuracy of 31%, our best result till now.

Three major improvements were carried out for this competition: first we eliminated the most time-consuming modules from the pre-processing step. Secondly, important improvements were made regarding the information retrieval module, where Lucene queries were built in a specific way for Definition questions. Thirdly, answer extraction was greatly improved by building very specific patterns. Also, we use a Romanian grammar in order to extract answers for Definition questions.

The significant improvements shown this year, combined with the major reduction in processing time, show promise with regards to our goal, which is to advance towards real-time good quality QA.

Acknowledgements

This paper presents the work of the Romanian team in the frame of the PNCDI II, SIR-RESDEC project number D1.1.0.0.0.7/18.09.2007.

References

1. Puşcaşu, G., Iftene, A., Pistol, I., Trandabăţ, D., Tufiş, D., Ceauşu, A., Stefănescu, D., Ion, R., Dornescu, I., Moruz, A., Cristea, D.: Cross-Lingual Romanian to English Question Answering at CLEF 2006. In: Peters, C., Clough, P., Gey, F.C., Karlgren, J., Magnini, B., Oard, D.W., de Rijke, M., Stempfhuber, M. (eds.) CLEF 2006. LNCS, vol. 4730, pp. 385–394. Springer, Heidelberg (2007)
2. Forner, P., Peñas, A., Alegria, I., Forăscu, C., Moreau, N., Osenova, P., Prokopidis, P., Rocha, P., Sacaleanu, B., Sutcliffe, R., Tjong, E., Sang, K.: Overview of the CLEF 2008 Multilingual Question Answering Track. In: Nardi, A., Peters, C. (eds.) Working Notes for the CLEF 2008 Workshop, Aarhus, Denmark, p. 39 (September 2008)
3. Iftene, A., Trandabăţ, D., Pistol, I., Moruz, A., Balahur-Dobrescu, A., Cotelea, D., Dornescu, I., Drăghici, I., Cristea, D.: UAIC Romanian Question Answering system for QA@CLEF. In: Peters, C., Jijkoun, V., Mandl, T., Müller, H., Oard, D.W., Peñas, A., Petras, V., Santos, D. (eds.) CLEF 2007. LNCS, vol. 5152, pp. 336–343. Springer, Heidelberg (2008)
4. Iftene, A., Trandabăţ, D., Pistol, I.: Grammar-based Automatic Extraction of Definitions and Applications for Romanian. In: Proc. of RANLP workshop "Natural Language Processing and Knowledge Representation for eLearning environments", Borovets, Bulgaria, September 26, pp. 19–25 (2007)

RACAI's QA System at the Romanian–Romanian QA@CLEF2008 Main Task

Radu Ion, Dan Ştefănescu, Alexandru Ceauşu, and Dan Tufiş

Research Institute for Artificial Intelligence, Romanian Academy
13, Calea 13 Septembrie, Bucharest 050711, Romania
{radu,danstef,aceausu,tufis}@racai.ro

Abstract. This paper describes the participation of the Research Institute for Artificial Intelligence of the Romanian Academy (RACAI) to the Multiple Language Question Answering Main Task at the CLEF 2008 competition. We present our Question Answering system answering Romanian questions from Romanian Wikipedia documents focusing on the implementation details. The presentation will also emphasize the fact that question analysis, snippet selection and ranking provide a useful basis of any answer extraction mechanism.

Keywords: Question Answering, query formulation, search engine, snippet selection, snippet ranking, question analysis, answer extraction, lexical chains.

1 Introduction

The Research Institute for Artificial Intelligence of the Romanian Academy (RACAI) is at the 3rd participation in the CLEF series of Question Answering competitions. This year (as in the previous one) we have focused on automatically answering questions in Romanian by searching their answers in Romanian Wikipedia documents. The classical requirement of the QA task is that the system must provide the shortest, syntactically well-formed string that completely answers the user's natural language question obeying the constraint that the string must be supported (contained) by a relevant text snippet which belongs to a document in the document collection. However, in the last two years, a new level of difficulty was added constraining the QA systems to resolve referential expressions and/or pronouns between questions topically grouped in a cluster in order to provide the answers. It is our firm belief that this added level of difficulty to the question analysis in QA@CLEF is counterproductive. If answering a question requires that you correctly answered *a previous one* (in the case of pronominal reference to a previous answer), the whole enterprise misses the point: the true ability of a QA system to answer natural language questions of a user which is mainly interested in finding the desired information and therefore, presumably, not inclined to "fool" the system by asking "complicated" questions (Tufiş and Popescu, 1991). In the light of these reflections, we have devised a second test set (from hereon called "the normalized test set" as opposed to "the official test set") in which we have manually identified and replaced all referents in all referential expressions/pronouns using the official Romanian-Romanian QA@CLEF2008 Gold Standard of Answers when we

C. Peters et al. (Eds.): CLEF 2008, LNCS 5706, pp. 393–400, 2009.

dealt with answers referents. We thus wanted to know how our QA system performs on the classical QA task in which the question itself supplies sufficient information so as to be able to identify the answer to it.

Our current system is based on the one that we have developed for the previous CLEF competition (Tufiş et al., 2008c). The main differences reside in an improved query formulation module and a completely redesigned answer extraction module which uses the results of a snippet selection and ranking component which did not exist in the 2007 version of the system. We have introduced this component because it is our belief that by incrementally improving every module of the system starting with the question analysis, we are able to eliminate cascading errors which obviously affect the most critical module which is the answer extractor. Thus, our current architecture consists of the following modules: question analysis, query formulation, information retrieval, snippet selection and ranking and answer extraction. These modules are pipelined from the first to the last so that the whole QA system receives as input a natural language question and outputs an ordered list of answers along with answer confidence and support snippet. It is our intention to implement all of these modules as Semantic Web web services so as to be able to **a)** develop each one independently of the others, **b)** easily measure the Mean Reciprocal Rank of the answer extraction module when modifications were made to the other dependencies and **c)** easily deploy the QA system itself as a web service and with it, a QA web application.

In what follows, we will briefly describe each of our QA system modules, concluding with the presentation of our results in the QA@CLEF2008 Romanian-Romanian Wikipedia QA competition. Part of the following presentation follows the one in (Ion et al., 2008).

2 The Search Engine

The document collection remained unchanged since the 2007 edition of the QA@CLEF. This collection is composed of 43486 Romanian language documents from Wikipedia (http://ro.wikipedia.org/). Each document has a title and several sections made up from paragraphs. All the logical sections of the documents were preprocessed with the TTL web service (Tufiş et al, 2008b) to obtain POS tagging, lemmatization and chunking of the text within.

The search engine is a C# port of the Apache Lucene (http://lucene.apache.org/) full-text searching engine. Lucene is a Java-based open source toolkit for text indexing and Boolean searching allowing queries formed with the usual Boolean operators such as AND, OR and NOT. Furthermore, it is capable to search for phrases (terms separated by spaces and enclosed in double quotes) and also to allow boosting for certain terms (the importance of a term is increased with the caret character '^' followed by an integer specifying the boost factor). We also used the field-specific term designation: a term may be prefixed by the field name to constrain the search to specific fields (such as title or text for instance) in the document index.

For the construction of the index, we considered that every document and every section within a document have different fields for the surface form of the words and their corresponding lemmas. This kind of distinction applies to titles and paragraph text resulting in four different index fields: title word form (**title**), title lemma (**ltitle**), paragraph word form (**text**) and paragraph lemma (**ltext**). We used the sentence and chunk annotation (from the TTL output) to insert phrase boundaries into our term index: a phrase query cannot match across different chunks or sentences. Thus, for instance, if we want to retrieve all documents about the TV series Twin Peaks, we would first like to search for the phrase "Twin Peaks" in the title field of the index (Lucene syntax **ltitle**:"Twin Peaks") and then, to increase our chance of obtaining some hits, to search in the word form field of the index for the same phrase (Lucene syntax **text**:"Twin Peaks"). Consequently, this Lucene query would look like this: **ltitle**:"Twin Peaks" **OR text**:"Twin Peaks".

The Romanian Wikipedia document retrieval mechanism is available as a web service (Tufiş et al., 2008b). The WSDL description can be found at http://nlp.racai.ro/ webservices/ SearchRoWikiWebService.asmx?WSDL.

3 Question Analysis and Query Formulation

The question analysis produces the focus and the topic of the question and was described in (Tufiş et al., 2008c). Basically, it uses the linkage of the question obtained with LexPar (Tufiş et al., 2008b) to identify linking patterns that describe the syntactic configuration of the focus, the main verb and the topic of the question.

The query formulation strategy improves the one described in (Tufiş et al., 2008c) which was successfully used in the Romanian-Romanian QA@CLEF2007 track. The input question must first be preprocessed with the TTL module to obtain POS tagging, lemmatization and chunking information. The CLEF 2007 version of the algorithm used to take into account all the word boundary substrings of each noun phrase regardless of their likeliness to appear. For instance, for the noun phrase "*cele mai avansate tehnologii ale armatei americane*"/"*the most advanced technologies of the US army*", terms like "*mai avansate tehnologii ale*" or "*cele mai avansate*" were valid. The present version of the question formulation algorithm fixes this aberration by constraining the substrings to be proper noun phrases themselves. In addition to that, the assignment of fields to each term was revised. Following, is the summary of modifications in four steps: 1. substring starting or ending with words of certain parts of speech are not considered terms; for instance substrings ending with adjectives or articles or beginning with adverbs; 2. substrings that do not contain a noun, a numeral or an abbreviation are not considered terms; 3. substrings starting with words other than nouns, numerals or abbreviations are not to be searched in the title field; 4. single word terms in occurrence form are not to be searched in the title field.

Table 1. Query formulation algorithm improvements onto the normalized test set

	MRR	Coverage
Initial	~0.7000	0.7500
Step 1	0.7366	0.7624
Step 2	0.7455	0.7715
Step 3	0.7689	0.8012
Step 4	0.7776	0.8092

We evaluated the retrieval accuracy of this query formulation algorithm onto *the normalized question* test set of the Romanian-Romanian QA@CLEF2008 track in which we participated this year. We used this test set in order to give a fair chance to the algorithm to discover the relevant terms that would produce the expected documents.

The query structure (its terms in our case) directly influences the *accuracy* and the *coverage* of the search engine. The accuracy is computed as a *Mean Reciprocal Rank* (MRR) score for documents (Voorhees, 1999), while the coverage is practically the recall score (coverage is the upper bound for the MRR). Although we primarily aim for covering all the questions (which means that we want to relax the queries in order to get documents/sections containing answers for as many questions as possible), a good MRR will ensure that these documents/sections will be among the top returned. Consequently, the detection of the exact answer should be facilitated if this procedure considers the ranks assigned by the search engine to the returned documents. The greater the MRR score, the better the improvement.

As Table 1 shows, starting from a MRR of around 0.7 and a coverage of 0.75 obtained with the 2007 version of the query formulation algorithm, the improved query formulation algorithms now achieves a MRR of 0.7776 and a coverage of 0.8092. The figures were computed using the reference Gold Standard of the Romanian-Romanian QA@CLEF2008 track in which for each question, the document identifier of the document containing the right answer is listed. We have not considered the questions which had a NIL answer and as such, no document identifier assigned.

The implementation of this query formulation algorithm is a web service (the WSDL description of which can be found at http://shadow.racai.ro/QADWebService/Service.asmx?WSDL) that takes the Romanian question as input and returns the query. To obtain POS tagging, lemma and chunking information, the web service uses the TTL web service.

4 Snippet Selection and Ranking

Snippet selection uses the question analysis to identify passages of text that, potentially, contain the answer to the question. For each section of each returned document, the snippet selection algorithm considers windows of at least N words at sentence boundary (that is, no window may have fewer than N words but it may have more than N words to enforce the sentence boundary condition). Each window is scored as follows (each word is searched in its lemma form): if the focus of the question is present, add 1; if the topic of the question is present, add 1; if both the topic and the focus of the question are present, add 10; if the k-th dependant of the focus/topic of the question is present add $1 / (k + 1)$ (the k-th dependent of a word a in the linkage of a sentence is the word b, if there exists a path of length k between a and b).

The above heuristics are simple and intuitive. Each window in which either focus or topic are found, receives one point. If both are to be found, a 10 point bonus is added because the window may contain the reformulation of the question into a statement which will thus resolve the value of the focus. The last heuristic aims at increasing the score of a window which contains dependents of the focus and/or topic but with a value which decreases with the distance (in links) between the two words in order to penalize snippets that contain long distance dependents of the focus/topic that may be irrelevant. The selection algorithm will retrieve at most *M* top-scored snippets from the documents returned by the search engine. A snippet may be added only if its selection score is greater than 0.

The snippet selection algorithm provides an initial ranking of the snippets. However, there are cases when the focus/topic is not present in its literal form but in a semantically related form like a synonym, hypernym, etc. This problem is known as the "lexical gap" between the question formulation and the text materialization of the answer. In these cases, our snippet selection procedure will assign lower scores to some of the important snippets because it will not find the literal representation of the words it looks for. To lighten the impact of this problem onto the snippets' scores, we developed a second ranking method which uses *lexical chains* to score the semantic relatedness of two different words.

Following (Moldovan and Novischi, 2002) we have developed a lexical chaining algorithm using the Romanian WordNet (RoWN) (Tufiş et al, 2008a) that for two words in lemma form along with their POS tags returns a list of lexical chains that exist between the meanings of the words in the Romanian WordNet. Each lexical chain is scored as to the type of semantic relations it contains. For instance, the synonymy relation receives a score of 1 and a hypernymy/hyponymy relation, a score of 0.85. Intuitively, if two words are synonymous, then their semantic similarity measure should have the highest value. The score of a lexical chain is obtained by summing the scores of the semantic relations that define it and dividing the sum to the number of semantic relations in the chain. All the semantic relations present in Romanian WordNet have been assigned scores (inspired by those in (Moldovan and Novischi, 2002)) between 0 and 1. Thus, the final score of a lexical chain may not exceed 1. Using the lexical chaining mechanism, we were able to re-rank the snippets that were selected with the previous procedure by computing the best lexical chain scores between focus, topic and their dependents and the words of the window.

As with the queries, we wanted to evaluate the two methods of snippet ranking individually and in combination using the normalized test set. We have set *N* (the number of words in a snippet) to 10 and 50 (these settings for *N* roughly correspond to 50-byte and 250-byte runs of the previous TREC QA competitions) and *M* (the number of retrieved snippets) to 20. We have also considered only the top 10 documents returned by the search engine. We counted a snippet (MRR style) only if it contains the answer *exactly* as it appears in the official Romanian-Romanian QA@CLEF2008 Gold Standard of Answers (no interest in NILs). Table 2 summarizes the results.

Table 2. MRR performance of the snippet selection and ranking algorithm on the normalized test set

N	Key word ranking	Lexical chain ranking	The combination	Coverage
10	**0.4372**	0.3371	0.4037	0.6894
50	0.4589	0.3679	**0.4747**	0.7

The combination of the two ranking methods consists in simply adding the scores provided for each snippet. When the snippet contains 10 words, the lexical chaining ranking method does not help the keyword ranking method because the semantic relatedness evidence is reduced by the short size of the snippet. When the snippet size increases (50 words), the contribution of the lexical chaining is more significant and this is reflected in the MRR score.

Unfortunately, the snippet selection and ranking module was developed and tested after the Romanian-Romanian QA@CLEF2008 track has ended. At the time of the writing, we didn't test this module onto the official test set but we are able to provide snippet MRR calculation from our official results where snippets were directly provided by the answer extraction procedure. Since only the first three answers were taken into consideration, it means that *M* equals 3 in this case. We count (MRR style) all the "YES" judgments from the `<answer-snippet>` element of the XML result file. This gives us a MRR of 0.2533 (coverage 0.325) for the first run and a MRR of 0.3041 (coverage 0.38) for the second run. All these figures are significantly lower than our current figures, using the normalized test set. This result shows that *if the use of anaphoric references would be really motivated in a practical natural language QA (NLQA) system, the anaphora resolution should be necessarily dealt with.* However, several user studies (e.g. (Slator, 1985), (Tufiş and Popescu, 1991)) brought evidence that, if a user is sincerely interested in getting the information he/she is looking for, in spite of the advertized understanding abilities of an artificial dialog system, the human language turns are unambiguous, direct and most of the time below the competence of the NLQA system.

5 The Answer Extraction Procedure

Answer extraction is responsible of extracting that syntactically well-formed substring that completely answers the user's question. Our answer extraction module relies on the question analysis and on the lexical chains algorithm and operates on the snippets provided by the snippet selection and ranking algorithm. Roughly, the extraction algorithm computes lexical chains scores between the question focus and words from a given snippet and then selects as answers those noun phrases whose heads are semantically close to the focus of the question. For instance, consider the question "*Câte* ***zile*** *avea aprilie înainte de 700 î.Hr.?*"/"*How many* ***days*** *did April have before 700 B.C.?*" and the text snippet:

```
...
luna/lună/Ncfsry/Np#2                   0.9
aprilie/aprilie/Ncms-n/Np#2             0.733
...
şi/şi/Crssp/                            0
avea/avea/Vmii3s/Vp#2                   0
29/29/Mc/Np#5                           0
de/de/Spsa/Pp#2                         0
zile/zi/Ncfp-n/Pp#2,Np#5                1
./././PERIOD/                           0
```

The left column contains the lemma, POS tagging and chunking analysis of the text snippet and the right column shows the lexical chains maximal scores between the focus of the question "*zi*"/"*day*" and the nouns of the snippet. Thus, competing for being the right answer to the user's question with other noun phrases from the snippet would be the string "*29 de zile*" (Np#5) as its head has a lexical chain score of 1 when linked with "*zi*".

As we have previously stated, when we submitted our results to the Romanian-Romanian QA@CLEF2008 track, we didn't have the snippet selection and ranking module and as such, our answer extraction module directly operated on the results of the search engine. Thus, the following accuracy figures are available (Table 3 summarizes the results): a) MRR and coverage using the official test set by counting the "R" (right answers) judgments from the XML official result file looking in the `<judgment>` element; b) MRR and coverage using the official test set by counting the "X" (inexact answers) judgments from the XML official result file looking in the `<judgment>` element; c) MRR and coverage using the normalized test set and the snippet selection and ranking module by counting (MRR style) the *exact* answers found in the official Romanian-Romanian QA@CLEF2008 Gold Standard of Answers (no interest in NILs).

Table 3. The answer extraction accuracy over the two test sets

Runs	MRR	Coverage
ICIA081RORO (official test set) Right (R)	0.0821	0.095
ICIA081RORO (official test set) ineXact (X)	0.0691	0.09
ICIA082RORO (official test set) Right (R)	0.1431	0.155
ICIA082RORO (official test set) ineXact (X)	0.0633	0.08
SSR (normalized test set) Right (R)	**0.1815**	**0.365**

The official test set contains 200 questions. We have submitted two runs: the first run, ICIA081RORO, was obtained by applying the answer extraction algorithm over the first 10 documents returned by the search engine when giving the query obtained from the first question in the cluster. All subsequent questions in the same cluster were answered from these 10 documents. The second run, ICIA082RORO, was the same as the first run except that: a) for definition type questions we have applied System A from (Tufiş et al, 2008c) which is specialized to answering definition type questions and b) for QA@CLEF2008 task, System A was modified slightly to answer some factoid questions and as such, any common factoid answers between this answer extraction algorithm and the modified System A was also preferred in the output.

6 Conclusions

As the Table 3 shows, the answer extraction procedure working on the output of the snippet selection and ranking module (SSR) and also on the 200 question normalized test set performs much better than applying the same answer extraction directly onto the results of the search engine and using ambiguous questions. Of course, this is to be expected but we want to emphasize that *these kinds of results need improving* and not the ones obtained from asking ambiguous questions. Also, we have shown that the same answer procedure greatly improves if an intermediate step selecting the snippets is involved. Our immediate goal is then to push the 0.1815 MRR figure to the possible achievable maximum which is the coverage of 0.365.

References

1. Ion, R., Ştefănescu, D., Ceauşu, A.: Important Practical Aspects of an Open-domain QA System Prototyping. In: Proceedings of the Romanian Academy, Series A, p. 6. The Publishing House of the Romanian Academy, Bucharest (2008)
2. Moldovan, D., Novischi, A.: Lexical Chains for Question Answering. In: Proceedings of COLING 2002, Taipei, Taiwan, pp. 674–680 (2002)
3. Slator, B.M., Anderson, M.P., Conley, W.: Pygmalion at the Interface. Commun. ACM 29(7), 599–604 (1986)
4. Tufiş, D., Ion, R., Bozianu, L., Ceauşu, A.l., Ştefănescu, D.: Romanian WordNet: Current State, New Applications and Prospects. In: Tanács, A., et al. (eds.) Proceedings of the Fourth Global WordNet Conference (GWC 2008), Szeged, Hungary, pp. 441–452 (2008a)
5. Tufiş, D., Ion, R., Ceauşu, A.l., Ştefănescu, D.: RACAI's Linguistic Web Services. In: Proceedings of the Sixth International Language Resources and Evaluation (LREC 2008), Marrakech, Morocco (2008b)
6. Tufiş, D., Popescu, O.: IURES2: Natural Language Environment and The Computer Speak Paradigm. In: Proceedings of the International Conference for Young Computer Scientists, Beijing, China (1991)
7. Tufiş, D., Ştefănescu, D., Ion, R., Ceauşu, A.l.: RACAI's Question Answering System at QA@CLEF2007. In: Peters, C., Jijkoun, V., Mandl, T., Müller, H., Oard, D.W., Peñas, A., Petras, V., Santos, D. (eds.) CLEF 2007. LNCS, vol. 5152, pp. 284–291. Springer, Heidelberg (2008c)
8. Voorhees, E.M.: The TREC-8 question answering track report. In: Proceedings of the 8th Text Retrieval Conference, Gaithersburg, Maryland, USA, pp. 77–82 (1999)

Combining Logic and Machine Learning for Answering Questions

Ingo Glöckner[1] and Björn Pelzer[2]

[1] Intelligent Information and Communication Systems Group (IICS), University of Hagen, 59084 Hagen, Germany
ingo.gloeckner@fernuni-hagen.de
[2] Department of Computer Science, Artificial Intelligence Research Group, University of Koblenz-Landau, 56070 Koblenz
bpelzer@uni-koblenz.de

Abstract. LogAnswer is a logic-oriented question answering system developed by the AI research group at the University of Koblenz-Landau and by the IICS at the University of Hagen. The system addresses two notorious problems of the logic-based approach: Achieving robustness and acceptable response times. Its main innovation is the use of logic for simultaneously extracting answer bindings and validating the corresponding answers. In this way the inefficiency of the classical answer extraction/answer validation pipeline is avoided. The prototype of the system, which can be tested on the web, demonstrates response times suitable for real-time querying. Robustness to gaps in the background knowledge and errors of linguistic analysis is achieved by combining the optimized deductive subsystem with shallow techniques by machine learning.

1 Introduction

The LogAnswer project (funded by the DFG - Deutsche Forschungsgemeinschaft - under contracts FU 263/12-1 and HE 2847/10-1) is aimed at investigating the opportunities of a logic-based approach for question answering (QA). Special emphasis is placed on two problems that still obstruct the successful application of logic in practical QA systems: achieving robustness (i.e., how can a logic-based QA system find useful results given that its background knowledge is necessarily incomplete?), and efficiency (i.e., how can answers be generated within a few seconds, given the computational effort of deductive reasoning?) The paper explains the design of the LogAnswer prototype that tries to overcome these problems by combining logic and machine learning. Based on an analysis of the results of the system in QA@CLEF 2008, the main shortcomings of the first prototype are identified. The results of this error analysis are instructive since they illustrate some general issues for the logic-based approach to question answering.

2 System Description

The system architecture of the LogAnswer QA system is shown in Fig. 1. In the following we describe the processing stages of the system.

C. Peters et al. (Eds.): CLEF 2008, LNCS 5706, pp. 401–408, 2009.

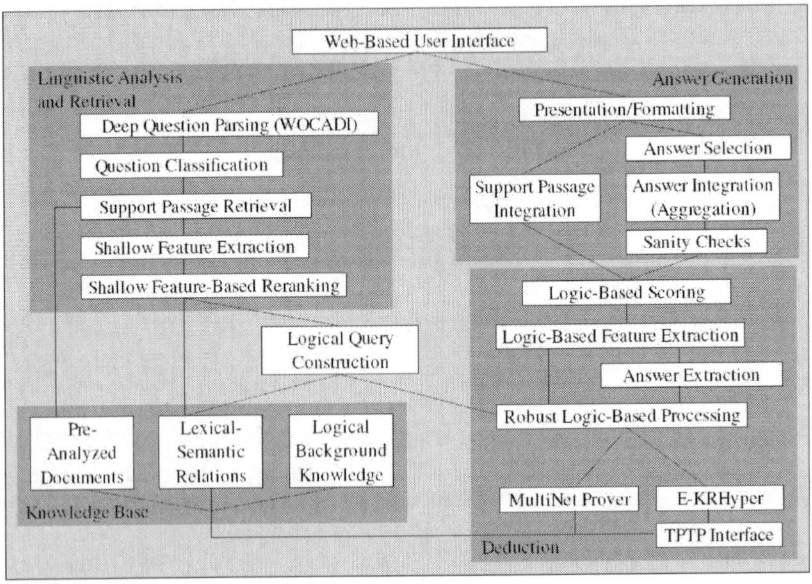

Fig. 1. System architecture of the LogAnswer prototype

Question Input. In normal operation, questions are entered into the LogAnswer web search box.[3] For QA@CLEF, a batch querying option was added.

Deep Linguistic Processing of the Question. The question is analyzed by the WOCADI parser [1], which generates a semantic representation in the MultiNet formalism [2]. The standard coreference resolution module of WOCADI is used for treating follow-up questions involving pronouns and nominal anaphora.

Question Classification. The category (*factual* vs. *definition*) and expected answer type (e.g. *PERSON*) of the question is identified by a system of 127 recognition rules, which also reduce the question to its descriptive core. Consider *Nennen Sie einige einfache Elementarteilchen!* (*'Name some elementary particles!'*). Then *nennen* (*'name'*) is not treated as part of the query content since it specifies what the system should do but does not describe the correct answers.

Support Passage Retrieval. The document collection of LogAnswer comprises the 11/2006 snapshot of the German Wikipedia; for QA@CLEF, the news collection of CLEF was added. All texts are parsed by WOCADI at indexing time. The resulting MultiNet representations are segmented into passages and stored in a Lucene-based retrieval module.[4] The following kinds of information are indexed:

[3] The system is available online at `www.loganswer.de`.

[4] Notice that at present, only single-sentence snippets are considered, but an extension to larger passages is planned for the future.

- The system uses lexical concepts (word senses) rather than word forms or word stems for indexing. However, there is no word sense disambiguation at the indexing level, i.e. all possible word senses for each word are indexed.
- Synonymy relationships are utilized for replacing all possible synonym variants by a canonical representation.[5] For example, *attacke.1.1* (attack) is replaced by the canonical *angriff.1.1* during indexing. Since all word senses of a word are normalized in this way, occurrences of other words that can have one of these meanings are also covered. A similar normalization at query time ensures that all synonym variants can be used for retrieval.
- Nominalizations are indexed. Thus, if the text contains *erfindung.1.1* (invention), then *erfinden.1.1* (invent) is also added to the index, and vice versa.
- Compound decompositions are indexed – *verteidigungsminister.1.1* (minister of defence) results in *minister.1.1* to be indexed as well.
- Adjective-attribute relationships are expanded. If the text contains *hoch.1.1* (high), then *höhe.1.1* (height) is indexed as well.

Moreover all answer types contained in a sentence are indexed. For answering definition questions, information about the containment of appositions, relative clauses, copula constructions, and defining verbs like *stehen für* (*'stand for'*), is also stored in the index. Notice that only sentences with a successful parse were indexed since the subsequent logic-based processing requires parsed sentences. The system was configured to retrieve 100 supporting snippets per question.

Shallow Feature Extraction and Reranking. In order to save processing time, LogAnswer normally restricts logical processing to a small number of most promising passages. To this end, the passages are reranked using shallow features that can be computed quickly without help of the prover. These features comprise: *failedMatch* (number of lexical concepts and numerals in the question which cannot be matched with the candidate document); *matchRatio* (relative proportion of lexical concepts and numerals in the question which find a match in the candidate document); *failedNames* (proper names mentioned in the question, but not in the passage); *containsBrackets* (the passage contains a pair of parentheses); *knownEat* (the expected answer type is known); *testableEat* (the expected answer type is fully supported by the current implementation of the answer type check); *eatFound* (an occurrence of the expected answer type has been found in the snippet); *isDefQuestion* (the question is a definition question). The *defLevel* feature is useful for definition questions. A value of *defLevel* = 2 indicates that the snippet contains a defining verb or apposition, and *defLevel* = 1 indicates a relative clause. Finally, the *irScore* feature provides the retrieval score determined by Lucene. The machine learning approach used for reranking the retrieved snippets based on the shallow features is the same as in [3,4]. It was implemented using the Weka toolbench [5]. The training data consisted of 17,350 annotated snippets retrieved in a run of LogAnswer on the QA@CLEF 2007 questions.

[5] The system uses 48,991 synsets (synonym sets) for 111,436 lexical constants.

Logical Query Construction. The parse of the question is turned into a conjunctive list of query literals. For example, *Wie hoch ist der chilenische Berg La Silla?* (*'How high is the Chilean mountain La Silla?'*) translates into the following logical query (with the $FOCUS$ variable representing the queried information):

$$\mathrm{modp}(X_1, \mathit{FOCUS}, \mathit{hoch.1.1}), \mathrm{sub}(X_2, \mathit{berg.1.1}), \mathrm{prop}(X_2, X_1), \mathrm{attr}(X_2, X_3),$$
$$\mathrm{prop}(X_2, \mathit{chilenisch.1.1}), \mathrm{val}(X_3, \mathit{la\_silla.0}), \mathrm{sub}(X_3, \mathit{name.1.1}).$$

During query construction, concept identifiers of synonyms are normalized by replacing the original concept identifiers with canonical synset representatives (however, no replacement occurs in the example).

Robust Logic-Based Processing. LogAnswer uses logic for simultaneously extracting and validating answers. To this end, the system tries to prove the logical representation of the question from the representation of the passage and the background knowledge (currently expressed in Horn logic).[6] Robustness is gained by using relaxation: if a proof is not found within a time limit, then query literals are skipped until a proof of the remaining query succeeds. The resulting skip count indicates (non-)entailment [6,4]. For efficiency reasons, relaxation is stopped before all literals are proved or skipped; a maximum of 3 relaxation cycles was chosen for the QA@CLEF runs. Notice that relaxation does not necessarily find the largest provable query fragment, since it only inspects a single sequence of simplification steps. Moreover the choice of skipped literals depends on factors like internal literal order of the prover which are arbitrary to some degree. Combining relaxation results of different provers can alleviate this problem. LogAnswer has interfaces to two provers in order to permit such a combination:

- The system includes a native prover for MultiNet representations, which is part of the MWR+ toolbench.[7] The MultiNet prover is very limited in expressive power (it only supports inferences over range-restricted Horn formulas), but its specialization to the task ensures high efficiency [7].
- E-KRHyper [8] is the latest version in the KRHyper-series of theorem provers and model generation systems for first-order logic with equality developed at the University Koblenz-Landau. It is an implementation of the E-*hyper tableau calculus* [9], which integrates a superposition-based handling of equality into the hyper tableau calculus [10]. E-KRHyper is capable of handling large sets of uniformly structured input facts, and it can rapidly switch and retract input clause sets for an efficient usage as a reasoning server. Embedded in the LogAnswer system, E-KRHyper is supplied with the MultiNet axioms transformed into first-order TPTP syntax [11]. The inference process operates on the axioms and the negated query literals, with a refutation result indicating a successful answer and providing the binding for the queried

---

6 The background knowledge of LogAnswer comprises 10,000 lexical-semantic facts (e.g. for nominalizations) and 109 logical rules, which define main characteristics of MultiNet relations and also handle meta verbs like 'stattfinden' (take place) [6].

7 See `http://pi7.fernuni-hagen.de/research/mwrplus`

variable. If the reasoning is interrupted due to exceeding the time limit, then partial results can be retrieved for guiding the relaxation process [12].

Answer Extraction. If a proof of the question from a passage succeeds, then LogAnswer obtains an answer binding which represents the queried information. In order to find more answers, LogAnswer also tries to determine a substitution when a strict proof of the query fails. The system then resorts to the intermediate substitution of the prover for the largest proven fragment of the query. LogAnswer uses word alignment information provided by WOCADI for extracting the corresponding answer string from the supporting text passage.

Logic-Based Feature Extraction. Based on the results of the relaxation proof and on the extracted answer, LogAnswer determines the following logic-oriented features: *skippedLitsLb* (number of literals skipped in the relaxation proof); *skippedLitsUb* (number of skipped literals, plus literals with unknown status); *litRatioLb* (relative proportion of actually proved literals compared to the total number of query literals, i.e. $1 - \mathit{skippedLitsUb}/\mathit{allLits}$); *litRatioUb* (relative proportion of potentially provable literals vs. all query literals, i.e. $1 - \mathit{skippedLitsLb}/\mathit{allLits}$); *npFocus* (the queried variable was bound to a constant which corresponds to a nominal phrase in the text); *focusEatMatch* (the answer type of the answer binding found by the prover matches the expected answer type). The *focusDefLevel* feature is relevant for definition questions. A value of $\mathit{focusDefLevel} = 2$ indicates that the answer binding found by the prover corresponds to an apposition, and $\mathit{focusDefLevel} = 1$ occurs if the answer binding corresponds to a noun phrase involving a relative clause.

Logic-Based Scoring. The logic-based answer scores are computed by the same ML approach also used for the shallow reranking. However, the shallow and logic-based features are now combined for better precision. In regular operation of LogAnswer, passages are considered in the order determined by the shallow feature-based ranking, and the logical processing is stopped after a pre-defined time limit. For QA@CLEF, no time limit was imposed, so every retrieved passage was subjected to deep processing and answer extraction.

Support Passage Selection. Depending on user preferences, the system answers the question either by presenting supporting text passages only, or alternatively, by presenting exact answers together with the supporting passage. For QA@CLEF, only the precise answer mode was relevant.

Sanity Checks. Two sanity checks are applied in order to eliminate false positives: A triviality check eliminates answers which only repeat contents of the question. For the question *'Who is Virginia Kelley?'*, this test rejects trivial answers like *'Virginia'* or *'Virginia Kelley'*. A special sanity check also rejects incomplete answers to definition questions. For example, *'the mother of Bill Clinton'* is a correct answer to the above question, but *'the mother'* must be rejected as incomplete. The compatibility of expected and found answer type is treated by answer-type related features passed to the machine learning method.

Table 1. Results of LogAnswer in QA@CLEF 2008

Run	#Right	#Unsupported	#Inexact	#Wrong	Accuracy	CWS	MRR
loga081dede	29	1	11	159	0.145	0.032	0.194
loga082dede	27	1	9	163	0.135	0.029	0.179

Aggregation and Answer Selection. The answer integration module computes a global score for each answer, based on the local score for each passage from which the answer was extracted [3]. The $k = 3$ distinct answers with the highest aggregated scores were chosen for the QA@CLEF runs. For each answer, the supporting passage with the highest score was selected as a justification.

3 Results on the QA@CLEF Test Set for German

The results of LogAnswer in the QA@CLEF 2008 task are shown in Table 1. The first run, *loga081dede*, used only the native prover of the MultiNet toolkit for logical processing. The second run, *loga082dede*, used the 'OPT' combination of the MultiNet prover and the E-KRHyper prover described in [4]. The motivation for using more than one prover is that following several relaxation paths by applying multiple provers might increase the chance of discovering a good provable query fragment. While the combination of the provers worked well in earlier experiments [4], results in the QA@CLEF 2008 task were slightly worse for the combined method compared to the first run which used only one prover.

4 Error Analysis and Discussion

An error analysis was made for the *loga081dede* run in order to identify the main deficits of the subsystems of LogAnswer. Concerning the linguistic processing stage, it was found that parsing of the question failed for 4 out of the 200 questions. Moreover the coreference resolution produced useless results (like unresolved pronouns) for 5 questions. Thus, the linguistic processing of the question was successful for 191 out of 200 questions in the QA@CLEF test set for German. Turning to the passage retrieval stage, the 19,064 retrieved supporting sentences (95.32 per question) were assessed for containment of a correct answer. The annotation revealed that for 119 of the questions, at least one passage which provides a correct answer was retrieved.

This means that, assuming perfect answer extraction and validation, the system can theoretically answer 119 non-NIL questions correctly. In order to extend this limit, the retrieval stage should be optimized. The following improvements are likely the most urgent:

- The retrieval module was configured to return only 100 sentences per question. Increasing this number will improve recall but incur more processing effort. A good trade-off for these factors should be assessed experimentally.

- The restriction to single-sentence snippets must be dropped. The analysis of the texts should be improved by resolving coreference. Deictic temporal expressions (like *'yesterday'*) should be resolved based on the document date.
- The system only indexes sentences with a full parse. This means that only about 60% of all sentences in the corpus are visible to LogAnswer. In order to improve recall, non-parseable answers should be indexed as well and a fallback method for answer extraction from such answers should be added.

Another significant source of errors is answer extraction: LogAnswer found 26 correct non-NIL answers in the *loga081dede* run. However, 46 of the supporting snippets for the top-1 answers actually contain a correct answer. Thus the success rate of answer extraction for sentences at top-1 position is 56.5%. For the top-3 results, the achieved MRR for 120 questions with multiple answers was 0.1944, compared to 0.3222 for perfect extraction. These problems of answer extraction are due to two phenomena not adequately treated in LogAnswer yet:

- The answer is often expressed by an apposition, as in *Albert Einstein, der Begründer der Relativitätstheorie* (*'Albert Einstein, the founder of the theory of relativity'*). In this case, the answer extractor must not return the full noun phrase which corresponds to the answer binding of the queried variable – it is necessary to split the extracted noun phrase and identify the relevant part.
- Copula constructions and constructions involving defining verbs also pose problems. For sentences like *'X is Y'* or *'X means Y'*, the logic-based answer extraction will often extract X even though the question targets at Y.

These problems result in wrong or inexact extractions, as shown by the relatively large number of 11 inexact answers of LogAnswer in the *loga081dede* run.

The chosen ML technique was also not very effective, which became clear when experimenting with refinements. While retaining bagging of decision trees as the basic method for probability estimation, the present version of LogAnswer no longer needs reweighting of training items.[8] This was made possible by changing the splitting criterion for decision tree induction in such a way that in each induction step, the selected split maximizes a generalized MRR metric over the training items, compared to all other nodes that currently await splitting:

$$\mathrm{k^*MRR} = \frac{1}{Q}\sum_{q=1}^{Q}\sum_{i=1}^{k_q^*}\frac{w_{i,k_q^*}}{\mathrm{rank}_{q,i} - i + 1} \quad \text{where} \quad w_{i,k_q^*} = \frac{2(k_q^* - i + 1)}{k_q^*\,(k_q^* + 1)},$$

$k_q^* = \min\{k, \mathrm{yes}_q\}$, yes_q is the number of correct results for question q, $\mathrm{rank}_{q,i}$ is the rank of the ith correct result for question q, Q is the number of questions, and k is the window size ($k = 3$ in the current system). The use of k*MRR-maximizing splits has a strong positive effect since the criterion is sensitive to the grouping of training items by question. Another refinement was permitting only those splits which fulfill *monotonicity constraints* on the estimated probabilities that can now be specified for each feature. For example, it makes sense to require

[8] Reweighting was necessary due to the low number of positives in the training set [7].

that (everything else being equal), the estimated correctness probability for a support passage must increase with the number of matched lexical concepts of the query. With these improvements, LogAnswer now finds 39 exact non-NIL answers, which means a 50% gain compared to the *loga081dede* run.

5 Conclusion

With LogAnswer, we have developed a prototype of a logic-based QA system suitable for real-time question answering. While the system works well when used for finding answer sentences [4,7], the naive solution for extracting exact answers that was added for QA@CLEF 2008 had problems with constructions involving appositions, copula, and defining verbs. Nevertheless, the simultaneous extraction of answer bindings and validation features from a relaxation proof of the question from the supporting snippet should be investigated further, since it avoids the extraction of a vast number of answer candidates from which the few correct ones must be selected by extensive validation. An intrinsic problem of logic-based answer extraction is the restriction to snippets with a full parse. Therefore the logic-based extraction should be complemented with a fallback extraction technique which covers sentences with a failed parse as well.

References

1. Hartrumpf, S.: Hybrid Disambiguation in Natural Language Analysis. Der Andere Verlag, Osnabrück (2003)
2. Helbig, H.: Knowledge Representation and the Semantics of Natural Language. Springer, Heidelberg (2006)
3. Glöckner, I.: University of Hagen at QA@CLEF 2008: Answer validation exercise. In: Working notes for the CLEF 2008 workshop, Århus, Denmark (2008)
4. Glöckner, I., Pelzer, B.: Exploring robustness enhancements for logic-based passage filtering. In: Lovrek, I., Howlett, R.J., Jain, L.C. (eds.) KES 2008, Part I. LNCS (LNAI), vol. 5177, pp. 606–614. Springer, Heidelberg (2008)
5. Witten, I.H., Frank, E.: Data Mining. Practical Machine Learning Tools and Techniques. Morgan Kaufmann, San Francisco (2005)
6. Glöckner, I.: University of Hagen at QA@CLEF 2007: Answer validation exercise. In: Working Notes for the CLEF 2007 Workshop, Budapest (2007)
7. Glöckner, I.: Towards logic-based question answering under time constraints. In: Proc. of ICAIA 2008, Hong Kong, pp. 13–18 (2008)
8. Pelzer, B., Wernhard, C.: System Description: E-KRHyper. In: Pfenning, F. (ed.) CADE 2007. LNCS (LNAI), vol. 4603, pp. 508–513. Springer, Heidelberg (2007)
9. Baumgartner, P., Furbach, U., Pelzer, B.: Hyper Tableaux with Equality. In: Pfenning, F. (ed.) CADE 2007. LNCS (LNAI), vol. 4603, pp. 492–507. Springer, Heidelberg (2007)
10. Baumgartner, P., Furbach, U., Niemelä, I.: Hyper Tableaux. In: JELIA 1996, Proceedings, pp. 1–17 (1996)
11. Sutcliffe, G., Suttner, C.: The TPTP Problem Library: CNF Release v1.2.1. Journal of Automated Reasoning 21(2), 177–203 (1998)
12. Pelzer, B., Glöckner, I.: Combining theorem proving with natural language processing. In: Proc. of the First Int. Workshop on Practical Aspects of Automated Reasoning (PAAR 2008), CEUR Workshop Proceedings, pp. 71–80 (2008)

The MIRACLE Team at the CLEF 2008 Multilingual Question Answering Track

Ángel Martínez-González[2], César de Pablo-Sánchez[1], Concepción Polo-Bayo[2], María Teresa Vicente-Díez[1], Paloma Martínez-Fernández[1], and José Luís Martínez-Fernández[1,2]

[1] Universidad Carlos III de Madrid
[2] DAEDALUS - Data, Decisions and Language, S.A.
{amartinez,cpolo,jmartinez}@daedalus.es
{Cdepablo,tvicente,pmf}@inf.uc3m.es

Abstract. The MIRACLE team participated in the monolingual Spanish and cross-language French to Spanish subtasks at QA@CLEF 2008. For the Spanish subtask, we used an almost completely rebuilt version of our system, designed with the aim of flexibly combining information sources and linguistic annotators for different languages. To allow easy development for new languages, most of the modules do not make any language dependent assumptions. The language dependent knowledge is encapsulated in a rule language developed within the MIRACLE team. By the time of submitting the runs, work on the new version was still ongoing, so we consider the results as a partial test of the possibilities of the new architecture. Subsystems for other languages were not yet available, so we tried a very simple approach for the French to Spanish subtask: questions were translated to Spanish with Babylon, and the output of this translation was fed into our system. The results had an accuracy of 16% for the monolingual Spanish task and 5% for the cross-language task.

1 Introduction

The MIRACLE team is a consortium formed by three universities from Madrid, (Universidad Politécnica de Madrid, Universidad Autónoma de Madrid and Universidad Carlos III de Madrid) and DAEDALUS, a small and medium size enterprise (SME). During the last year the MIRACLE QA system has gone through a major redesign and reimplementation that is still not finished. The main rationale of the new design is to flexibly combine heterogeneous information sources and linguistic annotation tools in a multilingual environment. In order to allow easy development for new languages, most of the modules do not make any language dependent assumptions. This language dependent knowledge is encapsulated in a rule language developed within the MIRACLE team.

By the time of submitting the runs, work on the new version implementation was still ongoing, so we consider it as a partial test of the possibilities of the new architecture. For the monolingual Spanish task, the MIRACLE team submitted two runs. We sent one main run with our system's best configuration, and another one where we

C. Peters et al. (Eds.): CLEF 2008, LNCS 5706, pp. 409–420, 2009.

test how the systems performs when varying the number of documents returned by the Information Retrieval subsystem. Although the modules for languages other than Spanish were not ready for this year's participation, we also took part in the French to Spanish task, with a very simple strategy. We just translated the questions from French to Spanish and fed our system with the results.

This paper is structured as follows. The next section describes the system architecture with special attention paid to three novelties: the rule language, the enhanced topic detection and the temporal expression analysis module. Section 3 presents and briefly analyzes the results. Conclusions and directions of future work follow in section 4.

2 System Architecture

This year the MIRACLE team took part in QA@CLEF with a system that has been rebuilt almost from scratch. The main objectives of the new design are:

- An architecture that can more easily be extended to Question Answering in languages other than Spanish. Though it was not ready for this year's CLEF, an English version of the system has been developed in parallel with the Spanish one.
- A system that can work simultaneously with several document collections in a flexible way. A new collection that has been processed offline can be added to the running system by just changing the parameters of a configuration file. Answers from different collections are not anymore extracted independently and then combined, but they are managed by the same extraction module. Nevertheless, collection specific extraction rules can also be formulated.
- Enhanced linguistic processing with a more detailed question analysis, the integration of a temporal expression analyzer and the use of a rule engine in all the modules that require some symbolic decision taking, so that more complex rules can be created and modified in a faster way.
- The system is required to be a web application that provides answers in real time.

The system architecture is presented in figure 1. It has a similar structure to most state-of-the-art QA systems. Several modules (Question Classification, Time Expression Analyzer, Answer Extraction and Topic Tracking) rely on a rule engine that we developed for this purpose. A brief introduction to the rule language is given in section 2.1. The strategy followed to deal with multilinguality was to gather all the language dependent knowledge in the rules, so that all the other modules (with the exception of the language processor) are language independent.

The main difference with previous versions of the system is that there are no separate streams for the EFE Newswire and Wikipedia Collection. Instead of that, the system is now ready for several collections to be considered jointly. Each source is assigned a confidence value and also source specific extraction rules can be added to the system.

The system modules are:

- *Linguistic Analysis:* the architecture allows several tokenizers and linguistic annotators for each language to be cascaded. For Spanish, we used the Daedalus STILUS [3] analyzer that provides tokenization, sentence detection and token analysis.

Token analyses included detailed part of speech analysis, lemmatization and semantic information. For English, a more heterogeneous set of tools was used that included Charniak parser [2], LingPipe Statistical Name Entity Recognizer [6], Wordnet [15] for lemmatization and self-created dictionaries for named entity annotation. Any other module that needs a linguistic processor for a language can get it without dealing with all those details using a Factory Design Pattern.

- *Time Expression Analyzer*: a component that analyzes and normalizes temporal expressions has also been integrated into the system. It is described in section 2.3.
- *Question Classification*: as an output of this module, the following values are determined for the question: focus, topic, question type, expected answer entity and a boolean feature telling whether the answer should be a list.

The value of the expected entity is taken quite directly from the entity tags used by the STILUS tokenizer. STILUS uses a multilevel named entity hierarchy, which in turn is inspired in the one developed by Sekine [11]. This hierarchy has been used to tag by hand a large number of entities in STILUS dictionaries. This hand tagging of the resources is a very labor-intensive task, which is still in process.

Table 1. Expected entity for sample questions

Question	Expected entity	Abstraction levels
¿Qué es Opel? (What is Opel?)	INDUSTRIAL_COMPANY	class or subclass
¿Cómo se llaman las líneas aéreas de Niki Lauda? (What is the name of Niki Lauda's airlines?)	SERVICE_COMPANY	instance
¿Qué empresa tiene a Bibendum como mascota? (Which company has Bibendum as mascot?)	COMPANY	instance

Though Sekine's hierarchy was originally thought for tagging instances (which roughly correspond to proper nouns), STILUS resources apply it in a novel way to tag common nouns, marking them as classes or subclasses. For example, "*Opel*" is tagged as an instance of "ORGANIZATION->COMPANY->INDUSTRIAL_COMPANY", while "*líneas_aéreas*" ("*airlines*") is tagged as a subclass of "ORGANIZATION->COMPANY->SERVICE_COMPANY" and the word "*empresa*" ("*company*") is tagged as the class "ORGANIZATION->COMPANY". These tags help the rules in the question classification module to determine the expected entity as shown in table 1. For example, in the case of the question "*¿Qué es Opel?*" ("*What is Opel?*"), the expected answer is a class or subclass of "INDUSTRIAL_COMPANY". And for the question "*¿Cómo se llaman las líneas aéreas de Niki Lauda?*" ("*What is the name of Niki Lauda's airlines?*") an instance of SERVICE_COMPANY is needed.

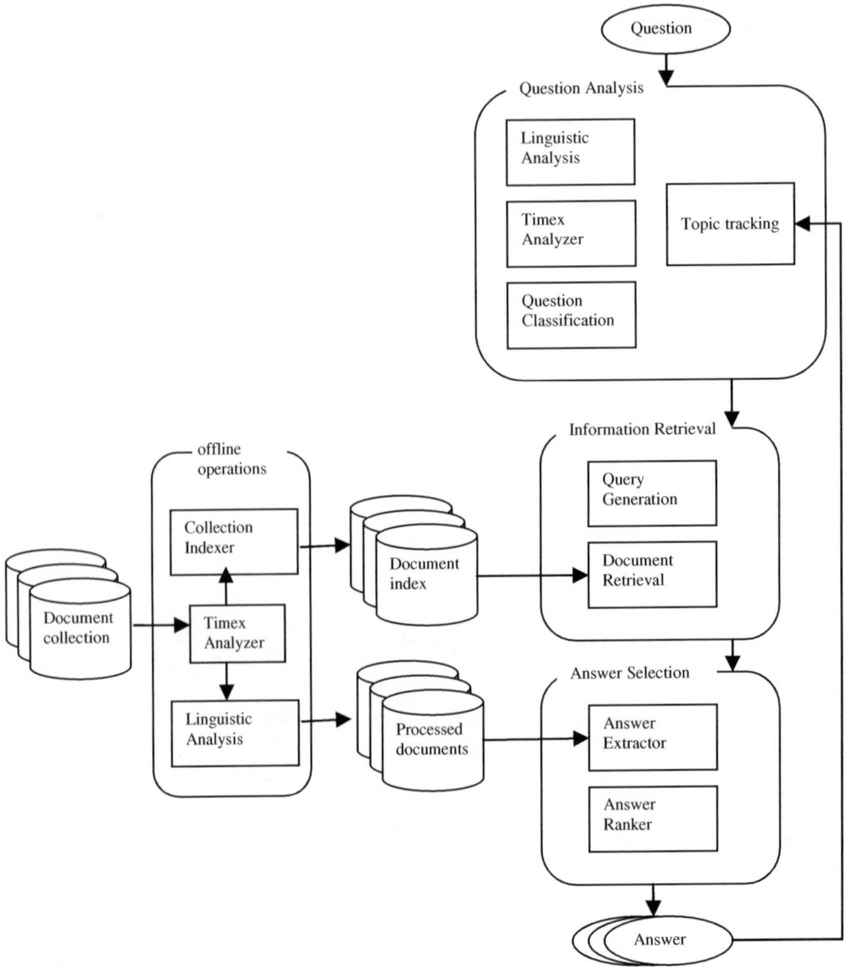

Fig. 1. MIRACLE 2008 system architecture

Note that the expected entity feature is used not only for factoid questions, but also for definition questions. In a question like "*¿Quién era Edgar P. Jacobs?*" ("*Who was Edgar P. Jacobs?*"), it is interesting for other modules to know that the answer will probably contain words such as "*writer*" or "*artist*", which are tagged as subclasses of "PERSON".

- *Topic and coreference tracker*: we have enhanced our topic candidate generator of previous years, by analyzing referring expressions in the follow-up questions, which often signal a change of topic inside the question group. This is explained in more detail in section 2.2.
- *Query generation and Document Retrieval*: Lucene was introduced for QA@CLEF 2007 as the Information Retrieval Engine and this year we kept it for this task. In

this module, a query in Lucene syntax is generated and the most relevant documents according to cosine similarity ranking are retrieved.

- *Answer Extractor*: a ruled based approach that detects patterns for each type of question is used. The Sentence Selection Module present just before the Answer Extractor in the pipeline of previous versions of the system was removed and its functionality has been merged into the Answer Extractor. The reason for this is that many rules of the Answer Extractor had to be duplicated in the Sentence Selector, to ensure that the sentence containing that pattern reached the input of the Extractor.

  One of the improvement needs identified after our last year's participation in CLEF [8] was in the extraction of definitions from Wikipedia. This year we made a special effort to write rules for definitions; some of the heuristics used to recognize definitions include selecting Wikipedia articles whose title matches the question focus, giving priority to the first sentences of each document and searching for patterns that include the focus, expressions such as "*es*" ("*is*"), "*son*" ("*are*"), "*se denomina*" ("*is called*") and entities of the expected type. As discussed in section 3 of this document, a substantial improvement for definition questions was achieved with this approach.
- *Answer Ranker*: sorts the answer candidates according to a ranking formula. Although this module ranks answers and not support sentences, other terms of the support sentence can also contribute to the answer's score. For this years runs, the following elements were taken into account:
  - Number of named entities in the support sentence that are compatible with the expected entity. An entity is said to be compatible with another if both are equal or the first is a child of the second in the hierarchy.
  - Number of term lemmas in the support sentence that also appear in the query.
  - Number of named entities in the support sentence that also appear in the question.

The values mentioned above are normalized in order not to favour answers contained in longer sentences. Besides, terms of the support sentence that are closer to the extracted answer are likely to be more related to it. To reflect this, the values are weighted with a factor that reflects term proximity in the support sentence and ranges between 0 (if the term is more than 9 words away from the extracted part of the sentence) and 10 (if the term belongs to the extracted answer). It can be the case that a sentence contains two answer candidates, and their scores will not necessarily be equal.

2.1 Rule Engine

In previous versions of the system a rule engine was used for question classification. This approach was found very useful to separate decisions related to symbolic linguistic knowledge from the rest of the code, therefore, rules were introduced in other parts of the system such as Answer Extraction, Temporal Expression Analysis and Topic Tracking modules. Rules in this language are preprocessed to generate Java code. New rule predicates and actions to expand the rule language can also be defined handily in Java. For the particular case of Answer Extraction, we have found the rules suitable to incrementally introduce the quite different heuristics necessary to cope with heterogeneous sources and different question types.

```
BEGIN_RULE
   WORD_FORM(0, "¿") AND
   EXISTENTIAL_LEMMA(
       POS_FIRST_EXISTENTIAL_ANALYSIS(Tag_WHPronoun),
       "quién")
THEN
   ADD_EXPECTED_ENTITY("PERSON")
END_RULE
```

Fig. 2. Example of a rule

The rules have a left part that expresses a pattern and a right part specifying the actions to be taken each time the pattern is found. The pattern is not necessarily only lexical; it can also have a syntactic or semantic component. While the rule engine is working, there is always a current sentence, for which all the patterns are tested. Depending on the module where the engine is running, this current sentence will simply be the question or each one of the selected document sentences. Figure 2 shows a simplified example of a rule. All it does is assign the expected entity "PERSON" to question starting with the interrogation sign "¿" and whose first interrogative pronoun is "*quién*". The rule language includes three types of constructs:

- *Predicates* (for example, EXISTENTIAL_LEMMA): return a boolean value and are used to check some linguistic feature such as word form, lemma, syntactical or semantic information. This predicates can be combined in the left part of the rule with boolean operators of conjunction, disjunction and negation. It has to be taken into account that in the case of ambiguity, a word might have more than one analysis, so predicates may require that all the analyses of a word satisfy a condition (universal test) o just that one of them does (existential text). Other context data such as the current document title or collection identifier can also be tested.
- *Actions* to be taken if the rule fires (such as ADD_EXPECTED_ENTITY): these actions form the right side of the rule. They assign a type to the question, extract a part of a sentence as an answer, calculate the normalized form of a date, etc.
- *Position Functions* (for example, POS_FIRST_EXISTENTIAL_ANALYSIS): these functions return a position in the current sentence and are used as arguments to predicates and to actions. They give the language a higher order of expressiveness. Positions can also be kept in temporary variables.

In our design, the rules are supposed to be the only language dependent part of the code. Another principle we have found useful about the rules, is that the right side of the rules should perform all the suitable actions considering the linguistic knowledge obtained by the tests on the left side of the rule. A different design, for example with only one action taken by each rule, would lead to more complexity and redundancy. This idea of avoiding redundancy is the reason why the Sentence Selector is not present in our new system as a module separated from the Answer Extractor (as explained above in the architecture outline).

2.2 Topic Detection and Coreference Tracking

In this year's evaluation, there were a large number of questions groups. In these groups the topic is presented in the first question and the following questions are related to this topic. The guidelines restrict the topic to any kind of entity or event introduced in the first question or the answer to this question. In contrast, an analysis of previous year's topics reveal that sometimes the topic can change within a group. In Spanish, a topic shift is naturally introduced by the use of a referring expression that recalls a different entity or event. This is a simplified view of the theory of centering (Grosz et al [4]). The example of table 2 is from group 2011 in CLEF 2007 topic set.

Table 2. Example of topic tracking for a question group

Question	Answer	Referent list	Referring Expressions	Topic
¿Quién fue Hermann Emil Fischer?	químico alemán			Hermann Emil Fischer
¿Que premio recibió en 1902?	Premio Nobel de Quimica	R1 = (Hermann Emil Fischer, químico alemán)	E1 (ellipsis)	E1=R1= (Hermann Emil Fischer, químico alemán)
¿Quién recibió el Premio Nobel de Literatura ese año?	Theodor Mommsen	R1 = (Herman Emil Fischer, químico alemán) R2 = Premio Nobel de Química R3 =1902	E2= ese año E3= Premio Nobel de Literatura	E2 = R3 = 1902

In our previous participation [8] we implemented a rule-based system for topic identification that considered candidates among the topic, the focus, the candidate answers and other constituents of the first question. The different candidates were selected based on syntactic heuristics and reordered depending on factors like the semantic type of the expected answer and the usual structure of an information seeking dialogue. For example, numbers, quantities and dates are uncommon as topics a priori when considered against persons or locations.

We have enhanced our topic tracking module by analyzing the follow-up question and tracking the use of referring expressions. Most referring relations in Spanish questions are realized by ellipsis and in those cases the a priori selection works well. In contrast when an explicit referring expression is introduced it usually signals a topic shift that reflects a reordering in the set of candidate referents. We have implemented rules that are able to track the most common cases in questions and answer dialogues: definite noun phrases (using determiners and articles), pronouns and named entities. Rare cases like epithets or verb nominalizations have been so far ignored. Once the candidate referring expressions have been selected the next step consists in solving their co-referent. For each candidate pair of referent and referring expressions we calculate if they satisfy a set of agreement constraints. We have implemented five different constraints based on the linguistic information that is available after analysis: number, genre, lemma, semantic type and acronym expansion. A candidate pair that satisfies more constraints than the a priori best rated candidate for co-reference could be promoted if the score is much higher. So far the weights have been adjusted manually using previous examples and counterexamples. This accounts for the most common co-reference phenomena in QA dialogues although some others are also feasible like partitive, meronymy or collective coreference and will be subject to future work ([12] and [5]).

2.3 Temporal Expression Analyzer

A precise analysis of temporal expressions is of vital importance both for questions about time ("*In what year did The Red Baron get the Blue Max?*") and for questions with some time restriction ("*Which city did Charlemagne capture in 778?*").

A temporal expression extractor and normalizer, which had been developed within the MIRACLE team ([13] and [14]), has been enhanced and integrated into our QA system for this year. The basis of the system is a set of recognition rules that defines a Finite State Grammar. For the definition of this grammar, an exhaustive study of the temporal expressions that appear in Spanish texts was necessary. The defined patterns include both absolute and relative expressions. Absolute expressions are completely defined by themselves, while relative expressions make reference to some other time that has to be known in order to be completely determined. Furthermore, both phrases that refer to time points or to intervals are considered. This latter classification is independent of the former, so we can have absolute time points ("*25/12/2007*"),

Table 3. Example of date recognition and normalization

Input	Description	Resolution rule	Reference date	Normalized output
El 31 de diciembre de 2005 ([the] 31th December 2005)	[ART|PREP]? DAY PREP MONTH_NAME PREP YYYY	Day =toDD (DAY) Month=toMM(MONTH_NAME) Year=YYYY	NA	2005-12-31
mañana (tomorrow)	DEICTIC_UNIT	Day=getDD(Creation_Time)+1 Month=getMM(Creation_Time) Year=getYYYY(Creation_Time)	2008-06-01	2008-06-02
Entre mayo y agosto (Between May and August)	PREP MONTH_NAME1 CONJ MONTH_NAME2	Year1=getYYYY(Creation_Time) Month1=getMM(MONTH_NAME1) Day1=1 Year2=getYYYY(Creation_Time) Month2=getMM(MONTH_NAME2) Day2=getLastDay(Month2)	2008-06-01	2008-05-01 2008-08-31

relative time points ("*ayer*"/ "*yesterday*"), absolute intervals ("*entre 2000 y 2003*"/ "*between 2000 and 2003*") and relative intervals ("desde mayo hasta junio"/ "*from May to June*"). To define the normalized output value the international standard ISO 8601 (2004) for representation of dates and times is used. For the resolution of relative temporal expressions, some reference date is necessary. Though in some cases the reference date should be deduced from the context, for the integration in the QA system a simpler approach was followed and the document's date of creation is always the reference.

The Temporal Expression Analyzer is integrated into the QA system at two levels:

- At the Information Retrieval level: Temporal expressions are normalized in the indexes generated from the document collections and in the queries generated from the questions. This allows an increase in recall.
- At a symbolic level: The rules for answer extraction can use predicates that check whether a given token is a time expression and, in that case, which normalized value it has. For time restriction checking, a basic temporal inference mechanism has been developed, based on the principle of inclusion of a time point or interval in another interval.

3 Results

3.1 Submitted Runs

Three runs were submitted by the MIRACLE team this year. Two monolingual runs for Spanish and a cross-language one for the French to Spanish subtask. For Spanish, we sent a main run using the system tuned as we thought it would yield best results. It is described in tables 4 and 5.

Table 4. Judged answers for the main Spanish run

Name	Right	Wrong	Inexact	Unsupp.
mira081eses	32	156	3	9

Table 5. Accuracy by question type for the main Spanish run

Factoids	Lists	Definitions	NIL Returned	Temporally Restricted
11.2 %	0.0 %	73.7 %	16.7 %	4.8 %

The results of the two other runs are summarized in table 6. Considering that two runs can be sent for each language, we wanted to employ the second Spanish run to test the variation in the system performance when changing a configuration parameter. We chose the maximum number of documents returned by the Information Retrieval module, setting it to 40 instead of the default value 20. As we expected, the result was worse, but not very significantly.

Finally, as explained in previous sections, the system for this year was developed with multilinguality in mind, but only the Spanish part was ready by the time of submitting the runs. Nevertheless, we decided to send a run with French as query language using a very simple approach: we translated the questions with Babylon [1] and just fed the Spanish system with the translation, with the non-surprising poor results shown in the second row of table 6.

Table 6. Results for runs other than the main Spanish run

Name	Right	Wrong	Inexact	Unsupp.
mira082eses	29	159	3	9
mira081fres	10	185	2	3

3.2 Error Analysis

In this section the results for our main Spanish task are analyzed. These results can be considered as disappointing, as they suppose a very small improvement from last year (from an accuracy of 15% to 16%). We consider that the main reason of this is the lack of time to complete the development and tuning of the system after fully rewriting it this year. For example, no rules for the extraction of lists were added to the system before the deadline, this explains that there are no right answers of this kind. We have dealt with list questions in previous years so there is no technical reason that explains this absence but for lack of time (or a failure in task planning). Therefore, we consider the results as a partial test of the possibilities of the new architecture at an intermediate stage of development.

The modest improvement can also be partially attributed to the greater difficulty of the question set. According to an evaluation we have done with this year's system on 2007 questions, a 22% accuracy was obtained, compared to 15% of last year's system on the same question set. The new set had many more group questions, 110 instead of 50. And also some rather tricky questions were included, for instance: "*¿Quién es el guardameta de la selección española de baloncesto?*" ("*Who is the goalkeeper of the Spanish basketball national team?*").

On the other hand, table 5 shows a great improvement for definition questions, with an accuracy of 73,684%. This was one of the main weaknesses in our last year's participation [8]. The number of definition questions fell from 32 to 19 from last year to this year.

After the submission of the runs, additional work was done to evaluate the impact of the Time Expression Analysis Module on the system [13]. Table 7 reflects the results of this evaluation, which was done using the CLEF questions but before the Golden Standard was available and therefore is based in our own criteria that might be slightly different from CLEF evaluators. The second row represents the results with time expression analysis applied both on the Lucene index and in the rules of the Answer Extraction module, as explained in section 2.3. It shows a moderate but positive influence of time normalization for temporally restricted questions.

Table 7. Evaluation of Time Analysis Module

System configuration	Temporally restricted	With timex answer	Temp. Restricted and with timex answer	Total
Without timex analysis	3,9%	21,1%	0%	10,9%
With timex analysis	11,5%	21,1%	0%	15,2%

4 Conclusions and Future Work

In the discussion about the results of the previous section, the Time Expression Analyzer was the only module whose influence was analyzed individually. We are currently working on an evaluation framework that will let as measure the performance of each module independently. This framework will also allow easily putting together different configurations of the system, with different implementation of one module or different setup parameters, and testing the overall performance.

The other main focus of work of the Miracle Team for next year is to introduce some more sophisticated logic description of the meaning of questions that will go beyond question focus and topic, probably using RDF. This semantic representation shall be compared with a similar analysis for document sentences, so that reasoning is possible with the aid of some of the available high-level open-domain ontologies ([10] and [16]).

Though this year's results in QA CLEF don't seem very promising, we consider it as an intermediate evaluation of an unfinished system. And we still keep our confidence the novelties we have introduced this year will yield fruit once we have the time to tune and debug the system.

Acknowledgements

This work has been partially supported by the Autonomous Community of Madrid by means of the project MAVIR (S-0505/TIC/000267) and by the Spanish Ministry of Education by means of the project BRAVO (TIN2007-67407-C3-01).

References

1. Babylon website, `http://www.babylon.com/` (visited 21/11/2008)
2. Charniak, E.: A maximum-entropy-inspired parser. In: Proceedings of the First Meeting of the North American Chapter of the Association for Computational Linguistics, pp. 132–139. Seattle, Washington (2000)
3. Daedalus. STILUS website, `http://www.daedalus.es` (visited 21/11/2008)
4. Grosz, B.J., Joshi, A.K., Weinstein, S.: Centering. A framework for modeling the local coherence of discourse. Computational Linguistics 21(2), 203–225 (1995)
5. Jurafsky, D., Martin, J.H.: Speech and Language Processing, ch. 21, 2nd edn (2008)
6. LingPipe (Java libraries for the linguistic analysis of human language): `http://www.alias-i.com/lingpipe/` (visited 21/11/2008)
7. Lucene webpage, `http://lucene.apache.org/` (visited 21/11/2008)
8. de Pablo-Sánchez, C., Martínez-Fernández, J.L., González-Ledesma, A., Samy, D., Moreno, A., Martínez, P., Al-Jumaily, H.: MIRACLE Question Answering System for Spanish at CLEF 2007. In: Working Notes of CLEF 2007, Budapest (2007)
9. de Pablo-Sánchez, C., Gonzalez-Ledesma, A., Moreno-Sandoval, A., Vicente-Díez, M.T.: MIRACLE experiments in QA@CLEF 2006 in spanish: main task, real-time QA and exploratory QA using wikipedia (WiQA). In: Peters, C., Clough, P., Gey, F.C., Karlgren, J., Magnini, B., Oard, D.W., de Rijke, M., Stempfhuber, M. (eds.) CLEF 2006. LNCS, vol. 4730, pp. 463–472. Springer, Heidelberg (2007)

10. Saias, J., Quaresma, P.: The senso question answering approach to Portuguese qa@clef-2007. In: Proceedings of CLEF – Cross Language Evaluation Forum, Budapest, Hungary (September 2007)
11. Sekine, S.: Sekine's extended named entity hierarchy, `http://nlp.cs.nyu.edu/ene/` (visited 21/11/2008)
12. Vicedo, J.L., Ferrández, A.: Correference in Q & A. In: Strzalkowski, T., Harabagiu, S. (eds.) Advances in Open Domain Question Answering. Text, Speech and Language Technology, vol. 32, pp. 71–96. Springer, Dordrecht (2006)
13. Vicente-Díez, M.T., de Pablo-Sánchez, C., Martinez, P.: In: Evaluacion de un Sistema de Reconocimiento y Normalización de Expresiones Temporales en Español. In: En Actas del XXIII Congreso de la Sociedad Española de Procesamiento de Lenguaje Natural (SEPLN 2007), páginas, Sevilla, Spain, September 2007, pp. 113–120 (2007)
14. Vicente-Díez, M.T., Samy, D., Martínez, P.: An Empirical Approach to a Preliminary Successful Identification and Resolution of Temporal Expressions in Spanish News Corpora. In: Proceedings of the Sixth International Language Resources and Evaluation (LREC 2008), Marrakech, Morocco (May 2008)
15. WordNet, a lexical database for the English language, `http://wordnet.princeton.edu/` (visited 21/11/2008)
16. Zajac, R.: Towards ontological question answering. In: Proceedings of the workshop on ARABIC language processing: status and prospects, Toulouse, France, July 06, pp. 1–7 (2001)

Efficient Question Answering with Question Decomposition and Multiple Answer Streams

Sven Hartrumpf[1], Ingo Glöckner[1], and Johannes Leveling[2]

[1] Intelligent Information and Communication Systems (IICS)
University of Hagen (FernUniversität in Hagen), 58084 Hagen, Germany
[2] Centre for Next Generation Localisation (CNGL)
Dublin City University, Dublin 9, Ireland

Abstract. The German question answering (QA) system IRSAW (formerly: InSicht) participated in QA@CLEF for the fifth time. IRSAW was introduced in 2007 by integrating the deep answer producer InSicht, several shallow answer producers, and a logical validator. InSicht builds on a deep QA approach: it transforms documents to semantic representations using a parser, draws inferences on semantic representations with rules, and matches semantic representations derived from questions and documents. InSicht was improved for QA@CLEF 2008 mainly in the following two areas. The coreference resolver was trained on question series instead of newspaper texts in order to be better applicable for follow-up questions. Questions are decomposed by several methods on the level of semantic representations. On the shallow processing side, the number of answer producers was increased from two to four by adding FACT, a fact index, and SHASE, a shallow semantic network matcher. The answer validator introduced in 2007 was replaced by the faster RAVE validator designed for logic-based answer validation under time constraints. Using RAVE for merging the results of the answer producers, monolingual German runs and bilingual runs with source language English and Spanish were produced by applying the machine translation web service Promt. An error analysis shows the main problems for the precision-oriented deep answer producer InSicht and the potential offered by the recall-oriented shallow answer producers.

1 Introduction

The German question answering (QA) system IRSAW (Intelligent Information Retrieval on the Basis of a Semantically Annotated Web) employs deep and shallow methods. The deep answer producer is InSicht, which transforms documents to semantic representations using a syntactico-semantic parser, draws inferences on semantic representations with rules, matches semantic representations derived from questions and documents, and generates natural language answers from the semantic representations of documents. Specialized modules refine the semantic representations in several directions: resolving coreferences in documents (and questions) and resolving temporal deixis in documents. To provide a robust strategy for difficult text passages or passages mixing text and

C. Peters et al. (Eds.): CLEF 2008, LNCS 5706, pp. 421–428, 2009.

other elements, four shallow[1] answer producers are employed. (Note that one of them, SHASE, is using the semantic representation in a simple way.) The resulting five streams of answer candidates, which are produced in parallel, are logically validated and merged by RAVE. Based on the results of validation, RAVE scores the answer candidates and selects the final results.

2 Changes of InSicht for QA@CLEF 2008

2.1 Improved Dialog Treatment

In contrast to QA@CLEF 2007, we trained the coreference resolver CORUDIS [1] on a dialog corpus with anaphors in questions, namely the test questions from QA@CLEF 2007. The training set was derived as follows. First, all coreferences (pronoun to NP, less specific NP to more specific NP) were annotated yielding 29 questions from 20 question series with a coreference. Second, as 20 training texts will not deliver good results, additional question series were created by taking every continuous sequence of 1 to 4 questions from the QA@CLEF 2007 questions. (A sequence is discarded for training if an anaphora leads outside the selected sequence.) Information about discourse boundaries (topic starts) was ignored because this kind of information is missing in many applications. Third, the resulting 462 question series were fed into the usual training process of CORUDIS. Note that also the answer to a question could be integrated as a possible antecedent, but as only two QA@CLEF 2007 questions show a coreference to the preceding answer, this was ignored. In 2008, the number of such cases increased to four so that this option has become more relevant.[2]

2.2 Question Decomposition

Question decomposition was systematically added to InSicht for QA@CLEF 2008. A decomposition method tries to simplify complex questions by first asking a *subquestion* whose answer is used to form a *revised question* which is often easier to answer than the original question.[3] For example, question decomposition for *Welches Metall wird zur Goldwäsche benutzt?*/*Which metal is used for washing gold?* (qa08_192) leads to the subquestion *Nenne Metalle*/*Name metals* with answers like *Eisen*/*iron* and *Quecksilber*/*quicksilver* and revised questions like *Wird Quecksilber zur Goldwäsche benutzt?*/*Is quicksilver used for washing gold?* Answers to original questions found by decomposition often require support for the answered subquestion and the revised question, i.e. the answer to the original question is supported by sentences from different documents.

To evaluate question decomposition after QA@CLEF 2008, we annotated all German QA@CLEF questions since 2003 with decomposition classes (see [3]

[1] i.e. not relying on semantic representations of sentences.

[2] For corpus documents, the statistical model trained on newspaper articles is chosen instead of the model from question series.

[3] The term *decomposition* is sometimes used in a different sense when a biographical question like *Who was Bernini?* is broken down into a set of standard questions [2].

for details on the annotation, the decomposition classes, and the decomposition methods). For 2008, 21 questions (10.5%) were annotated as decomposable. This percentage is lower than in previous years: from 2004 till 2007, the percentage was 17.1%. Examples from QA@CLEF 2008 are qa08_044 (*Wieviele Bundesländer hat Österreich?/How many states does Austria have?*) and qa08_192 as discussed above. As expected, some answers (e.g. for qa08_192) were not found when decomposition was turned off.

2.3 Performance Improvement

Adding features to the deep producer InSicht yields better results, but often with a longer runtime. Therefore, several performance improvements were tried. As query expansion by logical rules (applied in backward chaining) expands the search space dramatically, this expansion should be reduced by efficient heuristics that do not eliminate good answers. To this end, statistics on successful rule applications (i.e. combinations of logical rules that led to at least one correct answer) were collected from the test collections of QA@CLEF from 2003 to 2007 and some separate question collections. Restricting query expansion to successful rule combinations turned out to be very effective because results for the QA@CLEF 2008 questions stayed stable while runtime decreased by 56%.

3 Shallow QA Subsystems

In addition to the deep producer, IRSAW now employs four shallow producers of answer candidates: QAP [4], MIRA [5], FACT, and SHASE. The latter two have been added for QA@CLEF 2008. FACT employs a fact database, in which relational triples have been indexed, e.g. name2date_of_death("Galileo Galilei", "8. Januar 1642").[4] Relational triples take the same form as triples used in the MIRA producer. The relational triples have been extracted automatically from various sources, including the PND [6], the acronym database VERA, monetary names from ISO 4217, and appositions from the semantic network representation of the Wikipedia and CLEF-News corpora. To answer a question, the relational triple is determined for a question using a machine learning (ML) approach and keywords from the question are used to fill in one argument position of the triple. Answers are extracted from the other argument position of matching triples. Document sentences containing keywords from the question as well as the exact answer string are returned as support for the answer candidate.

SHASE uses the semantic network representation of both question and document sentences to produce answer candidates. The core node representing an answer node is identified in the question semantic network (i.e. the question focus node determined by the syntactico-semantic parser). To find answer candidates, the semantic relations to and from the core node, its semantic sort, and

[4] The relation type name2date_of_death is viewed as the first component of the triple. Variants of date formats (for the second argument) are explicitly generated and indexed as well because no normalization takes place at this level, yet.

its semantic entity are calculated; see [7] for details on the semantic hierarchies. These features are matched with corresponding features of nodes in the document networks. Matching nodes represent answer candidates: the answer string is generated from the semantic network and the document sentence is returned as answer support.

4 Merging Answer Streams by Validation

The answer candidates in the InSicht stream and the shallow QA streams are validated and merged by RAVE (Real-time Answer Validation Engine), a logic-based answer validator designed for real-time QA. It is crucial for the efficiency of RAVE that no answer must be parsed at query time – computing deep linguistic analyses for hundreds of extracted answer candidates during validation is not realistic in a real-time QA setting. The use of logic in RAVE is therefore restricted to validating support passages, i.e. deciding if a passage contains the requested information. This is the case if the logical representation of the question can be proved from the representation of the passage and from the available background knowledge, a criterion which can be checked independently of the answer candidates. Since the passage representations can be pre-computed, this eliminates the need for parsing during validation. Local validation scores are determined based on shallow and (if available) also logic-based features. Separate models were trained for each producer in order to tailor the validation criterion to the characteristics of each answer stream. Training data was obtained from a system run on the QA@CLEF 2007 questions. The resulting 21,447 answer candidates extracted from 27,919 retrieved passages were annotated for containment of a correct answer. Cross-validation experiments on the training set suggested that bagging of decision trees with reweighting of training examples is suited for the task. The local ML-based scores, which estimate the probability that an answer is correct judging from a specific supporting snippet, are aggregated in order to determine the total evidence for each answer. The aggregation model used by RAVE aims at robustness against duplicated information [8]. By pre-ranking arriving answers based on shallow features and computing improved logic-based scores for the most promising candidates until a given timeout is exceeded, RAVE implements an incremental, anytime validation technique [9]. Answer candidates from InSicht do not require logical validation since they result from a precision-oriented QA technique. Their validation rests on the self-assessment of InSicht and the number of alternative justifications found for the answer. Alternatively, the self-assessment can directly be used as the validation score.

5 Description of Runs

All runs with prefix *fuha081* were generated using the ML-based validation scores for InSicht, whereas the runs with prefix *fuha082* used the self-assessment of InSicht. For bilingual QA experiments, the Promt Online Translator (http://www.promt.com/) was employed to translate the questions from English or Spanish to German. Experience from previous CLEF campaigns suggested that Promt would return translations containing fewer errors than other web services for machine translation (MT), which becomes important when translated questions are parsed. However, we found that Promt employs a new MT service (in beta status) and experiments using translations from other web services had a higher performance [10].

6 Evaluation and Discussion

We submitted two runs for the German monolingual task in QA@CLEF 2008 and four bilingual runs with English and Spanish as source language and German as target language (see Table 1). The syntactico-semantic parser employed in InSicht was used to provide a complexity measure for the German questions by counting the semantic relations in parse results (after coreference resolution). This showed a decrease compared to previous years: 9.05 relations per question on average (2007: 11.41; 2006: 11.34; 2005: 11.33; 2004: 9.84). In the bilingual experiments with English and Spanish, about 60% and 40%, respectively, of the performance (measured in right answers) for monolingual German were achieved. Results may have been better with another MT service.

The evaluation of dialog treatment showed that the coreference resolver performed correctly. The only exceptions are the anaphors in the four questions that referred to the answer of the preceding question. These anaphors were incorrectly resolved because this case was not trained (see Sect. 2.1).

Table 2 shows an error analysis for the deep answer producer InSicht. The analysis is based on problem classes that lead to not finding a correct answer; the same classes were used for our participation in QA@CLEF 2004 [11], except that the new class q.incorrect_coreference (coreference resolution errors for questions) is needed for the question series introduced in QA@CLEF 2007. A random sample of 100 questions that InSicht answered incorrectly was investigated. For questions involving several problem classes, only the one that occurred in the earlier component of processing was annotated in order to avoid speculation about

Table 1. Results for the German question set from QA@CLEF 2008 (CWS: confidence-weighted score; MRR: mean reciprocal rank; R: right, U: unsupported, X: inexact, W: wrong). For accuracy, only first answers that are right or unsupported are counted as correct. Note that only 199 questions were assessed for *fuha081esde*.

Run	Results						
	#R	#U	#X	#W	Accuracy	CWS	MRR
fuha081dede	45	6	8	141	0.255	0.052	0.297
fuha082dede	46	4	11	139	0.250	0.049	0.296
fuha081ende	28	3	6	163	0.155	0.024	0.240
fuha082ende	28	6	6	160	0.170	0.020	0.226
fuha081esde	19	2	9	169	0.105	0.015	0.157
fuha082esde	17	5	5	173	0.110	0.049	0.296

Table 2. Problem classes and problem class frequencies for QA@CLEF 2008

Name	Description	%
q.error	error related to question side	
q.parse_error	question parse is not complete and correct	
q.no_parse	parse fails	3
q.chunk_parse	only chunk parse result	0
q.incorrect_coreference	a coreference is resolved incorrectly	4
q.incorrect_parse	parser generates full parse, but it contains errors	6
q.ungrammatical	question is ungrammatical	0
d.error	error related to document side	
d.parse_error	document sentence parse is not complete and correct	
d.no_parse	parse fails	12
d.chunk_parse	only chunk parse result	16
d.incorrect_parse	parser generates full parse, but it contains errors	16
d.ungrammatical	document sentence is ungrammatical	2
q-d.error	error in connecting question and document	
q-d.failed_generation	no answer string can be generated for a found answer	1
q-d.matching_error	match between semantic networks is incorrect	1
q-d.missing_cotext	answer is spread across several sentences	7
q-d.missing_inferences	inferential knowledge is missing	32

subsequent errors. Similar to our analysis for QA@CLEF 2004, parser errors on the document side and missing inferences between document and question representations are the two main problems for InSicht.

The performance of the shallow QA subsystem[5] has also been assessed. For the 200 questions, a total number of 36,757 distinct supporting passages was retrieved (183.8 per question). 1,264 of these passages contain a correct answer. For 165 of the questions, there is at least one passage that contains an answer to the question. Since these passages form the basis for answer extraction by the shallow producers, this means that for perfect answer extraction, it would theoretically be possible to answer 165 non-NIL questions correctly (or 175 questions including the NIL case). The extraction performance achieved by the answer producers of the shallow subsystem of IRSAW is shown in Table 3. The following labels are used in the table: *#candidates* (average number of extracted answer candidates per question), *#answers* (average number of right answers per question), *pass-rate* (fraction of the 1,264 correct passages from which a correct answer is extracted), *pass-prec* (precision of answer extraction for correct passages), *#answered* (number of questions for which at least one right answer is extracted), and *answer-rate* (answered questions divided by total number of

[5] This subsystem can be improved as follows. Most shallow producers used the semantic network representation for indexing, i.e. no stemming or stopword removal was applied, but full words were indexed. The tokenization and sentence segmentation underlying the semantic network representations often cause the answer extraction to fail. Finally, the shallow producers have not been trained on the Wikipedia.

Table 3. Extraction performance of shallow answer producers

Producer	Results					
	#Candidates	#Answers	Pass-rate	Pass-prec	#Answered	Answer-rate
FACT	14.38	1.43	0.19	0.57	34	0.21
MIRA	80.09	2.15	0.31	0.32	107	0.65
QAP	1.43	0.02	0.00	0.43	2	0.01
SHASE	80.89	1.15	0.16	0.16	81	0.49
all	176.79	4.74	0.50	0.29	132	0.80

questions with a correct supporting passage, i.e. *#answered*/165 in this case). As witnessed by the *answer-rate* of 0.8 for all shallow producers in combination, the answer candidates extracted by the shallow producers cover most of the correct answers contained in the retrieved passages. While perfect selection from the results of the shallow subsystem would answer 132 non-NIL questions correctly (or 142 including NIL questions),[6] RAVE only made 46 correct selections, which indicates that improvements are necessary:

- RAVE is good at identifying passages that contain an answer, but it often cannot discern right answer candidates found in such passages from wrong extractions. The validator needs better features for relating the answer candidate to the result of validating a supporting passage. Moreover, the rudimentary implementation of some existing features (like the answer type check) must be refined in order to achieve better performance.
- Due to technical problems when the training set was generated, the annotations cover only 151 questions of the 2007 test set and less than 30 definition questions. For better ML results, more questions must be annotated.
- The ML technique proved ineffective, but this problem has been addressed in the meantime: After modifying the induction of decision trees in such a way that the MRR on the training set is maximized, RAVE finds 60 correct answers and 102 correct support passages at top-1 position.

The average time for a complete logical validation, i.e. without a time limit, was 1.48 s per question.[7] Prior to the development of RAVE, logical validation used to be one of the most time-consuming stages of IRSAW, but now it no longer slows down the system response time (19.8 s on average).

7 Conclusion

The QA system IRSAW was successfully improved in several ways for QA@CLEF 2008. Coreference resolution for questions was strengthened by generating suitable training data. Question decomposition in the deep answer producer InSicht

[6] Including InSicht would further increase these numbers because often only a deep producer can deliver correct candidates for questions that require inferences.

[7] Times were measured on a standard PC (2.4 GHz CPU frequency).

opens interesting ways to a fusion of information from different documents or corpora. Adding two more shallow answer sources proved beneficial for robustness. With increasing system complexity, runtime performance becomes critical, but optimization techniques like parallelization and incremental processing help finding useful answers with response times acceptable for interactive querying. The RAVE prototype shows that applying logic-based validation techniques in a real-time QA setting is possible, but richer features and an improved training set must be provided in the next development phase.

References

1. Hartrumpf, S.: Coreference resolution with syntactico-semantic rules and corpus statistics. In: Proceedings of the Fifth Computational Natural Language Learning Workshop (CoNLL 2001), Toulouse, France, pp. 137–144 (2001)
2. Harabagiu, S.: Questions and intentions. In: Strzalkowski, T., Harabagiu, S. (eds.) Advances in Open Domain Question Answering. Text, Speech and Language Technology, vol. 32, pp. 99–147. Springer, Dordrecht (2006)
3. Hartrumpf, S.: Semantic decomposition for question answering. In: Ghallab, M., Spyropoulos, C.D., Fakotakis, N., Avouris, N. (eds.) Proceedings of the 18th European Conference on Artificial Intelligence (ECAI), Patras, Greece, pp. 313–317 (2008)
4. Leveling, J.: On the role of information retrieval in the question answering system IRSAW. In: Proceedings of the LWA 2006 (Learning, Knowledge, and Adaptability), Workshop Information Retrieval, pp. 119–125. Universität Hildesheim, Hildesheim (2006)
5. Leveling, J.: A modified information retrieval approach to produce answer candidates for question answering. In: Hinneburg, A. (ed.) Proceedings of the LWA 2007 (Lernen-Wissen-Adaption), Workshop FGIR. Gesellschaft für Informatik, Halle/Saale, Germany (2007)
6. Hengel, C., Pfeifer, B.: Kooperation der Personennamendatei (PND) mit Wikipedia. Dialog mit Bibliotheken 17(3), 18–24 (2005)
7. Helbig, H.: Knowledge Representation and the Semantics of Natural Language. Springer, Berlin (2006)
8. Glöckner, I.: University of Hagen at QA@CLEF 2008: Answer validation exercise. In: Results of the CLEF 2008 Cross-Language System Evaluation Campaign, Working Notes for the CLEF 2008 Workshop, Aarhus, Denmark (2008)
9. Hartrumpf, S., Glöckner, I., Leveling, J.: University of Hagen at QA@CLEF 2008: Efficient question answering with question decomposition and multiple answer streams. In: Results of the CLEF 2008 Cross-Language System Evaluation Campaign, Working Notes for the CLEF 2008 Workshop, Aarhus, Denmark (2008)
10. Leveling, J., Hartrumpf, S.: Integrating methods from IR and QA for geographic information retrieval. In: Peters, C., et al. (eds.) CLEF 2008. LNCS, vol. 5706, pp. 851–854. Springer, Heidelberg (2009)
11. Hartrumpf, S.: Question answering using sentence parsing and semantic network matching. In: Peters, C., Clough, P., Gonzalo, J., Jones, G.J.F., Kluck, M., Magnini, B. (eds.) CLEF 2004. LNCS, vol. 3491, pp. 512–521. Springer, Heidelberg (2005)

DFKI-LT at QA@CLEF 2008

Bogdan Sacaleanu, Günter Neumann, and Christian Spurk

LT-Lab, DFKI, Saarbrücken, Germany
{bogdan,neumann,cspurk}@dfki.de

Abstract. The paper describes QUANTICO, a cross-language open domain factoid question answering system for German and English document collections. The main features of the system are: use of preemptive off-line document annotation with information like Named Entities, abbreviation-extension pairs and appositional constructions; use of online translation services for the cross-language scenarios; use of redundancy as an indicator of good answer candidates; selection of the best answers based on distance metrics defined over graph representations. The results of evaluating the system's performance by QA@CLEF 2008 were as follows: for the German-German run we achieved a best overall accuracy (ACC) of 37%; for the English-German run 14.5% (ACC); and for the German-English run 14% (ACC).

1 Introduction

Though most of the research in Question Answering has been carried in monolingual settings, where the question and the answer-bearing documents share the same natural language, current approaches concentrate on cross-language scenarios, where the question and the documents are in different languages. Explored in this context and common with the Information Retrieval research are two methods of crossing the language barrier: by translating the question [2, 3, 7] or by translating the documents [1].

We present a cross-lingual English to German Question Answering system, QUANTICO, for both factoid and definition questions, using a German monolingual system and translating the questions from English to German. Two different techniques of translation are presented:

- direct translation of the English input question into German and
- transfer-based translation, using an intermediate representation that captures the "meaning" of the original German question and is translated into the target English language.

The intermediate representation captures the semantic of the question in terms of question type (*q-type*), expected answer type (*a-type*) and focus (*q-focus*), information that steers the workflow of the question answering process.

The German monolingual Question Answering system can answer both factoid and definition questions and is based on several premises:

- facts and definitions are usually expressed locally at the level of a sentence unit (Passage Retrieval)

C. Peters et al. (Eds.): CLEF 2008, LNCS 5706, pp. 429–437, 2009.

- for factoid questions redundancy of candidate answers is a good indicator for their suitability (Answer Extraction)
- definitions of concepts are expressed using fixed linguistic structures such as appositions, modifiers, abbreviation extensions (Answer Extraction)
- proximity of concepts within a sentence can be related to the semantic dependency of them (Answer Selection)

We will begin giving a short overview of the system and presenting its working for both factoid and definition questions in monolingual and cross-language scenarios. We will then continue with a short description of each component and close the paper with the presentation of the CLEF evaluation results and the error analysis outcome.

2 System Overview

QUANTICO uses a common framework for both monolingual and cross-language scenarios, with different workflow settings only for the translation component and different configurations of the extraction component for each type of question (definition or factoid).

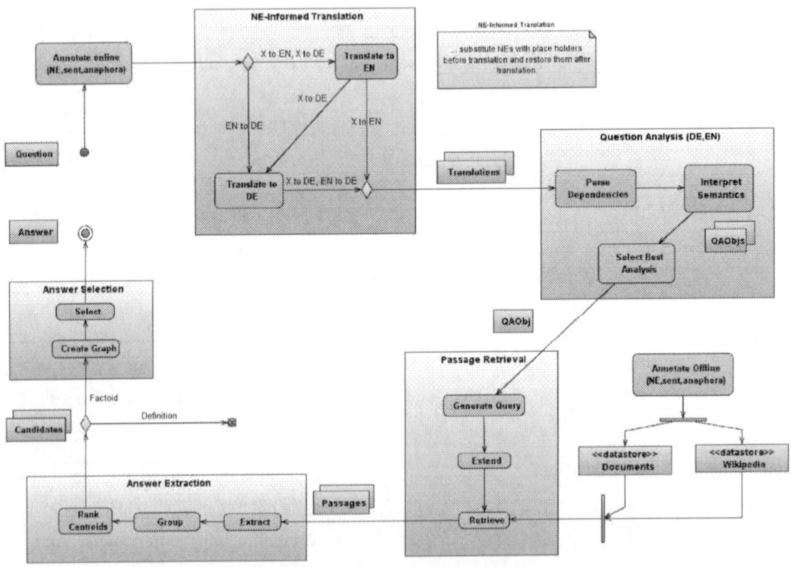

Fig. 1. System Architecture

Every question is translated into the target language resulting in a set of possible translations, which are individually interpreted. The outcome of the question analysis is ranked according to linguistic well-formedness and its completeness with respect to the query information (*q-type*, *q-focus*, *a–type*) and the best alternative is considered for further processing. Relevant passages are retrieved and possible answer candidates are extracted and ranked based on their redundancy. Finally, the best candidate is chosen based on a distance metric of the question's keywords and potential candidates.

The system is using online translation services[1] (AltaVista, FreeTranslation and VoilaTranslation) for crossing the language barrier from the source language of the question to the target language of the document collection.

Regarding the component configurations for each type of question (definition or factoid) the difference is to be noted only in the *Passage Retrieval* and *Answer Extraction* components. While the *Retrieve* process for the factoid questions builds on classic Information Retrieval methods, for definition questions it is merely a look-up procedure in a repository of offline extracted syntactic structures such as appositions, chunks and abbreviation-extension pairs. For the *Answer Extraction* the distinction consists in different methods of computing the clusters of candidate answers: for factoid question, where the candidates are usually named entities or chunks, is based on co-reference (*John ~ John Doe*) and stop-word removal (*of death ~ death*), while for definition questions, where candidates can vary from chunks to whole sentences, is based on topic similarity (*Italian designer ~ the designer of a new clothes collection*).

3 Component Description

3.1 NE-Informed Translation

Since named entities can pose some problems in translation, especially proper names, by being translated when they should not be, the translation component has been developed with a substitution module that replaces some types of named entities with place holders before translating the question. The process is being reversed after translation, resulting in more accurate results. The outcome of this module is highly dependent on the accuracy of the named entity (NE) recognizer, since an inaccurate mark-up of the NEs might prevent from translating semantically relevant information.

3.2 Question Analysis

In the context of a QA system we interpret the result of a NL question analysis as a *declarative description of search strategy and control information*, see [5]. Consider, for example, the NL question result for the question *"In welcher Stadt fanden 2002 die olympischen Winterspile statt?" (The Olympic winter games took place 2002 in which town?)*, where the value of the tag *q-type* represents the answer control strategy, *q-focus* and *q-scope* additional constraints for the search space:

```
<QA-control>
  <Q-FOCUS>Stadt</Q-FOCUS>
  <Q-SCOPE>stattfind_winter#spiel</Q-SCOPE>
  <Q-TYPE restriction="TEMP">C-COMPLETION</Q-TYPE>
  <A-TYPE type="atomic">LOCATION</A-TYPE>
</QA-control>
```

[1] http://babelfish.altavista.com, http:// ets.freetranslation.com, http:// trans.voila.fr

Parts of the information can already be determined on basis of local lexico-syntactic criteria (e.g., for the *wh*-phrase *where* we can simply infer that the expected answer type is *location*). However, in most cases we have to consider larger syntactic units in combination with the information extracted from external knowledge sources. For example for a definition question like "*What is a battery?*" we have to combine the syntactic and type information from the verb and the relevant NP in order to distinguish it from a description question like "*What is the name of the German Chancellor?*" We are doing this by following a two-step parsing schema:

- first a full syntactic analysis is performed using the robust parser SMES [4]
- second a question-specific semantic analysis takes place.

During the second step, the values for the question tags *a-type, q-type, q-focus* and *q-scope* are determined on basis of syntactic constraints applied on the dependency analysis of relevant NP and VP phrases and by taking into account information from two small knowledge bases. They basically perform a mapping from linguistic entities to values of the questions tags, e.g., trigger phrases like *name_of, type_of, abbreviation_of* or a mapping from lexical elements to expected answer types, like *town, person, and president*. For German, we additionally perform a *soft retrieval match* to the knowledge bases taking into account online compound analysis and string similarity tests. For example, assuming the lexical mapping *Stadt* → *LOCATION* for the lexeme *town*, then automatically we will also map the nominal compounds *Hauptstadt* (capital) and *Großstadt* (large city) to *LOCATION*.

3.3 Translation Services and Alignment

We have used two different methods for answering questions asked in a language different from the one of the answer-bearing documents. Both employ online translation services for crossing the language barrier, but at different processing steps, i.e. before and after formalizing the user information need into a *QAObj*.

The *a priori–method* translates the question string in an earlier step, resulting in several automatic translated strings, of which the best one is analyzed by the *Question Analysis* component and passed on to the *Passage Retrieval* component. This is the strategy we use in an English–German cross-lingual setting. To be more precise: the English source question is translated into several alternative German questions using online MT services. Each German question is then parsed with SMES. The resulting query object is then weighted according to its linguistic well–formedness and its completeness with respect to the query information (*q-type, q-focus, a-type*). The assumption behind this weighting scheme is that "a translated string is of greater utility, if its linguistic analysis is more complete or appropriate."

The *a posteriori–method* translates the formalized result of the *Query Analysis* component by using the question translations, a language modeling tool and a word alignment tool for creating a mapping of the formal information need from the source language into the target language. Translations returned by the on-line MT systems are first ranked according to a language model and those with a satisfactory degree of resemblance to a natural language utterance (i.e. linguistically well-formedness), given by a threshold on the language model ranking, are aligned based on several methods: MRD (machine readable dictionaries), statistical part-of-speech taggers and string similarity measures (dice coefficient, the lowest common substring ratio).

3.4 Passage Retrieval

The preemptive offline document annotation refers to the process of annotating the document collections with information that might be valuable during the retrieval process by increasing the accuracy of the hit list. Since the expected answer type for factoid questions is usually a named entity type, annotating the documents with named entities provides for an additional indexation unit that might help to scale down the range of retrieved passages only to those containing the searched answer type. The same practice applies for definition questions leveraging the fact that some structural linguistic patterns (appositions, abbreviation-extension pairs) are used with explanatory and descriptive purpose. Extracting these patterns in advance and looking up the definition term among them might return more accurate results.

The *Generate Query* process mediates between the question analysis result *QAObj* (*a-type*, *q-focus*, keywords) and the search engine (for factoid questions) or the repository of syntactic structures (for definition questions) serving the retrieval component with information units (passages). The *Generate Query* process builds on an abstract description of the processing method for every type of question to accordingly generate the *IRQuery* to make use of the advanced indexation units. For example given the question "*What is the capital of Germany?*", since named entities were annotated during the offline annotation and used as indexing units, the *Query Generator* adapts the *IRQuery* so as to restrict the search only to those passages having at least two locations: one as the possible answer (*Berlin*) and the other as the question's keyword (*Germany*), as the following example shows:

+capital^4 +Germany +neTypes:LOCATION +LOCATION:2.

It is often the case that the question has a semantic similarity with the passages containing the answer, but no lexical overlap. For example, for a question like "*Who is the French prime-minister?*", passages containing "*prime-minister X of France*", "*prime-minister X ... the Frenchman*" and "*the French leader of the government*" might be relevant for extracting the right answer. The *Extend* process accounts for bridging this gap at the lexical level, either through look-up of unambiguous resources or as a side-effect of the translation and alignment process (see [6]).

3.5 Answer Extraction

The *Answer Extraction* component is based on the assumption that the redundancy of information is a good indicator for its suitability. Based on the control information supplied by the *Analyse* component (*q-type*), different extraction strategies are being triggered (noun phrases, named entities, definitions) and even refined according to the *a-type* (definition as sentence in case of an OBJECT, definition as complex noun phrase in case of a PERSON).

Whereas the *Extract* process for definition questions is straightforward for cases in which the offline annotation repository lookup was successful, in other cases it implies an online extraction of those passages only that might bear a resemblance to a definition. The extraction of these passages is attained by matching them against a lexico-syntactic pattern of the form:

<Searched Concept> <definition verb> .+

whereby *<definition verb>* is being defined as a closed list of verbs like “is”, “means”, ”signify”, “stand for” and so on.

For factoid questions, with named entities or simple noun phrases as expected answer type, the *Group* (normalization) process consists in resolving cases of co-reference, while for definition questions, with complex phrases and sentences as possible answers, consists in finding out the focus of the explanatory sentence or the head of the considered phrase. Each cluster gets a weight assigned based solely on its size (definition questions) or using additional information like the average of the IR-scores and the document distribution for each of its members (factoid questions). For each of the clusters, the best scored member is considered to represent the cluster.

3.6 Answer Selection

Using the most representative sample (*centroid*) of the five best-weighed clusters of answer candidates, the *Answer Selection* component sorts out a list of top answers based on a distance metric defined over the answer's context. The context is first normalized by removing all functional words and then represented as a graph. The score of an answer is defined in terms of its proximity to the question concepts occurring in its context (*overlap weight*) and the proximity of those (*overlap cohesion*). The *overlap weight* is the average of proximity to all question concepts found to appear together with the answer, while the *overlap cohesion* is the minimum proximity between any two question concepts. Since we use proximity as a way of expressing semantic relatedness between concepts, we define it as:

$$proximity = \begin{cases} 1 - \exp(dist(C_i, C_j) - K), dist < K \\ 0.5, otherwise \end{cases}$$

where *exp* is the exponential function and $dist(C_i, C_j)$ is the shortest path between two concepts in the graph. The constant variable K highlights the concepts within a range of its value, making anything outside this range equal relevant/irrelevant.

Difference in vocabularies used by the question and by the documents can result in relevant documents not being matched and retrieved by the *Passage Retrieval*. This lowers the recall and therefore the performance of the whole system. In order to cope with this issue we have opted for using external general purpose lexical resources that provide semantically related concepts. For this purpose we employ the Wehrle-Eggers thesaurus (the German counterpart of ROGET's thesaurus) for extending the conceptual coverage of the question keywords with synonyms and related words.

4 Evaluation Results

We participated in three tasks: DE2DE (German to German), EN2DE (English to German) and DE2EN (German to English), with one run submitted for each of the cross-language tasks and two runs for the monolingual one. The second monolingual run submitted (*dfki082dede*) was distinct in that the question concepts were expanded with appropriate synonyms from Wehrle-Eggers Thesaurus during Answer Selection. A detailed description of the achieved results can be seen in Table 1.

Table 1. System Performance - Details

Run ID	Right		W	X	U	F	D	T	L	NIL	
	#	%	#	#	#	%	%	%	%	#	%
dfki081dede$_M$	73	36.5	119	2	6	30.62	80	44.44	0	0	0
dfki082dede$_M$	74	37	120	2	4	31.25	80	33.33	0	0	0
dfki081ende$_C$	29	14.5	164	2	5	10	43.3	0	0	0	0
dfki081deen$_C$	28	14	164	5	3	6.25	60	8.33	0	0	0

A preliminary error analysis of the results uncovered three weak places in our system:

- for the cross-language scenarios two of three translation services did not deliver any translations at all, so that we had to consider only one translation for further processing and in some cases no translation whatsoever,
- the use of named entities types during the unit retrieval improved the precision of the retrieval and assumes a high accuracy of the NE annotator at the document level; failure to correctly annotate the entities in the documents automatically brings along a lower recall during retrieval, that propagates on to the final results,
- using frequency as an indicator of answer suitability and distance among query keywords and answer candidates for selecting the right ones is a premise that does not hold when dealing with expected answer types other than named entities.

Translation Services

Failure to correctly translate the question has critical results when the information being erred on represents the focus or belongs to the scope of the question. Following are several examples of miss-translations that resulted in incorrect IR-queries generation and therefore wrong answers:

“states” → “Zustände, Staate” vs. „Bundesländer“
„Pointer Stick“ → „Zeigerstock“ vs. „Pointer Stick“
„Mt.“ (Mount) → „Millitorr“ vs. „Mt.“

Named Entity

The named entity tool used (LingPipe [8]), a statistical entity extractor, has a very good coverage and precision on annotating the documents, where lots of context data are available, but its performance drops when annotating short questions. Since our *Query Generator* uses named entities as mandatory items to restrain the amount of relevant passages retrieved, failure to consistently annotate entities on both question and document results in unusable units of information and therefore wrong answers.

Expected Answer Type

Our assumption of frequency being a good indicator for answer suitability does not hold for those cases when the expected answer type is an object. In such cases, our system extracts all nouns and noun phrases as possible answers and both experiments and results have shown a lot of noise being introduced this way. Not even the method of selecting the correct answer by considering its distance to the query keywords is efficient anymore, because of the fair amount of nouns and noun phrases targeted.

5 Conclusions

We have presented a framework for both monolingual and cross-lingual question answering for German/English factoid and definition questions. Based on a thorough analysis of the question, different strategies are considered and alternative work-flows and components are triggered depending on the question type.

Intuitive assumptions regarding the unit of retrieval granularity (i.e. sentence level) and the overlap of lexical information between the question and the relevant units have lead to promising results in the CLEF evaluation campaign, though the error analysis revealed some cases for which these premises do not hold.

Acknowledgments

The work presented in this paper was partly supported by the European Commission through the QALL-ME project (contract FP6-33860) and by a research grant from BMBF to the project HyLaP (FKZ: 01IWF02).

References

1. Bowden, M., Olteanu, M., Suriyentrakorn, P., Clark, J., Moldovan, D.: LCC's PowerAnswer at QA@CLEF 2006. In: Peters, C., Clough, P., Gey, F.C., Karlgren, J., Magnini, B., Oard, D.W., de Rijke, M., Stempfhuber, M. (eds.) CLEF 2006. LNCS, vol. 4730, pp. 310–317. Springer, Heidelberg (2007)
2. Echihabi, A., Oard, D.W., Marcu, D., Hermjakob, U.: Cross-Language QA at the USC Information Sciences Institute. In: Peters, C., Gonzalo, J., Braschler, M., Kluck, M. (eds.) CLEF 2003. LNCS, vol. 3237, pp. 514–522. Springer, Heidelberg (2004)
3. Lita, L.V., Rogati, M., Barbonell, J.: Cross Lingual QA: Modular Baseline. In: Peters, C., Gonzalo, J., Braschler, M., Kluck, M. (eds.) CLEF 2003. LNCS, vol. 3237, pp. 535–540. Springer, Heidelberg (2004)
4. Neumann, G., Piskorski, J.: A shallow text processing core engine. Computational Intelligence 18(3), 451–476 (2002)
5. Neumann, G., Sacaleanu, B.: Experiments on Robust NL Question Interpretation and Multi-layered Document Annotation for a Cross-Language Question/Answering System. In: Peters, C., Clough, P., Gonzalo, J., Jones, G.J.F., Kluck, M., Magnini, B. (eds.) CLEF 2004. LNCS, vol. 3491, pp. 411–422. Springer, Heidelberg (2005)

6. Sacaleanu, B., Neumann, G.: Cross-Cutting Aspects of Cross-Language Question Answering Systems. In: Proceedings of the EACL workshop on Multilingual Question Answering - MLQA 2006 (2006)
7. Sutcliffe, R., Gabbay, I., O'Gorman, A.: Cross-Language French-English QA using the DLT System at CLEF 2003. In: Peters, C., Gonzalo, J., Braschler, M., Kluck, M. (eds.) CLEF 2003. LNCS, vol. 3237, pp. 572–580. Springer, Heidelberg (2004)
8. Alias-i.2006. LingPipe1.7, http://www.alias-i.com/lingpipe

Integrating Logic Forms and Anaphora Resolution in the AliQAn System

Rafael Muñoz-Terol, Marcel Puchol-Blasco, María Pardiño, José Manuel Gómez, Sandra Roger, Katia Vila, Antonio Ferrández, Jesús Peral, and Patricio Martínez-Barco

University of Alicante, Department of Software and Computing Systems
San Vicente del Raspeig Road, 03690 Alicante, Spain
{rafamt,marcel,maria,jmgomez,sroger,
kvila,antonio,jperal,patricio}@dlsi.ua.es
http://www.dlsi.ua.es

Abstract. This paper deals with the AliQAn QA system in the multilingual (English - Spanish) task. It highlights the translation module of the QA system by applying two methods: the first one based on logic forms, and the other on machine translation techniques. Moreover, the system is able to solve the anaphora resolution problem by applying linguistic techniques. According to the results, machine translation techniques are a bit better than techniques based on logic forms in the performance of the question translation.

1 Introduction

The AliQAn system [1] applies two different methods for the translation of questions from one language to another. The first one consists in applying natural language processing techniques based on formal representation of questions, by using logic forms, and the second one, consists in using machine translation techniques. So, the main goal is to compare both methods of question translation. This fact implies that these two different translation methods have compared through the application of the QA process to both question translation set. A Spanish rule-based anaphora resolution system has also been developed to solve the problem of questions containing anaphoric expressions.

2 Description of the System

2.1 English-Spanish Translation Based on Logic Forms and Lexical Resources

Question translation from English into Spanish is performed by inferring the logic form of questions and using lexical resources to translate the logic form predicates. The technique applied to infer the logic forms of the questions is the one developed by [2]. Consequently, this translation technique is performed as follows:

C. Peters et al. (Eds.): CLEF 2008, LNCS 5706, pp. 438–441, 2009.

Table 1. Applied English-Spanish contrastive grammar rules

Rule Id.	English Structure	Spanish Translation
1	$JJ + NN$	$TR\ (NN) + TR\ (JJ)$
2	$JJ_1 + JJ_2 + NN$	$TR\ (JJ_1) + TR\ (NN) + \mathrm{TR}(JJ_2)$
3	$NN_1 + NN_2$	$TR\ (NN_1) + TR\ (NN_2)$
4	$NN + NNC$	$TR\ (NNC) + \text{"de"} + TR\ (NN)$
5	$JJ + NN_1 + NN_2$	$TR\ (NN_2) + TR\ (JJ) + \text{"de"} + TR\ (NN_1)$

- The predicates of types noun (NN) or verb (VB) are translated using the EuroWordNet [3] lexical resource. The connection between the synsets of the English and Spanish WordNets is performed in a similar way as treated by [4] using the Inter-Lingual-Index (ILI). Each one of the synsets that are mapped from English into Spanish towards ILI contains a set of synonym words. The process consists in counting the occurrences of the different synonym words that appear in the mapping process and, finally, the synonym word with the highest number of occurrences is chosen as the predicate translation.
- Predicates of type adjective (JJ), adverb (RB) or preposition (IN), and predicates treated in the previous step that are not translated by EuroWordNet are translated applying the English-Spanish Babylon dictionary[1]. As in the previous step, the dictionary can return a set of different translations grouped in synsets (different from WordNet synsets). Thus, the processing consists in counting the occurrences of the different translations that the dictionary returns and, finally, the translation with the highest number of occurrences is chosen as the predicate translation.
- Finally, the remaining predicates and the ones that were not translated in the two previous steps are definitely translated by using the Google Translation Toolkit[2].

Once the predicates of the logic form are translated according to the previously described rules, the last translation task consists in translating the question as a result of applying some English-Spanish contrastive grammar rules to the sequence of predicate translations of the logic form. The applied English-Spanish contrastive grammar rules are based on the ones derived from the previous study developed by [5] and [6] and are detailed in Table 1, where TR means translation.

Finally, the translation performed for the rest of predicates in the logic form, whose logic structure does not match with these English-Spanish contrastive grammar rules, consists in the concatenation of the sequence of translations of these predicates.

2.2 Anaphora Resolution

Since 2007 anaphora resolution has been part of the QA@CLEF challenge. Therefore, questions are grouped in topics. In these topics, the first question is

[1] http://www.babylon.com
[2] http://www.google.com/translate_t

anaphora-free, but the other questions may require information from data contained in the first question, or in the first answer. From our analysis of Spanish examples used last year, we have discovered three types of possible anaphora: anaphoric pronouns, definite descriptions, and zero anaphora. These types of anaphora have been analyzed and are the basis for our system, due to the fact that three different modules have been created for each type of anaphora. Our approach for anaphora resolution is rule-based. In short, it is based on the papers referenced in [7].

For all anaphora resolution modules, and for each topic group, noun phrases are extracted from the first question-answer pair as possible antecedent. Then, when we have an anaphoric pronoun, the possible anaphora is compared in gender and number with each possible antecedent. In definite descriptions and zero anaphora, for each noun phrase, a google search is launched joining the possible antecedent and the main words of the anaphora (nouns with semantic content), or main words contained in the noun phrases, in case of zero anaphora. Later, for each noun phrase in the possible anaphora, the relations between it and the main words contained in the antecedents are extracted using MultiWordNet[3]. In all these cases, a specific weight is assigned to each possible anaphora resolution module. Later, those weights are sorted, and the best case is selected as anaphora resolution for the related question. It is important to mention that if the noun phrase is in the answer, a greater weight is assigned to it, due to the analysis done on the basis of a corpus constructed of CLEF 2007 data.

3 Results and Conclusions

Table 2 shows the results obtained by the AliQAn system over the 200 questions treated in this task. The acronyms used in this table means: R (number of right answers), W (wrong answers), X (inexact answers), U (unsupported answers), MRR (Mean Reciprocal Rank), and CWS (Confident Weighted Score).

The scores obtained when applying the machine translation techniques are a bit better than the ones obtained when applying the techniques based on logic forms. This can be due to the fact that the use of logic forms is a good method to perform the language-independent knowledge representation, but this method must be improved to perform the translation of sentences from one language to another. For future work, the next research goal will be to improve the logic form processing methods in the translation process.

We have had some problems with the anaphora resolution system, because it is a rule-based system and it needs well-formed sentences to work. Due to the fact that translations offered by the system are not well-formed, in most cases, the accuracy obtained by the anaphora resolution system has decreased considerably, arriving at around 40% accuracy.

Finally, analyzing the results obtained by the English-Spanish QA systems [8], the best score of accuracy in this track was 42,5% and the average accuracy over

[3] `http://multiwordnet.itc.it/online/multiwordnet.php`

Table 2. Results obtained by our system at English-Spanish QA task 2008

Run_ID	#R	#W	#X	#U	Accuracy (%)	CWS	MRR
Run 1 (machine translation)	25	173	0	2	12.5	0.01114	0.17797
Run 2 (logic forms)	18	176	3	3	9.0	0.00626	0.11499

all the runs was 18%. Comparing all these scores with our best score (12,5%), we conclude that our AliQAn system must be improved.

Acknowledgments

This paper has been partially supported by the Spanish government, project TIN-2006-15265-C06-01 and project GV06/028, and by the framework of the project QALL-ME, which is a 6th Framework Research Programme of the European Union (EU), contract number: FP6-IST-033860, by the Spanish Government through the research grant AP2007-03072, by the Generalitat Valenciana through the research grant BFPI06/182 and by the University of Comahue under the project 04/E062.

References

1. Roger, S., Ferrández, S., Ferrández, A., Peral, J., Llopis, F., Aguilar, A., Tomás, D.: Aliqan, spanish qa system at clef-2005. In: Peters, C., Gey, F.C., Gonzalo, J., Müller, H., Jones, G.J.F., Kluck, M., Magnini, B., de Rijke, M., Giampiccolo, D. (eds.) CLEF 2005. LNCS, vol. 4022, pp. 457–466. Springer, Heidelberg (2006)
2. Terol, R., Martínez-Barco, P., Palomar, M.: A knowledge based method for the medical question answering problem. Computers in Biology and Medicine 37, 1511–1521 (2007)
3. Vossen, P.: EuroWordNet General Document. Part A. Final Document. EuroWordNet (LE2-4003, LE4-8328) (2002)
4. Ferrández, S., Ferrández, A., Roger, S., López-Moreno, P., Peral, J.: Brili, an english-spanish question answering system. In: Proceedings of the International Multiconference on Computer Science and Information Technology, pp. 23–29 (2006)
5. Fernández, F., Montero-Fleta, B.: La premodificación nominal en el ámbito de la informática. Estudio contrastivo inglés-español. Universidad de Valencia (2003)
6. Martínez-Vázquez, M.: Gramática contrastiva inglés-español. Servicio de publicaciones de la Universidad de Huelva (1996)
7. Mitkov, R.: Anaphora Resolution. Longman, London (2002)
8. Forner, P., Peñas, A.P., Aguirre, E., Alegria, I., Forascu, C., Moreau, N., Osenova, P., Prokopidis, P., Rocha, P., Sacaleanu, B., Sutcliffe, R., Sang, E.: Overview of the clef 2008 multilingual question answering track. In: Working Notes for the CLEF 2008 Workshop (2008)

Some Experiments in Question Answering with a Disambiguated Document Collection

Davide Buscaldi and Paolo Rosso

Natural Language Engineering Lab.,
Dpto. de Sistemas Informáticos y Computación (DSIC),
Universidad Politécnica de Valencia, Spain,
{dbuscaldi,prosso}@dsic.upv.es

Abstract. This paper describes our approach to the Question Answering - Word Sense Disambiguation task. This task consists in carrying out Question Answering over a disambiguated document collection. In our approach, disambiguated documents are used to improve the accuracy of the retrieval phase. In order to do this, we added a WordNet-expanded index to the document collection. The expanded index contains synonyms, hypernyms and holonyms of the words already in the documents. Question words are searched for in both the expanded WordNet index and the default index. The obtained results show that the system that exploited disambiguation obtained better precision than the non-WSD one.

1 Introduction

The evaluation of the impact of Word Sense Disambiguation (WSD) on Information Retrieval (IR) has been the object of many research efforts in the last decade [4,7,6]. One of the objectives of the QA-WSD and CLIR-WSD tasks at CLEF[1] 2008 was to attempt to find new evidence in favor of the utility of WSD in IR or not, by providing partecipants with disambiguated collections to perform tests on. The QA-WSD task put its focus on the Question Answering task, that can be seen as a specialized kind of IR.

The available collections were the one disambiguated using the method of the University of Basque Country (UBC) [1], and the one disambiguated by the method of the National University of Singapore (NUS) [8]. This is the first time that disambiguated collections of this size have been developed and released for a large-scale evaluation.

Our system is constituted by a modified version of the QUASAR system described in [2]. For this task the search engine (JIRS) has been replaced by Lucene[2], which can work with multiple indices. This capability was needed in order to put in different indices the terms extracted from the documents and the terms derived from WordNet [5]. The method we developed to take advantage of the disambiguated collection is similar to the "Index Term Expansion" method

[1] http://www.clef-campaign.org
[2] http://lucene.apache.org

C. Peters et al. (Eds.): CLEF 2008, LNCS 5706, pp. 442–447, 2009.

described in [3], in which the geographical terms in documents were expanded with their WordNet holonyms. In this case we added to the WordNet index all the hypernyms, holonyms, and synonyms of the disambiguated words in the document collection.

In the following section, we describe the system. In Section 3 we discuss the experiments carried out and the obtained results. Finally, in Section 4 we draw some conclusions.

2 WordNet-Based Index Expansion

Previous to the indexing phase, all documents are split into sentences. These are used later to form the passages. In the indexing phase, we create two indices: the first one (*text*) contains all the terms of the sentence; the second one (expanded index, or *wn* index) contains all the synonyms of the disambiguated words; in the case of nouns and verbs, it contains also their hypernyms. In the case of nouns, their holonyms (if present) are also added to the index. For instance, let us consider the following sentence from document `GH951115-000080-03`:

> Splitting the left from the Labour Party would weaken the battle for progressive policies inside the Labour Party.

The underlined words are those that have been disambiguated in the collection. For these words we can find their synonyms and related concepts in WordNet, as listed in Table 1.

Table 1. Expansion of terms of the example sentence. NA : not available (the relationship is not defined for the Part-Of-Speech of the related word).

lemma	ass. sense	synonyms	hypernyms	holonyms
split	4	separate part	move	NA
left	1	–	position place	–
Labour Party	2	labor party	political party party	–
weaken	1	–	change alter	NA
battle	1	conflict fight engagement	military action action	war warfare
progressive	2	reformist	NA	NA
policy	2	–	argumentation logical argument line of reasoning line	–

Therefore, the *wn* index will contain the following terms: *separate, part, move, position, place, labor party, political party, party, change, alter, conflict, fight, engagement, war, warfare, military action, action, reformist, argumentation, logical argument, line of reasoning, line.*

During the search phase, the *text* and *wn* indices are both searched for question terms. The top 20 sentences are returned for each question. The passages are built from these sentences, by appending them the previous and following sentences in the collection. For instance, if the above example was a retrieved sentence, the resulting passage would be composed by the sentences:

- `GH951115-000080-2` : "The real question is how these policies are best defeated and how the great mass of Labour voters can be won to see the need for a socialist alternative."
- `GH951115-000080-3` : "Splitting the left from the Labour Party would weaken the battle for progressive policies inside the Labour Party."
- `GH951115-000080-4` : "It would also make it easier for Tony Blair to cut the crucial links that remain with the trade-union movement."

Figure 1 shows the first 5 sentences returned for the question "What is the political party of Tony Blair?" using only the *text* index; in Figure 2 we show the first 5 sentences returned using the *wn* index.

It can be noted that sentences retrieved with the expanded WordNet index are shorter, because the keyword *political* was found only in the expanded index and not in the text.

Our system had some limitations on the type of questions it could answer. The reason is that the base system was developed for the 2006 edition of CLEF QA, which included guidelines that were different from the ones adopted in CLEF 2008. In 2006, questions did not include questions with references to a previous question (anaphora). Therefore, our system cannot solve anaphoras. We refer the reader to the description in [2] for a detailed description of the base system.

The Labour Party , under Tony Blair , is poised to achieve political power for the first time in 16 years	GH950821-000120-5
No Headline Present Political peace : A truce is agreed on the Westminster front as party leaders John Major , Paddy Ashdown , and Tony Blair celebrate VJ-Day	GH950821-000164-0
Blair puts the family centre stage LABOUR leader Tony Blair , in a further move to occupy the political centre ground , yesterday staked his claim for Labour to be the ' ' party of the family	GH950330-000184-0
Blair should beware I AM mystified by your political editor ' s comment about Tony Blair ' ' dispatching the Bennite Marxist left ' ' and rescuing the Labour Party from the ' ' false historical perspective ' ' of a ' ' Marxist intellectual analysis ' ' (March 23	GH950327-000057-0
Blair wrestles with age-old links TRADE union leaders of all political persuasions have confronted Labour leader Tony Blair with a demand that the unions retain a 50 % voice in the party they created	GH951002-000214-1

Fig. 1. Top 5 sentences retrieved with the standard Lucene search engine

The Labour Party , under Tony Blair , is poised to achieve political power for the first time in 16 years	GH950821-000120-5
The Labour Party has been set a simple test by Tony Blair	GH950310-000026-16
No Headline Present Political peace : A truce is agreed on the Westminster front as party leaders John Major , Paddy Ashdown , and Tony Blair celebrate VJ-Day	GH950821-000164-0
' The investigation is understood to have the full support of Labour Party leader Tony Blair	GH950227-000161-5
On the eve of the Labour party conference , there were also sharp words for Tony Blair	GH951002-000228-9

Fig. 2. Top 5 sentences retrieved with the WordNet extended index

3 Experiments

The participation at CLEF 2008 consisted in submitting two mandatory runs, one with the basic system (labeled as "*no WSD*" in Table 2) that does not use semantic information, and one with the system described above (*WSD-NUS* in Table 2), using as collection the NUS-disambiguated collection. Of the 200 questions in the test set, only 49 had an answer in the disambiguated collection (the other questions had their answers in Wikipedia, which was not featured for the QA-WSD track), according to the organisers. However, we manually checked the data and found that it was possible to find an answer to 25 of the Wikipedia questions, bringing the number of questions with an answer in the collection to 74.

In Table 2 we show the results obtained by the two mandatory runs and another run that used the UBC-disambiguated collection (*WSD-UBC*). The results of this last run are not official (i.e. we evaluated the run ourselves instead of the track organizers).

The table shows that, apart from the fact that the complete question set was not suitable for the evaluation, the runs that were carried out on the disambiguated collections obtained worse results. In order to understand the reason of this, we carried out an analysis of the average number of passages that contained the answer for each of the questions. Of the 49 questions, only three answers were present in more than nine passages. The average number of passages containing the answer for each question in the remaining 46 questions is 2.04. This number justifies the small differences between the WSD based system and the base one (the systems retrieve the same sets of relevant passages, independently from the method used).

Therefore, we carried out some additional experiments with the sets of questions from CLEF 2005 and 2006, in order to check what would be the results with questions that better fit the used collections. The questions were the same of the English-Spanish bilingual test sets, but in this case we employed them in a monolingual environment, with an English target collection. In Table 3 we show the results obtained with these questions. The evaluation was carried out taking into account the 2005 and 2006 guidelines.

From the results shown in Table 3 it can be observed that the average results are comparable to the ones obtained with the 49 questions of the CLEF 2008 test set: this confirms the fact that the whole question set included too many questions that the system could not answer. The results on the 2005 and 2006

Table 2. Results obtained with the three runs over the 49 questions that had (officially) an answer in the collection and all questions

	49 Questions				All Questions			
run ID	R	X	U	Accuracy	R	X	U	Accuracy
no WSD	8	0	0	16.32%	10	0	0	5.00%
WSD-NUS	7	0	0	14.29%	8	0	1	4.00%
WSD-UBC	6	0	0	12.24%	7	0	1	3.50%

Table 3. Results obtained with the CLEF QA 2005 and 2006 question sets, with the base system, the WSD-based system and the UBC collection, the WSD based system and the NUS collection

	CLEF 2005 questions				CLEF 2006 questions			
run ID	R	X	U	Accuracy	R	X	U	Accuracy
no WSD	30	6	0	15.00%	28	2	1	14.00%
WSD-NUS	36	5	0	18.00%	29	1	2	14.50%
WSD-UBC	37	5	1	18.50%	31	1	3	15.50%

test sets show also that with questions that present a higher redundancy of the answer the use of the disambiguated collection allowed to obtain a higher accuracy of the WSD system with respect to the non-WSD one. The answer redundancy for the 2005 collection was 29.28 answers per question, while in the 2006 collection was of 25.71 answers per question. There was no significant difference between the use of the NUS and the UBC document collections. Note that NIL questions were excluded from the computation of the results, since they were not taken into account in the evaluation at CLEF QA-WSD 2008.

4 Conclusions

The obtained results do not provide a decisive argument in favour of the utility of Word Sense Disambiguation in Information Retrieval. However, it is noteworthy that the WSD-based QA system performed better than the non-WSD one under two conditions: higher answer redundancy and the use of the disambiguated collection. We did not observe any significant difference on the smaller question set between the WordNet enhanced method and the base system. We believe that most errors were due to the poor performance of the QA system and not to the retrieval process. In the future, we will attempt to evaluate the impact of the use of the disambiguated collections only in passage retrieval, independently from the rest of the QA system.

Acknowledgements

We would like to thank the TIN2006-15265-C06-04 and the AECI-PCI A010317/07 research projects for partially supporting this work.

References

1. Agirre, E., Lopez de Lacalle, O.: UBC-ALM: Combnining k-NN with SVD for WSD. In: Proceedings of the 4th International Workshop on Semantic Evaluations (SemEval 2007), Prague, Czech republic, pp. 341–345 (2007)
2. Buscaldi, D., Gómez, J.M., Rosso, P., Sanchis, E.: N-gram vs. keyword-based passage retrieval for question answering. In: Peters, C., Clough, P., Gey, F.C., Karlgren, J., Magnini, B., Oard, D.W., de Rijke, M., Stempfhuber, M. (eds.) CLEF 2006. LNCS, vol. 4730, pp. 377–384. Springer, Heidelberg (2007)

3. Buscaldi, D., Rosso, P., Sanchis, E.: A wordnet-based indexing technique for geographical information retrieval. In: Peters, C., Clough, P., Gey, F.C., Karlgren, J., Magnini, B., Oard, D.W., de Rijke, M., Stempfhuber, M. (eds.) CLEF 2006. LNCS, vol. 4730, pp. 954–957. Springer, Heidelberg (2007)
4. Gonzalo, J., Verdejo, F., Chugur, I., Cigarrán, J.: Indexing with wordnet synsets can improve text retrieval. In: COLING/ACL 1998 workshop on the Usage of WordNet for NLP, Montreal, Canada, pp. 38–44 (1998)
5. Miller, G.A.: Wordnet: A lexical database for english. Communications of the ACM 38, 39–41 (1995)
6. Rosso, P., Ferretti, E., Vidal, V.: Text categorization and information retrieval using wordnet senses. In: 2nd Global WordNet Conference (GWC 2004), Brno, Czech Rep., pp. 299–304 (2004)
7. Sanderson, M.: Word Sense Disambiguation and Information Retrieval. PhD thesis, University of Glasgow, Glasgow, Scotland, UK (1996)
8. Chan, Y.S., Ng, H.T., Zhong, Z.: US-PT: Exploiting Parallel Texts for Word Sense Disambiguation in the English All-Words Tasks. In: Proceedings of the 4th International Workshop on Semantic Evaluations (SemEval 2007), Prague, Czech republic, pp. 253–256 (2007)

Answer Validation on English and Romanian Languages

Adrian Iftene[1] and Alexandra Balahur[1,2]

[1] UAIC: Faculty of Computer Science, "Alexandru Ioan Cuza" University, Romania
[2] University of Alicante, Department of Software and Computing Systems, Alicante, Spain
{adiftene,abalahur}@info.uaic.ro

Abstract. The present article describes the system built for the participation in the AVE 2008 track, stressing upon the new features added to the approach we had in the AVE 2007. The current version, while also based on the Textual Entailment system we built for the RTE-3 competition, adds and combines specific techniques used by Question Answering systems to improve answer classification. We outline the performance of this approach, presenting the high results obtained for both English and Romanian. Finally, we perform a critical analysis of the detected errors and propose the lines for future work.

1 Introduction

AVE[1] (Answer Validation Exercise) is a task in the QA@CLEF competition that evaluates subsystems assessing the correctness of the answers given by QA systems (Rodrigo et al, 2008), classifying them as SELECTED, VALIDATED or REJECTED.

This year, for our second participation in the AVE competition, we improved the system used last year and, additionally, introduced a question analysis part, which is specific to a Question Answering system. In this year's AVE competition we also participated with a system working in Romanian, using a Textual Entailment (TE) system working on Romanian (Iftene, Balahur-Dobrescu, 2007b). The latter is similar to the TE system working in English, with which we participated in the RTE 3 competition in 2007(Iftene, Balahur-Dobrescu, 2007a). For this reason, the present paper describes solely the AVE system working in English.

The following sections present the new functionalities that have been added to our English AVE system.

2 System Built for the AVE 2008 Competition

The main architecture of our AVE system is similar to the one used in the AVE 2007 competition (Iftene, Balahur-Dobrescu, 2008). The system is based on a Textual Entailment system we developed for the participation in the RTE-3 track in 2007. The steps executed by our AVE system are the following:

- Similarly to the approach we took in the AVE 2007 competition: (1) We build a pattern with variables for every question according to the question type; (2) Using

[1] AVE: http://nlp.uned.es/QA/ave/

C. Peters et al. (Eds.): CLEF 2008, LNCS 5706, pp. 448–451, 2009.

a pattern and all possible answers, we build a set of hypotheses for each of the questions: H_1, H_2, H_3 etc.; (3) We assign the justification snippet the role of text T and we run the TE system for all obtained pairs: (T_1, H_1), (T_2, H_2), (T_3, H_3), etc.
- Additionally, we perform the following new steps: (4) Identify the Answer Type (AT) for the answers; (5) Identify the question's Expected Answer Type (EAT). The two submitted runs are based on the following approaches:

1. In the first one we choose as correct answer for the current question the candidate from the hypothesis for which we obtained the greatest global fitness;
2. In the second one, we consider as correct answer for the current question the candidate with AT equal with EAT and for which we obtain the greatest global fitness.

The aim in determining the AT and the EAT is to eliminate the cases in which there are differences between these values. For example, in the case of question 13 in the test corpus (*What is the occupation of Richard Clayderman?*), since the expected answer type is JOB, seeking to identify the correct answer in the sub-set of answers of type JOB can obviously improve the probability to determine it.

In order to test the entailment relation between the possible hypotheses and the corresponding support snippets, we first transform the question into a statement with a variable of the expected answer type that will subsequently be replaced by each of the given answers to form the possible hypotheses. For question 13 in the corpus, the constructed statement is "*The occupation of Richard Clayderman is JOB.*", and the resulting hypotheses (with coinciding EAT and AT, as seen in Table 1) are "*The occupation of Richard Clayderman is pianist.*"; "*The occupation of Richard Clayderman is artist.*"; "*The occupation of Richard Clayderman is composer.*"; "*The occupation of Richard Clayderman is teachers.*".

The patterns used in the identification of the expected answer type (EAT) are specific to the following types of named entities: City, Count, Country, Date, Job, Measure, Location, Person, Organization, Year and Other. For this, we employ specific question answering techniques for question processing. Thus, for a question which starts with "When", "At which date", "In which year" etc. we consider the expected answer type as being of type "Date". Similar patterns are built for all cases.

Table 1: EAT and AT comparison

Pair	EAT	Answer	AT	Match score
13_1	JOB	Number	OTHER	0.25
13_2	JOB	teacher Qualifications	OTHER	0.25
13_3	JOB	Ways	OTHER	0.25
13_8	JOB	Pianist	JOB	1
13_11	JOB	Artist	JOB	1
13_12	JOB	Composer	JOB	1
13_13	JOB	teachers	JOB	1

For the identification of the answer type (AT), we use GATE[2] for the following types: Job, City, Country, Location, Person, Organization and we build specific patterns in order to identify the following types: Date, Measure, and Count. When an answer cannot be classified with GATE or with our patterns, it is considered with type Other. For question number 13, we correctly identify the EAT - which is JOB, and for all possible answers, we identify the AT as shown in the Table 1 (13_1 represent the answer 1 for question 13 with value "Number").

On the last column, we show the matching score between EAT and AT. Accordingly with this value, the last four answers will be preferred to the first three. In order to compute the match score value, we use a set of rules. The most important rules are:

Table 2. Rules for matching score calculation

Rule	Match score
AT = EAT	1
(EAT = "DEFINITION") and (AT = "OTHER")	1
EAT and AT are in the same class of entities: {City, Country, Region, Location} or {Year, Date} or {Count, Measure, Year}	0.5
(AT = "OTHER") or (EAT = "OTHER")	0.25
OTHERWISE	0

3 Results

We submit two runs on each of the languages (English and Romanian) according to the use or not of some system components. The systems are similar and only the external resources used by the TE system or by GATE are language-specific.

First run: is based on TE System output. The answers for which we have NE problems are considered as REJECTED, as in the system used for AVE 2007. Answers without NE problems are considered as VALIDATED and the answer with the highest global fitness is considered as SELECTED. If all answers contain NE problems, then all answers are considered REJECTED, except the answer with highest global fitness, which will be considered SELECTED.

Second run: in addition to the first run, we add the comparison between EAT and AT. In the cases where we have NE Problems, the answers are considered as REJECTED as well, and we also take into consideration if the matching score between EAT and AT is 0 (incompatible types). Of the remaining answers, if the matching score is not 0, then all answers are VALIDATED. For the identification of the SELECTED answer, we select the answers with the highest matching score (8, 11, 12) and the highest global fitness.

Our AVE 2008 systems obtained the following results: the best *qa_accuracy* (the number of correct SELECTED answers), for both Romanian and English systems, was 0.24 and the *estimated_qa_ performance* (estimated performance of a QA system

[2] GATE: http://www.gate.ac.uk/

that used for answers ranking the AVE system) was 0.24 for English and 0.25 for Romanian (Rodrigo et al., 2008). On English we obtained the highest *qa_accuracy*, equaled by the DFKI group, from seven participating groups. It is interesting to note that on Romanian, the 0.25 value is greater than the best value obtained by groups participating in the QA competition.

The incorrect classifications in our runs are regarding the *qa_rejected_accuracy* (the number of correct REJECTED questions), where results placed us fourth. The explanation is given by the fact that our AVE system tries to rank the answers in every situation and obtain the most probable answer for the current question, and does not use conditions for the identification of REJECT cases. The reason for this approach is that in the QA competition, the number of NIL questions is very low (8 out of 200) and therefore, we did not pay special attention to these cases. In the AVE competition, on the other hand, the number of rejected answers was so high because the test data was taken from answers given by QA participating systems and in many cases these answers were not correct.

4 Conclusions

In the AVE 2007 competition, we showed how the TE system we used in the RTE3 competition can successfully be employed as part of the AVE system to improve ranking between the possible answers, especially in the case of questions with answers of type Measure, Person, Location, Date and Organization.

This year, adding the question and answer type classification and the matching component, we showed how we improved, on the one hand, the correct classification of the answers, and on the other hand, the validation of more answers. Future work includes a more accurate identification of REJECTED answers, using thresholds estimated on the training data and changing the computation of match score between EAT and AT employing penalties instead of lower values.

References

Iftene, A., Balahur-Dobrescu, A.: Hypothesis Transformation and Semantic Variability Rules Used in Recognizing Textual Entailment. In: Proceedings of the ACL-PASCAL Workshop on Textual Entailment and Paraphrasing, Prague, Czech Republic, June 28-29, pp. 125–130 (2007a)

Iftene, A., Balahur-Dobrescu, A.: Improving a QA System for Romanian Using Textual Entailment. In: Proceedings of RANLP workshop "A Common Natural Language Processing Paradigm For Balkan Languages", Borovets, Bulgaria, September 26, pp. 7–14 (2007b)

Iftene, A., Balahur-Dobrescu, A.: UAIC Participation in AVE 2007. In: Peters, C., Jijkoun, V., Mandl, T., Müller, H., Oard, D.W., Peñas, A., Petras, V., Santos, D. (eds.) CLEF 2007. LNCS, vol. 5152, pp. 395–403. Springer, Heidelberg (2008)

Rodrigo, Á., Peñas, A., Verdejo, F.: Overview of the answer validation exercise 2008. In: Peters, C., et al. (eds.) CLEF 2008. LNCS, vol. 5706, pp. 296–313. Springer, Heidelberg (2008)

The Answer Validation System ProdicosAV Dedicated to French

Christine Jacquin, Laura Monceaux, and Emmanuel Desmontils

LINA, Laboratoire d'Informatique Nantes Atlantique
2 rue de la Houssinière
BP92208, 44322 NANTES Cedex 03
France
{christine.jacquin,laura.monceaux,emmanuel.desmontils}@univ-nantes.fr

Abstract. In this paper, we present the ProdicosAV answer validation system which was developed by the NLP team from the LINA institute. ProdicosAV system is based on the Prodicos System which participated two years ago in the Question Answering CLEF evaluation campaign for French. We firstly present the modifications made on Prodicos to improve it and to adapt it to a new kind of exercise. We present in details the ranking passage module and the temporal validator module. Secondly, the answer-validation module dedicated to the AVE task is presented. Finally, the evaluation is put forward to justify the results obtained.

1 Introduction

In this paper, we present the ProdicosAV system which was developed by the NLP team from the LINA institute and which participated in the Answer Validation Exercise for French. This system is based on the Prodicos System which participated two years ago in the Question Answering CLEF evaluation campaign for French. In this paper, we first present the modifications made on Prodicos QA system to improve it and to adapt it to a new kind of task. The main modification was made on the passage extraction module in order to transform it into a ranking passage module. In the same way, a new module dedicated to temporal validation was also developed. Secondly, the answer validation module devoted to the AVE task is presented. Its aim is to classify the proposed AVE answers as selected or rejected. Finally, the evaluation is put forward to justify the results obtained.

2 The Question Answering System Prodicos

In this section we will briefly describe the Prodicos system (more details are available in [3]). Prodicos system is divided into three parts (figure 1):

- The question analysis module: it extracts relevant features from questions that will make it possible to guide the passage extraction and the answer

C. Peters et al. (Eds.): CLEF 2008, LNCS 5706, pp. 452–459, 2009.

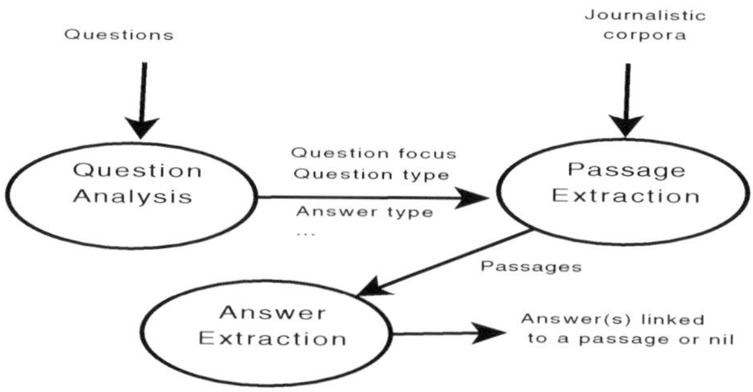

Fig. 1. The Prodicos System

search processes. These features are: question type, question focus, answer type and strategy. For example, let "*Qui est Abagelard de Paris ?*" (Who is Abagelard de Paris) be the question to be analysed; its analysis results are:
1. Question type: `WHO`
2. Answer type: `PERSON`, `ORGANIZATION`
3. Question Focus: *Abagelard de Paris*
4. Strategy: *named Entity*
5. Named Entity: "Abagelard de Paris"

- passage extraction module: for a particular request, this module provides a sorted list of passages which answer the question. The sort criterion is a confidence coefficient associated with each passage in the list. It is determined according to the number and the category of the question features which are found in passages.
- answer extraction module: it extracts the candidate answers from passages and ranks them. The question analysis step provides 4 different strategies to extract the answer (based on what kind of features the questions contain), namely a numerical entity based strategy, a named entity based strategy, an acronym definition based strategy and a pattern-based strategy.

3 Prodicos's System Modifications

The ProdicosAV system that is devoted to the answer validation task, is based on the Prodicos system. Our main idea was to operate the Question Answering System Prodicos on each question of the AVE task as well as on their related passages. Then, a decision module compared the results obtained by Prodicos with the AVE proposed answers and evaluated these last ones. Some modifications were made on Prodicos System in order to take into account AVE task particularities.

Firstly, the previous passage extraction module was adapted to a passage ranking module. Indeed, in the AVE campaign, the process consists in ranking a passage according to its ability to contain a correct answer. But in a QA campaign the aim is to provide passages which probably contain the answer. Moreover, we made an improvement to this module by using a density measure. Secondly, we also integrated a new process into Prodicos which validates (or not) a passage according to a date criterion. It determines the temporal adequacy between a question and a candidate passage.

3.1 Ranking Passage Module

The role of this module is to rank passages according to their ability to contain an answer to a question. It is based on the passage extraction module of the Prodicos system. But this one only uses the presence of certain question terms in passages in order to select or reject them. It never takes into account the distance between question features, which sometimes leads to incorrect results. [5] made a quantitative evaluation of passage retrieval algorithms for question answering and they showed that systems based on density measure perform better than the ones based on other techniques. The density measure approach rests on a scoring function based on the measure of query terms proximity to each other.

For each question, a particular request was built according to the data generated by the question analysis step. The request was composed of a combination of elements such as question focus, named entities, principal verbs, common nouns, adjectives, dates and other numerical entities. These elements were also weighted according to their importance for determining the potential answer. The weight of each element depends on the question types. For example for a question type equal to "date", the coefficient associated with a date element is greater than the one linked with a principal verb element. The density measure used was based on the one from [4] but some adjustments were made.

For all query passages, let m be the number of query terms belonging to the passage and let k be the number of words belonging to the passage. $wgt(qw_i)$ is the weight of query word i, $wgt(dw_i)$ is the weight of query word i with which document word j matches and $dist(j,questionFocus)$ is the distance beetween document word j and the question focus $questionFocus$.

$$score_{passage} = sum_{i=1}^{m} wgt(qw_i) + \left(\frac{\sum_{j=1}^{k-1} \frac{wgt(dw_j)+wgt(questionFocus)}{\alpha * dist(j,questionFocus)^2}}{k-1} * m\right) \quad (1)$$

The main adjustments made in comparaison with [4] was the introduction of the question focus in the calculus and the consideration of the whole passage instead of only selected sentences linked by anaphora. In fact, [4] did not consider the question focus as an important element but as an ordinary element. In our application, the density was computed by mainly taking into account the distance between the question focus and the other query terms in the passage. It must be noted that by experiment the alpha value was assigned to 0.5.

An other difference is that [4] computed the $score_{passage}$ coefficient for all sentences and they only gathered two sentences into a same passage if, for example, the second sentence contained an anaphora of a noun belonging to the first sentence. The aim of their system was to provide a passage which probably contained the answer. On the other hand, our application is of a different nature. The density measure was used in order to rank the given passages according to their ability to contain an answer to a question. So, we did not work at the sentence level but at the passage level.

3.2 Consideration of Dates

After having participated in several QA campaigns, we observed that certain answers given by our system corresponded to the expected answers type. In regard to temporal criteria, these answers became wrong.

The main objective of this module is on the one hand, to reject a passage or to decrease its confidence coefficient regarding temporal criteria. On the other hand, it might enhance the confidence coefficient of the passage if the temporal adequacy is good.

In order to implement this new module, we developed a date recognizer.

Dates recognition. The first stage was to locate the references of dates and times. Numerical values, integers, reals, and literals are annotated. Textual elements (days, month, etc) are also located. Then, the dates, hours and intervals of time were built. The date recognizer was developped on the French Corpus (le monde journal) using for previous clef campaigns.

Let's take the following passage: "En mars 1989, La Sept devient la Société européenne de programmes de télévision et reçoit du CSA l'autorisation d'émettre sur le satellite TDF 1 en avril 1989." After the labelling phase, this text becomes:

```
<duree type="date">
   <mod-pre type="eq">En</mod-pre> <mois type="car">mars</mois>
   <annee type="num">1989</annee>
</duree> , La <mois type="car">Sept</mois> devient la Société européenne de
programmes de télévision et
reçoit du CSA l' autorisation d' émettre sur le satellite TDF 1
<duree type="date">
   <mod-pre type="eq">en</mod-pre> <mois type="car">avril</mois>
    <annee type="num">1989</annee>
</duree> .  Elle commence à diffuser ses programmes
<date type="lin">
    <mod-pre type="eq">le</mod-pre> <no-jour type="num">30</no-jour>
<mois type="car">mai</mois>
    <annee type="num">1989</annee>
</date> ;
```

According to this result, we can make some remarks. First of all, not only the references including a day, a month and a year are regarded as "dates".

In the contrary case, this reference is labelled "duration" ("durée" in french). Moreover, we also labelled the article preceding this reference. This article gives information concerning the "direction of time" compared to the object of the sentence. Our temporal labelling system has still some imperfections. In this example, it labels "sept" as being September whereas it corresponds rather to the integer "7" (the name of the television channel ; "sept" means "seven" in french). This is produced by two different processes. First of all, we have a system allowing to recognize the numerical values in literal form. Then, the months whose name is long are often shortened. Also, we parameterized the system so that it recognizes "September" but also "sept." or "sept". Consequently "sept" indicates an integer but also September.

To facilitate the comparison of date, we chose to calculate the elements of date (and hour) in the ISO 8601 format. In this aim, we calculated the numerical values corresponding to the years, the months and the days. Then, we built the ISO form of the date. The same calculus was made with hours. For the AVE test set, the date extractor obtained a recall of 94% and a precision of 100%. It should be noted that we did not use the timex2 format to annotate the dates but our format is quiet close to this last (a simple XSLT transformation can be used).

Passages validation. The aim of this stage was to eliminate some proposed passages or to enhance other ones which had a good temporal adequacy with the question.

For this evaluation, we only took into account dates included into passages. In the future, we will consider other information, like metadata (the article date of issue where the passage comes from) and other temporal information like verb tense. The first step of the validation module aimed to compare the temporal elements of the question with the temporal elements of question passages.

For each passage, a temporal coefficient is calculated (between 0 to 1).

- If there is no temporal element into the question : score = 0.5 (nothing can be said)
- If there are temporal elements into the question :
  - If there is no temporal element into the passage : score = 0.5
  - If there are temporal elements but highly conflicting (not the same year into the question and into the passage): score = 0
  - If there are temporal elements but conflicting (year not specified into the question or the passage): score = 0.25
  - If there are certain joint temporal elements in the question and the passage (same year, but the days or months not specified into the question or the passage): score = 0.75
  - If there are exactly the same temporal elements into the question and into the passage : score = 1

We used the temporal coefficient to choose the best passages for the question (passages with temporal coefficient equal to 0.75 or 1)

Only 17 questions among the test set contained temporal elements in the questions and in the corresponding passages. They were not enough to evaluate our temporal approach.

3.3 Answer Validation

For each question, Prodicos system returned an answer. The answer validation module aimed at selecting for a question the correct answer among candidate answers of the test set according to the Prodicos's answer(figure 2).

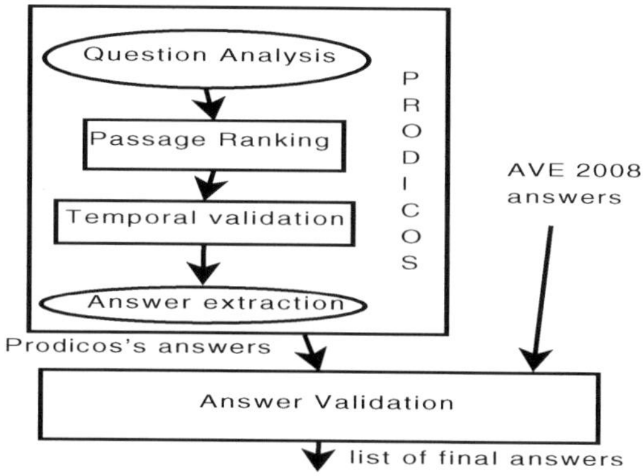

Fig. 2. The ProdicosAV System

The answer validation step is based on four rules. For a question:

- If Prodicos's answer was a candidate answer: candidate answer was validated
- If Prodicos's answer was included in a candidate answer: this candidate answer was validated but its confidence coefficient was lightly decreased
- If Prodicos's answer was partially included in a candidate answer (not all the Prodicos's answer words are present in a candidate answer) : the candidate answer was validated but the confidence coefficient was strongly decreased
- Otherwise the candidate answer was not validated

Finally, if only one answer was validated for a question, this one became the selected answer. Otherwise, if more than one answer was validated, the answer with the higher confidence coefficient became the selected answer.

4 Results Analysis

The French Answer Validation Exercise consisted of 108 questions which could each get one or more candidate answers. There were 199 candidate answers. All answers (199) given by the system were analysed by a human judge. It is worth noting that the "unknown" value given by a human expert to an answer was not taken into account in the evaluation. The evaluation of the results obtained by ProdicosAV are given in table 1. 51 answers were validated by ProdicosAV and among them 24 were validated too by human judges. 148 answers were rejected by our system and among them 109 were rejected too by human experts. Our system obtained a precision rate equal to 0.47 and a recall rate equal to 0.45. We made an other evaluation concerning the type of question; the results obtained are shown in table 1.

For the definitional questions, the test set contained 46 answers for 29 questions. For 16 answers validated by ProdicosAV only 5 of them did not correspond to the human judgment. Therefore, the system precision was high but its recall

Table 1. ProdicosAV evaluation

	Validated by ProdicosAV	**Validated by human**	**Validated by the two**	**Rejected by ProdicosAV**	**Rejected by human**
Definition	16	22	11	30	15
Numerical entity	7	8	3	21	20
Named entity	22	16	8	47	47
Other queries	6	7	2	50	44
Total	51	53	24	148	126

was worse. Indeed, only 11 answers were correctly validated by our system while 22 of them should have been. The problems mainly came from the question analysis problem (for example question 188: "Vasa" tagged as verb), from pattern extraction problems (for example question 159: "Jane Austen (16 décembre 1775, Steventon, Hampshire - 18 juillet 1817, Winchester) est une écrivain") or from acronym extraction problems (the meaning of the acronym is translated in the passage which implies that the acronym letters are completely independant of it). For the questions of numerical entity type, the test set contained 28 answers for 16 questions and for the questions of named entity type, the test set contained 69 answers for 36 questions. Only 11 answers among 29 validated by ProdicosAV corresponded to the human judgment. The system recall is also not close to being perfect. Indeed only 11 answers were validated by our system while 24 of them should have been. For other questions, the test set contained 56 answers for 27 questions. The system precision and recall are equivalent : only 2 answers among 6 validated by ProdicosAV were correctly annotated (7 answers were validated by the human judgment). The problems come mainly from the question analysis module or from the passage selection module. In fact, in this last case, the question focus was not always present into the passages which led to incorrect results for the density measure.

The second group of measures aimed at comparing the QA system's performance with the potential gain that the participant Answer Validation systems could add to them. The test set included 108 questions. Human experts found a response for 52 of them. Among them, our system gave 23 responses which were well validated. Human experts gave a negative response for 56 questions, among them, our system gave 38 negative responses. The *qa-accuracy* rate [7] obtained is equal to 22%. The maximum that the system should have obtained is 48% (according to the number of validated response the humans found). The *qa-rej-accuracy* rate [7] obtained is 35%. The maximum that the system could have obtained is 52% (according to the number of rejected response the humans found). Our system obtains a rather not so satisfactory *estimated-qa-performance* rate [7] equal to 29%. The maximum that it could have obtained is 73%. This shows that ProdicosAV System has not a good ability to acknowledge the identification of questions with a set of answers in which no correct one has been found.

5 Conclusion and Prospects

The results are not satisfactory, because we only recover 24 correct answers amoung 53 correct answers. The problems come mainly from the module of question analysis (specially word labeling), from the module of pattern extraction (specially due to the fact that semantics and coreference are not taken into consideration). Concerning temporal validation, the AVE task 2008 did not contain enough questions which involved temporal criteria. We will more thoroughly evaluate our system on a specific corpora containing temporal questions. Moreover, the temporal validation process could be improved by refining the decision criteria. We will also take into account in the validation step the date metadata related to document and other temporal information concerning question like verb tense.

References

1. Vossen, P.: EuroWordNet: A Multilingual Database with Lexical Semantic, editor Networks Piek Vossen, university of Amsterdam (1998)
2. Fourour, N.: Identification et catégorisation automatiques des entités nommées dans les textes français, PhD Thesis, Université de Nantes, LINA (2004)
3. Desmontils, E., Jacquin, C., Monceaux, L.: Question Types Specification for the Use of Specialized Patterns in Prodicos System. In: Peters, C., Clough, P., Gey, F.C., Karlgren, J., Magnini, B., Oard, D.W., de Rijke, M., Stempfhuber, M. (eds.) CLEF 2006. LNCS, vol. 4730, pp. 280–289. Springer, Heidelberg (2007)
4. Lee, G.G., Seo, J., Lee, S., Jung, H., Cho, B., Lee, C., Kwak, B., Cha, J., Kim, D., An, J., Kim, H., Kim, K.: SiteQ: Engineering High Performance QA System Using Lexico-Semantic Pattern Matching and Shallow NLP. In: Proceedings of tenth Text REtrieval Conference, TREC 2001 (2001)
5. Tellex, S., Katz, B., Lin, J., Fernandes, A., Marton, G.: Quantitative evaluation of passage retrieval algorithms for question answering. In: SIGIR conference on Research and development in information retrieval, pp. 41–47. ACM Press, New York (2003)
6. Morin, E.: Extraction de liens sémantiques entre termes á partir de corpus de textes techniques, PhD Thesis, Université de Nantes, LINA (1999)
7. Rodrigo, A., Penas, A., Verdejo, F.: Overview of the Answer Validation Exercise 2008. In: Working notes for the clef 2008 Workshop, Aahrus, Denmark, September 17-19 (2008), http://www.clef-campaign.org/2008/working_notes/

Studying the Influence of Semantic Constraints in AVE

Óscar Ferrández, Rafael Muñoz, and Manuel Palomar

Natural Language Processing and Information Systems Group
Department of Computing Languages and Systems
University of Alicante
{ofe,rafael,mpalomar}@dlsi.ua.es

Abstract. This paper discusses the participation of the University of Alicante in the Answer Validation Exercise (AVE) track. First, the proposed system uses a set of regular expressions in order to join the question and the answer into a declarative sentence, and afterwards applies several lexical-semantic inferences to attempt to detect whether the meaning of this sentence can be inferred by the meaning of the supporting text. Throughout the paper, we describe a basic system configuration and how it is enriched by the addition of semantic constraints. Moreover, we want to apply special emphasis to the language-independent capabilities of some system components. As a result, we were able to apply our techniques over both Spanish and English corpora achieving the first and second position in the AVE ranking.

1 Introduction

Question Answering (QA) appears with the aim of retrieving information required by natural language users' queries. The purpose of a QA system is to find the correct answers to user arbitrary questions in both non-structured and structured collection of digital data. Thus, the need to automatically extract knowledge from data has become acute with the dramatic growth of digital information. To overcome this problem, the three-year-old Cross-Language Evaluation Forum (CLEF) Answer Validation Exercise (AVE) track [1] provides an evaluation framework to consider appropriately those answers that are supported by the question and the passage from which they were extracted. This kind of inference will help QA systems to increase their performance as well as humans in the assessment of QA systems output. Traditionally this problem has been tackled by textual entailment techniques as shown in last AVE editions.

In our participation, we present a system that integrates several inferences from different knowledge sources. The base of the system is mainly based on lexical deductions, afterwards several modules have been added to the system in order to compute more sophisticated deductions (e.g. WordNet relations, named entities correspondences and verbs relations).

C. Peters et al. (Eds.): CLEF 2008, LNCS 5706, pp. 460–467, 2009.

2 Validating Answers

Aimed at achieving an approach that obtains promising results in a short lapse of time, we built a system using a reduced number of external resources. Figure 1 depicts the architecture of the system illustrating its modules and its workflow. It involves two main phases: (i) the preprocessing stage which is responsible for building a declarative sentence merging the question and the answer by means of a set of regular expressions, and (ii) the pure textual entailment component that detects lexical-semantic inferences between a pair of texts.

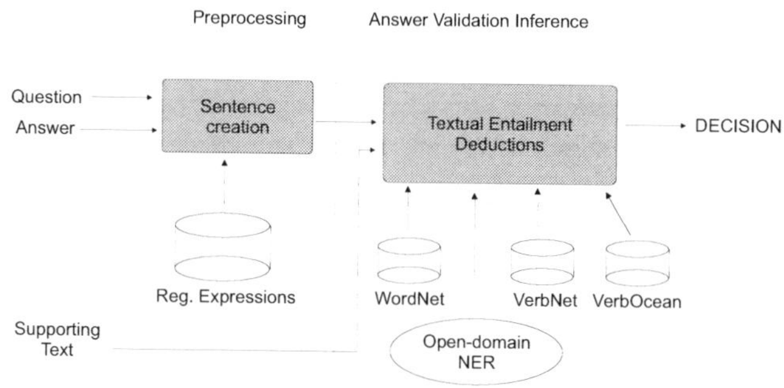

Fig. 1. System's Architecture

2.1 Building the Hypothesis

Each pair query-answer was preprocessed in order to obtain a declarative sentence (i.e. the *hypothesis*). For this purpose, we developed an extension of the set of regular expressions proposed in our previous participation in AVE [2]. It was done by analysing the questions within the development corpus and integrating new regular expressions capable of managing them. For both the development and test set every question is controlled by one regular expression, however it does not imply that the output sentence is grammatically well-formed. Obviously, it will also depend on the correctness of the answer.

2.2 The Textual Entailment Component

In order to tackle the AVE task, first we have created a base system making use of well-know techniques based on lexical inferences. These techniques have already been successfully applied by some research (including ourselves) in the task of recognising textual entailment [3,4,5]. Moreover, simple techniques based on word overlapping and shallow lexical inferences have obtained competitive results [6] being considered as a suitable starting point for further research.

Later on adjusting the system to the idiosyncrasies of the AVE task, we have generated some constraints that the candidate pairs of texts (hypothesis-supporting text) must fulfil.

The Base. Its performance is supported by the computation of a wide variety of lexical measures over the lemmas of the tokens (stop-words were discarded) that make up the texts. From the whole set of measures, we select those that are more significant according to the information gain that they provide to a machine learning classifier.[1] A Bayesian Net classifier implemented in Weka [2] was used for this issue, considering each measure as a feature. Next, brief descriptions of the most significant measures are listed[3]:

- **Levenshtein distance** [8]: it is a metric for measuring the amount of difference between two sequences. The Levenshtein distance between two strings is given by the minimum number of operations needed to transform one string into the other, where an operation is an insertion, deletion, or substitution of a single character. In our experiments the cost of an insertion, deletion or substitution is equal to one.
- **Smith-Waterman algorithm**: it is a well-known dynamic programming algorithm for performing local sequence alignment and determining similar regions between sequences. The algorithm was first proposed by [9] and consists of two steps: (i) calculate the similarity matrix score; and (ii) according to the dynamic programming method, trace back the similarity matrix to search for the optimal alignment.

  For two sequences SQ_1 and SQ_2, the optimal alignment score of two subsequence $SQ_1[1] \ldots SQ_1[i]$ and $SQ_2[1] \ldots SQ_2[j]$ is the calculation of $D(i,j)$ defined in formula 1:

$$D(i,j) = max \begin{cases} 0 & \text{start over,} \\ D(i-1,j-1) - f(SQ_1[j], SQ_2[j]) & \text{substitution or copy,} \\ D(i-1,j) - GAP & \text{insertion,} \\ D(i,j-1) - GAP & \text{deletion.} \end{cases} \quad (1)$$

  It permits two adjustable parameters regarding substitutions and copies for an alphabet mapping (the f function) and also allows costs to be attributed to a GAP for insertions or deletions. In our experiments we empirically set the values 0.3, -1 and 2 for a gap, copy and substitution respectively.
- **Cosine similarity**: it is a vector-based similarity. The input strings are transformed into vector space and it is computed as shown in equation 2:

$$\cos(\boldsymbol{x}, \boldsymbol{y}) = \frac{\boldsymbol{x} \cdot \boldsymbol{y}}{||\boldsymbol{x}|| \cdot ||\boldsymbol{y}||} \quad (2)$$

- **IDF specificity**: we determine the specificity of a word using the inverse document frequency (IDF) introduced in [10]. We derive the documents frequencies from the document collections used for the tracks reported within

---

[1] Unfortunately, due to space constraints just the most relevant measures are briefly detailed in this paper, we kindly redirect the reader to [7] for more details.

[2] http://www.cs.waikato.ac.nz/ml/weka/

[3] For some measures we use their implementation provided by the SimMetrics library (http://www.dcs.shef.ac.uk/~sam/simmetrics.html)

CLEF, in concrete the LA Times 94 and Glasgow Herald 95 collections, which contain a total number of 169,477 documents. The equation 3 shows how this measure is computed for two pair of texts:

$$ENT_{idf}(T,H) = \frac{\sum_{w \in H \wedge T} idf(w)}{\sum_{w \in H} idf(w)} \tag{3}$$

– **JWSL**: in order to discover word meaning relations that are not able to be detected directly from orthographic derivations we exploit WordNet. Relations such as synonymy, hypernyms, and semantic paths that connect two concepts can be found exploiting its taxonomy. Also, there are many implementations of similarity and relatedness measures between words based on WordNet. In our experiments, we have used the Java WordNet Similarity Library (JWSL[4]), which implements some of the most common semantic similarity measures. This feature automatically derives a score (the maximum score obtained from all similarity measures implemented in JWSL) that shows the similarity degree between the two texts.

The Constraints. Two constraints were added to the system prior to the computation of the aforementioned measures, they act as filters that eliminate some of the candidate pairs.

– **The Named Entities**: it is based on the detection, presence and absence of Named Entities (NEs). Despite the previous measures taken into account every token, even entities, these measures do not detect the importance of the presence or absence of an entity (e.g. when there is an entity in the hypothesis but the same entity is not present in the supporting text). This idea comes from the work presented in [11], where the authors successfully build their system only using the knowledge supplied by the recognition of NEs. In our case, we establish that "*in order to be considered as a candidate entailment pair, the hypothesis' entities must also appear within the supporting text*", so only pairs containing the same entities will be considered.

  In our experiments, we use NERUA system [12], an open domain NE recognizer which was trained by the corpus provided in CoNNL-2002 Share Task[5] and CoNLL-2003 Share Task[6] in order to recognise Spanish and English entities respectively. A partial entity matching was considered (i.e. "George Bush", "George Walker Bush", "G. Bush" and "Bush" are considered as the same entity). Unfortunately, reasoning about acronyms, date expansion, metonymy and location/demonymy was not developed at the current state of the system. Subsequent work on this area will be characterized by the addition of this sort of reasoning.

[4] http://grid.deis.unical.it/similarity/
[5] http://www.cnts.ua.ac.be/conll2002/ner/
[6] http://www.cnts.ua.ac.be/conll2003/ner/

– **The Verbs**: the other important particles in a sentence, apart from the NEs, are the verbs. Therefore, if we are able to detect whether the hypothesis' verbs are related to the supporting text's verbs, we could set another constraint showing this relatedness. To do this, we created two wrappers in Java for the VerbNet [13] and VerbOcean [14] resources. These wrappers allow us to detect semantic relationships between verbs.

  Thus, if every verb in the hypothesis (auxiliar verbs are not considered) can be related to one or more verbs in the supporting text, the pair will successfully pass this constraint. Two verbs are related whether: (i) they have the same lemma or are synonyms considering WordNet, (ii) they belong to the same VerbNet class or a subclass of their classes, and (iii) there is a relation in VerbOcean[7] that connects them.

One should note that the JWSL measure as well as the verb's constraint are language dependent, since they used resources specifically developed for English.

3 Experiments, Results and Discussion

Three experiments[8] were carried out: (1) *System Base (SB)*[9], which comprises the basic lexical measures together with the JWSL inference based on WordNet, (2) *SB+Entities Constraint (SB+EntC)*, which adds to SB the constraint about NEs, and (3) *SB+EntC+Verbs Constraint (SB+EntC+VerbC)*[9], which develops all the previous inferences including the constraint deduced by the relationships between verbs.

Table 1 shows the different experiments carried out and the results obtained by the system over the English corpora. The last column shows the AVE ranking of each run according to the F-measure. The proposed baselines were those provided by the AVE organizers.

As development corpora, we made several combinations from the corpora provided in the current and last AVE editions. The one that reached the best results (in a 10-cross fold validation test) was joining the development corpora of AVE'07 and AVE'08.

The results point out that a significant improvement is reached when the system considers the constraint about the NEs' inference. Unfortunately, although the constraint related to the verb's relationships considerably reduced the size of the corpus and consequently the processing time, it did not report any improvement except for the *estimated QA performance*. It reveals that complex treatment of verbs should be carried out, and the coverage of the resources used should be extended by means of other complementary knowledge sources (e.g. inferences about semantic frames rather than to only consider the verbs would improve these kinds of deductions).

[7] The VerbOcean's relations considered are: similarity, strength and happens-before.
[8] Some results presented in this section are not official due to the fact that some experiment runs were carried out after the deadline.
[9] These runs were developed after the CLEF Workshop deadline, therefore their ranking as well as the results shown in the tables are not official.

Table 1. English results obtained for the AVE 2008 track

Corpus	Run	Prec. YES	Rec. YES	F	QA acc.	estim. QA	rank
Dev.	SB	0.279	0.843	0.42	–	–	–
	SB+EntC	0.311	0.776	0.444	–	–	–
	SB+EntC+VerbC	0.307	0.748	0.436	–	–	–
Test	Baseline100	0.08	1	0.14	0.09	0.09	–
	Baseline50	0.08	0.5	0.13	–	–	–
	SB	0.23	0.92	0.37	0.19	0.23	4th
	SB+EntC	0.35	0.86	0.49	0.19	0.27	2nd
	SB+EntC+VerbC	0.35	0.78	0.48	0.19	0.28	3rd

Table 2. Spanish results obtained for the AVE 2008 track

Corpus	Run	Prec. YES	Rec. YES	F	QA acc.	estim. QA	rank
Development	SB_es	0.372	0.655	0.474	–	–	–
	SB+EntC_es	0.418	0.603	0.494	–	–	–
Test	Baseline100	0.10	1	0.18	0.11	0.11	–
	Baseline50	0.10	0.5	0.17	–	–	–
	SB_es	0.26	0.76	0.38	0.32	0.37	3rd
	SB+EntC_es	0.32	0.67	0.44	0.27	0.33	1st

Although the system makes use of language-dependent resources, its base as well as the NE recognizer components are language independent. It allowed us to apply the system over the Spanish corpora. However, this time just two experiments could be done: (1) *Spanish System Base (SB_es)*, similar to the one for English but without the JWSL measure[10], and (2) *SB_es+Entities Constraint (SB+EntC_es)*, which adds to SB_es the constraint about NEs using the Spanish configuration of NERUA.

Table 2 draws the results over the Spanish corpora. The system behaviour is somewhat similar. The entities constraint improves the system's performance and the system base configuration proves that it is a very good starting point for further language-independent research.

The SELECTED values were established to the pair that achieved the highest positive score among all pairs that belong to the same question. In the event that two or more pairs have the highest score, then one of them is randomly chosen.

Finally, we would like to show some statistics about the reduction of the corpora by our constraints. The development English corpora was reduced in a rate of 54% and 62% due to the application of the NEs and Verbs constraints respectively, whereas the test corpus reduction reached a rate of 58.7% and 66.7%. For Spanish, the NEs constraint helped the system to discard a rate of

[10] This is owing to JWSL works with English WordNet, and at present we do not have any implementation of these measures for the Spanish WordNet.

Table 3. Number of pairs considered for each experiment

Corpus	#Original	#SB	#SB+EntC	#SB+EntC+VerbC
train EN	1467	1467	674	557
test EN	1055	1055	435	351
train ES	2368	2368	1397	not applicable for Spanish
test ES	1528	1528	858	not applicable for Spanish

41% and 43.8% for the development and test corpora. Table 3 shows the number of pairs processed for each experiment.

4 Conclusions and Future Work

We present a basic configuration of a entailment system that, although using some language dependent resources, is easily portable to other languages. We proved it participating in both English and Spanish tracks. Moreover, we also describe another system configuration that deals with semantic inferences supported by two constraints: one based on the importance of the NE within the texts and the other based on discovering verb relations. Results show that our basic configuration is a good starting point and more importantly, the constraints apart from improving it, dramatically reduce the size of the processed corpora.

Future work can be related to the improvement in the treatment of verbs as well as the detection of NEs. For instance, some heuristics regarding semantic verb frames could help the system to extend the coverage of verb's relationships. Regarding the NE recognizer, currently we only detect a strict matching between the hypothesis and supporting text entities and whether an entity is contained by another. Subsequent work will be characterized by identifying deeper inference relations between entities such as acronyms, date expansion, etc.

Acknowledgements

This research has been partially funded by the QALL-ME consortium, which is a 6th Framework Research Programme of the European Union (EU), contract number FP6-IST-033860 and by the Spanish Government under the project CICyT number TIN2006-1526-C06-01.

References

1. Rodrigo, Á., Peñas, A., Verdejo, F.: Overview of the Answer Validation Exercise 2008. In: Peters, C., et al. (eds.) CLEF 2008. LNCS, vol. 5706, pp. 296–313. Springer, Heidelberg (2008)
2. Ferrández, Ó., Micol, D., Muñoz, R., Palomar, M.: On the Application of Lexical-Syntactic Knowledge to the Answer Validation Exercise. In: [15], pp. 377–380

3. Ferrández, Ó., Micol, D., Muñoz, R., Palomar, M.: A Perspective-Based Approach for Solving Textual Entailment Recognition. In: Proceedings of the ACL-PASCAL Workshop on Textual Entailment and Paraphrasing, Prague, Association for Computational Linguistics, June 2007, pp. 66–71 (2007)
4. Malakasiotis, P., Androutsopoulos, I.: Learning Textual Entailment using SVMs and String Similarity Measures. In: Proceedings of the ACL-PASCAL Workshop on Textual Entailment and Paraphrasing, Prague, Association for Computational Linguistics, June 2007, pp. 42–47 (2007)
5. Adams, R., Nicolae, G., Nicolae, C., Harabagiu, S.: Textual Entailment Through Extended Lexical Overlap and Lexico-Semantic Matching. In: Proceedings of the ACL-PASCAL Workshop on Textual Entailment and Paraphrasing, Prague, Association for Computational Linguistics, June 2007, pp. 119–124 (2007)
6. Giampiccolo, D., Magnini, B., Dagan, I., Dolan, B.: The Third PASCAL Recognizing Textual Entailment Challenge. In: Proceedings of the ACL-PASCAL Workshop on Textual Entailment and Paraphrasing, Prague, Czech Republic, Association for Computational Linguistics, June 2007, pp. 1–9 (2007)
7. Ferrández, Ó., Muñoz, R., Palomar, M.: A Lexical-Semantic approach to AVE. In: Working Notes of the CLEF 2008 Workshop, Aarhus, Denmark (September 2008)
8. Levenshtein, V.I.: Binary codes capable of correcting deletions, insertions, and reversals. Soviet Physics Doklady 10(8), 707–710 (1966)
9. Smith, T.F., Waterman, M.S.: Identification of common molecular subsequences. Journal of Molecular Biology 147, 195–197 (1981)
10. Sparck-Jones, K.: A statistical interpretation of term specificity and its application in retrieval. Journal of Documentation 28(1), 11–21 (1972)
11. Rodrigo, Á., Peñas, A., Verdejo, F.: UNED at Answer Validation Exercise 2007. In: [15], pp. 404–409 (2007)
12. Kozareva, Z., Ferrández, Ó., Montoyo, A., Muñoz, R.: Combining data-driven systems for improving Named Entity Recognition. Data and Knowledge Engineering 61(3), 449–466 (2007)
13. Kipper, K., Korhonen, A., Ryant, N., Palmer, M.: Extending VerbNet with Novel Verb Classes. In: Fifth International Conference on Language Resources and Evaluation (LREC 2006), Genova, Italy (June 2006)
14. Chklovski, T., Pantel, P.: VerbOcean: Mining the Web for Fine-Grained Semantic Verb Relations. In: Proceedings of Conference on Empirical Methods in Natural Language Processing (EMNLP 2004), Barcelona, Spain (2004)
15. Peters, C., Jijkoun, V., Mandl, T., Müller, H., Oard, D.W., Peñas, A., Petras, V., Santos, D. (eds.): CLEF 2007. LNCS, vol. 5152. Springer, Heidelberg (2008)

RAVE: A Fast Logic-Based Answer Validator

Ingo Glöckner

Intelligent Information and Communication Systems (IICS),
FernUniversität in Hagen, 58084 Hagen, Germany,
ingo.gloeckner@fernuni-hagen.de

Abstract. RAVE (Real-time Answer Validation Engine) is a logic-based answer validator/selector designed for real-time question answering. Instead of proving a hypothesis for each answer, RAVE uses logic only for checking if a considered passage supports a correct answer at all. In this way parsing of the answers is avoided, yielding low validation/selection times. Machine learning is used for assigning local validation scores based on logical and shallow features. The subsequent aggregation of these local scores strives to be robust to duplicated information in the support passages. To achieve this, the effect of aggregation is controlled by the lexical diversity of the support passages for a given answer.

1 Description of the Validation Task

The Answer Validation Exercise (AVE) [1] introduces a test set of validation items $i \in \mathcal{I}$ comprising the question q_i, answer candidate a_i and supporting snippet s_i. Let $\mathcal{Q} = \{q_i : i \in \mathcal{I}\}$ be the set of all questions and $\mathcal{I}_q = \{i \in \mathcal{I} : q_i = q\}$ the set of validation items for a question $q \in \mathcal{Q}$. The validator must assign a validation decision $v_i \in \{\mathit{REJECTED}, \mathit{SELECTED}, \mathit{VALIDATED}\}$ and confidence score $c_i \in [0, 1]$ to each $i \in \mathcal{I}$. At most one answer per question can be selected. Answers can only be validated if an answer was selected as best answer.

2 The RAVE Validator

The input to the validator comprises a question together with answer candidates for the question and the supporting text snippets, as represented by $\mathcal{I}_q$. Let $A_q = \{a_i : q_i = q\}$ denote the set of answer candidates for q in the test set. The AVE 2008 test set is redundancy-free, i.e. for each $a \in A_q$, there is only one item $i \in \mathcal{I}_q$ supporting a. As the basis for aggregation, the IRSAW system [2] was used to actively search for additional supporting snippets for each of the answer candidates. Answers were clustered into groups of minor variants with the same 'answer key' $\kappa(a_i)$ by applying a simplification function κ (which drops accents etc.) For example, $\kappa(\textit{in the year 2001}) = 2001$. The result is a set of auxiliary items $i \in \mathcal{I}'_q$ with $q_i = q$ and $\kappa(a_i) = \kappa(a_j)$ for some original answer $a_j \in A_q$, and supporting snippet s_i for a_i found by IRSAW. The original and auxiliary validation items are joined into the enhanced validation pool $\mathcal{I}^*_q = \mathcal{I}_q \cup \mathcal{I}'_q$.

C. Peters et al. (Eds.): CLEF 2008, LNCS 5706, pp. 468–471, 2009.

Validation starts with a *deep linguistic analysis* of the question, using the NLP toolchain of the IICS. For snippets, the parse is fetched from the pre-analyzed document collections. The *question classification* identifies the descriptive core of the question, the expected answer type and question category. *Sanity tests* eliminate trivial answers and non-informative answers to definition questions [3]. Failure of temporal restrictions and incompatibility of measurement units is also detected. For the remaining answers, *shallow features*[1] and *logic-based features* are computed. This involves proving the question from the snippet in a relaxation loop.[2] RAVE then extracts the number of proven literals and other features from the prover results; about half of all features are logic-based [3,5].

Machine learning is used to assign a local evidence score $\eta_i \in [0,1]$ to $i \in \mathcal{I}_q^*$, estimating the probability that answer a_i is correct judging from snippet s_i [5].

Intuitively, the plausibility of an answer candidate increases when multiple passages support the answer. But multiple copies of the same snippet should not increase the aggregated score $\gamma(a)$ since they do not provide independent evidence. Hence let $K_q = \{\kappa(a_i) : i \in \mathcal{I}_q\}$ be the set of answer keys (normalized answers) for a given question. Let $\mathcal{I}_{q,k} = \{i \in \mathcal{I}_q^* : \kappa(a_i) = k\}$ be the set of support items for answer key $k \in K_q$. For $i \in \mathcal{I}_q^*$, let Ω_i be the set of term occurrences[3] in snippet s_i. For a term occurrence $\omega \in \Omega_i$, let $t(\omega)$ be the corresponding term. Let $T_i = \{t(\omega) : \omega \in \Omega_i\}$ be the set of passage terms and $T^k = \bigcup\{T_i : i \in \mathcal{I}_{q,k}\}$ the set of all terms in any passage for $k \in K_q$. We abbreviate

$$\mu(k,t) = \min\{(1-\eta_i)^{\nu_{i,t}} : i \in \mathcal{I}_{q,k},\ t \in T_i\}, \quad \nu_{i,t} = \tfrac{\mathrm{occ}(t,i)}{|\Omega_i|},$$

where $\mathrm{occ}(t,i) = |\{\omega \in \Omega_i : t(\omega) = t\}|$ is the occurrence count of term t in snippet i, and η_i is the local evidence score. The aggregated support for $k \in K_q$ is then given by $\gamma(k) = 1 - \prod_{t \in T^k} \mu(k,t)$ and for extracted answers by $\gamma(a) = \gamma(\kappa(a))$.

Consider the answer $a = \kappa(a) = 42$ as to the age of death of Elvis, supported by *"Elvis died at 42."* $(i = 1)$, $\eta_1 = \frac{65}{81} \approx 0.802$ and *"Elvis (42) dead"* $(i = 2)$, score $\eta_2 = \frac{37}{64} \approx 0.578$. Then $T_1 = \{\mathit{Elvis, died, at, 42}\}$, $T_2 = \{\mathit{Elvis, 42, dead}\}$, and $|\Omega_1| = 4$, $|\Omega_2| = 3$ (number of words). Now $\mathrm{occ}(t,1) = 1$ and $\nu_{1,t} = \frac{1}{4}$ for all $t \in T_1$, $\nu_{1,t} = 0$ else; similarly $\mathrm{occ}(t,2) = 1$ and $\nu_{2,t} = \frac{1}{3}$ for $t \in T_2$, $\nu_{2,t} = 0$ else. Thus $\mu(\mathit{42, Elvis}) = \mu(\mathit{42, died}) = \mu(\mathit{42, at}) = \mu(\mathit{42, 42}) = (\frac{16}{81})^{\frac{1}{4}} = \frac{2}{3}$ and $\mu(\mathit{42, dead}) = (\frac{27}{64})^{\frac{1}{3}} = \frac{3}{4}$, i.e. $\gamma(42) = 1 - (\frac{2}{3})^4 \cdot \frac{3}{4} = \frac{23}{27} \approx 0.852$.

The effect of aggregation on $\gamma(a)$ is strongest if two aggregated passages have no terms in common; there is no increase at all if a passage is encountered repeatedly. After aggregation, the auxiliary support items in $\mathcal{I}_q'$ are dropped.

The final selection score depends on the aggregated score $\gamma(a_i)$ and the justification strength η_i of the snippet: $\sigma_i = \eta_i\,\gamma(a_i)/\max\{\eta_j : j \in \mathcal{I}_q, \kappa(a_i) = \kappa(a_j)\}$.[4] Based on the assignment of final selection scores σ_i, the system determines a choice of $i^{\mathrm{opt}} \in \mathcal{I}_q$ which maximizes σ_i, i.e. $\sigma_{i^{\mathrm{opt}}} = \max\{\sigma_i : i \in \mathcal{I}_q\}$. The chosen i^{opt} is marked as $v_{i^{\mathrm{opt}}} = \mathit{SELECTED}$ if $\sigma_{i^{\mathrm{opt}}} \geq \theta_{\mathrm{sel}}$, where $\theta_{\mathrm{sel}} \in [0,1]$

[1] Examples are lexical overlap and compatibility of expected/found answer types.
[2] RAVE uses the same background knowledge as its predecessor MAVE [4].
[3] A term occurrence is a pair (t,i) where t is a term and i the position in the passage.
[4] Due to a bug, the submitted runs used $j \in \mathcal{I}_q^*$ instead of the intended $j \in \mathcal{I}_q$.

Table 1. Results of RAVE in the AVE 2008 (plus additional experiments)

model	f-meas	prec	recall	qa-acc	s-rate	model	f-meas	prec	recall	qa-acc	s-rate
Run1	0.39	0.33	0.49	0.32	0.61	WF.75	0.47	0.45	0.50	0.25	0.48
Run2	0.29	0.25	0.34	0.23	0.44	WF1	0.47	0.44	0.50	0.26	0.50
RF	0.45	0.43	0.48	0.26	0.50	WQ.75	0.45	0.37	0.58	0.33	0.63
RQ	0.44	0.36	0.56	0.34	0.65	WQ1	0.45	0.36	0.58	0.34	0.65

is the selection threshold; otherwise i^{opt} is marked as $v_{i^{\mathrm{opt}}} = \mathit{REJECTED}$. If $\theta_{\mathrm{sel}} = 0$, then the best validation item for a question is always selected; this maximizes selection rate. In experiments aiming at high F-score, a threshold of $\theta_{\mathrm{sel}} = 0.23$ was chosen.[5] The non-best items $i \in \mathcal{I}_q \backslash \{i^{\mathrm{opt}}\}$ are classified as follows: if $v_{i^{\mathrm{opt}}} = \mathit{REJECTED}$, then $v_i = \mathit{REJECTED}$ for all $i \in \mathcal{I}_q \setminus \{i^{\mathrm{opt}}\}$ as well. If a selection has been made, i.e. $v_{i^{\mathrm{opt}}} = \mathit{SELECTED}$, then we set $v_i = \mathit{VALIDATED}$ if $\sigma_i \geq \theta_{\mathrm{val}}$ and $v_i = \mathit{REJECTED}$ otherwise, where $\theta_{\mathrm{val}} = 0.23$ is the decision threshold for non-best items. The confidence into this decision is $c_i = \sigma_i$ if $v_i = \mathit{SELECTED}$ or $v_i = \mathit{VALIDATED}$, and $c_i = 1 - \sigma_i$ if $v_i = \mathit{REJECTED}$.

3 Evaluation

The results of RAVE in the AVE 2008 (and further reference results) are shown in Table 1, using column labels *f-meas* (F-score), *prec* (precision), *recall*, *qa-acc* (qa-accuracy, i.e. correct selections divided by number of questions), and *s-rate* (selection rate, i.e. successful selections divided by optimal selections). Since it was not clear from the QA@CLEF 2008 guidelines if full-sentence answers to definition questions would be accepted or not, *Run1* was configured to accept such answers while *Run2* was configured to reject them; obviously the former policy was the intended one. The table also lists the results for the current validator after correcting minor bugs. The letter 'R' refers to the standard method of RAVE for combining scores using σ_i, 'F' means the use of $\theta_{\mathrm{sel}} = 0.23$ for F-score oriented runs, and 'Q' means use of $\theta_{\mathrm{sel}} = 0$ for runs aiming at qa-accuracy.

A weighted average $\sigma_i^{(\lambda)} = \lambda\gamma(a_i) + (1-\lambda)\eta_i$ for $\lambda \in \{0, 0.25, 0.5, 0.75, 1\}$ was also tried, see WFλ and WQλ runs for the best results so obtained. Some extra experiments were carried out with the aggregation model of RAVE replaced by 'best evidence' aggregation $\gamma'(k) = \max\{\eta_i : i \in \mathcal{I}_{q,k}\}$ (this decreased F-score by 5-6% and selection rate by 4%), or 'independent evidence' aggregation $\gamma''(k) = 1 - \prod_{i \in \mathcal{I}_{q,k}} (1 - \eta_i)$ (this method was only slightly worse, but it shows a spurious increase of aggregation for duplicated passages). As to support pool enhancement, experiments using $\mathcal{I}_q$ rather than $\mathcal{I}_q^*$ showed a drop of F-score by 5% compared to RF (6% for RQ); without any aggregation, the loss was 8%.

Finally the system was run with the prover switched off: selection rate of RF then dropped by 6% (RQ: 15%), but F-score increased by up to 5%. This contradicts experience from experiments on CLEF07 data [5]. A possible reason

[5] The threshold results from the parameters used for cost-sensitive learning.

is that 8 of the 11 runs for German were produced by QA systems that used RAVE for validation. The false positives that passed the validation by RAVE as part of these QA systems are likely to stay undetected when applying the validator again in the AVE. The use of a different classifier in the shallow-only run means an independence benefit that might explain the increased F-score.

Since RAVE is designed for real-time QA, processing times are also of interest. For the RQ method, validation took 126 ms per question (9.35 ms per validation item) on a standard PC. The extra effort for applying the prover to a validation item suitable for logical processing was an average 5.4 ms.

4 Conclusion

The main objective for the current work was that of making logic-based answer validation applicable in a real-time QA context. To achieve this, RAVE uses logic only for passage validation. The observed processing times confirm that this makes a fast validation possible. Sometimes a given snippet provides useful information but may not be disclosed for licensing reasons or because the user does not understand the language of the passage. The validator therefore supports 'auxiliary' passages which contribute to aggregation but are never shown. The aggregation model of RAVE is designed to be robust to replicated content. Experiments confirm its superiority over two alternative approaches.

RAVE achieved acceptable performance in the AVE (especially considering its departure from a full-fledged RTE-style answer validation), but it was outperformed by the best individual QA system with a selection rate of 0.73 (RAVE: 0.61 in *Run1*, and 0.65 in RQ after debugging). However, as a part of real QA systems, RAVE uses features not available in the AVE test set. For example, it normally considers the retrieval score of the passage retrieval system and a producer score assigned by each QA source [2]. Experiments on the CLEF07 data show that adding these features increases the F-score by up to 8%. Moreover, RAVE is normally customized by learning separate models for each QA source, which further improves results.

References

1. Rodrigo, A., Peñas, A., Verdejo, F.: Overview of the Answer Validation Exercise 2008. In: Peters, C., et al. (eds.) CLEF 2008. LNCS, vol. 5706, pp. 296–313. Springer, Heidelberg (2009)
2. Hartrumpf, S., Glöckner, I., Leveling, J.: Efficient question answering with question decomposition and multiple answer streams. In: Peters, C., et al. (eds.) CLEF 2008. LNCS, vol. 5706, pp. 421–428. Springer, Heidelberg (2009)
3. Glöckner, I., Pelzer, B.: Combining logic and machine learning for answering questions. In: Peters, C., et al. (eds.) CLEF 2008. LNCS, vol. 5706, pp. 401–408. Springer, Heidelberg (2009)
4. Glöckner, I.: Filtering and fusion of question-answering streams by robust textual inference. In: Proceedings of KRAQ 2007, Hyderabad, India (2007)
5. Glöckner, I.: Towards logic-based question answering under time constraints. In: Proc. 2008 IAENG Int. Conf. on Artificial Intelligence and Applications (2008)

Information Synthesis for Answer Validation

Rui Wang[1] and Günter Neumann[2]

[1] Saarland University
66123 Saarbrücken, Germany
rwang@coli.uni-sb.de
[2] LT-Lab, DFKI
Stuhlsatzenhausweg 3, 66123 Saarbrücken, Germany
neumann@dfki.de

Abstract. This paper proposes an integration of *Recognizing Textual Entailment* (RTE) with other additional information to deal with the *Answer Validation* task. The additional information used in our participation in the *Answer Validation Exercise* (AVE 2008) is from named-entity (NE) recognizer, question analysis component, etc. We have submitted two runs, one run for English and the other for German, achieving f-measures of 0.64 and 0.61 respectively. Compared with our system last year, which purely depends on the output of the RTE system, the extra information does show its effectiveness.

1 Introduction and Related Work

Using *Recognizing Textual Entailment* (RTE-1 – [3]; RTE-2 – [1]) to do *Answer Validation* has shown a great success [9]. We also developed our own RTE system and participated in AVE2007 [12]. The RTE system proposed a new sentence representation extracted from the dependency structure, and utilized the Subsequence Kernel method [2] to perform machine learning. We have achieved fairly high results on both the RTE-2 data set [10] and the RTE-3 data set [11], especially on *Information Extraction* (IE) and *Question Answering* (QA) pairs.

However, on the AVE data sets, we still found much space for the improvement. Therefore, based on the system we developed last year, our motivation this year is to see whether using extra information, e.g. *named-entity* (NE) recognition, question analysis, etc., can make further improvement on the final results.

2 The RTE System

The RTE system ([10]; [11]) is developed for the RTE-3 Challenge [5]. The system contains a main approach with two backup strategies. The main approach extracts parts of the dependency structures to form a new representation, named *Tree Skeleton*, as the feature space and then applies *Subsequence Kernels* to represent TSs and perform machine learning. The backup strategies will deal with the **T**-**H** pairs which cannot be solved by the main approach. One backup strategy is called *Triple Matcher*, as it calculates the overlapping ratio on top of the dependency structures in a triple

C. Peters et al. (Eds.): CLEF 2008, LNCS 5706, pp. 472–475, 2009.

representation; the other is simply a *Bag-of-Words* (BoW) method, which calculates the overlapping ratio of words in **T** and **H**.

3 The AVE System

3.1 Preprocessing and Post-processing

Since the input of the AVE task is a list of questions, their corresponding answers and the documents containing these answers, we need to adapt them into **T**-**H** pairs for the RTE system. In order to combine the question and the answer into a statement, manually construct some language patterns for the input questions (cf. [12] for more details). The constructed **T**-**H** pairs can be the input for any generic RTE systems. In practice, after applying our RTE system, if the **T**-**H** pairs are covered by our main approach, we will directly use the answers; if not, we will use a threshold to decide the answer based on the two similarity scores and together with other information (see the following subsection).

The post-processing is straightforward, the "YES" entailment cases will be validated answers and the "NO" entailment cases will be rejected answers. In addition, the selected answers (i.e. the best answers) will naturally be the pairs covered by our main approach or (if not,) with the highest similarity scores.

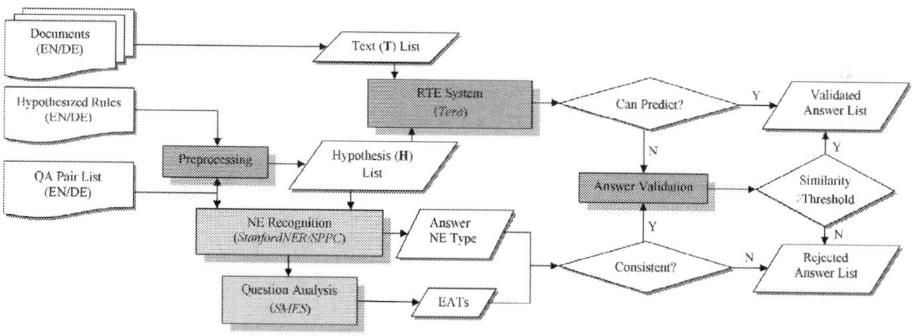

Fig. 1. Our AVE system uses the RTE system (**Tera** – *Textual Entailment Recognition for Application*) as a core component. The preprocessing module mainly adapts questions, their corresponding answers, and supporting documents into *Text* (**T**)-*Hypothesis* (**H**) pairs, assisted by some manually designed patterns. The post-processing module (i.e. the *Answer Validation* in the picture) will validate each answer and select a most proper one based on the output of the RTE system. The new modules added are the *NE Recognition* and *Question Analysis*. Thus, we will have extra information like NEs in the answers, *Expected Answer Types* (EATs).

3.2 Additional Components

The RTE system is used as a core component of the AVE system. Based on the error analysis of last year's results, this year, we use additional components to filter out noisy candidates. Therefore, two extra components are added to the architecture, the NE recognizer and the question analyzer. For NE recognition, we use StanfordNER

[4] for English and SMES [8] for German; and for question analysis, we use the SMES system [8]. The detailed workflow is as follows,

1. Annotate NEs in **H**, store them in an NE list; if the answer is an NE, store the NE type as A'_Type;
2. Analyze the question and obtain expected answer type, store it as A_Type;
3. Synthesize all the information, i.e. NE list, A_Type, A'_Type, BoW similarity, Triple similarity, etc.

Then, heuristic rules are straightforward to be applied, e.g. checking the consistence between A_Type and A'_Type, checking whether all (or how many of) the NEs also appear in the documents.

4 Results

We have submitted two runs, one for English and one for German.

Table 1. Results of our submissions compared with last year's

Submission Runs	Recall	Precision	F-measure	Estimated QA Performance	QA Accuracy
100% VALIDATED (EN)	1	0.08	0.14	N/A	N/A
50%VALIDATED (EN)	0.5	0.08	0.13	N/A	N/A
Perfect Selection (EN)	N/A	N/A	N/A	0.56	**0.34**
Best QA System (EN)	N/A	N/A	N/A	0.21	**0.21**
dfki07-run1 (EN)	0.62	0.37	**0.46**	N/A	0.16
dfki07-run2 (EN)	0.71	0.44	**0.55**	N/A	0.21
dfki08run1 (EN)	0.78	0.54	**0.64**	0.34	**0.24**
100% VALIDATED(DE)	1	0.12	0.21	N/A	N/A
50% VALIDATED (DE)	0.5	0.12	0.19	N/A	N/A
Perfect Selection (DE)	N/A	N/A	N/A	0.77	**0.52**
Best QA System (DE)	N/A	N/A	N/A	0.38	**0.38**
dfki08run1 (DE)	0.71	0.54	0.61	0.52	**0.43**

In the table, we notice that both for English and German, our validation system outperforms the best QA systems, which suggests the necessity of the validation step. Although there is a gap between the system performance and the perfect selection, the results are quite satisfactory. If we compare this year's results with last year's, the additional information does improve the results significantly. Comparing the recall and precision, for both languages, the latter is worse. After an error analysis (cf. the Working Notes), we find that to further synthesize the information we have, i.e. NE annotation and dependency parsing, might be more beneficial.

5 Conclusion and Future Work

To sum up, based on the experience of last year's participation, apart from the RTE core system, we add two extra components, NE recognizer and question analyzer, to further improve the results. The strategy is quite successful according to the comparison of system performances. However, the problem has not been fully solved. Filtering some documents in the preprocessing step could be even more effective than working on the post-processing phase; another direction considered by us is to take a closer look at the different performances between different languages.

References

1. Bar-Haim, R., Dagan, I., Dolan, B., Ferro, L., Giampiccolo, D., Magnini, B., Szpektor, I.: The Second PASCAL Recognising Textual Entailment Challenge. In: Proceedings of the Second PASCAL Challenges Workshop on Recognising Textual Entailment (2006)
2. Bunescu, R., Mooney, R.: Subsequence Kernels for Relation Extraction. In: Advances in Neural Information Processing Systems, vol. 18. MIT Press, Cambridge (2006)
3. Dagan, I., Glickman, O., Magnini, B.: The PASCAL Recognising Textual Entailment Challenge. In: Quiñonero-Candela, J., Dagan, I., Magnini, B., d'Alché-Buc, F. (eds.) MLCW 2005. LNCS (LNAI), vol. 3944, pp. 177–190. Springer, Heidelberg (2006)
4. Finkel, J.R., Grenager, T., Manning, C.: Incorporating Non-local Information into Information Extraction Systems by Gibbs Sampling. In: Proceedings of the 43rd Annual Meeting of the Association for Computational Linguistics, ACL 2005 (2005)
5. Giampiccolo, D., Magnini, B., Dagan, I., Dolan, B.: The Third PASCAL Recognizing Textual Entailment Challenge. In: Proceedings of the Workshop on Textual Entailment and Paraphrasing, Prague, June 2007, pp. 1–9 (2007)
6. Gildea, D., Palmer, M.: The Necessity of Parsing for Predicate Argument Recognition. In: Proceedings of the 40th Meeting of the Association for Computational Linguistics (ACL 2002), Philadelphia, PA, pp. 239–246 (2002)
7. Lin, D.: Dependency-based Evaluation of MINIPAR. In Workshop on the Evaluation of Parsing Systems (1998)
8. Neumann, G., Piskorski, J.: A Shallow Text Processing Core Engine. Journal of Computational Intelligence 18(3), 451–476 (2002)
9. Peñas, A., Rodrigo, Á., Verdejo, F.: Overview of the Answer Validation Exercise 2007. In: The CLEF 2007 Working Notes (2007)
10. Wang, R., Neumann, G.: Recognizing Textual Entailment Using a Subsequence Kernel Method. In: Proc. of AAAI 2007 (2007a)
11. Wang, R., Neumann, G.: Recognizing Textual Entailment Using Sentence Similarity based on Dependency Tree Skeletons. In: Proceedings of the Workshop on Textual Entailment and Paraphrasing, Prague, June 2007, pp. 36–41 (2007b)
12. Wang, R., Neumann, G.: DFKI–LT at AVE 2007: Using Recognizing Textual Entailment for Answer Validation. In: Online Proceedings of CLEF 2007 Working Notes, Budapest, Hungary, September 2007 (2007c) ISBN: 2-912335-31-0

Analyzing the Use of Non-overlap Features for Supervised Answer Validation

Alberto Téllez-Valero, Antonio Juárez-González,
Manuel Montes-y-Gómez, and Luis Villaseñor-Pineda

Laboratory of Language Technologies, INAOE, Mexico
{albertotellezv,antjug,mmontesg,villasen}@inaoep.mx

Abstract. This year we evaluated our supervised answer validation method at both, the Spanish Answer Validation Exercise (AVE) and the Spanish Question Answering Main Task. This paper describes and analyzes our evaluation results from both tracks. In resume, the F-measure of the proposed method outperformed the baseline result of the AVE 2008 task by more than 100%, and enhanced the performance of our question answering system, showing a gain in accuracy of 22% for answering factoid questions. A detailed analysis of the results shows that the proposed non–overlap features are most discriminative than the traditional overlap ones. In particular, these novel features allowed increasing the F-measure result of our method by 26%.

1 Introduction

An *answer validation* (AV) method try to determine if a specified answer is correct and supported. These methods are especially useful for filtering the best responses from question answering (QA) systems and for superficially combining them. In line with these efforts, we implemented a new AV method based on a supervised learning approach. In particular, our method implements a boosting ensemble —formed by ten decision tree classifiers— that decides whether to accept or reject each candidate answer based on the use of ninety-six attributes that characterize: (*i*) the compatibility between question and answer types; (*ii*) the redundancy of answers across streams; and (*iii*) the overlap and the non-overlap between the question-answer pair and the core fragment of the support text.

In order to evaluate the proposed method we considered two different scenarios: the Answer Validation Exercise (AVE 2008) and the Question Answering Main Task (QA@CLEF 2008). The objective of the first scenario was to evaluate the ability of our AV method to discriminate correct from incorrect answers as well as its capacity to combine the answers from several QA systems. In contrast, the goal of the second evaluation scenario was to measure the impact of including an answer validation module in our QA system [1].

The evaluation results were encouraging; the proposed method outperformed the F-measure result of the baseline at AVE 2008 task by more than 100%, and also enhanced the performance of our QA system producing a gain in accuracy of 22% for factoid questions. The analysis of the results showed the importance of

C. Peters et al. (Eds.): CLEF 2008, LNCS 5706, pp. 476–479, 2009.

the proposed non-overlap features, which increased our F-measure in 26%. This analysis also indicated that the redundancy features were not useful for the AVE 2008 test set; their elimination allowed achieving a gain in F-measure of 5%.

Due to space limitations, we decide to omit the description of our AV method (which can be found in [2]), and exclusively consider the analysis of the evaluation results. In particular, this paper is organized as follows. Section 2 resumes the results of the proposed method at CLEF. Section 3 presents the results analysis, and Section 4 exposes our conclusions and outlines some future work directions.

2 Evaluation Results

2.1 Spanish Answer Validation Exercise

Table 1 shows the answer validation results corresponding to our two submitted runs to Spanish AVE 2008 [3]. It also shows the results for the baseline (a 100% VALIDATED). The results indicate that relaxing the acceptance threshold over the answers's confidence value (RUN 1) our method achieved a high recall but a low precision. In contrast, the second run that maintained the default threshold (RUN 2) got a worst recall, but achieved a major precision overcoming the baseline F-measure result in more than 100%.

Complementary to the previous data, Table 2 shows the evaluation results for combining the answers from several QA systems. These results indicate that the QA-accuracy of RUN 1 is 19% better than the result of RUN 2. Given that RUN 2 outperformed the answer validation result of RUN 1 (see Table 1), these results confirm our observation in [2] that the best answer validation method (but not perfect) not necessary produces the best QA stream fusion performance. The results in Table 2 also indicate that, because of the better capacity of RUN 2 to rejected wrong answers, the estimated-QA-performance of both runs were very similar. Refer to [3] for a description of the evaluation measures at AVE.

Table 1. Results for the answer validation evaluation

	Precision	Recall	F-measure
RUN 1	0.13	0.86	0.23
RUN 2	0.30	0.59	0.39
100% VALIDATED	0.10	1.0	0.18

Table 2. Results for the QA stream fusion evaluation

	QA-accuracy	QA-reject-accuracy	estimated-QA-performance
RUN 1	0.32	0.06	0.34
RUN 2	0.27	0.22	0.33
PERFECT FUSION	0.62	0.38	0.85

Table 3. Results of the QA main task(It shows the accuracy, as well as the number of questions answered right (R), wrong (W), inexact (X), and unsupported (U))

	Factual				Definition				
	R	W	X	U	R	W	X	U	Accuracy
inaoe081eses (original QA system)	23	156	1	1	19	0	0	0	0.21
inaoe082eses (QA system with an AV module)	28	149	3	1	16	3	0	0	0.22
PERFECT VALIDATION	30	147	3	1	19	0	0	0	0.25

2.2 Spanish Question Answering Main Task

This year we submitted two different runs at the Main QA task [4]. The first run (inao081eses) was the original output of our QA system (refer to [1] for details), whereas the second run (inao082eses) was the result of applied the AV method over the set of candidate answers generated by the first run. Table 3 shows the evaluation results of both runs as well as a baseline result corresponding to a perfect validation of the output of our QA system.

Results from Table 3 indicate that the AV module helped increasing the number of right answers for factoid questions, improving the accuracy of our QA system by 22%. In contrast, the AV module damaged the treatment of definition question since it incorrectly rejected three right answers.

3 Results Analysis

In order to understand the behavior of the proposed AV method, we carried out a deep analysis of the usefulness of each one of the used characteristics over the AVE 2008 test set. The information gain (IG) values for the features used by our supervised AV method show two interesting facts. First, the proposed non–overlap features —with an average IG of 0.024— were more discriminative than the traditional overlap features (which got an average IG of 0.004). And second, in contrast to the high IG of the answer redundancy feature in the train set (a 0.184), in the test set this feature only reached a 0.015 of IG.

To evaluate the effects of these two facts over the performance of our AV method, we run two extra experiments: *i*) eliminating the attributes related to the non–overlap features; and *ii*) eliminating the attribute related to the answer redundancy. Taking as baseline the F-measure of our RUN 2 (see Table 2), the results of these extra experiments showed the following. First, the elimination of the non–overlap features decreased our baseline result in a 26% (getting a F-measure of 0.29), this result indicated the importance of these characteristics for answer validation. Second, the elimination of the answer redundancy feature allowed improving the baseline result by 5% (reaching a F-measure of 0.41), indicating that it is not relevant for this particular data set.

In relation to include the AV module into our QA system, the results in Table 3 indicate that it was only useful for factoid questions. However, due to the small number of extracted right answers for this kind of questions, it was impossible to obtain a better improvement. This fact was particularly evident for the questions answered from Wikipedia as described in [1]. Finally, taking into account that our QA system is very accurate for answering definition questions, we conclude that the use of an AV module is not convenient for this kind of questions.

4 Conclusions

This paper showed and discussed the evaluation results of our supervised AV system. The obtained results at two different scenarios (the Spanish AVE 2008 and the Spanish QA@CLEF 2008) were encouraging; the proposed method achieved a F-measure of 0.39 in the detection of correct answers, outperforming the baseline result by more than 100%. It also enhanced the performance of our Spanish QA system, producing a gain in accuracy of 22% for factoid questions. An analysis about of the results showed that the proposed non-overlap characteristics are more discriminative than the traditional overlap features, contributing in a 26% of the F-measure result.

Finally, it is important to comment that this year our best results in the AVE (a F-measure of 0.39 and a QA-accuracy of 0.32) were very distant from those corresponding to a perfect validation. We presume that this situation was caused by the decreasing number of right answers together with the increasing number of relevant support passages related to the wrong answers. In order to tackle these problems, and based on the fact that non-overlap attributes were the most discriminative, we plan to include more elements (such as prepositions, conjunctions, and some punctuation marks) for their computation.

Acknowledgments

This work was done under partial support of CONACYT (project grant 61335, and scholarships 171610 and 165499).

References

1. Téllez-Valero, A., et al.: INAOE's participation at QA@CLEF 2007. In: CLEF 2007 Working Notes, Budapest, Hungary (2007)
2. Téllez-Valero, A., et al.: INAOE at QA@CLEF 2008: Evaluating answer validation in spanish question answering. In: CLEF 2008 Working Notes, Denmark (2008)
3. Rodrigo, Á., Peñas, A., Verdejo, F.: Overview of the answer validation exercise 2008. In: CLEF 2008 Working Notes, Aarhus, Denmark (2008)
4. Forner, P., et al.: Overview of the CLEF 2008 multilingual question answering track. In: CLEF 2008 Working Notes, Aarhus, Denmark (2008)

The LIMSI Multilingual, Multitask QAst System

Sophie Rosset, Olivier Galibert, Guillaume Bernard, Eric Bilinski, and Gilles Adda

Spoken Language Processing Group, LIMSI-CNRS
{rosset,galibert,gbernard,bilinski,gadda}@limsi.fr

Abstract. In this paper, we present the LIMSI question-answering system which participated to the Question Answering on speech transcripts 2008 evaluation. This systems is based on a complete and multi-level analysis of both queries and documents. It uses an automatically generated research descriptor. A score based on those descriptors is used to select documents and snippets. The extraction and scoring of candidate answers is based on proximity measurements within the research descriptor elements and a number of secondary factors. We participated to all the subtasks and submitted 18 runs (for 16 sub-tasks). The evaluation results for manual transcripts range from 31% to 45% for accuracy depending on the task and from 16 to 41% for automatic transcripts.

1 Introduction

The Question Answering on Speech Transcripts track of the QA@CLEF task gives an opportunity to evaluate approaches able to handle speech transcriptions. In this paper, we present the architecture of the QA system developed at LIMSI for the QAst evaluation. This year 5 main tasks have been proposed: **QA in lectures** (T1, CHIL corpus), **QA in meetings** (T2, AMI corpus), **QA in broadcast news** for French (T3, ESTER corpus), **QA in European Parliament Plenary sessions** in English (T4, EPPS English corpus), and in Spanish (T5, EPPS Spanish corpus). Moreover, each task is subdivided in two sub-tasks, *a* for working on manual transcriptions and *b*, for working on one (T1b and T2b) or three (T3b, T4b and T5b) automatic transcriptions with varied word error rates (WER). We submitted 18 runs (2 runs for T3a and T5a tasks and one for each other tasks). We used the exact same system for each manual and ASR collection in order to be able to evaluate the impact of the WER on the overall system. For the different languages and tasks, we used basically the same system, the only changes were the analysis which is language dependant and the tuning parameters learned on the development data set.

The following section present the documents and queries pre-processing and the non-contextual analysis with the work carried out on the adaptation of our analysis system to Spanish. In section 3, we present the documents and snippets selection and the answer extraction and scoring. Section 4 finally presents the results for these two systems on both development and test data.

C. Peters et al. (Eds.): CLEF 2008, LNCS 5706, pp. 480–487, 2009.

2 Analysis of Documents and Queries

Our approach is to perform the same complete and multilevel analysis on both queries and documents. But first, we need to reduce the surface forms variations between the different modalities (text, manual transcripts, automatic transcripts) in order to have a common representation and use of words, sentences, case, etc. This process, a superset of tokenization, is called normalization.

2.1 Normalization

Normalization is the process by which *raw* texts are converted to a text form where words and numbers are unambiguously delimited, capitalization happens on proper nouns only, punctuation is separated from words, and the text is split into sentence-like segments (or as close to sentences as-is reasonably possible). Different normalization steps are applied, depending of the kind of input data; these steps are: **separating words and numbers** from punctuation, **reconstructing correct case** for the words, **adding punctuation** and **splitting into sentences** at period marks. Reconstructing the case and adding punctuation is done in the same process based on using a fully-cased, punctuated language model [1]. A word graph was built covering all the possible variants (all possible punctuations added between words, all possible word cases), and a 4-gram language model was used to select the most probable hypothesis. The language model was estimated on House of Commons Daily Debates, final edition of the European Parliament Proceedings and various newspapers archives. The final result, with uppercase only on proper nouns and words clearly separated by whitespace, is then passed to the non-contextual analysis.

2.2 Non Contextual Analysis Module

The *non-contextual analysis* [2] aims at extracting, from both user utterances and documents, what is considered to be *pertinent information*. The analysis covers multiple levels: Named entities detection, Linguistic chunking, Question words classification and Question topic detection. An example of an analysis result appears on figure 1. In that example, *New-York* is recognized as a named entity, specifically an organization. *municipal elections* is chunked together as a compound noun, which makes it available as a search key in the QA system. *who* is detection as a question word related to a person, and its combination with *won* allows to classify the question as one about someone's victory or achievement. We use a internal tool to write grammars based on regular expressions on words. Our tools allows the use of lists for initial detection, and the definition of local contexts and simple categorizations. This engine matches (and substitutes) regular expressions using words as the base unit This property enables the use of classes (lists of words) and macros (sub-expressions in-line in a larger expression).

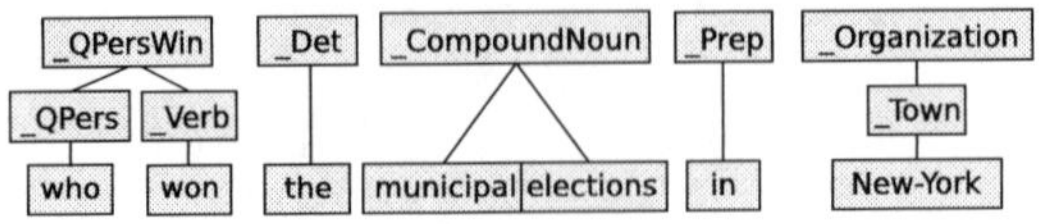

Fig. 1. Example of user utterance analysis

Adaptation to English and Spanish languages. This analysis is obviously language dependant. The French analyser detects about 300 types and constitutes the basis for the Spanish and English (T4 task only) analyzers adaptation. For T1 and T2 tasks, we used the same analyzer as in the 2007 evaluation (analyzer specifically developed for seminars and meetings). This year was our first attempt at working with Spanish. The Spanish analyser has been created as a simple adaptation of the French one where only the lexicons were adapted, and only around 50% of them. For the English a deeper adaptation is required, in particular the order in which the blocks of rules are applied is reversed. The English and Spanish analysers detect only about a hundred types. We now plan to use some aligned corpus in order to automatically acquire some specific lexicons.

3 Question-Answering System

3.1 Search Descriptor Generation

The input request takes the form of an analyzed question. From that information a *Search Descriptor* (SD) is built which is the basis of all the following search algorithms. This descriptor is structured in 3 parts: the elements of the input considered pertinent for the search, the expected type or types for the answer, and a number of tuning parameters. The types considered pertinent are the named entities and the linguistic chunks. Each entity also carries a weight, set by rules, and a critical/secondary flag. Critical entities must be present in a document near a candidate answer, secondary entities only give a bonus to the final score. This distinction aims at increasing the system precision. In practice, all named entities and some linguistic chunks are considered critical according to a set of rules. The expected answer types and their weights are decided using a 2-level rule-based classifier built by examining the development data and generalized by hand. The tuning parameters are set empirically by systematic trials on the development data. Moreover, as shown in Figure 2, possible transformations of the elements are described. These possible transformations are obtained from a few rules. This year, we used this concept to allow weighted morphological derivations and synonymic transformations. The lexicon used for morphological derivations have been built on our corpus using the analysis module to extract all values of the considered types (for example all adjectives and nouns) and to apply some derivational rules on these lists in order to built morphological correspondances.

Question: *when was Hans Krasa killed?*

- Critical element
  - 1,0 *pers* identity(Hans Krasa)
  - 0,2 *pers* expand(Hans Krasa)
- Secondary element
  - 1,0 *verb* identity(killed)
  - 0,7 *verb* lemma(killed)
  - 0,5 *verb* synonym(killed)
  - 0,5 *subs* verb_subs(killed)
- Answer types
  - 1,0 *full_date*
  - 0,9 *month_year*, *day_month*, *hour*
  - 0,7 *year*

Fig. 2. Example of a Search Descriptor: each element contains the list of triplets (type, transformation, value) under which it is expected to appear. Each triplet is weighted (*0,5* verb *synonym(killed)* a synonym of *killed* is accepted with a weight of 0.5); each possible answer type contains also a weight.

3.2 Documents Selection and Scoring

Once the SD is built, the next step is to generate a list of the n documents with the highest probability of containing the answer. The method is simple: give a score to all the documents that include at least one element of the SD and pick the n with the best scores. The score is based on the counts of occurrences of elements, ponderated by the SD weights. The tree structure is taken into account: the scores of elements in the same node are added, the scores for children have their geometric mean taken. The document score is the score of a virtual root node of all the top nodes. The index gives the raw occurrence counts for each of the elements. The analysis producing hierarchical annotations, the same instance of an element can appear under multiple types. For instance, *France* is typed as both country and location or organization each time it appears in a document. To compensate for that the counts are recomputed by subtracting the number of occurrences taken into account for the other elements of the same or upper nodes. In the specific case of QAst where the document count is very low, n is set high enough that all the documents with as least one element are picked.

3.3 Snippets Selection and Scoring

The snippet selection step aims at selecting in the documents blocks of lines with a high expectation of containing the answer. That action has a dual effect: faster answers by reducing the number of candidates to look at, and better precision of the answers given by reducing the noise introduced by faraway candidates. The idea of the method is that elements of the SD has a *distance of influence* or *range* which is counted in lines, that is sentences for text documents or utterances for spoken documents. The algorithm starts by extracting all the lines which have elements in range to satisfy all the critical elements of the SD, building that way a series of blocks. Too big blocks, i.e. above a critical *size*, are split up to try to push them under the critical size by temporarily promoting some of the secondary elements to critical status. Eventually all the blocks are small enough or all the elements have become critical and no more splitting is possible. We want these snippets to be self-contained for later candidate evaluation, which means that they must include all the elements found in the SD that made them pertinent.

3.4 Answers Selection and Scoring

The snippets are sorted by score and examined one by one independently. Every element in a snippet with a type found in the list of expected answer types is an answer candidate. Each candidate is given a score, which is the sum of the the distances between itself and the elements of the SD, each elevated to the power $-\alpha$, ponderated by the element weights. That score is smoothed with the snippet score through a δ-ponderated geometric mean. This extraction and scoring stops once a number m of candidates has been reached, once again to control the speed of the system. All the scores for the different instances of the same element are added together, and in order to compensate for the differencing natural frequencies of the entities in the documents the final score is divided by the occurence count in all the documents and in all the examined snippets, each elevated to the power β and γ respectively. The entities with the best scores then win. The tuning parameters α, β, γ, δ all come from the third part of the SD and have been selected by systematic trials on the develoment corpus. These parameters are set for each question class.

Our second approach for answer scoring is built upon the results of that first one. We compute a new ranking of the answers with a tree transformation method. For each candidate answer to a question, we transform the tree of the snippet from where the answer was extracted into the tree of the question. The sequence of operations used for the transformation gives us a transformation cost. The candidate answers are re-ranked using these costs. We applied this method as a second run for T3a and T5a tasks. The results do not yet show the expected improvement. But this work is still in progress and further analysis is needed. One positive aspect of these trials is that they show that this approach seems to be language independant (same results are obtained for French and Spanish languages).

4 Evaluation

4.1 Training and Development Data

The official development data consisted of 50 questions for each task. The development documents were 10 seminars for the T1 task, 50 meetings for the T2 task, 6 transcriptions for the T3 task, 4 for the T4 task and 1 for the T5

Table 1. The corpus: *Off. Dev.*: the official development data; *Ref. q.*: the reformulated questions based on the development documents; *Blind Corpus*: 2007 test data for T1 and T2 and new questions for T4, new questions and new documents for T3 and T5; Between parenthesis is the number of documents; *Test data*: data provided for the 2008 evaluation

	Off. Dev.	Ref. q.	Blind Corpus	Test data
T1	50 (10)	565 (10)	100 (15)	100 (15)
T2	50 (50)	587 (50)	100 (118)	100 (120)
T3	50 (6)	350 (6)	248 (3 new)	100 (12)
T4	50 (3)	277 (3)	186 (6)	100 (4)
T5	50 (1)	217 (1)	36 (1 + 1 new)	100 (4)

task. As we have observed last year, 50 questions are clearly not enough to correctly tune a system. We decided to hand-build and use a corpus of reformulated questions for each task and used them as training corpus. We built corpora of questions/answering/documents for the T3, T4 and T5 tasks and we used the 2007 evaluation data for the T1 and T2 tasks as blind development data. The Table 1 gives a general overview of the different corpus used and the test data.

4.2 Results

Table 2 presents the results and for each task the best system if applicable. In one task and its four associated sub-tasks, French Broadcast News, T3, we were the only one participant. Of the remaining 12 sub-tasks our systems reached top rank in 8 of them.

General results on manual transcripts. In Table 3, we compare the results obtained on our blind corpus and on the test data with the best tuning done on the on our reformulated questions corpus). The large drop in results between the blind corpus and the official test for the tasks T1 and T2 can be explained by a mismatch between the development and the test data. The blind corpus was creating following the question categories found in the development data,

Table 2. Official results of the QAst 2008 evaluation

Task		Acc.	Best	Task		Acc.	Best
T1	manual	41%	-	T4	manual	33%	34% UPC
	ASR	27%	31% UPC		ASR A	21%	30% INAOE
T2	manual	33%	-		ASR B	20%	-
	ASR	16%	18% UPC		ASR C	19%	-
T3	manual	45%	-	T5	manual	33%	-
	ASR A	41%	-		ASR A	24%	-
	ASR B	25%	-		ASR B	19%	-
	ASR C	21%	-		ASR C	23%	-

Table 3. Comparative results on blind corpus and test data

	T1		T2		T3		T4		T5	
	bc	test	bc	test	bc	test	bc	test	bc	test
Accuracy	64.3%	41%	44.8%	33%	41.5%	45%	26.9%	33%	36.1%	33%
MRR	0.71	0.45	0.52	0.40	0.50	0.49	0.31	0.42	0.45	0.36
Recall	80.6%	52%	61.5%	51%	61.3%	58%	38.7%	56%	61.1%	42%

Table 4. Comparison between Information Retrieval module and answer extraction and scoring module

	Information Retrieval			Answer Extraction		
Task	Acc.	MRR	Recall	Acc.	MRR	Recall
T1a	43%	0.50	58%	41%	0.45	52%
T2a	46%	0.53	62%	33%	0.40	51%
T3a	69%	0.75	84%	45%	0.49	58%
T4a	53%	0.62	73%	33%	0.42	56%
T5a	50%	0.56	65%	33%	0.36	42%

but this data was the same as for the 2007 evaluation. Meanwhile, the task had been redefined and new question categories were added. These changes were as a result missing from our blind corpus, which, while not directly used in training or tuning, was still an indicator on which changes were beneficial and as such partially directed our work. Meanwhile the results between blind corpus and test are consistant for the other tasks. We can even observe than on T3 and T4 the blind corpus is harder than the official evaluation, which made it a good guide for our work. The slight drop observed on T5 may be due to the small size of that particular corpus.

Table 4 gives the results for information retrieval and answer extraction allowing a direct comparison. A quick analysis of the problems have shown us that 3 main error sources were present: (i) Poor quality of the answer scoring. Intrinsically, working only with distances and redundancy is not enough, dependencies in particular would probably be a big help; (ii) For T1 and T2, large differences between the development and test data, in particular related to the definition questions, made for over-specialisation in some of the routing rules and poor tuning; And (iii) Some analysis errors, especially in Spanish and English, resulted in making some answers impossible to extract by the system. While the first and last point are entirely due to the system, the second one could have been avoided if the development data had been more representative of the test data.

General results on automatic transcripts. We did not do anything specific in order to handle recognition errors in the documents, the systems have been used as-is. As such our results show the loss due to the ASR. The T3b, T4b and T5b tasks provided three different ASR outputs allowing an analysis of

Table 5. Comparative results for T3b, T4b, T5b and corresponding manual data; *Acc.*: % correct answers in first rank; *WER*: Word Error Rate

	ASR_A		ASR_B		ASR_C		MAN
	Acc.	WER	Acc.	WER	Acc.	WER	Acc.
T3	41%	11%	25%	23.9%	21%	35.4%	45%
T4	21%	10.6%	20%	14%	19%	24.1%	33%
T5	24%	11.5%	19%	12.7%	23%	13.7%	33%

the impact of WER on the overall QA results. Table 5 gives the results on the ASR output depending on the task, the WER and the accuracy obtained on the respective manual transcriptions. The better quality, including robustness, on the French analysis shows up immediatly, the loss at equivalent error rate being roughly halved (5% instead of 10% at 11% WER). The loss rate does not seem to be easily predictable from the WER, but there are not enough data points to be sure. It may just be that 100 questions and a small number of documents is not enough to compute reliable statistics.

5 Conclusion

We presented the LIMSI question-answering systems on speech transcripts which participated to the QAst 2008 evaluation. These systems are based on a complete and multi-level language dependant analysis of both queries and documents followed by a language independant information retrieval and answer extraction and scoring. These systems obtained state-of-the-art results on the different tasks and languages.

References

1. Déchelotte, D., Schwenk, H., Adda, G., Gauvain, J.-L.: Improved machine translation of speech-to-text outputs, Antwerp. Belgium (2007)
2. Rosset, S., Galibert, O., Adda, G., Bilinski, E.: The limsi qast systems: comparison between human and automatic rules generation for question-answering on speech transcriptions. In: IEEE ASRU (December 2007)

IBQAst: A Question Answering System for Text Transcriptions*

María Pardiño, José M. Gómez, Héctor Llorens, Rafael Muñoz-Terol, Borja Navarro-Colorado, Estela Saquete, Patricio Martínez-Barco, Paloma Moreda, and Manuel Palomar

Natural Language Processing and Information Systems Group,
Department of Software and Computing Systems, University of Alicante, Spain
{maria,jmgomez,hllorens,rafamt,borja,stela,patricio,
moreda,mpalomar}@dlsi.ua.es
http://www.dlsi.ua.es/

Abstract. This paper shows the results of adapting a modular domain English QA system (called IBQAS, whose initials correspond to Interchangeable Blocks Question Answering System) to work with both manual and automatic text transcriptions. This system provides a generic and modular framework using an approach based on the recognition of named entities as a method of extracting answers.

1 Introduction

In this paper we are going to explain the adaptation of IBQAS, a QA system previously developed by the University of Alicante [1] [2], to the CLEF 2008 QAST (Question Answering on Speech Transcription) track. Moreover, we are going to report our official evaluation results in the frame of this CLEF 2008 QAST track.

The goal of the QAST process is to extract the correct answer to factual and definition questions over different corpus (CHIL corpus, AMI corpus, ESTER corpus, EPPS English corpus and EPPS Spanish corpus). Nevertheless, in order to perform the first participation of the University of Alicante in the CLEF 2008 QAST track, only the QA process over manual and automatic transcriptions of European Parliament Plenary sessions in English (EPPS English corpus) has been carried out.

The application of this QA system to the QAST process is explained in the following sections of the paper. Thus, next section presents a brief description of the system. Section 3 shows the results obtained by the system according to the CLEF 2008 QAST evaluation track. Finally, the last section details the conclusions and future works.

* This paper has been partially supported by the Spanish government, project TIN-2006-15265-C06-01 and project GV06028, and by the framework of the project QALL-ME, which is a 6th Framework Research Programme of the European Union (EU), contract number: FP6-IST-033860 and by the Spanish Government through the research grant AP2007-03072.

C. Peters et al. (Eds.): CLEF 2008, LNCS 5706, pp. 488–491, 2009.

2 Description of the System

This work shows the results of adapting IBQAS, a modular domain English QA system, based on the proposal of Pizzato [3] to work with text transcriptions both manual and automatic. This system provides a generic and modular framework using an approach based on the recognition of named entities as a method of extracting answers.

The system architecture follows the general methodology of QA systems incorporating the modules detailed below: question analysis, information retrieval and answer extraction. In the question analysis phase of the system, question type, keywords and focus are extracted. Question type represents the answer type and it is obtained by means of patterns that classify questions into the types of a previously defined taxonomy. In addition, keywords are extrated selecting main verb, nominal phrases nucleus, negations and their dependencies using the dependency analysis of the question. However, focus, the element related to interrogative particle what/which, is omitted from the keywords list because it rarely appears in the sentence with the answer. For the information retrieval process, we did not use the IR module incorporated in the original IBQAS because it performs searches on the Internet. Instead of it, JIRS [4] was used to search only in the predetermined corpus. Finally, relevant documents are filtered and candidate answers are extracted from them (using Lingpipe[1] to recognize location, person and organization entities; TERSEO [5] for temporal expressions and patterns to recognize other types of entities such as numeric entities, languages, ...). The last step consists in scoring and sorting the responses obtained in order to select several of them according to the distance between each response and the keywords, as well as the mutual information of the bigrams and trigrams of the passages.

3 Experiments and Results

Finally, we present the results obtained by our system in QAst task. We sent one manual run (for task T4a) and three automatic runs, one for each existing automatic transcriptions (for task T4b) all working with EPPS English corpus.

Such as we expected, the best results have been obtained with the manual transcription. This is due to the fact that this transcription has fewer errors than automatic transcriptions because most of the problems have been checked manually.

Our system, which is refered as "ua1" in result tables, has obtained results over average rates of correct answers to factual questions in the task T4a with manual transcriptions. Therefore, that satisfies us because we particularly focused on this specific part. Specifically, our system got 32 correct answers (43% of the factual questions) while the average number obtained by the set of the systems is 29,8 (40% of the factual questions). With this results, the MRR reached 0.30 and accuracy 21.3%. While the average number of successes with definitional

[1] http://alias-i.com/lingpipe/

Table 1. Results for task T4a, English EPPS, manual transcriptions (75 factual questions and 25 definitional ones

System	Factual			Definitional			All	
	# Correct	MRR	Acc(%)	# Correct	MRR	Acc(%)	MRR	Acc(%)
limsi1	44	0.43	33.3	12	0.39	32.0	0.42	33.0
inaoe1	41	0.43	37.3	6	0.21	20.0	0.38	33.0
upc1	38	0.44	40.0	4	0.16	16.0	0.37	34.0
ua1	32	0.30	21.3	4	0.16	16.0	0.27	20.0
cut1	12	0.16	16.0	9	0.36	36.0	0.21	21.0
cut2	12	0.16	16.0	11	0.39	36.0	0.22	21.0

Table 2. Results for task T4b, English EPPS, ASR transcriptions (75 factual questions and 25 definitional ones

ASR	System	Factual			Definitional			All	
		# Correct	MRR	Acc(%)	# Correct	MRR	Acc(%)	MRR	Acc(%)
a 10.6%	inaoe1	32	0.37	33.3	5	0.20	20.0	0.33	30.0
	inaoe2	34	0.38	32.0	5	0.20	20.0	0.33	29.0
	limsi1	24	0.23	18.7	9	0.31	28.0	0.25	21.0
	ua1	12	0.09	4.0	4	0.16	16.0	0.10	7.0
	upc1	18	0.22	20.0	4	0.17	16.7	0.21	19.0
	upc2	16	0.16	13.3	4	0.17	16.7	0.16	14.1
b 14.0%	limsi1	22	0.21	16.0	9	0.33	32.0	0.24	20.0
	ua1	12	0.11	8.0	4	0.16	16.0	0.12	10.0
	upc1	15	0.18	16.0	4	0.16	16.0	0.17	16.0
	upc2	14	0.16	13.3	4	0.16	16.0	0.16	14.0
c 24.1%	limsi1	21	0.21	16.0	8	0.30	28.0	0.23	19.0
	ua1	9	0.10	8.0	5	0.20	20.0	0.12	11.0
	upc1	11	0.11	9.3	5	0.20	20.0	0.14	12.0
	upc2	11	0.11	8.0	4	0.16	16.0	0.12	10.0

questions among all the system that have participated, was almost climb up to 0.31% of success (an average of 7,7 correct definitional answers), our system –as expected because none specific treatment was implemented for definitional questions– has obtained much lower performance getting only right answers in 16% of the cases (4 correct definitional answers). In this sense, the results obtained in connection with MRR and accuracy are 0.16 and 16.0% respectively, while the average values reach 0.28 for MRR and 26% of accuracy. In spite of the descense of performance due to not to deal with definitional questions, our results finally amount to 0.27 MRR and 20.0% of accuracy.

An important point is that, our system does not include any specific treatment to improve performance when working with automatic collections. In this regard, we believe it could be interesting to use the different automatic transcriptions simultaneously in the indexing process to compensate for the errors presented in each one.

With regard to the automatic transcriptions, the best results considering MRR and accuracy, were obtained with the transcription C versus transcription B (although these are very close to those of B) and transcription A (these are slightly smaller).

In addition, to explain the results obtained, we must not forget the problems arose in the development of this work. On the one hand, the small size of the corpus, and hence, the consequent low redundancy in them, made difficult to adapt our system. On the other hand, the existence of broad types of questions made not possible to cover them in our system.

4 Conclusions

In our first participation in QAst, we have adapted a generic and modular QA system to work with text transcriptions. We want to highlight that its results are above expectation because we did not use any specific resource to deal with automatic transcriptions. Despite using a generic system, the results are not discouraging.

In the future, we hope to obtain a better system capable of answering questions from the task in a more precise way. Moreover, we wish to measure the improvements we introduce in our system compared to the state-of-art at the moment. Emphasize that for more information about the proposal, QAst Overview [6] and Working Notes [7] are availabled.

References

1. Moreda, P., Llorens, H., Saquete, E., Palomar, M.: The influence of semantic roles in qa: A comparative analysis. In: Actas del XXIV Congreso de la SEPLN, vol. 41, pp. 55–62 (2008)
2. Moreda, P., Llorens, H., Saquete, E., Palomar, M.: Automatic generalization of a QA answer extraction module based on semantic roles. In: Geffner, H., Prada, R., Machado Alexandre, I., David, N. (eds.) IBERAMIA 2008. LNCS (LNAI), vol. 5290, pp. 233–242. Springer, Heidelberg (2008)
3. Pizzato, L.A., Molla, D.: Extracting exact answers using a meta question answering system. In: Proceedings of the Australasian Language Technology Workshop 2005, Sydney, Australia, December 2005, pp. 105–112 (2005)
4. Gómez-Soriano, J.M., Montes-y-Gómez, M., Sanchis-Arnal, E., Villaseñor-Pineda, L., Rosso, P.: Language independent passage retrieval for question answering. In: Gelbukh, A., de Albornoz, Á., Terashima-Marín, H. (eds.) MICAI 2005. LNCS (LNAI), vol. 3789, pp. 816–823. Springer, Heidelberg (2005)
5. Saquete, E.: Temporal information Resolution and its application to Temporal Question Answering. Phd, Departamento de Lenguages y Sistemas Informáticos. Universidad de Alicante (June 2005)
6. Turmo, J., Comas, P., Rosset, S., Lamel, L., Moreau, N., Mostefa, D.: Overview of qast 2008. In: QAst at CLEF (2008)
7. Pardiño, M., Gómez, J., Llorens, H., Muñoz-Terol, R., Navarro-Colorado, B., Saquete, E., Martínez-Barco, P., Moreda, P., Palomar, M.: Adapting ibqas to work with text transcriptions in qast task: Ibqast. In: QAst at CLEF Working Notes (2008)

Robust Question Answering for Speech Transcripts: UPC Experience in QAst 2008

Pere R. Comas and Jordi Turmo

TALP Research Center, Technical University of Catalonia (UPC)
pcomas@lsi.upc.edu, turmo@lsi.upc.edu

Abstract. This paper describes the participation of the Technical University of Catalonia in the CLEF 2008 Question Answering on Speech Transcripts track. We have participated in the English and Spanish scenarios of QAst. For the processing of manual transcripts we have deployed a robust factoid Question Answering that uses minimal syntactic information. For the handling of automatic transcripts we modify the QA system with a Passage Retrieval and Answer Extraction engine based on a sequence alignment algorithm that searches for "sounds like" sequences. We perform a detailed analysis of our results and draw conclusions relating QA performance to word error rate in transcripts.

Keywords: Question Answering, Spoken Document Retrieval, Evaluation.

1 Introduction

The CLEF 2008 Question Answering on Speech Transcripts (QAst) track [8] consists of five scenarios with several tasks: Question Answering (QA) in manual transcripts of recorded lectures (T1A) and their corresponding automatic transcripts (T1B), QA in manual transcripts of recorded meetings (T2A) and their corresponding automatic transcripts (T2B), QA in manual transcripts of French European Parliament Sessions (T3A) and three different automatic transcripts (T3B-A, T3B-B, T3B-C), QA in manual transcripts of English European Parliament Sessions (T4A) and three different automatic transcripts (T4B-A, T4B-B, T4B-C), QA in manual transcripts of Spanish European Parliament Sessions (T5A) and three different automatic transcripts (T5B-A, T5B-B, T5B-C). The automatic transcripts for tasks T3, T4 and T5 have different levels of word error rate (WER). WERs for T4 are 10.6%, 14%, and 24.1%. For T5 WERs are 11.5%, 12.7% and 13.7%. This paper summarizes our methods and results in QAst. We have participated in all the scenarios except the French language one (T3).

Our QA system is based on our previous work in [3,6] and [7]. We have used the same system architecture for all the tasks, having interchangeable language–dependant parts and different passage retrieval algorithms for automatic transcripts.

C. Peters et al. (Eds.): CLEF 2008, LNCS 5706, pp. 492–499, 2009.

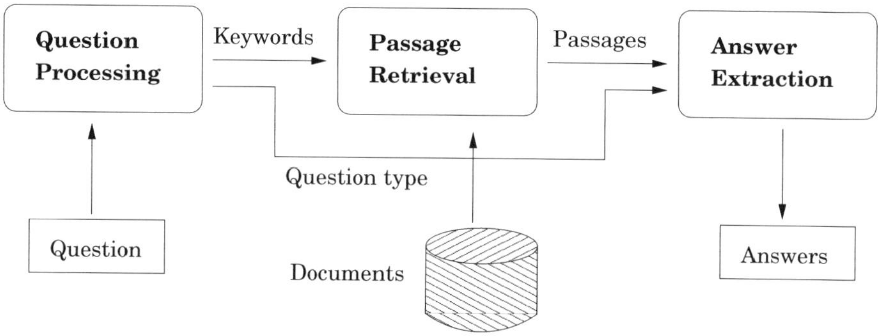

Fig. 1. Overview of QA architecture

2 Overview of the System Architecture

The architecture of our QA system follows a commonly-used schema which splits the process into three phases performed sequentially: Question Processing (QP), Passage Retrieval (PR), and Answer Extraction (AE), as shows Figure 1. These three phases are described in the following sections.

2.1 Question Processing and Classification

The main goal of this component is to detect the type of the expected answer. We currently recognize the 53 open-domain answer types from [4] plus 3 types specific to QAst corpora (i.e., `system/method`, `shape`, and `material`). The answer types are extracted using a multi-class Perceptron classifier and a rich set of lexical, semantic and syntactic features. This classifier obtains an accuracy of 88% on the corpus of [4]. Additionally, the QP component extracts and ranks relevant keywords from the question.

For scenario T5, he have developed an Spanish question classifier using human translated questions from the corpus of [4] following the same machine learning approach. This classifier obtains an accuracy of 74%.

2.2 Passage Retrieval

This component retrieves a set of relevant passages from the document collection, given the previously extracted question keywords. The PR algorithm uses a query relaxation procedure that iteratively adjusts the number of keywords used for retrieval and their proximity until the quality of the recovered information is satisfactory (see [6]). In each iteration a Document Retrieval application (Lucene IR engine) fetches the documents relevant for the current query and a subsequent passage construction module builds passages as segments where two consecutive keyword occurrences are separated by at most t words.

When dealing with automatic transcripts, you have to bear in mind that the state of the art in (Automatic Speech Recognizer) ASR technology is far from

1M: *"The host system it is a UNIX Sun workstation"*
1A: *"that of system it is a unique set some workstation"*
2M: *"Documents must be separated into relevant documents and irrelevant documents by a manual process, which is very time consuming."*
2A: *"documents must be separated into relevant documents and in relevant document by a manual process witches' of very time consuming"*

Fig. 2. Examples of manual (M) and automatic (A) transcripts

perfect. For example, the word error rate (WER) of the meetings automatic transcripts (T1B) is around 38% and the WER of the lectures (T2B) is over 20%, and from 10.6% to 24.1% for the T4B transcripts. Figure 2 shows two real examples of common errors when generating automatic transcripts. From the point of view of passage retrieval, imperfect transcripts create a new problem of incorrectly transcribed words that yield false positives and false negative for traditional search methods.

To overcome such drawbacks, we have used an IR engine relying on phonetic similarity for the automatic transcripts. This tool is called PHAST (after PHonetic Alignment Search Tool) and uses pattern matching algorithms to search for small sequences of phones (the keywords) into a larger sequence (the documents) using a measure of sound similarity. A detailed description of PHAST can be found in [2].

2.3 Answer Extraction

Identifies the exact answer to the given question within the retrieved passages. First, answer candidates are identified as the set of NEs that occur in these passages and have the same type as the answer type detected by QP. Then, these candidates are ranked using a scoring function based on a set of heuristics that measure keyword distance and density[5]. These heuristic measures use approximated matching for AE in automatic transcripts as shown in the passage retrieval module from the previous section.

The same measure is used for English and Spanish scenarios.

3 Named Entity Recognition and Classification

As described before, we extract candidate answers from the NEs that occur in the passages retrieved by the PR component.

We have used machine learning based Named Entity Recognizer and Classifier (NERC) for English tasks. We have learned different models for automatic and manual transcripts, including combination of data sources and atributes based on phonetic similarity. This NERC is close to the one used in our previous work for QAst 2007 [3].

For the Spanish track T5 we have used a previously developed NERC. It also uses a machine learning approach and it has been trained with the CoNLL

Spanish corpus. See details in [1]. Unfortunately, this NERC can recognize only person, location and organization Named Entity (NE) types. Thus only these types can be used as answer candidates. It supposes a serious shortcoming for QA performance as the results show in Section 4.

4 Experimental Results

UPC participated in 4 of the 5 scenarios, all but the French one (T3). We submitted two runs for the tasks on automatic transcripts, one run using the standard QA system for written text (QA_m) and another run using the system tailored for automatic transcripts (QA_a). See section 2 for the differences between both. Each scenario included 100 test questions, from which 10 do not have an answer in the corpora (these are *nil* questions). Around 75% of the questions are of factoid types and around 25% are definitional. Our QA system is designed to answer only factoid questions, therefore our experimental analysis will only refer to factoid questions.

We report two measures: (a) TOPk, which assigns to a question a score of 1 only if the system provided a correct answer in the top k returned; and (b) Mean Reciprocal Rank (MRR), which is the multiplicative inverse of the rank of the first correct answer returned. The official evaluation of QAst 2008 uses TOP1 and TOP5 measures [8]. An answer is considered correct by the human evaluators if it contains the complete answer and nothing more, and it is supported by the corresponding document. If an answer was incomplete or it included more information than necessary or the document did not provide the justification for the answer, the answer was considered incorrect.

Table 1 summarizes our overall results for factoid question only. The cost of moving from manual transcripts to automatic transcripts (i.e., the difference between TXA and TXB) is a loss in TOP1 score of at last 10% for T1, 43% for T2, 50% for T4 and 42% for T5. The performance of QA_a is very similar to QA_m. As shown in QAst 2008 Overview paper [8], UPC has ranked among the top teams in tasks T1, T2 and T4. Our team got the best TOP1 score in T1B,

Table 1. Overall results for our twenty QAst runs

Task, System	#Q	MRR	TOP1	TOP5	Task, System	#Q	MRR	TOP1	TOP5
T1A, QA_m	78	0.44	30	39	T2A, QA_m	74	0.35	23	29
T1B, QA_m	78	0.39	27	35	T2B, QA_m	74	0.20	13	19
T1B, QA_a	78	0.37	26	35	T2B, QA_a	74	0.16	8	16
T4A, QA_m	75	0.44	30	38	T5A, QA_m	75	0.11	7	9
T4B A, QA_m	75	0.22	15	18	T5B A, QA_m	75	0.05	3	5
T4B B, QA_m	75	0.18	12	15	T5B B, QA_m	75	0.06	4	5
T4B C, QA_m	75	0.11	7	11	T5B C, QA_m	75	0.03	2	2
T4B A, QA_a	75	0.16	10	16	T5B A, QA_a	75	0.06	4	5
T4B B, QA_a	75	0.16	10	14	T5B B, QA_a	75	0.06	4	5
T4B C, QA_a	75	0.11	6	11	T5B C, QA_a	75	0.03	2	3

Table 2. Distribution of correct answers (TOP5) according to answer type. Org = organization, Per = person, Tim = time, Mea = measure, Met/Sys = method/system, Mat = material, Col = color, Def = definitional.

Task, System	Org	Per	Loc	Tim	Mea	Sys	Lan	Sha	Mat	Col	Def
T1A, QA_m	4/8	8/9	1/2	3/5	13/19	4/5	6/10	0/8	0/3	0/9	4/22
T1B, QA_m	3/8	5/9	1/2	3/5	13/19	4/5	6/10	0/8	0/3	0/9	4/22
T1B, QA_a	4/8	4/9	2/2	2/5	14/19	3/5	6/10	0/8	0/3	0/9	4/22
T2A, QA_m	1/8	2/8	7/10	1/8	4/10	3/6	2/8	1/4	4/6	4/6	3/26
T2B, QA_m	3/8	2/8	1/10	0/8	3/10	1/8	1/8	1/4	4/6	3/6	5/26
T2B, QA_a	1/8	3/8	2/10	0/8	1/10	1/8	1/8	1/4	4/6	2/6	6/26
T4A, QA_m	7/14	9/14	6/15	9/15	7/15	0/2	-	-	-	-	4/25
T4B-A, QA_m	1/14	0/14	3/15	8/15	6/15	0/2	-	-	-	-	4/25
T4B-B, QA_m	1/14	0/14	2/15	8/15	4/15	0/2	-	-	-	-	5/25
T4B-C, QA_m	0/14	1/14	2/15	1/15	6/15	1/2	-	-	-	-	4/25
T4B-A, QA_a	1/14	0/14	2/15	7/15	6/15	0/2	-	-	-	-	4/25
T4B-B, QA_a	1/14	0/14	1/15	8/15	4/15	0/2	-	-	-	-	5/25
T4B-C, QA_a	0/14	1/14	2/15	1/15	6/15	1/2	-	-	-	-	4/25
T5A, QA_m	1/10	8/21	0/5	0/25	0/14	-	-	-	-	-	3/25
T5B-A, QA_m	1/10	3/21	1/5	0/25	0/14	-	-	-	-	-	0/25
T5B-B, QA_m	2/10	2/21	0/5	0/25	0/14	-	-	-	-	-	0/25
T5B-C, QA_m	0/10	3/21	0/5	0/25	0/14	-	-	-	-	-	2/25
T5B-A, QA_a	1/10	3/21	1/5	0/25	0/14	-	-	-	-	-	0/25
T5B-B, QA_a	2/10	3/21	0/5	0/25	0/14	-	-	-	-	-	2/25
T5B-C, QA_a	0/10	2/21	0/5	0/25	0/14	-	-	-	-	-	1/25

T2B and TA4 tracks, although the differences were not significant. For task T5 our results were far beyond other participants.

Table 2 shows the distribution of correct answers for all tasks according to the answer type. In scenario T4, a design error prevented our NERC from recognizing entity types `Sha`, `Mat` and `Col`. Therefore there are 20 unanswerable questions from the 78 factoid ones. Our system for the Spanish scenario (T5) is limited to answer types `Org`, `Per`, and `Loc`, so the real upper bound for factoid questions is 36 instead of 75.

Finally, Table 3 summarizes the error analysis of QP, PR, and AE parts. The meaning of each column is the following. Q: number of factoid question. QC: number of questions with answer type correctly detected by QP. PR: number of question where at least on passage with the correct answer war retrieved. C.NE: number of questions where the retrieved passages contain the correct answer tagged as a NE of the right type. U.NE: number of questions where the retrieved passages contain the correct answer but it remains undetected by the NERC. Er.NE: number of questions where the retrieved passages contain the correct answer tagged as a NE with an incorrect type. QC&PR: number of questions with correct answer type and correct passage retrieval. QC&NE: number of questions with correct answer type and correctly tagged answer in the passages. TOP5 non-nil: number of question with non-nil answer correctly

Table 3. Error analysis of the QA system components

Track	System	Q	QC	PR	C. NE	U. NE	Er. NE	QC& PR	QC& NE	TOP5 non-Null
T1A	QA_m	78	70	69	42	21	6	62	38	37
T1B	QA_m	78	70	61	39	20	2	55	34	33
	QA_a	78	70	59	36	22	1	53	33	33
T2A	QA_m	74	61	46	31	10	5	41	28	29
T4A	QA_m	75	62	60	41	6	13	60	41	37
T4B-A	QA_m	75	62	56	24	24	8	46	18	17
	QA_a	75	62	56	24	24	8	44	16	15
T4B-B	QA_m	75	62	55	21	26	8	45	16	14
	QA_a	75	62	57	21	28	8	46	15	14
T4B-C	QA_m	75	62	52	9	26	17	43	9	7
	QA_a	75	62	48	10	26	12	36	7	7
T5A	QA_m	75	18	55	21	31	3	21	8	9
T5B-A	QA_m	75	18	54	14	36	5	5	3	5
	QA_a	75	18	58	15	38	4	5	3	5
T5B-B	QA_m	75	18	58	14	40	4	6	2	5
	QA_a	75	18	60	15	39	6	6	2	5
T5B-C	QA_m	75	18	55	12	34	9	3	0	2
	QA_a	75	18	60	15	41	4	5	0	2

answered by our system in the TOP5 candidates. Due to technical reasons this analysis has not been performed on task T2B.

We can draw several important observations from this error analysis: Question classification performs better for T1 question set than T2 and T4 question sets. This suggests that in this evaluation T1 questions were more domain specific than the others. In T5, results are really disappointing and this suggests that our Spanish classifier may be too domain dependant since it achieves 74% accuracy in our test data. PR is specially degraded in task T4B-C, where we processed automatic transcripts with the highest WER (24.1%). This proves that passage retrieval is indeed affected by a high WER but is robust enough to be used with a *good* ASR. Passage retrieval using PHAST performed better than the passage retrieval with classical retrieval for tasks in T5 and worse for tasks in T4. Since both scenarios have similar domain, we think this difference is due to the nature of Spanish and English phonology. Further experiments in [2] show consistently that passage retrieval in Spanish is improved by using PHAST. As the table shows, the bad performance of NERC is the critical problem of our QA system. The difference between C.NE and PR values is much bigger than between PR and Q, thus the theoretical upper bound for answer extraction is limited specially by NERC performance. The average number of factoid questions in all runs is 75.3, the average value for PR is 56.61 and the average for C.NE is 22.44, so in less than 40% of the passages the answer is correctly tagged allowing its correct extraction in the answer extraction step. QC&NE is a theoretical upper bound of the total score of each task. We can see that the performance of our answer

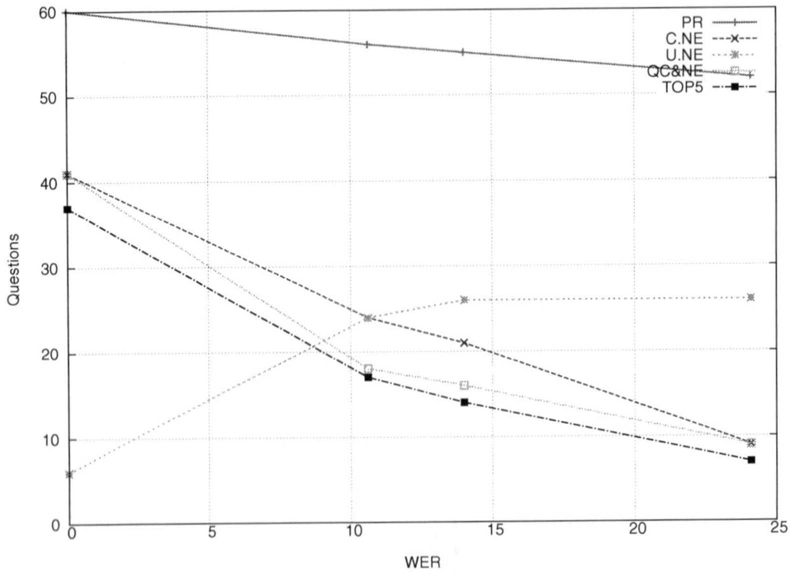

Fig. 3. Impact of ASR errors

extraction process is very good since TOP5 score is very near this upper bound in all tasks. As a remark, all of the scores in T5 are above the upper bound. This is due to the combination of two factors: first, a fall-back mechanism in our answer extraction process to help overcome the `PER/ORG` ambiguity[1] in question classification, this mechanism allows to answer misclassified questions. Second, a double–error situation when the question is misclassified and the answer is erroneously tagged but matches the question type.

The impact of transcription errors in QA can be analyzed in detail thanks to the three different automatic transcripts for task T4B (WERs of T5B have very close values and our overall performance is far too poor for this analysis). Figure 3 graphically shows the values in Table 3 for T4, QA_m. The WER percentage is on y axis and the lines show the evolution of variables PR, C.NE, U.NE, QC&NE and TOP5. The performance of passage retrieval decreases linearly as WER increases. The linear regression PR $= 59.78 - 0.33 \cdot$ WER fits the data with a Pearson coefficient $r = 0.99$. Other measures such as C.NE, QC&NE and TOP5 have a more pronounced diminishment. C.NE values fit also a linear regression C.NE $= 39.83 - 1.33 \cdot$ WER with a coefficient $r = 0.99$. The slope for C.NE is four times stepper that for PR, therefore we can conclude that the impact of WER on passage retrieval is much less severe than in NERC.

[1] In questions such as "*Who helped solving the packet loss problem?*" is impossible to know if the correct answer is a person name or an organization name.

5 Conclusions

This paper describes UPC's participation in the CLEF 2008 Question Answering on Speech Transcripts track. We submitted runs for all English and Spanish scenarios, obtaining the best results in some tasks. In this evaluation we analyzed the behavior of two systems differing in that one is tailored for manual transcripts while the other is tailored for automatic transcripts (uses approximate keyword search based on phonetic distances and a NERC enhanced with phonetic features).

Our approximated keyword search algorithm used for passage retrieval obtains mixed results. It can improve standard search for Spanish but makes little difference for English. We think this because in some document collections it may generated too many false-positive, introducing noise in sets of candidate passages and answers. Nevertheless, we believe that this approach is a good long-term research direction because it can truly address the phenomena specific to automatic transcripts.

Finally, our results show that automatic speech recognition has critical impact on the performance of NERC but its affect on passage retrieval is much less severe.

Acknowledgements

This work has been partially funded by the Spanish Ministry of Science and Technology (TEXTMESS project, TIN2006–15265–C06–05).

References

1. Carreras, X., Márquez, L., Padró, L.: Named entity extraction using adaboost. In: COLING 2002: proceedings of the 6th conference on Natural language learning (2002)
2. Comas, P.R., Turmo, J.: Spoken document retrieval based on approximated sequence alignment. In: Sojka, P., Horák, A., Kopeček, I., Pala, K. (eds.) TSD 2008. LNCS (LNAI), vol. 5246, pp. 285–292. Springer, Heidelberg (2008)
3. Comas, P.R., Turmo, J., Surdeanu, M.: Robust question answering for speech transcripts using minimal syntactic analysis. In: Peters, C., Jijkoun, V., Mandl, T., Müller, H., Oard, D.W., Peñas, A., Petras, V., Santos, D. (eds.) CLEF 2007. LNCS, vol. 5152, pp. 424–432. Springer, Heidelberg (2008)
4. Li, X., Roth, D.: Learning question classifiers: The role of semantic information. Journal of Natural Language Engineering (2005)
5. Paşca, M.: High-performance, open-domain question answering from large text collections. PhD thesis, Southern Methodist University, Dallas, TX (2001)
6. Surdeanu, M., Dominguez-Sal, D., Comas, P.R.: Design and performance analysis of a factoid question answering system for spontaneous speech transcriptions. In: INTERSPEECH 2006 (2006)
7. Surdeanu, M., Turmo, J., Comelles, E.: Named entity recognition from spontaneous open-domain speech. In: INTERSPEECH 2005 (2005)
8. Turmo, J., Comas, P.R., Rosset, S., Lamel, L., Moureau, N., Mostefa, D.: Overview of QAST 2008. In: Peters, C., et al. (eds.) CLEF 2008. LNCS, vol. 5706, pp. 314–324. Springer, Heidelberg (2009)

Overview of the ImageCLEFphoto 2008 Photographic Retrieval Task

Thomas Arni[1], Paul Clough[1], Mark Sanderson[1], and Michael Grubinger[2]

[1] Department of Information Studies, University of Sheffield, UK
[2] Victoria University, Melbourne, Australia

Abstract. ImageCLEFphoto 2008 is an ad-hoc photo retrieval task and part of the ImageCLEF evaluation campaign. This task provides both the resources and the framework necessary to perform comparative laboratory-style evaluation of visual information retrieval systems. In 2008, the evaluation task concentrated on promoting diversity within the top 20 results from a multilingual image collection. This new challenge attracted a record number of submissions: a total of 24 participating groups submitting 1,042 system runs. Some of the findings include that the choice of annotation language is almost negligible and the best runs are by combining concept and content-based retrieval methods.

Keywords: Performance Evaluation, IAPR TC-12 Benchmark, Image Retrieval, Diversity, Clustering.

1 Introduction

The evaluation of multilingual image retrieval systems (i.e. where associated texts are in languages different from written queries) has been the focus of ImageCLEF since its inception in 2003. The track has evolved over the years to address different domains (e.g. cultural heritage, medical imaging and Wikipedia), and different kinds of tasks (e.g. ad-hoc retrieval, automatic annotation and clustering). The focus of the ImageCLEFphoto task in 2008 has been to promote diversity in the top *n* results (see section 1.2). The resources provided enable system-centred evaluation for multilingual and diversity-based visual information retrieval based on a collection of "general" photographs (see section 2.1).

1.1 Evaluation Scenario

The evaluation scenario is similar to the classic TREC[1] ad-hoc retrieval task: a simulation of the situation in which a system knows the set of documents to be searched, but cannot anticipate the particular topic that will be investigated (i.e. the search topics are not known to the system in advance) [1]. The goal of the simulation is: given an alphanumeric statement (and/or sample images) describing a user's information need, find as many relevant images as possible from the given collection (with the

[1] http://trec.nist.gov/

C. Peters et al. (Eds.): CLEF 2008, LNCS 5706, pp. 500–511, 2009.

query language either being identical or different from that used to describe the images). For 2008, the scenario is slightly different in that systems must return relevant images from as many different sub-topics as possible (i.e. promote diversity) in the top *n* results.

1.2 Evaluation Objectives

The main objective of ImageCLEFphoto for 2008 comprised the evaluation of ad-hoc multilingual visual information retrieval systems from a general collection of annotated photographs (i.e. image with accompanying semi-structured captions such as the title, location, description, date or additional notes). However, this year focused on a particular aspect of retrieval: diversity of the results set (see section 1.3). More recently, research in image search has concentrated on ensuring that duplicate or near-duplicate documents retrieved in response to a query are hidden from the user. This should ideally lead to a ranked list where images are both relevant *and* diverse. In 2007, the task considered maximising the number of relevant documents in the resulting ranked list. In 2008, the task is to promote diversity in the top *n* results, which has been shown to better satisfy a user's information need [2, 3] (people often type in the same query but prefer to see results which represent different aspects of the results set). Hence, providing a diverse results list is especially important when a user types in a query that is either poorly specified or ambiguous.

This new challenge allows for the investigation of a number of research questions, including the following:

- Is it possible to promote diversity within the top *n* results?
- Which approaches work best at promoting diversity?
- Does promoting diversity reduce the number of relevant images in the top *n* results?
- Can "standard" text retrieval methods be used to promote diversity?
- How does the retrieval performance compare between bilingual and multilingual annotations?

One major goal of ImageCLEFphoto 2008 was to attract participants from various backgrounds and with different research interests. The collection developed for the 2008 task, in our view, provides a resource that can be used to evaluate both concept and content-based approaches for image retrieval. Further analysis of results can be found in [4].

2 Evaluation Framework

Similar to the 2006 and 2007 ImageCLEFphoto tasks [5, 6], we generated a subset of the IAPR TC-12 Benchmark as an evaluation resource for 2008. This section provides more information on these individual components: the document collection, the query topics, relevance judgements, cluster relevance judgements and performance indicators. More information on the design and implementation of the IAPR TC-12 Benchmark itself, created under Technical Committee 12 (TC-12) of the International Association of Pattern Recognition (IAPR[2]), can be found in [7].

[2] http://www.iapr.org/

2.1 Document Collection

The IAPR TC-12 Benchmark consists of 20,000 colour photographs taken from locations around the world and comprises a varying cross-section of still natural images. Figure 1 illustrates a number of sample images from a selection of categories.

Sports Landscapes Animals People

Fig. 1. Sample images from the IAPR TC-12 collection [7]

The majority of images have been provided by Viventura[3], an independent travel company that organises adventure and language trips to South America. Travel guides accompany the tourists and maintain a daily online diary including photographs of trips made and general pictures of each location including accommodation, facilities and ongoing social projects. In addition to these photos, a number of photos from a personal archive have also been added to form the collection used in ImageCLEF. The collection is publicly available for research purposes and, unlike many existing photographic collections, can be used to evaluate image retrieval systems. The collection is general in content with many different images of similar visual content, but varying illumination, viewing angle and background. This makes it a challenge for the successful application of techniques involving visual analysis.

Each image in the collection has a corresponding semi-structured caption consisting of the following six fields: (1) a unique identifier, (2) a title, (3) a free-text description of the semantic and visual contents of the image, (4) notes for additional information, (5) where and (6) when the photo was taken. Figure 2 shows a sample image with its corresponding textual annotation (in English). By using a custom-built application for managing the images, various subsets of the collection can be generated with respect to a variety of particular parameters (e.g. using a selected subset of caption fields). For 2008, the following data was provided:

- **Annotation (caption) language:** two sets of annotations in (1) English and (2) Random. In the random set, the annotation language was randomly selected from for each of the images (i.e. annotations are either German or English image captions).
- **Caption fields:** all caption fields were provided for the 2008 task.
- **Annotation completeness:** each image caption exhibited the same level of annotation completeness - there were no images without annotations (as experimented with in 2006). The participants were granted access to the data set on 22nd April 2008 and had exactly one month to familiarise themselves with the new subset. Most participants had to modify their standard retrieval systems in order to generate diverse results in the top *n*.

[3] http://www.viventura.de/

<**DOC**>
<**DOCNO**>annotations/16/16392.eng</**DOCNO**>
<**TITLE**>Sunset in Salvador</**TITLE**>
<**DESCRIPTION**>a sandy beach at the sea with dark rocks behind it; the setting sun in an orange sky in the background;
</**DESCRIPTION**>
<**NOTES**></**NOTES**>
<**LOCATION**>Salvador, Brazil</**LOCATION**>
<**DATE**>10 October 2004</**DATE**>
<**IMAGE**>images/16/16392.jpg</**IMAGE**>
<**THUMBNAIL**>thumbnails/16/16392.jpg</**THUMBNAIL**>
</**DOC**>

Fig. 2. Sample image and associated caption

2.2 Query Topics

From an existing set of 60 topics, 39 were selected and distributed to participants (Table 1) representing varying search requests (many of these are realistic and based on queries extracted during log file analysis – see [8] for more detailed information). We found that for the new retrieval challenge (promoting diversity), not all of the existing topics were suitable and therefore some were removed (see [9] for further details). Although 21 topics were removed, the remaining 39 topics are well-balanced, diverse and should present a retrieval challenge to participants wishing to use either text and/or low-level visual analysis techniques for creating clusters.

Similar to TREC, the query topics were provided as structured statements of user needs. The full description of a topic consists of (1) a topic titles (2) a topic narrative, (3) a newly added *cluster type* and (4) three example relevant images for that topic. An additional field was added called *cluster type*, which was augmented for easier assessment of the clusters as well as to facilitate the quantification of the result set diversity [9]. Below is an example augmented topic:

<top>
<num> Number: 48 </num>
<title> vehicle in South Korea </title>
<cluster> vehicle </cluster>
</top>

The cluster type in topic 48 is vehicle (in the <cluster> tag), which clearly defines how relevant images from this topic should be clustered. Different from previous years, topics were available in English only.

2.3 Relevance Assessments

The relevance assessments, with the exception of removing any additional images considered as non-relevant, are exactly the same as 2007 (no pooling of the images was carried out in 2008). Information about relevance assessments from previous years can be found in [6]. To enable diversity to be quantified, it was necessary to classify images relevant to a given topic to one or more sub-topics or clusters. This was performed by two assessors. In case of inconsistent judgements, a third assessor

Table 1. Topics selected for the ImageCLEFphoto 2008 task (from 2007 topics)

ID	Topic title	ID	Topic title
2	church with more than two towers	3	religious statue in the foreground
5	animal swimming	6	straight road in the USA
10	destinations in Venezuela	11	black and white photos of Russia
12	people observing football match	13	exterior view of school building
15	night shots of cathedrals	16	people in San Francisco
17	lighthouse at the sea	18	sport stadium outside Australia
19	exterior view of sport stadium	20	close-up photograph of an animal
21	accommodation provided by host families	23	sport photos from California
24	snowcapped building in Europe	28	cathedral in Ecuador
29	views of Sydney's world-famous landmarks	31	volcanoes around Quito
34	group picture on a beach	35	bird flying
37	sights along the Inka-Trail	39	people in bad weather
40	tourist destinations in bad weather	41	winter landscape in South America
43	sunset over water	44	mountains on mainland Australia
48	vehicle in South Korea	49	images of typical Australian animals
50	indoor photos of a church or cathedral	52	sports people with prizes
53	views of walls with unsymmetric stones	54	famous television (and telecommunication) towers
55	drawings in Peruvian deserts	56	photos of oxidised vehicles
58	seals near water	59	creative group pictures in Uyuni
60	salt heaps in salt pan		

was used to resolve the inconsistencies. The resulting cluster assessment judgements are then used in combination with the normal relevance assessment to determine the retrieval effectiveness of each submitted system run (for further details see [9]).

2.4 Generating the Results

Once the relevance judgments and the cluster relevance assessments were completed, the performance of individual systems and approaches can be evaluated. The results for submitted runs were computed using the latest version of trec eval (http://trec.nist.gov/trec_eval/trec_eval.8.1.tar.gz), as well as a custom-built tool to calculate diversity of the results set. Submissions were evaluated using two metrics: (1) precision at rank 20 (P20) and (2) cluster recall at rank 20 (CR20). Rank 20 was selected as the cut-off point to measure precision and cluster recall because most online image retrieval engines (e.g. Google, Yahoo! and AltaVista) display 18 to 20 images by default. Further measures considered included uninterpolated (arithmetic) Mean Average Precision (MAP), Geometric Mean Average Precision (GMAP) to test system robustness and binary preference (bpref), which is a good indicator of how complete relevance judgments are. To enable an absolute comparison between individual runs, a single metric is required: the F1-measure was used to combine scores from P20 and CR20 (representing the harmonic mean of P20 and CR20).

3 Participants and Submitted Runs

In 2008, 43 groups registered for ImageCLEFphoto (32 in 2007; 36 in 2006), with 24 groups eventually submitting a total of 1,042 runs (all of which were evaluated by the organisers). This is an increase in the number of runs from previous years (20 groups submitting 616 runs in 2007; 12 groups submitting 157 runs in 2006; 11 groups submitting 349 runs in 2005). The 24 participating groups are affiliated to 21 different institutions in 11 countries. New participants submitting in 2008 include joint work from four French labs (AVEIR), University of Waseda (GITS), Laboratory of Informatics of Grenoble (LIG), System and Information Science Lab (LSIS), Meiji University (Meiji), University of Ottawa (Ottawa), Telecom ParisTech (PTECH), University of Sheffield (Shef), University of Alicante (TEXTMESS) and Piere & Marie Curie University (UPMC). In total, 65% of the participants in 2007 returned and participated in 2008.

Increased participation might be an indicator of (1) the growing need for evaluation of visual information retrieval from more general photographic collections, (2) the growing need for comparative evaluation of diversity and/or (3) an interest by researchers world-wide to participate in evaluation events such as ImageCLEFphoto. Although the total number of runs rose, the geometric mean of runs per participating group was slightly lower than 2007 (12.4 in 2008; 13.8 in 2007). The reason for the increasing number of total runs is mainly due to the larger number of submissions from Dublin City University (DCU), who submitted a total of 733 runs (no upper limit was placed on the number of runs groups could submit).

3.1 Overview of Submissions

Overall, 1042 runs were submitted and categorised with respect to the following dimensions: (1) annotation language, (2) modality (text only, image only or combined) and (3) run type (automatic or manual). Table 2 provides an overview of all submitted runs according to these dimensions. Most submissions (96.8%) used the provided image annotations, with 22 groups submitting a total of 404 purely concept-based (textual) runs and 19 groups a total of 605 runs using a combination of content-based (visual) and concept-based features. A total of 11 groups submitted 33 purely content-based runs. Of all retrieval approaches, 61.2% involved the use of image retrieval (53.4% in 2007; 31% in 2006), 79% of all groups used content-based (i.e. visual) information in their runs (60% in 2007; 58% in 2006). Almost all of the runs (99.7%) were automatic (i.e. involving no human intervention); only 3 submitted runs were manual. Only one participating group made use of additional data, which was available from the Visual Concept Detection Task[4].

[4] http://www.imageclef.org/2008/iaprconcepts

Table 2. Overview of submissions categorized by run dimensions

Dimensions	Type	2008 Runs	2008 Groups	2007 Runs	2007 Groups	2006 Runs	2006 Groups
Annotation language	EN	514	24	271	17	137	2
	RND	495	2	32	2		
Modality	Text Only	404	22	167	15	121	2
	Mixed (text and image)	605	19	255	13	21	1
	Image Only	33	11	52	12		
Run type	Manual	3	1	19	3		
	Automatic	1039	25	455	19	142	2

4 Results

This section provides an overview of results with respect to the various submission dimensions (1) annotation language, (2) retrieval modality and (3) run type. The task for the participants was to maximise the number of relevant images in the top 20 results. At the same time the relevant images in the top 20 results should be from as many different sub-topics as possible. Simply getting lots of relevant images from one sub-topic or filling the ranking with diverse, but non-relevant images, results in a poor overall effectiveness score. Measures such as MAP are not suitable since it does not take into account diversity. To determine the diversity of a result set, S-Recall (sub-topic recall) proposed by Zhai et al [10] was used. S-recall at rank K is defined as the percentage of sub-topics covered by the first K documents in the list:

$$\textit{S-recall at K} \equiv \frac{\left| \cup_{i=1}^{K} \textit{subtopics} \left(d_i \right) \right|}{n_A}$$

where d_i represents the ith document, *subtopics*(d_i) the number of sub-topics d_i belongs to, and n_A the total number of sub-topics in a particular topic. Thus the evaluation is based on two measures: precision at 20 and cluster recall at rank 20 (S-recall). As previously mentioned, it was important to maximise both measures in order to get a high overall ranking. To provide a single measure of effectiveness, we used the F1-measure (harmonic mean) to combine P20 and CR20:

$$\text{F1-measure} = \frac{|2 \times (P20 \times CR20)|}{(P20 + CR20)}$$

The order of the diverse and relevant documents within the first top 20 result is not considered for the calculation of the cluster recall. This means that relevant documents from different sub-topics can be in a random order, without affecting the cluster recall score. A more detailed analysis of results can be found in [4].

4.1 Results by Annotation Language

Tables 3 and 4 show the runs which achieved highest F1-measure scores for the two annotation languages: ENG and RND. Taking into account that only two groups submitted 495 runs with a random annotation language, the result shows the same trend as in previous years: the highest monolingual run still outperforms the highest bilingual run, which consists of a random annotation language. However, as in previous years, the margin of difference is low and can be attributed to significant progress of the translation and retrieval methods using these languages. The best performing runs using random annotations performed with an F1-measure score at 97.4% of the highest monolingual run. Hence, the language barrier is no longer a critical factor in achieving good retrieval results.

Table 3. Systems with the highest F1-Measure for English

Query language	Caption language	Group	Run-ID	Run type	Modality	P20	CR20	F1-Measure
English	English	PTECH	PTECH-EN-EN-MAN-TXTIMG-MMBQI.run	MAN	TXTIMG	0.6885	0.6801	0.6843
English	English	PTECH	PTECH-EN-EN-MAN-TXTIMG-MMBMI.run	MAN	TXTIMG	0.6962	0.6719	0.6838
English	English	PTECH	PTECH-EN-EN-MAN-TXT-MTBTN.run	MAN	TXT	0.5756	0.5814	0.5785
English	English	XRCE	xrce_tilo_nbdiv_15	AUTO	TXTIMG	0.5115	0.4262	0.4650
English	English	DCU	DCU-EN-EN-AUTO-TXTIMG-qe.txt	AUTO	TXTIMG	0.4756	0.4542	0.4647
English	English	XRCE	xrce_tilo_nbdiv_10	AUTO	TXTIMG	0.5282	0.4146	0.4646
English	English	XRCE	xrce_cm_nbdiv_10	AUTO	TXTIMG	0.5269	0.4111	0.4619
English	English	DCU	DCU-EN-EN-AUTO-TXTIMG.txt	AUTO	TXTIMG	0.4628	0.4546	0.4587
English	English	XRCE	xrce_cm_mmr_07	AUTO	TXTIMG	0.5282	0.4015	0.4562
English	English	XRCE	xrce_tfidf_nbdiv_10	AUTO	TXTIMG	0.5115	0.4081	0.4540

XRCE - Xerox Research Centre Europe; PTECH – Institut TELECOM, TELECOM ParisTech, Paris, France; DCU – Dublin City University

4.2 Results by Retrieval Modality

In 2006 and 2007, the results showed that by combining visual features from the image and semantic knowledge derived from the captions offered optimum performance for retrieval from a general photographic collection with fully annotated images [5, 6]. As indicated in Table 5, the results of ImageCLEFphoto 2008 show that this also applies for our modified task, which promotes diversity in the results set. However, contrary to 2007 (24% MAP improvement over averages for combining techniques over solely text-based approaches), the improvement is not as clearly visible when combining visual features from the image and semantic information. The difference between "Mixed" and "Text Only" runs is across the averages from all runs, and differs only marginally. However, looking at the best runs in each modality, the "Mixed" runs (F1-Measure = 0.4650) outperform the "Text Only" runs by 16% (F1-measure = 0.4008). Purely content-based approaches still lag behind, although with a smaller gap than in previous years. The best "Image Only" runs (F1-Measure = 0.3396) is higher than both averages for the "Mixed" and "Text only" runs.

Table 4. Systems with the highest F1-Measure for Random annotations (German / English)

Query language	Caption language	Group	Run-ID	Run type	Modality	P20	CR20	F1-Measure
English	RND	DCU	DCU-EN-RND-AUTO-TXTIMG-tr.txt	AUTO	TXTIMG	0.4397	0.4673	0.4531
English	RND	DCU	DCU-EN-RND-AUTO-TXTIMG-tr-qe.txt	AUTO	TXTIMG	0.4423	0.4529	0.4475
English	RND	DCU	DCU-EN-RND-AUTO-TXTIMG-tr-d50-k40-tf-all.txt	AUTO	TXTIMG	0.4038	0.4967	0.4455
English	RND	DCU	DCU-EN-RND-AUTO-TXTIMG-tr-d50-k40-tfidf-all.txt	AUTO	TXTIMG	0.3974	0.4948	0.4408
English	RND	DCU	DCU-EN-RND-AUTO-TXTIMG-tr-d50-kx-tfidf-all.txt	AUTO	TXTIMG	0.3897	0.5049	0.4399
English	RND	DCU	DCU-EN-RND-AUTO-TXTIMG-tr-d50-kx-tf-qe-all.txt	AUTO	TXTIMG	0.4013	0.4806	0.4374
English	RND	DCU	DCU-EN-RND-AUTO-TXTIMG-tr-d50-kx-tf-all.txt	AUTO	TXTIMG	0.3910	0.4936	0.4363
English	RND	DCU	DCU-EN-RND-AUTO-TXTIMG-tr-d50-k40-tfidf-qe-all.txt	AUTO	TXTIMG	0.4013	0.4766	0.4357
English	RND	DCU	DCU-EN-RND-AUTO-TXTIMG-tr-d50-kx-tfidf-qe-all.txt	AUTO	TXTIMG	0.3897	0.4768	0.4289
English	RND	DCU	DCU-EN-RND-AUTO-TXTIMG-tr-d50-k40-tf-qe-all.txt	AUTO	TXTIMG	0.3897	0.4678	0.4252

DCU – Dublin City University

Table 5. Results by retrieval modality

Modality	Precision at 20		Cluster Recall at 20		F1-measure (P20/CR20)	
	Mean	SD	Mean	SD	Mean	SD
Mixed	0.2538	0.1023	0.3998	0.0977	0.3034	0.0932
Text Only	0.2431	0.0590	0.3915	0.0819	0.2957	0.0576
Image Only	0.1625	0.1138	0.2127	0.1244	0.1784	0.1170

4.3 Results by Run Type

Table 6 shows the average scores and the standard deviations across all systems runs with respect to the run type. Unsurprisingly, F1-Measure results of manual approaches are significantly higher than purely automatic runs. All submitted manual runs are done with English annotation, whereas the average of the automatic runs is both from English as well as Random annotation. However, as previously shown the translation does not have a big impact and can therefore be neglected. In case of the automatic runs the F1-measure is practically identical for the English (ENG) annotations and those with the language randomly selected (RND).

Table 6. Results by run type

Technique	Precision at 20		Cluster Recall at 20		F1-measure (P20/CR20)	
	Mean	SD	Mean	SD	Mean	SD
Manual	0.6534	0.0675	0.6445	0.0548	0.6489	0.0610
Automatic	0.2456	0.0873	0.3899	0.0975	0.2955	0.0829
Automatic RND Only	0.2353	0.0651	0.4191	0.0731	0.2992	0.0679
Automatic ENG Only	0.2609	0.0990	0.3731	0.1002	0.2994	0.0879
Automatic IMG Only	0.1625	0.1138	0.2127	0.1244	0.1784	0.1170

4.4 Approaches Used by Participants

Some of the participating groups started by using a baseline run, carried out using different weighting methods (e.g. BM25, DFR, LM), with or without query expansion (e.g. using Local Content Analysis, Pseudo Relevance Feedback, thesaurus-based query expansion, Conceptual Fuzzy Sets, using a location hierarchy, and using Wordnet), and using content- and/or concept-based retrieval methods. The aim of this initial step was obtaining the best possible ranking (i.e. maximising the number of relevant documents returned in the top *n*). The most common following step was to re-rank the initial baseline run in order to promote diversity. One approach of re-ranking is to cluster the top *n* documents into sub-topics or clusters and then select the highest ranked document in each cluster and promote higher in the ranked list (i.e. to the top *n*). Clustering was mostly based on the associated textual information using various clustering algorithms (e.g. k-means, k-medoids, knn-density, and latent dirichlet allocation) and different weighting parameters. Some groups also tried to re-rank results using Maximal Marginal Relevance. Other approaches included merging different kind of runs (e.g. calculating image ranking with average/min/mean) or combining scores (novelty/ranking score) to get a diverse and relevant results list. Overall, a majority of approaches applied post-processing methods in one way or another.

5 Conclusions

This paper has reported on the 2008 ImageCLEFphoto task, a general photographic ad-hoc retrieval task. The focus this year is different from this year and based on promoting diversity in the top *n* results. The challenge for participants was to maximise both the number of relevant images, as well as the number of sub-topics represented within the top 20 results. The 2008 task attracted a record number of submissions: 24 participating groups submitting a total of 1,042 system runs. The participants were provided with a subset of the IAPR TC-12 Benchmark: 20,000 colour photographs and two sets of semi-structured annotations in (1) English and (2) one set whereby the annotation language was randomly selected from English and German for each of the images. To measure the diversity of a ranked list, the existing collection was augmented with cluster assessments. Cluster assessments describe to which sub-topic a relevant image belongs to. Participants experimented with both content-and concept-based retrieval techniques. The main findings of this year include:

- Bilingual retrieval performs nearly as well as monolingual retrieval;
- Combining concept and content-based retrieval methods improves retrieval performance;
- A large number of participants used visual retrieval techniques (similar to previous years).

ImageCLEFphoto will continue to provide resources to the retrieval and computational vision communities to facilitate standardised laboratory-style testing of image retrieval systems. While these resources have predominately been used by systems applying a concept-based retrieval approach thus far, the number of participants who are using content-based retrieval techniques at ImageCLEFphoto is still increasing.

Acknowledgements

We would like to thank Michael Grubinger for providing the data collection and queries which formed the basis of the ImageCLEFphoto task for 2008. Work undertaken in this paper is supported by the EU-funded TrebleCLEF project (Grant agreement: 215231) and by the project Multimatch (contract IST-2005-2.5.10).

References

[1] Voorhees, E.M., Harman, D.: Overview of the Seventh Text REtrieval Conference (TREC–7). In: The Seventh Text Retrieval Conference, Gaithersburg, MD, USA, November 1998, pp. 1–23 (1998)

[2] Tian, S.K., Gao, Y., Huang, T.: Diversifying the image retrieval results. In: Proceedings of the 14th Annual ACM international Conference on Multimedia, MULTIMEDIA 2006, Santa Barbara, CA, USA, October 23-27, pp. 707–710. ACM, New York (2006)

[3] Chen, H., Karger, D.R.: Less is more: probabilistic models for retrieving fewer relevant documents. In: Proceedings of the 29th Annual international ACM SIGIR Conference on Research and Development in information Retrieval, SIGIR 2006, Seattle, Washington, USA, August 6-11, pp. 429–436. ACM, New York (2006)

[4] Sanderson, M., Tang, J., Arni, T., Clough, P.: What else is there? Search Diversity Examined. In: Boughanem, M., et al. (eds.) ECIR 2009. LNCS, vol. 5478, pp. 562–569. Springer, Heidelberg (2009)

[5] Clough, P., Grubinger, M., Deselaers, T., Hanbury, A., Müller, H.: Overview of the ImageCLEF 2006 photographic retrieval and object annotation tasks. In: Peters, C., Clough, P., Gey, F.C., Karlgren, J., Magnini, B., Oard, D.W., de Rijke, M., Stempfhuber, M. (eds.) CLEF 2006. LNCS, vol. 4730, pp. 579–594. Springer, Heidelberg (2007)

[6] Grubinger, M., Clough, P., Hanbury, A., Müller, H.: Overview of the ImageCLEFPhoto 2007 photographic retrieval task. In: Peters, C., Jijkoun, V., Mandl, T., Müller, H., Oard, D.W., Peñas, A., Petras, V., Santos, D. (eds.) CLEF 2007. LNCS, vol. 5152, pp. 433–444. Springer, Heidelberg (2008)

[7] Grubinger, M., Clough, P., Müller, H., Deselears, T.: The IAPR–TC12 Benchmark: A New Evaluation Resource for Visual Information Systems. In: International Workshop OntoImage 2006 Language Resources for Content-Based Image Retrieval, held in conjunction with LREC 2006, Genoa, Italy, May 22nd, pp. 13–23 (2006)
[8] Grubinger, M., Clough, P.: On the Creation of Query Topics for ImageCLEFPhoto. In: Proceedings of the third MUSCLE / ImageCLEF workshop on image and video retrieval evaluation, Budapest, Hungary, September 19-21 (2007)
[9] Arni, T., Tang, J., Sanderson, M., Clough, P.: Creating a test collection to evaluate diversity in image retrieval. In: Proceedings of the Workshop on Beyond Binary Relevance: Preferences, Diversity, and Set-Level Judgments, held at SIGIR 2008 (2008)
[10] Zhai, C., Cohen, W.W., Lafferty, J.: Beyond independent relevance: Methods and evaluation metrics for subtopic retrieval. In: Proceedings of ACM SIGIR 2003, pp. 10–17 (2003)

Overview of the ImageCLEFmed 2008 Medical Image Retrieval Task

Henning Müller[1,2], Jayashree Kalpathy-Cramer[3], Charles E. Kahn Jr.[4], William Hatt[3], Steven Bedrick[3], and William Hersh[3]

[1] Medical Informatics, University Hospitals and University of Geneva, Switzerland
[2] University of Applied Sciences Western Switzerland, Sierre, Switzerland
[3] Oregon Health and Science University (OHSU), Portland, OR, USA
[4] Department of Radiology, Medical College of Wisconsin, Milwaukee, WI, USA
henning.mueller@sim.hcuge.ch

Abstract. The medical image retrieval task of ImageCLEF is in its fifth year and participation continues to increase to a total of 37 registered research groups. About half the registered groups finally submit results. Main change in 2008 was the use of a new databases containing images of the medical scientific literature (articles from the Journals Radiology and Radiographics). Besides the images, the figure captions and the part of the caption referring to a particular sub–figure were supplied as well as access to the full text articles in html. All texts were in English and the topics were supplied in German, French, and English. 30 topics were made available, ten of each of the categories visual, mixed, semantic.

Most groups concentrated on fully automatic retrieval. Only three groups submitted a total of six manual or interactive runs not showing an increase of performance over automatic approaches. In previous years, multi–modal combinations were the most frequent submissions but in 2008 text only runs were clearly higher. Only very few fully visual runs were submitted and non of the fully visual runs had an extremely good performance. Part of these tendencies might be due to semantic topics and the extremely well annotated database. Best results regarding MAP were similar for textual and multi–modal approaches whereas early precision was better for some multi–modal approaches.

1 Introduction

ImageCLEF[1] [1,2,3] started within CLEF[2] (Cross Language Evaluation Forum, [5]) in 2003. A medical image retrieval task was added in 2004 to explore domain–specific multilingual visual information retrieval and also multi–modal retrieval by combining visual and textual features for retrieval. Medical image retrieval has been a very active domain over the past years [4]. Since 2005, a medical retrieval and a medical image annotation task are both presented as part of ImageCLEFmed [3].

[1] http://www.imageclef.org/
[2] http://www.clef-campaign.org/

C. Peters et al. (Eds.): CLEF 2008, LNCS 5706, pp. 512–522, 2009.

This paper reports on the medical retrieval task whereas additional papers describe the four other tasks of ImageCLEF. More detailed information can also be found on the task web pages for ImageCLEFmed. A detailed analysis of previous medical image retrieval tasks is available in [6].

2 The Medical Retrieval Task in 2008

The main change in the medical retrieval task in 2008 was the use of a new database. The search tasks remained essentially the same as in the previous years. The collection distributed to the participants included the images and the captions, as published in the medical journals. URLs to access the full text of the journal articles were also made available to the participants.

2.1 Registration and Participation

Registration has continued to rise for the medical retrieval task in 2008 as in previous years, albeit slowly. In total, 37 research groups registered for this task and obtained the dataset. Several of the groups registered solely to obtain the test collection in order to use it as training or test data data for their algorithms, rather than actually participating in the competition. In the end, 15 research groups submitted a total of 130 runs. Groups were asked to not submit more than ten runs in 2008 (different from previous years) so as not to bias the pools too much towards any single group or approach. There were significant problems with many of the 130 initial runs: some were submitted in incorrect formats; several runs were duplicated; and there were runs that covered only a part of the topics. These problems were corrected in collaboration with the authors as much as was possible, resulting in 111 valid runs that were used to generate the pools that were finally judged for relevance. All these runs were included in the official evaluation. The following groups submitted valid runs:

- Hungarian Acadamy of Sciences, Budapest, Hungary;
- National Library of Medicine (NLM), National Institutes of Health NIH, Bethesda, MD, USA;
- Bania Luka University, Bosnia-Hercegovina;
- MedGIFT group, University of Geneva, Switzerland;
- Natural Language Processing group, University Hospitals of Geneva, Switzerland;
- GPLSI group, University of Alicante, Spain;
- Multimedia Modelling Group, LIG, Grenoble, France;
- Natural Language Processing at UNED. Madrid, Spain;
- Miracle group, Spain;
- Oregon Health and Science University (OHSU), Portland, OR, USA;
- IRIT Toulouse, France;
- University of Jaen, Spain;
- Tel Aviv University, Israel;

- National University of Bogota, Colombia;
- TextMess group, University of Alicante, Spain.

A total of 15 groups from eight countries and four continents thus submitted results that are presented in the following chapters.

2.2 Database

The database used for the task in 2008 was made available by the Radiological Society of North America (RSNA[3]). The database contains in total slightly more than 66,000 images taken from the radiological journals *Radiology* and *Radiographics*. The images are original figures used in published articles. The collection is a subset of a larger database that is also available via the Goldminer[4] image search interface. For each image, the text of the figure caption was supplied as free text. However, this caption was sometimes associated with a multi–part image. In over 90% of the images the part of the caption actually referring to this sub–image was also provided. Additionally, links to HTML versions of the full–text articles were provided along with the relevant PubMed[5] accession ID numbers. These PubMed identifications also gave acces to the MeSH (Medical Subject Headings) terms, that are manually added to all references added to PubMed. Both the full–size images as well as thumbnails were available to the participants. All texts of the collection were in English.

The contents of this database represent a broad and significant body of medical knowledge, which makes this year's competition a potentially realistic scenario for how clinicians might use image retrieval systems in the future.

2.3 Query Topics

The query topics in 2008 were a selection of 30 topics from the previous three years of ImageCLEFmed [7]. Training data in the form of the 2005–2007 database with images, annotations, topics, sample query images and qrel files (for trec_eval) were made available to participants. All topics were supposed to cover at least two of the following axes:

- Anatomic region shown in the image;
- Image modality (x–ray, CT, MRI, gross pathology, ...);
- View (frontal, sagittal,...);
- Pathology or disease shown in the image;
- abnormal visual observation (eg. enlarged heart).

From the 85 possible topics of the past three years, similar topics were removed to cover a wide range of different modalities and anatomic regions. A visual and textual check was then performed to make sure that at least a few relevant images

[3] http://www.rsna.org/
[4] http://goldminer.arrs.org/
[5] http://www.pubmed.gov/

exist in the dataset. Since the databases of 2008 and 2007 were very different, we wanted to ensure that each topic had more than one relevant image existing.

Each query topic consists of the information need in three languages (English, French, German) and at least two example images. Groups could decide which language and media to use for the query processing and also which part of the text to use.

2.4 Relevance Judgments

A new system for relevance judgments was introduced in 2008 building on a Ruby for Rails framework and allowing for simple judgments via a web interface for all judges. The first 35 images of every run were combined into "pools" with an average size of around 900 images. Such pooling is necessary to reduce the amount of data to judge, and the bias can be regarded as very limited [8]. Medical Doctors who are also students of biomedical informatics at OHSU were hired for the judgment process and paid by the hour for the judgments.

A ternary judgment scheme was used, wherein each image in each pool was judged to be "relevant", "partly relevant", or "non–relevant". Images clearly corresponding to all criteria were judged as "relevant", images whose relevance could not be safely confirmed but could still be possible were marked as "partly relevant", and images for which one or more criteria of the topic were not met were marked as "non–relevant". Judges were instructed in these criteria and results were manually controlled during the judgment process.

During the judging, the new system exhibited a minor problem that resulted in certain images losing their judgments. This resulted in a short delay in the judging process, after which the affected images were re–judged by the same persons.

3 Submissions and Results

This section details the submissions for the tasks and a first brief evaluation.

3.1 Submissions

A total of 130 runs were submitted via the electronic submission system. Scripts to check the validity of the runs were made available to participants ahead of the submission phase, but even so, almost half of the submitted runs contained errors in either content or format and required changes. Common mistakes included a wrong trec_eval format, use of only a subset of the topics and incorrect image identifiers. In collaboration with the authors a large number of runs were repaired, resulting in 111 valid runs taken into account for the pools.

In total, only seven runs were "manual" or "interactive". There were also fewer "visual–only" runs than in all previous years, with only 8 such runs being submitted. The large majority were text–only runs, with 65 submissions. Mixed automatic runs had 31 submissions.

Groups subsequently had the chance to evaluate additional runs themselves as the qrels were made available to participants two weeks ahead of the submission deadline for these working notes.

3.2 Visual Retrieval

The number of visual runs in 2008 was much lower than in previous years, and the evolution is not as fast as with textual retrieval techniques. Five groups submitted a total of eight runs in 2008. Performance as measured in MAP is very low for all these runs, reaching a maximum of 0.04 for the best run. Early precision averaged over all topics reaches around 0.2, which is absolutely acceptable. When taking into account only the visual topics these results are much better, whereas the purely semantic topics obtained extremely poor results.

Table 1 shows the results and particularly the large differences between the runs. Some runs managed to retrieve a larger part of the relevant images (809) but with a fairly low MAP, whereas some results with a higher MAP only found a very small number of relevant images in the first 1000 results. A higher bpref in this context can mean that a larger number of images from these runs were not judged for relevance. This might also be due to the fact that only very few visual runs were submitted and thus only few visually retrieved documents were finally judged.

Results of GIFT were available to all the participants for combinations of visual and textual runs.

3.3 Textual Retrieval

Purely automatic textual retrieval had by far the largest number of runs in 2008 with 65, more than half of all submitted runs. Table 2 shows the results for all submitted automatic text runs, ordered by MAP. Most performance measures such as bpref and early precision are similar in order. Only early precision sometimes has significant differences from the ranking with MAP.

University of Alicante (Textmess), University of Jaen (SINAI), and LIG Grenoble obtain the best results, mainly with using ontologies such as MeSH (Medical Subject Headings) to code the documents. A MAP of 0.29 could be obtained and several systems have a high score very close to this. A more detailed analysis is required with the exact techniques applied for each of the runs.

Table 1. Results of the automatic runs using only visual information

Run	run_type	MAP	bpref	P5	P10	P30	num_rel
TAU_MIPLAB-TAU_norm	Visual Automatic	0.04	0.09	0.22	0.17	0.15	568
UNAL-W+QE+JS	Visual Automatic	0.04	0.06	0.13	0.13	0.11	297
GE_GIFT8	Visual Automatic	0.03	0.09	0.17	0.17	0.15	809
MIPLAB-TAU_orig	Visual Automatic	0.03	0.08	0.16	0.14	0.11	519
etfbl-max11111	Visual Automatic	0.03	0.04	0.15	0.13	0.11	212
etfbl-sum11111	Visual Automatic	0.03	0.04	0.12	0.10	0.12	194
GE_GIFT16	Visual Automatic	0.03	0.07	0.13	0.13	0.11	670
LSI_UNED	Visual Automatic	0.02	0.03	0.11	0.11	0.08	94
CEB_Image	Visual Automatic	0.01	0.04	0.03	0.04	0.05	390

Table 2. Results of the automatic runs using only text

Run	run_type	MAP	bpref	P5	P10	P30	num_rel
EXPPRFNegativaMesh	Text Automatic	0.29	0.35	0.49	0.46	0.41	2165
sinai_CT_Mesh	Text Automatic	0.28	0.33	0.44	0.41	0.37	2106
LIG_COS0506_MPTT_Emi	Text Automatic	0.28	0.34	0.51	0.47	0.43	2224
LIG-LIG_MPTT_Emix	Text Automatic	0.28	0.34	0.43	0.45	0.43	2138
TEXTMESSmeshType_CT	Text Automatic	0.28	0.32	0.43	0.41	0.37	2106
IRn2baseline	Text Automatic	0.28	0.33	0.48	0.42	0.35	1986
IRn2ExpNeg	Text Automatic	0.28	0.33	0.45	0.40	0.34	2006
LIG_RET_MPTT_Emix	Text Automatic	0.27	0.34	0.46	0.45	0.41	2129
LIG_COS_MPTT_Emix	Text Automatic	0.27	0.33	0.47	0.47	0.43	2275
LIG_CR_MPTT_Emix	Text Automatic	0.27	0.33	0.48	0.47	0.41	2265
IRn2ExpNegMesh	Text Automatic	0.27	0.32	0.45	0.42	0.36	2038
MirBaselineEN	Text Automatic	0.27	0.32	0.51	0.47	0.39	1861
IRn2Explca	Text Automatic	0.26	0.33	0.45	0.41	0.35	2096
LIG_RET_MP_Emix	Text Automatic	0.26	0.32	0.47	0.45	0.42	1979
IRn2ExpPRF	Text Automatic	0.26	0.32	0.47	0.41	0.36	1980
LIG_MP_Emix	Text Automatic	0.25	0.33	0.45	0.42	0.43	2007
MirAPEN	Text Automatic	0.25	0.31	0.49	0.46	0.39	1773
sinai_CT_Base	Text Automatic	0.25	0.31	0.32	0.35	0.33	2030
MirTaxEN	Text Automatic	0.25	0.32	0.38	0.37	0.37	1867
LIG-LIG_COS_MP_Emix	Text Automatic	0.24	0.31	0.45	0.41	0.41	2120
LIG-LIG_CR_MP_Emix	Text Automatic	0.24	0.31	0.47	0.40	0.39	2108
sinai_CT_Umls	Text Automatic	0.23	0.27	0.37	0.35	0.30	1927
bp_acad_textonly	Text Automatic	0.22	0.28	0.49	0.43	0.35	1726
Ssinai_CTA_Mesh	Text Automatic	0.21	0.27	0.46	0.40	0.29	1683
ohsu_text_umls_4	Text Automatic	0.20	0.30	0.31	0.29	0.25	1973
sinai_CTA_Base	Text Automatic	0.20	0.27	0.41	0.36	0.30	1702
LIG-LIG_MPadd_Emix	Text Automatic	0.19	0.29	0.34	0.37	0.34	2032
sinai_CTS_Base	Text Automatic	0.18	0.25	0.33	0.31	0.31	1790
sinai_CTA_Umls	Text Automatic	0.18	0.25	0.35	0.32	0.32	1553
HUG-MH-EN	Text Automatic	0.18	0.24	0.34	0.30	0.22	1957
HUG-MHnOVID-EN	Text Automatic	0.18	0.24	0.34	0.30	0.22	1957
sinai_CTS_Mesh	Text Automatic	0.16	0.24	0.32	0.29	0.27	1828
HUG-ltc-EN	Text Automatic	0.16	0.23	0.31	0.28	0.20	1713
HUG-mixPapers_EN	Text Automatic	0.15	0.21	0.33	0.27	0.20	1883
ohsu_text_3	Text Automatic	0.15	0.23	0.39	0.31	0.22	1786
sinai_CTS_Umls	Text Automatic	0.14	0.21	0.23	0.21	0.21	1558
TEXTMESSumlsType_CT	Text Automatic	0.14	0.17	0.33	0.32	0.25	1045
sigRunTxt	Text Automatic	0.14	0.19	0.29	0.24	0.22	858
HUG-BL_EN	Text Automatic	0.14	0.21	0.31	0.26	0.24	1615
HUG-HUG-BL HUG-BL	Text Automatic	0.14	0.21	0.31	0.26	0.24	1615
HUG-capMH_EN	Text Automatic	0.13	0.19	0.33	0.28	0.24	1499
HUG-capMH_EN	Text Automatic	0.13	0.19	0.33	0.28	0.24	1499
OHSU-text_or_1	Text Automatic	0.11	0.18	0.31	0.26	0.24	1420
HUG-ltc-FR	Text Automatic	0.11	0.18	0.19	0.20	0.16	1218
HUG-MH-FR	Text Automatic	0.11	0.17	0.19	0.17	0.16	1419
HUG-MHnOVID-FR	Text Automatic	0.11	0.17	0.19	0.17	0.16	1419
MirRF0505EN	Text Automatic	0.11	0.18	0.28	0.24	0.24	1372
HUG-MHnOVID-GE	Text Automatic	0.10	0.14	0.21	0.19	0.17	894
TEXTMESSmeshType_CTS	Text Automatic	0.10	0.18	0.23	0.23	0.15	1828
HUG-ltc-GE	Text Automatic	0.09	0.14	0.17	0.16	0.13	869
HUG-capMH_FR	Text Automatic	0.09	0.16	0.23	0.20	0.17	1364
TEXTMESSumlsType_CTS	Text Automatic	0.09	0.14	0.23	0.23	0.16	933
CEB_BaseC_QE	Text Automatic	0.08	0.14	0.33	0.29	0.23	887
CCEB_BaseC_QE	Text Automatic	0.08	0.14	0.35	0.28	0.23	887
CEB_BaseC	Text Automatic	0.08	0.14	0.31	0.28	0.22	893
MirRF1005EN	Text Automatic	0.07	0.15	0.22	0.16	0.15	1248
HUG-MH-GE	Text Automatic	0.07	0.11	0.17	0.15	0.14	866
HUG-BL-FR	Text Automatic	0.07	0.11	0.17	0.16	0.15	942
MirRFTax1005EN	Text Automatic	0.07	0.14	0.15	0.13	0.14	1260
MirRFTax1005FR	Text Automatic	0.07	0.11	0.13	0.11	0.09	823
MirRFTax1005DE	Text Automatic	0.05	0.08	0.09	0.09	0.06	461
CEB_BaseM	Text Automatic	0.04	0.09	0.20	0.17	0.15	532
HUG-BL-GE	Text Automatic	0.03	0.05	0.07	0.06	0.06	432
HUG-capMH_GE	Text Automatic	0.03	0.05	0.07	0.06	0.06	432
CEB_BaseC_QE	Text Automatic	0.02	0.03	0.06	0.05	0.04	182

Using various languages for the retrieval. Unfortunately, only little information was available on which languages the groups used for the retrieval. It can be assumed that most groups used English as this promises the best results. It was also possible to use all three query languages together, for example, for extracting MeSH terms. While this multi–lingual approach is not necessarily a realistic scenario, it can lead to interesting results.

The HUG group used the same techniques with several languages and showed that English obtained by far the best results, better than either French or German. The technique they applied was to map the MeSH terms form the text and queries in various languages. Through the PMIDs, the officially (manually) assigned MeSH terms of the articles were also available. The MeSH terms extracted from the article and query text performed worse for retrieval than the officially assigned terms.

Additional resources used for the retrieval. Groups could also state which additional resources were used for retrieval. The goal of this was to make a collection of available resources that can potentially be shared among participants to improve performance in future challenges. A large variety of resources were used, in large part for the combination of visual and textual runs, but also for purely textual runs. Many of the best runs used the ImageCLEFmed 2005-2007 data for training. Official MeSH terms manually assigned by the National Library of Medicine could be used through the PMIDs of the articles.

Most commonly used resources were the training data sets of ImageCLEF 2005-2007. The data itself was fairly different as the annotation was of much poorer quality and the images were significantly different. Still, topics were a subset of those from the past and so the scenario was very realistic with respect to the training data.

3.4 Mixed Retrieval

The promotion of mixed–media retrieval has always been one of the main goals of ImageCLEF. In past years mixed–media retrieval had the highest submission rate but in 2008 only half as many mixed runs were submitted as purely textual runs.

Table 3 shows the results for all submitted runs. It is clear that, for a large number of the runs, the MAP results for the mixed retrieval submissions were very similar to those from the purely textual retrieval systems. An interesting observation is that the mixed-media submissions often have higher early precision than the purely textual retrieval submissions. This confirms what has been previously observed. In the mixed media runs the ranking between bpref and MAP was only more different than for the purely textual approaches, meaning that the variety of techniques used might be larger and that there are more non–judged images.

When comparing results with textual retrieval it also becomes clear that mixed retrieval can obtain very low results. Particularly results with known text runs obtain often lower results than the text alone stressing the fragility of such combination methods.

Table 3. Results of the automatic runs mixing text and visual information

Run	run_type	MAP	bpref	P5	P10	P30	num_rel
sinai_CT_Mesh_Fire20	Mixed Automatic	0.29	0.33	0.45	0.43	0.40	2132
TEXTMESSmeshTypeFIREidf_CT	Mixed Automatic	0.28	0.32	0.43	0.41	0.37	2106
IRn2ExpNegRRIDF	Mixed Automatic	0.28	0.33	0.45	0.40	0.34	2006
IRn2ExpNegMeshRRIDF	Mixed Automatic	0.27	0.32	0.45	0.42	0.36	2038
ohsu_vis_mod_umls_4	Mixed Automatic	0.23	0.35	0.41	0.37	0.28	2052
ohsu_vis_mod_5	Mixed Automatic	0.23	0.33	0.41	0.38	0.29	1995
EXTMESSmeshTypeFIRE_CT	Mixed Automatic	0.22	0.27	0.30	0.30	0.30	2106
ohsu_mod_pars2_sp	Mixed Automatic	0.21	0.30	0.58	0.55	0.46	1561
OHSU_vis_mod_3	Mixed Automatic	0.15	0.25	0.41	0.32	0.24	1829
TEXTMESSumlsTypeFIREidf_CT	Mixed Automatic	0.14	0.17	0.33	0.32	0.25	1045
TEXTMESSumlsTypeFIRE_CT	Mixed Automatic	0.13	0.18	0.25	0.22	0.23	1045
TEXTMESSmeshTypeFIRE_CTS	Mixed Automatic	0.12	0.19	0.25	0.25	0.20	1828
SIG_IRIT-SigRunMixt	Mixed Automatic	0.11	0.16	0.30	0.29	0.23	859
TEXTMESSumlsTypeFIRE_CTS	Mixed Automatic	0.09	0.15	0.21	0.22	0.21	928
GE_GIFT8_EN_0.5	Mixed Automatic	0.08	0.19	0.27	0.24	0.24	1835
GE_EN_reGIFT8	Mixed Automatic	0.08	0.19	0.24	0.23	0.23	1957
GE_EN_GIFT8_mix	Mixed Automatic	0.08	0.19	0.28	0.24	0.25	1610
GE_GIFT8_EN_0.9	Mixed Automatic	0.07	0.12	0.31	0.27	0.25	812
GE_GIFT8_reEN	Mixed Automatic	0.07	0.12	0.29	0.24	0.25	812
IRn2ExpNegGiftRR	Mixed Automatic	0.05	0.11	0.13	0.12	0.11	830
IRIT-SigRunComb5	Mixed Automatic	0.05	0.10	0.28	0.24	0.17	793
IRIT-SigRunComb1	Mixed Automatic	0.05	0.10	0.28	0.25	0.17	791
IRIT-SigRunComb2	Mixed Automatic	0.05	0.10	0.28	0.24	0.16	789
IRIT-SigRunComb3	Mixed Automatic	0.05	0.10	0.27	0.24	0.16	782
IRIT-SigRunComb7	Mixed Automatic	0.04	0.09	0.25	0.22	0.16	805
IRIT-SigRunComb4	Mixed Automatic	0.04	0.09	0.25	0.22	0.16	770
IRIT-SigRunComb6	Mixed Automatic	0.04	0.09	0.25	0.22	0.16	771
IRIT-SigRunComb8	Mixed Automatic	0.04	0.09	0.24	0.22	0.16	817
CEB_IBaseC	Mixed Automatic	0.04	0.13	0.17	0.15	0.10	893
CEB_ITD3	Mixed Automatic	0.03	0.10	0.07	0.11	0.10	945
IRn2ExpNegMeshGiftRR	Mixed Automatic	0.03	0.08	0.11	0.11	0.09	662

Table 4. Results of the interactive and manual runs

Run	run_type	MAP	bpref	P5	P10	P30	num_rel
ohsu_int_2	Mixed Interactive	0.22	0.31	0.57	0.49	0.39	1580
ohsu_sdb_full_inter.	Mixed Interactive	0.18	0.29	0.53	0.46	0.33	1626
ohsu_sdb_lsa	Mixed Interactive	0.10	0.20	0.27	0.27	0.27	1601
CEB_ITD_ALL	Mixed Manual	0.03	0.11	0.08	0.11	0.11	964
CEB_IBaseM	Mixed Manual	0.02	0.10	0.08	0.08	0.06	532
CEB_TD_ALL	Text Manual	0.08	0.16	0.24	0.27	0.25	1198
CEB_TD3	Text Manual	0.08	0.16	0.24	0.27	0.25	1189

3.5 Interactive Retrieval

This year, as in previous years, interactive retrieval was only used by a very small number of participants. Interactive retrieval is extremely important, and it is a pity that it is hard to motivate groups to anything else than pure automatic technology assessment.

Table 4 shows the results of all manual and interactive runs submitted. Only two runs from OHSU had fairly good results, the other runs were not competitive in either the MAP or early precision categories compared to the fully automatic runs.

Table 5. Best results and average for all topics, showing the significant differences between topics

Topic	Topic	Ave. MAP	Max. MAP	no. rel.
1.	Show me photographs of benign or malignant skin lesions.	0.04	0.29	2
2.	Show me images containing one or several full-body scintigraphies.	0.02	0.61	10
3.	Show me Doppler ultrasound images (colored).	0.24	0.50	284
4.	Show me photographs showing an entire fetus.	0.04	0.26	5
5.	Show me chest CT images with emphysema.	0.16	0.58	69
6.	Show me images of a frontal head MRI.	0.01	0.08	27
7.	Show me images of a knee x-ray.	0.07	0.40	137
8.	Show me x-ray images of a hip joint with prosthesis.	0.07	0.38	28
9.	Show me images of PowerPoint slides.	0.32	1.00	17
10.	mediastinal CT	0.23	0.52	358
11.	Show me abdominal CT images showing liver blood vessels.	0.05	0.21	331
12.	Show me microscopic pathology images of the kidney.	0.04	0.47	51
13.	Show me gross pathologies of myocardial infarction.	0.08	0.35	10
14.	Show me chest CT images showing micro nodules.	0.06	0.22	71
15.	Show me chest x-ray images of cases with tuberculosis.	0.07	0.33	204
16.	Show me all x-ray images containing one or more fractures.	0.04	0.27	218
17.	Show me MRI images of the brain with a blood clot.	0.01	0.09	11
18.	gastrointestinal endoscopy with polyp	0.08	0.35	46
19.	CT liver abscess	0.24	0.76	101
20.	MRI or CT of colonoscopy	0.20	0.60	306
21.	Show me photographs of tumours.	0.11	0.39	334
22.	Show me images of muscle cells.	0.13	0.50	90
23.	Show me x-ray images of bone cysts .	0.05	0.29	17
24.	Show me images containing a Budd-Chiari malformation.	0.38	0.94	74
25.	Merkel cell carcinoma	0.40	1.00	24
26.	gastrointestinal neoplasm	0.13	0.37	279
27.	tuberous sclerosis	0.34	0.77	52
28.	mitral valve prolapse	0.14	0.53	3
29.	pulmonary embolism all modalities	0.26	0.55	237
30.	microscopic giant cell	0.13	0.50	39

3.6 Topic Analysis

Overall, most groups performed significantly better on the semantic topics than on the mixed or visual topics, as can be seen in the table 5. Topics 6 and 11–18 were quite difficult for many participants. Table 5 gives an overview of the best and average perform per topic. Some topics with a small number of relevant images have a particularly low performance.

The fact that many of the visual topics obtain poorer performance than the semantic topics also shows that groups have much more experience working on semantic topics and that visual retrieval currently has much more difficulty obtaining good results. Still, visual retrieval can have an important positive influence and it seems necessary to promote it further by having potentially a larger number of visual topics to push groups towards using visual techniques.

4 Conclusions

The focus of many participants in this year's ImageCLEF 2008 has been text–based retrieval. The increasingly semantic topics combined with a database containing high–quality annotations in 2008 may have resulted in less impact of using visual techniques as compared to previous years. This tendency is also shown when looking at the performance per topic where visual topics had significantly lower results than the semantic topics. Our goal in the upcoming ImageCLEF medical retrieval task is to increase visual runs. We hope to modify the task to favor more integrated approaches. The interactive approaches with the use of relevance feedback Another important aspect is interactive retrieval that has always had a poor participation and definitely needs to be regarded more strongly. Relevance feedback and query modifications have a potential to really improve results but of course research favors laboratory style evaluations.

Visual runs were rare and had no single run with a very convincing performance as for example was the case in 2007 where the best visual runs had an extremely good performance. Mixed–media runs were very similar in performance to textual runs when looking at MAP. The only difference was that mixed–media runs obtained better early precision in general. Several mixed–media runs were also broken, resulting in a very poor performance. This highlights that the combination is still not very stable.

A per topic analysis shows that visual topics obtained lower average results than semantic topics. The analysis also shows that several runs with very few relevant images have a very low average performance, whereas topics with a larger number seem to perform better.

Acknowledgements

We would like to thank the CLEF campaign for supporting the ImageCLEF initiative. The images for the 2008 ImageCLEFmed challenge were contributed by the Radiological Society of North America (RSNA) and were published in the journals Radiology and Radiographics. This work was partially funded by the Swiss National Science Foundation (FNS) under contract 205321–109304/1, the American National Science Foundation (NSF) with grant ITR–0325160, and by the University of Applied Sciences Western Switzerland (HES SO) in the context of the BeMeVIS project.

References

1. Clough, P., Müller, H., Sanderson, M.: The CLEF cross–language image retrieval track (ImageCLEF) 2004. In: Peters, C., Clough, P., Gonzalo, J., Jones, G.J.F., Kluck, M., Magnini, B. (eds.) CLEF 2004. LNCS, vol. 3491, pp. 597–613. Springer, Heidelberg (2005)
2. Clough, P., Grubinger, M., Deselaers, T., Hanbury, A., Müller, H.: Overview of the ImageCLEF 2006 photo retrieval and object annotation tasks. In: Peters, C., Clough, P., Gey, F.C., Karlgren, J., Magnini, B., Oard, D.W., de Rijke, M., Stempfhuber, M. (eds.) CLEF 2006. LNCS, vol. 4730, pp. 579–594. Springer, Heidelberg (2007)
3. Müller, H., Deselaers, T., Kim, E., Kalpathy-Cramer, J., Deserno, T.M., Clough, P., Hersh, W.: Overview of the ImageCLEFmed 2007 medical retrieval and annotation tasks. In: Peters, C., Jijkoun, V., Mandl, T., Müller, H., Oard, D.W., Peñas, A., Petras, V., Santos, D. (eds.) CLEF 2007. LNCS, vol. 5152, pp. 472–491. Springer, Heidelberg (2008)

4. Müller, H., Michoux, N., Bandon, D., Geissbuhler, A.: A review of content–based image retrieval systems in medicine – clinical benefits and future directions. International Journal of Medical Informatics 73, 1–23 (2004)
5. Savoy, J.: Report on CLEF–2001 experiments. In: Peters, C., Braschler, M., Gonzalo, J., Kluck, M. (eds.) CLEF 2001. LNCS, vol. 2406, pp. 27–43. Springer, Heidelberg (2002)
6. Hersh, W., Müller, H., Jensen, J., Yang, J., Gorman, P., Ruch, P.: Advancing biomedical image retrieval: Development and analysis of a test collection. Journal of the American Medical Informatics Association, 488–496 (September/October 2006)
7. Hersh, W., Müller, H., Kalpathy-Cramer, J.: The imageclefmed medical image retrieval task test collection. Journal of Digital Imaging (2008)
8. Zobel, J.: How reliable are the results of large–scale information retrieval experiments? In: Croft, W.B., Moffat, A., van Rijsbergen, C.J., Wilkinson, R., Zobel, J. (eds.) Proceedings of the 21st Annual International ACM SIGIR Conference on Research and Development in Information Retrieval, Melbourne, Australia, pp. 307–314. ACM Press, New York (1998)

Medical Image Annotation in ImageCLEF 2008

Thomas Deselaers[1] and Thomas M. Deserno[2]

[1] RWTH Aachen University, Computer Science Department, Aachen, Germany
deselaers@cs.rwth-aachen.de
[2] RWTH Aachen University, Dept. of Medical Informatics, Aachen, Germany
deserno@ieee.org

Abstract. The ImageCLEF 2008 medical image annotation task is designed to assess the quality of content-based image retrieval and image classification by means of global signatures. In contrast to the previous years, the 2008 task was designed such that the hierarchy of reference IRMA code classifications is essential for good performance. In total, 12,076 images were used, and 24 runs of 6 groups were submitted. Multi-class classification schemes for support vector machines outperformed the other methods. A scoring scheme was defined to penalise wrong classification in early code positions over those in later branches of the code hierarchy, and to penalise false category association over the assignment of a "not known" code. The obtained scores rage from 74.92 over 182.77 to 313.01 for best, baseline and worst results, respectively.

1 Introduction

From the first introduction of the medical image annotation task in ImageCLEF to now this task evolved form a simple classification task with only about 60 classes [3] to a task with nearly 120 classes [6] and further to a task where a complex class hierarchy of potentially several thousand classes had to be considered [4].

In 2005, the aim of the medical image annotation task was defined as exploring and promoting the use of automatic annotation techniques to for extracting semantic information from little-annotated medical images. Therefore a new database of 10,000 images from 57 classes was created. This database was extended each year by adding at least 1,000 images. Furthermore the difficulty of the classification was increased by first increasing the number of classes and later including a complex hierarchical class structure: the Image Retrieval in Medical Applications (IRMA) code [5]. However, even the 2007 task could be solved using flat classification hierarchies since large parts of the hierarchy were unused and the effective number of classes was only slightly higher than in 2006.

With the 2008 task, we have achieved the goals that were set out initially: an image annotation task which requires the explicit use of the class-hierarchy in order to achieve good results and a wide variety of different methods has been systematically evaluated by the participating groups.

C. Peters et al. (Eds.): CLEF 2008, LNCS 5706, pp. 523–530, 2009.

Other tracks in ImageCLEF 2008 were the photo retrieval task [1], the medical retrieval task [7], the Wikipedia multimedia retrieval task [8], and the visual concept detection task [2].

2 Materials and Methods

The aim of the 2008 medical image annotation task was to promote the use of hierarchical classification techniques and foster the use of the prior knowledge encoded into the hierarchy of classes. Thus, the task was similar to the task of 2007 in that the classes were based on the IRMA code [5]. The main difference this year was that the prior distribution of the classes in the test data differed strongly from the prior distribution of the training data and that thus in particular classes which were badly represented in the training data were present in the test data to encourage the use of the hierarchy and the placement of wild card operators.

2.1 Database and Task Description

The training data of this year consisted of 12,076 images (10,000 training images from last year + 1,000 development images from last year + 1,000 test images from last year + 76 new images) and the test data consisted of 1,000 new images. In total 196 unique codes were present in the training images and 187 of these were present in the test images. The most frequent class in the training data consisted of more than 2,300 images, but the test data had only one example from this class. In Figure 1, the frequency of classes in the training and in the test data is shown. It can be seen that the classes in the test data were nearly uniformly distributed, but, in the training data, some classes were far more frequent than others.

Each of the radiographs is annotated with its complete IRMA code (see Sec. 2.2). In total, 196 different IRMA codes occurred in the database. Example

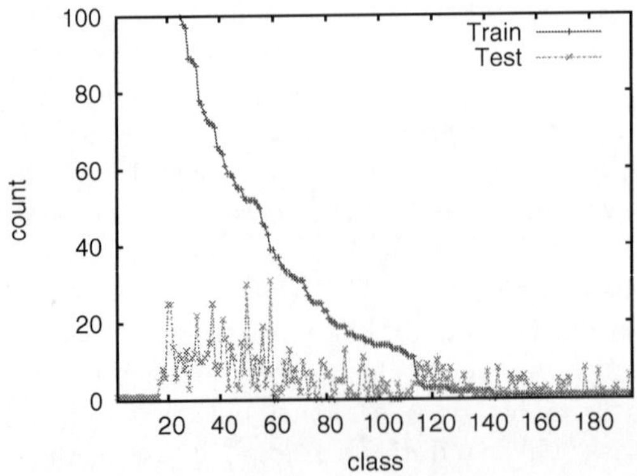

Fig. 1. Frequency of images in the training and test data

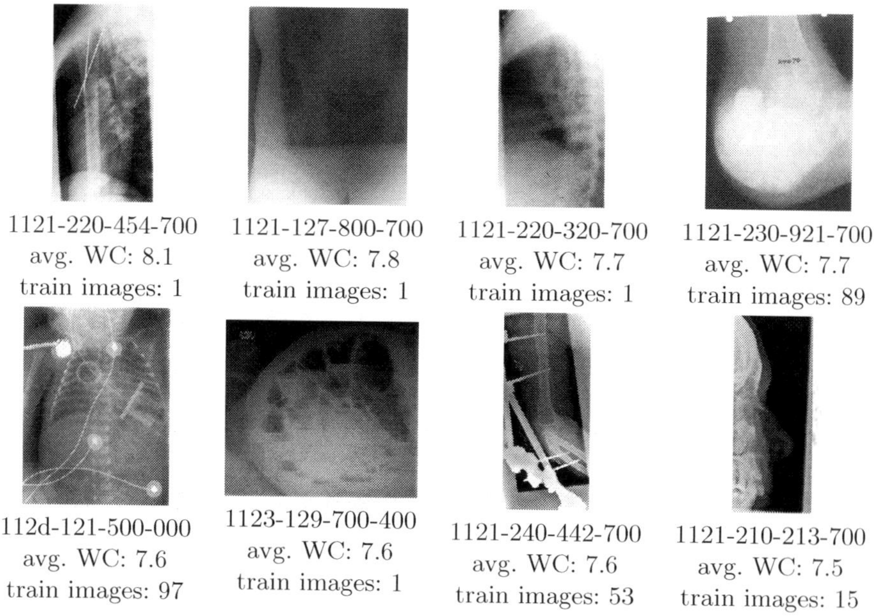

Fig. 2. The test images from the database where most wildcards were used with their full IRMA code and the average number of wildcards over all runs

images from the database together with textual labels and their complete code are given in Figure 2 and 3.

2.2 IRMA Code

Existing medical terminologies such as the MeSH thesaurus are poly-hierarchical, i.e., a code entity can be reached over several paths. However, in the field of content-based image retrieval, we frequently find class-subclass relations. The mono-hierarchical multi-axial IRMA code strictly relies on such part-of hierarchies and, therefore, avoids ambiguities in textual classification [5]. In particular, the IRMA code is composed from four axes having three to four positions, each in $\{0, \dots 9, a, \dots z\}$, where "0" denotes "not further specified". More precisely,

- the technical code (T) describes the imaging modality;
- the directional code (D) models body orientations;
- the anatomical code (A) refers to the body region examined; and
- the biological code (B) describes the biological system examined.

This results in a string of 13 characters (IRMA: TTTT – DDD – AAA – BBB). A small exemplary excerpt from the anatomy axis of the IRMA code is given in Table 1.

The IRMA code can be easily extended by introducing characters in a certain code position, e.g., if new imaging modalities are introduced. Based on the hierarchy, the more code position differ from "0", the more detailed is the description.

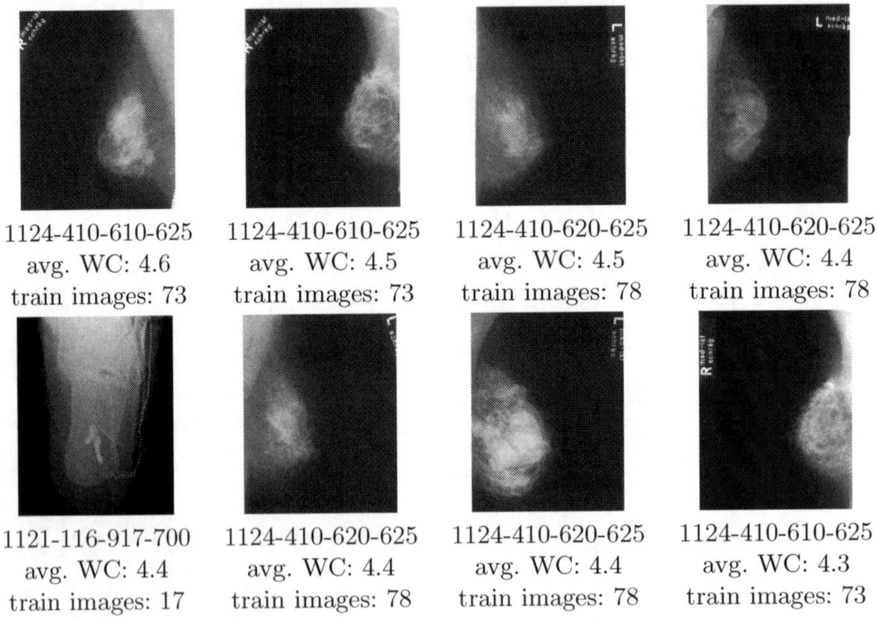

Fig. 3. The test images from the database where fewest wildcards were used with their full IRMA code and the average number of wildcards over all runs

Table 1. Examples from the IRMA code

AAA code	textual description
000	not further specified
...	
400	upper extremity (arm)
410	upper extremity (arm); hand
411	upper extremity (arm); hand; finger
412	upper extremity (arm); hand; middle hand
413	upper extremity (arm); hand; carpal bones
420	upper extremity (arm); radio carpal join ...
430	upper extremity (arm); forearm
431	upper extremity (arm); forearm; distal forearm
432	upper extremity (arm); forearm; proximal forearm
440	upper extremity (arm); ellbow
...	

2.3 Hierarchical Classification

Let an image be coded by the above 4 *independent* axes, such that we can consider the axes independently and just sum up the errors for each axis independently:

- let $l_1^I = l_1, l_2, \ldots, l_i, \ldots, l_I$ be the *correct* code (for one axis) of an image;
- let $\hat{l}_1^I = \hat{l}_1, \hat{l}_2, \ldots, \hat{l}_i, \ldots, \hat{l}_I$ be the *classified* code (for one axis) of an image;

where l_i is specified precisely for each position, and in $\hat{l}_i$ it is allowed to say "*don't know*", which is encoded by $*$. Note that I (the depth of the tree to which the classification is specified) may be different for different images.

Given an incorrect classification at position $\hat{l}_i$ we consider all succeeding decisions to be wrong and given a not specified position, we consider all succeeding decisions to be not specified. Furthermore, we do not count any error if the correct code is unspecified and the predicted code is a wildcard. In that case, we do consider all remaining positions to be not specified.

Since we want to penalise wrong decisions that are easy (fewer possible choices at that node) over wrong decisions that are difficult (many possible choices at that node), a decision at position l_i is considered to be correct by chance with a probability of $\frac{1}{b_i}$, if b_i is the number of possible labels for position i. This assumes equal priors for each class at each position.

Furthermore, we want to penalise wrong decisions at an early stage in the code (higher up in the hierarchy) over wrong decisions at a later stage in the code (lower down on the hierarchy) (i.e. l_i is more important than l_{i+1}).

Putting this together yields:

$$\sum_{i=1}^{I} \underbrace{\frac{1}{b_i}}_{(a)} \underbrace{\frac{1}{i}}_{(b)} \underbrace{\delta(l_i, \hat{l}_i)}_{(c)} \tag{1}$$

with

$$\delta(l_i, \hat{l}_i) = \begin{cases} 0 & \text{if } l_j = \hat{l}_j \quad \forall j \leq i \\ 0.5 & \text{if } l_j = * \quad \exists j \leq i \\ 1 & \text{if } l_j \neq \hat{l}_j \quad \exists j \leq i \end{cases}$$

where the parts of the equation account for

Table 2. Example for different errors in the hierarchical classification scheme. Assuming the code 318a is correct.

predicted code	error score
318a	0.0
318*	0.0
3187	0.0
31*a	0.1
31**	0.1
3177	0.2
3***	0.3
32**	0.7
1000	1.0

(a) difficulty of the decision at position i (branching factor);
(b) the level in the hierarchy (position in the string); and
(c) the correct/not specified/wrong labelling, respectively.

In addition, for each axis, the maximal possible error is calculated and the errors are normed such that a completely wrong decision (i.e. all positions for that axis are wrong) gets an error count of 0.25, and a completely correctly predicted axis has an error of 0. Thus, an image where all positions in all axes are wrong has an error count of 1, and an image where all positions in all axes are correct has an error count of 0. An example of this scheme is given in Table 2.

3 Results from the Evaluation

In 2008, 6 groups participated in the medical annotation task submitting 24 runs in total. In the following, we briefly describe the methods applied by the participating groups.

FEIT. The Faculty of Electrical Engineering and Information Technologies from the University of Skopje in Macedonia submitted two runs using global and local image descriptors, which are classified using bagging and random forests.

medGIFT. The medical Gnu Image Finding Tool (medGIFT) group from University Hospitals of Geneva in Switzerland submitted four runs using different descriptors and voting schemes in the medGIFT image retrieval system.

Miracle. The Miracle group from Daedalus University in Spain submitted four runs using different global and local image descriptors in a nearest neighbour classifier.

TAU-BIOMED. The Biomedical Image Processing Lab from Tel Aviv University in Israel submitted four runs using a bag-of-visual words approach with dense sampling and support vector machines for classification.

IDIAP. The "Institut Dalle Molle d'Intelligence Artificielle Perceptive" (IDIAP) Research Institute from Switzerland submitted nine runs using different multi-class classification schemes for support vector machines and different image descriptors.

RWTH-MI. The Image Retrieval in Medical Applications (IRMA) group at the Department of Medical Informatics, RWTH Aachen University in Aachen, Germany, provides a baseline-run that was computed using Tamura Texture Measures and the Image Distortion Model. Since 2004, the parameterisation remains unchanged, and, therefore, the hierarchy was disregarded.

The results from the evaluation are given in Table 3 sorted by error score. It can be seen that the classification accuracy varies strongly from 74.9 to 313 error points according to the above described error measurement. Also, the number of wildcards used varies very strongly between 0 in the model free approach from the IRMA group up to about 7,000, which means that almost seven wildcards

Table 3. Results from the medical image annotation task

group	run	error score	wildcards
idiap	LOW_MULT_2MARG	74.92	4148
idiap	LOW_MULT	83.45	3154
idiap	LOW_2MARG	83.79	4353
idiap	MCK_MULT_2MARG	85.91	4655
idiap	LOW_lbp_siftnew	93.20	3157
idiap	SIFTnew	100.27	3144
TAU	BIOMED-svm_full	105.75	1000
TAU	BIOMED-svm_prob	105.86	4868
TAU	BIOMED-svm_vote	109.37	1000
TAU	BIOMED-svm_small	117.17	1000
idiap	LBP	128.58	3173
rwth_mi	baseline	182.77	0
MIRACLE	MIRACLE-3I-OF	187.90	4426
MIRACLE	MIRACLE-2I-OF	190.38	3194
MIRACLE	MIRACLE-2I-2F	190.38	3194
MIRACLE	MIRACLE-3I-2F	194.26	3871
GE	GIFT0.9_0.5_vcad_5	210.93	2146
GE	GIFT0.9_0.5_vca_5	217.34	2466
idiap	MCK_pix_sift_2MARG	227.82	6994
GE	GIFT0.9_akNN_2	241.11	1000
GE	GIFT0.9_kNN_2	251.97	1000
FEIT	1	286.48	1117
FEIT	2	290.50	1024
idiap	MCK_pix_sift	313.01	3420

per image were used on the average, i.e. more than half of the positions for the images are undefined.

In general, it an be seen that the discriminative models using local descriptors from the IDIAP group outperform the other approaches.

In Figures 2 and 3, some example test images are given along with their full IRMA code. The number of wildcards used by the submitted runs on average and the number of training images from this particular class. The top and the bottom parts of the figure show the images where, on the average, the most and the fewest wildcards were used, respectively. It can be observed that for classes with bad support in the training data far more wildcards were used.

4 Discussion and Conclusion

We have presented the ImageCLEF 2008 medical image annotation task. In contrast to previous years, the distribution of training and test images was chosen such that using the hierarchy of the IRMA code was necessary to obtain good results. For classes with very few training images, the submitted runs employed up to more than eight wildcards out of thirteen code positions per image to express their uncertainty about the classifications. Multi-class classification schemes for support vector machines, as used by the IDIAP Research Institute of Switzerland, outperformed the other methods. The obtained scores rage from 74.92 over 182.77 to 313.01 for best, baseline and worst, respectively.

In total the goals initially setup for the medical image annotation task were achieved: techniques for the annotation of medical images were systematically evaluated on a series of tasks of gradually increasing difficulty and still the results of the best system was improved over the years. The medical image annotation will not be continued in ImageCLEF in its current form but hopefully new and challenging tasks will be proposed and offered.

References

1. Arni, T., Clough, P., Sanderson, M., Grubinger, M.: Overview of the ImageCLEF-photo 2008 photographic retrieval task. In: Peters, C., et al. (eds.) CLEF 2008. LNCS, vol. 5706, pp. 500–511. Springer, Heidelberg (2009)
2. Deselaers, T., Hanbury, A.: The visual concept detection task in ImageCLEF 2008. In: Peters, C., et al. (eds.) CLEF 2008. LNCS, vol. 5706, pp. 531–538. Springer, Heidelberg (2009)
3. Deselaers, T., Müller, H., Clogh, P., Ney, H., Lehmann, T.M.: The CLEF 2005 automatic medical image annotation task. International Journal of Computer Vision 74(1), 51–58 (2007)
4. Deselaers, T., Müller, H., Deserno, T.M.: Automatic medical image annotation in ImageCLEF 2007: Overview, results, and discussion. Pattern Recognition Letters (2008) (page in press)
5. Lehmann, T.M., Schubert, H., Keysers, D., Kohnen, M., Wein, B.B.: The IRMA code for unique classification of medical images. In: Proceedings SPIE, vol. 5033, pp. 440–451 (2003)
6. Müller, H., Deselaers, T., Lehmann, T.M., Clough, P., Hersh, W.: Overview of the ImageCLEFmed 2006 medical retrieval and annotation tasks. In: Peters, C., Clough, P., Gey, F.C., Karlgren, J., Magnini, B., Oard, D.W., de Rijke, M., Stempfhuber, M. (eds.) CLEF 2006. LNCS, vol. 4730, pp. 579–594. Springer, Heidelberg (2007)
7. Müller, H., Kalpathy-Cramer, J., Kahn Jr., C.E., Hatt, W., Bedrick, S., Hersh, W.: Overview of the ImageCLEFmed 2008 medical image retrieval task. In: Peters, C., et al. (eds.) CLEF 2008. LNCS, vol. 5706, pp. 512–522. Springer, Heidelberg (2009)
8. Tsikrika, T., Kludas, J.: Overview of the wikipediaMM task at ImageCLEF 2008. In: Peters, C., et al. (eds.) CLEF 2008. LNCS, vol. 5706, pp. 539–550. Springer, Heidelberg (2009)

The Visual Concept Detection Task in ImageCLEF 2008

Thomas Deselaers[1] and Allan Hanbury[2,3]

[1] RWTH Aachen University, Computer Science Department, Aachen, Germany
deselaers@cs.rwth-aachen.de
[2] PRIP, Inst. of Computer-Aided Automation, Vienna Univ. of Technology, Austria
hanbury@prip.tuwien.ac.at
[3] CogVis GmbH, Vienna, Austria

Abstract. The Visual Concept Detection Task (VCDT) of ImageCLEF 2008 is described. A database of 2,827 images were manually annotated with 17 concepts. Of these, 1,827 were used for training and 1,000 for testing the automated assignment of categories. In total 11 groups participated and submitted 53 runs. The runs were evaluated using ROC curves, from which the Area Under the Curve (AUC) and Equal Error Rate (EER) were calculated. For each concept, the best runs obtained an AUC of 80% or above.

1 Introduction

Searching for images is, despite intensive research on alternative methods in the last 20 years, still a task that is mainly done based on textual information. For a long time, searching for images based on text was the most feasible method because on the one hand, the number of images to be searched was rather restricted, and on the other hand, only few people needed to access huge repositories of images. Both of these conditions have changed. The number of available images is growing more rapidly than ever due to the falling prices of high-end imaging equipment for professional use and of digital cameras for consumer use. Publicly available image databases such as Google picassa and Flickr have become major sites of interest on the Internet.

Nevertheless, accessing images is still a tedious task because sites such as Flickr do not allow images to be accessed based on their content but only based on the annotations that users create. These annotations are commonly disorganised, not very precise, and multilingual. Access problems can be addressed by improving the textual access methods, but none of these improvements can ever be perfect as long as the users do not annotate their images perfectly, which is very unlikely. Therefore, content-based methods have to be employed to improve access methods to digitally stored images.

A problem with content-based methods is that they are often computationally costly and cannot be applied in real-time. An intermediate step is to automatically create textual labels based on the images' content. To make these labels

C. Peters et al. (Eds.): CLEF 2008, LNCS 5706, pp. 531–538, 2009.

as useful as possible, frequently occurring visual concepts should be annotated in a standard manner.

In the visual concept detection task (VCDT) of ImageCLEF 2008, the aim was to apply labels of frequent categories in the photo retrieval task to the images and evaluate how well automated visual concept annotation algorithms function. Additionally, participants of the VCDT could create annotations for all images used in the photo retrieval task, which were provided to the participants of this task. In the following, we describe the visual concept detection task of ImageCLEF 2008, the database used, the methods of the participating groups, and the results.

Other tracks in ImageCLEF 2008 were the photo retrieval task [1], the medical retrieval task [2], the Wikipedia multimedia retrieval task [3], and the medical image annotation task [4].

2 Database and Task Description

As database for the ImageCLEF 2008 visual concept detection task, a total of 2,827 images were used. These are taken from the same pool of images used to create the IAPR-TC12 database [5], but are not included in the IAPR-TC12 database used in the ImageCLEF photo retrieval task.

The visual concepts were chosen based on concepts used in previous work on visual concept annotation. In particular they are an extension of the hierarchy used in the attribute recognition task of the ImagEVAL 2006 campaign [6]. They are organised hierarchically, as shown in Figure 1. An image can be labelled by a group of concepts. The hierarchy demonstrates the interdependency between

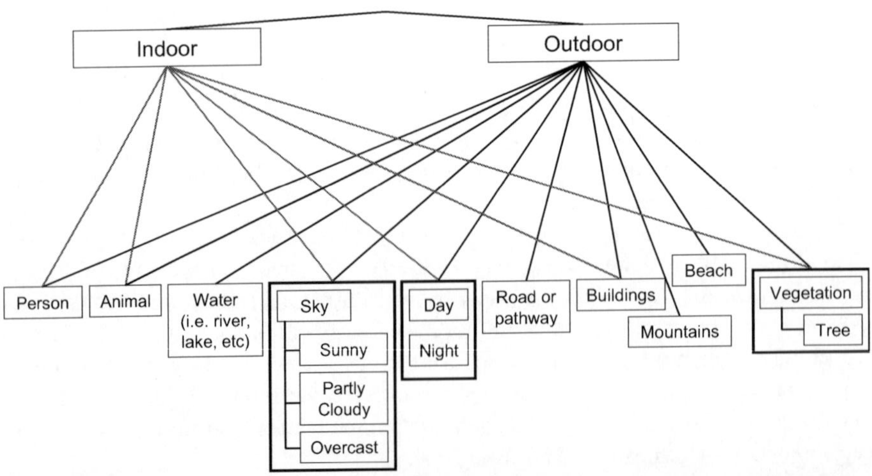

Fig. 1. Visual concept hierarchy used in the visual concept detection task of ImageCLEF 2008

Fig. 2. Example images for each of the concepts

some concepts, e.g. if the *sunny*, *partly cloudy* or *overcast* concept applies to an image, then the *sky* concept must apply too.

As for the ImageCLEF object detection task in 2007 [7], a web interface was created for manual annotation of the images by the concepts. Annotation was mainly carried out by undergraduate students at the RWTH Aachen University and by the track coordinators. A general opinion expressed by the annotators was that the concept annotation required more time than the object annotation of 2007. The number of images that were voluntarily annotated this year was also significantly less than the 20,000 images annotated by object labels in 2007.

Of the 2,827 manually annotated images, 1,827 were distributed with annotations to the participants as training data. The remaining 1,000 images were provided without labels as test data. The participants' task was to apply labels to these 1,000 images. Table 1 gives an overview of the frequency of the 17 visual concepts in the training data and in the test data and Figure 2 gives an example image for each of the categories.

Table 1. Statistics on the frequency of concepts in the training and test data

number	category	train [%]	test [%]
00	indoor	9.9	10.2
01	outdoor	88.0	88.1
02	person	43.8	44.9
03	day	82.0	81.9
04	night	3.7	2.3
05	water	23.1	21.7
06	road or pathway	20.0	19.4
07	vegetation	52.5	51.7
08	tree	29.3	30.8
09	mountains	14.3	13.8
10	beach	4.4	3.7
11	buildings	45.5	43.6
12	sky	66.9	69.3
13	sunny	12.3	13.1
14	partly cloudy	22.7	22.2
15	overcast	19.6	21.4
16	animal	5.6	5.8

3 Results from the Evaluation

In total 11 groups participated and submitted 53 runs. For each run, results for each concept were evaluated by plotting ROC curves. The results for each concept were summarised by two values: the area under the ROC curve (AUC) and the Equal Error Rate (EER). The latter is the error rate at which the false positive rate is equal to the false negative rate. Furthermore, for each run, the average AUC and average EER over all concepts were calculated.

Below, we briefly describe the methods employed by each group:

CEA-LIST. The Lab of Applied Research on Software-Intensive Technologies of the CEA, France submitted 3 runs using image features accounting for color and spatial layout with nearest neighbour and SVM classifiers.

HJ FA. The Microsoft Key Laboratory of Multimedia Computing and Communication of the University of Science and Technology, China submitted one run using color and SIFT descriptors which are combined and classified using a nearest neighbour classifier.

IPAL I2R. The IPAL French-Singaporean Joint Lab of the Institute for Infocomm Research in Singapore submitted 8 runs using a variety of different image descriptors.

LSIS. The Laboratory of Information Science and Systems, France submitted 7 runs using a structural feature combined with several other features using multi-layer perceptrons.

MMIS. The Multimedia and Information Systems Group of the Open University, UK submitted 4 runs using CIELAB and Tamura features and combinations of these.

Table 2. Summary of the results of the VCDT in ImageCLEF 2008

		best run			average		
	runs	rank	EER	AUC	rank	EER	AUC
XRCE	2	1	16.7	90.7	1.5	18.0	89.7
RWTH	1	3	20.5	86.2	3.0	20.5	86.2
UPMC	6	4	24.6	82.7	11.0	27.2	65.2
LSIS	7	5	25.9	80.5	20.3	32.8	71.8
MMIS	4	13	28.4	77.9	23.3	32.6	73.0
CEA-LIST	3	17	29.0	73.4	26.3	33.4	59.7
IPAL-I2R	8	19	29.7	76.4	32.1	36.0	68.3
budapest	13	20	31.1	74.9	31.8	35.2	68.6
TIA	7	24	32.1	55.6	39.6	39.9	36.3
HJ-FA	1	47	45.1	20.0	47.0	45.1	20.0
Makere	1	51	49.3	30.8	51.0	49.3	30.8

Makerere. The Faculty of Computing and Information Technology, Makerere University, Uganda submitted one run using luminance, dominant colors, and texture and shape features classified using a nearest neighbour classifier.

RWTH. The Human Language Technology and Pattern Recognition Group from RWTH Aachen University, Germany submitted one run using a patch-based bag-of-visual words approach using a log-linear classifier.

TIA. The Group for Machine Learning for Image Processing and Information Retrieval from the National Institute of Astrophysics, Optics and Electronics, Mexico submitted 7 runs using global and local features with SVMs and random forest classifiers.

UPMC. The University Pierre et Marie Curie in Paris, France submitted 5 runs using fuzzy decision forests.

XRCE. The Textual and Visual Pattern Analysis group from the Xerox Research Center Europe in France submitted two runs using multi-scale, regular grid, patch-based image features and a Fisher-Kernel Vector classifier.

budapest. The Datamining and Websearch Research Group, Hungarian Academy of Sciences, Hungary submitted 13 runs using a wide variety of features, classifiers, and combinations.

Table 2 gives an overview of the submissions and results for the task. The table is ranked by the performance of the best run submitted by the groups. It can be seen that the XRCE runs perform best.

Table 3 shows a breakdown of the results per concept. For each concept, the best and worst EER and AUC are shown, along with the average EER and AUC over all runs submitted. The best results were obtained for all concepts by XRCE, with budapest doing equally well on the *night* concept. The AUC per concept for all the best runs is 80.0% or above. Among the best results, the concepts having the highest scores are *indoor* and *night*. The concept with the worst score among the best results is *road or pathway*, most likely due to the

Table 3. Overview of the results per concept

# concept	best			average		worst	
	EER	AUC	group	EER	AUC	EER	AUC
00 indoor	8.9	97.4	XRCE	28.0	67.6	46.8	2.0
01 outdoor	9.2	96.6	XRCE	30.6	70.5	54.6	13.3
02 person	17.8	89.7	XRCE	35.9	62.2	53.0	0.4
03 day	21.0	85.7	XRCE	35.4	64.9	52.5	9.7
04 night	8.7	97.4	XRCE/budapest	27.6	72.5	73.3	0.0
05 water	23.8	84.6	XRCE	38.1	57.8	53.0	3.2
06 road/pathway	28.8	80.0	XRCE	42.6	50.7	56.8	0.0
07 vegetation	17.6	89.9	XRCE	33.9	67.4	49.7	30.7
08 tree	18.9	88.3	XRCE	36.1	62.8	59.5	1.0
09 mountains	15.3	93.8	XRCE	33.1	61.2	55.8	0.0
10 beach	21.7	86.8	XRCE	35.8	57.6	51.4	0.0
11 buildings	17.0	89.7	XRCE	37.4	60.8	64.0	0.5
12 sky	10.4	95.7	XRCE	24.0	78.6	50.8	37.3
13 sunny	9.2	96.4	XRCE	30.3	66.5	55.4	0.0
14 partly cloudy	15.4	92.1	XRCE	37.5	58.9	55.5	0.0
15 overcast	14.1	93.7	XRCE	32.1	67.6	61.5	0.0
16 animal	20.7	85.7	XRCE	38.2	54.2	58.4	0.0

high variability in the appearance of this concept. The concept with the highest average score, in other words, the concept that was detected best in most runs is *sky*. Again, the concept with the worst average score is *road or pathway*.

Two automatic runs provided by participants of the VCDT were made available to ImageCLEF participants. These provide annotations of the 20,000 ImageCLEF photo images with the VCDT concepts. Two groups participating in the photo retrieval task of ImageCLEF made use of these annotations, while one group used VCDT annotations provided by their own algorithm.

The group from Université Pierre et Marie Curie in Paris, France made use of their own VCDT algorithm to provide concepts for the photo retrieval task. They used the detected visual concepts to re-rank the first 50 results returned using text retrieval approaches. The concepts to use for the re-ranking were chosen using two approaches: (i) the concept word appears in the query text and (ii) the concept word appears in the list of synonyms (obtained using WordNet) of the words in the query text. The first approach improved the results of all the queries for which it was applicable, while the second resulted in worse results for some topics. Both approaches resulted in better overall performance than using text alone: the F-measure for the best text only run (using TF-IDF) is 0.273, while the F-measure for the run re-ranked using the first approach is 0.289.

The group from the National Institute of Informatics in Tokyo, Japan made use of both provided VCDT concept annotations. They also used the concepts to re-rank results returned by a text retrieval approach, where the best results were obtained by re-ranking based on a hierarchical clustering using distances between vectors encoding the VCDT concepts. This re-ranking decreased the P20 metric while increasing the CR20 metric, resulting in an increase of the F-measure from 0.224 for text only to 0.230 after the re-ranking.

Although the TIA-INAOE group in Puebla, Mexico also made use of one of the provided VCDT concept annotations, this was as part of a group of visual retrieval algorithms whose results were used in a late fusion process. It is therefore not possible to determine the effect of only the VCDT concepts on the results.

4 Conclusion

This paper summarises the ImageCLEF 2008 Visual Concept Detection Task. The aim was to automatically annotate images with concepts, with a list of 17 hierarchically organised concepts provided. The results demonstrate that this task can be solved reasonably well, with the best run having an average AUC over all concepts of 90.66%. Six further runs obtained AUCs between 80% and 90%. When evaluating the runs on a per concept basis, the best run also obtained an AUC of 80% or above for every concept. Concepts for which automatic detection was particularly successful are: *indoor/outdoor*, *night*, and *sky*. The worst results were obtained for the concept *road or pathway*.

References

1. Arni, T., Clough, P., Sanderson, M., Grubinger, M.: Overview of the ImageCLEFphoto 2008 photographic retrieval task. In: Peters, C., et al. (eds.) CLEF 2008. LNCS, vol. 5706, pp. 500–511. Springer, Heidelberg (2009)
2. Müller, H., Kalpathy-Cramer, J., Kahn Jr., C.E., Hatt, W., Bedrick, S., Hersh, W.: Overview of the ImageCLEFmed 2008 medical image retrieval task. In: Peters, C., et al. (eds.) CLEF 2008. LNCS, vol. 5706, pp. 512–522. Springer, Heidelberg (2009)
3. Tsikrika, T., Kludas, J.: Overview of the wikipediaMM task at ImageCLEF 2008. In: Peters, C., et al. (eds.) CLEF 2008. LNCS, vol. 5706, pp. 539–550. Springer, Heidelberg (2009)
4. Deselaers, T., Deserno, T.M.: Medical image annotation in ImageCLEF 2008. In: Peters, C., et al. (eds.) CLEF 2008. LNCS, vol. 5706, pp. 523–530. Springer, Heidelberg (2009)
5. Grubinger, M., Clough, P., Müller, H., Deselaers, T.: The IAPR TC-12 benchmark - a new evaluation resource for visual information systems. In: Proceedings of the International Workshop OntoImage 2006, pp. 13–23 (2006)
6. Fluhr, C., Moëllic, P.A., Hède, P.: ImagEVAL: Usage-oriented multimedia information retrieval evaluation. In: Proceedings of the second MUSCLE/ImageCLEF Workshop on Image and Video Retrieval Evaluation, Alicante, Spain, pp. 3–8 (2006)
7. Deselaers, T., Hanbury, A., Viitaniemi, V., Benczúr, A., Brendel, M., Daróczy, B., Balderas, H.E., Gevers, T., Gracidas, C.H., Hoi, S.C.H., Laaksonen, J., Li, M., Castro, H.M., Ney, H., Rui, X., Sebe, N., Stöttinger, J., Wu, L.: Overview of the image-CLEF 2007 object retrieval task. In: Peters, C., Jijkoun, V., Mandl, T., Müller, H., Oard, D.W., Peñas, A., Petras, V., Santos, D. (eds.) CLEF 2007. LNCS, vol. 5152, pp. 445–471. Springer, Heidelberg (2008)

A Results for All Submissions

The results for each submitted run are given in Table 4.

Table 4. Average EER and Average AUC over all concepts for all runs of all participating groups

group	run	EER [%]	AUC [%]
CEA_LIST	CEA_LIST_2	29.71	71.44
CEA_LIST	CEA_LIST_3	41.43	34.25
CEA_LIST	CEA_LIST_4	29.04	73.40
HJ_FA	HJ_Result	45.07	19.96
IPAL_I2R	I2R_IPAL_Cor_Run1	40.02	62.62
IPAL_I2R	I2R_IPAL_Edge_Run2	45.71	55.79
IPAL_I2R	I2R_IPAL_HIST_Run4	31.83	73.80
IPAL_I2R	I2R_IPAL_Linear_Run5	36.09	68.65
IPAL_I2R	I2R_IPAL_Texture_Run	39.22	62.93
IPAL_I2R	I2R_IPAL_model_Run6	33.93	72.01
IPAL_I2R	IPAL_I2R_FuseMCE_R7	31.17	74.05
IPAL_I2R	IPAL_I2R_FuseNMCE_R8	29.71	76.44
LSIS	GLOT-methode23_LSIS_evaOK	26.56	79.92
LSIS	new_kda_results.txt	25.88	80.51
LSIS	FusionA_LSIS.txt	49.29	50.84
LSIS	FusionH_LSIS.txt	49.38	50.20
LSIS	MLP1_LSIS_GLOT	25.95	80.67
LSIS	MLP1_vcdt_LSIS	25.95	80.67
LSIS	method2_LSIS	26.61	79.75
MMIS	MMIS_Ruihu	41.05	62.50
MMIS	ainhoa	28.44	77.94
MMIS	alexei	28.82	77.65
MMIS	combinedREPLACEMENT	31.90	73.69
Makerere	MAK	49.25	30.83
RWTH	PHME	20.45	86.19
TIA	INAOE-kr_00_HJ_TIA	42.93	28.90
TIA	INAOE-kr_04_HJ_TIA	47.12	17.58
TIA	INAOE-lb_01_HJ_TIA	39.12	42.15
TIA	INAOE-psms_00_HJ_TIA	32.09	55.64
TIA	INAOE-psms_02_HJ_TIA	35.90	47.07
TIA	INAOE-rf_00_HJ_TIA	39.29	36.11
TIA	INAOE-rf_03_HJ_TIA	42.64	26.37
UPMC	LIP6-B50trees100C5N5	27.32	71.98
UPMC	LIP6-B50trees100C5N5T25	28.93	53.78
UPMC	LIP6-B50trees100COOC5T25	28.83	54.19
UPMC	LIP6-B50trees100pc	24.55	82.74
UPMC	LIP6-B50trees100pc_COOC5	27.37	71.58
UPMC	LIP6-B50trees100pc_T25	26.20	57.09
XRCE	TVPA-XRCE_KNN	16.65	90.66
XRCE	TVPA-XRCE_LIN	19.29	88.73
budapest	acad-acad-logreg1	37.36	66.39
budapest	acad-acad-logreg2	37.12	66.53
budapest	acad-acad-lowppnn	36.07	67.15
budapest	acad-acad-lowppnpnn	32.46	73.05
budapest	acad-acad-medfi	32.47	73.57
budapest	acad-acad-mednofi	32.10	74.18
budapest	acad-acad-medppnn	37.01	59.30
budapest	acad-acad-medppnpnn	32.47	73.61
budapest	acad-acad-mixed	38.34	63.80
budapest	acad-budapest-acad-glob1	45.72	52.78
budapest	acad-budapest-acad-glob2	31.14	74.90
budapest	acad-budapest-acad-lowfi	32.48	73.03
budapest	acad-budapest-acad-lownfi	32.44	73.32

Overview of the WikipediaMM Task at ImageCLEF 2008

Theodora Tsikrika[1] and Jana Kludas[2]

[1] CWI, Amsterdam, The Netherlands
Theodora.Tsikrika@cwi.nl
[2] CUI, University of Geneva, Switzerland
jana.kludas@unige.ch

Abstract. The wikipediaMM task provides a testbed for the system-oriented evaluation of ad-hoc retrieval from a large collection of Wikipedia images. It became a part of the ImageCLEF evaluation campaign in 2008 with the aim of investigating the use of visual and textual sources in combination for improving the retrieval performance. This paper presents an overview of the task's resources, topics, assessments, participants' approaches, and main results.

1 Introduction

The wikipediaMM task provides a testbed for the system-oriented evaluation of multimedia information retrieval from a collection of Wikipedia (http://www.wikipedia.org/) images. This collection has been previously used in the INEX 2006-2007 Multimedia track [5,4]. The aim is to investigate mono-media and cross-media retrieval approaches in the context of a large and heterogeneous collection of images (similar to those encountered on the Web) that are accompanied by unstructured and noisy textual annotations in English, and are searched for by users with diverse information needs.

It is an ad-hoc image retrieval task with an evaluation scenario similar to the classic TREC ad-hoc retrieval task and the ImageCLEFphoto task: simulation of the situation in which a system knows the set of documents to be searched, but cannot anticipate the particular topic that will be investigated (i.e., the topics are not known to the system in advance). Given a textual query (and/or sample images and/or concepts) describing a user's multimedia information need, the aim is to find as many relevant images as possible from the Wikipedia image collection. A multimedia retrieval approach in that case should aim at combining the evidence of relevance from different media types into a single ranking that is presented to the user. In this first year of the task, the focus in on monolingual retrieval.

The paper is organised as follows. First, we introduce the task resources (the image collection and other available resources), the topics, and assessments (Sections 2–4). Section 5 presents the approaches employed by the different participants, while Section 6 summarises their main results. Section 7 concludes the paper and provides an outlook on next year's task.

C. Peters et al. (Eds.): CLEF 2008, LNCS 5706, pp. 539–550, 2009.

2 Task Resources

The resources used for the wikipediaMM task are based on Wikipedia data. The following resources were made available to the participants:

(INEX MM) wikipedia image collection: The collection consists of approximately 150,000 JPEG and PNG images provided by Wikipedia's users. Each image is associated with user generated, alphanumeric, unstructured metadata in English. These metadata typically contain a brief caption or description of the image, the Wikipedia user who uploaded the image, and the copyright information (see Figure 1 for an example). These descriptions are highly heterogeneous and of varying length. Further information about the image collection can be found in [5].

Image classification scores: For each image, the classification scores for the 101 different MediaMill concepts were provided by UvA [3]. The UvA classifier had been trained on manually annotated TRECVID video data and the concepts were selected for the broadcast news domain.

Image features: For each image, the set of the 120D feature vectors used to derive the above image classification scores [1] was also made available. Participants could use these feature vectors to custom-build a content-based image retrieval (CBIR) system, without having to pre-process the image collection.

These resources had also been provided in the INEX 2006-2007 Multimedia track. The latter two resources are beneficial to researchers who wish to exploit visual evidence without performing image analysis.

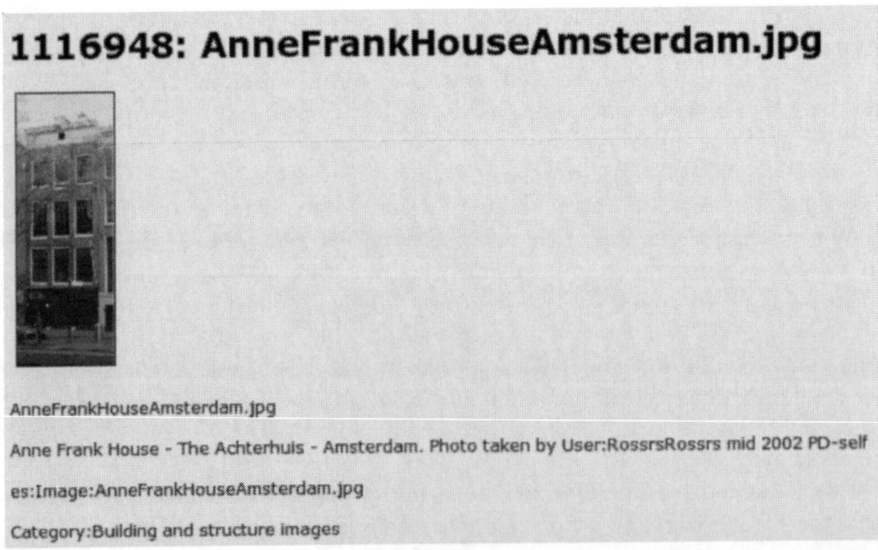

Fig. 1. Example Wikipedia image+metadata from the (INEX MM) wikipedia image collection

3 Topics

The topics for the ImageCLEF 2008 wikipediaMM task include (i) topics previously used in INEX 2006-2007 Multimedia track, and (ii) topics created by this year's task participants. They are descriptions of multimedia information needs that may contain not only textual, but also visual evidence, in the form of sample images and concepts.

3.1 Topic Format

The wikipediaMM topics are multimedia queries that can consist of a textual, visual, and conceptual part, with the latter two parts being optional.

<title> query by keywords
<concept> query by concept (optional)
<image> query by image content (optional)
<narrative> description in which the definition of relevance and irrelevance are given

<title> The topic <title> simulates a user who does not have (or does not want to use) example images or other visual information. The query expressed in the topic <title> is therefore a text-only query. This profile is likely to fit most users searching digital libraries.

Upon discovering that a <title>-only query returns many irrelevant hits, users might decide to reformulate it by adding visual information.

<concept> This field is directly related to the concepts for which classification results are provided as an additional source of information (see Section 2), i.e., they are restricted to the 101 MediaMill concepts.

<image> The second type of visual evidence are example images, which can be taken from outside or inside Wikipedia and can be of any common format.

<narrative> A clear and precise description of the information need is required in order to unambiguously determine whether or not a given image fulfils the given need. In a test collection setting, this description is known as the narrative. It is the only true and accurate interpretation of a user's need. Precise recording of the narrative is important for scientific repeatability - there must exist a definitive description of what is and is not relevant to the user. To aid this, the <narrative> should explain not only what information is being sought, but also the context and motivation of the information need, i.e., why the information is being sought and what work-task it might help to solve.

The three different types of information sources (textual terms, visual examples, and concepts) can be used in any combination. For each field more than one entry can be specified. It is up to the systems how to use, combine or ignore this information; the relevance of a result item does not directly depend on them, but it is decided by manual assessments based on the <narrative>.

Table 1. Topics for the ImageCLEF 2008 wikipediaMM task: their IDs, titles, and whether they include visual information (Yes/No) in the form of image examples (IMG) and concepts (CON)

ID	Topic title	IMG	CON	ID	Topic title	IMG	CON
1	blue flower	Y	Y	2	sea sunset	N	N
3	ferrari red	Y	Y	4	white cat	Y	Y
5	silver race car	N	Y	6	potato chips	N	N
7	spider web	Y	N	8	beach volleyball	Y	Y
9	surfing	Y	Y	10	portrait of Jintao Hu	Y	Y
11	map of the United States	N	Y	12	rabbit in cartoons	Y	Y
13	DNA helix	Y	Y	14	people playing guitar	Y	Y
15	sars china	Y	N	16	Roads in California	Y	N
17	race car	N	Y	18	can or bottle of beer	N	N
19	war with guns	N	N	20	hunting dog	N	Y
21	oak tree	Y	Y	22	car game covers	Y	N
23	british trains	Y	N	24	peace anti-war protest	Y	Y
25	daily show	N	Y	26	house architecture	N	Y
27	baseball game	Y	N	28	cactus in desert	Y	Y
29	pyramid	Y	Y	30	video games	N	N
31	bridges	N	N	32	mickey mouse	Y	N
33	Big Ben	N	N	34	polar bear	N	N
35	George W Bush	Y	N	36	Eiffel tower	N	N
37	Golden gate bridge	Y	N	38	Da Vinci paintings	Y	N
39	skyscraper	Y	Y	40	saturn	Y	N
41	ice hockey players	N	Y	42	labor demonstrations	N	Y
43	mountains under sky	N	Y	44	graph of a convex function	Y	Y
45	paintings related to cubism	Y	Y	46	London parks in daylight	Y	Y
47	maple leaf	Y	Y	48	a white house with a garden	Y	Y
49	plant	Y	Y	50	stars and nebulae in the dark sky	Y	N
51	views of Scottish lochs	Y	N	52	Cambridge university buildings	Y	N
53	military aircraft	N	Y	54	winter landscape	N	N
55	animated cartoon	N	Y	56	London city palaces	N	Y
57	people riding bicycles	Y	N	58	sail boat	Y	Y
59	dancing couple	N	Y	60	atomic bomb	Y	Y
61	Singapore	N	N	62	cities by night	Y	Y
63	star galaxy	N	N	64	football stadium	N	N
65	famous buildings of Paris	N	Y	66	historic castle	N	N
67	bees with flowers	Y	Y	68	pyramids in Egypt	Y	Y
69	mountains with snow under sky	N	Y	70	female players beach volleyball	Y	Y
71	children riding bicycles	N	N	72	civil aircraft	N	Y
73	bridges at night	Y	Y	74	gothic cathedral	Y	Y
75	manga female character	N	Y				

3.2 Topic Development

The topics in the wikipediaMM task have been mainly developed by the participants. Altogether, 12 of the participating groups submitted 70 candidate topics. The 35 topics used in INEX 2006-2007 Multimedia were also added to the candidate topic pool. The task organisers judged all topics in the pool in terms of their "visuality" as proposed in [2] (replacing though the so-called "neutral" option with a "textual" one). This led to the following classification of candidate topics:

visual: topics that have visual properties that are highly discriminating for the problem (e.g., "blue flower"). Therefore, it is highly likely that CBIR systems would be able to deal with them.
textual: topics that often consist of proper nouns of persons, buildings, locations, etc. (e.g., "Da Vinci paintings"). As long as the images are correctly annotated, text-only approaches are likely to suffice.
semantic: topics that have a complex set of constraints, need world knowledge, or contain ambiguous terms (e.g., "labor demonstrations"). It is highly likely that no modality alone is effective.

The candidate topics were classified by the organisers; for the old INEX topics, the results of the INEX participants' runs were also used to aid this classification. The final topic set is listed in Table 1 and consists of 75 topics: 5 visual (topic IDs: 1-5), 35 textual (topic IDs: 6-40), and 35 semantic (topic IDs: 41-75). Table 2 shows some statistics on the topics. Not all topics contain visual/multimedia information (i.e., image examples or concepts); this corresponds well with realistic scenarios, since users who express multimedia information needs do not necessarily want to employ visual evidence.

Table 2. ImageCLEF 2008 wikipediaMM topics

	all	visual	textual	semantic
Number of topics	75	5	35	35
Average number of terms in title	2.64	2.2	2.3	2.9
Number of topics with image(s)	43	3	22	18
Number of topics with concept(s)	45	4	16	25
Number of topics with both image and concept	28	3	11	14
Number of topics with text only	15	1	8	6

4 Assessments

The wikipediaMM task is an image retrieval task, where an image with its metadata is either relevant or not (binary relevance). We adopted TREC-style pooling of the retrieved images with a pool depth of 100, resulting in pools of between 753 and 1850 images with a mean and median both around 1290. The evaluation was performed by the participants of the task within a period of 4 weeks after the submission of runs. The 13 groups that participated in the evaluation process

Table 3. Resources used by the 77 submitted runs

Resource modality		# runs using it
textual	Txt	35
visual	Img	5
concept	Con	0
textual/visual	TxtImg	22
textual/concept	TxtCon	13
textual/visual/concept	TxtImgCon	2

used a web-based interface previously employed in the INEX Multimedia and TREC Enterprise tracks. Each participating group was assigned 4-5 topics and an effort was made to ensure that most of the topics were assigned to their creators. This was achieved in 76% of the assignments for the topics created by this year's participants.

5 Participants

A total of 12 groups submitted 77 runs. Table 3 gives an overview of the resources used by the submitted runs[1]. Most of the runs are textual only approaches, but compared to the INEX Multimedia track, there is a rise in fusion approaches that combine text and images, text and concepts, and all three modalities.

Below we briefly describe the approaches investigated by the participating groups:

Digital Media Institute, Peking University, China (7 runs). They investigated the following three approaches: (i) a text-based approach with query expansion where the expansion terms are automatically selected from a knowledge base that is (semi-)automatically constructed from Wikipedia, (ii) a content-based visual approach, where they first train 1-vs-all classifiers for all queries by using the training images obtained by Yahoo! search, and then treat the retrieval task as a visual concept detection in the given Wikipedia image set, and (iii) a cross-media approach that combines and reranks the text- and content-based retrieval results.

CEA LIST, France (2 runs). Their approach was based on query reformulation using concepts considered to be semantically related to those in the initial query. For each interesting entity in the query, they used Wikipedia and WordNet to extract related concepts, which were further ranked based on their perceived usefulness. A small list of visual concepts was used to rerank the answers to queries that included them, implicitly or explicitly. They submitted two automatic runs, one based on query reformulation only, and one combining query reformulation and visual concepts.

NLP and Information Systems, University of Alicante, Spain (24 runs). They employed their textual passage-based retrieval system as their

[1] Our analysis is based on the runs' descriptions given by the participants themselves.

baseline approach, which was enhanced by a module that decomposes the (compound) file names in camel case notation into single terms, and by a module that performs geographical query expansion. They also investigated Probabilistic Relevance Feedback and Local Context Analysis techniques.

Data Mining and Web Search, Hungarian Academy of Sciences (8 runs). They used their own retrieval system to experiment with a text-based approach, that uses BM25 and query expansion based on blind relevance feedback, and its combination with a segment-based visual approach.

Database Architectures and Information Access, CWI, Netherlands (2 runs). They used a language modelling approach based on purely textual evidence and also incorporated a length prior to bias retrieval towards images with longer descriptions than the ones retrieved by the language model.

Laboratoire Hubert Curien, Université Jean Monnet, Saint-Etienne, France (6 runs). They used a vector space model to compute similarities between vectors of both textual and visual terms; the textual part is a term vector of the terms' BM25 weights and the visual part a 6-dimensional vector of clusters of colour features. The applied both manual and blind relevance feedback to a text-based run in order to expand the query with visual terms.

Dept. of Computer Science and Media, Chemnitz University of Technology (4 runs). They used their Xtrieval framework based on both textual and visual features, and also made use of the provided resources (concepts and features). They experimented with text-based retrieval, its combination with a visual approach, the combination of all three modalities, and thesaurus-based query expansion. They also investigated the efficiency of the employed approaches.

Multimedia and Information Systems, Imperial College, UK (6 runs). They examined textual features, visual features, and their combination. Their text-based approach was combined with evidence derived from a geographic co-occurrence model mined from Wikipedia which aimed at disambiguating geographic references either in a context-dependent or a context-independent manner. Their visual-based approach, employing Gabor texture features and the Cityblock distance as a similarity measure, was combined with the text-based approach and with blind relevance feedback using a convex combination of ranks.

SIG-IRIT, Toulouse, France (4 runs). They explored the use of images' names as evidence in text-based image retrieval. They used them in isolation, by computing a similarity score between the query and the name of images using the vector space model, and in combination with textual evidence, either by fusing the ranking of a text-based approach (based on the XFIRM retrieval system) with the ranking produced by the name-based technique, or by using the text-based approach with an increase in the weight of terms in the image name.

Computer Vision and Multimedia, Université de Genève, Switzerland (2 runs). Their approach was based on the preference ranking option of the SVM light library developed by Cornell University. Their first run employed a text-based retrieval approach. Their second run applied a feature selection to the high dimensional textual feature vector based on the features relevant to each query.

UPMC/LIP6 - Computer Science Lab, Paris, France (7 runs). They investigated (i) text-based retrieval, using a *tf.idf* approach, a language modelling framework, and their combination based on the ranks of retrieved images, and (ii) the combination of textual and visual evidence, by reranking the text-based results using visual similarity scores (Euclidean distance and a manifold-based technique, both based on HSV features).

LSIS, UMR CNRS & Université Sud Toulon-Var, France (5 runs). They applied the same techniques they used for the Visual Concept Detection task at ImageCLEF 2008, by relating each of the wikipediaMM topics to one or more visual concepts from that task. They also fused these visual-based rankings with the results of a text-based approach.

6 Results

The analysis presented in this section takes into account only the top 75% of all submitted runs that perform best in terms of Mean Average Precision (MAP), so as to exclude noise by removing erroneous and buggy runs. Table 4 shows the top 30 runs ranked by their MAP; the complete result list of results can be found at: `http://www.imageclef.org/2008/wikimm-results`.

We first analyse the performance per modality by comparing the average performance over runs that employ the same type of resources. Table 5 shows the average performance and standard deviation with respect to modality. There were no automatic image-only runs in the top 75% of the runs. According to the MAP measure, the best performing runs fuse text and concepts, followed by the runs that fuse text, concepts and images, and the text-only baseline. The latter two perform almost identically. A similar result is obtained with the precision after R (= number of relevant) documents are retrieved (R-prec.). Again, the TxtCon runs outperform all others. The runs that fuse textual and visual information (TxtImg) perform worst for all measures, so this remains an open research issue. But, in general, it can be said that this year the fusion approaches perform surprisingly well, mostly due to the incorporation of concepts.

Next, we analyse the performance per topic. Figure 2 shows for each topic the average MAP over all (top 75% of) runs, over only the text-based ones among them, and over only the text/concept-based ones among them. The text/concept-based runs outperform the text-based ones for 64% of the topics with a maximum improvement of 21% in a single topic. Table 6 presents the top 10 topics, where the text/concept runs outperform the textual. The list includes a mixture of topics with proper nouns and complex concepts. Oddly, this top 10 also includes topics that do not have an example concept (see Table 1), which suggests that some conceptual query expansion or a concept schema different from the provided one has been used by some of the participants. These could also be topics that are not well annotated in the collection.

Table 4. Results for the top 30 submitted runs ranked by their MAP

Group	Run	Modality	FB/QE	MAP	P@10	P@20	R-prec.
1 upeking	zzhou3	Txt	QE	0.3444	0.4760	0.3993	0.3794
2 cea	ceaTxtCon	TxtCon	QE	0.2735	0.4653	0.3840	0.3225
3 ualicante	IRnNoCamel	Txt	NOFB	0.2700	0.3893	0.3040	0.3075
4 cea	ceaTxt	Txt	QE	0.2632	0.4427	0.3673	0.3080
5 ualicante	IRnNoCamelLca	Txt	FB	0.2614	0.3587	0.3167	0.2950
6 ualicante	IRnNoCamelGeog	Txt	QE	0.2605	0.3640	0.2913	0.3000
7 ualicante	IRnConcSinCamLca	TxtCon	FB	0.2593	0.3493	0.2900	0.3016
8 ualicante	IRnConcSinCam	TxtCon	NOFB	0.2587	0.3627	0.2900	0.2975
9 ualicante	IRnNoCamelLcaGeog	Txt	FBQE	0.2583	0.3613	0.3140	0.2922
10 sztaki	bp_acad_avg5	TxtImg	NOFB	0.2551	0.3653	0.2773	0.3020
11 sztaki	bp_acad_textonly_qe	Txt	QE	0.2546	0.3720	0.2907	0.2993
12 ualicante	IRnConcSinCamLcaGeog	TxtCon	FBQE	0.2537	0.3440	0.2853	0.2940
13 cwi	cwi_lm_txt	Txt	NOFB	0.2528	0.3427	0.2833	0.3080
14 sztaki	bp_acad_avgw_glob10	TxtImg	NOFB	0.2526	0.3640	0.2793	0.2955
15 sztaki	bp_acad_avgw_glob10_qe	TxtImg	QE	0.2514	0.3693	0.2833	0.2939
16 ualicante	IRnConcSinCamGeog	TxtCon	QE	0.2509	0.3427	0.2787	0.2924
17 sztaki	bp_acad_glob10_qe	TxtImg	QE	0.2502	0.3653	0.2833	0.2905
18 sztaki	bp_acad_glob10	TxtImg	NOFB	0.2497	0.3627	0.2780	0.2955
19 cwi	cwi_lm_lprior_txt	Txt	NOFB	0.2493	0.3467	0.2787	0.2965
20 sztaki	bp_acad_avg5	TxtImg	NOFB	0.2491	0.3640	0.2773	0.2970
21 sztaki	bp_acad_avgw_qe	TxtImg	QE	0.2465	0.3640	0.2780	0.2887
22 curien	LaHC_run01	Txt	NOFB	0.2453	0.3680	0.2860	0.2905
23 ualicante	IRnConcSinCamPrf	TxtCon	FB	0.2326	0.2840	0.2700	0.2673
24 ualicante	IRnNoCamelPrf	Txt	FB	0.2321	0.3107	0.2800	0.2665
25 ualicante	IRnNoCamelPrfGeog	Txt	FBQE	0.2287	0.3120	0.2787	0.2611
26 ualicante	IRnConcSinCamPrfGeog	TxtCon	FBQE	0.2238	0.2853	0.2673	0.2561
27 chemnitz	cut-mix-qe	TxtImgCon	QE	0.2195	0.3627	0.2747	0.2734
28 ualicante	IRnConcepto	TxtCon	NOFB	0.2183	0.3213	0.2520	0.2574
29 ualicante	IRn	Txt	NOFB	0.2178	0.3760	0.2507	0.2569
30 chemnitz	cut-txt-a	Txt	NOFB	0.2166	0.3440	0.2833	0.2695

Table 5. Results per modality over all topics

Modality	MAP		P@20		R-prec.	
	Mean	SD	Mean	SD	Mean	SD
All top 75% runs	0.2149	0.049	0.2676	0.047	0.2566	0.050
Txt in top 75% runs	0.2104	0.053	0.2643	0.052	0.2510	0.055
Img in top 75% runs	–	–	–	–	–	–
TxtCon in top 75% runs	**0.2316**	0.025	**0.2874**	0.033	**0.2742**	0.026
TxtImg in top 75% runs	0.2078	0.061	0.2522	0.047	0.2516	0.060
TxtImgCon in top 75% runs	0.2122	0.010	0.2683	0.014	0.2559	0.012

Table 6. Top 10 topics for which text/concept runs outperform text-only runs

text/concept
(23) british trains
(63) star galaxy
(38) Da Vinci Paintings
(50) stars/nebulae dark sky
(10) portrait of Jintao Hu
(67) bees with flowers
(34) polar bear
(43) mountains under sky
(51) Views of Scottish lochs
(3) ferrari red

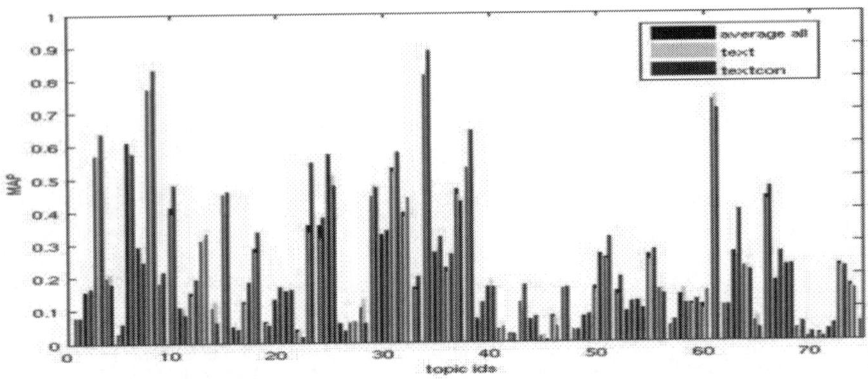

Fig. 2. Performance of different modalities per topic

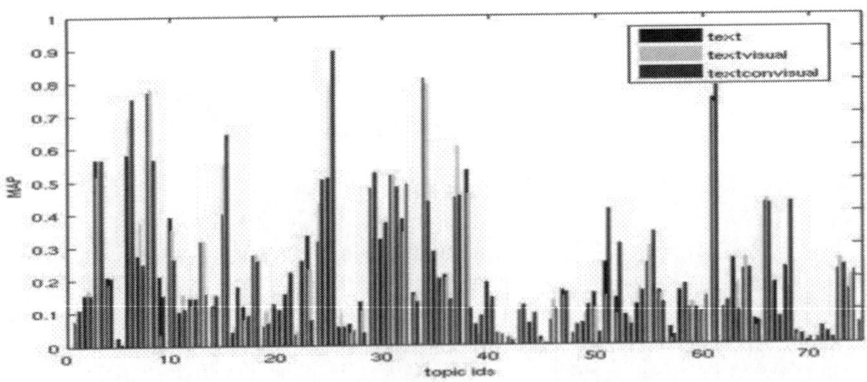

Fig. 3. Performance of different fusion approaches per topic

Table 7. Top 10 topics for which fusion runs outperform textl-only runs

text/visual	text/concept/visual
(25) daily show	(25) daily show
(37) Golden gate bridge	(15) sars china
(15) sars china	(22) car game covers
(6) potato chips	(68) pyramids in egypt
(24) peace anti-war protest	(24) peace anti-war protest
(7) spider web	(6) potato chips
(46) London parks in daylight	(52) Cambridge University Buildings
(11) map of the united states	(51) Views of Scottish lochs
(55) animated cartoon	(16) Roads in California
(54) winter landscape	(32) Mickey Mouse

Figure 3 compares the text baseline with the other 2 fusion approaches: (1) text/visual and (2) text/concept/visual. The fusion of text and images outperforms the text-based runs only for less than half of the topics (44%) and the fusion of all 3 modalities for 56%. The maximum improvement of a topic over the baseline is 29% for TextImg and 38% for TextConImg respectively. We again create a list of the top 10 topics for which each fusion approach outperforms the text-only baseline. The lists for the fusion of text/visual and text/concept/visual have many entries in common, and contain mostly topics with complex concepts that have an image and/or a concept defined.

In summary, the results indicate that fusion approaches are catching up with the text-based approaches. Especially the text/concept fusion approaches perform particularly well. The visual hints help mainly for topics that incorporate an image example, but can also improve the overall performance.

7 Conclusion and Outlook

Our debut in ImageCLEF 2008 attracted much interest from groups researching multimedia retrieval and significantly more participants than the INEX 2006-2007 Multimedia track. With the help of our participants, we both developed and assessed a diverse set of 75 multimedia topics. The results indicate that the dominance of text-based image retrieval is coming to an end; multi-modal fusion approaches help to improve the retrieval performance in this domain.

Our main focus for next year remains the same: researching the combination of evidence from different modalities in a "standard" ad-hoc image retrieval task. Possible new directions for 2009 include the addition of multilinguality in form of multi-lingual topics (and if possible annotations), and access to the context of the images, i.e., the Wikipedia web pages that contain them. We also aim at providing new sets of classification scores and low-level features, so that participants can concentrate their research effort on information fusion.

Acknowledgements

Theodora Tsikrika was supported by the European Union via the European Commission project VITALAS (contract no. 045389). Jana Kludas was funded by the European project MultiMATCH (EU-IST-STREP#033104). The authors would also like to thank Thomas Deselaers for invaluable technical support and all the groups participating in the relevance assessment process.

References

1. van Gemert, J.C., Geusebroek, J.-M., Veenman, C.J., Snoek, C.G.M., Smeulders, A.W.M.: Robust scene categorization by learning image statistics in context. In: Proceedings of the 2006 Conference on Computer Vision and Pattern Recognition Workshop, Washington, DC, USA, p. 105. IEEE Computer Society, Los Alamitos (2006)
2. Grubinger, M., Clough, P.D.: On the creation of query topics for ImageCLEFphoto. In: Proceedings of the Third MUSCLE / ImageCLEF Workshop on Image and Video Retrieval Evaluation (2007)
3. Snoek, C.G.M., Worring, M., van Gemert, J.C., Geusebroek, J.-M., Smeulders, A.W.M.: The challenge problem for automated detection of 101 semantic concepts in multimedia. In: Proceedings of the 14th annual ACM international conference on Multimedia, pp. 421–430. ACM Press, New York (2006)
4. Tsikrika, T., Westerveld, T.: The INEX 2007 multimedia track. In: Fuhr, N., Kamps, J., Lalmas, M., Trotman, A. (eds.) INEX 2007. LNCS, vol. 4862, pp. 440–453. Springer, Heidelberg (2008)
5. Westerveld, T., van Zwol, R.: The INEX 2006 multimedia track. In: Fuhr, N., Lalmas, M., Trotman, A. (eds.) INEX 2006. LNCS, vol. 4518, pp. 331–344. Springer, Heidelberg (2007)

Meiji University at ImageCLEF2008 Photo Retrieval Task: Evaluation of Image Retrieval Methods Integrating Different Media

Kosuke Yamauchi, Takuya Nomura, Keiko Usui, Yusuke Kamoi, and Tomohiro Takagi

Department of Computer Science
Meiji University
1-1-1 Higashimita, Tama-ku, Kawasaki-shi, Kanagawa, 214-8571, Japan
{yamauchi,takagi}@cs.meiji.ac.jp

Abstract. This paper describes the participation of the Human Interface Laboratory of Meiji University in the ImageCLEF2008 photo retrieval task. We submitted eight retrieval runs taking two main approaches. The first approach combined Text-Based Image Retrieval (TBIR) and Context-Based Image Retrieval (CBIR). The second approach applied query expansion using conceptual fuzzy sets (CFS). A CFS is a method that uses the expression of meaning depending on the context, which an ordinary fuzzy set does not recognize. A conceptual dictionary is necessary to perform query expansion using CFS, and this is constructed by clustering. We propose here the use of query expansion with CFS and other techniques, for image retrieval that integrates different media, and we verify the utility of the system by explaining our experimental results. This time, *TBIR+CFS* in the system which we proposed is selected No.1 with "Text Only" runs, and we demonstrated that question expansion with CFS produced higher search results.

Keywords: Information Retrieval, Image Retrieval, Query Expansion, Conceptual Fuzzy Sets, Fuzzy Clustering.

1 Introduction

We present here our participation in the ImageCLEF2008 photo retrieval task [1]. The task deals with answering 39 queries of variable complexity from a repository of 20,000 photographic images in the IAPR TC-12 photographic collection. The details of this task can be found in ImageCLEFphoto 2008 [2].

We submitted eight retrieval runs this year taking two main approaches.

The first approach is an image retrieval system that integrates Text-Based Image Retrieval (TBIR) and Context-Based Image Retrieval (CBIR). We are taking this approach in order to overcome the difficulties that arise with these methods individually. For example, using TBIR, a person's subjectivity can easily be introduced, and furthermore, images without text cannot be retrieved. Using CBIR with the present

C. Peters et al. (Eds.): CLEF 2008, LNCS 5706, pp. 551–559, 2009.

technology, it is very difficult to see only the contents of the image and to make the computer understand what the image is. This method is also less accurate than TBIR. We expect that these difficulties can be avoided by combining TBIR and CBIR, and the accuracy of the image retrieval can be improved by the synergistic effect of different media used to solve these problems.

The second approach is query expansion using Conceptual Fuzzy Sets (CFS). Current search engines are powerful. However, several words are difficult to search. Takagi et al. [3] proposed resolving this problem by query expansion depending on the context using CFS. A CFS is a method that uses meaning expression depending on the context, which a fuzzy set does not recognize. We propose a system that depends more on query context than on query expansion for improving the packaging method of a conceptual fuzzy set.

This paper is organized as follows. In section 2, CFS and a method of query expansion using CFS are presented. Section 3 describes the details of a good system of the precision in all eight retrieval systems that we submitted. Section 4 describes the results of our experiment. In Section 5, we consider the results and discuss our study.

2 Conceptual Fuzzy Sets

In this section, we explain what the conceptual fuzzy sets (CFS) is. Next, the method necessary to construct a conceptual dictionary in order to perform a query expansion using CFS is described, and finally, the query expansion technique is described.

2.1 Conceptual Fuzzy Sets

A CFS is based on the use theory of meaning propounded by Wittgenstein for the expression of meaning of a concept. According to this theory, the meaning of a word can be expressed by another word. Thus, the meaning of a word can be expressed by other words associated with one another. Also, the expression of the meaning of a word by another word makes a closed circular system. In CFS, the meaning of a word is expressed by the set of words and their activity values.

General fuzzy sets express phenomena without a clear boundary. However, when applying them to various realistic problems, they are not situation-dependent. The meaning of a concept changes in situation-dependent phenomena, and cannot correspond to a fixed expression. This problem occurs when the generality of knowledge is not obtained because the degree the ambiguity of a fuzzy set is fixed, and the mechanism including background is not given. The essence of this problem is the expression of meaning.

Conceptual fuzzy sets solve situation-dependent problems by a recall mechanism (i.e., combined fuzzy sets) and the general versatility of knowledge by making a closed circular system based on the use theory of meaning.

In our system, we created a conceptual fuzzy set by the superpositioning of concepts. The concept component of a fuzzy set is called a prototype concept, and a set of prototype concepts is called a concept dictionary. The concept depending on the input context can be generated by superpositioning a prototype concept that is similar to the input.

2.2 Conceptual Dictionary

We constructed a conceptual dictionary to perform query expansion using conceptual fuzzy sets. The following methods are commonly used to construct a conceptual dictionary.

i) Manual construction.
ii) Clustering corpora, assuming one cluster is one concept.
iii) Using an existing edifice of knowledge (e.g., Wordnet), and making a concept from words included in each directory, etc.

However, the first method is a very tedious task, and the third method cannot be used to make a concept if the existing edifice of knowledge cannot be found. Thus, in our experiment, we constructed a concept dictionary by clustering.

We compared the precision of query expansion by constructing two concept dictionaries. We show each method for constructing these concept dictionaries below.

2.2.1 Conceptual Dictionary Using Fuzzy C-Means

We used the fuzzy C-means method to construct the conceptual dictionary. The construction process was as follows.

1. From the data included in the corpus, make a word vector from the annotation text and a region ID vector from an image. Make a feature vector by combining the region ID vector and the word vector.
2. Cluster the feature vectors using the fuzzy C-means method; in this case, we set the number of clusters to be 1,000.
3. The fuzzy C-means method was used for clustering, and all vectors belonged to all clusters. Therefore, we set the threshold based on the membership value and limit the number of vectors that belong to one cluster. In addition, in a reorganization of the Term Frequency-Inverse Document Frequency (TF-IDF) method, we weight the feature vectors taking into consideration the degree of belonging to a cluster. First, we define E_j as a clustered element (i.e., feature vector in our experiment), and j expresses an identifier. Here, E_j has the same value of the degree of belonging to each cluster I (in our experiment, it was the same as the membership value), we define this as $Belong(E_j, I)$, and we define TF as the multiplication of feature vectors taking into consideration $Belong(E_j, I)$. TF is obtained with Eq. (1).

$$TF = \sum_{E_j \in I} Belong\ (E_j, I) \times E_j \tag{1}$$

4. We define the word set from 3. As the prototype concept.

2.3 Query Expansion Using CFS

The technique of query expansion using CFS is described here. The flow of query expansion is shown below.

1. Extract a word vector from <title> and <narr> of topic, and extract a region ID vector from the image. Then, make a feature vector by combining the word vector and the region ID vector.
2. Calculate the degree of similarity of the feature vector of the query and each prototype concept in a concept dictionary using the cosine measure.
3. Make the query's concept by overlapping a number of similar prototype concepts *N*. The overlap of the prototype concept is calculated by Eq. (2) Here, *Similarity* is the cosine measure, C_i is the number of prototype concept *i*, and we set $N = 5$.

$$NewConcept = \sum_{i}^{N} Similarity\ (C_i, Query) \times C_i \tag{2}$$

4. Extract words that have a high score, and append this to the query. In our experiment, we extracted 15 words and appended these words to the query.

3 System Description

In this section, details of the retrieval system that we submitted are described.

3.1 Integration System

This system is the bare system used as the base of all submitted systems, and it is considered to be the final retrieval result that integrates the retrieval results of TBIR and CBIR. This system was built based on the system of CINDY [4], which participated in the ImageCLEF2006 photo retrieval task. First, the details of TBIR and CBIR are explained, and then the method of integration is described.

3.1.1 Text-Based Image Retrieval

We used Apache Lucene [5] as the TBIR method. Apache Lucene is an all-text retrieval engine developed as open source software. In Lucene, we can use text retrieval with TF-IDF, but we built Okapi BM25 [6] into Lucene for our system.

3.1.2 Content-Based Image Retrieval

CBIR consists of three retrieval modules called global retrieval, grid retrieval, and region retrieval. The retrieval results of the three retrieval modules are integrated by Eq. (3), and this integration result is considered to be the final retrieval result of CBIR.

$$\begin{aligned} CBIR\ result &= global\_result * global\_weight \\ &+ grid\_result * grid\_weight + region\_result * region\_weight \end{aligned} \tag{3}$$

where *result* indicates the retrieval result of each retrieval module, and *weight* indicates weight. In this system, each weight is as follows: *global_weight* = 0.3, *grid_weight* = 0.6, *region_weight* = 0.1.

The details of the three retrieval modules are described as follows.

- Global Retrieval
  In the global retrieval module, first, the feature values of color (lab) and texture (wavelet transform) are extracted from the whole image. The distance scale between the images uses Earth Mover's Distance for the color and Euclidean distance for the texture. In this module, the weights of color and texture are as follows: Color_Weight = 0.9, Texture_Weight = 0.1.

- Grid Retrieval
  In the grid retrieval module, first, an image is divided into a 3x3 block. And, feature values of color (lab average, standard deviation) and texture (wavelet transform) are extracted for each block. The distance scale between the images uses Euclidean distance for both color and texture. In this module, the weight of color and texture are as follows: Color_Weight = 1.0, Texture_Weight = 0.0.

- Region Retrieval
  The region retrieval module first segments an image with the JSEG algorithm [7]. Then feature values are extracted from all regions. These features are normalized into Z-scores and combined into a 47-dimensional vector. The distance scale between the images uses the cosine measure. Table 1 lists the adopted features and their dimensions.

Table 1. Feature values of region (numbers in parentheses indicate number of dimensions)

Color	RGB average (3), standard deviation (3)
	Lab average (3), standard deviation (3)
Texture	Wavelet transform (24)
Shape	Z-Fourier descriptors (8)
Position	X and Y coordinates of median point (2)
Size	Number of pixels (1)

3.1.3 Integration System

Here, an integrated part of the retrieval result is described. Figure 1 outlines the flow of the Integration System.

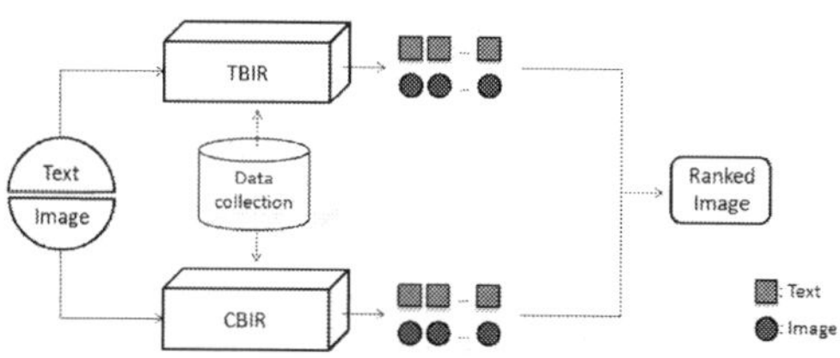

Fig. 1. Integration System

If a query is input, it is divided into an annotation and three images, and the annotation is passed to the TBIR, and the images are passed to the CBIR. TBIR described in 3.1.1 is executed using <title> and <narr> from the query annotation. At the same time, CBIR described in 3.1.2, is executed. Finally, the retrieval results of TBIR and CBIR are integrated by Eq. (4), and the final retrieval result is obtained.

$$\begin{aligned} Final\ retrieval\ result &= TBIR\_result * TBIR\_weight \\ &+ CBIR\_result * CBIR\_weight \end{aligned} \tag{4}$$

where *result* indicates the retrieval result of each retrieval system, and *weight* indicates the weight. In this system, each weight is as follows: *TBIR_weight* = 0.5, *CBIR_weight* = 0.5.

3.2 Integration System + CFS

This system combined CFS with the Integration System.

If a query is input, query expansion using CFS is performed when TBIR is executed. The details of query expansion using CFS are described in 2.3 of Section 2. The result of TBIR using query expansion is integrated with the result of CBIR, and the final retrieval result is obtained. In this system, each weight is as follows: *TBIR_Weight* = 0.5, *CBIR_Weight* = 0.5.

3.3 Inter Media Pseudo Relevance Feedback

Inter Media Pseudo Relevance Feedback (IMPRF) is a query expansion system that is performed using an annotation of the image obtained through the retrieval result of CBIR, and that also executes TBIR. This system is based on the system of IPAL [8], which participated in the ImageCLEF2006 photo retrieval task. Figure 2 outlines the flow of IMPRF.

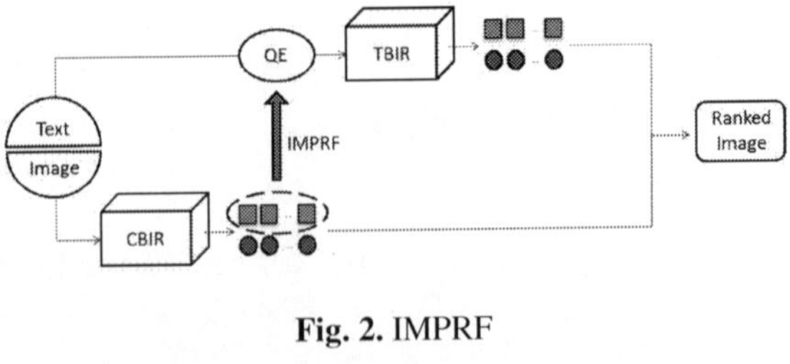

Fig. 2. IMPRF

When a query is input, first CBIR is executed. Next, query expansion is performed using the retrieval result of CBIR. Query expansion is performed by extracting 15 words from <TITLE>, <DESCRIPTION>, <NOTES> and <LOCATION> of the annotations of the top 5 images of the retrieval result of CBIR. TBIR is executed using the words obtained by the query expansion. The retrieval results of TBIR and CBIR are integrated, and the final retrieval result is obtained. In this system, each weight is as follows: *TBIR_weight* = 0.5, *CBIR_weight* = 0.5.

3.4 IMPRF + CFS

This system combines CFS with IMPRF.

If a query is input, first CBIR is executed. Next, query expansion is performed using the retrieval result of CBIR. Query expansion is carried out by extracting 15 words from annotations of the top 5 images obtained in the retrieval result of CBIR. Moreover, query expansion is performed using CFS for the words obtained by query expansion, and TBIR is executed. Finally, the retrieval results of TBIR and CBIR are integrated, and the final retrieval result is obtained. In this system, each weight is as follows: *TBIR_weight* = 0.5, *CBIR_weight* = 0.5.

3.5 TBIR + CFS

This system combines CFS with TBIR and uses a text only approach. In section 2.2.1, the use of word vector and region ID vector was explained in terms of constructing the conceptual dictionary. However, the conceptual dictionary is constructed using only a word vector because this system uses text only. In addition, when a query expansion is performed, a word vector is generated from <title> and <narr> of the query, and this is assumed to be a feature vector. TBIR is executed as explained in 3.1.1.

4 Experimental Results

Table 2 lists the experimental results of the systems submitted.

Table 2. Experimental results of photo retrieval task

RunID	*Modality*	*P@20*	*CR20*	*MAP*
Integration System	TXTIMG	0.4282	0.3975	0.2940
Integration System + CFS	TXTIMG	0.4218	0.3297	0.2966
IMPRF	TXTIMG	0.3949	0.3223	0.2711
IMPRF + CFS	TXTIMG	0.4231	0.3226	0.3032
TBIR + CFS	TXT	0.4051	0.3966	0.3015

The Integration System had the highest accuracy in *P@20* and *CR20*. As for this result, it was understood that integrating different media, i.e., TBIR and CBIR, was conducive to the higher retrieval result. Moreover, when looking at the mean average precision (*MAP*), which can comprehensively assess how well the retrieval system performs, the top three systems perform query expansion using CFS. This indicates that performing query expansion using CFS produces a higher retrieval result. Here, the notable conclusion is TBIR + CFS. Although this system approach uses text only, it retrieves with very high accuracy on the whole. However, it is clear that of the systems we submitted, the best accuracy is obtained using IMPRF + CFS. Although an approach using image only is not very accurate, it is understood that accuracy improves by integrating different media.

5 Conclusion

We demonstrated that retrieval accuracy improved by performing query expansion using CFS in image retrieval that integrates different media.

However, future tasks remain, and these are as follows.

It is necessary to improve the accuracy because the accuracy of CBIR is low. One factor is thought of as the distance scale between the images. We used Euclidean distance as the dissimilarity measure on several occasions. Haiming et al [9] systematically investigated 14 typical dissimilarity measures in CBIR, and showed that Euclidean distance was not effective. Consequently, it is thought that it is necessary to change into another dissimilarity measure. Because it is clear that accuracy improves by integrating TBIR and CBIR, it is assumed that higher accuracy of CBIR will lead to a further rise in overall accuracy. In addition, it is thought that accuracy will increase by revising the method of integration because simply adding TBIR and CBIR is not sufficient (e.g., Eq. 4). In other words, the method of integration plays a very important role when TBIR and CBIR are integrated.

Also, another problem is how to determine the initial point of clustering when the conceptual dictionary is constructed. This system sets the initial point at random. It is thought that accuracy will fall outside the accuracy range if another corpus is used, but there is the possibility that accuracy will be higher or lower than the results reported here when the conceptual dictionary is constructed anew. It is necessary to consider using another fuzzy clustering technique if the accuracy changes each time the conceptual dictionary is constructed.

References

[1] Arni, T., Clough, P., Sanderson, M., Grubinger, M.: Overview of the ImageCLEFphoto2008 Photographic Retrieval Task. In: Peters, C., et al. (eds.) CLEF 2008. LNCS, vol. 5706, pp. 500–511. Springer, Heidelberg (2009)

[2] ImageCLEFphoto2008, `http://www.imageclef.org/2008/photo`

[3] Chen, Y.-C., Sekiya, H., Takagi, T.: Conceptualized Queries for Information Retrieval. In: Proceedings of North American Fuzzy Information Processing Society Annual Conference, NAFIPS (2007)

[4] Rahman, M.M., Sood, V.K., Desai, B.C., Bhattacharya, P.: CINDI at imageCLEF 2006: Image retrieval & annotation tasks for the general photographic and medical image collections. In: Peters, C., Clough, P., Gey, F.C., Karlgren, J., Magnini, B., Oard, D.W., de Rijke, M., Stempfhuber, M. (eds.) CLEF 2006. LNCS, vol. 4730, pp. 715–724. Springer, Heidelberg (2007)

[5] Apache Lucene, `http://lucene.apache.org/`

[6] Robertson, S.E., Walker, S., Jones, S.: Micheline Hancock-Beaulieu, and Mike Gatford, "Okapi at TREC-3". In: Proceedings of the Third Text Retrieval Conference, TREC (1994)

[7] Deng, Y., Manjunath, B.S.: Unsupervised segmentation of color-texture regions in images and video. Transactions of Pattern Analysis and Machine Intelligence 2001 (2001)

[8] Maillot, N., Chevallet, J.-P., Valea, V., Lim, J.H.: IPAL Inter-Media Pseudo-Relevance Feedback Approach to ImageCLEF 2006 Photo Retrieval. In: Proceedings of the Cross Language Evaluation Forum (CLEF) 2006 Workshop, Alicante, Spain (2006)

[9] Liu, H., Song, D., Ruger, S., Hu, R., Uren, V.: Comparing Dissimilarity Measures for Content-Based Image Retrieval. In: Li, H., Liu, T., Ma, W.-Y., Sakai, T., Wong, K.-F., Zhou, G. (eds.) AIRS 2008. LNCS, vol. 4993, pp. 44–50. Springer, Heidelberg (2008)

Building a Diversity Featured Search System by Fusing Existing Tools

Jiayu Tang, Thomas Arni, Mark Sanderson, and Paul Clough

Department of Information Studies, University of Sheffield, UK,
{j.tang,t.arni,m.sanderson,p.d.clough}@sheffield.ac.uk

Abstract. This paper describes a diversity featured retrieval system which is built for the task of ImageCLEFPhoto 2008. Two existing tools are used: Solr and Carrot^2. We have experimented with different settings of the system to see how the performance changes. The results suggest that the system can indeed increase diversity of the retrieved results, without sacrificing too much of the precision.

1 Introduction

In this paper, we describe how to quickly set up a diversity featured search system by combining existing tools that are available to the public, namely Solr [2] and Carrot^2 [3]. We further describe how to tune the system for better performance for the task of ImageCLEFPhoto 2008 [4].

2 Tools

We use Solr for text search and Carrot^2 for increasing diversity of results.

Solr is a text search server based on the popular search library Lucene [1]. Solr provides some very useful and convenient features. The web-services like API allows users to index documents via XML over HTTP, query it via HTTP GET and receive results in XML format. The fields and field types of documents can be easily defined in the schema.xml file. In the same file, users can also specify the Solr "out of the box" tokenizers and token filters to be used for indexing and query. In addition, the HTML administration interface gives users comprehensive insight into the system.

Carrot^2 is a open source search results clustering engine. It provides five different algorithms for automatically organising search results into thematic categories. Carrot^2 works as a pipeline of three kinds of components: input components, filter components and visualisation components. Input components obtain search results from a source of choice (e.g. YahooAPI, GoogleAPI, Lucene, Solr, etc.), then filter components apply clustering algorithms to the search results, and finally visualisation components render the clustered results to the user.

C. Peters et al. (Eds.): CLEF 2008, LNCS 5706, pp. 560–567, 2009.

3 System Setup

In ImageCLEFPhoto 2008, each image comes with 9 fields of data: DOCNO, TITLE, DESCRIPTION, NOTES, LOCATION, DATE, IMAGE, THUMBNAIL, and TOPIC. We decided that TITLE, DESCRIPTION, NOTES, LOCATION are the fields that would provide useful information for text based image retrieval. Therefore, we constructed a new field named TEXT by combining the text from the four fields, and specified it as the default search field in Solr. Solr's default configuration of tokenizers and token filters is used in our experiments, namely WhitespaceTokenizer, SynonymFilter, StopFilter, WordDelimiterFilter, LowerCaseFilter, EnglishPorterFilter and RemoveDuplicatesTokenFilter. More details on each tokenizer and token filter can be found on [2]. After feeding the Solr server with all the 20,000 documents in XML format, we have a running search engine that is able to return a ranked list of documents based on the query submitted by the user. Note that all the fields are indexed by Solr.

In Carrot^2, we construct an input component for acquiring results from the Solr server. Then, a filter component is used for clustering the results, and a visualisation component is used for displaying the clusters and their members. Since ImageCLEFPhoto 2008 assesses the S-recall [6] of the top 20 results of the list to be submitted, we use the following procedure to find the best 20 results from the input/ranked list generated by Solr:

1. Denote the ranked list from the search engine as L, a temporary list as T, and a final re-ranked list as R. T and R are empty.
2. Add first document in L to group T, and remove the document from L.
3. Find the document in L which has the highest rank in L and belongs to a cluster that does not exist in T. Add the document to T and remove it from L.
4. If L is empty, append T to R and exit the procedure.
5. If the number of document in T equals to the total number of clusters, append T to R, and empty T.
6. Do process 3, 4 and 5.

Basically, the above procedure chooses documents based on two criteria: 1. appear as early as possible in the input/ranked list, 2. cover as many different clusters as possible. The 20 documents chosen by the above procedure form the list to be submitted for assessment.

4 Experiments

We have submitted 32 runs, all of which are EN-EN-AUTO-TXT (see [4] for details), meaning that all the submissions are English-English monolingual runs using fully automated text clustering methods. Among them, there are 8 groups, each of which includes 4 runs with the same system configuration except that different number of documents are used for clustering. In other words, for each group, we vary the number of documents (40, 60, 80 and 100) that are used from the top of the input list for clustering by Carrot^2, in order to get different output lists. For example, a run with top 40 documents applies clustering on the 40 documents and chooses 20 for submission based on the procedure in Section 3.

We increasingly change another 3 kinds of settings of the system to examine how the performances would change. Firstly, for topics whose CLUSTER field is Country, City, State or Location, we specify Carrot^2 to cluster the documents by the LOCATION filed, otherwise by the TEXT field. The field CLUSTER is generated by the organizers of ImageCLEFPhoto 2008 to indicate the cluster type based on which cluster recall will be evaluated (refer to [4] for more details), and is included as part of each topic description. For example, if the CLUSTER field is Country, cluster recall will evaluate how many different countries are returned in the results. Secondly, we change the parameters of the clustering algorithm used in Carrot^2. Finally, we apply expansion to the indexing and query stage. In the following, we describe each group of runs.

Baseline. This group uses the default settings of Solr and Carrot^2. In terms of text retrieval, expansion is not applied during indexing or query. Clustering is applied to the TEXT field for all topics. Carrot^2's default clustering algorithm (Lingo [5]) and default parameters (0.150, 0.775) are used. This group is used as a comparison baseline for other groups.

Clustering by Location. It seems that for topics that have been specified to be clustered by Country, City, State and Location (as indicated in the CLUSTER field), the LOCATION filed in each document contains the essential information for clustering. Therefore, such topics are clustered based on the LOCATION field. This group is different from the "Baseline" group in that topics whose CLUSTER field are Country, City, State and Location, are clustered based on the location field of images. Other topics are clustered based on the TEXT field.

Parameters of Lingo set to (0.05, 0.95). Lingo [5] is a singular value decomposition based clustering algorithm that has been implemented in Carrot^2. The first parameter 0.05 is the Cluster Assignment Threshold, determining how precise the assignment of documents to clusters should be. Lower threshold assign more documents to clusters and less to "Other Topics", which contains unclassified documents. With a low threshold, more irrelevant documents are also assigned to the clusters. The second parameter 0.95 is the Candidate Cluster Threshold, determining how many clusters Lingo will try to create. Higher values give more clusters. This group is based on group "Clustering by Location", but uses (0.05, 0.95) as the parameters.

Parameters of Lingo set to (0.10, 0.90). This group is the same as group "Parameters of Lingo set to (0.05, 0.95)" except that it uses (0.10, 0.90).

Parameters of Lingo set to (0.15, 0.85). This group is the same as group "Parameters of Lingo set to (0.05, 0.95)" except that it uses (0.15, 0.85).

Indexing and Query Expansion with (0.05, 0.95). This group is based on group "Parameters of Lingo set to (0.05, 0.95)", but applies indexing and query expansion in Solr. Specific domain ontologies have been a popular choice for expansion. Cyclopedia websites such as Wikipedia have also been adopted for expansion. Due to limited time, we used neither of the approaches. Instead, we examined the topics and data-set, and then manually built an indexing expansion list and query expansion list, as shown in Appendix. These lists are used as the synonym lists in Solr for indexing and query. In the expansion lists, for lines containing "=>", any of the words before "=>" are replaced by words after "=>" during expansion. For example, "ship, ships => ship, vehicle" will replace any "ship" or "ships" by "ship, vehicle". For lines without "=>", any word from the line will be replaced by the whole set of words from the same line. For example, "USA" or "United States of America" or "US" will be replaced by "USA, United States of America, US".

Indexing and Query Expansion with (0.10, 0.90). This group is the same as group "Indexing and Query Expansion with (0.05, 0.95)" except that it uses (0.10, 0.90) in Lingo.

Indexing and Query Expansion with (0.15, 0.85). This group is the same as group "Indexing and Query Expansion with (0.05, 0.95)" except that it uses (0.15, 0.85) in Lingo.

5 Results and Discussions

Figure 1(a) and 1(b) depict the precision and cluster recall at 20 of all submitted runs. For example, in Figure 1(b), group 1 corresponds to the results of cluster recall generated by the runs from group "Baseline" described in the previous Section.

As can be seen in Figure 1(a), group 6, 7 and 8 have clearly higher precisions than the other groups. This is due to indexing and query expansion using the expansion lists in Appendix. As have been mentioned, the expansion lists are built based on an examination of the data-set. For example, in Topic 48 "vehicle in South Korea", "South Korea" normally means the country, so there is

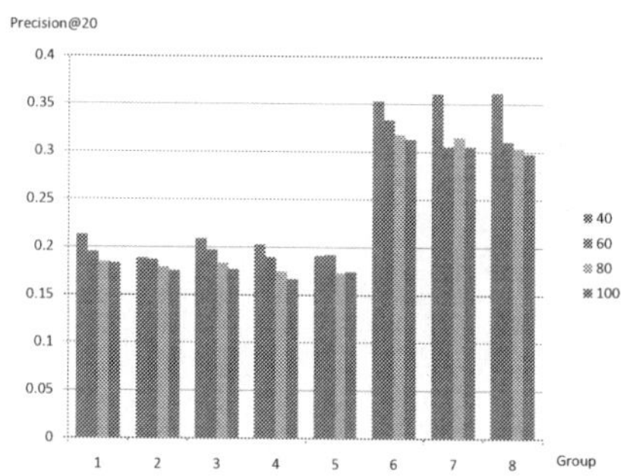

(a) Precision at 20 for all groups of runs.

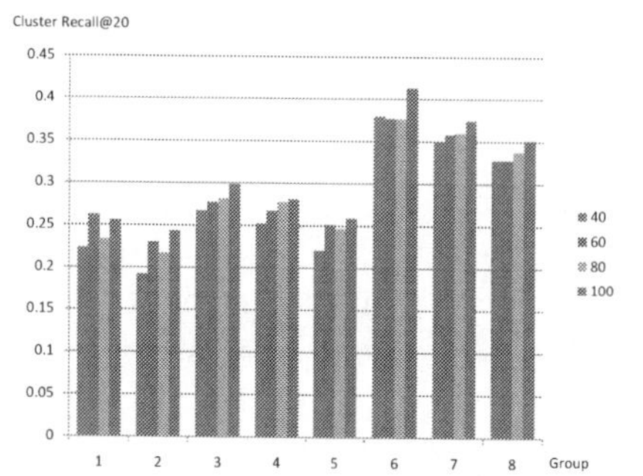

(b) Cluster recall at 20 for all groups of runs.

Fig. 1. Results

not much ambiguity. However, "vehicle" can mean many things, e.g. car, bus, boat. Intuitively, expanding names of different types of vehicle with the word "vehicle" during indexing will boost the precision, because many images are only annotated with specific vehicle names rather than the word "vehicle". Therefore, after indexing expansion, "car" becomes "car vehicle". Similarly, we expanded specific animal names with the word "animal", so "fish" becomes "fish animal".

In terms of cluster recall, we can see in Figure 1(b) that different parameters of the clustering algorithm (Lingo) have led to different performances. It is a little surprising that group 2 ("Clustering by Location") performed worse than group 1 ("Baseline"), the reason of which needs to be examined. In addition, it seems that low Cluster Assignment Threshold (i.e. more documents are clustered) and high Candidate Cluster Threshold (i.e. more clusters are created) give better cluster recall. In our experiments, (0.05, 0.95) gives the best results: group 3 is better than group 4 and 5; group 6 is better than group 7 and 8.

By comparing the two charts, it can also be noticed that groups with the same settings of Solr have very similar precisions, no matter what settings of $Carrot^2$ were used. For example, with different parameters of Lingo, the precisions of group 6, 7 and 8 are relative stable, but the values of cluster recall vary. This can be seen as an evidence that the diversity featured retrieval system can make the results more diverse while maintaining the precision.

6 Conclusions and Future Work

This paper has described our submissions to ImageCLEFPhoto 2008. We have changed 4 kinds of settings: the field used for clustering, the number of images used for clustering by $Carrot^2$, indexing and query expansion, and parameters of the clustering algorithm. The results suggest that indexing and query expansion can fairly improve precision. Appropriately chosen clustering method can increase diversity of the results while keeping precision almost the same.

As we have mentioned, the performance of group 2 is a little out of our initial expectation. It would be interesting to find out why. On the other hand, we plan to build an automatic expansion approach using resources such as ontologies, rather than using the manually built expansion lists.

Acknowledgements

Work undertaken in this paper is supported by the EU-funded TrebleCLEF project (Grant agreement: 215231) and by Tripod (Contract No. 045335).

References

1. Apache lucene project, `http://lucene.apache.org` (Visited 23/07/08)
2. Apache solr project, `http://lucene.apache.org/solr/index.html` (Visited 23/07/08)

3. Carrot^2 project, `http://project.carrot2.org/` (Visited 23/07/08)
4. Arni, T., Clough, P., Sanderson, M., Grubinger, M.: Overview of the ImageCLEF-photo 2008 photographic retrieval task. In: Peters, C., et al. (eds.) CLEF 2008. LNCS, vol. 5706, pp. 500–511. Springer, Heidelberg (2009)
5. Osinski, S., Stefanowski, J., Weiss, D.: Lingo: Search results clustering algorithm based on singular value decomposition. In: Proceedings of the International Conference on Intelligent Information Systems, Zakopane, Poland, pp. 359–368 (2004)
6. Zhai, C.X., Cohen, W.W., Lafferty, J.: Beyond independent relevance: methods and evaluation metrics for subtopic retrieval. In: Proceedings of the 26th annual international ACM SIGIR conference on Research and development in informaion retrieval, Toronto, Canada, pp. 10–17 (2003)

A Appendix

A.1 Indexing Expansion List

vehicles
ship, ships => ship, vehicle
cutter, cutters => cutter, vehicle
train, trains => train, vehicle
rail, rails => rail, vehicle
locomotives, locomotive => locomotive, vehicle
wagon, wagons => wagon, vehicle
tractor, tractors => tractor, vehicle
bus, buses => bus, vehicle
car, cars => car, vehicle
forklift, forklifts => forklift, vehicle
boat, boats => boat, vehicle

animals
Alligator, Alligators => Alligator, animal
Turtle, Turtles => Turtles, animal
Ducks, Duck => Duck, animal
Dolphins, Dolphin => Dolphins, animal
Fish => Fish, animal
Whale, Whales => Whale, animal
Pelicans, Pelican => Pelicans, animal
blowfish => blowfish, animal
shoal => shoal, animal
Shark, Sharks => Shark, animal
Crocodile => Crocodile, animal
orcas => orcas, animal
Angelfish => Angelfish, animal
kangaroo, kangaroos => kangaroo, animal
wallaby, wallabies => wallaby, animal
koalas, koala => koala, animal

wombats, wombat => wombat, animal
quokkas, quokka => quokka, animal
platypuses, platypus => platypus, animal
possums, possum => possum, animal
Tasmanian devils, Tasmanian devil => Tasmanian devil, animal
gull => gull, animal
flamingo, flamingos => flamingo, animal
anacondas, anaconda => anacondas, animal
ocelot, ocelots => ocelot, animal
Penguin, Penguins => penguin, animal
condors, condor => condor, animal
monkey, monkeys => monkey, animal
bird, birds => bird, animal
iguana, iguanas => iguana, animal
snail, snails => snail, animal
toucan, toucans => toucan, animal
lion, lions => lion, animal
llama, llamas => llama, animal
snake, snakes => snakes, animal
Tortoise, Tortoises => Tortoise, animal
parrot, parrots => parrot, animal
Booby, Boobies => Booby, animal
horse, horses => horse, animal
sea lion, sea lions => seal, animal

\# sports
football => football, sport
surf => surf, sport
motorcycle => motorcycle, sport
race => race, sport

\# people
fans, fan => fan, people

\# water
river => river, water
lake => lake, water

\# stone, rock
rock => rock, stone
brick, bricks => brick, stone

A.2 Query Expansion List

countries
USA, United States of America, US

synonym
church, churches, cathedral, cathedrals
oxidised, rusty
observing, watch, see
football, soccer
match, game
accommodation, room

# common sense
swimming => swimming in the water
drawings, drawing => drawing, line, petroglyph
prize, prizes => prize, medal
water => water, sea

Some Results Using Different Approaches to Merge Visual and Text-Based Features in CLEF'08 Photo Collection

Ana García-Serrano[1], Xaro Benavent[2], Ruben Granados[3], and José Miguel Goñi-Menoyo[3]

[1] Universidad Nacional de Educación a Distancia, UNED
[2] Universidad de Valencia
[3] Universidad Politécnica de Madrid
agarcia@lsi.uned.es, xaro.benavent@uv.es,
rgranados@fi.upm.es, josemiguel.goni@upm.es

Abstract. This paper describes the participation of the MIRACLE team[1] at the ImageCLEF Photographic Retrieval task of CLEF 2008. We succeeded in submitting 41 runs. Obtained results from text-based retrieval are better than content-based as previous experiments in the MIRACLE team campaigns [5, 6] using different software. Our main aim was to experiment with several merging approaches to fuse text-based retrieval and content-based retrieval results, and it happened that we improve the text-based baseline when applying one of the three merging algorithms, although visual results are lower than textual ones.

Keywords: Information Retrieval, Text-based image Retrieval and Content-based image Retrieval, Visual features, Textual features, Merge result lists, Indexing.

1 Introduction

MIRACLE is a consortium formed by research groups from different universities in Madrid, Universidad Politécnica (UPM), Universidad Autónoma and Universidad Carlos III, along with DAEDALUS, a SME spin-off of UPM. This paper describes our participation at the ImageCLEF Photographic Retrieval task of CLEF 2008, fully described in [1, 2]. This campaign Mir-FI team (MIRACLE at UPM) joined the Vision-Team at the University of Valencia (UV) who has developed a Content-Based retrieval system (CBIR) [4], in which the low-level features have been adapted to be used at the ImageCLEFphoto.

We succeeded in submitting 41 runs with results obtained by using (1) our re-implemented module for textual retrieval based on the classical vector model (VSM) in Information Retrieval, (2) the content-based image module (developed by UV)

[1] This work has been supported by the Spanish R+D National Plan, projects TIN2007-67407-C03-03, TIN2006-10134 and by Madrid R+D Regional Plan, project MAVIR, S-0505/TIC/0 00267 (www. mavir.net).

C. Peters et al. (Eds.): CLEF 2008, LNCS 5706, pp. 568–571, 2009.

with five different methods for aggregation, and (3) the three new merging algorithms using textual and visual features.

Obtained results from text-based retrieval are better than content-based ones. By merging both textual and visual retrieval we improve the text-based one when applying one of the merging algorithms implemented (the so-called ENRICH).

2 Detailed Description of Experiments

This year Mir-FI system allows executing the different configurations that are explained in the following.

MIRACLE-FI textual retrieval is based on the VSM approach using weighted vectors based on the TF-IDF weight. The implemented components are: (1) The Text Extractor, (2) the Preprocessor (to special characters deletion and Stop-word detection), the (3) Annotations/Topics Tags Selector, to select tags from the annotations files (TITLE, DESCRIPTION, NOTES and LOCATION), and from the topics (TITLE and NARR), (3 and 4) MirFi-VSM Indexer and MirFi-VSM Searcher (applied on the associated text to the images and using a slightly different cosine measure).

The CBIR use different low-level features describing (a) Color information with a histogram of the HS (Hue, Saturation) values of the image pixels (quantization of the HSV space into 30 color bins) and also describing (b) Texture information using different feature textures (Gabor Convolution Energies, Gray Level Coocurrence Matrix also known as Spatial Gray Level Dependence, Gaussian Random Markov Fields, the granulometric distribution function and the spatial distribution).

Secondly the CBIR module calculates the similarity distance between the feature vectors from each image on the database to the three topic images. The two distance metrics used are: the Euclidean and the Mahalanobis. The so-called OWA [4] operators have been used to aggregate the three low-level feature vectors of the topic images.

Finally, textual and image results lists are merged in three different ways:

FILTER-N. This way of merging the image and textual results lists consists on checking which results in the textual results list are also included in between the N first results of the visual results list. The value of N indicates the number of results taken into account from the visual list when narrowing down the textual one. The resulting merged list will have a maximum of 1000 results for each query.

ENRICH. Also uses two results lists, the main and the support list. If a concrete result appears in both lists for the same query, the relevance of this result in the merged list will be increased in the following way:

$$new\,\mathrm{Re}\,l = main\,\mathrm{Re}\,l + \frac{\sup \mathrm{Re}\,l}{\left(pos\,\mathrm{Re}\,l + 1\right)} \tag{1}$$

where newRel is the relevance value in the merged list, mainRel is the relevance value in the main list, supRel in the support list, posRel is the position in the support list. Relevance values will be then normalized from 0 to 1.

Every results appearing in the support list but not in the main one (for each query), will be added at the end of the results for each query (a maximum of 1000 results per query). In this case, relevance values will be normalized according with the lower value in this moment.

TEXT-FILTER. In this case, the textual module is applied to the complete database and only those images that have a relevance value above zero are passed to the CBIR. Then, the CBIR calculates the distance similarity to each of the three query and these values are merged with the different OWA aggregation operators.

Table 1. Our best results for each of the different runs we participated

Run Identifier	Textual Retrieval	Visual Retrieval			P20	MAP
		Distance	Merge topics	Merge		
TXT-baseline	SVM	--	--	--	0.2253	0.2846
IMG-mahamin	--	maha	min	--	0.0213	0.0679
TXTIMG-merge06mahamin	SVM	maha	min	ENRICH	0.3090	0.2401
TXTIMG-criba10000mahamin	SVM	maha	o3	FILTER-10000	0.3179	0.1936
TXTIMG-merge06mahamin	SVM	maha	min	ENRICH	0.2401	0.3090

3 Runs and Results

Finally, it was submitted one text-based run, 10 content-based runs and 30 mixed runs using a combination of both. The details can be found in [2].

In the general classification with all the automatic runs (1039) from all the participant groups (25), our best position was obtained is 306, corresponding to the English automatic textual and visual retrieval run using the ENRICH merge, being our best values not far away from the bests values from all the automatic experiments; we were even better if taking the best 4 runs from each participating group. The English automatic textual retrieval module with no linguistic processes, that is our "baseline" run, appears in the position 185 (over 399). The content-based image module results show that Mahalanobis distance outperforms the Euclidean one, and the best aggregation method in both metrics is the minimum (AND), followed by the orness(W)_0.3 that is a smoothed AND. Our best result for this group of experiments was the combination of Mahalanobis metric with orness(W)_0.3 aggregation method, and was considerably lower than the best results. The merge results in which we experimented with the 3 different algorithms, are: FILTER-10000 that improved the text-based baseline in low precision values (P5, P10 and P20), but never MAP neither number of relevant images retrieved. The run English automatic textual and visual retrieval with Mahalanobis metric and the min operator, obtained the best P20 value from all ours experiments (190th in the general P20 classification, over 1039).

Experiments applying ENRICH improved the baseline in MAP and in number of relevant images retrieved. Our best MAP was achieved merging the textual results with the visuals ones obtained using the Mahalanobis metric and the AND operator which is in the 91st position in the general MAP classification (over 1039). This value is higher than the average MAP taken from the best 4 runs from each participating group (0.2187).

Both FILTER-10000 and ENRICH algorithms worked with textual results as primary list, and merge it with all the visual results lists (secondary). TEXT-FILTER uses visual lists as primaries and merges all of them with the textual one (secondary). The bests results corresponded to the experiments which use the Mahalanobis distance and the AND operator.

4 Conclusions and Future Work

In this participation the MAP value obtained for the text-based baseline experiments was 0.2253, higher than the average MAP (0.2187) calculated from the best 4 runs from each participating group.

For the content-based image retrieval, the results have not been very successful. Our results are lower than the best top ten. The most interesting conclusion in that the Mahalanobis distance works better than the Euclidean one, and the best aggregation method is the AND operator. For following editions more low-level features based on local color descriptors and shape descriptors will be included.

Merged results show that the ENRICH algorithm improves very lightly the baseline. This is important taken into account the poor results obtained from the visual retrieval. FILTER-10000 algorithm improves the textual baseline results in terms of precision at low values.

References

1. Arni, T., Clough, P., Sandersin, M., Grubinger, M.: Overview of the ImageCLEFphoto 2008 Photographic Retrieval Task. In: Peters, C., et al. (eds.) CLEF 2008. LNCS, vol. 5706, pp. 500–511. Springer, Heidelberg (2008)
2. Granados, R., Benavent, X., García-Serrano, A., Goñi, J.M.: MIRACLE-FI at ImageCLEFphoto 2008: Experiences in merging Text-based and Content-based Retrievals. In: Working Notes of the 2008 CLEF Workshop, Aarhust, Denmark (September 2008)
3. Grubinger, M., Clough, P., Müller, H., Deselaers, T.: The IAPR-TC12 benchmark: A new evaluation resource for visual information systems. In: International Workshop OntoImage'2006 Language Resources for Content-Based Image Retrieval, held in conjunction with LREC 2006, Genoa, Italy, May 2006, pp. 13–23 (2006)
4. Leon, T., Zuccarello, P., Ayala, G., de Ves, E., Domingo, J.: Applying logistic regression to relevance feedback in image retrieval systems. Pattern Recognition 40, 2621–2632 (2007)
5. Martínez-Fernández, J.L., Villena-Román, J., García-Serrano, A., González-Cristóbal, J.C.: Combining Textual and Visual Features for Cross-Language Medical Image Retrieval. In: Peters, C., Gey, F.C., Gonzalo, J., Müller, H., Jones, G.J.F., Kluck, M., Magnini, B., de Rijke, M., Giampiccolo, D. (eds.) CLEF 2005. LNCS, vol. 4022, pp. 712–723. Springer, Heidelberg (2006)
6. Villena-Román, J., Lana-Serrano, S., Martínez-Fernández, J.L., González-Cristóbal, J.-C.: MIRACLE at ImageCLEFphoto 2007: Evaluation of Merging Strategies for Multilingual and Multimedia Information Retrieval. In: Peters, C., Jijkoun, V., Mandl, T., Müller, H., Oard, D.W., Peñas, A., Petras, V., Santos, D. (eds.) CLEF 2007. LNCS, vol. 5152, pp. 500–503. Springer, Heidelberg (2008)

MIRACLE-GSI at ImageCLEFphoto 2008: Different Strategies for Automatic Topic Expansion

Julio Villena-Román[1,3], Sara Lana-Serrano[2,3], and José Carlos González-Cristóbal[2,3]

[1] Universidad Carlos III de Madrid
[2] Universidad Politécnica de Madrid
[3] DAEDALUS - Data, Decisions and Language, S.A.
jvillena@it.uc3m.es, slana@diatel.upm.es,
josecarlos.gonzalez@upm.es

Abstract. This paper describes the participation of MIRACLE-GSI research consortium at the ImageCLEFphoto task of ImageCLEF 2008. For this campaign, the main purpose of our experiments was to evaluate different strategies for topic expansion in a pure textual retrieval context. Two approaches were used: methods based on linguistic information such as thesauri, and statistical methods that use term frequency. First a common baseline algorithm was used in all experiments to process the document collection. Then different expansion techniques are applied. For the semantic expansion, we used WordNet to expand topic terms with related terms. The statistical method consisted of expanding the topics using Agrawal's apriori algorithm. Relevance-feedback techniques were also used. Last, the result list is reranked using an implementation of k-Medoids clustering algorithm with the target number of clusters set to 20. 14 fully-automatic runs were finally submitted. MAP values achieved are on the average, comparing to other groups. However, results show a significant improvement in cluster precision (6% at CR10, 12% at CR20, for runs in English) when clustering is applied, thus proving to be valuable.

Keywords. Image retrieval, domain-specific vocabulary, thesaurus, linguistic engineering, information retrieval, indexing, relevance feedback, topic expansion, ImageCLEF Photographical Retrieval task, ImageCLEF, CLEF, 2008.

1 Introduction

The MIRACLE team is a research consortium formed by research groups of three different Spanish universities (UPM, UAM and UC3M) along with DAEDALUS, a spin-off company of these groups. To participate at the ImageCLEF Photographic Retrieval task [1] of ImageCLEF 2008, we decided to split into two subgroups, MIRACLE-GSI (Grupo de Sistemas Inteligentes – Intelligent System Group) in charge of purely textual runs and MIRACLE-FI (Facultad de Informática, Computer Science Faculty) in charge of visual and mixed runs. This paper reviews the participation of MIRACLE-GSI [2] and the participation of the other subgroup is described in another paper [3].

C. Peters et al. (Eds.): CLEF 2008, LNCS 5706, pp. 572–576, 2009.

Our main goal this year was to evaluate different strategies for topic expansion in a pure textual context. Two approaches were used: methods based on linguistic information such as thesauri, and statistical methods that use term frequency. Finally 14 fully-automatic runs were submitted.

2 Description of the System

The system is composed of five different modules, as shown in Figure 1: 1) the textual (text-based) retrieval module, which indexes image annotations in order to find the most relevant ones to the text of the topic; 2) the expander module, which expands annotations and/or topics with additional related terms; 3) the relevance-feedback module, which allows the execution of reformulated queries that include the results of previous queries; 4) the result combination module, which merges the results of the previous subsystems; and, finally, 5) a clustering module that reranks the result list to provide cluster diversity.

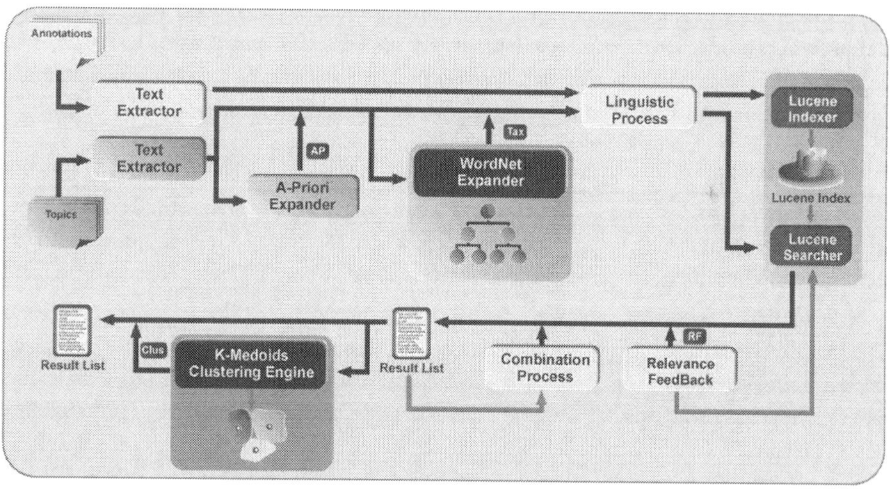

Fig. 1. Overview of the system architecture

A common baseline algorithm was used in all experiments to process the document collection. First, image annotations are extracted from the XML files and parsed to detect basic textual components, specifically common single words, numbers and tagged entities. Then stopwords are filtered out and a stemming process (based on Porter's stemming algorithm) is applied to all terms, which are then indexed with Lucene [4].

Then expansion techniques are applied to enrich the original contents of both topics and image annotations. We studied and compared a semantic- versus a statistical-based technique. For the semantic expansion, WordNet [5] was used to expand topic terms with related terms corresponding to a variety of semantic relationships (mainly synonyms and hyponyms). The statistical method consisted of expanding the topics using the Agrawal's apriori algorithm [6]. First, a term-document matrix is built using the terms in the document corpus. Then apriori algorithm is used to discover out rules having the topic terms as antecedent and a confidence value greater than 0.5. Finally the topic is expanded with the (one-term) consequent of those rules.

Additionally, relevance-feedback techniques were also used. The top M indexing terms of each of the top N result documents were extracted and weighted by a factor that is proportional to their document frequency to reformulate the query. Values for N and M were arbitrarily chosen to limit the number of runs (5 and 10, in both cases).

To provide cluster diversity, the last step of the process is to rerank the result list, moving the discovered cluster prototypes to the top positions. An implementation of k-Medoids clustering algorithm [7] was used, with k (the target number of clusters) equal to 20. For each resulting cluster, the element with higher relevance in the baseline result list is selected as the class prototype, and reranked to the top of the final result list.

3 Experiments and Results

Experiments are defined by the different combinations of modules, topic expansion techniques and relevance-feedback. Results for the best runs are shown in Table 1.

Table 1. Results for both English and Random language

Name[1]	RelRet	MAP	P10	P20	CR10	CR20
EN_TitleBaseline	1406	**0.1802**	0.2513	**0.2090**	0.2216	0.2697
EN_TitleTagClus	**1812**	0.1748	**0.2564**	0.1756	**0.2366**	**0.3029**
EN_TitleBaselineClus	1406	0.1662	0.2333	0.1782	0.2150	0.2787
EN_TitleAPClus	1550	0.1551	0.2385	0.1590	0.2323	0.2670
RND_TitleTagClus	*1270*	*0.1048*	*0.2154*	0.1449	*0.2133*	*0.2758*
RND_TitleBaseline	900	0.0995	0.1692	*0.1692*	0.1858	0.2398
RND_TitleBaselineClus	900	0.0954	0.1872	0.1295	0.1797	0.2393
RND_TitleAPClus	984	0.0892	0.1897	0.1192	0.1786	0.2110
EN_TitleRF1005Clus	1333	0.0873	0.1051	0.0859	0.1087	0.1546
EN_TitleTagRF1005Clus	1047	0.0795	0.1333	0.0846	0.1263	0.1625
EN_TitleAPRF1005Clus	1414	0.0722	0.1359	0.1077	0.1393	0.2037
RND_TitleTagRF1005Clus	724	0.0537	0.1000	0.0667	0.1234	0.1406
RND_TitleRF1005Clus	801	0.0536	0.0949	0.0654	0.1114	0.1456

[1] Baseline (stem+stopwords), Clus (k-Medoids clustering), AP (apriori topic expansion), Tag (WordNet topic expansion), RF<N><M> (relevance feedback), EN (captions in English), RND (captions in "Random" language)

For English, the baseline experiment achieves the best result in terms of MAP. However, the best cluster precision (CR), which was the variable to maximize in this task, is achieved when k-Medoids algorithm is applied, thus proving to be valuable. There is a noticeable improvement in cluster precision over 6% at CR10 and 12% at CR20, though no statistical test has been carried out to prove its significance. For the "Random" (mixed) language, the best results in terms of CR are also achieved with k-Medoids clustering (16% increment at CR20). These are average results, compared to other groups.

4 Conclusions and Future Work

MAP values are similar in practice for experiments using topic expansion (Tag and AP) and significantly worse (0.08 against 0.18) in the case of relevance-feedback (RF). Thus, apparently no strategy for topic expansion or especially relevance-feedback has proven to be useful. In the case of the topic expansion techniques, the reason is that the OR operator was used to build the reformulated query, i.e., both the original terms and the expanded terms were combined with the OR operator. This implies that documents that contain any of those terms are considered as relevant, no matter if the term belongs to the original topic or it is included in the expansion process. A combination of OR and AND operators should have been used to ensure that documents contain the original topic terms and, optionally, any of the expanded terms: "($original_1$ OR $expanded_1$) AND ($original_2$ OR $expanded_2$)".

On the other hand, we found that the reranking algorithm used for combining the different results list is the reason for the low precision values obtained in the experiments that make use of the relevance-feedback methods. Many interesting documents returned after the initial query were reassigned a very low relevance value and thus pushed down in the result list. To avoid this issue, other combination operators must be studied, especially those that assign a higher weight to documents that correspond to the initial query and a lower weight to documents found by the relevance feedback query.

The last conclusion is that the application of clustering techniques smoothes the negative effect of the expansion processes, showing quite promising results.

Acknowledgements. This work has been partially supported by the Spanish R+D National Plan, project BRAVO (Multilingual and Multimodal Answers Advanced Search–Information Retrieval), TIN2007-67407-C03-03, and by the R+D Regional Plan of Madrid, project MAVIR (Enhancing the Access and the Visibility of Networked Multilingual Information), S-0505/TIC/000267.

References

1. Arni, T., Clough, P., Sanderson, M., Grubinger, M.: Overview of the ImageCLEFphoto 2008 Photographic Retrieval Task. In: Peters, C., et al. (eds.) CLEF 2008. LNCS, vol. 5706, pp. 500–511. Springer, Heidelberg (2009)
2. Villena-Román, J., Lana-Serrano, S., González-Cristóbal, J.C.: MIRACLE-GSI at Image-CLEFphoto 2008: Experiments on Semantic and Statistical Topic Expansion. In: Working Notes of the 2008 CLEF Workshop, Aarhus, Denmark (2008)
3. García-Serrano, A., Benavent, X., Granados, R., Goñi, J.M.: Some results using different approaches to merge visual and text-based features in CLEF 2008 Photo Collection. In: Peters, C., et al. (eds.) CLEF 2008. LNCS, vol. 5706, pp. 568–571. Springer, Heidelberg (2009)
4. Apache Lucene project, `http://lucene.apache.org` (Visited 09/11/2008)
5. Eurowordnet: Building a Multilingual Database with Wordnets for several European Languages (March 1996), `http://www.illc.uva.nl/EuroWordNet/` (Visited 09/11/2008)

6. Agrawal, R., Srikan, R.: Fast algorithms for mining association rules. In: Proceedings of the International Conference on Very Large Data Bases, pp. 407–419 (1994)
7. Park, H.-s., Lee, J.-s., Jun, C.-h.: A K-means-like Algorithm for K-medoids Clustering and Its Performance. In: Proceedings of the 36th CIE Conference on Computers & Industrial Engineering, Taipei, Taiwan, June 20-23, pp. 1222–1231 (2006)

Using Visual Concepts and Fast Visual Diversity to Improve Image Retrieval

Sabrina Tollari, Marcin Detyniecki, Ali Fakeri-Tabrizi, Christophe Marsala, Massih-Reza Amini, and Patrick Gallinari

Université Pierre et Marie Curie - Paris 6
Laboratoire d'Informatique de Paris 6 - UMR CNRS 7606
104 avenue du président Kennedy, 75016 Paris, France
firstname.lastname@lip6.fr

Abstract. In this article, we focus our efforts (i) on the study of how to automatically extract and exploit visual concepts and (ii) on fast visual diversity. First, in the Visual Concept Detection Task (VCDT), we look at the mutual exclusion and implication relations between VCDT concepts in order to improve the automatic image annotation by Forest of Fuzzy Decision Trees (FFDTs). Second, in the ImageCLEFphoto task, we use the FFDTs learnt in VCDT task and WordNet to improve image retrieval. Third, we apply a fast visual diversity method based on space clustering to improve the cluster recall score. This study shows that there is a clear improvement, in terms of precision or cluster recall at 20, when using the visual concepts explicitly appearing in the query and that space clustering can be efficiently used to improve cluster recall.

1 Introduction

Automatic image annotation is an important issue to improve image retrieval. In fact, users prefer to use words to express their need of information. The ImageCLEF track of the 2008 CLEF campaign permits us to study image annotation in the same context as image retrieval: the Visual Concept Detection Task (VCDT) [2] allows us to study how to extract visual concepts, and then in the Photo Retrieval task (ImageCLEFphoto) [1], we use the visual concept to filter a text based query. In the other hand, the particularity of the 2008 ImageCLEFphoto edition was its focus on diversity. Most of the diversity methods propose to apply the diversification of the results after retrieving the images. This means that the diversification must be done on line and so must be very fast. So we proposed to use visual space clustering which is well known to be a fast clustering technique.

In Section 2, we present our Forests of Fuzzy Decision Trees methods and the cooccurrences analysis applied in the VCDT task. In Section 3, we describe the techniques we use in the ImageCLEFphoto task, especially how we use the VCDT concepts in this task and our diversification method. Finally, in the last section, we conclude.

C. Peters et al. (Eds.): CLEF 2008, LNCS 5706, pp. 577–584, 2009.

2 The Visual Concept Detection Task (VCDT)

2.1 Forests of Fuzzy Decision Trees (FFDTs)

Automatic image annotation is a typical inductive machine learning approach. One of the most common methods in this research topic is the decision tree approach (DT). In fact, recently, this approach (based on random decisions) has obtained great interest as a tool for tackling this challenge [6]. One limitation when considering classical DTs is their robustness and threshold problems when dealing with numerical or imprecisely defined data. The introduction of fuzzy set theory smoothes out these negative effects. In general, inductive learning consists in raising from the *particular* to the *general*. A tree is built, from the root to the leaves, by successively partitioning the training set into subsets. Each partition is done by means of a test on an attribute and leads to the definition of a node of the tree [4]. In [5] was shown that, when addressing unbalanced and large (in terms of dimension and size) data sets, it is interesting to combine several DTs, obtaining a Forest of Fuzzy Decision Trees (FFDTs). Moreover, when combining the results provided by several DTs the overall score becomes a degree of confidence in the classification.

During the learning step, a FFDT of n trees is constructed for each concept C. Each tree F_j of the forest is constructed based on a training set T_j, each being a balanced random sample of the whole training set.

During the classification step, each image I is classified by means of each tree F_j. We obtain a degree $d_j \in [0,1]$ representing the degree to which concept C is present on the image I. Thus, for each I, n degrees d_j, $j = 1 \ldots n$ are obtained from the forest. Then all these degrees are aggregated by a weighted vote, which mathematically corresponds to the sum of all the degrees: $d = \sum_{j=1}^{n} d_j$. Finally, to decide if an image presents a concept or not, we use a threshold value $t \leq n$.

2.2 Cooccurrences Analysis

DTs learn each concept independently, but concepts can be related. For instance, a scene cannot be simultaneously *indoor* and *outdoor*, furthermore if we observe that it is *overcast*, we can imply that the concept *sky* is present. Here, we propose to use cooccurrence analysis to automatically find these relations. Once we have discovered the relations, we need a rule to resolve the conflicting annotations. In fact, each concept is annotated by a FFDT with a certain confidence degree. For instance, for each image we will have a degree of having the concept *outdoor* and a certain degree of having *indoor*. We know that both can not appear simultaneously, something has to be done. We propose to use simple rules. In this paper, we study two type of relations between concepts: exclusion and implication.

Exclusion Discovery and Rule. To discover the *exclusions*, we look at which concepts *never* appear together. Therefore, we calculate a cooccurrence matrix COOC. Since there may be some noise (e.g. wrong annotation), we use a threshold α to decide which pair of concepts never appears together. Once we know which concepts are related, we apply a resolution rule to the scores provided by the FFDT. We choose the rule that, for mutually excluding concepts, eliminates (i.e. gives a confidence of zero) the label having the lowest confidence. For instance, if we have *outdoor* with a degree of confidence of 42/50 and *indoor* with a degree of 20/50 then we will say that it is certainly not *indoor* and its degree should equal 0. For each test image I, let d(I,C) be the FFDT degree of I for concept C, we then apply the following algorithm:

for each couple of concepts (A,B) where $COOC(A, B) \leq \alpha$ (*discovery*)
if d(I,A) > d(I,B) then d(I,A)=0 else d(I,B)=0 (*resolution rule*)

where COOC is the concept cooccurrence matrix.

Implication Discovery and Rule. To discover *implications*, we look, by definition of the implication, at the cooccurrence of the absence of concepts and of the presence of concepts. The resulting cooccurrence matrix COOCNEG is non symmetric, which reflects the fact that one concept may imply another one, but the reciprocal may not be true. The resolution rule says that if a concept implies another one, the confidence degree of the latter should be at least equal to the former. Since there may be some noise, we use a threshold β to decide which concepts imply other ones. For each test image I, let d(I,C) be the FFDT degree of I for concept C, we then apply the following algorithm:

for each couple of concepts (A,B) where $COOCNEG(A, B) \leq \beta$ (*discovery*)
d(I,B)=max(d(I,A),d(I,B)) (*resolution rule*)

where COOCNEG is the concept cooccurrence asymmetric matrix between a concept and the negation of an other concept.

2.3 VCDT Experiments

Visual Descriptors. The visual descriptors used in this paper are exclusively color based. In order to obtain spatial-related information, the images were segmented into 9 overlapping regions. For each region, we compute a color histogram in the HSV space. The number of bins of the histogram (i.e. numbers of colors) reflects the importance of the region. The large central region (the image without borders) represents the purpose of the picture. Two other regions, top and bottom, correspond to a spatial focus of these areas. We believe that they are particularly interesting for general concepts (i.e. not objects), as for instance: sky, sunny, vegetation, etc. The remaining regions (left and right top, left and right middle, left and right bottom) are described in terms of color difference between the right and the left. The idea is to make explicit any *systematic* symmetries. In fact, objects can appear on either side. Moreover, decision trees are not able to automatically discover this type of relations.

Corpus. The VCDT corpus contains 1827 training images and 1000 test images. There are 17 concepts. A training image is labelled by 5.4 concepts on average (standard deviation=2.0, between 0 (2 images) to 11 concepts per image). A concept label in average 584 training images (standard deviation=490, between 68 to 1607 training images by concept). This task corresponds to a multi-class multi-label image classification.

Table 1. Results of VCDT task (EER: Equal Error Rate - AUC: Area under ROC)

	Excl. rule	Impl. rule	Without class decision EER(AUC)	EER(AUC) gains %	With class decision (t=25) EER(AUC)	EER(AUC) gains %
FFDT			24.55 (82.74)	-	26.20 (57.09)	-
FFDT	X		27.37 (71.58)	-11 (-13)	28.83 (54.19)	-10 (-5)
FFDT		X	25.66 (82.48)	-5 (0)	27.51 (54.89)	-5 (-4)
FFDT	X	X	27.32 (71.98)	-11 (-13)	28.93 (53.78)	-10 (-6)
Random			50.17(49.68)	-104(-40)	50.26 (24.89)	-48(-56)

Exclusive and Implication Relations. A preliminary step before extracting visual concepts is to study cooccurrence values to discover exclusions and implications. For the 17 concepts, there are 136 cooccurrences values. Those values vary from 0 to 1443 (there are 1827 training images). Since there may be some noise (e.g. wrong annotation), we set $\alpha = 5$ (two concepts are considered exclusive if at the maximum 5 of the 1827 training images were annotated as presenting the two concepts in the training sets). For the same reason, we set $\beta = 5$ (a concept implies an other concept if at the maximum 5 training images are not annoted by the first concept, but annoted by the second one). Our system automatically discovered 25 exclusive relations and 12 implication relations. We found not only most of the relations suggested in the schema describing the training data, but also several other ones. For the latter, some are logic and some are the result of the fact that some labels are not very frequent. We notice, for instance, that *sunny* and *night* never appear together, but also that there is never a *beach* and a *road* together.

In order to appreciate the effect of the implication and exclusion rules, we compare, in Table 1, the scores obtained by the FFDTs composed of 50 trees (first line) and the scores obtained using implication and/or exclusion rules. Based on these scores, the exclusion and implication rules seem to worsen the results provided by the FFDTs. We believe that this is due to the fact that these scores are not adapted to boolean classification (and our rules provide boolean decisions). The area under the curve and the equal error rate are interesting when the classification is accompanied by a degree of confidence. Moreover, this measure penalizes boolean decision over degrees.

3 The Photo Retrieval Task 2008

3.1 Using VCDT Concepts in ImageCLEFphoto

Previous works show that combining text and visual information improves image retrieval, but most of this work use an early or late fusion of visual and textual modality. Following the idea of VCDT and ImageCLEFphoto tasks, we propose to use VCDT visual concepts to filter ImageCLEFphoto text runs in order to answer if visual concept filtering can improve text only retrieval.

The difficulty is to determine how to use the visual concepts of VCDT in ImageCLEFphoto 2008. In the VCDT task, we have obtained a FFDT per concept (see Section 2). Each of these FFDTs can give a degree that the corresponding visual concept appears in a new image. In order to make a decision, we put a threshold t to determine if an image contains the given concept according to the corresponding FFDT. First, if the name of a concept appears in the <title> element (VCDT filtering), we propose to filter the rank images list according to the FFDT of this concept. Second, if the name of a concept appears in the <title> element or in the list of synonyms (according to WordNet [3]) of the words in the <title> element (VCDTWN filtering), we also propose to filter the rank images list according to the FFDT of this concept. For example, the <title> of topic 5 is "animal swimming". Using only VCDT filtering, the system automatically determine that it must use the FFDT of the concept *animal*. If, in addition, we use WordNet (VCDTWN filtering), the system automatically determine that it must use the FFDT of the concept *animal* and of the concept *water* (because according to WordNet, the synonym of "swimming" is: "water sport, aquatics"). For each query, we obtain a list of images ranked by their text relevance according to a language model (LM) or TF-IDF text models. Then, using the decision of the FFDTs, we rerank the first 50 ranked images: the system browses the retrieves images from rank 1 to rank 50. If the degree of an image is lower than the threshold t, then this image is reranked to the end of the current 50 images list.

3.2 Promote Diversity by Fast Clustering Visual Space

For a given query *similar* documents are naturally similarly ranked. When a user makes a query, he should want that the first relevant documents are as diverse as possible. So the ImageCLEFphoto 2008 task is very interesting to improve image retrieval, but the definition of diversity in the ImageCLEFphoto 2008 task is not very clear, in particular in term of granularity. In most cases, it is strongly related to the text. For us, there are two kinds of diversification in the ImageCLEFphoto 2008. The first one is knowledge based: *city, state, country, venue, landmark....* The second one is based on visual information: *weather condition, group composition, statue....* For these clusters, visual diversification should improve results. As in real applications, it is not obvious to determine automatically which kind of diversification applying for a given query [7], we choose to apply, for all query (even if it is suboptimal), the same diversification technique (the visual one) by clustering the visual space.

Visual clustering has been studied for a long time. Two approaches are generally proposed: data clustering and space clustering. The first one requires lots of computation time and should be adapted to distribution of the first images ranked by a given query. The second approach, since it is computed independently of the data, is often less effective, but can be applied extremely fast. We choose to cluster the visual space based on the hue dimension of the HSV space. For each image, we binarize its associated 8 bin hue histogram. Each

Table 2. Comparison of VCDT and VCDTWN filtering. For VCDT filtering, only 11 topics are modified. For VCDTWN, only 25 topics are modified.

Text	Visual concept filtering	All 39 topics P20 (gain %)	All 39 topics CR20 (gain %)	Topics modified by filtering Nb topics	P20 (gain %)	CR20 (gain %)
LM	-	0.185 (-)	0.247 (-)	11	0.041 (-)	0.090 (-)
				25	0.148 (-)	0.254 (-)
	VCDT	0.195 (+6)	0.257 (+4)	11	0.077 (+88)	0.126 (+40)
	VCDTWN	0.176 (-5)	0.248 (+1)	25	0.134 (-9)	0.257 (+1)
TF-IDF	-	0.250 (-)	0.300 (-)	11	0.155 (-)	0.161 (-)
				25	0.210 (-)	0.305 (-)
	VCDT	0.269 (+8)	0.313 (+5)	11	0.223 (+44)	0.209 (+30)
	VCDTWN	0.260 (+4)	0.293 (-2)	25	0.226 (+8)	0.294 (-4)

binary vector correspond to a cluster. The number of clusters is 256 (not all are instantiated), a reasonable number for a re-ranking at P20.

We use the visual space clusters to rerank the 50 retrieve images. For each query, the system browses the retrieves images from rank 1 to rank 50. If an image has the same visual space cluster as an image of highest rank, then this image is reranked to the end of the current 50 images list. In this way, if in the 50 first images, there are n different visual space clusters, then at the end of the rerank process, the first n images correspond to strictly different visual space clusters. We call this diversification method: DIVVISU. In order to have a point of comparison, we also propose to randomly permute the first 40 retrieve images. We call this naive method of diversification: DIVALEA.

3.3 ImageCLEFphoto Experiments and Results

The ImageCLEFphoto2008 corpus contains 20k images and 39 topics. Each image is associated with a caption stored in a semi-structured format. These captions include the title of the image, its creation date, the location at which the photograph was taken, the name of the photographer, a semantic description of the contents of the image (as determined by the photographer) and additional notes. In the text retrieval, we use all these elements. We build 18 runs: on the beginning, we build two runs based on classical text models (language model and TF-IDF), then we apply, on each of these runs, VCDT filtering or VCDTWN filtering, and finally we apply DIVVISU and DIVALEA diversity methods.

VCDT and VCDTWN Filtering. To determine if an image contains a visual concept, we choose to set the threshold t to the median of all the degrees values for a given concept (this value varies from 7.3 (*overcast*) to 28.8 (*outdoor*)). We do not use cooccurrence analysis (neither exclusion nor implication rules) in the ImageCLEFphoto task because it was not conclusive in the VCDT task. Table 2 shows that, for all topics, VCDT filtering improves P20 by 8% and VCDTWN filtering improves P20 by 4% in comparison to TF-IDF P20. Since

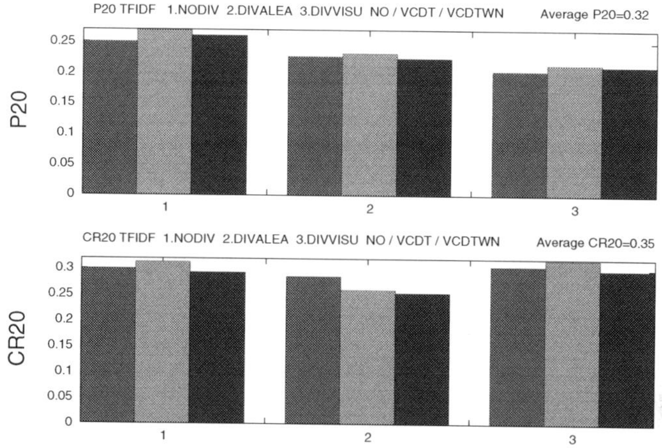

Fig. 1. Comparison of diversification methods 1. no diversification, 2. random diversification (DIVALEA) 3. diversification by visual space clustering (DIVVISU). For each diversification method, scores for TF-IDF only (1st bar), TF-IDF+VCDT (2nd bar) and TF-IDF+VCDTWN filtering (3rd bar) are given.

our method depends on the presence of a concept in the textual query, it does not apply to every topic. Using VCDT filtering, only 11 topics were filtered. Using VCDTWN filtering, 25 topics were modified. For the other topics, result images from text retrieval remain unchanged. Thus, we separate the study into three groups: all the topics, the 11 topics modified by VCDT filtering and the 25 topics for which we applied VCDTWN filtering. On Table 2, we observe an improvement on TF-IDF scores of +44% for P20 and +30% for the 11 topics modified by VCDT filtering, but not by VCDTWN filtering (+8% for P20 and -4% for CR20). Using VCDT filtering, all the modified topics are improved, but using VCDTWN filtering, some topics are improved and others are worsened. Then, we conclude that the way we use WordNet is not adapted for this task. Further study is needed.

Diversification. Figure 1 compares diversification method scores. DIVALEA and DIVVISU give lower P20 than no diversification, but DIVVISU slightly improves CR20 (in average +2%). So our DIVVISU diversification method works slightly better for diversification, but lowers precision as many others diversity methods (see [8]).

4 Conclusion

In this article, we focus our efforts (i) on the study of how to automatically extract and exploit visual concepts and (ii) on fast visual diversity. First, in VCDT task, we look at the mutual exclusion and implication relations between the concepts, in order to improve the automatic labelling. Our best VCDT run is the 4th ones under 53 submitted runs (3rd team under 11 teams). In our experiments, the use of the relations do not improve nor worsen the quality of the labeling. Second, in ImageCLEFphoto task, we analyse the influence of extracted visual concepts models to the diversity and precision, in a text retrieval context. This study shows that there is a clear improvement, in terms of precision or cluster recall at 20, when using the visual concepts explicitly appearing in the query. Third, we show that our fast visual diversity method based on fast clustering improved the cluster recall at 20. In our future researches, we will focus on how using image query to improve image retrieval using concept.

Acknowledgment. This work was partially supported by the French National Agency of Research (ANR-06-MDCA-002 AVEIR project).

References

1. Arni, T., Clough, P., Sanderson, M., Grubinger, M.: Overview of the ImageCLEF-photo 2008 photographic retrieval task. In: Peters, C., et al. (eds.) CLEF 2008. LNCS, vol. 5706, pp. 500–511. Springer, Heidelberg (2009)
2. Deselaers, T., Deserno, T.M.: The visual concept detection task in ImageCLEF 2008. In: Peters, C., et al. (eds.) CLEF 2008. LNCS, vol. 5706, pp. 531–538. Springer, Heidelberg (2009)
3. Fellbaum, C. (ed.): WordNet - An Electronic Lexical Database. Bradford Books (1998)
4. Marsala, C., Bouchon-Meunier, B.: Forest of fuzzy decision trees. In: Proceedings of the Seventh International Fuzzy Systems Association World Congress, vol. 1, pp. 369–374 (1997)
5. Marsala, C., Detyniecki, M.: Trecvid 2006: Forests of fuzzy decision trees for high-level feature extraction. In: TREC Video Retrieval Evaluation Online Proceedings (2006)
6. Shotton, J., Johnson, M., Cipolla, R.: Semantic texton forests for image categorization and segmentation. In: IEEE Conference on Computer Vision and Pattern Recognition (CVPR), pp. 1–8 (2008)
7. Tollari, S., Glotin, H.: Web image retrieval on ImagEVAL: Evidences on visualness and textualness concept dependency in fusion model. In: ACM Conference on Image and Video Retrieval (CIVR), pp. 65–72 (2007)
8. Tollari, S., Mulhem, P., Ferecatu, M., Glotin, H., Detyniecki, M., Gallinari, P., Sahbi, H., Zhao, Z.-Q.: A comparative study of diversity methods for hybrid text and image retrieval approaches. In: Peters, C., et al. (eds.) CLEF 2008. LNCS, vol. 5706, pp. 585–592. Springer, Heidelberg (2009)

A Comparative Study of Diversity Methods for Hybrid Text and Image Retrieval Approaches

Sabrina Tollari[1], Philippe Mulhem[2], Marin Ferecatu[3], Hervé Glotin[4], Marcin Detyniecki[1], Patrick Gallinari[1], Hichem Sahbi[3], and Zhong-Qiu Zhao[4,5]

[1] Université Pierre et Marie Curie - Paris 6, UMR CNRS 7606 LIP6, Paris
firstname.lastname@lip6.fr
[2] Université Joseph Fourier, UMR CNRS 5217 LIG, Grenoble
first.lastname@imag.fr
[3] TELECOM ParisTech, UMR CNRS 5141 LTCI, Paris
firstname.lastname@telecom-paristech.fr
[4] Université du Sud Toulon-Var, UMR CNRS 6168 LSIS, Toulon
name@univ-tln.fr
[5] Computer and Information School, Hefei University of Technology, China

Abstract. This article compares eight different diversity methods: 3 based on visual information, 1 based on date information, 3 adapted to each topic based on location and visual information; finally, for completeness, 1 based on random permutation. To compare the effectiveness of these methods, we apply them on 26 runs obtained with varied methods from different research teams and based on different modalities. We then discuss the results of the more than 200 obtained runs. The results show that query-adapted methods are more efficient than non-adapted method, that visual only runs are more difficult to diversify than text only and text-image runs, and finally that only few methods maximize both the precision and the cluster recall at 20 documents.

1 Introduction

Information retrieval is generally based on the computation of similarity with the query. Thus very similar images appearing in almost identical documents are retrieved with comparable degrees, producing clusters of alike images alike in the final ranking. In order to reduce this redundancy, several research teams [1,2,3,5,6,7] propose to apply, after retrieving the images, a diversity method.

The 2008 ImageCLEFphoto task [1] was focused on diversity. The evaluation was based on two measures: precision at rank 20 (P20) and instance recall at rank 20 (also called cluster recall (CR20) or S-recall), which calculates the percentage of *different* classes or clusters represented in the top 20 results. The idea behind these measures was to focus on relevant but at the same time diverse - in terms of clusters - images. Since it is important to maximize simultaneously both measures, for the overall ranking, the F1-measure (harmonic mean of P20 and CR20) was used.

The number of parameters in play during a retrieval process makes the study of the diversification approaches complex. For instance, the diversity (measured

C. Peters et al. (Eds.): CLEF 2008, LNCS 5706, pp. 585–592, 2009.

by the cluster recall) depends on the classical recall of the different underlying retrieving methods. To successfully compare the effectiveness of diversification methods, it is important to compare them using different multimodal non-diversified runs. In this paper we propose to study eight diversification methods applied to 26 ranked lists obtained with varied methods from different research teams [2,3,5,6] and based on different modalities (text, image, both...).

In Section 2, we briefly describe the characteristics of the eight diversity methods. In Section 3, we first present the 26 non-diversified runs used, then we discuss the results of the application of the diversity methods on the non-diversified runs. Finally we conclude the paper in Section 4.

2 The Diversity Methods

We study two kinds of diversity methods: (i) those where we apply the same diversity criterion for each topic, and (ii) those where the diversity criterion is adapted to the topic in function of the <cluster> field. For more details, please refer to the specific papers listed below.

2.1 The Non-adapted Diversity Methods

We study three kinds of non-adapted diversity methods: one based on the <date> field, three based on visual information only and one based on random permutation.

`ClustDMY`[5]. It uses an approach in which all images taken at the same date (day-month-year) are grouped together. All the images that have the same month and year but do not have a day specified are grouped together. All the images that have the same year, but no month and no day are grouped together.

`AffProp`[3]. First, a clustering on the top 1000 images is performed using affinity propagation and setting the parameters for 20 clusters. Then, images with the lowest rank are selected in each cluster, the remaining images are put in their original rank order.

`DIVVISU`[6]. A visual space partition of 256 clusters is obtained by binarization of the 8-bin Hue histogram (Hue from HSV space). Images are reranked in order to have each image of the top 20 belonging to a different cluster.

`VisKmeans`[5]. A KMeans clustering based on the visual description of the images (4608 dimension histograms). The number of clusters is 500. There are on average 40 images per cluster.

`DIVALEA`[6]. In order to have a point of comparison, this naive method proposes to randomly permute the first 40 retrieve images.

2.2 The Query-Adapted Diversity Methods

The <cluster> field of ImageCLEFphoto 2008 is related to the location of the pictures in 26 out of 39 topics. For the following methods, the diversity is based on text field for those 26 topics, and based on visual information for the 13 remaining.

Table 1. Comparison of the information used for each diversity method

	Diversity Method	Text	Visual	Other
1	`ClustDMY`	<DATE>	-	-
2	`AffProp`	-	20 visual clusters built by affinity propagation	-
3	`DIVVISU`	-	256 clusters (Hue)	-
4	`VisKmeans`	-	500 clusters (RGB)	-
5	`DIVALEA`	-	-	random
6	`Kmeans`	<LOCATION>	500 clusters (RGB)	-
7	`MAXMIN`	<LOCATION>	maximising the visual distances	-
8	`QT`	<LOCATION>	20 visual clusters built iteratively	-

`Kmeans`[5]**.** Queries having a cluster *city* are diversified using a clustering based on the city name coming from the <LOCATION> field of the image descriptions. Queries having a cluster *country* are diversified using a clustering based on the country name coming from the <LOCATION> of image descriptions. Queries having another cluster name are diversified using the `VisKmeans` diversity method (see before).

`MAXMIN`[2]**.** For the 26 topics related to location, text clustering is used. Otherwise, the diversity method based on visual descriptors is used. The `MAXMIN` diversity algorithm is based on the maximization of the smallest visual distance of a given document with respect to the so far selected results. Let E be the candidate set (the 40 best elements in our experiments) and consider $\mathcal{C} \subset \mathcal{S}$ as the set of already selected examples, the next document is then chosen as $x = \arg\max_{x_k \in \mathcal{S} \setminus \mathcal{C}} \min_{x_i \in \mathcal{C}} d(x_k, x_i)$ This procedure generates a permutation of the relevant query results such as its prefix corresponds to the most diversified results regarding the distance defined in the description space.

`QT`[2]**.** Text clustering is the same as the `MAXMIN` method. Visual diversity is inspired both by the Quality Threshold clustering and Voronoi algorithms. Let $s = N - n_{\mathcal{C}}$ be the cluster size, where $N, n_{\mathcal{C}}$ are respectively the number of images and the expected number of clusters. The algorithm iteratively updates a list of Voronoi cell prototypes by minimizing the following criterion $y_l = \arg\min_{x_i} \mathcal{R}(KNN_s(x_i; \mathcal{S}_t))$ where $KNN_s(x_i; \mathcal{S}_t$ denotes the s nearest neighbors of x_i in $\mathcal{S}_t$ ($\mathcal{S}_1 = \{x_1, \cdots, x_N\}$) and $\mathcal{R}$ the radius of the smallest ball enclosing $KNN_s(x_i; \mathcal{S}_t$. The new generated Voronoi cells (denoted $\mathcal{C}_l$) are removed and the process is iterated on the remaining data $\mathcal{S}_{l+1} - \mathcal{S}_l \setminus \mathcal{C}_l$. The final result is a partition of Voronoi cells and their prototypes which corresponds to the most diverse results.

The `AffProp`, `ClustDMY`, `DIVVISU`, `VisKmeans`, `Kmeans` and `QT` methods are based on clustering contrary to `DIVALEA` and `MAXMIN` which are based on permutation.

3 Experiments

3.1 The Non-diversified Runs

In order to make a true comparison of the diversity methods, we propose to apply these methods on 26 different non-diversified runs[1].

Automatic Text Only Runs. There are 3 text only runs, each from a different team. The first is based on language modelling [5], the second on tf-idf [6], and the last one uses a cosine measure between topics and matching candidate documents [2].

Automatic Visual Only Runs. There are also 3 visual runs from three different teams. The first is based on grid segmentation and Jensen-Shannon divergence [5], the second on entropic visual features and a 2-class SVM learning machine trained with the Gaussian kernel [3], and the last one on global color, texture and shape visual descriptors and a 2-class SVM learning machine trained with the Laplacian kernel [2].

Automatic Text-Image Runs. There are 9 automatic text-image runs. Four runs are from each of the different teams, and the last five ones are the AVEIR fusion runs (see also [7] for more details). The four text-image individual runs are:

LIG: The first run is based on the linear combination of the scores provided by a language model using Dirichlet smoothing on the text and by a Jeffrey-Divergence correspondence on the images [5].

LIP6: In the second run, text processing is based on standard TF-IDF with cosine similarity. Forest of Fuzzy Decision Trees (FFDT) trained on VCDT ImageCLEF task 2008 are used for a visual concept filtering of the textual results. The matching of the concepts and the topics text used WordNet [6].

LSIS: In the third run, the visual features are entropic features. Lots of SVMs are trained and generated with different parameters using the sample images provided. Then the first 20 images of the LIG run are used as the positive samples for each topic, and the others as the negative samples to construct the validation set for selecting the best one among the generated SVMs [3].

PTECH: The last run uses a combination of text and image descriptors. For a given topic, a separate query is performed for each modality (text and image). The results are merged by a minimum rank criterion: each image keeps the best rank [2].

The five AVEIR runs [7] correspond to different fusion strategies applied to the four text-image individual runs described before. Those runs were well ranked in the ImageCLEFphoto 2008 competition[2].

[1] The 26 non-diversified runs and the 208 runs generated by the proposed diversity methods are available at `http://aveir.lip6.fr/diversity`

[2] Particularly, the MEAN run is ranked 18 under 1042 submitted runs (P20=0.43, CR20=0.46 and F1-measure=0.45) (see [7] for more details).

MIN: for each image, the fusion-rank corresponds to the minimum rank observed on each of the 4 team's runs. This strategy corresponds to creating a rank by alternatively choosing an image from each of the teams' runs. The first image of the fusion rank corresponds to the first image of the first team; the second image corresponds to the first image of the second team; the fifth corresponds to the second image of the first team, and so on.

MEAN: for each image, the fusion-rank corresponds to the average rank observed on each of the 4 team's runs. This strategy corresponds to a compromise taking into account all the systems. Images not present in one of the ranked lists are considered as having rank 1001.

MEAN1on4: here only images that are ranked by at least 1 teams were considered. The fusion-rank correspond to the average of the available ranks.

MEAN2on4: same as MEAN1on4, but only images that are ranked by at least two teams were considered. The idea behind this strategy is to avoid fusion of images returned only by one team.

MEAN3on4: same as MEAN1on4, but only images that are ranked by at least three teams were considered. The idea behind this strategy is to avoid fusion of images returned only by two teams.

Manual Text-Image Run. There is only one manual run proposed in [2]. For this run, text descriptions are built using the vector space model, while topics are represented as boolean queries built manually. The visual and text results are combined by intersecting the underlying results, so this will guarantee that the output is consistent with respect to both modalities.

Ideal Runs. To measure to what extent our diversity methods are efficient independently of the runs there are applied on, we also propose to build ideal runs. Using the ground truth, we build the set of relevant images for each topic, then we randomly permute those images to obtain the ideal runs. We reiterate the process 10 times in order to obtain 10 ideal runs. Obviously, the precision at 20 documents will be closer to 1, because all the retrieved documents are relevant, but the cluster recall at 20 documents will not be optimal.

3.2 Results

When averaging on the 10 ideal runs (see Figure 1(a)), `Kmeans`, `MAXMIN` and `QT` give the best results for the CR20 (above 0.84) and for the F1-measure (above 0.90). For ideal runs, we consider that checking the precision at 20 documents is not relevant (P20=0.993 ± 0.001), because reordering only relevant documents lead to the same value. It is worth noting that, on average on the 10 ideal runs, the random reranking `DIVALEA` increases the CR20 results, and such reranking also outperforms for the CR20 value the `ClustDMY` and `AffProp`, for which we do not have yet a clear explanation.

When considering only the text only runs (see Figure 1(b)), `AffProp`, `VisKmeans` and `QT` achieves a F1-measure value as high as the non diversified runs. An important point to notice here is that all the diversification schemes

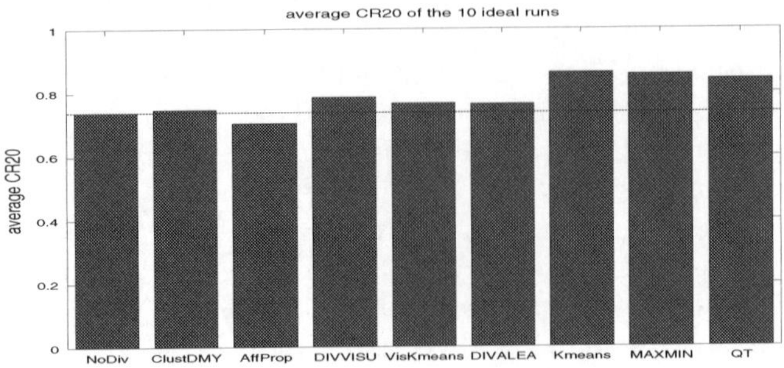

(a) Average CR20 on the 10 ideal runs (P20=0.993 ± 0.001)

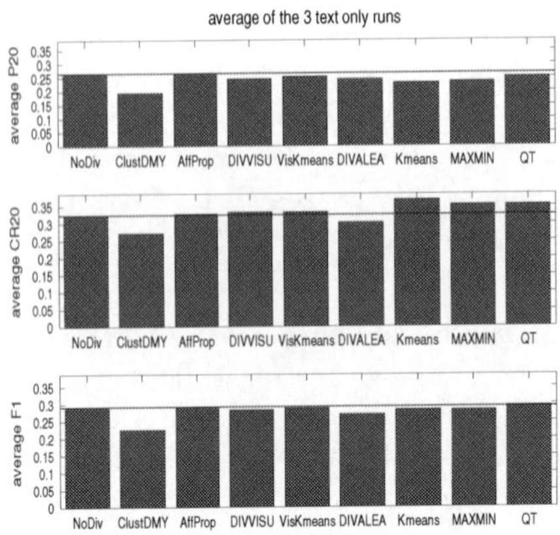

(b) Average P20, average CR20 and average F1 measure of the 3 text only runs

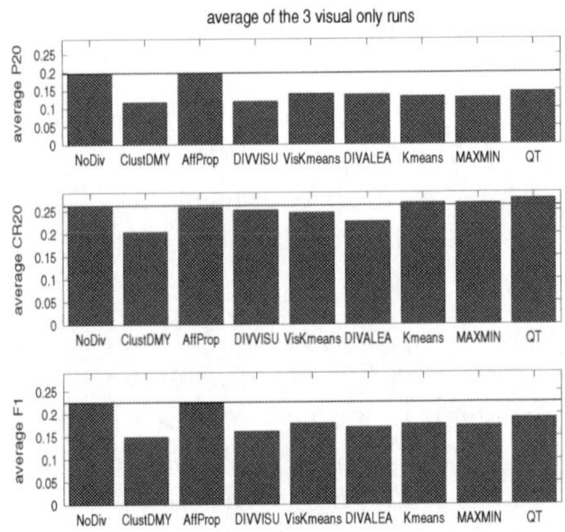

(c) Average P20, average CR20 and average F1 measure of the 3 visual only runs

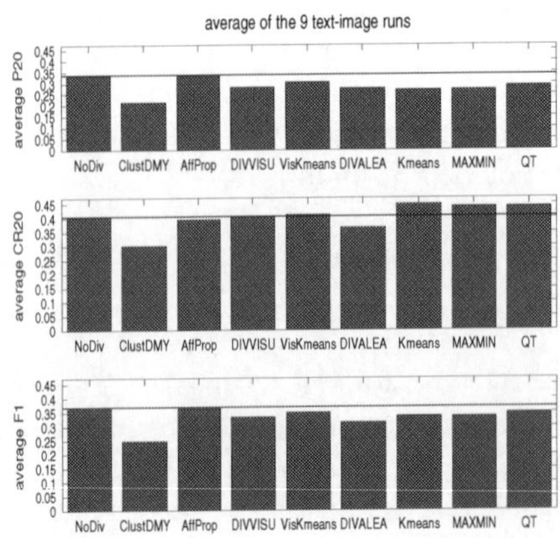

(d) Average P20, average CR20 and average F1 measure of 9 text-image runs

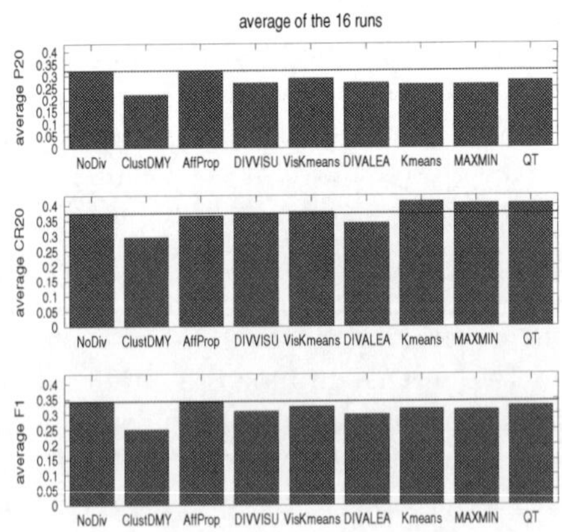

(e) Average P20, average CR20 and average F1 measure of the 16 runs

Fig. 1. Comparison of the diversity methods in function of the modality of the runs. First bar (and line) corresponds to the value without diversity.

(except ClustDMY and DIVALEA) outperform the non diversified result of 0.325 for the CR20, with values higher than 0.35 for Kmeans, QT and MAXMIN.

When considering the three visual only runs (see Figure 1(c)), the P20 scores strongly decrease except for AffPropwhere the P20 score is similar with the non-diversified run. Even if the the P20 scores strongly decrease, the CR20 scores for Kmeans, MAXMIN and QT are above the CR20 of the non-diversified run. But finally the F1-measures are very low. This leads to the fact that diversifying visual only runs is much more difficult to achieve.

For the automatic text-image runs (see Figure 1(d)), the non diversified run outperforms the diversified runs for the F1-measure. One again, the Kmeans, QT and MAXMIN diversifications give better results that the non diversified runs for the CR20 value. The same conclusion are inferred from the 5 AVEIR runs, from the 4 individual text-image runs and from the manual run.

For all the runs (image only, text only and image-text) without the ideal runs (see Figure 1(e)), the Kmeans, MAXMIN and QT diversity methods outperforms the CR20 of the non diversified run; AffProp, DIVVISU and VisKmeans give similar CR20 than the non diversified run, we conclude from these results that diversity based only on visual information is not effective; DIVALEA decreases the CR20; clustering on day/month/year ClustDMY gives also bad results, this may be due to the fact that all the images do not have the full date given. Finally, if we compare the F1-measure of the 16 runs (see Figure 1(e)), we noted that the F1-measure values of the diversified runs are always below the F1-measure value of the non-diversified run. We conclude that it is very hard to have a high CR20 score and at the same time a high P20 score. We only achieve this goal with the query-adapted diversity methods and the ideal runs (see Figure 1(a)).

4 Conclusion

In this article, we compare eight diversity methods applied on 26 varied runs from different teams. This study shows that for all the diversity methods, the precision at 20 always decreases compared to the run without diversification (except - of course - for ideal runs).

The results first show that query-adapted methods (Kmeans, MAXMIN and QT) are more efficient than non-adapted methods. But, in the case of real web search engines, it is difficult for a user to choose the right diversity criterion to apply to a query, so even if non-adapted methods are less efficient they must also be considered. The diversity based on the date ClustDMY appears to be non efficient and the visual-only diversity methods (AffProp, VisKmeans and DIVVISU) gives similar cluster recall than the non diversified results. Diversifying a ranked list of image results when no diversity cluster is given is always an open question. Second, because visual only runs are more difficult to diversify than text only and text-image runs, diversifying content based image retrieval results are more difficult than diversifying text-based image retrieval. Finally, we unfortunately notice that, in the case of 16 non ideal runs, none of the eight diversity methods obtains a better F1-measure value than the non-diversified run. It is only in the case of the 10 ideal runs that some of the diversity methods gives a better F1-measure value. In future work, we will also compare the effect of diversity based on visual concept [4].

Acknowledgment

This work was supported by the French National Agency of Research (ANR-06-MDCA-002). Zhong-Qiu Zhao is also supported by Research Fund for the Doctoral Program of Higher Education of China (200803591024).

References

1. Arni, T., Clough, P., Sanderson, M., Grubinger, M.: Overview of the ImageCLEF-photo 2008 photographic retrieval task. In: Peters, C., et al. (eds.) CLEF 2008. LNCS, vol. 5706, pp. 500–511. Springer, Heidelberg (2009)
2. Ferecatu, M., Sahbi, H.: TELECOM ParisTech at ImageClefphoto 2008: Bi-modal text and image retrieval with diversity enhancement. In: Working Notes for the CLEF 2008 workshop (2008)
3. Glotin, H., Zhao, Z.: Visual-only affinity propagation promoting diversity for CLEF2008 photo retrieval task. In: Peters, C., et al. (eds.) CLEF 2008. LNCS, vol. 5706, pp. 628–631. Springer, Heidelberg (2009)
4. Inoue, M., Grover, P.: Effects of visual concept-based post-retrieval clustering in imageclefphoto 2008. In: Working Notes for the CLEF 2008 workshop (2008)
5. Maisonnasse, L., Mulhem, P., Gaussier, E., Chevallet, J.-P.: LIG at ImageCLEF. In: Peters, C., et al. (eds.) CLEF 2008. LNCS, vol. 5706, pp. 704–711. Springer, Heidelberg (2009)
6. Tollari, S., Detyniecki, M., Fakeri-Tabrizi, A., Amini, M.-R., Gallinari, P.: Using visual concepts and fast visual diversity to improve image retrieval. In: Peters, C., et al. (eds.) CLEF 2008. LNCS, vol. 5706, pp. 577–584. Springer, Heidelberg (2009)
7. Tollari, S., Detyniecki, M., Ferecatu, M., Glotin, H., Mulhem, P., Amini, M.-R., Fakeri-Tabrizi, A., Gallinari, P., Sahbi, H., Zhao, Z.-Q.: Consortium AVEIR at ImageCLEFphoto 2008: on the fusion of runs. In: Working Notes for the CLEF 2008 workshop (2008)

University of Jaén at ImagePhoto 2008: Filtering the Results with the Cluster Term

Miguel Angel García-Cumbreras, Manuel Carlos Díaz-Galiano, María Teresa Martín-Valdivia, and L. Alfonso Ureña-López

SINAI Research Group*, Computer Science Department, University of Jaén, Spain
{magc,mcdiaz,maite,laurena}@ujaen.es
http://sinai.ujaen.es

Abstract. This paper describes the University of Jan system presented at ImagePhoto CLEF 2008. Previous systems used translation approaches and different information retrieval systems to obtain good results. The queries used are monolingual, so translation methods are not necessary. The new system uses the parameters that obtain the best results in the past. The novelty of our method consists of some filtered methods that are used to improve the results, with the cluster terms and its WordNet synonyms. The combination of different weighting functions (Okapi and Tfidf), the results obtained by the information retrieval systems (Lemur and Jirs), and the use or not of automatic feedback complete the experimentation.

1 Introduction

In this paper our system has been tested with the framework provided by ImagePhoto CLEF organization[1].

Our system only uses textual information, not visual information, to improve the retrieval methods. Two Information Retrieval (IR) systems have been run, and the experiments test the use of automatic feedback and different weighting functions (Okapi and Tfidf). A simple method has been developed to filter the results with the cluster term and its WordNet[1] synonyms. It has been applied in some experiments.

Section 2 describes the complete system. In Section 3 the experiments and results are shown. Finally, Section 4 contains the analysis of the results and the main conclusions.

2 System Description

In our system there is not user interaction, and we have used the English textual information (not visual information).

* http://sinai.ujaen.es
[1] Available at http://wordnet.princeton.edu

C. Peters et al. (Eds.): CLEF 2008, LNCS 5706, pp. 593–596, 2009.

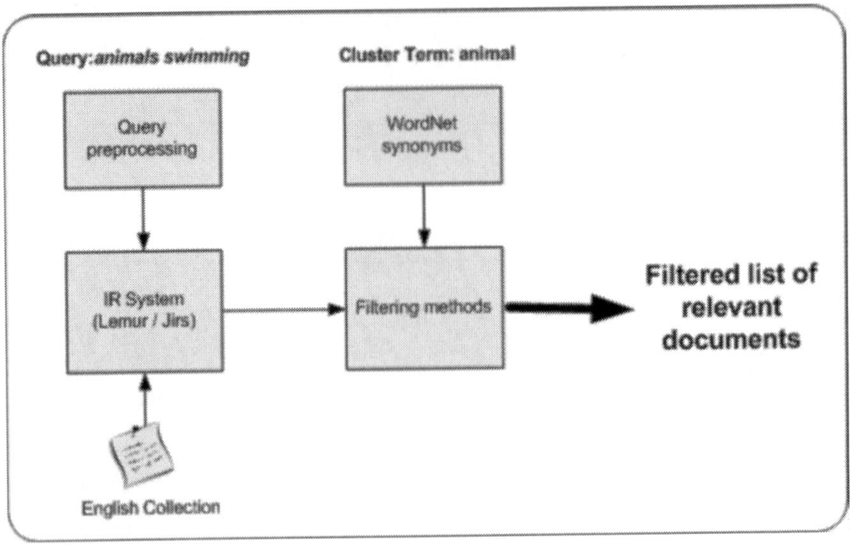

Fig. 1. General scheme of the SINAI system at ImagePhoto 2008

The English collection documents have been preprocessed as usual (English stopwords removal and the Porter's stemmer[2]). Then, the documents have been indexed using as IR systems: Lemur[2] and Jirs[3].

Previous results of our system shown that precision results obtained with both IR systems were very similar. Only the results with Italian queries were quite different[4]. A simple combination method with both IR results was developed, and the evaluation of the combined list of relevant documents fix the parameter that weight each list in 0.8 for Lemur documents and 0.2 for Jirs documents. Using the same combination parameters, the new system try to improve the results with different combinations of methods and the application of a filter with the cluster term. The weighting function of the IR systems is a parameter changed to test the results.

The use of Psedo-Relevance Feedback (PRF) to improve the retrieval process is not conclusive, but in general the precision is increased in past experiments, so it is used always with Lemur.

After the retrieval process, the documents or passages marked as relevant are filtered, using the cluster terms, as follows:

1. The cluster term is expanded with its WordNet synonyms (the first sense).
2. The list of relevant documents generated by the IR system is filtered. If the relevant document contains the cluster term or a synonym its docid (the identifier of the document) is written in another list.
3. Finally, the new list with the filtered documents is combined with the original ones (Lemur and Jirs) in order to improve them. A simple method to do this

[2] Available at `http://www.lemurproject.org/`

was to duplicate the score value of the documents in the filtered list and to add them to the original ones.

Figure 2 shows a general architecture of the system delevoped.

3 Experiments Description and Results

We have made the following experiments:

1. **SINAI exp1 Baseline**. It is the baseline. Lemur is used as IR system with automatic feedback. The weighting function applied was Okapi. There was no combination of results, nor filtering method with the cluster term.
2. **SINAI exp2 LemurJirs**. This experiment combines the IR lists of relevant documents. Lemur also uses Okapi as weighting function and PRF. Before the combination of results Lemur and Jirs lists are filtered, only with the cluster term.
3. **SINAI exp3 Lemur fb okapi**. The Lemur list of relevant documents is filtered with the cluster term and its WordNet synonyms. Okapi is used as weighting funcion, and PRF is applied automatically.
4. **SINAI exp4 Lemur fb tfidf**. It is the same experiment as before, but in this case the weighting function used was Tfidf.
5. **SINAI exp5 Lemur simple okapi**. Lemur IR system has been run with Okapi as weighting function and without feedback. The list of relevant documents has been filtered with the cluster term and its WordNet synonyms.
6. **SINAI exp6 Lemur simple tfidf**. Lemur IR system has been used with Tfidf as weighting function and without feedback. The list of relevant documents has not been filtered.

Table 1 presents the results, with and without filtering. All the results are based on textual information of the English queries. The last column shows the mean F1-Measure obtained in ImagePhoto 2008, with an automatic system and only text.

Table 1. Results

Id	Filtering	Modality	FB	Expansion	MAP	P@5	P@10	Best F1
(1)	No	Text	Yes	No	**0.2125**	**0.3744**	**0.3308**	0.2957
(6)	No	Text	No	No	0.2016	0.3077	0.2872	0.2957
(2)	Yes	Text	Yes	No	0.2063	0.3385	0.2949	0.2957
(3)	Yes	Text	Yes	No	**0.2089**	**0.3538**	0.3128	0.2957
(4)	Yes	Text	Yes	No	0.2043	0.2872	0.2949	0.2957
(5)	Yes	Text	No	No	0.1972	0.3385	**0.3179**	0.2957

4 Discussion and Conclusions

In this paper we have presented the results of our architecture to retrieve information from a multimedia corpus, presented in the ImageCLEF 2008 Photo task. We have experimented with two major variables, a filtering process that used the cluster term, and its synonyms in one case, and some changes in the retrieval parameters, such as the weighting function or the use or automatic feedback.

The results show that a filtering method is not useful if the cluster term or related words are used to filter the IR retrieved documents, because some good documents are deleted and none of non retrieved relevant documents are included in the second step. In general, the results in term of MAP or other precision values are not so different. Between the best MAP and the worse one the difference is less than 8%. Filtering methods have not improved the baseline cases. After an analysis of the performance we can write some reasons:

- Some relevant documents that appear in the first retrieval phase have been deleted because they not contain the cluster term, so the cluster term is not useful in a filtering process.
- Other documents retrieved by the IR, that are not relevant, contains synonyms of the cluster term, so they are not deleted and the precision decrease.

The final conclusion is that this filtering process is not good with the cluster term to improved the results. As future work a clustering or classifying method will be developed, working with textual information, to classify and improved the baseline results.

Acknowledgements

This project has been partially supported by a grant from the Spanish Government, project TIMOM (TIN2006-15265-C06-03).

References

1. Arni, T., Clough, P., Sanderson, M., Grubinger, M.: Overview of the ImageCLEF-photo 2008 Photographic Retrieval Task. In: Working Notes of the 2008 CLEF Workshop, Aarhus, Denmark (September 2008)
2. Porter, M.F.: An algorithm for suffix stripping. In Readings in information retrieval, pp. 313–316. Morgan Kaufmann Publishers Inc., San Francisco (1997)
3. Gómez-Soriano, J.M., Montes-y-Gómez, M., Sanchis-Arnal, E., Rosso, P.: A Passage Retrieval System for Multilingual Question Answering. In: Matoušek, V., Mautner, P., Pavelka, T. (eds.) TSD 2005. LNCS (LNAI), vol. 3658, pp. 443–450. Springer, Heidelberg (2005)
4. Díaz-Galiano, M.C., García-Cumbreras, M.A., Martín-Valdivia, M.T., Montejo-Raez, A., Ureña-López, L.A.: Integrating MeSH Ontology to Improve Medical Information Retrieval. In: Peters, C., Jijkoun, V., Mandl, T., Müller, H., Oard, D.W., Peñas, A., Petras, V., Santos, D. (eds.) CLEF 2007. LNCS, vol. 5152, pp. 601–606. Springer, Heidelberg (2008)

Combining TEXT-MESS Systems at ImageCLEF 2008

Sergio Navarro[1], Miguel Angel García-Cumbreras[2], Fernando Llopis[1], Manuel Carlos Díaz-Galiano[2], Rafael Muñoz[1], María Teresa Martín-Valdivia[2], L. Alfonso Ureña-López[2], and Arturo Montejo-Ráez[2]

[1] Natural Language Processing and Information Systems Group, University of Alicante, Spain
{snavarro,llopis,rafael}@dlsi.ua.es

[2] SINAI Research Group, Computer Science Department, University of Jaén, Spain
{magc,mcdiaz,maite,laurena,amontejo}@ujaen.es

Abstract. This paper describes the joint work of two teams belonging to the TEXT-MESS project. The system presented at ImageCLEFPhoto task combines one module based on filtering and other based on clustering. The main objective was to study the behavior of these methods with a large number of configurations in order to increase our chances of success. The system presented at ImageCLEFmed task uses the IR-n system with a negative query expansion based on the acquisition type of the image mixed with the SINAI system with a MeSH based query expansion.

1 Introduction

This paper describes the joint work of two teams belonging to the TEXT-MESS project, a complete Information Retrieval (IR) system which works in a multi-modal environment, with textual information and medical or general images.

1.1 ImageCLEFphoto

The goal of the photographic task is, given a query, to retrieve a diverse, yet relevant set of images at the top of a ranked list [1]. Text and visual information can be used to improve the retrieval methods, and the main evaluation point is to study how the use of clustering or filtering methods affects to the precision and diversity of the results achieved by two different IR systems.

1.2 ImageCLEFmed

The goal of the medical task is to retrieve relevant medical images from a query based on one or several medical images and a textual query [2]. The collection contains images from articles published in Radiology and Radiographics including the text of the captions and a link to the HTML of the full text articles.

C. Peters et al. (Eds.): CLEF 2008, LNCS 5706, pp. 597–604, 2009.

We downloaded articles from the web and constructed a new textual collection including the text of the article section where the image appears. Besides, for the experiments, we worked on the combination of two IR systems, SINAI and IR-n. Both systems use adaptations to the medical domain, and their major difference is that SINAI is based on documents while IR-n is based on passages.

2 ImageCLEFphoto System Description

The complete system is composed by an IR retrieval system and two modules that work in a serial mode, a filtering module and a clustering module. The output of the IR system is the input of the first module and its output is the input of the second one. For our participation we worked with two different IR systems which participated individually in this edition of the ImageCLEFphoto task, SINAI [3] and IR-n [4].

The SINAI system combines the output of two IR systems (Lemur[1] and Jirs [5]), is automatic (without user interaction), works with English text information (not visual information), and its blind feedback algorithm is based on the Probabilistic Relevance Feedback formula (PRF) [6].

IR-n is an IR system based on passages. It allows to select between two different blind feedback algorithms, PRF and Local Context Analysis (LCA) [7]. Also it uses different mixing strategies to perform multimodal retrieval using a provided CBIR ranked list. [4].

2.1 Filtering Module

The use of the cluster term is oriented in a filtering way. After the retrieval process the documents or passages marked as relevant are filtered as follows:

1. The cluster term is expanded with its WordNet synonyms (the first sense).
2. The list of relevant documents generated by the IR system is filtered. If the relevant document contains the cluster term or a synonym its docid (the identifier of the document) is written in another list.
3. Finally, the new list with the filtered documents is combined with the original one (Lemur and Jirs) in order to improve them. A simple method to do this was to duplicate the score value of the documents in the filtered list and to add them to the original ones.

A similar filtering method is applied in the SINAI system that works with geographical information [8].

2.2 Clustering Module

The clustering strategy seeks to promote the diversity among the 20 top ranked documents returned by the system. It is based on the relevance assigned by

[1] Available at http://www.lemurproject.org/

the IR system for each document and on the clusters of documents created by Carrot2 [2], an open source clustering engine. This module rises to the top of the ranking the most relevant document within each cluster and a number of documents without cluster assigned equal to the remaining number of documents until 20. The clustering can use only the image annotations or the annotations enriched with visual concepts extracted from their related images.

3 ImageCLEFmed System Description

The complete system is composed by two systems working in a parallel mode, IR-n and SINAI systems. Both use their adaptations to the IR medical domain developed for their individual participation in this edition of the task [9] [10]. The major difference between these systems is that while SINAI is a system based on documents IR-n is based on passages.

The ranking returned by each one of the systems is merged following a standard Reranking (RR) method. Finally, the result of this step is merged with the output of a CBIR system using a rerankig strategy. We tested two different multimodal RR strategies. In the following lines we describe the main modules of these systems.

3.1 Query Expansion with MeSH and UMLS Ontologies

The SINAI expansion method using MeSH ontology [3] is the same as we carried out in the past, which obtained good results [11].

Moreover, to expand the queries with UMLS [4] , SINAI group used MetaMap program [12]. In order to reduce the number of terms that could expand the query, to make it equal to that of MeSH expansion, we restricted the semantic types in the mapped terms [9] as follows: *bpoc* (Body Part, Organ, or Organ Component), *diap* (Diagnostic Procedure), *dsyn* (Disease or Syndrome) and *neop* (Neoplastic Process). Therefore, for this expansion we used the Meta Candidate terms, because these terms provide similar terms with differences in the words. For a detailed view of the process and some examples, see [9].

3.2 Negative Query Expansion Based on the Acquisition Type of the Image

IR-n system uses a query expansion method based on the acquisition type of the image to penalize those images which do not pertain to the same acquisition type found in the query. It is based on [13], but in our approach we neither used visual features nor filtering strategy but we used a modified version of the classification proposed at that work. In order to only retrieve images of the desired type in the query we used the textual query and the text annotations for the retrieval [10].

[2] http://www.carrot2.org
[3] http://www.nlm.nih.gov/mesh/
[4] http://www.nlm.nih.gov/research/umls/

3.3 TF-IDF Multimodal RR Strategy

IR-n system allows us to use an alternative RR strategy for merging a text based list and a list returned by a CBIR system.

This approach is based on two assumptions: on the one hand the textual list is more confident than the list based on images and on the other hand the TF-IDF formula is a suitable way to measure the quantity and the quality of a text. Thus, the system only uses the relevance value returned by a CBIR system for those documents which have a TF-IDF value under an established TF-IDF threshold. Further information about the reranking formula used can be found in [10].

4 Experiments Description and Results

4.1 ImageCLEFphoto

The dataset is the collection IAPR TC-12 image collection, which consists of 20,000 images taken from different locations around the world and comprises a varying cross-section of still natural images. It includes pictures of a range of sports and actions, photographs of people, animals, cities, landscapes and many others of contemporary life. Each image is associated with alphanumeric captions stored in a semi-structured format (title, creation date, location, name of the photographer, description and additional notes). The topics statements also have a semi-structured format which includes the query, a cluster tag and a narrative tag.

We used the SINAI system for our experiments in the following configurations:

- **LemurJirs:** This experiment combines the IR lists of relevant documents. Lemur also uses Okapi as weighting function and PRF. Before the combination of results Lemur and Jirs lists are filtered, only with the cluster term.
- **Lemur fb okapi:** The Lemur list of relevant documents is filtered with the cluster term and its WordNet synonyms. Okapi is used as weighting function, and PRF is applied automatically.
- **Lemur fb tfidf:** It is the same experiment as before, but in this case the weighting function used was TF-IDF.
- **Lemur simple okapi:** Lemur IR system uses Okapi as weighting function and without feedback. The list of relevant documents is filtered with the cluster term and its WordNet synonyms.
- **Lemur simple tfidf:** Lemur IR system is used with TF-IDF as weighting function and without feedback. The list of relevant documents is not filtered.

The following configurations were used for the IR-n system experiments:

- **IRnExp:** This experiment uses PRF as relevance feedback strategy.
- **IRnExpClust:** It uses the annotations related to IRnExp output as input for the clustering module.
- **IRnFBFIRE:** It uses a baseline experiment of the FIRE [14] system and LCA as a multimodal relevance feedback strategy.

Table 1. ImageCLEFphoto Official Textual Results

run name	Standalone Run				Official Run			
	MAP	P20	CR20	FMea	MAP	P20	CR20	FMea
IRnExp**Filt**	0.2699	**0.3244**	0.2816	0.3015	**0.2671**	0.3154	0.2875	0.3008
IRnExpClust**Filt**	0.2699	**0.3244**	0.2816	0.3015	0.2287	0.2090	0.3011	0.2467
LemurSimpleOkapi Filt**Clust**	0.1972	0.2795	0.2930	0.2861	0.1750	0.1987	**0.3241**	0.2464
LemurFbOkapiFilt**Clust**	0.2089	0.2808	0.2682	0.2744	0.1804	0.1897	0.2764	0.2250
LemurJirs**Clust**	0.2063	0.2769	0.2900	0.2833	0.1840	0.2051	0.2815	0.2373
LemurFbTfidfFilt**Clust**	0.2043	0.2679	0.2704	0.2691	0.1786	0.1974	0.3185	0.2437

- **IRnFBFIREClustC:** It uses the IRnFBFIRE output run to perform a clustering based on the image annotations and the visual concepts extracted from their related images.
- **IRnConcepFBFIRE:** The image annotations indexed by IR-n are previously enriched with visual concepts extracted from the image. For the retrieval phase, the system uses a baseline run of the FIRE system and LCA as a multimodal relevance feedback strategy.
- **IRnConcepFBFIREClustC:** It uses the IRnConcepFBFIREClustC run output to perform a clustering based on the image annotations and the visual concepts extracted from their related images.

Our aim is to analyze the effect of adding to the work flow of the two IR systems the SINAI filtering method and the IR-n clustering module in order to improve their performance.

Table 1 and Table 2 show the results of the textual runs and the mixed runs respectively. Furthermore, we can see the results previously obtained by the standalone runs, without adding the external filtering or clustering module, in order to observe the improvement or worsening obtained with the added module. For each run name we show a term in bold letters which identifies the external module which has been added to that base configuration, **Filt** or **Clust**.

We can observe in the Table 1 that the CR20 value has increased its value for almost all the experiments which have used the clustering module. Indeed the best CR20 value for the textual runs has been obtained using the clustering module with SINAI system.

4.2 ImageCLEFmed

In the training phase, for each IR system, we worked in the selection of those runs which better results obtained. Next, in order to figure out which are the most suitable weighting values for the reranking module, we carried out a training phase with each pair of selected runs in the previous step. The runs submitted to the competition (**TEXTMESS** runs) were the followings:

- **meshType_CT:** Standard RR strategy which fuses the output of the SINAI system with query expansion based on MeSH ontology and the output of the

Table 2. ImageCLEFphoto Official Mixed Results (Image + Text)

run name	Standalone Run MAP	P20	CR20	FMea	Official Run MAP	P20	CR20	FMea
IRnFBFIRE**Filt**	**0.3436**	**0.4564**	0.3119	0.3706	0.3354	0.4333	0.3041	0.3574
IRnFBFIREClustC**Filt**	0.3032	0.3782	0.3483	0.3626	0.3183	0.3808	0.3178	0.3465
IRnFBFIRE**Filt**ClustC	0.3032	0.3782	0.3483	0.3626	0.3097	0.3564	0.3223	0.3385
IRnConcepFBFIRE**Filt**	0.3333	0.4333	0.3316	**0.3757**	0.3272	0.4115	0.3311	0.3669
IRnConcepFBFIRE **Filt**ClustC	0.3032	0.3782	0.3483	0.3626	0.2917	0.3410	**0.3483**	0.3446
IRnConcepFBFIRE **Filt**ClustC	0.3032	0.3782	0.3483	0.3626	0.2973	0.3603	0.3446	0.3523

IR-n system with the negative expansion. Both IR systems use the image captions and article titles of the image collection.

- **umlsType_CT:** meshType_CT configuration but the SINAI system works with the query expansion based on UMLS metathesaurus using MetaMap program instead of use MeSH ontology.
- **meshType_CTS:** meshType_CT configuration but both IR systems use captions, titles and texts of the sections where the images appear.
- **umlsType_CTS:** umlsType_CT configuration but both IR systems use captions, titles and texts of the sections where the images appear.
- **meshTypeFIREidf_CT:** TF-IDF RR strategy fusing the meshType_CT output and the FIRE system output.
- **meshTypeFIRE_CT:** Standard RR strategy fusing the meshType_CT output and the FIRE system output.
- **umlsTypeFIREidf_CT:** TF-IDF RR strategy fusing the umlsType_CT output and the FIRE system output.
- **umlsTypeFIRE_CT:** Standard RR strategy fusing the umlsType_CT run output and the FIRE system output.
- **meshTypeFIRE_CTS:** Standard RR strategy fusing the meshType_CTS output and the FIRE system output.
- **umlsTypeFIRE_CTS:** Standard RR strategy fusing the umlsType_CTS output run and a CBIR system output.

Table 3. ImageCLEFMed Official Textual Runs Results

run name	map	txt rk	CLEF rk
TEXTMESSmeshType_CT	0.2777	5	6
TEXTMESSumlsType_CT	0.1413	37	49
TEXTMESSmeshType_CTS	0.1026	49	65
TEXTMESSumlsType_CTS	0.0858	52	70

Table 4. ImageCLEFmed Official Mixed Runs Results (Image + Text)

run name	map	txt rk	CLEF rk
TEXTMESSmeshTypeFIREidf_CT	**0.2777**	**2**	**6**
TEXTMESSmeshTypeFIRE_CT	0.2223	7	29
TEXTMESSumlsTypeFIREidf_CT	0.1412	10	50
TEXTMESSumlsTypeFIRE_CT	0.1325	11	56
TEXTMESSmeshTypeFIRE_CTS	0.1188	12	57
TEXTMESSumlsTypeFIRE_CTS	0.0887	14	69

Tables 3 and 4 show the MAP results, the ranking position within the textual and the mixed modality respectively and the ranking position within all the participant runs.

In the Table 3 we can see the best TEXT-MESS run in the textual modality is the one which uses IR-n with negative expansion based on the acquisition type of the image and SINAI with MeSH based expansion.

5 Conclusion and Future Work

On the one hand after the analysis of the ImageCLEFphoto results we can see the following conclusions: The filtering method is not useful when we use the cluster term or related words to filter retrieved documents, because some relevant documents are deleted and none of non retrieved relevant documents are included in the second step. The clustering method, without using the cluster term, can improve the results of cluster detection, although at the expense of a decrease in precision of the results that is greater than the gain obtained for the CR20.

On the other hand, we can observe in the results of the ImageCLEFmed task, that the precision values reached by our standard multimodal RR runs has gone down in the task ranking, while the TF-IDF multimodal RR runs are in the top positions of this ranking, reaching the same MAP values achieved by the runs with the same configuration but without multimodal RR (textual runs). It is explained by the low threshold used for TF-IDF RR strategy which made that the system handles all the images retrieved by the CBIR system as images with enough textual information to perform a suitable retrieval only using the relevance returned by the textual IR system. The reasons to obtain this low threshold in the tuning phase were the use of a different collection and the use of a different CBIR system from the one which was used in the test phase. These problems affected negatively to the performance of this strategy with the test collection.

As future work we plan to improve the combination of the filtering and clustering methods only applying the filter when we predict the results obtained by the IR system will be poor. Moreover, we are planning on to work on finding an alternative method to establish the TF-IDF RR threshold.

Acknowledgements

This research has been partially funded by the Spanish Government within the framework of the TEXT-MESS (TIN-2006-15265-C06-01 and TIN-2006-15265-C06-03) project.

References

1. Arni, T., Clough, P., Sanderson, M., Grubinger, M.: Overview of the ImageCLEFphoto 2008 Photographic Retrieval Task. In: Peters, C., et al. (eds.) CLEF 2008. LNCS, vol. 5706, pp. 500–511. Springer, Heidelberg (2009)
2. Müller, H., Kalpathy-Cramer, J., Kahn, C.E., Hatt, W., Bedrick, S., Hersh, W.: Overview of the ImageCLEFmed 2008 medical image retrieval task. In: Peters, C., et al. (eds.) CLEF 2008. LNCS, vol. 5706, pp. 512–522. Springer, Heidelberg (2009)
3. García-Cumbreras, M.A., Díaz-Galiano, M.C., Martín-Valdivia, M.T., Ureña López, L.A.: SINAI at ImageCLEFphoto 2008. In: On-line Working Notes, CLEF 2008 (2008)
4. Navarro, S., Llopis, F., Muñoz, R.: Different Multimodal Approaches using IR-n in ImageCLEFphoto 2008. In: On-line Working Notes, CLEF 2008 (2008)
5. Gómez-Soriano, J.M., Montes-y-Gómez, M., Sanchis-Arnal, E., Rosso, P.: A Passage Retrieval System for Multilingual Question Answering. In: Matoušek, V., Mautner, P., Pavelka, T. (eds.) TSD 2005. LNCS (LNAI), vol. 3658, pp. 443–450. Springer, Heidelberg (2005)
6. Robertson, S.E., Sparck Jones, K.: Relevance weighting of search terms. Journal of the American Society for Information Science 27(3), 129–146 (1976)
7. Xu, J., Croft, W.B.: Improving the efectiveness of information retrieval with local context analysis. ACM Trans. Inf. Syst. 18(1), 79–112 (2000)
8. Perea-Ortega, J.M., García-Cumbreras, M.A., García-Vega, M., Montejo-Raez, A.: Filtering for Improving the Geographic Information Search. In: Peters, C., et al. (eds.) CLEF 2007. LNCS, vol. 5152, pp. 823–829. Springer, Heidelberg (2008)
9. Díaz-Galiano, M.C., García-Cumbreras, M.A., Martín-Valdivia, M.T., Ureña López, L.A., Montejo-Ráez, A.: SINAI at ImageCLEFmed 2008. In: On-line Working Notes, CLEF 2008 (2008)
10. Navarro, S., Muñoz, R., Llopis, F.: A Multimodal Approach to the Medical Retrieval Task using IR-n. In: On-line Working Notes, CLEF 2008 (2008)
11. Díaz-Galiano, M.C., García-Cumbreras, M.A., Martín-Valdivia, M.T., Montejo-Raez, A., Ureña-López, L.A.: Integrating MeSH Ontology to Improve Medical Information Retrieval. In: Peters, C., Jijkoun, V., Mandl, T., Müller, H., Oard, D.W., Peñas, A., Petras, V., Santos, D. (eds.) CLEF 2007. LNCS, vol. 5152, pp. 601–606. Springer, Heidelberg (2008)
12. Aronson, A.R.: Effective Mapping of Biomedical text to the UMLS Metathesaurus: the Meta Map Program. In: Proc. of the AMIA Symposium, Washington, DC, November 3-7, pp. 17–21 (2001)
13. Kalpathy-Cramera, J., Hersh, W.: Automatic Image Modality Based Classification and Annotation to Improve Medical Image Retrieval. In: MEDINFO 2007: Proceedings of the 12th World Congress on Health (Medical) Informatics; Building Sustainable Health Systems, Brisbane, Australia, pp. 1334–1338 (2007)
14. Deselaers, T., Keysers, D., Ney, H.: Features for Image Retrieval: An Experimental Comparison. Information Retrieval 11(77), 77–107 (2008)

Image Retrieval by Inter-media Fusion and Pseudo-relevance Feedback

Osama El Demerdash, Leila Kosseim, and Sabine Bergler

CLaC Laboratory
Department of Computer Science & Software Engineering
Concordia University
{osama_el,kosseim,bergler}@cse.concordia.ca

Abstract. This paper presents our participation at the ImageCLEF Photo 2008 task. We submitted six runs, experimenting with our own block-based visual retrieval as well as with query expansion. The results we obtained show that despite the poor performance of the visual and text retrieval components, better results can be obtained through pseudo-relevance feedback and the inter-media fusion of the results.

1 Introduction

This paper presents our participation at the ImageCLEFPhoto 2008 task. ImageCLEFPhoto is a benchmark for open-domain image retrieval involving 39 queries on 20,000 images in 2008. A full description of the track as well as the data used can be found in [1]. We approach photo image retrieval in a hybrid fashion. The text portions of the image annotations are very short and do not give convincing retrieval results in isolation. Given the low performance of image retrieval algorithms and the maturity of text retrieval, we combine the two in a hybrid system. Mixed systems have been successful in the past ImageCLEFPhoto editions. We demonstrate here that combining two low-performing algorithms with relevance feedback results in an ensemble with much better performance.

This is our second participation in imageCLEFPhoto (see [2]). We used a new image retrieval system, as well as applied new methods for data fusion. While the 2008 task introduced a focused clustering theme for the first time, we did not attempt to use the cluster target information. The main purpose of our experiments was to maximize the Mean Average Precision (MAP) of the results. Therefore, we did not attempt the clustering task. We submitted six runs experimenting with our own block-based visual retrieval as well as with query expansion.

Our resources comprised a text search engine and a content-based retrieval system that we developed. The results we obtained show that despite the poor performance of the visual and text retrieval components, better results can be obtained through pseudo-relevance feedback and the inter-media fusion of the results.

C. Peters et al. (Eds.): CLEF 2008, LNCS 5706, pp. 605–611, 2009.

2 Text and Visual Retrieval

For text retrieval, we used the Apache Lucene engine [3], a Java-based text search engine available at http://lucene.apache.org. Stemming was done using the Snow-ball stemmer, based on the Porter algorithm and available separately under the BSD license at http://snowball.tartarus.org. For image analysis, we developed our own system utilizing the Java Advanced Imaging (JAI) API v.1.1.3, which provides a set of object-oriented interfaces supporting a high-level programming model for image manipulation. JAI is available at http://www.java.net.

2.1 Text Retrieval

As mentioned earlier, for text retrieval, we used the Apache Lucene engine, which implements a TF-IDF paradigm. Stop-words were removed and the rest of the terms were stemmed using the Snow-ball stemmer. The documents were then indexed as *field data* retaining only the *title*, *notes* and *location* fields, all of which were concatenated into a single field. All text query terms were joined using the *OR* operator in order to increase recall. This year's collection was also annotated with a *description* field. However, we inadvertently skipped this field by reusing the index from the previous year which did not include it. As will be illustrated in section 4, not using the *description* field had a significant negative impact on the precision. Indeed, the text index we used contained only 7577 terms of the 28,087 officially provided ones. Results obtained with the *description* field are labeled in this paper as *unofficial* results.

When searching the text, the query was also stemmed and stop words were removed. We found that using the first sentence of the *narrative* field which expresses positive examples, in addition to the title field, improves the result (see section 4). By contrast, the rest of the narrative, like negative examples, needs semantic processing to avoid introducing noise. Hence, we only used the first sentence of the *narrative* field.

2.2 Visual Retrieval

For visual retrieval, we implemented a system based on unsupervised analysis of the image. We sought to capture basic global and local color, texture and shape information utilizing a block-based method. To achieve this, we employed block-based techniques, which have been used extensively in image retrieval. Examples can be found in [4] and [5]. These techniques are based on partitioning the image into blocks, then performing feature extraction on each block independently. This approach is suitable to general, non domain-specific photographic databases, such as the one used in ImageCLEFPhoto, where there is not enough information to correctly and meaningfully segment the images.

In order to capture different levels of detail, we divided the image into 2X2, 3X3, 4X4 and 5X5 blocks yielding 4, 9, 16 and 25 equal partitions respectively. Using finer granularity for partitioning is possible and might slightly improve the precision, although at the cost of execution time and storage space. We also used the image as a whole, as well as a center block occupying half the image dimensions. Figure 1 shows the different regional divisions used to analyze the image. The image was first converted to the Intensity/Hue/Saturation (IHS) color space, a perceptual color space which is more intuitive and reflective of human color perception than the RGB color space, then the following features were extracted:

- A three-band color histogram for each of the image divisions
- A histogram of the grey-level image
- A histogram of the gradient magnitude image for each of the divisions of the grey-level image
- A three-band color histogram of the thumbnail of the image

The first feature captures the color characteristics of the image, while the grey-level histogram conveys some texture information. The gradient magnitude adds the outline of the shapes in the image. Finally, the thumbnail represents a visual summary of the image.

For retrieval, the different partitions are compared to their counter parts in the query images. Although this simple method does not account for translations and rotations in the image, we found it a reasonable choice, especially in the case of outdoor images which account for a significant proportion of the data. We did not resort to assigning different weights to features to avoid over-fitting the data. After experimenting with several measures including the Euclidean and the Mahalanobis distances, the Manhattan distance (L^1 Norm) was chosen as the distance measure. Since all features were represented as histograms with the same number of bins (256), no normalization was necessary. The images in the

Fig. 1. Partitioning the Image for Visual Retrieval

database were ranked according to their highest proximity to any of the three query images.

3 Query Expansion and Fusion of the Results

The next step in processing the query is the text query expansion involving a pseudo-relevance feedback mechanism and the fusion of the text and visual search results. Figure 2 shows an overview of the method we used. First the visual query is executed, then the highest results obtained are used to expand the text query. A late fusion is performed on the results obtained from both engines. [6] investigated pseudo-relevance feedback and fusion methods on the same dataset used in our experiments, the IAPR TC-12 collection. They reported a precision gain with feedback but not with fusion.

3.1 Query Expansion

We experimented with several ways of query expansion:

1 The highest ranked n results from the text search engine were passed as additional example images to the visual search. We experimented with values of n from 1-5.
2 All terms in the metadata of the highest ranked visual results were added to the text query.
3 The highest ranked text search results were used to expand the text query.
4 Noun synonyms from WordNet were added to the query.

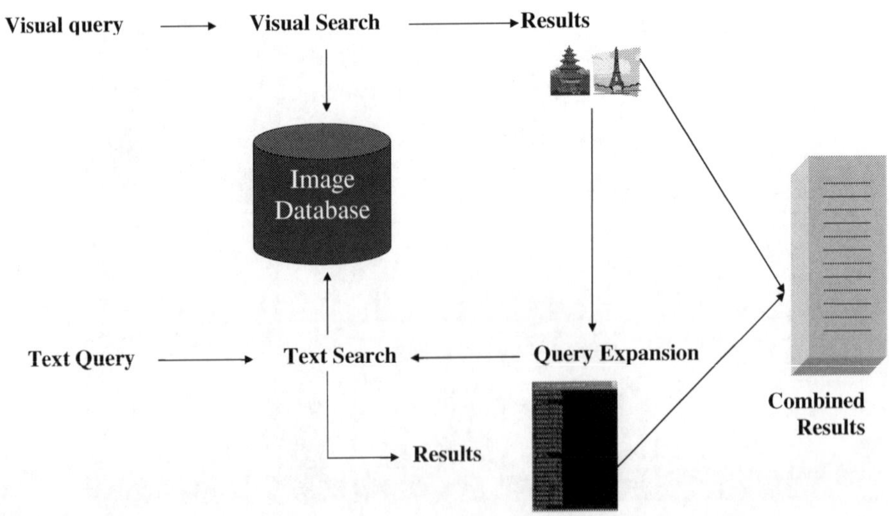

Fig. 2. Overview of Our System

Our best run uses pseudo-relevance feedback for query expansion using the second method listed above, as will be shown in section 4. While the MAP of the visual only run is only 0.055, its precision at five retrieved documents (0.328) is significantly higher than that of the text only run (0.236). For this reason, we use the highest ranked document for expansion of the text query. This is only done if the document meets a confidence level that we determined empirically. The confidence score is assigned based on the proximity score to the query image.

3.2 Fusion of Image and Text Search Results

To combine the results from the two media searches, we again took into consideration the confidence level in the visual results (i.e. the level of proximity from the query images). A maximum of three highest ranked images is taken from the visual query result depending on the confidence score, followed by the text results after query expansion. This simple re-ranking method only improved a little on the run that utilized only pseudo-relevance feedback in the official results. When adding the *description* field, it lowered considerably the precision. We deduce from this result that post-fusion methods could be useful on top of pseudo-relevance feedback in case of the availability of little textual data or text retrieval with low-precision.

4 Results

We submitted six runs at ImageCLEFPhoto 2008. In addition, we later computed corrected results taking into account the *description* field:

1 clacTX: Uses text search only on *title* field
2 clacIR: Uses only visual search
3 clacTxNr: Combines text search on *title* field and the first sentence of *narrative* field with the text from the first result of the visual search(Pseudo-relevance feedback)
4 clacIRTX: Combines the results from clacTxNR and clacIR
5 clacNoQE: Uses text search on *title* and *narrative* fields
6 clacNoQEMX: Same as clacNoQE combined with clacIR
7 clac(*unofficial* run): same as clacTxNR but including the *description* field

Table 1 shows the results we obtained in comparison to the mean, median and best runs of the track, taken from the best four runs from each participating group (25 groups and 100 runs in total) . As we expected, we obtained our highest Mean Average Precision (MAP) for the runs that utilized the maximum of our resources and methods.

Despite the weak results of the visual-only run (clacIR), we believe that the block-based method we used was appropriate for the data set and that the low MAP score is due to the simple features chosen. We made a conscious trade-off between precision and execution time.

Figure 3 shows the break down of the *MAP* by topic for five of the official six runs we submitted sorted by the precision per topic of our best run IRTX.

Table 1. Results at ImageCLEFPhoto 2008

Run ID	Modality	**MAP**	P10	P20	P30	GMAP	Rel	F-measure
clacTX	Text	0.1201	0.1872	0.1487	0.1462	0.019	1155	0.1741
clacTxNr	Text	0.2577	0.4103	0.3449	0.3085	0.1081	1859	0.3290
clacIR	Visual	0.0552	0.2282	0.1615	0.1214	0.0268	629	0.1877
clacIRTX	Mixed	0.2622	0.4359	0.3744	0.3308	0.1551	1630	0.3546
clacNoQE	Mixed	0.2034	0.3205	0.2705	0.2487	0.078	1701	0.2875
clacNoQEMX	Mixed	0.218	0.4026	0.3269	0.2855	0.129	1546	0.3384
clacUnofficial	Mixed	**0.3419**	0.5051	0.4256	0.3726	0.1794	2401	
ImageCLEF best three runs/team								
Average run	N/A	0.2187		0.3203				
Median run	N/A	0.2096		0.3203				
Best run(Manual)	N/A	0.4288		0.6962				
Best run(Automatic)	N/A	0.4105		0.5731				

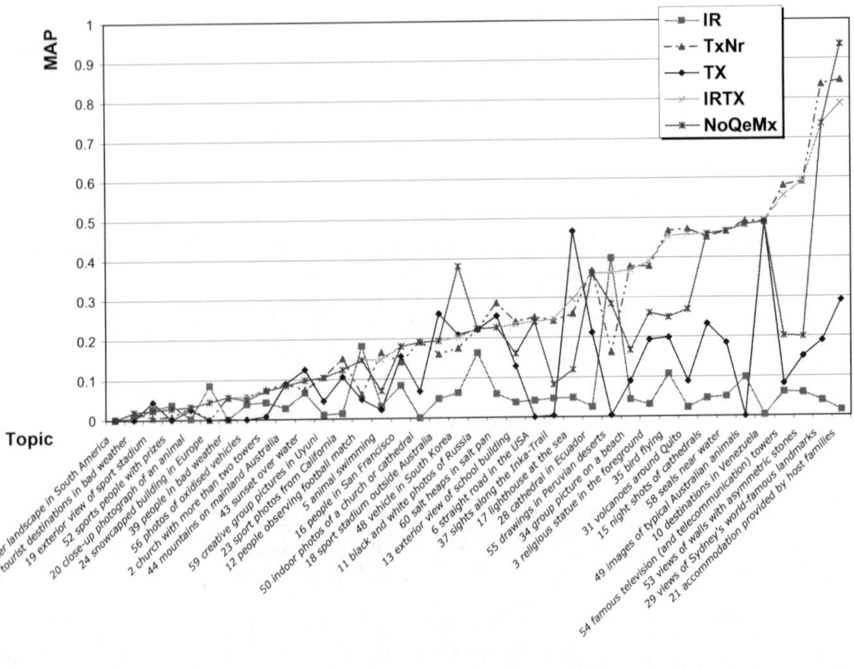

Fig. 3. Comparison of Runs by MAP of Each Topic

We note from this figure that the runs with feedback, (TxNR) and (IRTX), performed consistently better than the single media runs, (TX) and (IR), as well as the combined run without feedback (NoQEMX), except in cases where there was a significant divergence between the visual and text search results. The relevance feedback mechanism tends to average between these diverging results.

5 Conclusion

We have illustrated on ImageCLEFPhoto 2008 data how the use of simple, lightweight, low-cost and relatively lower-precision retrieval systems can be significantly improved through the use of pseudo-relevance feedback. We believe that there was little correlation between the visual descriptors we chose and the annotation of the images. It is possible in this case to have confidence in the top results only. A supervised-training-based visual search system would likely have a much higher correlation. While this leads to higher precision, more overlap with the text results would render pseudo-relevance feedback less useful. In the future, we intend to test this system on another data set of ImageCLEF which is the wikipedia multimedia task set [7].

References

1. Arni, T., Clough, P., Sanderson, M., Grubinger, M.: Overview of the ImageCLEFphoto 2008 photographic retrieval task. In: Peters, C., et al. (eds.) CLEF 2008. LNCS, vol. 5706, pp. 500–511. Springer, Heidelberg (2009)
2. El Demerdash, O., Kosseim, L., Bergler, S.: Text-based clustering of the imageclef-photo collection for augmenting the retrieved results, pp. 562–568 (2008)
3. Hatcher, E., Gospodnetic, O.: Lucene in Action (In Action series). Manning Publications Co., Greenwich (2004)
4. Han, J.H., Huang, D.S.: A novel BP-Based Image Retrieval System. In: International Symposium on Circuits and Systems (ISCAS 2005), Kobe, Japan, May 23-26, pp. 1557–1560. IEEE, Los Alamitos (2005)
5. Takala, V., Ahonen, T., Pietikäinen, M.: Block-based methods for image retrieval using local binary patterns. In: Kalviainen, H., Parkkinen, J., Kaarna, A. (eds.) SCIA 2005. LNCS, vol. 3540, pp. 882–891. Springer, Heidelberg (2005)
6. Maillot, N., Chevallet, J.P., Lim, J.H.: Inter-media pseudo-relevance feedback application to imageclef 2006 photo retrieval. In: Peters, C., Clough, P., Gey, F.C., Karlgren, J., Magnini, B., Oard, D.W., de Rijke, M., Stempfhuber, M. (eds.) CLEF 2006. LNCS, vol. 4730, pp. 735–738. Springer, Heidelberg (2007)
7. Tsikrika, T., Kludas, J.: Overview of the wikipediaMM task at ImageCLEF 2008. In: Peters, C., et al. (eds.) CLEF 2008. LNCS, vol. 5706, pp. 539–550. Springer, Heidelberg (2008)

Increasing Precision and Diversity in Photo Retrieval by Result Fusion

Yih-Chen Chang and Hsin-Hsi Chen*

Department of Computer Science and Information Engineering
National Taiwan University
Taipei, Taiwan
ycchang@nlg.csie.ntu.edu.tw, hhchen@csie.ntu.edu.tw

Abstract. This paper considers the strategies of query expansion, relevance feedback and result fusion to increase both precision and diversity in photo retrieval. In the text-based retrieval only experiments, the run with query expansion has better MAP and P20 than that without query expansion, and only has 0.85% decrease in CR20. Although relevance feedback run increases both MAP and P20, its CR20 decreases 10.18% compared with the non-feedback run. It shows that relevance feedback brings in relevant but similar images, thus diversity may be decreased. The run with both query expansion and relevance feedback is the best in the four text-based runs. Its F1-measure is 0.2791, which has 20.8% increase to the baseline model. In the content-based retrieval only experiments, the run without feedback outperforms the run with feedback. The latter has 10.84%, 9.13%, 20.46%, and 16.7% performance decrease in MAP, P20, CR20, and F1-measure. In the fusion experiment, integrating text-based and content-based retrieval not only reports more relevant images, but also more diverse ones. Its F1-measure is 0.3189.

1 Introduction

In the photo retrieval task of ImageCLEF 2008 [1], the focus has shifted from cross language image retrieval [2][3][4][5] to promoting diversity. Besides precision, retrieving diverse items representing different subtopics is also of concern. How to balance precision and diversity is challenging.

These papers [6][7] explore the uses of both textual information and visual features for cross-language image retrieval. The trans-media dictionary approach [6] shows that textual and low-level visual features have different semantic levels. Textual features are highly semantic, while low-level visual features are less semantic and are more emotive. These two types of features are complementary and provide different aspects of information about images. Chen and Chang [7] transform a query in one medium into a query in another medium by referencing to an aligned trans-media corpus. From the counterpart of results of an initial retrieval, they generate a new query in different medium. Such a media-mapping approach derives image query to capture extra information other than text query does, and generates an expanded text

* Corresponding author.

C. Peters et al. (Eds.): CLEF 2008, LNCS 5706, pp. 612–619, 2009.

query for more reliable text-based retrieval. We will examine their effects on balancing precision and diversity.

Arni, et al. [1] employ F1-measure in terms of precision and cluster recall to evaluate the performance of photographic retrieval in ImageCLEF 2008. Clarke et al. [8] deal with evaluation of novelty and diversity in information retrieval. The measure α-nDCG (α-Normalized Discounted Cumulative Gain) is proposed. We adopt the measure in [1] to evaluate the methods proposed in this paper.

This paper studies the strategies of query expansion and relevance feedback in text-based and content-based retrieval, and shows how to merge the results of text and image queries to increase both precision and diversity. It is organized as follows. Sections 2 presents text-based retrieval, content-based retrieval and a combination of both methods. Section 3 shows the runs submmited for formal evaluation in the photo retrieval task, and discsses the effects of different retrieval and fusion strategies. Section 4 provides a summary of the paper.

2 Three Retrieval Models

In this section, we will present three retrieval methods, including text-based retrieval method, content-based retrieval method, and a combination of both methods, to deal with balancing precision and diversity in photo retrieval.

2.1 Text-Based Retrieval

In text-based retrieval, we consider the strategies of query expansion and relevance feedback. Assuming the text corpus T is composed of n terms, t_1, t_2, ..., t_n, and a query Q contains m query terms, q_1, q_2, ..., q_m. We expand Q in the following way:

(1) For each corpus term t_i and query term q_j, compute $P(q_j \mid t_i)=P(t_i,q_j)/ P(t_i)$.

(2) For each corpus term t_i, compute $OverlapNum(t_i,Q)$ defined below.

$$OverlapNum(t_i,Q) = cardinality\{q \mid q \in Q, P(q \mid t_i)>0\}$$

(3) For all $t_i \in T$, if

$$\sum_{j=1}^{m} P(q_j \mid t_i) \times OverlapNum(t_i,Q) > thd$$

then t_i will be added to new query Q'. In the experiments, thd is set to 1. In other words, the original query terms which also appear in the corpus will be added to Q'.

We adopt Lemur as our text IR system. During indexing, stop words are removed; the remaining words are converted to root forms by using Porter stemmer; and the weighting function is BM25 with parameters (K1=1.2, B=0.75, K3=7). For relevance feedback, we select the top-10 terms of the highest BM25 scores from the top-5 retrieved documents, and add them to the query. The expanded terms have 1/2 weight of the original query terms.

2.2 Content-Based Retrieval

In content-based retrieval, we adopt a simple approach. For each image g_i, we extract two kinds of features: *SizeFeature*(g_i) and *ColorFeature*(g_i). *SizeFeature* is the size of an image. We postulate that portrait images tend to be vertical photos. In other words, the height of a portrait is larger than its width. Basically this feature is used to detect lanscape versus portrait images. We will assign higher weight to those query and target images having the same size. Besides the size of an image, we also consider its color. We divide an image into equal number of blocks, compute their RGB values, and count the number of blocks with similar colors.

SizeFeature(g_i) and *ColorFeature*(g_i) are defined below.

(1) *SizeFeature*(g_i) = 0, if *height*(g_i)>*width*(g_i)

= 1, if *height*(g_i)<=*width*(g_i)

(2) *ColorFeature*(g_i): divide g_i into 32×32 blocks, and extract their RGB values.

The similarity of two images, g_i and g_j, is computed as follows.

(1) Compute the color similarity of g_i and g_j based on their color features. *ColorSimilar*(g_i, g_j) = number of blocks in g_i and g_j, whose R, G and B value differences are not larger than 10.
(2) Compute the size similarity of g_i and g_j based on their size features. If *SizeFeature*(g_i) and *SizeFeature*(g_j) are the same, then *SizeSimilar*(g_i, g_j)=1.5. Otherwise, *SizeSimilar*(g_i, g_j)=1.0.
(3) The similarity of g_i and g_j is in terms of *SizeSimilar* and *ColorSimilar* as follows:

$$Similar(g_i, g_j) = SizeSimilar(g_i, g_j) \times ColorSimilar(g_i, g_j)$$

The collection of images along with their text descriptions forms an image-text-aligned trans-media corpus in the media mapping approach [8, 9]. In the initial content-based image retrieval, we compute the similarities of the query images and all the images in the data set with the method specified above, and select the most similar image for media mapping. That is, the corresponding text description of the reported image is regarded as a text query for further retrieval. To avoid noise introduced by media mapping in our simple content-based retrieval, only the top-1 retrieved image is considered.

2.3 Combining Text-Based and Content-Based Retrieval

We postulate that text-based retrieval and content-based retrieval have their own specific features and may contribute images from different view points. Therefore, the results of text-based and content-based retrieval are merged. Several heuristic merging methods like raw scoring merging, round-robin merging, normalized-scoring, and so on, have been proposed for multilingual information retrieval to fuse with the results lists retrieved from collections of different languages [10].

Round robin merging interleaves the retrieved documents by their ranks only to produce a result list. Normalized-score merging uses the score of the top one document to normalize the other documents in the same list, and then sorts the normalized scores to obtain the final list. We adopt the normalized-by-top-1 approach in this paper. At first, we normalize the scores of the result lists of text-based and content-based retrieval by the corresponding top-1 scores, in other words, the normalized scores will be within 0 and 1, and merge the lists with the same weights by their normalized scores. The same image is proposed only once in the final result list to avoid duplication.

3 Experiments and Discussion

We submit 7 runs shown below for the formal evaluation:

(1) NTU-EN-EN-AUTO-NOFB-TXT
This run is a baseline. We employ Lemur for text-based retrieval without query expansion and relevance feedback.

(2) NTU-EN-EN-AUTO-FB-TXT
This run employs Lemur for text-based retrieval with relevance feedback.

(3) NTU-EN-EN-AUTO-QE-NOFB-TXT
This run employs Lemur for text-based retrieval with query expansion.

(4) NTU-EN-EN-AUTO-QE-FB-TXT
This run employs Lemur for text-based retrieval with query expansion and relevance feedback.

(5) NTU-IMG-EN-AUTO-NOFB-TXTIMG
This run employs content-based retrieval first, then adopts media mapping to transform the image query to text query, and employs Lemur for text-based retrieval without relevance feedback.

(6) NTU-IMG-EN-AUTO-FB-TXTIMG
This run is similar to NTU-IMG-EN-AUTO-NOFB-TXTIMG except that relevance feedback is done.

(7) NTU-EN-EN-AUTO-QE-FB-TXTIMG
This run merges the results of NTU-IMG-EN-AUTO-FB-TXTIMG and NTU-EN-EN-AUTO-QE-FB-TXT.

The evaluation of the formal runs is based on mean average precision (MAP), precision at 20 (P20) and cluster recall at 20 (CR20), which calculates the percentage of different clusters represented in the top 20. The F1-measure in terms of P20 and CR20 is defined as follows.

$$\text{F1 - measure} = \frac{2 \times \text{P20} \times \text{CR20}}{\text{P20} + \text{CR20}}$$

Table 1 lists the experimental results of employing text query only. The run with query expansion has better MAP and P20 than that without query expansion, and only has 0.85% decrease in CR20. Although relevance feedback increases both MAP and P20 in EN-EN-AUTO-FB-TXT run, its CR20 decreases 10.18% compared with EN-EN-AUTO-NOFB-TXT. It shows that relevance feedback brings in relevant but similar images, thus diversity may be decreased. The run with both query expansion and relevance feedback is better than the other three runs. Compared with the baseline model, it has 33.79%, 44.44%, 0.27% and 20.80% increase in MAP, P20, CR20, and F1-measure, respectively.

Table 1. Comparisons of runs employing text query only

Runs	Feedback	Expansion	MAP	P20	CR20	F1
EN-EN-AUTO-NOFB-TXT	No	No	0.1790	0.2077	0.2602	0.2310
EN-EN-AUTO-QE-NOFB-TXT	No	Yes	0.1967 +9.88%	0.2244 +8.04%	0.2580 -0.85%	0.2400 +3.9%
EN-EN-AUTO-FB-TXT	Yes	No	0.2122 +18.54%	0.2692 29.61%	0.2337 -10.18%	0.2502 +8.3%
EN-EN-AUTO-QE-FB-TXT	Yes	Yes	0.2395 +33.79%	0.3000 +44.44%	0.2609 +0.27%	0.2791 +20.8%

Table 2 lists the experimental results of employing sample images. In the experiments, 3 example images are considered. The run without feedback outperforms the run with feedback. The latter has 10.84%, 9.13%, 20.46%, and 16.70% performance decrease in MAP, P20, CR20, and F1-measure, respectively.

Table 2. Comparisons of runs employing image query only

Runs	Feedback	MAP	P20	CR20	F1
IMG-EN-AUTO-NOFB-TXTIMG	No	0.2103	0.3090	0.1779	0.2258
IMG-EN-AUTO-FB-TXTIMG	Yes	0.1875 -10.84%	0.2808 -9.13%	0.1415 -20.46%	0.1882 -16.70%

The possible reason for the drop in precision is that the top-5 retrieved images for feedback may be very specific. This may introduce noises. For example, topic 43, *sunset over water*. The correct image should contain both *sunset* and *water*. In the top-5 retrieved images, only one contains both scenes, but all of them contain *sunset* scene. Figure 1 lists the top-5 images retrieved by these two methods.

Fig. 1. Top-5 returned images for runs without feedback (the 1st row) and with feedback (the 2nd row)

These images are the same, but retrieved in different order. Only the first image in the first row and the fourth image in the second row contain both concepts. The following shows the text descriptions of these 5 images.

(1) Image 1: the dark outlines of a mountain in the foreground; the sun is rising over the sea behind it; a light orange sky in the background.
(2) Image 2; the dark outlines of a mountain range in the foreground; dark grey and orange clouds, illuminated by the sun, in a light blue sky in the background.
(3) Image 3: a road and the dark outlines of trees in the foreground; the setting sun and an orange sky in the background.
(4) Image 4: the dark outline of a buggy in a dark, flat landscape in the foreground; the setting sun in dark orange sky in the background.
(5) Image5: the dark outline of trees and bushes in the foreground; the setting sun and an orange sky in the background.

In reality, only 5 contains both *sunset* and *water* in the top-20 images. There are 34 relevant images in the result list before feedback, and only 25 relevant images after feedback. This is due to that the occurrences of *sunset* increase the weight of the images related to *sunset*. The MAP decreases from 0.1776 to 0.0535 after feedback. CR20 decreases more than MAP and P20. The F1-measure of the run with feedback is 16.70% decrease relative to the run without feedback. It shows that pure relevance feedback is harmful to diversity.

Table 3 compares the performance of employing text query only, image query only, and both. The fusion run, which achieves MAP 0.2809, P20 0.3769, CR20 0.2763 and F1-measure 0.3189, is the best of our 7 submitted runs in the formal evaluation. It shows that integrating text-based and content-based retrieval not only finds more relevant images, but also more diverse ones. It confirms our expectation of semantic diversity in the text and image queries.

Table 3. Comparisons of runs employing text query, image query and both

Run	Feedback	MAP	P20	CR20	F1
EN-EN-AUTO-QE-FB-TXT	Yes	0.2395	0.3000	0.2609	0.2791
IMG-EN-AUTO-FB-TXTIMG	Yes	0.1875	0.2808	0.1415	0.1882
EN-EN-AUTO-QE-FB-TXTIMG	Yes	0.2809	0.3769	0.2763	0.3189

Compared Table 3 with the official results by retrieval modality [1], we find that the F1 of our mixed method (0.3189) is above the mean (0.3034), the P20 of our text only, image only and mixed methods (i.e., 0.3000, 0.2808, and 0.3769) is also above the respective mean (i.e., 0.2431, 0.1625, and 0.2538), but the CR20 of all of our methods (i.e., 0.2609, 0.1415, and 0.2763) is below the mean (i.e., 0.3915, 0.2127, and 0.3998).

4 Conclusion

This paper considers query expansion, relevance feedback and result fusion to deal with precision and diversity in image retrieval. We adopt a media mapping approach to transform an image query to a text one. Query expansion is useful to increase the precsion in text-based retrieval, but has a small negative effect on diversity. Relevance feedback is harmful to diversity when this strategy is used independently of query expansion or image only strategy. Text-based and content-based retrievals have their own special capability, so that both precision and diversity are improved.

How to merge the retrieval results is crucial for balancing precision and diversity. A simple normalized-by-top-1 approach is adopted and then equal weighting for text and image retrieval is used for merge in this paper. Such a heuristic merging strategy does not reflect topic diversity efficiently, so that cluster recall of our methods in different retrieval modality is below the mean of the official results. A learning to merge model beyond heuristic approach [11] may be considered in future work.

Acknowledgments. Research of this paper was partially supported by Excellent Research Projects of National Taiwan University (97R0062-04).

References

1. Arni, T., Clough, P., Sanderson, M., Grubinger, M.: Overview of the ImageCLEFphoto 2008 Photographic Retrieval Task. In: Peters, C., et al. (eds.) CLEF 2008. LNCS, vol. 5706, pp. 500–511. Springer, Heidelberg (2009)
2. Clough, P., Sanderson, M., Müller, H.: The CLEF 2004 Cross-Language Image Retrieval Track. In: Peters, C., Clough, P., Gonzalo, J., Jones, G.J.F., Kluck, M., Magnini, B. (eds.) CLEF 2004. LNCS, vol. 3491, pp. 597–613. Springer, Heidelberg (2005)
3. Clough, P., Müller, H., Deselaers, T., Grubinger, M., Lehmann, T.M., Jensen, J., Hersh, W.: The CLEF 2005 Cross-Language Image Retrieval Track. In: Peters, C., Gey, F.C., Gonzalo, J., Müller, H., Jones, G.J.F., Kluck, M., Magnini, B., de Rijke, M., Giampiccolo, D. (eds.) CLEF 2005. LNCS, vol. 4022, pp. 535–557. Springer, Heidelberg (2006)
4. Clough, P., Grubinger, M., Deselaers, T., Hanbury, A., Müller, H.: Overview of the ImageCLEF 2006 Photographic Retrieval and Object Annotation Tasks. In: Peters, C., Clough, P., Gey, F.C., Karlgren, J., Magnini, B., Oard, D.W., de Rijke, M., Stempfhuber, M. (eds.) CLEF 2006. LNCS, vol. 4730, pp. 579–594. Springer, Heidelberg (2007)
5. Grubinger, M., Clough, P., Hanbury, A., Müller, H.: Overview of the ImageCLEFphoto 2007 Photographic Retrieval Task. In: Peters, C., Jijkoun, V., Mandl, T., Müller, H., Oard, D.W., Peñas, A., Petras, V., Santos, D. (eds.) CLEF 2007. LNCS, vol. 5152, pp. 433–444. Springer, Heidelberg (2008)
6. Lin, W.C., Chang, Y.C., Chen, H.H.: Integrating Textual and Visual Information for Cross-Language Image Retrieval: A Trans-Media Dictionary Approach. Information Processing and Management 43, 488–502 (2007)
7. Chen, H.H., Chang, Y.C.: Language Translation and Media Transformation in Cross-Language Image Retrieval. In: Sugimoto, S., Hunter, J., Rauber, A., Morishima, A. (eds.) ICADL 2006. LNCS, vol. 4312, pp. 350–359. Springer, Heidelberg (2006)
8. Clarke, C., Kolla, M., Cormack, G., Vechtomova, O., Ashkan, A., Büttcher, S., MacKinnon, I.: Novelty and Diversity in Information Retrieval Evaluation. In: 31st ACM SIGIR, Singapore, pp. 659–666 (2008)

9. Chang, Y.C., Chen, H.H.: Approaches of Using a Word-Image Ontology and an Annotated Image Corpus as Intermedia for Cross-Language Image Retrieval. In: Peters, C., Clough, P., Gey, F.C., Karlgren, J., Magnini, B., Oard, D.W., de Rijke, M., Stempfhuber, M. (eds.) CLEF 2006. LNCS, vol. 4730, pp. 625–632. Springer, Heidelberg (2007)
10. Lin, W.C., Chen, H.H.: Merging Results Using Predicted Retrieval Effectiveness. In: Peters, C., Gonzalo, J., Braschler, M., Kluck, M. (eds.) CLEF 2003. LNCS, vol. 3237, pp. 202–209. Springer, Heidelberg (2004)
11. Tsai, M.F., Wang, Y.T., Chen, H.H.: A Study of Learning a Merge Model for Multilingual Information Retrieval. In: 31st ACM SIGIR, Singapore, pp. 195–202 (2008)

Diversity in Image Retrieval: DCU at ImageCLEFPhoto 2008

Neil O'Hare[1], Peter Wilkins[1], Cathal Gurrin[1,2], Eamonn Newman[1], Gareth J.F. Jones[1], and Alan F. Smeaton[1,2]

[1] Centre For Digital Video Processing, Dublin City University, Ireland
[2] CLARITY: Centre for Sensor Web Technologies
nohare@computing.dcu.ie
http://www.cdvp.dcu.ie

Abstract. DCU participated in the ImageCLEF 2008 photo retrieval task, which aimed to evaluate diversity in Image Retrieval, submitting runs for both the English and Random language annotation conditions. Our approaches used text-based and image-based retrieval to give baseline runs, with the the highest-ranked images from these baseline runs clustered using K-Means clustering of the text annotations, with representative images from each cluster ranked for the final submission. For random language annotations, we compared results from translated runs with untranslated runs. Our results show that combining image and text outperforms text alone and image alone, both for general retrieval performance and for diversity. Our baseline image and text runs give our best overall balance between retrieval and diversity; indeed, our baseline text and image run was the 2nd best automatic run for ImageCLEF 2008 Photographic Retrieval task. We found that clustering consistently gives a large improvement in diversity performance over the baseline, unclustered results, while degrading retrieval performance. Pseudo relevance feedback consistently improved retrieval, but always at the cost of diversity. We also found that the diversity of untranslated random runs was quite close to that of translated random runs, indicating that for this dataset at least, if diversity is our main concern it may not be necessary to translate the image annotations.

Keywords: Content-Based Image Retrieval, Data Fusion, Clustering.

1 Introduction

The CLEF 2008 ImageCLEF photo retrieval task was concerned with evaluating diversity in image retrieval [1]. For our participation in this task we used standard text retrieval, with and without pseudo relevance feedback, and content-based image retrieval (CBIR) based on MPEG-7 low level visual features, and a combination of text retrieval and CBIR. K-Means clustering was run on the outputs from these retrieval approaches to create a more diverse set of images at the top of the result list. For cross-language information retrieval (i.e. random language runs), we classified documents as English or German, and then translated

C. Peters et al. (Eds.): CLEF 2008, LNCS 5706, pp. 620–627, 2009.

German documents to English using machine translation. We also submitted runs that did not translate the random language documents, to explore whether it is necessary to translate non-English annotations in order to achieve diversity.

The remainder of this paper is organised as follows: Section 2 outlines the approaches that we used for both retrieval and clustering and details our submitted runs; Section 3 gives our results, along with some preliminary analysis of them, and finally Section 4 concludes the paper.

2 System Description

Our approach for the ImageCLEF photo retrieval task can be broken down into 3 main phases, which are described in more detail below.

- **Retrieval.** We first use text-based and image-based retrieval algorithms to create a traditional ranked list of images ordered by relevance to the query.
- **Clustering.** To improve the diversity of the results, the images towards the top of the result list are clustered, which will give us groups of similar images.
- **Cluster Representative selection and Final Ranking.** The clusters are then ranked in order of relevance to the query, and one representative image from each cluster is output to the final result list.

2.1 Retrieval

Since the topic set for CLEF 2008 consists of a subset of 39 of the 60 topics used in ImageCLEFPhoto 2006 and 2007, we used the remaining 21 topics as a training set for system development. We used the retrieval ground truth for these topics to guide development of our baseline retrieval systems.

Text Retrieval. We index the Title, Description, Notes and Location fields from the annotation of each photo, and use these for text-based retrieval. The location field is matched to a world gazetteer based on freely available resources[1], expanding the Town and Country location information to Town, State/County, Country and Continent. We construct text queries using the Title and Narr fields from the topics; since the Narr field often includes information about non-relevant documents, we remove any sentences containing the phrase 'not relevant' from this field. We use the BM25 ranking algorithm [9], as implemented in the Terrier search engine [8], for text retrieval. For pseudo relevance feedback (PRF) we use the diversion from randomness approach [8], using the top 10 terms from the top 3 documents for query expansion.

For random annotation language runs the annotation documents were processed using TextCat[2], an implementation of the text categorization algorithm proposed by Cavnar & Trenkle [3]. After identifying the German documents, we translated them from German to English using Systran Version:3.0 Machine

[1] http://nhd.usgs.gov/gnis.html, http://earth-info.nga.mil/gns/html/index.html
[2] http://odur.let.rug.nl/vannoord/textcat

Translator[3]. The translated documents were then indexed by the search engine identically to the English documents.

We used 3 language conditions (English, translated random and untranslated random), with and without PRF, giving 6 distinct baseline text retrieval runs.

Image Retrieval. For content-based image retrieval we make use of the following six global visual features defined in the MPEG-7 specification [7]:

- **Scalable Colour (SC)** is a Haar transform encoded colour histogram defined in the HSV colour space.
- **Colour Structure (CS)** represents an image by both the colour distribution (similar to a colour histogram) and the local spatial structure of the colour.
- **Colour Layout (CL)** is a compact descriptor that captures the spatial layout of the representative colours on a grid superimposed on an image.
- **Colour Moments (CM)** is similar to Colour Layout, it divides an image into 4x4 subimages and for each subimage the mean and the variance on each LUV colour space component is computed.
- **Edge Histogram (EH)** represents the spatial distribution of edges in an image, with edges categorized as vertical, horizontal, 45 degrees diagonal, 135 degrees diagonal or non-directional.
- **Homogeneous Texture (HT)** is a quantitative representation consisting of the mean energy and the energy deviation from a set of frequency channels.

To create a visual query we take the topic images and extract the six Query-Terms from each (i.e. a representation of the image by each of the six features above). For each Query-Term we query its associated retrieval expert (i.e. visual index and ranking function) to produce a ranked list. The ranking metric for each feature is as specified by MPEG-7 and is typically a variation on Euclidian distance. For our experiments we kept the top 1000 results per Query-Term. Each ranked list was then weighted and the results from all ranked lists are normalized using MinMax [5], then linearly combined using CombSUM [5].

We used a query-dependent weighting scheme for combining visual experts using an approach that requires no training data. This approach is based on the observation that if one was to plot the normalized scores of an expert against that of scores of other experts used for a particular query, then the expert whose scores showed the greatest initial change tends to be the best performer for that query. While we acknowledge this observation is not universal, it has been shown empirically to improve retrieval performance [10]; we also used this technique for our participation in ImageCLEFPhoto 2007 [6].

So, if a topic has three query images for example, we will extract six features per image, resulting in the generation of 18 Query-Terms. Each of these is then queried against its respective retrieval expert to produce 18 ranked lists, each ranked list is then individually weighted and the lists linearly combined.

[3] http://www.systran.co.uk

Combination of Image and Text Retrieval. As with the combination of visual features, image and text results are combined using weighted CombSUM and MinMax normalisation [5]. Based on experiments on the set of 21 training topics we used global weights of 0.7 for text and 0.3 for image, as this outperformed the query-dependant weighting approach described in Section 2.1.

2.2 Clustering

The baseline retrieval results, whether text-based, image-based, or a combination of the two, are clustered into groups in an attempt to increase the diversity of the results. All of our clustering approaches use text information exclusively; we do not perform clustering on visual features. The topic description for the ImageCLEF Photo task in 2008 includes a 'cluster' tag, which defined what criteria would be used to create the ground truth for diversity evaluation [1]. To avoid confusion with the K-Means clustering algorithm that we use in this work, we will refer to this cluster tag as 'diversity criteria'. Since it was permitted to manually inspect this diversity criteria from the topic, we classified the diversity criteria into 3 categories: 'location', 'non-location' or 'general'. The 39 topics include 17 unique entries for this diversity criteria tag. After classifying them into the 3 categories, we use a different subset of the fields from the structured annotation as input into our text clustering algorithm, as follows:

- **Location:** Topics for which only the location field is used as input to the clustering algorithm, corresponding to the diversity criteria 'city', 'state', 'location', 'country', 'city national park' and 'venue'.
- **Non-location:** Topics for which the location field is ignored for clustering, corresponding to the diversity criteria 'animal', 'sport', 'bird', 'weather condition', 'vehicle type', 'composition' and 'group composition'.
- **General:** Topics for which all fields used for retrieval are also used for clustering: 'statue', 'venue', 'landmark', 'volcano' and 'tourist attraction'.

Apart from using a different subset of the annotation fields, each of these types is treated identically in our subsequent clustering. We also submitted runs that did not classify the diversity criteria and treated all topics the same. We use the K-Means clustering algorithm, as implemented in the Text Clustering Toolkit[4]. Using annotation fields from one of the 3 classes defined above, we take the top X documents from our baseline retrieval algorithms and cluster them using K-Means; we varied the parameter X in a number of runs, using values of 50, 100 and 150. We also varied k, the number of clusters, using 20, 30 and 40 clusters. An additional variant used the the Calinski-Harabasz index to automatically estimate the optimum number of clusters [2]. Since we are clustering a small number of documents (150 or less), the tf-idf weighting scheme may not have enough documents to calculate reliable inverse document frequency scores, so we use two separate approaches to term normalisation for clustering: term frequency (tf) and term frequency * inverse document frequency (tf-idf).

[4] http://mlg.ucd.ie/content/view/20/

2.3 Cluster Representative Selection and Final Ranking

Finally, we rank all clusters in order of relevance and select a representative image for each cluster for the final ranked list. We use the maximum individual image ranking score within the cluster as the overall cluster score, using the same maximum image as the cluster representative, and our final output is k images, corresponding to the most relevant image from each cluster.

2.4 Submitted Runs

We created 13 baseline retrieval runs as follows: 3 language conditions (English, translated and untranslated random) with and without pseudo relevance feedback; each of these 6 text-only baselines was combined with image retrieval to give 6 text-image baselines; additionally, we had 1 image-only baseline. These 13 baseline runs were used as input into clustering using a number of parameter variations, creating a number of different runs. The parameters were: X, the number of documents to cluster; k, the number of clusters; term normalisation method; diversity criteria classification. This gives a total of 48 variations of clustering for each baseline submission. We cluster the image-only baseline using each of the 3 language conditions, meaning we cluster 13 baselines plus two additional language variants for the image baseline, we have $15x48 = 720$ clustered runs and 13 baseline runs, giving a total of 733 runs submitted.

3 Results

Our results are summarised in Table 1. This shows our baseline unclustered text and text-image results along with the best clustered variation for each baseline. As one would expect, runs that combine text and image retrieval always give the best performance. It is noteworthy, however, that there is no tradeoff involved: general retrieval (measured by MAP or P@20) and diversity (CR@20) are both improved simultaneously by combining text and image. For English language retrieval with PRF but without clustering, for example, P@20 is improved from 0.405 to 0.476, and CR@20 is improved from 0.348 to 0.454, by combining text retrieval with image retrieval. Similar improvements can be observed for all comparable configurations when text retrieval and image retrieval are combined. This makes intuitive sense because these different modalities will retrieve different relevant documents, and so combining them will improve both retrieval and diversity. Since there is no tradeoff here, it suggests that one very effective way to improve diversity is to use evidence from independent modalities, and we expect that we could further improve our results by using automatically extracted visual concepts such as those extracted as part of the ImageCLEF Visual Concept Detection Task [4].

Using K-Means Clustering gives a large improvement in diversity, but this comes at the cost of degraded retrieval performance. Our best CR@20 score of 0.552 on English language text and image with clustering and without PRF, for example, gives a 21% improvement over the unclustered equivalent, but P@20

Table 1. DCU Results for ImageCLEFPhoto 2008

Language	Translated	Modality	Clustered	PRF	MAP	P@20	CR@20
English	-	Txt	No	No	0.312	0.376	0.407
English	-	Txt	No	Yes	**0.351**	**0.405**	0.348
English	-	Txt	Yes	No	0.070	0.232	**0.514**
English	-	Txt	Yes	Yes	0.092	0.294	0.50
English	-	TxtImg	No	No	0.352	0.463	0.455
English	-	TxtImg	No	Yes	**0.354**	**0.476**	0.454
English	-	TxtImg	Yes	No	0.095	0.265	**0.552**
English	-	TxtImg	Yes	Yes	0.097	0.262	0.525
Random	Yes	Txt	No	No	0.258	0.339	0.406
Random	Yes	Txt	No	Yes	**0.279**	**0.345**	0.353
Random	Yes	Txt	Yes	No	0.081	0.246	**0.472**
Random	Yes	Txt	Yes	Yes	0.073	0.231	0.464
Random	No	Txt	No	No	0.169	0.283	0.404
Random	No	Txt	No	Yes	**0.173**	**0.289**	0.381
Random	No	Txt	Yes	No	0.053	0.214	**0.488**
Random	No	Txt	Yes	Yes	0.059	0.209	0.473
Random	Yes	TxtImg	No	No	0.309	0.440	0.467
Random	Yes	TxtImg	No	Yes	**0.309**	**0.442**	0.453
Random	Yes	TxtImg	Yes	No	0.1063	0.332	**0.536**
Random	Yes	TxtImg	Yes	Yes	0.101	0.283	0.513
Random	No	TxtImg	No	No	**0.225**	**0.381**	0.455
Random	No	TxtImg	No	Yes	0.222	0.372	0.400
Random	No	TxtImg	Yes	No	0.081	0.264	**0.518**
Random	No	TxtImg	Yes	Yes	0.077	0.247	0.491

for this run falls by 43%% to 0.265; a similar tradeoff can be seen with all comparable clustered and unclustered runs.

While PRF leads to consistently better retrieval performance in terms of MAP and P@20, it also consistently harms diversity. Runs without feedback consistently perform better for CR@20, and this pattern can be observed both in clustered and unclustered runs. This result is not particularly surprising as PRF uses that top retrieved images to expand the query, meaning that the results will be dominated by images similar to these.

Comparing Random language runs with English language runs, the best text and image Random runs in terms of diversity perform quite close to the English runs, achieving a CR@20 score of 0.536, only a 3% decrease from best English score at 0.552, although this run is 7% worse than English for P@20. Comparing English text-only runs with Random text-only runs, the best Random runs for diversity are 5% worse for CR@20 and 15% worse for CR@20. The fact that this difference is much smaller for text and image retrieval shows that image retrieval can is particularly helpful for cross-lingual retrieval, where it can help to close the gap between mono-lingual retrieval and cross-lingual retrieval. Comparing the best overall (ie. F1-measure, P@20 and CR@20) translated Random runs, again the text and image unclustered run, at 0.4531 performs within 3% of the best English run for this measure (0.4647).

Our untranslated runs show that by essentially discarding 50% of the documents in the collection (although, for the text and image runs, some of these 'discarded' documents may be recovered if their image score is high enough), we can still maintain a similar level of diversity with a best CR@20 score of 0.518 for the clustered run, only 3% below the score achieved if we translate the annotation documents. The untranslated runs perform much more poorly for P@20 and MAP, but for scenarios where we consider diversity to be our main concern this result suggests that it is not necessary to translate non-English documents, particularly when we are combining text retrieval with image retrieval. It is unclear whether this is an effect of this particular test collection or if this conclusion would be valid in other scenarios.

Comparing our results with those of other participants [1], DCU had the 2nd best automatic run for the English language condition, with an F1-measure (P@20 and CR@20) score of 0.4647. This run is only 0.0003 behind the best automatic run, submitted by Xerox Research Centre Europe, a difference small enough to suggest there is no clear difference between our best run and the best overall run for the task in 2008. In fact, our system performs better in terms of diversity than their system (0.4542 CR@20 compared with 0.4262) and worse in terms of retrieval (0.5115 P@20 compared with 0.4756), so we would argue that our system would be preferable if the focus is on diversity. Due to the small number of submissions from other groups for the Random language condition, it is not possible to fruitfully compare our results with other participants.

While we submitted an order of magnitude more runs than any other participant it would be incorrect to suggest that we were simply searching for good runs among these submissions. The purpose of the multiple runs was to find good parameters for our clustered runs (48 runs per baseline); our best performing runs, however, were from unclustered, baseline runs. Also, we feel that by conducting this search of the clustering parameter space, we can be confident that the clustering parameters are close to optimal. This means that we can be confident that the comparison between clustered and unclustered runs is valid, and is not biased by a poor choice of clustering parameters.

4 Conclusions

Our paritipation in ImageCLEF Photo 2008 has allowed us to draw a number of conclusions about diversity in image retrieval. PRF improves performance for standard retrieval measures, but at the cost of less diversity. Clustering the results of the baseline retrieval algorithms gives a large improvement in diversity, while harming retrieval performance. Combining image with text retrieval gives a large improvement in diversity and retrieval over text alone, and we found this to be the most effective way of improving diversity. For cross-lingual information retrieval we have shown that it is possible to maintain diversity without translating the German annotations into English; our cross-lingual runs have also shown that using image retrieval in combination with text retrieval narrows the gap in performance between cross-lingual and mono-lingual information retrieval.

Acknowledgements

This work is supported by Science Foundation Ireland under grant number 07/CE/I1147.

References

1. Arni, T., Clough, P., Sanderson, M., Grubinger, M.: Overview of the ImageCLEFphoto 2008 Photographic Retrieval Task. In: Peters, C., et al. (eds.) CLEF 2008. LNCS, vol. 5706, pp. 500–511. Springer, Heidelberg (2009)
2. Calinski, T., Harabasz, J.: A Dendrite Method for Cluster Analysis. Communications in Statistica 3, 1–27 (1974)
3. Cavnar, W.B., Trenkle, J.M.: N-Gram-Based Text Categorization. In: Cavnar, W.B., Trenkle, J.M. (eds.) Proceedings of Third Annual Symposium on Document Analysis and Information Retrieval, Las Vegas, NV, April 11-13, pp. 161–175. UNLV Publications/Reprographic (1994)
4. Deselaers, T., Hanbury, A.: The Visual Concept Detection Task in ImageCLEF 2008. In: Peters, C., Giampiccol, D., Ferro, N., Petras, V., Gonzalo, J., Peñas, A., Deselaers, T., Mandl, T., Jones, G.J.F., Kurimo, M. (eds.) Evaluating Systems for Multilingual and Multimodal Information Access – 9th Workshop of the Cross-Language Evaluation Forum, Aarhus, Denmark. LNCS. Springer, Heidelberg (2008) (printed in, 2009)
5. Fox, E.A., Shaw, J.A.: Combination of Multiple Searches. In: Proceedings of the Third Text REtreival Conference (TREC 1994), Gaithersburg, MD, pp. 243–252 (1994)
6. Jarvelin, A., Wilkins, P., Adamek, T., Airio, E., Jones, G.J.F., Smeaton, A.F., Sormunen, E.: DCU and UTA at ImageCLEFPhoto 2007. In: Peters, C., Jijkoun, V., Mandl, T., Müller, H., Oard, D.W., Peñas, A., Petras, V., Santos, D. (eds.) CLEF 2007. LNCS, vol. 5152, pp. 530–537. Springer, Heidelberg (2008)
7. Manjunath, B.S., Salembier, P., Sikora, T. (eds.): Introduction to MPEG-7: Multimedia Content Description Language. Wiley, Chichester (2002)
8. Ounis, I., Lioma, C., Macdonald, C., Plachouras, V.: Research Directions in Terrier: a Search Engine for Advanced Retrieval on the Web. In: Novatica/UPGRADE Special Issue on Web Information Access (2007)
9. Robertson, S.E., Walker, S., Jones, S., Hancock-Beaulieu, M.M., Gatford, M.: Okapi at TREC-3. In: Proceedings of the Third Text Retrieval Conference (TREC-3), Gaithersburg, MD, pp. 109–126 (1995)
10. Wilkins, P., Ferguson, P., Smeaton, A.F.: Using Score Distributions for Querytime Fusion in Multimedia Retrieval. In: MIR 2006 - 8th ACM SIGMM International Workshop on Multimedia Information Retrieval, Santa Barbara, CA (2006)

Visual Affinity Propagation Improves Sub-topics Diversity without Loss of Precision in Web Photo Retrieval*

Hervé Glotin[1] and Zhong-Qiu Zhao[1,2]

[1] Systems & Information Sciences Lab., UMR CNRS 6168
& Univ. Sud Toulon-Var, France
glotin@univ-tln.fr, zhongqiuzhao@gmail.com
[2] Computer & Information School, Hefei Univ. of Technology, China

Abstract. This paper demonstrates that Affinity Propagation (AP) outperforms Kmeans for sub-topic clustering of web image retrieval. A SVM visual images retrieval system is built, and then clustering is performed on the results of each topic. Then we heighten the diversity of the 20 top results, by moving into the top the image with the lowest rank in each cluster. Using 45 dimensions Profile Entropy visual Features, we show for the 39 topics of the imageCLEF08 web image retrieval clustering campaign on 20K IAPR images, that the Cluster-Recall (CR) after AP is 13% better than the baseline without clustering, while the Precision stays almost the same. Moreover, CR and Precision without clustering are altered by Kmeans. We finally discuss that some high-level topics require text information for good CR, and that more discriminant visual features would also allow Precision enhancement after AP.

Keywords: Web Image Retrieval, Diversity, Clustering, Profile Entropy Features, Affinity Propagation, XML, SVM.

1 Introduction and Profile Entropy Features Extraction

ImageCLEF08 Photo Retrieval Task [5] takes both retrieval accuracy and diversity to evaluate image retrieval system. The considered 39 semantic topics are difficult to translate with low level visual features, however local features like interest points are too much costly for scaled search engine. Thus we use our global descriptor called "Profile Entropy Feature" (PEF). It is equal, for each color, to the entropy of the distribution of te pixels' projection (arithmetic mean in horizontal (resp. vertical) direction) [1]. Computed in 3 horizontal image subbands, PEF combines raw shape and texture informations. It is processed on an image of 250K pixels in 1/10 sec. on a Pentium IV. It includes mean and STD of the color of each image subbands, yielding to 45 dimensions. Our system using

* Work supported by the French National Agency of Research (ANR-06-MDCA-002) & Research Fund for the Dr. Program of Higher Education of China (200803591024). We thank P. Mulhem (LIG) for providing the training labels and Kmeans results.

C. Peters et al. (Eds.): CLEF 2008, LNCS 5706, pp. 628–631, 2009.

PEF had the 2nd rank in the web images retrieval campaing ImagEval[1] [3]; and a medium rank in the NIST TRECVIDEOO8 High Level Features task [2].

2 Affinity Propagation Clustering (AP)

We think that AP [7], may improve the image retrieval diversity because it is not initialization dependant, contrary to some other clustering like Kmeans. In AP, the similarity between the points i and k is $s(i,k) = -\|x_i - x_k\|^2$; but $s(k,k)$ is an input parameter, influencing the number of identified clusters. It is set [7] as the median of the input similarities (resulting in a moderate number of clusters), or to their minimum (resulting in a small one). Here we set the initialized value of $s(k,k)$ varying from $\min_{i,j} s(i,j)$ to $\max_{i,j} s(i,j)$, namely : $s(k,k) = \min_{i,j} s(i,j) + \alpha(\max_{i,j} s(i,j) - \min_{i,j} s(i,j))$, where $\alpha \in [0,1]$. In this paper we set α to get 20 clusters by topics, as required in the task.

3 Experiments

The IAPR-TC12 [6] data consists of 20K images associated with a set of XML text descriptions, location tags, title and date. The challenge proposes 39 query topics, each is defined by a set of three image queries and a XML block indicating the search target and the diversity criterion.

Nevertheless, we buid here a visual-only system (IMG type), without using any text information during testing (the bi-modal textuo-visual system version is in [4]).Thus we train several visual-only SVMs models, using positive images labeled after a Porter process [8]. Each model ranks the test images for each topic. Then we make 20 clusters using AP, or Kmeans (K=20, processed by [8]), on the top 1000 images for each of the 39 considered topics. We place the best image (the one with the lowest rank) of each cluster into the top 20 ranks, and we keep the rest of the list as before this clustering. Evaluation are based on two measures : Precision at 20 (P20) and instance recall at rank 20 (CR20=percentage of different clusters represented in the top 20) [5].

4 Results and Conclusion

Our "IMG" type retrieval system is one of the 5 best among the 25 teams[2], but using only 45 dimensions. Figure 1 shows that AP outperforms the Kmeans clustering : P20 is not affected by AP, while the CR20 increases using AP on the contrary using Kmeans. This may be due to the stability of AP [7].

Table 1 gives, for 4 topics, the P20 and CR20 after AP or of the system without clustering. AP doesn't improve CR20 for "Black & White photos of

[1] www.imageval.org

[2] DCU was the 1st in the official campaign with P20=0.237, CR20=0.325; http://imageclef.org/2008/results-photo

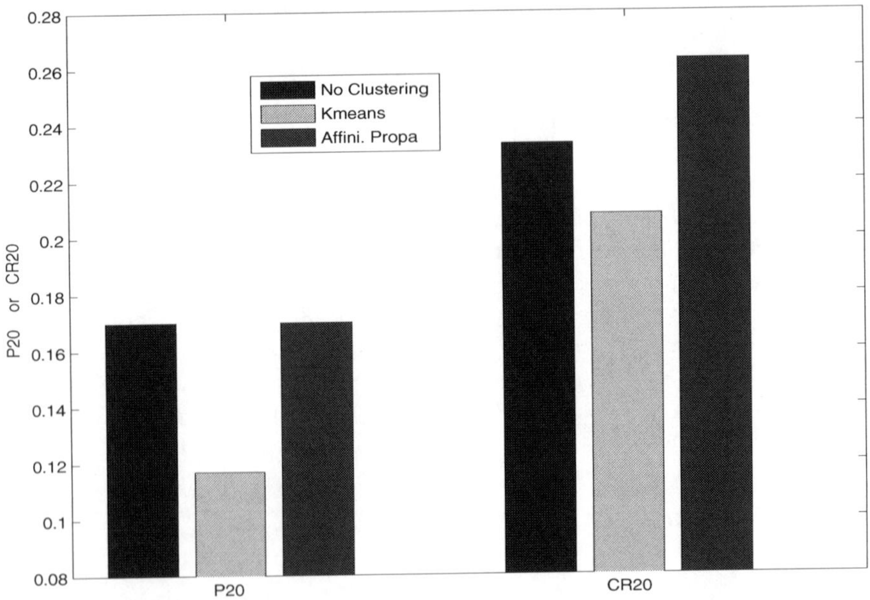

Fig. 1. Average on 39 topics of the P20 or CR20, without clustering ('No Clustering'), with Kmeans, or with AP. P20 is not affected by AP, while the CR20 significantly increases using AP, on the contrary when using Kmeans.

Table 1. Comparisons between AP and the without clustering system, for two topics where AP doesn't improve CR20, versus two easier topics where AP doubles it

Topic	P20 No Clustering = P20 AP	No Clustering CR20	AP CR20	relative gain using AP
Black & white photos of Russia	1.00	0.87	0.87	0 %
Sights along the Inka-Trail	0.50	0.37	0.37	0 %
Australian animals	0.25	0.16	0.34	+112 %
Seals near water	0.10	0.20	0.40	+100 %
Average on the 39 topics	0.17	0.23	0.26	+13 %

Russia", maybe because two third of the PEF are useless on B&W images, resulting in a too small visual space (15 dim) for visual clustering. The "Sights along the Inka Trail" is a too high-level topic : Clusters must show both natural and man-made sights along the Inka-Trail on the way to the ruins of Machu Picchu. This includes waterfalls as well as ruins, but not the Machu Picchu. It is clear then that text features are required. The other two topics are more simple, then AP generates good clusters. The "Australian animals" topic consists in clustering kangaroos, wallabies, koalas, wombats, quokkas, platypuses,... We can assume that AP performs well because their shape and biotop differ in textures and colors. For the topic "Seals near water", clusters must show seals at different body of water (sea, lake, etc.). Here again, the various water colors and textures are fundamental.

For any of the 39 topics, Kmeans generates always worst results than the without clustering system, which may be due to the fact that kmeans clustering is sentive to initialization. Remarkably, AP shows better CR20 performance than visual Kmeans (up to a factor 2 for some topics).

Moreover, we show that AP never reduces Precision, even using low cost PEF. We tested AP clustering on the 100 dimensions medium-level visual features computed in [10] : The relative gain against the baseline system is smaller than using PEF. Thus AP may be quickly affected by the curse of dimensionality. Further research will consist in getting more discriminant low dimensional features, such that P20 and CR20 could be both enhanced by visual-only clustering. Comparing visual only CR20 to other runs using text features (see [4] for TXT-IMG results, and AVEIR runs in [9]), we conclude that the text information is necessary for diversifying the high-level sub-topics . Thus, we will also try to estimate for which topic text features are necessary for promotting diversity. Actually, future image retrieval systems will tend to such joint maximization of P20 and CR20, getting few relevant images from each sub-topic cluster, and filling the top ranks with relevant but diverse images.

References

1. Glotin, H.: Robust Information Retrieval and Perception for a Scaled Lego-Audio-Video Multi-structuration. Pr. Habilitation Thesis, Univ. Toulon (2007)
2. Glotin, Zhao: LSIS TREC VIDEO 2008 High Level Feature Shot Segmentation using Compact Profil Entropy. In: NIST TRECVIDEO 2008 notebook (2008)
3. Tollari, Glotin: Learning Optimal Visual Features from Web Sampling in Online Image Retrieval. In: IEEE Conf. Acoustics Speech Signal Image Proc. (2008)
4. Glotin, Zhao: LSIS Imageclef Photo: combining text with entropic pixel features for texto-visual photo retrieval. CLEF keynotes (2008)
5. Thomas, A., Paul, C., Mark, S., Michael, G.: Overview of the ImageCLEFphoto 2008 Photo Retrieval Task Eval. Systems for Multilingual and Multimodal Information Access. In: 9th Wkp of the Cross-Language Eval. (2008)
6. Grubinger, Clough, Muller, Deselaers: The IAPR TC-12 benchmark: A new evaluation resource for visual information systems. In: Proc. OntoImage Language Resources for Content-Based Image Retrieval Wkp, with LREC (2006)
7. Frey, B., Dueck, D.: Clustering by Passing Messages Between Data Points. Science 315, 972–976 (2007)
8. Mulhem, et al.: LIG working notes on ImageCLEFphoto. CLEF keynotes (2008)
9. Tollari, S., Mulhem, P., Ferecatu, M., Glotin, H., Detyniecki, M., Gallinari, P., Sahbi, H., Zhao, Z.-Q.: A comparative study of diversity methods for different text and image retrieval approaches. In: Peters, C., et al. (eds.) CLEF 2008. LNCS, vol. 5706, pp. 585–592. Springer, Heidelberg (2009)
10. Ferecatu, M., Sahbi, H.: Bi-Modal Text and Image Retrieval with Diversity Enhancement. CLEF keynotes (2008)

Exploiting Term Co-occurrence for Enhancing Automated Image Annotation

Ainhoa Llorente[1,2], Simon Overell[3], Haiming Liu[1], Rui Hu[1], Adam Rae[1], Jianhan Zhu[4], Dawei Song[5], and Stefan Rüger[1]

[1] Knowledge Media Institute, The Open University
Walton Hall, Milton Keynes, MK7 6AA, United Kingdom
[2] INFOTECH Unit, ROBOTIKER-TECNALIA
Parque Tecnológico, Edificio 202, Zamudio, E-48170, Bizkaia, Spain
[3] Department of Computing, Imperial College London
London, SW7 2AZ, United Kingdom
[4] University College London
Adastral Campus, Ipswich, Suffolk, IP5 3RE, United Kingdom
[5] School of Computing, The Robert Gordon University
Andrew Street, Aberdeen, AB25 1HG, United Kingdom
{a.llorente,h.liu,r.hu,a.rae,s.rueger}@open.ac.uk
seo01@doc.ic.ac.uk,j.zhu@adastral.ucl.ac.uk,d.song@rgu.ac.uk

Abstract. This paper describes an application of statistical co-occurrence techniques that built on top of a probabilistic image annotation framework is able to increase the precision of an image annotation system. We observe that probabilistic image analysis by itself is not enough to describe the rich semantics of an image. Our hypothesis is that more accurate annotations can be produced by introducing additional knowledge in the form of statistical co-occurrence of terms. This is provided by the context of images that otherwise independent keyword generation would miss. We applied our algorithm to the dataset provided by ImageCLEF 2008 for the Visual Concept Detection Task (VCDT). Our algorithm not only obtained better results but also it appeared in the top quartile of all methods submitted in ImageCLEF 2008.

Keywords: automated image annotation, statistical co-occurrence, image analysis, semantic similarity.

1 Introduction

Most of the approaches in automated image annotation use machine learning techniques to learn statistical models from a training set of pre-annotated images and apply them to generate annotations for unseen images using visual feature extracting technology. Thus, given an image x, the *posterior probability distribution* $p(\omega|x)$ for each annotation keyword ω can be estimated under Bayes's law as:

$$p(\omega|x) = \frac{p(x|\omega).p(\omega)}{p(x)}, \tag{1}$$

C. Peters et al. (Eds.): CLEF 2008, LNCS 5706, pp. 632–639, 2009.

where $p(x|\omega)$ is the *conditional probability* of seeing the image x if the hypothesis ω happens to be true, $p(\omega)$ is the *prior probability* inferred before x is available and $p(x)$ is the *marginal probability* of x, the *a priori probability* of witnessing the new evidence x under all possible hypotheses.

Therefore, the ultimate goal of any generative automated annotation system is to model $p(x|\omega)$ for each annotation keyword. The resulting algorithms differ mainly on the way $p(x|\omega)$ is computed, the way the image x is treated, as a whole or as segments, how the feature information is extracted and on the model used for representing this information.

After analysing some examples of these probabilistic annotation systems we detect that sometimes keywords are generated independently from others, without considering that they represent objects that co-occur in the same scene. For example, let us consider an image of the North Pole that represents the figure of an animal surrounded by a landscape covered by ice and snow. Clearly, the depicted animal can not be a "camel", no matter how high the probability value associated with it. Consequently, having a basic understanding of the scene represented in an image, or at least a certain knowledge of other objects contained there, may actually help to recognise an object.

Our work attempts to overcome the limitations of the probabilistic annotation systems, that generate keywords independently, by applying statistical analysis techniques. The final goal is to increase the accuracy of annotations after removing keywords that are incoherent with the rest.

The rest of this paper is organised as follows: Section 2 introduces the baseline probabilistic annotation system. Section 3 discusses the two semantic similarity measures used. Section 4 explains our algorithm, while Section 5 provides results for the submitted runs. Finally, Section 6 contains our conclusions and plans for future work.

2 Non-parametric Density Estimation

The baseline system used is the probabilistic framework developed by Yavlinsky et al. [1] who use global features together with a non-parametric density estimation. The starting point of this approach is the Bayes rule defined in Equation 1 whereas the final goal is to model $f(x|\omega)$ for each annotation keyword. They used a non-parametric approach because the distributions of image features have irregular shapes that do not resemble *a priori* any simple parametric form. The function $f(x|\omega)$ is estimated following a kernel k based approach as represented in:

$$f(x|\omega) = \frac{1}{nC} \sum_{i=1}^{n} k \frac{x - x_{\omega}^{(i)}}{h}, \tag{2}$$

where $x_{\omega}^{(1)}$, $x_{\omega}^{(2)}$,..., $x_{\omega}^{(n)}$, is a sample of feature vectors from the training set labelled with the keyword ω and $x = (x_1, ..., x_d)$ is a vector of real-valued image features. Then, a d-dimensional Gaussian kernel is placed over each point $x^{(i)}$ as expressed in:

$$k_G(t;h) = \prod_{l=1}^{d} \frac{1}{\sqrt{2\pi h_l}} e^{-\frac{1}{2}(\frac{t_l}{h_l})^2}, \tag{3}$$

where h is the bandwidth of the kernel and $t = x - x^{(i)}$.

Once the model is determined, the annotation process can be described as follows. First, images are segmented into nine equal tiles, and then, low-level features are extracted. The feature vectors, CIELAB [2] colour and Tamura [3] texture, from the nine tiles are concatenated to form a global image feature vector. The next step is to extract the same feature information from an unseen image in order to compare it with all the previously created models. The result of this comparison yields a probability value of each keyword ω being present in each image, $p(\omega|x)$. The final annotations are obtained after selecting five of the keywords with the highest probability. Finally, the system performance is evaluated.

3 Semantic Similarity Measures

The context of the images is computed using statistical co-occurrence of pairs of keywords appearing together in the training set. This information is represented in the form of a co-occurrence matrix. The starting point for computing it is an image-keyword matrix A, where each row represents an image of the training set and each column a word of the vocabulary. Each cell indicates the presence or absence of a keyword in the image. The co-occurrence matrix B is obtained after multiplying the image-word matrix A by its transpose A^T. The resulting co-occurrence matrix ($B = A^T \cdot A$) is a symmetric matrix where each entry b_{jk} contains the number of times the keyword w_j co-occurs with the keyword w_k. The elements in the diagonal b_{jj} represent the number of images of the training set annotated by the keyword w_j.

The dataset provided by ImageCLEF 2008 [4] for the Visual Concept Detection Task (VCDT) comes with a vocabulary of 17 visual concepts with a hierarchical structure that represent general concepts such as "indoor", "outdoor", "person", "day", "night", "water", "road or pathway", "vegetation", "tree", "beach", "mountains", "buildings", "sky", "sunny", "partly cloudy", "overcast" and "animal". In the first level of the hierarchy, we find two general concepts like "indoor" and "outdoor", which are mutually exclusive while in lower levels more specific concepts subclasses of the previous ones appear. Some concepts may belong to more than one class, for instance, a "person" can be part of an "indoor" or an "outdoor" scene but others are mutually exclusive, e.g., a scene can not represent "day" and "night" at the same time.

Authors like Pedersen et al. [5] and more recently, Gracia and Mena [6] make a clear distinction between the term "semantic relatedness" and "semantic similarity" based on the fact that two concepts can be related without being similar. Therefore, "relatedness" should be seen as a more general notion than "similarity". At first glance, it seems we should adopt "semantic relatedness" measures as a way to estimate the distance between two words that define the context of an image. However, due to the fact that antagonism is considered a relation that may appear between two related terms and it is something that we would like to avoid by all means, we adopt finally "semantic similarity" measures.

Consequently, we refer to the definition of "semantic similarity" by Miller and Charles [7] who consider it as *the degree of contextual interchangeability or the degree to which one word can be replaced by another in a certain context.* As a result, two keywords are similar if they refer to entities that are likely to co-occur together like "mountains" and "vegetation", "beach" and "water", "buildings" and "road", etc. On the contrary, dissimilarity occurs between conflicting keywords such as "outdoor" and "indoor", "day" and "night", etc.

In our experiments, we apply two different semantic measures to determine the distance between two annotation keywords: one defined by Manning and Schütze and the other, the cosine distance.

The process starts by assigning a vector to each annotation keyword that represents all the co-occurrences of that keyword in the keyword-keyword space. Thus, the vector $\boldsymbol{w_j}$ corresponds to the row j of the co-occurrence matrix and it is represented by the following elements: $\boldsymbol{w_j} = \{b_{j1}, b_{j2}, ..., b_{jn}\}$ with n being the number of keywords in the vocabulary.

Manning and Schütze [8] propose that a good estimation of the semantic similarity between two keywords can be obtained after dividing each entry of the co-occurrence matrix by the norm of its associated vector as represented in Equation 4. In this way, the co-occurrence matrix is transformed into a conditional probability distribution. After performing several experiments, we finally select the Euclidean norm as opposed to the Manhattan because of the better performance obtained:

$$d(\omega_j, \omega_k) = \frac{b_{jk}}{\sqrt{b_{j1}^2 + b_{j2}^2 + ... + b_{jk}^2 + ... + b_{jn}^2}} \tag{4}$$

The other approach used, the cosine distance, is estimated as follows:

$$d(\omega_j, \omega_k) = \cos\theta = \frac{\boldsymbol{w_j}.\boldsymbol{w_k}}{|\boldsymbol{w_j}|.|\boldsymbol{w_k}|} \tag{5}$$

We consider that two keywords ω_j and ω_k are semantically similar if the distance $d(\omega_j, \omega_k)$ is greater than a parameter β. On the contrary, they are dissimilar if this value is lower than γ. The initial value of these parameters β and γ are obtained after taking into account some considerations derived from the linear algebra.

Thus, if the angle between two vectors is zero it means that the vectors are equal or parallel. This reflects the similarity case. The value of cosine of the angle between them is one. Then, the cosine starts decreasing, reaching its minimum value zero when the two vectors are perpendicular — dissimilarity case. The situation when the angle between the two vectors is 45° is considered the inflection point and corresponds to a cosine value of 0.70, which is the initial value of β and γ before adjusting them empirically. These parameters depend on the dataset provided for training.

4 Algorithm Description

The underlying idea of our algorithm is to detect incoherence among the annotation keywords. Once we have detected incoherence between two keywords, the probability of the keyword with the lowest probability will be lowered, as well as all the keywords that are semantically similar. The input for our algorithm are the top five keywords w_j and their associated probability $p(w_j|i_i)$ generated by the probabilistic framework described in Section 2.

Let *AnnoSet* be the annotations assigned to an image i_i ordered according to the decreasing probability:

$$\text{AnnoSet}(i_i)=\{(w_1,p(w_1|i_i));(w_2,p(w_2|i_i));(w_3,p(w_3|i_i));(w_4,p(w_4|i_i));(w_5,p(w_5|i_i))\}$$

A schema of the algorithm follows:

```
For each image i_i in testSet:
  if (max{p(w_j|i_i) with j = 1..5} > threshold α):
    for all pairs of keywords (w_j, w_k) in AnnoSet(i_i):
      if incoherence(w_j, w_k):
        lowerProbability(w_k)
        for each keyword w_l in vocabulary V:
          if not incoherence(w_k, w_l):
            lowerProbability(w_l)
Generate(AnnoSet(i_i))
```

Our algorithm only works with a selection of images from the test set for which the underlining system is *"confident enough"* i.e. at least one of the keywords has greater probability than a threshold α which is estimated empirically. Once an image i_i is selected the objective is to prune the keywords that are incoherent with the rest. The function $incoherence(w_j, w_k)$ with $j, k = 1...5$ will detect whether a pair of keywords are semantically dissimilar or not. We applied the semantic similarity measures of Section 3. If the system finds that the keywords w_j and w_k are incoherent, the function $lowerProbability(w_k)$ will lower the probability of the keyword associated to the lowest probability w_k. Furthermore, the probability of each keyword w_l semantically similar to w_k is also lowered. This is done in order to ensure that all words incoherent with the context are removed from the annotation set. After modifying the probability values of these keywords, the function $generate(AnnoSet(i_i))$ sorts the keywords according to their probability and by selecting the five highest, new and more precise annotations are produced.

5 Experimental Runs and Results

In Table 1, we present the results obtained for ImageCLEF 2008 dataset for three different algorithms, the baseline algorithm explained in Section 2, the enhanced version of Section 4 using the semantic similarity of Manning and Schütze (Enhanced 1) and the one based on the cosine distance (Enhanced 2).

Table 1. Performance of the algorithm under EER metric

Keyword	Baseline	Enhanced 1	Enhanced 2
indoor	0.1674	0.1830	0.1674
outdoor	0.4444	0.4444	0.4274
person	0.3297	0.3242	0.3333
day	0.4358	0.4022	0.4078
night	0.1426	0.1692	0.1426
water	0.3329	0.3278	0.3329
road_or_pathway	0.3507	0.3507	0.3532
vegetation	0.3098	0.2994	0.3056
tree	0.3188	0.3188	0.3188
mountains	0.2547	0.2477	0.2535
beach	0.2706	0.2706	0.2706
buildings	0.2722	0.2722	0.2740
sky	0.1672	0.1607	0.1607
sunny	0.2215	0.2215	0.2215
partly_cloudy	0.3093	0.3067	0.3067
overcast	0.2245	0.2219	0.2245
animal	0.3457	0.3457	0.3447

Table 2. Average performance under EER and MAP

Metric	Baseline	Enhanced 1	Enhanced 2
EER	0.2881	0.2863	0.2850
MAP	0.5885	0.5885	0.5911

The performance of our results is measured under the metric adopted by the ImageCLEF organisation based on ROC curves and under the ranked retrieval metric, mean average precision (MAP). ROC curves represent the fraction of true positives (TP) against the fraction of false positives (FP) in a binary classifier. The Equal Error Rate (EER) is the error rate at the threshold where FP=FN. Mean average precision measures the average precision, over all queries, at the ranks where recall changes where relevant items occur. The queries are the 17 keywords from the vocabulary. Notice in Table 2, where we have averaged the values, we obtain significantly better results with our enhanced algorithm when using the cosine similarity. Regarding our performance with respect to the rest of the ImageCLEF 2008 participants, we have represented in Figure 1 the % EER values corresponding to the best algorithm (in black), our best algorithm (in white) and the average value for all the participants (in grey). Our best performance is for "night", "sky" and "indoor", while for "outdoor" and "day" the results were lower than the average. It is worth noting that our results were in the top quartile of all methods submitted.

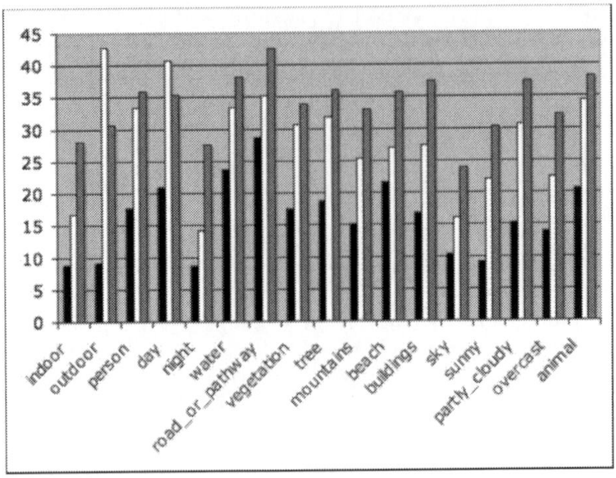

Fig. 1. Comparison with the rest of participants (%EER)

6 Conclusions and Future Work

The main goal of this work is to improve the accuracy of a traditional automated image annotation system based on a machine learning method. We have demonstrated that building a system that models the image context on top of another that is able to accomplish the initial identification of the objects increases significantly the precision of the automated annotation system.

Our algorithm has shown that modelling a scene using co-occurrence values between pairs of words and using this information appropriately helps to achieve better accuracy. Experiments have been carried out with the ImageCLEF 2008 dataset using two similarity measures. The best results were obtained with cosine similarity although we obtained statistically significant better results with the Corel 5k dataset [9] in the past.

It is worth pointing out that another participant of ImageCLEF 2008, the LIP6 group, has deployed a similar framework to ours [10]. The underlying probabilistic method they used is based on *Forest of Fuzzy Decision Trees.* They undertook a co-occurrence analysis in order to find the relationships among the annotation keywords and they implemented some "resolution rules" in order to resolve the conflicting annotations. They considered two types of relations between concepts: exclusion and implication. By exclusion they mean concepts that never appear together and by implication they refer to relationships between concepts. Despite the fact that their baseline algorithm performs quite well, the fourth best of all submitted runs, their results become worse after using a co-occurrence approach. However, their best performance is achieved when they use the exclusion and implication rules together.

Regarding future work, we want to improve the encouraging results shown in this paper by using different kinds of background knowledge such as WordNet, Wikipedia and Web search engines.

Acknowledgments. This work was partially funded by the Pharos project sponsored by the Commission of the European Communities as part of the Information Society Technologies programme under grant number IST-FP6-45035.

References

1. Yavlinsky, A., Schofield, E., Rüger, S.: Automated image annotation using global features and robust nonparametric density estimation. In: Proceedings of the International ACM Conference on Image and Video Retrieval, pp. 507–517 (2005)
2. Hanbury, A., Serra, J.: Mathematical morphology in the CIELAB space. Image Analysis & Stereology 21, 201–206 (2002)
3. Tamura, H., Mori, T., Yamawaki, T.: Textural features corresponding to visual perception. IEEE Transactions on Systems, Man and Cybernetics 8(6), 460–473 (1978)
4. Deselaers, T., Hanbury, A.: The visual concept detection task in ImageCLEF 2008. In: Peters, C., et al. (eds.) CLEF 2008. LNCS, vol. 5706, pp. 531–538. Springer, Heidelberg (2009)
5. Pedersen, Banerjee, Patwardhan: Maximizing semantic relatedness to perform word sense disambiguation. Technical report, University of Minnesota (2003)
6. Gracia, J., Mena, E.: Web-based measure of semantic relatedness. In: Bailey, J., Maier, D., Schewe, K.-D., Thalheim, B., Wang, X.S. (eds.) WISE 2008. LNCS, vol. 5175, pp. 136–150. Springer, Heidelberg (2008)
7. Miller, G.A., Charles, W.G.: Contextual correlates of semantic similarity. Journal of Language and Cognitive Processes 6, 1–28 (1991)
8. Manning, C.D., Schütze, H.: Foundations of statistical natural language processing. MIT Press, Cambridge (1999)
9. Llorente, A., Rüger, S.: Can a probabilistic image annotation system be improved using a co-occurrence approach? In: Workshop on Cross-Media Information Analysis, Extraction and Management at the 3rd International Conference on Semantic and Digital Media Technologies (2008)
10. Tollari, S., Detyniecki, M., Fakeri-Tabrizi, A., Amini, M.R., Gallinari, P.: UPMC/LIP6 at ImageCLEFphoto 2008: On the exploitation of visual concepts (VCDT). In: Evaluating Systems for Multilingual and Multimodal Information Access – 9th Workshop of the Cross-Language Evaluation Forum (2008)

Enhancing Visual Concept Detection by a Novel Matrix Modular Scheme on SVM*

Zhong-Qiu Zhao[1,2] and Hervé Glotin[1]

[1] Systems and Information Sciences Lab. (LSIS), UMR CNRS & Univ. Sud Toulon-Var, France
[2] Computer and Information School, Hefei Univ. of Technology, China
glotin@univ-tln.fr, zhongqiuzhao@gmail.com

Abstract. A novel Matrix Modular Support Vector Machine(MMSVM) classifier is proposed to partition a visual concept problem into many easier two-class problems.This MMSVM shows significant detection improvements on the ImageClef2008 VCDT task, with a relative reduction of 15% of the classification error, compared with usual SVMs.

1 Introduction

The huge amount of available video underlines the necessity of designing more and more efficient pattern classification schemes. Recent trends show increasing uses of statistical machine learning that provides a computational framework for mapping low level features to high level semantic concepts. In this paper we expose how to improve the performance of general classifiers on Visual Concept Detection Task (VCDT).

It has been proven that generalization capability can be improved using modular classifiers [1], which are composed of several classifiers and an integration machine. The voting machine has been improved with the new modular approach called the Matrix Modular (MM) classifier [2], where we consider smaller two-class subtasks between clusters of different classes and the classifiers are set as Neural Networks. In this paper, we extend MM-NN to MM-SVM for improving the performance on VCDT, where the classifiers are set as support vector machines (SVMs).

2 Matrix Modular Support Vector Machines

The Matrix Modular strategy divides the complex visual concept detection problem into several much easier subtasks : each visual concept space is divided into several clusters. The main steps to construct an MMSVM system for VCDT are presented below.

Task Decomposition. The detection for one visual concept among K concepts is a 2-class problem. The divide-and-conquer strategy is dividing it into $(K-1)$

* Work supported by French National Agency of Research ANR-06-MDCA-002, and Research Fund for the Doctoral Program of Higher Education of China 200803591024.

C. Peters et al. (Eds.): CLEF 2008, LNCS 5706, pp. 640–643, 2009.

smaller two-class subtasks, each of which is to distinguish between the retrieval topic and each of the other topics. Then all the $(K-1)$ pairwise decisions are combined to form the final decision as in [2]. Thereby, for topic i, we contruct $K-1$ classifiers P_{ij} $(j=1,...,K, j \neq i)$, each of which is to distinguish class i from class j. Thus, assuming that χ_k denotes the positive input set for topic k, $\chi_k = \{X_k^l\}_{l=1}^{N_k}, k = 1,2,...K$, where X_k^l is the input values for the positive samples of topic k. Further, using clustering methods, we divide the input set of class c_k, χ_k, into D_k subsets as, $\chi_k^d, d = 1,2,...D_k$. Then for retrieval topic i we construct $D_i(D-D_i)$ classifiers, $P_{i^{di}j^{dj}}$, where $D = \sum_{k=1}^{K} D_k, di = 1,...,D_i, j = 1,...,K, j \neq i, dj = 1,...,D_j$.

The Matrix of SVMs. For retrieving each topic i, we design two SVM matrices

$$\mathbb{M}_i = (M_{i1},...,M_{i(i-1)}, \Phi, M_{i(i+1)},...,M_{iK}),$$

$$\mathbb{M}_i' = ({M_{1i}}^T,...,{M_{(i-1)i}}^T, \Phi, {M_{(i+1)i}}^T,...{M_{Ki}}^T)^T,$$

where M_{ij} is a $D_i * D_j$ SVM matrix in charge of distinguishing class i from class j, and

$$M_{ij} = \begin{pmatrix} P_{i^1j^1} & P_{i^1j^2} & ... & P_{i^1j^{D_j}} \\ P_{i^2j^1} & P_{i^2j^2} & ... & P_{i^2j^{D_j}} \\ & ... & ... & \\ P_{i^{D_i}j^1} & P_{i^{D_i}j^2} & ... & P_{i^{D_i}j^{D_j}} \end{pmatrix} = (P_{i^{di}j^{dj}}).$$

Here $P_{i^{di}j^{dj}}$ is a SVM with only one output node, which generates the output $o_{i^{di}j^{dj}}$ $(o_{i^{di}j^{dj}} \in [0,1])$. The module of $P_{i^{di}j^{dj}}$ undertakes the subtask of distinguishing the subset χ_i^{di} of from that of χ_j^{dj}. Then we get the output matrices $\mathbb{O}_i$ and $\mathbb{O}_i'$, yielded by the SVM matrices as:

$$\mathbb{O}_i = (O_{i1},...,O_{i(i-1)}, \Phi, O_{i(i+1)},...,O_{iK}),$$

$$\mathbb{O}_i' = ({O_{1i}}^T,...,{O_{(i-1)i}}^T, \Phi, {O_{(i+1)i}}^T,...{O_{Ki}}^T)^T,$$

where O_{ij} and O_{ji} are $D_i * D_j$ and $D_j * D_i$ output matrix corresponding to the matrices M_{ij} and M_{ji} respectively, and

$$O_{ij} = \begin{pmatrix} o_{i^1j^1} & o_{i^1j^2} & ... & o_{i^1j^{D_j}} \\ o_{i^2j^1} & o_{i^2j^2} & ... & o_{i^2j^{D_j}} \\ & ... & ... & \\ o_{i^{D_i}j^1} & o_{i^{D_i}j^2} & ... & o_{i^{D_i}j^{D_j}} \end{pmatrix} = (o_{i^{di}j^{dj}}).$$

Integration Machine. The averaging approach [3] used in general modular classifiers system adopts the average of the results of all modules as the basis of the final classification decision. In our MMSVMs architecture, the averaging machine is modified as follows. Since the desired outputs of $P_{i^{di}j^{dj}}$ for the samples from χ_i^{di} and those from χ_j^{dj} are set to 1 and 0, respectively, the values of $o_{i^{di}j^{dj}}$

can be regarded as a conditional posterior-probability with which the input belongs to χ_i^{di}, namely

$$Prob(x \in \chi_i^{di} \mid P_{i^{di}j^{dj}}) = o_{i^{di}j^{dj}}.$$

In the matrix M_{ij}, all of the elements which have the desired outputs of 1 for the inputs from χ_i^{di}, denoted by the $\mathcal{P}_i^{di}$ row of the matrix $\mathbb{M}_i$ are in charge of distinguishing χ_i^{di} from all $D - D_i$ subsets of the classes except i. While in the matrix M_{ji}, all of the elements which have the desired outputs of 1 for the inputs from χ_j^{dj}, denoted by the $\mathcal{P}_j^{dj}$ row of the matrix $\mathbb{M}_i^{'}$ are in charge of distinguishing χ_j^{dj} from all D_i subsets of the class i. So our averaging approach is to use the average output of the $\mathcal{P}_i^{di}$ row of $\mathbb{M}_i$ as a new estimate of the posterior-probability with which x belongs to χ_i^{di}:

$$Prob_{average}(x \in \chi_i^{di} \mid \mathcal{P}_i^{di}) = \frac{1}{D - D_i} \sum_{j=1, j \neq i}^{K} \sum_{dj=1}^{D_j} o_{i^{di}j^{dj}},$$

where $\mathcal{P}_i^{di}$ denotes the set of the SVMs $P_{i^{di}j^{dj}}$(for all $j = 1, ..., K, j \neq i, dj = 1, ..., D_j$). We use the average output of the $\mathcal{P}_j^{dj}$ row of $\mathbb{M}_i^{'}$ as a new estimate of the posterior-probability with which x belongs to χ_j^{dj}:

$$Prob_{average}(x \in \chi_j^{dj} \mid \mathcal{P}_j^{dj}) = \frac{1}{D_i} \sum_{di=1}^{D_i} o_{j^{dj}i^{di}},$$

where $\mathcal{P}_j^{dj}$ denotes the set of the SVMs $P_{j^{dj}i^{di}}$(for all $di = 1, ..., D_i$). Then the final decision is for all $k = 1, ..., K, dk = 1, ..., D_k$:

$$x \longrightarrow Topic \quad i, \quad if \quad i == \arg max\{Prob_{average}(x \in \chi_k^{dk})\}.$$

Clustering for Subset Divisions. The simple Kmeans clustering method is used for the subset division. We use the MSE criterior to determine the number of clusters. The MSE decreases with the increase of the number of cluster. A turning point can be found, where the number of clusters is the proper one.

3 Visual Concept Detection Results and Conclusions

Our MMSVM is compared with the single SVM on VCDT task [4]. Firstly, 45 profile entropy features (PEF)[5,6] are extracted for each image. Then we construct an MMSVM and perform it on the features. In our experiments, we use RBF kernels and cross validation to tune SVMs parameters. Finally, the average classification accuracy of the MMSVM is 83.8%, while the accuracy of the baseline SVM is 81.0% (which equals to AUC = 81.13% , the 4th rank system in the official test [5]). Figure 1 details that the MMSVM does better than the SVM for concepts 'animals', 'mountains', 'road' and 'path'. One reason could be

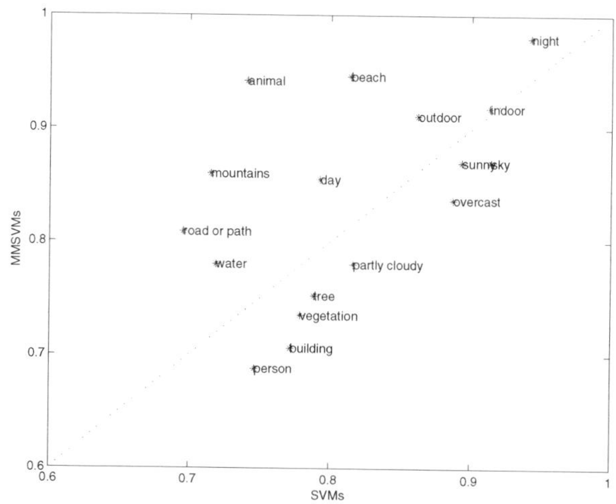

Fig. 1. Classification accuracies of SVM vs MMSVMs for the 17 topics of the CLEF08 VCD Task. The 9 concepts above the diagonal are better detected using the MMSVM than the SVM. The average classification error equals 19% using SVM, but 16.25% using MMSVM, yielding to a relative error reduction of 15%.

that these polymorphic topics are well clusterized by Kmeans in this PEF visual space, while other topics such as 'person' or 'building' are not.

We are not sure that the same topics would be enhanced by MMSVM using other kind of features. However, we demonstrated in this paper that our Matrix Modular SVM can enhance SVM when the visual space of is well clusterized. Other clustering algorithms will be integrated in our further works, since the Kmeans is sensitive to its initialization. Moreover, AUC evaluations will be conducted after transformation of the binary outputs of the MMSVM into soft ones.

References

1. Nilsson, N.J.: Learning Machines: Foundations of Trainable Pattern-Classifying Systems. McGraw-Hill, New York (1965)
2. Zhao, Z.Q., Huang, D.S., Jia, W.: Palmprint recognition with 2DPCA+PCA based on modular neural networks. Neurocomputing 71, 448–454 (2007)
3. Xu, L., Krzyzak, A., Suen, C.Y.: Methods of Combining Multiple Classifiers and Their Applications to Handwriting Recognition. IEEE Trans. Sys. Man and Cybernetics. 22(3), 418–433 (1992)
4. Thomas, D., Allan, H.: The Visual Concept Detection Task in ImageCLEF 2008. In: Evaluating Systems for Multilingual and Multimodal Information Access (2008)
5. Glotin, H., Zhao, Z.Q.: Profile Entropic visual Features for VCDT. In: Working Notes CLEF 2008, Danmark, in conjunction with ECDL 2008 (2008)
6. Glotin, H.: Robust Information Retrieval and perception for a scaled Lego-Audio-Video multi-structuration, Thesis of habilitation for research direction, University Sud Toulon-Var (2007)

SZTAKI @ ImageCLEF 2008: Visual Feature Analysis in Segmented Images*

Bálint Daróczy, Zsolt Fekete, Mátyás Brendel, Simon Rácz, András Benczúr, Dávid Siklósi, and Attila Pereszlényi

Data Mining and Web search Research Group, Informatics Laboratory
Computer and Automation Research Institute of the Hungarian Academy of Sciences
{daroczyb,zsfekete,mbrendel,sracz,benczur,peresz,sdavid}@ilab.sztaki.hu

Abstract. We describe our image processing system used in the ImageCLEF 2008 Photo Retrieval and Visual Concept Detection tasks. Our method consists of image segmentation followed by feature generation over the segments based on color, shape and texture. In the paper we elaborate on the importance of choices in the segmentation procedure with emphasis on edge detection. We also measure the relative importance of the visual features as well as the right choice of the distance function. Finally, given a very large number of parameters in our image processing system, we give a method for parameter optimization by measuring how well the similarity measures separate sample images of the same topic from those of different topics.

1 Introduction

The ImageCLEF 2008 Photo Retrieval [1] and Visual Concept Detection tasks [2] both targeted towards image processing and visual feature generation over the IAPR TC-12 benchmark collection [3]. While the actual systems we used in the ImageCLEF 2008 campaign are described in the Working Notes [4,5], in this paper we concentrate on the main lessons we have learned considering the strength of various visual processing elements in image categorization and similarity search.

Our image processing system common to both tasks is based on image segmentation and then feature generation for the individual image segments, typically around 100 for each image in the corpus. The segmentation procedure consist of a novel combination of the Felzenszwalb–Huttenlocher graph cut method [6] with smoothing over the Gaussian-Laplacian Pyramid [7]. We map all image segments into a roughly 400-dimensional space with features describing the color, shape and texture of the segment. While in the image categorization task we can learn the relative importance of the feature classes, the similarity search procedure used in our content based retrieval system is sensible to the weight. We make an

* This work was supported by the EU FP7 project JUMAS – Judicial Management by Digital Libraries Semantics and by grants OTKA NK 72845 and NKFP-07-A2 *TEXTREND*.

C. Peters et al. (Eds.): CLEF 2008, LNCS 5706, pp. 644–651, 2009.

Table 1. Description and number of visual features used to characterize a single image segment

dimensions	description
3	Mean HSV (or RGB)
60	RGB histogram, 20 bins each
30	Hue histogram
15	Saturation histogram
15	Value histogram
210	Zig-Zag Fourier amplitude (105) and absolute phase (105) low frequency components
1	Size
1	Aspect ratio
64	Shape: density in 8x8 regions

excessive analysis of the feature weights as well as give a novel method to learn these weights based solely on the sample images of the photo retrieval topics.

We briefly describe our text IR system; for more details we refer to [4]. We use the Hungarian Academy of Sciences search engine as our information retrieval system that is based on Okapi BM25 with the proximity of query terms taken into account. We used the original automatic query expansion formula of [8] that, in our implementation, turned out to give minor improvement only. While we show results with and without query expansion, the improvement is minor and hence we omit detailed description and analysis from this report. We also omit details on cluster recall as clusters were typically organized based on the location of the photograph and, in our opinion, image processing could not assist in identifying images of the same cluster.

2 The Image Processing System

In our system we segment images by a novel combination of the graph based image segmentation method of Felzenszwalb and Huttenlocher [6] with the Gaussian-Laplacian Pyramid. While the pyramid is used with success in the ImageCLEF campaign for example in combination with the region of interest method [9], we find other elements of the segmentation procedure of more importance.

After segmentation we map each segment into a feature space characterizing its color, shape and texture with description and dimensionality shown in Table 1. These features are used directly for image classification in Section 4. Their use in the content-based retrieval system (CBIR) of Section 3 is via the distance from sample images. Given a pair of a sample and a target image, for each sample segment we compute the distance of the closest segment in the target image. The final (asymmetric) distance arises by simply averaging over all sample image segments.

Next we describe the details of the segmentation (Section 2.1) procedure and in Section 2.2 we give a novel method to learn the weight of the feature groups based solely on the sample images of the photo retrieval queries. The effect of various settings on the image processing quality is analyzed over the Visual Concept Detection (Section 4) and Photo Retrieval (Section 3) tasks of ImageCLEF 2008.

2.1 Segmentation

Our segmentation procedure is based on a multilevel Gaussian-Laplacian pyramid [7] that enables a gradual refinement of the segments starting out from a coarse segmentation on the top level of the pyramid. Given a coarser segmentation on a higher level, we first try to replace each segment pixel by pixel with the four lower level pixels if their similarity in the RGB space is within a threshold. If the four pixels of the finer resolution are dissimilar, we remove those pixels from the segment. The remaining segments are kept together as starting segments for the lower level procedure while the removed pixels can join existing segments or form new ones.

On the top level of the pyramid we use a modified Felzenszwalb–Huttenlocher graph cut method [6] that, on lower levels, simply continues the growth of the segments obtained on the higher level. Our main improvement over the original method is the use of Canny edge detection [10] values to weight the connection between neighboring pixels. The original method only uses distance in the RGB space as weight that we add to the edge detection weight.

We also require a similar number of segments in the images that are large enough to be meaningful for retrieval or classification purposes. The original Felzenszwalb–Huttenlocher method builds a minimum spanning forest where the addition of a new pixel to the component is constrained by the weight of the connection with the next pixel and the size of the existing component. We test two post-processing rules that reject the smallest segments. The pixels of rejected segments are then redistributed by the same minimum spanning forest method but now without any further restriction on the growth of the existing large segments. The two different rules are as follows:

- Segments of size below a threshold are rejected.
- All segments are rejected except for the prescribed number of largest ones.

2.2 Learning Feature Weights for Image Similarity Search

Our CBIR ranks images based on the distance of the target image segments with the sample image segments. Unlike image classification where classifiers may be capable of learning the relative importance of the features, when considering distances in the feature space, we cannot distinguish between directions relevant or irrelevant with respect to image retrieval.

When applying feature weight optimization for the Photo Retrieval task, we face several problems. First, training data consists solely of the three sample images of the topics. Second, relevance to certain Photo Retrieval topics are based on aspects other than image similarity such as the location of the scene. Third, the three sample images of the same topic are sometimes not even similar.

Our method for training the image processing weights is based on a test for topic separation. We select those topics manually where the three sample images are similar to one another. For ImageCLEF Photo 2008 the list of the selected topics (some of which are ImageCLEF 2007 only) is as follows: 01, 02, 04, 07, 14, 15, 17, 22, 24, 27, 33, 36, 41, 43, 45, 51, 53, 55, 58, 60.

The training data consists of image pairs with an identical number of pairs from the same topic and from different topics. Since our distance is asymmetric, we have six pairs for one topic that results in 120 positive pairs. The negative pairs are formed by selecting two random pairs from a different topic for each of the 60 sample images.

We optimize weights for the AUC value of the two-class classification. Since the task at hand is computationally very inexpensive, we simply performed a brute force parameter search. For larger problems we could choose from logistic regression (if we only train linear weights), simulated annealing or genetic algorithms to name a few.

Given the post-campaign evaluation data, we could perform another manual parameter search to find the best performing weights in terms of the MAP of the retrieval system. As shown in Section 3 we could reach very close to the best settings we found manually, a result that is in fact overtrained due to the use of all evaluation data.

3 The Photo Retrieval Task

For the Photo Retrieval task we combine the scores of our text retrieval system (with or without query expansion) with the following visual relevance score. For a target image to be ranked we take each segment of a given topic sample image and find the closest segment in the target image. We average distances over all these segments. Finally among the three sample images we use the smallest value that corresponds to the closest, most similar one. Since we compute distance instead of similarity, we simply negate the values.

When combining the much lower quality visual scores with the text retrieval scores, we use a method that basically optimizes for early precision but reaches very good improvement in MAP as well. Due to the low quality of the visual scores, low ranked images carry little information and act as noise when combining with text retrieval. Hence we replace all except the highest scores by the same largest value among them, i.e. after some position i, for all $j > i$ we let $\text{score}_j = \text{score}_i$. In our experiment we choose i to be the first value where $\text{score}_i = \text{score}_{i+1}$.

Our results are summarized in Table 2 for a choice of 100 segments with the best segmentation method that uses a 7-level Gaussian-Laplacian pyramid and Canny edge detection. We observe the following behavior. First, ℓ_1 distance outperforms ℓ_2 in all cases. Second, better CBIR scores translate into better combined scores. Third, the test for topic separation (method Section 2.2) finds weights that perform nearly as well as the overtrained best weight setting that we were only able to compute given all relevance assessment data and by far

Table 2. Photo Retrieval performance of different methods (left) with explanation on the right

	MAP	P5	P20
ℓ_1 w1.0	0.0835	0.3795	0.2026
ℓ_1 w1.0	0.0615	0.3231	0.1372
ℓ_1 TST	0.0970	0.4564	0.2103
ℓ_2 TST	0.0800	0.4051	0.1808
ℓ_1 best	0.0985	0.4821	0.2192
ℓ_2 best	0.0813	0.4256	0.1833
txt	0.2956	0.4462	0.3769
txt+qe	0.2999	0.4821	0.3731
txt+ℓ_1 w1.0	0.3279	0.5846	0.4500
txt+ℓ_2 w1.0	0.3130	0.5538	0.4154
txt+TST ℓ_1	0.3344	0.6103	0.4487
txt+TST ℓ_2	0.3206	0.5795	0.4295
txt+best ℓ_1	0.3416	0.6359	0.4603
txt+best ℓ_2	0.3200	0.5538	0.4321
txt+qe+TST ℓ_1	0.3363	0.6103	0.4436
txt+qe+best ℓ_1	0.3444	0.6359	0.4615

ℓ_1	ℓ_1 norm is used
ℓ_2	ℓ_2 norm is used
TST	visual feature weights optimized with test for topic separation (Section 2.2)
w1.0	all 1.0 weights
best	weights hand picked based on the evaluation data
txt	text based information retrieval
qe	query expansion

Table 3. Performance of the variants method evaluated by different measures

	MAP	P5	P20
RGB	0.0278	0.1877	0.0795
RGB + Canny	0.0289	0.1877	0.0795
RGB + pyramid	0.0286	0.1846	0.0846
RGB + Canny + pyramid	0.0314	0.2103	0.1000
RGB+HSV	0.0538	0.3077	0.1308
RGB+HSV + Canny	0.0549	0.3077	0.1308
RGB+HSV + pyramid	0.0521	0.2974	0.1308
RGB+HSV + Canny + pyramid	0.0572	0.3026	0.1410

outperforms the all-1.0 weight case. Unfortunately query expansion gives only very minor improvement and we will have to revise this component of our system.

Figure 1 shows the performance of our best methods on the different topics. Topics are sorted by the MAP of the pure visual result. As it can be seen, the visual result improves text result in most of the topics with the exception of four topics (31, 60, 17 and 15) only. Interestingly, for five topics (23, 59, 50 and 53) the MAP improvement is higher than the visual MAP itself.

The main factors in our CBIR performance consist of ℓ_1 distance in the HSV and DFT feature space as well as our home grown parameter optimization method. Table 3 compares some variations. In general the HSV space is better than RGB but RGB yields additional improvement in combination. The table also justifies the use of both the Gaussian-Laplacian pyramid and the Canny edge weight in the Felzenszwalb–Huttenlocher segmentation algorithm. Finally

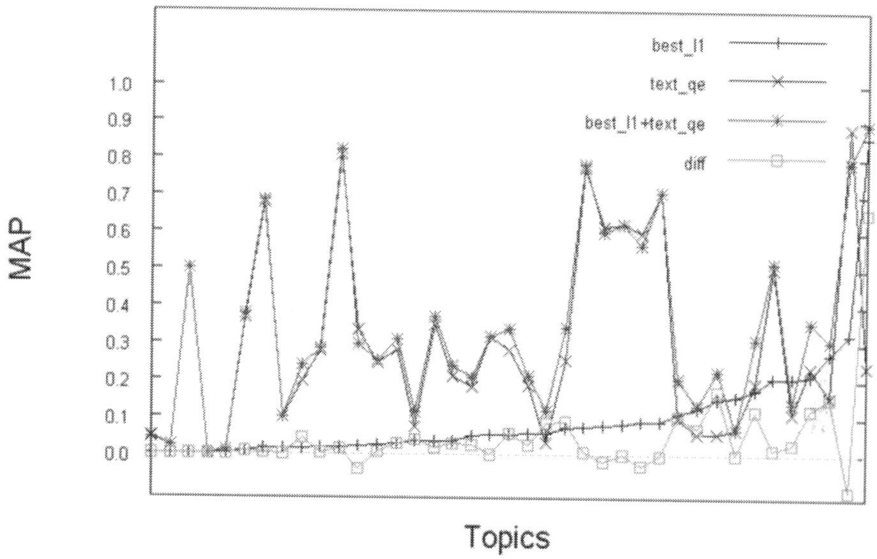

Fig. 1. Performance of different methods by topic. The diff line denotes the improvement of the CBIR over text retrieval with query expansion.

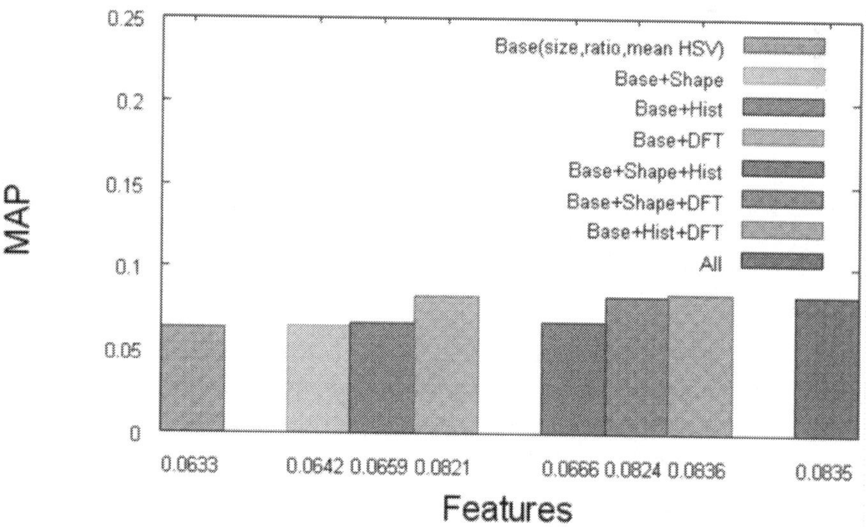

Fig. 2. Performance of different feature combinations

in Fig. 2 we compare the relative strength of the features. Five features, size, aspect ratio and the three mean HSV values themselves form a strong similarity space. This fact is due to the large number of segments so that these features

Table 4. Left: Performance of the three basic methods and their combination, evaluated by different measures. **Right:** Examples of global and local types of concepts with performance given in the form of EER/AUC.

	Glob1	Large	Small	Mixed	Logreg
EER	30.02	32.22	32.37	28.83	26.74
AUC	75.80	74.09	73.21	77.05	79.86

	Glob1	Large	Small
Night	**8.72/94.63**	22.17/78.85	30.49/79.72
Overcast	**19.28/88.92**	28.73/80.43	29.24/78.85
Vegetation	34.85/71.44	33.19/75.19	**30.91/77.88**
Buildings	36.12/68.55	37.18/67.49	**30.96/74.03**

act as histograms. Over this feature set DFT gives the largest additional improvement while histograms and shape add very little, though positive, increase in MAP.

4 The Visual Concept Detection Task

For the Visual Concept Detection Task we used our image processing system with three main settings:

global: features computed for the whole image: mean color, histogram and DFT;
medium: 50 segments, features: size, ratio, mean color, histogram, shape and DFT;
small: 100 segments, features: size, ratio, mean color, histogram, shape and DFT;

Logistic regression was used for classification with the global or segment features as input. For a single image we averaged the segment based predictions, which turned out more accurate than either the minimum or the maximum. We note that we did not use the class hierarchy information. Our main classification results summarized in Table 4 where, in addition to the three above settings for image processing, we give two additional combinations:

Logreg: The output of the classifiers are combined by logistic regression on the 1/4 random fraction training data as heldout set. For the rest of the training set predictions are generated in a 3-fold crossvalidation.
Mixed: For each class the method performing best on the above defined heldout set was selected.

As seen in Table 4, left, best overall performance is attained with a high dimensional global feature space, closely followed by the medium resolution segmentation. We find a clear distinction between concepts that give an overall characterization of the image (day, night, overcast) and those that describe objects in the image (people, vegatation, buildings). The former concepts are best classified in a global while the latter in a segmentwise local feature space (Table 4, right).

5 Conclusion and Future Work

We have demonstrated that image segmentation based retrieval and categorization systems perform well and analyzed the right choice for the segmenter and the visual features. In future work we will conduct a more thorough investigation of possible features and using of more sophisticated methods for computing image distances such as the mixture of Gaussian models. We also plan to strengthen our results by improving our query expansion procedure and using more sophisticated methods for text and image retrieval fusion as well as utilize visual concepts for retrieval.

References

1. Arni, T., Clough, P., Sanderson, M., Grubinger, M.: Overview of the ImageCLEFphoto 2008 photographic retrieval task. In: Peters, C., et al. (eds.) CLEF 2008. LNCS, vol. 5706, pp. 500–511. Springer, Heidelberg (2009)
2. Deselaers, T., Hanbury, A.: The visual concept detection task in ImageCLEF 2008. In: Peters, C., et al. (eds.) CLEF 2008. LNCS, vol. 5706, pp. 531–538. Springer, Heidelberg (2009)
3. Grubinger, M., Clough, P., Müller, H., Deselears, T.: The IAPR TC-12 benchmark - a new evaluation resource for visual information systems. OntoImage, 13–23 (2006)
4. Rácz, S., Daróczy, B., Siklósi, D., Pereszlényi, A., Brendel, M., Benczúr, A.: Increasing cluster recall of cross-modal image retrieval. In: Working Notes for the CLEF 2008 Workshop, Aarhus, Denmark (September 2008)
5. Daróczy, B., Fekete, Z., Brendel, M.: Sztaki @ imageclef 2008 visual concept detection. In: Working Notes for the CLEF 2008 Workshop, Aarhus, Denmark (September 2008)
6. Felzenszwalb, P.F., Huttenlocher, D.P.: Efficient graph-based image segmentation. International Journal of Computer Vision 59 (2004)
7. Burt, P., Adelson, E.: The Laplacian Pyramid as a Compact Image Code. IEEE Transactions on Communications 31(4), 532–540 (1983)
8. Xu, J., Croft, W.: Query expansion using local and global document analysis. In: Proceedings of the 19th annual international ACM SIGIR conference on Research and development in information retrieval, pp. 4–11 (1996)
9. Ah-Pine, J., Cifarelli, C., Clinchant, S., Csurka, G., Renders, J.: XRCE's Participation to ImageCLEF 2008. In: Working Notes of the 2008 CLEF Workshop (2008)
10. Canny, J.: A computational approach to edge detection. IEEE Trans. Pattern Anal. Mach. Intell. 8(6), 679–698 (1986)

THESEUS Meets ImageCLEF: Combining Evaluation Strategies for a New Visual Concept Detection Task 2009*

Stefanie Nowak, Peter Dunker, and Ronny Paduschek

Fraunhofer Institute for Digital Media Technology IDMT
Ehrenbergstrasse 31, 98693 Ilmenau, Germany
{nwk,dkr,pdk}@idmt.fraunhofer.de

Abstract. Automatic methods for archiving, indexing and retrieving multimedia content become more and more important through the steadily increasing amount of digital data in the web and at home. THESEUS, a German research program, focuses on developing sophisticated algorithms and evaluation strategies for the automated processing of digital data. In this paper we present how evaluation is performed in THESEUS and introduce a generic framework for the evaluation of various video and image analysis algorithms. Besides, evaluation campaigns like the Cross Evaluation Language Forum (CLEF) and subprojects like ImageCLEF deal with the evaluation of such algorithms and provide an objective comparison of their performance. We relate the THESEUS tasks to the work done in ImageCLEF and propose a new task for ImageCLEF 2009.

1 Introduction

THESEUS is a German research program that focuses on the development of an Internet-based infrastructure that provides better access to knowledge stored in the World Wide Web. The program is divided into the core technology cluster (CTC), responsible for developing core algorithms, and the Use Cases that transfer the know-how to real world application scenarios. The CTC technologies focus on e.g. image and video analysis, ontologies, user interfaces and evaluation strategies. The work package dealing with evaluation strategies, is lead by the Fraunhofer Institute for Digital Media Technology (IDMT) and is responsible for the quality assessment of the core technologies.

In this paper, we introduce the concept and realization of a generic evaluation framework, present the evaluation procedure of the first evaluations in THESEUS and point out the relationship to the ImageCLEF activities. The evaluation framework concentrates on key features such as easy extension to new formats and measures, storing of previous test results for comparison and measurement of improvement, sophisticated visualizations for interactive reviewing and the generation of descriptive result documents. To compare the results

* This work has been supported by the German THESEUS program.

C. Peters et al. (Eds.): CLEF 2008, LNCS 5706, pp. 652–660, 2009.

of developments in the THESEUS program with algorithms of other research groups, a public benchmark such as ImageCLEF [1] offers a good platform.

The paper is structured as followed. First, we discuss characteristics of multimedia databases and outline benchmarking activities in image and video processing. Next, the concept of the evaluation framework is presented taking into account the different needs from the THESEUS program. The results of the framework and the first evaluation cycle are presented. Finally the concept for a new task in ImageCLEF 2009 is presented in which algorithms from THESEUS will be benchmarked in comparison to the performance of other algorithms from the community.

1.1 Multimedia Databases and Benchmarking Contests

Multimedia databases are generally characterized through a variety in their characteristics, be it the size of items or the way they include metadata like class labels or segment information. In video retrieval the TRECVid databases originated by the TRECVid benchmark [2], are established as a gold standard and widely accepted in the research community. For image retrieval, no database has been accepted as standard for now. An often used commercial database is the Corel Database or subsets of it, although this database was criticized in the community (see e.g., [3]). Collecting databases and defining a ground truth is mainly hindered by the high ambiguity in annotating images and the amount of time needed. Also the requirements concerning the characteristics of the databases are commonly high and strongly task dependent and range from representativeness, availability and size to variety of media items (compare also [4]). This leads to the utilization and collection of a large amount of diverse databases specific for the retrieval task.

A new database with 25000 freely distributable images collected from Flickr was presented on the ACM MIR 2008 for image classification and retrieval [5]. Large scale collaboratively tagged image sets can be found at LabelMe[1]. Popular datasets for image retrieval, object detection or segmentation are also the IAPR TC12 dataset[2], the Caltech datasets[3], the PASCAL VOC classes[4] or the Berkeley Segmentation Dataset[5]. One comprehensive overview of available multimedia databases was summarized by the MUSCLE (Multimedia Understanding through Semantics, Computation and Learning) project[6].

Due to the variety of database characteristics, results of retrieval algorithms are often incomparable and it is hardly possible to decide, whether one approach outperforms another or not. As a result, several contests for different tasks in image and video analysis became popular. Contests define challenging tasks with

[1] `http://labelme.csail.mit.edu/`
[2] `http://eureka.vu.edu.au/~grubinger/IAPR/TC12_Benchmark.html`
[3] `http://www.vision.caltech.edu/Image_Datasets`
[4] `http://www.pascal-network.org/challenges/VOC/voc2007/`
[5] `http://www.eecs.berkeley.edu/Research/Projects/CS/vision/grouping/segbench/`
[6] `http://muscle.prip.tuwien.ac.at/data_links.php`

the goal to objectively measure the performance of algorithms and to establish a baseline for comparing systems. When defining the tasks of a contest, the following criterions are important to consider: objectivity, scalability, processing times, user's interest and expected real-world scenarios. Video retrieval evaluation is for example performed in TRECVid[7]or VideoOlympics[8]. In 2008, TRECVid focused on the evaluation of five categories of video retrieval tasks like surveillance event detection or high-level feature extraction. Image-based evaluation is, amongst others, performed in ImageCLEF[9] or PASCAL VOC Challenge[10]. ImageCLEF focuses on the multimodal and multilingual evaluation taking also text into account while the VOC Challenge deals mainly with the evaluation of object detection and segmentation strategies. An overview of existing IR evaluation campaigns with a special focus on TRECVid can be found in Smeaton et al [2].

2 Evaluation Concept

The evaluation concept has to cover evaluation methodologies for a variety of image and video retrieval algorithms. The aim is to develop one generic evaluation framework with which the evaluation of all image and video analysis tasks in THESEUS can be performed. In the next section a brief presentation of the research on image and video analysis in THESEUS is given, followed by the concept of the framework.

2.1 Image and Video Analysis in THESEUS

In THESEUS, research is performed in the field of fast image and video identification or content-based copy detection based on perceptual hashing. Another focus is the automatic quality assessment of image and video documents and their restoration. Further investigation in the video domain concentrates on temporal shot, subshot and scene change detection, on video genre estimation and video analysis for event detection. The video event detection technologies focus on a flexible framework for a fast and easy integration and optimization (e.g., for different surveillance tasks). Next to image segmentation algorithms, partners research on spatio-temporal segmentation algorithms which incorporate the time dimension to analyse moving regions. Research in the area of image classification and annotation focuses on new visual features, image and object representations, new classification approaches and fast indexing methods. These methods concentrate on generic approaches (e.g., to be agnostic for use with medical images or user generated content). A specific image classification task focuses on a robust face detection. Finally, partners research on different machine learning algorithms (e.g., for parameter learning or statistical modelling) that can be applied to image and video analysis tasks.

[7] http://www-nlpir.nist.gov/projects/trecvid/
[8] http://staff.science.uva.nl/~cgmsnoek/VideOlympics/index.php
[9] http://www.imageclef.org/
[10] http://www.pascal-network.org/challenges/VOC/

Especially in the area of image classification, intersections with the ImageCLEF visual concept detection task, photographic or medical retrieval tasks or PASCAL VOC classification task are existent. This leads to the motivation of organizing an evaluation task that poses a challenge not only for THESEUS partners.

2.2 Concept

Abstract **test cases** were defined that cover the evaluation of all THESEUS developments in the area of image and video analysis. Test cases can be understood as concepts that encapsulate similar multimedia retrieval procedures and are used to generalize the evaluation framework for different evaluation needs at the conceptual level. Altogether we focus on three test cases: Retrieval, keyword or segment indexing and multimedia enhancement.

1. Retrieval:
   Retrieval describes the scenario where one multimedia document serves as input into the analysis application and a list of similar documents is the output. This list can be further enriched with holistic annotations or segments and segment-based annotations of the single documents. Applications are low- or high-level based search scenarios.
2. Keyword or Segment Indexing:
   The test case keyword or segment indexing covers all scenarios, where one media item is the input for an application and a description of this item is computed as output. These descriptions are holistic annotations, segment information or segment-based annotations. This case is applied for the evaluation of face or object detection as well as classification algorithms.
3. Multimedia Enhancement:
   Multimedia enhancement deals with all cases where the input multimedia document is processed and an enhanced version of this document serves as output like in automatic distortion corrections in images or videos.

Depending on the current test case, adapted evaluation measures are chosen and different views for the visualization and interpretation of the results are available. Besides, the evaluation framework contains modules for the generation of query topics, for carrying out performance tests, to provide result visualizations and to track the performance improvement.

3 Results

In this section the results of the first evaluation cycle in THESEUS are presented. The proposed concept for the evaluation framework was realized and is introduced in the following subsection. Subsequently the evaluation process of the first cycle, the image segmentation and the face detection, is illustrated.

3.1 Evaluation Framework

The evaluation framework consists of the Evaluation Toolbox itself, a Graphical User Interface (GUI) to visualize and invoke the evaluation processes and an Input Interface that converts the output of any evaluated system to an internal file format. The Evaluation Toolbox covers the test cases (compare Sec. 2.2), the evaluation measures and the performance tests. Additionally a binding to a database that holds the ground truth data and saves all evaluation results will be established. So it can be used for comparisons and to judge the improvement over time in the prospective evaluation cycles.

In the overall workflow, the system from the CTC task is executed and its results are saved on the file system. A loader module reads the files and converts them to the internal processing format. In dependence from the data and provided annotations, an evaluation process is defined and computed. The results of the evaluation are displayed in the GUI (see Fig. 1) depicting the example of image segmentation evaluation. The segmentation results are presented as crosses in a Precision-Recall Plot. The mean segmentation performance is denoted by a filled circle. The images can be enlarged in a separate window and the boundaries of the computed segments and the ground truth are marked in red and green respectively. Overlapping boundaries are drawn in yellow to show the correspondences of both segmentations.

Image Segmentation. For the evaluation of image segmentation, the Berkeley Segmentation Dataset provided by Martin et. al. [6] was utilized as test corpus. The public part consists of 300 images with 5 to 10 human segmentations from

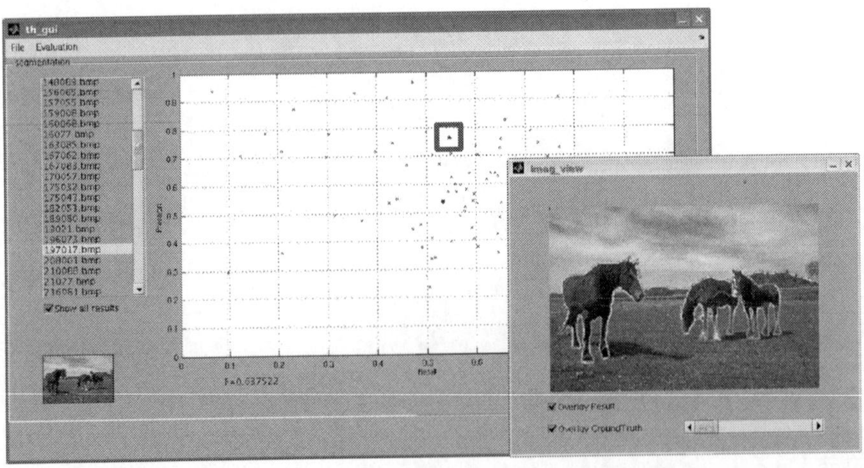

Fig. 1. Result presentation in the GUI of the evaluation framework using the segmentation results as example

different persons for each image. Also, Martin et. al. set up a public segmentation benchmark[11]. Its results allow for a comparison with the THESEUS results.

The evaluation framework incorporates two measures for segmentation evaluation, one boundary-based and one region-based measure. The boundary-based measure is the F-Measure that is computed for corresponding boundaries of the segmented and ground truth images for each boundary map. The region-based measure is the *normalized Hamming Distance* proposed by Huang and Dom in [7]. It is computed as the sum of the *directional Hamming Distance* between the ground truth regions and the segmented regions and vice versa divided by two times the image size. It associates every region of the first segmentation with a region of the second segmentation where the overlapping pixels are maximal and sums up the number of unassociated pixels.

Face Detection. For the evaluation of face detection systems, a ground truth set consisting of more than 350 images with 1000 faces was collected and manually annotated with bounding boxes. To relate a detected face to a ground truth face, it has to fulfil two constraints which refer to the position and size of a bounding box. The Euclidean distance between the centres of the bounding boxes has to be smaller than half of the side length of bounding box 1. Also a tolerance value was defined as the difference between the side length of both bounding boxes divided through the side length of the subtrahend, which has to be smaller than a certain threshold to cope with different sizes of bounding boxes. The sum of the criterions over all images are used to calculate the measurements precision and recall.

4 Visual Concept Detection in ImageCLEF 2009

In THESEUS it is not important to merely monitor the performance improvement of the developed algorithms over time, but additionally to compare them to other state-of-the-art systems. So, carrying out a task in an evaluation campaign like ImageCLEF allows for a comparison of the program results on one hand and offers other participants the opportunity to submit their work to a new challenging task on the other hand. The presented evaluation framework can be used for the evaluation of the submitted results and further extended to additional needs. Besides one can benefit from the result visualizations and the intuitive integration of different evaluation measures to provide a comprehensive and efficient insight into the algorithms performance and gather new knowledge for research. In THESEUS, research is performed on image classification and machine learning algorithms as already pointed out in Sec. 2.1. Especially the scalability of these algorithms to a large amount of data and image concepts plays a crucial rule. In the next section we introduce the new task for ImageCLEF 2009: *Large Scale Visual Concept Detection and Annotation*.

[11] `http://www.eecs.berkeley.edu/Research/Projects/CS/vision/grouping/segbench/`

4.1 Large Scale Visual Concept Detection and Annotation Task

This image annotation and classification task is an extension of the visual concept detection task (VCDT) of 2008 [8]. One focus lies on an enlargement concerning the amount of data and the amount of concepts to be annotated. In 2008, there were about 1800 images for training and 1000 images for testing available. This year the training and test set will consist of several thousand images, respectively, from the MIR Flickr 25.000 image dataset [5] with multiple annotations. This dataset fulfills the needs for a benchmark dataset (compare Sec. 1.1) as it is publicly available, has an acceptable size, is representative because 9862 Flickr users contributed at least one image to the dataset and contains a variety of image contents and camera models used.

Most of the visual concepts refer to a holistic impression of the image and are annotated on an image-based level. Categories for the visual concepts will be e.g., Abstract Categories (Landscape, Family&Friends, Partylife etc.), Seasons, Persons (no, single, big groups) or Quality (blurred, underexposed etc.). Altogether we think about providing ca. 50 concepts, depending on the annotation effort that is feasible and the occurrence of the concepts in the dataset. While most of the holistic concepts can be objectively determined (e.g., the presence or absence of objects) there are also some concepts that are influenced through the subjective impression of the annotators. During groundtruthing a description of what is relevant for a specific visual concept will be provided. Altough it is interesting if a concordant opinion can be achieved throughout annotators for concepts like *aesthetic image*. A challenge is posed through the unbalanced number of photos per concept.

The second focus lies on the structuring of the concepts. All visual concepts are provided in a small ontology. Besides the hierachical structure of the visual concepts, the ontology additionally offers relations between concepts that can help in annotation. For example, for the visual class `Portrait`, there exists the condition that it only can be a portrait if the relation `hasPersons` or `hasAnimals` is fulfilled. Another example is the exclusion of concepts if one concept is present. This information can be useful to build robust concept detectors. Participants may use the hierarchical order and the relations between concepts for solving the annotation task. In summary this task poses two main challenges:

1. Can image classifiers scale to the large amount of concepts and data?
2. Can an ontology (hierarchy and relations) help in large scale annotations?

The different requirements for a task in a benchmark as defined in Sec. 1.1 are partly fulfilled. The task mainly focuses on scalability. Objectivity will be adressed by the groundtruthing through several annotators (at least for a part of the dataset), but especially for some concepts this will be hard to achieve. Processing times can only be determined on mutual trust. We suppose that this task is of great user's interest as image annotation via ontologies are potentially interesting for many applications.

There are related tasks in already existing benchmarks like PASCAL VOC or TRECVid, but with a different focus. PASCAL VOC image detection and classification task from 2008 focuses on the detection of 20 object classes in contrast to the detection of holistic concepts in the proposed task. The concepts therefore do not overlap aside the concept *people*. Additionally the object categories are not structured in a hierarchy or ontology. In the TRECVid High-level feature extraction task from 2008, there were 20 semantic classes that should be detected in videos. As the task focuses on videos, the participants had the possibility to use features that take the time, motion or flow into consideration. The semantic classes have some (potential) overlap like cityscape, nighttime or classroom. This task is also limited to 20 classes without any structural information of the concepts.

5 Conclusion and Future Work

All in all we presented a generic evaluation framework that encapsulates methods for the evaluation of a large variety of image and video analysis algorithms. We introduced three types of test cases in which all developments of the CTC partners in the area of image and video analysis in THESEUS can be categorized in. Two initial evaluations already took place, namely image segmentation and face detection, and were briefly presented here. We highlighted the overlaps between research in the THESEUS program and the ImageCLEF tasks and proposed a concept for a new challenging task for ImageCLEF 2009. This task fits to the objective of ImageCLEF, is not covered yet by a comparable evaluation campaign and suits to developments in THESEUS. The proposed evaluation framework can be applied to the examination of the submitted results.

References

1. Müller, H., Geissbuhler, A., Marchand-Maillet, S., Clough, P.: Benchmarking image retrieval applications. In: Proc. of the 10th Intern. Conf. Distributed Multimedia Systems, Workshop on Visual Information Systems, San Francisco (2004)
2. Smeaton, A.F., Over, P., Kraaij, W.: Evaluation campaigns and TRECVid. In: MIR 2006: Proceedings of the 8th ACM International Workshop on Multimedia Information Retrieval, pp. 321–330. ACM Press, New York (2006)
3. Müller, H., Marchand-Maillet, S., Pun, T.: The Truth about Corel-Evaluation in Image Retrieval. In: Lew, M., Sebe, N., Eakins, J.P. (eds.) CIVR 2002. LNCS, vol. 2383, pp. 38–49. Springer, Heidelberg (2002)
4. Datta, R., Joshi, D., Li, J., Wang, J., Surveys, A.: Image Retrieval: Ideas, Influences, and Trends of the New Age. ACM Computing Surveys 40(2) (2008)
5. Huiskes, M.J., Lew, M.S.: The MIR Flickr Retrieval Evaluation. In: MIR 2008: Proceedings of the 2008 ACM International Conference on Multimedia Information Retrieval. ACM, New York (2008)
6. Martin, D., Fowlkes, C., Tal, D., Malik, J.: A Database of Human Segmented Natural Images and its Application to Evaluating Segmentation Algorithms and Measuring Ecological Statistics. In: Proc. 8th Int. Conf. Computer Vision, vol. 2, pp. 416–423 (2001)

7. Huang, Q., Dom, B.: Quantitative methods of evaluating image segmentation. In: IEEE International Conference on Image Processing, vol. 3, pp. 53–56 (1995)
8. Deselaers, T., Hanbury, A.: The Visual Concept Detection Task in ImageCLEF 2008. In: Peters, C., et al. (eds.) CLEF 2008. LNCS, vol. 5706, pp. 531–538. Springer, Heidelberg (2009)

Query Types and Visual Concept-Based Post-retrieval Clustering

Masashi Inoue[1] and Piyush Grover[2,*]

[1] National Institute of Informatics
m.inoue@acm.org
[2] Indian Institute of Technology, Kharagpur
pgrover@cse.iitkgp.ernet.in

Abstract. In the photo retrieval task of ImageCLEF 2008, we examined the influence of image representations, clustering methods, and query types in enhancing result diversity. Two types of visual concept vectors and hierarchical and partitioning clustering as post-retrieval clustering methods were compared. We used the title fields in the search topics, and either only the title field or both the title and description fields of the annotations were in English. The experimental results showed that one type of visual concept representation dominated the other except under one condition. Also, it was found that hierarchical clustering can enhance instance recall while preserving the precision when the threshold parameters are appropriately set. In contrast, partitioning clustering degraded the results. We also categorized the queries into geographical and non-geographical, and found that the geographical queries are relatively easy in terms of the precision of retrieval results and post-retrieval clustering also works better for them.

1 Introduction

The target of this year's ImageCLEFphoto ad hoc task is to enhance the topical diversity in retrieved results. Usually, instance recall is measured by counting the number of correctly retrieved relevant documents. To measure the topical diversity of retrieved images, the instances are assumed to be the topic. This change in measurement is intended to partially reflect a user's potential needs, which is that many users look at many different choices in terms of the objects or topics given in the retrieval results. For example, if a search topic is associated with the <city> criterion, all the images of the same city in the retrieval results are considered the same in terms of value for the user. Similarly, all images whose subjects are the same species of animal, they are treated as the same if <animal> is the criterion for the search topic. Assuming this model of a user's preference is true, the results should be diverse, which includes as many different objects or topics as possible. To address this problem, we examined the utility of

* This work was conducted while the author was with the National Institute of Informatics.

C. Peters et al. (Eds.): CLEF 2008, LNCS 5706, pp. 661–668, 2009.

clustering techniques that are based on visual content after acquiring the initial ranking that was based solely on textual annotations. We can assume that the topical diversity of the images in the top range of the ranked list will increase by using only representative images from the clusters. The experimental procedures, algorithms used, and experimental results, will be explained and discussed in the following sections.

2 Experimental Setup

2.1 Initial Retrieval

We used the ImageCLEFphoto 2008 ad hoc test collection that consists of 39 search topics, and 20,000 images with structured annotations for this research. The design of the task is explained in [1]. It consists of a monolingual collection in English and a mixed language collection in English and German. We used only the monolingual English collection and all of our queries were in English.

As the retrieval engine, we used the Terrier Information Retrieval Platform[1] for all the textual processing including the pre-processing of the image annotations, indexing, and the matching between queries and indexes. As for the pre-processing, the default stop-word word list and the Porter's stemming in the Terrier toolkit were used.

We tested two variations in the indexing: First, we indexed only the <TITLE> field of the image annotations. In the second case we used both the <TITLE> and <DESCRIPTION> fields for the indexing. All the retrieval experiments were performed on both indexes. In the indexes, the words are assigned weights. The weights are determined by the retrieval model used. The retrieval models also specify the scoring of a particular document when given the query. In the Terrier toolkit, the ranking of documents follows the framework of divergence from randomness (DFR).

The Terrier IR platform offers a variety of retrieval models. To obtain a reasonable baseline retrieval system, we selected the models and their parameters based on our pilot runs. In the pilot runs, we did not use any formal training collection, but we compared the retrieval results on the test collection through the manual inspection of the relevance regarding the several top ranked images. In this process, all topics provided for 2008 were used, but we did not tune the models for each query but the same models and parameter values were used throughout the queries. Therefore, these retrieval models and parameter values returns some reasonable results but not optimal for the test collection. When we constructed indices for the collection using only the <TITLE> field of the annotations, we used the following **IFB2 DFR** model.

$$w(t,d) = \frac{F+1}{n_t \cdot (tfn+1)}(tfn \cdot \log_2 \frac{N+1}{F+0.5}) \tag{1}$$

where tf is the within-document frequency of t in d, N is the number of documents in the entire collection, F is the term frequency of t in the entire collection,

[1] http://ir.dcs.gla.ac.uk/terrier/

and n_t is the document frequency of t. tfn is the normalised term frequency. This is given by normalisation 2:

$$tfn = tf \cdot \log_2(1 + c \cdot \frac{\bar{l}}{l_d}),$$

where l_d is the document length of d, which is the number of tokens in d, $\bar{l}$ is the average document length in the collection, and c is a tuning parameter. We set the parameter to $c = 2.5$.

When we used the <TITLE> and <DESCRIPTION> fields of the image annotations for the indexing, we used the following **In_expC2 DFR** model with $c = 1.1$.

$$w(t,d) = \frac{F+1}{n_t \cdot (tfn_e + 1)}(tfn_e \cdot \log_2 \frac{N+1}{n_e + 0.5}) \tag{2}$$

Our retrieval task consists of two main stages. In the first stage we obtained the retrieval results by using only the indexed data, which is the text retrieval, and the <TITLE> field of the queries in the topic file. The submitted runs corresponding to the text only retrieval were named as follows:

1. EN-EN-TXT-TITLE-AUTO.res
2. EN-EN-TXT-TITDESC-AUTO.res

where TITLE means only the <TITLE> fields were used and TITDESC corresponds to the runs in which both the <TITLE> and <DESCRIPTION> fields were used. Both of them were automatic runs with automatic query expansion by the BE1 model. For the former run, the IFB2 model was used and, the In_expC2 model was used for the latter run. These runs correspond to the baseline conditions for our experiments.

2.2 Post-retrieval Clustering

Diversification by clustering. The initial ranking obtained using only the text contains many duplicate or near duplicate images in terms of their topics. Thus, the retrieved images were clustered to include diverse image sets in the limited window size of the retrieval results, which was 20 in our case. Topically similar images in clusters were represented by the most representative image and did not appear in the final ranked list. As a result, we were able to include diverse types of images on the screen.

Different features can be used in determining the clusters. We used the visual concept vectors that were the semantic concepts extracted from the raw visual signals of the images. These concepts were prepared for the VCDT 2008 task [3]. Although the appearance of the images does not directly correspond to the clustering topical criteria, as we have already used text features in obtaining the initial scores for the documents, we may use another feature of the documents to compensate for the lack of detail in the ranking. We applied two simple clustering approaches to the results obtained from the text retrieval to diversify the final results.

visual concept vectors. visual concept vectors are different from raw visual signals, but they are the semantic entities represented by word tokens that correspond to the visual content in images. Therefore, later on, they can be used as an extra vocabulary. The concepts are extracted using various image processing and pattern recognition techniques. We used two visual concept vectors files:

1. *DISC*—annotations created by Thomas Deselaers from RWTH Aachen University following the described method [2]
2. *CONT*—annotations created by Jean-Michel Renders from XEROX Europe following the method in [6]

The first concept set is labeled DISC because their values are discrete and each image contains concepts represented as binary values. The second concept set is labeled CONT because their values are continuous and each image contains concepts probabilistically. Since automatic image annotation is a difficult task, it contains some errors. We use them with inherent noise.

Hierarchical clustering approach. The first approach is based on a hierarchical clustering in which we produced a dendrogram using the visual concept vectors of the initial ranking given a particular query. Here, we explain the clustering process. All retrieved images that have some relevance scores are clustered. The process is further explained in [5] using an example. Let the number of images in the initial ranking be N; then, each image is represented by its rank from 1 to N. The Euclidean distance between two images represented by concept vectors was used to create the image pairs regardless of the initial index. In the next step, this cluster forms a new higher level cluster with another individual node or cluster. Cluster centers are defined as the mean value of the concept features for member images. The new distance is calculated between the new cluster center and the neighboring new cluster center.

Once the dendrogram has been constructed, we have to decide which granularity we should use to constitute a new ranked list. The dendrogram was sliced at a certain distance level. For both indexing and both visual concept vectors, we changed the distance values for the threshold value from 1.6 to 0.7 at a step size of 0.1. We select the representative images in the clusters at these 9 different levels from the higher values to the lower ones. These parameter values were selected based on the manual inspection of the retrieval results for the 2008 queries. We fixed the the values that returns seemingly reasonable results for all queries. Once we have set the threshold, in the final clusters, images with the smaller index number are regarded as the representative images because the smaller index number indicates a higher original relevance score. In our example, since we start this merging process from a distance level of 1.6 and come down to 0.7, we first make clusters and obtain the representative images for all the clusters at a distance level of 1.6. They will be included in the modified rankings, but their positions have not yet been determined at this point. In the next step, as we come down to a distance level of 1.5, we select the representative images at this distance level. If they are not chosen already, we modify this new image score to the initial retrieval score divided by *level*, which is the step number the process has passed through (here it is 2). This score adjustment is made because we want to topically shuffle the new ranked list. The representative images of the clusters in the lower levels that are visually quite similar to the images that are already placed in the new ranked list have smaller scores and are placed in the lower rankings. Similarly, we continue going down until we reach a distance level of 0.7. After getting all the representative images up to the last level (here the 9th level) and their scores have all been modified, we sort the list according to the new scores and obtain the final modified ranked list for a particular query. We used a threshold value ranging *0.7-1.6* for all our experimental runs. The step size and ranges were determined by conducting a manual inspection of the clustered results.

K-means clustering approach. As a second approach, we applied k-means clustering to the visual concept vectors of the all resulting images obtained by the text retrieval of a particular query. Our clustering process itself is the same as an ordinary k-means clustering. If we randomly assign the initial K means, the final result will also contain randomness and then it becomes difficult to compare the differing conditions. To avoid such randomness, a modification of initialization of k-means clustering was made. K initial cluster centers were evenly allocated in the initial ranked list. Another modification lies in the representative image selection process. We use the densities of the clusters. If a cluster is dense, we assume that the cluster contains near identical images homogeneously; thus, only representative images are included in the final ranking. On the other hand, if clusters are sparse, they likely contain different concepts; therefore, we include all the diverse images in the cluster. In the k-means method, original scores are used in sorting candidate representative images for the final ranking. The details of these procedures are explained by using the pseudo codes in [5].

3 Experimental Results

The two evaluation measures for our submitted runs that were used were precision at the 20th document (P@20) and cluster recall at the 20th document (CR@20). The goal of post-retrieval clustering is to enhance cluster recall. Therefore, a small drop in precision is acceptable as long as we can sufficiently enhance the cluster recall. Degradation may happen because very relevant images of the same categories are removed from the ranked list. To summarize this, we want to improve CR@20 while minimizing the degradation of the precision.

Table 1 shows the results of the two measures. A clear difference in the upper half of the table (<TITLE> only) and the lower half of it (<TITLE> and <DESCRIPTION>) can be seen. More information given in the description fields resulted in better P@20 and CR@20 scores. Also, between the two clustering methods, the modified k-means algorithm was not effective. Although it is not systematic, the difference between the title field only runs and the title and description field runs suggest that a good initial performance may lead to bigger improvement when clustering is used.

Table 1. Precision at 20, Cluster Recall at 20, and F-measure are shown. The cluster recall scores for both media that are better than the text-only runs are marked with boldface.

Run Name	P@20	CR@20	F-measure
EN-EN-TXT-TITLE-AUTO	0.1397	0.1858	0.1620
EN-EN-TXTIMG-TITLE-CONT-Kmeans-AUTO	0.0654	0.1201	0.0858
EN-EN-TXTIMG-TITLE-DISC-Kmeans-AUTO	0.0859	0.1431	0.1063
EN-EN-TXTIMG-TITLE-CONT-0.7-1.6-AUTO	0.1372	**0.1941**	0.1599
EN-EN-TXTIMG-TITLE-DISC-0.7-1.6-AUTO	0.1090	0.1827	0.1365
EN-EN-TXT-TITDESC-AUTO	0.2090	0.2409	0.2238
EN-EN-TXTIMG-TITDESC-CONT-Kmeans-AUTO	0.1115	0.2062	0.1447
EN-EN-TXTIMG-TITDESC-DISC-Kmeans-AUTO	0.1090	0.1730	0.1337
EN-EN-TXTIMG-TITDESC-CONT-0.7-1.6-AUTO	0.1859	**0.3027**	0.2303
EN-EN-TXTIMG-TITDESC-DISC-0.7-1.6-AUTO	0.1590	**0.2703**	0.2002

4 Discussion

4.1 Query and Cluster Topic Dependency

The clustering criteria used to calculate the instance recall can be divided into two groups: geographical criteria, such as the country or city, and others such as the objects. The geographical categorization is based on the official clustering criteria. The geographical criteria include the name of country, name of city, or just location. Geographical criteria dominate about 60% of criteria among all 39 topics. The query numbers for each category are listed in Table 2. The topic dependencies may influence the effectiveness of the post-clustering. Table 3 shows the difference in precision at 20 values for different categorizations. Since the CONT feature usually works better than the DISC feature and only hierarchical clustering could enhance the instance recall as discussed in Sec. 3, we only examined the CONT-0.7-1.6 conditions here. The queries that are associated with the geographical clustering criteria achieved a higher precision in the initial retrieval and after clustering. A similar tendency was observed in the cluster recall values. Actually, in non-geographical topics, clustering damaged the cluster recall scores but enhanced the precision scores for the TITLE only condition. When both TITLE and DESCRIPTION fields were used, cluster recall had been improved in both geographical and non-geographical queries; however, compared with the notable gain in geographical queries, the change in non-graphical ones can be considered marginal. The reasons why images of geographical topics can

Table 2. Categorization of queries based on clustering criteria

	Query Number
Geographical:	2 6 10 11 12 13 15 17 18 19 21 24 28 34 40 41 43 44 50 53 54 55 58
Non-geographical:	3 5 16 20 23 29 31 35 37 39 48 49 52 56 59 60

Table 3. This table shows a performance comparison among the query groups defined in Table 2 under the CONT-0.7-1.6 condition. The changes in retrieval effectiveness before (text-only: t/o) and after clustering (clstd) are shown in terms of the precision (PR) and cluster recall (CR) values at 20th rankings. The scores after clustering that are better than the text-only runs are marked with boldface.

TITLE only

Query groups	P@20 (t/o)	P@20 (clstd)	CR@20 (t/o)	CR@20 (clstd)
All queries	0.1397	0.1372	0.1858	**0.1941**
Geographical queries	0.1500	0.1413	0.1878	**0.2080**
Non-geographical queries	0.1250	**0.1313**	0.1828	0.1742

TITLE & DESCRIPTION

Query groups	P@20 (t/o)	P@20 (clstd)	CR@20 (t/o)	CR@20 (clstd)
All queries	0.2090	0.1859	0.2409	**0.3027**
Geographical queries	0.2130	0.1983	0.2522	**0.3523**
Non-geographical queries	0.2031	0.1488	0.2247	**0.2315**

be clustered well by visual content should be examined in the future. The higher initial precision due to the existence of proper names for geographical queries may explain part of this phenomenon. Another possible hypothesis is that the geographical topics are associated with landmarks that are easier to identify visually.

4.2 Multilingual Retrieval

In our experiment, we used only a monolingual corpus. When the target collection images are annotated in different languages, the initial ranked list given by the text retrieval contains few relevant images. The post-retrieval clustering methods used here eliminate any redundancy found in the top region of the ranked list, but do not actively search for lower ranked hidden relevant images. If our method is used in the multilingual setting, some new methods are needed to enhance the initial relevant retrieved set. Existing techniques for multilingual image retrieval that rely on visual near-identity such as [4] can be used together with this post-retrieval clustering approach because they use the visual similarity in opposite ways.

4.3 Evaluation Measures

The new evaluation measure used in this year's experiments is a cluster recall whose relevance to the ad hoc tourist photo retrieval task has not yet been clarified. The relationship between the utility that users may choose and the increase in cluster recall should be examined. Also, the conventional P@20 measure and the cluster recall are not orthogonal in evaluating ranked lists. Both of them count the number of relevant images in the top region of the ranked lists.

5 Conclusion

We have experimentally compared two post-retrieval clustering methods relying on two types of visual concept vectors that were derived from the images. The experimental results of a monolingual retrieval showed that the use of hierarchical clustering can enhance the instance recall such that the top ranked images are diverse in terms of the topics. Also, we found that the clustering criteria that are assigned to search topics influence the improvement of scores. Generally, the benefit of post-clustering is observed when images are clustered with geographical perspectives. To make our results more reliable, we should further examine the following points: the use of perfectly created visual concept vectors based on the ground truth data, and a comparison between the extracted high-level visual concept vectors and the low-level feature values themselves in the clustering. Future research topics may include the automation of thresholding in the clustering methods that is now manually set by results inspection. The categorization of queries in other criteria such as whether they are context-oriented or content-oriented might be interesting.

References

1. Arni, T., Clough, P., Sanderson, M., Grubinger, M.: Overview of the ImageCLEFphoto 2008 photographic retrieval task. In: Peters, C., et al. (eds.) CLEF 2008. LNCS, vol. 5706, pp. 500–511. Springer, Heidelberg (2009)
2. Deselaers, T., Keysers, D., Ney, H.: Discriminative training for object recognition using image patches. In: CVPR, San Diego, CA, USA, June 2005, vol. 2, pp. 157–162 (2005)
3. Deselaers, T., Hanbury, A.: The visual concept detection task in ImageCLEF 2008. In: Peters, C., et al. (eds.) CLEF 2008. LNCS, vol. 5706, pp. 531–538. Springer, Heidelberg (2009)
4. Inoue, M.: Mining visual knowledge for multi-lingual image retrieval. In: DMIR 2007, Niagara Falls, Ontario, Canada, May 21-23, vol. 1, pp. 307–312 (2007)
5. Inoue, M., Grover, P.: Effects of visual concept-based post-retrieval clustering in ImageCLEFphoto 2008. In: 9th Workshop of the Cross-Language Evaluation Forum (2008)
6. Perronin, F., Dance, C.: Fisher kernels on visual vocabularies for image categorization. In: CVPR, Minneapolis, Minnesota, US, June 18-23 (2007)

Annotation-Based Expansion and Late Fusion of Mixed Methods for Multimedia Image Retrieval

Hugo Jair Escalante, Jesús A. Gonzalez, Carlos A. Hernández, Aurelio López, Manuel Montes, Eduardo Morales, Luis E. Sucar, and Luis Villaseñor-Pineda

Instituto Nacional de Astrofísica, Óptica y Electrónica
Luis Enrique Erro No. 1, 72840, Puebla, México
{hugojair,jagonzalez,carloshg,allopez,
mmontesg,emorales,esucar,villasen}@inaoep.mx

Abstract. This paper describes experimental results of two approaches to multimedia image retrieval: *annotation-based expansion* and *late fusion of mixed methods*. The former formulation consists of expanding manual annotations with labels generated by automatic annotation methods. Experimental results show that the performance of text-based methods can be improved with this strategy, specially, for visual topics; motivating further research in several directions. The second approach consists of combining the outputs of diverse image retrieval models based on different information. Experimental results show that competitive performance, in both retrieval and results diversification, can be obtained with this simple strategy. It is interesting that, contrary to previous work, the best results of the fusion were obtained by assigning a high weight to visual methods. Furthermore, a probabilistic modeling approach to result-diversification is proposed; experimental results reveal that some modifications are needed to achieve satisfactory results with this method.

1 Introduction

Multimedia image retrieval (MIR) is a problem that has been attracting the interest from diverse communities during the last decade [9]. The interest is increasing because of the availability of cheap devices (e.g. cell phones) able to generate large amounts of images everyday. MIR is more challenging than text-based and content-based image retrieval (TBIR and CBIR, respectively) because two modalities must be handled; the problem is further complicated because of the lack of correspondence between low-level image features (e.g. color and texture) and high-level semantics (e.g. named entities like locations or names). Nevertheless, the availability of information from different modalities, yet making reference to a common document, motivate the development of methods able to exploit the diversity, redundancy and complementariness of information. This paper describes experimental results on two novel formulations that follow this line of thinking: *annotation-based expansion* (ABE) and *late fusion of heterogeneous methods* (LFHM); these approaches were developed and evaluated in the context of the photographic retrieval task at ImageCLEF2008, which is described in detail by Arni et al. [2].

C. Peters et al. (Eds.): CLEF 2008, LNCS 5706, pp. 669–676, 2009.

The rest of this document is organized as follows. Next section describes the annotation-based approach. Section 3 presents the LFHM formulation; note that since LFHM is described in detail elsewhere [5], Section 3 is brief in its description; in this section it is also described a probabilistic modeling approach to result-diversification. Section 4 describes and analyzes experimental results. Finally, Section 5, summarizes the findings and contributions of this paper and outlines future work directions.

2 Annotation-Based Document Expansion

Automatic image annotation (AIA) is the task of assigning semantic labels to images [9]; it has been recognized as one of the *hot topics* on MIR. The ultimate goal of AIA is to allow un-annotated image collections to be searched by keywords; however, the usefulness of AIA methods should not be limited to un-annotated collections as shown in this paper and in previous work by the authors [4]. In this work region-level AIA methods were used to expand the manual annotations of images. The underlying idea is to represent documents by considering both their high-level (given by manual annotations of images) and low-level (given by labels automatically assigned to images) semantics; and then using this representation for MIR. The ABE approach is depicted in Figure 1.

ABE was first proposed by Escalante et al. in the framework of ImageCLEF2007[1] [4]; however, the size and quality of the training data used for annotation prevented the authors of deriving concluding facts on the usefulness of automatic labels on image retrieval. In this paper we consider a much better collection to train AIA methods: a subset[2] of *the segmented and annotated IAPR-TC12 benchmark* [3]. This subset is composed of about 7,000 manually segmented and annotated images from the IAPR-TC12 collection; a sample image is shown in Figure 1. Only the regions annotated with the 100 most common labels were considered; resulting in about 37,000 regions that are described by the following features: area, boundary/area, width and height of the region, average and standard deviation in x and y, convexity, average, standard deviation and skewness in both color spaces RGB and CIE-Lab.

The 20,000 images in the IAPR-TC12 benchmark [2] were automatically segmented using the normalized nuts algorithm and the above features were extracted from each region. Using the subset of annotated regions together with a classifier all of the regions in the segmented collection were automatically labeled. For annotation a simple knn classifier was used. Additionally, we considered a method for improving the quality of the knn annotations. This postprocessing method (referred to as MRFS) is based on a Markov random field that uses spatial relationships between connected regions for maximizing the annotation coherence for each image [8]. For each image, the generated labels (manual labels

[1] Note that some participants at ImageCLEF2008 adopted a similar approach for expanding manual annotations with visual concepts [1].

[2] This is an extension to the IAPR-TC12 benchmark that will allow to study the impact of AIA methods on MIR [3]; see, http://ccc.inaoep.mx/~tia/saiapr

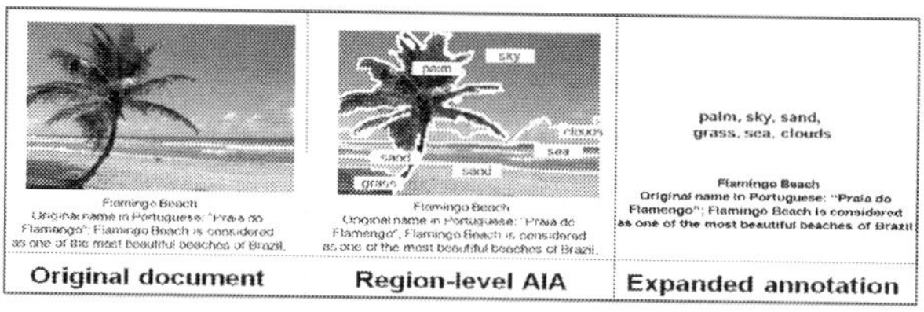

Fig. 1. Diagram of the ABE approach

were used for images in the training subset) were used as expansion of the original annotation, see Figure 1. The expanded annotation was considered a textual document and a text-based retrieval model was used for indexing the documents; the textual statement in each topic was used as query for the retrieval model. Based on previous work we selected as retrieval engine a vector space model (VSM) with a combination of augmented-normalized term-frequency and entropy for indexing/weighting documents [4,5]. The TMG-Matlab^R toolbox was used for the implementation of all of the text-based methods we considered [11].

3 Late Fusion of Heterogeneous Retrieval Methods

Late fusion of independent retrieval methods is one of the simplest and most widely used approaches to combine visual and textual information for MIR [7,5]. The approach consists of building several retrieval systems (i. e. independent retrieval models, hereafter IRMs) based on different information from the same collection of documents. At querying time, each IRM returns a list of documents relevant to a given query. The output of the different IRMs is combined to obtain a single list of ranked documents, see Figure 2 left.

This work considered the combination of multiple heterogeneous IRMs through the late fusion fusion approach (i.e. LFHM). Heterogeneity in IRMs has proved to be important for improving the fusion results by providing complementary and

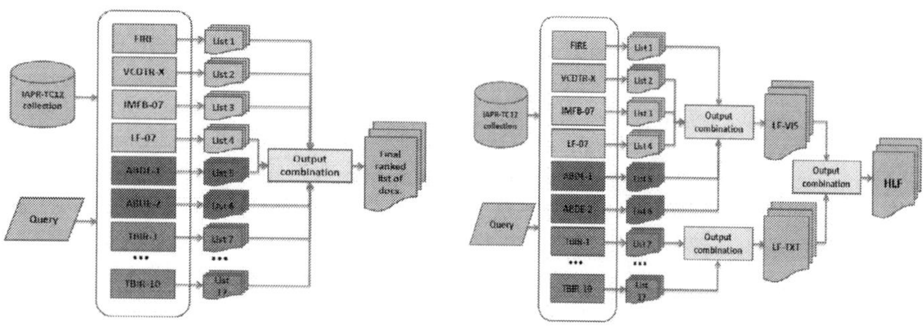

Fig. 2. Configurations considered with LFHM. Left: straight fusion. Right: per-modality and hierarchical LFHM, see text and [5].

Table 1. IRMs considered in this work . From rows 7 and on, column 4 describes the local/global weighting schemas used. Abbreviations are as follows: t, term-frequency; f, inverse document-frequency; n, augmented normalized t; e, entropy; a, alternate log; g, global-frequency/t; l, logarithmic frequency, see [5] for details. For rows 1,3 and 4 it is specified the rank (rk.) position of the respective entry at ImageCLEF2007 [7].

ID	Name	Modality	Description
1	FIRE	IMG	The FIRE CBIR system [6] (rk. 377/474 [7])
2	VCDTR-X	IMG	Boolean TBIR build on the visual concepts provided by XRCE [1]
3	IMFB-07	TXT+IMG	Our best entry, in MAP, at ImageCLEF2007, see [4] (rk. 41/474 [7])
4	LF-07	TXT+IMG	Our best entry, in recall, at ImageCLEF2007, see [4] (rk. 82/474 [7])
5	ABDE-1	TXT+IMG	A TBIR that implements ABE as described in Section 2 (knn)
6	ABDE-2	TXT+IMG	A TBIR that implements ABE as described in Section 2 (MRFS)
7	TBIR-1	TXT	TBIR model based on the VSM representation and t/f weighting
8	TBIR-2	TXT	TBIR model based on the VSM representation and n/e weighting
9	TBIR-3	TXT	TBIR model based on the VSM representation and a/g weighting
10	TBIR-4	TXT	TBIR model based on the VSM representation and a/e weighting
11	TBIR-5	TXT	TBIR model based on the VSM representation and n/g weighting
12	TBIR-6	TXT	TBIR model based on the VSM representation and t/g weighting
13	TBIR-7	TXT	TBIR model based on the VSM representation and n/f weighting
14	TBIR-8	TXT	TBIR model based on the VSM representation and a/f weighting
15	TBIR-9	TXT	TBIR model based on the VSM representation and t/e weighting
17	TBIR-10	TXT	TBIR model based on the VSM representation and t/g weighting

diverse, yet redundant, lists of documents to the fusion [5]; the inclusion of many IRMs (the largest number of IRMs considered so far in late fusion for MIR, to the best of our knowledge) contributed in the same directions as well, although mostly in redundancy. The lists of ranked documents are combined by assigning a score W to each document d_j as follows:

$$W(d_j) = (\sum_{i=1}^{N} \mathbf{1}_{d_j \in L_i}) \times \sum_{i=1}^{N} (\alpha_i \times \frac{1}{\psi(d_j, L_i)}) \tag{1}$$

where i indexes the N available lists of documents $L_{\{1,\dots,N\}}$; $\psi(x, H)$ is the position of document x in ranked list H; $\mathbf{1}_y$ is an indicator function that takes the unit value when y is true and α_i ($\sum_{k=1}^{N} \alpha_k = 1$) is the relevance weighting for IRM i, when using hierarchical LFHM (the α_i values were fixed manually, see [5], in the future we could learn these values from data). Each list L_i is the output of one of the IRMs we considered, these are shown in Table 1, we considered the top 1000 documents from each IRM. Documents are re-ranked in descending order of this score, and the top$-x$ documents are kept.

Different configurations for LFHM were considered: *simple* is the straight fusion of IRMs as depicted at the left of Figure 2; *per-modality* is the combination of IRMs based on the same modality; specifically, IRMs that use text only (LF-TXT) and IRMs that use images (LF-VIS) were fused separately, both configurations are shown at the right of Figure 2; finally, *hierarchical* LFHM (abbreviated HLF), is the fusion of the already fused lists LF-TXT and LF-VIS, as shown at the right of Figure 2; for HLF different weights were assigned to the textual (LF-TXT) and visual lists (LF-VIS), see Section 4.

In order to diversify the results of LFHM, an approach based on latent Dirichlet allocation (LDA) [10] was developed. LDA is a probabilistic modeling tool widely used in text analysis; it assumes documents are mixtures of unknown LDA-topics, which are nothing but (learned) probability distributions of words over documents, characterizing semantic themes. Since documents are mixtures of topics one can calculate the probability of each document given an LDA-topic $P(\mathbf{w}|z_i)$; thus, we associate each document $\mathbf{w}$ to the topic that maximizes the latter probability. In this way, each document is associated to a single LDA-topic, which can be considered a cluster. To diversify retrieval results it was considered the set of documents returned by LFHM (any configuration) to a query-topic. We used the toolbox of Steyvers et al. to obtain k LDA-topics from such set [10]; k was fixed to 20 because diversification at 20 documents was evaluated in ImageCLEF2008. Documents were grouped according the LDA-topic they belong to and a single document was selected from each LDA-topic as representative of it. The representative document was selected according its relevance weight in the list of ranked documents returned by the LFHM method. The k representative documents were placed at the top of a new list and the rest of documents were placed below them according to their initial relevance weight.

4 Experimental Results

This section describes results of ABE and LFHM in the photographic retrieval task at ImageCLEF2008; the following evaluation measures were considered: precision (**p20**) and cluster-recall (**c20**) at 20 documents retrieved, mean average precision (**MAP**) and number of relevant documents retrieved (**Rel-Ret**).

4.1 Annotation-Based Expansion

Results with different configurations of ABE are shown in Table 2. In order to illustrate the advantages of this approach, results are shown over all topics (as provided by organizers [2]) and over visual/textual[3] topics (as categorized in ImageCLEF2006). *Baseline* is a TBIR that uses the original annotations; *Manual* uses annotations expanded with the labels from our training set, since not all images have been manually annotated, not all images have their annotations expanded for this run; *KNN* uses annotations expanded with labels from the training set plus labels assigned with knn (for those images that are not manually annotated); *KNN-MRFS* is the same as KNN though labels assigned by knn were improved with MRFS.

From this table one can see that, as expected, the *Baseline* method obtained the best results over textual topics; although over all and, mainly, over visual topics, ABE configurations consistently outperformed the baseline. Among ABE runs the best **MAP** is achieved with *Manual*; this can be due to the fact that for this setting the expanded labels were always correct. The best results on **P20**

[3] Visual topics are those queries well suited to be answered by using information from images content (e.g. *"night shots of cathedrals"*); textual topics, on the other hand, are those that require using textual information to effectively retrieve documents (e.g. *"Destinations in Venezuela"*).

Table 2. Performance of different configurations with ABE evaluated in ImageCLEF2008. It is shown the performance over all topics (**All**); over visual topics (**Visual**) and over textual topics (**Textual**); the best results are shown in **bold**.

Method	All			Visual			Textual		
	MAP	P20	C20	MAP	P20	C20	MAP	P20	C20
Baseline	0.2625	0.3295	0.3493	0.2916	0.3500	0.3378	**0.2334**	**0.3090**	**0.3608**
ABE-Manual	**0.2648**	0.3333	0.3510	**0.3085**	0.3700	0.3574	0.2211	0.2966	0.3446
ABE-KNN	0.2554	**0.3397**	0.3582	0.2943	**0.3775**	0.3648	0.2165	0.3019	0.3516
ABE-KNN+MRFS	0.2546	0.3295	**0.3733**	0.2971	0.3750	**0.3942**	0.2121	0.2840	0.3524

Fig. 3. Relevant-retrieved images for the Baseline and ABE runs at the top-20 positions for the topic #15 ("*night shots of cathedrals*"). For each run it is shown the [**MAP** / **P20** / **c20**] for this topic.

were obtained with the *KNN* technique; this means that *KNN*-labels resulted more helpful for placing relevant documents at the first positions. Finally, the best result in **c20** were obtained with the *KNN+MRFS* approach. Note that the differences may appear small in number, however, ABE can provide significant advantages to users of MIR systems. Figure 3 shows the relevant-retrieved images to a visual-topic in the first 20 positions for the *Baseline* and ABE runs; this figure illustrates the advantages offered by using the ABE approach. It can be seen that more relevant-images were retrieved with ABE methods and that the performance is improved in all measures; results were likewise for all visual topics. Results shown in Table 2 give evidence that the use of labels generated with AIA methods can be helpful to improve the performance of TBIR methods on both retrieval and diversification of results; specially for visual topics. Note that ABE is the simplest way of taking advantage of automatic labels; therefore, better results are expected by using more sophisticated strategies. Also note that ABE runs resulted very useful for the LFHM approach (see next section) [5].

4.2 Late Fusion of Mixed Methods

Results obtained with different configurations of LFHM are shown in Table 3. It can be seen that performance of all of the configurations is quite competitive with respect to the best and average runs. Averaging over all measures the

Table 3. Performance of different settings with LFHM; rows 5-7 in column 1 show the weights α_1/α_2 assigned to visual and textual lists, respectively; column 7 shows the **c20** performance after applying the LDA technique. For reference, the average over runs evaluated at ImageCLEF2008 [2] and the best run overall [1] are shown.

Run	p20	MAP	c20	Avg.	Rel-Ret	+LDA
Simple	0.3782	0.3001	0.4058	0.3613	1946	0.4291
LF-TXT	0.341	0.2706	0.3815	0.3311	1885	0.3335
LF-VIS	**0.4141**	0.2923	0.3864	0.3642	1966	0.3941
HLF-0.5/0.5	0.3795	0.303	0.3906	0.3577	1970	0.3721
HLF-0.8/0.2	0.391	**0.3066**	0.4033	**0.3667**	**1978**	0.3976
HLF-0.2/0.8	0.3731	0.2949	**0.4175**	0.3619	1964	0.4132
Avg. ImageCLEF	0.2460	0.1091	0.3900	0.2383	543.9	-
Best ImageCLEF XRCE	0.5423	0.3797	0.4146	0.4455	1896	-

best result was obtained by using the HLF approach assigning a weight of 0.8 to visual methods and of 0.2 to textual ones (row 6). This is a very interesting result, contrary to previous work where higher weight to textual methods results on improved performance; this is due to the fact that visual methods are indeed a mixture of CBIR and MIR strategies. It was also interesting that the inclusion of low-performance IRMs (e.g. FIRE and VCDTR-X) resulted beneficial to the LFHM approach, see [5] for further details. Note that the recall of *HLF-0.8/0.2* was among the top-3 over all (1042) ImageCLEF2008 runs; giving evidence of the potential advantages offered by LFHM and that a better strategy for re-ranking documents is required. We would like to emphasize that the considered IRMs (shown in Table 1) are not the best methods one can try and better results are expected by using IRMs of better individual performance.

Column 7 in Table 3 shows the **c20** performance after applying the LDA diversification strategy. The application of such technique improved the **c20** performance of *Simple* and *LF-VIS*, although it decreased the performance of the rest. However, it is important to mention that the **MAP** and **P20** of all of the results was significantly decreased by using the LDA method. This can be due to several factors that motivate further research with this approach; namely, the top 1000 results were considered for clustering, which introduced too much noise; the ranking of LFHM is not the best approach to select representative documents; the initial list of documents may not be correct enough; and the restriction of generating k=20 clusters may be inappropriate.

5 Conclusions

We have described experimental results on two novel approaches to MIR: ABE and LFHM. Experimental results with ABE provide evidence that indicates the use of AIA labels can be helpful to improve the performance of TBIR methods. This is an interesting result, because even with a very simplistic approach we were able to improve both retrieval performance and diversification of results. As expected, the use of labels resulted particularly helpful for visual topics. On the other hand, results obtained with LFHM show that competitive performance can be obtained with this method; even when late fusion is the simplest approach one may try to MIR and when the considered IRMs were not the best retrieval methods one can try. Both formulations motivate further research in several aspects; namely, studying different strategies to combine manual and automatic annotations; improving the performance of AIA methods; applying LFHM with better IRMs and different fusion strategies.

Acknowledgements. We would like to thank the organizers of *ImageCLEF2008* for their support and the reviewers for their useful suggestions. This work was supported by CONACyT under project grant 61335.

References

1. Ah-Pine, J., Cifarelli, C., Clinchant, S., Csurka, G., Renders, J.M.: Xrce's participation to imageclef 2008. In: Peters, C., et al. (eds.) Working Notes of the CLEF, Aarhus, Denmark (September 2008)
2. Arni, T., Sanderson, M., Clough, P., Grubinger, M.: Overview of the 2008 photographic retrieval task. In: Peters, C., et al. (eds.) CLEF 2008. LNCS, vol. 5706, pp. 500–511. Springer, Heidelberg (2009)
3. Escalante, H.J., Hernandez, C., Gonzalez, J., Lopez, A., Montes, M., Morales, E., Sucar, E., Villasenor, L.: The segmented and annotated IAPR-TC12 benchmark. Computer Vision and Image Understanding (in press, 2009), doi:10.1016/j.cviu.2009.03.008
4. Escalante, H.J., Hernandez, C., Lopez, A., Marin, H., Montes, M., Morales, E., Sucar, E., Villasenor, L.: Towards annotation-based query and document expansion for image retrieval. In: Peters, C., Jijkoun, V., Mandl, T., Müller, H., Oard, D.W., Peñas, A., Petras, V., Santos, D. (eds.) CLEF 2007. LNCS, vol. 5152, pp. 546–553. Springer, Heidelberg (2008)
5. Escalante, H.J., Hernandez, C., Sucar, E., Montes, M.: Late fusion of heterogeneous methods for multimedia image retrieval. In: Proc. of MIR 2008, Vancouver, Canada, pp. 172–179. ACM, New York (2008)
6. Gass, T., Weyand, T., Deselaers, T., Ney, H.: Fire in imageclef 2007: Support vector machines and logistic regression to fuse image descriptors for photo retrieval. In: Peters, C., Jijkoun, V., Mandl, T., Müller, H., Oard, D.W., Peñas, A., Petras, V., Santos, D. (eds.) CLEF 2007. LNCS, vol. 5152, pp. 492–499. Springer, Heidelberg (2008)
7. Grubinger, M., Clough, P., Hanbury, A., Müller, H.: Overview of the ImageCLEF 2007 photographic retrieval task. In: Peters, C., Jijkoun, V., Mandl, T., Müller, H., Oard, D.W., Peñas, A., Petras, V., Santos, D. (eds.) CLEF 2007. LNCS, vol. 5152, pp. 433–444. Springer, Heidelberg (2008)
8. Hernández, C., Sucar, E.: Mrfs and spatial information to improve automatic image annotation. In: Mery, D., Rueda, L. (eds.) PSIVT 2007. LNCS, vol. 4872, pp. 879–892. Springer, Heidelberg (2007)
9. Li, J., Datta, R., Joshi, D., Wang, J.Z.: Image retrieval: Ideas, influences, and trends of the new age. ACM Computing Surveys 40(2) (2008)
10. Steyvers, M., Griffiths, T.: Probabilistic topic models. In: Latent Semantic Analysis: A Road to Meaning. Laurence Erlbaum, Mahwah (2007)
11. Zeimpekis, D., Gallopoulos, E.: Tmg: A matlab toolbox for generating term-document matrices from text collections. In: Kogan, J., et al. (eds.) Grouping Multidimensional Data: Recent Adv. in Clustering, pp. 187–210. Springer, Heidelberg (2005)

Evaluation of Diversity-Focused Strategies for Multimedia Retrieval

Julien Ah-Pine, Gabriela Csurka, and Jean-Michel Renders

Xerox Research Centre Europe
6 ch. de Maupertuis
38240 Meylan, France
FirstName.LastName@xrce.xerox.com

Abstract. In this paper, we propose and evaluate different strategies to promote diversity in the top results of multimedia retrieval systems. These strategies consist in clustering, explicitly or implictly, the elements of the top list of some initial ranking and produce a re-ranking that favours elements belonging to different clusters. We evaluate these strategies in the particular case of ImageCLEFPhoto 2008 Collection. Results show that most of these strategies succeed in increasing a diversity performance measure, while keeping or slightly degrading precision of the top list and, more interestingly, they achieve this in complementary ways.

1 Introduction

Up to now, many multi-modal retrieval systems have focused on addressing the semantic gap between textual and visual contents of multimedia documents. The scientific challenges are traditionnaly oriented towards exploiting in an efficient manner these two types of information in order to improve the search results. In ImageCLEFPhoto 2008, however, the main goal is to promote diversity among the first search results. We address this issue using a two-step approach. In the first step, we ignore the question of diversity. In other words, we first try to find the most relevant objects using the material introduced in [1]. Then, in a second step, we re-rank the first relevant objects by taking into account their mutual similarities in order to avoid redundancy and thus to promote diversity. These different approaches are described in details in [2]. Here, we will just briefly remind them, as the main goal of this paper is to perform a deep analysis of the results, as well as to compare and to discuss the strengths and weaknesses of the proposed approaches.

2 Incorporating Diversity in Retrieval Ranking

As mentioned previously, one of the aims of ImageCLEFPhoto 2008 compared to previous sessions, is to promote diversity in the search results so that the first retrieved elements are not redundant. Nevertheless, let us begin by recalling our basic multimedia retrieval strategy, that does not take into account the redundancy in

C. Peters et al. (Eds.): CLEF 2008, LNCS 5706, pp. 677–684, 2009.

the returned ranked list. Complete description and rationale can be found in [2]. Remind that diversity-based extensions will be built on top of this method.

As textual similarity measures, we used the cross-entropy between the language models of two objects [3], the documents of the collection being expanded by thesaurus enrichment[1], while the query was expanded by pseudo-relevance feedback. As visual similarity measures, a cosine-like norm between the "Fisher Vector" representation of two images was employed [4]. Based on these mono-modal similarities, we developed an intermediate fusion method which efficiently combine them in the context of multimedia retrieval. This kind of fusion operator can be understood as query score regularization through a two-step diffusion process, the first step being performed in one mode and the second step being performed in the other one [1]. The ranking obtained with the basis method will be denoted by *xrce_cm_best_basic*.

Let us go back to the objective of promoting diversity in the first elements of the ranked retrieval list. To this end, we investigate two main families of methods: implicit and explicit clustering-based approaches. Both approaches use a pair-wise similarity matrix (designated by *Sim*) between objects of the collection in order to model the diversity. In ImageCLEFPhoto 2008, we tested two kinds of similarities. The first one, designated by *tilo* is purely textual and is computed by the cross-entropy measures between pairs of smoothed language models of the documents restricted to their *ti*tle and *lo*cation fields. The second one, denoted by *cm*, is a pair-wise cross-media similarity matrix that takes into account both textual and visual content in the same way that we did for computing the similarity between a query and an element of the collection.

The first method to avoid redundancy is commonly known as "Maximal Margin Relevance" (MMR) [5]. It amounts to re-rank the search results so that the element chosen at rank j has to be dissimilar to elements that were already selected at ranks $j' < j$. More concretely, given an initial relevance score vector $Score(q)$ (for a given query q, this is typically our *xrce_cm_best_basic* ranking), as well as a pair-wise similarity matrix Sim, the MMR framework supposes that the elements should be ranked by greedily choosing at each step (rank) j the element o^i that maximizes the following re-ranking criterion:

$$MMR(o^i) = \beta(j) Score(q, o^i) - (1 - \beta(j)) \max_{o^{i'} \in P_j} Sim(o^i, o^{i'}) \tag{1}$$

where $\beta(j)$ is a mixture parameter[2] (between 0 and 1) depending on the rank and P_j is the set of objects already selected (ranks lower than j).

The second family of methods are based on an explicit clustering of the first k elements, followed by a strategy designed to re-rank the elements so that many

[1] Using the English Open Office thesaurus available on *http://wiki.services.openoffice.org/wiki/Dictionaries*, giving 15 times more weights to the original words.

[2] Traditionally, β is kept constant, but we propose here a more efficient variant, where $\beta(j)$ linearly increases between $\beta(1) = \alpha$ (< 1) and $\beta(k)$=1 for some k (typically k=100), before saturating at value $\beta = 1$.

different clusters are represented among the first elements of the re-ranked list. In this method, for a topic q, we start by selecting the first k elements according to $Score(q)$, which we denote by P_k. Next, we cluster objects within P_k in order to find different themes, using the Sim pair-wise similarity matrix as basic material. This is realized using the Relational Analysis approach [6,7], which has particular advantages in this framework because it does not impose to fix a priori the number of clusters and because it can deal with strongly unbalanced cluster size distributions (by isolating very small, but significant clusters).

After having clustered the elements within P_k, the top elements of the initial list are re-ranked by the following strategy: we go top down through the elements of this list and, if the cluster id of the current object is not yet represented by other objects previously considered, it is put in the priority list; otherwise, it is put in the non-priority list; we iterate this operation until the number of different clusters represented in the priority list reaches *nbdiv*; after this, we naturally build a new list by taking the priority list, followed by the non-priority list (in each list, objects are ranked following their original relevance scores).

As far as nomenclature is concerned, to designate our ranking methods, the *tilo* suffix indicates that we used the purely textual (title + location) similarity measure to cluster the objects of the collection. The *cm* suffix refers to the use of the pair-wise cross-media similarity measure. Implicit and explicit clustering are differentiated by the *mmr* and *nbdiv* suffixes respectively; in the latter case, the maximum number of clusters represented in the top list is indicated just after.

3 Analysis of Experimental Results

3.1 Analysis of Retrieval Results

For details about the task and the collection, we refer to the overview paper [8]. Before analysing the effects of the diversity-focused strategies, it is interesting to analyse the basic ranking algorithm (*xrce_cm_best_basic*), in order to understand the strengths and weaknesses of the underlying method. That is the goal of this sub-section. Clearly, this algorithm outperforms very significantly the mono-modal strategies, as well as the late fusion approach (see our results in [1]). Let us have a deeper look on results query by query (see Figure 5).

First, we recall (see also [2]) that the only linguistic analyses made on the text or the query were lemmatization and stop-word removal.

Even if the lemmatization was in most cases beneficial, (e.g. for topic[3] 55/35 Peru and Peruvian), we also noticed a negative effect in the case of the topics 39/24 and 40/25 where *"bad weather"* and the adjective "weathered" were considered as identical words after lemmatization. For this reason, texts containing words like "rain, hail, fog, wind, storm, etc" (making reference to bad weather, but not using explicitely the word "weather") had much lower relevance scores than images containing "weathered sandstone", even if the word "weather" was added by

[3] We will refer to topics by t/n, where t refers to the topic number provided and n to its order in the topic list. In the figures the topics' orders appear from 1 to 39.

Fig. 1. We can obviously see the visual similarity (apart weather conditions) between the query images of topic 39/24 (first three images) and three of the four top images retrieved by our system. The missing top retrieved image contained was one of the top ranked 'weathered sandstone" images.

Fig. 2. Top 10 images for the topic topic 2/1 with the xrce_cm_best_basic (top row) and xrce_cm_mmr_07 (bottom row)

thesaurus enrichment (but with a low weight by construction). Moreover, from a purely visual viewpoint, *"bad weather"* is a concept rather difficult to be detected at least with our visual similarity measures as we can see in Figure 1. Indeed, top images look very similar to the query images apart from some small umbrella's that could be interpreted as a sign of bad weather.

The lack of semantic or linguistic analyses of queries made some of the queries quite difficult to be handled by our approach. We can mention here the topic 2/1, *"church with more than two towers"* where the notion of "more than two" would require a semantic analysis and a possible transformation to "three", "four", etc. Furthermore, visually, even if we retrieve the same building as present in a given query (e.g. the St Patrick Cathedral in Melbourne as shown in Figure 2[4] from different points of view), it is difficult to ensure that all towers are visible in the image (the assessor appeared to rely more on the image than on her prior knowledge about the building). For buildings that are not in the query images, the task becomes even more difficult.

Similarly, for topic 18/12 entitled *"sport stadium outside Australia"*, as we removed "outside" as a stopword and did no semantic analysis of the query, texts mentioning "Australian sport stadiums" get higher textual ranks than any "sport stadiums" from other countries.

The failures in the above examples were rather due to the lack of semantic analysis or logical reasoning about the query content. Another cause of failure

[4] In all figures, a colored box around an image means that the image is relevant and different colors make reference to different ground truth clusters.

Fig. 3. The first line represents the top 10 results provided by the xrce_cm_best_basic method for the query 20/14. The second line shows the top 10 results of the diversity-focused approach xrce_cm_mmr_07. Compared to the former, the latter was able to bring a relevant image to the 5th position from 70th (in the first case).

Fig. 4. The method xrce_tilo_nbdiv_10 introduced diversity for the query 6/4 within the top 10 results (second line) compared to the xrce_cm_best_basic method keeping the same P10 performance

stems from the cross-media pseudo-relevance feedback mechanism that is the basis for our cross-media similarity measure. The real limitation (and therefore risk) of the proposed cross-media similarity can be seen through the analysis of the topic 20/14, *"close-up of animals"*. As shown in Figure 3, the results were really poor and suprising. Trying to understand why, we realized that due to high visual similarity between the "golden eagle" close-up query image and some close-up "mummies" images (top 2), the textual query was enriched by the words related to "Nazca grave's mummies" images. This resulted in pushing at the top similar "mummies" images. This was possible because there were no stronger candidates, as the original query virtually shared no textual content (for most animal images, the word "animal" does not appear explicitly) and no visual content with the documents of the collection.

Fortunately, this was a rare case, and in most cases the system was able to take advantage either from textual or from visual similarities (or both) and to successfully combine them. We can mention the excellent results obtained for queries 6/4 or 44/28 (see Figure 4) where the top results show alternatively documents that are visually similar or textually similar.

3.2 Analysis of Diversity Results

In this sub-section we focus our analysis on the different strategies we proposed for avoiding redundancy in the top lists. These comments are made on the basis of the Figure 5 given previously and the Table 1. The latter represents some

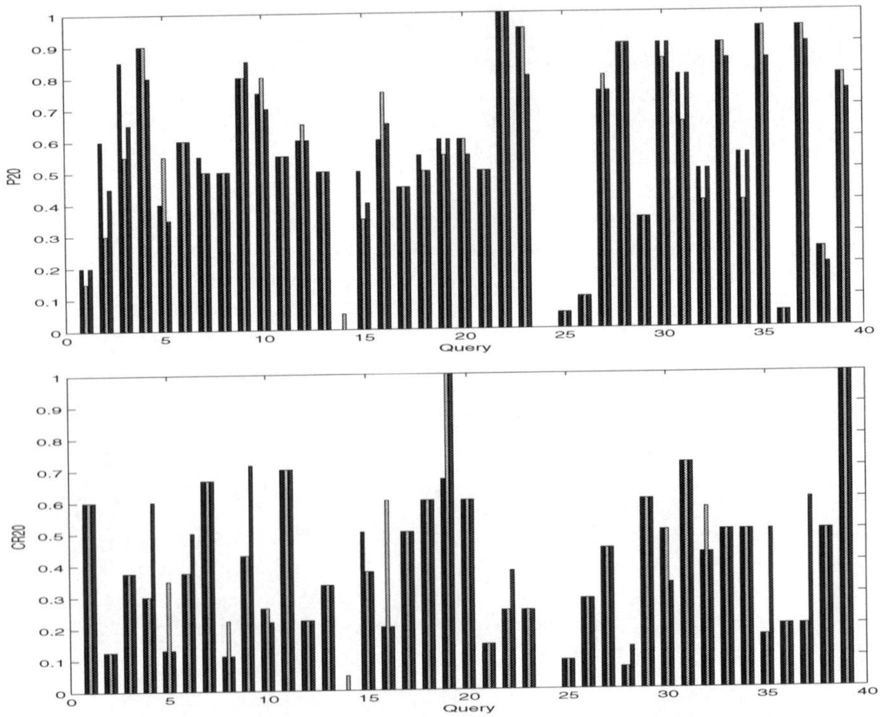

Fig. 5. P20 and CR20 results per query for xrce_cm_best_basic (blue), xrce_cm_mmr_07 (green) and xrce_tilo_nbdiv_10 (red)

Table 1. Some Best results among XRCE's runs

Rank	Run	P10	CR10	P20	CR20	P100	CR100
1	xrce_tilo_nbdiv_10	0.5103	0.3636	0.5423	0.4146	0.2831	0.6374
2	xrce_cm_nbdiv_10	0.4923	0.3608	0.5474	0.4111	0.2831	0.6374
4	xrce_cm_mmr_07	0.6538	0.3012	0.5500	0.4015	0.2790	0.6407
7	xrce_tilo_mmr_07	0.6462	0.3076	0.5385	0.4006	0.2751	0.6492
12	xrce_cm_best_basic	0.6846	0.2816	0.5731	0.3727	0.2831	0.6374

of the best results we obtained for ImageCLEFPhoto 2008 (incidentally, they were also the best ones when compared to the other participants). Particularly, we will pay more attention to the methods denoted by *xrce_tilo_nbdiv_10* and *xrce_cm_mmr_07* that are respectively the best results we obtained with explicit and implicit clustering approaches.

The analysis we give below takes into account the P20 and CR20 measures, following the criteria that were chosen to evaluate the results during ImageCLEFPhoto 2008.

Fig. 6. The method xrce_tilo_nbdiv_10 introduced diversity for the query 3/2 within the top 10 results (second line) compared to the xrce_cm_best_basic method but decreasing the P10 performance

When seeking to promote diversity departing from the basic run *xrce_cm_best_basic*, we can observe in Table 1, that any diversity-focused method fails to increase, on average, the P20 measure. However, all methods perform better than the basic run regarding the CR20 measure. In other words, by trying to eliminate redundancy among the first retrieved objets, unfortunately, we might push relevant objects out of the 20 first re-ranked elements and on the contrary, we might put into this final top list some irrelevant objects. Nevertheless, this is an average trend. Indeed, according to Figure 5, it happens, that seeking for diversity also leads to improved precision.

Next, if we compare the results given by the two kinds of method (implicit and explicit clustering), we can notice, according to Table 1, that the former has, on average, a better P20 than the latter. On the contrary, with respect to CR20, explicit clustering method generally performs better.

Then, let us look at Figure 5 more closely. The explicit clustering strategy exhibits a consistent, stable behaviour, where it systematically gives slightly lower or equal P20 performance than the basic ranking, while offering CR20 performance that are superior or equal to the baseline. In that context, Figure 4 that illustrates the results of topic 11/6 is a good example that demonstrates the benefits of the explicit clustering method for avoiding redundancy.

Implicit clustering does not offer such a stability in its behaviour. In fact, the implicit clustering method seems to "take more risk" in the re-ranking process with a diversity seeking goal than the explicit clustering method, with a consequence of increased variance in the performance. In that context, the topic 20/14 illustrated in Figure 3, is a good example that shows the benefits of the implicit clustering. Indeed, in that case neither the basic run nor the clustering explicit runs allow to find a single relevant object in their top lists. Despite this fact, the implicit clustering had allowed to find a relevant objects in its top 10.

Finally, we illustrate in Figure 6 another issue, related to the fact that the clustering process can result in a totally different thematic than the one intended by the organizers. For the topic 3/2, entitled *"religious statue in the foreground"*, the ranking *xrce_tilo_nbdiv_10* (based on purely textual information) offers diversity (in terms of location), but different from what was expected (types of religious statues).

4 Conclusion

From a pure relevance-based assessment, our method based on cross-media relevance feedback was able to adequately combine and leverage textual and visual information in the query and the documents. Incapacity to give good results was in general due to the lack of deeper logical and semantic analysis of the query content. When considering diversity-focused strategies, three main lessons can be learned. First of all, these strategies are very dependent on the results of the basic ranking (they often fail to capture relevant elements that are low in the list). Secondly, implicit and explicit clustering strategies are often complementary (they are not beneficial for the same type of queries). Finally, implicit clustering strategies favour more exploration than explicit clustering strategies, with the consequence of exhibiting larger variance in the performance results. Clearly, deeper linguistic processings that lead to better document or query enrichment and understanding would certainly allow to leverage the results both from a purely textual viewpoint as well as from a cross-media viewpoint.

Acknowledgments. This work was partly funded by the French National Agency under the *OMNIA* project (OMNIA ANR-07-MDCO-009-01) and under the *Infomagic* project, part of the Pole CAP DIGITAL (IMVN) de Paris.

References

1. Clinchant, S., Renders, J.M., Csurka, G.: Trans-media pseudo-relevance feedback methods in multimedia retrieval. In: Peters, C., Jijkoun, V., Mandl, T., Müller, H., Oard, D.W., Peñas, A., Petras, V., Santos, D. (eds.) CLEF 2007. LNCS, vol. 5152, pp. 569–576. Springer, Heidelberg (2008)
2. Ah-Pine, J., Cifarelli, C., Clinchant, S., Csurka, G., Renders, J.: XRCE's participation to imageCLEF 2008. In: Working Notes of the 2008 CLEF Workshop, Aarhus, Denmark (2008)
3. Ponte, J., Croft, W.: A language modelling approach to information retrieval. In: Proceedings of the 21st Annual International ACM SIGIR Conference on Information Retrieval, Melbourne, Australia, pp. 275–281. ACM, New York (1998)
4. Perronnin, F., Dance, C.: Fisher kernels on visual vocabularies for image categorization. In: Proceedings of IEEE CVPR Computer Vision and Pattern Recognition, Minneapolis, Minnesota, USA (2007)
5. Carbonell, J., Goldstein, J.: The use of MMR, diversity-based reranking for reordering documents and producing summaries. In: SIGIR 1998: Proceedings of the 21st annual international ACM SIGIR conference on Research and development in information retrieval, Melbourne, Australia, pp. 335–336. ACM, New York (1998)
6. Michaud, P., Marcotorchino, F.: Modeles d'optimisation en analyse des données relationnelles. Math. Sci. Hum. 67, 7–38 (1979)
7. Marcotorchino, J., Michaud, P.: Heuristic approach of the similarity aggregation problem. Methods of Operation Research 43, 395–404 (1981)
8. Arni, T., Clough, P., Sanderson, M., Grubinger, M.: Overview of the ImageCLEFphoto 2008 photographic retrieval task. In: Peters, C., et al. (eds.) CLEF 2008. LNCS, vol. 5706, pp. 500–511. Springer, Heidelberg (2009)

Clustering for Photo Retrieval at Image CLEF 2008

Diana Inkpen, Marc Stogaitis, François DeGuire, and Muath Alzghool

School of Information Technology and Engineering,
University of Ottawa
diana@site.uottawa.ca, mstog024@uottawa.ca,
fdegu079@uottawa.ca, alzghool@site.uottawa.ca

Abstract. This paper presents the first participation of the University of Ottawa group in the Photo Retrieval task at Image CLEF 2008. Our system uses Lucene for text indexing and LIRE for image indexing. We experiment with several clustering methods in order to retrieve images from diverse clusters. The clustering methods are: k-means clustering, hierarchical clustering, and our own method based on WordNet. We present results for thirteen runs, in order to compare retrieval based on text description, to image-only retrieval, and to merged retrieval, and to compare results for the different clustering methods.

Keywords: Information retrieval, image retrieval, photographs, text retrieval, k-means clustering, agglomerative clustering, WordNet-based clustering.

1 Introduction

This paper presents the first participation of the University of Ottawa group in the photo retrieval track at Image CLEF 2008. This year's task focused on clustering images in order to retrieve images from different clusters. We present our system, followed by results for the several runs. We worked only with the English part of the collection (English captions and queries). The research questions that we are investigating include: what happens if we index only the text captions, only the images, or the captions and the images; what is the performance of the system with and without clustering. We investigate different types of clustering. First, the k-means clustering algorithm, then hierarchical clustering in three variants: based on average link similarity, complete link, and single link. Then we try our own clustering method, based on searching words from the query and from the text caption in the WordNet lexical knowledge base [1]. We present four versions of this algorithm.

2 System Description

The University of Ottawa's Image Retrieval system was built with off-the-shelf components. For text retrieval we used Lucene[1] and for image retrieval we use LIRE[2].

[1] Lucene text search engine library, http://lucene.apache.org/java/docs/
[2] LIRE open source java content-based image retrieval library, http://www.semanticmetadata.net/lire/

C. Peters et al. (Eds.): CLEF 2008, LNCS 5706, pp. 685–690, 2009.

Lucene is an open-source text search engine that we have used in order to match the text captions of the queries with text description of the images from the collection. This tool provided us with a wide variety of options that we were able to use to quickly create our index and, later on, to perform the appropriate queries on it. It also provided us with parsers that we used in order to automatically remove the stopwords from the provided queries. We added our own parsing component that removed phrases from the narrative fields such as: "the relevant images should not include …".

LIRE is the tool that we have used in order to perform the content-based image search. The main advantage of this tool for us is that it is compatible with Lucene, thus providing a good solution for the problem at hand. Lucene and LIRE were designed to work together. Furthermore, these tools are also fairly simple to use, also providing straightforward methods for indexing and searching.

We have used a data fusion technique to merge the text retrieval with the image retrieval based on the method proposed in [3]. Their method, called combMNZ, sums up all the scores of a document multiplied by the number of non-zero scores of the document, as in formula 1:

$$combMNZ = \sum_{i \in IR\ schemes} score_i * n \tag{1}$$

where $score_i$ is the similarity score of the document for the indexing technique used to retrieve this document (text or image indexing), and n is the number of non-zero scores of the document.

Since there are there are two indexing techniques based on text and image retrieval from different systems, these systems will generate different ranges of similarity scores, so it is necessary to normalize the similarity scores of the document. Lee [2] proposed a normalization method by utilizing the maximum and minimum scores for each weighting scheme as defined by formula 2.

$$NormalizedScore = \frac{score - MinScore}{MaxScore - MinScore} \tag{2}$$

We have adapted combMNZ to carry a weight for each indexing technique. Our data fusion model uses a fusion formula that we call WCombMNZ represented by formula 3.

$$WCombMNZ = \sum_{i \in indexing\ schemes} W_i * NormalizedScore_i * n \tag{3}$$

where W_i is a predefinded weight associated with each indexing technique's results, n is the number of non-zero scores of the document, and the $NormalizedScore_i$ is calculated by formula 2 as described before.

We have tried different weights (W_i) for each indexing scheme, as shown in the next section, usually giving more weight to the text retrieval.

We added a **clustering component** that clusters the text captions. Our text clustering component was implemented with the use of the Dragon3 Toolkit. Our text clustering

[3] Dragon Toolkit http://www.dragontoolkit.org/textcluster.asp

works as follows: using the Dragon Toolkit, a second index of all image annotations, indexed by the doc number field, was created. When performing a clustering operation during a search, the program goes through the Lucene results, extracts their doc number, and uses them to generate a list of Dragon Toolkit documents. The Dragon Toolkit algorithm is then called, which clusters the data based on its own index (the second index described above).

The way we integrated the results of the clustering component into the system was by post processing the search results for the test queries. After the clusters were formed, the results were re-ranked so that as many as possible distinct clusters were retrieved in the first 20 results, keeping only the first image from each cluster. By the first image from a cluster we mean the image with the highest similarity score. The other images from the same cluster were re-arranged below the limit of first 20, in the order of their scores.

Our clustering component used the k-means clustering algorithm, a hierarchical agglomerative clustering algorithm, and a new clustering method that we designed based on WordNet search.

The WordNet-based runs work as follows. The algorithm takes as input the ranked results list that was created by our standard text/image search. It then cycles through each word of the first document, looking for words that match the current clustering criteria. For example, if the query indicates that we should cluster based on animals, then the algorithm will try to spot a word in the document that it recognizes as an animal (for example, the word "dog"). This is accomplished by performing a recursive search up the WordNet semantic relation called hypernym ("is a" relation). We also included the "is an instance of" relation together with the hypermyms. Here is an example of this process: suppose we have the sentence "Run fast fox and don't look back". Also assume that our clustering criterion is "animal". The algorithm would first take the word "run" and, for each of the possible word senses, recursively look through its hypernym ancestry to see if it can find the word "animal". In this case, it would not find it. It would then move on to the next word, "fast". After not finding it for this word, it would move on to the word "fox". In this case, it would find "animal" somewhere in its hypernym tree and would therefore be able to return the word "fox" as being related to the cluster. Once the algorithm has identified the relevant word which describes the current document in relation to the cluster, it then checks if we have already entered a result in the top 20 results with that word. If we have not, it adds this result to the current list of top results. If it is already in the list, it holds on to this result and will eventually append it lower in the list (after the first 20 results). The performance of the algorithm seems pretty good. It is particularly good at detecting locations, countries, states etc. To help enhance this capability, the algorithm looks at the <location> field of the documents first.

3 Experimental Results

Table 1 shows the results of the thirteen submitted results on the test topics/queries. The evaluation measure we report are standard measures computed with the trec_eval script: MAP (Mean Average Precision) and the precisions at 20 documents. In addition, we report the number of distinct clusters included in the first 20 documents, which was the main focus of this year's task.

Next, we describe the first 9 runs from Table 1.

Run1 is a text only search followed by re-arranging using the k-means clustering algorithm.

Run2 merged the results of four different retrieval runs, one with text only and one for each of the three sample images (similarity with that image). The results were merged with the following weights: 70% for the text retrieval, and 10% for each image retrieval.

Run3 is the same as Run2, but with the following weights: 85% for the text retrieval, and 5% for each image retrieval.

Run4 is the same as Run3 but with the following weights: 55% for the text retrieval, and 15% for each image retrieval.

Run5 was an image only run. It was done by assigning the following weights: 0% for the text retrieval, and 33% for each image retrieval.

Run6 is a run in which the clustering step was disabled so that the order that we present is the same as from a text & image retrieval only run. The weights that were used are: 55% for the text retrieval, and 15% for each image retrieval.

Run7 uses the Hierarchical Clustering algorithm instead of the K-Means. It uses Average Linkage to measure similarity. The weights that were used are: 55% for the text retrieval, and 15% for each image retrieval.

Run8 uses the Hierarchical Clustering algorithm instead of the K-Means. It uses Complete Linkage to measure similarity. The weights that were used are: 55% for the text retrieval, and 15% for each image retrieval.

Run9 uses the Hierarchical Clustering algorithm instead of the K-Means. It uses Single Linkage to measure similarity. The weights that were used are: 55% for the text retrieval, 15% for each image retrieval.

Three more runs use a different way of re-arranging the data that is based on WordNet, as explained at the end of the previous section. Below is an explanation of the next three runs that we performed using our WordNet-based clustering algorithm.

Run10 was a basic WordNet run where the algorithm functions as mentioned above.

Run11 was the same as Run10, except that instead of using the clustering field from the query directly, we process it a bit (basically, we only take the first word of the clustering criterion). This helped fix a few exceptions that were found in the clustering queries, such as a cluster called "vehicle type", which would not be recognized by WordNet. WordNet will however recognize vehicle and should be able to cluster properly with that term.

Run12 is the same as Run11 with the added constraint that we only take the first 3 definitions of a word sense when looking for hypernym links. This helped fixed problems with things like the word "blue" being recognized as an animal since its 7th word sense according to WordNet is a "butterfly".

In the previous algorithm, there were cases when none of the terms of a document were found to match the category, according to WordNet. In the previous runs, we just kept the result where it was. However, another option was to only put results that had matches in WordNet in the 20 first documents. Run 13 tries this second option.

Therefore, Run13 is similar to Run12 except that we now make sure that the first 20 results returned contain words that are found to match the clustering criterion according to WordNet.

Table 1. Results of the thirteen submitted runs. All of them are for English queries and English captions of the images (EN-EN). They are all automatic, no manual feedback was involved (AUTO). TXT indicates that the image annotation text was used during the search. IMG indicates that the images themselves were used during the search. The names of the runs also contain a code for the clustering algorithm (KM for k-means clustering; HA – hierarchical clustering, average linkage; HC – hierarchical clustering, complete linkage; HS – hierarchical clustering, single linkage; WN – WordNet-based clustering; NO – no clustering).

Run name	CR@20	P@20	MAP
UOt01_EN_EN_AUTO_TXT_KM	0.3684	0.3333	0.2314
UOt02_EN_EN_AUTO_TXTIMG_KM	0.3767	0.3474	0.2429
UOt03_EN_EN_AUTO_TXTIMG_KM	0.3716	0.3436	0.2384
UOt04_EN_EN_AUTO_TXTIMG_KM	0.3970	0.3615	0.2479
UOt05_EN_EN_AUTO_IMG_KM	0.2694	0.1590	0.0693
UOt06_EN_EN_AUTO_TXTIMG_NO	0.3970	**0.3654**	**0.2490**
UOt07_EN_EN_AUTO_TXTIMG_HA	0.1488	0.0705	0.1665
UOt08_EN_EN_AUTO_TXTIMG_HC	0.3149	0.1333	0.1869
UOt09_EN_EN_AUTO_TXTIMG_HS	0.1204	0.0590	0.1625
UOt10_EN_EN_AUTO_TXTIMG_WN	**0.4102**	0.2756	0.1382
UOt11_EN_EN_AUTO_TXTIMG_WN	0.4038	0.2679	0.1391
UOt12_EN_EN_AUTO_TXTIMG_WN	0.4027	0.2705	0.1372
UOt13_EN_EN_AUTO_TXTIMG_WN	0.4069	0.2026	0.1965

4 Discussion of the Results

The first five runs used the k-means clustering algorithm, Run1 used only the text, and Run5 used only the images. We can see that using only the images the performance is the lowest. Text-only retrieval was pretty good, and using a combination of text and image retrieval brings a small improvement over text-only. Among the first 5 runs, the best weights for merging text and image retrieval results were for Run4, namely 55% for the text retrieval, and 15% for each image retrieval. These weights for merging results were kept for the remaining runs, with good results.

Run6, with no clustering achieved the best P@20 and MAP core, and a score of 0.3970 from CR@20. Adding clustering we reduces P@20 and MAP score, but some of the methods improved the number of distinct clusters retrieved in the first 20.

The four WordNet-based clustering methods worked best in terms of retrieving many relevant clusters, especially the version from Run 10, which archived 0.4102 for CR@20. The hierarchical clustering was the worst, especially the complete link and single link.

5 Conclusion

Our system used merged results from Lucene for text retrieval and Lire for images search. We incorporated a clustering component. We experimented with the k-means algorithm, agglomerative clustering, and a WordNet-based clustering algorithm.

Our experiments showed that text retrieval works well, and adding image similarity brings a bit of improvement. In terms of retrieving many different clusters, our WordNet-based algorithm worked best.

References

1. Fellbaum, C. (ed.): WordNet, An Electronic Lexical Database. MIT Press, Cambridge (1998)
2. Lee, J.H.: Combining multiple evidence from different properties of weighting schemes. In: Proceedings of the Proceedings of the 18th annual international ACM SIGIR conference on Research and development in information retrieval, Seattle, Washington, United States. ACM, New York (1995)
3. Shaw, J.A., Fox, E.A.: Combination of Multiple Searches. National Institute of Standards and Technology Special Publication (1994)

Methods for Combining Content-Based and Textual-Based Approaches in Medical Image Retrieval

Mouna Torjmen, Karen Pinel-Sauvagnat, and Mohand Boughanem

SIG-RFI, IRIT, Toulouse, France

Abstract. This paper describes our participation to the Medical Image Retrieval task of Image CLEF 2008. Our aim was to evaluate different combination approaches for context-based and content-base image retrieval. Our test set is composed of 30 queries, which has been classified by organizers into three categories: visual, textual (semantic) and mixed.

Our most interesting conclusion is that combining results provided by both methods using a classical combination function on all query types, obtains higher retrieval accuracy than combining according to query type. Moreover, it is more successful than using only textual retrieval or using only visual retrieval.

Keywords: Contextual image retrieval, content-based image retrieval, combination, query classification.

1 Introduction

In Image Retrieval, two main approaches are distinguished [6]: Context Based and Content Based Image Retrieval (CBIR). In the first approach, the context of an image, i.e. all information about the image coming from other sources than the image itself, is used to search the images that are likely to be relevant for a textual query. It is often reduced to textual information. The main problem of this approach comes from the classical textual information retrieval problems: vocabulary mismatch. Indeed, documents may use different words to describe the same image or they can use the same words to describe different concepts. In Content Based Image Retrieval, low-level image features such as color and texture, are used to return images similar to an image used as query example. However, this visual similarity may do not correspond to semantic similarity, for instance, a CBIR system can return a picture of blue sky while the example image is a blue car.

In order to take advantages of both approaches, we evaluated combination methods in the Medical retrieval task of CLEF 2008. Our aim was to compare classical combination using a linear combination function and a combination method that took into account the query type: visual, textual and mixed. The systems used for content-based image retrieval and text-based image retrieval are respectively *GIFT* [4] and *XFIRM* [5].

C. Peters et al. (Eds.): CLEF 2008, LNCS 5706, pp. 691–695, 2009.

The rest of the paper is organized as follows. Section 2 describes our approach to evaluate Medical Retrieval queries. In section 3, we present an empirical evaluation of the proposed methods carried out in the Medical Retrieval Task [3] of Image CLEF 2008. We conclude in section 4 with a discussion on our findings and suggestions for future work.

2 Retrieval Approaches

2.1 The XFIRM Model

We used the *XFIRM* XML search engine [5] as a baseline model for the textual query processing. The model is based on a relevance propagation method. Scores are propagated from the leaf nodes to the inner nodes, each node is assigned a score.

In the indexing phase, only two fields of documents of the medical textual collection are indexed ("caption" and "title") as they are the only ones that contain significant textual information.

As the aim of the task is to return image identifier (document), we ask the XFIRM system to only return the whole document, and not parts of the document. Document relevance is then considered as equivalent to image relevance.

2.2 The GIFT System

GIFT or GNU Image Finding Tool [4] is a free CBIR system released under GNU After license. It processes Query By Example (QBE[1]) on images, with the opportunity to improve query results by relevance feedback. In the experiments presented here, we only used the *GIFT* results provided by the organizers, with no further processing.

2.3 Combination of *XFIRM* and *GIFT* Systems

Classical combination. In this approach, we used the two aforementioned systems on the whole set of queries and merged their results to obtain a single result list. The overall structure of this approach is depicted in part (a) of figure 1.

To merge the two result lists into a single list of ranked results, we first normalized scores obtained by the two systems, and then used a simple and classical linear combination of evidences:

$$FS(image) = \alpha \cdot S_{XFIRM}(image) + (1 - \alpha) \cdot S_{GIFT}(image) \quad (1)$$

where α is a decay parameter $\in [0..1]$, $S_{XFIRM}(image)$ and $S_{GIFT}(image)$ represent the image score obtained respectively using the *XFIRM* system and the *GIFT* system.

[1] http://en.wikipedia.org/wiki/QBE

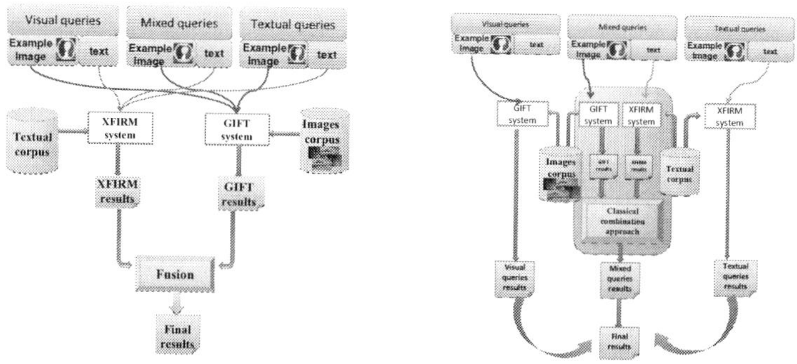

(a) Classical combination

(b) Combination according to query type

Fig. 1. Overview of our approaches

Combination according to query type. In ImageCLEFmed task 2008 [3], there were 30 topics composed of both textual and example image queries, classified in three categories: Visual, Mixed, and Textual. We evaluated the processing of each category with a specific system. We thus used the *GIFT* system to evaluate visual topics, the *XFIRM* system to evaluate textual (semantic) topics, and a classic combination function (Equation 1) of the two systems to evaluate mixed topics. Part (b) of figure 1 illustrates this second approach.

3 Runs and Results

Results are listed in Table 1. Our official runs are in grayed boxes. In these runs, the number of returned results by the XFIRM system was 250 while the number of returned results could be up to 1000. We corrected these runs by returning the correct number of results. Equivalent runs are marked with the same symbol.

XFIRMRun is the complete run of *SigRunText*. It used textual parts of queries and was processed using the XFIRM system. *GiftRun* is the run obtained by only using example image of queries and was processed using the GIFT system. Comparing both runs, we notice that textual based retrieval is significantly better.

RunComb01 to *RunComb09* are obtained with the method exposed in paragraph 2.3 (Part (a) of figure 1). The optimal parameter of the linear combination function used in the classical combination (Figure 1, part(a)) was found for $\alpha = 0.9$ (*RunComb09*). This means that the two information sources are usefull, but as results are improved when α increases, textual information should be considered as the main source of evidence.

RunMix09 is obtained using the approach illustrated in part (b) of figure 1. For the ten mixed queries, we set the combination parameter $\alpha = 0.9$ (the best run obtained using the classical combination). *RunMix1* is obtained by processing textual and mixed queries using XFIRM system and visual queries using GIFT system. We can conclude that processing mixed queries combining

Table 1. Results of Clef Medical Retrieval Task

Runs	α	MAP	bpref	P@10	p@30	Symbol
GiftRun	0	0.0349	0.0898	0.1700	0.1511	
SigRunText	1	0.1410	0.1851	0.2430	0.2244	*
XFIRMRun	1	0.1524	0.2454	0.2600	0.2278	*
SigCombAlpha01	0.1	0.0427	0.0929	0.2200	0.1600	**
RunComb01	0.1	0.0483	0.1021	0.2200	0.1800	**
RunComb02	0.2	0.0572	0.1098	0.2700	0.2111	
RunComb03	0.3	0.0658	0.1170	0.3167	0.2311	
RunComb04	0.4	0.0749	0.1190	0.3000	0.2452	
SigCombAlpha05	0.5	0.0432	0.0948	0.2200	0.1600	***
RunComb05	0.5	0.0820	0.1376	**0.3733**	0.2833	***
RunComb06	0.6	0.0909	0.1462	**0.3733**	0.2889	
RunComb07	0.7	0.1014	0.1591	0.3433	**0.2989**	
RunComb08	0.8	0.1409	0.2167	0.3100	0.2933	
SigCombAlpha09	0.9	0.0415	0.0947	0.2170	0.1611	****
RunComb09	0.9	**0.1705**	**0.2614**	0.2900	0.2611	****
SigMix	0.5	0.1113	0.1637	0.2870	0.2311	
RunMix1	1	0.1043	0.1906	0.2067	0.1678	
RunMix09	0.9	0.1101	0.1914	0.2133	0.1756	

Table 2. Some Results of Clef Medical Retrieval Task by query category

	Runs	MAP	bpref	P@10	p@30
Visual Queries (10)	RunVisXFIRM	0.1658	**0.2256**	0.2600	0.2600
	RunVisGift	0.0159	0.0615	0.1000	0.0800
	RunVisComb09	**0.1667**	0.2199	**0.2800**	**0.2633**
Mixed Queries (10)	RunMixXFIRM	0.0630	0.1609	0.1400	0.1033
	RunMixGift	0.0613	0.1302	**0.3200**	**0.2800**
	RunMixComb09	**0.0805**	**0.1632**	0.1600	0.1267
Textual Queries (10)	RunTexXFIRM	0.2341	0.3496	**0.3800**	**0.3200**
	RunTexGift	0.0275	0.0778	0.0900	0.0933
	RunTexComb09	**0.2643**	**0.4041**	0.2800	0.2633

textual and visual features (MAP=0.1101) is better than processing them using only textual data (MAP=0.1043).

More experiments have to be conducted to show the difference and the effectiveness of each approach. We plan in future work to evaluate our approaches using other systems as results depend also of the systems used for retrieval.

Table 2 shows results for each query category using the XFIRM system, the GIFT system and the best classical combination of both systems (with $\alpha = 0.9$). For all query categories, the best MAP is obtained using classical combination with $\alpha = 0.9$. The intuition that visual queries must be processed with a CBIR system, textual queries must be processed with textual based retrieval system and mixed queries must be processed with a combination of both is thus not validated in these results. This shows that the classification of queries into visual, mixed and textual categories is not adequate for improving results, and consequently, using score classical combination remains the best way to improve performance.

Our proposed approach described in part (b) of figure 1 is a preliminary method as it used query classification made by the organizers. We expect that retrieval results will be improved if we apply techniques of automatic query classification using other features. In fact, query classification is an active research field where Good [2] and Fairthorne [1] were among the first to recommend automatic query classification to improve document retrieval.

4 Conclusion and Future Work

In this paper, we evaluated a classical score combination approach and an approach based on query classification in image retrieval. Best results are obtained using the classical combination method, by assigning higher weight to the textual information than the visual one. This could be explained by the fact that classifying queries into visual, textual and mixed queries is not an adequate way of classification. In addition, results depend also of the used systems.

This is our first study of query classification conducted with queries already classified by CLEFMED organisers, and hence in the future, we plan to apply automatic query classification to effectively study its efficiency in image retrieval.

References

1. R. FAIRTHORNE. The mathematics of classification (1961)
2. Good, I.J.: Speculations concerning information retrieval. Technical report, Research Report PC-78, IBM Research Centre, Yorktown Heights, New York (1958)
3. Müller, H., Kalpathy-Cramer, J., Kahn Jr., C.E., Hatt, W., Bedrick, S., Hersh, W.: Overview of the ImageCLEFmed 2008 medical image retrieval task. In: Peters, C., et al. (eds.) CLEF 2008. LNCS, vol. 5706, pp. 512–522. Springer, Heidelberg (2009)
4. Müller, H., Müller, W., Squire, D.M., Pecenovic, Z., Marchand-Maillet, S., Pun, T.: An open framework for distributed multimedia retrieval
5. Pinel-Sauvagnat, K., Boughanem, M., Chrisment, C.: Searching XML documents using relevance propagation. In: Apostolico, A., Melucci, M. (eds.) SPIRE 2004. LNCS, vol. 3246, pp. 242–254. Springer, Heidelberg (2004)
6. Westerveld, T.: Image retrieval: Content versus context. In: Content-Based Multimedia Information Access, RIAO 2000 Conference Proceedings, pp. 276–284 (2000)

An SVM Confidence-Based Approach to Medical Image Annotation

Tatiana Tommasi, Francesco Orabona, and Barbara Caputo*

Idiap Research Institute
Centre Du Parc, Rue Marconi 19
P.O. Box 592, CH-1920 Martigny, Switzerland
{ttommasi,forabona,bcaputo}@idiap.ch

Abstract. This paper presents the algorithms and results of the "idiap" team participation to the ImageCLEFmed annotation task in 2008. On the basis of our successful experience in 2007 we decided to integrate two different local structural and textural descriptors. Cues are combined through concatenation of feature vectors and through the Multi-Cue Kernel. The challenge this year was to annotate images coming mainly from classes with only few training examples. We tackled the problem on two fronts: (1) we introduced a further integration strategy using SVM as an opinion maker; (2) we enriched the poorly populated classes adding virtual examples. We submitted several runs considering different combinations of the proposed techniques. The run jointly using the feature concatenation, the confidence-based opinion fusion and the virtual examples ranked first among all submissions.

1 Introduction

The rapid development of new medical image acquisition techniques and the widespread use of computerized equipment to save, transfer, and store medical imagery in digital format have led to the need for new methods to manage and archive this data. Automatic image annotation systems turn out to be important tools to manage big databases, in avoiding manual classification errors and helping in image retrieval. In 2008 the ImageCLEFmed annotation task provided participants with 12076 x-ray images as training data spread across 197 classes. The task consisted in assigning the correct label to 1000 test images. To recognize these images, an automatic annotation system has to face two major problems: intra-class variability vs inter-class similarity, and data imbalance. The ImageCLEFmed organizers decided to focus on this second problem introducing in the training set 82 classes with a maximum of 6 images each and preparing a test set mainly with images from these low populated classes.

* This work was supported by the EMMA project (B.C. and F.O.) thanks to the Hasler foundation (www.haslerstiftung.ch) and by the Blanceflor Boncompagni Ludovisi foundation (T.T., www.blanceflor.se).

C. Peters et al. (Eds.): CLEF 2008, LNCS 5706, pp. 696–703, 2009.

This paper describes the algorithms submitted by the "idiap" team as its second participation to the CLEF benchmark competition[1]. Last year we proposed different cue-integration approaches based on Support Vector Machine (SVM, [1]), using global and local features. They proved robust and able to tackle the inter-vs-intra class variability problem. Our run based on the use of the Multi-Cue Kernel (MCK, [2]) ranked first in 2007. After the competition we compared the results obtained by MCK with a scheme consistent in concatenating the different feature vectors. The benchmark showed that the two methods do not produce significatively different results [2]. This year we decided to reuse both the above described methods changing the selected features into two different types of local descriptors: Scale Invariant Feature Transform (SIFT, [3]) and Local Binary Pattern (LBP, [4]). We also propose two strategies to tackle the imbalancing problem. On one hand we explore a technique to estimate the confidence of the classifier's decision. When it is not considered reliable, a soft decision is made using SVM as an opinion maker and combining its first two opinions to produce a less specific label. On the other hand we created examples for the classes with few images to enrich them. The new images were produced as slightly modified copies of the original ones through translation, rotation and brightness changes. We submitted several runs. The one which combines feature concatenation with confidence based opinion fusion and introduction of virtual examples ranked first among all submissions.

2 Cue Integration

In the previous editions of the challenge, top-performing methods were based on local features which thus seem to be the most discriminative cues for medical image annotation [5,6]. Our past experience confirms this assumption, so this year we decided to explore two local approaches. We considered them separated and combined through two different SVM-based integration schemes.

2.1 Feature Extraction

In 2007 for the medical annotation task we defined a modified version of the classical SIFT descriptor that we called modSIFT. We used it through a "bag of words" approach [2]. The two runs based on this feature ranked third and fourth in 2007, so we decided to reuse it doing only a slight modification, inspired by the approach in [7]. We added to the original feature vector the histogram obtained extracting modSIFT from the entire image producing a vector of 2500 elements. In this way we are considering the image at two different space levels: in our preliminary tests this simple method brought a gain of approximately 2 score points.

As second local descriptor we chose the LBP operator, a powerful method well known in face recognition, object classification [8,9] and also in the medical area [10,11]. The LBP basic idea is to build a a binary code that describes the local

[1] In 2007 the name was "BLOOM" due to our sponsors.

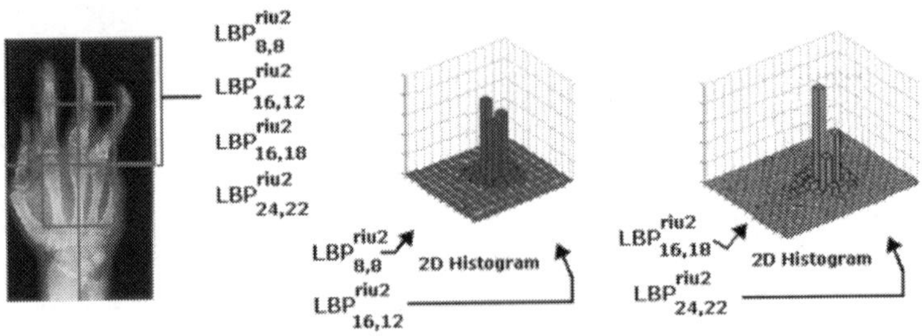

Fig. 1. A schematic drawing which shows how we built the texture feature vector combining the 1-dimensional histograms produced by the LBP operators in 2-dimensional histograms

texture pattern in a circular region thresholding each neighborhood on the circle by the gray value of its center. After choosing the dimension of the radius R and the number of points P to be considered on each circle, the images are scanned with the LBP operator pixel by pixel and the outputs are accumulated into a discrete histogram [4]. The operator is gray-scale invariant, moreover we used the $riu2$ rotational invariant LBP version which considers the uniform patterns with two spatial transitions ($\mathrm{LBP}^{riu2}_{P,R}$, [4]).

Our preliminary results on a validation set showed that the best way to use LBP on the medical image database at hand was combining in a two dimensional histogram $\mathrm{LBP}^{riu2}_{8,8}$ together with $\mathrm{LBP}^{riu2}_{16,12}$ and concatenating it with the two dimensional histogram made by $\mathrm{LBP}^{riu2}_{16,18}$ together with $\mathrm{LBP}^{riu2}_{24,22}$. In this way a feature vector of 648 elements is obtained. Each image is divided in four parts, one vector is extracted from each subimage and from the central area and then they are concatenated producing a vector of 3240 elements (see Figure 1).

2.2 Low and Mid Level Integration Schemes

In the computer vision and pattern recognition literature some authors suggested different methods to combine information derived from different cues. They can all be reconducted to one of these three approaches: high-level, mid-level and low-level integration [12]. Considered our results in the ImageCLEF 2007 [2], we decided to use again the Multi-Cue Kernel as mid-level integration scheme and the concatenation of feature vectors as low-level integration.

The Multi-Cue Kernel is a linear combination of kernels each dealing with a single feature. Suppose that for each image I_i, we extract a set of P different cues, $T_p(I_i)$, $p = 1 \dots P$. Hence we have P different training sets and a corresponding set of P kernels K_p, $p = 1 \dots P$. The Multi-Cue Kernel between two images, I_i and I_j, is defined as

$$K_{MC}(I_i, I_j) = \sum_{p=1}^{P} a_p K_p(T_p(I_i), T_p(I_j)) \qquad (1)$$

where $a_p \in \Re^+$ are weighting factors found through cross validation while determining the optimal separating hyperplane.

On the other hand, in the low-level scheme, the single features vectors are combined in a unique vector, which is normalized to have sum equal to one.

2.3 Classification

For the classification step we used an SVM with an exponential χ^2 as kernel [13], for both the local structural and textural approaches and the cue-integration methods:

$$K(X,Y) = \exp\left(-\gamma \sum_{i=1}^{N} \frac{(X_i - Y_i)^2}{X_i + Y_i}\right) . \tag{2}$$

The parameter γ was tuned through cross-validation. In our experiments we used also the linear, RBF and histogram intersection kernel but all of them gave worse results than the χ^2.

Even if the labels are hierarchical, we have chosen to use the standard multi-class approaches. This choice is motivated by the finding that, with our features, the error score was higher using an axis-wise classification.

3 Confidence Based Opinion Fusion

The evaluation scheme for the medical image annotation task addresses the hierarchical structure of the IRMA code by allowing the classifier to decide a "don't know" at any level of the code, independently for each of the four axes [14]. To effectively support this scheme, models which estimate the classifier's confidence in its decision could be useful. Discriminative classifiers usually do not provide any out-of-the-box solution for estimating confidence of the decision, but in some cases they can be transformed in opinion makers on the basis of the value of the used discriminative function. In case of SVM, it can be done considering the distances between the test samples and the hyperplanes. This approach turns out to be very efficient due to the use of kernel functions and does not require additional processing in the training phase. In the One-vs-All multiclass extension of SVM, if M is the number of classes, M SVMs are trained each separating a single class from all remaining ones. The decision is then based on the distances of the test sample, $\boldsymbol{x}$, to the M hyperplanes, $D_j(\boldsymbol{x})$, $j = 1 \dots M$. The final output is the class corresponding to the hyperplane for which the distance is largest:

$$j^* = \underset{j=1\dots M}{\operatorname{argmax}} D_j(\boldsymbol{x}) . \tag{3}$$

If now we think at the confidence as a measure of unambiguity of the decision, we can define it as the difference between the maximal and the next largest distance:

$$C(\boldsymbol{x}) = D_{j^*}(\boldsymbol{x}) - \max_{j=1\dots M, j \neq j^*} D_j(\boldsymbol{x}) . \tag{4}$$

The value $C(\boldsymbol{x})$ can be thresholded to obtain a binary confidence information. Confidence is then assumed if $C(\boldsymbol{x}) > \tau$ for threshold τ. In the cases in which the decision is not confident, we decided to compare the labels corresponding to the first two margins and to put a "don't know" term in the points of the code in which they differ.

4 Adding Virtual Examples

An SVM, working with classes very sparsely populated is not able to produce reliable results. To create the models it is forced to individuate the best hyperplane which separates classes with few examples, to all the rest of the training set. To improve the classification reliability, we enriched the poorly populated classes. In [15] the creators of the IRMA corpus describe that small transformation of the images do not alter the class membership. So we produced modified copies of the training images increasing and decreasing each side (100, 50 pixels); rotating them right and left (20,40 degrees); shifting right, left, up, down and in the four diagonal directions (50 pixels); increasing and decreasing brightness (add and subtract 20 to the original gray level). Thus for each of the images belonging to poorly populated classes we produced 17 different versions.

5 Experiments

Before starting our validation experiments, we studied in-depth how to divide the released database to consider the high imbalancing between classes. We decided to separate the training images in:

- rich_set: images belonging to classes with more than 10 elements. A total of 11947 images divided in 115 classes. From this group we built 5 disjoint sets, $\text{rich\_train}_i/\text{rich\_test}_i$, each with of 11372/575 images, where the test sets were created randomly extracting five images for each of the 115 classes. Note that in this way we are automatically considering a normalization on the classes.
- poor_set: images belonging to classes with less than 10 elements. A total of 129 images divided in 82 classes. We used the whole poor_set as a second test set.

We trained the classifier on the rich_train_i set and tested both on the rich_test_i and on the poor_set, for each of the 5 splits. The error score was evaluated using the program released by the ImageCLEF organizers. The score values were normalized by the number of images in the corresponding test set, producing two average error scores. They were then multiplied by 500 and summed together to produce the value of the score on the test set of the challenge as if it was constituted half by images from the rich_set and half by images form the poor_set. The expected value of the score is then defined as the average of the scores obtained on the 5 splits. Each parameter in our methods was found optimizing this expected score.

Cross validation was done considering for LBP SVM_C=[50 **100** 150 200], γ=[0.5 1.5 **2.5** 3.5] and for modSIFT SVM_C=[80 120 **160** 200], γ=[0.01 0.025 **0.05** 0.1]. For the two cue integration schemes we used: low level feature concatenation SVM_C=[50 **100** 150 200], γ=[0.5 **1** 2 4]; MCK SVM_C=[50 **100** 150 200], γ_{LBP}=[0.25 0.5 **1** 1.5], $\gamma_{modSIFT}$=[0.25 **0.5** 1 1.5], a_{LBP}=[0.4 **0.3** 0.2 0.1], $a_{modSIFT}$=[0.6 **0.7** 0.8 0.9]. The best parameters are in bold.

On top of these preliminary experiments we applied, the confidence based opinion fusion technique described in Section 3. Both the single-cue and the multiple-cue runs were executed using the One-vs-All SVM multiclass extension. The first two higher margins for every test images were subtracted and the difference compared to the threshold τ varying in $[0.1, 0.2, \ldots 0.9]$. The best threshold led to the lowest expected score.

To evaluate the effect of introducing virtual examples in the poor_set we extracted from it only images belonging to classes with more than one element. We called this set poor_more, it contained a total of 76 images from 29 classes. From it we created 6 poor_more_train$_j$/poor_more_test$_j$ splits of 29/47 images, where the train sets were defined extracting one image from each of the 29 classes. We also introduced virtual examples as described in Section 4 such that each poor_more_train set was enriched with 29*17=493 images. Then we combined these sets joining rich_train$_i$ and poor_more_train$_j$ to build the training set and testing separately on rich_test$_i$ and poor_more_test$_j$. We run experiments with this setup and the best kernel parameters obtained form the previous single and multiple-cue experiments. The described procedure, for each i, j couple produced again two classification outputs. The error scores were normalized and combined as described above. We also repeated this group of experiments without introducing the virtual examples and the score resulted lower of approximately 4 points on average showing that the addition of virtual elements is useful for the classification task.

Finally we applied the confidence based decision fusion on the output of the just presented experiments with the virtual examples in the training set. Independently of the selected feature or combination of features, applying together our two proposed methods improved the score.

Even if the cross validation experiments required a preliminary effort in computational resources and time to select the best parameters, the subsequent confidence based opinion fusion, introduction of virtual examples and the combination of these two strategies turned out to be very fast.

All the parameters of the validation phase were then used to run our submission experiments on the 1000 unlabelled images of the challenge test set using all the 12076 images of the original dataset as training. We submitted 9 runs. One of them (idiap-MCK_pix_sift) consisted simply in repeating our 2007 winner run, that is combining modSIFT and pixel features through MCK using exactly the same parameters of last year [2]. As expected, this run ranked last this year, due to the fact that the dataset varied a lot respect of 2007 and a new search for all the parameters was needed. It is interesting to note that simply applying the confidence based opinion fusion on the this run (idiap-MCK_pix_sift_2MARG) we have a gain in score of 85.19.

Table 1. Ranking of our submitted runs, name, score and gain respect to the best run of the other participants. The extension MULT stands for image multiplication, that is the use of virtual examples. 2MARG stands for the combination of the first two SVM margins for the confidence based opinion fusion.

Rank	Name	Score	Gain
1	idiap-LOW_MULT_2MARG	74.92	30.83
2	idiap-LOW_MULT	83.45	22.30
3	idiap-LOW_2MARG	83.79	21.96
4	idiap-MCK_MULT_2MARG	85.91	19.84
5	idiap-LOW_lbp_siftnew	93.20	12.55
6	idiap-SIFTnew	100.27	5.48
7	TAU-BIOMED-svm_full	105.75	0
11	idiap-LBP	128.58	−22.83
19	idiap-MCK_pix_sift_2MARG	227.82	−122.07
24	idiap-MCK_pix_sift	313.01	−207.26

Considering that our validation results did not show great differences between the low-level and the mid-level integration scheme we decided to use just the low-level cue-integration scheme for sake of simplicity. We submitted only one MCK run using both the confidence based opinion fusion and the virtual examples. Hence the remaining runs consisted in:

- using the two new cues separately (idiap-SIFTnew, idiap-LBP);
- applying cue-integration (idiap-LOW_lbp_siftnew);
- combining cue-integration with the confidence based opinion fusion (idiap-LOW_2MARG);
- combining cue-integration with the introduction of virtual examples in the training set (idiap-LOW_MULT);
- combining cue-integration with the confidence based opinion fusion and the introduction of virtual examples in the training set (idiap-LOW_MULT_2MARG, idiap-MCK_MULT_2MARG).

The ranking, name and score of our submitted runs together with the score gain respect to the best run of other participants are listed in Table 1.

6 Conclusions

This paper presents a combination of three different strategies to face the medical image annotation in a highly imbalanced database with great inter-vs-intra class variability. The first consists in combining cues through two different SVM approaches. The second allows to estimate the confidence of the classifier decision and, on this basis, to assign to a test image the class label corresponding to the hard decision of the classifier, or to a combination of the labels related to the first two produced opinions. The third consists in enlarging the training set through virtual examples. The method obtained combining the low-level cue-integration scheme together with the confidence based opinion fusion and the introduction of virtual examples obtained a score of 74.92 ranking first among all submissions.

This work can be extended in many ways. First, it could be interesting to understand if the low-level cue-integration scheme results still better then the mid-level one when the number of combined cues grows. Second, we would like to integrate the confidence estimation and the cue integration in a unique strategy. The classifier should measure its own level of confidence and, in case of uncertainty, to seek for extra information considering multiple cues, so to increase its own knowledge only when necessary. Future work will explore these directions.

References

1. Cristianini, N., Shawe-Taylor, J.: An Introduction to Support Vector Machines (and Other Kernel-Based Learning Methods). Cambridge University Press, Cambridge (2004)
2. Tommasi, T., Orabona, F., Caputo, B.: Discriminative cue integration for medical image annotation. PRL (2008) (in Press)
3. Lowe, D.G.: Object Recognition from Local Scale-Invariant Features. In: ICCV (1999)
4. Ojala, T., Pietikainen, M., Maenpaa, T.: Multiresolution Gray-Scale and Rotation Invariant Texture Classification with Local Binary Patterns. PAMI (2002)
5. Shyu, C.R., Brodley, C.E., Kak, A.C., Kosaka, A., Aisen, A., Broderick, L.: Local versus global features for content-based image retrieval. CBAIVL (1998)
6. Müller, H., Deselaers, T., Deserno, T.M., Clough, P., Kim, E., Hersh, W.R.: Overview of the ImageCLEFmed 2006 Medical Retrieval and Medical Annotation Tasks. In: Peters, C., Clough, P., Gey, F.C., Karlgren, J., Magnini, B., Oard, D.W., de Rijke, M., Stempfhuber, M. (eds.) CLEF 2006. LNCS, vol. 4730, pp. 595–608. Springer, Heidelberg (2007)
7. Lazebnik, S., Schmid, C., Ponce, J.: Beyond Bags of Features: Spatial Pyramid Matching for Recognizing Natural Scene Categories. In: CVPR (2006)
8. Ahonen, T., Hadid, A., Pietikinen, M.: Face description with local binary patterns: application to face recognition. PAMI (2006)
9. Zhang, L., Li, S.Z., Yuan, X.T., Xiang, S.M.: Real-time Object Classification in Video Surveillance Based on Appearance Learning. In: CVPR (2007)
10. Unay, D., Ekin, A., Cetin, M., Jasinschi, R., Ercil, A.: Robustness of Local Binary Patterns in Brain MR Image Analysis. In: EMBS (2007)
11. Oliver, A., Lladó, X., Freixenet, J., Martí, J.: False Positive Reduction in Mammographic Mass Detection Using Local Binary Patterns. In: Ayache, N., Ourselin, S., Maeder, A. (eds.) MICCAI 2007, Part I. LNCS, vol. 4791, pp. 286–293. Springer, Heidelberg (2007)
12. Sanderson, C., Paliwal, K.K.: Identity Verification Using Speech and Face Information. Digital Signal Processing 14, 449–480 (2004)
13. Fowlkes, C., Belongie, S., Chung, F., Malik, J.: Spectral Grouping Using the Nyström Method. PAMI (2004)
14. Lehmann, T.M., Schubert, H., Keysers, D., Kohnen, M., Wein, B.B.: The IRMA code for unique classification of medical images. In: SPIE (2003)
15. Keysers, D., Dahmen, J., Ney, H., Wein, B.B., Lehmann, T.M.: A Statistical Framework for Model-Based Image Retrieval in Medical Applications. J. Electronic Imaging 12, 59–68 (2003)

LIG at ImageCLEF 2008

Loic Maisonnasse[1], Philippe Mulhem[2], Eric Gaussier[2], and Jean Pierre Chevallet[2]

[1] Université de Lyon, INSA-Lyon, LIRIS
loic.maisonnasse@insa-lyon.fr
[2] Grenoble University, LIG
Philippe.Mulhem@imag.fr, Eric.Gaussier@imag.fr,
Jean-Pierre.Chevallet@imag.fr

Abstract. This paper describes the work of the LIG for ImageCLEF 2008. For ImageCLEFPhoto, two non diversified runs (text only and text + image), and two diversified runs were officially submitted. We add in this paper results on image only runs. The text retrieval part is based on a language model of Information Retrieval, and the image part uses RGB histograms. Text+image results are obtained by late fusion, by merging text and image results. We tested three strategies for promoting diversity using date/location or visual features. Diversification on image only runs does not perform well. Diversification on image and text+image outperforms non diversified runs. In a second part, this paper describes the runs and results obtained by the LIG at ImageCLEFmed 2008. This contribution incorporates knowledge in the language modeling approach to information retrieval (IR) through the graph modeling approach proposed in [4]. Our model makes use of the textual part of the corpus and of the medical knowledge found in the Unified Medical Language System (UMLS) knowledge sources. And the model is extended to combine different graph detection methods on queries and documents. The results show that detection combination improves the performances.

1 Introduction

This paper describes the work of the LIG for ImageCLEF 2008. The aim of our work was to study language model approaches for image+text indexing and retrieval. Section 2 describes runs submitted to ImageCLEFphoto 2008 by the LIG (Laboratoire d'Informatique de Grenoble) as well as non officially submitted ones: runs on image only, text only and text+image. We compare here three different diversification processes based on three different clustering methods. We show that diversification impact negatively the precision at 20 documents but positively the cluster recall, and that diversification processes based on simple features have potential interests. Section 3 describes LIG contribution to the ImageCLEFmed 2008 task. This contribution evaluates the combination of graph models with different processing on queries and documents, and also evaluates the extent of textual descriptions for a given image.

C. Peters et al. (Eds.): CLEF 2008, LNCS 5706, pp. 704–711, 2009.

2 LIG Runs on ImageCLEFPhoto Collection

2.1 Text Processing

When considering text retrieval, it has been shown in [5] that language models (LM) of Information Retrieval, inspired by speech recognition, give results that are close to, or outperform, existing approaches (like Vector Space Model [7] or probabilistic models based on BM25 for instance [6]). We used a language model that expresses the probability for the query Q to be generated from a document model D by : P(Q|D)=$Sim_t(Q, D)$. Such a probability is computed using the probability of any term t_i to be generated by the document, P(t_i|D). One strong aspect of language models is the use of smoothing methods leading to consider that a document that does not contain a term does not have a zero probability of generating this term. The smoothing used in our experiments is the Dirichlet smoothing, that has been shown in [9] to be effective. The results are ranked according to the decreasing order of this probability.

2.2 Image Processing

The processing of the images computes histograms on image blocks. Results obtained on 9 blocks gave better results than taking the whole image in our previous studies on person photographs. Each image of the corpus is split into a 3x3 regular grid. For each of the blocks $b_{I,i}, 1 \leq i \leq 9$ of an image I, we extract an RGB histogram, $H_{I,i}, 1 \leq i \leq 9$. The choice of the RGB color space is due to the fact that it gives similar results than other color spaces that require more processing time to be computed. Each histogram has 512 bins, according to a 8×8×8 regular split of the RGB cube (a tradeof between effectiveness, storage usage and processing speed). Then, one global histogram GH_I with 4602 bins is created for one image, by concatenating the 9 $H_{I,i}$ histograms, and then normalized to 1. The similarity $Sim_v(I, J)$ between two images I and J is defined as $1 - JS(GH_I||GH_J)$, with $JS(GH_I||GH_J)$ the Jensen-Shannon divergence (a symmetrical version of the Kullback-Leibler divergence). The visual part of a query is composed of 3 images, each image representing one sample of what is expected. We consider that the relevance of images in the corpus depends on the best relevance value wrt each of the query images. The similarity between a set of images $IS_q = \{I_{q,i}\}$ and one image J is then defined as the maximum of the similarity between J and each of the images of IS_q. The results are then ranked according to the decreasing order of this similarity.

2.3 Image+Text Processing

Several ways may be used to represent mixed textual and visual data for retrieval. We consider that even if the internal representation for these two media are somewhat similar (distribution of probabilities), their inner nature refrains us from applying early fusion by concatenating the two distributions. Another reason is that "adding" two distributions does not necessarily create another one.

That is why, in our run, we consider a late fusion defined as a linear combination of the two results (text and image) obtained. In fact, the linear distribution is applied on normalized results, using for the text (resp. the image) the minimum and the maximum matching values to ensure the normalized results to be in [0, 1]. This linear combination is refined using the following heuristics: we assume that the best visual results (i.e. very similar to one of the query images) are often relevant, which is not the case for the textual ones. This heuristics is consistent with partial results we obtained previously. So, we extend the matching function using a condition on the matching:

$$Sim(Q, D) = \begin{cases} 1, & \text{if } Sim_v(Q_v, D_v) > t_v \\ \alpha.Sim_t(Q_t, D_t) + (1 - \alpha).Sim_v(Q_v, D_v) & \text{otherwise} \end{cases} \quad (1)$$

with Q_v (resp. Q_t) the visual (resp. textual) part of the query, and D_v (resp. D_t) the visual (resp. textual) part of the document. The results are then ranked according to the decreasing order of this similarity.

2.4 Diversification Processing

One task in ImageCLEFphoto this year was dedicated to study the capacity of systems to provide diverse results for a query, and not only near duplicate results. To achieve this goal, we defined clusters according to several criteria:

- $C_{country}$: based on the `LOCATION` field of the image description, we generated a set of clusters $C_{country}$, that groups images by country name. $C_{country}$ contains 23 classes, with on average 952 images per cluster.
- C_{city}: same as $C_{country}$ but based on city names. One cluster of C_{city} contains all the images with no city. C_{city} contains 511 classes, with on average 38 images per cluster.
- C_{dmy} : all images taken at the same date (day-month-year, dmy) are grouped. All the images that have the same month and year but do not have a day specified are grouped together. All the images that have the same year, but no month and no day are grouped together.
- C_{vis}: k-means clustering, namely C_{visual}, based on the visual description of the images (4608 dimension histograms) was applied. The number of cluster is 500. There are on average 40 images per cluster.

We experimented with three kinds of diversification, on our three runs (text only, visual only and text+image). The first diversification, Div_{dmy}, uses the C_{dmy} clustering. The second diversification, namely Div_{vis}, uses the C_{vis} clustering. The third diversification, namely Div_{Qclust}, depends on the cluster description of the query: a) Queries having a cluster name `city` are diversified using the C_{city} clustering, b) Queries having a cluster name `state` or `country` are diversified using the $C_{country}$ clustering, and c) Queries having another cluster name are diversified using visual clusters C_{visual}. The diversification process according to a clustering C_x is straightforward: rerank the images from the top non diversified results so as to obtain 20 images from different clusters of C_x.

2.5 Results

The official results submitted are for text only and text+image runs, undiversified and with the Div_{Qclust} diversification. All the results are presented in the first four lines of Table 1 for the precision at 20, the second part (between line 5 to line 8) of this table for the cluster recall at 20, and the F measure in the last part (between lines 9 and 12) of the table. In each part of this table, the values underlined are the official results we obtained, the values in bold are the best result for one run (with our without diversification), and the results in italic are the best diversification results per run. The parameters are: λ, text language model, is set to 1500, using the Zettair system [8]; threshold t_v is set to 0.99 and α, used in the linear combination, to 0.55 .

From the first part of table 1, we notice that for the precision, image and image+text runs using no diversification outperform greatly any diversification. For these runs, the diversification based on visual features is above the two other diversification schemes. Diversity impacts to a lesser extent the text only results. The reason is that the results given for the text are already diversified, so the diversification processes do not modify the ranking. The text only run and the visual diversification have a higher precision at 20 than the non diversified results, but this improvement is only around 3%. From these results, we conclude that the diversification processes degrade the results, by shifting non relevant documents to the top 20 for image and image+text runs. For the text, the diversification has less impact because the documents that are shift up are also relevant. In any cases, text+image runs outperform text only runs, and text only runs outperform image runs.

We study now the impact of the diversification processes on the cluster recall. From the second part of table 1, we see that the diversification process dependent on the query type provides consistently higher cluster recall values than the non diversified runs. Visual based diversity outperforms also slightly the non diversified results. In both cases however, the image only runs do not benefit

Table 1. P20 Results for LIG with 3 runs and 3 diversifications

Evaluation	IMG	TXT	TXTIMG
P20 NoDiv	**0.1795**	0.2026	**0.3218**
P20 Div_{dmy}	0.1064 (-59%)	0.1897 (-6%)	0.2167 (-33%)
P20 Div_{vis}	textit0.1269 (-29%)	***0.2077*** (+3%)	*0.2833* (-12%)
P20 Div_{Qclust}	0.1218 (-32%)	0.1897 (-6%)	0.2462 (-23%)
CR20 NoDiv	0.2518	0.2764	0.3795
CR20 Div_{dmy}	0.1907 (-24%)	0.2315 (-16%)	0.3086 (-19%)
CR20 Div_{vis}	0.2439 (-3%)	0.2857 (+3%)	0.3927 (+3%)
CR20 Div_{Qclust}	***0.2551*** (+1%)	***0.3431*** (+24%)	***0.4209*** (+11%)
F1 NoDiv	**0.2096**	0.2338	**0.3483**
F1 Div_{dmy}	0.1366 (-35%)	0.2085 (-11%)	0.2546 (-27%)
F1 Div_{vis}	*0.1670* (-20%)	0.2405 (+3%)	*0.3291* (-6%)
F1 Div_{Qclust}	0.1648 (-21%)	***0.2443*** (+4%)	0.3106 (-11%)

from the diversification, because the images that are shift up do not belong to "relevant" clusters. For the cluster recall also, text+image runs outperform text only runs, and text only runs outperform image runs.

Regarding the third part of table 1 corresponding to the official evaluation of ImageCLEFphoto, the best results are obtained without diversification for image only and image+text runs. For text only runs however, the visual and the query based diversifications achieve slightly better results than no diversification. If we compare to the official results, our best image only run should have been at position 16 on 34, our best text only run is at position 347 on 400, and the image+text run is at position 225 on 609 runs.

2.6 Discussion

When comparing our different runs, we found that text+image runs consistently outperform text only runs. When considering diversification processes, we show that the diversification lower the results for the F1 measure for image only and image+test runs. These results come from the fact that the precision that we loose when diversifying is not balanced by the gain coming from the cluster recall. For the precision values, non diversified results outperfom diversified runs. For the cluster recall, diversified runs outperfom non diversified results. So, we conclude that precision and recall do not benefit both from the diversification process. Such an antagonism between recall and precision is not new in IR, but we need to study in detail the effect of our diversification process.

3 LIG Runs on ImageCLEFmed Collection

In previous ImageCLEFmed editions, conceptual indexing based on UMLS provided some of the best systems on text, significantly outperforming standard keyword indexing (cf. [4]). Similar results have also been obtained on TREC genomics by [10]. Some works, as [4], go beyond the use of concepts by exploiting relations between them as a way to better capture the content of queries and documents and allow a matching at an abstract semantic level. Along this line, [2] and [4] have proposed extensions of the language modeling approach that can deal with dependencies, syntactic ones in the case of [2], either syntactic or semantic in the case of [4]. if such models allow one to take in account advanced representations in an efficient IR model, they do not solve the problems associated with the difficulty of detecting such representations in text. We address here this problem by combining different graph extraction methods, on both queries and documents. We first present an overview of the graph model and how we combine different models. We then describe the graph extraction process and the results obtained on the CLEF 2008 medical retrieval task.

3.1 Graph Model

We used the graph model proposed in [4]. In this model query and documents are represented as graphs $G = (C, E)$, where C represents the node set of the graph and E the relation set, that they assumed labeled. E is defined by an application that indicates the labels associated to relations. The probability that the query graph Gq is generated by the document model M_d is:

$$P(G_q|M_d) = P(C|M_d)\,P(E|C,M_d) \tag{2}$$

where $P(C|M_d)$ corresponds to the node contribution and $P(E|c, M_d)$ to the edge contribution. This model assumes that, conditioned on M_d, query concepts are independent of one another. Thus the node contribution is computed through a standard unigram model based on a Jelinek-Mercer smoothing with a parameter λ_u.

For the relations, the model assumes that E is an application from $C \times C$ in $P(\mathcal{L})$[1] that associates to each relation a set of labels. Thus the edge contribution can be decomposed as:

$$\begin{aligned} P(E|C,M_d) &= \textstyle\prod_{i,j\in C, i\leq j} P(E(c_i,c_j) = \mathcal{L}|c_i,c_j,M_d) \\ &= \textstyle\prod_{i,j\in C, i\leq j} \prod_{label\in\mathcal{L}_{ij}} P(e(c_i,c_j) = label|c_i,c_j,M_d) \end{aligned} \tag{3}$$

where $E(c_i, c_j) = \mathcal{L}$ indicates that a relation exists between c_i and c_j and that this relation is associated to the label set $\mathcal{L}$ and $e(q_i, q_j) = label$ indicates that there is a relation between q_i and q_j, the label set of which contains *label*. $P(e(c_i, c_j) = label|c_i, c_j, M_d)$ is computed through a Jelinek-Mercer smoothing, using a parameter λ_r.

We present now the methods used to combine different graphs (i.e. different structures obtained from different analyses of the queries and/or documents) in the model presented above. First, we group the different analysis of a query. To do so, we assume that a query is represented by a set of graphs $Q = G_q$ and that the probability of a set of graphs assuming a document model is computed by the product of the probability of each query graph:

$$P(Q = \{G_q\}|M_d) = \prod_{G_q} P(G_q|M_d) \tag{4}$$

Thus, this model considers that a relevant document model must generate all the possible analyses of a query Q.

Second, we group the different analyses of a document. To do so, we assume that a query can be generated by different models of the same document M_d^* (i.e. a set of models). As a result of this generation process, we keep the higher probability among the different models of the document:

$$P(G_q|M_d^*) = argmax_{M_d\in M_d^*} P(G_q|M_d) \tag{5}$$

3.2 Evaluation

Graph Detections. UMLS is a good candidate as a knowledge source for medical text indexing. The large set of concepts associated to term variants allows

[1] $\mathcal{L}$ is the set of all possible labels for a relationship and $\mathcal{P}(\mathcal{L})$ is the set of sets of $\mathcal{L}$.

Table 2. Results for mean average precision (MAP) and precision at five documents (P@5) with combination of models and extended text descriptions

	2006-2007[4]		2008		2008 with citation parragraph		2008 with citation line	
model	MP	MP TT	MP	MP TT	MP	MP TT	MP	MP TT
MAP								
UNI	0.298	0.306	0.254	**0.28**	0.222	0.240	0.265	0.276
RET	0.304	0.313	0.248	0.270	0.252	277	0.252	277
P@5								
UNI	0.487	0.490	0.447	0.444	0.373	0.373	0.460	**0.467**
RET	0.531	0.516	0.453	0.453	0.447	0.453	**0.467**	0.453

one to build on top of it a full scale conceptual indexing system. In UMLS, all concepts are assigned to at least one semantic type from the Semantic Network. This enables to detect semantic relation between concepts. With UMLS, we produce graphs from a text in two steps: first by detecting concepts, second by detecting relations between detected concepts. These two steps are detailed in [3]. With variations on concept detection we obtained two graph detection methods: the first one (MP) uses a term mapping over MiniPar[2] analysis and the second one (TT) uses a term mapping over TreeTagger[3] analysis.

Experiments. We present here the results obtain on CLEFmed 2008 [1]. Last year results showed that merging queries improves the results. As a consequence, we use here the two graphs detected on a query (MP and TT analysis). We test two model variations: a first one (UNI) that only uses the node contribution and a second one (RET) that uses both nodes and relations. For each model, we learn its parameter with the MAP on the previous ImageCLEFmed corpus. We then test the parameters on the collection analyzed with MiniPar (MP) and on the collection analyzed with both MiniPar and TreeTagger (MPTT) using the combinations proposed in this paper. We also test the use of part of the article that contains the images. To do so, we add to the image description the text article that corresponds to the paragraph or the line where the image is cited. The results are presented on table 3.2. The best results on the 2008 collection are those obtained with the unigram model over both MiniPar and TreeTagger. On this collection, integrating relations only improves the P@5. The results also show that adding citations does not improve the system.

3.3 Discussion

The use of graphs allows one to attain a good precision. Combining models on documents provides a better coverage of the documents but does not help to

[2] http://www.cs.ualberta.ca/ lindek/minipar.htm

[3] http://www.ims.uni-stuttgart.de/projekte/corplex/TreeTagger/DecisionTreeTagger.html

improve the selection of concepts. The combination thus mainly improves the MAP. Extending description of images, using references, lowers the results. This may come from the fact that we consider here the added text, potentially less precise, similar to the description. In the future, we will incorporate this text directly in our model.

Acknowledgement

This work was partially supported by the French National Agency of Research (ANR-06-MDCA-002).

References

[1] Müller, H., Kalpathy-Cramer, J., Kahn Jr., C., Hatt, W., Bedrick, S., Hersh, W.: Overview of the ImageCLEFmed 2008 medical image retrieval task. In: Peters, C., et al. (eds.) CLEF 2008. LNCS, vol. 5706, pp. 512–522. Springer, Heidelberg (2009)
[2] Jianfeng, G., Jian-Yun, N., Guangyuan, W., Guihong, C.: Dependence language model for information retrieval. In: SIGIR 2004 (2004)
[3] Maisonnasse, L., Gaussier, E., Chevallet, J.P.: Multi-relation modeling on multi concept extraction lig participation at imageclefmed. In: Working Notes for the CLEF 2008 Workshop, Aarhus, Denmark, September 17-19 (2008)
[4] Maisonnasse, L., Gaussier, E., Chevallet, J.P.: Multiplying concept sources for graph modeling. In: Peters, C., Jijkoun, V., Mandl, T., Müller, H., Oard, D.W., Peñas, A., Petras, V., Santos, D. (eds.) CLEF 2007. LNCS, vol. 5152, pp. 585–592. Springer, Heidelberg (2008)
[5] Ponte, J.M., Croft, W.B.: A language modeling approach to information retrieval. In: Proceedings of the ACM SIGIR, pp. 275–281 (1998)
[6] Robertson, S.E., Walker, S., Jones, S., Hancock-beaulieu, M.M., Gatford, M.: Okapi at trec-3, pp. 109–126 (1995)
[7] Salton, G., McGill, M.J.: Introduction to modern information retrieval. McGraw-Hill, New York (1983)
[8] The Zettair search engine, `http://www.seg.rmit.edu.au/zettair/`
[9] Zhai, C., Lafferty, J.: A study of smoothing methods for language models applied to information retrieval. ACM Trans. Inf. Syst. 22(2), 179–214 (2004)
[10] Zhou, W., Yu, C., Smalheiser, N., Torvik, V., Hong, J.: Knowledge-intensive conceptual retrieval and passage extraction of biomedical literature. In: Research and Development in Information Retrieval (2007)

The MedGIFT Group at ImageCLEF 2008

Xin Zhou[1], Julien Gobeill[1], and Henning Müller[1,2]

[1] Geneva University Hospitals and University of Geneva, Switzerland
[2] University of Applied Sciences Western Switzerland, Sierre, Switzerland
henning.mueller@sim.hcuge.ch

Abstract. This article describes the participation of the MedGIFT research group at the 2008 ImageCLEFmed image retrieval benchmark. We concentrated on the two tasks concerning medical imaging. The visual information analysis is mainly based on the GNU Image Finding Tool (GIFT). Other information such as textual information and aspect ratio were integrated to improve our results. The main techniques are similar to past years, with tuning a few parameters to improve results.

For the visual tasks it becomes clear that the baseline GIFT runs do not have the same performance as some more sophisticated and more modern techniques. GIFT can be seen as a baseline for the visual retrieval as it has been used for the past five years in ImageCLEF. Due to time constraints not all optimizations could be performed and no relevance feedback was used, one of the strong points of GIFT. Still, a clear difference in performance can be observed depending on the various optimizations applied, and the difference with the best groups is smaller than in past years.

1 Introduction

The MedGIFT group of the Geneva University Hospitals and the University of Geneva contributes regularly to ImageCLEF[1]. The principle domains of interest are medical retrieval and medical image annotation [1]. More details on the ImageCLEF databases, topics, and a comparison of all medical retrieval results can be found in [2]. In [10] the medical classification is detailed.

2 Basic Retrieval Strategies

This section describes the basic technologies that were used for the retrieval by the medGIFT group. More details on optimizations per task are given in the results section.

2.1 Text Retrieval Approach

The text retrieval approach used in 2008 is detailed in a paper of the text retrieval group of the Geneva University Hospitals [3]. It is similar to approaches in past years, where queries and documents were translated into MeSH (Medical Subject Heading) terms.

[1] http://www.imageclef.org/

C. Peters et al. (Eds.): CLEF 2008, LNCS 5706, pp. 712–718, 2009.

2.2 Visual Retrieval Techniques

The technology used for the visual retrieval is mainly taken from the *Viper*[2] (Visual Information Processing for Enhanced Retrieval) project [4]. Outcome of the *Viper* project is the GNU Image Finding Tool, *GIFT*[3]. This tool is open source and can be used by other participants of ImageCLEF as well. A ranked list of visually similar images for all query topics was made available for participants and serves as baseline to measure the quality of submissions. Feature sets used by *GIFT* are:

- Local color features at different scales by partitioning the images successively into four equally sized regions (four times) and taking the mode color of each region as a descriptor;
- global color features in the form of a color histogram, compared by a simple histogram intersection;
- local texture features by partitioning the image and applying Gabor filters in various scales and directions, quantized into 10 strengths;
- global texture features represented as a simple histogram of responses of the local Gabor filters in various directions and scales.

A particularity of *GIFT* is that it uses many techniques well–known from text retrieval. Most visual features are quantized and the feature space is similar to the distribution of words in texts. A standard *tf/idf* weighting is used and the query weights are normalized by the results of the query itself. The histogram features are compared based on a histogram intersection [5].

3 Results

In this section, the results and technical details for the two medical tasks of ImageCLEF 2008 are detailed.

3.1 Medical Image Retrieval

Results of our runs for the medical retrieval task are shown in Table 1 highlighting the most important performance measures such as MAP (Mean Average Precision), Bpref, and early precision. 3 purely visual retrieval runs using GIFT with 4 gray levels (*GIFT4*), 8 gray levels (*GIFT8*), and 16 gray levels (*GIFT16*) were submitted for evaluation. Using GIFT with 8 gray levels gives the best result for purely visual retrieval. Increasing the number of gray levels further decreases basically all performance measures.

Purely visual retrieval results in past years were often not robust [6]. Thus, more effort was invested into mixing visual retrieval and textual retrieval. The textual retrieval run (*HUG–BL–EN*) was provided by the text retrieval group of the Geneva University Hospitals [3]. This text retrieval run was used for several

[2] `http://viper.unige.ch/`

[3] `http://www.gnu.org/software/gift/`

Table 1. Results of the runs submitted to the medical retrieval task

Run	run_type	MAP	bpref	P10	P30	num_ret
best system	Mixed	0.2908	0.327	0.4267	0.3956	30000
HUG–BL–EN	Textual	0.1365	0.2053	0.26	0.24	28095
GE–GE_GIFT8_EN0.5	Mixed	0.0848	0.1927	0.2433	0.2378	29999
GE–GE_EN_reGIFT8	Mixed	0.0815	0.1896	0.2267	0.2267	29452
GE–GE_EN_GIFT8_mix	Mixed	0.0812	0.1867	0.24	0.2467	29999
GE–GE_GIFT8_EN0.9	Mixed	0.0731	0.1248	0.2733	0.25	30000
GE–GE_GIFT8_reEN	Mixed	0.0724	0.1244	0.2433	0.2544	30000
GE–GE_GIFT4	Visual	0.0315	0.0901	0.1433	0.12	30000
GE–GE_GIFT8	Visual	0.0349	0.0898	0.17	0.1511	30000
GE–GE_GIFT16	Visual	0.0255	0.0715	0.1333	0.1111	30000

combinations with our best–performing visual run (*GIFT8*). In total, 5 mixed–media automatic runs were generated based on these runs with the following combination strategies:

- combination of textual and visual runs with equal weight (*GIFT8_EN0.5*);
- reordering of the ranked lists of the textual run based on the visual run (*EN_reGIFT8*);
- mixing visual and textual retrieval by giving varying weights based on the kind of topic: for visual topics the visual run is at 90%, for textual topics the visual run is at 10%, for mixed topics the visual run is at 50% (*EN_GIFT8_mix*);
- combining textual and visual runs but favoring the text (90%) over the visual information (10%) (*GIFT8_EN0.9*);
- reordering the visual run based on the textual run (*GIFT8_reEN*).

Mixing two runs with varying weights based on the topic type (*EN_GIFT8_mix*) gives second best early precision (P30), and third best MAP among the 5 runs. The best MAP is reached by simply combining textual and visual runs with equal weight (*GIFT8_EN0.5*). Favoring the textual run (*GIFT8_EN0.9*) gives best early precision, but surprisingly poor MAP. Compared to the original text runs, the combination with our visual run improves early precision slightly, but reduces MAP significantly.

3.2 Medical Image Annotation

For the medical image annotation task, the basic GIFT system was used for the feature extraction as in previous years but with significant changes [7]. Aspect ratio as feature and annotation by axis were again used for our participation in 2008. Main new approaches for 2008 were a modified classification strategy and changed parameter settings.

The annotation is based on the known labels of similar images of the training set retrieved by GIFT. In [7], the classification strategies were regrouped around a kNN (k Nearest Neighbor) approach and a voting–based approach. The voting–based approach takes into account the n most similar images. In 2008, we took into account two other factors: the frequency of images of each class in the training data and the hierarchy information inside each axis of the IRMA (Image Retrieval in Medical Applications) code.

One problem of classifying images with training data is that the classification strategy most often favors large classes in the training data and punishes small ones, as images of large classes have a higher chance to be selected. The frequency of each class in the training data is analyzed to avoid this bias. Such a dynamic kNN approach is then used instead of a standard kNN approach to give a different k value for each class. The disadvantages for the smaller classes are thus reduced. In previous years, the distribution of classes in the test data was the same as in the training data, which is not the case in 2008. Thus, using a dynamic kNN approach to avoid the bias is even more necessary.

Another useful information is the hierarchy information inside each code axis (the IRMA code in total contains four). The output of the classification per axis is usually an entire axis or a wild card for the entire axis. Another possibility is to chop only the lowest level (the last letter) of each axis. The remainder can then be used for a second round of classification. This additional step allows to use less wild cards in the classification process and thus can potentially improve the score.

The results of our basic runs and the best overall system are presented in Table 2. Three submitted runs use the kNN approach with classification for the entire code (*kNN*), classification per axis (*akNN*), and dynamic kNN classification per axis (*adkNN*). Dynamic kNN obtains the best result of these three approaches. Three other runs use a voting–based approach described in [7]: per axis with descending vote (*vad*), per axis with chopping letter by letter with a descending vote (*vcad*), and per axis with chopping letter by letter using equal weights (*vca*). The confidence thresholds were all set to 0.5 (as this obtained good results in past years) and we submitted the runs that take into account the first 5 similar images, only. In tests this lead to good results and no optimization for this parameter was tried. The best results among these runs are obtained using the voting strategy per axis with descending vote(*vad*). Surprisingly, chopping the lowest level and redoing the classification for the rest gives slightly worse results. To detail

Table 2. Results of the main runs submitted by MedGIFT to the medical image annotation task

run ID	score
best system	74.92
GE–GIFT0.9_0.5_vad_5.run	209.70
GE–GIFT0.9_0.5_vcad_5.run	210.93
GE–GIFT0.9_0.5_vca_5.run	217.34
GE–GIFT0.9_adkNN_2.run	233.02
GE–GIFT0.9_akNN_2.run	241.11
GE–GIFT0.9_kNN_2.run	251.97

Table 3. Classification per axis with and without a chopping strategy

run ID	score
GE–GIFT0.9_0.5_vad_5.run	209.70
GE–GIFT0.9_0.6_vad_5.run	198.79
GE–GIFT0.9_0.7_vad_5.run	198.79
GE–GIFT0.9_0.8_vad_5.run	198.79
GE–GIFT0.9_0.9_vad_5.run	208.23
GE–GIFT0.9_0.5_vcad_5.run	210.93
GE–GIFT0.9_0.6_vcad_5.run	191.53
GE–GIFT0.9_0.7_vcad_5.run	191.53
GE–GIFT0.9_0.8_vcad_5.run	191.53
GE–GIFT0.9_0.9_vcad_5.run	181.17

the two best–performing techniques and optimize results a further comparison is performed with varying parameters and presented in Table 3.

Chopping at the lowest level and re–classification performs better but only when using a high threshold.

The two best groups (IDIAP and MIPLAB) in the classification competition in 2008 both use a similar approach for their visual characteristics. This *bag of features* approach is based on neighborhoods of interest points randomly selected from the image, followed by a Support Vector Machine (*SVM*)–based classification approach [8,9]. Both use a large number of features (1'000–5'000 patches per image) and Principle Component Analysis (*PCA*) to reduce the dimensionality. The IDIAP group duplicated the instances of small classes in the training data in order to reduce the possibilities that the large classes mask the small ones [8].

An important aspect of the evaluation is to understand how much of the performance is based on the visual features used, and how much based on the machine learning techniques. Table 4 shows a comparison to have an idea about the influence of the visual features only. To minimize the impact of machine learning techniques, classifiers for patch–based approaches (such as SVM) will be replaced by a simple Euclidean distance, which translates an annotation approach into a retrieval one. GIFT is used as it is to give a baseline. The presumption is: appropriate features should rank images of the same class as "close" without the help from machine learning algorithm. The evaluation is based on the 1000 images in test dataset. For each of them, with selected feature and distance function, 100 nearest images were extracted from the training dataset. The goal is to know among the 1000 images in test dataset, how many of them have found at least one image of the same class. Results were obtained with 100, 30 and 10 nearest images. The comparison shows that particularly the axis anatomy and thus also for the full code, the patch based features work significantly better.

On the other hand we can also see, that a large number of images has no correspondence in the top N=10 results for neither of the two feature sets. This means that a large part of the higher performance of these approaches is not due to the features alone but to a combination of features, distance measures and the learning approach.

Table 4. Comparison of GIFT features with a patch–based approach with respect to the number of test images that have at least one exact correspondence in the top N results of the system (on the axis level and for the full code; T=type, modality, D=direction, A=Anatomy, B=Bio system)

Feature	entire code	axis T	axis D	axis A	axis B
in the 100 most similar images					
GIFT	736	996	949	798	987
random patches	821	994	972	882	984
in the 30 most similar images					
GIFT	691	993	919	754	976
random patches	752	990	923	821	980
in the 10 most similar images					
GIFT	621	985	847	682	966
random patches	682	982	870	743	967

4 Conclusions

For the medical retrieval task only very few purely visual runs (8 runs among 111) were submitted by the participants. The pools of the relevance judgments can thus be slightly biased and even further worsen results such as MAP for these runs. All visual approaches obtain poor scores underlying the high–quality annotations, and tasks that are much more oriented towards text–based approaches. The use of text alone is in our test even better than the combinations with visual retrieval. Few groups actually manage to increase performance with a visual approach over purely textual retrieval. Only early precision can be improved through the combination of textual runs with visual runs. The visual baseline seems to be of insufficient quality for really improving the combined runs significantly and better visual approaches seem necessary. A small number of gray levels still gives best results in our tests.

Differently from previous years, the training dataset and the test dataset do not have the same distribution of classes. Goal of this was to force participants to use the supplied hierarchy for classification including wild cards [10]. An analysis on the wild card frequency of participants is also given in the overview article, indicating a relationship between the wild card frequency and the number of training images available. The difference between our runs and the best techniques was reduced compared to previous years. The voting–based approaches perform generally better than the simple kNN approaches. Classifying each axis separately with a suitable threshold gives best results in our tests. When the threshold cannot be reached in the first step, chopping the lowest level and repeating the classification for the remaining levels can improve the result slightly. The advantage of the chopping strategy is that the classification is repeated iteratively. High threshold values increase the confidence without totally blocking the classification. The idea of the IDIAP group of oversampling the small classes in the training data is easy to implement and considerably increases performance.

Acknowledgments

This study was partially supported by the Swiss National Science Foundation (Grant 200020–118638/1), the HES SO with the BeMeVIS project, and the European Union in the 6th Framework Program through the KnowARC project (Grant IST 032691).

References

1. Müller, H., Deselaers, T., Kim, E., Kalpathy-Cramer, J., Deserno, T.M., Clough, P., Hersh, W.: Overview of the imageCLEFmed 2007 medical retrieval and medical annotation tasks. In: Peters, C., Jijkoun, V., Mandl, T., Müller, H., Oard, D.W., Peñas, A., Petras, V., Santos, D. (eds.) CLEF 2007. LNCS, vol. 5152, pp. 472–491. Springer, Heidelberg (2008)
2. Müller, H., Kalpathy-Cramer, J., Kahn Jr., C.E., Hatt, W., Bedrick, S., Hersh, W.: Overview of the ImageCLEFmed 2008 medical image retrieval task. In: Peters, C., et al. (eds.) CLEF 2008. LNCS, vol. 5706, pp. 512–522. Springer, Heidelberg (2009)
3. Gobeill, J., Ruch, P., Zhou, X.: Text-only cross language image search at medical ImageCLEF 2008. In: Working Notes of the 2008 CLEF Workshop, Aarhus, Denmark (September 2008)
4. Squire, D.M., Müller, W., Müller, H., Pun, T.: Content-based query of image databases: inspirations from text retrieval. In: Ersboll, B.K., Johansen, P. (eds.) Pattern Recognition Letters (Selected Papers from The 11th Scandinavian Conference on Image Analysis SCIA 1999), vol. 21(13-14), pp. 1193–1198 (2000)
5. Swain, M.J., Ballard, D.H.: Color indexing. International Journal of Computer Vision 7(1), 11–32 (1991)
6. Zhou, X., Gobeill, J., Ruch, P., Müller, H.: University and hospitals of geneva participating at imageCLEF 2007. In: Peters, C., Jijkoun, V., Mandl, T., Müller, H., Oard, D.W., Peñas, A., Petras, V., Santos, D. (eds.) CLEF 2007. LNCS, vol. 5152, pp. 649–656. Springer, Heidelberg (2008)
7. Zhou, X., Depeursinge, A., Müller, H.: Hierarchial classification using a frequency–based weighting and simple visual features. Pattern Recognition Letters 29(15), 2011–2017 (2008)
8. Tommasi, T., Orabona, F., Caputo, B.: CLEF2008 image annotation task: an SVM confidence-based approach. In: Working Notes of the 2008 CLEF Workshop, Aarhus, Denmark (September 2008)
9. Avni, U., Goldberger, J., Greenspan, H.: TAU MIPLAB at ImageClef 2008. In: Working Notes of the 2008 CLEF Workshop, Aarhus, Denmark (September 2008)
10. Deselaers, T., Deserno, T.M.: Medical image annotation in ImageCLEF 2008. In: Peters, C., et al. (eds.) CLEF 2008. LNCS, vol. 5706, pp. 523–530. Springer, Heidelberg (2009)

MIRACLE at ImageCLEFmed 2008: Semantic vs. Statistical Strategies for Topic Expansion

Sara Lana-Serrano[1,3], Julio Villena-Román[2,3], and José Carlos González-Cristóbal[1,3]

[1] Universidad Politécnica de Madrid
[2] Universidad Carlos III de Madrid
[3] DAEDALUS - Data, Decisions and Language, S.A.
slana@diatel.upm.es, jvillena@it.uc3m.es,
josecarlos.gonzalez@upm.es

Abstract. This paper describes the participation of MIRACLE research consortium at the ImageCLEFmed task of ImageCLEF 2008. The main goal of our participation this year is to evaluate different text-based topic expansion approaches: methods based on linguistic information such as thesauri or knowledge bases, and statistical techniques based mainly on term frequency. First a common baseline algorithm is used to process the document collection: text extraction, medical-vocabulary recognition, tokenization, conversion to lowercase, filtering, stemming and indexing and retrieval. Then different expansion techniques are applied. For the semantic expansion, the MeSH concept hierarchy using UMLS entities as basic root elements was used. The statistical method expanded the topics using the apriori algorithm. Relevance-feedback techniques were also used.

Keywords: Image retrieval, medical domain-specific vocabulary, thesaurus, linguistic engineering, information retrieval, indexing, topic expansion, relevance feedback, ImageCLEF Medical Retrieval Task, ImageCLEF, CLEF, 2008.

1 Introduction

MIRACLE team is a research consortium formed by research groups of three different Spanish universities (UPM, UAM and UC3M) along with the private company DAEDALUS. This paper reviews our participation [1] at the ImageCLEFmed task of ImageCLEF 2008 [2]. The main goal of our participation this year was to evaluate different text-based query expansion techniques using different approaches: methods based on linguistic information such as thesauri or knowledge bases, and statistical techniques based mainly on term frequency. All experiments were fully automatic, with no manual intervention.

2 Description of the System

Figure 1 gives an overview of the system architecture. It is composed of four different modules: the textual (text-based) retrieval module, which indexes medical case descriptions

C. Peters et al. (Eds.): CLEF 2008, LNCS 5706, pp. 719–723, 2009.

in order to search and find the most relevant ones to the text of the topic; the expander module, which performs the expansion of the content of documents and/or topics with related terms using textual or statistical algorithms; the relevance-feedback module, which allows to execute reformulated queries that include the results of an initial seed query; and, finally, the result combination module, which merges, if necessary, the result lists provided by the previous subsystems.

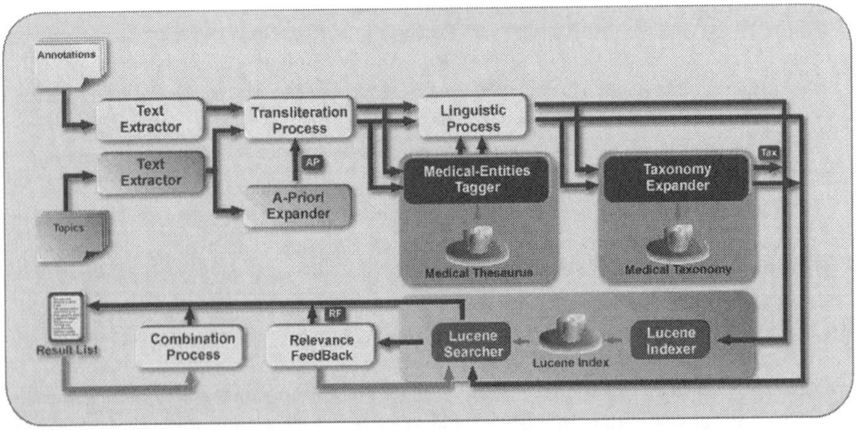

Fig. 1. Overview of the system architecture

Both topics and documents are parsed and tagged to unify all terms into concepts of medical entities, i.e., concept identifiers [3] are used instead of terms in the text-based information retrieval. For this purpose, a terminological dictionary was created by using a subset of the Unified Medical Language System (UMLS) metathesaurus [4] containing 3,211,169 entries matching 1,215,749 medical concepts in English, French and German.

A common baseline algorithm was used in all experiments to process the document collection. First, annotations and metadata are extracted from the medical cases and then case descriptions and topics are parsed and tagged using the UMLS-based terminological dictionary to identify and disambiguate medical terms. Next, the tokenization process extracts the basic textual components, specifically common single words, numbers and tagged entities. Then all terms are converted to lowercase, stopwords are filtered out [5] and a stemming process is applied to each term to be indexed or used for retrieval. Standard Porter stemmers [6] for each considered language have been used. Last, Lucene system [7] was used as the information retrieval engine for the whole retrieval task.

Then expansion techniques are applied to enrich the original contents of both topics and documents. We studied and compared a semantic- versus a statistical-based technique. For the semantic expansion, MeSH concept hierarchy [8] was used on the UMLS entities detected in document and topics as basic root elements to expand with their hyponyms (i.e., other entities whose semantic range is included within that of the root entity). The statistical method consisted of expanding the topics using the Agrawal's apriori algorithm [9]. First, a term-document matrix is built using the UMLS entities found in the document corpus. Then the classical apriori algorithm is applied in order to discover out association rules among UMLS entities in the corpus. The idea is to find rules whose antecedent contains any of the UMLS entities that have been identified in the topic, because the consequent of those rules would be a UMLS entity that is related to that topic. These related entities are used to expand the topic. As a threshold, rules with a confidence value greater than 0.5 are selected.

In addition, relevance-feedback techniques were also used. The top *M* UMLS entities of each of the top *N* result documents were extracted and weighted by a factor that is proportional to their document frequency to reformulate a new query that is executed once again to get the final result list.

3 Experiments and Results

Results are presented in Table 1, which shows the number of relevant documents retrieved, the mean average precision and the precision at several top results. Our best results are highlighted in bold. The table also includes the results of the best run, shown in italics.

Table 1. Results of experiments

Run Identifier[1]	RelRet	MAP	P5	P10	P30
MirBaselineEN	1861	**0.27**	**0.51**	**0.47**	0.39
MirAPEN	1773	0.25	0.49	0.46	**0.39**
MirTaxEN	**1867**	0.25	0.38	0.37	0.37
MirRF0505EN	1372	0.11	0.28	0.24	0.24
MirRFTax1005EN	1260	0.07	0.15	0.13	0.14
MirRF1005EN	1248	0.07	0.22	0.16	0.15
MirRFTax1005DE	461	0.05	0.09	0.09	0.06
MirRFTax1005FR	823	0.07	0.13	0.11	0.09
EXPPRFNegativaMesh	*2165*	*0.29*	*0.49*	*0.46*	*0.41*

[1] Baseline (stem+stopwords+UMLS tagging), AP (apriori topic expansion), Tax (MeSH topic expansion), RF<N><M> (relevance feedback), EN (English), FR (French), DE (German)

Our baseline experiment in English achieves the best result in terms of MAP and early precision. Moreover, MAP values are similar in practice for experiments using topic expansion, and noticeably worse (0.105 against 0.266) in the case of relevance-feedback. Overall, this run ranks 12th, compared to all groups, which is rather average.

4 Conclusions and Future Work

Apparently, no strategy for either topic expansion or specially relevance-feedback has proved to be useful. However, the post-workshop analysis showed that the main reason for the low precision values obtained in the experiments that included topic expansion techniques was that, in all cases, the OR operator was used to build the reformulated query, i.e., both the original terms and the expanded terms were combined with the OR operator. This implies that documents that contain any of those terms are considered as relevant, no matter if the term belongs to the original topic or it is included in the expansion process. We think that a combination of OR and AND operators should have been used to be sure that documents do contain the original topic terms and, optionally, any of the expanded terms: "($original_1$ OR $expanded_1$) AND ($original_2$ OR $expanded_2$)".

In addition, we found that the reranking algorithm used for combining the different results list is the reason for the low precision values obtained in the experiments that make use of the relevance-feedback methods. Other combination operators must be studied, in special those that assign a higher weight to documents that correspond to the initial query and a lower weight to documents found by the relevance feedback query.

There is a significant difference in the number of terms among languages. This might bias the results towards the best covered language, English in this case. A possible explanation is that the process of entity unification for the other languages is poor, due to the reduced coverage of the knowledge base.

Finally, as in previous participation, the value for early precisions (P5, P10) quickly decreases as more documents are considered for the calculation and therefore decreasing the final MAP value. Although the first results may be appropriate, we probably fail to filter non-relevant documents out of the result list, or perhaps to sort out relevant documents that are "more difficult" to find. More effort will be invested on this issue.

Acknowledgements. This work has been partially supported by the Spanish R+D National Plan, project BRAVO (Multilingual and Multimodal Answers Advanced Search–Information Retrieval), TIN2007-67407-C03-03, and by the R+D Regional Plan of Madrid, project MAVIR (Enhancing the Access and the Visibility of Networked Multilingual Information), S-0505/TIC/000267.

References

1. Lana-Serrano, S., Villena-Román, J., González-Cristóbal, J.C.: MIRACLE at ImageCLEFmed 2008: Evaluating Strategies for Automatic Topic Expansion. In: Working Notes of the 2008 CLEF Workshop, Aarhus, Denmark (2008)
2. Müller, H., Kalpathy-Cramer, J., Kahn Jr., C.E., Hatt, W., Bedrick, S., Hersh, W.: Overview of the ImageCLEFmed 2008 Medical Image Retrieval Task. In: Peters, C., et al. (eds.) CLEF 2008. LNCS, vol. 5706, pp. 512–522. Springer, Heidelberg (2009)
3. González, J.C., Villena, J., Moreno, C., Martínez, J.L.: Semiautomatic Extraction of Thesauri and Semantic Search in a Digital Image Archive. In: Integrating Technology and Culture: 10th International Conference on Electronic Publishing, ELPUB 2006, Bulgaria (2006)
4. U.S. National Library of Medicine. National Institutes of Health. Unified Medical Language System, `http://www.nlm.nih.gov/research/umls/`

5. University of Neuchatel. Page of resources for CLEF, http://www.unine.ch/info/clef
6. Porter, M.: Snowball stemmers and resources page, http://www.snowball.tartarus.org
7. Apache Lucene project, http://lucene.apache.org
8. U.S. National Library of Medicine. National Institutes of Health. Medical Subject Headings, http://www.nlm.nih.gov/mesh/
9. Agrawal, R., Srikan, R.: Fast algorithms for mining association rules. In: Proceedings of the International Conference on Very Large Data Bases, pp. 407–419 (1994)

Experiments in Calibration and Validation for Medical Content-Based Images Retrieval

Jose L. Delgado, Covadonga Rodrigo, and Gonzalo León

Dpto. Lenguajes y Sistemas Informáticos - E.T.S.I. Informática - U.N.E.D.
{jdelgado,covadonga}@lsi.uned.es, gleon1@alumno.uned.es

Abstract. We present a CBIR system (Content-based Image Retrieval). The system establishes a set of visual features which will be automatically generated. The sort of features is diverse and they are related to various concepts. After visual features calculation, a calibration process is performed whereby the system estimates the best weight for each feature. It uses a calibration algorithm (an iterative process) and a set of experiments, and the result is the influence of each feature in the main function that is used for the retrieval process. In image validation, the modifications to the main function are verified so as to ensure that the new function is better than the preceding one. Finally, the image retrieval process is performed according to the ImageCLEFmed rules, fully described in [2, 5]. The retrieval results have not been the expected ones, but they are a good starting for the future.

Keywords: CBIR, calibration, validation, JAI.

1 Introduction

Image characterization has been studied for a long time. If we focus on a graphical way (based upon the image observation and its properties), the work entails the extraction of sufficient significant data from these images to identify the images. This significant data (from now on we will call them only visual features) is a set of short-scope attributes that allows identification, or at least characterization, within a certain group. The sort of features is very diverse.

There are several image processing systems which pursue this aim (for instance, the GIFT[10]). This work is aimed at developing a system that is able to extract a set of visual features which will be further used for cataloguing and comparing purposes. The visual features are not based in textual metadata.

2 The System at a Glance

We introduced the concept of visual feature as an attribute that allows, alone or in combination with other attributes, establishing the membership of an image to a certain group. Obtaining visual features and its later use in the images identification has been applied in many fields, for instance in the biometric identification [7, 3], with great results. So there is a need in investigating the visual features which are likely to

C. Peters et al. (Eds.): CLEF 2008, LNCS 5706, pp. 724–727, 2009.

be extracted from an image, and to what extent they are suitable for their comparison. Another goal is to plan the system in a scalable way in terms of the visual features extraction software. In the same way, the system developed is able to:

- Store a collection of images (corpus) and their related visual features.
- Store a collection of classes/groups of visual features which can be applied to images and their influence in later comparisons.
- Compare the image's features with those belonging to the ones within the corpus.
- Combine the comparison's results provided by the different features when obtaining a unified measure of similarity, for its later sorting.

We have chosen a Java development starting from scratch, mainly due to the great number of libraries which are available for Java. In this sense, we can find JAI (Java Advanced Imaging), that is an optional library (not included in the standard distribution J2SE) allowing the reading and processing of images for extracting their visual features. Further-more, the use of the concurrent paradigm by means of threads was also considered to choose Java as the selected technology.

3 Selection of Visual Features

The number and source/nature of the features available was very diverse, so there are some visual features more connected to the image's source, whereas there are some other features obtained through a mathematical algorithm upon the image and its later processing requires a better knowledge. Hence, it is necessary to perform an analysis of these features and evaluate whether they are useful or not.

After performing some experimental assessments and addressing some indications shown in several articles (see [4, 6, 8, 1]), the number of features was fixed to 9 (Euclidean distance between two histograms, Arithmetic average from Histogram, Entropy, Local Entropy, Features related to Gabor's filters, Colour Bits, Aspect Ratio, Image Miniaturization and Size). The comparison procedure used was to give a value of 0 for very low similarity and of 1 when the similarity was higher. In all the cases, the JAI library has been used for the image reading and other related information.

4 Calibration and Validation

The calibration process leads to modifications of each feature's weight to increase the hit ratio in the group's prediction. It is possible for one feature to be better than any other one to classify an image and worse to classify other one. For this reason there is a need in finding an average weight of influence in terms of final similarity.

The calibration process is a non trivial task because it is necessary to select the best (possible) weight that improves our classification process for the same collection of images. For this reason, the calibration requires a later process named validation. The validation allows the system to decide whether the modification of the weights enclosed in the calibration enhances or not the results. To achieve this goal, the validation process utilizes a collection of images not used in the calibration, and a sort of validity is obtained for the weights calculated in the previous step.

To carry out the calibration process it was used the ImageCLEF information from the previous years. Specifically the database used was Casimage [9], that was divided in disjoint groups. A total of 27 groups were extracted from the 57.

The goal was to obtain the visual features' weights that make the similarity value, for each image within the group, as high as possible. A secondary objective is to get a similarity value for a certain image belonging to a group as low as possible when comparing to images belonging to other groups to prevent a wrong classification.

The calibration algorithm is a very time-consuming task. However, the algorithm has demonstrated to be the most suitable to obtain the best results compared to other algorithms evaluated. For a collection of images outside the corpus to calibrate the system, the hit ratio was increased from the 56% (that gave the same weight to all the visual features) to the 86% with the weights coming from the algorithm.

The calibration process is required since it is reasonable that the system improves future classifications when the weights for the right classifications are previously extracted (feedback). Table 1 illustrates the values obtained through several iterations:

Table 1. Calibration Algorithm results

Visual Feature	Weight
Euclidean Distance between Histograms	0.04474
Entropy	0.04517
Local Entropy	0.17401
Image Miniaturization	0.30959
Colour Bits	0.06239
Aspect Ratio	0.04784
Size	0.08883
Typical Desviation of Gabor Transform (0)	0.03907
Typical Desviation of Gabor Transform (-(1/2)π)	0.04099
Arithmetic Averages of Gabor Transform (0)	0.03919
Arithmetic Averages of Gabor Transform (-(1/2)π)	0.05582
Arithmetic Averages of the Histogram	0.05232

5 Analysis of the Results and Conclusions

This is our first participation with the ImageCLEF and we can say that we have not achieved the expected results. We have got the last position in most of the cases. The reason lies on the number of extracted images for each topic, always under the number of images provided by the rest of participants (dozens versus a thousand).

This fact can be explained from the point of view of our system. We defined some criteria to decide whether an image was relevant or not: for each topic, a list of images was generated, filtered by similarity and sorted descendantly. Afterwards, for each image within the list, the typical desviation of each image and its previous ten similar ones was calculated. When this value was under 0.01, i.e. when the similarity of this image with the previous ten was very close, the list was "cut" in that point, and the images provided were the preceding ones. The reason lies on the observation of the similarities in such ordered list and the fact that the similarity within the images at the first position decreased significantly, and after these first position it decreased softly.

The use of a lower value would have been more suitable. Above all, perhaps we have not taken into account of the fact of that some "good" images, due to the calibration, may fall to the latest position of the list and be discarded. Besides, it seems to be logical that scanning the entire list, the number of images retrieved would be bigger; this possibility was discarded due to the lack of efficiency and speed of the overall system. In any case, it will be a very relevant issue to consider for the future.

In the same way, the visual features used in the development should be revised (more visual features should be desirable and even required), as well as a significant improvement of the efficiency in the calibration process so as to get the best fits in less time; and finally, to make the process to be adapted depending on the source of the images, making the algorithm execute in a different and optimum fashion.

References

1. Rahman, M.M., Sood, V., Desai, B.C., Bhattacharya, P.: Cindi at Image CLEF 2006. In: Image Retrieval & Annotation Tasks for the General Photographic and Medical Image Collections. Dept. of Computer Science & Software Engineering. Concordia University (2006)
2. Deselaers, T., Deserno, T.M.: Medical Image Annotation in Image CLEF 2008. In: Peters, C., et al. (eds.) CLEF 2008. LNCS, vol. 5706, pp. 523–530. Springer, Heidelberg (2009)
3. Martín Rodríguez, F., Suárez López, F.J.: Identificación Dactilar Basada en Filtros de Gabor. Departamento de Teoría de la Señal y Comunicaciones. Universidad de Vigo
4. Besançon, R., Mollet, C.: Using Text and Image Retrieval Systems: Lic2m Experiments at Image CLEF 2006. In: Grupo CEA-LIST/LI2CM en Fontenay-aux-Roses - CEDEX (2006)
5. León, G., Delgado, J.L., Rodrigo, C., López, F., Sama, V.: Automatic system for extraction of content-based characteristics from digital images. In: Working Notes of the 2008 CLEF Workshop, Aarhus, Denmark (September 2008)
6. Clough, P., Müller, H., Sanderson, M.: The CLEF 2004 Cross-Language Image Retrieval Track. In: Peters, C., Clough, P., Gonzalo, J., Jones, G.J.F., Kluck, M., Magnini, B. (eds.) CLEF 2004. LNCS, vol. 3491, pp. 597–613. Springer, Heidelberg (2005)
7. Sánchez Ávila, C., Sánchez-Reíllo, R.: Sistemas de Identificación Biométrica mediante Patrón de Iris utilizando Representación Multiescala e Información de Fase. (1) Dpto. de Matemática Aplicada a las Tecnologías de la Información. E.T.S.I. Telecomunicación – UPM (2) Dpto. de Ingeniería Eléctrica, Electrónica y Automática. Universidad Carlos III de Madrid (UC3M)
8. Lacoste, C., Lim, J., Wei, X., Raccoceanu, D., Le Thi Hoang, D., Teodorescu, R., Vuillemot, N.: Ipal knowledge-base medical image retrieval in ImagcCLEFMed 2006. IPAL French-Singaporean Joint Lab (2006)
9. Casimage Corpus, `http://pubimage.hcuge.ch`
10. GIFT – The GNU Image-Finding Tool, `http://www.gnu.org/software/gift/`

MIRACLE at ImageCLEFannot 2008: Nearest Neighbour Classification of Image Feature Vectors for Medical Image Annotation

Sara Lana-Serrano[1,3], Julio Villena-Román[2,3]
José Carlos González-Cristóbal[1,3], and José Miguel Goñi-Menoyo[1]

[1] Universidad Politécnica de Madrid
[2] Universidad Carlos III de Madrid
[3] DAEDALUS - Data, Decisions and Language, S.A.
slana@diatel.upm.es, jvillena@it.uc3m.es
josecarlos.gonzalez@upm.es, josemiguel.goni@upm.es

Abstract. This paper describes the participation of MIRACLE research consortium at the ImageCLEF Medical Image Annotation task of ImageCLEF 2008. During the last year, our own image analysis system was developed, based on MATLAB. This system extracts a variety of global and local features including histogram, image statistics, Gabor features, fractal dimension, DCT and DWT coefficients, Tamura features and co-occurrence matrix statistics. A classifier based on the k-Nearest Neighbour algorithm is trained on the extracted image feature vectors to determine the IRMA code associated to a given image. The focus of our participation was mainly to test and evaluate this system in-depth and to compare among diverse configuration parameters such as number of images for the relevance feedback to use in the classification module.

Keywords: Information Retrieval, medical image, image annotation, classification, IRMA code, axis, learning algorithms, nearest-neighbour, machine learning, ImageCLEF Medical Automatic Image Annotation task, CLEF, 2008.

1 Introduction

The MIRACLE team is a research consortium formed by research groups of three different Spanish universities (Universidad Politécnica de Madrid, Universidad Autónoma de Madrid and Universidad Carlos III de Madrid) along with DAEDALUS, a private company founded as a spin-off of these groups and a leading company in the field of linguistic technologies in Spain.

This paper reviews our participation [1] at the Medical Image Annotation task of ImageCLEF 2008 [2]. While in previous participations [3] [4] we approached this task as a domain-independent machine learning problem, as our areas of expertise did not include image analysis research, a lot of effort was invested during the last year to develop our own image analysis system. Thus, the main purpose of our participation in this task was to test and evaluate this system in-depth and determine the optimum settings to use in the classification module. In the following sections, we will give an overview of our approach and describe and analyze the results achieved.

C. Peters et al. (Eds.): CLEF 2008, LNCS 5706, pp. 728–731, 2009.

2 Description of the System

Our system is composed of two different functional modules. First, the Feature Extraction module is in charge of the extraction of a variety of feature descriptors for a given image. It has been entirely developed during the last year, using MATLAB.

Each image is first converted to gray-scale and rescaled to 256x256 pixels. Then, the following feature descriptors are extracted: gray histogram (128 levels of gray), image statistics (mean, median, variance, maximum singular value, skewness and kurtosis), Gabor features (4 scales, 6 filter orientations), fractal dimension, Discrete Cosine Transform, Discrete Wavelet Transform, Tamura features (coarseness, contrast, directionality), and co-occurrence matrix statistics (energy, entropy, contrast, homogeneity, correlation). Both global features for the whole image and local features for 64x64 blocks are obtained. All features are linearly combined (weight=1), and no feature selection is carried out. The final feature vector contains 3,741 descriptors.

On the other hand, the Classification module determines the IRMA code associated to a given image, basically comparing its feature vector and the feature matrix of the training set. The classifier is internally composed of two blocks: an initial module in charge of selecting those images in the training set whose vectors are at a distance lower than a given threshold from the vector associated to the image to classify, and then a second module that actually generates the IRMA code, depending on the codes and similarity of nearby images.

3 Experimental Results and Discussion

Four runs were submitted. All of them are based on the classical k-Nearest Neighbour algorithm [5] with a specific adaptation to generate the output class. The IRMA code is generated from the combination of the codes of the first k images in the training set that are most similar to the image to classify. The combination consists of a simple "addition" of code strings characters in which, if both characters are different, the result is the wildcard "*" representing the ambiguity ("hesitation" to choose).

Additionally, two runs use relevance feedback (RF) with the first n images in the training set that are at a lowest distance. Feature vectors of those images are added and averaged to build a new feature vector that is used for querying the system again.

Results are shown in Table 1. The "Error score" is defined [2] so as to penalize wrong decisions in which there are few possible choices over wrong decisions with many possible choices. Furthermore, it also penalizes wrong decisions higher up in the IRMA code hierarchy over wrong decisions lower down in the hierarchy. The "Well Classified" column shows the images with complete correct predicted code.

Table 1. Results of experiments

Run[1]	Error Score	Well Classified
2I-0F	190.38	**219**
3I-0F	**187.90**	144
2I-2F	190.38	**219**
3I-2F	194.26	167

[1] Run identifiers hold k (neighbours) and n (images for relevance feedback)

The best score is achieved by the run that combines the codes of the first 3 images, with no relevance feedback. Moreover, runs using the codes of the first 2 images seem to get the same final score no matter if relevance feedback is considered or not. Although it is not shown in the table, no image code was completely wrong, i.e., there was at least one valid code character for every annotated image.

One important issue related to the classification algorithm is the fact that, as the cost of making an incorrect decision is higher than the cost of not actually making a decision, the choice criteria of the system is biased for "hesitation", i.e., the system is very cautious and assigns a wildcard "*" if there is any kind of ambiguity. As confirmed by results shown in Table 1, when the number of codes taken for generating the final IRMA code increases, so ambiguity does, thus the number of complete correct predictions decreases and also the error score.

Strategies for improving this decision criterion must be found for next years. One possible strategy is to assign a different relevance to each result according to its position P in the list, with different weighting factors, such as (1/P), (1-P), etc.

Comparing to other groups, we achieve average results and rank 4th (out of 6).

The analysis axis-by-axis shows interesting results. For each of the four axis of the IRMA code, the "Error Score" shown in Table 2 is calculated as the sum of the errors made for each image, and the number of images in which the full prediction of the axis (i.e., no wildcards in the output) is correct.

Table 2. Axis-by-axis analysis

Run	T-Axis		D-Axis		B-Axis		A-Axis	
	Error Score	Well Classified	Error Score	Well Classified	Error Score	Well Classified	Error Score	Well Classified
2I-0F	**5.24**	**852**	318.04	**381**	**362.56**	**283**	75.6	**789**
3I-0F	6.32	808	**309.78**	293	367.82	206	**67.67**	735
2I-2F	**5.24**	**852**	318.04	**381**	**362.56**	**283**	75.67	**789**
3I-2F	6.12	818	322.09	318	374.74	235	74.07	744

The Technical axis achieves the best error score. However, this is somehow misleading, as the whole image database holds only 4 possible values (codes) for this axis ("1121", "1123", "1124" and "112d"), thus making the decision easier than for the other axes. In fact, 93% of the images have the value "1121", i.e., it is possible to achieve a good error score even with a simple majority classifier (such as ZeroR [5]). Obviously, these different distribution and frequency of code values for each axis will be taken into account in future participations.

Moreover, the prediction for the Anatomical (A) axis shows a significant difference with respect to the Direction (D) and Biological (B) axes. We must conclude that the chosen features are not useful for modelling the image concepts that are intrinsic to those axes. Particularly, in the case of the D-axis the differences among images are very subtle and strongly rely on different brightness or contrast areas in the left or right side (or top or bottom) of the image. A possible approach for improving the classification of those axes is to give more weight to local features with respect to the global analysis, as local feature can model the differences among image regions.

4 Conclusions and Future Work

Based on the analysis performed over each axis, the first conclusion to be drawn is that the first weak point of our experiments is the prediction of the Direction (D) and Biological (B) axis. Some extra effort must be invested on determining which image features could be most useful to predict those axis.

Another aspect is that the calculation of the distance among vectors assigns the same weight to every dimension of the vectors, regardless of the nature of the feature to which this component belongs and/or the number of components belonging to that feature. This was actually our mistake when carrying out the experiments and the feature matrix should have been divided into the different feature sub-matrixes that employ different distances for calculating similarity and are combined to each other using different weight strategies. This fact will be taken into account for next years.

Acknowledgements. This work has been partially supported by the Spanish R+D National Plan, project BRAVO (Multilingual and Multimodal Answers Advanced Search–Information Retrieval), TIN2007-67407-C03-03, and by the R+D Regional Plan of Madrid, project MAVIR (Enhancing the Access and the Visibility of Networked Multilingual Information), S-0505/TIC/000267.

References

1. Lana-Serrano, S., Villena-Román, J., González-Cristóbal, J.C., Goñi-Menoyo, J.M.: MIRACLE at ImageCLEFannot 2008: Classification of Image Features for Medical Image Annotation. In: Working Notes of the 2008 CLEF Workshop, Aarhus, Denmark (2008)
2. Deselaers, T., Deserno, T.M.: Medical Image Annotation in ImageCLEF 2008. In: Peters, C., et al. (eds.) CLEF 2008. LNCS, vol. 5706, pp. 523–530. Springer, Heidelberg (2009)
3. Lana-Serrano, S., Villena-Román, J., González-Cristóbal, J.-C., Goñi-Menoyo, J.M.: MIRACLE at ImageCLEFanot 2007: Machine Learning Experiments on Medical Image Annotation. In: Peters, C., Jijkoun, V., Mandl, T., Müller, H., Oard, D.W., Peñas, A., Petras, V., Santos, D. (eds.) CLEF 2007. LNCS, vol. 5152, pp. 597–600. Springer, Heidelberg (2008)
4. Villena-Román, J., González-Cristóbal, J.C., Goñi-Menoyo, J.M., Martínez Fernández, J.L.: MIRACLE's Naive Approach to Medical Images Annotation. In: Working Notes for the CLEF 2005 Workshop, Vienna, Austria (2005)
5. Witten, I.H., Frank, E.: Data Mining: Practical machine learning tools and techniques, 2nd edn. Morgan Kaufmann, San Francisco (2005)

Query Expansion on Medical Image Retrieval: MeSH vs. UMLS

Manuel Carlos Díaz-Galiano, Miguel Angel García-Cumbreras, María Teresa Martín-Valdivia, L. Alfonso Ureña-López, and Arturo Montejo-Ráez

University of Jaén, Computer Science Department
Grupo Sistemas Inteligentes de Acceso a la Información
Campus Las Lagunillas, Ed. A3, E-23071, Jaén, Spain
{mcdiaz,magc,maite,laurena,amontejo}@ujaen.es

Abstract. In this paper we explain experiments in the medical information retrieval task (ImageCLEFmed). We experimented with query expansion and the amount of textual information obtained from the collection. For expansion, we carried out experiments using MeSH ontology and UMLS separately. With respect to textual collection, we produced three different collections, the first one with caption and title, the second one with caption, title and the text of the section where the image appears, and the third one with the full text article. Moreover, we experimented with textual and visual search, along with the combination of these two results. For image retrieval we used the results generated by the FIRE software. The best results were obtained using MeSH query expansion on shortest textual collection (only caption and title) merging with the FIRE results.

1 Introduction

This is the fourth participation of the SINAI research group in the medical task of the ImageCLEF campaign [7]. This year, the organizers released a new corpus, that we have enriched with the full text of the referenced articles. Thus, new textual collections have been built by combining different sections [4]. We also created two groups of expanded queries, a group expanded with MeSH ontology[1] and another group expanded with UMLS[2].

For the experiment on mixed results, we used the list of images retrieved by FIRE[3] [2], which was supplied by the organizers of this task.

The following section describes the different textual collections generated, the expansion of the queries and the experiments carried out. Finally, conclusions and further work are presented in Section 3.

[1] http://www.nlm.nih.gov/mesh/
[2] http://www.nlm.nih.gov/research/umls/
[3] http://www-i6.informatik.rwth-aachen.de/~deselaers/fire.html

C. Peters et al. (Eds.): CLEF 2008, LNCS 5706, pp. 732–735, 2009.

2 System Description and Experiments

To create the different textual collections, we first obtained the textual information by dowloading full text articles from the web and combining certain sections, as explained in [4]. Therefore, three different collections were generated: with *caption* of image and *title* of the article (CT), with caption, title and text of the *section* where the image appear (CTS), and with caption, title and text of the full *article* (CTA).

One of the purposes of these experiments was to compare the performance of query expansion using two different ontologies: MeSH and UMLS. Therefore we used three sets of queries: original (base), expanded with MeSH and expanded with UMLS. The expansion using MeSH follows the same method as presented in CLEF 2007 [3]. To expand the queries with UMLS information we used the MetaMap program [1]. In order to reduce the number of terms that could expand the query, to make it equal to that of MeSH expansion, we restricted the semantic types in MetaMap to the following: *bpoc* (Body Part, Organ, or Organ Component), *diap* (Diagnostic Procedure), *dsyn* (Disease or Syndrome), and *neop* (Neoplastic Process). For our expansion we used the Meta Candidate terms because these terms provide similar terms with differences in the words. Prior to the inclusion of Meta Candidates terms in the queries, the term words were added to a set where repeated words are deleted. All words in this set were included in the query. For a detailed view of the this process and some examples, see [4].

The results of textual experiments are shown in Table 1:

Moreover, we experimented with the influence of mixing visual information with our results. We mixed our textual results with the visual results given by the organizers [7] in order to improve the baseline results. The visual results were obtained with the FIRE software. To mix textual and visual results, we used the same algorithm as applied in 2007 [3]. The results of this aproach with the CT collection are shown in Table 2.

Figure 1 shown a graphical representation of our mixed experiments results. In general, the experiments with the CT collection obtained best results (Figure 1.a) over the other collections. Therefore, we only compared the influence of different expansions using the CT collection in mixed retrieval approach as shown in Figure 1.b.

Our baseline result with only text was 0.2480 of MAP (Mean Average Precision). With MeSH query expansion and using the CT collection we obtained an improvement of about 12%. Our best result in mixed experiment was 0.2929 with MeSH query expansion and a weight value of 0.3 for the visual results.

Table 1. MAP values of official textual experiments

Expansion	CT	CTS	CTA
Base	0.2480	0.1784	0.1982
MeSH	**0.2792**	0.1582	0.2057
UMLS	0.2275	0.1429	0.1781

Table 2. MAP values of mixed experiments (textual and visual fusion) in CT collection

Expansion	0.1	0.2	0.3	0.4	0.5	0.6	0.7	0.8	0.9	1.0
Base	0.2581	0.2616	0.2602	0.2404	0.1822	0.1158	0.0821	0.0588	0.0459	0.0305
MeSH	0.2846	0.2908	**0.2929**	0.2771	0.2045	0.1325	0.0934	0.0661	0.0486	0.0305
UMLS	0.2387	0.2442	0.2491	0.2369	0.1753	0.1106	0.0798	0.0575	0.0450	0.0306

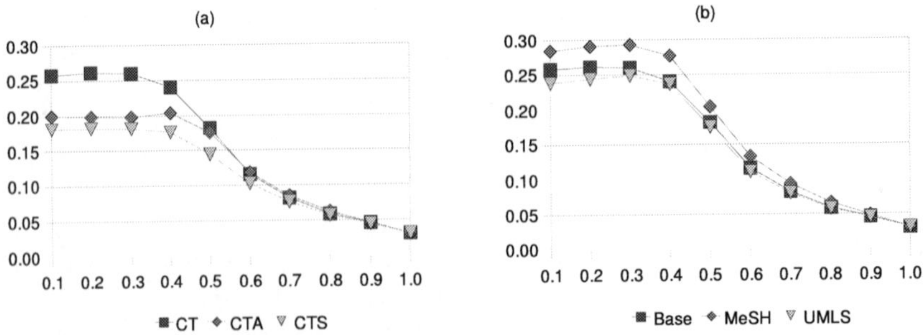

Fig. 1. Results of mixed experiments. (a) With different collections and queries without expansion. (b) With different expansions and CT collection.

3 Conclusions and Further Work

Similarly to previous years, the use of textual information improved the results of baseline visual results. In this case, the use of FIRE and MeSH expansion with the minimal collection (only caption and title) obtained the best results.

The MeSH expansion obtained better results than no expasion or UMLS expansion. The group of University of Alicante obtained the bests results [8] in only textual task using a similar MeSH expansion and negative feedback. The Miracle group performed a MeSH expansion in documents and topics using the hyponyms of UMLS entities [6], but the results obtained are worse than the baseline results.

The use of UMLS expansion obtained worse results than the baseline. Although the UMLS Metathesaurus includes the MeSH ontology in the source vocabularies, MetaMap added, in general, more terms in the queries. The MetaMap mapping was different from MeSH mapping, therefore the terms selected to expand were not the same. The OHSU group used UMLS synonyms expansion in theirs experiments [5] obtaining lower results than baseline results.

In short, groups that used UMLS expansion obtain worse results that other types of expansion. A conclusion is that it is better to have less but more specific textual information. Including the whole section where the image appears was not a good approach. Sometimes, a section contains several images, therefore the same information references different images.

As further work we plan to obtain a more precise textual information by finding the phrase where a reference to the image exists, that is, by finding HTML tags that reference to the image locally (for example: A HREF="#F1") or syntactic structures of type *"in Figure 1 ..."*. Moreover, we want to expand the queries with UMLS Metathesaurus using an algorithm different to that used by the MetaMap tool. We are approaching new methods to expand with UMLS similar to the expansion performed by MeSH, so less terms but more specific ones are proposed.

Acknowledgements

This project has been partially supported by a grant from the Spanish Government, project TIMOM (TIN2006-15265-C06-03), the RFC/PP2006/Id_514 granted by the University of Jaén and the project FFIEXP06-TU2301-2007/000024 granted by the Andalusian Government.

References

1. Aronson, A.R.: Effective Mapping of Biomedical Text to the UMLS Metathesaurus: the MetaMap Program. In: Proc. of the AMIA Symposium, pp. 17–21 (2001)
2. Deselaers, T., Weyand, T., Keysers, D., Macherey, W., Ney, H.: FIRE in ImageCLEF 2005: Combining Content-Based Image Retrieval with Textual Information Retrieval. In: Peters, C., Gey, F.C., Gonzalo, J., Müller, H., Jones, G.J.F., Kluck, M., Magnini, B., de Rijke, M., Giampiccolo, D. (eds.) CLEF 2005. LNCS, vol. 4022, pp. 652–661. Springer, Heidelberg (2006)
3. Díaz-Galiano, M.C., García-Cumbreras, M.A., Martín-Valdivia, M.T., Montejo-Ráez, A., Ureña-López, L.A.: Integrating meSH ontology to improve medical information retrieval. In: Peters, C., Jijkoun, V., Mandl, T., Müller, H., Oard, D.W., Peñas, A., Petras, V., Santos, D. (eds.) CLEF 2007. LNCS, vol. 5152, pp. 601–606. Springer, Heidelberg (2008)
4. Díaz-Galiano, M.C., García-Cumbreras, M.A., Martín-Valdivia, M.T., Ureña López, L.A., Montejo-Ráez, A.: SINAI at ImageCLEFmed 2008. In: On-line Working Notes, CLEF 2008 (2008)
5. Kalpathy-Cramer, J., Bedrick, S., Hatt, W., Hersh, W.: Multimodal Medical Image Retrieval: OHSU at ImageCLEF 2008. In: On-line Working Notes, CLEF 2008 (2008)
6. Lana-Serrano, S., Villena-Román, J., González-Cristóbal, J.C.: MIRACLE at ImageCLEFmed 2008: Evaluating Strategies for Automatic Topic Expansion. In: On-line Working Notes, CLEF 2008 (2008)
7. Müller, H., Kalpathy-Cramer, J., Kahn, C.E., Hatt, W., Bedrick, S., Hersh, W.: Overview of the ImageCLEFmed 2008 medical image retrieval task. In: Peters, C., et al. (eds.) CLEF 2008. LNCS, vol. 5706, pp. 512–522. Springer, Heidelberg (2009)
8. Navarro, S., Llopis, F., Muñoz, R.: Different Multimodal Approaches using IR-n in ImageCLEFphoto 2008. In: On-line Working Notes, CLEF 2008 (2008)

Query and Document Expansion with Medical Subject Headings Terms at Medical Imageclef 2008

Julien Gobeill, Patrick Ruch, and Xin Zhou

University and Hospitals of Geneva, Service d'Informatique Médicale,
2 rue des Acacias 1211 Genève, Switzerland
{Julien.Gobeill,Patrick.Ruch,Xin.Zhou}@sim.hcuge.ch

Abstract. In this paper, we report on query and document expansion using Medical Subject Headings (MeSH) terms designed for medical ImageCLEF 2008. In this collection, MeSH terms describing an image could be obtained in two different ways: either being collected with the associated MEDLINE's paper, or being extracted from the associated caption. We compared document expansion using both. From a baseline of 0.136 for Mean Average Precision (MAP), we reached a MAP of respectively 0.176 (+29%) with the first method, and 0.154 (+13%) with the second. In-depth analyses show how both strategies were beneficial, as they covered different aspects of the image. Finally, we combined them in order to produce a significantly better run (0.254 MAP, +86%). Combining the MeSH terms using both methods gives hence a better representation of the images, in order to perform document expansion.

Keywords: Cross-Language Image Retrieval, Text Categorization, Query Expansion, Document Expansion.

1 Introduction

Cross-Language Information Retrieval (CLIR) is becoming progressively more relevant as network-based resources become commonplace. In the medical domain, CLIR is of strategic importance in order to fill the gap between clinical records, usually written in national languages, and research reports mostly written in English. Images are also getting more important and varied in the medical domain, as they become available in digital form. Despite the fact that images are language-independent, they are often accompanied by textual notes in various languages. These textual notes can strongly improve the retrieval quality [1].

The Cross Language Evaluation Forum (CLEF) is a challenge that has been occurring yearly since 2000. The goal of CLEF is to evaluate systems on a common multilingual task, to establish a state of the art in techniques used in a domain, and to build a benchmark for future evaluations. The medical ImageCLEF challenge started in 2004 aiming at retrieving relevant medical images in a multilingual document collection, using visual features (images) or textual features (associated captions, titles and articles).

In this challenge, we applied textual strategies based on picture's metadata and use of Medical Subject Headings (MeSH) in order to perform query and document expansion

C. Peters et al. (Eds.): CLEF 2008, LNCS 5706, pp. 736–743, 2009.

[2] [3]. In medical ImageCLEF 2008, groups that obtained the best results tried to perform query or document expansion with MeSH terms, often without obtaining significant improvements. The SINAI group performed query expansion with MeSH terms, extracted using MetaMap [4], however they observed a small gain. The University of Alicante performed the same query expansion as the SINAI group [5], observing this time no significant improvement. The MIRACLE group performed query and document expansion with MeSH terms, extracted with strict matching [6], but their baseline run obtained better results than the run using document and query expansion. So, query and document expansion is a difficult process, which needs to be efficiently designed in order to be productive.

2 Data and Strategy

The medical ImageCLEF 2008 collection was entirely new. The organizers obtained the new images from two imaging and radiology journals using the GoldMiner system [7]. Since images were linked to MEDLINE's papers, MeSH terms describing the image in order to perform document expansion could be obtained by different ways.

2.1 Data

The data collection consisted of a set of 67'115 images, along with their captions, article titles, and linkage to PubMed and the full text of the associated article. In addition to the images, XML files containing metadata were distributed. A detailed description of the protocol can be found in [8].

Contrary to previous medical ImageCLEF collections, 65697 images or 98% of the collection were linked to MEDLINE through their PMID. There were 4961 different PMIDs for these 65697 images (around 13 images per PMID). In the context of document expansion, this means that MeSH terms could be obtained from 2 different origins: either being collected with the associated MEDLINE's paper, or being extracted from the associated caption. The first one is supposed to be more accurate because it is manually added to each MEDLINE's paper. Nevertheless, as these MEDLINE's papers concerned a set of images – around 13 – these MeSH terms could not be relevant to each image. On the other side, MeSH terms extracted from the caption are supposed to be less relevant because they were automatically extracted, but we are sure that the accurate ones concerned the associated image.

2.2 Strategy

The keystone of our strategy for participating in medical ImageCLEF was focused on associating MeSH terms to any textual components – documents or queries – in order to expand the text with language-independent descriptors, and then to perform a standard Information Retrieval process. The core technical component of our cross-language search engine was an automatic text categorizer, which associated a set of MeSH terms to any input text. The precision at high ranks of this engine for MeSH terms was evaluated above 90% for MEDLINE's abstracts [9].

For multilingual purpose, we merged three versions of MeSH (English, German and French) in order to expand each document with several MeSH terms – between 3 and 8 in 2006, 15 in 2007 – and their unique identifier, making them efficient regardless of the original language of the document. The expanded documents were then indexed.

The same MeSH categorization was then performed on queries; according to past studies, associating three terms by query is the most efficient [2]. For instance, if a German query dealt with "magen-darm-endoskopie", this concept had great chances to be mapped by our categorizer, and to expand the query with this German form and the MeSH id: then, even in an English collection, the MeSH id D016099 was a strongly discriminant feature.

In medical ImageCLEF 2008, MeSH terms could also be collected via PubMed, thanks to the associated PMID. In this case, the MeSH terms were obtained via the PubMed e-utilities [10].

MeSH was hence considered not only as a metadata vocabulary, but also as a intermediate language [2], as MeSH unique identifiers are the same in all languages.

2.3 MeSH Distribution in Data

Since MeSH has a hierarchical structure, we could observe in which aspect a given MeSH term was associated to an image, both those extracted from the caption and those collected with the associated MEDLINE's paper. The MeSH structures can be consulted on [11]. In Table 1, we show the distribution of the MeSH terms associated with the images in the medical ImageCLEF 2008 collection. The column "*Caption*" concerns the MeSH terms extracted by our automatic categorizer from caption, keeping 12 terms per caption. The column "*Paper*" concerns the MeSH terms collected with the associated MEDLINE's paper, which are 12 terms in mean.

Table 1. The different MeSH subtrees, covering different aspects of the medical sciences. Column 3 and 4 shows the distribution of MeSH terms respectively extracted from the associated caption and collected with the associated paper

#	Name	Caption	Paper
A	Anatomy	13.1%	5.0%
B	Organisms	3.1%	4.8%
C	Diseases	27.3%	16.9%
D	Chemicals and Drugs	15.3%	5.3%
E	Analytical, Diagnostic and Therapeutic Techniques and Equipment	14.1%	29.2%
F	Psychiatry and Psychology	2.4%	0.3%
G	Biological Sciences	6.3%	8.3%
H	Natural Sciences	3.6%	6.1%
I	Anthropology, Education, Sociology and Social Phenomena	2.9%	0.2%
J	Technology, Industry, Agriculture	1.1%	0.3%
K	Humanities	0.4%	0.1%
L	Information Science	1.6%	2.3%
M	Named Groups	2.2%	8.2%
N	Health Care	5.2%	7.7%
V	Publication Characteristics	1.1%	0.0%
Z	Geographicals	0.2%	0.2%

The differences between the two origins are significant. Anatomical parts concepts (A subtree) were more frequently found in the captions than in the papers. The B subtree deals with organisms, and is not very relevant to the task. Disease concepts (C subtree) were twice more frequent on the captions than on the papers. The D subtree deals with chemical products. Terms belonging to the D subtree were 3 times more present on the captions than on the papers. Actually, a deeper analysis shows that these terms often referred to chemical products used for the protocol, for instance for scintigraphies. It seems that this information was more present in the captions than in the paper's MeSH terms. Concepts describing the type of image (E subtree) were more frequent on the papers than on the captions. MeSH terms belonging to the G, H and N subtrees often deals with the scientific description of the study, and are often not relevant. For instance, the MeSH term "Sensitivity and Specificity" was present in 17% of terms collected with the associated paper, and "Retrospective Studies" in 15%. Nevertheless, some terms – especially for the G subtree – could be extremely relevant, like "Blood Coagulation" (G subtree), which is critical for the topic 17 "Show me MRI images of the brain with a blood clot". MeSH terms belonging to the M subtree dealt with the patient, like "Adult" or "Adolescent", and were not relevant for this task. Other MeSH subtrees were considered as not relevant for this task.

In this paper, we focus on the three most important aspects that appeared in the topics: the type of the image, the possible anatomical parts concerned, and the possible disease concerned. Each of them belongs to a MeSH subtree.

Table 2. Percentage of images having at least one MeSH term belonging to the specified MeSH subtree in their MeSH terms set, and average number of MeSH terms belonging to this subtree for these images

	Papers		Captions	
MeSH subtree	%	Average number	%	Average number
E_{DiagIm}	86	1.7	32	1.7
A	54	1.7	80	3
C	86	2.4	92	4

Type of image. The type of image concepts belong to the E subtree. More precisely, as the E subtree contains more terms than only those dealing with medical images, we can reduce the subtree to all the children of the term "Diagnostic Imaging". We call this new subtree E_{DiagIm}. As shown in Table 2, for the 65'697 images having an associated MEDLINE's paper, by collecting the MeSH terms from the paper, we could obtain for 56'555 (86%) at least one MeSH term belonging to the E_{DiagIm} subtree, each of them having 1.7 terms in mean. Table 3 shows the distribution of the 5 most frequent terms belonging to E_{DiagIm} in this collection.

It appears that the types of images were well indicated and standardized in the MeSH terms collected from their associated paper. By comparison, with our automatic extraction on captions, for 65'510 images having a caption, we could obtain only 21'152 (32%) with at least one MeSH term belonging to EDiagIm. For this type of image, the use of MeSH terms collected from the associated paper seems preferable.

Table 3. Distribution of the MeSH terms belonging to E_{DiagIm} in the sets of MeSH terms collected with the associated MEDLINE's paper

MeSH term	Frequency in the "paper" terms
Magnetic Resonance Imaging	31%
Tomography, X-Ray Computed	31%
Magnetic Resonance Angiography	5%
Mammography	5%
Angiography	4 %

Anatomical Parts. The anatomical parts concepts belong to the A subtree. In the same process, for the 65'697 images having an associated MEDLINE's paper, we could obtain 35'822 (54%) with at least one MeSH term belonging to the A subtree, each of them having 1.7 terms in mean. By comparison, with our automatic extraction on captions, for 65'510 images having a caption, we could obtain 52'302 (80%) with at least one MeSH term belonging to the A subtree, each of them having 3 terms in mean. For the anatomical parts, this information was not well indicated in the associated paper, whereas we could extract more often concepts from the captions.

Diseases. The diseases belong to the C subtree. In the same process, for the 65'697 images having an associated MEDLINE's paper, we could obtain 56'211 (86%) with at least one MeSH term belonging to the C subtree, each of them having 2.4 terms in mean. By comparison, with our automatic extraction on captions, for 65'510 images having a caption, we could obtain 60'269 (92%) with at least one MeSH term belonging to the C subtree, each of them having 4 terms on mean. For the diseases, by analyzing the distribution, no origin of MeSH terms showed a significant superiority.

3 Methods

Both the text categorization – for query and document expansion – and the Information Retrieval process were performed with our generic toolkit, EasyIR. EasyIR is available at [12]. More information on the EasyIR toolkit and the methods can be obtained in our previous proceedings [2]. As for the official results, the measure used in this paper is Mean Average Precision (MAP).

4 Results and Discussion

We firstly describe the Baseline run, computed without any query or document expansion. Then, we describe several runs using the query and document expansion's strategy. We finally quickly describe several other official runs.

4.1 Baseline Run

The Baseline run (HUG-BL) was submitted in order to evaluate the performance of our Information Retrieval engine. The strategy was simple: for each image, the caption and the title were indexed. Then, queries were submitted in the three languages, without any expansion.

As captions and titles were in English. It is not stunning that the English run was the best one. Nevertheless, French and German runs, without any strategy, had an acceptable performance.

4.2 Runs with Document and Query Expansion

For the following runs, queries were expended with 3 MeSH terms, automatically extracted by our categorizer.

The HUG-MH run was finally the best official run. For each image, the caption, the title and the MeSH terms collected from the associated MEDLINE's paper – MeSH term + MeSH id – were indexed.

The HUG-cap-MH run was submitted too. For this run, documents were expanded with 15 MeSH terms extracted from the captions.

The best result for document expansion with MeSH terms extracted from captions was 0.154 (+13 %) – see Table 4. Document expansion with MeSH terms collected with MEDLINE's paper performed better – MAP of 0.176 (+29%).

In additional experiments, we created a new document expansion, by concatenating both origins of MeSH terms, captions and papers. We expanded each document with the MeSH terms collected with the MEDLINE's paper, and with several numbers of MeSH terms extracted from the caption.

Table 4. MAP for three official runs. Baseline run (HUG-BL) was performed without any document and query expansion. HUG-MH run was performed with document expansion, using MeSH terms collected with the associated MEDLINE's paper. HUG-cap-MH run was performed with document expansion, using MeSH terms automatically extracted from the captions.

MAP	HUG-BL	HUG-MH	HUG-cap-MH
EN	0.136	0.176	0.154
FR	0.069	0.105	0.093
GE	0.07	0.076	0.073

Table 5. MAP for different document expansion strategies in English. Documents were expanded with 3, 5, 10 or 15 MeSH terms extracted from the captions, plus the MeSH terms collected with the associated MEDLINE's paper.

# of extracted MeSH terms	3	5	10	15
only captions MeSH terms	0.137	0.165	0.162	0.154
captions + papers MeSH terms	0.213	0.254	0.245	0.23

When we merged both origins, MAP achieved 0.254 (+ 86%) for English queries. The same experiments let to a MAP of 0.141 (+101%) for German queries, and 0.109 (+57%) for French queries.

4.3 Mix with a Visual Run

To obtain these last runs, we supplied our best run (HUG-MH) to another team of the University and Hospitals of Geneva, Xin Zhou and Henning Muller, who are specialized in Visual Information Retrieval. MAP for the two combined runs (GE-GE_GIFT8 and GE-GE_GIFT8_EN0.5) were respectively 0.035 and 0.085. Zhou and Muller performed runs that relied mainly on GIFT [3], which meant to be baseline runs rather than candidates for the high ranks. Since visual strategies led to poor performances this year, it was not surprising that the combination with a visual run led to no improvements.

5 Conclusion and Future Work

The keystone of our strategy for medical ImageCLEF was to expand documents and queries with MeSH terms. In this year's collection, MeSH terms for document expansion could be either collected with the associated MEDLINE's paper, or extracted from the associated caption. We showed that both origins covered different aspects of the concepts searched in topics, and that both origins were complementary, as their combination for document expansion led to great improvements for our system (+86% in English, + 101% in German, + 57% in French). Particularly, the MeSH terms in MEDLINE better indicated and standardized the type of image, while anatomical parts concepts were better extracted from captions. Combining the MeSH terms from both origins gave a better representation of the images, in order to perform document expansion.

Obviously, in order to benefit from these metadata, the query expansion has to be consistent with these representations. In this way, it is essential to apply a set of rules in order to expand the query with the most appropriate and standard MeSH terms. These rules can be derived from a thesaurus manually built. They can also exploit the hierarchical structure of the MeSH.

References

1. Müller, H., Michoux, N., Bandon, D., Geissbuhler, A.: A review of content-based image retrieval systems in medicine - clinical benefits and future directions. International Journal of Medical Informatics 73, 1–23 (2004)
2. Gobeill, J., Müller, H., Rucj, P.: Query and Document Translation by Automatic Text Categorization: A Simple Approach to Establish a String Textual Baseline for ImageCLEFmed 2006. In: ImageCLEF Working Notes (2006)
3. Zhou, X., Gobeill, J., Ruch, P., Müller, H.: University and Hospitals of Geneva at ImageCLEF 2007. In: Zhou, X., Gobeill, J., Ruch, P., Muller, H. (eds.) ImageCLEF Working Notes (2007)
4. Diaz-Galiano, M.C., Garcia-Cumbreras, M.A., Mertin-Valdidia, M.T., Urena-Lopez, L.A., Montejo-Raez, A.: SINAI at ImageCLEFmed 2008. In: ImageCLEF Working Notes (2008)
5. Navarro, S., Munoz, R., Llopis, F.: A Multimodal Approach to the Medical Retrieval Task using IR-n. In: ImageCLEF Working Notes (2008)

6. Lana-Serrano, S., Villena-Roman, J., Gonzalez-Cristobal, J.C.: MIRACLE at ImageCLEF med 2008: Evaluating Strategies for Automatic Expansion. In: ImageCLEF Working Notes (2008)
7. Kahn, C., Thao Jr., C.: GoldMiner: a radiology image search engine. American Journal of Roentgenology, 1475–1478 (2007)
8. ImageCLEF 2008 protocole, http://ir.ohsu.edu/image/2008protocol.html
9. Ruch, P.: Automatic Assignment of Biomedical Categories: Toward a Generic Approach. Bioinformatics 22(6), 658–664 (2006)
10. PubMed e-utilities, http://www.ncbi.nlm.nih.gov/entrez/query/static/eutils_help.html
11. Medical Subject Headings trees, http://www.nlm.nih.gov/mesh/2008/MeSHtree.html
12. University and Hospitals of Geneva, http://www.natlang.hcuge.ch/Resources/easyIR.zip

Multimodal Medical Image Retrieval OHSU at ImageCLEF 2008

Jayashree Kalpathy-Cramer, Steven Bedrick, William Hatt, and William Hersh

Department of Medical Informatics & Clinical Epidemiology
Oregon Health and Science University (OHSU)
Portland, OR, USA
kalpathy@ohsu.edu

Abstract. We present results from the Oregon Health & Science University's participation in the medical retrieval task of ImageCLEF 2008. Our web-based retrieval system was built using a Ruby on Rails framework. Ferret, a Ruby port of Lucene was used to create the full-text based index and search engine. In addition to the textual index of annotations, supervised machine learning techniques using visual features were used to classify the images based on image acquisition modality. Our system provides the user with a number of search options including the ability to limit their search by modality, UMLS-based query expansion, and Natural Language Processing-based techniques. Purely textual runs as well as mixed runs using the purported modality were submitted. We also submitted interactive runs using user specified search options. Although the use of the UMLS metathesaurus increased our recall, our system is geared towards early precision. Consequently, many of our multimodal automatic runs using the custom parser as well as interactive runs had high early precision including the highest P10 and P30 among the official runs. Our runs also performed well using the `bpref` metric, a measure that is more robust in the case of incomplete judgments.

1 Introduction

Advances in digital imaging technologies and the increasing prevalence of Picture Archival and Communication Systems (PACS) have led to a substantial growth in the number of digital images stored in hospitals and medical systems in recent years. Medical images can form an essential component of a patient's health records. The ability to retrieve relevant images can be useful in diagnosis, treatment, education and research.

Image retrieval systems do not currently perform as well as their text counterparts [1]. Medical and other image retrieval systems have historically relied on annotations or captions associated with the images for indexing the retrieval system. The last few decades, however, have seen advancements in the area of content-based image retrieval (CBIR) [2, 3]. Although CBIR systems have demonstrated success in fairly constrained medical domains including pathology, dermatology, chest radiology, and mammography, they have demonstrated

C. Peters et al. (Eds.): CLEF 2008, LNCS 5706, pp. 744–751, 2009.

poor performance when applied to databases with a wide spectrum of imaging modalities, anatomies and pathologies [1, 4, 5, 6].

Retrieval performance, especially early precision, has been improved demonstrably by fusing the results of textual and visual search techniques. [7, 8] The medical image retrieval task within 2008 ImageCLEF 2008 campaign is TREC-style [9] and provides a forum and set of test collections for the medical image retrieval community to use to benchmark their algorithms on a set of queries. The ImageCLEF campaign has, since 2003, been a part of the Cross Language Evaluation Forum (CLEF) [9, 10] which is an offshoot of the Text Retrieval Conference (TREC[1]).

2 System Description of Our Adaptive Medical Image Retrieval System

The 2008 ImageCLEF collection consists of 67,115 medical images and annotations associated with them [11]. The collection contains images and captions from *Radiology* and *Radiographics*, two Radiological Society of North America (RSNA) journals. We created a flexible database schema that allows us to easily incorporate new collections while facilitating retrieval using both text and visual techniques. The captions and titles in the collection are currently indexed and we continue to add indexable fields for incorporating visual information.

2.1 Database and Web Application

The data distribution included an xml file with the image id, the captions of the images, the titles of the journal articles in which the image had appeared and the PubMed ID of the journal article. In addition, a compressed file containing the approximately 67,000 images was provided.

We used the Ruby programming language [2], with the open source Ruby On Rails[3] web application framework. The PostgreSQL[4] relational database was used to store mapping between the images and the various fields associated with the image. The title, full caption and precise caption, as provided in the data distribution, were indexed.

2.2 Image Processing and Analysis

The image itself has important visual characteristics such as color and texture that can help in the retrieval process. We created additional tables in the database to store image information that was created using a variety of image processing techniques in MATLAB[5]. These include color and intensity histograms as well as measures of texture using gray-level co-occurrence matrices

[1] `http://trec.nist.gov/`
[2] `http://www.ruby-lang.org/`
[3] `http://www.rubyonrails.org/`
[4] `http://www.postgresql.org/`
[5] `http://www.mathworks.com/`

and discrete cosine transforms. These features can be used to find images that are visually similar to the query image. We used this in the interactive, mixed mode to reorder the images obtained from the textual search such that images that are visually similar to an image marked relevant by the user are returned at the top of the list.

Medical images often begin life with rich metadata in the form of DICOM headers describing their imaging modality or anatomy. However, since most teaching or on-line image collections are made up of compressed standalone JPEG files, it is very common for medical images to exist *sans* metadata. In previous work [8], we described a modality classifier that could identify the imaging modality for medical images using supervised machine learning. We extended that work to the new dataset used for ImageCLEF 2008.

One of the biggest challenges in creating such a classifier is creating a labeled training dataset of sufficient size and quality. Our system, as previously described [8], relied on a external training set of modality-labeled images for its supervised learning. In 2008, we did not use any external databases for training the modality classifier. Instead, a Ruby text parser was written to extract the modality from the image captions for all images in the collection using regular expressions as well as a simple Bayesian classifier.

Note that the quality and accuracy of these labels are not as good as in case of the external training set used in previous experiments. Images where a unique modality could be identified based on the caption, were used for training the modality classifier. Grey scale images were classified into a set of modalities including x-rays, CT, MRI, ultrasound and nuclear medicine. Color image classes include gross pathology, microscopy, and endoscopy. The rest of the dataset (i.e., images for which zero or more than one modalities were parsed) was classified using the above classifier. We created two fields in the database for the modality that were indexed by our search engine. The first field contained the modality as extracted by the text parser, and the second contained the modality resulting from the classification process using visual features.

2.3 Query Parser and Search Engine

The system presents a variety of search options to the user including Boolean OR, AND, and "exact match". There are also options to perform fuzzy searches, as well as a custom query parser. A critical aspect of our system is the query parser, written in Ruby. Ferret, a Ruby port of the popular Lucene system, was used in our system as the underlying search engine. The custom query parser performs stop-word removal using a specially-constructed list of stop-words. The custom query parser is highly customizable, and the user has several configuration options from which to choose. The first such option is modality limitation. If the user selects this option, the query is parsed to extract the desired modality, if available. Using the modality fields described in the previous section, only those images that are of the desired modality are returned. This is expected to improve the precision, as only images of the desired modality would be included within the result set. However, there could be a loss in recall if the process of modality extraction and classification is inaccurate.

The system is linked to the UMLS Metathesaurus; the user may choose to perform manual or automatic query expansion using synonyms from the Metathesaurus. In the manual mode, a list of synonyms is presented to the user, which the user can choose to add to the query. In the automatic mode, all synonyms of the UMLS preferred term are added to the query. Another configuration option is the "stem and star" option, in which all the terms in the query are first stemmed. A wildcard (*) is then appended to the word to allow the search of words containing the desired root. The last option allows the user to only send unique terms to the search engine. This can be useful when using the UMLS option, as many of the synonyms have a lot of overlap in the preferred terms.

2.4 Interactive Mode

In addition to user-selectable search engine configuration options described above, our system provides users with other interactive features. Once a user has submitted a query using the above-described query parser, they have the option to improve the precision of their results by using an interactive re-ordering system. In this year's system, users select what they feel to be a visually representative image from their search's results. The system then attempts to re-rank the search results according to their degree of visual similarity with the "probe image" that the user selected. If the user is not satisfied with the re-ordering produced by their choice if image, they may repeat the process by selecting different probe images until they arrive at a satisfactory sorting.

To assess the visual similarity of the images within a result set, the system uses a relatively straightforward approach derived from Latent Semantic Analysis [12]. In this approach, each image in the result set is abstracted into a feature vector, which thereafter plays the same role that a document's "term vector" would play in classical LSA. We have experimented with sets of features derived from image color, texture, and frequency attributes; in our final system, the user is able to select which combinations of features they wish to use.

3 Runs Submitted

We submitted a total of 10 offical runs. The search options for the different runs are provided in 1. These runs included textual and mixed, automatic and interactive options. Although the ImageCLEF2005-2007 collection with qrels and topics were available, we did not use any external training data. Three automatic text-based runs were submitted with different custom parsing options including the use of UMLS term expansion. We also submitted four mixed, automatic runs. The modality classification based on the text parsing of the caption and the classification based on visual features was used to improve the precision of the search.

Table 1. Description of OHSU runs

Run Name	Text/ Visual/ Mixed	Automatic/ Manual/ Interactive	Data used	Parsing options
text_or_1	text	automatic	full caption	none
text_3	text	automatic	full caption, title	custom
text_umls_4	text	automatic	full caption, title	custom, umls, unique
vis_mod_3	mixed	automatic	full caption	custom, modality
mod_pars2_sp	mixed	automatic	full caption	custom, modality
vis_mod_5	mixed	automatic	full caption, title	custom, modality
vis_mod_umls_4	mixed	automatic	full caption	custom, modality, umls
sdb_lsa	mixed	interactive	full caption	custom, modality
sdb_full_interactive	mixed	interactive	full caption, title	custom, modality
int_2	mixed	interactive	precisise caption, title	custom, modality, umls

While the majority of our runs were automatic in nature, three of ours were interactive. In the first such run (ohsu_int_2), the user chose different combinations of options for each topic and added terms based on the list provided using the UMLS query expansion option. Two runs using the interactive result sorting system were submitted. The first such run, "ohsu_sdb_lsa", used the result sorting system on every topic. The second run, "ohsu_sdb_full_interactive", only used the result sorting system on topics where the user thought that it would be beneficial to the run's precision. This second run also featured much more intervention on the part of the user, who took full advantage of our retrieval system's interactive nature and enabled or disabled options and features as needed.

4 Results and Discussion

Table 2 contains a subset of the official performance metrics for the OHSU runs. We have also included the average of these metrics for all runs, the highest measure in each category as well as data from the best run (based on MAP) in the 2008 campaign. Table 2 shows the results and particularly the large differences between the runs.

OHSU performed well, especially among the runs that did not use any external training data. One of our runs (mod_pars2_sp) had the highest early precision including P10 and P30 among all 111 official runs. All but two of our runs performed better than the average for all measures. As described in the previous section, our systems have been designed to improve precision, perhaps at the expense of recall. Our custom parsing improved the mean average precision as well as the early precision, as can be seen in the text runs. The use of modality

Table 2. Results of the automatic runs using only visual information

Run Name	MAP	bpref	P10	P30	Recall
text_or_1	0.11	0.18	0.26	0.24	0.41
text_3	0.15	0.23	0.31	0.22	0.52
text_umls_4	0.20	0.30	0.29	0.25	0.57
vis_mod_3	0.15	0.25	0.32	0.24	0.53
mod_pars2_sp	0.21	0.30	0.55	0.46	0.45
vis_mod_5	0.23	0.33	0.38	0.29	0.58
vis_mod_umls_4	0.23	0.35	0.37	0.28	0.60
sdb_lsa	0.10	0.20	0.27	0.27	0.47
sdb_full_interactive	0.18	0.29	0.46	0.33	0.47
int_2	0.22	0.31	0.49	0.39	0.46
SINAI-sinai_CT_Mesh_Fire20	*0.29*	*0.33*	*0.43*	*0.40*	*0.62*
average	*0.14*	*0.20*	*0.28*	*0.24*	*0.40*
best in category	*0.29*	*0.35*	*0.55*	*0.46*	*0.66*

parsing and detection improved the MAP as well as the early precision. All our mixed runs performed better than the corresponding text runs.

OHSU had submitted four of the top ten mixed runs, as sorted using the precision at 10. The use of term expansion with UMLS increased the recall. We had submitted runs after the creation of the pools but prior to the official deadline. This potentially biases the results against these runs as potentially fewer of the images in these runs are judged. Two of these runs (vis_mod5) and (vis_mod_umls4) had relatively high recall but moderate MAP. One of these runs (vis_mod_umls4) had the highest bpref. Bpref is a measure that is robust in the case of incomplete judgments and one that would not penalize the fact that the images in this run were not used in creating the pools as much. Some of our runs managed to find a larger part of the relevant images (809) but with a fairly low MAP, whereas some results with a higher MAP only find very small relevant images in the first 1000 results.

The performance of the first LSA run (ohsu_sdb_lsa) was unsatisfactory: there are many situations in which the original result sorting provided by our textual search engine was adequate, and changing it by means of our interactive visual re-sorting system damaged a topic's precision. The second LSA run, "ohsu_sdb_full_interactive", performed much better. In fact, its p10 was greater than that of the overall competition winner's (0.46 for "ohsu_sdb_full_interactive" vs 0.43 for "SINAI-sinai_CT_Mesh_Fire20").

Obviously, the utility of the interactive portion of our system is variable, and depends highly on the contents of the initial result set. In the case of a set where the desired images are simultaneously visually similar to one another and distinct from the rest of the images in the set, this visual re-sorting system works quite well. However, in the case where the desired images are visually different from one another, or where all of the results (including the non-relevant ones) are visually similar, this re-sorting system is not very useful.

For example, a result set consisting entirely of ultrasound images will not be improved very much by re-sorting. In fact, in this particular case, resorting the result set may hurt its precision, as any ordering imposed by our textual search engine will be lost. On the other hand, a result set in which most of the relevant images are ultrasounds and most of the non-relevant images are x-rays could benefit from being re-ordered based on visual similarity to a user-selected probe image.

The third interactive run (int_2), where the parsing mode and UMLS term expansion was performed interactively also performed quite well.

5 Conclusions and Future Work

Our image retrieval system built using open-source tools is a flexible framework for evaluating various tools and techniques in image processing as well as natural language processing for medical image retrieval. Combining visual and textual information, use of UMLS-based term expansion and query parsing all add value over the basic Ferret (Lucene) search engine. The use of visual information to automatically extract the imaging modality is a promising approach for the ImageCLEFmed campaign and can greatly improve early precision. The captions can be used to create labels for the purposes of creating a training set of the supervised learning process. We will continue to improve our image retrieval system by adding more image tags using automatic visual feature extraction, including anatomical location and view attributes.

Acknowledgements

We acknowledge the support of NLM Training Grant 2T15LM007088 and NSF Grant ITR-0325160.

References

1. Hersh, W.R., Müller, H., Jensen, J.R., Yang, J., Gorman, P.N., Ruch, P.: Advancing biomedical image retrieval: Development and analysis of a test collection. J. Am. Med. Inform. Assoc. (June 2006); M2082
2. Smeulders, A., Worring, M., Santini, S., Gupta, A., Jain, R.: Content-based image retrieval at the end of the early years. IEEE Transactions on Pattern Analysis and Machine Intelligence 22(12), 1349–1380 (2000)
3. Tagare, H.D., Jaffe, C.C., Duncan, J.: Medical image databases: A content-based retrieval approach. J. Am. Med. Inform. Assoc. 4(3), 184–198 (1997)
4. Aisen, A.M., Broderick, L.S., Winer-Muram, H., Brodley, C.E., Kak, A.C., Pavlopoulou, C., Dy, J., Shyu, C.R., Marchiori, A.: Automated storage and retrieval of thin-section ct images to assist diagnosis: System description and preliminary assessment. Radiology 228(1), 265–270 (2003)
5. Schmid-Saugeona, P., Guillodb, J., Thirana, J.P.: Towards a computer-aided diagnosis system for pigmented skin lesions. Computerized Medical Imaging and Graphics: The Official Journal of the Computerized Medical Imaging Society 27(1), 65–78 (2003); PMID: 12573891

6. Müller, H., Michoux, N., Bandon, D., Geissbuhler, A.: A review of content-based image retrieval systems in medical applications–clinical benefits and future directions. International Journal of Medical Informatics 73(1), 1–23 (2004)
7. Hersh, W.R., Kalpathy-Cramer, J., Jensen, J.: Medical image retrieval and automated annotation: OHSU at imageCLEF 2006. In: Peters, C., Clough, P., Gey, F.C., Karlgren, J., Magnini, B., Oard, D.W., de Rijke, M., Stempfhuber, M. (eds.) CLEF 2006. LNCS, vol. 4730, pp. 660–669. Springer, Heidelberg (2007)
8. Kalpathy-Cramer, J., Hersh, W.: Automatic image modality based classification and annotation to improve medical image retrieval. Studies in Health Technology and Informatics 129(Pt 2), 1334–1338 (2007); PMID: 17911931
9. Braschler, M., Peters, C.: Cross-language evaluation forum: Objectives, results, achievements. Information Retrieval 7(1), 7–31 (2004)
10. Müller, H., Clough, P., Hersh, W., Deselaers, T., Lehmann, T., Geissbuhler, A.: Evaluation axes for medical image retrieval systems: the imageclef experience. In: Proceedings of the 13th annual ACM international conference on Multimedia, Hilton, Singapore, pp. 1014–1022. ACM, New York (2005)
11. Müller, H., Kalpathy-Cramer, J., Kahn Jr., C.E., Hatt, W., Bedrick, S., Hersh, W.: Overview of the ImageCLEFmed 2008 medical image retrieval task. In: Peters, C., et al. (eds.) CLEF 2008. LNCS, vol. 5706, pp. 512–522. Springer, Heidelberg (2009)
12. Furnas, G.W., Deerwester, S., Dumais, S.T., Landauer, T.K., Harshman, R.A., Streeter, L.A., Lochbaum, K.E.: Information retrieval using a singular value decomposition model of latent semantic structure. In: Proceedings of the 11th annual international ACM SIGIR conference on Research and development in information retrieval, Grenoble, France, pp. 465–480. ACM, New York (1988)

Baseline Results for the ImageCLEF 2008 Medical Automatic Annotation Task in Comparison over the Years

Mark O. Güld, Petra Welter, and Thomas M. Deserno

Department of Medical Informatics, RWTH Aachen University, Aachen, Germany
mgueld@mi.rwth-aachen.de, pwelter@mi.rwth-aachen.de, deserno@ieee.org

Abstract. This work reports baseline results for the CLEF 2008 Medical Automatic Annotation Task (MAAT) by applying a classifier with a fixed parameter set to all tasks 2005 – 2008. A nearest-neighbor (NN) classifier is used, which uses a weighted combination of three distance and similarity measures operating on global image features: Scaled-down representations of the images are compared using models for the typical variability in the image data, mainly translation, local deformation, and radiation dose. In addition, a distance measure based on texture features is used. In 2008, the baseline classifier yields error scores of 170.34 and 182.77 for $k = 1$ and $k = 5$ when the full code is reported, which corresponds to error rates of 51.3% and 52.8% for 1-NN and 5-NN, respectively. Judging the relative increases of the number of classes and the error rates over the years, MAAT 2008 is estimated to be the most difficult in the four years.

1 Introduction

In 2008, the Medical Automatic Annotation Task (MAAT) [1] is held for the fourth time as part of the annual challenge issued by the Cross-Language Evaluation Forum (CLEF). It demands the non-interactive classification of a set of 1,000 radiographs according to a hierarchical, multi-axial code [2]. For training, a separate set of radiographs is given along with their code, which was defined by expert physicians. Over the four years, the task difficulty changed: the challenge in 2005 used a grouping based on the code hierarchy, whereas the later challenges use the full code. In addition, a modified error counting scheme is employed in 2007 and 2008 in order to address the severity of classification errors. It penalizes misclassification in upper (broader) hierarchy levels more than errors on lower, more detailed levels. The participants in the task also varied over the years.

It is therefore desirable to have baseline results for the CLEF MAATs, which allow a rough estimation of the task difficulties. Based on the Image Retrieval in Medical Applications (IRMA) framework [3,4], we provide this baseline computations.

C. Peters et al. (Eds.): CLEF 2008, LNCS 5706, pp. 752–755, 2009.

2 Methods

The content of each radiograph is represented by TAMURA's texture measures (TTM) proposed in [5] and down-scaled representations of the original images, 32×32 and $X \times 32$ pixels disregarding and according to the original aspect ratio, respectively. Since these image icons maintain the spatial intensity information, variabilities that are commonly found in a medical imagery are modeled by the distance measure. These include radiation dose, global translation, and local deformation. In particular, the cross-correlation function (CCF) that is based on SHANNON, and the image distortion model (IDM) from [6] are used.

The single classifiers are combined within a parallel scheme, which performs a weighting of the normalized distances obtained from the single classifiers C_i, and applies the NN decision function C to the resulting distances:

$$d_{\text{combined}}(q,r) = \sum_i \lambda_i \cdot d'_i(q,r), \tag{1}$$

$$d'_i(q,r) = \frac{d_i(q,r)}{\sum_{r' \in R} d_i(q,r')} \tag{2}$$

where $0 \leq \lambda_i \leq 1$, $\sum_i \lambda_i = 1$ denotes the weight for the normalized distance $d_i(q,r)$ obtained from classifier C_i for a sample q and a reference r from the set of reference images, R. Values $0 \leq s_i(q,r) \leq 1$ obtained from similarity measures are transformed via $d_i(q,r) = 1 - s_i(q,r)$.

The three content descriptors and their distance measures use the following parameters:

- TTM: texture histograms from down-scaled image (256×256), 384 bins, Jensen-Shannon divergence as a distance measure;
- CCF: 32×32 icon, 9×9 translation window; and
- IDM: $X \times 32$ icon, gradients, 5×5 window, 3×3 context

The weighting coefficients were set empirically during CLEF MAAT 2005:

$$\begin{aligned} \lambda_{\text{IDM}} &= 0.42, \\ \lambda_{\text{CCF}} &= 0.18, \quad \text{and} \\ \lambda_{\text{TTM}} &= 0.40. \end{aligned}$$

3 Results

Tab. 1 lists the baseline results for the four years [7,8,9]. Runs which were not submitted are marked with asterisks, along with their hypothetic rank. In 2007 and 2008, the evaluation was not based on the error rate. Therefore, the table lists the rank based on the modified evaluation scheme on full codes. In average, the $k = 1$ NN classifier is better than $k = 5$. Disregarding the hierarchical information, which was made essential to solve the 2008 task, we obtain an error rate of 51,3%.

Table 1. Baseline error rates (ER) and ranks among submissions

			$k = 1$		$k = 5$	
Year	References	Classes	ER	Rank	ER	Rank
2005	9,000	57	13.3%	2/42	14.8%	*7/42
2006	10,000	116	21.7%	13/28	22.0%	*13/28
2007	11,000	116	20.0%	*17/68	18.0%	18/68
2008	12,089	197	51.3%	*12/24	52.8%	12/24

4 Discussion

The baseline error rates allow a rough estimation of the task difficulty: Comparing 2005 and 2006, the number of classes increased by 103%, while the error rates only increased by 63% and 48% for 1-NN and 5-NN, respectively. This suggests that the task in 2006 was easier than in 2005. Since the challenges in 2006 and 2007 use the same class definitions, the obtained error rates are directly comparable and show a slightly reduced task difficulty in 2007. In 2008, the number of classes increased by 70% compared to 2007, while the error rate increased by 157% and 193%, respectively. The 2008 task is therefore considered to be more difficult than the 2007 task. With similar estimation, the 2008 task is also found to be more difficult than the 2005 task, as the number of classes increased by 246%, but the error rate increased by 286% and 257%, respectively.

Acknowledgment

This work is part of the IRMA project, which is funded by the German Research Foundation (DFG), grants Le 1108/4, Le 1108/6, and Le 1108/9.

References

1. Deselaers, T., Deserno, T.M.: Medical Image Annotation in ImageCLEF 2008. In: Peters, C., et al. (eds.) Evaluating Systems for Multilingual and Multimodal Information Access – 9th Workshop of the Cross-Language Evaluation Forum. LNCS, vol. 5706, pp. 523–530. Springer, Heidelberg (2009)
2. Lehmann, T.M., Schubert, H., Keysers, D., Kohnen, M., Wein, B.B.: The IRMA code for unique classification of medical images. In: Proceedings SPIE, vol. 5033, pp. 109–117 (2003)
3. Lehmann, T.M., Güld, M.O., Thies, C., Fischer, B., Spitzer, K., Keysers, D., Ney, H., Kohnen, M., Schubert, H., Wein, B.B.: Content-based image retrieval in medical applications. Methods of Information in Medicine 43(4), 354–361 (2004)
4. Güld, M.O., Thies, C., Fischer, B., Lehmann, T.M.: A generic concept for the implementation of medical image retrieval systems. International Journal of Medical Informatics 76(2-3), 252–259 (2007)
5. Tamura, H., Mori, S., Yamawaki, T.: Textural features corresponding to visual perception. IEEE Transactions on Systems, Man, and Cybernetics, B 8(6), 460–473 (1978)

6. Keysers, D., Dahmen, J., Ney, H., Wein, B.B., Lehmann, T.M.: A statistical framework for model-based image retrieval in medical applications. Journal of Electronic Imaging 12(1), 59–68 (2003)
7. Clough, P., Müller, H., Deselaers, T., Grubinger, M., Lehmann, T.M., Jensen, J., Hersh, W.: The CLEF 2005 cross-language image retrieval track. In: Peters, C., Gey, F.C., Gonzalo, J., Müller, H., Jones, G.J.F., Kluck, M., Magnini, B., de Rijke, M., Giampiccolo, D. (eds.) CLEF 2005. LNCS, vol. 4022, pp. 535–557. Springer, Heidelberg (2006)
8. Müller, H., Deselaers, T., Lehmann, T.M., Clough, P., Kim, E., Hersh, W.: Overview of the ImageCLEFmed 2006 medical retrieval and medical annotation tasks. In: Peters, C., Clough, P., Gey, F.C., Karlgren, J., Magnini, B., Oard, D.W., de Rijke, M., Stempfhuber, M. (eds.) CLEF 2006. LNCS, vol. 4730, pp. 595–608. Springer, Heidelberg (2007)
9. Müller, H., Deselaers, T., Kim, E., Kalpathy-Cramer, J., Deserno, T.M., Clough, P., Hersh, W.: Overview of the ImageCLEFmed 2007 medical retrieval and annotation tasks. In: Peters, C., Jijkoun, V., Mandl, T., Müller, H., Oard, D.W., Peñas, A., Petras, V., Santos, D. (eds.) CLEF 2007. LNCS, vol. 5152, pp. 472–491. Springer, Heidelberg (2008)

Evaluating the Impact of Image Names in Context-Based Image Retrieval

Mouna Torjmen, Karen Pinel-Sauvagnat, and Mohand Boughanem

SIG-RFI, IRIT, Toulouse, France

Abstract. This paper describes our work at the CLEF 2008 WikipediaMM Task. We study the use of image name in a context-based image retrieval approach. This factor is evaluated in three manners. The first one consists of using image names explicitly: we computed a similarity score between the query and the name of images using the vector space model. The second one consists of combining results obtained using the textual content of documents and results obtained using the first method. Finally, in our last approach, image names are used less explicitly: we proposed to use all the textual content of image annotations, but we increased the weight of terms in the image name. Results show that the image name can be an interesting factor to improve image retrieval results.

Keywords: Contextual image retrieval, image name, annotation.

1 Introduction

The textual content of multimedia documents[1] can be used as potential high-level semantic features to represent images. This technique , called context-based image retrieval, uses evidences like the surrounding text of the image, titles, hyperlinks or the image name to represent images.

In this paper, we study the impact of one of these factors, the *name of images*, in the retrieval process. This study was carried out within the WikipediaMM task of CLEF 2008 [6].

We evaluated three algorithms that take into account the name of images in the retrieval process. In the first one, we calculate a score for each image by only using its name. This score is combined in the second approach with the score of the textual content[2] of the document containing the image. Finally, the third approach evaluates the impact of the image name implicitly: we increase the weight of terms in the image name. Consequently, documents having a relevant image name will be top ranked.

The rest of the paper is organized as follows: section 2 describes our motivation and related works. In section 3, we present our approaches to study the impact

[1] A multimedia document is a document containing textual information and at least one multimedia object as an image for example.

[2] In this paper, the textual content is the image metadata, i.e. the image name and the image annotation. In fact each document of the used collection (MMWikipedia) are composed of one image and its metadata.

C. Peters et al. (Eds.): CLEF 2008, LNCS 5706, pp. 756–762, 2009.

of the image name in the retrieval process. In section 4, we detail our empirical evaluation of the proposed approaches and discuss our results. Finally, section 5 concludes with possible directions for future work.

2 Motivation and Related Works

In previous works on the WikipediaMM collection, we evaluated some contextual components in the image retrieval as the document structure [3] and the concept classification scores [5] to estimate the relevance scores of images.

In this paper, we aim to study the impact of another contextual component in the image retrieval process which is the *image name*. Many works [2], [1] and many web search engines as Google Images use the name of image as a contextual element of the image itself to participate in its relevance estimation. However, to our knowledge, there is no real and explicit study concerning this factor.

Our intuition behind this study is that the name of an image, if it is significant, describes well the image content, and consequently plays an important role in estimating the image relevance. To be significant, the image name should be be composed of real terms. For example, 'Barcelona.jpg' is a significant image name, whereas this is not the case for 'DSCN0075.jpg'.

Some issues arise when image names are composed of concatenated terms. Let us consider the following image name: "CityBarcelona.jpg". This name is considered as a single term, while in reality it is composed of two terms: city and Barcelona.

To solve this problem, all image names containing at least one query term are considered as significant. For example, if our query term is "Barcelona", it will be identified in the image name "CityBarcelona.jpg".

However, this solution partially solves the problem, since some cases where a relevant image name cannot be identified still exist. For example, if the query is "city", the query term, thanks to the Porter algorithm, will for example be indexed as "citi", and cannot be found in our image name "CityBarcelona.jpg".

In the WikipediaMM collection, the metadata of images (textual content) are formatted in XML. An XML document (part (a) of figure 1) can be represented as a hierarchical tree (part (b) of figure 1), composed of a root (document), simple nodes (element and/or attributes) and leaf nodes (values as text and images). An inner node is any node of the tree that has child nodes (i.e a non-leaf node). The hierarchical relationships between elements can be used to improve retriaval results, but it is not our aim in this paper.

As the element name where we can find image names can change from a collection to another, we identified the names of images in documents using their extension. Any token that ends by an image extension such as ".jpeg" and ".gif" for example is considered as an image name (like "Barcelona.jpg" in figure 1). In the indexing phase, we thus index image names in a separate index.

In our study, we only use the metadata of images in the MMWikipedia collection. Our study focusses on several issues: do the names of the images have an impact on the search results? If it is the case, how this source of information should be exploited?

```
<article>
  <title> Barcelona </title>
  <figure>
    <caption>A picture of Barcelona city ... </caption>
    <image> Barcelona.jpg</image>
  </figure>
  <text>
    <wikilink>Sagrada Familia</wikilink>
    Barcelona is...
  </text>
</article>
```

(a) XML document

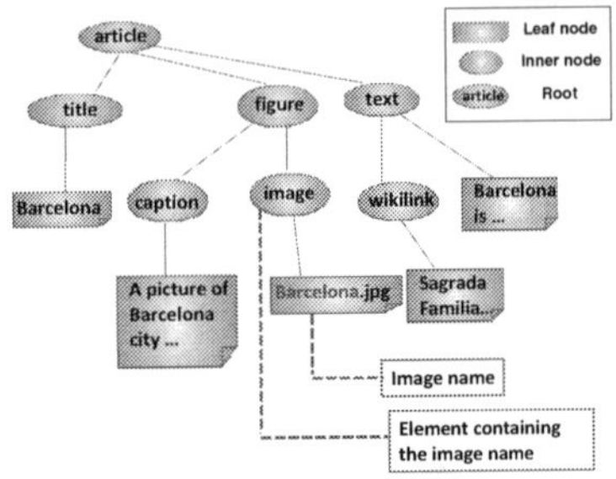

(b) Associated tree

Fig. 1. Example of an image name identification in an XML document tree

3 Retrieval Approaches

In this section, we describe briefly our baseline model in the first part, and then in the second part, we describe the three approaches used for the image name evaluation.

3.1 Baseline Model: The XFIRM Model

As the image metadata are formatted in XML, we use an XML search engine, the XFIRM system [4], as a baseline model for retrieval.

In CLEF 2008 WikipediaMM task, image annotations have approximately the same simple structure in the collection (the average document depth is low, about 6.60). We thus used a simplified version of the XFIRM model, which is based on a relevance propagation method. During query processing, relevance scores are computed at leaf nodes level and then at inner nodes level thanks to a propagation of leaf nodes scores through the document tree.

As the aim of the task is to return relevant images with their metadata (more precisely the names of files containing images and metadata), we ask the XFIRM system to only return whole documents and not parts of documents as an image or a part of annotation.

3.2 Algorithms Used to Evaluate the Impact of Image Names

Using only the image name terms. To explicitly use the terms composing the image name and study their importance, we first proposed to only use the image name terms to retrieve relevant images. We computed a score for each image using the vector space model. We evaluated three similarity measures.

Let us consider a space of N dimensions, with N the number of terms in the collection, a query q, and an image name $ImName$.

The first similarity measure used is the Cosine Similarity (Equation 1).

$$W_{ImName}(Im) = cos(q, ImName) = \frac{\sum_{j=1}^{N} w_j^q w_j^{ImName}}{\sqrt{\sum_{j=1}^{N} (w_j^q)^2}\sqrt{\sum_{j=1}^{N} (w_j^{ImName})^2}} \quad (1)$$

The second one is the Dice Coefficient presented in Equation 2.

$$W_{ImName}(Im) = sim(q, ImName) = \frac{2\sum_{j=1}^{N} w_j^q w_j^{ImName}}{\sum_{j=1}^{N}(w_j^q)^2 + \sum_{j=1}^{N}(w_j^{ImName})^2} \quad (2)$$

The last one is the Inner Product measure, expressed in Equation 3.

$$W_{ImName}(Im) = sim(q, ImName) = \sum_{j=1}^{N} w_j^q w_j^{ImName} \quad (3)$$

In the three formula, $w_q^j \in [0..1]$ is the weight of term t_j in the query, and $w_j^{ImName_i} \in [0..1]$ is the weight of t_j in the image name. These weights are evaluated using a tf*idf formula.

We return to users all file names containing an image having a positive weight, in decreasing order of image weights.

Combining image name scores to document scores. The score of the image obtained thanks to its name $W_{ImName}(Im)$ can be combined with the score of document textual content (image metadata) obtained by the XFIRM system $W_{XFIRM}(Text)$. The combination function between the two scores is presented in equation 4.

$$W(Im) = \lambda W_{XFIRM}(txt) + (1 - \lambda) W_{ImName}(Im) \quad (4)$$

where λ is a pivot parameter $\in [0..1]$.

We also return to users all file names containing an image having a positive weight, in decreasing order of image weights.

Implicit use of the image name terms. We propose in this section to modify the term weighting formula used in the XFIRM model, by increasing the score of terms belonging to the image name. The new formula is as follows:

$$RSV(q, ln) = \sum_{i=1}^{n} w_i^q * w_i^{ln},$$

where q is the query, ln a leaf node, $w_i^q = tf_i^q * idf_i$ and

$$w_i^{ln} = \begin{cases} K * tf_i^{ln} * idf_i \quad if\ t_i\ is\ an\ image\ name\ term \\ tf_i^{ln} * idf_i\ otherwise \end{cases} \quad (5)$$

$K > 1$ is a constant used to increase the weight of image name terms.

In the example of figure 2, without the implicit specification of the image name, the weight of the relevant term in the image name is computed like the other terms in the metadata. So, the scores of document 1 and document 2 are equal for query "Barcelona" (tf=2 in the two documents). However, when applying our algorithm, document 1 becomes more relevant than document 2 as the weight of the relevant term in the image name will be increased. This seems coherant as the image of document 1 concerns more the query than the image of document 2.

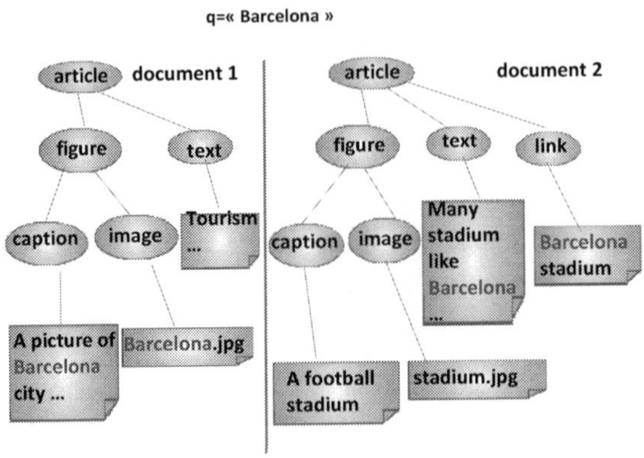

Fig. 2. Importance of image name terms

4 Runs and Results

Before giving and discussing results, we will mention some details about the applicability of our study on the used collection. In fact, analysis of the image names in the collection shows that about 47% of image name are significant in the collection and can be used in the evaluation of our methods.

An other analysis of the image names in the assessment files demonstrates that about 43% of image names of relevant documents are significant.

In the following, our official runs are in grayed boxes.

Results obtained using our first algorithm (equation 1, 2 and 3) are presented in table 1.

Table 1. Explicit use of image name

Runs	Similarity Formula	MAP	P@5	P@10	R-prec	BPREF
SigRunImageName	Cosine	0.0743	**0.1813**	**0.1573**	0.1146	0.0918
SigRunImageNameDice	Dice	0.0416	0.1200	0.1040	0.0745	0.0563
SigRunImageNameInner	Inner Product	**0.0757**	0.1760	0.1493	**0.1238**	**0.1078**

The Cosine (*SigRunImageName*) and Inner Product (*SigRunImageNameInner*) measures give approximately the same results, which are better than those using the Dice coefficient.

Table 2 shows results obtained when combining image scores evaluated using the Cosine measure (equation 1) with the whole document score (equation 4).

Run *SigRunText* only uses the document textual content (image annotations) evaluated with the XFIRM system. The other runs, obtained by combining image score evaluated with the cosine measure and image score evaluated with all textual content of documents, shows that this combination leads to better performance. The best run obtained is the one evaluated with $\lambda = 0.9$. This means

Table 2. Combination of image name results and XFIRM system results using Equation 4

Runs	λ	MAP	P@5	P@10	R-prec	BPREF
SigRunComb0	0	0.0743	0.1813	0.1573	0.1146	0.0918
SigRunComb01	0.1	0.1318	0.2560	0.2147	0.1782	0.1458
SigRunComb02	0.2	0.1380	0.2880	0.2320	0.1857	0.1537
SigRunComb03	0.3	0.1446	0.2987	0.2493	0.1919	0.1609
SigRunComb04	0.4	0.1472	0.3067	0.2600	0.1961	0.1645
SigRunComb05	0.5	0.1537	0.3227	0.2747	0.1979	0.1684
SigRunComb06	0.6	0.1572	**0.3253**	0.2720	0.2043	0.1755
SigRunComb07	0.7	0.1614	0.2327	0.2773	0.2123	0.1817
SigRunComb08	0.8	0.1610	0.3093	0.2773	0.2215	0.1825
SigRunComb09	0.9	**0.1681**	0.3093	0.2800	**0.2270**	**0.1876**
SigRunText	1	0.1652	0.3067	**0.2880**	0.2148	0.1773

that the image name can improve results as an additionnal source of evidence but not as tha main one: the combination parameter of textual content is 0.9 versus to 0.1 for the image name.

We analysed results query by query and we notice that the performance improvement concerns about 41% of queries, while a degration of performance affects 32% of queries, according to the MAP measure.

Runs obtained by implicit use of the image name (equation 5) are detailed in table 3.

Comparing the best run obtained by implicit use of image name ($K = 1.1$, *RunImage11*) to the run obtained using document textual content (*SigRunText*), we notice that the MAP measure of the first one is better.

Query by query analysis shows that 36% of queries are improved (and 44% degraded) according to the MAP measure.

Table 3. Implicit use of image name

Runs	K	MAP	P@5	P@10	R-prec	BPREF
SigRunText	1	0.1652	0.3067	0.2880	0.2148	0.1773
RunImage11	1.1	**0.1724**	**0.3227**	**0.2867**	**0.2329**	**0.1874**
RunImage12	1.2	0.1701	0.3200	0.2787	0.2271	0.1845
RunImage13	1.3	0.1721	0.3093	0.2760	0.2267	0.1854
RunImage14	1.4	0.1714	0.3093	0.2720	0.2258	0.1855
RunImage15	1.5	0.1686	0.3040	0.2760	0.2247	0.1858
RunImage16	1.6	0.1681	0.3040	0.2720	0.2252	0.1859
RunImage17	1.7	0.1665	0.3093	0.2733	0.2238	0.1841
RunImage18	1.8	0.1667	0.3067	0.2707	0.2253	0.1841
RunImage19	1.9	0.1658	0.3093	0.2760	0.2261	0.1839
SigRunImage	2	0.1595	0.3147	0.2787	0.2211	0.1798
SigRunImage	5	0.1481	0.2960	0.2547	0.2130	0.1716
SigRunImage	10	0.1410	0.2933	0.2573	0.2011	0.1615

As a main conclusion about the use of image names in contextual image retrieval, we notice that the image name can be in some cases a relevant contextual element in image retrieval.

However, obtained results cannot definitively confirm our intuition that the image name is an important factor in image retrieval, since results are lower for a siginificant part of queries.

5 Conclusion and Future Work

In this paper, we presented a study about the impact of the use of image name terms in contextual image retrieval. We evaluated the explicit and the implicit use of these terms. Results in Wikipedia Clef 2008 showed that this factor leads to performance improvement in some cases and to performance degradation in other cases.

Our approaches cannot have been applied in the whole collection, as some image names have some problems related to the index phase (some terms are concatenated). In future work, we will try to find a solution of the image name indexing problem. We also plan to evaluate our methods on other collections having significant image names.

Moreover, we plan to add to our model the process of concept clauses of queries.

References

1. Aslandogan, Y.A., Yu, C.T.: Diogenes: a web search agent for person images. In: MULTIMEDIA 2000: Proceedings of the eighth ACM international conference on Multimedia, pp. 481–482 (2000)
2. Frankel, C., Swain, M.J., Athitsos, V.: Webseer: An image search engine for the world wide web. Technical Report TR-96-14, 31 (1996)
3. Hlaoua, L., Torjmen, M., Pinel-Sauvagnat, K., Boughanem, M.: XFIRM at INEX 2006. Ad-hoc, Relevance Feedback and MultiMedia tracks. In: Fuhr, N., Lalmas, M., Trotman, A. (eds.) INEX 2006. LNCS, vol. 4518, pp. 373–386. Springer, Heidelberg (2007)
4. Pinel-Sauvagnat, K., Boughanem, M., Chrisment, C.: Searching XML documents using relevance propagation. In: Apostolico, A., Melucci, M. (eds.) SPIRE 2004. LNCS, vol. 3246, pp. 242–254. Springer, Heidelberg (2004)
5. Torjmen, M., Pinel-Sauvagnat, K., Boughanem, M.: Mm-xfirm at inex multimedia track 2007. In: Pre-Proceedings of INEX 2007 Workshop, Dagstuhl, Germany (2007)
6. Tsikrika, T., Kludas, J.: Overview of the wikipediaMM task at ImageCLEF 2008. In: Peters, C., et al. (eds.) CLEF 2008. LNCS, vol. 5706, pp. 539–550. Springer, Heidelberg (2009)

Large-Scale Cross-Media Retrieval of WikipediaMM Images with Textual and Visual Query Expansion

Zhi Zhou[1,2,3,*], Yonghong Tian[3,*], Yuanning Li[1,2,3], Tiejun Huang[3], and Wen Gao[3]

[1] Key Laboratory of Intelligent Information Processing, Institute of Computing Technology, Chinese Academy of Sciences, Beijing 100190, China
[2] Graduate University of Chinese Academy of Sciences, Beijing 100049, China
[3] Institute of Digital Media, School of EE & CS, Peking University, Beijing 100871, China
{zzhou,ynli}@jdl.ac.cn, {yhtian,tjhuang,wgao}@pku.edu.cn

Abstract. In this paper, we present our approaches for the WikipediaMM task at ImageCLEF 2008. We first experimented with a text-based image retrieval approach with query expansion, where the extension terms were automatically selected from a knowledge base that was semi-automatically constructed from Wikipedia. Encouragingly, the experimental results rank in the first place among all submitted runs. We also implemented a content-based image retrieval approach with query-dependent visual concept detection. Then cross-media retrieval was successfully carried out by independently applying the two meta-search tools and then combining the results through a weighted summation of scores. Though not submitted, this approach outperforms our text-based and content-based approaches remarkably.

Keywords: Image retrieval, textual query expansion, query-dependent visual concept detection, cross-media re-ranking.

1 Introduction

The WikipediaMM task at ImageCLEF 2008 aims to investigate effective retrieval approaches in a large-scale collection of Wikipedia images. In the task, participants need to deal with searching 75 topics from approximately 150,000 images. Search over such a large-scale image collection offers many challenges. Among them, the most glaring challenge is the so-called semantic gap [8]. Even in the situation where images are associated with some textual descriptions, this semantic gap is still present since they do not fully capture all the subtleties of the semantics of the images.

To address the semantic gap issue, we experimented with several image retrieval approaches on the WikipediaMM dataset. A retrieve engine was implemented in this participation, which consists of four components respectively for data pre-processing, text-based image retrieval (TBIR), content-based image retrieval (CBIR), and cross-media retrieval. In TBIR, textual query expansion technique is used where the extension terms are automatically selected from a knowledge base (KB) that is semi-automatically constructed from the online large-scale encyclopedia — Wikipedia.

* Corresponding author.

C. Peters et al. (Eds.): CLEF 2008, LNCS 5706, pp. 763–770, 2009.

Encouragingly, the experimental results rank in the first place among all submitted runs. For CBIR, visual query expansion is employed through query-dependent visual concept detection to semantically annotate images or augment their rough semantics gathered from related text. By comparison, this approach performs better than the other submitted CBIR runs. Then cross-media retrieval is performed by independently applying the two meta-search tools and then combining the results through a weighted summation of scores. Though not submitted, this approach outperforms our text-based or content-based approaches remarkably.

The rest of this paper is organized as follows. Textual and visual query expansion approaches for two meta-search tools are described respectively in Section 2 and 3. Then the cross-media re-ranking approach is presented in Section 4. The experimental results are shown in Section 5. Finally we draw a conclusion in Section 6.

2 Textual Query Expansion for TBIR

A natural solution for WikipediaMM 2008 task is to use TBIR method. To help the retrieval system get close to users' real intent, query expansion techniques are often used by adding terms to queries or modifying preliminary queries. In this participation, we focus on how to automatically extract the expansion terms from a KB that is semi-automatically constructed from Wikipedia. Organized with concepts identified by URLs and links between concepts and external nodes, Wikipedia is not only a Web collection but also an online knowledge center which assembles all users' intelligences. Therefore, it is naturally attractive and promising that this open, and constantly evolving encyclopedia can yield inexpensive knowledge structures that can be exploited to enhance the semantics of queries.

Recently, "Wikipedia mining" has been addressed as a new research topic. WikiRelate [2] used link-based path length for computing relatedness for given concepts; Nakayama *et al.* [3] proposed the PFIBF (Path Frequency – Inversed Backward link Frequency) algorithm for Web thesaurus construction. However, none of work is made on using Wikipedia as the KB in information retrieval.

In Wikipedia, each non-administrative page is used as a term/concept describing individuals (e.g., *Jingtao Hu*), concepts (e.g., *Emissions trading*), locations (e.g., *Big Ben*), events (e.g., *collapse of the World Trade Center*), and categories (e.g., *microbiology*). For a given term, the related terms can be easily extracted from the corresponding Wikipedia pages, and then used to extend the query when this term is used as the query input. Finally, the extended query is fed into the retrieval engine to generate the final search results. In our implementation, we use the TF-IDF paradigm for text retrieval which has been widely used in text mining and information retrieval.

As shown in Fig. 1, three steps are used to construct the KB from Wikipedia:

(1) ***Near Pages Selection.*** We first download and index all Wikipedia pages with TF-IDF model. Only pages with a similarity score higher than threshold θ (θ is set to be 0.9 in our experiments) are chosen as the related pages of the input query.

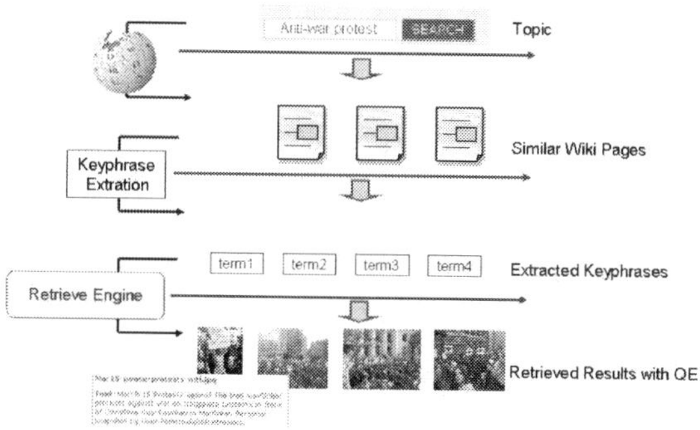

Fig. 1. Textual query expansion using the KB constructed from Wikipedia

(2) ***Page Keyphrase Extraction.*** In a Wikipeida page, keyphrases or keywords briefly describe the content of a concept. Thus they can be used to enhance the semantics of that concept. In our system, we employ an unsupervised keyphrase extraction algorithm presented in our previous work [4]. By treating text in a page as a semantic network, this algorithm computes several structure variables of Small-World Network (SWN) to select key nodes as keyphrases $K = \{(t_k, P(t_k))\}$, each with a probability score $P(t_k)$ indicating the importance of the extracted keyphrase t_k.

(3) ***Term Selection for Query Expansion.*** In practice, the top-ranked keyphrases cannot be directly used for query expansion. For instance, when searching "*saturn*", term "*moon*" is extracted as the keyphrase with a high score, but "*moon*" may appear on many pages and should be considered more *general*. To address this problem, a statistical feature *Inverse Backward link Frequency* (*ibf*) [3] is calculated as:

$$ibf = \log(\frac{N}{bf(t) + \beta}), \tag{1}$$

where $bf(t)$ is the number of backward links in which the link text contains term *t*, *N* denotes the total number of pages and β is a parameter in case $bf(t)$ is zero. Therefore, the final weight of a keyphrase can be computed as:

$$w_{t_k \in K} = P(t_k) \cdot ibf(t_k). \tag{2}$$

Then the keyphrases with their normalized weights are combined with the original query to construct an extended query to be fed into the retrieve engine.

3 Query-Dependent Visual Concept Detection for CBIR

In the WikipediaMM dataset, some images have few or even no descriptive texts. To address this problem, *query-dependent visual concept detection* can be used to semantically annotate images or augment their rough semantics gathered from related text. Given the pre-defined query concepts, *1-vs-all* visual concept detectors are trained for all these concepts by using the training images obtained by Yahoo! search. Clearly, these training images can be used for visual query expansion to enhance the CBIR task. As shown in Fig 2, the training process includes the following three steps:

(1) ***Building the training set.*** For each query concept, top k (k=30 in our system) images are clawed from Yahoo! image search engine. Then some unrelated images with respect to the concept are manually filtered out whichforms a positive training set. Negative images for each concept are randomly selected from positive images of the other concepts.

(2) ***Building Bag of Words (BOW) representation.*** SIFT [5], Dense-SIFT [6] and Color-Dense-SIFT are extracted from the training sets of all concepts. Then k-means algorithm is employed to quantize different types of features and create a combined visual codebook. All images are represented by a set of tokens of the visual words.

(3) ***Supervised training for each topic.*** Unsupervised probabilistic latent semantic analysis (pLSA) [7] is utilized to infer the latent topic distribution of the training images based on the BOW representation. Then support vector machine (SVM) is used to train a *one-class classifier* for each concept in the latent topic space.

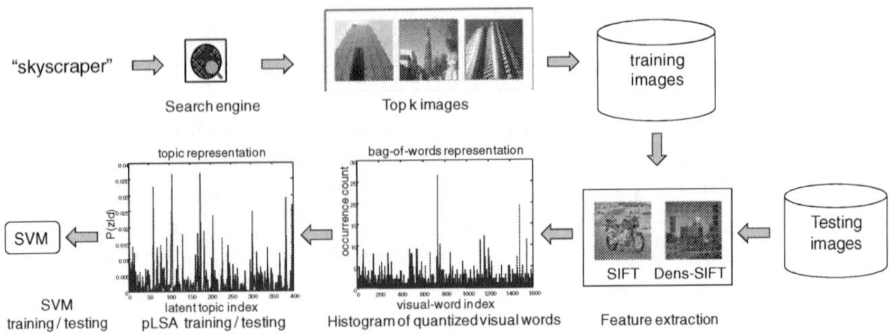

Fig. 2. Query-dependent visual concept detection for CBIR

Given the trained *1-vs-all* visual concept detectors for all query topics, we can perform the concept detection for each test image by firstly representing it with the visual words from the trained codebook, inferring its latent topic distribution based on the trained pLSA model, and finally computing the responds of the trained SVMs for different concepts. Concept is detected only when the corresponding respond is above a given threshold. For CBIR, test images are finally ranked according to their responds with respect to the query concept.

4 Query-Independent Cross-Media Re-ranking

For better retrieval performance, we study cross-media image retrieval by combining both TBIR and CBIR methodologies. In our implementation, cross-media retrieval is performed by independently applying the two meta-search tools and then combining the results through a weighted summation of scores. Here the weights are query-independent, say, identical for all queries. Then the re-ranking score is computed as:

$$WeightedScore(q,d) = w_1 * Score(q_{text}, d_{text}) + w_2 * Score(q_{visual}, d_{visual}) \quad (3)$$

A key point here is to compare the overlap of the results returned by different retrieve engines. Let $\mathbf{R}_1$ and $\mathbf{R}_2$ respectively denote the result sets of TBIR-based and CBIR-based retrieval engines, and M_1 and M_2 be their sizes. Let image $d_i^1 \in \mathbf{R}_1, i < M_1$, and $d_j^2 \in \mathbf{R}_2, j < M_2$, then an overlap set $\mathbf{G}$ can be obtained:

$$\mathbf{G} = \{(d_i^1, d_j^2) : d_i^1 = d_j^2, i < M_1, j < M_2\}, \quad (4)$$

where (d_i^1, d_j^2) stands for an image both returned by the two engines. Let H_1 and H_2 be the numbers of overlap images in Top N ranked images, $H_1 = \#\{d_i^1 : d_i^1 \in \mathbf{G}, i < N\}, H_2 = \#\{d_j^2 : d_j^2 \in \mathbf{G}, j < N\}$, then the weight of each engine can be calculated as:

$$w_l = \frac{\sigma/2 + H_l/N}{\sigma + \sum_l H_l/N}, \quad (5)$$

where l is the engine identifier and σ (we set $\sigma = 0.1$) is an adjusting parameter.

5 Experiments

This section describes our experiments for the WikipediaMM task. Note that some of the experimental results reported here were not submitted before the deadline.

The experiments are evaluated by *MAP* (Mean Average Precision), *P@N* (precision of top *N* images), and *R-precision*. The ground-truth results are given in the evaluation phase of the WikipediaMM task.

5.1 Experiments with TBIR

The first set of experiments is to evaluate the performance of TBIR approach with different query expansion methods.

Query expansion by using the automatically-constructed KB. Different methods are used to automatically construct KB from Wikipedia for query expansion, by using different text sources (e.g., titles, links or fulltext of Wikipedia articles) and different term selection algorithms (e.g., TFIDF-based, Small-World (SW)-based, SWIBF-based). Therefore, four *automatic query expansion* methods were evaluated in our experiments, respectively denoted by *QE-Title-TFIDF*, *QE-Link-TFIDF*, *QE-Fulltext-SW*, and *QE-Fulltext-SWIBF*. We also use *NO-QE* to denote TBIR without query expansion. In all experiments, only top 20 terms are used.

Surprisingly, all these automatic query expansion methods can not significantly improve the TBIR performance, compared with *NO-QE* (See Table 1). Thus we should consider how to improve the quality of the constructed KB.

Query expansion by using the semi-automatically-constructed KB. After the KB was automatically constructed from Wikipedia, we then performed some manual confirmations. Here we use *QE-Fulltext-Semi* to denote this query expansion method. Note that in this case, the query expansion method still automatically selects terms from the KB to semantically expand a given query term. From Table 1, we can see that this *QE-Fulltext-Semi* method performs much better than all other models.

Table 1. The experimental results of different textual query expansion methods

Run ID	QE	Modality	MAP	P@5	P@10	R-Prec
NO-QE	without	TXT	0.2565	**0.4427**	**0.3747**	0.2929
QE-Title-TFIDF	with	TXT	0.2566	0.4187	0.3627	**0.2967**
QE-Link-TFIDF	with	TXT	0.2271	0.376	0.3147	0.2533
QE-Fulltext-SW	with	TXT	0.2365	0.3733	0.336	0.2618
QE- Fulltext-SWIBF	with	TXT	**0.2609**	0.44	0.3693	0.2859
QE- Fulltext-SEMI	with	TXT	**0.3444**	**0.5733**	**0.476**	**0.3794**

5.2 Experiments with CBIR

Compared with TBIR, our CBIR obtained a comparable precision in the top-ranked images (P@5=0.5307 and P@10= 0.4507 of CBIR *vs.* P@5=0. 5733 and P@10=0.476 of TBIR), but much lower MAP (0.1928 of CBIR *vs.* 0.3444 of TBIR) and R-Prec (0.2295 of CBIR *vs.* 0.3794 of TBIR). Although visual content ambiguity reduces the overall performance (MAP) by returning images with similar low-level features, the experimental results show that learning visual models from Web images (e.g., from Yahoo! search) do help to rank the content-relevant images higher. It also should be noted that, our CBIR approach performs best among all submitted CBIR runs in WikipediaMM 2008 task.

Table 2. The experimental results of CBIR

Run ID	QE	Modality	MAP	P@5	P@10	R-Prec
CBIR run1	with	IMG	0.1912	**0.5333**	0.4427	**0.2929**
CBIR run2	with	IMG	**0.1928**	0.5307	**0.4507**	0.2295

5.3 Experiments with Cross-Media Retrieval

In the last set of experiments, cross-media retrieval approach is used to achieve better performance by combining text-based and content-based retrieval results. In the experiments, we set M_2 smaller than M_1. This means that only the *top-ranked* images returned by CBIR are included in the re-ranking phase since the lower-ranked images may have much higher probabilities to be noises. Table 3 shows the experimental results, where *ReRank-Text-Visual-N* denotes the combination of CBIR and TBIR without query expansion, and *ReRank-Semi-Visual-N* denotes the combination of CBIR and TBIR with semi-automatic query expansion, and *N* denotes the corresponding parameter in Eq. (5).

Table 3. Some experimental results of cross-media retrieval

Run ID	QE	Modality	MAP	P@5	P@10	R-Prec
NO-QE	without	TXT	0.2565	0.4427	0.3747	0.2929
CBIR run2	without	IMG	**0.1928**	0.5307	**0.4507**	0.2295
ReRank-Text-Visual-10	without	TXTIMG	**0.3099**	**0.608**	**0.5213**	0.3387
ReRank-Text-Visual-20	without	TXTIMG	0.3035	0.6027	0.512	**0.3420**
ReRank-Text-Visual-40	without	TXTIMG	0.2972	0.584	0.4893	0.3393
ReRank-Text-Visual-60	without	TXTIMG	0.2928	0.5547	0.4733	0.3366
ReRank-Text-Visual-80	without	TXTIMG	0.2910	0.5387	0.4693	0.3349
QE- Fulltext-SEMI	with	TXT	0.3444	0.5733	0.476	0.3794
CBIR run2	with	IMG	0.1928	0.5307	0.4507	0.2295
ReRank-Semi-Visual-10	with	TXTIMG	**0.3584**	**0.6293**	**0.5147**	**0.3993**
ReRank-Semi-Visual-20	with	TXTIMG	0.3568	0.6187	0.5147	0.3974
ReRank-Semi-Visual-40	with	TXTIMG	0.3519	0.5867	0.5013	0.3988
ReRank-Semi-Visual-60	with	TXTIMG	0.3487	0.568	0.492	0.3988
ReRank-Semi-Visual-80	with	TXTIMG	0.3483	0.5653	0.4907	0.3988

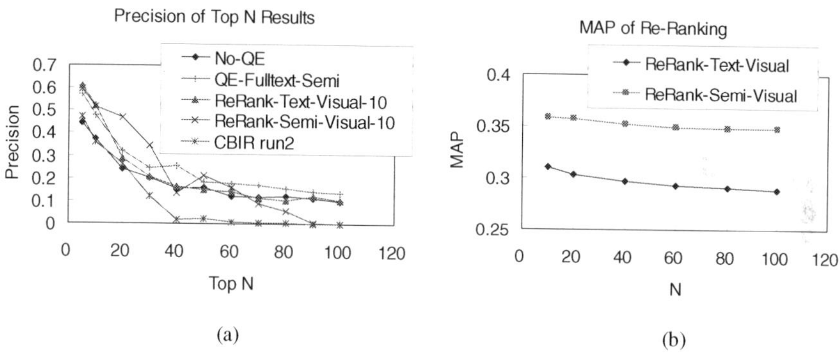

Fig. 3. Performance of cross-media retrieval: (a) P@N and (b) MAP results with different values of N in Eq. (5)

From Table 3 and Fig. 3, it's interesting to find that when N increases, the preliminary result of each system is more likely to be equally treated and the overall performance decreases. For the combination of CBIR and text-based retrieval without query expansion, the average improvement of all the queries in *ReRank-Text-Visual-10* is around 5.34% over the single text-based retrieval approach (25.65% of MAP). While for the combination of CBIR and text-based retrieval with semi-automatic query expansion, the average improvement for all the queries in *ReRank-Semi-Visual-10* is around 1.4% over the single text-based retrieval approach (34.44% of MAP).

We also observed that the cross-media retrieval results have much higher precision of top-ranked images than both text-based retrieval or CBIR results. Generally speaking, text-based retrieval can return more relevant images by searching keywords with image descriptions, while CBIR can obtain high precision of top-ranked images but too many noises in lower-ranked images. Thus combining CBIR with text-based retrieval can help increase the precision of top-ranked images.

In conclusion, the cross-media retrieval approach performs remarkably well. This indicates that cross-media fusion is definitely a promising direction to investigate effective retrieval approaches in the context of a large-scale and heterogeneous collection of images.

6 Conclusion and Future Work

This paper reported our approaches for the WikipediaMM task at ImageCLEF 2008. We experimented with TBIR, CBIR and cross-media image retrieval approaches with query expansion. Encouragingly, the experimental results of our TBIR approach rank in the first place among all submitted runs. Despite not submitted, the cross-media approach performs much better than the single TBIR or CBIR approaches. Further experiments will be done by optimizing the KB construction procedure and taking better cross-media re-ranking strategies into account.

Acknowledgement

The work is supported by grants from Chinese NSF under contract No. 60605020 and No. 90820003, National Hi-Tech R&D Program (863) of China under contract No. 2006AA01Z320 and No. 2006AA010105, and National Basic Research Program of China under contract No. 2009CB320906.

References

1. Tian, Y.H., Huang, T.J., Gao, W.: Exploiting multi-context analysis in semantic image classification. J. Zhejiang Univ. SCI. 6A(11), 1268–1283 (2005)
2. Strube, M., Ponzetto, S.: WikiRelate! Computing semantic relatedness using Wikipedia. In: Proc. of National Conference on Artificial Intelligence (AAAI 2006), Boston, Mass, pp. 1419–1424 (2006)
3. Nakayama, K., Hara, T., Nishio, S.: A thesaurus construction method from large scale web dictionaries. In: Proc. of IEEE International Conference on Advanced Information Networking and Applications (AINA 2007), pp. 932–939 (2007)
4. Huang, C., Tian, Y.H., Zhou, Z., Ling, C.X., Huang, T.J.: Keyphrase extraction using Semantic Networks Structure Analysis. In: Proc. of the sixth IEEE Int'l. Conf. on Data Mining (ICDM 2006), pp. 275–284. IEEE press, Hong Kong (2006)
5. Lowe, D.: Object recognition from local scale-invariant feature. In: Proc. Int'l Conf. Computer Vision (ICCV 1999), pp. 1150–1157 (1999)
6. Lazebnik, S., Schmid, C., Ponce, J.: Beyond Bags of Features: Spatial Pyramid Matching for Recognizing Natural Scene Categories. In: Proceedings of the IEEE CVPR 2006, pp. 2169–2178 (2006)
7. Hofmann, T.: Unsupervised learning by probabilistic latent semantic analysis. Machine Learning 41, 177–196 (2001)
8. Smeulders, A.W.M., Worring, M., Santini, S., Gupta, A., Jain, R.: Content-based image retrieval at the end of the early years. IEEE Trans. Pattern Anal. Mach Intell. 22(12), 1349–1380 (2000)

Conceptual Image Retrieval over a Large Scale Database

Adrian Popescu*, Hervé Le Borgne, and Pierre-Alain Moëllic

CEA, LIST, Laboratoire d'ingénierie de la connaissance multimédia et multilingue
F-92265 Fontenay-aux-Roses, France
adrian.popescu@telecom-bretagne.eu,
{herve.le-borgne,pierre-alain.moellic}@cea.fr

Abstract. Image retrieval in large-scale databases is currently based on a textual chains matching procedure. However, this approach requires an accurate annotation of images, which is not the case on the Web. To tackle this issue, we propose a reformulation method that reduces the influence of noisy image annotations. We extract a ranked list of related concepts for terms in the query from WordNet and Wikipedia, and use them to expand the initial query. Then some visual concepts are used to re-rank the results for queries containing, explicitly or implicitly, visual cues. First evaluations on a diversified corpus of 150000 images were convincing since the proposed system was ranked 4th and 2nd at the WikipediaMM task of the ImageCLEF 2008 campaign [1].

Keywords: image retrieval, large-scale database, query reformulation.

1 Introduction

Existing Web-scale image search engines consider the text found *around* the images (caption, HTML tags...) as a relevant description to describe them, and thus match the query to those terms to propose results. The main advantages of this approach are its computational tractability and its applicability to large volume of data. Unfortunately, the descriptive text is often unrelated to image content and leads to an important imprecision of results. Query ambiguity is another important noise source. For instance, the word *bridge* can refer to the structure or to the card game, and the expected results are completely different for the two meanings. The use of semantic structures is a possible solution to cope with such problems, as long as these structures can cover the query space. We propose to expand the queries using conceptual relations from a prebuilt large-scale semantic structure, a process that enhances results and requires little computational overload.

Semantic structures, such as WordNet [2] were already used in image retrieval [3] but they do not ensure a sufficient coverage of the query space. For instance WordNet includes only few artefact instances for each concept (e.g. there is

* Adrian Popescu is currently with the Computer Science Dept., Télécom Bretagne.

C. Peters et al. (Eds.): CLEF 2008, LNCS 5706, pp. 771–778, 2009.

no WordNet entry for *Peugeot*) and these instances are popular Web queries. Wikipedia is a rich source of semi-structured information and has already been used to structure large quantities of knowledge [4,5]. [5] proposed a method to clean the categorical tree of Wikipedia in order to obtain a sound taxonomy. Kazama et al. [6] successfully extracted *IsA* relations from the first sentence of articles using a syntaxic analysis. [7] explored the automatic enrichment of WordNet using Wikipedia content. They extract hyponymy, hyperonymy, holonymy and meronymy relations based on lexical patterns learned from a text corpus. The overall precision of the extraction process exceeds 50%, leaving a lot of incorrect relations in the extracted structure. DBPedia [4] is a translation of parts of Wikipedia articles to a database format, enabling structured queries over the content of the encyclopaedia. It parses structured parts of the articles (such as info boxes, tables, or categories), which contain a fairly detailed description of the concepts presented in the article.

Content based image retrieval (CBIR) is an alternative to text-based search, but it suffers from important drawbacks, such as the semantic gap [8] and its poor scalability. As a consequence, the use of image processing techniques in Web-scale image retrieval is currently limited to face detection (proposed by Google or Exalead). Previous works [9,3] advocate that a combination of CBIR and text-based retrieval improves the quality of results. WordNet was exploited in CBIR applications [3], to create multimodal similarity vectors for the visual description of the images [10] or to limit the conceptual neighbourhood where visually similar images are searched [11]. Wang et al. [9] enriched an existing taxonomy of animals (620 terms) with visual information about animal's color and image properties (in/outdoor, photo/graph). The resulting structure outperformed Google Image and a purely textual version of the taxonomy when retrieving images from 20 animal species. However, this interesting approach was limited to a specific domain with quite stable visual properties (the colors of animals). Here we investigate a late fusion scheme of textual information and low level image descriptions, applied to diversified queries.

Image queries reformulation based on semantic resources has already been experimented. In [12], the authors compare a WordNet based query expansion to a ConceptNet based one and conclude that both semantic structures are complementary. The use of WordNet provides a better discrimination of the expanded queries whereas the use of ConceptNet supports better diversity. This was expectable since ConceptNet includes a larger number of inter-conceptual relations. In this paper, we advocate that only parts of the query should be reformulated. We consider that nouns are the most important part of image queries and focus the query expansion on them. For mono-conceptual queries, if knowledge exists about that particular concept, we should use it to expand the query. However, the reformulation is harder for more complex queries because the number of reformulations becomes rapidly unmanageable. This case is thus out of the scope of this work.

Section 2 presents our method based on conceptual structures reformulation. It is experimentally validated in section 3 and discussed in section 4.

2 System Description

Our approach integrates a textual query reformulation using automatically mined conceptual structures and a visual reformulation based on a list of visual concepts that can be automatically detected using image processing. In our approach, we distinguish a knowledge base building and a retrieval phase. The first, which aims at associating precise subtypes to nominal concepts, is performed off-line and its results are exploited during the retrieval, which has to be realized under real time constraints. In retrieval mode, a user request is analyzed and the system separates nominal and visual concepts which will be processed separately, leaving the rest of the query untouched. Each nominal concept in an initial query is reformulated using subtypes or synonyms in the knowledge base and the expanded query is probed against the textual descriptions of the image database. We consider the chance to mistake the annotation of an image is higher when the number of concept is low. Therefore, the images containing the largest number of terms are ranked better. The visual analysis consists in the detection of several visual concepts (from an existing list) and a classification of images with respect to these concepts. The multimedia reformulation of queries consists in re-ranking the text-based reformulation using the visual classification of images.

2.1 Automatic Building of Conceptual Structures

Building automatically conceptual neighbourhood of a good quality for nominal concepts is crucial for our approach. We first draw up a comprehensive list of terms that are to be probed against WordNet and Wikipedia in order to extract and rank their subtypes and synonyms. WordNet is used because it contains good quality structured knowledge, providing at low cost some lists of subtypes and synonyms as well as sense separation for ambiguous concepts. Unfortunately, WordNet has little information related to named entities (which often appear in Web queries) and is less complete than Wikipedia (for instance, there are just over 100 dog races in WordNet and around 600 in Wikipedia). The English version of the collaborative encyclopaedia currently includes over two million articles and, since its content is semi-structured, allows to extract good quality nominal hierarchies [5]. In order to increase the number of discovered subtypes, we first perform a WordNet-based concept expansion, then we reuse the subtypes to match the Wikipedia articles. For instance, when the system looks for subconcepts of *building*, it exploits the *isA* relation between *skyscraper* or *hotel* and *building* and therefore retains these subtypes as representative for *building*.

The concept matching procedure (table 1) relies on the analysis of the first sentence and of the "Categories" box of the articles. As illustrated in table 1, the information of the first sentence and that of the categories box is often complementary. We can extract *skyscraper* as parent concept from both parts of the article for *Empire State Building* and *Transamerica Pyramid*, but only from the Categories box for *50 California Street*. Nominal concepts often have a high number of subtypes and it is necessary to order them so as to favour

Table 1. Concept matching in Wikipedia. We present a ranked list of subtypes for *skyscraper*.

Concept	First sentence	Categories	Article length
Skyscaper	The *Empire State Building* is a 102-story Art Deco *skyscraper...*	*Skyscrapers in New York City*	165510
Skyscaper	The Transamerica Pyramid is the tallest and most recognizable *skyscraper...*	*Skyscrapers* in San Francisco	76403
Skyscraper	50 California Street is a massive *office tower...*	*Skyscrapers* in San Francisco	41049

those that are the most representative. Here we used the length of Wikipedia articles as a simple ranking measure, considering that subtypes described in more detail tend to be more representative. We illustrate the results of the ranking process in table 1, where the presented subtypes of *skyscraper* are ranked (first *Empire State Building* (165510), then *Transamerica Pyramid* (76403) and finally *50 California Street*). With the joint use of WordNet and Wikipedia, we obtain a large scale knowledge base, including good quality conceptual relations, which is usable during the retrieval phase.

2.2 Image Retrieval Phase

The query analysis is the key element of our image retrieval scheme. It separates the user requests in atomic parts, which can be one of the elements presented in table 2. This separation is necessary in order to process each query component adequately. For instance, we attempt a textual reformulation only for *nominal concepts* (NC) and a part of *named entities* (NE) but not for *visual concepts* (VIS), *modifiers* (MOD) and *others* (OTH). It is performed using existing lists of VISs, NCs, NEs and MODs and considers everything that is not in a list as being something else (OTH). At the end of the analysis, we remove stop words from the query. The list of visual concepts is arbitrary determined according to the hierarchy proposed by [13], corresponding to some concepts that can be processed by image processing algorithms. *Nominal concepts* and *named entities* are extracted from WordNet and Wikipedia, while *modifiers* are WordNet adjectives. In table 3, we present two examples of textual query reformulation using our technique. The textual reformulation works for concepts existing in the

Table 2. Type of elements that can be identified within a query

Element	Short denomination	possible instances
visual concepts	VIS	sky, night, day, portrait etc.
nominal concepts	NC	skyscraper, building, dog etc.
named entities	NE	Eiffel Tower, Ferrari, George W. Bush...
modifiers	MOD	white, red, gothic, historic
others	OTH	

Table 3. Examples of query reformulations

Initial query	Query analysis	Reformulated query
skyscraper	NC(skyscraper)	skyscraper + Empire State Building
		skyscraper + Transamerica Pyramid
bridges by night	NC(bridges) by VIS(night)	Golden Gate Bridge + bridge + night
		Pont Alexandre III + bridge + night

knowledge base only. If the query is composed of unknown concepts, it will not be reformulated and the results will be identical to a chain matching retrieval. During queries analysis, we chose to consider multiwords (such as *hunting dog* or *White House*) as single concepts because they refer to a single entity. For short queries, which are often ambiguous, we retain the default WordNet or Wikipedia sense of the concept. This choice is made because of the lack of information on the user's intent: we thus consider the most common sense of a term as the most adequate to answer the user need. If additional information is provided, we try to match the query to most appropriate word meaning.

The reformulated queries are compared to the textual descriptions of the images and the results are ranked to favour those images that are described by the highest number of concepts. The rank of a result is given by:

$$\begin{aligned} Rank = \alpha \times (\mathcal{N}_{NCinit} + \mathcal{N}_{NEinit} + \mathcal{N}_{VISinit}) + \\ \beta \times (\mathcal{N}_{NCrefo} + \mathcal{N}_{NErefo}) + \\ \gamma \times (\mathcal{N}_{MOD} + \mathcal{N}_{OTH}) \end{aligned} \tag{1}$$

where $\mathcal{N}_{Xy}$ is the number of concepts of a certain type (see table 2) appearing in the user query ($y = init$) or in its reformulated version ($y = refo$). Equation 1 gives a first ranking of results, favouring those results that are described by a high number of query related concepts. We studied different results configuration and decided that NCs, NEs and VISs in the initial queries should be given the highest weight, followed by NCs and NEs obtained after the query reformulation and by MODs and OTHs ($\alpha > \beta > \gamma$). Equation 1 differentiates between answers that are described by a different number of concepts or by different types of concepts. For instance, a picture annotated with *skyscraper* and *Empire State Building* is ranked higher than a second one annotated with *skyscraper* only, which is ranked higher than a third picture annotated with *Empire State Building* only. Equation 1 fails to separate queries having the same quantity and type of concepts (for instance, two pictures annotated with *Empire State Building*, respectively with *Transamerica Pyramid*). To discriminate these last types of answers, we use the subtypes based on Wikipedia articles length.

The proposed retrieval scheme is flexible and is able to retrieve results that are described by the initial query and expanded concepts, by the initial query or the expanded concepts only or by parts of the initial query. Equation 1 and the use of the subtypes ranking order answers considering their closeness to the query.

2.3 Multimedia Query Reformulation and Matching

This section describes the visual analysis of queries that aims at (possibly) rearranging the order of the answers returned by the textual reformulation with respect to the visual concepts in the query.

We used two systems to detect visual concepts within the images. The first one is the Viola-Jones face detector that is based on the boosting of Haar wavelets [14]. The second system [13] is a set of SVM-based classifiers learnt (RBF kernel) to determine the *type* of an image (clipart, map, painting or photo). In this last case (if the image is a photo), other sets of SVM determine whether the image is *indoor* or *outdoor*, *day* or *night*, as well as whether it is a *urban* or a *natural* scene. The multi-class classification scheme is solved using a one-versus-one approach. For each classifier, the images of the learning databases were chosen separately of the wikipedia corpus used in the experimental evaluation.

The queries were analysed to detect those containing (explicitly or implicitly) visual cues that can be detected using the visual analysis described above. Each visual concept was linked to a pre-defined list of textual concept that triggers its use. For instance, the presence of a person name (such as *Georges W Bush*) will trigger the use of the face detector. The presence of the word *map* in the query will claim for the use of the *image type detector* and favour the images tagged as *maps*; the word *cartoon* will trigger a search for images classified as *cliparts*. When a list of answers coming from the two first layers is reordered, the images detected as relevant according to the visual concept associated to the query are put at the head of the list without changing their relative order.

3 Experimental Validation

Our method has been evaluated in the context of the wikipedia MM task at ImageCLEF 2008 [1]. We submitted two runs, in order to compare our method to the state-of-the-art on the one hand, and to evaluate more specifically the influence of the multimedia query reformulation on the other hand.

Our system returns 170 documents to each query on average. Over the 75 queries to process, only 33 were reformulated with respect to the visual concepts. We quantified the change this brought about with the Levenshtein distance[15] between the index of the lists of results before and after this multimedia reformulation. The Levenshtein distance is a classic metric to measure the distance between two strings (so called "edit distance"), given as the minimum number of operations needed to transform one string into the other. In our case, we found an average Levenshtein distance of 98.1. The average "rank change", defined as the difference of rank within the lists before and after the multimedia reformulation, is 37.6.

Table 4 reports the main results of the two runs we submitted. The run *ceaTxt* is the output of the textual query reformulation and matching only, whereas the run *ceaConTxt* is the output of the full system including the multimedia

Table 4. Performances of our method at ImageCLEF wikipedia task. The results are given in terms of Mean Average Precision, and precision at ranks five and ten.

Run	MAP	P@5	P@10
ceaConTxt	0.2735	0.5467	0.4653
ceaTxt	0.2632	0.52	0.4427

query reformulation and matching. The textual reformulation is effective since our system is ranked 4th (MAP - 0.2632, P@10 - 0.4427) and the first purely textual approach (no reformulation and no feedback) is only ranked 10th (MAP - 0.2551, P@10 - 0.44). The difference between our two runs shows an interest for the multimedia reformulation and rearrangement that led to an improvement of one point in terms of MAP (from 0.263 to 0.273). It is worth noting that about half of the images were judged as relevant among the ten first answers returned by our system, demonstrating a practical interest for a real user.

4 Conclusions and Perspectives

We proposed a new image retrieval scheme that exploits both textual and visual information. The approach is based on a query reformulation using concepts that are semantically related to those in the initial query. We used Wikipedia and WordNet to extract a ranked list of related concepts for a large number of concepts and reformulate text queries. We also added an image processing which exploits visual cues in queries.

The results submitted at ImageCLEF 2008 were ranked 4th and 2nd with a mean average precision of 0.2632 and 0.2735. The small difference between the two submitted runs shows that the greater contribution to the final results was probably due to the use of conceptual structures, although a rigorous comparison would have required submitting a run with the third layer (visual concept detection) only. Nevertheless, the improvement of the results' precision accounts for the interest of introducing visual concept detection in the retrieval schema.

Number of features of our system are currently still under investigation. The detection of associated concepts is currently limited to the use of Wikipedia and WordNet. We plan to extend our approach so as to exploit search engine snippets, in order to improve the coverage of the resources. As well, while simple and generally effective, the current ranking procedure can certainly be improved if, for instance, we favour unambiguous hyponyms over ambiguous ones. Finally, we are currently exploring a finer grained filtering of visual concepts.

Acknowledgments. We thank the Direction Générale des Entreprises for funding us through the regional business cluster Systematic (project POPS) and Cap Digital (project Mediatic).

References

1. Tsikrika, T., Kludas, J.: Overview of the WikipediaMM task at ImageCLEF 2008. In: Peters, C., et al. (eds.) CLEF 2008. LNCS, vol. 5706, pp. 539–550. Springer, Heidelberg (2009)
2. Fellbaum, C.: WordNet: an electronic lexical database. MIT press, Cambridge (1998)
3. Yang, J., Wenyin, L., Zhang, H., Zhuang, Y.: Thesaurus-aided approach for image browsing and retrieval. In: IEEE Intl. Conference on Multimedia and Expo, 2001. ICME 2001, pp. 1135–1138 (2001)
4. Auer, S., Bizer, C., Kobilarov, G., Lehmann, J., Cyganiak, R., Ives, Z.: DBpedia: A nucleus for a web of open data. In: Aberer, K., Choi, K.-S., Noy, N., Allemang, D., Lee, K.-I., Nixon, L.J.B., Golbeck, J., Mika, P., Maynard, D., Mizoguchi, R., Schreiber, G., Cudré-Mauroux, P. (eds.) ASWC 2007 and ISWC 2007. LNCS, vol. 4825, pp. 722–735. Springer, Heidelberg (2008)
5. Ponzetto, S., Strube, M.: Deriving a large scale taxonomy from wikipedia. In: Proc. of the 22nd National Conference on Artificial Intelligence (AAAI 2007), Vancouver, B.C, pp. 1440–1447 (2007)
6. Kazama, J., Torisawa, K.: Exploiting wikipedia as external knowledge for named entity recognition. In: Joint Conference on Empirical Methods in Natural Language Processing and Computational Natural Language Learning, pp. 698–707 (2007)
7. Ruiz-Casado, M., Alfonseca, E., Castells, P.: Automatising the learning of lexical patterns: An application to the enrichment of wordnet by extracting semantic relationships from wikipedia. Data Knowl. Eng. 61, 484–499 (2007)
8. Datta, R., Joshi, D., Li, J., Wang, J.: Image retrieval: Ideas, influences and trends of the new age. ACM Transactions on Computing Surveys (2008)
9. Wang, H., Liu, S., Chia, L.T.: Does ontology help in image retrieval?: a comparison between keyword, text ontology and multi-modality ontology approaches. In: Proc. of the 14th ACM Intl. Conference on Multimedia, pp. 109–112. ACM, New York (2006)
10. Ferecatu, M., Boujemaa, N., Crucianu, M.: Semantic interactive image retrieval combining visual and conceptual content description. Multimedia systems 13, 309–322 (2007)
11. Popescu, A., Millet, C., Moëllic, P.A.: Ontology driven content based image retrieval. In: Proc. of the 6th ACM Intl. Conference on Image and Video Retrieval, pp. 387–394. ACM, New York (2007)
12. Hsu, M.H., Tsai, M.F., Chen, H.H.: Query expansion with conceptnet and wordnet: An intrinsic comparison. In: Ng, H.T., Leong, M.-K., Kan, M.-Y., Ji, D. (eds.) AIRS 2006. LNCS, vol. 4182, pp. 1–13. Springer, Heidelberg (2006)
13. Millet, C.: Automatic image annotation: consistent annotation, and creating automatically a learning database. ENST, Paris, PhD thesis (2008)
14. Viola, P., Jones, M.: Rapid object detection using a boosted cascade of simple features. In: Proc. of the 2001 IEEE Computer Society Conference on Computer Vision and Pattern Recognition, vol. 1, pp. I-511–I-518 (2001)
15. Levenshtein, V.I.: Binary codes capable of correcting deletions, insertions, and reversals. Soviet Physics Doklady 10, 707–710 (1966)

UJM at ImageCLEFwiki 2008

Christophe Moulin, Cécile Barat, Mathias Géry, Christophe Ducottet, and Christine Largeron

Université de Lyon, F-69003, Lyon, France
Université de Saint-Étienne, F-42000, Saint-Étienne, France
CNRS UMR5516, Laboratoire Hubert Curien
{christophe.moulin,cecile.barat,mathias.gery,ducottet, largeron}@univ-st-etienne.fr

Abstract. This paper reports our multimedia information retrieval experiments carried out for the ImageCLEF track (ImageCLEFwiki[10]). We propose a new multimedia model combining textual and/or visual information which enables to perform textual, visual, or multimedia queries. We experiment the model on ImageCLEF data and we compare the results obtained using the different modalities.

Our multimedia document model is based on a vector of textual and visual terms. Textual terms correspond to textual words while the visual ones are computed using local colour features. We obtain good results using only the textual part and we show that the visual information is useful in some particular cases.

1 Introduction

The capacity of data storage increases constantly, making possible the collection of large amount of information of all kinds, as texts, images, videos or combinations of them. In order to retrieve documents in such amount of data, information retrieval techniques tailored for the data types are required.

First early methods in multimedia retrieval only considered the textual part of documents [2]. In order to take into account as much information as possible, other methods began to exploit the image names. The content of image is more and more used to improve results. Different features can be used such as colour [9], texture[5] or shape [1] information.

The ImageCLEF collection consists of multimedia documents made up of text and images. In this paper, we present our participation to the ImageCLEFwiki[10] task. Our research goals are twofold: First, we aim to propose a multimedia document model combining text and image modalities adapted for multimedia retrieval. Second, we study the performance of our model compared to a text retrieval approach. In order to benefit from our experience with textual model, we develop a vector-based model composed of textual and visual terms. The textual terms correspond to words of the text. The visual terms are obtained through a bag of words approach. Local colour descriptors are extracted from images and quantized by k-means leading to an image vocabulary.

C. Peters et al. (Eds.): CLEF 2008, LNCS 5706, pp. 779–786, 2009.

After presenting our model, we describe the submitted runs. Then, we comment on the results we obtained and conclude.

2 Visual and Textual Document Model

ImageCLEFwiki is a multimedia collection where each document is composed of text and one image. User needs are represented by queries ("topics"), which are also multimedia (text, image and concept). Hence a multimedia document model is necessary to handle such a collection. We focus our work on combining textual and visual information without using the concept field of the topics. Figure 1 shows our system architecture detailed in the following paragraphs.

2.1 Textual Representation Model

One of the most known document model in textual information retrieval is the vector space model introduced by Salton and al. [8]. This model is based on a textual vocabulary $T = \{t_1, ..., t_j, ...t_{|T|}\}$ where $|\ |$ denotes the cardinal of the set T. The document i, denoted d_i, is represented as a vector of weights $w_{i,j}, j \in 1...|T|$ where $w_{i,j}$ is the weight of the term t_j in the document d_i: $\boldsymbol{d_i} = (w_{i,1}, ..., w_{i,j}, ..., w_{i,|T|})$. In order to calculate the weight of a term t_j in a document d_i, a $tf.idf$ formula is usually applied. The term frequency $tf_{i,j}$ measures the relative frequency of a term t_j in a document d_i. We use the one defined in the Okapi formula from Robertson and Jones [7]:

$$tf_{i,j} = \frac{(k_1 + 1) * n_{i,j}}{n_{i,j} + k_1 * (1 - b + b * \frac{|d_i|}{d_{avg}})}$$

where $k_1 = 1.2$ and $b = 0.75$ are two constants empirically defined in the Okapi formula, $n_{i,j}$ is the occurrence of the term t_j in the document d_i, $|d_i|$ is the size of the document d_i and d_{avg} is the average size of all documents in the corpus. The size of document corresponds to the number of terms in this document.

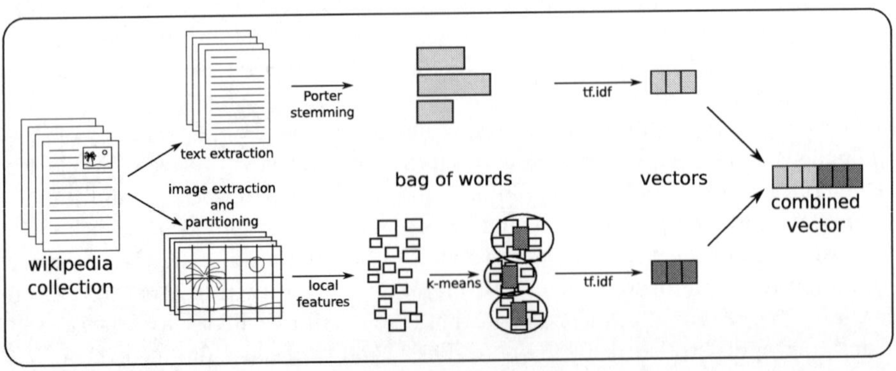

Fig. 1. System architecture based on a vector of textual and visual terms

The inverse document frequency idf_j measures the discriminatory power of a term t_j and is defined as [7]:

$$idf_j = \log \frac{|D| - df_j + 0.5}{df_j + 0.5}$$

where $|D|$ is the number of documents in the collection and df_j is the number of documents in which the term t_j occurs at least one time.

The weight $w_{i,j}$ is then obtained by multiplying $tf_{i,j}$ and idf_j. This weight is high when the term t_j is frequent in the document d_i but rare in the others. In our case, the number of terms in the vocabulary T is 217'323 after applying a Porter stemming[6]. The indexing has been performed with the Lemur software[1].

2.2 Visual Representation Model

In order to combine the visual information with the textual one, we also represent images with a vector of visual words. Using these visual words, it is possible to use the $tf.idf$ formula in the same way as in textual model. It is therefore necessary to create a visual vocabulary $V = \{v_1, ..., v_j, ..., v_{|V|}\}$ as in [4]. Our method consists in partitioning all images into 16x16 grids, a minimum of 8x8 pixels being required for each cell. It leads to about 256 cells per image, or about 38 million over all images.

For each cell, we compute a feature vector containing the colour properties of the region. The vector has 6 dimensions which correspond to the mean and the standard deviation for $\frac{R}{R+G+B}$, $\frac{G}{R+G+B}$ and $\frac{R+G+B}{3*255}$ where R, G and B are the red, green and blue components of the cell.

We apply a k-means algorithm [3] over 4 millions of cells randomly selected within the 38 millions of cells to obtain 2'000 visual terms, which correspond to our visual vocabulary V. 2'000 for k has been chosen arbitrarily while 4 millions correspond to the maximum number of cells we could compute due to the ressources required by the k-means computation. Each visual term represents a cluster of feature vectors.

Then, each new image can be represented using a vector of visual terms. It is decomposed into a 16x16 grid and the local features are computed. Each cell is then assigned to the closest visual term from our visual vocabulary V, using the euclidean distance.

In the same way as for textual words, the weight of each visual term is computed using a $tf.idf$ approach.

3 Experiments

Using the model described in the previous section, we present our approach for multimedia document retrieval from multimedia queries. Then we describe the submitted runs to ImageCLEFwiki in order to evaluate our model.

[1] Lemur project : http://www.lemurproject.com

3.1 Queries and Matching

As mentioned before, the ImageCLEFwiki topics are composed of text, image and concept modalities. However, our model is designed to only take into account text and image modalities. Our retrieval approach consists in computing a similarity score between each document d_i and a query k, denoted q_k, using the Okapi method. Documents are then ranked according to their scores. The following expression is used to compute the score:

$$score(q_k, d_i) = \sum_{u_j \in q_k} tf_{i,j} * idf_j * qtw_{k,j}$$

where $qtw_{k,j}$ is defined as:

$$qtw_{k,j} = \frac{(k_3 + 1) * n_{k,j}}{k_3 + n_{k,j}}$$

where $k_3 = 7$ is a constant defined in the okapi formula, u_j represents a textual term t_j or a visual term v_j and where $n_{k,j}$ represents the occurrence of the term u_j in the query q_k.

Let us insist on the fact that the term u_j can be either a textual term t_j or a visual term v_j and that queries can be composed of textual terms only, visual terms only, or both (textual and visual terms) which allows to perform text only queries, image only queries or multimedia queries.

The textual terms used for queries are those provided with topics. When visual terms are used, they are extracted either from the topic images or from the collection images as detailed in the following paragraph.

3.2 Submitted Runs

We have submitted 6 runs to ImageCLEFwiki 2008, labelled from LaHC_run01 to LaHC_run06. In order to simplify the notation, we use run 01 instead of LaHC_run01. All these runs are presented and summarised with their characteristics in Figure 2. Part of them are automatic (*auto*), others required manual selection of some relevant documents (*manual*). Thanks to these 6 runs, we aim to study several aspects of our model, as the choice of the visual words and local features, the combination of textual and visual words for a query and the performance improvements obtained when adding visual information to a pure textual model.

We define a baseline, run 01, that corresponds to a pure text model. It uses only textual terms for the query and scoring of documents. Its results are noted R_1. We do not use neither feedback nor query expansion for this automatic run.

All other runs exploit both textual and visual information of documents. They consist of two successive queries: first one, Q_1, is textual and corresponds to the baseline (R_1), while the second one is a visual or textual and visual query (Q_2). As the user did not always provide an image for the query, we decided to build the visual information from the baseline results (R_1) either automatically or

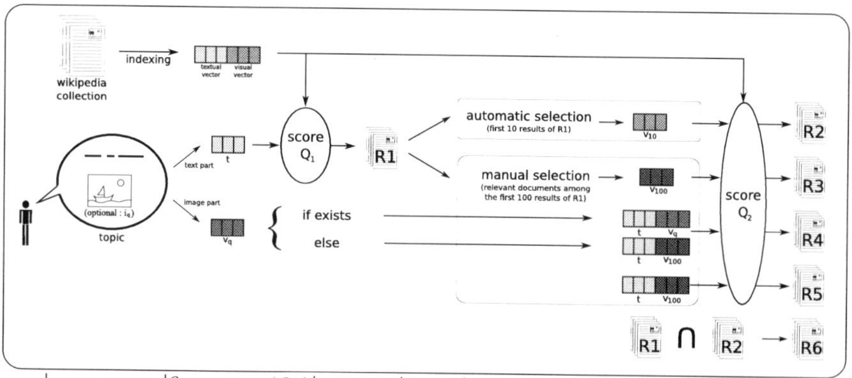

run name	first query (Q_1)	run type	Q_1 use	second query (Q_2)	results
LaHC_run01	t	*auto*	-	-	R_1
LaHC_run02	t	*auto*	v_{10}	v_{10}	R_2
LaHC_run03	t	*manual*	v_{100}	v_{100}	R_3
LaHC_run04	t	*auto* *manual*	- v_{100}	$t + \begin{cases} v_q & \text{if } i_q \text{ exists} \\ v_{100} & \text{else} \end{cases}$	R_4
LaHC_run05	t	*manual*	v_{100}	$t{+}v_{100}$	R_5
LaHC_run06	-	*auto*	-	-	$R_6 = R_1 \cap R_2$

with t: text only query: $u_j \in q \cap T$, R_i : results of run_i, i_q: query image, v_{10}: automatic selection of the first 10 results from R_1, v_{100}: manual selection of the relevant documents in the first 100 results from R_1, v_q: visual words extracted from the query image

Fig. 2. Presentation of the runs: run 01 is the text run (baseline). run 02 to run 06 consist of two successive queries: the first one Q_1 correspond to a textual query while the second one Q_2 is a visual or textual and visual query. Visual words are selected from query images, by an automatic or manual selection.

manually. Runs are automatic when we select all the visual words of the top 10 retrieved documents issued from the baseline (v_{10}), assuming these results as relevant. Runs are manual when the user is asked to select relevant documents among the first 100 results of the baseline. (v_{100}). There is no limit in the number of selected documents as the user has to choose all relevant documents over the first 100 results. In both cases, automatic and manual runs, the visual words of the selected documents are chosen to build the second query (Q_2) except for the run 04 where the topic image is used when it is available.

Run 02 and run 06 are automatic. Run 02 only uses visual information v_{10} for the second query and its results are noted R_2. For run 06, an intersection is performed between the results of the baseline and those of run 02 ($R_1 \cap R_2$). Results of this run are noted R_6. This intersection is interesting as it emphasises the gain of the visual information use. When a document is retrieved with run 02, and not with run 06, it means that only the visual information lets us find this document. The higher is the number of relevant retrieved documents with run 02 and not with run 06, the more efficient is the visual information.

Run 03, run 04 and run 05 are manual. The second query for Run 03 is only visual, while run 04 and run 05 use multimedia queries. For run 03, all the visual words of the selected images are used for the second query. For run 04 and run 05, we perform a query expansion in order to analyse the combination of textual and visual information (t+ v_{100}). We keep the textual words of the initial query and add some visual words. For run 05, these words come from the manual selected images. Run 04 proceeds as run 05, unless a query image is provided for the considered topic. In that case, the second query is composed of the visual words extracted from the query image (v_q). Thanks to these two runs, we can study the influence of the number of relevant images used for queries.

4 Results

All our results are summed up in Table 1. The comparison of our textual results with other participants is presented in Table 2. Concerning visual approaches, we did not compare our runs to those of other participants as they are too different from each other. We comment the results below according to the information used for the query.

Table 1. Summary of our results

Rank	Run	MAP	P@10	Number of retrieved documents	Number of relevant retrieved documents
22	LaHC_run01	0.2453	0.3680	54638	3467
57	LaHC_run03	0.1174	0.2613	74986	1004
58	LaHC_run05	0.1161	0.2600	74986	987
61	LaHC_run06	0.1067	0.3280	1741	429
65	LaHC_run04	0.0760	0.1813	74986	822
69	LaHC_run02	0.0577	0.1613	74989	643

Table 2. Best textual baseline runs of each participant

Rank	Participant	Run	MAP	P@10
11	sztaki	bp_acad_textonly_qe	0.2546	0.3720
13	cwi	cwi_lm_txt	0.2528	0.3427
22	curien	LaHC_run01	0.2453	0.3680
29	ualicante	IRn	0.2178	0.3200
30	chemnitz	cut-txt-a	0.2166	0.3440
44	imperial	SimpleText	0.1918	0.3240
48	irit	SigRunText	0.1652	0.2880
50	upeking	zhou1	0.1525	0.2573
52	ugeneva	unige_text_baseline	0.1440	0.2053
56	upmc-lip6	TFUSION_TFIDF_LM	0.1193	0.2160
70	utoulon	LSIS_TXT_method1	0.0399	0.0467

Text-based retrieval. The text only run (run 01) corresponds to our best run. This run is ranked 22 out of 77 runs. If we consider every baseline runs (text only using neither feedback nor query expansion) from other participants, this run is ranked 3 out of 11 (Table 2). Our textual model is quite good and let us retrieve 3'467 relevant documents out of 5'593 provided by the ground truth.

Image-based retrieval. Concerning the visual runs (run 02 and run 03), the automatic run (run 02) is the worst of our results while the manual run (run 03) is our second one. As the precision at ten documents retrieved (P@10) for run 01 is low (0.3680 (see table 1)), it is not surprising that run 02 is our worst run. Indeed, only 36,80% of the visual information used for the query is meaningful. For run 03, using only the visual words, we are able to retrieve more relevant documents than the number of relevant documents selected by the user and used for the query. For example, we retrieved 42 relevant documents for the query "blue flowers" whereas we had just selected 9 images manually. Unfortunately, there are some topics for which the results are bad and only a third of the topics leads to improvements. This is due to the visualness of the query. It is obvious that for a query like "blue flower", the visual information is more useful than for a query like "peace anti-war protest". If we consider all the relevant results from run 02 and run 03, we are able to find 1'222 relevant documents. This means that using only the visual information we are able to find a fifth of the relevant documents which is quite encouraging.

Improvements with visual information. Even if using the visual only model leads to worse results than the text only model, the visual information brings complementary relevant documents that are not found with the text query. The comparison between run 01, 02 and 06 informs us that over all topics, 214 new relevant documents are retrieved with only the visual query. Indeed, as we can see on Table 1, 643 relevant documents are found with run 02 and 429 with run 06. As run 06 corresponds to the intersection between the results of run 01 and the results of run 02, this means that $643 - 429 = 214$ relevant documents are found using only the visual information. To take an example with the "blue flower" topic, 32 relevant documents are found with run 01 and 30 with run 02. The intersection of both results coming from run 06 gives 13 shared documents. Thus, 17 relevant documents are retrieved with only the visual information. Performing the same experiment between run 01 and run 03, we find that 351 relevant documents are found with the visual information and not with the textual one.

Text and image combination. The conclusion about the combination of textual and visual words did not improve the results as expected. From Table 1, we can see that run 01 leads to 3'467 relevant documents. Using only visual information, we find 351 relevant documents with run 03. If we had perfectly combined the textual and the visual information, we should have found 3'818 relevant documents for run 05. However, we obtain less results with only 987 relevant retrieved documents out of the 3'818 expected. The comparison between run 03 and run 05 tells us that 92% of documents are shared between these two runs. Thus adding directly 2 or 3 textual words to the visual words in the query is not efficient as it gives too many importance to visuals words.

Image topic use. The difference between run 04 and run 05 shows us that one image is not enough efficient to represent the visual information. Indeed, for topics with one image, the results (run 04) are always worse than results obtained when several images are selected (run 05). This can be explained by the fact that topic images were not representative enough. Furthermore, it is obvious that one image can not be enough expressive compared to several relevant images.

5 Conclusion

We proposed a vector based model for multimedia documents. Thanks to the ImageCLEFwiki[10] collection, we were able to test our model as this collection provides visual and textual information. We obtained encouraging results with visual words only based on coloured features.

For future work, concerning the textual part, we could exploit additional information such as image names. For example, TheWhiteHouse.jpg could be replace by The White House. For the visual part, as our local image features are basic, other features such as texture and edge information would surely lead to performance improvements. We also plan to automatically select the number of visual words using machine learning approaches. Finally, we aim to combine more efficiently textual and visual information.

Acknowledgements

This work was partly supported by the LIMA project[2] and the Web Intelligence project[3] which are 2 Rhône-Alpes region projects of ISLE cluster[4].

References

1. Ferrari, V., Fevrier, L., Jurie, F., Schmid, C.: Groups of adjacent contour segments for object detection. IEEE Transactions on Pattern Analysis and Machine Intelligence 30(1), 36–51 (2008)
2. Goodrum, A.A.: Image information retrieval: An overview of current research. Informing Science 3, 2000 (2000)
3. Gordon, A.D.: Classification. Chapman & Hall, Boca Raton (1981)
4. Jurie, F., Triggs, B.: Creating efficient codebooks for visual recognition. In: International Conference on Computer Vision (2005)
5. Lowe, D.G.: Object recognition from local scale-invariant features. In: Proceedings of the International Conference on Computer Vision ICCV, Corfu, pp. 1150–1157 (1999)
6. Porter, M.F.: An algorithm for suffix stripping. Program 14(3), 130–137 (1980)
7. Robertson, S.E., Walker, S., Hancock-Beaulieu, M., Gull, A., Lau, M.: Okapi at trec-3. In: Text REtrieval Conference, pp. 21–30 (1994)
8. Salton, G., Wong, A., Yang, C.S.: A vector space model for automatic indexing. Communations of the ACM 18(11), 613–620 (1975)
9. Tollari, S., Glotin, H.: Wisti: a simple efficient textuo-visual web image retrieval model - specifications and benchmarks. In: ImagEval (2006)
10. Tsikrika, T., Kludas, J.: Overview of the wikipediaMM task at ImageCLEF 2008. In: Peters, C., et al. (eds.) CLEF 2008. LNCS, vol. 5706, pp. 539–550. Springer, Heidelberg (2009)

[2] LIMA project: `http://liris.cnrs.fr/lima/`
[3] WI project: `http://www.web-intelligence-rhone-alpes.org/`
[4] ISLE cluster: `http://ksup-gu.grenet.fr/isle`

Overview of WebCLEF 2008

Valentin Jijkoun and Maarten de Rijke

ISLA, University of Amsterdam
Science Park 107, 1098 XG Amsterdam, The Netherlands
{jijkoun,mdr}@science.uva.nl

Abstract. We describe the WebCLEF 2008 task. Similarly to the 2007 edition of WebCLEF, the 2008 edition implements a multilingual "information synthesis" task, where, for a given topic, participating systems have to extract important snippets from web pages. We detail the task, the assessment procedure, the evaluation measures and results.

The WebCLEF 2008 task is based on its 2007 predecessor [4]: for a given topic (undirected information need of the type *"Tell me all about X"*) automatic systems need to compile a set of snippets, extracting them from web pages found using Google. Thus, WebCLEF 2008 has similarities with (topic-oriented) multi-document summarization.

In the remainder of the paper we describe the task, the submissions, the assessment procedure and the results. We also give an analysis of the evaluation measures and differences between the participating systems.

1 Task Description

The user model for WebCLEF 2008 is the same as in the 2007 task definition [4]. Specifically, in our task model, our hypothetical user is a knowledgeable person writing a survey article on a specific topic with a clear goal and audience (e.g., a Wikipedia article, or a state of the art survey, or an article in a scientific journal). She needs to locate items of information to be included in the article and wants to use an automatic system for this purpose. The user only uses online sources found via a Web search engine.

The user information needs (operationalized as WebCLEF 2008 topics) are specified as follows:

- a short *topic title* (e.g., the title of the survey article),
- a free text *description* of the goals and the intended audience of the article,
- a list of *languages* in which the user is willing to accept the information found,
- an optional list of *known sources*: online resources (URLs of web pages) that the user considers to be relevant to the topic and information from which might already have been included in the article, and
- an optional list of *Google retrieval queries* that can be used to locate the relevant information; each query specifies the expected language of the documents it is supposed to locate.

C. Peters et al. (Eds.): CLEF 2008, LNCS 5706, pp. 787–793, 2009.

Below is an example of an information need:

- topic title: *Paul Verhoeven*
- description: I'm looking for information on similarities, differences, connections, influences between Paul Verhoeven's movies of his Dutch period and his American period.
- language: English, Dutch
- known source(s): `http://en.wikipedia.org/wiki/Paul_Verhoeven`, `http://nl.wikipedia.org/wiki/Paul_Verhoeven`
- retrieval queries: "`paul verhoeven (dutch AND american)`", "`paul verhoeven (nederlandse AND amerikaanse OR hollywood OR VS)`"

Each participating team was asked to develop 10 topics and subsequently assess responses of all participating systems for the created topics. In total, 61 multilingual topics were created, of which 48 were bilingual and 13 trilingual; specifically:

- 21 English-Spanish topics
- 21 English-Dutch topics;
- 10 English-Romanian-Spanish topics;
- 6 Russian-English topics;
- 2 English-German-Dutch topics; and
- 1 Russian-English-Dutch topic.

1.1 Data Collection

The test collection consists of the web documents found using Google with the queries provided by the topic creators. For each topic the collection includes the following documents along with their URLs:

- all "known" sources specified for the topic;
- the top 100 (or less, depending on the actual availability) hits from Google for each of the retrieval queries; in the 2007 edition of the task the test collection included up to 1000 documents per query;
- for each online document included in the collection, its URL, the original content retrieved from the URL and the plain text conversion of the content are provided. The plain text (UTF-8) conversion is only available for HTML, PDF and Postscript documents. For each document, the collection also provides its origin: which query or queries were used to locate it and at which rank(s) in the Google result list it was found.

1.2 System Response

For each topic, a response of an automatic system consists of a ranked list of plain text snippets extracted from the test collection. Each snippet should indicate what document in the collection it comes from.

2 Assessment

The assessment procedure was a simplification of the procedure from 2007. The assessment was blind. For a given topic, all responses of all systems were pooled into an anonymized randomized sequence of text segments. To limit the amount of assessments required, for each topic only the first 7,000 characters of each response were included (according to the ranking of the snippets in the response); this is also similar to the procedure used at WebCLEF 2007. For the pool created in this way for each topic, the assessors were asked to mark text spans that either (1) repeat the information already present in the known sources, or (2) contain new important information. Unlike the 2007 tasks, assessors were not asked to group such text snippets into subtopics (by using *nuggets*), as the 2007 assessment results proved inconsistent with respect to nuggets. The assessors used a GUI to mark character spans in the responses.

Similar to INEX [3] and to some tasks at TREC (i.e., the 2006 Expert Finding task [8]) assessment was carried out by the topic developer, i.e., by the participants themselves.

Out of the total 61 developed topics, 51 topics were actually assessed. For two of these 51 topics assessors did not find any relevant information beside the information from the known sources: topic 30 ("*Thomas Bernhard*") and topic 53 ("*Canned food in Soviet Union*"). Systems were evaluated on the remaining 49 topics.

3 Evaluation Measures

Submissions were evaluated using the following measures:

- *Average character precision (AP)*: the fraction of a system's response that matches at least one of the spans identified by assessors as relevant in the pool of all responses for a given topic; we only used alpha-numerical characters when determining substring matches, but included all characters when computing precision values;
- *Average character recall (AR)*: the sum of the character lengths of the relevant spans that are present in the system's response, divided by the total length of the relevant spans; like for precision, only alpha-numerical characters were used for substring matching, but all characters were used for computing the recall values;
- *ROUGE-1* and *ROUGE-1-2*: the values of the ROUGE evaluation metric [5] computed on word unigrams (ROUGE-1) and word unigrams and bigrams, (ROUGE 1-2); in a nutshell, ROUGE-*n* measures *n*-gram recall: the fraction of *n*-grams of the relevant spans that were found by a system; we excluded stopwords from the ROUGE evaluation.

Similarly to the WebCLEF 2007 task, for a system's response for a given topic, we computed all measures on the first 7,000 bytes of the response.

4 Approaches and Evaluation Results

In total, 9 runs were submitted by 3 research groups, the University of Twente, UNED, and the University of Salamanca. For reference and comparison, we also included a run generated by the best system participating in WebCLEF 2007.[1]

The University of Twente [6] developed three modifications of the baseline, including bugfixes in the baseline's software (namely, in stopword removal). The University of Salamanca [2] implemented three versions of query formulation for estimating query relevance: using only the topic description, using terms extracted from the known sources of the topic, and using only English words from known sources. Finally, UNED [1] extended the baseline with a key term extraction, relevance-based document re-ranking and a method for eliminating cross-lingual redundancy.

Table 1 shows the submitted runs with the basic statistics: the average length (the number of bytes) of the snippets in the run, the average number of snippets in the response for one topic, and the average total length of response per topic; we also show the four evaluation measures for the runs: average precision, average recall, ROUGE-1 and ROUGE-1-2.

Table 1. Simple statistics for the baseline (one of the systems from WebCLEF 2007) and the 9 submitted runs

Participant	Run	Average snippet length	Average snippets per topic	Average response length	AP	AR	ROUGE 1	ROUGE 1-2
	baseline 2007	286	20	5,861	0.08	0.07	0.14	0.05
U. Twente	ip2008	450	32	14,580	0.23	0.23	**0.20**	0.07
	ipt2008	464	31	14,678	**0.24**	**0.24**	0.19	**0.08**
	ipu2008	439	33	14,607	0.21	0.21	0.17	0.07
UNED	Uned RUN1	594	24	14,817	0.23	0.21	0.18	0.06
	Uned RUN2	577	25	14,879	0.18	0.18	0.18	0.05
	Uned RUN3	596	24	14,861	0.21	0.19	0.18	0.05
U. Samalanca	usal 0	851	91	77,668	0.21	0.23	0.17	0.06
	usal 1	1,494	86	129,803	0.11	0.09	0.16	0.06
	usal 2	1,427	88	126,708	0.09	0.09	0.15	0.05

We see that the best performing runs improve substantially over the baseline run (which was the best performing system in 2007), according to all measures. We looked at the statistical significance of the differences in precision (AP) using the paired two-tailed t-test with $p = 0.05$. There are two groups of statistically indistinguishable runs (when considering AP): {baseline, usal 1, usal 2} and {ip2008, ipt2008, ipu2008, Uned RUN1, Uned RUN2, Uned RUN3, usal 0}. Although all systems improve over the baseline, it is impossible to tell reliably which individual approach gives the best performance.

When we look at the per topic breakdown of the (precision) scores, we see a mixed story. Figure 1 shows the precision scores of the submitted runs for individual topics. On

[1] The source code of the system is publicly available at
`http://ilps.science.uva.nl/WebCLEF/WebCLEF2008/Resources.`

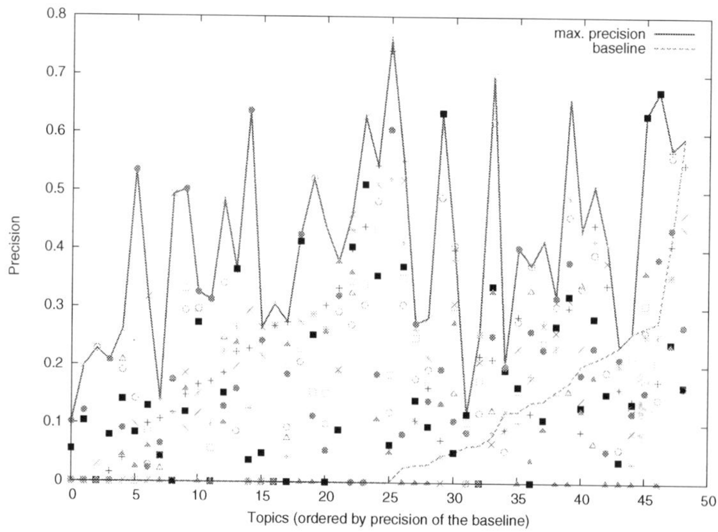

Fig. 1. Precision for the 49 non-empty topics. Points give precision values for the 9 submitted runs; lines show the maximum precision value for each topic and the precision of the baseline.

many topics, runs that perform poorly on average outperform runs that perform best (on average). Also, there is no run that outperforms all other runs on all (or even on most) topics—this is in line with our observation of a large set of statistically indistinguishable runs.

5 Discussion

In this section we take a brief look at the evaluation measures used at WebCLEF 2008. Figure 2 shows the values of the four evaluation measures for all runs. Clearly, the correlation between different measures is far from perfect. The measures generally agree on the best and worst runs, but the ranking of the runs in the middle is less unanimous.

Table 2 shows Kendall's rank correlation coefficient for the pairs of measures (values close to 1 mean that the two measures rank the runs similarly, values close to 0 indicate no correlation between measures). Note the relatively low correlation between the two ROUGE measures and between precision/recall and ROUGE. Since precision and

Table 2. Kendall's rank correlation coefficient for agreement between evaluation measures

	AP	AR	ROUGE 1	ROUGE 1-2
AP	–	0.82	0.73	0.69
AR	0.82	–	0.56	0.69
ROUGE 1	0.73	0.56	–	0.51
ROUGE 1-2	0.69	0.69	0.51	–

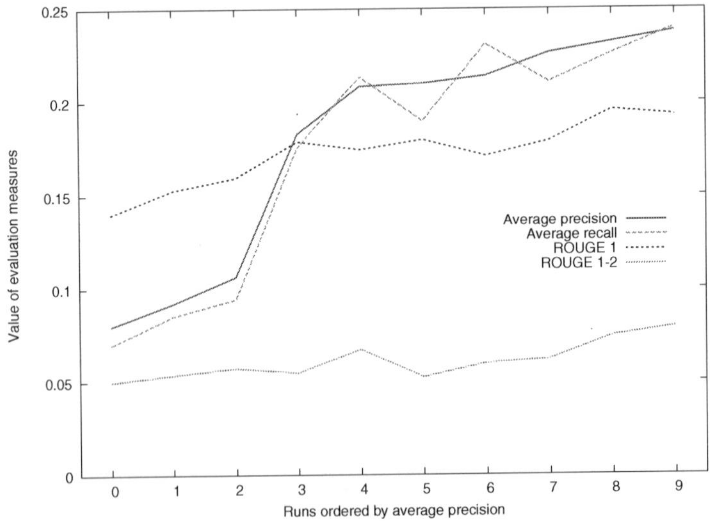

Fig. 2. Values of the evaluation measures for the baseline and the 9 submitted runs (runs ordered by the average precision)

recall are computed straightforwardly from human assessments (in every run, assessors mark up relevant character spans), we conclude that while ROUGE is successfully used in tasks such as summarization or machine translation, it is not fully appropriate for evaluating the WebCLEF task. This is unfortunate, because, as [6] argues, the strict precision/recall-based evaluation of the task does not allow us to reuse the human judgements for evaluating runs that humans have not assessed directly. As a consequence, it is virtually impossible to create a proper test collection for the task.

6 Conclusions

We detailed the task description and evaluation procedure for the 2008 edition of WebCLEF, the multilingual web retrieval task at CLEF. In 2008, participating systems showed substantial improvements over the best system from 2007 (that was used a baseline). For the best 2008 system, on average 24% of its output is judged relevant by human assessors (compared to 8% for the 2007 baseline). However, all runs with a reasonable performance are statistically indistinguishable from each other. Moreover, we found that the ROUGE measure, often used in machines translation and summarization, is not directly applicable for the evaluating the task.

Unfortunately, 2008 was the last year in which WebCLEF was run. The track is now being retired, due to a lack interest from the CLEF research community.

Acknowledgments

Valentin Jijkoun was supported by the STEVIN programme funded by the Dutch and Flemish Governments (`http://www.stevin-tst.org`). Maarten de Rijke was supported by the Netherlands Organisation for Scientific Research (NWO) under project numbers 017.001.190, 640.001.501, 640.002.501, 612.066.512, STE-07-012, 612.061.-814, 612.061.815, 640.004.802.

References

1. Amigo, E., Martinez-Romo, J., Araujo, L., Peinado, V.: UNED at WebCLEF 2008: Applying High Restrictive Summarization, Low Restrictive Information Retrieval and Multilingual Techniques. In: Peters, et al [7]
2. Figuerola, C., Berrocal, J., Rodriguez, A., Mateos, M.: Retrieval of snippets of Web pages converted to plain text. More questions than answers. In: Peters, et al. [7]
3. Fuhr, N., Lalmas, M., Trotman, A. (eds.): INEX 2006. LNCS, vol. 4518. Springer, Heidelberg (2007)
4. Jijkoun, V., de Rijke, M.: Overview of WebCLEF 2007. In: Peters, C., Jijkoun, V., Mandl, T., Müller, H., Oard, D.W., Peñas, A., Petras, V., Santos, D. (eds.) CLEF 2007. LNCS, vol. 5152, pp. 725–731. Springer, Heidelberg (2008)
5. Lin, C.-Y.: ROUGE: A Package for Automatic Evaluation of Summaries. In: Proceedings of Workshop on Text Summarization Branches Out, WAS 2004 (2004)
6. Overwijk, A., Nguyen, D., Hauff, C., Trieschnigg, R., Hiemstra, D., de Jong, F.: On the Evaluation of Snippet Selection for WebCLEF. In: Peters, et al. [7]
7. Peters, C., et al.: CLEF 2008. LNCS, vol. 5706. Springer, Heidelberg (2009)
8. Soboroff, I., de Vries, A., Craswell, N.: Overview of the TREC 2006 Enterprise Track. In: The Fifteenth Text REtrieval Conference, TREC 2006 (2007)

On the Evaluation of Snippet Selection for WebCLEF

Arnold Overwijk, Dong Nguyen, Claudia Hauff, Dolf Trieschnigg, Djoerd Hiemstra, and Franciska de Jong

University of Twente,
The Netherlands
arnold.overwijk@gmail.com, dong.p.ng@gmail.com,
c.hauff@ewi.utwente.nl, trieschn@ewi.utwente.nl,
hiemstra@cs.utwente.nl, f.m.g.dejong@ewi.utwente.nl

Abstract. WebCLEF is about supporting a user who is an expert in writing a survey article on a specific topic with a clear goal and audience by generating a ranked list with relevant snippets. This paper focuses on the evaluation methodology of WebCLEF. We show that the evaluation method and test set used for WebCLEF 2007 cannot be used to evaluate new systems and give recommendations how to improve the evaluation.

Keywords: Measurement, Performance, Experimentation.

1 Introduction

WebCLEF is about supporting a user who is writing an article and therefore wants to know more about a certain topic (i.e. undirected information search), which is the most common search goal [1]. This support consists of a list with relevant snippets. The degree to which the user's information need is satisfied is measured by the number of distinct atomic facts that the user includes in the article after analyzing the top snippets returned by the system.

The evaluation method should give insight into the parameters of the system and the performance of both participating and non-participating systems. In this paper we investigate the usefulness of the evaluation method of WebCLEF 2007 [2].

First, a brief overview of WebCLEF 2007's evaluation method is given, followed by a description of the experimental setup and the results. Based on the results, we propose a number of alternative evaluation methods. We finish with conclusions and possible future work.

2 Evaluation Method of WebCLEF 2007

The evaluation of WebCLEF relies on manual assessments created by the participants, who have manually selected the most relevant snippets from snippets delivered by the participating systems. The measures currently employed in the WebCLEF evaluation are *recall* and *precision*. Here, *recall* is defined as the sum of character lengths of all spans in the response of the system linked to nuggets (i.e. an aspect the

C. Peters et al. (Eds.): CLEF 2008, LNCS 5706, pp. 794–797, 2009.

user includes in his article), divided by the total sum of span lengths in the responses for a topic in all submitted runs. *Precision* is defined as the number of characters that belong to at least one span linked to a nugget, divided by the total character length of the system's response. More details about these measures as well as the data provided by WebCLEF can be found in the overview paper [2].

3 Experimental Setup

We investigate the evaluation method by creating several experimental systems. The general idea of our experiment is that if we can reason that a system is worse, almost equal or better than another system, this should also be reflected in the performance indicated by the evaluation method. As a baseline we use last year's best performing system, S_{base} [3]. We create three experimental systems that we argue to perform *worse*, *very similar* and *better* than this baseline, named S_{worse}, $S_{similar}$ and S_{better} respectively.

S_{worse} performs no sophisticated snippet selection. It simply delivers the snippets (i.e. paragraphs as in S_{base}) in order of occurrence; the first snippet is the first paragraph of the first document, etc. Therefore this system does much less than S_{base}, which orders snippets by relevance, removes redundant snippets, etc.

$S_{similar}$ gives almost identical output as S_{base}: it removes the last word of every snippet in the output of S_{base}. The amount of information returned to the user is almost the same when a snippet lacks only the last word, since the average length of a snippet is over 40 words. Obviously, $S_{similar}$ is not a realistic system but since it almost returns the same output as S_{base}, we argue that the evaluation metrics should return similar performance scores.

Initial experiments showed that S_{base}, performing best last year, actually contained a small programming error: only half of the intended stop word list was removed during a preprocessing step. Since it is not certain that the removal of the error leads to a better performing system, we compared S_{base} to two other systems, a system that filters all stop words and one that does not filter any stop words at all. One of these systems should perform better, whether filtering stop words is a good approach or not.

4 Results and Discussion

The measured performance of the evaluated systems are given in table 1.

Table 1. Performance of the experimental systems compared to the baseline

System	Precision	Recall	Rank
S_{base}	0.2018	0.2561	1
S_{worse}	0.0536	0.0680	5
$S_{similar}$	0.0597	0.0758	4
$S_{better - filtering\ stop\ words}$	0.1328	0.1685	2
$S_{better - not\ filtering\ stopwords}$	0.1087	0.1380	3

It is notable that the metric indicates that all systems perform worse than the baseline. Only S_{worse} meets our expectations; however, a more in dept analysis of the results tells us that simply returning the snippets in order of their occurrence results in the same performance as the baseline for six (i.e. topic 17, 18, 21, 23, 25, 26) out of thirty topics (20%). Moreover, the metric shows only a small performance difference between S_{worse} and $S_{similar}$. These results indicate that the available relevance judgments in combination with the evaluation methodology cannot be used to evaluate new systems.

An important problem of the evaluation metric is its strictness. According to the evaluation script a snippet from the manual assessments should exactly occur in the output of the system, otherwise there is no match at all. This explains why $S_{similar}$ has much lower performance scores. A slight change to the output of a perfect system results in a strong decrease of the measured performance.

Additionally, the pool of snippets to create relevance judgments was not very large, since there were only three participating systems. There might be snippets that are relevant to the user, but which are not delivered by one of the participants, resulting in incomplete relevance judgments. Such a setup gives a disadvantage to non-participating systems, since they might deliver such a snippet. This in combination with the strictness of the evaluation explains why S_{better} has lower performance scores. Notice that according to the evaluation metric, filtering only half of the intended stop word list performs better than filtering all stop words as well as not filtering any stop words at all. Again, the evaluation metric does not reflect the quality of the systems in its scores.

Furthermore, we noticed that some of the relevance judgments were not carefully created, which might influence the evaluation of new systems. For example some topics only contain non-relevant snippets (e.g. topic 14) and other topics do not contain any snippets at all (e.g. topic 12), which automatically results in a precision and recall value of zero. In topic 14 for example the user wants to find out if there are any blog search engines in Europe that are not subsidiaries of the big three search engines (Google, Yahoo! and Microsoft). Here the assessments file contains snippets like "blog search engines are hardly usable so far", which is not relevant to the user at all. This in combination with the strictness problem explains why the evaluation metric indicates that S_{worse} performs almost the same as $S_{similar}$. To be more precise, S_{base} provided for six topics exactly the same output as S_{worse}. Due to an error in the ranking algorithm no ranking could be determined for some topics and snippets were delivered in order of occurrence.

The pool problem can be solved with a larger number of participants. The problem with the manual assessments can also be solved with some effort, namely with multiple assessors per topic, which is already done in some other tracks (e.g. [4]). Unfortunately the strictness problem is not as easily solved, since the same information can be represented in several ways. The TREC QA task also has to deal with this problem [4]. However there are some existing evaluation methods that are less strict by calculating the amount of overlap.

One of them that is close to the current one, and therefore a reasonable solution, is already used in XML Retrieval [5]. In this approach the systems provide the offsets (i.e. the start and end of a passage in the document) of the delivered snippets from which the amount of overlap can be calculated to get an indication of the performance.

Another more common, approach for evaluating extractive summaries, which is the case in WebCLEF, is automatic comparison between reference and system summaries using n-grams. Originally this approach was applied to machine translation, but it has been developed in the ROUGE program for summary evaluation as well [6].

5 Conclusion and Future Work

For developers it is important to measure the system performance, especially in a task where it is hard to measure the quality of the output (i.e. WebCLEF). We explored several weaknesses in the evaluation method and the dataset of WebCLEF 2007. Unfortunately the evaluation does not provide information that is of the developers' interest nor does it reflect the performance of the system in a correct way. We showed that the manual assessments were not carefully created, which is mainly caused by the fact that it most of the times is very hard to judge whether a snippet is relevant to the user. Moreover we have shown that the measurement in general is not appropriate. With the current evaluation method a snippet in the assessments must occur exactly in the system's output. This is not realistic, since the same information can be variably expressed. A possible solution to this problem can be found in using n-grams (e.g. ROUGE [6]), because it is likely that the same information makes use of the same words. In addition it might be even better to combine this approach with TF.IDF measures to give different values to different n-grams. With such an approach words that occur less frequent, which are probably more specific and therefore contain more information, are given a higher value. We leave this question for future work.

Acknowledgments. This paper is based on research partly funded by IST project MESH (http://www.mesh-ip.eu) and by bsik program MultimediaN (http://www.multimedian.nl).

References

1. Rose, D.E., Levinson, D.: Understanding user goals in web search. In: Proceedings of the 13th international conference on World Wide Web. ACM, New York (2004)
2. Jijkoun, V., de Rijke, M.: Overview of WebCLEF 2007. In: Peters, C., Jijkoun, V., Mandl, T., Müller, H., Oard, D.W., Peñas, A., Petras, V., Santos, D. (eds.) CLEF 2007. LNCS, vol. 5152, pp. 725–731. Springer, Heidelberg (2008)
3. Jijkoun, V., de Rijke, M.: The University of Amsterdam at Web CLEF 2007: Using Centrality to Rank Web Snippets. In: CLEF 2007, Budapest, Hungary (2007)
4. Voorhees, E.M., Tice, D.M.: The TREC-8 question answering track evaluation. In: Text Retrieval Conference TREC-8, pp. 83–105 (1999)
5. Pehcevski, J., Thom, J.A.: HiXEval: Highlighting XML Retrieval Evaluation. In: Fuhr, N., Lalmas, M., Malik, S., Kazai, G. (eds.) INEX 2005. LNCS, vol. 3977, pp. 43–57. Springer, Heidelberg (2006)
6. Lin, C.-Y.: ROUGE: a Package for Automatic Evaluation of Summaries. In: Proceedings of Workshop on Text Summarization, Barcelona, Spain (2004)

UNED at WebCLEF 2008: Applying High Restrictive Summarization, Low Restrictive Information Retrieval and Multilingual Techniques

Enrique Amigó, Juan Martinez-Romo, Lourdes Araujo, and Víctor Peinado

NLP & IR Group at UNED, ETSI Informática UNED
c/ Juan del Rosal, 16. E-28040 Madrid, Spain
{enrique,juaner,lurdes,victor}@lsi.uned.es

Abstract. This paper describes our participation in the WebCLEF 2008 task, targeted at snippet retrieval from new data. Our system assumes that the task can be tackled as a summarization problem and that the document retrieval and multilinguism treatment steps can be ignored. Our approach assumes also that the redundancy of information in the Web allows the system to be very restrictive when picking information pieces. Our evaluation results suggest that, while the first assumption is feasible, the second one is not always true.

1 Introduction

The WebCLEF 2008 task has been defined in a similar way to the previous edition. Systems are asked to return a ranked list of snippets extracted from the 1000 web documents identified using the Google web search engine. Multiple languages are covered by the queries and retrieved documents. This task inherits several aspect from Information Retrieval, Summarization and Question Answering tasks. Our approach, as we will describe, is oriented to summarization strategies.

2 Assumptions

Participants are provided with a topic title, a description of the information need, the languages in which the information must be returned, a set of known sources, and a set of queries and their relevant web pages retrieved using Google. The snippets returned by the system must cover the information need without introducing any redundant information already included in the known sources or in other retrieved snippets. Our approach makes the some assumptions that will be tested in the following sections, namely:

1. The terms included in the queries are unambiguous. For instance, "machine translation" (topic 41) refers to systems that translate text from one language into another.

C. Peters et al. (Eds.): CLEF 2008, LNCS 5706, pp. 798–801, 2009.

2. Snippets written in different languages tend to contain non redundant information. This assumption avoids the management of multilingual texts that would require additional processing time and linguistic resources.
3. It is possible to find enough information in the Web to build a report containing only sentences that satisfy all requirements established by all basic summarization techniques.
4. The information needs described in topics correspond to the most frequent information returned by Google. This assumption is feasible when queries are defined manually in order to obtain a relatively clean initial ranking.

3 System Architecture

Our system has been implemented over the system described in [1]. In our approach the set of candidate snippets are re-ranked as they are added to the solution. The considered features are:

Noise elimination. Sentences containing more than 5% of words with non-alphabetical characters are discarded. This step removes noisy snippets from the sources.

Snippet length. Sentences containing less that 50 words or more than 200 bytes are removed.[1]

Query terms. The system awards snippets containing query terms, specially when they appear at the beginning of the snippets.[2]

Document relevance. We consider the relevance of the document from which the snippet has been extracted. Initially, we model the document relevance counting the number of query terms appearing in the document.

Centrality. We compute the *vector similarity* described in [1] between the candidate snippet and the rest of candidates. The centrality is the averaged similarity to all candidates.

Redundancy. In a first step, as in [1], we do not consider snippets exceeding a certain similarity threshold with respect to any other snippet in the known sources. In addition, a quantitative redundancy measure is computed by considering the maximum similarity with respect to previously picked snippets.

Key terms contribution. [2] showed that the distribution of key terms has a relevant role in Information Synthesis tasks. Following the approach described in [3], for each topic, we have produced a list of 100 key terms by considering words located immediately before a verb. In order to cover all languages without requiring linguistic processing or big lexicons, we have considered just auxiliary or common verbs such as "is" or "has". After testing several configurations, we

[1] Our exploratory studies showed that a minimum length reduces the number of non informative snippets and the maximum length awards the recall of different contents.
[2] The exploratory tentatives have suggested that snippets containing a query term at the beginning are usually more focused on it.

have included in the list only those key terms that appear before a verb more than 10 times in the document ranking and in more than 10% of the cases. Finally, we have consider the number of key terms appearing in the sentences that didn't appear in previously selected snippets.

In order to compute the snippet score, we calculate the harmonic mean (Rijsbergen's F measure) over Centrality, Redundancy, Key Term Contribution, Document Relevance and Query Terms. Each feature is previously normalized for all the candidate snippets. The motivation for using the harmonic mean rather than other combining criterion is that it is very sensitive to decreases in any of the features, because we expect to find snippets satisfying all requirements at the same time.

3.1 First Variant: More Sophisticated Document Retrieval Step

In order to to test the validity of the assumptions described in Section 2, we have applied information retrieval techniques to select a subset of documents from which our system extracts the snippets. The idea is to select the more relevant documents with respect to several queries composed of terms obtained from different sources. These sources depend on the languages in which the topic is described. We construct an *extended query* for each language in which the system provides some query for the topic. There is always an English extended query composed of terms extracted from the English title and description. This extended query is expanded with terms obtained from the English queries of the topic, if there are any.

For other languages, we translate the topic description from English to the corresponding language[3] and we extract the query terms from this translation and from the queries provided in the considered language. The document relevance is computed following the traditional vector space model which computes the relevance as the minimum cosine distance. The proposed document selection has been implemented using Lucene. For each query we only take the first 50 retrieved documents, which are expected to be more relevant.

3.2 Second Variant: Eliminating Cross-Lingual Redundancy

The second variant consists of a slight modification of our original proposal: we have included a filter in order to eliminate cross-lingual redundancy over pairs of snippets. This filter also uses Google's translation tools and proceeds as follows: 1) automatically detect the language of both input snippets; 2) when necessary, translate each snippet into English (we use machine-translated English as a kind of interlingua to easily compare snippets); 3) remove stop words and; 4) compute words overlap between the two resulting snippets. If the overlap between a candidate snippet and any previously added snippet exceeds a given threshold, it is discarded.

[3] We used Google's Language API services. See `http://code.google.com/apis/ajaxlanguage/documentation` for further details.

Table 1. Results

System	Character Precision	Character Recall
Original approach	0.22	0.21
First variant	0.18	0.17
Second variant	0.21	0.20

4 Conclusions

Analyzing the failures across topics, we have seen that: 1) there are not ambiguous query terms that could affect the results; 2) avoiding redundant information among snippets written in different languages does not contribute to the results; 3) for all topics, the system has found snippets that satisfy all summarization restrictions. These observations suggest that our first three assumptions are correct.

However, the analysis of results across topics suggests that assuming the most frequent information in documents is correlated with the information needs is not applicable to this corpus. In fact not all queries are designed to produce a clean initial set of relevant documents for a given information need. On one hand, some information needs are scattered in two queries: e.g. the "Vanellus" and "collect lapwing eggs" are two independent queries launched to Google to generate the initial ranking for topic 1, and the relevant documents are actually associated to both queries simultaneously. On the other hand, the initial Google queries are not sufficiently precise producing non-relevant documents. For instance, "Algorithms, Data structures and Complexity" do not appear in the general query "computer algorithms contest". In addition, in some cases a bag of words is not enough for detecting the information need, as in "causes of the schizophrenia" vs. "schizophrenia causes".

Our first system tried to solve these problems by including some additional information retrieval techniques but obtained worse results. It seems that it is necessary to analyze more deeply the information need by applying e.g. Question Answering techniques to match the snippet content with the information needs, and to make use of more sophisticated Information Retrieval techniques to tackle the deficiencies of the user query.

References

1. Jijkoun, V., de Rijke, M.: Using Centrality to Rank Web Snippets. In: Peters, C., Jijkoun, V., Mandl, T., Müller, H., Oard, D.W., Peñas, A., Petras, V., Santos, D. (eds.) CLEF 2007. LNCS, vol. 5152, pp. 737–741. Springer, Heidelberg (2008)
2. Amigó, E., Gonzalo, J., Peinado, V., Peñas, A., Verdejo, F.: An empirical study of information synthesis tasks. In: Proceedings of the 42nd Annual Meeting on Association for Computational Linguistics, ACL 2004 (2004)
3. Amigó, E., Gonzalo, J., Peinado, V., Peñas, A., Verdejo, F.: Using syntactic information to extract relevant terms for multi-document summarization. In: Proceedings of the 20th international conference on Computational Linguistics, COLING 2004 (2004)

Retrieval of Snippets of Web Pages Converted to Plain Text. More Questions Than Answers

Carlos G. Figuerola, José Luis Alonso Berrocal, Ángel F. Zazo Rodríguez, and Montserrat Mateos

University of Salamanca, REINA Research Group
c/ Fco. de Vitoria, 6-16, 37008 Salamanca, Spain
reina@usal.es
http://reina.usal.es

Abstract. This year's WebCLEF task was to retrieve snippets and pieces from documents on various topics. The extraction and the choice of the most widely used snippets can be carried out using various methods. However, the way in which web pages are usually converted to plain text introduces a series of problems that cause inefficiency in the retrieval. Duplicate information, absolutely irrelevants snippets or even meaningless, are some of these problems. Also, it is intended in this paper to explore the real impact of the use of several languages in obtaining relevant fragments.

1 Introduction

This year, the WebCLEF track is similar to the 2007 edition, namely retrieving text snippets or fragments of web pages which bring up information about a topic [1]; additionally, snippets must be in a language from a set of accepted ones. As in 2007, we have a set of topics, each with a title and a short description, as well as several documents or *known sources* about the topic. Additionally, for each topic, we have one or several searches in Google, with the first 1000 documents retrieved.

The system used is basically the same as last year [2], for each topic we considered all documents retrieved after queries to Google as the collection of documents with which to work. These documents are to be fragmented into pieces, each of whom will be treated as a separate document.

For the queries, we use the description that we have for each topic. This query can be enriched with more terms from the *known sources*. So, the task can be approached like a classic problem of retrieval, and apply, consequently, conventional techniques.

2 Segmentation of Web Pages

Task organizers provide the translation to plain text of documents retrieved by Google. We have assigned equal values to all Google searches for the same topic.

C. Peters et al. (Eds.): CLEF 2008, LNCS 5706, pp. 802–807, 2009.

UTF-8 worked fine in almost all cases, something important as there were documents in several languages and with different alphabets (including Cyrillic, for example). That freed us of the many problems experienced in previous editions with the detection of the coding system of each document [3]. So, for each document translated to plain text, we have to segment it in fragments, to obtain the terms of each fragment and to calculate their weights.

To segment documents and to obtain fragments or short text passages diverse techniques can be applied. Basically, some are based on the size in bytes, or words; and others are oriented in the separation in phrases or paragraphs [4]. The former techniques produce, of course, pieces wich are more homogenous in size, but often devoid of sense, as the partition point is blind. The latter tend to produce fragments of very different sizes. In addition, its application is not always simple; in many cases the conversion of a web document to plain text loses the separations between paragraphs, does not distinguish between soft and hard line feeds, or blurs structural elements, like the tables [5].

A simplistic approach, like the election of an orthographic character, such as the period (.) as a reference to fragment the text [6], tends to produce too short passages and, therefore, of little use for the objectives of this task. In our case, we adopted a mixed approach. After several tests, we decided that the suitable size for each fragment was around the 1500 bytes, but as we wanted fragments that had informative sense, our fragmenter looks for the period closest the 1500 bytes, and parts by that point.

Some other transformations were carried out: lowercase conversion, accent and stopwords removal(with a long list of stop words for all the accepted languages), application of a simple s-stemmer [7].

Each fragment thus obtained and transformed was considered an independent document. Terms were extracted and they were weighed according to scheme ATU (slope=0.2) [8], applied to the good well-known vector space retrieval model.

3 Formation of Queries

From the document collection formed with snippets, we must select those that are more useful for each topic. The key is composing suitable queries that can produce this selection. As sources of information to compose those queries, we have topics with a short title and a brief description. Additionally, we also have, for each topic, a few documents denominated *known sources*, in full text.

So we can use topics (title and description) as the core of each query, and refine it with terms coming from the *known sources*. The *known sources* are complete documents, which can contain many terms.

4 Runs Carried Out and Results

As last year, we worked on the formation of the queries. The basic dilemma was whether the core of each topic (title and description) could be enriched by the terms of the *known sources*, and to what extent the use of these terms adds useful information to the query.

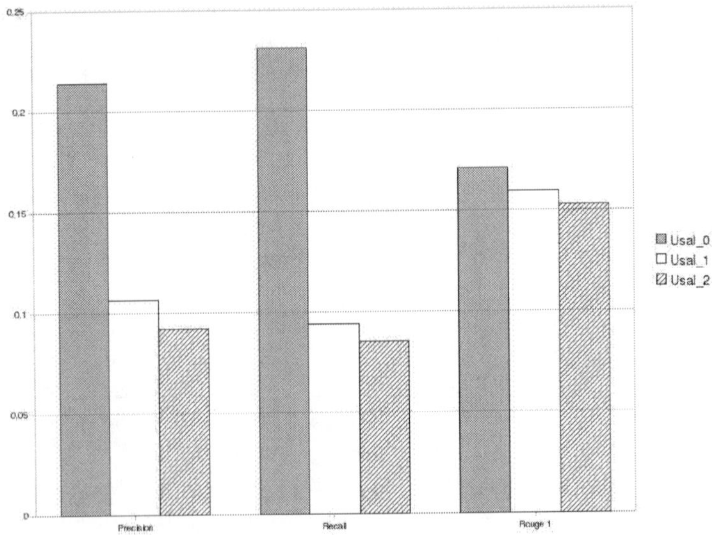

Fig. 1. Results of submitted runs

So, we have sent three official runs. The first one (USAL_0) uses only the topic terms and their descriptions. The second (USAL_1) also uses the topics, but enriched with the terms extracted from *known sources*, the latter were weighted to a lesser extent with the aim of preventing or reducing the noise that those terms might introduce.

Finally, we made a third run similar to the second, but using only words in English. One may wonder whether the retrieval of fragments in different languages provides more relevant information, and to what extent. All topics allow at least English fragments as a useful response and, additionally, in other languages. It is expected that these fragments in other languages are derived from queries which include terms in those other languages. We made a run using queries in English only, which should allow us to compare results and assess to what extent the use of other languages aid in retrieval. Our idea was to test the hypothesis that working only in English was possible to resolve satisfactorily the needs of information, in other words, assess the extent of the retrieval in other languages makes improvements on the final outcome.

Results are showed in the graph. Precison and recall, in WebCLEF, are defined in a different way, explained in [1]; the evaluation procedure on the other hand, has been criticized by [9]. Average precision and recall show that retrieval with only the terms of topics is much better. Even the measure ROUGE, although it tends to smooth differences mark clearly the trend. These results are different from those achieved last year. Then, it seemed that enrich the queries with the terms of the *known sources* improves the results of the retrieval. However, this improvement was very small; there was little difference in using terms of *known sources* or not.

5 More Questions Than Answers

Anyway, with or without term from the *known sources*, results of runs USAL_1 and USAL_2 are below the average of runs submitted by all partipants.Run USAL_0, although with better results, was the sixth of ten runs from all participants.

Thus, the experience of both years showed other interesting things. Web pages are not conventional documents; in addition to hyperlinks and hypermedia elements, they have a structure that is not always sequential. Many web pages are viewed by the user as a set of visual blocks that have different functions and containing different types of information [10]. From the standpoint of obtaining this information, some blocks are more useful than others. The conventional tools of conversion to plain text are not able to reproduce this visual structure, the result is that many of the fragments that we get are meaningless. Others contain information not relevant to our purposes: navigational aids, copyright notices, advertising, etc..

Unfortunately, this visual structure can not be obtained from the elements of HTML and this is a challenging research area which has recently started to be addressed [11].

We tried a very naive approach, filtering and dropping snippets based on a simple heuristics: fragments with too many blank lines, with very short lines, with a few words in relation to the size of the fragment, and so on. So, from 639,215 snippets obtained from documents , our filter deleted 165,442 (=25.88 %). This would suggest that we work with a database with a lot of noise, a more refined way of extracting fragments could possibly improve results in a substantial way.

In a similar way, last year we observed a lot of duplicated snippets. Information is replicated across the web, and so we have fragments of different pages that have the same information. However, as visual presentation is not always the same, the results of the conversion to plain text produces different strings. We used the Dice Coefficient as a measure to compare snippets and discover duplicates and near duplicates. In this case, we applied detection of duplicates on the retrieved snippets for each topic. So, if we consider snippets with Dice similarity greater than 0.7, we found that 11.08 % are duplicates ones.

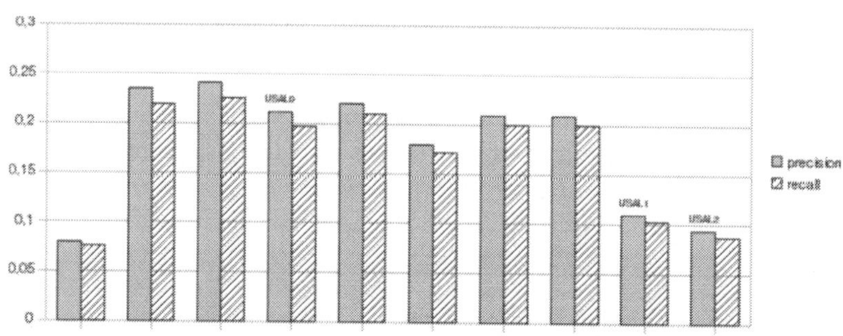

Fig. 2. Results of runs submitted by all participants

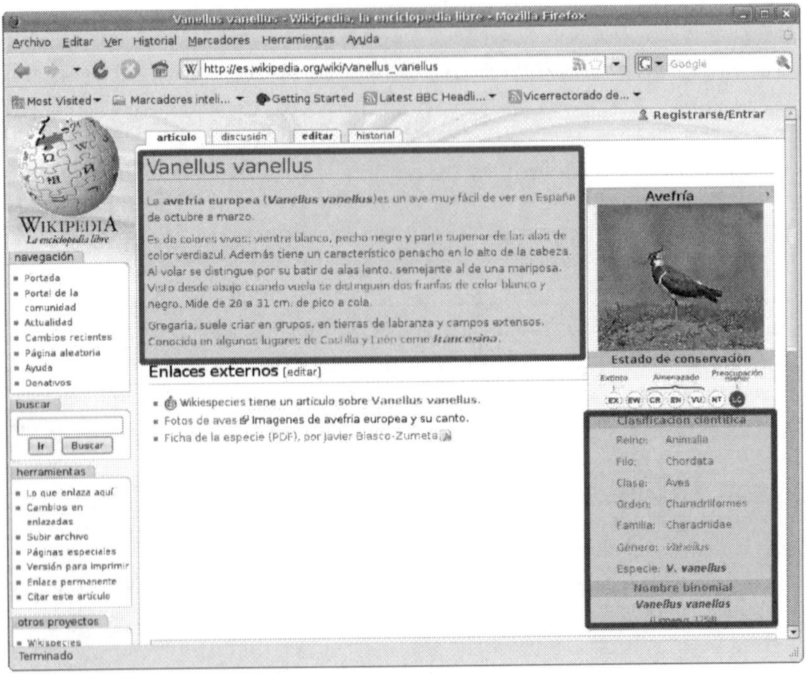

Fig. 3. Only some parts of the web page are relevant

On the other hand, queries in English only worked worse than multilingual ones. This was, in some way, the expected result; but there is not a remarkable difference. Is it really worth working in several languages? Probably the answer depends on the type of information need, on the semantic content of the queries. It is true that the dominant language in the web is English, but is it true for all information needs? Given that the overall results are not very good, it is difficult to draw a clear conclusion.

6 Conclusions

We have described our approach to WebCLEF task, similar to last year, but incorporating the experience of the last edition. The results are not conclusive and provide more questions than answers. The questions focus on how we get the text of Web pages, the amount of noise that contain fragments; the effect of filtering meaningless fragments; the amount of duplicate information we find on the web.

References

1. Jijkoun, V., de Rijke, M.: Overview of webclef 2007. In: Peters, C., Jijkoun, V., Mandl, T., Müller, H., Oard, D.W., Peñas, A., Petras, V., Santos, D. (eds.) CLEF 2007. LNCS, vol. 5152, pp. 725–731. Springer, Heidelberg (2008)

2. Figuerola, C.G., Alonso Berrocal, J., Zazo Rodrguez, A.: Reina at webclef 2007. selecting useful snippets. In: Nardi, A., Peters, C. (eds.) Working Notes for the CLEF 2007 Workshop (2007)
3. Li, S., Momoi, K.: A composite approach to language/encoding detection. In: 19th International Conference on Unicode (2001)
4. Zazo, Á.F., Figuerola, C.G., Alonso Berrocal, J.L., Rodríguez, E.: Reformulation of queries using similarity thesauri. Information Processing & Management 41(5), 1163–1173 (2005)
5. Yu, S., Cai, D., Wen, J.R., Ma, W.Y.: Improving pseudo-relevance feedback in web information retrieval using web page segmentation. In: Proceedings of the Twelfth International World Wide Web Conference, WWW 2003, Budapest, Hungary, May 20-24, pp. 11–18. ACM, New York (2003)
6. Mikheev, A.: Tagging sentence boundaries. In: Proceedings of the First Meeting of the North American Chapter of the Computational Linguistics (NAACL 2000), pp. 264–271. Morgan Kaufmann, San Francisco (2000)
7. Figuerola, C.G., Zazo, Á.F., Rodríguez Vázquez de Aldana, E., Alonso Berrocal, J.L.: La recuperación de información en español y la normalización de términos. Inteligencia Artificial. Revista Iberoamericana de Inteligencia Artificial 8(22), 135–145 (2004), `http://tornado.dia.fi.upm.es/caepia/numeros/22/raepiaF08.pdf`
8. Singhal, A., Buckley, C., Mitra, M.: Pivoted document length normalization. In: Proceedings of the 19th Annual International ACM SIGIR Conference on Research and Development in Information Retrieval (Special Issue of the SIGIR Forum), Zurich, Switzerland, August 18–22, pp. 21–29. ACM, New York (1996)
9. Overwijk, A., Nguyen, D., Hauff, C., Trieschnigg, R., Hiemstra, D., Jong, F.d.: On the evaluation of snippet selection for information retrieval. In: Working Notes for the CLEF 2008 Workshop (2008)
10. Yang, Y., Zhang, H.: Html page analysis based on visual cues. In: ICDAR, pp. 859–864. IEEE Computer Society, Los Alamitos (2001)
11. Kang, J., Choi, J.: A preliminary report for an information extraction system based on visual block segmentation. Technical Report TR-IS-2007-1, Hanyang University, Intelligent Systems Laboratory (2007)

GeoCLEF 2008: The CLEF 2008 Cross-Language Geographic Information Retrieval Track Overview

Thomas Mandl[1], Paula Carvalho[2], Giorgio Maria Di Nunzio[3], Fredric Gey[4], Ray R. Larson[4], Diana Santos[5], and Christa Womser-Hacker[1]

[1] Information Science, University of Hildesheim, Germany
mandl@uni-hildesheim.de, womser@uni-hildesheim.de
[2] University of Lisbon, DI, LasiGE, XLDB
Linguateca, Portugal
pqfcarvalho@gmail.com
[3] Department of Information Engineering, University of Padua, Italy
dinunzio@dei.unipd.it
[4] University of California, Berkeley, CA, USA
gey@berkeley.edu, ray@sims.berkeley.edu
[5] Linguateca, SINTEF ICT, Norway
Diana.Santos@sintef.no

Abstract. GeoCLEF is an evaluation task running under the scope of the Cross Language Evaluation Forum (CLEF). The purpose of GeoCLEF is to test and evaluate cross-language geographic information retrieval (GIR). The GeoCLEF 2008 task presented twenty-five geographically challenging search topics for English, German and Portuguese. Eleven participants submitted 131 runs, based on a variety of approaches, including sample documents, named entity extraction and ontology based retrieval. The evaluation methodology and results are presented in the paper.

1 Introduction

The Internet propelled a variety of geographic services that range from map services to route planning and hotel reservation systems. Many queries for search engines involve some sort of geographic processing and reasoning. Therefore, the development and evaluation of information retrieval systems that optimize the geographically oriented access to information is very important.

Geographical Information Retrieval (GIR) concerns the retrieval of information involving some kind of spatial awareness. Many documents containing spatial references are important to retrieve, rank and visualize information needs, such as "find me news stories about riots near Paris and their consequences".

GeoCLEF is the first track of an evaluation campaign dedicated to evaluating geographic information retrieval systems. The aim of GeoCLEF is to provide the framework for evaluating GIR systems, in both a spatial and a multilingual dimension. Participants were presented with a TREC style ad hoc retrieval task, based on the existing CLEF newspaper collections.

C. Peters et al. (Eds.): CLEF 2008, LNCS 5706, pp. 808–821, 2009.

GeoCLEF was a pilot track in 2005 and, since then, it was a regular track. It evaluates document retrieval with an emphasis on geographic information text retrieval. Spatial reasoning is often necessary to solve the search tasks.

Eleven research groups (thirteen in 2007) from different backgrounds and nationalities submitted 131 runs (108 in 2007) to GeoCLEF 2008.

Portuguese, German and English were available as document and topic languages. As in previous editions, there were two Geographic Information Retrieval tasks: monolingual (English to English, German to German and Portuguese to Portuguese) and bilingual (language X to language Y, where X and Y correspond to one of the above mentioned languages).

GeoCLEF developed a standard evaluation collection which supports long-term research. Altogether, 100 topics including relevance assessments have been developed over the last four years (one pilot run and three regular tracks). Additionally, a set of 26 CLEF ad-hoc topics with spatial restrictions has been identified and can be used as a benchmark. Topics and the relevance judgment files will be publicly available on the GeoCLEF website[1].

Table 1. GeoCLEF test collection – collection and topic languages

GeoCLEF Year	Collection Languages	Topic Languages
2005 (pilot)	English, German	English, German
2006	English, German, Portuguese, Spanish	English, German, Portuguese, Spanish, Japanese
2007	English, German, Portuguese	English, German, Portuguese
2008	English, German, Portuguese	English, German, Portuguese

Geographic IR is a challenging task, namely because it deals with geographical references which are often vague, ambiguous and multilingually challenging. Multilingual retrieval requires systems matching references to a place from one language to another, which may have different correspondents (e.g. *Athens*, *Athen*, *Atenas*, *Atina*). Spatial reasoning is usually mandatory to solve information needs, such as "demonstrations in cities in *Northern Germany*", where the geographic term corresponds to a selection of places and locations that are not explicitly specified in the topic.

The GeoCLEF track comprises two sub tasks. The main task is described in the following sections. The GikiP task[2] which evaluates searches for Wikipedia entries that require some geographical processing, is described in a separate overview paper [5].

2 GeoCLEF 2008 Search Task

The geographic search task is the main task of GeoCLEF and it is developed following the general framework underlying the CLEF ad-hoc task. The following sections describe the test design.

1 http://www.uni-hildesheim.de/geoclef

2 http://www.linguateca.pt/GikiP

2.1 Document Collections Used in GeoCLEF 2008

The document collections used in the third GeoCLEF edition are the same as the ones used in GeoCLEF 2007, and in previous CLEF ad-hoc evaluations [1]. They are newspaper and newswire stories, from 1994 to 1995, covering international and national news and events that mention a wide variety of geographical entities. The English collection contains 169,477 documents, which are made out of stories from the British newspaper *The Glasgow Herald* (1995) and the American newspaper *The Los Angeles Times* (1994). The German collection contains 294,809 documents from the German magazine *Der Spiegel* (1994/95), the German newspaper *Frankfurter Rundschau* (1994) and the Swiss newswire agency *Schweizer Depeschen Agentur* (SDA, 1994/95). The Portuguese collection is made out of two major daily newspapers, namely the Portuguese newspaper *Público* (106,821 documents) and the Brazilian newspaper *Folha de São Paulo* (103,913 documents). The Portuguese collections are distributed by Linguateca as the CHAVE collection[3].

Table 2. GeoCLEF 2008 test collection size

Language	English	German	Portuguese
Number of documents	169,477	294,809	210,734

The documents have a common structure in the three language collections: newspaper-specific information, like date, (optionally) page, issue, special filing numbers and often one or more titles, a by-line and the actual text. Geographic entities were not previously recognized and none semantic location-specific information was added to the documents.

2.2 Generating Search Topics

A total of 25 topics were created for this year's GeoCLEF (GC76 - GC100). Topics express a natural information need that a user of the collection might have. Topic creation was a shared task between the Portuguese and the German groups. The task was supported, by the use of the DIRECT System, provided by the University of Padua. This system includes a search utility for handling the collections.

Topic creation was performed in two stages. First, each group devised a set of candidate topics in their own language, whose appropriateness was checked in the text collection available for that language. Topic candidates were subsequently checked for relevant documents in the other collections. Sometimes, it is difficult to find geographically interesting topics below the granularity of a country. Regional events with a wide coverage in one country do not often correspond to many newspaper articles in other countries. As a consequence, some topics needed to be partially modified or refined, by relaxing or tightening the content or the geographic focus.

Other reasons driving this process were the absence of relevant documents in one of the languages, the complexity of topic interpretation and the translation into the other languages. For example, a candidate topic on fish living in the Iberian Peninsula

[3] http://www.linguateca.pt/CHAVE

had relevant matches in the Portuguese collection. However, this topic was not mentioned in the other newspapers. Moreover, some of the species described in the "narrative" (e.g. *"saramugo"*, a species which lives only in Spanish and Portuguese rivers) were difficult to translate into German and English. The spatial parameter (*Iberian Peninsula*) remained in a topic, but the subject was replaced by a matter that potentially interests the international mass media, namely, the state of agriculture in the Iberian Peninsula. In most cases, the changes were not radical. For example, the initial candidate topic "Nobel Prize winners in Physics from Northern European countries" was replaced by a more general one: "Nobel prize winners from Northern European countries". In other cases, the geographic term was replaced by other(s) involving a more difficult but interesting exercise of geographic reasoning and processing. For example, "Most visited sights in the capital of France" was changed to: "Most visited sights in the capital of France and its vicinity", which is more challenging from the geographic point of view. The new form involves the processing of relative proximity and neighborhood concepts.

The final topic set was agreed upon after intensive discussion. All missing topics were translated into Portuguese and German and all translations were checked. The next section discusses the creation of topics with spatial parameters for the track.

2.3 Spatial Parameters

One goal of GeoCLEF is the creation of a geographically challenging topic set. Geographic knowledge is necessary to successfully retrieve relevant documents for most documents. While many geographic searches may be reasonably satisfied by keyword approaches, others require geographic reasoning. Most systems, especially keyword based systems, might perform better on average with a realistic topic set, where these difficulties occur less frequently.

To increase the difficulty of the topic set, the following issues were explicitly included in the topics of GeoCLEF 2008:

- imprecise /vague geographic regions (Sub-Saharan Africa, Western Europe)
- geographical relations beyond IN (forest fires on Spanish islands)
- granularity below the country level (fairs in Lower Saxony)
- terms which are not explicitly mentioned in documents (Portuguese communities in other countries)

We tried to create a set of topics representing different kinds of geographic queries. These queries present different levels of complexity and may require different approaches to process them adequately, and successfully retrieve relevant documents. Instead of privileging specific geographical places, such as a country or city, preference was given to reference geographical regions, comprehending more than one physical or administrative place. Different kinds of regions were, then, considered, which may correspond, for instance, to a delimited geographical area of a given continent (e.g. Sub-Saharan Africa, Southeast Asia, Northern Africa, Western Europe) or country (e.g. Western USA, Lower Saxony, Spanish islands). Other interesting geo-economic-political terms, such as OECD countries, were also considered in the topic creation.

The majority of the GeoCLEF 2008 topics specify complex (multiply defined) geographical relations, a property introduced in the GeoCLEF 2007 [8], kept in this evaluation. Such geographical relations, which can be explicitly or implicitly mentioned in the topic, may represent:

- Proximity (e.g. Most visited sights in the capital of France and its vicinity);
- Inclusion (e.g. Attacks in Japanese subways);
- Exclusion (e.g. Portuguese immigrant communities in the world).

The example illustrating proximity also presents a relation of inclusion (between sights and capital of France), explicitly formalized by the preposition "in". That relation can also be inferred in the phrase “Japanese subways” occurring in the topic illustrating inclusion, which can be paraphrased by the expression "subways in Japan".

Different from the GeoCLEF 2007 topics, which might represent explicit relations of exclusion (e.g. Europe excluding the Alps), such relations were only implicitly represented in the topics of GeoCLEF 2008, as illustrated above. This topic has the particularity of presenting simultaneously a relation of inclusion (communities from Portugal in the world) and exclusion (in this context, world represents any country except Portugal).

Just as in previous GeoCLEF editions, vague geographic designations were introduced for certain topics. For example, in the topic: “Nuclear tests in the South Pacific”, the geographical term South Pacific may refer to both Australasia ("an area including Australia, New Zealand, New Guinea and other islands including the eastern part of Indonesia") and Oceania ("a geographical (often geopolitical) region of many countries/territories (mostly islands) in the southern Pacific Ocean"). The interpretation of this geographical term (ambiguous between and an ocean and the islands within it) is only possible if the full topic content is considered.

A similar situation is observed in the topic "American troops in the Persian Gulf". In this case, the Persian Gulf does not stand for the gulf itself but for a Southwest Asian region, which is an extension of the Indian Ocean located between Iran and the Arabian Peninsula. Once again, the adequate processing of the information in the topic requires term disambiguation.

Another case of vagueness can be observed in the topic “Environmental pollution in European waters”, where the term waters can refer to rivers, lakes or the sea.

2.4 Approaches to Geographic Information Retrieval

The format of GeoCLEF 2008 is identical to that of GeoCLEF 2006 and 2007. Table 3 illustrates the syntax of two different topics, the one on the left hand side in English and the one on the right side in Portuguese. As it can be observed, the topics do not contain any geographic tag.

The table shows, the short topic description, within the title and description tags, is followed by the narrative tag, which contains a detailed description of the geographic requirements and the relevance criteria. In some topics, relevant geographic names are listed in the narrative.

Table 3. Two examples from the Topics: 10.2452/89-GC and 10.2452/84-GC

<num>10.2452/89-GC</num> <title>Trade fairs in Lower Saxony </title> <desc>Documents reporting about industrial or cultural fairs in Lower Saxony. </desc> <narr>Relevant documents should contain information about trade or industrial fairs which take place in the German federal state of Lower Saxony, i.e. name, type and place of the fair. The capital of Lower Saxony is Hanover. Other cities include Braunschweig, Osnabrück, Oldenburg and Göttingen. </narr> </top>	<num>10.2452/84-GC</num> <title>Atentados à bomba na Irlanda do Norte </title> <desc>Os documentos relevantes mencionam atentados bombistas em localidades da Irlanda do Norte </desc> <narr>Documentos relevantes devem mencionar atentados à bomba na Irlanda do Norte, indicando a localização do atentado. </narr> </top>

2.5 Approaches to Geographic Information Retrieval

In the last three editions of GeoCLEF, traditional ad-hoc retrieval approaches and specific geographic reasoning systems have been explored in parallel. Successful results have often been achieved by ad-hoc techniques without any specific geographic knowledge or processing. These approaches have sometimes been developed as a baseline for more sophisticated systems. Some of the traditional techniques may have beneficial effects for geographic search tasks. Blind relevance feedback can lead to a geographic term expansion necessary to solve a search problem. For example, a query for riots in German cities does not contain the name of any German city. A query including the term German may lead to documents containing the word German and the names of some cities which can be included in subsequent optimized queries. As a result, geographic term expansion has been achieved without proper geographic knowledge being available to the system. This form of pseudo-geographic processing is not very reliable, but the specific components often have a high error rate or introduce significant noise. In GeoCLEF 2007, some systems tried combinations of both approaches and the dedicated geographic systems have further matured. In 2008, new ideas were introduced. For example, an ontology based approach presented by the DFKI was successful for the most competitive task: monolingual English. The University of Berkeley implemented a system designed like an ad-hoc system without any geographic components.

The participants used varied approaches to the GeoCLEF tasks, ranging from basic IR approaches to deep natural language processing (NLP). The approaches include the use of full documents for ranking the result set, map based techniques and Wikipedia as a knowledge source. For details, the reader can consult the description of the systems in the papers of the participants (in this volume).

2.6 Relevance Assessment

The English assessment was shared by Berkeley and Hildesheim Universities. The German assessment was done by the University of Hildesheim and the Portuguese assessment by Linguateca. The DIRECT System, used for topic development, was also used for relevance assessment. The system provided by the University of Padua

Table 4. GeoCLEF 2008 Size of Pools

Language	# Documents
English	14.528
German	15.081
Portuguese	14.780

Table 5. GeoCLEF 2008 Relevant Documents per Topic

Language	Minimum	Maximum
English	0	109
German	1	146
Portuguese	2	158

allowed the automatic submission of runs by participating groups and supported the GeoCLEF assessment pools by language. All runs were included in the pool. The size of the pool is shown in table 4 and the distribution of relevant documents over topics is given in table 5.

During the assessment process, the assessor tried to find the best collection of keywords – namely, based on the detailed information described in the narrative, using the DIRECT system. The following subsections report some of the issues concerning the relevance assessment for each language.

Some topics caused assessment difficulties, especially when the narrative required specific information, not expressed in the text. For example, from the sentence: Bonn ... former chancellor Willy Brandt ... Nobel Peace prize winner... is it possible to infer that Willy Brandt was German?

In assessments, topic drifts typically occur. GeoCLEF 2008 assessment was not an exception. Is a document about kidnapping of a French aid worker in Kenya relevant for "foreign aid in Sub-Saharan Africa"? The kidnapping of an aid worker implies the existence of foreign aid in Kenya, but a kidnapping is not related in any sense to foreign aid.

The assessment usually provides hints on why systems failed. The German topic about “fairs in Lower Saxony” points to inappropriate stemming rules or to high n-gram similarity.. The German word for fairs (*Messe*) was matched against similar words with a different meaning (e.g. *angemessen* -> appropriate, *Messer* -> knife).

The English document pool also led to borderline cases that needed to be discussed among the assessors. One topic required documents on “natural disasters in Western states of the USA”. Some documents only reported the insurance costs caused by natural disaster overall. In such cases, it was decided to consider relevant the documents mentioning a geographically relevant place (for example, *Los Angeles*) even when they did not mention the disaster explicitly and directly.

3 GeoCLEF 2008 Results

The results of the participating groups are reported in the following sections.

3.1 Participants and Experiments

As shown in Table 6, a total of eleven groups from seven different countries participated in one or more GeoCLEF tasks. A total of 131 experiments (runs) were submitted. Five of these groups participated in GeoCLEF for the first time.

Table 6. GeoCLEF 2008 participants – new groups are indicated by *

Participant	Institution	Country
Alivale*	U.Jaén & U.Politecnica Valencia	Spain
Cheshire	U.C.Berkeley	United States
Csusm	Cal. State U.- San Marcos	United States
dfki*	German Research Center for AI	Germany
Hagen	U.Hagen-Comp.Science	Germany
icl	Imperial College London	United Kingdom
Inaoe*	Lab. Tecnologıas del Lenguaje	Mexico
jaen*	U.Jaén	Spain
pittsburgh*	U.Chengdu & U.Pittsburgh	China & United States
Valencia	U.Politecnica Valencia	Spain
xldb	U.Lisbon	Portugal

Table 7 provides an overview of the experiments submitted per task and participant. Three different topic languages were allowed for the GeoCLEF bilingual experiments. Again, the most popular language for queries was English; German took the second place. The number of bilingual runs by topic language is shown in Table 8.

Table 7. GeoCLEF 2008 experiments by task

Participant	Monolingual Tasks			Bilingual Tasks			TOTAL
	DE	EN	PT	X2DE	X2EN	X2PT	
alivale*		9					**9**
cheshire	3	3	3	6	6	6	**27**
csusm	1	1	2	1	1	1	**7**
Dfki*		5					**5**
hagen	5			10			**15**
icl		9					**9**
inaoe*		12					**12**
Jaen*		7			6		**13**
pittsburgh*		4					**4**
valencia		6					**6**
xldb		12	12				**24**
TOTAL	**9**	**68**	**17**	**17**	**13**	**7**	**131**

Table 8. Bilingual experiments by topic language

Track	Source Language			TOTAL
	DE	EN	PT	
Bilingual X2DE		10	7	**17**
Bilingual X2EN	4		3	**7**
Bilingual X2PT	7	6		**13**
TOTAL	**11**	**16**	**10**	**27**

3.2 Monolingual Experiments

Monolingual retrieval was offered for the following target collections: English, German, and Portuguese. Figures 1 to 3 show the interpolated recall vs. average precision for the top participants of the monolingual tasks.

The most competitive task was the monolingual English task with half of all GeoCLEF runs. The DFKI submitted the best run based on ontology processing but the results of the other participants are very close. The University of California at Berkeley applied no geographic processing and is not only in the top group for monolingual English but also for the bilingual experiments.

3.3 Bilingual Experiments

The bilingual task was structured in four subtasks (X → DE, EN or PT target collection). The best performing system for each of the three bilingual sub-tasks was

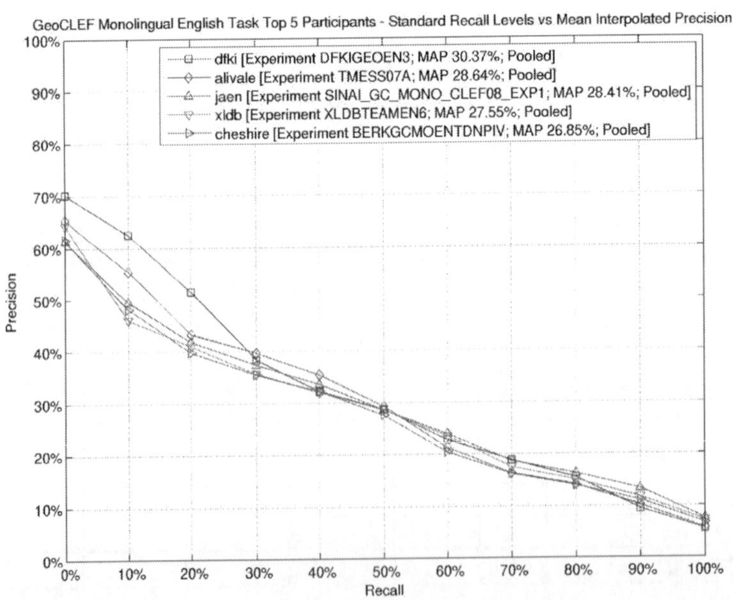

Fig. 1. Monolingual English top participants. Interpolated Recall vs. Average Precision.

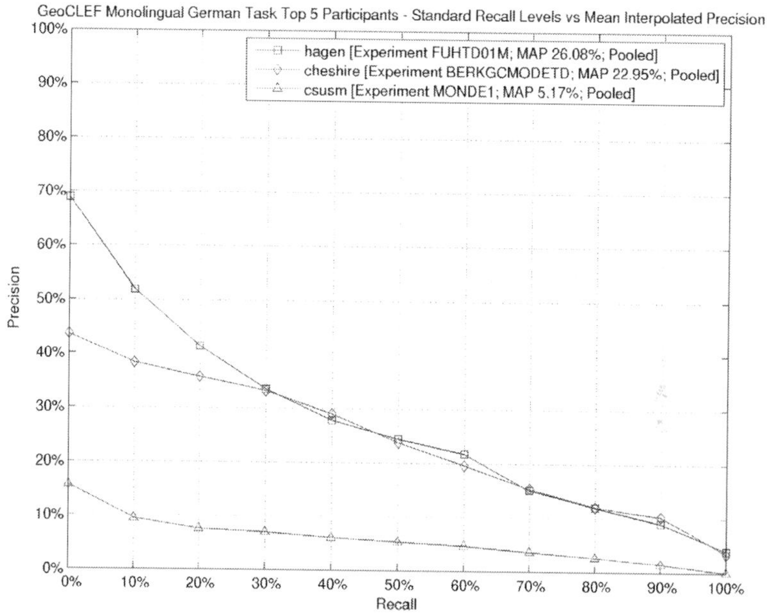

Fig. 2. Monolingual German top participants. Interpolated Recall vs. Average Precision.

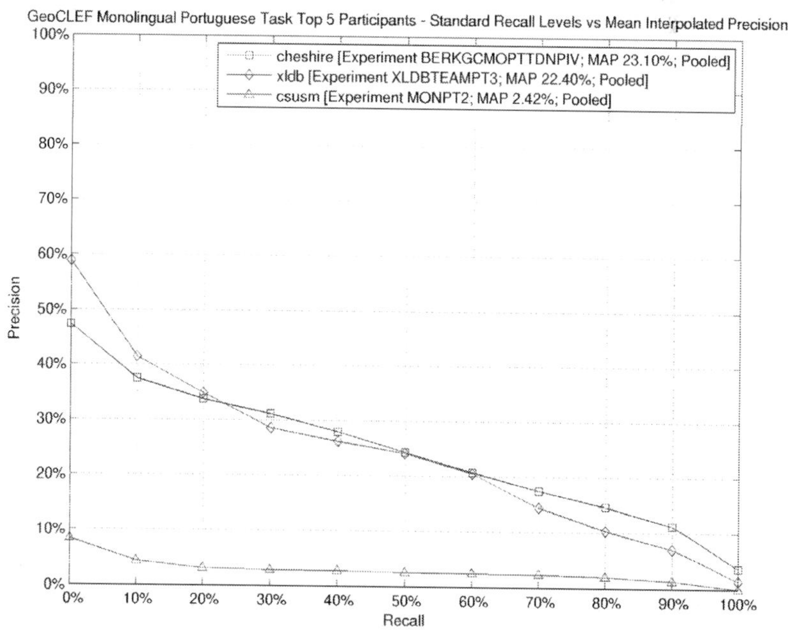

Fig. 3. Monolingual Portuguese top participants. Interpolated Recall vs. Average Precision.

presented by the University of California at Berkeley. This system did not use any specific geographic reasoning or knowledge source. Figures 4 to 6 show the interpolated recall vs. average precision graph for the top participants of the different bilingual tasks.

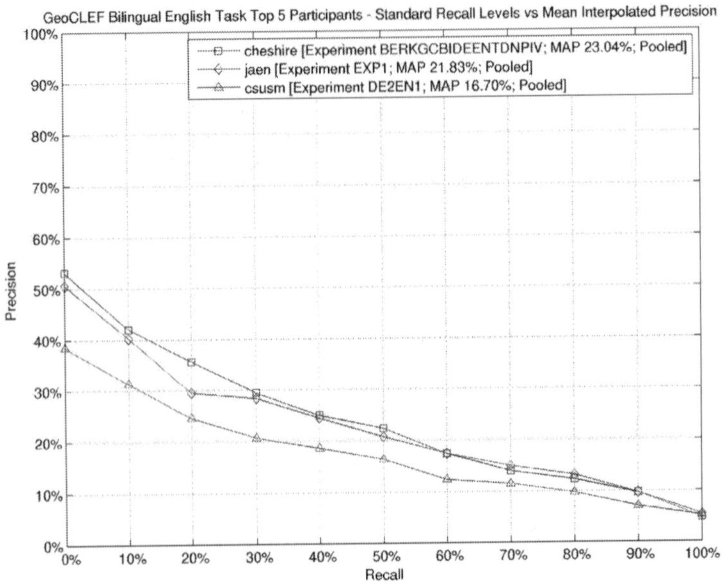

Fig. 4. Bilingual English top participants. Interpolated Recall vs Average Precision.

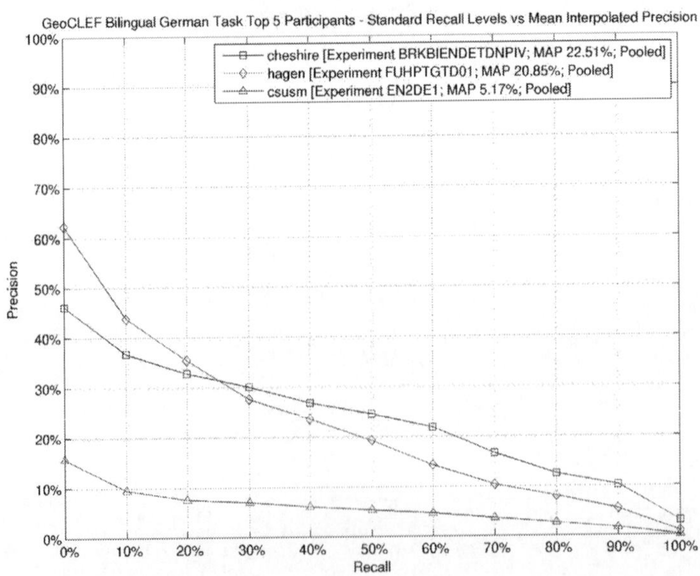

Fig. 5. Bilingual German top participants. Interpolated Recall vs Average Precision.

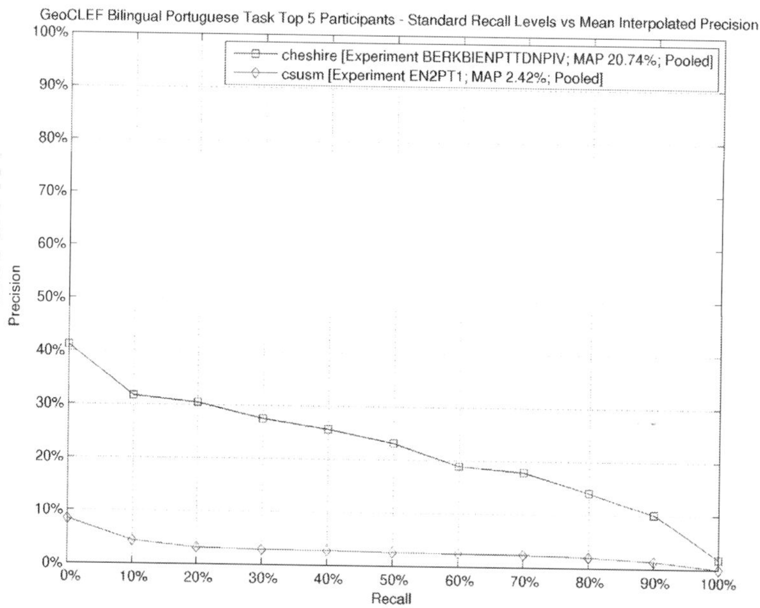

Fig. 6. Bilingual Portuguese top participants. Interpolated Recall vs Average Precision.

4 Result Analysis

The test collection of GeoCLEF increased 25 topics each year. Statistical testing and further analysis were performed to assess the validity of the results obtained.

Statistical testing for retrieval tests is used to determine whether the order of the systems which results from the evaluation reliably measures the quality of the systems in a reliable manner [2]. In most cases, the statistical analysis gives an approximate conservative estimate of the upper level of significance. The MATLAB Statistics Toolbox and the *ANalysis Of VAriance* (ANOVA) test were used for statistical testing. In all the experiments a value of alpha = 0.05 has been used to determine when to accept or reject the null hypothesis.

4.1 Monolingual vs. Bilingual Retrieval

In order to evaluate bilingual retrieval experiments, a common method is to compare results against monolingual baselines, which is comparing the best monolingual experiment vs. the best bilingual experiment and transform the ratio into a percentage:

- X → DE: 86 % of best monolingual German IR system
- X → EN: 76 % of best monolingual English IR system
- X → PT: 90 % of best monolingual Portuguese IR system

Note that there is an almost constant proportion in this result since CLEF 2006: Portuguese is usually the best performer. German and surprisingly English are last, even though there are several geographical and linguistic resources for these languages.

It is possible to run another kind of statistical analysis for a comparison between bilingual and monolingual performance which is not based on the comparison of the single best experiments, but on the average performance of each topic on the monolingual and bilingual task [6]. The results of this analysis are as follows:

- Monolingual German performs better than bilingual German. The mean average precision per topic of the monolingual task is significantly higher than the mean average precision per topic of the bilingual task;
- Monolingual English performs significantly better than bilingual English;
- Monolingual Portuguese performance is not significantly different from bilingual Portuguese.

That means, even though the performance difference between 86% and 90% of the German and Portuguese tasks presented above seems to be small, only for Portuguese, the difference between monolingual and bilingual performance is not significant.

4.2 Grouped Analysis

When the goal is to validate how well results can be expected to hold beyond a particular set of queries, statistical testing can help to determine what differences between runs appear to be real as opposed to differences that are due to sampling issues. For this purpose, a Tukey t-test was performed in order to study the groups of experiments which performed equally or significantly different [7].

There was an interesting result: the performance of all the experiments were statistically not different except for one participant, California State University San Marcos (CSUSM) who performed significantly worse compared to all other experiments. This experiment is an important baseline for comparison with all the approaches because the experiments sent by CSUMS were:

- automatic (no manual processing),
- without any query expansion,
- using only title and description (without narrative),
- without any translation in the bilingual task (no translation module at all),
- without removing diacritic marks in the collection.

This shows that if a cross-lingual system is designed with the basic functionalities, the performances of this system will be significantly worse compared to systems with advanced components. For the other systems, different optimization approaches can lead to optimal performances and no approach can be considered superior yet.

5 Conclusions and Future Work

GeoCLEF developed 100 topics and relevance judgments for geographic information retrieval. Another 26 topics with geographic specification were selected out of previous ad-hoc CLEF topics. This test collection is the first GIR test collection available for the research community and it will be a benchmark for future research.

GIR is receiving increased attention both through the GeoCLEF effort and through scientific workshops on the topic. The wide availability of geographic systems on the Internet will further increase the demand for and the interest in geographic information retrieval.

For GeoCLEF 2009, a new GikIP track is again planned, [5]. A query parsing and classification task is again planned for CLEF 2009. It requires the participants to identify geographic queries within a large set of queries from a search engine log.

Acknowledgments

The organization of GeoCLEF was mainly volunteer work. The English assessment was performed by Fredric Gey and Ray Larson from the University of California at Berkeley and Samaneh Beheshti-Kashi and Wiebke Alscher from the University of Hildesheim. The German assessment was carried out by a group of people from the University of Hildesheim: Lea Drolshagen, Kathrin Stackmann,, Julia Schulz, Nadine Mahrholz, Daniela Wilczek and Ralph Kölle. The Portuguese documents were assessed by Ana Frankenberg, Cláudia Freitas, Cristina Mota, David Cruz, Diana Santos, Hugo Oliveira, Luís Costa, Luís Miguel Cabral, Marcirio Chaves, Paula Carvalho, Paulo Rocha, Pedro Martins, Rosário Silva, Sérgio Matos and Susana Inácio, all from Linguateca (the Linguateca project is jointly funded by the Portuguese Government and the European Union (FEDER and FSE) under contract ref. POSC/339/1.3/C/NAC.). The topics were checked by Sven Hartrumpf from the University of Hagen (Germany).

References

1. Braschler, M., Peters, C.: Cross-Language Evaluation Forum: Objectives, Results, Achievements. Information Retrieval 7(1-2), 7–31
2. Buckley, C., Voorhees, E.: Retrieval System Evaluation. In: TREC: Experiment and Evaluation in Information Retrieval, pp. 53–75. MIT Press, Cambridge (2005)
3. Gey, F., Larson, R., Sanderson, M., Bishoff, K., Mandl, T., Womser-Hacker, C., Santos, D., Rocha, P., Di Nunzio, G.M., Ferro, N.: GeoCLEF 2006: the CLEF 2006 Cross-Language Geographic Information Retrieval Track Overview. In: Peters, C., Clough, P., Gey, F.C., Karlgren, J., Magnini, B., Oard, D.W., de Rijke, M., Stempfhuber, M. (eds.) CLEF 2006. LNCS, vol. 4730, pp. 852–876. Springer, Heidelberg (2007)
4. Mandl, T., Gey, F., Di Nunzio, G.M., Ferro, N., Larson, R., Sanderson, M., Santos, D., Womser-Hacker, C., Xing, X.: GeoCLEF 2007: the CLEF 2007 Cross-Language Geographic Information Retrieval Track Overview. In: Peters, C., Jijkoun, V., Mandl, T., Müller, H., Oard, D.W., Peñas, A., Petras, V., Santos, D. (eds.) CLEF 2007. LNCS, vol. 5152, pp. 745–772. Springer, Heidelberg (2008)
5. Santos, D., Cardoso, N., Carvalho, P., Dornescu, I., Hartrumpf, S., Leveling, J., Skalban, Y.: GikiP at GeoCLEF 2008: Joining GIR and QA forces for querying Wikipedia. In: Peters, C., et al. (eds.) CLEF 2008. LNCS, vol. 5706, pp. 894–905. Springer, Heidelberg (2009)
6. Crivellari, F., Di Nunzio, G.M., Ferro, N.: A Statistical and Graphical Methodology for Comparing Bilingual to Mono-lingual Cross-Language Information Retrieval. In: Agosti, M. (ed.) Information Access through Search Engines and Digital Libraries. Information Retrieval Series, vol. 22, pp. 171–188. Springer, Heidelberg (2008)
7. Gey, F., Larson, R., Mandl, T., Womser-Hacker, C., Santos, D., Carvalho, P., Di Nunzio, G.M., Ferro, N.: GeoCLEF 2008: the CLEF 2008 Cross-Language Geographic Information Retrieval Track Overview. In: Carol, P., Borri, F. (eds.) Working Notes for the CLEF 2008 Workshop, Aarhus, Danemark (September 2008), `http://www.clef-campaign.org/2008/working_notes/GeoCLEF-2008-overview-notebook-paperWNfinal.pdf`

GIR with Language Modeling and DFR Using Terrier

Rocio Guillén

California State University San Marcos, San Marcos, CA 92096, USA
rguillen@csusm.edu

Abstract. This paper reports on additional experiments in the Monolingual English, German and Portuguese collections tasks to those described in CLEF2008 Working Notes. Experiments were performed using the language modeling approach and the Divergence From Randomness (DFR) InL2 model as implemented in Terrier (TERabyte RetrIEveR) version 2.1. The main purpose was twofold: 1) to compare these approaches to determine their impact on performance retrieval and 2) to compare results from these experiments with the results generated in the first set of experiments to determine whether query expansion and the presence or absence of diacritic marks have an impact on performance retrieval. The stopword list provided by Terrier was used to index all the collections. We removed diacritic marks from the topics and collections for German and Portuguese before indexing and retrieval. Topics were processed automatically and the query tags specified were the title and the description. Query expansion was included using the 20 top ranked documents and 40 terms. These parameters were selected arbitrarily. Results show that the DFR InL2 model outperformed language modeling for all the languages. Results of the new experiments with query expansion show an improvement in performance retrieval for all the languages. They also suggest that removing diacritic marks may also have an impact in the case of German and Portuguese.

1 Introduction

Geographic Information Retrieval (GIR) is aimed at the retrieval of geographic data based not only on conceptual keywords, but also on spatial information. Building GIR systems with such capabilities requires research on diverse areas such as information extraction of geographic terms from structured and unstructured data; word sense disambiguation, which is geographically relevant; ontology creation; combination of geographical and contextual relevance; and geographic term translation among others.

Research efforts on GIR are addressing issues such as access to multilingual documents, techniques for information mining (i.e., extraction, exploration and visualization of geo-referenced information), investigation of spatial representations and ranking methods for different representations, application of machine learning techniques for place name recognition, development of datasets containing annotated geographic entities, among others [2,1]. Other researchers are

C. Peters et al. (Eds.): CLEF 2008, LNCS 5706, pp. 822–829, 2009.

exploring the usage of the World Wide Web as the largest collection of geospatial data.

The focus of the task was on experimenting with and evaluating the performance of GIR systems when topics include geographic references. Collections of documents and topics in different languages were available to carry out monolingual and bilingual experiments. We ran monolingual experiments in English, German, and Portuguese; for bilingual retrieval, we worked with topics in German and English and collections in English, German and Portuguese. We used Terrier platform to perform indexing and retrieval. The queries included only the title or a combination of the title and narrative; seven runs were submitted for evaluation. Results showed that no pre-processing of queries and data in the collection affects recall and precision.

Further experiments have been performed using the language modeling approach [3] and the Divergence From Randomness (DFR) [9] framework as implemented in Terrier (TERabyte RetrIEveR). The aim was to compare language modeling with InL2, a DFR model to determine their impact on performance retrieval, and to compare results from these experiments with the results generated in the first set of experiments to determine whether query expansion and the presence or absence of diacritic marks have an impact on performance retrieval. Additionally, we carried the following steps before retrieval: a) query expansion, and b) removal of diacritic marks in the collection and the topics. Results were evaluated with TRECEVAL, which is a program to evaluate TREC [7] results using the standard, NIST evaluation procedures.

The paper is organized as follows. In Section 2 we review the language modeling approach, the DFR framework and their implementation in Terrier. In Section 3 we describe the new set of experiments in the monolingual task. Section 4 describes our setting and experiments in the bilingual task. In Section 5 we discuss the language modeling approach and the DFR framework results. Finally, we present conclusions in Section 6.

2 Indexing and Retrieval Approaches

In this section we review the Ponte and Croft language modeling approach, the DFR framework and their implementation in Terrier 2.1. We used the Terrier information retrieval platform for both indexing and retrieval.

2.1 Language Modeling

Language modeling refers to a probability distribution that captures the statistical regularities of the generation of language, in speech recognition. Language models for speech attempt to predict the probability of the next word in an ordered sequence.

In information retrieval, the generation of queries can be treated as a random process modeled by a probability distribution. Ponte and Croft's [3] approach to retrieval is to infer a language model for each document and to estimate the probability of generating the query according to each of these models. The probability of the query given the language model of document d is $\hat{p}(Q \mid M_d)$. Documents are then ranked according to these probabilities.

The maximum likelihood estimate of the probability of term t under the term distribution for document d is:

$$\hat{p}_{ml}(t \mid M_d) = \frac{tf_{(t,d)}}{dl_d}$$

where $tf_(t,d)$ is the raw term frequency of term t in document d and dl_d is the total number of tokens in document d. It is assumed that query terms occur independently given a particular language model. Thus, the ranking formula is $\prod_{t \in Q} \hat{p}_{ml}(t,d)$ for each document.

There are two main problems with this formula: 1) there is a risk of assigning a probability of zero to a document that is missing one or more terms, and 2) there is only a document sized sample rather than an arbitrary sized sample data from M_d to be reasonably confident in the maximum likelihood estimator. To solve these problems, an estimate for a larger amount of data is the mean probability of term t in documents containing it. Additionally, the authors use a risk function as a mixing parameter to calculate $\hat{p}(Q \mid M_d)$. The estimate of the probability of producing the query for a given document model is calculated based on the following: 1) the probability of producing the terms in the query and 2) the probability of not producing other terms.

2.2 Divergence from Randomness

Like the language model approach of Ponte and Croft, a nonparametric model is derived as a combination of different probability distributions. The DFR paradigm is a generalization of Harter's 2-poisson indexing model [8]. In the DFR approach, a query term is weighted by how different its term distribution in the document d is, compared to the whole collection. The more divergence of the within document term frequency from its frequency within the collection, the more the information carried by term t in document d.

The weighting formulae is calculated in sequential steps. The first step is to calculate a probability that measures the informative content of the term in the document. Amati and Van Rijsbergen [9] introduce five basic models for this measure. Two basic models are approximated by two formulae each, resulting in seven weighting formulae: a) inverse term frequency, b) inverse document frequency, c) inverse expected document frequency, d) two approximations for the binomial distribution, divergence and Poisson, and e) two approximations for the Bose-Einstein statistics, geometric and Bose-Einstein. In the second step a first normalization calculates the information gain when accepting the term in the observed document as a good document descriptor. There are two first normalization formulae: L derived from Laplace's law of succession, and B obtained by a ratio of two Bernoulli processes. The third and last step recalculate the term frequency based on the length of the document. A second normalization calculates the uniform distribution of the term frequency and the term frequency density, which is inversely related to the length.

2.3 Terrier

Terrier [4,5,6] is a high performance and scalable search engine platform for the rapid development of large-scale retrieval applications. It offers a variety of IR models based on the Divergence from Randomness (DFR) framework and supports classic retrieval models like the Ponte-Croft language model. Among the DFR models included in Terrier is the Inverse Document Frequency (InL2) with Laplace after-effect and normalization 2. The InL2 model has been used in experiments in the past, GeoCLEF2007, GeoCLEF2006 and GeoCLEF2005[11,12,13], successfully.

3 Experiments

We start by providing an overview of our first set of experiments for the monolingual task [14].

The document collections indexed were the LA Times (American) 1994 and the Glasgow Herald (British) 1995 for English, publico94, publico95, folha94 and folha95 for Portuguese, and der_spiegel, frankfurter and fr_rundschau for German. There were 25 topics for each of the languages tested. Documents and topics were processed using the English stopword list and the Porter stemmer provided by Terrier. No stopword lists for German and Portuguese were used. The query tags specified were title and description for English, German and Portuguese, and title only for Portuguese.

The setup for the new experiments was as follows:

- Diacritic marks both from the German and Portuguese collections and from the German and Portuguese topics were removed.
- English, German and Portuguese collections were indexed using the Ponte-Croft language model.
- Documents in English, German and Portuguese were retrieved and ranked using the Ponte-Croft model with query expansion of 20 documents and 40 terms.
- Results for each language were evaluated using TRECEVAL.
- English, German and Portuguese collections were indexed using the InL2 DFR model.
- Documents in English, German and Portuguese were retrieved and ranked using the InL2 DFR model with query expansion of 20 documents and 40 terms.
- Results for each language were evaluated using TRECEVAL.

3.1 Evaluation Results

We show the evaluation results of the runs submitted (RunS) and the evaluation results of the new experiments (RunN) for the monolingual task in English, German and Portuguese in Table 1. Language modeling was outperformed by

Table 1. English, German and Portuguese Monolingual Retrieval Performance

Language	Run	IR Model	Topic Fields	MAP	R. Precision
English	RunS	InL2	title and description	0.167	0.193
	RunN	InL2	title and description	**0.2356**	**0.2359**
	RunN	Language Modeling	title and description	0.0394	0.0305
German	RunS	InL2	title and description	0.051	0.058
	RunN	InL2	title and description	**0.2059**	**0.2098**
	RunN	Language Modeling	title and description	0.0014	0.0020
Portuguese	RunS	InL2	title and description	0.024	0.03
	RunN	InL2	title and description	**0.2164**	**0.2257**
	RunN	Language Modeling	title and description	0.0025	0.0032

InL2 for all languages. One reason may be the way topics were designed. Language modeling estimates the likelihood that a query and a document could have been generated by the same language model, given the language model of the document.

For instance, there were two relevant documents for the English topic 91 with title and description as follows:

"<title>Forest fires on Spanish islands</title>"
"<description>Documents mentioning forest fires on Spanish islands</description>"

No documents were retrieved using language modeling, whereas the two documents judged as relevant were retrieved using InL2 with query expansion. Only one relevant document was retrieved without query expansion. One reason could be that the information the n-gram "Spanish islands" does not appear as such in the documents.

In the case of Portuguese, for the same topic, no documents were retrieved with language modeling, whereas the four documents considered relevant were retrieved using InL2 with query expansion. No documents were retrieved without query expansion.

Analysis of the same topic in German look similar to those of Portuguese. No documents were retrieved with language modeling, whereas the 22 documents selected to be relevant were retrieved using InL2 with query expansion. Ten documents were retrieved without query expansion.

A comparison in terms of the mean average precision (MAP) of the new experiments using InL2 with the overall best results of the monolingual task for English, German and Portuguese [10] is presented in Table 2.

The MAP for the new experiments was below the best results, however the improvement was significant compared to our first set of experiments [14].

Diacritic marks had no influence in the results for English because we did not remove them from the collection. We assumed that because English has very few words which contain diacritics the results were not going to vary significantly. The only difference between English RunS InL2 and English RunN InL2 was query expansion.

Table 2. Best Overall Map vs. New Experiments Map

Language	Run	MAP
English	DFKIGEOEN3	30.37
	RunN-InL2	23.56
German	FUHTD01M	26.08
	RunN-InL2	20.59
Portuguese	BERKGCMOPTTDNPIV	23.10
	RunN-InL2	21.64

Table 3. German Relevant Documents vs. Relevant Documents Retrieved

Topic	Relevant	Relevant Retrieved RunN	Relevant Retrieved RunS
76	31	12	0
78	1	1	1
79	49	49	48
81	65	47	51
82	13	11	1
84	62	62	9
85	109	108	52
88	19	5	1
89	28	18	0
91	22	22	10
92	56	3	3
93	133	133	4
94	146	59	61
95	41	41	3
98	22	22	1
99	128	120	0

Results for German in terms of relevant documents and relevant documents retrieved improved for most topics, in very few cases the numbers are the same and in two cases the numbers are lower. In Table 3 we present for some topics the number of relevant documents, the number of relevant documents retrieved without diacritics and the number of relevant documents retrieved with the original collection. For instance, for Topic 99 no relevant documents were retrieved with the original collection, whereas 120 out of 128 relevant documents were retrieved after removing diacritics from the collection. Fewer documents were retrieved for Topics 81 and 94 without diacritics. There were no changes in performance retrieval for topics 78 and 92.

In contrast, for Portuguese results improved just for nine topics (76, 80, 83, 90, 92, 94, 96, 97, 99), there were no changes for nine topics (79, 84, 85, 86, 89, 91, 93, 95, 98) and results were worse for seven topics (77, 78, 81, 82, 87, 88, 100) . In Table 4 we present for some topics the number of relevant documents, the number of relevant documents retrieved without diacritics and the number of relevant documents retrieved with the original collection.

Table 4. Portuguese Relevant Documents vs. Relevant Documents Retrieved

Topic	Relevant	Relevant Retrieved RunN	Relevant Retrieved RunS
76	96	2	4
78	49	46	44
79	99	99	99
81	50	49	31
82	73	66	25
83	6	1	6
85	87	87	87
87	158	150	137
90	80	33	74
93	138	138	138
96	40	22	28
97	56	31	38
99	34	3	32

The main difference between German and Portuguese seems to be the number of diacritical marks in each language and their usage, in terms of frequency, in the language. Further experiments would allow us to reach a more informed conclusion.

4 Conclusions

In this paper we presented new experimental results on monolingual geographical information retrieval. We used Terrier to run our experiments using the InL2 model and the Ponte-Croft language model. Language modeling was outperformed by InL2 for all languages. Comparison of results between new experiments using query expansion and those submitted to GeoCLEF 2008 show improved performance retrieval for English, German and Portuguese. This is specially true in the case of English which is a language with no diacritic marks. Analysis and evaluation of results suggest that removing diacritic marks have an impact on performance retrieval for German and Portuguese. However the impact differs depending on the language. We obtained better results for German than Portuguese. We have concluded that the number of diacritic marks and their usage in each language might be other factors that need to be further studied. In general, precision was improved, but it was still below the overall best results for all the participating teams.

References

1. Larson, R.R.: Geographic Information Retrieval and Spatial Browsing. In: Smith, L., Gluck, M. (eds.) GIS and Libraries: Patrons, Maps and Spatial Information, pp. 81–124. University of Illinois (1996)
2. Purves, R., Jones, C. (eds.): SIGIR 2004: Workshop on Geographic Information Retrieval, Sheffield, UK (2004)

3. Ponte, J.M., Croft, W.B.: A Language Modeling Approach to Information Retrieval. In: Proceedings of the 21st Annual International ACM SIGIR Conference on Research and Development in Information Retrieval, pp. 275–281 (1998)
4. Ounis, I., Liorna, C., Macdonald, C., Plachouras, V.: Research Directions in Terrier: a Search Engine for Advanced Retrieval on the Web. Novatica/UPGRADE Special Issue on Next Generation Web Search 8(1), 49–56 (2007)
5. Ounis, I., Amati, G., Plachouras, V., He, B., Macdonald, C., Liorna, C.: Performance and Scalable Information Retrieval Platform. In: Proceedings of ACM SIGIR 2006 Workshop on Open Source Retrieval (OSIR 2006), Seattle, Washington, USA, August 10 (2006)
6. Ounis, I., Amati, G., Plachouras, V., He, B., Macdonald, C., Johnson, D.: Terrier Information Retrieval Platform. In: Losada, D.E., Fernández-Luna, J.M. (eds.) ECIR 2005. LNCS, vol. 3408, pp. 517–519. Springer, Heidelberg (2005)
7. `http://trec.nist.gov/trec_eval/`
8. Harter, S.P.: A probabilistic approach to automatic keyword indexing. Journal of the American Society for Information Science 26, 197–206, 280-289 (1975)
9. Amati, G., Van Rijsbergen, C.J.: Probabilistic models of information retrieval based on measuring divergence from randomness. ACM Transactions on Information Systems (TOIS) 20(4), 357–389 (2002)
10. Mandl, T., Carvalho, P., Gey, F., Larson, R., Santos, D., Womser-Hacker, C., Nunzio, G.D., Ferro, N.: GeoCLEF 2008: the CLEF2008 Cross-Language Geographic Information Retrieval Track Overview. In: Nardi, A., Peters, C. (eds.) Working Notes for the CLEF 2008 Workshop, Aarhus, Denmark (2008)
11. Guillén, R.: CSUSM Experiments at GeoCLEF2005: 6th Workshop of the Cross-Language Evaluation Forum. In: Peters, C., Gey, F.C., Gonzalo, J., Müller, H., Jones, G.J.F., Kluck, M., Magnini, B., de Rijke, M., Giampiccolo, D. (eds.) CLEF 2005. LNCS, vol. 4022, pp. 956–962. Springer, Heidelberg (2006)
12. Guillén, R.: Monolingual and Bilingual Experiments in GeoCLEF 2006: Evaluation of Multilingual and Multi-modal Information Retrieval Cross-Language Information Forum. In: Peters, C., Clough, P., Gey, F.C., Karlgren, J., Magnini, B., Oard, D.W., de Rijke, M., Stempfhuber, M. (eds.) CLEF 2006. LNCS, vol. 4730, pp. 893–900. Springer, Heidelberg (2007)
13. Guillén, R.: GeoCLEF 2007 Experiments in Query Parsing and Cross-language GIR: CLEF 2007 Working Notes. In: Nardi, A., Peters, C. (eds.) ISSN per Working Notes and CD: 1818-8044 (2007), ISBN Abstracts: 2-912335-31-0
14. Guillén, R.: Cross-lingual Geographical Information Retrieval: CLEF 2008 Working Notes. In: Peters, C. (ed.), Aarhus, Denmark (2008)

Cheshire at GeoCLEF 2008: Text and Fusion Approaches for GIR

Ray R. Larson

School of Information
University of California, Berkeley, USA
ray@ischoolbib.berkeley.edu

Abstract. In this paper we will briefly describe the approaches taken by the Berkeley Cheshire group for the main GeoCLEF 2008 tasks (Mono and Bilingual retrieval), and present some analyses of the fusion approach used. This year our submissions used probabilistic text retrieval based on logistic regression and incorporating blind relevance feedback for all of the runs and in addition we ran a number of tests combining this type of search with OKAPI BM25 searches using a fusion approach. We did not, however, use any explicit geographic processing. All translation for bilingual tasks was performed using the LEC Power Translator PC-based MT system.

1 Introduction

Geographic Information Retrieval (GIR) as it was originally defined was concerned with providing access to georeferenced information resources using a combination of Information Retrieval (IR) and Data Retrieval (DR) (or database) methods[2]. In GeoCLEF the nature of the topics has tended to emphasize the IR aspects of GIR, largely because no explicit georeferencing of documents or topics has been supplied to the experimenter.

Without the explicit georeferencing of documents and/or topics the experimenter (or searcher) is faced with attempting to provide such georeferencing and therefore solving all of the attendent problems of ambiguity and multiplicity of toponyms and the issues of name polysemy that explicit georeferencing is intended to alleviate. An alternative approach is to, in effect, ignore geographic clues by treating them like any other term in a normal IR search process. This year we took this latter approach, and use only text retrieval methods on the provided topics with no explicit identification or treatment of toponyms.

This paper describes the retrieval algorithms and evaluation results for the Cheshire group's official submissions for the GeoCLEF 2008 track. All of the submitted runs were automatic without manual intervention in the queries (or translations). We submitted nine Monolingual runs (three German, three English, and three Portuguese) and eighteen Bilingual runs (three runs for each of the three languages to each other language). The runs varied in the topic elements used, and whether or not a fusion approach (described below) was used.

C. Peters et al. (Eds.): CLEF 2008, LNCS 5706, pp. 830–837, 2009.

This paper first describes the retrieval algorithms and fusion operations used for our submissions, followed by a discussion of the processing used for the runs. We then examine the results obtained for our officially submitted runs, and finally present conclusions and future directions.

2 The Retrieval Algorithms and Fusion Operators

The basic form and variables of the *Logistic Regression* (LR) algorithm used for all of our submissions was originally developed by Cooper, et al. [1] and our use of them has been described in detail in previous CLEF proceedings[6] and the blind feedback approach used with it also described[4]. The basic form used estimates the "log odds" of relevance ($\log O(R \mid Q, D)$) for a given query and document, which can be simply converted to an estimated probability of relevance:

$$\begin{aligned}
\log O(R|C,Q) = log\frac{p(R|C,Q)}{1-p(R|C,Q)} &= log\frac{p(R|C,Q)}{p(\overline{R}|C,Q)} \\
&= c_0 + c_1 * \frac{1}{\sqrt{|Q_c|}+1}\sum_{i=1}^{|Q_c|}\frac{qtf_i}{ql+35} \\
&+ c_2 * \frac{1}{\sqrt{|Q_c|}+1}\sum_{i=1}^{|Q_c|}\log\frac{tf_i}{cl+80} \\
&- c_3 * \frac{1}{\sqrt{|Q_c|}+1}\sum_{i=1}^{|Q_c|}\log\frac{ctf_i}{N_t} \\
&+ c_4 * |Q_c|
\end{aligned} \tag{1}$$

where C denotes a document component (i.e., an indexed part of a document which may be the entire document) and Q a query, R is a relevance variable,

$p(R|C,Q)$ is the probability that document component C is relevant to query Q,
$p(\overline{R}|C,Q)$ the probability that document component C is *not relevant* to query Q, which is 1.0 - $p(R|C,Q)$
$|Q_c|$ is the number of matching terms between a document component and a query,
qtf_i is the within-query frequency of the ith matching term,
tf_i is the within-document frequency of the ith matching term,
ctf_i is the occurrence frequency in a collection of the ith matching term,
ql is query length (i.e., number of terms in a query like $|Q|$ for non-feedback situations),
cl is component length (i.e., number of terms in a component), and
N_t is collection length (i.e., number of terms in a test collection).
c_k are the k coefficients obtained though the regression analysis.

The coefficients, c_k, used for our official runs are the same as those described in [6], i.e.: $c_0 = -3.51$, $c_1 = 37.4$, $c_2 = 0.330$, $c_3 = 0.1937$ and $c_4 = 0.0929$.

Our "blind feedback" approach assumes that the top 13 documents in each result set are relevant, and then uses them to calculate the Robertson-Sparck Jones term weights for each term in the documents, then takes the 16 top ranked terms and either adds them to the initial query (or augments their weights if already present in the query). The new query is submitted using the same LR algorithm shown above.

This LR algorithm with blind feedback was used alone in our submitted runs as well as in combination with the Okapi BM-25 algorithm. The version of the Okapi BM-25 algorithm used in these experiments is based on the description of the algorithm in Robertson [8], and in TREC notebook proceedings [9]. These algorithms and our implementations of them are described in more detail in the conference working notes[5].

2.1 Fusion Operators

The Cheshire II system used in this evaluation provides a number of operators to combine the intermediate results of a search from different components or indexes. The basic fusion approaches with the Cheshire II system are described in[3]. With these fusion operators we have available an entire spectrum of combination methods ranging from strict Boolean operations to fuzzy Boolean and normalized score combinations for probabilistic and Boolean results. These operators are the means available for performing fusion operations between the results for different retrieval algorithms and the search results from different different components of a document.

The MERGE_PIVOT operator was the only one of these algorithms tested in this year's GeoCLEF. It is used primarily to adjust the probability of relevance for one search result based on matching elements in another search result. It was developed primarily to adjust the probabilities of a search result consisting of sub-elements of a document (such as titles or paragraphs) based on the probability obtained for the same search over the entire document. It is basically a weighted combination of the probabilities based on a "DocPivot" fraction, such that:

$$P_n = DocPivot * P_d + (1 - DocPivot) * P_s \tag{2}$$

where P_d represents the document-level probability of relevance, P_s represents the subelement probability, and P_n represents the resulting new probability estimate.

For all of our fusion experiments this year, the P_s was the estimated probability of relevance for a document obtained by a TREC2 with blind feedback search using the topic title and description (as described above) normalized using MINMAX normalization, and P_d was an OKAPI BM-25 search using the topic title, description, and narrative, also normalized to 0-1 range using MINMAX normalization. The "*DocPivot*" value used for all of the runs submitted was 0.29. This value was chosen by numerous experiments using GeoCLEF 2007 queries, collections and relevance data for pivot values ranging from .01 to .99, and 0.29 was found to give the best results on the 2007 data.

Note that this is not the first time we have used some fusion approaches in GeoCLEF. In the first GeoCLEF (2005) we also employed fusion approaches in some of our runs, but these did not use the TREC2 with blind feedback algorithm[7], which appears to make an important difference.

3 Approaches for GeoCLEF

In this section we describe the specific approaches taken for our submitted runs for the GeoCLEF tasks. First we describe the indexing and term extraction methods used, and then the search features we used for the submitted runs.

3.1 Indexing and Term Extraction

The Cheshire II system uses the XML structure of the documents to extract selected portions for indexing and retrieval. Any combination of tags can be used to define the index contents. Although we extracted a number of indexes for the collection, we only used a single index in the submitted runs that included all of the text and title elements from the various collections. For our official submissions we did not use any explicit georeferencing of the text (although the Cheshire II system has that capability and it was used in earlier GeoCLEF evaluations).

For all indexing we used language-specific stoplists to exclude function words and very common words from the indexing and searching. The German language runs *did not* use decompounding in the indexing and querying processes to generate simple word forms from compounds. This was due to a failure of our decompounding software (no compound terms were correctly matched). The Snowball stemmer was used by Cheshire II for language-specific stemming.

3.2 Search Processing

Searching the GeoCLEF collection using the Cheshire II system involved using TCL scripts to parse the topics and submit the title and description or the title, description, and narrative from the topics. For monolingual search tasks we used the topics in the appropriate language (English, German, and Portuguese), for bilingual tasks the topics were translated from the source language to the target language using the LEC Power Translator PC-based machine translation system (http://www.lec.com/). Table 1 shows the runs submitted and the characteristics of each run, including which task it was submitted for, and the topic elements used in searching (this are indicated by using the "T" for the "title" element, "D" for the "description" element and "N" for the "narrative" element. The topic elements used were combined into a single probabilistic query. For those runs including the term "fusion" in the "Type" column, both TREC2 with blind

Table 1. Submitted GeoCLEF Runs

Description	TD MAP	TDN MAP	TD-fusion MAP
DE	0.2295 *	0.2050	0.2292
EN	0.2652	0.2001	0.2685 *
PT	0.2170	0.1741	0.2310 *
EN⇒DE	0.2150	0.1682	0.2251 *
PT⇒DE	0.1950	0.1108	0.1912
DE⇒EN	0.2274	0.1894	0.2304 *
PT⇒EN	0.1886	0.1540	0.2101
DE⇒PT	0.1346	0.1260	0.1488
EN⇒PT	0.1913	0.1762	0.2074 *

feedback algorithm results (using only topic title and description) and Okapi BM-25 results (using title description and narrative) were combined using the MERGE_PIVOT fusion operator described above with a pivot value of 0.29.

4 Results for Submitted Runs

The summary results (as Mean Average Precision) for our submitted monolingual and bilingual runs for both English, German and Portuguese are shown in Table 1. The precision-recall curves for these runs are also shown in Figures 1 and 2, for monolingual and bilingual runs respectively. Table 1 indicates the runs that had the highest overall MAP for the task and language by asterisks next to the run name.

Once again we found some rather interesting results among the official runs. For example, it seems clear that using topic title and description alone is a much better approach with our algorithms than using title description and narrative. In most cases the fusion approach either exceeds, or is very close to the performance of the TREC2 with blind feedback search with title and description due to "supporting evidence" for relevance from the Okapi BM-25 algorithm.

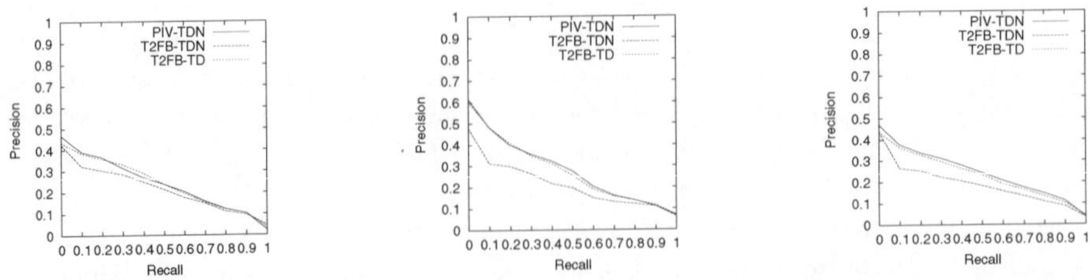

Fig. 1. Monolingual Runs – German (left), English (center), and Portuguese (right)

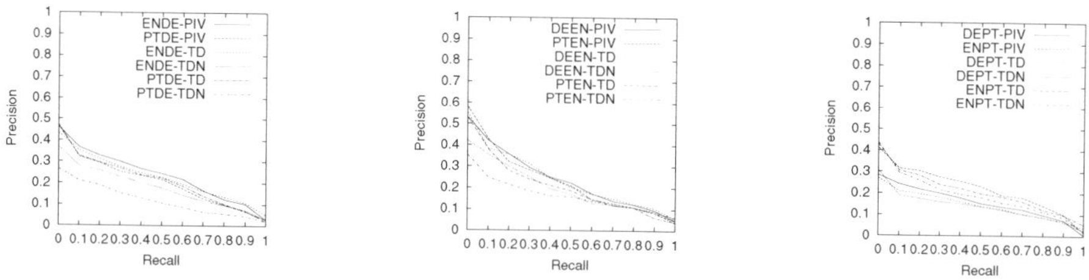

Fig. 2. Bilingual Runs – To German (left), To English (center) and to Portuguese (right)

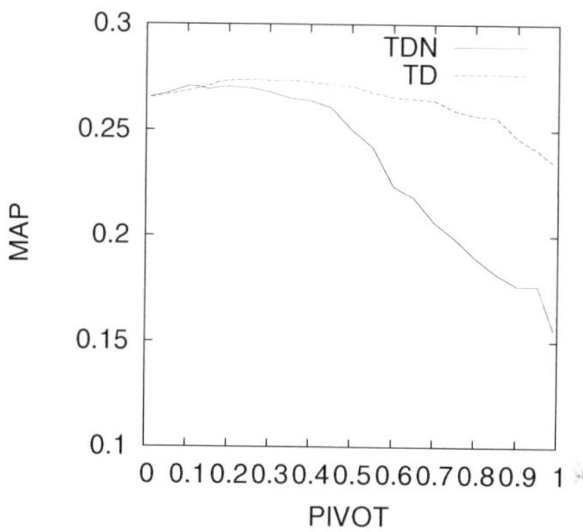

Fig. 3. Mean Average Precision for variations in the "pivot" value for Monolingual English

5 Additional Analysis

In Table 2 the MAP for our best runs from 2006, 2007, and 2008 for various tasks are shown, along with the percentage difference between each pair of years. Overall, we have seen some distinct improvements in the text-based approaches this year over those used in 2006 and 2007. The large differences in Monolingual and Bilingual German were between 2007 and the other years were caused by an undiscovered failure to index all of the GeoCLEF German data for that year.

Based on the summary data across participants for GeoCLEF available on the DIRECT system, it is apparent that the text-based and fusion approaches that we used this year are quite effective relative to some other approaches. In all of the six main GeoCLEF tasks there was a Cheshire group run in the top five participants listed. In Monolingual Portuguese and each Bilingual task (German, English, and Portuguese) one of the Cheshire group runs was ranked highest in MAP over all participants.

Because we have previously observed significant variation in fusion results when differing "pivot" values are used, we conducted a post-hoc analysis of our

Table 2. Comparison of Berkeley's best 2006, 2007 and 2008 runs for English, German and Portuguese

TASK	MAP 2006	MAP 2007	MAP 2008	Diff. '06-'07	Diff. '07-'08	Diff. '06-'08
Monolingual English	0.250	0.264	0.268	5.303	1.493	6.716
Monolingual German	0.215	0.139	0.230	-54.676	39.565	6.522
Monolingual Portuguese	0.162	0.174	0.231	6.897	24.675	29.870
Bilingual English⇒German	0.156	0.090	0.225	-73.333	60.00	30.667
Bilingual English⇒Portuguese	0.126	0.201	0.207	37.313	2.899	39.130

English Monolingual run. For these experiments we used the same script as for the official runs and only varied the value used for the "pivot" parameter. We performed 48 runs with pivot values ranging from 0.01 to 0.99 (in 0.05 increments, with a few extra runs within the 0.10 - 0.30 range, with half using only the Title and Description (TD) topic elements for both the Logistic Regression and OKAPI algorithms, and the other half using just Title and Description for the Logistic Regression and Title, Description and Narrative(TDN). Figure 3 shows the results for this variation in pivot value in terms of Mean Average Precision, and elements used. Once again this analysis shows the relative importance of using only the title and description elements instead of title, description, and narrative. The value of the "pivot" indicates the relative weight given to the OKAPI algorithm versus the Logistic Regression algorithm (i.e., if the pivot value is 0.10, only ten percent of the fused weight is coming from OKAPI and ninety percent from LR). As Figure 3 indicates, when using TD alone, taking about thirty percent of the fusion score from OKAPI produces the best results. In our submitted runs we used only the TDN approach, and thus missed the improvement shown by the TD approach. In both cases, however the fusion approaches marginally out-perform either single algorithm (whose performance can be seen as the end-points of the curves in Figure 3. This analysis confirms that the pivot value chosen for our official runs (0.29), which was based on similar analysis of GeoCLEF 2007 queries, data, and relevance judgements, seems to be fairly stable and seems to represent the optimum for combining these algorithms when searching this collection.

6 Conclusion

The GeoCLEF track is being replaced by the GikiCLEF track using multilingual versions of Wikipedia for geographical queries. Our challenge for next year is to reintroduce actual geographic elements into our mix of processing approaches to see if, for example, automatic expansion of toponyms in the topic texts will enhance or degrade performance over the purely textual approach for a different collection. Such expansion was done explicitly in many of the topic narratives this year, but we found that using those narratives in queries proved counterproductive. We suspect that this kind of query expansion may be a source of noise instead of fostering improved results. In previous years it appeared that implicit or explicit toponym inclusion in queries led to better performance when compared to using titles and descriptions alone in retrieval. But given the results this time, some doubt has been cast over that assumption, at least for the algorithms that we have been using.

Although we did not do any explicit geographic processing other than in indexing for this year, we plan to do so in the future, because we still believe that use of geographical knowledge and evidence in topics and documents *should* improve performance over purely text-based methods. However, given the results reported above, this belief is still unsupported in our experiments.

References

1. Cooper, W.S., Gey, F.C., Dabney, D.P.: Probabilistic retrieval based on staged logistic regression. In: 15th Annual International ACM SIGIR Conference on Research and Development in Information Retrieval, June 21-24, pp. 198–210. ACM, New York (1992)
2. Larson, R.R.: Geographic information retrieval and spatial browsing. In: Smith, L., Gluck, M. (eds.) GIS and Libraries: Patrons, Maps and Spatial Information, pp. 81–124. University of Illinois at Urbana-Champaign, GSLIS, Urbana-Champaign (1996)
3. Larson, R.R.: A fusion approach to XML structured document retrieval. Information Retrieval 8, 601–629 (2005)
4. Larson, R.R.: Cheshire at geoclef 2007: Retesting text retrieval baselines. In: Peters, C., Jijkoun, V., Mandl, T., Müller, H., Oard, D.W., Peñas, A., Petras, V., Santos, D. (eds.) CLEF 2007. LNCS, vol. 5152, pp. 811–814. Springer, Heidelberg (2008)
5. Larson, R.R.: Cheshire at GeoCLEF 2008: Text and fusion approaches for GIR: CLEF working notes (2008), `http://www.clef-campaign.org/2008/working_notes/larson_GeoCLEF.pdf`
6. Larson, R.R.: Experiments in classification clustering and thesaurus expansion for domain specific cross-language retrieval. In: Peters, C., Jijkoun, V., Mandl, T., Müller, H., Oard, D.W., Peñas, A., Petras, V., Santos, D. (eds.) CLEF 2007. LNCS, vol. 5152, pp. 188–195. Springer, Heidelberg (2008)
7. Larson, R.R., Gey, F.C., Petras, V.: Berkeley at GeoCLEF: Logistic regression and fusion for geographic information retrieval. In: Peters, C., Gey, F.C., Gonzalo, J., Müller, H., Jones, G.J.F., Kluck, M., Magnini, B., de Rijke, M., Giampiccolo, D. (eds.) CLEF 2005. LNCS, vol. 4022, pp. 963–976. Springer, Heidelberg (2006)
8. Robertson, S.E., Walker, S.: On relevance weights with little relevance information. In: Proceedings of the 20th annual international ACM SIGIR conference on Research and development in information retrieval, pp. 16–24. ACM Press, New York (1997)
9. Robertson, S.E., Walker, S., Hancock-Beauliee, M.M.: OKAPI at TREC-7: ad hoc, filtering, vlc and interactive track. In: Text Retrieval Conference (TREC-7), November 9-1, pp. 152–164 (1998) (Notebook)

Geographic and Textual Data Fusion in Forostar

Simon Overell[1], Adam Rae[2], and Stefan Rüger[2,1]

[1] Department of Computing, Imperial College London, London SW7 2AZ, UK
[2] Knowledge Media Institute, The Open University Walton Hall, Milton Keynes, MK7 6AA, UK
simon.overell@imperial.ac.uk, {a.rae,s.rueger}@open.ac.uk

Abstract. In this paper we provide some analysis of data fusion techniques employed at GeoCLEF 2008 to merge textual and geographic relevance. These methods are compared to our own experiments, where using our GIR system, *Forostar*, we show that an aggressive filter-based data fusion method can outperform a more sophisticated penalisation method.

1 Introduction

In 2005 when the first GeoCLEF track was run the structuring of the queries implied one was expected to build a textual retrieval engine, a geographic retrieval engine, and then fuse the results. In fact the first query set was split along these lines into concept, location and spatial relation [1]. In recent years opinion has become divided whether this is the most appropriate approach and whether all evidence needs to be considered simultaneously [2]. The results presented in the experiments part of this paper follow the methods found best in [3], where the unprocessed query is submitted to a text retrieval engine and fused with the results of a geographic retrieval engine (effectively doubling up on geographic evidence).

2 Data Fusion

There are many ways of including geographic relevance in a GIR system. One of the simpler ways is through query or document expansion. Strictly speaking this is pre-processing rather than data fusion but it is a method often employed [2,4]. Placenames are expanded either at query time or indexing time to include other locations where relevance is implied.

Traditional data fusion combines a number of ranks or filters at query time. The difference between a rank and a filter is that a rank is ordered (assumes scored retrieval), while all documents contained in a filter are considered of equal relevance (assumes binary retrieval). Textual retrieval generally produces results ranked according to a relevance score. Geographic relevance ranking is a less mature and less studied area than text retrieval. The results of a text rank are generally considered more reliable than the results found through geographic

C. Peters et al. (Eds.): CLEF 2008, LNCS 5706, pp. 838–842, 2009.

relevance. Therefore, the geographic relevance results are often used to re-rank [5] or filter [6,7] the text results, or combined with the text results at a lower priority [8]. Martins et al. [9] outlines four ways of combining text and geographic relevance: linear combination of similarity, product of similarity, maximum similarity or a step-linear function (equivalent to filtering). The most common ways of combining a geographic and text rank is either as a convex combination of ranks [10] or scores [8,2]. The disadvantage of using scores rather than ranks is that the distribution of scores produced by different relevance methods may differ greatly. This problem can be mitigated by normalising the scores [11]. Using ranks rather than scores has the disadvantage that information is discarded.

3 Forostar

Forostar is described in detail in our GeoCLEF working notes paper [10]. Essentially placenames are extracted from the GeoCLEF corpus and grounded to a unique location ID. Multiple disambiguation methods were tested, these will not be described here and we would like to refer the reader to [10] for more details. The disambiguation methods used in this paper are MR, where placenames are disambiguated as the *most referred* to location ID, and NoDis, where for each ambiguous placename multiple location IDs are indexed. The location IDs form a geographic index that is queried in parallel with a text index to produce a text rank and a geographic filter. The text rank consists of a ranked list of documents ordered by relevance to the query; the geographic rank consists of an unordered set of documents that refer to a location mentioned in the query. These are merged by the data fusion module, which is of particular interest to this paper, to produce a single rank. We compared two different data fusion methods:

Penalisation. The penalisation method multiplies the rank r, of each element in the text rank that is not in the geographic filter by a penalisation value p, to give an intermediate rank $r\prime$. The intermediate rank is sorted by $r\prime$ to give the final returned rank. The penalisation value p is found using a brute force search using the 75 queries and relevance judgements from GeoCLEF 2005-07 as training data. The search finds the value of p that maximises mean average precision (MAP). The p values found for each disambiguation method are shown in Table 1.

Filtering. The filtering method reorders the text rank in a more aggressive way than the penalisation method. All the results of the text rank that are also contained in the geographic filter are returned first, followed by the text results

Table 1. Penalisation values

Disambig. Method	Penalisation Values
MR	2.0
NoDis.	3.0

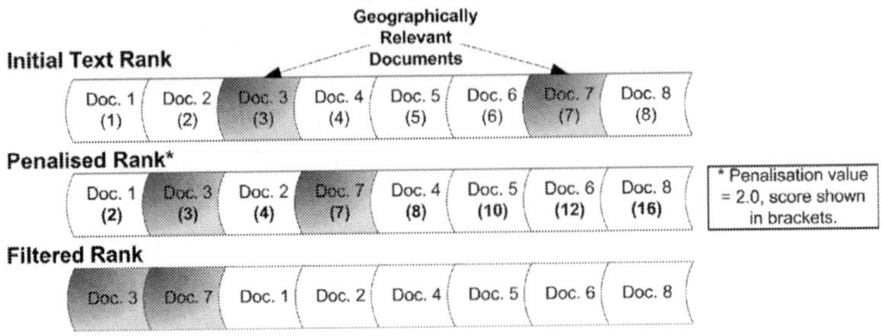

Fig. 1. Rank Example

that are not in the geographic filter. The filter method and text baseline are both equivalent to the penalisation method with p values of a high value (e.g. 1000) and 1 respectively.

An example of these two methods are shown in Figure 1. It shows a hypothetical text rank containing two entries also in the geographic filter. The penalisation method calculates $r\prime$, shown in brackets, to re-order the results, while the filter method simply promotes all the documents also in the geographic filter to the top of the rank.

Our experiments are all monolingual. We use the 25 English Language GeoCLEF queries and the English Language corpus consisting of $\approx 170,000$ documents from the Glasgow Herald and Los Angeles Times [1]. GeoCLEF topics contain title, description and narrative fields. We only use the title field as this field bears most similarity to the types of geographic query submitted to search engines. The experiments presented here hope to test whether a brute force search with 75 training queries is sufficient not to over fit the p value. We believe as the training queries are not substantially different from the test queries, this should be the case.

The results of five of our runs are displayed in Table 2. For comparison with the rest of the participants at GeoCLEF 2008, alongside our result table the quartile ranges of all submitted results are shown. MAP is the metric primarily used in GeoCLEF, however Geometric Average Precision is also shown to give an indication of how consistent the methods are. Notice that combining the geographic information using the trained penalisation value actually gives worse results than the text baseline. Our conclusion here is that the penalisation training is significantly over fitting. On the other hand, the filter method outperforms the baseline in every case showing it to be more robust.

We performed pairwise statistical significance testing of each method with the baseline using the Wilcoxon signed rank test rejecting the null hypothesis only when $p < 5\%$ [12]. We found that all the penalisation results were statistically significantly worse than the baseline, and only the NoDis-Filter method was statistically significantly better. We also performed pairwise significance testing between the penalisation method and filter method runs with the same disambiguation method and found that in every case the filter method was statistically

Table 2. Results

Disambig.	Fusion	MAP (%)	GeoAP (%)
Text Baseline		24.1	6.52
MR	Penalis.	18.9	5.12
NoDis.	Penalis.	18.9	5.17
MR	Filter	24.5	11.22
NoDis.	Filter	26.4	10.98

Quartile	MAP (%)
Best	30.4
Q3	26.1
Median	23.7
Q1	21.4
Worst	16.1

significantly better. Our best result, NoDis-Filter, occurs in the top quartile of all submitted results. The other filtered results and the baseline occur between the Median and Q3. The penalisation results occur in the lower quartile.

4 Conclusions

We have shown that brute force training of a penalisation value to combine text and geographic data is highly sensitive to over fitting. In fact this resulted in a MAP on the test data statistically significantly worse than the baseline or filter methods. These results concur with other participants [8].

Due to the current immaturity of geographic relevance ranking techniques it is our conclusion that more robust training methods or methods such as filtering are the most appropriate for GIR systems. Alternatively post-hoc query analysis could be performed to assess which queries are causing this dramatic over fitting. Wilkins et al. [13] propose a method of adjusting the weightings given to different features based on the distribution of scores at query time. Such a method may also be appropriate for geographic information retrieval and could help with the search for synergy between textual and geographic evidence.

References

1. Mandl, T., Carvalho, P., Gey, F., Larson, R., Santos, D., Womser-Hacker, C.: GeoCLEF 2008 overview. In: Working Notes from the Cross Language Evaluation Forum (2008)
2. Cardoso, N., Sousa, P., Silva, M.: The University of Lisbon at GeoCLEF 2008. In: Working Notes from the Cross Language Evaluation Forum (2008)
3. Overell, S., Magalhães, J., Rüger, S.: Forostar: A System for GIR. In: Peters, C., Clough, P., Gey, F.C., Karlgren, J., Magnini, B., Oard, D.W., de Rijke, M., Stempfhuber, M. (eds.) CLEF 2006. LNCS, vol. 4730, pp. 930–937. Springer, Heidelberg (2007)
4. Li, Y., Stokes, N., Cavedon, L., Moffat, A.: NICTA I2D2 group at GeoCLEF 2006. In: Working Notes from the Cross Language Evaluation Forum (2006)
5. Buscaldi, D., Rosso, P.: The UPV at GeoCLEF 2008: The GeoWorSE system. In: Working Notes from the Cross Language Evaluation Forum (2008)
6. Ferrés, D., Rodríguez, H.: TALP at GeoCLEF 2007: Using Terrier with geographical knowledge filtering. In: Working Notes from the Cross Language Evaluation Forum (2007)

7. Hauff, C., Trieschnigg, D., Rode, H.: University of Twente at GeoCLEF 2006. In: Working Notes from the Cross Language Evaluation Forum (2006)
8. Cardoso, N., Cruz, D., Chaves, M., Silva, M.: The University of Lisbon at GeoCLEF 2007. In: Working Notes from the Cross Language Evaluation Forum (2007)
9. Martins, B., Silva, M., Andrade, L.: Indexing and ranking in Geo-IR systems. In: CIKM Workshop on GIR, pp. 31–34 (2005)
10. Overell, S., Rae, A., Rüger, S.: MMIS at GeoCLEF 2008: Experiments in GIR. In: Working Notes from the Cross Language Evaluation Forum (2008)
11. Martins, B., Cardoso, N., Chaves, M., Andrade, L., Silva, M.: The University of Lisbon at GeoCLEF 2006. In: Working Notes from the Cross Language Evaluation Forum (2006)
12. Hull, D.: Using statistical testing in the evaluation of retrieval experiments. In: SIGIR, pp. 329–338 (1993)
13. Wilkins, P., Ferguson, P., Smeaton, A.: Using score distributions for querytime fusion in multimedia retrieval. In: SIGMM workshop on MIR (2006)

Query Expansion for Effective Geographic Information Retrieval

Qiang Pu[1], Daqing He[2], and Qi Li[2]

[1] School of Computer Science and Engineering,
University of Electronic Science and Technology of China, Chengdu, China
puqiang@uestc.edu.cn
[2] School of Information Sciences, University of Pittsburgh, Pittsburgh, PA, USA
{dah44,qil14}@pitt.edu

Abstract. We developed two methods for monolingual Geo-CLEF 2008 task. The GCEC method aims to test the effectiveness of our online geographic coordinates extraction and clustering algorithm, and the WIKIGEO method wants to examine the usefulness of using the geographic coordinates information in Wikipedia for identifying geo-locations. We proposed a measure of topic distance to evaluate these two methods. The experiments results show that: 1) our online geographic coordinates extraction and clustering algorithm is useful for the type of locations that do not have clear corresponding coordinates; 2) the expansion based on the geo-locations generated by GCEC is effective in improving geographic retrieval; 3) Wikipedia can help in finding the coordinates for many geo-locations, but its usage for query expansion still needs further study; 4) query expansion based on title only obtained better results than that on the title and narrative parts, even though the latter contains more related geographic information. Further study is needed for this part.

1 Introduction

Along with the rapidly developed Web technologies and services, Web users' queries increasingly contain geographic information. It is, therefore, important for Web search engines to be able to recognize the geographic information, and expand it with more concrete locations if the initial geographic information is inaccurate. This is the motivation that our research team participated in GeoCLEF 2008.

We propose two different methods for extracting geographic location information for query expansion. The first one is Geographic Information Retrieval with Geographic Coordinates Extraction and Clustering (GCEC). Its basic idea is that those locations grouped in the same geographic area with the original geographic location should be treated as the geographic approximations of the original location which can be used for geographic query expansion. The second method is Wikipedia-based Geographic Information Retrieval (WIKIGEO). Geographic locations are mined from Wikipedia - the online encyclopedia which provides abundant types of knowledge. We also assume that a query in our task can be segmented into a topic part, a geo part and the relation part that separates the topic part from the geo part.

In the remainder of the paper, Section 2 presents the two methods we developed. We will propose a concept of topic distance as a measure to evaluate these two methods in

C. Peters et al. (Eds.): CLEF 2008, LNCS 5706, pp. 843–850, 2009.

Section 3. Then we talk about the experiments and analyze in detail the results in Section 4. Section 5 is the conclusions.

2 Two Systems for Geographic Information Retrieval

2.1 GeoIR with Geographic Coordinates Extraction/Clustering

We built a system, called GCEC, by utilizing geographic coordinates and clustering method. This system consists of four main functional modules as shown in Figure 1.

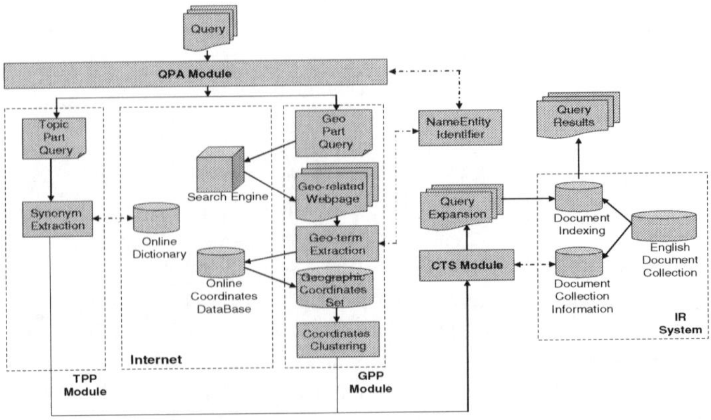

Fig. 1. The architecture of GCEC system

(1) *Query Pattern Analysis (QPA) Module*
We assume that a query contains three distinctive segments: the topic part indicates what a user wants to know; the geo part explains a geographic location the user wants to search for; and the third segment expresses a relationship, which is often a preposition, acting as the boundary that separates the topic part from the geo part. A set of such prepositions can be obtained from [3]. This is similar to the segmentation of query into "what", "relation-type", and "where" part. We acknowledge that this view of queries could be too rigid, but it does help parsing queries. We hope that our remaining modules could compensate for the insufficiency in query parsing.

Some pre-processing steps were performed before query parsing, and a set of heuristic rules were developed for fine tuning the two parts, more details can be found in our working notes paper [7].

(2) *Topic Part Processing (TPP) Module*
This module extracts synonyms for the terms in the topic part from an online dictionary or from the Web. To obtain synonyms for English words, we borrowed the back translation idea from CLIR [1]. By using an English-Chinese dictionary and a Chinese-English dictionary, an English word such as "prison" can use its translation "监狱" as the bridge to bring back the synonymous English words like "jail, jailhouse, Job's pound, penitentiary, quod, pokey" and so on. Of course, it is clear that,

from the example, some noise would be introduced via this method. We will talk about how to select terms from a synonym set for query expansion in CTS module.

(3) *Geo Part Processing (GPP) Module*
This module expands geographic query by utilizing geographic coordinates extraction and clustering. We assume that if those locations are grouped in the same geographic cluster with the original location in the geo part by our clustering algorithm, we can treat them as geographic approximation of the original location, thus we can use them for expansion of the geo part. The concrete process of geographic coordinates extraction and clustering can refer to [7].

We used K-means algorithm to clustering geographic locations, and our clustering algorithm used the geographic coordinates instead of term co-occurrence. The reason is because some geographic locations such as "United States, Germany" often co-occur with unrelated geographic location, such as "former Yugoslavia" [2].

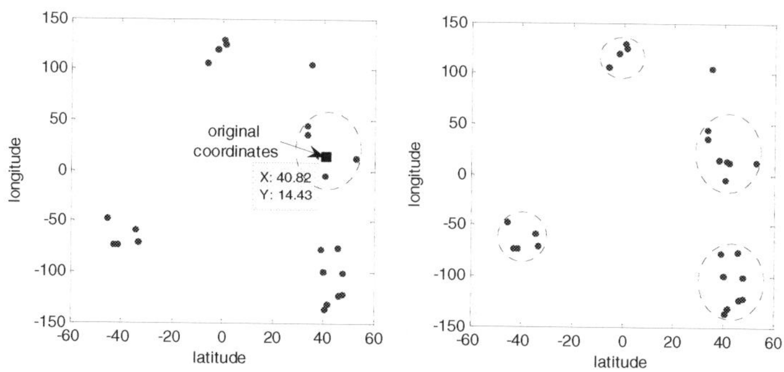

Fig. 2. The left figure illustrates that only one geographic cluster is considered if the location in geo part has definite geographic coordinates as the cluster's center, which is the black rectangle, those points in the circle can be viewed as the geographic approximation of the location in the geo part. The right one has several clusters if the location in geo part has not definite geographic coordinates.

During location clustering, when the location in geo part has concrete geographic coordinates, the generated cluster uses this location as the center (see the left figure of Figure 2). However, not all geographic locations have definite coordinates, which make the geographic cluster center undetermined. Therefore, we grouped those locations into several clusters according to their own geographic coordinates (see the right figure in Figure 2). We selected the location with the shortest distance (d_{min}) to its temporary cluster center (formed by K-means algorithm) as the actual cluster center. If the ratio ($(d_i$-$d_{min})/ d_{min}$) of other location with its distance d_i to the actual center is less than a threshold (set to 10 in this paper), we treat such a location as geographic approximation for the geo part.

(4) *Candidate Term Selection (CTS) Module*
This module uses global collection statistics to filter out noisy candidate terms. Its output can be used for query expansion. Some other researches on the TREC robust

track [5] [6] give us the motivations of using global collection statistics for query expansion. We calculated the collection weight $w_{t,c}$ for a term t from the candidate set based on a variant form of the tf-idf model [4]:

$$w_{t,c} = \left(\frac{tf_{t,c}}{df_t} \right) \times \log\left(1 + \frac{cdf}{df_t} \right) \quad (1)$$

where $tf_{t,c}$ is the number of t occurring in collection c, df_t is the number of document containing t, cdf is the total number of documents in collection. Our experiments showed that retrieval results were improved if terms with weight, $1.6 < w_{t,c} < 3.5$, were chosen as expanded terms.

2.2 Wikipedia-Based Geographic Information Retrieval

The second system is called WIKIGEO. Many pages on Wikipedia contain geographic locations, along with corresponding geographic coordinates, which motivates us to identify and mine geographic locations from Wikipedia.

Our mining method is as follows: for each Wikipedia webpage, we checked if it contains the keyword "Coordinates" with digit number surrounding it. We then extracted potential geographic location with its coordinates. If a webpage has too many contents, we used the first paragraph of the page. Some redundant geographic locations were removed. In the calculation of duplicate locations, we used a heuristic rule saying that if the distance between two locations is less than three kilometers, we treat the two locations as the same place.

Table 1. The evaluation of geographic locations extracted from Wikipedia using 2007 geographic query parsing task. * means that there are mistakes in the ground truth.

Total Query	Correctly Identified	Ambiguous locations	Incorrectly identified	Organization name	Special geo name
500	389	24	73*	11	3

We totally extracted 370,787 geographic locations from Wikipedia, which were used as candidate geographic terms for query expansion. The coverage is much more comprehensive than the world gazetteer which only contains 172,076 locations. We also evaluated the extracted results on geographic query parsing task of 2007. The evaluation results are shown in Table 1. The correct identification ratio 83% ((389+11+3)/500) proves our geographic locations extracted from Wikipedia is useful for building a comprehensive gazetteer for geographic information retrieval.

3 Topic Distance

To help intrinsically examining the quality of the expansion terms, we propose "*topic distance*" as a measure. Given two probability mass functions $p(Q|\theta_{Topic})$ and $p(Q|\theta_{Coll})$, we define a concept of *topic distance* by the Kullback-Leibler divergence between the two probabilities, which is also used as the query clarity in [5]:

$$topic\ dist = \sum_{t \in Q} p(t \mid \theta_{Topic}) \log \frac{p(t \mid \theta_{Topic})}{p(t \mid \theta_{Coll})} \tag{2}$$

where t is a term in query Q, the topic model θ_{Topic} can be estimated by the expansion part. The collection model θ_{Coll} can be estimated by two different collections: one is the whole collection which stands for a random non-topical model, denoted as θ_{Coll_all}, the other is the relevance collection that contains all the true relevant documents for a query topic, denoted as θ_{Coll_rel}. We assume that a good query should be topically closer to the true topic θ_{Coll_rel} and further away from the random topic θ_{Coll_all}.

4 Experimental Results and Analysis

Our experiments were at monolingual English retrieval. The English collection contains documents from Glasgow Herald in 1995 (GH95) and documents from Los Angeles Times in 1994 (LA94). Our retrieval system was Indri.

4.1 Experimental Results from GCEC and WIKIGEO System

Our research questions in the experiments were: 1) is the geo information mined from the Web effective for query expansion in GeoIR? 2) will geographic information manually selected from narrative part as geographic query expansion give better result? 3) is the geo-coordinates information from Wikipedia useful for GeoIR?

Table 2. Comparison of Mean Average Precision (MAP) and R-Precision (RP) between GCEC system and WIKIGEO system. PITTQP1: only title part of a topic was used for automatic query expansion as mentioned in Section 2. PITTQP2: both title part and narrative part were used. Those geographic locations in the narrative part were manually selected. The expansion was still automatic. PITTQI1: the title, description and narrative part were used. PITTQI2: the title and narrative part were used.

Experiments	MAP	RP
PITTQP1 (*Title Only*)	0.2624	0.2805
PITTQP2 (*Title+Narrative*)	0.2623	0.2706
PITTQI1 (*Title+Narrative+Description*)	0.1857	0.1935
PITTQI2 (*Title+Narrative*)	0.1719	0.1799
BASELINE (*Title without expansion*)	0.2091	0.2298

Each method had two submission runs (i.e., PITTQP1 and PITTQP2 for GCEC, PITTQI1 and PITTQI2 for WIKIGEO). As shown in Table 2, the mining on the Web does provide improvement in the retrieval effectiveness, but manually selected geographic information from narrative part hurt both MAP and RP when used in the GCEC system whereas helped in the WIKIGEO system. WIKIGEO performed inferior to GCEC, which indicates that, though we can use Wikipedia to find the coordinates for many geo-locations, its usage for query expansion needs further study.

4.2 Experimental Analysis

Using topic distance we analyzed intrinsically the expansion terms obtained from GCEC and WIKIGEO. This may give us more insights on how different sources for extracting geographic information should be used in query expansion. Table 3 shows the comparison of topic distance between the two methods. The Coll_all and Coll_rel in column 1 indicate the random non-topical model and relevant topic model respectively. We observed that the two GCEC runs have a significant larger distance to Coll_all than the two WIKIGEO runs. That means the query topics of PITTQP1 and PITTQP2 are far away from the random topics. Meanwhile, both two GCEC runs have closer distance to Coll_rel than the two WIKIGEO runs. In the case of PITTQP1 against the two WIKIGEO runs, the difference is significant. This indicates that the query topic of PITTQP1 is significantly closer to the true relevant topics.

Table 3. Comparison of Topic Distance between GCEC and WIKIGEO to Coll_all and Coll_rel. The left values in columns PITTQI1 and PITTQI2 are the topic distance of GCEC system to the two collections, the values in parenthesis are the topic distance of WIKIGEO system to the two collections. Column 4 and 6 are the p value of the Wilcoxon test. The "*" indicates statistically significant difference in the topic distance between GCEC and WIKIGEO with a 95% confidence by the Wilcoxon test.

		PITTQI1	Wilc.	PITTQI2	Wilc.
Coll_all	PITTQP1	17.11 (15.42)	0.0012*	17.11 (15.12)	0.0004*
	PITTQP2	16.88 (15.42)	0.0032*	16.88 (15.12)	0.0011*
Coll_rel	PITTQP1	10.58 (11.58)	0.0037*	10.58 (11.56)	0.0037*
	PITTQP2	10.97 (11.58)	0.1229	10.97 (11.56)	0.1161

When studying the expansion terms from individual topics, we find that it may be restricted in WIKIGEO's usage of the first paragraph of a Wikipedia webpage in extracting geographic information. This simplification may work with concrete geographic locations, but this year's topics contain many broad geographic terms, like "South America". Using the simplification, we only obtained those geographic terms like "America", "Pacific Ocean", or "Atlantic Ocean" for expanding the "South America". Experiments showed that such expansion hurt the retrieval effectiveness.

Table 4 shows the comparison of retrieval performance between PITTQP1 and PITTQP2. We observe that, though there had been a large improvement on the precision of MAP, RP and 11-point average when query expansion was performed on either title only part or title and narrative parts , the improvement was not statistically significant. There is no statistical difference between the results of PITTQP1 and that of PITTQP2 too. It is worth noting that though PITTQP1 got better precisions than BL, its total number of retrieved relevant documents is actually less (597 vs 618) . Comparing to BL, PITTQP2 got better precisions and more retrieved relevant documents (621vs 618).

When plotted into precision-recall curve (see Figure 3), we observe that both PITTQP1 and PITTQP2 consistently outperformed the baseline except PITTQP1 at low recall end.

Table 4. Comparison between PITTQP1 (denoted as T) and PITTQP2 (denoted as TN). BL is the baseline. Column Imp1 indicates the relative improvement of T over BL, Imp2 indicates the relative improvement of TN over BL, Imp3 indicates the relative improvement of TN over T. Columns Wilc. are the p values of the Wilcoxon test.

	BL	T	Imp1	Wilc1.	TN	Imp2	Wilc2.	Imp3	Wilc3.
Rel	747	747			747				
Rret	618	597	-3.4%	0.5694	621	+4.9%	0.6863	+3.5%	0.7031
MAP	0.2091	0.2624	+25.5%	0.6373	0.2623	+25.4%	0.5019	-0.0%	0.9176
RP	0.2298	0.2805	+22.1%	0.5862	0.2706	+17.8%	0.7771	-3.5%	1.0000
11Pt	0.2278	0.2796	+22.7%	0.6682	0.2809	+23.3%	0.3758	+0.5%	0.8767

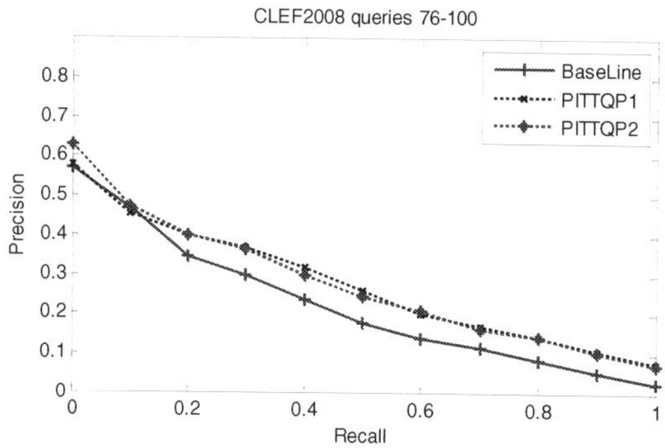

Fig. 3. Comparison of retrieval performance among the Baseline, the PITTQP1 (Title-only) and the PITTQP2 (Title+Narra)

Based on the above result analysis, we have the following observations:

- Without statistically significant differences from the baseline, further study is needed for more effective query expansion. How to correctly and effectively use the geographic information extracted online, how to effectively filter the noise are important questions to be answered.
- Narrative part of topic statements is helpful in query expansion especially at low recall end. Our results show that narrative parts of most queries do provide more geographical information. This is why the total number of retrieved relevant documents is larger than that of using title only for query expansion. But this approach also brings lots of noise so that the final performance is actually worse in MAP and RP. How to effectively integrate narrative part is still a question.
- It should be noted that though GCEC is useful for the type of locations that do not have clear corresponding coordinates, the coordinates does not work all the time. For example, in query 97 "Foreign aid in Sub-Saharan Africa", our calcuation viewed "Barcelona Spain" as reasonable close to "Morocco Africa" so that it used Barcelona for expansion, which hurt the results.

5 Conclusions

In this paper, we presented our participation in GeoCLEF 2008. As first time participant, we only worked on the monolingual GeoCLEF evaluation and submitted four runs under two different methods. Our GCEC method aimed to test the effectiveness of our online geographic coordinates extraction and clustering algorithm, and our WIKIGEO method wanted to examine the usefulness of using the geographic coordinates information from Wikipedia for identifying geo-locations.

Our experiments results show that: 1) our online geographic coordinates extraction and clustering algorithm is useful for the type of locations that do not have clear corresponding coordinates; 2) the expansion based on the geo-locations generated by GCEC is effectiveness in improving geographic retrieval; 3) Using Wikipedia we can find the coordinates for many geo-locations, but its usage for query expansion still needs further study; 4) query expansion based on title only obtained better results than using the combination of title and narrative parts which are thought to contain more related geographic information. Further study is needed for this part too.

Our future work will move in several directions, which includes better method for locating related geo-location information, determining when it is appropriate to perform query expansion, and the parameters for effectiveness of query expansion.

Acknowledgments. This work was partially supported by China Scholarship Council and the University of Pittsburgh.

References

1. He, D., Oard, D.W., Wang, J., Luo, J., Demner-Fushman, D., Darwish, K., Resnik, P., Khudanpur, S., Nossal, M., Subotin, M., Leuski, A.: Making MIRACLEs: Interactive Translingual Search for Cebuano and Hindi. ACM Transactions on Asian Language Information Processing 2(3), 219–244 (2003)
2. Li, Z.S., Wang, C., Xie, X., Ma, W.Y.: MSRA Columbus at GeoCLEF 2006. In: Peters, C., Clough, P., Gey, F.C., Karlgren, J., Magnini, B., Oard, D.W., de Rijke, M., Stempfhuber, M. (eds.) CLEF 2006. LNCS, vol. 4730, pp. 926–929. Springer, Heidelberg (2007)
3. Li, Z.S., Wang, C., Xie, X., Ma, W.Y.: Query Parsing Task for GeoCLEF 2007 Report. In: Working Notes of the Cross Language Evaluation Forum (CLEF) 2007 Workshop, Budapest, Hungary (2007)
4. Salton, G., Wong, A., Yang, C.S.A.: Vector Space Model for Automatic Indexing. Communication of the ACM 18(11), 613–620 (1975)
5. Cronen-Townsend, S., Zhou, Y., Croft, W.B.: Predicting Query Performance. In: Proceedings of the 25th annual international ACM SIGIR conference on Research and development in information retrieval, pp. 299–306. ACM Press, New York (2002)
6. Amati, G., Carpineto, C., Romano, G.: Query Difficulty, Robustness, and Selective Application of Query Expansion. In: McDonald, S., Tait, J.I. (eds.) ECIR 2004. LNCS, vol. 2997, pp. 127–137. Springer, Heidelberg (2004)
7. Pu, Q., He, D., Li, Q.: University of Pittsburgh at GeoCLEF 2008: Towards Effective Geographic Information Retrieval. In: Working Notes of the Cross Language Evaluation Forum (CLEF) 2008 Workshop, Aarhus, Denmark (2008)

Integrating Methods from IR and QA for Geographic Information Retrieval

Johannes Leveling[1] and Sven Hartrumpf[2]

[1] Centre for Next Generation Localisation (CNGL),
Dublin City University, Dublin 9, Ireland
johannes.leveling@computing.dcu.ie
[2] Intelligent Information and Communication Systems (IICS),
University of Hagen, 58084 Hagen, Germany
sven.hartrumpf@fernuni-hagen.de

Abstract. This paper describes the participation of GIRSA at GeoCLEF 2008, the geographic information retrieval task at CLEF. GIRSA combines information retrieval (IR) on geographically annotated data and question answering (QA) employing query decomposition.

For the monolingual German experiments, several parameter settings were varied: using a single index or separate indexes for content and geographic annotation, using complex term weighting, adding location names from the topic narrative, and merging results from IR and QA, which yields the highest mean average precision (0.2608 MAP).

For bilingual experiments, English and Portuguese topics were translated via the web services Applied Language Solutions, Google Translate, and Promt Online Translator. For both source languages, Google Translate seems to return the best translations. For English (Portuguese) topics, 60.2% (80.0%) of the maximum MAP for monolingual German experiments, or 0.1571 MAP (0.2085 MAP), is achieved.

As a post-official experiment, translations of English topics were analysed with a parser. The results were employed to select the best translation for topic titles and descriptions. The corresponding retrieval experiment achieved 69.7% of the MAP of the best monolingual experiment.

1 Introduction

GeoCLEF is the geographic information retrieval (GIR) task at CLEF. In recent years, we have developed GIRSA (GIR by Semantic Annotation), a system for exploring novel approaches at GIR. GIRSA supports methods to improve precision (e.g. annotation of metonymic uses of location names [1]) and methods to improve recall (e.g. normalisation of location names [2] and decompounding). For GeoCLEF 2008, the major improvement lies in the combination of results from information retrieval (IR) and question answering (QA).

2 System Description and Experimental Setup

GIRSA is a system for the evaluation of novel indexing and retrieval methods for GIR. Basically, the GIRSA setup introduced at GeoCLEF 2007 was used

C. Peters et al. (Eds.): CLEF 2008, LNCS 5706, pp. 851–854, 2009.

Table 1. Selected results for retrieval experiments on German GeoCLEF documents

Run	Parameters					Results				
	lang.	transl.	fields	index	comb.	MAP	rel_ret	P@5	P@10	P@20
FUHtd01	DE	-	TD	A	N	0.2420	977	0.39	0.37	0.31
FUHtd01m	DE	-	TD	A	Y	0.2608	1028	0.38	0.37	0.35
FUHtd20	DE	-	TD	B	N	0.1719	914	0.20	0.29	0.27
FUHtd20m	DE	-	TD	B	Y	0.2211	998	0.36	0.35	0.34
FUHtdn20	DE	-	TDN	B	N	0.1478	834	0.17	0.24	0.20
FUHENAtd20	EN	A	TD	B	N	0.1076	644	0.18	0.17	0.17
FUHENAtdn20	EN	A	TDN	B	N	0.0962	610	0.14	0.15	0.13
FUHENGtdn20	EN	G	TDN	B	N	0.1571	800	0.21	0.21	0.21
FUHENOtd20	EN	O	TD	B	N	0.1179	703	0.23	0.23	0.21
FUHENOtdn20	EN	O	TDN	B	N	0.1146	699	0.21	0.21	0.19
FUHENVtd20	EN	V	TD	B	N	0.1817	808	0.32	0.32	0.29
FUHENVtd20m	EN	V	TD	B	Y	0.1857	877	0.33	0.31	0.29

for GeoCLEF 2008, too. This setup involves the identification and normalisation of location indicators, i.e. text segments from which a geographic scope can be inferred. Location adjectives, names for inhabitants of a place, geographic codes, orthographic variants, acronyms, and abbreviations are mapped to location names. For details on the system's improvements, see [3].

GIRSA was employed to produce results for a number of monolingual and bilingual experiments. The following parameter settings were varied in different retrieval experiments (see Table 1): the *topic source language* (lang.): German (DE) or English (EN) serves as topic source language; the *translation service* (transl.): Applied Language Solutions (A, http://www.appliedlanguage.com/free_translation.shtml), Google Translate (G, http://translate.google.com/), or Promt Online Translator (O, http://www.online-translator.com/), and – in post-official experiments – a combination of translations (V); the *content fields*: content words and location indicators are extracted from the topic title and description: with location names from the topic narrative (TDN) or without (TD); the *index type*: all words are stemmed and a single index is produced (A), content words are decompounded and stemmed, location names are identified, both are indexed separately (B); the *combination* (comb.): results from IR and QA are combined (Y) or not (N). Results are merged and the top 1000 documents are returned. Five metrics are employed to measure retrieval performance: mean average precision (MAP), the number of relevant and retrieved documents (rel_ret), and precision at N documents (P@N).

3 Results and Discussion

The following four hypotheses were formulated before the experiments and investigated after the experiments as follows.

Experiments using additional location names from the narrative part of the topics will achieve a higher MAP than experiments that do not (to confirm results from GeoCLEF 2007). This turned out to be false. The MAP for experiments with additional location names from the topic narrative is lower than for the experiments using title and description only (e.g. FUHtd20 vs. FUHtdn20).

The MAP for experiments adding results from the QA subsystem will be somewhat higher than for experiments with pure GIR. This is also not true: performance is considerably higher due to the improvements (query decomposition, less strict matching) in InSicht, the QA subsystem. The MAP for merged runs is higher in all cases. FUHtd01m shows a relative improvement of 7.8% in MAP compared to FUHtd01, FUHtd20m shows an improvement of 28.6% compared to FUHtd20; also, more relevant documents are retrieved in both cases. InSicht found documents for 13 (of the 25) topics, which is much better than last year. These results alone are not sufficient for GIR, but due to their high complementarity merging these results improves GIRSA significantly.

Topic translations with the Promt Online Translator web service will be better (e.g. containing less untranslated words) than those from the other web services tested. The corresponding results will therefore have a higher MAP. The MAP for the best bilingual English-German experiment is 0.1571 (about 60.2% of the best MAP for monolingual German); the MAP for the best bilingual Portuguese-German experiment is 0.2085 (about 80.0% compared to monolingual German). The experiments with topics translated by Google Translate returned the best results. Promt offers a web service (in beta status) different from previous years, which may be a reason why topics could not be translated well enough.

Applying the weighting from QA (for all experiments), merging results from IR and QA, and combining indexes for location names and content words will result in a higher initial MAP. In comparison, the initial MAP was quite high: GIRSA returned 69% MAP at 0% recall for monolingual German experiments (experiment FUHtd01m), other participants achieved 43% and 16%, respectively (see [4]).

A result analysis for the QA subsystem InSicht showed that query decomposition was vital: With decomposition, 1238 documents (232 assessed as relevant) were retrieved; only 125 documents (77 assessed as relevant) without decomposition. InSicht's orientation towards precision was confirmed because if documents were retrieved for a topic, also relevant documents were retrieved.

To find the causes of low performance for the bilingual experiments, we analysed the topic titles and descriptions translated by the MT web services into German. The topics show many types of errors: ending with a wrong translation (using a different word sense, e.g. *schießen*/*'shoot'* instead of *Feuer*/*'fire'* in topic GC88), using uncommon translations, using a wrong preposition, generating a translation with incongruence between words, using a wrong verb position, and untranslated words. Except for getting wrong prepositions, these errors do not seem to ultimately have much impact on the performance of a GIR system. Prepositions will become important in a GIR system which is capable of interpreting the prepositions as geographic semantic relations.

Analysing the translations per web service used, the following errors were observed for translations from English: Applied Language Solutions returns untranslated words or completely untranslated title and description fields for 4 topics. The translations also include uncommon words in 3 topics and wrong translations in 3 topics. Google Translate returns two untranslated words only, *'resons'* in GC99 and *'Douments'* in GC100, both spelling errors. The Promt Online Translator returns uncommon translations for 3 topics and wrong translations for 6 topics. The Promt translator added translation alternatives in brackets, which might have caused a topic shift if translations with different senses were added. The performance of these machine translation web services is reflected in the performance results for bilingual experiments: translations with Google Translate show the least number of errors and the corresponding experiments return the best performance.

As the three translators presented quite diverse translation mistakes, a virtual translator was implemented (after the official experiments) that picks one of the translations for a given sentence t using the scoring function q:

$$q(t) := w_1 \cdot \text{parse\_quality}(t) - w_2 \cdot |\text{unknown\_words}(t)| \text{ with } w_1 = 1.0 \text{ and } w_2 = 0.1$$

The parse quality is a real number between 0 and 1 obtained from analysing the topics with InSicht's syntactico-semantic parser. The virtual English translator returned an acceptable translation for 92% of the topic titles and for 76% of the topic descriptions. Selection with the virtual translator gives much better results than using translations from the best single translator and allows better retrieval results: e.g., InSicht lost only 11 relevant documents compared to the monolingual run. GIRSA achieved 0.1817 MAP and 0.1857 MAP and a much higher initial precision when using the virtual translator (see FUHENVtd20 and FUHENVtd20m in Table 1).

Future work will continue in the field of integrating methods from information retrieval and question answering for geographic information retrieval, evaluating GIRSA in the GikiCLEF task planned for CLEF 2009.

References

1. Leveling, J., Hartrumpf, S.: On metonymy recognition for geographic information retrieval. IJGIS 22(3), 289–299 (2008)
2. Leveling, J., Hartrumpf, S.: Inferring location names for geographic information retrieval. In: Peters, C., Jijkoun, V., Mandl, T., Müller, H., Oard, D.W., Peñas, A., Petras, V., Santos, D. (eds.) CLEF 2007. LNCS, vol. 5152, pp. 773–780. Springer, Heidelberg (2008)
3. Leveling, J., Hartrumpf, S.: University of Hagen at GeoCLEF 2008: Combining IR and QA for geographic information retrieval. In: Results of the CLEF 2008 Cross-Language System Evaluation Campaign, Working Notes for the CLEF 2008 Workshop, Aarhus, Denmark (2008)
4. Mandl, T., Carvalho, P., Gey, F., Larson, R., Santos, D., Womser-Hacker, C.: GeoCLEF 2008: the CLEF 2008 cross-language geographic information retrieval track overview. In: Peters, C., et al. (eds.) CLEF 2008. LNCS, vol. 5706, pp. 808–821. Springer, Heidelberg (2008)

Using Query Reformulation and Keywords in the Geographic Information Retrieval Task

José Manuel Perea-Ortega, L. Alfonso Ureña-López, Manuel García-Vega, and Miguel Angel García-Cumbreras

SINAI Research Group*, Computer Science Department, University of Jaén, Spain
{jmperea,laurena,mgarcia,magc}@ujaen.es

Abstract. This paper describes the use of query reformulation to improve the Geographic Information Retrieval (GIR) task. This technique also includes the geographic expansion of the topics. Moreover, several experiments related to the use of keywords and hyponyms in the filtering process are performed. We also use a new approach in the re-ranking process based on the original position of each document in the ranking. The results obtained show that our query reformulation sometimes retrieves valid documents that the default query is not able to find, but on average it does not improve the baseline case. The best result is obtained considering the geographic entities in the traditional retrieval process.

1 Introduction

In this paper we describe our system to resolve the Geographic Information Retrieval task. GeoCLEF is a cross-language GIR task, whose aim is to evaluate GIR systems. It belongs to the Cross-Language Evaluation Forum[1] (CLEF) campaign since 2005. GIR is concerned with improving the quality of geographically specific information retrieval with a focus on access to unstructured documents [4].

For GeoCLEF 2006 [3], we studied the behavior of query expansion using a gazetteer and a thesaurus. Those experiments showed us that the method we used to make the query expansion was not very good. For GeoCLEF 2007 [6], we changed the approach and we applied a filtering process to the documents retrieved by the IR subsystem without using any query expansion. The results for this approach were better than those achieved in 2006 using query expansion.

In the new system, we have added some Natural Language Processing (NLP) techniques such as query reformulation, keywords and hyponyms extraction and even query geo-expansion. In the next section we describe the general architecture. In Section 3 we present the experiments and results and finally we expound the conclusions and future work.

2 SINAI-GIR System Overview

As we can see in Figure 1, our GIR system is made up of five main subsystems: *Translator*, *Collection Preprocessing*, *Query Analyzer*, *Information Retrieval* and

* http://sinai.ujaen.es
[1] http://www.clef-campaign.org/

C. Peters et al. (Eds.): CLEF 2008, LNCS 5706, pp. 855–862, 2009.

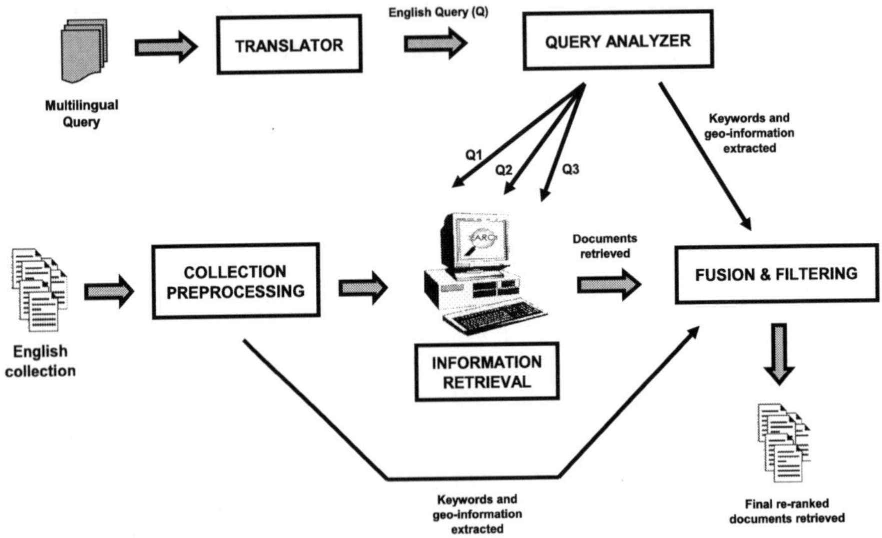

Fig. 1. SINAI-GIR system architecture

Fusion and Filtering. We make use of the *GeoNames*[2] gazetteer as geographic knowledge base for the whole system. As IR index-search engine we have used Lemur[3].

Each translated query is preprocessed and analyzed by the *Query Analyzer*, identifying their geo-entities and spatial relations. This module also applies query reformulation, generating several independent queries which will be run against the IR subsystem. On the other hand, the collection is preprocessed by the *Collection Preprocessing* module and finally the documents recovered by the IR subsystem are filtered and re-ranked by means of the *Fusion and Filtering* subsystem.

2.1 Translator

As translation module, we have used SINTRAM (SINai TRAnslation Module), our Machine Translation system which works with different online machine translators and implements several heuristics to combine different translations [2]. This module translates the topics from other languages into English.

2.2 Collection Preprocessing Subsystem

In our architecture we only work with an English document collection[4] and we have applied an off-line preprocess to it, using the Porter stemmer [7], the Brill

[2] http://www.geonames.org/

[3] http://www.lemurproject.org/

[4] The collection consists of 169,477 documents composed of stories from the British newspaper *Glasgow Herald* (1995) and the American newspaper *Los Angeles Times* (1994).

POS tagger [1] and LingPipe[5] as Named Entity Recognizer (NER). We also remove the meaningless terms using the predefined list of English *stop-words* from the SMART System [8]. This list contains 571 words.

During the collection preprocessing, two indexes are generated:

- The **locations index**. It stores all location entities detected and recognized by the NER in each document of the collection. All entities typified as *LOC* (location) by the NER are checked using the GeoNames gazetteer.
- The **keywords index**. It stores the *keywords* of each document (nouns that appear more than once in the document, because they have more meaning than verbs or adjectives).

2.3 Query Analyzer

This module preprocesses the English topics and it generates different query reformulations. It is one of the most important modules in our architecture because its aim is to perform a thorough analysis of the query. In the field of GIR, a query can be characterized as a triplet of <theme><spatial_relationship><location> composed of a topic of interest in combination with a place name qualified by a spatial preposition [4].

The *Query Analyzer* is made up of several components:

Preprocessing module. The guidance information and irrelevant descriptions from the topics are removed. We have analyzed manually all the GeoCLEF topics from 2005 to 2008, recognizing these unnecessary phrases. This module simply removes them from the topics. Some examples of unnecessary phrases are: "*Relevant documents contain information about*", "*Find documents describing*", "*To be relevant, documents must describe*", etc. This component also preprocesses the terms of the topics (the Porter stemmer is used and the *stop-words* are removed).

NER module. This module recognizes the locations in the topics. As in the *Collection Preprocessing* subsystem, we have used LingPipe with its default training model. All locations detected are also verified using the GeoNames gazetteer.

Geo-Relation Finder module. This module is used to find the spatial relations in the topics. It is based on manual rules and the entities detected by the NER module. Some examples of spatial relations that can be detected are: *in, of, near, north of, next to, in or around, in the west of*, etc.

Query Reformulation module. This module parses only the *title* of the query, detecting the three components: "*what*", "*geo-relation*" and "*where*". Before the query parsing, we apply a particular *translation* of sentences like "*capital of <entity>*" or "*<entity> 's capital*", replacing the entire sentence by the corrected location using the GeoNames gazetteer. For instance, the sentence "*the capital of France*" would be replaced by the word "*Paris*". This module also generates three types of query reformulation:

[5] http://alias-i.com/lingpipe/

- Q_1: it is formed by the preprocessed content of the topic labels, depending on the experiment. Some experiments consider only the content of the *title* and *description* from the topics. Others consider the content of all labels (*title*, *description* and *narrative*).
- Q_2: it is formed only by the concatenation of "*what*" and "*where*" components detected in the *title* of the topic.
- Q_3: this is the query expansion using geographic terms. It is formed by the concatenation of "*what*" and "*where*" components, in addition to the expanded locations using the "*where*" component. For instance, if the title of the query is "*Riots in South American prisons*", the Q_3 query reformulation would be "*prisons riots South America Brazil Colombia Argentina*" (we expand the query with the three most important countries in South America).

Keywords-Hyponyms extractor module. The aim of this module is to detect the *keywords* only in the *title* of the topics. We have considered as *keywords* only the nouns (detected by the Brill POS tagger) because they have more importance than verbs or adjectives in the re-ranking process. In addition, for each keyword recognized, this module extracts its hyponyms using WordNet[6]. These extracted hyponyms will also be used later in the re-ranking process by the *Fusion and Filtering* subsystem for some experiments.

2.4 Information Retrieval Subsystem

The index-search engine used in the experiments is Lemur. It supports several weighting functions such as *Okapi*, *TF·IDF* and the use of *Pseudo-Relevant Feedback* (PRF).

This module retrieves the most relevant documents for each query reformulation (Q_1, Q_2 and Q_3) based on the *Okapi with PRF* weighting schema.

2.5 Fusion and Filtering Subsystem

The *Fusion and Filtering* process decides which of the retrieved documents by the IR subsystem are valid. The process will consider a document as valid when their geo-references are related to the "*where*" component and the spatial relationship detected in the query. On the other hand, this subsystem also determines the final ranking of documents, based on manual rules and the initial position of the document in the ranking.

We have used a particular approach to make the **fusion** of the lists of documents returned for each query reformulation. We have considered as baseline the list of documents recovered for the Q_1 query and we have added the documents of Q_2 and Q_3 that are not on the Q_1 list, using a normalized value of the Retrieval Status Value (RSV).

In order to **filter** each document recovered, this subsystem applies different manual rules, making use of geographic data detected in the topic. These manual

[6] http://wordnet.princeton.edu/

rules have been developed based on the observation of documents and topics following a simple geographic reasoning. Some examples of these manual rules are:

- If the entity of the topic is a *country* and its *geo-relation* associated is "*north of*", this subsystem will accept the document if it has any location whose latitude is greater than the mid-latitude of that country. This is a rule not very accurate to limit the northern part of a region.
- If the entity of the topic is a *city* and its *geo-relation* associated is "*near to*", the module will consider that a location is *near to* another one when it is at a geographic distance of less than 50 kilometers. We have tried several thresholds in the experiments and the distance of 50 kilometers provided us the best results. In order to measure the geographic distance between two locations we have used the *Great-Circle formula*:

$$D = arcos((\sin a)(\sin b) + (\cos a)(\cos b)(\cos P))$$

  where:
  a is the latitude of point A
  b is the latitude of point B
  P is the longitudinal difference between points A and B
- If the entity of the topic is a *continent* or a *country* and its *geo-relation* associated is "*in*", "*of*", "*at*", "*on*", "*from*" or "*along*", the module will accept the document recovered if a location exists in the document that belongs to that continent or country.

In the **re-ranking** process, this subsystem calculates the new RSV of the valid documents using a formula based on the score obtained by the document in the old list:

$$RSV_{new} = \log(RSV_{old} + 1)$$

3 Experiments and Results

We have used the GeoCLEF 2008 framework [5] for the experiments described in this paper. Topics are textual descriptions with three fields: *title* (T), *description* (D) and *narrative* (N). In some experiments we have used the text contained in all labels (TDN). In others, we have only used TD labels. For all experiments we have used the *Okapi with PRF* as weighting function with the following parameters:

- Okapi: $b = 0.75$, $k1 = 1.2$, $k3 = 7$
- PRF: $DocCount = 5$, $TermCount = 20$

We have considered two baseline cases: removing the geo-entities from the pre-processed query and without removing them (Q_1). In both experiments, we do not apply any filtering or re-ranking process. We have also retrieved the query reformulation (Q_2) and the query expansion (Q_3) for each topic without applying

Table 1. Experiments without applying any filtering and re-ranking process

ID Exp	Topic Labels	Description	MAP
TD	TD	Without removing geo-entities (Q_1)	**0.2841**
TD-GeoEnt	TD	Removing geo-entities	0.2362
TDN	TDN	Without removing geo-entities	0.2258
TDN-GeoEnt	TDN	Removing geo-entities	0.2353
QR	T	Query reformulation (Q_2)	0.2831
QRGE	T	Query geo-expansion (Q_3)	0.1346
FusionTD	TD	Fusion list (Q_1-Q_2-Q_3) with normalized RSV	0.2838
FusionTDN	TDN	Fusion list (Q_1-Q_2-Q_3) with normalized RSV	0.2250

Table 2. Experiments applying the filtering and re-ranking process

Original ID Exp	Filtering and Re-ranking process	MAP
TD	Without using keywords or hyponyms	0.2746
TD-GeoEnt	Without using keywords or hyponyms	0.2470
TDN	Without using keywords or hyponyms	0.2119
FusionTDN	Without using keywords or hyponyms	0.1960
TD	Using only keywords	**0.2790**
TDN	Using only keywords	0.2260
TDN	Using keywords and hyponyms	0.2221
FusionTDN	Using only keywords	0.2122

any filtering or re-ranking process either. Moreover, we have tried to evaluate the fusion of three query reformulations, following different approaches. The results of these experiments are shown in Table 1.

Other experiments described in this paper have been performed by applying the filtering and re-ranking process to the list returned for each experiment before. In these experiments we have combined the use of keywords and hyponyms in the re-ranking process. For that case, our approach has been to extract the nouns of the topics and to obtain their hyponyms from WordNet. In order to raise a document in the final list, we searched if the keyword or the hyponym appeared in each document retrieved by the IR subsystem. In Table 2 are shown these experiments applying the filtering and re-ranking process.

4 Conclusions and Future Work

In this paper we have presented some approaches to resolve the GIR task. We have tried some NLP techniques such as query reformulation and geographic term expansion. In relation to the filtering and re-ranking process of the documents recovered, we have tried the use of keywords and hyponyms.

Analyzing the results, the best result is obtained considering geo-entities in the topics and without applying any filtering or re-ranking process to the list of documents recovered by the IR subsystem. The main reason is that we have not used an optimal method to raise valid documents in the final ranking. However, if we do not consider geo-entities in the topics for the text retrieval, the use of the filtering and re-ranking process improves the results.

In relation to the use of *keywords* in the re-ranking process, it seems to improve slightly the filtering results for some experiments. Instead, the use of *hyponyms* does not improve the results. This is because we have tried to find each keyword (or hyponym) from the topics, into each document retrieved by the IR subsystem through a simple matching. It is difficult that appears the keyword or hyponyms exactly in the document. Nevertheless, the proper use of *keywords* in the re-ranking process could be interesting for the future.

On the other hand, the type of query reformulations we have used in the experiments does not seem to work well, although in some topics the Q_2 and Q_3 query types add valid documents to the final list which have not been found using the default query (Q_1). The main reason to explain the low results obtained with query geo-expansion is that we expand the topics with all geo-entities related to the "*where*" component, for the same query reformulation, so it is introducing a lot of noise in the retrieval process. For the future, we will generate a query geo-expansion for each geo-term related to the "*where*" component detected in the topic and we will retrieve them separately. In addition, we have tried to improve the query reformulation and the geographic expansion analyzing when they should be expanded and how.

Acknowledgments

This work has been supported by a grant from the Spanish Government, project TIMOM (TIN2006-15265-C06-03), and the RFC/PP2006/Id_514 granted by the University of Jaén (Spain).

References

1. Brill, E.: A simple rule-based part-of-speech tagger. In: Proceedings of the third Conference on Applied Natural Language Processing (ANLP 1992), Trento, Italy, pp. 152–155 (1992)
2. García-Cumbreras, M.A., Ureña-López, L.A., Martínez-Santiago, F., Perea-Ortega, J.M.: BRUJA System. In: The University of Jaén at the Spanish task of QA@CLEF 2006. LNCS, vol. 4730, pp. 328–338. Springer, Heidelberg (2007)
3. García-Vega, M., García-Cumbreras, M.A., Ureña-López, L.A., Perea-Ortega, J.M.: GEOUJA System. In: The first participation of the University of Jaén at GEOCLEF 2006. LNCS, vol. 4730, pp. 913–917. Springer, Heidelberg (2007)
4. Jones, C.B., Purves, R.S.: Geographical Information Retrieval. International Journal of Geographical Information Science 22, 1365–8816, 219–228 (2008)
5. Mandl, T., Carvalho, P., Gey, F., Larson, R., Santos, D., Womser-Hacker, C.: GeoCLEF 2008: the CLEF 2008 Cross-Language Geographic Information Retrieval Track Overview. In: Peters, C., et al. (eds.) CLEF 2008. LNCS, vol. 5706, pp. 808–821. Springer, Heidelberg (2009)

6. Perea-Ortega, J.M., García-Cumbreras, M.A., García-Vega, M., Ureña-López, L.A.: Filtering for Improving the Geographic Information Search. In: Peters, C., Jijkoun, V., Mandl, T., Müller, H., Oard, D.W., Peñas, A., Petras, V., Santos, D. (eds.) CLEF 2007. LNCS, vol. 5152, pp. 823–829. Springer, Heidelberg (2008)
7. Porter, M.F.: An algorithm for suffix stripping. Program 14, 130–137 (1980)
8. Salton, G.: The SMART Retrieval System: Experiments in Automatic Document Processing. Prentice-Hall, Englewood Cliffs (1971)

Using GeoWordNet for Geographical Information Retrieval*

Davide Buscaldi and Paolo Rosso

Natural Language Engineering Lab., ELiRF Research Group,
Dpto. de Sistemas Informáticos y Computación (DSIC),
Universidad Politécnica de Valencia, Spain
{dbuscaldi,prosso}@dsic.upv.es

Abstract. We present a method that uses GeoWordNet for Geographical Information Retrieval. During the indexing phase, all places are disambiguated and assigned their coordinates on the world map. Documents are first searched for by means of a term-based search method, and then re-ranked according to the geographical information. The results show that map-based re-ranking allows to improve the results obtained by the base system, which relies only on textual information.

1 Introduction

One of the main issues in Geographical Information Retrieval (GIR) consists in finding the perfect balance between the thematic part and the geographical part in queries [1,2]. Currently available GIR systems are not able to perform significantly better than standard keyword-based IR systems. In our past participations at GeoCLEF we attempted to integrate geographical knowledge at keyword level in the Lucene[1] search engine, focusing on the use of the WordNet [3] ontology for both query reformulation and index term expansion.

Ferres and Rodríguez [4] obtained good results at GeoCLEF 2007 by combining textual retrieval with map-based filtering and ranking. This kind of integration between geographical knowledge and term-based ranking was previously introduced by [5] in 2006, but it did not demonstrate useful. However, we attempted to introduce a similar feature in our system. The main obstacle was determined by the fact that we use WordNet, which did not provide us with geographical coordinates for toponyms. Therefore, we first had to develop GeoWordNet[2], a georeferenced version of WordNet [6]. By combining this resource with the WordNet-based toponym disambiguation algorithm presented in [7], we were able to assign to the place names in the collection their actual geographical coordinates and to perform some geographical reasoning. We named the resulting system GeoWorSE (an acronym for *Geographical Wordnet Search Engine*). This is the first time that GeoWordNet is used for IR.

* We would like to thank the TIN2006-15265-C06-04 research project for partially supporting this work.

1 http://lucene.apache.org/

2 http://www.dsic.upv.es/grupos/nle/resources/geo-wn/download.html

C. Peters et al. (Eds.): CLEF 2008, LNCS 5706, pp. 863–866, 2009.

2 The GeoWorSE GIR System

During the indexing phase the documents are examined in order to find location names (*toponyms*) by means of the Stanford NER system [8]. When a toponym is found, the disambiguator determines the correct reference for the toponym. Then, the system adds the toponym coordinates (retrieved from GeoWordNet) to the *geo* index and stores in the *wn* index the toponym together with its holonyms and synonyms. All document terms are stored in the *text* index.

The topic text is split into "content" terms, which are searched in the *text* index, and the "geo" part, constituted by toponyms extracted by the Stanford NER. The "geo" terms are searched for in the *wn* index with a weight 0.25 with respect to the content terms. The result of the search is a list of documents ranked using Lucene's weighting scheme. At the same time, the toponyms are analyzed in order to find a geographical constraint that can be of the following two types:

- a *distance* constraint, corresponding to a point in the map: documents that contain locations closer to this point will be ranked higher;
- an *area* constraint, corresponding to a polygon in the map: documents that contain locations included in the polygon will be ranked higher;

The nature of the constraint is determined automatically, on the basis of the data contained in GeoWordNet (that is, whether the toponym can be expanded to an area by means of its meronyms or not). For instance, topic 10.2452/58 – *GC* contains a distance constraint: "Travel problems at major airports near to *London*". Topic 10.2452/76 – *GC* contains an area constraint: "Riots in *South American* prisons". The GeoAnalyzer expands *South America* to its meronyms: *Argentina, Bolivia, Brazil, Chile, Colombia, Ecuador, Guyana, Paraguay, Peru, Uruguay, Venezuela.* The area is obtained by calculating the convex hull of the points associated to the meronyms using the Graham algorithm [9].

If the constraint extracted from the topic is a *distance* constraint, the weights of the documents are modified according to the following formula:

$$w(doc) = w_{Lucene}(doc) * (1 + \exp(-\min_{p \in P} d(q,p))) \tag{1}$$

Where w_{Lucene} is the weight returned by Lucene for the document *doc*, P is the set of points in the document, and q is the point extracted from the topic.

If the constraint extracted from the topic is an *area* constraint, the weights of the documents are modified according to Formula 2:

$$w(doc) = w_{Lucene}(doc) * \left(1 + \frac{|P_q|}{|P|}\right) \tag{2}$$

where P_q is the set of points in the document that are contained in the area extracted from the topic.

3 Experiments

We compared the results obtained with the system using three configurations:

- The Lucene system, without WordNet expansion neither the map-based reranking (label: Luc)
- The system with WordNet expansion but without the map-based reranking (label: L+WN)
- The system with WordNet expansion and map-based reranking (label: GWN)

The results were calculated over all the topics of the GeoCLEF since 2005.

In Table 1 we show the obtained results.

Table 1. Mean Average Precision (MAP) and R-Precision obtained for all topics, using TD (topic and description) and TDN (topic, description and narrative) fields

		TD		TDN	
system	year	MAP	R-Prec	MAP	R-Prec
Luc	2005	0.311	0.340	0.321	0.333
	2006	**0.251**	**0.242**	**0.274**	**0.265**
	2007	0.228	0.245	0.249	0.268
	2008	0.224	0.248	0.210	0.223
average		0.253	0.269	0.263	0.272
L+WN	2005	**0.328**	**0.362**	0.324	0.339
	2006	0.245	0.236	0.261	0.252
	2007	0.242	**0.252**	**0.264**	**0.272**
	2008	**0.269**	**0.277**	**0.216**	**0.226**
average		**0.271**	**0.282**	**0.266**	0.272
GWN	2005	0.320	0.352	**0.326**	**0.347**
	2006	0.247	0.239	0.263	0.261
	2007	0.242	0.247	0.253	0.263
	2008	0.264	0.267	0.204	0.211
average		0.268	0.276	0.262	0.271

The results show that there is no significant difference between the use of the map-based re-ranking and the use of the WordNet-enhanced method. We believe that there are two reasons for this behaviour: the first one is the presence of errors in toponym disambiguation, the second one the fact that in the re-ranking phase the rank of documents is not taken into account. In both cases further work is needed in order to estimate how these features may affect the results.

4 Conclusions and Further Work

We introduced a map-based filtering method in our WordNet-based GIR system. The obtained results do not show any significant improvement over the previous method. We will carry out a study of the weights and the formulae that are used to re-rank documents. As a future work, we would like to implement a dynamical ranking scheme, such as the one proposed by [10], based on *geographic specificity*.

References

1. Kornai, A.: Evaluating Geographic Information Retrieval. In: Peters, C., Gey, F.C., Gonzalo, J., Müller, H., Jones, G.J.F., Kluck, M., Magnini, B., de Rijke, M., Giampiccolo, D. (eds.) CLEF 2005. LNCS, vol. 4022, pp. 928–938. Springer, Heidelberg (2006)
2. Buscaldi, D., Rosso, P.: On the Relative Importance of Toponyms in GeoCLEF. In: Peters, C., Jijkoun, V., Mandl, T., Müller, H., Oard, D.W., Peñas, A., Petras, V., Santos, D. (eds.) CLEF 2007. LNCS, vol. 5152, pp. 815–822. Springer, Heidelberg (2008)
3. Miller, G.A.: WordNet: A Lexical Database for English. Communications of the ACM 38, 39–41 (1995)
4. Ferrés, D., Rodríguez, H.: TALP at GeoCLEF 2007: Results of a Geographical Knowledge Filtering Approach with Terrier. In: Peters, C., Jijkoun, V., Mandl, T., Müller, H., Oard, D.W., Peñas, A., Petras, V., Santos, D. (eds.) CLEF 2007. LNCS, vol. 5152, pp. 830–833. Springer, Heidelberg (2008)
5. Martins, B., Cardoso, N., Silveira Chaves, M., Andrade, L., Silva, M.J.: The University of Lisbon at GeoCLEF 2006. In: Peters, C., Clough, P., Gey, F.C., Karlgren, J., Magnini, B., Oard, D.W., de Rijke, M., Stempfhuber, M. (eds.) CLEF 2006. LNCS, vol. 4730, pp. 986–994. Springer, Heidelberg (2007)
6. Buscaldi, D., Rosso, P.: Geo-WordNet: Automatic Georeferencing of WordNet. In: Proc. 5th Int. Conf. on Language Resources and Evaluation, LREC-2008, Marrakech, Morocco (2008)
7. Buscaldi, D., Rosso, P.: A Conceptual Density-based Approach for the Disambiguation of Toponyms. International Journal of Geographical Information Systems 22(3), 301–313 (2008)
8. Finkel, J.R., Grenager, T., Manning, C.: Incorporating Non-local Information into Information Extraction Systems by Gibbs Sampling. In: Proceedings of the 43rd Annual Meeting of the Association for Computational Linguistics (ACL 2005), U. of Michigan - Ann Arbor, ACL, pp. 363–370 (2005)
9. Graham, R.L.: An Efficient Algorithm for Determining the Convex Hull of a Finite Planar Set. Information Processing Letters 1(4), 132–133 (1972)
10. Yu, B., Cai, G.: A Query-aware Document Ranking Method for Geographic Information Retrieval. In: Purves, R., Jones, C. (eds.) Proceedings of the 4th ACM Workshop On Geographic Information Retrieval, GIR 2007, Lisbon, Portugal, November 9, pp. 49–54. ACM, New York (2007)

GeoTextMESS: Result Fusion with Fuzzy Borda Ranking in Geographical Information Retrieval

Davide Buscaldi[1], José Manuel Perea Ortega[2], Paolo Rosso[1], L. Alfonso Ureña López[2], Daniel Ferrés[3], and Horacio Rodríguez[3]

[1] Natural Language Engineering Lab,
ELiRF Research Group,
Dpto. de Sistemas Informáticos y Computación,
Universidad Politécnica de Valencia
{dbuscaldi,prosso}@dsic.upv.es
[2] SINAI Research Group,
Computer Science Department,
Universidad de Jaén
{jmperea,laurena}@ujaen.es
[3] TALP Research Center,
Universitat Politècnica de Catalunya
{dferres,horacio}@lsi.upc.edu

Abstract. In this paper we discuss the integration of different GIR systems by means of a fuzzy Borda method for result fusion. Two of the systems, the one by the Universidad Politécnica de Valencia and the one of the Universidad of Jaén participated to the GeoCLEF task under the name TextMess. The proposed result fusion method takes as input the document lists returned by the different systems and returns a document list where the documents are ranked according to the fuzzy Borda voting scheme. The obtained results show that the fusion method allows to improve the results of the component systems, although the fusion is not optimal, because it is effective only if the components return a similar set of relevant documents.

1 Introduction

Result fusion has been studied as an option for obtaining better results in Information Retrieval (IR) by taking advantage from the combination of existing methods. Many fusion method have been proposed, such as linear combinations [1,2] and voting schemes like the Condorcet [3] and the Borda [4] schemes. Aslam and Montague [4] concluded that the Borda fusion is a simple, unsupervised method that is capable to exceed the performance of the best component system. The fuzzy Borda voting scheme is an improvement of the standard Borda voting scheme that was introduced by [5,6]. This is the first time it is used in the IR task, although it has been used in the Word Sense Disambiguation task at Semeval[1] with good results [7].

[1] http://nlp.cs.swarthmore.edu/semeval

C. Peters et al. (Eds.): CLEF 2008, LNCS 5706, pp. 867–874, 2009.

In Sections 2, 3 and 4 we describe briefly the systems of each group. In Section 5 we describe the fuzzy Borda ranking method, in Section 6 we present the experiments carried out and the obtained results. Finally, in Section 7 we draw some conclusions.

2 The SINAI-GIR System

The SINAI-GIR system is composed of the following subsystems: the *Collection Preprocessing subsystem*, the *Query Analyzer*, the *Information Retrieval subsystem* and the *Validator*. Each query is preprocessed and analyzed by the *Query Analyzer*, identifying its geo-entities and spatial relations and making use of the Geonames gazetteer[2]. This module also applies *query reformulation* based on the query parsing sub-task, generating several independent queries which will be indexed and searched by means of the IR subsystem. On the other hand, the collection is pre-processed by the *Collection Preprocessing* module and finally the documents retrieved by the IR subsystem are filtered and re-ranked by means of the *Validator* subsystem.

The features of each subsystem are:

- *Collection Preprocessing Subsystem.* During the collection preprocessing, two indexes are generated (*locations* and *keywords* indexes). We apply the Porter *stemmer*, the Brill POS tagger and a the LingPipe[3] Named Entity Recognizer (NER). We also discard the English *stop-words.*
- *Query Analyzer.* It is responsible for the preprocessing of English queries as well as the generation of different query reformulations.
- *Information Retrieval Subsystem.* As IR index-search engine we have used Lemur[4].
- *Validator.* The aim of this subsystem is to filter the lists of documents recovered by the IR subsystem, establishing which of them are valid, depending on the locations and the *geo-relations* detected in the query. Another important function is to establish the final ranking of documents, based on manual rules and predefined weights.

3 The NLEL GeoWorSE System

The system is built around the Lucene[5] open source search engine, version 2.1. The Stanford NER system based on Conditional Random Fields [8] is used for Named Entity Recognition and classification. The toponym disambiguator is based on the method presented in [9].

During the indexing phase, the documents are examined in order to find location names (*toponyms*) by means of the Stanford NER system. When a toponym

[2] http://www.geonames.org
[3] http://alias-i.com/lingpipe
[4] http://www.lemurproject.org
[5] http://lucene.apache.org/

is found, the disambiguator determines the correct reference for the toponym. Then, the toponym coordinates are added to the *geo* index, and the toponym is stored in the *wn* index together with its holonyms and synonyms. All document terms are stored in the *text* index.

The search phase starts with the search of the topic keywords in the *text* index. The toponyms extracted by the Stanford NER are searched for in the *wn* index with a weight 0.25 with respect to the content terms. The result of the search is a list of documents ranked using the Lucene's weighting scheme. At the same time, the toponyms are used to define a geographical constraint that is used to re-rank the document list. There are two types of geographical constraints:

- a *distance* constraint, corresponding to a point in the map: documents that contain locations closer to this point will be ranked higher;
- an *area* constraint, corresponding to a polygon in the map: documents that contain locations included in the polygon will be ranked higher.

Finally, the documents retrieved by Lucene are re-ranked depending on the geographical constraints.

4 The TALP GeoIR System

The TALPGeoIR system [10] has five phases performed sequentially: collection processing and indexing, linguistic and geographical analysis of the topics, textual IR with Terrier, Geographical Retrieval with Geographical Knowledge Bases (GKBs), and geographical document re-ranking.

The collection is processed and indexed in two different indexes: a geographical index with geographical information extracted from the documents and enriched with the aid of GKBs and a textual index with the lemmatized content of the documents.

The linguistic analysis uses the following Natural Language Processing tools: *TnT* , a statistical POS tagger, the *WordNet lemmatizer 2.0*, and a *Maximum Entropy-based NERC* system, trained with the CONLL-2003 shared task English data set.

The retrieval system is a textual IR system based on Terrier [11]. Terrier configuration includes a TF-IDF schema, lemmatized query topics, Porter Stemmer, and Relevance Feedback using 10 top documents and 40 top terms.

The Geographical Retrieval uses geographical terms and/or geographical feature types appearing in the topics to retrieve documents from the geographical index. The geographical search allows to retrieve documents with geographical terms that are included in the sub-ontological path of the query terms (e.g. documents containing *Alaska* are retrieved from a query *United States*).

Finally, a geographical re-ranking is performed using the set of documents retrieved by Terrier. From this set of documents those that have been also retrieved in the Geographical Retrieval set are re-ranked giving them more weight than the other ones.

5 Fuzzy Borda Fusion

5.1 Fuzzy Borda Count

In the classical (discrete) Borda count, each expert gives a mark to each alternative. The mark is given by the number of alternatives worse than it. The fuzzy variant [5,6] allows the experts to show numerically how much alternatives are preferred over others, expressing their preference intensities from 0 to 1.

Let $R^1, R^2, \ldots, R^m$ be the fuzzy preference relations of m experts over n alternatives $x_1, x_2, \ldots, x_n$. Each expert k expresses its preferences by means of a matrix of preference intensities:

$$\begin{pmatrix} r^k_{11} & r^k_{12} & \ldots & r^k_{1n} \\ r^k_{21} & r^k_{22} & \ldots & r^k_{2n} \\ \ldots & \ldots & \ldots & \ldots \\ r^k_{n1} & r^k_{n2} & \ldots & r^k_{nn} \end{pmatrix}$$

where each $r^k_{ij} = \mu_{R^k}(x_i, x_j)$, with $\mu_{R^k} : X \times X \rightarrow [0,1]$ is the membership function of R^k. The number $r^k_{ij} \in [0,1]$ is considered as the degree of confidence with which the expert k prefers x_i over x_j. The final value assigned by the expert k to each alternative x_i is the sum by row of the entries greater than 0.5 in the preference matrix, or, formally:

$$r_k(x_i) = \sum_{j=1, r^k_{ij} > 0.5}^{n} r^k_{ij} \tag{1}$$

The threshold 0.5 ensure the relation $R^k$to be an ordinary preference relation.

The fuzzy Borda count for an alternative x_i is obtained as the sum of the values assigned by each expert to that alternative: $\mathbf{r}(x_i) = \sum_{k=1}^{m} r_k(x_i)$.

5.2 Application of Fuzzy Borda Count to Result Merging

In our approach each system is an expert: therefore, for m systems, there are m preference matrices. The size of these matrices is variable: the reason is that the document list is not the same for all the systems. The size of a preference matrix is $N_t \times N_t$, where N_t is the number of unique documents retrieved by the systems (i.e. the number of documents that appear at least in one of the lists returned by the systems) for topic t.

Each system may rank the documents using weights that are not in the same range of the other ones. Therefore, the output weights $w_1, w_2, \ldots, w_n$ of each expert k are transformed to fuzzy confidence values by means of the following transformation:

$$r^k_{ij} = \frac{w_i}{w_i + w_j} \tag{2}$$

This transformation ensures that the preference values are in the range $[0,1]$. In order to adapt the fuzzy Borda count to the merging of the results of IR systems, we had to determine the preference values in all the cases where one of the systems does not retrieve a document that has been retrieved by another one. The preference values of these documents were set to 0.5, corresponding to the idea that the expert is presented with an option on which it cannot express a preference.

6 Experiments and Results

In Tables 1 and 2 we show the detail of each run in terms of the component systems and the topic fields used. "Official" runs (i.e., the ones submitted to GeoCLEF) are labeled with `TMESS02-08` and `TMESS07A`.

In order to evaluate the contribution of each system to the final result, we calculated the overlap rate O of the documents retrieved by the systems: $O = \frac{|D_1 \cap \ldots \cap D_m|}{|D_1 \cup \ldots \cup D_m|}$, where m is the number of systems that have been combined together and $D_i, 0 < i \leq m$ is the set of documents retrieved by the i-th system. The obtained value measures how different are the sets of documents retrieved by each system.

The R-overlap and N-overlap coefficients introduced by [12] are used to calculate the degree of overlap of relevant and non-relevant documents in the results of different systems. R-overlap is defined as $R_{overlap} = \frac{m \cdot |R_1| \cdot \ldots \cdot |R_m|}{|R_1| + \ldots + |R_m|}$, where $R_i, 0 < i \leq m$ is the set of relevant documents retrieved by the system i. N-overlap is calculated in the same way, where each R_i has been substituted by N_i, the set of the *non*-relevant documents retrieved by the system i.

In Table 3 we show the Mean Average Precision (MAP) obtained for each run and its composing runs, together with the average MAP calculated over the composing runs.

The obtained results show that the fuzzy Borda merging method always allows to improve the average of the results of the components, and only in two cases it cannot improve the best component result (TMESS13 and TMESS14). The results in Table 4 show that the best results are obtained if the systems returns a

Table 1. Description of the runs of each system

run ID	description
	NLEL
NLEL0802	base system (only text index, no wordnet, no map filtering)
NLEL0803	2007 system (no map filtering)
NLEL0804	base system, title and description only
NLEL0505	2008 system, all indices and map filtering enabled
NLEL01	complete 2008 system, title and description
	SINAI
EXP1	base system, title and description only
EXP2	base system, all fields
EXP4	filtering system, title and description only
EXP5	filtering system (rule-based)
	TALP
TALP01	system without GeoKB, title and description only
TALP02	complete system, including GeoKB, title and description

Table 2. Details of the composition of all the evaluated runs

run ID	fields	NLEL run ID	SINAI run ID	TALP run ID
		Officially evaluated runs		
TMESS02	TDN	NLEL0802	EXP2	
TMESS03	TDN	NLEL0802	EXP5	
TMESS05	TDN	NLEL0803	EXP2	
TMESS06	TDN	NLEL0803	EXP5	
TMESS07A	TD	NLEL0804	EXP1	
TMESS08	TDN	NLEL0505	EXP5	
		Non-official runs		
TMESS10	TD		EXP1	TALP01
TMESS11	TD	NLEL01	EXP1	
TMESS12	TD	NLEL01		TALP01
TMESS13	TD	NLEL0804		TALP01
TMESS14	TD	NLEL0804	EXP1	TALP01
TMESS15	TD	NLEL01	EXP1	TALP01

Table 3. Results obtained for the various system combinations

run ID	MAP	MAP_{NLEL}	MAP_{SINAI}	MAP_{TALP}	avg. MAP
TMESS02	0.227	0.201	0.226		0.213
TMESS03	0.219	0.201	0.212		0.206
TMESS05	0.235	0.216	0.226		0.221
TMESS06	0.226	0.216	0.212		0.214
TMESS07A	0.286	0.256	0.284		0.270
TMESS08	0.216	0.203	0.212		0.207
TMESS10	0.289		0.284	0.280	0.282
TMESS11	0.285	0.254		0.280	0.267
TMESS12	0.287	0.254	0.284		0.269
TMESS13	0.271	0.256		0.280	0.268
TMESS14	0.282	0.256	0.284	0.280	0.273
TMESS15	0.289	0.254	0.284	0.280	0.273

similar set of relevant documents (TMESS10 and TMESS12). In order to better understand this result, we calculated the results that would have been obtained by calculating the fusion over different configurations of each group's system. These results are shown in Table 5.

The fuzzy Borda method allowed also in the case of the fusion of two configurations of the same system to improve the results of the component runs. O, $R_{overlap}$ and $N_{overlap}$ values for same-group fusions are well above the O values obtained in the case of different systems (more than 0.73 with respect to $0.31 - 0.47$). However, the obtained results show that the method is not able to combine in an optimal way the systems that returns different sets of relevant documents. This is due to the fact that a relevant document that is retrieved by a system and not by another one has a 0.5 weight in the preference matrix, making that its ranking will be worse than a non-relevant document retrieved by both systems.

Table 4. O, $R_{overlap}$, $N_{overlap}$ coefficients, difference from the best system (*diff. best*) and difference from the average of the systems (*diff. avg.*) for all runs

run ID	MAP	diff. best	diff. avg.	O	$R_{overlap}$	$N_{overlap}$
TMESS01	0.226	0.001	0.013	0.315	0.698	0.459
TMESS02	0.227	0.001	0.014	0.346	0.692	0.496
TMESS03	0.219	0.007	0.013	0.317	0.693	0.465
TMESS05	0.235	0.009	0.014	0.358	0.692	0.508
TMESS06	0.226	**0.010**	0.012	0.334	0.693	0.484
TMESS07A	0.286	0.002	0.016	0.356	0.775	0.563
TMESS08	0.216	0.004	0.013	0.326	0.690	0.475
TMESS10	**0.289**	0.005	0.007	0.485	**0.854**	0.625
TMESS11	0.285	0.005	**0.018**	0.475	0.796	0.626
TMESS12	0.287	0.003	**0.018**	0.356	0.822	**0.356**
TMESS13	0.271	−0.009	0.003	0.475	0.796	0.626
TMESS14	0.282	−0.002	0.009	0.284	0.751	0.429
TMESS15	**0.289**	0.005	0.016	0.277	0.790	0.429

Table 5. Results obtained with the fusion of systems from the same participant. M_1: MAP of the system in the first configuration, M_2: MAP of the system in the second configuration.

run ID	MAP	M_1	M_2	O	$R_{overlap}$	$N_{overlap}$
EXP1+EXP4	0.289	0.284	0.275	0.792	0.904	0.852
NLEL0804+NLEL01	0.261	0.254	0.256	0.736	0.850	0.828
TALP01+TALP02	0.283	0.280	0.272	0.792	0.904	0.852

7 Conclusions and Further Work

We combined different systems by means of the fuzzy Borda voting scheme. The implemented method allowed to improve in most cases the results of the combined systems, although the improvement was limited. The best results with this method were obtained when the systems returned a similar set of relevant documents, which means that the method needs to be improved in order to better combine sets of different relevant results. This could be done by assigning to the unknown documents a weight different from 0.5, calculating the similarity of these documents with the ones that have been retrieved by the system. This will be the focus of future research efforts.

Acknowledgements

We would like to thank the TextMESS TIN2006-15265-C06 research project and its subprojects TIN2006-15265-C06-04 MiDeS, TIN2006-15265-C06-04-05 and TIN2006-15265-C06-03 TIMOM for partially supporting this work. Daniel Ferrés is supported by a UPC-Recerca grant.

References

1. Bartell, B.T., Cottrell, G.W., Belew, R.K.: Automatic combination of multiple ranked retrieval systems. In: SIGIR 1994: Proceedings of the 17th annual international ACM SIGIR conference on Research and development in information retrieval, pp. 173–181. Springer, New York (1994)
2. Vogt, C.C., Cottrell, G.W.: Fusion via a linear combination of scores. Information Retrieval 1(3), 151–173 (1999)
3. Montague, M., Aslam, J.A.: Condorcet fusion for improved retrieval. In: CIKM 2002: Proceedings of the eleventh international conference on Information and knowledge management, pp. 538–548. ACM, New York (2002)
4. Aslam, J.A., Montague, M.: Models for metasearch. In: SIGIR 2001: Proceedings of the 24th annual international ACM SIGIR conference on Research and development in information retrieval, pp. 276–284. ACM, New York (2001)
5. Nurmi, H.: Resolving Group Choice Paradoxes Using Probabilistic and Fuzzy Concepts. Group Decision and Negotiation 10(2), 177–199 (2001)
6. García Lapresta, J., Martínez Panero, M.: Borda Count Versus Approval Voting: A Fuzzy Approach. Public Choice 112(1-2), 167–184 (2002)
7. Buscaldi, D., Rosso, P.: Upv-wsd: Combining different wsd methods by means of fuzzy borda voting. In: Fourth International Workshop on Semantic Evaluations (SemEval-2007), ACL, pp. 434–437 (2007)
8. Finkel, J.R., Grenager, T., Manning, C.: Incorporating non-local information into information extraction systems by gibbs sampling. In: Proceedings of the 43rd Annual Meeting of the Association for Computational Linguistics (ACL 2005), U. of Michigan - Ann Arbor, ACL, pp. 363–370 (2005)
9. Buscaldi, D., Rosso, P.: A conceptual density-based approach for the disambiguation of toponyms. International Journal of Geographical Information Systems 22(3), 301–313 (2008)
10. Ferrés, D., Rodríguez, H.: TALP at GeoCLEF 2007: Results of a Geographical Knowledge Filtering Approach with Terrier. In: Peters, C., Jijkoun, V., Mandl, T., Müller, H., Oard, D.W., Peñas, A., Petras, V., Santos, D. (eds.) CLEF 2007. LNCS, vol. 5152, pp. 830–833. Springer, Heidelberg (2008)
11. Ounis, I., Amati, G., Plachouras, V., He, B., Macdonald, C., Lioma, C.: Terrier: A High Performance and Scalable Information Retrieval Platform. In: Proceedings of ACM SIGIR 2006 Workshop on Open Source Information Retrieval, OSIR 2006 (2006)
12. Lee, J.H.: Analyses of multiple evidence combination. In: SIGIR 1997: Proceedings of the 20th annual international ACM SIGIR conference on Research and development in information retrieval, pp. 267–276. ACM, New York (1997)

A Ranking Approach Based on Example Texts for Geographic Information Retrieval*

Esaú Villatoro-Tello, Manuel Montes-y-Gómez, and Luis Villaseñor-Pineda

Language Technologies Group, Computer Science Department,
National Institute of Astrophysics, Optics and Electronics (INAOE), Mexico
{villatoroe,mmontesg,villasen}@inaoep.mx

Abstract. This paper focuses on the problem of ranking documents for Geographic Information Retrieval. It aims to demonstrate that by using some query-related *example texts* it is possible to improve the final ranking of the retrieved documents. Experimental results indicated that our approach could improve the MAP of some sets of retrieved documents using only two *example texts*.

1 Introduction

Geographic Information Retrieval (GIR) considers the search of documents based not only on conceptual keywords, but also on spatial information (i.e. geographical references) [1,2]. Recent development of GIR systems [3] evidence that: *i)* traditional IR machines are able to retrieve the majority of the relevant documents for most queries, but that, *ii)* they have severe difficulties to generate a pertinent ranking of them. Based on these facts, we designed a new GIR method that aims to improve the ranking of retrieved documents by considering information from some query-related *example texts*.

The proposed method was evaluated in the Monolingual English exercise of the 2008 GeoCLEF task [4]. In particular, the purpose of our experiments was two-fold: first, to confirm that traditional IR machines can achieve high recall levels, and second, to probe that using some query-related *example texts* allow improving the original ranking of the retrieved documents.

2 Proposed Method

Our method is divided in two main stages: the *retrieval stage* and the *ranking stage*. The goal of the first is to retrieve as many as possible relevant documents for a given query, whereas, the function of the second is to improve the ranking of the retrieved documents.

* This work was done under partial support of CONACyT (scholarship 165545), SNI and INAOE.

C. Peters et al. (Eds.): CLEF 2008, LNCS 5706, pp. 875–879, 2009.

2.1 The Retrieval Stage

The core module of our method is the information retrieval (IR) machine. It is used two times: in a *first iteration*, it retrieves a set of relevant documents using the original query; then, in a *second iteration*, it retrieves a larger set of relevant documents considering an expanded query. The IR machine was implemented using LEMUR[1].

2.2 The Ranking Stage

Feedback Module. This module selects some "presumably relevant" items from the set of retrieved documents generated in the first iteration of the IR process. We call these items *example texts*, and use them for two different purposes: *i)* to modify the original query and perform the second iteration of the IR process, and *ii)* to re-rank the set of retrieved documents. The implementation of this module was based on the blind relevance feedback (BRF) technique.

Query Expansion Module. This module takes as input the set of *example texts* and extracts from them a set of relevant terms. Then, it uses these terms to expand the original query. The expanded query is sent to the IR machine, and a new set of documents is retrieved. Finally, this new set of documents is analyzed by the re-ranking module and the output of the system is generated.

Re-ranking Module. This module is the main contribution of our system. Its goal is to re-rank the set of retrieved documents using the information contained in the query-related *example texts.* This process considers the following steps:

1. **Geo-Expansion.** Expands all geographical terms contained in the *example texts.* It adds to each term its two nearest ancestors (e.g., *Madrid* $\rightarrow$ *Spain,Europe*). For this process we employed the Geonames database[2].
2. **Similarity Calculation.** Compares the retrieved documents against each *example text*, generating this way several different ranking proposals (one for each *example text*). The comparison of documents considers their *thematic* and *geographic* information. In particular, the similarity is computed as follows.

$$SQ(s,r) = (\lambda \times SQ_{thematic}(s,r)) + ((1-\lambda) \times SQ_{geographic}(s,r)) \qquad (1)$$

   where s represents an *example text*, r represents a document from the set of retrieved documents, and λ is a weighting value.
3. **Information Fusion.** Combines, into one single result list, all the information from the different ranking proposals. For this process we employed the well-known Round Robin technique.

---
[1] http://www.lemurproject.org/
[2] http://www.geonames.org/

3 Experiments and Results

This section describes the results from a subset of our experiments evaluated at GeoCLEF 2008. Table 1 shows the results from our four baseline runs. The first two rows correspond to the results of the first IR iteration. In this case, the run inaoe-BASELINE1 employed the title and description fields, whereas, the run inaoe-BASELINE2 used all available information: title, description, and narrative. As it can be noticed, the inclusion of the narrative did not improve the IR performance.

The third and fourth rows of the table show the results achieved in the second IR iteration, after the query expansion process. For these two experiments, we expanded the original query using the K most frequent terms from the top N documents retrieved by the inaoe-BASELINE1 run. We named these experiments as inaoe-BRF-N-K. As expected, the query expansion process allowed both configurations to obtain better results than the BASELINE1, especially for the case of the recall rate.

Table 2, under the column "submitted runs", shows the results achieved by the proposed method. In these experiments, we used the same N *example texts* for query expansion and for re-ranking the output of the *2nd iteration*. In particular, we considered $N = 5$ (i.e., inaoe-BRF-5-5) because we wanted to provide the greatest information to the re-ranking method. The results correspond to the following configurations:

1. RRBF: retrieved documents in the *2nd iteration* were re-ranked making **no** distinction between the *thematic* and *geographic* parts, i.e., similarity was computed using entire documents.
2. RRGeo: retrieved documents in the *2nd iteration* were re-ranked considering both *thematic* and *geographic* parts separately, i.e., applying Formula 1.
3. RRGeoExp: retrieved documents in the *2nd iteration* were re-ranked making distinction between *thematic* and *geographic* parts (applying Formula 1), and considering the Geo-Expansion process.

From Table 2, we can observe that the distinction of the thematic and geographic parts allowed obtaining the best performance. It is also possible to notice that the MAP difference between the experiments RRGeo-5-5 and RRGeoExp-5-5 was not very significant. We believe this performance was consequence of the noise introduced by our naïve geo-expansion process, which does not consider

Table 1. Baseline results

	Experiment ID	MAP	R-Prec	P@5	Recall
1st	inaoe-BASELINE1	0.234	0.261	0.384	0.835
iteration	inaoe-BASELINE2	0.201	0.226	0.272	0.815
2nd	inaoe-BRF-5-2	**0.258**	**0.267**	**0.344**	**0.863**
iteration	inaoe-BRF-5-5	0.246	0.264	0.328	**0.863**

Table 2. Results of the proposed approach

Submitted Runs				Additional Experiments			
Experiment ID	**MAP**	**R-Prec**	**P@5**	**Experiment ID**	**MAP**	**R-Prec**	**P@5**
inaoe-RRBF-5-5	0.241	0.268	**0.384**	inaoe-RRBF‡	0.306	0.304	0.496
inaoe-RRGeo-5-5	0.244	0.266	**0.384**	inaoe-RRGeo‡	0.315	0.307	0.520
inaoe-RRGeoExp-5-5	**0.246**	**0.270**	**0.384**	inaoe-RRGeoExp‡	**0.318**	**0.310**	**0.536**

the disambiguation of geographical terms. That is, it can not distinguish between *Cordoba-Spain* and *Cordoba-Mexico.*

On the other hand, Table 2, under the column "additional experiments", shows the results of our method when there were used only truly relevant *example texts*. These experiments considered the manual selection of the *example texts*, and, somehow, they aimed to determine the best-possible result of our method. In all cases we used, at most, two *example texts*. Therefore, these results can be interpreted as: "By determining only two relevant *example texts*, we could reach a MAP of 0.318". These results show that the proposed method works well, but also indicate that it is very sensitive to the presence of incorrect *example texts*.

Finally, in order to support this conclusion, we made some significant tests. In particular, we employed the well-known Wilcoxon test. As noticed in Table 2(‡), only when we used manually-selected *example texts* we could obtain a significant improvement over the baseline result.

4 Conclusions

The results from our participation at GeoCLEF 2008 showed that the use of query-related *example texts* allows improving the original ranking of the retrieved documents. Nevertheless, they also showed that the proposed method is very sensitive to the presence of incorrect *example texts*, and that it is also affected by the incorrect expansion of the geographical terms. Our current work is mainly focused on tackling these drawbacks. In particular, due to our interest for having a fully automatic GIR process, we are working in a new example-text selection method based on machine learning techniques. On the other hand, we are also working in a better strategy for geographic query expansion.

References

1. Purves, R., Jones, C. (eds.): SIGIR 2004 Workshop on Geographic Information Retrieval, Sheffield, UK (2004)
2. Henrich, A., Lüdecke, V.: Characteristics of Geographic Information needs. In: Proceedings of Workshop on Geographic Information Retrieval, GIR 2007, Lisbon, Portugal. ACM Press, New York (2007)

3. Mandl, T., Gey, F., Di Nunzio, G., Ferro, N., Larson, R., Sanderson, M., Santos, D., Womser-Hacker, C., Xie, X.: GeoCLEF 2007: the CLEF 2007 Cross-Language Geographic Information Retrieval Track Overview. In: Peters, C., Jijkoun, V., Mandl, T., Müller, H., Oard, D.W., Peñas, A., Petras, V., Santos, D. (eds.) CLEF 2007. LNCS, vol. 5152, pp. 745–772. Springer, Heidelberg (2008)
4. Villatoro-Tello, E., Montes-y-Gómez, M., Villaseñor-Pineda, L.: INAOE at GeoCLEF 2008: A Ranking Approach based on Sample Documents. In: Working Notes for the CLEF 2008 Workshop, Aarhus, Denmark (2008)

Ontology-Based Query Construction for GeoCLEF

Rui Wang[1] and Günter Neumann[2]

[1] Saarland University
66123 Saarbrücken, Germany
rwang@coli.uni-sb.de
[2] LT-Lab, DFKI
Stuhlsatzenhausweg 3, 66123 Saarbrücken, Germany
neumann@dfki.de

Abstract. This paper describes experiments with geographical information retrieval (GIR). Being different from the traditional information IR, we focus more on the query expansion instead of document ranking. We parse each topic into the *event* part and the *geographic* part and use different ontologies to expand both parts respectively. The results show promising results of our strategy for this task.

1 Introduction

The goal of geographic information retrieval (GIR) is to retrieve documents for topics with a geographic specification [2]. For example, given the query "*riots in South American prisons*", the system is asked to retrieve all the relevant documents about these *events* (i.e. "*riots*") happening at those *places* (i.e. "*South American prisons*").

Traditional information retrieval consists of three main components: query expansion, document retrieval, and document ranking, of which the last component attracts the most attention [4]. As for GIR, since geographic variation is an important criterion for evaluating such systems, we assume that the query processing will have more impact on the final results. Furthermore, we show that ontologies both for events and geographic terms can improve the results greatly.

2 System Description

Our system is a pipeline consisting of query processing, document indexing, and document ranking. Since we focus mainly on the first component, we will not talk about the rest two in this report, which is a straightforward use of Lucene[1]. The query processing module can be further divided into three submodules: topic parsing, keywords expansion, and query construction. We preprocess the input topics and documents with named-entity (NE) recognition[2]. The documents are indexed after that; and the topics with NE annotations are sent to later processing stages. The following picture shows the workflow.

[1] http://lucene.apache.org/
[2] We use Stanford NER [1].

C. Peters et al. (Eds.): CLEF 2008, LNCS 5706, pp. 880–884, 2009.

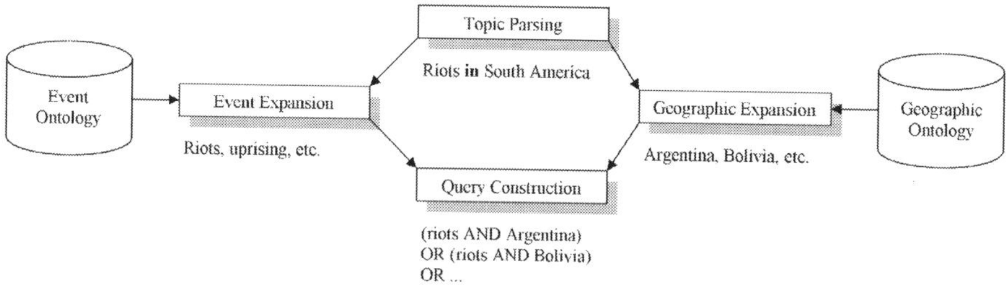

Fig. 1. Topic Parsing splits each topic into two parts, the *Event* part and the *Geographic* part, and send them to Event Expansion and Geographic Expansion components. These two components are assisted by Event Ontology and Geographic Ontology respectively. After the expansion, the query for the indexed documents will be constructed by Query Construction.

2.1 Topic Parsing

As mentioned before, we preprocess the input topics with NE recognition and identify the two parts of each topic, i.e. the *Event* part and the *Geographic* part. By doing this, we use prepositions as indicators for the division. Some topics are listed as follows,

Riots ***in*** *South American prisons*
Most visited sights ***in*** *the capital* ***of*** *France and its vicinity*

In most cases, the prepositions are effective as in the first example. Together with the NE information (i.e. location names), the two corresponding parts will be identified out. However, there are some cases, like the last example, which consist of several parts, if they are divided by prepositions. In practice, we take location names as the *Geographic* part (marked with double underline) and all the rest as the *Event* part (marked with underline).

2.2 Ontology-Based Keywords Expansion

In this step, the *Event* part and the *Geographic* part will be tackled separately, assisted by two ontologies,

Geographic Ontology. After referring several geographic taxonomies (Geonames[3], WorldGazetteer[4], etc.), we construct a geographic ontology using geographic terms and two relations. The backbone taxonomy of the ontology is as follows,

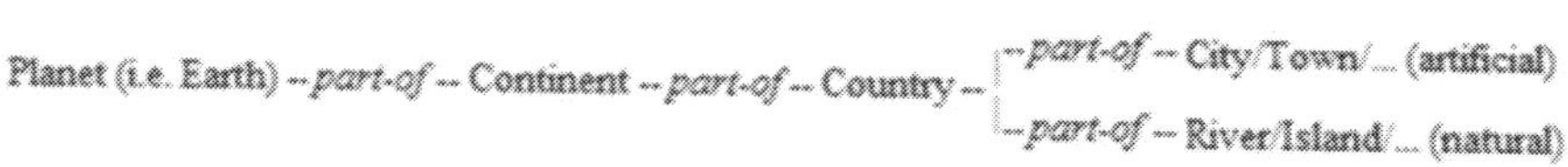

Fig. 2. The basic structure of the geographic ontology consists of geographic terms referring different granularities of areas. The basic relation in-between is the directional *part-of* relation, which means the geographic area on the left side contains the area on the right side.

---

[3] Geonames geo coding web service: http://www.geonames.org/
[4] WorldGazetteer: http://www.world-gazetteer.com

In addition, extra geographic areas are connected with these basic terms using the same *part-of* relation. For example, the following geographic areas consist of the basic terms above,

Subcontinent: *the Indian subcontinent, the Persian Gulf*, etc.

Organization: *the Organization for Economic Co-operation and Development (OECD)*, etc.

An additional *equal* relation is utilized for synonyms and abbreviations of the same geographic area, e.g. *the United Kingdom, the UK, Great Britain*, etc.

Event Ontology. The event ontology is constructed using Wikipedia as an extra resource. Unlike the linguistic classification of events, we consider this ontology as a rather flat structure of two main categories, natural events and human activities. The first category mainly contains natural disasters, e.g. floods, earthquakes, etc; the second category takes all the rest, e.g. meetings, sports, wars, etc. Two examples are,

Earthquakes: *San Francisco Earthquake (1906), Good Friday Earthquake Earthquake (1964)*, etc.

Nobel Prize winners: *Marie Curie (Russian Poland, Physics, 1903), Albert Einstein(Germany, Physics, 1921), Mother Teresa (Albania, Peace, 1979)*, etc.

Keywords Expansion. The population of the ontologies is done with either the narratives given or Wikipedia. The former can be done automatically from the texts after NE recognition; the latter has to be done manually. The usage of the event ontology is to take all the terms contained in that category; the use of the geographic ontology follows the rule: if the geographic part contains the granularity of the basic terms, the ontology will provide all the geographic terms at that level; otherwise, the ontology will provide all the geographic terms below the level of that term.

2.3 Query Construction

After the expansion of both the events and the geographic terms, the query can be constructed using Boolean operators. In order to achieve both high precision and recall, we setup four levels of queries, giving different weights (the numbers in the front) for the retrieved documents. The higher levels of queries aim to obtain accurate results, while the lower levels for the recall. The four levels are as follows,

Level 4 (1000): the event ontology **AND** the geographic ontology
Level 3 (100): the event terms **AND** the geographic ontology
Level 2 (10): the event terms **AND** the geographic terms
Level 1 (1): the event terms **OR** the geographic terms

Here, *event terms* and *geographic terms* mean those words appearing in the topics but not the narratives. In fact, both the event ontology and the geographic ontology can be further divided into two cases, the *automatic* meaning the ontology is constructed automatically using the narratives and the *manual* meaning the ontology is constructed also with Wikipedia information.

3 Submissions and Results

In the GeoCLEF track, we submitted 5 runs for the monolingual task of English. Different runs were constructed from combinations of different levels of queries,

Run1 (M): Use queries from Level 1~4 and both ontologies are constructed with Wikipedia information
Run2 (A): Similar to Run1, but both ontologies are constructed with narratives
Run3 (M): Use queries from Level 1~3 and the ontology is constructed with Wikipedia information
Run4 (A): Similar to Run3, but the ontology is constructed with narratives
Run5 (A): Use queries from Level 1~2

Since we consider the ontologies constructed from Wikipedia are manual work, Run1 and Run3 are Manual (M) submissions and the other three are Automatic (A) submissions. The following table shows the final results of our five submissions,

Table 1. Results of our five submissions

Submissions	R-Prec	MAP
Run1 (M)	33.38% (1/68[5])	29.18% (3/68)
Run2 (A)	33.19% (2/68)	29.24% (2/68)
Run3 (M)	31.70% (3/68)	30.37% (1/68)
Run4 (A)	31.41% (4/68)	27.73% (6/68)
Run5 (A)	20.95% (58/68)	16.07% (68/68)

The results suggest the impact of focusing on ontology-based query expansion for GIR. The best automatic submission will be Run2, which has both high R-Prec and MAP scores. For the best manual submissions, Run1 and Run3 have the best R-Prec and MAP scores respectively. Comparing automatic and manual submissions, the R-Prec has a slight difference, while for MAP, the difference is bigger. Consequently, the manual work of populating the ontology with Wikipedia information does help to improve the precision. At last, only using the terms in the topics without any help from the narratives or Wikipedia, the results are quite poor (Run5).

Taking a closer look at the results, we find that the system has increased performance in some topics, but decreased in some others. This may be because the improvement from the ontology is not stable, since different topics contain various events, which cannot be treated uniformly.

Additionally, since the only language dependent components of our system are the NE recognizer and the ontologies, we also did experiments on the German data sets. The SPPC system [3] was used for German NE recognition and the ontologies were constructed with the help of German Wikipedia. The preliminary evaluation was not so satisfactory, so that we did not make submissions, but our approach can be easily adapted to other languages.

4 Conclusion and Future Work

In this paper, we showed our approach of GIR, focusing on the query processing part instead of the document ranking as in traditional IR systems. In particular, we analyzed the topics and applied ontologies to expand the keywords in both the geographic

[5] The rank of the corresponding submission among all the 68 submissions.

part and the event part. We also setup four levels of queries in order to achieve both high precision and recall. The results suggest the success of our strategy.

In the future, we will take into account the document ranking part as well. One direction could be to use a context window to control the distance between the event and the geographic term in order to filter out some documents. More experiments on other languages are also considered by us in the near future.

References

1. Finkel, J.R., Grenager, T., Manning, C.: Incorporating Non-local Information into Information Extraction Systems by Gibbs Sampling. In: Proceedings of the 43nd Annual Meeting of the Association for Computational Linguistics, ACL 2005 (2005)
2. Mandl, T., Gey, F., Nunzio, G.D., Ferro, N., Larson, R., Sanderson, M., Santos, D., Womser-Hacker, C., Xie, X.: GeoCLEF 2007: the CLEF 2007 Cross-Language Geographic Information Retrieval Track Overview. In: Peters, C., Jijkoun, V., Mandl, T., Müller, H., Oard, D.W., Peñas, A., Petras, V., Santos, D. (eds.) CLEF 2007. LNCS, vol. 5152, pp. 745–772. Springer, Heidelberg (2008)
3. Neumann, G., Piskorski, J.: A Shallow Text Processing Core Engine. Journal of Computational Intelligence 18(3), 451–476 (2002)
4. Singhal, A.: Modern Information Retrieval: A Brief Overview. Bulletin of the IEEE Computer Society Technical Committee on Data Engineering 24(4), 35–43 (2001)

Experiments with Geographic Evidence Extracted from Documents

Nuno Cardoso, Patrícia Sousa, and Mário J. Silva

Faculty of Sciences, University of Lisbon, LASIGE
{ncardoso,csousa,mjs}@xldb.di.fc.ul.pt

Abstract. For the 2008 participation at GeoCLEF, we focused on improving the extraction of geographic signatures from documents and optimising their use for GIR. The results show that the detection of explicit geographic named entities for including their terms in a tuned weighted index field significantly improves retrieval performance when compared to classic text retrieval.

1 Introduction

This paper presents the participation of the XLDB Team from the University of Lisbon at the 2008 GeoCLEF task. Following a thorough analysis of the results achieved on the 2007 participation [1], we identified the following improvement points:

Experiment with new ways to handle thematic and geographic criteria. Our previous methodology was moulded on the assumption that the thematic and geographic facets of documents and queries were complementary and non-redundant [2]. Previous GIR prototypes handled thematic and geographic subspaces in separate pipelines. As the evaluation results did not show significative improvements compared to classic IR, an alternative GIR methodology should be tested.

Capture more geographic evidence from documents. The text mining module, based on shallow pattern matching of placenames, often failed on the extraction of essential geographic evidence for geo-referencing many relevant documents [1]. We therefore considered reformulating our text annotation tools, making them capable of capturing more geographic evidence from the documents. As people describe sought places in several other ways other than providing explicit placenames (e.g., “Big Apple”, “Kremlin” or “UE Headquarters”), these named entities (NEs) can be captured and grounded to their locations, having an important role on defining the geographic area of interest of documents.

Smooth query expansion. Query expansion (QE) is known to improve IR performance in most queries, but often at the cost of degrading the performance of other queries. We do not assign weights to query terms, so the expanded terms have the same weight as the initial query terms. This means that we do not control the impact of QE in some topics, which causes query drifting [3]. This year, we wanted to use QE with automatic re-weighting of text and geographic terms, to soften the undesired effect of query drifting.

C. Peters et al. (Eds.): CLEF 2008, LNCS 5706, pp. 885–893, 2009.

To address these topics, we made the following improvements to our GIR prototype:

Query Processing. We now handle placenames both as geographic terms and plain query terms. In fact, placenames revealed to be good retrieval terms, and were frequently ranked at the top on the query expansion step [1]. While placenames may be used in other unrelated contexts, such as proper names, they seem to help recall when used as plain terms. In addition, their geographic content can be used afterwards to refine the ranking scores, promoting documents with placenames referred in a geographic context.

Text mining. We developed REMBRANDT, a new named entity recognition module that identifies and classifies all kinds of named entities in the CLEF collection [4]. Used as a text annotation tool, REMBRANDT generates more comprehensive geographic document signatures (D_{sig}). We first introduced D_{sig} on last year's participation [1] as a means to capture the geographic scope of documents as lists of geographic concepts corresponding to the grounded names in the documents used for computing the geographic similarity of documents and queries. The D_{sig} comprise two kinds of geographic evidence: i) *explicit* geographic evidence, consisting of grounded placenames that designate locations, and ii) *implicit* geographic evidence, consisting of other grounded entities that do not designate explicitly geographic locations but are strongly related to a geographic location (e.g., monuments, summits or buildings).

Document Processing. To cope with the new query processing approach, we needed a simple ranking model that elegantly combined the text and geographic similarity models, eliminating the need for merging text and geographic ranking scores, while still allowing us to assign a weight to each term. We extended MG4J [5] to suit our requirements for this year's experiments, and we chose the BM25 [6] weighting scheme to compute a single ranking score for documents using three index fields: `text` field, for standard term indexes, `explicit local` field, for geographic terms labeled as explicit geographic evidence, and `implicit local` field, for geographic terms associated to implicit geographic evidence.

2 System Description

Figure 1 presents the architecture of the assembled GIR prototype. In a nutshell, the CLEF topics are pre-processed by the QE module, QuerCol, which generates term-weighted query strings in MG4J syntax. The CLEF collection is annotated by REMBRANDT, which generates the geographic signatures of the documents (D_{sig}). The text and D_{sig} of the documents are indexed by MG4J, which uses an optimised BM25 weighting scheme. For the retrieval, MG4J receives query strings and generates results in the `trec_eval` format. A geographic ontology assists QuerCol in its geographic term expansion. REMBRANDT and MG4J use other geographic knowledge resources, as described further in this section.

2.1 REMBRANDT

REMBRANDT is a language-dependent named-entity recognition (NER) system that uses Wikipedia as a knowledge resource, and explores its document structure to classify

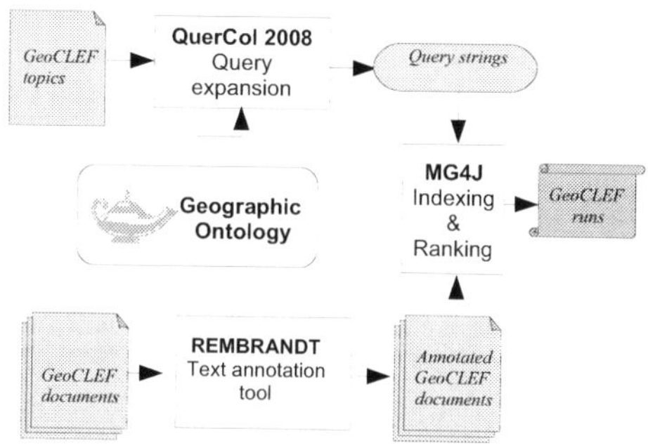

Fig. 1. Architecture of the GIR prototype used in GeoCLEF 2008

all kinds of named entities (NE) in the text. Through Wikipedia, REMBRANDT obtains additional knowledge on every NE that is also a Wikipedia entry, which can be useful for understanding the context, detecting relationships with other NEs, and contextualise and classify surrounding NEs in the text. One example of use of this additional knowledge is deriving implicit geographic evidence for each NE from Wikipedia's page categories. REMBRANDT handles category strings as text sentences and searches for place names in a similar way as it is performed on normal texts, generating a list of captured place names that are considered as implicit geographic evidence for the given NE.

REMBRANDT currently classifies NEs using the categorization defined by HAREM, a NER evaluation contest for Portuguese [7,8]. The main categories of HAREM are: `PERSON`, `ORGANIZATION`, `PLACE`, `DATETIME`, `VALUE`, `ABSTRACTION`, `EVENT`, `THING` and `MASTERPIECE`. REMBRANDT can handle vagueness and ambiguity by tagging a NE with more than one category or sub-category. REMBRANDT's strategy relies on mapping each NE to a Wikipedia page and subsequently analysing its structure, links and categories, searching for suggestive evidences. REMBRANDT also uses manually crafted rules for capturing NE internal and external evidences, classifying the NEs that were not mapped to a Wikipedia page or mapped to a page with insufficient information, and contextualising NEs that have a different meaning (for example, in "I live in Portugal Street", "Portugal" designates a street, not a country).

The classification is best illustrated by following how an example NE, "Empire State Building," is handled: the English Wikipedia page of the Empire State Building (`en.wikipedia.org/wiki/Empire_State_Building`) is labelled with 10 categories, such as "Skyscrapers in New York City" and "Office buildings in the United States." With this information, REMBRANDT classifies the NE as a `PLACE/HUMAN/CONSTRUCTION`. In the hypothetical case that this NE could not be mapped to a Wikipedia page, the presence of the term "Building" in the end (internal evidence) gives a hint for `PLACE/HUMAN/CONSTRUCTION`. External evidence rules check the context of the NE, ensuring that it is not referred in a different context (for example, as a hypothetical movie, street or restaurant name). Finally, the categories "Skyscrapers in New York City" and "Office buildings in the United States" are handled by REMBRANDT as additional text, and the placenames "New York City" and "United States" are treated as implicit geographic evidence associated to the NE.

Table 1. Classification of HAREM categories and sub-categories as having explicit, implicit or no geographic evidence for the generation of D_{sig}

Explicit geographic evidence	No geographic evidence
PLACE/PHYSICAL: {ISLAND, WATERCOURSE, WATERMASS, MOUNTAIN, REGION, PLANET}	**THING**: {CLASS, CLASSMEMBER, OBJECT, SUBSTANCE}
PLACE/HUMAN: {REGION, DIVISION, STREET, COUNTRY}	**PLACE**/VIRTUAL: {MEDIA, ARTICLE, SITE}
Implicit geographic evidence	**PERSON**: {POSITION, INDIVIDUAL, PEOPLE, INDIV.GROUP, POSIT.GROUP, MEMBER, MEMBERGROUP}
EVENT: {PASTEVENT, ORGANIZED, HAPPENING}	**VALUE**: {CURRENCY, CLASSIFICATION, QUANTITY}
PLACE/HUMAN: {CONSTRUCTION}	**ABSTRACTION**: {DISCIPLINE, STATE, IDEA, NAME}
ORGANIZATION: {ADMINISTRATION, INSTITUTION, COMPANY}	**MASTERPIECE**: {WORKOFART, REPRODUCED, PLAN}
	{GENERIC, DURATION, FREQUENCY, HOUR,
	TIME: INTERVAL, DATE}

From REMBRANDT annotations to geographic document signatures

Each document annotated with REMBRANDT contains a list of NEs that might convey explicit or implicit geographic evidence. We can now generate rich geographic document signatures D_{sig} by adding NEs that have explicit geographic evidence, together with the placenames that are associated as implicit geographic evidence for other NEs. We divide the 47 NE sub-categories into 3 levels of eligibility, as depicted in Table 1:

1. **Sub-categories having explicit geographic evidence:** all sub-categories under the main category `PLACE`, with the exception of the sub-categories `PLACE/HUMAN/CONSTRUCTION` and `PLACE/VIRTUAL/*`. The category `PLACE` mostly spans the administrative domain and physical domain, but the `PLACE/VIRTUAL/*`Âăsub-categories span virtual places, such as web sites or TV programs, and therefore are not eligible. In HAREM, the subcategory `PLACE/HUMAN/CONSTRUCTION` is included in the `PLACE` main category, precisely because of its strong geographic connotation. However, since it is not an explicit geographic entity, it is included in the next level.
2. **Sub-categories having implicit geographic evidence:** the main categories `ORGANIZATION`, `EVENTS` and sub-category `PLACE/HUMAN/CONSTRUCTION`. The category `ORGANIZATION` spans institutions and corporations, such as city halls, schools or companies, which are normally related to a defined geographic area of interest. The category `EVENTS` spans organised events that normally take place in a defined place, such as tournaments, concerts or conferences.
3. **Sub-categories not having geographic evidence:** the remaining categories.

This eligibility table of NE classifications into D_{sig} signatures is a far from consensual simplification. It is questionable whether categories such as `PERSON` can also convey a significative geographic evidence for document signatures. For instance, the NE "Nelson Mandela" is associated by REMBRANDT to "South Africa" as an implicit geographic evidence because the Wikipedia page of Nelson Mandela (`en.wikipedia.org/wiki/Nelson_Mandela`) contains the category "Presidents of South Africa." Yet, since not all documents mentioning "Nelson Mandela" have the South African territory as their geographic scope, adding this geographic evidence may produce noisy document scopes. On the other hand, we are assuming that all captured geographic evidence is relevant for the D_{sig}, but this is neither always true. For example, the NE "Empire State Building" conveys an implicit location when it is addressed

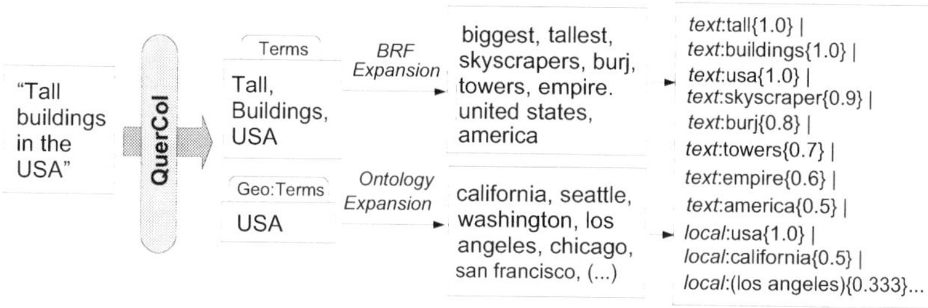

Fig. 2. QuerCol's query reformulation strategy

in a context of office headquarters, but is not relevant as a geographic scope when it is addressing an architectural style.

2.2 QuerCol

QuerCol's query reformulation has two different procedures, illustrated in Figure 2 for the example query "Tall buildings in the USA." First, it uses blind relevance feedback (BRF) to expand the non-stopwords *tall*, *buildings* and *usa*, and weights the expanded terms with the $w_t(p_t\text{-}q_t)$ algorithm [9]. Secondly, it performs geographic query expansion for geographic terms, by exploring their relationships as described in a geographic ontology [10]. As such, QuerCol recognises the geographic term "USA" with the help of REMBRANDT, and grounds it to the geographic concept 'United States of America (country)', triggering the ontology-driven geographic query expansion that searches for other geographic concepts known to be contained within the USA territory. The expanded geographic terms are then re-weighted according to the graph distance in the ontology between the node associated to the expanded concept and descendent nodes by the formula $\frac{1}{distance-1}$. For the given example, USA generates 50 states with a weight of $\frac{1}{2}$ and several cities with weight $\frac{1}{3}$ (considering that the node distance in the ontology between states and countries is 1, and between cities and countries is 2). The final geographic terms represent the query geographic signature(Q_{sig}). In the end, all text and geographic terms are labelled with their targeted index field, and assembled in a final query string connecting them with OR operators (|).

2.3 Indexing and Ranking in MG4J

MG4J is responsible for the indexing and retrieval of documents. MG4J indexes the text of CLEF documents into a `text` index field, while the D_{sig} of the documents is divided in two geographic indexes: the `explicit local` and `implicit local` index fields, according to each type of geographic evidence. Figure 2.3 presents an example of REMBRANDT's annotation and subsequent MG4J indexing steps.

We define *term similarity* as the similarity between query terms and document terms computed with BM25 on the `text` index field only. *Geographic similarity* is the similarity between the geographic signatures of queries and documents (Q_{sig} and D_{sig}) computed by BM25 on the `explicit local` and `implicit local` index fields. MG4J

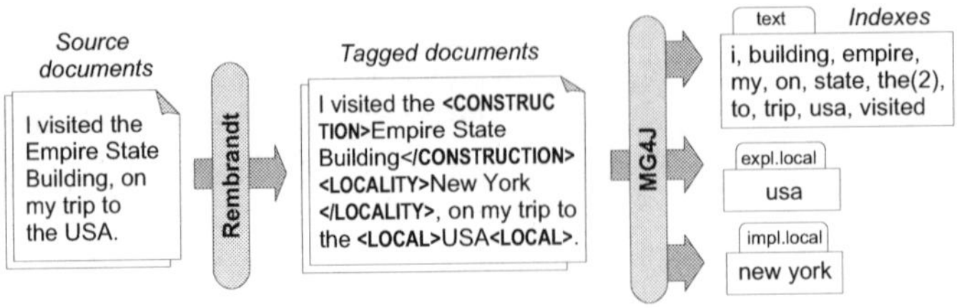

Fig. 3. REMBRANDT's text annotation and MG4J's indexing steps

enables the dynamic selection of the indexes to be used in the retrieval, and changing the weight of a field before retrieval. Unfortunately, the BM25 implementation included in MG4J does not support term weights, so all the terms weights were set to the default value of 1 for all the generated runs.

3 Runs

Before run generation, the BM25 parameters and the index field weights were tweaked to fit the GeoCLEF collection. Using the 2007 GeoCLEF topics and relevance judgements, we generated several runs with different parameters and weights and then selected the run with the highest MAP value, the *initial run*. Afterwards, we generated several final query strings with different blind relevance feedback (BRF) parameters (the number of top-ranked expanded terms, *top-k terms*, and the number of top-ranked documents, *top-k docs*) from the initial run, and again generated several runs with different parameters and weights. The run with the highest MAP value, the *final run*, corresponded therefore to the best BM25 parameters and index field weights for the GeoCLEF collection with the 2007 topics and relevance judgements.

We submitted a total of 12 runs for each subtask, using slight variations of the parameter values from the best optimised runs. Table 2 gives the parameter values used for the official runs. The official runs are composed by initial runs (i.e., runs before BRF, #1 to #3) and final runs (i.e., runs after BRF, #4 to #12). We experimented different ratios of `text` / `explicit local` index weights, by increasing and decreasing the `text` index weight by 0.5.

Table 2. The configuration parameters used for the official runs

	Initial Run					BRF		Final Run				
Run number	BM25 opt.		Index field weight			top-k	top-k	BM25 opt.		Index field weight		
Portuguese	b	k_1	text	exp.l.	imp.l.	terms	docs	b	k_1	text	exp.l.	imp.l.
#1, #2, #3	0.4	0.9	{2.0, 2.5, 3.0}	0.25	0.0	-	-	-	-	-	-	-
#4, #5, #6	0.4	0.9	2.5	0.25	0.0	8	5	0.95	0.3	{2.0, 2.5, 3.0}	0.25	0.0
#7, #8, #9	0.4	0.9	2.5	0.25	0.0	8	5	0.65	0.35	{2.0, 2.5, 3.0}	0.25	0.0
#10,#11, #12	0.4	0.9	2.5	0.25	0.0	8	5	0.65	0.5	{2.0, 2.5, 3.0}	0.25	0.0
English												
#1, #2, #3	0.65	1.4	{1.5, 2.0, 2.5}	0.5	0.0	-	-	-	-	-	-	-
#4, #5, #6	0.65	1.4	2.0	0.5	0.0	8	15	0.65	1.4	{1.5, 2.0, 2.5}	0.5	0.0
#7, #8, #9	0.4	0.9	2.5	0.25	0.0	8	10	0.65	0.35	{1.5, 2.0, 2.5}	0.5	0.0
#10,#11,#12	0.4	0.9	2.5	0.25	0.0	8	5	0.65	0.5	{1.5, 2.0, 2.5}	0.5	0.0

We observed that the BM25 optimisation for the Portuguese subtask presents many local optimal MAP values, so we used three BM25 configurations for the official runs, to increase the odds of standing near a global optimal BM25 parameter. For the English subtask, on the other hand, the BRF parameters were more influent for the optimal MAP values than the BM25 parameters, so its runs have different BRF parameter values. The `implicit local` index field did not improve MAP values in any optimisation scenario, and thus it was turned off on the submitted runs.

4 Results

Table 3 presents the official GeoCLEF 2008 results (top part) and the results for previous GeoCLEF evaluations with optimised parameters (bottom part). We observe that our best Portuguese run was in fact an initial run (with a MAP of 0.2234). The post-hoc optimisation corroborates the observation that the best MAP values for Portuguese are achieved by initial runs (with the best MAP value of 0.2301), which is somewhat unexpected. For the English subtask, the best run was a final run. It achieved a MAP value of 0.2755, which could be pushed further up to 0.2814 with optimised parameters.

The results show that the use of `explicit local` index field on the retrieval process improves the results in all GeoCLEF evaluations, while the `implicit local` index field does not contribute at all to the improvement of the retrieval results. This means that the GIR prototype outperforms classic IR system consistently, but also contradicts the initial assumption that implicit geographic evidence would improve searches. In fact, we observe that the `implicit local` index field takes part only on the best MAP values for GeoCLEF 2006 topics. We believe that this can be explained by the fact that the implicit geographic evidence captured by REMBRANDT is grounded to countries and continents, given that in GeoCLEF 2006 the geographic scopes of topics were mostly about countries and continents.

A group of statistical significance tests (Wilcoxon signed-rank, Student's t and randomization tests) comparing the 2008 official runs and the 2008 post-hoc runs show that their differences in MAP values are not statistically significant, meaning that the best official runs were obtained with near optimal BM25 parameter values and index field weights. This also shows that the best parameter values from the GeoCLEF 2007

Table 3. MAP values and optimising parameters for all GeoCLEF evaluations

	Initial Run							BRF		Final Run						
	BM25 parameters			Index field weight				top-k	top-k	BM25 parameters			Index field weight			
	b	k_1	MAP	text	exp.l.	imp.l.	MAP	terms	docs	b	k_1	MAP	text	exp.l.	imp.l.	MAP
Best GeoCLEF 2008 runs																
PT3	0.4	0.9	0.2222	2.5	0.25	0.0	**0.2234**	-	-	-	-	-	-	-	-	-
EN6	0.65	1,4	0.2519	2.0	0.5	0.0	0.2332	8	15	0.65	1.4	-	2.5	0.5	0.0	**0.2755**
Past GeoCLEF evaluations																
PT																
2006	0.4	0.4	0.1613	2.0	0.25	0.0	0.1810	16	5	0.55	0.9	0.1967	2.0	1.25	0.5	**0.2082**
2007	0.4	0.9	0.273	2.5	0.25	0.0	0.3037	8	5	0.3	0.95	**0.3310**	2.5	0.25	0.0	**0.3310**
2008	0.35	1.2	0.2233	4.0	0.25	0.0	**0.2301**	12	15	0.5	1.0	0.2069	1.5	0.25	0.0	0.2089
EN																
2006	0.3	1.6	0.2158	2.25	0.5	0.25	0.2442	16	5	0.8	0.2	0.2704	0.75	0.25	0.5	**0.2714**
2007	0.65	1.4	0.2238	2.0	0.5	0.0	0.2713	8	15	0.65	1.4	**0.2758**	2.0	0.5	0.0	**0.2758**
2008	0.65	1.6	0.2528	3.5	0.25	0.25	0.2641	12	10	0.75	0.6	0.2809	2.0	0.25	0.0	**0.2814**

optimisation were also good parameter values for GeoCLEF 2008, as they were not over-fitted to the GeoCLEF 2007 data.

5 Conclusions

We participated in GeoCLEF 2008 with the purpose of maturing the ideas that were introduced for the 2007 participation, namely the use of signatures for representing the geographic scopes of queries and documents. We focused on generating more comprehensive document geographic signatures, by capturing and using both explicit and implicit geographic evidence.

The results showed that our GIR prototype is consistently better when using the geographic indexes on the retrieval, meaning that our GIR approach outperforms a classic IR retrieval in every GeoCLEF evaluation scenario since 2006. For future work, we plan to improve REMBRANDT's strategy for capturing implicit geographic evidence. We believe that its naïve approach generated noisy signatures and was responsible for the futility of the `implicit local` index field. We also want to develop a new adaptive strategy for QuerCol, as the optimal QE parameters vary for each topic, and using the same configuration set for all topics generates sub-optimal expanded queries. Finally, we intend to optimise term weights on the BM25 implementation available in MG4J for document retrieval.

Acknowledgements

We thank David Cruz and Sebastiano Vigna for the modifications made to MG4J for the experiments, and to Marcirio Chaves for updating the geographic ontology. Our participation was jointly funded by the Portuguese government and the European Union (FEDER and FSE) under contract ref. POSC/339/1.3/C/NAC (Linguateca), and partially supported by grants SFRH/BD/29817/2006 and PTDC/EIA/73614/2006 (GREASE II) from FCT (Portugal), co-financed by POSI.

References

1. Cardoso, N., Cruz, D., Chaves, M., Silva, M.J.: Using Geographic Signatures as Query and Document Scopes in Geographic IR. In: Peters, C., Jijkoun, V., Mandl, T., Müller, H., Oard, D.W., Peñas, A., Petras, V., Santos, D. (eds.) CLEF 2007. LNCS, vol. 5152, pp. 802–810. Springer, Heidelberg (2008)
2. Cai, G.: GeoVSM: An Integrated Retrieval Model for Geographic Information. In: Egenhofer, M.J., Mark, D.M. (eds.) GIScience 2002. LNCS, vol. 2478, pp. 65–79. Springer, Heidelberg (2002)
3. Mitra, M., Singhal, A., Buckley, C.: Improving Automatic Query Expansion. In: Proceedings of SIGIR 1998, Melbourne, Australia, pp. 206–214. ACM, New York (1998)
4. Cardoso, N.: REMBRANDT - Reconhecimento de Entidades Mencionadas Baseado em Relações e Análise Detalhada do Texto. In: Mota, C., Santos, D. (eds.) Desafios na avaliação conjunta do reconhecimento de entidades mencionadas: O Segundo HAREM. Linguateca (2009)

5. Boldi, P., Vigna, S.: MG4J at TREC 2005. In: Proceedings of TREC 2005, NIST (2005), http://mg4j.dsi.unimi.it
6. Robertson, S.E., Walker, S., Hancock-Beaulieu, M., Gull, A., Lau, M.: Okapi at TREC-3. In: Proceedings of TREC-3, Gaithersburg, MD, USA, pp. 21–30 (1992)
7. Santos, D., Seco, N., Cardoso, N., Vilela, R.: HAREM: An Advanced NER Evaluation Contest for Portuguese. In: Proceedings of LREC 2006, Genoa, Italy, pp. 1986–1991 (2006)
8. Santos, D., Carvalho, P., Oliveira, H., Freitas, C.: Second HAREM: new challenges and old wisdom. In: Teixeira, A., de Lima, V.L.S., de Oliveira, L.C., Quaresma, P. (eds.) PROPOR 2008. LNCS (LNAI), vol. 5190, pp. 212–215. Springer, Heidelberg (2008)
9. Efthimiadis, E.N.: A user-centered evaluation of ranking algorithms for interactive query expansion. In: Proceedings of SIGIR 1993, pp. 146–159. ACM, Pittsburgh (1993)
10. Cardoso, N., Silva, M.J.: Query Expansion through Geographical Feature Types. In: 4th Workshop on Geographic Information Retrieval, Lisbon, Portugal. ACM, New York (2007)

GikiP at GeoCLEF 2008: Joining GIR and QA Forces for Querying Wikipedia

Diana Santos[1], Nuno Cardoso[1,2], Paula Carvalho[2], Iustin Dornescu[3], Sven Hartrumpf[4], Johannes Leveling[5], and Yvonne Skalban[3]

[1] Linguateca, SINTEF ICT, Norway
[2] University of Lisbon, DI, LasiGE, XLDB, Linguateca, Portugal
[3] Research Group in Computational Linguistics (CLG) at the University of Wolverhampton, UK
[4] Intelligent Information and Communication Systems (IICS), University of Hagen (FernUniversität in Hagen), Germany
[5] Centre for Next Generation Localisation (CNGL), Dublin City University, Ireland

Diana.Santos@sintef.no, ncardoso@xldb.di.fc.ul.pt, pqfcarvalho@gmail.com, i.dornescu2@wlv.ac.uk, Sven.Hartrumpf@fernuni-hagen.de, Johannes.Leveling@computing.dcu.ie, yvonne.skalban@wlv.ac.uk

Abstract. This paper reports on the GikiP pilot that took place in 2008 in GeoCLEF. This pilot task requires a combination of methods from geographical information retrieval and question answering to answer queries to the Wikipedia. We start by the task description, providing details on topic choice and evaluation measures. Then we offer a brief motivation from several perspectives, and we present results in detail. A comparison of participants' approaches is then presented, and the paper concludes with improvements for the next edition.

1 Introduction

This paper introduces GikiP, an evaluation contest on retrieving geographically-related information from Wikipedia in the form of a list of answers (corresponding to articles). Or, as stated on the website[1]: *Find Wikipedia entries (i.e. articles) that answer a particular information need which requires geographical reasoning of some sort.*

To guarantee a common evaluation ground, participants were requested to use the Wikipedia collection(s) already used in the QA@CLEF main track (2007 and 2008), dating from the end of 2006. Fifteen topics (see Table 1) were released on the 2nd of June 2008 in English, German, and Portuguese (eight example topics had already been published).

In order to conform to expectations of both the question answering (QA) and the geographical information retrieval (GIR) practitioners, topic titles were in a

[1] http://www.linguateca.pt/GikiP/

C. Peters et al. (Eds.): CLEF 2008, LNCS 5706, pp. 894–905, 2009.

Table 1. Topic titles in GikiP 2008. "Lang." stands for the language biases in topic choice. There were three English, three Portuguese, four German and five other topics.

ID	English topic title	Lang.
GP1	Which waterfalls are used in the film "The Last of the Mohicans"?	EN
GP2	Which Vienna circle members or visitors were born outside the Austria-Hungarian empire or Germany?	DE
GP3	Portuguese rivers that flow through cities with more than 150,000 inhabitants	PT
GP4	Which Swiss cantons border Germany?	DE
GP5	Name all wars that occurred on Greek soil.	other
GP6	Which Australian mountains are higher than 2000 m?	EN
GP7	African capitals with a population of two million inhabitants or more	other
GP8	Suspension bridges in Brazil	PT
GP9	Composers of Renaissance music born in Germany	DE
GP10	Polynesian islands with more than 5,000 inhabitants	other
GP11	Which plays of Shakespeare take place in an Italian setting?	EN
GP12	Places where Goethe lived	DE
GP13	Which navigable rivers in Afghanistan are longer than 1000 km?	other
GP14	Brazilian architects who designed buildings in Europe	PT
GP15	French bridges which were in construction between 1980 and 1990	other

QA format, while the topic description was generally a less condensed and more verbose version of the topic title, but would not add crucial information. See for example topic GP5 in the English version:

```
<top lang="en">
<num>GP5</num>
<title>Name all wars that occurred on Greek soil.</title>
<description>Wars that took place in (ancient or modern) Greece
are relevant.</description> </top>
```

Participants had ten days to return the results as a list of titles of Wikipedia pages. The maximum number of documents returned per topic was set to 100, but the topics chosen by the organizers had typically considerably fewer hits.

Only answers / documents of the right type were considered correct. In other words, if a topic concerned painters, the result should be a list of names of painters, and not names of boats or countries. That is, it was not enough that the answer were found in a particular Wikipedia document: it had to be its title. GikiP expected only precise and concise answers that must have been mapped to the correct Wikipedia article name (no homographs).

Evaluation was devised in order to emphasize diversity and multilinguality. Systems able to retrieve a higher number of answers and in more languages should be considered better, so we introduced a simple bonus in order to reward multilinguality, *mult*, being 1, 2 or 3 depending on the number of languages tried out by the systems. More precisely, the score for each topic was calculated according to the following formula: $mult \cdot N \cdot N/total$, where N is the number of correct answers found, and $N/total$ is the precision. The system's final score was defined as the average of its scores per topic.

For further discussion on topic choice, as well as interesting problems on topic translation and assessment, see our longer paper in the CLEF 2008 Working Notes [1], where for example, we point out that different questions are easier to answer (and more natural to pose) in different languages.

2 Motivation

A number of different motivations led us to organize or participate in GikiP:

- the wish to innovate and add difficulty to both QA and GIR (in its GeoCLEF variety), given that both tasks have been quite stable in the last 3–4 years,
- the fact that Wikipedia has established itself as the main/largest multilingual resource for NLP in the last years,
- the belief that asking list questions to an encyclopaedia is very useful for a wide variety of people,
- the need to devise truly crosslingual and multilingual evaluation set-ups so that it makes sense to harvest information in more than one language (as previously pointed out e.g. by [2]),
- the hope that merging QA and IR is a fruitful path for information access in general, and
- the interest in geographical information in natural language, which led us to explore different and more complex ways of encoding place in texts than those ordinarily coped for by GeoCLEF (see e.g. [3,4]).

GikiP is closely related to WiQA [5], a challenging task devised to assist creators of new Wikipedia pages with multilingual data. GikiP has a more general user model, and a more modest requirement: we simply expect systems to answer open list questions with a list of factoids.

From an IR perspective, we believe that IR research has been well aligned with the real demand for bigger and better retrieval models. Nowadays, another trend on the user's demand on search tools can be observed: users start to expect an IR system to understand better the topics addressed and to reason over the answers, instead of just computing and ranking results according to simple term similarity approaches. Therefore, new IR approaches require a good understanding of the context of the user's query to capture the real information need behind it, and must therefore turn to knowledge extraction and use what traditionally is considered an NLP approach to improve their retrieval and ranking based on more than bag-of-words approaches. Geographic IR is a good place to start working on these new semantic IR approaches, as it focuses on a specific angle (geographic) and has as obvious application to endorse IR systems with geographic reasoning capabilities.

One of the bottlenecks that limit the performance of QA systems is the fact that they focus on extracting the answer from plain text, using several specific

NLP tools that may offer extra information (e.g. named entity recognition, parsing, semantic role labeling) at the cost of adding their own biases and limitations.

One property of GikiP that may put QA systems on the right track is that it encourages systems to exploit hyperdata, not just plain text, and also process semi-structured contexts. In fact, to answer GikiP questions, information had to be extracted from several articles, links between articles had to be analyzed, and often the category hierarchy and the infoboxes of Wikipedia had to be used. This is a property that we believe realistic QA systems should have: process several kinds of information sources and strategies and merge that information in a coherent and more informative answer.

Another issue in which we believe GikiP represents considerable progress compared to usual QA contests (at least QA@CLEF) is the fact that these restrict list questions to what we consider the rarest and least interesting kind of lists, dubbed closed lists, such as "Name the seven hills of Rome" or "Name the four Beatles members". In fact, it seems that most participating systems in QA@CLEF even ignore this kind of questions altogether (probably because of the rigid evaluation and their little number – 10 out of 200 in 2008). Our opinion is that a realistic QA system should not be restricted to closed list questions. From a user point of view, any list question makes sense and, in fact, the user may not even know the right number of answers from the start.

The strongest point for GikiP is that it definitely calls for an integration perspective between QA and IR. In fact, there are several arguments for combining approaches from QA and IR for a successful solution to the GikiP task.

Taking a look at the topics, one finds that several require an interpretation of geographic relations (e.g. GP2, GP4), some include measurable properties of locations (e.g. GP3, GP6, GP7), others aim at resolving temporal constraints (e.g. GP2, GP9, GP15), and still others include words with an irregular morphology which are derived from location names (e.g. GP3, GP4, GP5, GP11, GP15).

While IR methods will be useful to provide an initial result set for these topics, methods like inferences or semantic processing will be required to ensure high precision. Interpreting the task as a kind of QA, on the other hand, QA often provides high precision originating from methods that can deal successfully with these kinds of problems. However, some QA methods lack robustness to provide each question with a correct answer, as already pointed out.

On the document level, information required to find matching Wikipedia articles can be contained in either the textual part or in other parts like tables. Measurable properties of entities are typically listed as attribute-value pairs in tables. Thus, processing the GikiP topics benefits from approaches analyzing the textual information in the Wikipedia articles (e.g. methods from QA) and approaches employing structured information (e.g. methods from IR or information extraction). Furthermore, the Wikipedia corpus offers a wealth of additional resources which are useful for both QA and IR approaches (e.g. Wikipedia categories, Wikipedia links between articles, and inter-language links).

Finally, a result for a GikiP topic is a small set of answers to an open list question. The size of result sets for GikiP lies between fixed-size result sets of

Table 2. Topic size of GikiP 2008, only automatic runs. "Unique correct" stands for number of correct hits, removing duplicates in other languages.

Topic	Results	Correct	Accuracy (%)	Unique correct
GP1	5	1	20.00	1
GP2	31	7	22.58	4
GP3	28	8	28.57	5
GP4	79	19	24.05	6
GP5	69	19	27.54	15
GP6	36	7	19.44	4
GP7	90	33	36.67	14
GP8	49	2	4.08	1
GP9	49	15	30.61	15
GP10	53	2	3.77	1
GP11	35	24	68.57	12
GP12	51	25	49.02	10
GP13	9	4	44.44	2
GP14	60	6	10.00	4
GP15	18	2	11.11	1
Total	662	174	26.28	95

1000 ranked documents found in IR tasks and the single-answer result set for QA tasks. In contrast to QA@CLEF, no redundancy in candidate answers can be exploited (because there are no duplicate Wikipedia articles); furthermore, no answer extraction from the articles is necessary for GikiP.

In summary, a successful solution to the GikiP task calls for a combination of approaches from IR and QA, either in sequence (e.g. filtering candidates and applying semantic filters) or in parallel (e.g. using IR as a fall-back to brittle QA methods).

3 Global Results

Before comparing the performance of the actual participating systems, we wanted to assess the task's feasibility, and investigate whether the results were of interest from a crosslingual point of view. We therefore pooled all results obtained, listing, for each topic, the total number of answers, as well as the number of correct answers found, displayed in Table 2. The answers themselves may not be different, they just need to correspond to a different Wikipedia article, so we had to also compute the number of **distinct** correct answers (Unique correct). If we group the answers by closest language (the one described in Table 1), no group seems to stand out: for *correct/total* we get 33.2% for German, 34.2% for Portuguese, 35.0% for English and 25.2% for the remaining topics.

A different way of investigating language weight is presented in Table 3(a) which counts the cases where answers were found in all three languages as well as in only one of them. This is not, however, a reliable measure of Wikipedia

Table 3. An investigation of the weight of languages in GikiP 2008

(a) Number of answers per language in GikiP 2008.

Kind	**DE**	**EN**	**PT**
Total	233	255	174
Correct (174)	32	84	58
Unique correct	0	32	11

(b) Crosslingual boost by simply following the direct links.

From/To	**DE**	**EN**	**PT**
DE	-	32C	29C, 3M
EN	71C, 13M	-	45C, 35M, 1W
PT	52C, 6M	51C, 7M	-

contents, due to the very different approaches (and success rate) of the different participants, to which we turn in the following section.

Finally, we investigated the issue of, departing from each language in turn, how many other language hits were possible to recover using only direct translation links. In Table 3(b) we present the results by manually following the links present in the correct answers of each language, and classifying them into C(orrect): the answer arrived at by following the translation link is correct; M(issing): there is no translation link, so one would not arrive at this answer by simply following the translation link; or W(rong): the answer arrived at by following the translation link is wrong.

4 Overview of Participation and System Results

These are the participating systems: one participant per country where one of the three languages is spoken, and participation was divided equally between GeoCLEFers and QA@CLEFers (given that the IICS group is known to participate in both):

- **GIRSA-WP**, represented by Sven Hartrumpf and Johannes Leveling, Intelligent Information and Communication Systems (IICS) at the FernUniversität in Hagen (Germany), submitted six fully automatic runs with results in the three languages
- **RENOIR** (acronym for REMBRANDT's Extended NER On Interactive Retrievals), represented by Nuno Cardoso, University of Lisbon, Faculty of Sciences, LaSIGE, XLDB (Portugal), submitted one semi-automatic run only in English and Portuguese
- **WikipediaListQA@wlv**, represented by Iustin Dornescu, Research Group in Computational Linguistics (CLG) at the University of Wolverhampton (UK), submitted a fully automatic run in the three languages

We have also requested manual runs (based on the current on-line Wikipedia), for two reasons: we wanted to compare human performance to automatic answers (after the evaluation), and we also wanted to assess how much the information in Wikipedia had changed regarding the particular topics, from the official collections dating from November 2006 to the June 2008 date.

The results obtained by the systems, together with a fully manual run based on the on-line Wikipedia, can be found in Table 4. It is interesting to note that the human participant was not able to find any results for topics 2 and 5, contrary to the automatic systems, which together managed to find 7 and 5 correct hits, respectively.

Turning now to a short description of the participating systems, as can be seen in further detail in [1], they took a wide range of approaches, as well as different starting collections: GIRSA-WP used the German corpus, WikipediaListQA@wlv the English corpus, RENOIR the English and the Portuguese corpus. RENOIR is currently only semi-automatic, the other two systems are fully automatic.

4.1 GIRSA-WP

GIRSA-WP (GIRSA for Wikipedia) is a fully automatic, hybrid system combining methods from QA and GIR. In particular, it merges results from InSicht, an open-domain QA system [6], and GIRSA, a system for textual GIR [7]. In comparison with the two underlying systems, GIRSA-WP applies a semantic filter on the article titles (which are encoded in the answers in GikiP) to increase precision. This semantic filter ensures that the expected answer type (EAT) of the topic and the title of a Wikipedia article are compatible. This technique is widely known from QA for typical answer types such as PERSON, ORGANIZATION, or LOCATION. In the GIRSA-WP system, a concept (a disambiguated word) corresponding to the EAT is extracted from the topic title or description. Then, this concept and the title of a candidate article are parsed by WOCADI [8], a syntactico-semantic parser for German text. The semantic representations (more specifically, the ontological sort and the semantic features, see [9] for details) of the semantic heads are unified. If this unification succeeds, the candidate article is kept; otherwise it is discarded.

The major differences to InSicht and GIRSA are that GIRSA-WP does not merge streams of answers and does not include a logical answer validation. In contrast to GIRSA, the retrieval is based on documents indexed on a per-sentence basis of Wikipedia articles. In addition, the documents from Wikipedia had not been geographically annotated at all.

For the GikiP experiments, the topic title and description were analyzed and sent to GIRSA and InSicht. In GIRSA, the top 1000 results were retrieved and scores were normalized in the interval from 0 to 1. For results returned by both

Table 4. GikiP results in 2008

Run	Answers	Correct	Avg. Prec.	Score
GIRSA-WP (best)	79	9	0.107	0.704
GIRSA-WP (all runs merged)	372	11	0.038	0.286
RENOIR	218	120	0.551	10.706
WikipediaListQA@wlv	123	94	0.634	16.143

GIRSA and InSicht, the maximum score was chosen. Results whose score was below a given threshold were discarded and the semantic filter was applied to the remaining results. To obtain multilingual results, the German article names were translated to English and Portuguese using the Wikipedia linking between languages. Note that this linking was the only non-textual information GIRSA-WP used from Wikipedia; for example, categories and inter-wiki links were completely ignored.

In InSicht, the semantic representation of the query and the semantic representations of document sentences are compared. To go beyond perfect matches, InSicht uses many techniques, for example intratextual coreference resolution, query expansion by inference rules and lexico-semantic relations, and splitting the query semantic network at certain semantic relations. InSicht employed a special technique called *query decomposition* (first tried in GeoCLEF in 2007 [7]) or *question decomposition* in the context of QA [10]. Among the different decomposition classes described in the latter paper, only meronymy decomposition and description decomposition are promising for current queries in GikiP.

The results during the evaluation were somewhat disappointing. There are several reasons for this. Due to time constraints, the Wikipedia articles had not been fully processed for GIRSA and some methods have been applied to the topics only although they should have been applied to the documents, too. For InSicht, the main problems were (1) that important information is given in tables (like inhabitant numbers), but the syntactico-semantic parser ignores these parts of articles and (2) that the semantic matching approach forming the basis of QA is still too strict for the IR-oriented parts of GikiP queries (similar problems occurred for GeoCLEF experiments).

Future work will include enabling the annotation of geographic entities and geo-inferences, and preferring special regions of Wikipedia articles (for example, the introductory sentences).

4.2 RENOIR

The goal of RENOIR's participation in GikiP was to explore new ways of doing GIR, specially for those kinds of geographic queries that cannot be correctly handled by just naïvely expanding the query terms and hoping that an IR system with some sort of geographic reasoning capabilities would capture the full meaning of the topic at stake, as XLDB does for GeoCLEF [11].

As such, XLDB chose to participate with one semi-automatic run using *query procedures* as retrieval input, instead of query terms, defining query procedures as a group of pipelined actions that express each GikiP topic. The selection of query procedures for a given topic was entirely manual, and the execution varied between automatic, semi-automatic and manual.

RENOIR is an interactive tool where query procedures are executed, generating partial and final results for each GikiP topic. RENOIR makes extensive use of REMBRANDT [12], a named entity recognition module which explores the Wikipedia document structure, links and categories, to identify and classify named entities (NEs) in Portuguese and English texts.

Table 5. Query procedures in RENOIR (A: automatic, M: manual, A/M: semi-automatic)

`SEARCH TERM`	A	Performs a simple term query search in the GikiP 2008 collection, and returns a list of Wikipedia documents.
`SEARCH CATEGORY`	A	Searches the Wikipedia dumps for documents with the given Wikipedia category, and returns a list of Wikipedia documents.
`SEARCH INLINKS`	A	Searches the Wikipedia dumps for documents that link to a given Wikipedia document.
`MAP DOC`	A/M	Maps a document from the Wikipedia dump to its counterpart in the GikiP 2008 collection.
`MAP NE`	A/M	Maps a NE to its corresponding document in the GikiP 2008 collection.
`REMBRANDT`	A	Annotates selected Wikipedia document(s) with REMBRANDT, generating lists of NEs for each document.
`REMB. DOC TO NE`	A	Invokes REMBRANDT to classify the title of a given Wikipedia document, generating the respective NE.
`FILT. NE BY TYPE`	A	Filters a list of NEs of a given classification category, generating a subset of NEs.
`FILT. DOC BY TERM`	A	Filters a list of Wikipedia documents by having (or not) a given term/pattern
`FILT. DOC BY EVAL`	M	Filters a list of Wikipedia document by evaluating a condition for a given subset of NEs. For instance, if the document has a number NE greater than 1000, or if it has a place name NE within Europe.

The GikiP 2008 collection was indexed with MG4J [13], which was used for basic document retrieval. For retrievals involving Wikipedia categories and links, different snapshots of Wikipedia (namely the Portuguese and English static SQL dumps from April 2008) were used, since the information regarding Wikipedia categories, redirections and page links was already available in SQL databases.

The RENOIR actions used for the query procedures are described in Table 5. The query procedures were formulated in a simple modular and pipelined approach, to "divide and conquer" the complex task of translating the GikiP topics into a machine-understandable way. So, the actions that could be made automatically were therefore implemented, while the more complex actions performed in GikiP with human intervention (so far) were also kept simple in order to be possible to extend RENOIR to perform them automatically in the future.

The next obvious step is to implement the automatic generator of query procedures, dealing with the problems that were mitigated by using human reasoning. At the same time, future work in RENOIR includes the improvement of the Wikipedia mining approaches, namely extracting information from infoboxes.

4.3 WikipediaListQA@wlv

The participation in this pilot task was motivated by CLG's interest in using Wikipedia as a backbone in QA. In addition, the task required the system to rely on the information inherent in the Wikipedia article link graph and the relation between entities, rather than developing accurate textual answer extractors. For example, in GP4, the fact that a river flows through a city with a high population is not stated in any Wikipedia article. However, given an article that describes a river (e.g. Douro), all the out links can be extracted and by just examining the category assignments and the infoboxes of the corresponding articles, the list of cities that have a population higher than the given threshold can be obtained.

WikipediaListQA@wlv proposes a simple model for topic interpretation that uses few language dependent resources. It exploits relationships between entities that may not be expressed in the article text, but are implied by the links between the articles. In order to navigate the Wikipedia link graph, the Wikipedia SQL dump was used and the articles' text was indexed with Lucene [14]. The system starts by identifying a domain category that comprises candidate articles, and then removes the ones that do not correspond to the topic filter. Thus two parts are identified in each topic: a) the *domain* of the candidate answers, and b) the *filters* to apply in order to select the correct ones.

In order to identify a Wikipedia category that would describe the domain of candidate articles, a parser [15] was used to extract the first noun phrase of the topic which was then matched to a category, by querying the Lucene index. For the purpose of this pilot only very simple filters were implemented: entity filters and factoid filters. The entity filters match documents that mention or have a link to a given entity (8 of the 15 topics). The factoid filters try to extract a particular fact, and the value is then compared to the selection criterion. The facts were: *population* (GP3, GP7, GP10), *nationality* (GP2, GP9), *height* (GP6) and *length* (GP13). Articles from which the fact could not be extracted were dismissed.

The accuracy of the system is limited due to the ambiguity of links. Category relations are not classified: hypernymy vs. meronymy vs. similarity, thus very large article sets might be extracted (the system did not return any results for GP5,GP9 and GP10 because of ambiguity). This might be avoided, in further work, by using resources that map Wikipedia articles to WordNet and disambiguate the type of the entity described in each article (e.g. Yago [16] and DBpedia [17]).

The main advantage of the system is that – using a small set of filters – very complex data can be extracted from Wikipedia. Its disadvantage lies in the complexity of correctly identifying (combined) filters in natural language questions.

5 Concluding Remarks

The results presented by this pilot are encouraging. We believe to have demonstrated the possibility of automating the particular task at hand, and that there are interesting kinds of non-trivial questions that have a retrievable answer in Wikipedia. Furthermore, answer correctness can be quickly assessed by the users, often without even having to visit the page.

But before these systems can reach the general public, much work still remains to be done: for example, dealing with issues of redundancy removal, choice of which language / answer to present first (or only), how to present a compound set of pages to justify a particular answer, and so on.

In particular, we found that mixing answers of different granularity should be properly dealt with. (A list of places containing cities, castles, and countries is not a pleasing outcome for most users.) More often than expected, answers to (apparently simple) questions required clarification: Temporal aspects are important and cannot be overlooked, and even definitional aspects pop up: if you are looking for wars, do you want also battles?

A new edition of this task with more languages (Bulgarian, Dutch, Italian, Norwegian, Romanian and Spanish), more (50) topics, and with a stronger focus on cross-cultural issues, is currently being organized, named GikiCLEF.[2] Some of the improvements are (i) a more appropriate scoring function rewarding multilinguality, which distinguishes, on a per topic basis, whether there were answers at all in a particular language, (ii) reducing the value of (trivial) direct translation links compared to radically different items in another language, and (iii) allowing for more complex justifications if needed.

Acknowledgements. The GikiP organizers are grateful to Ross Purves for checking the English topics, to Sven Hartrumpf's WOCADI parser for debugging the German and English topics, to Anselmo Peñas for supplying the Wikipedia collections and to Paulo Rocha for providing a manual run in two days.

Organization was done in the scope of the Linguateca project, jointly funded by the Portuguese Government and the European Union (FEDER and FSE) under contract ref. POSC/339/1.3/C/NAC. Development of the WikipediaListQA@wlv system was partly supported by the EU funded project QALL-ME (FP6 IST-033860). Grant SFRH/BD/29817/2006 from FCT (Portugal) supported the development of RENOIR.

References

1. Santos, D., Cardoso, N., Carvalho, P., Dornescu, I., Hartrumpf, S., Leveling, J., Skalban, Y.: Getting geographical answers from Wikipedia: the GikiP pilot at CLEF. In: Borri, F., Nardi, A., Peters, C. (eds.) Working Notes for the CLEF 2008 Workshop, Aarhus, Denmark, September 17-19 (2008)
2. Santos, D., Rocha, P.: The key to the first CLEF in Portuguese: Topics, questions and answers in CHAVE. In: [18], pp. 821–832
3. Santos, D., Chaves, M.: The place of place in geographical IR. In: Proceedings of GIR 2006, the 3rd Workshop on Geographic Information Retrieval (GIR 2006), Seattle, August 10, pp. 5–8 (2006)
4. Gey, F., Larson, R., Sanderson, M., Bishoff, K., Mandl, T., Womser-Hacker, C., Santos, D., Rocha, P., Nunzio, G.D., Ferro, N.: GeoCLEF 2006: the CLEF 2006 Cross-Language Geographic Information Retrieval Track Overview. In: [19], pp. 852–876

[2] http://www.linguateca.pt/GikiCLEF/

5. Jijkoun, V., de Rijke, M.: Overview of the WiQA Task at CLEF 2006. In: [19], pp. 265–274
6. Hartrumpf, S.: Question answering using sentence parsing and semantic network matching. In: [18], pp. 512–521
7. Leveling, J., Hartrumpf, S.: Inferring location names for geographic information retrieval. In: Peters, C., Jijkoun, V., Mandl, T., Müller, H., Oard, D.W., Peñas, A., Petras, V., Santos, D. (eds.) CLEF 2007. LNCS, vol. 5152, pp. 773–780. Springer, Heidelberg (2008)
8. Hartrumpf, S.: Hybrid Disambiguation in Natural Language Analysis. Der Andere Verlag, Osnabrück (2003)
9. Helbig, H.: Knowledge Representation and the Semantics of Natural Language. Springer, Berlin (2006)
10. Hartrumpf, S.: Semantic decomposition for question answering. In: Ghallab, M., Spyropoulos, C.D., Fakotakis, N., Avouris, N. (eds.) Proceedings of the 18th European Conference on Artificial Intelligence (ECAI), Patras, Greece, pp. 313–317 (2008)
11. Cardoso, N., Sousa, P., Silva, M.J.: The University of Lisbon at GeoCLEF 2008. In: Borri, F., Nardi, A., Peters, C. (eds.) Working Notes for the CLEF 2008 Workshop, Aarhus, Denmark, September 17-19 (2008)
12. Cardoso, N.: REMBRANDT - Reconhecimento de Entidades Mencionadas Baseado em Relações e Análise Detalhada do Texto. In: Mota, C., Santos, D. (eds.) Desafios na avaliação conjunta do reconhecimento de entidades mencionadas, Linguateca, pp. 187–204 (2008)
13. Boldi, P., Vigna, S.: MG4J at TREC 2005. In: Proceedings of the 14th Text REtrieval Conference, TREC 2005 (2005)
14. Hatcher, E., Gospodnetic, O.: Lucene in Action (In Action series), Manning, Greenwich, CT, USA (2004)
15. Tapanainen, P., Järvinen, T.: A non-projective dependency parser. In: Proceedings of the 5th Conference of Applied Natural Language Processing, Washington D.C., USA, ACL, pp. 64–71 (1997)
16. Suchanek, F.M., Kasneci, G., Weikum, G.: Yago: a core of semantic knowledge. In: WWW 2007: Proceedings of the 16th international conference on World Wide Web, pp. 697–706. ACM Press, New York (2007)
17. Auer, S., Bizer, C., Kobilarov, G., Lehmann, J., Cyganiak, R., Ives, Z.: DBpedia: A Nucleus for a Web of Open Data. In: Aberer, K., Choi, K.-S., Noy, N., Allemang, D., Lee, K.-I., Nixon, L.J.B., Golbeck, J., Mika, P., Maynard, D., Mizoguchi, R., Schreiber, G., Cudré-Mauroux, P. (eds.) ASWC 2007 and ISWC 2007. LNCS, vol. 4825, pp. 722–735. Springer, Heidelberg (2007), http://www.springerlink.com/content/rm32474088w54378/
18. Peters, C., Clough, P., Gonzalo, J., Jones, G.J.F., Kluck, M., Magnini, B. (eds.): CLEF 2004. LNCS, vol. 3491. Springer, Heidelberg (2005)
19. Peters, C., Clough, P., Gey, F.C., Karlgren, J., Magnini, B., Oard, D.W., de Rijke, M., Stempfhuber, M. (eds.): CLEF 2006. LNCS, vol. 4730. Springer, Heidelberg (2007)

Overview of VideoCLEF 2008: Automatic Generation of Topic-Based Feeds for Dual Language Audio-Visual Content

Martha Larson[1], Eamonn Newman[2], and Gareth J.F. Jones[2]

[1] EEMCS, Delft University of Technology, 2628 CD Delft, Netherlands
[2] Centre for Digital Video Processing, Dublin City University, Dublin 9, Ireland
m.a.larson@tudelft.nl, {enewman,gjones}@computing.dcu.ie

Abstract. The VideoCLEF track, introduced in 2008, aims to develop and evaluate tasks related to analysis of and access to multilingual multimedia content. In its first year, VideoCLEF piloted the Vid2RSS task, whose main subtask was the classification of dual language video (Dutch-language television content featuring English-speaking experts and studio guests). The task offered two additional discretionary subtasks: feed translation and automatic keyframe extraction. Task participants were supplied with Dutch archival metadata, Dutch speech transcripts, English speech transcripts and ten thematic category labels, which they were required to assign to the test set videos. The videos were grouped by class label into topic-based RSS-feeds, displaying title, description and keyframe for each video.

Five groups participated in the 2008 VideoCLEF track. Participants were required to collect their own training data; both Wikipedia and general web content were used. Groups deployed various classifiers (SVM, Naive Bayes and k-NN) or treated the problem as an information retrieval task. Both the Dutch speech transcripts and the archival metadata performed well as sources of indexing features, but no group succeeded in exploiting combinations of feature sources to significantly enhance performance. A small scale fluency/adequacy evaluation of the translation task output revealed the translation to be of sufficient quality to make it valuable to a non-Dutch speaking English speaker. For keyframe extraction, the strategy chosen was to select the keyframe from the shot with the most representative speech transcript content. The automatically selected shots were shown, with a small user study, to be competitive with manually selected shots. Future years of VideoCLEF will aim to expand the corpus and the class label list, as well as to extend the track to additional tasks.

Keywords: Classification, Translation, Keyframe Extraction, Speech Recognition, Evaluation, Benchmark, Video.

C. Peters et al. (Eds.): CLEF 2008, LNCS 5706, pp. 906–917, 2009.

1 Introduction

VideoCLEF was a new track piloted at CLEF 2008.[1] The goal of the track is to develop and evaluate tasks involving the analysis of multilingual video content. In particular, we are interested in *dual language* video. Dual language video is video content in which two languages are spoken, but the content of one does not duplicate (i.e., is not a translation of) the content of the other. Prime examples of dual language video content are documentaries and talk shows where interviewees and studio guests do not speak the dominant language of the show (referred to as the *matrix language*), but rather speak another language (referred to as the *embedded language*). The VideoCLEF task was introduced as the successor to the Cross-Language Speech Retrieval (CL-SR) run at CLEF from 2005 to 2007 [6]. The goal is to extend the achievements of CL-SR to the broader challenge of search for video data. VideoCLEF is intended to complement the TRECVid benchmark [7] by emphasizing the exploitation of spoken content (via speech recognition transcripts) and also of archival metadata associated with videos. While TRECVid concerns itself with what is depicted in a video, VideoCLEF focuses on what is described in a video, in other words, what a video is about. VideoCLEF participants are free to use features derived from the visual track of the video, but it is not a required aspect of the task.

1.1 Data

The video data for VideoCLEF 2008 was supplied by the Netherlands Institute of Sound and Vision[2] (called *Beeld & Geluid* in Dutch), one of the largest audio/video archives in Europe. The dual language content contained in the Sound and Vision archives provided the initial inspiration for the VideoCLEF 2008 task. Although the dominant spoken language of much Dutch television programming is, not surprisingly, Dutch, many other languages are spoken. Dutch television is subtitled rather than dubbed. The extensive use of English and other languages in interviews and studio discussions in Dutch television programming means that Dutch media archives are a rich source of spoken content in languages other than Dutch.

Dual language content is an interesting subject of research investigation for two reasons. First, as mentioned above, in dual language content, two or more languages exist side by side. The languages are intertwined, but not duplicated. Each spoken language represents a separate source of evidence for semantic analysis, classification and retrieval of video. Although we limited VideoCLEF 2008 to two languages, the natural extension of the task is to involve all languages present in the video content as information sources. In the Sound and Vision archive, additional languages include not only other European languages, but also a mixture of languages from the other continents. Further, dual language video also implies the presence of subtitles, which (again, this was not yet done

[1] `http://www.clef-campaign.org`

[2] `http://www.beeldengeluid.nl`

in the pilot year 2008) are a valuable further source of semantic evidence. Second, dual language content is useful to information seekers who do not speak the dominant language of the archive. Dutch documentaries are of high quality and media archives contain valuable information nuggets in the form of interviews with historically significant figures. VideoCLEF 2008 aimed to take a first step towards providing access to non-Dutch content hidden within a predominantly Dutch language video collection.

1.2 Tasks

In 2008, VideoCLEF consisted of one task, called *Vid2RSS*.[3] A supplementary description of the Vid2RSS task can be found in [4]. The main subtask was a classification task involving automatically assigning thematic subject category labels to dual language video. This classification task was chosen since it is a straightforward classic video analysis task with high potential for application in real world systems. Thematic subject labels can be understood to be high-level semantic features. Such features can be applied directly in a faceted browsing system or they can be used to support retrieval or other video analysis tasks downstream. The subject labels used for Vid2RSS have known utility for multimedia search. They are a subset of classes used by archive staff for archival and retrieval at Sound and Vision. The creation of groups of resources related to one topic is a familiar task to the staff of large archives, who are often called upon to create a dossier on a particular topic for use in production of new content for broadcast. The choice of the classification task as a task for VideoCLEF was also influenced by an important practical consideration — archivist assigned subject labels are available for the test data and provide the gold standard for task performance evaluation.

Participants submitted their Vid2RSS results as a series of topic-based RSS-feeds. The feeds are trivial to generate. Generation involves concatenating feed item elements corresponding to the videos that have been assigned a certain class label. The feed item elements were supplied with the test data and contain the title of the video, a short description and a representative keyframe. The purpose of requiring output in RSS-feed format was to make the results of the runs submitted by the different sites easily visualizable. RSS-feeds can be displayed in a feedreader and can be easily assessed by end users, for example archive staff. By using RSS as the output format, we hope that we can narrow the distance that must be traversed between experimental runs in a benchmark campaign and exploitation of results achieved in a real-world application.

In addition to the classification subtask, which was mandatory, participants could also carry out two additional subtasks, a translation subtask and a keyframe extraction subtask. The following sections of this paper describe each of the tasks in turn, summarizing the approaches chosen by the individual participants and the task results. The paper finishes with a conclusion and outlook.

[3] `http://ilps.science.uva.nl/Vid2RSS`

2 Classification Task

The goal of this task was to reproduce the subject labels that were hand assigned to the test set videos by archivists at Sound and Vision. Ten thematic categories were chosen, representing a small subset of the subject labels in use at Sound and Vision: Archeology (archeologie), Architecture (architectuur), Chemistry (chemie), Dance (dansen), Film (film), History (geschiedenis), Music (muziek), Paintings (schilderijen), Scientific research (wetenschappelijk onderzoek) and Visual arts (beeldende kunst).

For each video, the task participants were provided with archival metadata including the description and title of the video. As mentioned above, subject labels were removed from this archival metadata record. Participants received speech transcripts from both languages. The speech transcripts included the first best hypothesis of the speech recognition system and were encoded in MPEG-7 format. The transcripts were generated by the University of Twente [2]. No language detection was used, so both the Dutch and the English transcripts reflect a recognition of the video in its entirety. The required task was to perform classification making use of the speech recognition transcripts only.

2.1 Techniques

Chemnitz University of Technology (CUT). The Chemnitz University of Technology (CUT) team chose to carry out the task using a Naive-Bayes and a k-nearest neighbor (k-NN) classifier. They derived training data for the classifiers from identically or similarly named categories in the English and the Dutch Wikipedia. In their experiments, they varied the composition of the feature set (i.e., the vocabulary of terms) used for classification. Stemming and stopword removal were applied. The results suggest that it is helpful to eliminate terms that occur in multiple classes. Also, the depth to which they descended into the Wikipedia category while gathering data impacted results. The performance achieved by their method on the development data unfortunately did not transfer to the test data. In particular, the CUT team notes that classification performance did not improve when the archival metadata was added to the mix. Performance on the combination of archival metadata and transcripts remained comparable to performance on transcripts alone.

Dublin City University (DCU). The Dublin City University (DCU) team approached the task as an information retrieval problem and used an off-the-shelf information retrieval system implementing the vector space model. Both stopword removal and stemming were applied in the feature extraction step. The label of each subject category was used as a query. The DCU team experimented with two dimensions: (1) limiting the recall of the task by labeling a video only with the most specific category label that retrieved it, and, (2) using blind relevance feedback to expand the label of the subject category into a richer query. The Dutch speech transcripts used alone were more useful than the English speech transcripts used alone. Using metadata alone allowed the system to achieve high precision, but did not out-perform the run using Dutch speech transcripts alone.

MIRACLE Research Consortium (MIRACLE). The MIRACLE Research Consortium (MIRACLE) chose a classifier based on the k-nearest neighbor algorithm. Representations of the video episodes were used as queries to perform retrieval on a knowledge base containing Wikipedia articles. Each episode was assigned the label that was associated with the most retrieved Wikipedia articles. In the experiments, the length of the results list was set to ten. Stopword removal and stemming were applied. The MIRACLE team hypothesized that performance is improved in cases where there are a larger number of Wikipedia articles of the appropriate class available in the knowledge base.

University of Amsterdam (UAms). The University of Amsterdam (UAms) team picked a Support Vector Machine with a linear kernel to use as the classifier. They applied χ^2 feature selection; no stopword removal or stemming was performed. In order to collect training data, the class labels were submitted as a query to the Dutch and English Wikipedia and the articles returned were used as the training set. Experimentation was performed adding archival metadata to speech transcripts for the representation of test documents (which improved performance) and combining Dutch and English speech transcripts (which did not outperform use of Dutch speech transcripts by themselves).

University of Jaén (SINAI). The SINAI team from the University of Jaén collected topical data from the internet by submitting the thematic class labels as queries to Google and harvesting the top ten documents returned, which were amalgamated into a single document. One such document from each class was indexed. Stopword removal and stemming were applied. Retrieval was performed on this collection using the language modeling framework. The queries were derived from the speech transcripts and from the archival metadata. A video was assigned the label corresponding to the top ranked document.

2.2 Results

This section reports the results achieved on the classification task by all participating sites and comments on the techniques used and the trends observed. Results are reported in terms of micro-averaged f-scores and macro-averaged f-scores [3]. The f-score is the harmonic mean between precision and recall. The micro-average reflects a document-centric system performance and is calculated directly with respect to the entire collection. The macro-average reflects class-centric performance and is calculated by first computing the f-score for each individual topic class and then averaging over all classes.

The results of all runs are presented in Table 1. The top micro-averaged f-score was 0.53, achieved by SINAI with run SINAI-JEAN-Class-II and the top macro-averaged f-score was 0.59, achieved by DCU with run dcu_run4. It should be noted that good micro-averages and macro-averages might not reflect the

Table 1. Evaluation results for all runs from all participants (nl = Dutch; en = English, asr = Automatic Speech Recognition transcripts; md= archival metadata; test doc. rep. = source of the features for the test document representation). Runs with a statistically significant improvement over at least one other competitor run are indicated by ▲ (Wilcoxon signed rank test; $p <= 0.05$).

RunID	micro-averaged f-score	macro-averaged f-score	feature language	test doc rep	site
CUT-C1R1▲	0.15	0.27	en/nl	asr	CUT
CUT-C1R2	0.11	0.14	en/nl	asr	CUT
CUT-C2R1	0.13	0.26	en/nl	asr/md	CUT
CUT-C2R2	0.13	0.17	en/nl	asr/md	CUT
dcu_run1▲	0.41	0.54	nl	asr	DCU
dcu_run2▲	0.25	0.47	en	asr	DCU
dcu_run3▲	0.28	0.58	nl	asr	DCU
dcu_run4	0.28	**0.59**	en	asr	DCU
dcu_run5	0.29	0.43	nl	md	DCU
MIRACLE-CNL	0.46	0.49	nl	asr	MIRACLE
MIRACLE-CNLEN	0.39	0.27	nl/en	asr	MIRACLE
MIRACLE-CNLMeta▲	0.47	0.47	nl	asr/md	MIRACLE
uams08m	0.18	0.17	nl	md	UAms
uams08asrd	0.10	0.41	nl	asr	UAms
uams08masrd	0.15	0.45	nl	asr/md	UAms
uams08asrde	0.09	0.14	nl/en	asr	UAms
uams08masrde	0.09	0.33	nl/en	asr/md	UAms
SINAI-Class-I	0.51	0.49	nl	asr	SINAI
SINAI-Class-II	**0.53**	0.51	en	asr	SINAI
SINAI-Class-I-Trans	0.10	0.40	nl	md	SINAI

type of performance that humans intuitively feel is best. Run dcu_run4 has a high macro-average since it sacrifices precision for recall in six of the ten classes and sacrifices recall for perfect precision in the other four. Run SINAI-JEAN-Class-II has a high micro-precision due to the fact that it assigns class labels in only three of the ten classes and in these three it performs well. In both cases the performance scores are high, but it can be argued that such a classification strategy might not appeal to a human user who would like to have a chance to find videos in all ten topic categories.

Humans might actually find the runs dcu_run1 and MIRACLE-CNLMeta to yield more usable performance. Here, the macro-precision and the micro-precision are better balanced. Note that these are two categories in which the improvement in system performance over multiple (although not all) competitor runs can be shown to be statistically significant at the $p <= 0.05$ level according to the Wilcoxon signed rank test.

Runs using speech recognition transcripts only. Runs that used speech transcripts alone were competitive with runs that used archival metadata alone or combined archival metadata with speech transcripts. These results indicate that there is potential for automatically assigning thematic category labels for videos that lack archival metadata.

Runs integrating English speech transcripts with Dutch information sources. No participant was able to exploit the speech recognition transcripts for the English language in order to improve performance. Participants conjectured that this might have been due to the fact that there was more Dutch spoken content in the documentaries than there was English spoken content, or that the English speech recognition transcripts had a higher word error rate than the Dutch speech recognition transcripts. In the future, we would like to try segmenting the videos using a language detector so that transcripts for English are generated only where English is spoken in the video.

Top performing classes. The breakdown of the performance of the runs over the individual thematic categories is shown in Table 2. It can be seen that *Music* is the class for which the best performance was achieved, cf., SINAI-Class-II, MIRACLE-CNLEN, MIRACLE-CNLMeta. It should be noted that this is also the class with the highest number of videos in the test corpus. The fact that relatively high performance levels could be attained for individual classes suggests that progress can still be achieved on the classification task if more research and development effort is devoted to it in the future.

Comments on the evaluation metric. We would like to mention here why the metrics chosen for VideoCLEF 2008 may not be adequate to reflect all relevant

Table 2. F-scores of each run reported for each individual class. Runs with a statistically significant improvement over at least one other competitor run are indicated by ▲ (Wilcoxon signed rank test; $p <= 0.05$). Full names of the thematic categories are: Archeology, Architecture, Chemistry, Dance, Film, History, Music, Paintings, Scientific research and Visual arts.

RunID	Arche	Archi	Chem	Dance	Film	Hist	Mus	Paint	Sci	Arts
Raw count correct videos	**7**	**0**	**0**	**3**	**3**	**10**	**22**	**3**	**4**	**5**
CUT-C1R1▲	0.44	0.00	0.00	0.00	0.20	0.14	0.16	0.20	0.00	0.14
CUT-C1R2	0.15	0.00	0.00	0.00	0.00	0.10	0.21	0.17	0.00	0.15
CUT-C2R1	0.44	0.00	0.00	0.00	0.20	0.14	0.08	0.20	0.00	0.13
CUT-C2R2	0.15	0.00	0.00	0.18	0.00	0.10	0.21	0.22	0.00	0.17
dcu_run1▲	0.60	1.00	0.00	0.50	0.46	0.24	0.47	0.57	0.40	0.00
dcu_run2▲	0.25	0.00	1.00	0.18	0.15	0.50	0.17	0.40	0.00	0.00
dcu_run3▲	0.30	0.00	1.00	1.00	0.14	0.40	0.71	0.14	0.18	0.00
dcu_run4	0.00	0.00	1.00	0.14	0.14	0.40	0.71	0.14	0.00	0.00
dcu_run5	0.00	1.00	1.00	0.00	0.33	0.18	0.53	0.00	0.00	0.00
MIRACLE-CNL	0.18	1.00	0.00	0.00	0.00	0.27	0.76	0.00	0.44	0.00
MIRACLE-CNLEN	0.18	0.00	0.00	0.00	0.29	0.34	0.79	0.00	0.35	0.27
MIRACLE-CNLMeta▲	0.33	0.00	0.00	0.22	0.00	0.46	0.79	0.00	0.35	0.17
uams08m	0.00	0.00	0.00	0.00	0.09	0.38	0.44	0.00	0.15	0.00
uams08asrd	0.25	0.00	0.00	0.00	0.06	0.18	0.00	0.00	0.33	0.00
uams08masrd	0.26	1.00	1.00	0.11	0.00	0.00	0.00	0.00	0.18	0.14
uams08asrde	0.17	0.00	0.00	0.00	0.00	0.27	0.00	0.11	0.11	0.22
uams08masrde	0.26	0.00	0.00	0.00	0.00	0.00	0.00	0.00	0.14	0.00
SINAI-Class-I	0.00	1.00	1.00	0.00	0.00	0.18	0.79	0.00	0.33	0.00
SINAI-Class-II	0.00	1.00	1.00	0.00	0.00	0.17	0.81	0.00	0.57	0.00
SINAI-Class-I-Trans	0.00	1.00	1.00	0.00	0.00	0.00	0.16	0.00	0.16	0.00

aspects of system performance. A high micro-average does not reflect how system performance is distributed over classes. However, the macro-average also has its shortcomings. In order to calculate the macro-average, it is necessary to define its behavior in cases where, for a given class, there are no videos in the collection that belong to that class. This case leads to division by zero when calculating the recall. We defined recall for these classes to be 1.0, since there exists no video of this class in the collection to which the classifier has failed to assign the correct class label. It is also necessary to define behavior in cases where a system fails to assign any videos to a particular class. This case leads to division by zero when calculating the precision. We defined precision for these classes to be 1.0, since there are no videos in the collection to which the classifier has erroneously assigned this class label. An alternative solution would be to average precision and recall only over classes that don't give rise to a division by zero problem. This solution might encourage systems to artificially inflate precision by ignoring difficult classes. We did not adopt this solution. An advantageous aspect of the Vid2RSS task is that the visualization of the results as RSS-feeds makes it easy for humans to grasp differences in run performance and to understand the mismatch between runs that might be most useful for real world applications and runs that achieve high performance. The visualization serves to compensate for evaluation metrics which do not present a well-rounded picture of system performance.

3 Translation Task

The translation task, which was a discretionary task, required participants to translate topic-based feeds from Dutch into a target language. The feeds consist of concatenations of feed items, each describing a video with that video's title, a small description derived from the archival metadata and a keyframe representing the video's content. One participant, CUT, carried out the translation task. CUT chose to translate the feeds into English and to use Google's AJAX language API.

Evaluation of the feeds was carried out using human assessment of adequacy and fluency performed by three assessors. All assessors had high-level mastery of both the source and target language. The assessment procedure was adapted from the TIDES *Specifications for human assessment of translation quality.*[4] Assessors were asked to assess fluency and adequacy of the translation of the feed item metadata (title and description) for each video on a five point scale. For fluency, they were asked to answer the question *How do you judge the fluency of this translation?* and assign points on the basis of the following answers: 5 = Flawless English, 4 = Good English, 3 = Non-native English, 2 = Disfluent English, 1 = Incomprehensible. For adequacy, they were asked to answer the questions *How much of the meaning expressed in the original Dutch version of the video title and description is also expressed in the English translation?* and assign points on the basis of the following answers: 5 = All, 4 = Most, 3 = Much,

[4] `http://projects.ldc.upenn.edu/TIDES/Translation/TransAssess04.pdf`

2 = Little, 1 = None. On average, assessors gave feed items a score of 2.82 for fluency and 3.49 for adequacy.

One of the main problems with the translation is that compound words often failed to be translated. For example the Dutch word "tiendelige," which is a compound that means consisting of ten parts, is written simply as "tiendelige" in the English translation. The word "concertpianist" meaning concert pianist, is translated as only "concert" with mention of pianist dropped. Another problem is that proper names were translated in cases when then are homonyms with other words. Despite these glitches, on the whole the translation was satisfactory and certainly demonstrated potential to allow non-Dutch speakers to understand the contents of the topic-based feeds.

4 Keyframe Extraction Task

Participants were provided with a segmentation of the videos into shots and a set of keyframes, one keyframe per shot. The segmentation and the shot level keyframe data was provided by Dublin City University [1]. The Vid2RSS keyframe extraction task required the participant to pick the keyframe from the provided set that best represented the semantic content of the video. Note that the task of automatically extracting a keyframe to represent a shot was not evaluated in VideoCLEF 2008. The set of keyframe level shots was taken as a given, and participants were required to chose the most appropriate keyframe from this set.

4.1 Keyframe Extraction Experiments and Results

Only one participant, MIRACLE, participated in the keyframe extraction task, which was discretionary. MIRACLE chose the keyframes based on the content of the speech recognition transcript associated with the shot. The MIRACLE team based their approach on the assumption that a representative shot for the video is a shot for which the spoken content is the least different from the spoken content of the video as a whole. They selected the keyframe of the shot whose speech recognition transcript vector has the closest cosine distance to the speech recognition transcript vector of the video as a whole.

Keyframe extraction was tested in a small scale user study in which subjects were given the title and description of a video and asked to chose between two candidates for a keyframe to represent the semantic content of that video. One candidate was the baseline human selected keyframe and the other was the keyframe automatically selected by MIRACLE. On an average 44% of the videos had automatic keyframes that were well selected, meaning that they were either identical to the manually selected keyframes (two cases), or that the subjects preferred the automatically selected keyframe to the manual one. These results suggest that the automatic keyframe extraction is a very viable competitor with manual keyframe selection.

4.2 Keyframe Extraction Evaluation

During the course of the user study, several important trends emerged that should be mentioned here since they serve to illustrate how challenging the keyframe extraction task actually is and the limited ability of the evaluation score to reflect the level of challenge. First, subjects often have a very mild preference for one keyframe over the other, implying that when a subject chooses the manually selected keyframe over the automatically extracted keyframe, it does not mean that the automatically extracted keyframe was inappropriate. Second, subjects' preference of keyframe was dependent on their knowledge of the topic of the video. In the case of a documentary about Frank Zappa, subjects who could recognize Frank Zappa by sight chose the keyframe picturing him, and rejected the other keyframe, which was technically a much clearer picture. The same phenomenon was observed for a video about World War II. Subjects that could identify Churchill by sight preferred the keyframe picturing him. Third, the comments of the experimental subjects reflected that their picks were dependent on whether they felt that the keyframe should depict the genre (documentary) or particular television series, or whether they felt it should depict the novel content of the particular video episode. In some cases, familiarity with the television series to which the video belonged impacted the subject's decision on which keyframe to pick. Subjects commented on some occasions that they simply preferred the "prettier" keyframe. One subject preferred the keyframe that made the video seem more enticing. Finally, there were a lot of details that subjects paid attention to. For example, in one case one keyboard shot was preferred above another because it was slightly shifted revealing knobs that showed the keyboard to be an electric one. Taking such details into account will probably remain a challenge for semantic keyframe extraction until far into the future.

5 Conclusions and Future Plans

The Vid2RSS task in the VideoCLEF 2008 track involved classification, translation and keyframe extraction performed on dual language video. All in all, evaluation of the classification runs demonstrated that there is quite a bit of improvement left to be achieved on this task. However, strong performance by classifiers for particular thematic categories, especially classifiers for videos treating the topic of *Music*, leads us to believe that improved performance can be achieved in the future. Further, the classification task demonstrated both Wikipedia and the Web at large to be promising sources of training data. Finally, simple approaches that recast the classification problem as an information retrieval task, yielded strong results.

The results of the discretionary translation task were satisfactory, although they would have been more revealing had more than a single site participated. A further comment should be made at this juncture concerning translation in the Vid2RSS task. Recall that the Vid2RSS task was originally motivated by the idea the multimedia archives contain multilingual content that is of high informational value if it can be made available to users who do not master the dominant language of the archive. The results of the translation task strongly suggest that the impasse for providing non-Dutch speakers access to usable content in a predominantly Dutch archive does not lie in the problem of providing usable translations of video titles and descriptions.

The results of the discretionary keyframe selection task were very encouraging. Here again, more elaborate conclusions would be supported had more than a single site participated. However, the small scale user study did demonstrate that automatically selected keyframes are competitive with manually selected keyframes. This result confirms the usefulness of the speech transcript associated with the video as a source of features for selecting a keyframe capable of semantically representing the video.

We were pleased with the success of the idea of having participants deliver their results as topic-based RSS-feeds. Feeds for the same class from different runs can be compared graphically in a feed reader with very minimal effort. Such a visualization makes it easier to get feedback on the usability of task results from potential end users who can gain a quick impression of the potential utility of the classification. The visualization aspect proved to be particularly important since, as mentioned above, we were not particularly convinced that the evaluation metrics chosen for this year's task truly reflected the potential usefulness of the results in an application.

We consider the VideoCLEF pilot track to have successfully demonstrated that the classification of dual language television documentaries into subject classes is a challenging and interesting task. In particular, we would like to note that the experiences of the pilot year of Vid2RSS strongly suggest that classification of video content is not always as easy as classification of broadcast news content, for which reasonable performance can be achieved in a relatively straightforward fashion [5]. We believe that a significant source of challenge lies in the fact that the videos contain a high proportion of unscripted speech in the form of interviews and discussions. Associated with such speech, which can be characterized as conversational, is a wide vocabulary, potentially sparse in on-topic words, and an informal style including disfluencies and sentence fragments. The combination of features derived from multiple sources (speech transcripts of both matrix and embedded language and metadata, where available) seems to offer a line of investigation with solid potential to improve classification performance, although such improvement was not realized in the initial year of the VideoCLEF track.

In future years, VideoCLEF plans to expand its data set and continue the classification task using a larger number of classes. We would also like to provide participants with training data in order to compare classification approaches that collect their own training data with approaches that use training data from the same domain. We plan to continue the keyframe selection task, most probably admitting the possibility of choosing more than a single keyframe to represent a video. Two new tasks are in planning for introduction in 2009. First, a task called *Affect and Appeal*, whose focus is classification of videos according to characteristics reaching beyond their informational content. For this task, participants will be asked to automatically predict which videos users find most "boring" or "outdated." Second, a task called *Finding Related Resources* which requires participants to identify English-language resources that will support information seekers in their understanding of Dutch language video.

References

1. Calic, J., Sav, S., Izquierdo, E., Marlow, S., Murphy, N., O'Connor, N.: Temporal video segmentation for real-time key frame extraction. In: Proceedings of the International Conference on Acoustics, Speech and Signal Processing, ICASSP 2002, Orlando, Florida (2002)
2. Huijbregts, M., Ordelman, R., de Jong, F.: Annotation of heterogeneous multimedia content using automatic speech recognition. In: Falcidieno, B., Spagnuolo, M., Avrithis, Y., Kompatsiaris, I., Buitelaar, P. (eds.) SAMT 2007. LNCS, vol. 4816, pp. 78–90. Springer, Heidelberg (2007)
3. Jackson, P.: Natural Language Processing for Online Applications. Natural Language Processing. John Benjamins, Philadelphia (2002)
4. Larson, M., Newman, E., Jones, G.: Classification of dual language audio-visual content: Introduction to the VideoCLEF 2008 pilot benchmark evaluation task. In: Proceedings of the SIGIR 2008 Workshop on Searching Spontaneous Conversational Speech, pp. 71–72 (2008)
5. Paass, G., Leopold, E., Larson, M., Kindermann, J., Eickeler, S.: SVM classification using sequences of phonemes and syllables. In: Elomaa, T., Mannila, H., Toivonen, H. (eds.) PKDD 2002. LNCS (LNAI), vol. 2431, pp. 373–384. Springer, Heidelberg (2002)
6. Pecina, P., Hoffmannova, P., Jones, G.J.F., Zhang, Y., Oard, D.W.: Overview of the CLEF 2007 cross-language speech retrieval. In: Peters, C., Jijkoun, V., Mandl, T., Müller, H., Oard, D.W., Peñas, A., Petras, V., Santos, D. (eds.) CLEF 2007. LNCS, vol. 5152, pp. 674–686. Springer, Heidelberg (2008)
7. Smeaton, A.F., Over, P., Kraaij, W.: Evaluation campaigns and TRECVid. In: MIR 2006: Proceedings of the 8th ACM International Workshop on Multimedia Information Retrieval, pp. 321–330. ACM, New York (2006)

MIRACLE at VideoCLEF 2008: Topic Identification and Keyframe Extraction in Dual Language Videos

Julio Villena-Román[1,3] and Sara Lana-Serrano[2,3]

[1] Universidad Carlos III de Madrid
[2] Universidad Politécnica de Madrid
[3] DAEDALUS - Data, Decisions and Language, S.A.
jvillena@it.uc3m.es, slana@diatel.upm.es

Abstract. This paper describes the participation of MIRACLE research consortium at the VideoCLEF track at CLEF 2008. We took part in both the main mandatory Classification task (classify videos of television episodes using speech transcripts and metadata) and the Keyframe Extraction task (select keyframes that represent individual episodes from a set of supplied keyframes). Our system for the first task is composed of two main blocks: the core system knowledge base and the set of operational elements that are needed to classify the speech transcripts of the topic episodes and generate the output in RSS format. For the second task, our approach is based on the assumption that the most representative fragment (shot) of each episode is the one with the lowest distance to the whole episode, considering a vector space model. In the classification task, our runs ranked 3rd (out of 6 participants) in terms of precision.

Keywords: Video retrieval, domain-specific vocabulary, thesaurus, linguistic engineering, information retrieval, indexing, relevance feedback, multilingual speech transcripts. VideoCLEF, VID2RSS, CLEF, 2008.

1 Introduction

MIRACLE team is a research consortium formed by research groups of three different Spanish universities (Universidad Politécnica de Madrid, Universidad Autónoma de Madrid and Universidad Carlos III de Madrid) along with DAEDALUS, a private company founded as a spin-off of these groups and a leading company in the field of linguistic technologies in Spain. MIRACLE has taken part in CLEF since 2003 in main tracks, including the main bilingual, monolingual and cross lingual tasks as well as ImageCLEF, WebCLEF, GeoCLEF and Question Answering activities.

This paper reviews our participation [1] at the VideoCLEF task [2], a new track for CLEF 2008. The goal of this track is focused on topic classification performed on dual language videos, i.e., assigning topic class labels to videos of television episodes, using speech recognition transcripts and, optionally, metadata records (title and description). We participated in the Classification and the Keyframe Extraction tasks. In

C. Peters et al. (Eds.): CLEF 2008, LNCS 5706, pp. 918–922, 2009.

the following sections, we will give an overview of our approach in both tasks and then the results of our experiments will be presented and analyzed.

2 Approach to the Classification Task

The architecture of the system for the Classification task is shown in Figure 1. The objective of the Training Set Extractor is to build a corpus that can be used as the core system knowledge base. The Classifier includes the set of operational elements that are needed to classify the speech transcripts of the topic episodes and generate the output in RSS format. Those operational elements include a search engine and a classifier, as well as auxiliary modules for text processing and RSS generation.

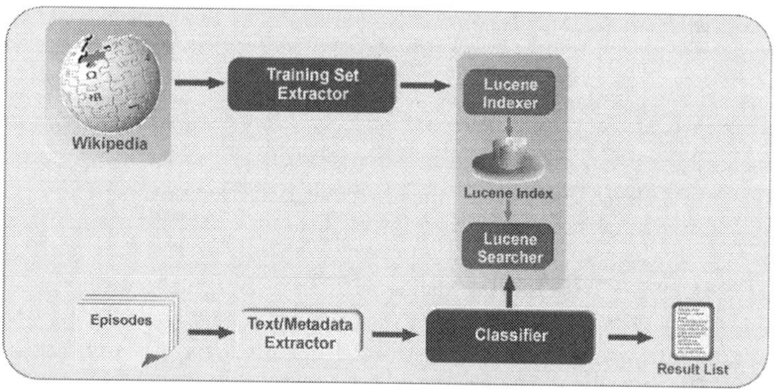

Fig. 1. Overview of the classification system

The first step is to gather the necessary data to train the classifier. The knowledge base was generated from Wikipedia articles. First, a matching was established between the topic classes provided for the task and the classification topics that Wikipedia uses for articles (encoded in metadata). Then a corpus of Wikipedia articles belonging to each of the 10 topic class was obtained for both task languages.

Each document is processed with the usual sequence of operations: text extraction, diacritics removal and conversion to lowercase, stopword filtering [3] and stemming (using standard stemmers from Porter [4]). The processed corpus is indexed with Lucene engine [5], building two different indexes, one for each language.

The classifier is based on the k-Nearest Neighbour algorithm [6]. To find the class for a given episode, the content is first processed as explained before. Then the set of resulting terms is used to build a query that is used to find the list of the top k most relevant (i.e., most similar) articles in the Wikipedia-based corpus. Finally, the class of the given episode is the most frequent class in the top k results.

3 Approach to the Keyframe Extraction Task

Our approach is that, in the context of a vector space model representation [7], the most representative fragment (shot) of each episode (represented by a vector) is the one with the lowest distance to the whole episode (also a vector). An overview of the system architecture is shown in Figure 2. The contents of both each shot and the whole episode are first processed the same way as training corpus, after extracting the text from the speech transcription. Based on the vector space model, a weighted vector is built for each episode and set of shots, representing the term frequency of the most significant terms in the episode. Finally the extraction module selects the keyframe belonging to the most representative shot in the episode whose vector is nearest to the vector of the whole episode, using the cosine distance [7].

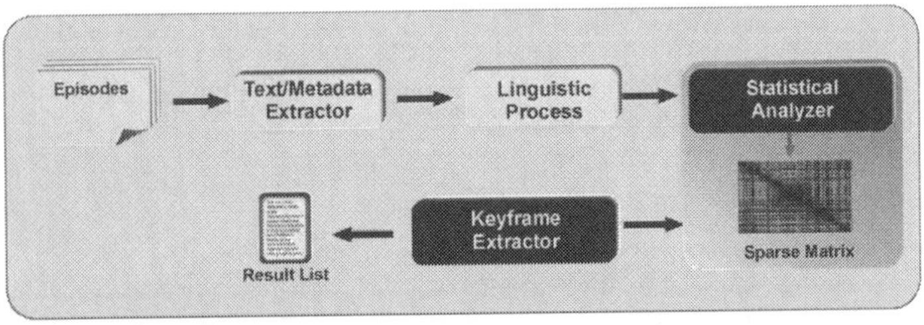

Fig. 2. Overview of the keyframe extraction system

4 Experiments and Results

Three runs were submitted. In short, "CNL" run only uses the index for the Dutch corpus, "CNLEN" uses both indexes and gathers together results from any of them, and "CNLMeta" uses the Dutch index but also includes the episode metadata to build the query for the retrieval engine.

Table 1. Classification task results

	Precision			Recall		
	CNL	CNLEN	CNLMeta	CNL	CNLEN	CNLMeta
Archaeology	0.25	0.25	0.40	0.14	0.14	0.29
Architecture	0.00	0.00	0.00	1.00	1.00	1.00
Chemistry	0.00	0.00	0.00	1.00	1.00	1.00
Dance	0.00	0.00	0.13	0.00	0.00	0.67
Film	1.00	0.25	1.00	0.00	0.33	0.00
History	0.25	0.26	0.38	0.30	0.50	0.60
Music	**0.64**	**0.65**	**0.65**	**0.95**	**1.00**	**1.00**
Paintings	1.00	0.00	1.00	0.00	0.00	0.00
Scientific Res.	0.29	0.21	0.21	1.00	1.00	1.00
Visual Arts	1.00	0.20	0.14	0.00	0.40	0.20
Microaverage	**0.43**	0.29	0.37	0.51	0.61	**0.65**

Table 1 shows the values of precision and recall obtained by the different runs in the classification task. The best precision is achieved when only the Dutch transcription is used. When the knowledge base and the transcription in English are involved, results are noticeably and significantly worse. We verified that this is directly motivated by the fact that the dominant language of the episodes is Dutch. Comparing to other groups, we successfully ranked 3rd out of 6 participants in terms of precision, 2nd in terms of recall and also F-score (not shown in the table).

Regarding the keyframe extraction task, we were the only group that submitted results [2]. Thus, the evaluation was manually made. Five native Dutch speakers were presented with the title and the description of each video episode along with two keyframes, one manually extracted and one automatically extracted provided by us. They were asked to choose which keyframe they preferred. On average, the subjects chose the automatic over the manually selected keyframe in 15.2 cases (41.08%). These promising figures indicate that the automatically extracted keyframes may be strong competitors with the manual ones in the short- or middle-term future.

5. Conclusions and Future Work

As shown in Table 1, the best modelled class is "Music", which is the class that owns the higher number of Wikipedia articles in the training set. This fact may indicate a certain relationship between the size of the knowledge base associated to a given class and results achieved for it, although this conclusion has to be further analyzed. This can also explain why results for Dutch are considerably better (0.43 vs. 0.29), as the knowledge base for English is smaller than the one available for Dutch. This fact indicates that the architecture of the system and provided algorithms are useful, but more effort must be invested to improve the knowledge base, both in its volume (coverage) and the pre-processing activities.

Despite the subjectivity of the keyframe extraction task and lack of any reference experiment to which compare our own system, these results are promising and encourage us to keep on this line of research for future participations.

Acknowledgements. This work has been partially supported by the Spanish R+D National Plan, project BRAVO (Multilingual and Multimodal Answers Advanced Search–Information Retrieval), TIN2007-67407-C03-03, and by the R+D Regional Plan of Madrid, project MAVIR (Enhancing the Access and the Visibility of Networked Multilingual Information), S-0505/TIC/000267.

References

1. Villena-Román, J., Lana-Serrano, S.: MIRACLE at VideoCLEF 2008: Classification of Multilingual Speech Transcripts. In: Working Notes of the 2008 CLEF Workshop, Aarhus, Denmark (2008)
2. Larson, M., Newman, E., Jones, G.J.F.: Overview of VideoCLEF 2008: Automatic Generation of Topic-based Feeds for Dual Language Audio-Visual Content. In: Peters, C., et al. (eds.) CLEF 2008. LNCS, vol. 5706, pp. 906–917. Springer, Heidelberg (2009)

3. University of Neuchatel. Page of resources for CLEF, http://www.unine.ch/info/clef (Visited 09/11/2008)
4. Porter, M.: Snowball stemmers and resources page, http://www.snowball.tartarus.org (Visited 09/11/2008)
5. Apache Lucene project, http://lucene.apache.org (Visited 09/11/2008)
6. Witten, I.H., Frank, E.: Data Mining: Practical machine learning tools and techniques, 2nd edn. Morgan Kaufmann, San Francisco (2005)
7. Baeza-Yates, R., Ribeiro-Prieto, B.: Modern Information Retrieval. Addison Wesley, Reading (1999)

DCU at VideoClef 2008

Eamonn Newman and Gareth J.F. Jones

Centre for Digital Video Processing, Dublin City University, Dublin 9, Ireland
{enewman,gjones}@computing.dcu.ie
http://www.cdvp.dcu.ie

Abstract. We describe a baseline system for the VideoCLEF Vid2RSS task in which videos are to be classified into thematic categories based on their content. The system uses an off-the-shelf Information Retrieval system. Speech transcripts generated using automated speech recognition are indexed using default stemming and stopping methods. The categories are populated by using the category theme (or label) as a query on the collection, and assigning the retrieved items to that particular category. Run 4 of our system achieved the highest f-score in the task by maximising recall. We discuss this in terms of the primary aims of the task, i.e., automating video classification.

Keywords: Classification, Information Retrieval, Automatic Speech Recognition.

1 Introduction

The VideoClef Vid2RSS task required users to classify videos into one (or more) of a set of categories. Audio content consists primarily of Dutch with some embedded English content. The data provided consists of automatic speech recognition (ASR) transcripts (generated independently using Dutch and English ASR systems), shot boundary keyframes, and catalogue metadata (in Dutch). Each category is then published as an RSS feed. The system described in this paper is based on an Information Retrieval approach. We built a standard free text index using the ASR transcripts and associated metadata as the content.

2 System Description

We used the open source Lucene Search Engine technology [1] as the base technology for our system. Dutch-language content was stopped, stemmed and tokenised using Lucene's built-in Dutch analyser, `DutchAnalyzer`[1]. English-language content was stopped and tokenised by the Lucene default tokeniser, `StandardAnalyzer`[2]. The `StandardAnalyzer` does not perform any stemming of tokens.

[1] org.apache.lucene.analysis.nl.DutchAnalyzer
[2] org.apache.lucene.analysis.standard.StandardAnalyzer

C. Peters et al. (Eds.): CLEF 2008, LNCS 5706, pp. 923–926, 2009.

2.1 Run Configurations

Five separate runs were prepared and submitted to the task. The runs varied in terms of both system configuration and the data which was used.

1. **Dutch ASR transcripts:** In this run, we indexed the entire set of Dutch ASR transcripts (the `FreeTextAnnotation` elements). The index was queried with the labels in the order given in Table 1 and each item was classified into a single category.
2. **English ASR transcripts:** This is identical to Run 1, using English ASR transcripts and translations of the category labels as queries.
3. **Dutch ASR with query expansion:** We ran the same queries as Run 1, but added an additional step of query expansion in order to improve the recall of certain categories. Some categories in earlier runs had nothing assigned to them. Because of this, we chose to allow items to be assigned to multiple categories. Queries were expanded by performing an initial query which consisted of just the category label. We take the first 10 retrieved documents and extract the 5 most frequently occuring terms in each. We process this set of 50 terms to remove any duplicates. The remaining terms are combined with the original query to form the expanded query.
4. **English ASR with query expansion:** This is identical in method to Run 3, but the data now consists of the English ASR transcripts, rather than the Dutch.
5. **Dutch metadata:** We indexed the catalogue metadata which were supplied in the data sets. Specifically, we used the `description` elements from the metadata documents. Once again, the Dutch category term labels were used as queries, and the items were restricted to appear in one feed only.

2.2 Category Order

The categories were ordered from most specific to least specific, as in Table 1. For each category, a query was made to the IR system using the category name as the query keyword. All retrieved items were labelled as belonging to that category.

Table 1. Category Label Order

Dutch	English
archeologie	archeology
architectuur	architecture
chemie	chemistry
dansen	dance
schilderijen	paintings
wetenschappelijk onderzoek	scientific research
beeldende kunst	visual arts
geschiedenis	history
film	film
muziek	music

The ordering meant that when an item was retrieved, it was placed into the most specific category possible. For our submitted runs, in Runs 1, 3, and 5, a retrieved item was placed only into the first category for which it was retrieved. In Runs 2 and 4, it was placed in all categories for which it was retrieved. This restriction was imposed to improve the precision of the classification task, since labels such as "film" were very general and tended to capture most, if not all, of the items.

3 Results

In Table 2 we present the retrieval scores attained by our system runs. The metrics are defined in Section 2.2 of the Track Overview paper [2]. A direct comparison of Runs 1 and 2 suggests that the Dutch transcripts were more useful in identifying the subject categories than the English ones. Indeed, the English transcripts had the poorest f-scores at both micro and macro level. This is most likely attributable to the fact that the majority of the dialogue was in Dutch and so contained less "noise" than the English counterparts. Processing of the ASR transcripts to identify the points at which the language changed would allow for the combination of transcripts (or the removal of erroneous segments) which would improve classification performance.

Runs 3 and 4 used query expansion to add new keywords to the queries and allowed items to be placed in multiple categories. As we can see, this relaxation resulted in a large drop in the micro-average precision scores of these systems; conversely, the micro-average recall is much higher in these runs. As items were placed in multiple categories the chances of an item being correctly classified were much greater, however the number of false positives also increased.

As can be seen from the results, Run 5 performed particularly well in terms of precision, and relatively well (when compared to our other runs) in terms of recall. However, since this was on the metadata and not on the ASR transcripts, it cannot be directly compared to the others. The higher precision scores do suggest that there may be merit in combining the different data sets available.

One drawback that is immediately obvious with this system is that it is not possible to guarantee that all items will be classified. If an item is not retrieved for any of the queries, then it will not be placed in any of the category feeds. As it happens, this was not the case for any of the runs with this particular data set (probably due to the presence of highly generic labels such as "film" and "music").

Table 2. Vid2RSS Scores for Runs 1 to 5

metric	Run 1	Run 2	Run 3	Run 4	Run 5
micro-average precision	0.50	0.32	0.16	0.17	0.83
micro-average recall	0.35	0.21	0.91	0.72	0.18
f-score micro-average	0.41	0.25	0.28	0.28	0.29
macro-average precision	0.54	0.62	0.42	0.50	0.93
macro-average recall	0.55	0.38	0.90	0.70	0.28
f-score macro-average	0.54	0.47	0.58	0.59	0.43

Additionally, the number of terms added in the query expansion phase could be reduced. The maximum for this was 50, but elimination of duplicates meant that the size of the set was generally much smaller. Nevertheless, it seems that too many terms were added to the queries, and this is supported by the difference in micro-average precision between Runs 1 and 3 and Runs 2 and 4. To overcome this, we plan to implement a standard query expansion method where this can be controlled.

4 Conclusions

On comparing the results of our runs with those of other participants (see Track Overview [2] for full comparative analysis), Run 4 was shown to have the highest f-score of all systems when averaged over the individual f-scores for each of the topic classes (macro-average). This was attained by deliberately promoting recall over precision, in allowing videos to be classified under multiple topics. Furthermore, as mentioned in the Overview paper, the imbalance between precision and recall may not result in a particularly useful system. From this point of view, Run 1, which has the closest balance between precision and recall, may be seen as most useful to a human evaluator.

The results suggest that there is room for improvement in our system. The precision scores could be improved by finer-grained query expansion, which will be examined in future experiments. Additionally, the performance on the English-language content could be improved by use of a stemming algorithm, such as Porter [3].

Acknowledgements

This work was supported by the EU IST programme of the 6th FP for RTD under project MultiMATCH contract IST-033104.

References

1. Apache Software Foundation. Lucene: Java-based Indexing and Searching technology, `http://lucene.apache.org/`
2. Larson, M., Newman, E., Jones, G.J.F.: Overview of VideoCLEF2008: Automatic Generation of Topic-based Feeds for Dual Language Audio-Visual Content. In: Peters, C., et al. (eds.) CLEF 2008. LNCS, vol. 5706, pp. 906–917. Springer, Heidelberg (2009)
3. Porter, M.: An algorithm for suffix stripping. Program (July 1980)

Using an Information Retrieval System for Video Classification

José Manuel Perea-Ortega, Arturo Montejo-Ráez, Manuel Carlos Díaz-Galiano, María Teresa Martín-Valdivia, and L. Alfonso Ureña-López

SINAI Research Group*, Computer Science Department, University of Jaén, Spain
{jmperea,amontejo,mcdiaz,maite,laurena}@ujaen.es

Abstract. This paper describes a simple approach to resolve the video classification task. This approach consists in applying an Information Retrieval (IR) system as classifier. We have generated a document collection for each topic class predefined. This collection has been composed of documents retrieved using the Google[1] search engine. Following the IR strategy, we have used the speech transcriptions of the videos as textual queries. The results obtained show that an IR system can perform well as video classifier if the speech transcriptions of the videos to classify have good quality.

1 Introduction

In this paper we describe a simple approach to resolve the video classification task. Multimedia content-based retrieval is a challenging research field that has drawn significant attention in the multimedia research community. With the rapid growth of multimedia data, methods for effective indexing and search of visual content are decisive.

The video classification can be considered a subtask of the multimedia content-based retrieval. For instance, one of its applications is the automatic generation of RSS feeds specific to a particular information need. In addition, they could be personalized to a particular language preference. Therefore, the aim of the video classification task is to assign specific class labels to videos.

The aim of this work has been the study of the problem of the video classification task, and the development of a basic architecture which approaches it. We have some experience in the field of multimedia video retrieval [3] and in image retrieval [2][1][5].

This paper is organized as follows: Section 2 describes the whole system. Then, in the Section 3, experiments and results are described. Finally, in the Section 4, the conclusions are presented.

* http://sinai.ujaen.es
[1] http://www.google.com

C. Peters et al. (Eds.): CLEF 2008, LNCS 5706, pp. 927–930, 2009.

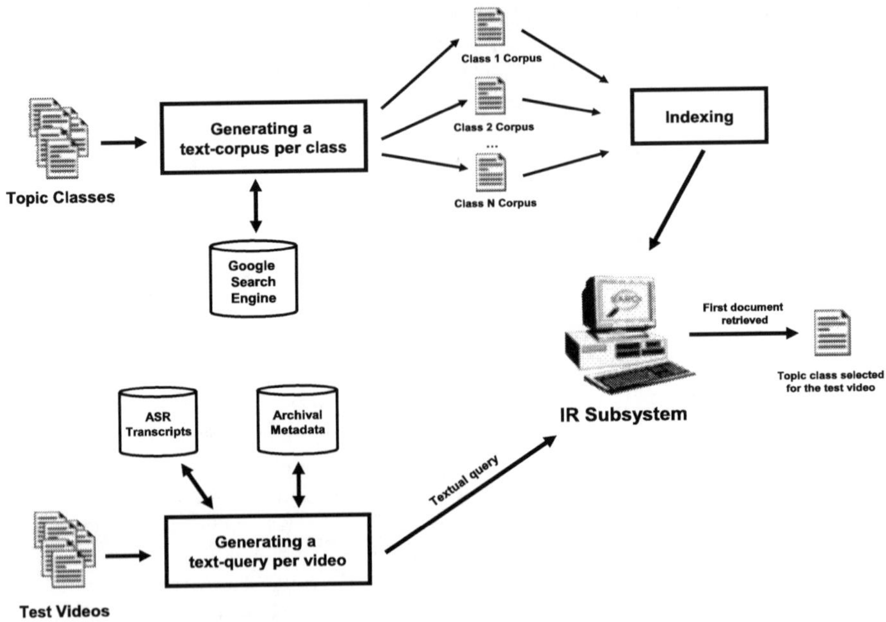

Fig. 1. Basic architecture of the SINAI video classifier

2 System Overview

The overall architecture of our automatic video classifier is based on the use of a particular IR system as text-classifier. In our experiments we have used Lemur[2] as IR system.

In our approach we have two main processes:

- **Generating a text-corpus per class.** We have generated a textual corpus per topic class. This corpus corresponds to the ten top retrieved results by Google, using the topic word (e.g. *Architecture*) as search query. The use of Google has been found useful in other tasks like Robust Retrieval[6]. These results have been combined into one single document. Therefore, a document per class is obtained after this process. Each document per class is indexed by means of IR system. This index will be used for retrieving the speech transcriptions preprocessed of each test video (textual queries).
- **Generating a text-query per video.** We have used the textual information available from Automatic Speech Recognition (ASR) output in order to generate the textual queries for each test video. This data has been preprocessed using the *Dutch stemmer* from Snowball[3] for Dutch language and *Porter stemmer*[7] for English. We have also discarded the *stop-words* for both languages.

[2] http://www.lemurproject.org

[3] http://snowball.tartarus.org/algorithms/dutch/stemmer.html

For each textual query generated the IR subsystem retrieves the document class more relevant, using the standard *TF·IDF* weighting schema. Figure 1 shows the basic architecture of our approach.

3 Experiments and Results

In order to evaluate our video classifier, we have used the video files provided by the organization of VideoCLEF [4]. This video data are Dutch television documentaries and contain Dutch as dominant language, but also contain a high proportion of spoken English, such as interviewed guests. On the other hand, we have the speech transcriptions of the videos and *metadata* files, which contain the program titles and short descriptions of the content of each video file.

In the experiments carried out in this paper, we have studied the difference between using speech transcriptions only and adding to them the information of *metadata* files. In addition, we have used the speech transcriptions for each language supplied (Dutch and English), but the *metadata* files have only been provided in Dutch language.

The results using micro-averaged and macro-averaged measures are shown in Tables 1 and 2.

Analyzing the results obtained, we can observe that English language leads to worse results than Dutch. This could be an expected behavior, due to the lower relevance of this language in the corpus compared to Dutch. In micro-averaged values, the use of *metadata* brings slightly better precision and recall measurements, although in macro-averaged results this improvement is only present in recall.

Table 1. Micro-averaging results

Experiment Description	Language	P	R	F1
Using speech transcriptions only	Dutch	0.65	0.42	**0.51**
Using speech transcriptions only	English	0.13	0.09	0.10
Using speech transcriptions and *metadata*	Dutch	0.68	0.44	0.53

Table 2. Macro-averaging results

Experiment Description	Language	P	R	F1
Using speech transcriptions only	Dutch	0.91	0.34	0.49
Using speech transcriptions only	English	0.68	0.28	0.40
Using speech transcriptions and *metadata*	Dutch	0.89	0.36	**0.51**

4 Conclusions and Future Work

In this paper we have presented a simple approach to resolve the video classification task. This approach consists in applying an Information Retrieval (IR) system as content-based classifier. The main processes of our approach are the generation of a document collection per topic class and the generation of a text-query per video to classify. Then, for each textual query generated, the IR system retrieves the document class more relevant.

Our results show that, despite the simplicity of our system, transcriptions are a good source of information for video classification. In other hand, the use of metadata from videos improves the results. In any case, some enhancements on the system can be performed, by selecting additional sources of learning data: for the future, we will work on a system that uses Wikipedia articles too.

Acknowledgments

This work has been supported by a grant from the Spanish Government, project TIMOM (TIN2006-15265-C06-03), and the RFC/PP2006/Id_514 granted by the University of Jaén.

References

1. Díaz-Galiano, M.C., García-Cumbreras, M.A., Martín-Valdivia, M.T., Montejo-Ráez, A., Ureña-López, L.A.: SINAI at ImageCLEF 2006. In: Peters, C., Clough, P., Gey, F.C., Karlgren, J., Magnini, B., Oard, D.W., de Rijke, M., Stempfhuber, M. (eds.) CLEF 2006. LNCS, vol. 4730, pp. 119–126. Springer, Heidelberg (2007)
2. Díaz-Galiano, M.C., García-Cumbreras, M.A., Martín-Valdivia, M.T., Montejo-Ráez, A., Ureña-López, L.A.: SINAI at ImageCLEF 2007. In: Peters, C., Jijkoun, V., Mandl, T., Müller, H., Oard, D.W., Peñas, A., Petras, V., Santos, D. (eds.) CLEF 2007. LNCS, vol. 5152, pp. 137–142. Springer, Heidelberg (2008)
3. Díaz-Galiano, M.C., Perea-Ortega, J.M., Martín-Valdivia, M.T., Montejo-Ráez, A., Ureña-López, L.A.: SINAI at TRECVID 2007. In: Proceedings of the TRECVID 2007 Workshop, TRECVID 2007 (2007)
4. Larson, M., Newman, E., Jones, G.J.F.: Overview of VideoCLEF 2008: Automatic Generation of Topic-based Feeds for Dual Language Audio-Visual Content. In: Peters, C., et al. (eds.) CLEF 2008. LNCS, vol. 5706, pp. 906–917. Springer, Heidelberg (2009)
5. Martín-Valdivia, M.T., García-Cumbreras, M.A., Díaz-Galiano, M.C., Ureña-López, L.A., Montejo-Ráez, A.: SINAI at ImageCLEF 2005. In: Peters, C., Gey, F.C., Gonzalo, J., Müller, H., Jones, G.J.F., Kluck, M., Magnini, B., de Rijke, M., Giampiccolo, D. (eds.) CLEF 2005. LNCS, vol. 4022, pp. 113–120. Springer, Heidelberg (2006)
6. Martínez-Santiago, F., García-Cumbreras, M.A., Montejo-Ráez, A.: SINAI at CLEF 2007 Ad Hoc Robust track 2007: Applying google search engine for robust cross-lingual retrieval. In: Peters, C., Jijkoun, V., Mandl, T., Müller, H., Oard, D.W., Peñas, A., Petras, V., Santos, D. (eds.) CLEF 2007. LNCS, vol. 5152, pp. 137–142. Springer, Heidelberg (2008)
7. Porter, M.F.: An algorithm for suffix stripping. Program 14, 130–137 (1980)

VideoCLEF 2008: ASR Classification with Wikipedia Categories

Jens Kürsten, Daniel Richter, and Maximilian Eibl

Chemnitz University of Technology
Faculty of Computer Science, Chair Computer Science and Media
Straße der Nationen 62
09111 Chemnitz, Germany
{jens.kuersten,daniel.richter,eibl}@cs.tu-chemnitz.de

Abstract. This article describes our participation at the *VideoCLEF track*. We designed and implemented a prototype for the classification of the Video ASR data. Our approach was to regard the task as text classification problem. We used terms from Wikipedia categories as training data for our text classifiers. For the text classification the Naive-Bayes and kNN classifier from the WEKA toolkit were used. We submitted experiments for classification task 1 and 2. For the translation of the feeds to English (translation task) Google's AJAX language API was used. Although our experiments achieved only low precision of 10 to 15 percent, we assume those results will be useful in a combined setting with the retrieval approach that was widely used. Interestingly, we could not improve the quality of the classification by using the provided metadata.

Keywords: Evaluation, Experimentation, Automatic Speech Transcripts, Video Classification.

1 Introduction

In this article we describe the general architecture of a system for the participation at the *VideoCLEF track*. The task was to categorize dual-language video into 10 given classes based on provided ASR transcripts [3]. The participants had to generate RSS Feeds that contain the videos for each of the 10 categories. The content of the RSS items for each of the videos was also given[1].

Our approach to solve the problem mainly relies on the application of a text classifier. We use the textual content of Wikipedia[2] categories that are identical or at least highly related to the 10 given categories. The classification of the ASR transcripts will be done by classifiers from the WEKA toolkit [4].

The remainder of the article is organized as follows. In sections 2 and 3 we describe the architecture of our system and present the results of our experiments. A summary of the result analysis is given in section 4. Finally, we conclude our experimental results and give and outlook to future work.

[1] http://ilps.science.uva.nl/Vid2RSS/Vid2RSS08/Vid2RSS08.html
[2] http://en.wikipedia.org

C. Peters et al. (Eds.): CLEF 2008, LNCS 5706, pp. 931–934, 2009.

2 System Architecture

The general architecture of the system we used is illustrated in figure 1. Besides the given input data (archival metadata, ASR transcripts and RSS items) we used a snapshot of the English and the Dutch Wikipedia as training data. We extracted terms related to the given categories by using the JWPL library [5] and applied a category mapping. These extracted terms were later used to train our text classifiers. A detailed description of all operational steps is given in [1].

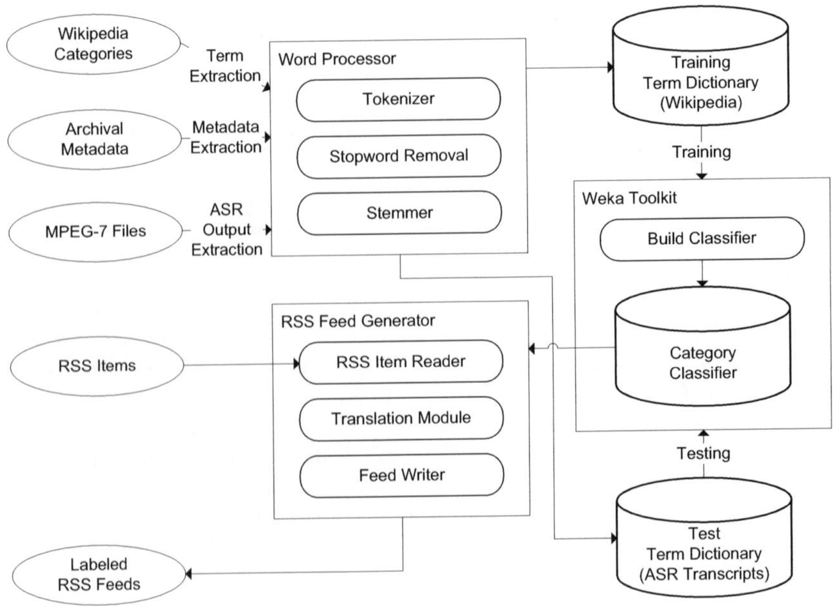

Fig. 1. General System Architecture

3 Experimental Setup and Results

We submitted two experiments for each of the two classification tasks. The results of the evaluation are presented in table 1. For our experiments we alternated three out of six essential parameters of the classification:

- Depth of Wikipedia Category Extraction (D)
- Maximum Number of Training Terms (TMAX)
- Training Term Duplicate Deletion (WT); 0.5 for deletion of terms that appear in at least 50 percent of the training categories

The remaining three parameters were kept fixed, i.e. the frequency-based selection of the extracted terms by using only the most frequent terms, removed duplicates from the extracted test data, when it appeared in at least 50 percent of the test samples and we used a combination of the Naive-Bayes and kNN (k=4) classifiers from the WEKA toolkit.

Table 1. Experimental Results based on the Evaluation Data

TMAX	*D*	*WT*	*P*	*R*
3000	3	0.2	0.15	0.14
5000	4	0.5	0.10	0.12
3000	3	0.2	0.13	0.12
5000	4	0.5	0.12	0.14

The performance of our experiments was not very good and did not meet our expectations and observations on the development data. Interestingly, using the metadata in classification task 2 did not improve the classification performance in either case. The main reason for that was the classification on a term-by-term basis, i.e. each term of the document was classified separately and the final category of the document was set by determining the category with the highest occurrence frequency. Due to the ratio of metadata and transcribed document terms this approach results in a very low weight of the extracted archival metadata terms.

Table 2. Assessment of the Translation

Criterion	*Ass. 1*	*Ass. 2*	*Ass. 3*	*Average*
fluency	2.88	2.65	2.93	2.82
adequacy	3.53	3.15	3.80	3.49

Additionally, we submitted a translation of the RSS Feeds. The translation was evaluated by three assessors in terms of fluency (1-5) and adequacy (1-5). The results are summarized in table 2.

4 Result Analysis - Summary

The following items conclude our observations of the experimental evaluation:

- *Classification task 1:* The quality of the video classification was not as good as expected, both in terms of precision and in terms of recall.
- *Classification task 2:* Surprisingly, the quality of the video classification could not be improved by utilizing the given metadata. The reason for that was our approach to classify the video documents on a term-by-term basis.
- *Translation task:* The translation of the RSS Feeds was quite good, but there is also room for improvement, especially in terms of fluency.

5 Conclusion and Future Work

The experiments showed that the classification of dual-language video based on ASR transcripts is a quite hard task. Nevertheless, we presented an idea to tackle the problem. But there are a numerous points to improve the system. The two most important problems are the size of the training data on the one hand and the balance of the categories on the other hand. We consider omitting the balancing step of the training data and shrink its size in further experiments. Another approach to improve the classification rate might be weighting the test data based on an approximated distribution of the categories in the video collection. This could be a good indicator on how to find the correct classes for a given video. Finally, we intend to integrate the video classification prototype into our Xtrieval framework [2].

Acknowledgments

This work was partially accomplished in conjunction with the project *sachsMedia*, which is funded by the *Entrepreneurial Regions*[3] program of the German Federal Ministry of Education and Research.

References

1. Kürsten, J., Richter, D., Eibl, M.: VideoCLEF 2008: ASR Classification based on Wikipedia Categories. In: Working Notes for the CLEF 2008 Workshop, Aarhus, Denmark, September 17-19 (2008)
2. Kürsten, J., Wilhelm, T., Eibl, M.: Extensible Retrieval and Evaluation Framework: Xtrieval. In: LWA 2008: Lernen - Wissen - Adaption, Workshop Proceedings, Würzburg (October 2008)
3. Larson, M., Newman, E., Jones, G.: Overview of VideoCLEF 2008: Automatic Generation of Topic-based Feeds for Dual Language Audio-Visual Content. In: Peters, C., et al. (eds.) CLEF 2008. LNCS, vol. 5706, pp. 906–917. Springer, Heidelberg (2009)
4. Witten, I.H., Frank, E.: Data mining: practical machine learning tools and techniques, 2nd edn. Elsevier, Morgan Kaufman, Amsterdam (2005)
5. Zesch, T., Müller, C., Gurevych, I.: Extracting Lexical Semantic Knowledge from Wikipedia and Wiktionary. In: Proceedings of the Sixth International Language Resources and Evaluation, LREC 2008 (2008)

[3] The Innovation Initiative for the New German Federal States.

Metadata and Multilinguality in Video Classification

Jiyin He, Xu Zhang, Wouter Weerkamp, and Martha Larson*

ISLA, University of Amsterdam, 1098 SJ Amsterdam
{j.he,x.zhang,w.weerkamp,m.a.larson}@uva.nl

Abstract. The VideoCLEF 2008 Vid2RSS task involves the assignment of thematic category labels to dual language (Dutch/English) television episode videos. The University of Amsterdam chose to focus on exploiting archival metadata and speech transcripts generated by both Dutch and English speech recognizers. A Support Vector Machine (SVM) classifier was trained on training data collected from Wikipedia. The results provide evidence that combining archival metadata with speech transcripts can improve classification performance, but that adding speech transcripts in an additional language does not yield performance gains.

Keywords: Video classification, SVM, speech recognition.

1 Introduction

The University of Amsterdam participated in the Vid2RSS,[1] task of the VideoCLEF track in the 2008 Cross Language Evaluation Forum (CLEF) campaign.[2] The task involves classification on a video corpus containing episodes of dual language television programs. Dutch is the main language, which we call the *matrix* language, and English, which we call the *embedded* language, is spoken during interviews. The classification task is formulated as follows: given a set of thematic categories, the participants should assign the right category label or labels to each video. The data set includes automatic speech recognition (ASR) transcripts (both languages) and archival metadata (Dutch only). A final feature of the task is the lack of training data: participants should develop methods to collect their own training data. Further details concerning the task can be found in [1]. We experimented with (i) the impact of metadata on classification, and (ii) the multilinguality of the data set. Our larger goal is to develop and refine techniques for classification of multimedia content, and especially of speech content, in a multilingual setting.

We chose to use Support Vector Machine (SVM) classifiers for all our runs. We opt for the SVM since it has been reported to perform well in general on text classification problems [2,3] and this performance has been demonstrated

* Currently at ICT, Delft University of Technology, 2628 CD Delft, Netherlands.
[1] http://ilps.science.UvA.nl/Vid2RSS
[2] http://www.clef-campaign.org

C. Peters et al. (Eds.): CLEF 2008, LNCS 5706, pp. 935–938, 2009.

Table 1. Submitted runs and their metadata and multilingual dimensions

Settings	uams08m	uams08asrd	uams08masrd	uams08asrde	uams08masrde
Dutch transcripts		X	X	X	X
English transcripts				X	X
metadata	X		X		X

to transfer to ASR transcripts that contain a significant level of noise [4]. Also, a linear kernel is selected based on the results of the exploratory experiments. These results were consistent with previous work [4].

Test data was not provided for VideoCLEF 2008. Instead, we trained our classifiers with data collected from Wikipedia.[3] The data from this source is freely obtainable, treats the right topics, and is available in multiple languages (including Dutch and English). The main disadvantage of using Wikipedia is its dissimilarity to the test data: Wikipedia content is created to be read and is much more structured and uses a language style more formal than that of the conversational speech present in the task video.

We set our experimental focus on two dimensions of the task: *metadata* and *language*. Exploring the *metadata* dimension involves attempting to understand the impact of using the archival metadata associated with the video as a source of features for classification. In contrast to the metadata, which has high informational content and is relatively noise free, the ASR transcripts can be expected to contain relatively fewer informational terms and a high level of recognition errors. Although terms that reflect the video topic are without doubt present, speech transcripts are diluted with the kinds of vocabulary typical of conversational speech, namely reflecting social convention, expressing feelings and opinions and drawing connections between entities and concepts not always explicitly mentioned. Exploring the *language* dimension involves investigating the contribution of ASR transcripts generated by a speech recognizer that transcribes the embedded language. We select combinations of metadata-derived and ASR-derived features for a total of five runs, as shown in Table 1.

2 Experimental Setup

For the implementation of the SVM classifier, we use the Least Square-SVM (LS-SVM) toolbox.[4] The LS-SVM is similar to Vapnik's SVM formulation, but instead of solving the Quadratic Programming (QP) problem, it solves a set of linear equations. For tuning the parameters, we do a grid search with leave-one-out cross-validation on the training data. Since we are using the linear kernel, the only parameter that needs to be estimated is C, which controls error rate.

We select the training data by performing retrieval in the Wikipedia collection using the class label as query. For each class, we collect the top 200 relevant Wikipedia pages as positive examples. We use the "one-against-all" strategy to

[3] http://www.wikipedia.org
[4] http://www.esat.kuleuven.ac.be/sista/lssvmlab/

construct the training set, i.e., the target class supplies the positive examples and all the rest of the classes supply the negative examples. A classifier is trained for each class separately. We use the χ^2 statistic to select the relevant features for each class, which has proven useful in previous work [5]. For our experiments, we heuristically select the 80 top features given the χ^2 values.

3 Results and Observations

The results of the official runs are listed in Table 2. Below, we discuss our observation concerning the use of metadata and of ASR transcripts from multiple languages as sources of classification features.

3.1 Impact of Metadata

The use of archival metdata improves performance in terms of micro-averaged f-score (i.e., cf. uams08m vs. uams08asrd). The classifiers tend to assign their thematic category labels to too many documents, reflected in the relatively high recall compared to the precision. The picture is different, when results are evaluated using macro-averaging. Here, results are best when metadata is combined with speech transcripts. The reason for the difference between macro-averaging and micro-averaging is that macro-averaging can assign disproportionately large weights to thematic categories whose classifiers perform well, even though these classifiers are relevant to only a small number of videos. We would like to note some of our best performing single classifiers are the "music" classifier (precision: 0.57; recall: 0.36) and the "history" classifier (precision: 0.36; recall: 0.40) for uams08m. We believe that this relatively high performance is due to the satisfactory match between the vocabulary used in the archival metadata and that used in Wikipedia to describe both topics. Both with respect to micro-averaging and to macro-averaging, adding metadata to the ASR transcripts improves, or at least does not hurt results (i.e., uams08asrd vs. uams08masrd, and uams08asrde vs. uams08masrde).

3.2 Impact of Speech Transcripts from Two Languages

The run using Dutch ASR transcripts alone (uams08asrd) is not improved by the addition of English ASR transcripts (cf. uams08asrde), nor is the run using Dutch ASR transcripts with archival metadata (uams08masrd) improved by the addition of English ASR transcripts (cf. uams08masrde).

Table 2. Classification results for various runs

measure	uams08m	uams08asrd	uams08masrd	uams08asrde	uams08masrde
micro average precision	**0.13**	0.08	0.11	0.07	0.07
micro average recall	**0.28**	0.16	0.23	0.14	0.14
micro f-score	**0.18**	0.10	0.15	0.09	0.09
macro average precision	0.11	**0.44**	**0.44**	0.09	0.32
macro average recall	0.38	0.38	**0.46**	0.35	0.35
macro f-score	0.17	0.41	**0.45**	0.14	0.33

4 Conclusions

In this year's VideoCLEF classification task, we explored the use of archival metadata and the use of speech transcripts from both the matrix and the embedded language in the video. We drew our training data from the Dutch and the English editions of Wikipedia and used an SVM classifier with linear kernel to carry out classification.

The results show that metadata is very useful in video classification: highest scores on both macro and micro measures are achieved by runs using metadata. Regarding the use of one (matrix) or two (matrix and embedded) languages, we conclude that adding an extra language does not lead to improved results. Performance of runs using only the matrix language are consistently higher than using two languages.

Overall, the results achieved on this task fell short of being satisfactory. However, individual classifiers in individual runs proved promising (e.g., "music" and "history" in the metadata only run), suggesting that further development of our methods could be successful in generalizing this performance to more topic classes. In the future, we would like to perform per class failure analysis. We suspect that the root of the problem lies in the mismatch between the training data and the test data. Additionally, we would also like to experiment with methods to filter the ASR transcripts and discard those portions where the Dutch recognizer is producing output while English is being spoken and vice versa. We believe that a more judicious selection of speech-based features from the transcripts will serve to make them useful to the classification process.

Acknowledgements. This research was supported by the E.U. IST programme of the 6th FP for RTD under project MultiMATCH contract IST-033104.

References

1. Larson, M., Newman, E., Jones, G.: CLEF 2008 working notes. In: Borri, F., Nardi, A., Peters, C. (eds.) Overview of VideoCLEF 2008: Automatic generation of topic-based feeds for dual language audio-visual content (2008)
2. Joachims, T.: Text categorization with support vector machines: Learning with many relevant features. In: Nédellec, C., Rouveirol, C. (eds.) ECML 1998. LNCS, vol. 1398, pp. 137–142. Springer, Heidelberg (1998)
3. Drucker, H., Wu, D., Vapnik, V.N.: Support vector machines for spam categorization. IEEE Transactions on Neural Networks (5), 1048–1054 (1999)
4. Paass, G., Leopold, E., Larson, M., Kindermann, J., Eickeler, S.: SVM classification using sequences of phonemes and syllables. In: Elomaa, T., Mannila, H., Toivonen, H. (eds.) PKDD 2002. LNCS (LNAI), vol. 2431, pp. 373–384. Springer, Heidelberg (2002)
5. Leopold, E., Kindermann, J., Paass, G., Volmer, S., Cavet, R., Larson, M., Eickeler, S., Kastner, T.: Integrated classification of audio, video and speech using partitions of low-level features. In: Proceedings of the Workshop on Multimedia Discovery and Mining (2003)

Overview of CLEF 2008 INFILE Pilot Track

Romaric Besançon[1], Stéphane Chaudiron[2], Djamel Mostefa[3], Olivier Hamon[3,4], Ismaïl Timimi[2], and Khalid Choukri[3]

[1] CEA LIST
18 route du Panorama
BP 6 92265 Fontenay aux Roses France
romaric.besancon@cea.fr

[2] Université de Lille 3 - GERiiCO
Domaine univ. du Pont de Bois
BP 60149 - 59653
Villeneuve d'Ascq cedex France
{ismail.timimi,stephane.chaudiron}@univ-lille3.fr

[3] Evaluations and Language Distribution Agency (ELDA)
55-57 rue Brillat Savarin
75013 Paris France
{choukri,hamon,mostefa}@elda.org
http://www.elda.org

[4] LIPN - Université Paris 13 & CNRS
99 avenue J.-B. Clément
93430 Villetaneuse France

Abstract. The INFILE campaign was run for the first time as a pilot track in CLEF 2008. Its purpose was the evaluation of cross-language adaptive filtering systems. It used a corpus of 300,000 newswires from Agence France Presse (AFP) in three languages: Arabic, English and French, and a set of 50 topics in general and specific domain (scientific and technological information). Due to delays in the organization of the task, the campaign only had 3 submissions (from one participant) which are presented in this article.

1 Introduction

The purpose of the INFILE (INformation FILtering Evaluation) evaluation campaign[1] was to evaluate cross-language adaptive filtering systems, i.e. the ability of automated systems to successfully separate relevant and non-relevant documents in an incoming stream of textual information with respect to a given profile. The document and profile were possibly written in different languages.

The INFILE campaign was a pilot track in CLEF 2008 campaigns and is funded by the French National Research Agency (ANR) and co-organized by the CEA LIST, ELDA and the University of Lille3-GERiiCO.

Information filtering has many applications (routing, categorization, email filtering, anti-spamming). In the INFILE campaign, we considered the context of

[1] ANR-06 MDCA-011, http://www.infile.org

C. Peters et al. (Eds.): CLEF 2008, LNCS 5706, pp. 939–946, 2009.

competitive intelligence: in this context, the evaluation protocol of the campaign was designed with a particular attention to the context of use of filtering systems by real professional users. Even if the campaign was mainly a technological oriented evaluation process, we adapted the protocol and the metrics, as close as possible, to how a normal user would proceed, including through some interaction and adaptation of his system.

The INFILE campaign could mainly be seen as a cross-lingual pursuit of the TREC 2002 Adaptive Filtering task [1] (adaptive filtering track has been run from 2000 to 2002), with a particular interest in the correspondence of the protocol with the ground truth of competitive intelligence (CI) professionals. In this goal, we asked CI professionals to write the topics according to their experience in the domain.

Other related campaigns were the Topic Detection and Tracking (TDT) campaigns from 1998 to 2004 [2]. However, in the TDT campaigns, focus was mainly on topics defined as "events", with a fine granularity level, and often temporally restricted, whereas in INFILE (similar to TREC 2002) topics were of long-term interest and supposed to be stable, which could induce different techniques, even if some studies showed that some models could be efficiently trained to have good performance on both tasks [3].

2 Description of the Task

The main features of the INFILE evaluation campaign are summarized here:

- Crosslingual: English, French and Arabic were concerned by the process but participants could be evaluated on mono or bilingual runs.
- A newswire corpus provided by the Agence France Presse (AFP) and covering recent years.
- The topic set was composed of two different kinds of profiles, one concerning general news and events, and a second one on scientific and technological subjects.
- The evaluation was performed using an automatic interactive process for the participating systems to get documents and filter them, with a simulated user feedback.
- Systems were allowed to use the feedback at any time to increase performance.
- Systems provided a boolean decision for each document according to each profile.
- Relevance judgments were performed by human assessors.
- Participants were asked to fill a form to specify the languages used, the fields used in the profiles and a summary of the technology used.

We used an automatic process for the submission protocol. Indeed, the protocol of the INFILE campaign was designed to be a realist task for a filtering system. In particular, the idea was to avoid making the whole corpus available to the participants before the campaign, but to make it available one document at a time, simulating the behavior of the newswire service. The protocol then forced participating systems to be evaluated in a one-pass test.

The protocol was interactive and evaluation works as follows:

- a document server was started at the beginning of the campaign, initialized with the document collection: documents were retrieved from this server and filtering results were sent back by the participants to the server;
- the participant systems communicated with this server using a web service protocol (web services had been chosen to be able to bypass possible corporate firewalls of the participants):
  1. a participant system connected to the server from which it got a run identifier: if a participant wanted to submit several runs, the system had to connect several times to get different run identifiers;
  2. the system retrieved one document;
  3. the system filtered the document, i.e. it associated the document with one or several profiles, or discarded it;
  4. for adaptive systems, a relevance feedback was provided for filtered documents;
  5. the system retrieved a new document (back to step 2) that could only be retrieved when the previous document had been filtered;

A simulated relevance feedback was provided for adaptive systems: the idea was again to have a simulation of a realist behavior of the CI professional. In a real process, the CI professional receives the documents found relevant to a profile in a corresponding mailbox or directory and he can read the document and decide to remove it if it was a filtering error. In the INFILE automated process, it was also the only feedback authorized: relevance feedback could only be asked on a document associated with a profile by the system, there was no relevance feedback on discarded documents.

Furthermore, we assumed that a CI professional would not have an infinite patience: the number of feedbacks was then limited to 50, from the advice taken from CI professionals. This tended to give more interest to systems with quick adaptivity, than to systems that needed a large amount of data to be trained, but it seemed right for the organizers to put systems in a the context of a realistic task.

A dry run was organized from June 26th to July 3rd to check the technical viability of the protocol. The official campaign was run from July 7th to July 26th.

3 Test Collections

3.1 The Topics

A set of 50 profiles was prepared covering two different categories. The first group (30 topics) dealt with general news and events concerning national and international affairs, sports, politics, etc. The second one (20 topics) dealt with scientific and technological subjects. The scientific topics were developed by competitive intelligence professionals from INIST[2], ARIST Nord Pas de Calais[3], Digiport[4] and OTO Research[5]. The topics were developped in both English and French. The Arabic version was translated from French by native speakers.

Topics were defined with the following structure:

- a unique identifier;
- a title (6 words max.) describing the topic in a few words;
- a description (20 words max.) corresponding to a sentence-long description;
- a narrative (60 words max.) corresponding to the description of what should be considered a relevant document and possibly what should not;
- up to 5 keywords allowing to characterize the profile;
- an example of relevant text (120 words max.) taken from a document that is not in the collection (typically from the web).

Each record of the structure in the different languages correspond to translations, except for the samples which need to be extracted from real documents.

3.2 The Document Collection

The INFILE corpus was provided by the Agence France Presse (AFP) for research purpose. AFP is the oldest news agency in the world and one of the three largest with Associated Press and Reuters. Although AFP is the largest French news agency, it transmits news in other languages such as English, Arabic, Spanish, German and Portuguese. Newswires are available in different languages but are not necessarily translations from a language to another, since the same information is generally completely rewritten from one language to another to match the interest of the audience in the corresponding country.

For INFILE, we selected 3 languages (Arabic, English and French) and a 3 years period (2004-2006) which represented a collection of about one and half millions newswires for around 10 GB, from which 100,000 documents of each language were selected to be used for the filtering test. News articles were encoded in XML format and follow the News Markup Language (NewsML) specifications[6].

Since we provided a real-time simulated feedback to the participants, we needed to have the identification of relevant documents prior to the campaign, as in [4]. For each language, the 100,000 documents were selected in the following way:

---

[2] The French Institute for Scientific and Technical Information Center, http://international.inist.fr

[3] Agence Régionale d'Information Stratégique et Technologique, http://www.aristnpdc.org

[4] http://www.digiport.org

[5] http://www.otoresearch.fr

[6] NewsML is an XML standard designed to provide a media-independent, structural framework for multi-media news. NewsML was developed by the International Press Telecommunications Council. see http://www.newsml.org

- The whole collection was indexed with 4 different search engines: Lucene[7], Indri[8], Zettair[9] and our own search engine developed at CEA LIST. Zettair was originally only working in English, but was modified to also deal with French. The three other engines worked in the three languages (English, French, Arabic).
- Each search engine was queried independently using the 5 different fields of the topics, plus one query taking all fields and one query taking all fields but the sample (considering that the sample may introduce more noise than other fields). This gave a pool of 28 runs.
- The relevance of retrieved documents was judged by human assessors[10], two criteria being used: relevant or not relevant. The assessment process was performed using a *Mixture of Experts* model: the first 10 documents of each run were taken as first pool and assessed. Then, a score was computed for each run and each topic according to the current assessments and a next pool is created by merging the runs using a weighted sum of scores (where weights were proportional to the score)[11].
- The document collection was built by taking:
  - all documents that were relevant to at least one topic;
  - all documents that were assessed and judged not relevant: these documents form a set of difficult documents (not relevant, but which shared something in common with at least one topic, because they were retrieved by a search engine);
  - a set of documents taken randomly in the rest of the collection (i.e. from documents that were not retrieved by any search engines for any topic; this should limit the number of relevant documents in the corpus that were not assessed).

4 Metrics

The results returned by the participants were binary decisions on the association of a document with a profile. The results, for a given profile, can then be summarized in a contingency table of the form:

	Relevant	Not Relevant
Retrieved	a	b
Not Retrieved	c	d

---

[7] http://lucene.apache.org

[8] http://www.lemurproject.org/indri

[9] http://www.seg.rmit.edu.au/zettair

[10] Assessments were performed on a subset of the topics by 5 assessors, showing an inter-annotator agreement of 81% (kappa=0.7). Given this good agreement, the rest of the documents were judged by 2 assessors, and the documents for which the assessors did not agree were submitted to a 3rd one.

[11] Due to a lack of time and resources, this iterative process was not used for all assessments: for some of the queries, we used only the first pool.

On these data, a set of standard evaluation measures was computed:

- Precision, defined as $P = \frac{a}{a+b}$
- Recall , defined as $R = \frac{a}{a+c}$
- F-measure, which is a standard combination of precision and recall [5] depending on a parameter α, and defined as

$$F = \frac{1}{\alpha \frac{1}{P} + (1-\alpha)\frac{1}{R}}$$

We used the standard value $\alpha = 0.5$, which gave the same importance to precision and recall (F-measure is then the harmonic mean of the two values).

Following the TREC Filtering tracks [6,1] and the TDT 2004 Adaptive tracking task [2], we also considered the linear utility, defined as

$$u = w_1 \times a - w_2 \times b$$

where w_1 is the importance given to a relevant document retrieved and w_2 is the cost of a not relevant document retrieved.

Linear utility is bounded positively (to 1 for a perfect filtering), but unbounded negatively (negative values depend on the number of relevant documents for a profile). Hence, the average value on all profiles would give too much importance to the few profiles on which a systems would perform poorly. To be able to average the value, the measure is scaled as follows:

$$u_n = \frac{\max(\frac{u}{u_{max}}, u_{min}) - u_{min}}{1 - u_{min}}$$

where u_{max} is the maximum value of the utility and u_{min} a parameter considered to be the minimum utility value under which a user would not even consider the following documents for the profile.

In the INFILE campaign, we used the values $w_1 = 1$, $w_2 = 0.5$, $u_{min} = -0.5$, $u_{max} = a + c$ (same as in TREC 2002).

From the Topic Detection and Tracking campaigns [7], other measures were also considered:

- The estimated probability of missing a relevant document, defined as $P_{miss} = \frac{c}{a+c}$
- The estimated probability of raising a false alarm on a non-relevant document defined as $P_{false} = \frac{b}{b+d}$
- The detection cost, defined as

$$c_{det} = c_{miss} \times P_{miss} \times P_{topic} + c_{false} \times P_{false} \times (1 - P_{topic})$$

where

- c_{miss} if the cost of a missed document
- c_{false} is the cost of a false alarm

- P_{topic} is the *a priori* probability that a document is relevant to a given profile.

In the INFILE campaign, we used the values $c_{miss} = 10$, $c_{false} = 0.1$ and $P_{topic} = 0.001$ (according to an estimation of the average ratio of relevant documents in the corpus).

To compute average scores, the values were first computed for each profile and then averaged. Another way of averaging would be to sum up the values for all profiles in each cell of the contingency table and compute the scores on the resulting table. The first method was preferred because it allows equalizing the contribution of the profiles, whose differences are supposed to be the main source of variance in measures.

In order to measure the adaptivity of the systems, the measures were also computed at different times in the process, each 10,000 documents, and an evolution curve of the different values across time was presented.

Additionally, we proposed two following experimental measures. The first one was an originality measure, defined as a comparative measure corresponding to the number of relevant documents the system uniquely retrieved (among participants). It gave more importance to systems that use innovative and promising technologies that retrieved "difficult" documents. Since we only had too few runs, this measure was not really relevant.

The second one was an anticipation measure, designed to give more interest to systems that can find the first document in a given profile. This measure was motivated in competitive intelligence by the interest of being at the cutting edge of a domain, and not missing the first information to be reactive. It was measured by the inverse rank of the first relevant document detected (in the list of the documents), averaged on all profiles. The measure was similar to the mean reciprocal rank (MRR) used for instance in Question Answering Evaluation [8], but was not computed on the ranked list of retrieved documents but on the chronological list of the relevant documents.

5 Overview of the Results

During the development of the campaign, around 10 teams indicated their intent to participate to the INFILE track. Unfortunately, only one participant actually submitted runs, the IMAG team, which submitted 3 runs, in monolingual English filtering. Table 1 presents the runs and Table 2 presents the results on the runs, using the metrics described in previous section, averaged on all queries. More precise results are available in individual results.

Table 1. Submitted runs in the INFILE campaign

run identifier	team	language pair	topic fields used
run2G	IMAG	eng-eng	all
run5G	IMAG	eng-eng	all
runname	IMAG	eng-eng	all

Table 2. Results of the INFILE campaign

results	prec	recall	F_0.5	util_1_0.5_-0.5	cdet_10_0.1
run2G.eval	0.298	0.056	0.082	0.300	0.009
run5G.eval	0.298	0.324	0.231	0.362	0.006
runname.eval	0.362	0.052	0.071	0.307	0.009

6 Conclusion

The INFILE campaign was organized for the first time this year as a pilot track of CLEF, to evaluate cross-language adaptive filtering systems. The campaign followed the TREC 2002 Adaptive Filtering track, in a cross-language environment. An original setup was proposed to simulate the incoming of newswires documents and the interaction of a user, with a simulated feedback. Due to delays in the implementation of this setup, the campaign was postponed in July. Only one team participated in the campaign, which at least validated the viability of the interactive approach chosen. For the future of this track, it has to be verified if the complexity of the protocol is the element that has discouraged participants, or if it was the lack of information or communication around this evaluation, or the lack of interest in the subject.

References

1. Robertson, S., Soboroff, I.: The trec 2002 filtering track report. In: Proceedings of The Eleventh Text Retrieval Conference (TREC 2002), NIST (2002)
2. Fiscus, J., Wheatley, B.: Overview of the tdt 2004 evaluation and results. In: TDT 2002, NIST (2004)
3. Yang, Y., Yoo, S., Zhang, J., Kisiel, B.: Robustness of adaptive filtering methods in a cross-benchmark evaluation. In: Proceedings of the 28th annual international ACM SIGIR conference on Research and development in information retrieval, Salvador, Brazil, pp. 98–105 (2005)
4. Soboroff, I., Robertson, S.: Building a filtering test collection for trec 2002. In: Proceedings of The Eleventh Text Retrieval Conference (TREC 2002), NIST (2002)
5. Van Rijsbergen, C.: Information Retrieval. Butterworths, London (1979)
6. Hull, D., Roberston, S.: The trec-8 filtering track final report. In: Proceedings of the Eighth Text REtrieval Conference (TREC-8), NIST (1999)
7. NIST: The topic detection and tracking phase 2 (tdt2) evaluation plan (1998), `http://www.nist.gov/speech/tests/tdt/1998/doc/tdt2.eval.plan.98.v3.7.pdf`
8. Voorhees, E.: The trec-8 question answering track report. In: Proceedings of the Eighth Text REtrieval Conference (TREC-8), NIST (1999)

Online Document Filtering Using Adaptive k-NN

Vincent Bodinier, Ali Mustafa Qamar, and Eric Gaussier

Laboratoire d'Informatique de Grenoble (LIG)
Université Joseph Fourier
{vincent.bodinier,ali-mustafa.qamar,eric.gaussier}@imag.fr

Abstract. We propose in this paper an adaptation of the k-Nearest Neighbor (k-NN) algorithm using category specific thresholds in a multiclass environment where a document can belong to more than one class. Our method uses feedback to tune the thresholds and in turn the classification performance over time. The experiments were run on the InFile data, comprising 100,000 English documents and 50 topics.

Keywords: Filtering, on-line algorithms, k-nearest neighbors.

1 Adaptive k-NN for the InFile Campaign

The goal of the InFile campaign [1] is to filter 100,000 documents into 50 topics (plus a category "other") in an online fashion. 30 topics are related to general news and events, while 20 concern scientific and technical subjects. A document can belong to zero, one or more topics, each topic being described by a set of sentences. In addition, a certain number of feedbacks (50) was allowed for tuning (however limited) the participating systems.

To address this problem, we rely on a similarity measure between new documents and topics, and a set of thresholds on this similarity which evolve over time. The similarity we retained between a document d, to be filtered, and a topic t_i is defined by:

$$\mathrm{sim}(t_i, d) = \alpha * \underbrace{cos(t_i, d)}_{s_1(t_i,d)} + (1-\alpha)\underbrace{max_{(d' \in t_i)} cos(d, d')}_{s_2(t_i,d)} \tag{1}$$

where $\alpha \in [0, 1]$. This similarity is based on (a) a direct similarity between the new document and the topic ($s_1(t_i, d)$), and (b) the similarity between the new document and the documents already assigned to the topic ($s_2(t_i, d)$). α controls the importance given to these two similarities. At first, when no, or few documents have been assigned to topic t_i, only s_1 is taken into account for computing the similarity between the document and the topic.

For each topic, we then introduced two thresholds: the first one (θ_i^1) allows filtering out documents in the early stages of the process (i.e. when few documents are assigned to the topic) and operates on s_1, while the second one (θ_i^2)

C. Peters et al. (Eds.): CLEF 2008, LNCS 5706, pp. 947–950, 2009.

operates on the global similarity, after a certain number of documents have been assigned to the topic. θ_i^2 (see algorithm description below) accounts for the fact that new information has been incorporated in the topic.

Lastly, we make use of the possible feedbacks (50 in total for the whole collection) to try and ensure that only relevant documents are placed in the various topics. The general algorithm we used is summarized below:

if ($l_i < 10$)
    if ($s_1(t_i, d) > \theta_i^1$)
        Ask for feedback (if possible) and assign d to t_i if feedback positive
else if ($\text{sim}(t_i, d) > \theta_i^2$)
        assign d to t_i
where $\theta_i^2 = min_{d \in t_i} \text{sim}(t_i, d)$

The parameter α and the threshold θ_i^1 were tuned during the dry run phase.

Simplification of the general algorithm

We also investigated a simplified version of the above, general algorithm, which does not use any feedback and does not update the threshold θ_i^2. In this version, a threshold θ is derived from θ_i^1 and θ_i^2 according to equation 1, which integrates the two similarities θ_i^1 and θ_i^2 operate upon:

$$\theta = \alpha * \theta_i^1 + (1 - \alpha) * \theta_i^2$$

Documents are then filtered according to the following, simple algorithm:

for each new document d
    for each topic i ($i \in \{101,102,...,150\}$)
        if ($\text{sim}(t_i, d) \geq \theta$) {Assign d to topic i}
        else {Do not assign d to topic i}

Here again, values for the different parameters were tuned during the dry run phase. We then tested slight modifications of these values in the final experiments. A complete description of the algorithms and experiments can be found in [2].

2 Experiments

For each new document retrieved, first of all stemming is performed using Porter's algorithm. This is followed by the removal of stop-words, XML tags skipping and the building of a document vector (which associates each term with its frequency)

Table 1. Run Features

	Name	Algorithm	Parameters
Run 1	run5G	General	$\alpha = 0.7;\ \theta_i^1 = 0.42$
Run 2	runname	Simplified	$\alpha = 0.7;\ \theta_i^2 = 0.8;\ \theta_i^1 = 0.45$
Run 3	run2G	Simplified	$\alpha = 0.7;\ \theta_i^2 = 0.7;\ \theta_i^1 = 0.4$

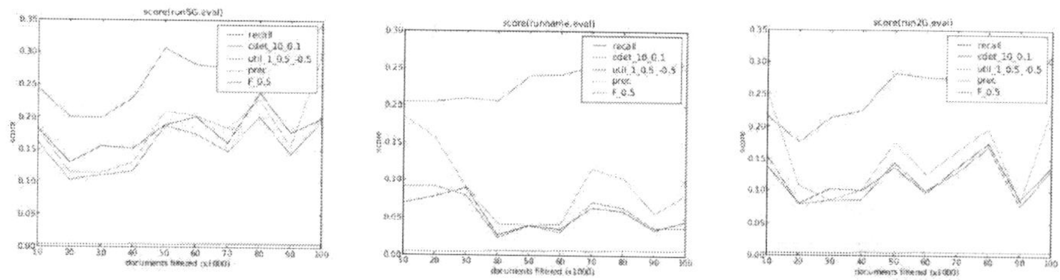

Fig. 1. Score Evolution for the 3 runs

Table 2. Run Scores

	Precision	Recall	F-measure	Linear Utility	Detection Cost
Run 1	0.306	0.260	0.209	0.351	0.007
Run 2	0.366	0.068	0.086	0.311	0.009
Run 3	0.357	0.165	0.165	0.335	0.008

using rainbow [3]. During the InFile campaign evaluation, three runs were submitted, which are summarized in Table 1.

There was a total of 1597 documents relevant to one or more topics in the InFile data. All the methods were evaluated according to 5 measures: precision, recall, F-measure, linear utility and detection cost (see [1]). Figure 1 depicts the evolution of the different measures for the 3 runs we submitted.

Finally, Table 2 contains the average scores for the three runs (the average are taken over the whole profiles).

As one can note, run 2 scores are rather low when compared to the other two runs. In particular, the average recall value is very low whereas the precision is around 0.36, which shows that this run is precision-oriented (only a small fraction of the relevant documents are retrieved in this run). The precision value for run 3 is close to the one of run 2. The recall and the F-measure are however slightly better. This run is also precision-oriented with a precision value clearly better than the recall one. If we consider the overall scores, run 1 is better than the other two. Although the precision is slightly lower, the recall score attains 0.26 while the F-measure reaches 0.2. The overall detection cost is very low during the runs (less than 0.01). This a strong point for our algorithms. We can also notice that the linear utility progressively increases between 0.2 and 0.3. Run 2 is a more conservative method compared to run 3, the thresholds used are higher. Run

3 thus assigns more documents to the topics than run 2. Regarding evolution measures, for run 2, precision, recall and F-measure tend to decrease slightly. For run 3, they randomly vary but remain the same at the end, whereas they increase slightly during run 1, corresponding to the general algorithm. Clearly, the values obtained for the different evaluation measures show that our runs are rather precision-oriented. Different settings of the thresholds need be tested in the future, hopefully with additional feedback information.

3 Conclusion

We have presented here a simple extension of k-NN using thresholds. A similar extension was proposed in [4], however based on a direct similarity measure whereas our algorithm relies on the combination of two similarity measures. The first results we obtained are encouraging as the F-measure roughly equals 20%, for a collection of 100,000 documents and 50 topics (and bearing in mind as well that, overall, there are only 1597 relevant documents in the collection). It is difficult to really make use of the feedback information, as it is really limited wrt the size of the collection. Because of this particularity, the InFile collection is not a standard filtering collection, and it is difficult to fully deploy machine learning techniques on it. In the future, we plan to investigate different settings of the parameters of our method, and to find ways of learning them more adequately.

References

1. Besancon, R., Chaudiron, S., Mostefa, D., Timimi, I., Choukri, K.: The infile project: a crosslingual filtering systems evaluation campaign. In: ELRA (ed.) Proceedings of LREC 2008, Morocco (May 2008)
2. Bodinier, V., Qamar, A.M., Gaussier, E.: Working notes for the infile campaign: Online document filtering using 1 nearest neighbor. In: Workshop CLEF 2008, Aarhus, Denmark, September 17-19 (2008)
3. McCallum, A.K.: Bow: A toolkit for statistical language modeling, text retrieval, classification and clustering (1996)
4. Yang, Y., Liu, X.: A re-examination of text categorization methods. In: SIGIR 1999, USA, pp. 42–49. ACM Press, New York (1999)

Overview of Morpho Challenge 2008

Mikko Kurimo, Ville Turunen, and Matti Varjokallio

Adaptive Informatics Research Centre, Helsinki University of Technology,
P.O. Box 5400, FIN-02015 TKK, Finland
{mikko.kurimo,ville.t.turunen,matti.varjokallio}@tkk.fi
http://www.cis.hut.fi/morphochallenge2008/

Abstract. This paper gives an overview of Morpho Challenge 2008 competition and results. The goal of the challenge was to evaluate unsupervised algorithms that provide morpheme analyses for words in different languages. For morphologically complex languages, such as Finnish, Turkish and Arabic, morpheme analysis is particularly important for lexical modeling of words in speech recognition, information retrieval and machine translation. The evaluation in Morpho Challenge competitions consisted of both a linguistic and an application oriented performance analysis. In addition to the Finnish, Turkish, German and English evaluations performed in Morpho Challenge 2007, the competition this year had an additional evaluation for Arabic. The results in linguistic evaluation in 2008 show that although the level of precision and recall varies substantially between the tasks in different languages, the best methods seem to deal quite well with all languages involved. The results in information retrieval evaluation indicate that the morpheme analysis has a significant effect in all the tested languages (Finnish, English and German). The best unsupervised and language-independent morpheme analysis methods can also rival the best language-dependent word normalization methods. The Morpho Challenge was part of the EU Network of Excellence PASCAL Challenge Program and organized in collaboration with CLEF.

Keywords: Morphological analysis, Machine learning.

1 Introduction

The goal of Morpho Challenge 2008 was to evaluate proposed unsupervised machine learning algorithms for the task of morpheme analysis for words in different languages. The evaluation consisted of both a linguistic and an application oriented performance analysis. The linguistic evaluation, called *Competition 1*, was based on a comparison of the suggested morpheme analysis to a linguistic morpheme analysis gold standard. The practical application oriented evaluation, called *Competition 2*, contained information retrieval (IR) experiments from CLEF, where all words in queries and text corpus were replaced by the results of their morpheme analysis.

The Morpho Challenge 2008 tasks and training corpora were the same as those for the previous year [5], except that another morphologically complex

C. Peters et al. (Eds.): CLEF 2008, LNCS 5706, pp. 951–966, 2009.

language, Arabic, was added. There was also an optional evaluation of the IR performance using the morpheme analysis of word forms in their full text context. Its main difference to our first Morpho Challenge 2005 [1], which focused on the segmentation of words into morphologically meaningful units, is that the goal was set to find units that reflect abstract classes of morphemes. For example, an analysis could find a link between the word forms “foot” and “feet”. In Morpho Challenge 2005 [1], there were two speech recognition tasks, but no IR, and morpheme segmentations were utilized to train language models.

For morphologically complex languages, such as Finnish, Turkish and Arabic, the morpheme analysis is particularly important in lexical modeling of words for speech recognition [2,1], information retrieval [4,6] and machine translation [3,17]. Due to the high level of agglutination, inflection, and compounding, there are millions of different word forms, which is clearly too much for building an effective vocabulary and training probabilistic models for the relations between words.

There exist carefully constructed linguistic tools for morphological analysis, but only for a few languages. Even in these cases statistical machine learning methods can discover interesting alternatives to rival even the most sophisticated linguistically designed morphologies. The IR tasks attempted by the Morpho Challenge participants' morpheme analyses were also tested by a number of reference methods to verify the usefulness of the unsupervised morpheme analysis. These references included the unsupervised baseline algorithms Morfessor Categories-Map [9] and Morfessor Baseline [11,12], the rule-based grammatical morpheme analysis based on linguistic gold standards [13], a commercial word normalization tool (TWOL) and the traditional stemming approaches for different languages based on the Porter stemming [16]. The IR results were also provided for the case when all words were left unprocessed.

The scientific objectives of the Morpho Challenge competitions are: to learn about word construction in natural languages, to advance machine learning methodology, and to explore suitable approaches for various languages. Portability to different languages is very important, because language technology often needs to be quickly extended to various new languages for which only limited amount of resources are available. Unsupervised learning is then the most attractive approach, because the majority of available data is unannotated and human annotation work is expensive.

2 Tasks and Data

The task of the participants in Morpho Challenge 2008 was to extend the given list of words in each language by a morpheme analysis of each word form. Examples of reference analyses are shown in Table 3. The results of morpheme analyses was expected to be obtained by an unsupervised learning algorithm that would preferably be as language independent as possible. As the learning is unsupervised, the returned morpheme labels can be arbitrary and be listed in any order. Several interpretations for the same word can also be supplied, and it was left to the participants to decide whether they would be useful in the task, or not.

In each language, the participants were pointed to a training corpus in which all the words occur (in a sentence), so that the algorithms may also utilize information about the word context. The tasks were the same as in the Morpho Challenge 2007 last year with the addition of one new language, Arabic.

The training corpora were the same as in the Morpho Challenge 2007, except for Arabic: 3 million sentences for English, Finnish and German, and 1 million sentences for Turkish in plain unannotated text files that were all downloadable from the Wortschatz collection[1] at the University of Leipzig (Germany). The corpora were specially preprocessed for the Morpho Challenge (tokenized, lower-cased, some conversion of character encodings).

The Arabic text data (135K sentences with 3.9M words) is the same as used by Habash and Sadat [15]. Because this text data is unfortunately not freely available, only a list of word forms was provided, so if the participants wanted to use typical word contexts in training their models in Arabic, they had to find their own text corpus. All words in the Arabic data were presented in Buckwalter transliteration[2]. In other languages the lists of word forms to be analyzed were extracted from the Wortschatz corpora and included all the different word forms existing there and their frequencies in the corpora. The total amount of word types were 2,206,719 (Finnish), 617,298 (Turkish), 1,266,159 (German), 384,903 (English), and 143,966 (Arabic).

In Competition 2, the Morpho Challenge organizers performed IR experiments based on the morpheme analyses submitted by the participants for the given word lists. Two word lists in each language were provided for analysis, the first for Competition 1 and then another which included the same words plus the word forms that occurred in the IR tasks. For the IR experiments both the words in the documents and in the test queries were then replaced by their proposed morpheme representations and the search was based on morphemes instead of words. Three tasks were provided for three different languages: Finnish, German and English, and the participants were encouraged to use the same algorithms for all of them.

The data sets for testing the IR performance were exactly the same as in the previous Morpho Challenge 2007. In each language there were newspaper articles, test queries and the binary relevance judgments regarding to the queries. Because the organizers performed the IR experiments based on the morpheme analyses submitted by the participants, it was not necessary for the participants to get these data sets. However, all the data was available for registered participants in the Cross-Language Evaluation Forum (CLEF)[3], so that it was possible to use the full text corpora for preparing the morpheme analyses. In Morpho Challenge 2008 IR evaluation, an option was also given to use the full text corpora to get information and train models for morpheme analysis using the context in which the words occur. It was also possible to submit the morpheme analysis for the full text corpora, instead of just for the word lists, to disambiguate the alternative analyses.

[1] `http://corpora.informatik.uni-leipzig.de/`
[2] `http://www.qamus.org/transliteration.htm`
[3] `http://www.clef-campaign.org/`

The source documents were news articles collected from different news papers selected as follows:

- In Finnish: 55K documents from short articles in Aamulehti 1994-95, 50 test queries on specific news topics and 23K binary relevance assessments (CLEF 2004)
- In English: 170K documents from short articles in Los Angeles Times 1994 and Glasgow Herald 1995, 50 test queries on specific news topics and 20K binary relevance assessments (CLEF 2005).
- In German: 300K documents from short articles in Frankfurter Rundschau 1994, Der Spiegel 1994-95 and SDA German 1994-95, 60 test queries with 23K binary relevance assessments (CLEF 2003).

3 Participants and the Submissions

By the submission deadline at the end of June, 2008, four research groups had submitted nine different algorithms which were then evaluated by the organizers. After the deadline, more submissions were received from another author (Goodman), which were evaluated separately outside the Competition 1. One group (Can) decided not to submit the final wordlists that could be evaluated and one (McNamee) wanted only to participate in Competition 2. The algorithm submissions and their authors are listed in Table 1.

Some characteristics of morpheme analyses proposed by the unsupervised algorithms together with the gold standard analyses are briefly presented in Table 2. The statistics of each submission include the average amount of alternative analyses per word, the average amount of morphemes per analysis, and the total amount of morpheme types. The "Allomorfessor" by Kohonen et al. is an extension to the

Table 1. The submitted algorithms. "C1" shows which were evaluated in Competition 1 and "C2" shows which were evaluated in Competition 2. "1" in C2 means that only analyses of Competition 1 words were used in Competition 2.

Algorithm	Author	Affiliation	C1	C2
"Can (no wordlists)"	Burcu Can	Univ. York, UK	no	no
"Goodman (late submission)"	Sarah A. Goodman	Univ. Maryland, USA	yes	no
"Kohonen Allomorfessor"	Oskar Kohonen et al.	Helsinki U. Tech, FI	yes	1
"McNamee five"	Paul McNamee	JHU, USA	no	yes
"McNamee four"	Paul McNamee	JHU, USA	no	yes
"McNamee lcn5"	Paul McNamee	JHU, USA	no	yes
"Monson Morfessor"	Christian Monson et al.	CMU, USA	yes	yes
"Monson ParaMor"	Christian Monson et al.	CMU, USA	yes	yes
"Monson ParaMor-Morfessor"	Christian Monson et al.	CMU, USA	yes	yes
"Zeman 1"	Daniel Zeman	Karlova Univ., CZ	yes	1
"Zeman 3"	Daniel Zeman	Karlova Univ., CZ	yes	1

Table 2. Some statistics of the compared morpheme analyses in Competition 1. #a is the average amount of analyses per word, #m the average amount of morphemes per analysis, and lexicon the total amount of morpheme types.

	Finnish			**Turkish**			**Arabic**		
	#a	#m	lexicon	#a	#m	lexicon	#a	#m	lexicon
Kohonen	1	1.86	486096	1	1.76	183297			
Monson paramor	1	2.62	1123572	1	2.89	245737	1	1.72	81978
Monson morfessor	1	2.83	223412	1	2.76	107431	1	2.03	46526
Monson p+m	2	2.72	1359325	2	2.83	354280	2	1.87	133309
Zeman 1	3.61	1.81	5379817	3.24	1.76	1205970	2.24	1.65	217232
Zeman 3	1.21	1.62	1830751	1.14	1.52	501154	1.23	1.61	106378
Morfessor baseline	1	2.21	149417	1	2.14	53473	1	2.45	16735
Morfessor catmap	1	2.94	217001	1	2.64	114834	1	2.04	46789
Gold Standard	1.16	3.29	33754	1.99	3.36	21163	1.78	3.39	43914

	German			**English**		
	#a	#m	lexicon	#a	#m	lexicon
Kohonen	1	1.83	334851	1	1.62	180813
Monson paramor	1.25	1.65	908556	1.27	1.75	252997
Monson morfessor	1	3.10	166963	1	2.07	137973
Monson p+m	2.25	2.30	1094322	2.27	1.89	378364
Zeman 1	4.11	1.80	4054397	3.18	1.74	905251
Zeman 3	1.12	1.43	1053275	1.08	1.37	319982
Morfessor baseline	1	2.30	90009	1	2.32	40293
Morfessor catmap	1	3.06	172907	1	2.12	132086
Gold Standard	1.30	2.97	14298	1.10	2.13	16902

"Morfessor Baseline" that attempts to discover common baseforms for the different surface forms that are likely to represent the same morpheme. The "ParaMor" by Monson et al. is another algorithm for segmenting words into morphemes which, after improvements from the previous Morpho Challenge, was submitted also as a combination with the publicly available "Morfessor CATMAP". The "Zeman 1" is a resubmission from the previous Morpho Challenge which, after attempts to include a new treatment of prefix, was submitted as the "Zeman 3". It is interesting to note that this year all the algorithms resulted in a very large lexicon, usually much larger than the reference methods did.

In the IR task (Competition 2), totally nine algorithms were evaluated in all three languages. For six of those, the morpheme analyses were available for all the words in the IR text corpus. For the remaining three only those words were analyzed that existed in the text corpus for Competition 1 and the others were indexed without analysis. In the Morpho Challenge 2007 [6] experiments were made to compare the IR performance with and without the analysis of these "new" words. The results indicated that in the Finnish task the extra analyses were helpful for almost all participants, but in the German and English task they did not seem to affect the results.

Unlike the others, the algorithms by McNamee were no real attempts to find morphemes, but rather focused directly on extracting substrings from words that would be suitable for IR.

4 Reference Analysis for Competition 1

4.1 Linguistic Gold Standard

In Competition 1 the proposed unsupervised morpheme analyses were compared to the correct grammatical morpheme analyses called here the linguistic gold standard. The gold standard morpheme analyses were prepared in exactly the same format as the result file the participants were asked to submit, alternative analyses separated by commas. See Table 3 for examples.

The gold standard reference analyses were the same as in the Morpho Challenge 2007 [5], except in Arabic. The Arabic gold standard analyses are based on the representation of lexeme and features used in the Aragen system (a wrapper using publicly available BAMA-1 databases) [14]. The first part of an analysis is a lexeme followed by a list of features. The original features were here modified to connect the POS label to the root of the word, e.g. "Algbn = gabon_POS:N Al+ +SG". In addition, the gender morphemes were removed (e.g. the German gold standard doesn't contain these either). This did not affect the ranking of the submissions, but made the evaluation resemble more the other tested languages.

In the word lists described in the previous section, the gold standard analyses were available for 650,169 (Finnish), 214,818 (Turkish), 125,641 (German), 63,225 (English), and 141,876 (Arabic) word types.

4.2 Morfessor

As baseline results for unsupervised morpheme analysis, the organizers provided morpheme analysis by a publicly available unsupervised algorithm called

Table 3. Examples of gold standard morpheme analyses

Language	Examples	
English	baby-sitters	baby_N sit_V er_s +PL
	indoctrinated	in_p doctrine_N ate_s +PAST
Finnish	linuxiin	linux_N +ILL
	makaronia	makaroni_N +PTV
German	choreographische	choreographie_N isch +ADJ-e
	zurueckzubehalten	zurueck_B zu be halt_V +INF
Turkish	kontrole	kontrol +DAT
	popUlerliGini	popUler +DER_lHg +POS2S +ACC, popUler +DER_lHg +POS3 +ACC3
Arabic	Algbn	gabon_POS:N Al+ +SG
	AlmtHdp	mut aHidap_POS:PN Al+ +SG, mut aHid_POS:AJ Al+ +SG

"Morfessor Categories-MAP" developed at Helsinki University of Technology [9] (or here "Morfessor catmap" or "Morfessor MAP", for short as in [5]). Analysis by the original Morfessor [11,12] (or here "Morfessor baseline"), which provides only a surface-level segmentation, was also provided for reference.

5 Reference Methods for Competition 2

In addition to the participating algorithms, a number of different reference methods were evaluated for the same tasks. The purpose of these methods was to provide views on the difficulty and various characteristics of these tasks and on the usefulness of the unsupervised morpheme analysis in the IR tasks.

1. *Morfessor Categories-Map*: The same Morfessor Categories-Map (or here just "catmap", for short) as described in Competition 1 was used for the unsupervised morpheme analysis. The stem vs. suffix tags were kept, but did not receive any special treatment in the indexing as we wanted to keep the IR evaluation as unsupervised as possible.
2. *Morfessor Baseline*: All the words were simply split into smaller pieces without any morpheme analysis. This means that the obtained subword units were directly used as index terms. This was performed using the Morfessor Baseline algorithm as in Morpho Challenge 2005 [1].
3. *dummy*: No words were split nor any morpheme analysis provided except hyphens were replaced by spaces so that hyphenated words were indexed as separate words (changed from last year). This means that words were directly used as index terms as such without any stemming or tags. We expected that although the morpheme analysis should provide helpful information for IR, all the submissions would not probably be able to beat this brute force baseline. However, if some morpheme analysis method would consistently beat this baseline in all languages and task, it would mean that the method would probably be useful in a language and task independent way.
4. *grammatical*: The words were analyzed using the same gold standard analyses in each language that were utilized as the "ground truth" in the Competition 1. Besides the stems and suffixes, the gold standard analyses typically consist of all kinds of grammatical tags which we decided to simply include as index terms, as well. For many words the gold standard analyses included several alternative interpretations that were all included in the indexing. However, we decided to also try the method adopted in the morpheme segmentation for Morpho Challenge 2005 [1] that only the first interpretation of each word is applied. This was here called "grammatical first" whereas the default was called "grammatical all". Words that were not in the gold standard segmentation were indexed as such. Because our gold standards are quite small, 60k (English) - 600k (Finnish), compared to the amount of words that the unsupervised methods can analyze, we did not expect "grammatical" to perform particularly well, even though it would probably capture some useful indexing features to beat the "dummy", at least.

5. *snowball*: No real morpheme analysis was performed, but the words were stemmed by stemming algorithms provided by Snowball libstemmer library[4]. Porter stemming algorithm was used for English. Finnish and German stemmers were used for the other languages. Hyphenated words were first split to parts that were then stemmed separately. Stemming is expected to perform very well for English but not necessarily for the other languages because it is harder to find good stems.
6. *TWOL*: Two-level morphological analyzer TWOL from Lingsoft[5] Inc. was used to find the normalized forms of the words. These forms were then used as index terms. Some words may have several alternative normalized forms and two cases were studied similarly to the grammatical case. Either all alternatives were used ("all") or only the first one ("first"). Compound words were split to parts. Words not recognized by the analyzer were indexed as such. German analyzer was not available for the organizers.
7. *best 2007*: This is the algorithm in each task that provided the highest average precision in Morpho Challenge 2007.

6 Evaluation in Competition 1

The evaluation of Competition 1 in Morpho Challenge 2008 was similar as in Morpho Challenge 2007 except that there was one new language, Arabic. The full description of the method to compare the submitted unsupervised morpheme analyses were to the linguistic gold standard analyses is in [5]. In the current paper we just remind the main points and obtained performance measures.

Because the morpheme analysis candidates are achieved by unsupervised learning, the morpheme labels can be arbitrary and different from the ones designed by linguists. The basis of the evaluation is, thus, to compare whether any two word forms that contain the same morpheme according to the participants' algorithm also has a morpheme in common according to the gold standard and vice versa. In practice, the evaluation is performed by randomly sampling a large number of morpheme sharing word pairs from the compared analyses. Then the *precision* is calculated as the proportion of morpheme sharing word pairs in the participant's sample that really has a morpheme in common according to the gold standard. Correspondingly, the *recall* is calculated as the proportion of morpheme sharing word pairs in the gold standard sample that also exist in the participant's submission. The sample size in different languages varied depending on the size of the word lists and gold standard: 200,000 (Finnish), 50,000 (Turkish), 50,000 (German), 10,000 (English), and 20,000 (Arabic) word pairs.

The *F-measure*, which is the harmonic mean of *Precision* and *Recall*, was selected as the final evaluation measure:

$$\text{F-measure} = 1/(1/\text{Precision} + 1/\text{Recall})\,. \tag{1}$$

[4] `http://snowball.tartarus.org/`

[5] `http://www.lingsoft.fi/`

7 Evaluation in Competition 2

The submitted morpheme analyses were evaluated by IR experiments in three different tasks: one in Finnish, one in German and one in English. It would have been interesting to evaluate also the performance in Turkish and Arabic, but unfortunately no IR tasks in these languages were available to the organizers. In the IR corpora the words were replaced by the provided morpheme analyses both in the text and the queries, and then the search was performed based on morphemes instead of full words. Any word without morpheme analysis was left un-replaced and indexed as if it were just a single morpheme on its own.

Those participants who only provided morpheme analyses for words that exist in the text corpus for Competition 1 had a slight disadvantage, because then the "new" words in the IR task were indexed and searched without splitting. However, the experiments in the Morpho Challenge 2007 [6] revealed that the extra analyses were helpful only in the Finnish task. In the German and English task they did not seem to affect the results.

In Morpho Challenge 2008 we provided the participants an option to use the full text corpora in order to get information and train models using the context in which the different words occur and, for the first time, also to submit morpheme analysis for words in their actual context. However, none of the participants dared to go for this even more challenging option.

In practice, the IR evaluation was performed using the latest version of the freely available LEMUR toolkit[6]. Okapi (BM25) term weighting was used for all index terms excluding an automatic stoplist. The automatic stoplist was separately determined for each morpheme analysis run by extracting the morphemes that have a collection frequency higher than 75000 (Finnish) or 150000 (German and English). The stoplist was used with the Okapi weighting, because in the previous Morpho Challenge [6] it was observed that the performance of indexes that have many very common terms was poor. The optimal frequency threshold for the stoplists may vary depending on the size of the lexicon, but in initial experiments it was also observed that changing the threshold between 50000 and 200000 did not affect the performance. A corresponding automatic stoplist was also determined and applied for every reference method. The evaluation criterion for all IR experiments was the Uninterpolated Average Precision.

8 Results in Competition 1

The results of the linguistic evaluation are presented in Tables 4 and 5. The tasks in Competition 1 were the same as in Morpho Challenge 2007, so it is possible to directly compare the improvements made over the previous algorithms. However, direct comparisons between the evaluation measures in different languages are not valid, because the corpora and gold standards are different. In all tasks except the English one, improvements were made in 2008 and the best obtained F-measure was now higher. As clearly seen in Tables 4 and 5, this is mainly due to

[6] `http://www.lemurproject.org/`

Table 4. The submitted unsupervised morpheme analyses compared to the gold standard in **Finnish**, **Turkish** and **Arabic**. The Competition 1 participants are shown in bold and the various reference methods in normal font.

Finnish	PRECISION	RECALL	F-MEASURE
Monson p+m	49.76%	47.25%	48.47%
reference Morfessor catmap	76.83%	27.54%	40.55%
Monson paramor	46.40%	34.44%	39.53%
best 2007 Bernhard 1	75.99%	25.01%	37.63%
Monson morfessor	77.40%	21.52%	33.68%
Zeman 1	58.51%	20.47%	30.33%
reference Morfessor baseline	88.12%	12.01%	21.16%
Goodman methodB.deduped	62.19%	7.71%	13.71%
Kohonen allomorfessor	92.55%	6.89%	12.82%
Zeman 3	72.41%	3.42%	6.54%

Turkish	PRECISION	RECALL	F-MEASURE
Monson p+m	51.88%	52.10%	51.99%
Monson paramor	56.67%	39.42%	46.50%
Monson morfessor	73.92%	26.06%	38.53%
reference Morfessor catmap	76.36%	24.50%	37.10%
Zeman 1	65.81%	18.79%	29.23%
best 2007 Zeman	65.81%	18.79%	29.23%
reference Morfessor baseline	89.20%	11.32%	20.08%
Goodman pruned	69.96%	8.42%	15.04%
Kohonen allomorfessor	93.25%	6.15%	11.53%
Zeman 3	73.30%	3.01%	5.79%

Arabic	PRECISION	RECALL	F-MEASURE
Monson p+m	79.77%	27.47%	40.87%
reference Morfessor baseline	78.16%	23.74%	36.41%
reference Morfessor catmap	90.17%	20.97%	34.03%
Monson morfessor	90.35%	20.95%	34.01%
Zeman 1	77.24%	12.73%	21.86%
Monson paramor	78.58%	8.52%	15.37%
Zeman 3	89.62%	5.18%	9.79%

the improved version of "Monson paramor+morfessor" that dominated all tasks. The difference is especially clear in the recall statistics where the performance of the "Monson paramor+morfessor" is superior. Behind Monson's algorithms, the "Zeman 1" that is a re-submission from last year, was better than the rest of the algorithms, which all suffered from a very low recall. It is worth noting that the "Kohonen allomorfessor" algorithm achieved clearly the highest precision of all algorithms in all tasks, but due to the low recall, or undersegmentation, it got rather low F-measure values.

From the Competition 1 in Morpho Challenge 2007 [5], only the winner "best 2007" in each task was chosen in Tables 4 and 5 for reference. The "Monson paramor+morfessor" was able to clearly beat the publicly available reference

Table 5. The submitted unsupervised morpheme analyses compared to the gold standard in **German** and **English**. The Competition 1 participants are shown in bold and the various reference methods in normal font.

German	PRECISION	RECALL	F-MEASURE
Monson p+m	49.53%	59.51%	54.06%
best 2007 Monson p+m	51.45%	55.55%	53.42%
reference Morfessor catmap	67.56%	36.92%	47.75%
Monson morfessor	67.16%	36.83%	47.57%
Monson paramor	53.42%	38.15%	44.51%
Zeman 1	53.12%	28.37%	36.98%
reference Morfessor baseline	80.23%	19.22%	31.01%
Goodman methodB.deduped	54.53%	12.70%	20.60%
Kohonen allomorfessor	87.92%	7.44%	13.71%
Zeman 3	72.27%	7.15%	13.01%

English	PRECISION	RECALL	F-MEASURE
best 2007 Bernhard 2	61.63%	60.01%	60.81%
Monson p+m	50.64%	63.30%	56.26%
reference Morfessor baseline	71.93%	43.27%	54.04%
Monson paramor	58.50%	48.10%	52.79%
reference Morfessor catmap	82.17%	33.08%	47.17%
Monson morfessor	77.22%	33.95%	47.16%
Zeman 1	52.98%	42.07%	46.90%
Goodman methodB.deduped	66.19%	16.51%	26.43%
Kohonen allomorfessor	83.39%	13.43%	23.13%
Zeman 3	76.92%	8.47%	15.27%

methods “Morfessor baseline” and “Morfessor catmap” in all tasks. It is interesting to note that the “Morfessor baseline”, which is the original simpler Morfessor version and only attempts to split words into morphemes without any further analysis, actually beats the more sophisticated “Morfessor catmap”, as well as “Monson morfessor” and “Zeman 1”, in English and Arabic. Otherwise, the ranking between the different 2008 algorithms remains the same in all tasks.

9 Results in Competition 2

Table 6 presents the IR evaluation results. The algorithms had been improved from the previous competition, and in all tasks there was a new winner. The highest average precision in the Finnish task was, slightly surprisingly, achieved by the character 4-gram approach “McNamee four” that was equal in performance to last year’s winner, but clearly beat the other 2008 competitors.

In the English and German tasks the winner was “Monson Paramor+Morfessor” that also won the Competition 1 in all languages. The marginal to the best 2007 results was very tight, but clear to the other 2008 competitors. In both English and German tasks the “McNamee four” was second after Monson’s algorithms.

Table 6. The obtained average precision (AP%) in the three different IR tasks. The Competition 2 participants are shown in bold and the various reference methods in normal font. (a) the IR tasks are the same as in Morpho Challenge 2007, but because some values in the word frequency statistics provided for the participants differed slightly, the 2007 results may not be exactly comparable. (b) some participants provided morpheme analyses only for words that existed also in the text corpus for Competition 1.

Finnish IR task	AP%	English IR task	AP%
TWOL first	0.4976	snowball porter	0.4081
McNamee four	0.4918	**Monson Paramor+Morfessor**	0.3989
best 2007 Bernhard 2	0.4915a	TWOL first	0.3957
TWOL all	0.4845	best 2007 Bernhard 2	0.3943a
Monson Morfessor	0.4679	**Monson Paramor**	0.3928
Monson Paramor+Morfessor	0.4673	TWOL all	0.3922
McNamee five	0.4515	Morfessor baseline	0.3861
Morfessor catmap	0.4441	grammatical first	0.3734
Morfessor baseline	0.4425	Morfessor catmap	0.3713
grammatical first	0.4312	**Monson Morfessor**	0.3637
snowball finnish	0.4275	**McNamee five**	0.3630
grammatical all	0.4090	**McNamee four**	0.3566
Monson Paramor	0.3965	**McNamee lcn5**	0.3563
McNamee lcn5	0.3688	grammatical all	0.3542
Kohonen	0.3548b	**Kohonen**	0.3342b
dummy	0.3519	dummy	0.3293
Zeman 3	0.3282b	**Zeman 3**	0.3125b
Zeman 1	0.2627b	**Zeman 1**	0.2631b

German IR task	AP%
Monson Paramor+Morfessor	0.4734
best 2007 Bernhard 1	0.4729a
Monson Morfessor	0.4671
Morfessor baseline	0.4656
Morfessor catmap	0.4642
McNamee four	0.4388
McNamee five	0.4331
snowball german	0.3865
Kohonen	0.3671b
Monson Paramor	0.3631
dummy	0.3509
grammatical first	0.3353
McNamee lcn5	0.3276
Zeman 3	0.3206b
grammatical all	0.3014
Zeman 1	0.2343b

The "Monson Paramor+Morfessor" which was built by combining the publicly available Morfessor algorithm and the "Monson Paramor" managed to improve both of them, except in the Finnish task, where it was very close to "Monson

Morfessor". It is interesting to note that while being far behind Morfessor in both Finnish and German, the "Monson Paramor" does a very good job in English being close to the combined version "Monson Paramor+Morfessor".

The new rule-based reference method "TWOL" that was evaluated this year in the Finnish and English task, was unbeatable in Finnish and only narrowly beaten in English by the best unsupervised algorithm and the traditional "Snowball Porter" stemmer. In Finnish and German the "Snowball" stemmers did not perform very well and had clearly lower average precision than the best unsupervised algorithms and "TWOL". The performance of the "grammatical" references based on the linguistic gold standards were not very high, which is not surprising given that the gold standards are relatively small.

The algorithms by Kohonen and Zeman that did not have morpheme analyses for all the words in the IR corpora were left behind Monson and McNamee. This may partly be due to those words that were not split in the morphemes, but as the importance of the analysis of those relatively rare words has not generally been very large in the previous tests, the performance gap may also be due to the morpheme analyses the algorithms provide.

10 Conclusions and Discussion

The Morpho Challenge 2008 was a successful follow-up to our previous Morpho Challenges 2005 and 2007. Since the main tasks were unchanged, the participants of the previous challenges were able to track improvements of their algorithms. It also gave a possibility for the new participants and those who missed the previous deadlines to try more established benchmark tasks. This year the evaluation was performed also in Arabic, and despite the relatively small wordlist and the inability to distribute a relevant text corpus, this task was again successful in finding significant differences between the submitted algorithms. The new IR task which allowed full text context to be used in the unsupervised morpheme analysis was not yet attempted by anyone. However, as it seems like a natural way to improve the models, it may be included in the next Morpho Challenge as well, giving participants more time to develop the new kinds of models and learning algorithms needed.

The significance of the differences in F-measure in Competition 1 was analyzed for all algorithm pairs in all tasks using the t-test. The analysis was performed by splitting the data into several partitions and comparing the results in each independent partition separately. The results of the tests show that all differences were statistically significant, except "Zeman 1" vs "Morfessor catmap" in the English task.

As already noted in the previous section, the ranking of the algorithms would have been very different, if only the precision measure was utilized in Competition 1. Some of the methods, especially "Kohonen allomorfessor" undersegmented the word forms heavily, which produced high precision but low recall. However, because it is difficult to estimate the relative weight of precision against recall in different applications, it remains for the application based evaluations in different tasks to show which algorithms are most useful. Many of the grammatical morphemes (such as +PL and +PAST in Table 3) are very common and may not be very relevant in IR, for example, compared to recognizing the right stem.

The use of unweighted F-measure in Competition 1 influences the results in several ways as discussed in the Morpho Challenge 2008 workshop and after it. Because there is a clear trade-off between precision and recall caused by favoring the under- or oversegmentation, F-measure improvements can be obtained by optimizing this. For example, the baseline Morfessor can be tuned to obtain a significantly higher F-measure than the other algorithms. Its recall can be increased with respect to the precision, for example, by discarding the rare words in training. This produces a smaller morpheme lexicon and, in average, shorter morphemes, which leads to a higher recall and lower precision [18]. However, this tuning does not seem to lead to significant improvements in the practical Competition 2 task. Direct recall improvement can be also obtained by combining the output of different algorithms as if they were alternative morpheme analysis of the same word [19]. In this combination the final precision will be roughly the average of the two alternatives and the final recall the sum of the two. In IR this combination does not give much help either, although the results seem to be at least as good as the better of the two algorithms.

As future work there remains the need to develop better methods to combine the different existing algorithms and to cluster the different surface forms produced by the morphemes. This might also somewhat improve the relatively low recall that several algorithms suffered in the Competition 1. New IR tasks should also be included and languages like Arabic which pose new kinds of morphological problems. To better serve the goal of producing a general purpose morpheme-based vocabulary that would be useful for several applications where large vocabulary is needed, we should also target new evaluation applications, e.g. in machine translation, text understanding and speech recognition.

Acknowledgments

We thank all the participants for their submissions and enthusiasm. We owe great thanks as well to the organizers of the PASCAL Challenge Program and CLEF who helped us organize this challenge and the challenge workshop. We are most grateful to the University of Leipzig for making the training data resources available to the Challenge, and in particular we thank Stefan Bordag for his kind assistance. We are indebted to Ebru Arisoy for making the Turkish gold standard available to us. We are most grateful to the Nizar Habash from the University of Columbia for his kind assistance and making the Arabic word frequency list and reference analyses available to the Challenge. Our work was supported by the Academy of Finland in the projects *Adaptive Informatics* and *New adaptive and learning methods in speech recognition*. This work was supported in part by the IST Programme of the European Community, under the PASCAL Network of Excellence, IST-2002-506778. This publication only reflects the authors' views. We acknowledge that access rights to data and other materials are restricted due to other commitments.

References

1. Kurimo, M., Creutz, M., Varjokallio, M., Arisoy, E., Saraclar, M.: Unsupervised segmentation of words into morphemes - Challenge 2005, an introduction and evaluation report. In: PASCAL Challenge Workshop on Unsupervised segmentation of words into morphemes, Venice, Italy (2006)
2. Bilmes, J.A., Kirchhoff, K.: Factored language models and generalized parallel backoff. In: Proceedings of HLT-NAACL, Edmonton, Canada, pp. 4–6 (2003)
3. Lee, Y.S.: Morphological analysis for statistical machine translation. In: Proceedings of HLT-NAACL, Boston, MA, USA (2004)
4. Zieman, Y., Bleich, H.: Conceptual mapping of user's queries to medical subject headings. In: Proceedings of the 1997 American Medical Informatics Association (AMIA) Annual Fall Symposium (October 1997)
5. Kurimo, M., Creutz, M., Varjokallio, M.: Morpho Challenge evaluation using a linguistic Gold Standard. In: Peters, C., Jijkoun, V., Mandl, T., Müller, H., Oard, D.W., Peñas, A., Petras, V., Santos, D. (eds.) CLEF 2007. LNCS, vol. 5152, pp. 864–872. Springer, Heidelberg (2008)
6. Kurimo, M., Creutz, M., Turunen, V.: Morpho Challenge evaluation by IR experiments. In: Peters, C., et al. (eds.) CLEF 2008. LNCS, vol. 5706, pp. 991–998. Springer, Heidelberg (2009)
7. Cetinoglu, O.: Prolog based natural language processing infrastructure for Turkish. M.Sc. thesis, Bogazici University, Istanbul, Turkey (2000)
8. Dutagaci, H.: Statistical language models for large vocabulary continuous speech recognition of Turkish. M.Sc. thesis, Bogazici University, Istanbul, Turkey (2002)
9. Creutz, M., Lagus, K.: Inducing the morphological lexicon of a natural language from unannotated text. In: Proceedings of the International and Interdisciplinary Conference on Adaptive Knowledge Representation and Reasoning (AKRR 2005), Espoo, Finland, pp. 106–113 (2005)
10. Creutz, M., Lagus, K.: Morfessor in the Morpho Challenge. In: PASCAL Challenge Workshop on Unsupervised segmentation of words into morphemes, Venice, Italy (2006)
11. Creutz, M., Lagus, K.: Unsupervised discovery of morphemes. In: Proceedings of the Workshop on Morphological and Phonological Learning of ACL 2002, pp. 21–30 (2002)
12. Creutz, M., Lagus, K.: Unsupervised morpheme segmentation and morphology induction from text corpora using Morfessor. Technical Report A81, Publications in Computer and Information Science, Helsinki University of Technology (2005), `http://www.cis.hut.fi/projects/morpho/`
13. Creutz, M., Linden, K.: Morpheme segmentation gold standards for finnish and english. Technical Report A77, Publications in Computer and Information Science, Helsinki University of Technology (2004), `http://www.cis.hut.fi/projects/morpho/`
14. Habash, N.: Large scale lexeme based arabic morphological generation. In: Proceedings of Traitement Automatique du Langage Naturel (TALN 2004), Fez, Morocco (2004)
15. Habash, N., Sadat, F.: Arabic preprocessing schemes for statistical machine translation. In: Proceedings of the Human Language Technology, Conference of the North American Chapter of the Association for Computational Linguistics (HLT-NAACL), New York, USA (2006)
16. Porter, M.: An algorithm for suffix stripping. Program 14(3), 130–137 (1980)

17. Virpioja, S., Väyrynen, J.J., Creutz, M., Sadeniemi, M.: Morphology-aware statistical machine translation based on morphs induced in an unsupervised manner. In: Proceedings of Machine Translation Summit XI, Copenhagen, Denmark (2007)
18. Virpioja, S.: Private communication (2008)
19. Monson, C., Carbonell, J., Lavie, A., Levin, L.: ParaMor and Morpho Challenge 2008. In: Peters, C., et al. (eds.) CLEF 2008. LNCS, vol. 5706, pp. 967–974. Springer, Heidelberg (2009)

ParaMor and Morpho Challenge 2008

Christian Monson[1], Jaime Carbonell[2], Alon Lavie[2], and Lori Levin[2]

[1] Center for Spoken Language Understanding, Oregon Health and Science University
20000 NW Walker Rd., Beaverton, OR, 97006, USA
[2] Language Technologies Institute, Carnegie Mellon University
5000 Forbes Ave., Pittsburgh, PA, 15213, USA
monsonc@ohsu.edu, jgc@cs.cmu.edu,
alavie@cs.cmu.edu, lsl@cs.cmu.edu

Abstract. We summarize the strong performance of ParaMor, an unsupervised morphology induction system, at Morpho Challenge 2008. When ParaMor's morphological analyses, which specialize at identifying inflectional morphology, are added to the analyses from the general-purpose unsupervised morphology induction system, Morfessor, the combined system identifies the morphemes of all five Morpho Challenge languages at recall scores higher than those of any other system which competed in the Challenge. These strong recall scores lead to F_1 values for morpheme identification as high as or higher than those of any competing system for all the competition languages but English.

Categories and Subject Descriptors: I.2 [Artificial Intelligence]: I.2.7 Natural Language Processing.

General Terms: Experimentation.

Keywords: Natural language morphology, Unsupervised learning, morphology induction.

1 Introduction

This paper describes the performance of the unsupervised morphology induction algorithm ParaMor in Morpho Challenge 2008. Morpho Challenge is a series of peer-operated competitions for algorithms designed to discover the morphological structure of individual natural languages in an unsupervised fashion. Two major considerations motivate our work on unsupervised morphology induction. First, we are interested in developing methods to quickly bring a morphology analysis system online for a new language. Of the more than 5000 languages in the world [4], only a handful have freely available computational morphological analysis systems. Unsupervised morphology induction promises to significantly decrease the time and expertise needed to build a morphology system for the many remaining languages. Second, we are interested in this problem from a theoretical standpoint: We would like to know how much of the morphology of a language it is possible to learn from nothing but raw text. Although we do not claim that our unsupervised morphology induction algorithm mimics how a human learns morphology, we hope that our work can place constraints on the range of strategies that humans might use.

C. Peters et al. (Eds.): CLEF 2008, LNCS 5706, pp. 967–974, 2009.

1.1 Related Work

To place the ParaMor algorithm in perspective, this section summarizes and categorizes proposed approaches to unsupervised morphology induction into four categories. The first category comprises systems that examine word-internal character transitions for probabilistic evidence of a morpheme boundary. The character transition approach is exemplified by [8], which builds forward and backward character tries and takes trie locations with significant branching factors as likely morpheme boundaries. More recently, Bernhard, in [1], measures the probabilities of word-internal character sequences without fragmenting a vocabulary down the leaves of a trie.

A second category of unsupervised morphology induction system treats morphology as a minimum description length (MDL) problem. This approach views morphemes as a compact representation of natural language words: If a system can identify the morphemes of a language, then that system could efficiently encode that language. Systems that employ this MDL approach include [3] and [6], as well as the Morfessor algorithm described in [5].

A third category of unsupervised morphology induction algorithm brings to bear the larger context in which a word occurs. The works in [12] and [16] exploit the idea that morphologically distinct surface forms of the same lexeme will often occur in the context of similar surrounding words. These two systems use combinations of word edit-distance heuristics and latent semantic analysis of word contexts to identify morphologically related words.

A fourth category of unsupervised morphology induction system purposefully models the paradigmatic structure of morphology. A morphological paradigm is a mutually substitutable set of morphological operations. In particular, conjugation and declension tables, as commonly found in language text books, are paradigms. Systems that appeal to the paradigmatic structure of morphology include [2], [6], [7], [13], [14], and [17], as well as the ParaMor algorithm, whose performance at Morpho Challenge 2008 is the topic of this paper. ParaMor differs from all paradigm-learning algorithms but the small-scale system described in [13] by directly modeling agglutinative sequences of affixes.

2 The ParaMor Algorithm

An unsupervised morphology induction algorithm, such as ParaMor, discovers the morphology of a language from nothing more than raw text. Although the Morpho Challenge 2008 competition evaluated morphology induction systems against full morphological analyses of words, the ParaMor algorithm only identifies purely concatenative sequences of suffixes. The section gives a high-level description of the ParaMor algorithm as outlined in Fig. 1. For a more in-depth description of ParaMor, please reference [11].

Using whitespace, the ParaMor algorithm first reduces an input text to a list of unique word types. A priori, ParaMor does not know where the morpheme boundaries fall in any given word, and so ParaMor proposes, for each word, a separate analysis that hypothesizes a morpheme boundary at each character boundary of that word. Whenever two or more corpus types end in the same word-final string, ParaMor constructs a paradigm seed. This paradigm seed contains the word-final string together with all word-initial strings that allow the word-final string to attach.

Phase 1: Paradigm Discovery (Only words above a threshold character length participate*).
Step 1: Search for candidate paradigms in a greedy recall-centric fashion
Step 2: Cluster candidate paradigms with a bottom-up agglomerative algorithm
Step 3: Filter paradigm clusters
A) Filter for the number of suffixes and stems in a paradigm
B) Filter for unlikely character transitions at morpheme boundaries*
Phase 2: Word-to-Morpheme Segmentation* (All input words participate).
For each word, *w*, and for all possible segmentations of *w*=*t.f*, where *f* is a suffix in a discovered paradigm *P*: If *f′*≠*f* is also a suffix in *P* and if *t.f′* is also a word in the vocabulary, then place a morpheme boundary in *w* between *t* and *f*.

Fig 1. The ParaMor unsupervised morphology induction algorithm. The three pieces of the ParaMor algorithm that have been introduced or modified since Morpho Challenge 2007 are marked with a star (*).

The ParaMor algorithm then proceeds in two main phases (see Fig. 1). In the first phase, Paradigm Discovery, ParaMor searches for sets of word-final strings which likely represent the suffixes of a paradigm. In the second phase, ParaMor segments word forms exactly where the discovered paradigms suggest a morpheme boundary. The Paradigm Discovery stage consists of three steps. Step 1 is a recall centric search that greedily expands ParaMor's paradigm seeds into full candidate paradigms by successively adding additional suffixes. Step 2 clusters initially selected candidate paradigms, which likely model the same underlying paradigm of a language. And Step 3 applies a series of filters to weed out paradigm candidates which likely do not model true paradigms of a language. When deciding to keep or discard a candidate paradigm, ParaMor's filters consider the number of suffixes and stems in the candidate paradigm cluster as well as the character transition probabilities that surround the morpheme boundary that the paradigm candidate hypothesizes, in the style of [8].

In Phase 2, ParaMor's word-to-morpheme segmentation algorithm looks for paradigmatic evidence of a morpheme boundary: When a word, *w*, belongs to a paradigm, then the stem of *w* can combine with other suffixes from that paradigm to form new words which may have occurred in the text corpus. Hence, to segment *w*, ParaMor matches each word-final string, *f*, of *w* against the suffixes in each discovered paradigm, *P*. ParaMor then substitutes in for *f*, one at a time, the other suffixes in *P*. If at least one of the substituted suffixes builds a word type from the corpus, then ParaMor segments *w* immediately before *f*.

2.1 Changes to ParaMor Since Morpho Challenge 2007

Morpho Challenge 2008 is the second Morpho Challenge competition in which ParaMor has taken part. In Morpho Challenge 2007, ParaMor participated in the English and German competitions. This year, ParaMor again analyzed English and German morphology, but also participated in the Finnish, Turkish, and Arabic tracks. Three major additions and adaptations to the ParaMor algorithm made participation in these morphologically more challenging language tracks practical. These three adaptations are described in detail in [11], while here they are only briefly summarized. Fig. 1 marks each of the three adaptations to ParaMor with a star (*).

The first two adaptations extend techniques to ParaMor that have been developed for the unsupervised morphology induction algorithms described in [8] and [6]. These first two adaptations are designed to improve the precision of ParaMor's discovered paradigms and resulting word-to-morpheme segmentations. The first adaptation restricts the set of word types which participate in ParaMor's paradigm discovery phase. Because, combinatorially, there are fewer possible short strings than there are long strings, words that consist of just a few characters are more likely than longer strings to suggest spurious morphological relationships. Hence, the first adaptation excludes short types from the Paradigm Discovery phase.

The second adaptation to ParaMor since the 2007 Challenge improves paradigm precision by removing initially induced paradigms that incorrectly hypothesize a morpheme boundary internal to a true suffix. At morpheme-internal character boundaries, the variety of characters that can extend a set of words is severely limited. Our adaptation measures the entropy in the distribution of stem-final characters in each candidate paradigm. ParaMor discards candidates with an entropy below a parameterized threshold.

The final adaptation to the ParaMor system from the 2007 Challenge acknowledges the agglutinative structure of natural language morphology: Many natural languages, including the Challenge languages of Turkish and Finnish, form surface words from several suffixes in sequence. But any individual candidate paradigm that ParaMor constructs during the Paradigm Discovery phase proposes at most a single morpheme boundary per word. Our third adaptation straightforwardly merges the separate morpheme boundaries, which arise from distinct discovered paradigms, into a single morphological segmentation containing multiple morpheme boundaries.

2.2 Combining ParaMor with Morfessor

ParaMor is designed to identify the sets of suffixes that form the morphological paradigms of inflectional morphology. But Morpho Challenge 2008 specifically evaluates morphology analysis systems on not only inflectional but also derivational morphology. In inflectional morphology the vast majority of lexemes that adhere to a paradigm form separate surface forms with each and every suffix of the paradigm. On the other hand, derivational morphology is more idiosyncratic [15]: Although productive derivational suffixes exist, often a particular stem may or may not form a new word with a particular derivational suffix.

To more practically compete in Morpho Challenge, we add to ParaMor's morphological analyses the morphological analyses suggested by the unsupervised morphology induction system Morfessor (Creutz, 2006). Morfessor is designed to identify all concatenative morphology, whether inflectional or derivational. Because a single word may have multiple legitimate morphological analyses, Morpho Challenge permits participants to submit multiple analyses of each particular word. In our combined ParaMor-Morfessor system, we submit the ParaMor and the Morfessor segmentations of each word as separate analyses of that word—as if each word were ambiguous between a ParaMor and a Morfessor analysis. Additional discussion of ParaMor's performance on inflectional and derivational morphology can be found in [11].

3 Results

Morpho Challenge 2008 evaluated unsupervised morphology induction systems in two ways [10]. First, systems competed in a linguistic evaluation that measured precision, recall, and F_1 at morpheme identification. And second, Morpho Challenge evaluated competing systems by measuring improvement on an information retrieval (IR) task. Of the three language tracks of the IR evaluation of Morpho Challenge, the combined ParaMor-Morfessor system placed first at average precision in the English and German tracks. Space limitations dictate, however, that the remainder of this paper focus on ParaMor's success in the linguistic evaluation of Morpho Challenge.

Table 1 summarizes the results of the linguistic competition. Systems competed in up to five languages in the linguistic evaluation: English, German, Finnish, Turkish, and Arabic. Table 1 contains the scores of nine individual unsupervised morphology induction algorithms. Six of these nine systems competed in Morpho Challenge 2008, while three systems participated in the 2007 Challenge. The scores from the 2007 competition are directly comparable to scores from the 2008 challenge because:

1. The linguistic evaluation of Morpho Challenge 2007 used the same evaluation methodology as the 2008 challenge; and moreover,
2. The 2007 challenge scored systems over the same corpora and against the same answer key as the more recent 2008 competition.

Of the six systems which competed in the 2008 challenge that appear in Table 1, three are systems we, Monson et al., submitted, while three are systems submitted by others. The three systems which we entered in Morpho Challenge 2008 are:

1. The ParaMor system alone,
2. A version of Morfessor, [5], which we trained ourselves, and
3. Our ParaMor and Morfessor analyses submitted as alternate, ambiguous, analyses.

Both the ParaMor and the Morfessor algorithms have free parameters. ParaMor's parameters, which control the Paradigm Discovery phase, were set to values that produced reasonable Spanish paradigms. ParaMor's parameters were then frozen before running over the five languages of Morpho Challenge. In the case of the Morfessor algorithm's single free parameter, which controls a perplexity threshold, we optimized the threshold for English, German, and Turkish against morphological answer keys we built ourselves. In Finnish and Arabic, we used that parameter setting which performed best against Turkish, a setting of 80.

The six systems in Table 1, which were prepared by others, are the systems with the top performance in the linguistic evaluation of Morpho Challenge 2007/2008. The final five systems found in Table 1 bear the names of their principle authors, while the system labeled Morf. MAP is the same Morfessor algorithm as the Morf. system which we submitted, but with different parameter settings. A change in parameter setting can sometimes result in quite different performance for Morfessor, c.f. Finnish. The specific parameter settings used to generate the results from the Morf. MAP column of Table 1 have not been reported.

Table 1. Results from the linguistic evaluation of Morpho Challenge. Systems participated in up to five language tracks. In each language track all participating systems were scored at precision (P), recall (R), and F_1 of morpheme identification. For each language track, the system or systems which place first at F_1 by a statistically significant margin appear in **bold**.

		Monson et al. 2008			Other Authors					
					2008			2007		
		ParaMor + Morf.	ParaMor	Morf.	Morf. MAP	Zeman	Kohonen	Bernhard	Bordag	Pitler
Eng	P	50.6	58.5	77.2	82.2	53.0	83.4	61.6	59.7	74.7
	R	63.3	48.1	34.0	33.1	42.1	13.4	60.0	32.1	40.6
	F_1	56.3	52.8	47.2	47.2	46.9	23.1	**60.8**	41.8	52.6
Ger	P	49.5	53.4	67.2	67.6	53.1	87.9	49.1	60.5	-
	R	59.5	38.2	36.8	36.9	28.4	7.4	57.4	41.6	-
	F_1	**54.1**	44.5	47.6	47.8	37.0	13.7	52.9	49.3	-
Finn	P	49.8	46.4	77.4	76.8	58.5	92.6	59.7	71.3	-
	R	47.3	34.4	21.5	27.5	20.5	6.9	40.4	24.4	-
	F_1	**48.5**	39.5	33.7	40.6	30.3	12.8	**48.2**	36.4	-
Tur	P	51.9	56.7	73.9	76.4	65.8	93.3	73.7	81.3	-
	R	52.1	39.4	26.1	24.5	18.8	6.2	14.8	17.6	-
	F_1	**52.0**	46.5	38.5	37.1	29.2	11.5	24.7	28.9	-
Arab	P	79.8	78.6	90.4	90.2	77.2	-	-	-	-
	R	27.5	8.5	21.0	21.0	12.7	-	-	-	-
	F_1	**40.9**	15.4	34.0	34.0	21.9	-	-	-	-

Although ParaMor alone performs respectably, it is when ParaMor's analyses are combined with Morfessor's that ParaMor shines. The combined system achieves the highest F_1 of any system which competed in the 2007 or 2008 Challenges in all language tracks but English (where the combined ParaMor-Morfessor system placed a strong second). In general, the ParaMor-Morfessor system attains this higher F_1 by balancing precision and recall. Where other systems are biased toward cautious precision, the ParaMor-Morfessor system trades high precision for improved recall in two ways. First, the search step in ParaMor's Paradigm Discovery phase (see Fig. 1) is intentionally recall-centric. Second, proposing morphological analyses from both ParaMor and Morfessor increases the likelihood of finding a particular morpheme.

The language ParaMor performs most poorly at is Arabic. New to Morpho Challenge in 2008, Arabic's morphology is distinctly different from that of the other four languages in the challenge. Arabic morphology differs most notably in possessing templatic morphology, where a consonantal root is interleaved with vowels to produce specific surface forms. Equally important, from ParaMor's perspective, is that Arabic is the only language in Morpho Challenge with significant prefixation. Arabic verbal morphology includes inflectional prefixes. In addition, Arabic orthography attaches a number of common determiners and prepositions directly onto the written form of the following word. These attached function words act as prepositions in text. As ParaMor is limited to looking for suffixes, both the templatic morphology and the prefixational morphology lower ParaMor's morpheme recall. In the near term, since prefixes are the mirror image of suffixes, a simple augmentation could allow ParaMor to analyze prefixation.

In general, all the systems that competed identified less than a third of the morphemes of Arabic. Interestingly, when ParaMor's Arabic analyses are presented in combination with Morfessor's, the increase in recall between the two systems is practically additive—implying very little overlap between the morphemes which the two systems identify. When recall scores are depressed across the board, any increase in recall implies an increase in F_1. And indeed, the ParaMor-Morfessor system receives the highest F_1 of any system which analyzed Arabic morphology.

4 Conclusions

The premise that the paradigmatic structure of morphology can be leveraged toward unsupervised morphology induction is clearly justified by the state-of-the-art performance of the ParaMor algorithm in Morpho Challenge 2008. In addition, the improved performance that results from joining the morphological analyses of the ParaMor and Morfessor systems demonstrates that current unsupervised morphology algorithms are highly complementary, and have much to gain from uniting their unique strengths.

While we are pleased with ParaMor's performance in Morpho Challenge 2008, we also see significant room for improvement on ParaMor's morphology induction algorithms. A careful examination of the paradigms which ParaMor produces over Spanish data identifies two major error classes. The first class of erroneous candidate paradigm results from inadequate clustering of initially selected candidate paradigms. We would like to more tightly integrate the search and clustering phases of ParaMor to enable more complete clustering of the initially selected partial paradigms. The second major class of erroneous paradigm is a consequence of morphophonology. Specifically, a stem or a suffix may appear in different surface forms conditioned on the morphemes with which it occurs. To conflate the varied surface forms of a single underlying morpheme we believe we will need to look at evidence outside the word as Schone (2001) and Wicentowski (2002) do.

Acknowledgements. We thank the organizing committee of Morpho Challenge 2008 for running the logistics of the Morpho Challenge competition. We are particularly grateful for the individualized attention Mikko Kurimo gave our specific situation. The research described in this paper was supported by NSF grants IIS-0121631 (AVENUE) and IIS-0534217 (LETRAS), with supplemental funding from NSF's Office of Polar Programs and Office of International Science and Education.

References

1. Bernhard, D.: Simple Morpheme Labeling in Unsupervised Morpheme Analysis. In: Peters, C., Jijkoun, V., Mandl, T., Müller, H., Oard, D.W., Peñas, A., Petras, V., Santos, D. (eds.) CLEF 2007. LNCS, vol. 5152, pp. 873–880. Springer, Heidelberg (2008)
2. Bernhard, D.: Apprentissage non supervisé de familles morphologiques par classifi-cation ascendante hiérarchique. In: Actes de la 14e conférence sur le Traitement Automati-que des Langues Naturelles, TALN (2007)

3. Brent, M.R., Murthy, S.K., Lundberg, A.: Discovering Morphemic Suffixes: A Case Study in MDL Induction. In: The Fifth International Workshop on Artificial Intelligence and Statistics (1995)
4. Comrie, B., Dryer, M.S., Gil, D., Haspelmath, M.: Introduction. In: Haspelmath, M., Dryer, M.S., Gil, D., Comrie, B. (eds.) The World Atlas of Language Structures. Oxford University Press, Oxford (2005)
5. Creutz, M.: Induction of the Morphology of Natural Language: Unsupervised Mor-pheme Segmentation with Application to Automatic Speech Recognition. Ph.D. Thesis. Computer and Information Science, Report D13. Helsinki: University of Technology, Es-poo, Finland (2006)
6. Goldsmith, J.: An Algorithm for the Unsupervised Learning of Morphology. Natural Language Engineering 12.4, 335–351 (2006)
7. Hammarström, H.: Unsupervised Learning of Morphology: Survey, Model, Algorithm and Experiments. Thesis for the Degree of Licentiate of Engineering. Chalmers, Göteborg University (2007)
8. Harris, Z.: From Phoneme to Morpheme. Language 31.2, 190–222 (1955)
9. Harris, Z.: Papers in Structural and Transformational Linguists. D. Reidel, Dordrecht (1970)
10. Kurimo, M., Turunen, V., Varjokallio, M.: Unsupervised Morpheme Analysis – Morpho Challenge (2008), `http://www.cis.hut.fi/morphochallenge-2008/` (accessed on: August 11, 2008)
11. Monson, C.: ParaMor: From Paradigm Structure to Natural Language Morphology Induction. Ph.D. Thesis. Language Technologies Institute, Carnegie Mellon University, Pittsburgh, Pennsylvania (2009)
12. Schone, P., Jurafsky, D.: Knowledge-Free Induction of Inflectional Morphologies. In: North American Chapter of the Association for Computational Linguistics, NAACL (2001)
13. Sharma, U., Kalita, J., Das, R.: Unsupervised Learning of Morphology for Building Lexicon for a Highly Inflectional Language. In: The Sixth Workshop of the ACL Special Interest Group in Computational Phonology, SIGPHON (2002)
14. Snover, M.G.: An Unsupervised Knowledge Free Algorithm for the Learning of Morphology in Natural Languages. M.S. Thesis. Computer Science, Sever Institute of Technology, Washington University, Saint Louis, Missouri (2002)
15. Stump, G.T.: Inflection. In: Spencer, A., Zwicky, A.M. (eds.) The Handbook of Morphology, Blackwell Publishers Inc., Malden (2001)
16. Wicentowski, R.: Modeling and Learning Multilingual Inflectional Morphology in a Minimally Supervised Framework. Ph.D. Thesis. Johns Hopkins University, Baltimore, Maryland (2002)
17. Zeman, D.: Using Unsupervised Paradigm Acquisition for Prefixes. In: Working Notes for the CLEF 2008 Workshop (2008)

Allomorfessor: Towards Unsupervised Morpheme Analysis

Oskar Kohonen, Sami Virpioja, and Mikaela Klami

Adaptive Informatics Research Centre,
Helsinki University of Technology
{oskar.kohonen,sami.virpioja,mikaela.klami}@tkk.fi

Abstract. We extend the unsupervised morpheme segmentation method Morfessor Baseline to account for the linguistic phenomenon of allomorphy, where one morpheme has several different surface forms. Our method discovers common base forms for allomorphs from an unannotated corpus. We evaluate the method by participating in the Morpho Challenge 2008 competition 1, where inferred analyses are compared against a linguistic gold standard. While our competition entry achieves high precision, but low recall, and therefore low F-measure scores, we show that a small model change gives state-of-the-art results.

1 Introduction

Morphological analysis is crucial to many modern natural language processing applications, especially when dealing with morphologically rich languages where the enormous number of inflected word forms lead to severe problems with data sparsity and computational efficiency. There are several successful methods for unsupervised segmentation of word forms into smaller, morpheme-like units [2,3]. The phenomenon of allomorphy limits the quality of morpheme analysis achievable by segmentation alone. Allomorphy is defined in linguistics as when an underlying morpheme-level unit has two or more morph-level surface realizations which only occur in a complementary distribution: only one of the different allomorphs of a given morpheme appear may appear in a certain morpho- and phonotactical context. For example, in Finnish, the singular genitive case is marked with a suffix `n`, e.g. `auto` (car) – `auton` (car's). Many Finnish nouns undergo a stem change when producing the genitive: `kenkä` (shoe) – `kengän` (shoe's), `pappi` (priest) – `papin` (priest's), `tapa` (habit) – `tavan` (habit's). A segmentation based approach models changed stems as distinct morphemes.

There are two main tasks in literature on learning allomorphy: finding morphologically related words (e.g. [9,1]), and learning a morphological analyzer (e.g. [10,4]). In our contribution to Morpho Challenge 2008 [7], we present an analyzer that is similar to Morfessor Baseline [3], but in addition finds common base forms for the inflected forms that derive from the same root word. We currently ignore allomorphic variation in suffixes. Information sources used in literature are orthographic similarity, word frequencies [10] and similar word contexts [9,1].

C. Peters et al. (Eds.): CLEF 2008, LNCS 5706, pp. 975–982, 2009.

We currently use only orthographic features. They are used in a similar manner in [10], but our model needs less supervision and allows concatenative morphology, rather than only stem-suffix pairs. Maybe the closest work to ours is presented in [4]. They study more general orthographic rewrite rules than we do, but the algorithm includes several phases and many heuristics. They also allow concatenative morphology, but the approach is not as general and cannot find, e.g., suffixes between stems. By embedding allomorphy learning into the Morfessor framework, we keep the algorithm flexible and conceptually simple.

2 Allomorfessor Model

In this section, we describe *Allomorfessor*, a morphological model that takes allomorphic variation into account. We start by defining a probabilistic generative model $\mathcal{M}$ for a text corpus. With *Maximum a Posteriori* (MAP) estimation, we try to find the model that is the most probable given the training corpus:

$$\mathcal{M}_{\mathrm{MAP}} = \arg\max_{\mathcal{M}} P(\mathcal{M}|\mathrm{corpus}) = \arg\max_{\mathcal{M}} P(\mathcal{M})P(\mathrm{corpus}|\mathcal{M}) \tag{1}$$

$P(\mathcal{M})$ is the Bayesian prior probability for the model and $P(\mathrm{corpus}|\mathcal{M})$ is the likelihood of the training corpus. Compared to Maximum Likelihood estimation, MAP provides a systematic way of balancing the model complexity and accuracy, and thus helps with the problem of overlearning (see, e.g., Ch. 3 in [5]).

Modeling a corpus with a morphological model is not straightforward. As occurrences of words in a corpus follow power law distributions (Zipf's law), any realistic model should abide by that phenomenon. Instead of using an explicit model for the corpus, as in, e.g., [6], we separate word-level and morpheme-level models, and concentrate only on the latter. We set the word-level model $\mathcal{M}_W$ to be a constant given a word lexicon $\mathcal{L}_W$, which contains all the word forms in the corpus, and try to find only the morpheme-level model $\mathcal{M}_M$. In addition, we divide $\mathcal{M}_M$ into two parts: morpheme lexicon $\mathcal{L}_M$ and morpheme grammar $\mathcal{G}_M$. The former models word-internal syntax and the latter provides the morphemes that from which the words are constructed. The optimization task is thus:

$$\mathcal{M}_{\mathrm{MAP}} = \arg\max_{\mathcal{G}_M, \mathcal{L}_M} P(\mathcal{L}_W|\mathcal{G}_M, \mathcal{L}_M)P(\mathcal{G}_M)P(\mathcal{L}_M). \tag{2}$$

This is equivalent to the approach used in Morfessor [3], but instead of modeling the original corpus, we are now modeling a lexicon of the words in the corpus.[1]

Our morpheme-level model resembles Morfessor Baseline, but it has a hierarchical structure, whereas Morfessor Baseline models words as sequences of morphs. At its core, our model is a probabilistic context-free grammar. Terminals of the grammar are strings resembling linguistic morphemes, specifically root stems and affixes. Non-terminals μ are morphs or their combinations. There

[1] This has been recommended to be done also with Morfessor by setting all the word counts to one. Otherwise, frequent word forms are often undersegmented.

are only two kinds of rules: μ is either replaced by a terminal (string), or two non-terminals, a *prefix* and a *suffix morph*, with a *mutation* terminal. In the former case, μ is a single real morph (root stem or affix). In the latter case, it is a *virtual morph*, which has substructure. Prefix and suffix morphs of a virtual morph can be either real or virtual morphs. The mutation is a special kind of terminal which modifies the virtual prefix. The mutation may be *empty*, which corresponds to a regular inflection or compound word, where the previous morph does not undergo any changes. For an illustration, see Fig. 1.

When designing the mutation model for allomorphy we strive to: (1) Make wrong analyses costly by favoring mutations close to the suffix. E.g., the edit distance between `blue` and `glue` is only one, but they are not allomorphs of the same morpheme. (2) Use mutation types general enough to allow statistical analysis. I.e., similar variations in different words should be modeled with the same mutation. The mutation type used in Allomorfessor is a special case of the standard edit distance (see, e.g., [8]). We allow only substitution and deletion operations, and make the mutation position independent. The affected position is found by matching to k:th instance of a target letter, that is scanned for starting from the end of the virtual prefix (or previous operation). Examples are shown in Table 1.

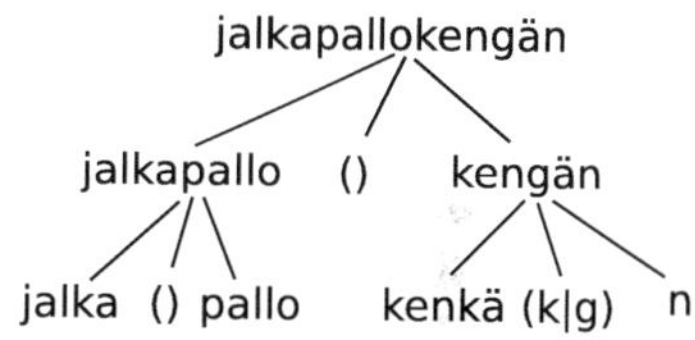

Fig. 1. An example analysis of the Finnish word `jalkapallokengän` (football shoe's). First, the word is split in two with an empty mutation denoted as `()`, then the virtual prefix `jalkapallo` is further split into the stems `jalka` and `pallo`. The virtual suffix `kengän` is split into the stem `kenkä`, the mutation `(k|g)` which transforms it into `kengä`, and the suffix `n`.

To calculate the smallest mutation of this kind between two arbitrary strings we apply the dynamic programming based algorithm for minimum edit distance (see, e.g., [8]), which can be modified to return also the edit operations needed. We want the optimal path not containing insertions, so we set the cost of insertions to be larger than what the other operations may yield for the given

Table 1. The allowed operations in mutations and some examples in Finnish

Operation	*Notation*	*Description*
substitution	k`x|y`	Change k:th `x` to `y`
deletion	-k`x`	Remove k:th `x`

(k is omitted when $k = 1$)

Source	*Mutation*	*Target*
`kenkä` (shoe)	`(k|g)`	`kengä` (e.g. `kengä+ssä`, in shoe)
`tanko` (pole)	`(k|g)`	`tango` (e.g. `tango+t`, poles)
`ranta` (shore)	`(-a t|n)`	`rann` (e.g. `rann+oi+lla`, on shores)
`ihminen` (human)	`(2n|s)`	`ihmisen` (human's)

string lengths. In this way we always find alternative paths without insertions if possible, by discarding candidates with too high costs. It is trivial to transform the edit operations into the Allomorfessor mutation format.

2.1 Model Probabilities

Next we give a formal description of the probabilities of Equation 2 for the Allomorfessor model. The formulation follows the work by Creutz and Lagus [3], with a few changes. First, every word form in the word lexicon is represented by one real or virtual morph μ_j. Thus the likelihood of the word lexicon is simply

$$P(\mathcal{L}_W|\mathcal{G}_M, \mathcal{L}_M) = \prod_{j=1}^{M_W} P(\mu_j), \tag{3}$$

where M_W is the number of words in the lexicon. The probability of the morph μ is estimated from the number of references to it from the word lexicon and virtual morphs.

The morph lexicon $\mathcal{L}_M$ consists of the real and virtual morphs. The probability of the morph lexicon is based on the properties of the morphs:

$$P(\mathcal{L}_M) = P(\text{size}(\mathcal{L}_M) = M)P(\text{properties}(\mu_1)\dots\text{properties}(\mu_M))M! \tag{4}$$

If a non-informative prior is used for the probability of the lexicon size M, its effect is minimal and it can be neglected. The factor $M!$ is explained by the fact that there are $M!$ possible orderings of M items, and the lexicon is the same regardless of the order in which the morphs are discovered.

The properties of the morphs are divided into two parts, usage and form. The usage includes properties of the morph itself and the properties of its context. In this model, we use only morph frequencies. For the probability of the frequency distribution, we use a non-informative, implicit frequency prior

$$P(\text{usage}(\mu_1)\dots\text{usage}(\mu_M)) = P(\text{freq}(\mu_1)\dots\text{freq}(\mu_M)) = 1/\binom{N-1}{M-1}, \tag{5}$$

where N is the sum of the counts of the morphs.

The form of a morph is its representation in the model. Forms of the morphs are assumed to be independent. As described before, μ_i is either a real morph represented by a string of letters, or a virtual morph consisting of prefix (μ_{pre}) and suffix (μ_{suf}) morphs and a (possibly empty) mutation δ_k. The probabilities are defined as:

$$P(\text{form}(\mu_i)) = \begin{cases} P(\text{sub})P(\mu_{\text{pre}})P(\delta_k)P(\mu_{\text{suf}}), & \text{if } \mu_i \text{ is virtual;} \\ [1 - P(\text{sub})]\,P(\text{len}(\mu_i)) \prod_{j=1}^{\text{len}(\mu_i)} P(\hat{c}_{ij}), & \text{otherwise.} \end{cases} \tag{6}$$

$P(\text{sub})$ is the probability that a morph has substructure, and for real morphs, $\hat{c}_{ij}$ is the jth character of the morph. The lengths of the real morphs are modeled explicitly using a gamma distribution with shape a and scale b:

$$P(\text{len}(\mu_i)) = \frac{1}{\Gamma(a)b^a}\text{len}(\mu_i)^{a-1}e^{-\text{len}(\mu_i)/b}. \tag{7}$$

Grammar $\mathcal{G}_M$ of the model contains the set of mutations Δ. Similarly to the lexicons,

$$P(\mathcal{G}_M) = P(\text{size}(\Delta) = M_\delta)P(\text{properties}(\delta_1)\ldots\text{properties}(\delta_{M_\delta}))M_\delta!, \quad (8)$$

and properties can be divided into usage and form. Usage features include only the frequencies; the non-informative prior is applied (cf. Equation 5). The prior probability for the form of a mutation δ_i with $\text{len}(\delta_i)$ operations is given by:

$$P(\text{form}(\delta_i)) = P(\text{len}(\delta_i)) \prod_{j=1}^{\text{len}(\delta_i)} P(k_{ij})P(\text{op}_{ij}) \quad (9)$$

$$P(\text{op}_{ij}) = \begin{cases} P(\text{del})\frac{1}{\Sigma} & \text{if op}_{ij} \text{ is a deletion} \\ P(\text{sub})\frac{1}{\Sigma^2} & \text{if op}_{ij} \text{ is a substitution} \end{cases} \quad (10)$$

For the weights we use $P(\text{del}) = P(\text{sub}) = 0.5$, Σ is the alphabet size, and k_{ij} tells which instance of the target letter of the operation op_{ij} is matched. $P(\text{len}(\delta_i))$ and $P(k_{ij})$ are taken from Gamma distributions.

2.2 Learning the Model

The model is learned by iteratively improving the model posterior $P(\mathcal{M}|\text{corpus})$, processing one word at a time and selecting the analysis of that word that maximizes the probability, as shown in Algorithm 1. Note that A_w is a list and we use + to denote the append operation. The algorithm considers analyzing the word w (1) without splits (2) with all possible splits of w and an empty mutation (3) with all possible splits and a base form similar to the virtual prefix and the required mutation. The two former ones are the same as in Morfessor Baseline and the third is our extension, with details shown in Algorithm 2.

Since each word has $2^{(\text{len}(w)-1)}$ possible analyses without considering mutations, we search greedily for the best split at any time, reducing the search space to $O(\text{len}(w)^2)$. When considering mutations, any word w could potentially be the base form for any other word w^*. This would lead naturally to a $O(N^2)$ algorithm. This is unfeasible for large datasets, and therefore we constrain the candidates in heuristic ways, such as limiting the number of analyses to K per morph and iteration, as can be seen in Algorithm 2. Since finding the $baseforms$ can be done as a range search it requires $O(K\log(N))$ time, and thus the time complexity for the whole learning algorithm is $O(NK\log(N))$.

3 Experiments

The model was evaluated in Morpho Challenge 2008 competition 1 [7]. The following parameter settings are used: Morph lexicon length distribution in Equation 7: shape $a = 5$ and scale $b = 1$. The number of candidates considered for each virtual morph $K = 20$. For the mutation lengths and k_{ij} in Equation 9, we used parameters $a = 1$ and $b = 1$ of the gamma prior to prefer short mutations.

Algorithm 1. The learning algorithm

while $P(\mathcal{M} \,|\, \text{corpus})$ increases **do**
  for $w \in \mathcal{L}_W$ in random order **do** optimize(w,len(w))
end while
function optimize(w,n)
  $A_w \leftarrow [w] + [(w_{1..i}, w_{(i+1)..n}) : i \in 1, ..., n-1] + \text{mutated\_analyses}(w, n)$
  Apply the analysis a^*_w of the first K elements of A_w that maximizes $P(\mathcal{M} \,|\, \text{corpus})$

  if a^*_w involved a split **then** optimize($w_{1..i}, i$); optimize($w_{(i+1)..n}, n-i$)

Algorithm 2. mutated_analyses(w, n)

for $i \in 1, ..., n-1$ **do**
  if $n >= 4 \wedge \text{len}(w_{(i+1)..n}) <= 5 \wedge w_{(i+1)..n} \in \mathcal{L}_M$ **then**
    if $n > 6$ **then** *difflen* $\leftarrow 4$ **else** *difflen* $\leftarrow 3$
    $\textit{baseforms} \leftarrow \{v \in \mathcal{L}_W : v_{1..(n-\textit{difflen})} = w_{1..(n-\textit{difflen})}\}$
    Calculate mutations δ_j between each $\textit{baseforms}_j$ and $w_{(i+1)..n}$
    $A_w \leftarrow A_w + [(v_j, w_{(i+1)..n}, \delta_j) : v_j \in \textit{baseforms}]$
  end if
end for
return A_w sorted by i and descending len(v_j)

The Morpho Challenge results are summarized in Table 2. The most striking figures are our very low recall numbers. Low recall means that the model undersegments heavily, i.e., the algorithm should find more morphemes per word (e.g. `kengän` and `papin` are both unsegmented). The precisions are quite good, especially for Turkish and Finnish, but are explained by the low recall.

Mutations were not used very frequently in the analyses. Where substructure was found in the word form, 98% of the mutations were empty for English and 96% for Finnish. The algorithm often favors using new base forms over using mutations, e.g. `prettier` is analysed as `pretti () er`, not `pretty -y er`. The five most common mutations for English and Finnish are shown in Table 3. Some of the example analyses shown are desired (e.g., `-e` in `abjure`, `-a` in `haljeta`), but in many cases the mutation is clearly unnecessary. E.g., a simpler analysis for `suspicions` would be `suspicion () s`. Mutations are also used commonly in misspelled words. E.g., both `contructed` and `contructive` exist in the English corpus, and mutation `-d-e` is used to get the missing base form `contruct`.

Table 2. Results from the Morpho Challenge evaluation. See [7] for details.

Language	*Precision*	*Recall*	*F-Measure*	*F/Winner*	*F/Morf.Baseline*
English	83.39%	13.43%	23.13%	56.26%	54.04%
German	87.92%	7.44%	13.71%	54.06%	31.01%
Turkish	93.25%	6.15%	11.53%	51.99%	20.08%
Finnish	92.55%	6.89%	12.82%	48.47%	21.16%

Table 3. The five most frequent mutations found by the algorithm for English (left side) and Finnish (right side)

Mutation	*Freq.*	*Example*	*Mutation*	*Freq.*	*Example*
(-e)	2033	abjure (-e) ed	(-n)	27510	antiikin (-n) lle
(-s)	537	actress (-s) s'	(-n -e)	15830	edustajien (-n-e) esi
(-y)	386	inequity (-y) able	(-a)	6241	haljeta (-a) essa
(-n)	243	suspicion (-n) ns	(-i)	4203	kliimaksi (-i) in
(-d -e)	183	contructed (-d-e) ive	(-a -t)	2792	alokkaita (-a-t) lle

4 Discussion

Our model gave poor results in Morpho Challenge even compared to Morfessor Baseline (see Table 2). Afterwards, we have found out the reasons for the undersegmentation and implemented a new version that solves the problems. The main reasons for the undersegmentation are hierarchical model structure and context independent mutations, which both result in increased cost of data. Compare, e.g., the following analyses of English word "mispronouncing":

$$\begin{aligned} P(\text{mispronouncing}|\mathcal{M}) &= P(\text{mis})P(\epsilon)P(\text{pronouncing}) \\ P(\text{pronouncing}|\mathcal{M}) &= P(\text{pronounce})P(\text{-e})P(\text{ing}) \\ &\text{vs.} \\ P(\text{mispronouncing}|\mathcal{M}) &= P(\text{mis})P(\text{pronounc})P(\text{ing}), \end{aligned}$$

where ϵ is an empty mutation. The former analysis may save one lexical item due to the use of the mutation -e, but the data cost will have six probabilities compared to three in the latter analysis. If "mispronouncing" were analyzed as a single morph, the data probabilities of the two model would be equal. Thus complex segmentations are penalized more in our model.

The problem of mutation costs can be solved by conditioning the mutations by the following morph (suffix). In the previous example, we can get:

$$\begin{aligned} P(\text{mispronouncing}|\mathcal{M}) = P(\text{mis})P(\epsilon|\text{mis}) \times P(\text{pronounce})P(\epsilon|\text{pronounce}) \times \\ P(\text{ing})P(\text{-e}|\text{ing}). \end{aligned}$$

Note that most of the morphs are stems, and occur only with the empty mutation. Then $P(\epsilon \mid \mu) = 1$, and the data cost does not increase. Using a flat structure and conditioning the mutation probabilities on suffixes do not require complicated changes to the Allomorfessor model. The most relevant change is that the word lexicon is represented by a sequence of morphs and mutations:

$$P(\mathcal{L}_W|\mathcal{G}_M, \mathcal{L}_M) = \prod_{j=1}^{M_W} \prod_{k=1}^{n_j} P(\mu_{jk})P(\delta_{jk}|\mu_{jk}), \tag{11}$$

where n_j is the number of morphs in word j. As morphs will not have any substructure, $P(\text{sub})$ is zero in Equation 6. Probabilities $P(\delta \mid \mu)$ are estimated from the observed frequencies. Non-informative prior probabilities of the co-occurrences of mutations and morphs are added to the usage properties of the morphs.

We have made preliminary test with the new version using the English task of competition 1. To speed up the computation and reduce misspellings, the word forms that occurred only once were excluded from training. With this setting, F-measure was 57.12% (precision 65.26%, recall 50.79%). This is significant improvement over the submitted version, and shows that the general framework is working. Part of the improvement in recall and F-measure was due to the pruned training data: with the same data, F-measure for Morfessor Baseline was 56.14% (precision 64.49%, recall 49.69%). However, the improvement over Morfessor Baseline was statistically significant for both precision and recall.

We conclude that our framework for learning allomorphy seems to be promising. In addition to more extensive testing and error analysis with the new version, future work will include using context and frequency information to both limit and weight the potential allomorphs.

References

1. Baroni, M., Matiasek, J., Trost, H.: Unsupervised discovery of morphologically related words based on orthographic and semantic similarity. In: Proceedings of the ACL 2002 workshop on Morphological and phonological learning, Morristown, NJ, USA, pp. 48–57. ACL (2002)
2. Bernhard, D.: Simple morpheme labelling in unsupervised morpheme analysis. In: Peters, C., Jijkoun, V., Mandl, T., Müller, H., Oard, D.W., Peñas, A., Petras, V., Santos, D. (eds.) CLEF 2007. LNCS, vol. 5152, pp. 873–880. Springer, Heidelberg (2008)
3. Creutz, M., Lagus, K.: Unsupervised models for morpheme segmentation and morphology learning. ACM Transactions on Speech and Language Processing 4(1) (January 2007)
4. Dasgupta, S., Ng, V.: High-performance, language-independent morphological segmentation. In: The annual conference of the North American Chapter of the ACL, NAACL-HLT (2007)
5. de Marcken, C.G.: Unsupervised Language Acquisition. PhD thesis, MIT (1996)
6. Goldwater, S., Griffiths, T.L., Johnson, M.: Interpolating between types and tokens by estimating power-law generators. In: Advances in Neural Information Processing Systems (NIPS), p. 18 (2006)
7. Kurimo, M., Turunen, V., Varjokallio, M.: Overview of Morpho Challenge 2008. In: Peters, C., et al. (eds.) CLEF 2008. LNCS, vol. 5706, pp. 951–966. Springer, Heidelberg (2009)
8. Navarro, G.: A guided tour to approximate string matching. ACM Comput. Surv. 33(1), 31–88 (2001)
9. Schone, P., Jurafsky, D.: Knowledge-free induction of morphology using latent semantic analysis. In: Proceedings of the 2nd workshop on Learning language in logic and the 4th conference on Computational natural language learning, Morristown, NJ, USA, pp. 67–72. ACL (2000)
10. Yarowsky, D., Wicentowski, R.: Minimally supervised morphological analysis by multimodal alignment. In: Proceedings of the 38th Meeting of the ACL, pp. 207–216 (2000)

Using Unsupervised Paradigm Acquisition for Prefixes

Daniel Zeman

Ústav formální a aplikované lingvistiky
Univerzita Karlova
Malostranské náměstí 25
CZ-11800 Praha, Czechia
zeman@ufal.mff.cuni.cz

Abstract. We describe a simple method of unsupervised morpheme segmentation of words in an unknown language. All that is needed is a raw text corpus (or a list of words) in the given language. The algorithm identifies word parts occurring in many words and interprets them as morpheme candidates (prefixes, stems and suffixes). New treatment of prefixes is the main innovation in comparison to [1]. After filtering out spurious hypotheses, the list of morphemes is applied to segment input words. Official Morpho Challenge 2008 evaluation is given together with some additional experiments. Processing of prefixes improved the F-score by 5 to 11 points for German, Finnish and Turkish, while it failed to improve English and Arabic. We also analyze and discuss errors with respect to the evaluation method.

1 Introduction

Morphological analysis (MA) is an important step in natural language processing, needed by subsequent processes such as parsing or translation. Unsupervised approaches to MA are important in that they help process understudied (and corpus-poor) languages, for which we have inadequate machine-readable dictionaries and tools. Ideally, an unsupervised morphological analyzer (UMA) would learn to analyze a language by looking at a large text in that language, without any additional resources, not to mention an expert or speaker of the language.

Supervised approaches to MA can provide us with three types of information: 1. segmentation of a word into *morphemes,* i.e. the smallest units bearing lexical or grammatical meaning; 2. functional explanation of grammatical morphemes, often expressed in a *morphological tag* (e.g. the PDT [2] tag `AAMS2----3N----` would mean that all the grammatical morphemes together set the features gender = masculine, number = singular, case = genitive (2), degree = superlative (3) and negation = negative); 3. lexical anchoring of morphemes – a *lemma* (part-of-speech information, although lexical, is usually encoded in the tag). Unsupervised approaches cannot use dictionaries nor do they know about labels such as *singular* or *genitive* that people have attached to morphemes. Thus, neither lexical nor grammatical explanation is possible for any morpheme. What unsupervised methods can attempt is morpheme segmentation. Algorithms for such segmentation may not know if a particular morpheme denotes plural,

C. Peters et al. (Eds.): CLEF 2008, LNCS 5706, pp. 983–990, 2009.

they may even disagree where exactly a morpheme boundary lies. But they can figure out that a particular morpheme occurs at the end of a certain set of words (a linguist would say that all these words are plural), and that it can be optional, i.e. a word may occur without it (identified as singular by the linguist).

In many languages, morphemes are classified as *stems* and *affixes* with latter being further subclassified as *prefixes* (preceding stems) and *suffixes* (following stems). A common word pattern consists of a stem (bearing the lexical meaning) and, optionally, some prefixes (bearing lexical or grammatical meaning) and/or suffixes (often bearing grammatical meaning). In a language such as German, *compound words* containing more than one stem are quite frequent. While a stem can appear without any affix, affixes hardly appear on their own. In this paper, a morphological *paradigm* is a collection of affixes that can be attached to the same group of stems, plus the set of affected stems.

Although the segmentation of a word does not provide any linguistically justified explanation for any of its components, the output can still be useful for further processing. Having got a paradigm, we can generate all unseen morpheme combinations within that paradigm. We can recognize stems of new words and then group all words with the same stem. The hope is that a stem may have a lexical meaning. All words in a group will share this lexical meaning and differ grammatically. By dropping some part of the meaning (hopefully the less important part) we reduce the data sparseness of more complex models for syntactic parsing, machine translation, and information retrieval.

A comprehensive overview of related work is given in [1]. In addition, there are several new papers on UMA describing results of the previous Morpho Challenge. The approaches in [3], [4] and [7] utilize segment predictability and transition probabilities. [5] models character N-grams to find word stems. [8] is a semi-supervised method using a limited set of hand-written rules. [6] seems to be the most similar approach to ours, which is a direct extension of [1].

The rest of the paper is organized as follows: in Sections 2 and 3, we review the method of [1] for stem-suffix learning. The main innovation in learning and identifying prefixes using two different methods is described in Section 4. The rest of the paper presents our experiments and discusses their results.

2 Learning Stems and Suffixes

We begin with reviewing the paradigm acquisition that was first described in [1]. The algorithm searches for positions to cut words each into two parts: the *stem* and the *suffix*. An important feature of such a split is that a stem occurs in training data with an arbitrary number of suffixes and a suffix occurs with multiple stems. Otherwise, the algorithm would just collect words with coincidentally identical parts.

In the first step, all possible segmentations of every word are generated. For instance, the word *bank* can be segmented as *bank, ban+k, ba+nk, b+ank*. We collect all stems and suffixes. For each stem, we store all co-occurring suffixes, and for each suffix we keep all co-occurring stems.

Various techniques are applied to filter out spurious paradigm candidates (see [1] for more details and examples):

1. If there are more suffixes than stems in a paradigm, the paradigm is removed.
2. If all suffixes in a paradigm begin with the same letter, there is another paradigm where the letter is part of the stem. The rule is to prefer longer stems and shorter suffixes. It means that the paradigm in which the border letter is part of stems will be preserved, and the other will be removed.
3. If the suffixes of paradigm B form a subset of suffixes of paradigm A ($A \supset B$) and there is no C, different from A, such that B is also subset of C ($\forall C \neq A : (B \not\subset C)$), we add the stems of B to the stems of A, and remove B. A subset paradigm is merged with its superset, as long as there is only one superset candidate. As mentioned in [1], the process of identifying subsets is computationally quite expensive. We replaced the algorithm used in [1] by a new one based on dynamic programming. Starting with the longest suffix sets, we gradually identify their possible subsets by dropping one element at a time. The resulting graph of superset-subset relations contains suffix sets that do not occur in any paradigm. However, building it is linear in original number of paradigms and their mean size. Traversing the graph is trivial and relatively few steps are required to find the closest real superset paradigms. In the case of approx. 69,000 English paradigms, the old approach required billions of hash queries, whereas now we need only about 600,000 steps together for constructing and querying the graph. The new algorithm makes the method capable of processing more data in less time, allowing for morphologically more complex languages and/or more benevolent filtering in the preceding steps.
4. Paradigms with only one suffix are removed.

The final set of paradigms yields the lists of known stems and suffixes. The information what stem can occur with what suffix is also available. Note however, that due to subset merging, the expression "can occur" is no longer equivalent to "has occurred in training data". Also, some words are covered by no known stem and/or suffix because all their segmentations were removed during the paradigm filtering.

The three lists (known stems, known suffixes, and known stem-suffix pairs) are the output of the learning phase. They are now used to identify morphemes in new words.

3 Morphemic Segmentation

Given the lists obtained during training, we want to find the stem-suffix boundary in a word of the same language. We can also find out that the word does not have any suffix.

Again, we consider all possible segmentations of each analyzed word. For each stem-suffix pair, we look up the table of learned stems and suffixes. The following cases are possible:

1. Both stem and suffix are known and allowed to occur together.
2. Both parts are known but they are not known to occur together.
3. Only the stem is known.
4. Only the suffix is known.
5. Neither the stem nor the suffix is known.

If both parts were known (case 2), the procedure from [1] marked the segmentation as *certain*. If only one part was known (cases 3 and 4), the segmentation was labeled as *possible*. If there were certain segmentations, they were returned as competing analyses of the word. Otherwise, the possible segmentations were returned. If there was no possible segmentation either, the whole word was returned as a single morpheme.

4 Learning and Detecting Prefixes

The main weakness of the approach in [1] is that it assumes one or two morphemes per word. There is no means of correctly segmenting words that contain both prefixes and suffixes, and compound words. In the present work, we explored two ways of identifying prefixes in addition to the stem-suffix splitting. Both methods work separately from the stem-suffix learning, so after learning a separate list of prefixes, modified segmentation algorithm has to be employed.

4.1 Reversed Word Method

The least expensive prefix-learning approach seems to be to take the whole apparatus and apply it on reversed words (right-to-left). The strings that the system marks as suffixes are reversed again and marked as prefixes. The problem is that we now have two sets of stems: one from the suffix learning, one from the learning of prefixes. For instance, the English word *un+beat+able* yields the stems *unbeat* and *beatable*, respectively. The following algorithm uses the four lists (prefixes, prefix-stems, suffix-stems and suffixes) to find one or more segmentations of a word:

1. For all split points, check whether the left part is a known prefix and the right part is a known prefixed stem. (This corresponds to *certain* segmentations of [1].) If so, remember the prefix as applicable.
2. For all split points, check whether the left part is a known suffixed stem and the right part is a known suffix. (This corresponds to *certain* segmentations of [1].) If so, remember the suffix as applicable.
3. Remember also the empty prefix and the empty suffix as applicable.
4. Loop over all combinations of applicable prefixes and suffixes. Make sure that they do not overlap and that at least one character of the word remains to play the role of stem. Save segmentations found this way.
5. If at least one morpheme boundary has been found, remove the "dummy" segmentation (which marks the whole word as a stem).

4.2 Rule-Based Method

The other approach we tested defines a prefix using the following set of parameterized rules. The values of the four parameters are estimates based on a few experiments. We wanted to keep the approach language-independent, so we did the experiments with English data only, and used the same values for all languages. We set $K = 5$, $L = 2$, $M = 5$ and $N = 100$.

1. A prefix is formed by 1 to K word-initial characters.
2. Minimal length of the stem (the remainder of the word after removing the prefix) is L.
3. The prefix occurs at least with N stems for which the following condition holds.
4. The stems with which the prefix occurs also occur without the prefix or with another prefix. The number of different prefixes (including the empty one) seen with the common stem must be at least M.

The algorithm to find prefixes is simple. First split each word in up to K positions, observing conditions 1 and 2, and generate the initial set of prefix candidates. Then loop over them and discard those not complying with conditions 3 and 4. This process typically needs to be repeated iteratively because discarding a prefix decreases the number of prefixes at other stems, which in turn could invalidate another prefix, although it already passed the first check-up. Note that the prefixes counted in condition 4 have to conform to the definition themselves. We observed that the process converged rather quickly. For instance, the English data generated 119,000 first-round candidates. Only 678 candidates passed to the second round, and the set converged to 665 prefixes after four rounds.

We would prefer to get even smaller set of prefixes but were unable to find better setup. Either the system discarded all candidates, or at least the real prefixes did not survive (while some garbage did). For more details, see Section 6.

A few other comments on the method: We further revised the morphemic segmentation based on prefixes. We ignored the stems found with prefixes. We took the stem-suffix segmentation found by [1] and just looked for a known prefix in the beginning of the stem. If we found it, the prefix was made a separate morpheme (regardless whether the stem was actually seen with this prefix).

5 Results

Results are compared to gold standard segmentation created by a supervised morphological analyzer. The evaluation method must reflect the limitations of unsupervised approaches, thus the only information being compared with the gold standard is the fact that two words share a morpheme with the same label. Even the order of the morphemes is not significant, although [1] stated the contrary. List of pairs of words and the morphemes they share is created first for both the gold standard and the output of the unsupervised analyzer. The next step is computing **precision** (what portion of the morphemes shared in system output were shared correctly, i.e. were also shared in the gold standard?) and **recall** (what portion of the morphemes shared in gold standard were also shared in system output?) The **F** score is then computed the usual way, i.e. $F = (P + R) / 2PR$. For more details on the evaluation procedure, see [9].

The main result of the experiment is the comparison of the two methods for finding prefixes. The reversed word method generally brings very high precision and very low recall and F-score. The rule-based method improves results for three languages. It generally decreases precision as a price for improving both recall and F-score. It is not clear what kind of damage the method caused on English and Arabic. This calls for further investigation because previous evaluation on smaller data suggested improvement on all the languages [10].

Table 1. Results. Next to the language name is the best F-score of the other participants. "Stem+suff" is equivalent to [1] (submitted to MC2008 as "method 1"). "Rev strict" adds the reversed prefix method (submitted as "method 3"). "Rule weak" is unofficial post-deadline result (but evaluated on the same data).

	English (56.26)		
	P	R	F
stem+suff	52.98	42.07	**46.90**
rev strict	76.92	8.47	15.27
rule weak	27.72	62.47	38.40

German (54.06)		
P	R	F
53.12	28.37	36.98
72.27	7.15	13.01
41.75	41.97	**41.86**

	Finnish (48.47)		
	P	R	F
stem+suff	58.51	20.47	30.33
rev strict	72.41	3.42	6.54
rule weak	50.12	35.85	**41.80**

Turkish (51.99)		
P	R	F
65.81	18.79	29.23
73.30	3.01	5.79
52.54	33.43	**40.86**

	Arabic (40.87)		
	P	R	F
stem+suff	77.24	12.73	**21.86**
rev strict	89.62	5.18	9.79
rule weak	68.96	11.20	19.27

Vocabulary sizes:	Ar	144 K
	En	385 K
	Tr	617 K
	De	1.3 M
	Fi	2.2 M

The processing of one language took from about 5 minutes (Arabic) to almost 1 hour (Finnish) on a 64bit AMD Opteron.

6 Error Analysis

The training data is noisy and contains many typos. Our system does not use up the word frequency statistics that could help to filter out noise. The damaging impact that noise can cause can be illustrated on the group of English words *abrupt – abruptly – abruptness – *abrupty*. The first three words form a well understood pattern that occurs with a few hundred other stems *(absent-minded, aimless, anxious, artless, assertive... etc.)* The fourth word, *abrupty,* is in fact a misspelled version of *abruptly*. The typo in the suffix prevented this group from being included in the main paradigm. Typos are infrequent (compared to correct material) and there was only one other stem *(explicit)* that occurred with the same kind of error. As a result, the group formed a separate paradigm and got filtered out on the rule 1 (more suffixes than stems).

Since the rule was introduced to exclude spurious paradigms with a one-letter stem and thousands of suffixes, we tried to weaken it and allow paradigms that have N times more suffixes than stems for a small N (≤ 5). However, the change decreased recall as well as the overall F-score, so we did not include it in our final experiments.

Two-way checking of morphemes in the reversed word method is the probable cause for the high precision and extremely low recall for all languages. The learned paradigms look reasonable, although they are not too large. Some examples are: (English) *three-, two-, four-, 0 + decade-old, foot-thick, fifths, hour-plus; 0, re, re-, over +*

capitalise, stimulate, tighten, commit; (German) *südo, nordo, o, südwe, nordwe, we + stprovinz, sthorizont, stchinesischen, stpolnischen, stafrikanischen, stzipfel, stdeutsche, stengland, stchina; 0, ab, ein, aus, zu + gewanderter, gewanderten, wanderungsdruck, gewanderte, wanderungswelle.* The first German example well illustrates that it would make more sense to put the border characters to the prefix instead of the stem. The reversed word method found 3,279 English prefixes, 12,585 German, 20,537 Finnish, 5,127 Turkish and 261 Arabic.

The rule-based prefix method is generally more restrictive in learning prefixes, although the segmentation approach we combined with it is more benevolent. The rule-based algorithm learned 665 English, 1,890 German, 4,628 Finnish, 2,178 Turkish and 331 Arabic prefixes. There are three kinds of prefixes: 1. very short strings that are probably not true prefixes but are too frequent to be filtered out *(aa, abf, abg, ac, ag, ah, ai, ak...)*; 2. real prefixes (English *anti, anti-, auto, by, co, co-, dis, ex, mis, re, un, ...;* German *ab, an, anti, anti-, anzu, auf, aufge, aufzu, aus, ausge, be, dar, ...)*; 3. first parts of compounds (English *ash, back, bank, bell, down, five-, half, ...;* German *abend, acht, aids, aids-, akten, alarm, alpen, ...)*.

7 Ways to Improve

Some options have already been mentioned: using word frequencies to eliminate typos, more or less strict stem matching in the segmentation phase. There are alternatives to the strictest matching (stem and suffix must have occurred together). For instance, instead of requiring that the stem and the suffix occur together, we can ask whether the stem occurred with *N* other suffixes that co-occur with the tried suffix in at least one paradigm. Another option is to try to figure out whether the word can belong to a paradigm that allows for such suffix. For example, the strictest method did not split *a-com's* to *a-com* and *'s*, although the word *a-com* was in training data and *'s* is part of many paradigms. However, this particular segmentation got discarded in the filtering phase.

A better approach to compound words is needed. The prefix processing is able to separate first parts of compounds in many instances. However, there are many other compounds that are not solved satisfactorily. Either their first part is longer than the threshold, or they have more than two parts.

The naming of morphemes matters! Even when the way how the evaluation works is designed not to depend on the labels, we are still responsible for giving same labels to same things. If we fail to recognize that "-d" and "-ed" is essentially the same English morpheme (which can easily happen if they came from different paradigms), we will be penalized in F-score.

The border letters that occur in all words of a paradigm can trouble subset merging mechanisms. For instance, the largest German paradigm has suffixes *0, m, n, r, re, rem, ren, rer, res, s*. All stems of this paradigm end in *e*. There is another paradigm with suffixes *0, e, em, en, er, ere, es*. Here the *e* must remain in suffixes because of the only exception, the empty suffix *0*. The two paradigms cannot be merged. However, if the *e* in the former paradigm was shifted to the suffixes, merging would be trivial and immediate. The question is, how do we know that in this particular case the bordering letters should be treated differently?

8 Conclusion

Our method can be used for unsupervised segmentation of words into morphemes. The main improvement over [1] is the unsupervised learning of prefixes. Compounds and typos are the most important yet-to-be-addressed issues.

Acknowledgements. My thanks go to the anonymous reviewers for their helpful feedback. Funding: Czech Academy of Sciences, project No. 1ET101470416, and Czech Ministry of Education, project MSM0021620838.

References

1. Zeman, D.: Unsupervised Acquiring of Morphological Paradigms from Tokenized Text. In: Peters, C., Jijkoun, V., Mandl, T., Müller, H., Oard, D.W., Peñas, A., Petras, V., Santos, D. (eds.) CLEF 2007. LNCS, vol. 5152, pp. 892–899. Springer, Heidelberg (2008)
2. Böhmová, A., Hajič, J., Hajičová, E., Hladká, B.: The Prague Dep. Treebank: A Three-Level Annotation Scenario. In: Treebanks: Building and Using.... Kluwer, Dordrecht (2003)
3. Bernhard, D.: Simple Morpheme Labeling in Unsupervised Morpheme Analysis. In: Peters, C., Jijkoun, V., Mandl, T., Müller, H., Oard, D.W., Peñas, A., Petras, V., Santos, D. (eds.) CLEF 2007. LNCS, vol. 5152, pp. 873–880. Springer, Heidelberg (2008)
4. Bordag, S.: Unsupervised and Knowledge-free Morpheme Segmentation and Analysis. In: Peters, C., Jijkoun, V., Mandl, T., Müller, H., Oard, D.W., Peñas, A., Petras, V., Santos, D. (eds.) CLEF 2007. LNCS, vol. 5152, pp. 881–891. Springer, Heidelberg (2008)
5. McNamee, P., Mayfield, J.: N-Gram Morphemes for Retrieval. In: Working Notes for the CLEF Worksh., Budapest, Hungary (2007)
6. Monson, C., Carbonell, J., Lavie, A., Levin, L.: ParaMor: Finding Paradigms across Morphology. In: Peters, C., Jijkoun, V., Mandl, T., Müller, H., Oard, D.W., Peñas, A., Petras, V., Santos, D. (eds.) CLEF 2007. LNCS, vol. 5152, pp. 900–907. Springer, Heidelberg (2008)
7. Pitler, E., Keshava, S.: A Segmentation Approach to Morpheme Analysis. In: Working Notes for the CLEF Worksh., Budapest, Hungary (2007)
8. Tepper, M.A.: Using Hand-Written Rewrite Rules to Induce Underlying Morphology. In: Working Notes for the CLEF Worksh., Budapest, Hungary (2007)
9. Kurimo, M., Turunen, V., Varjokallio, M.: Overview of Morpho Challenge 2008. In: Peters, C., et al. (eds.) CLEF 2008. LNCS, vol. 5706, pp. 951–966. Springer, Heidelberg (2009)
10. Zeman, D.: Using Unsupervised Paradigm Acquisition for Prefixes. In: Working Notes for the CLEF Worksh., Århus, Denmark (2008)

Morpho Challenge Evaluation by Information Retrieval Experiments

Mikko Kurimo, Mathias Creutz, and Ville Turunen

Adaptive Informatics Research Centre, Helsinki University of Technology,
P.O. Box 5400, FIN-02015 TKK, Finland
{mikko.kurimo,mathias.creutz,ville.t.turunen}@tkk.fi
http://www.cis.hut.fi/morphochallenge2007/

Abstract. In Morpho Challenge competitions, the objective has been to design statistical machine learning algorithms that discover which morphemes (smallest individually meaningful units of language) words consist of. Ideally, these are basic vocabulary units suitable for different tasks, such as text understanding, machine translation, information retrieval (IR), and statistical language modeling. In this paper, we propose to evaluate the morpheme analyses by performing IR experiments, where the words in the documents and queries are replaced by their proposed morpheme representations and the search is based on morphemes instead of words. In this paper, the evaluations are run for three languages: Finnish, German, and English using the queries, texts, and relevance judgments available in CLEF forum. The results show that the morpheme analysis has a significant effect in IR performance in all languages, and that the performance of the best unsupervised methods can be superior to the supervised reference methods.

Keywords: Morphological analysis, Machine learning.

1 Introduction

The scientific objectives of Morpho Challenge are: to learn the phenomena underlying word construction in natural languages, to advance machine learning methodology, and to discover approaches that work well for a wide range of languages. The suitability for a wide range of languages is becoming increasingly important, because language technology methods need to be quickly and as automatically as possible extended to new languages that have limited previous resources. That is why learning morpheme analysis directly from large text corpora using unsupervised machine learning algorithms is such an attractive approach and a very relevant research topic today.

The problem of learning morphemes directly from large text corpora using an unsupervised machine learning algorithm is clearly a difficult one. First the words should be somehow segmented into meaningful parts, and then these parts should be clustered to the abstract classes of morphemes that would be useful for modeling. It is also challenging to learn to generalize the analysis to rare

C. Peters et al. (Eds.): CLEF 2008, LNCS 5706, pp. 991–998, 2009.

words, because even the largest text corpora are very sparse, a significant portion of the words may occur only once. Many important words, for example proper names and their inflections or some forms of long compound words, may also not exist in the training material, and their analysis is often even more challenging. However, benefits for successful morpheme analysis, in addition to obtaining a set of basic vocabulary units for modeling, can be seen for many important tasks in language technology. The additional information included in the units can provide support for building more sophisticated language models, for example, in speech recognition [1], machine translation [2], and IR [3].

In Morpho Challenge two complementary evaluations are performed for unsupervised morpheme analysis. The first one, described in [4], compares the analysis to a linguistic morpheme analysis gold standard. The second one, described in this paper, performs a practical real-world task where morpheme analysis might be useful. This application developed originally for Morpho Challenge 2007 and continued in Morpho Challenge 2008, was to find useful index terms for IR tasks in multiple languages using the queries, texts, and relevance judgments available in the CLEF forum and morpheme analysis methods submitted by the challenge participants.

Traditionally, and especially in processing English texts, stemming algorithms have been used to reduce the different inflected word forms into common roots or stems for indexing. However, to achieve best results when ported to new languages the development of stemming algorithms requires a considerable amount of special development work. In many highly-inflecting, compounding, and agglutinative European languages the amount of different word forms is huge and the task of extracting the useful index terms becomes both more complex and more important.

To evaluate the performance level of the best unsupervised morpheme analysis, the same IR tasks are also run by a number of reference methods. These references included the organizers' public Morfessor Categories-Map [5] and Morfessor Baseline [6,7], the Morfessor analysis improved by a hybrid method [8], grammatical morpheme analysis based on a linguistic gold standard [9], the traditional Porter stemming [10] of words, and also by the words as such without any processing.

2 Task and Data Sets

In the first Morpho Challenge 2005 (Unsupervised Segmentation of Words into Morphemes) [11] the focus was in the segmentation of data into useful statistical modeling units. The specific task for the competition was to design an unsupervised statistical machine learning algorithm that segments words into the smallest meaning-bearing units of language, morphemes. In addition to comparing the obtained morphemes to a linguistic gold standard, their usefulness was evaluated by using them for training statistical language models for speech recognition.

In Morpho Challenge 2007 a more general focus was chosen to not only to segment words into smaller units, but also to perform *morpheme analysis* of the word forms in the data. The specific task was to return the unsupervised morpheme analysis of every word form contained in a long word list supplied by the organizers for each test language [4]. The participants were pointed to corpora [4] in which the words occur, so that the algorithms could utilize information about word context. The IR experiments were performed by the organizers based on the morpheme analysis submitted by the participants.

The source documents were articles collected from different news papers in Finnish, English and German, which were all available for registered participants in the Cross-Language Evaluation Forum (CLEF)[1]:

- In Finnish: 55K documents from short articles in Aamulehti 1994-95, 50 test queries on specific news topics and 23K binary relevance assessments (CLEF 2004)
- In English: 170K documents from short articles in Los Angeles Times 1994 and Glasgow Herald 1995, 50 test queries on specific news topics and 20K binary relevance assessments (CLEF 2005).
- In German: 300K documents from short articles in Frankfurter Rundschau 1994, Der Spiegel 1994-95 and SDA German 1994-95, 60 test queries with 23K binary relevance assessments (CLEF 2003).

When performing the IR experiments, it turned out that the test data contained quite many new words in addition to those that were provided as training data for the analysis algorithms [4]. Thus, the participants were offered a chance to improve the retrieval results of their morpheme analyses by providing them a list of the new words found in all test languages. The participants then had the choice to either run their algorithms to analyze as many of the new words as they could or liked, or to provide no extra analyses.

3 Participants' Submissions and the Reference Methods

The participants and their submissions are analyzed closer in [4]. From the IR point of view it is interesting to note that only Monson and Zeman decided to provide several alternative analyses for most words instead of just the most likely one. McNamee's algorithms did not attempt to provide a real morpheme analysis, but focused directly on finding a representative substring for each word type that would be likely to perform well in the IR evaluation.

In general, the submissions were all interesting and relevant and all of them met the exact specifications given and could be properly evaluated. In addition to the participants' 12 morpheme analysis algorithms, we evaluated a number of reference methods:

1. Public baseline methods called "Morfessor Baseline" and "Morfessor Categories-MAP" (or here just "Morfessor MAP") developed by the organizers [5].

[1] `http://www.clef-campaign.org/`

2. No words were split nor any morpheme analysis provided, "dummy".
3. The words were analyzed using the gold standard in each language utilized as the "ground truth" in [4]. Besides the stems and suffixes, the gold standard analyses typically consist of all kinds of grammatical tags which we decided to simply include as index terms. "grammatical first" uses only the first interpretation of each word whereas "grammatical all" use all.
4. *Porter*: No real morpheme analysis was performed, but the words were stemmed using the Porter stemming option provided by the Lemur toolkit.
5. *Tepper*: A hybrid method developed by Michael Tepper [8] was utilized to improve the morpheme analysis reference obtained by our Morfessor Categories-MAP.

4 IR Evaluation

In this evaluation, the organizers applied the analyses provided by the participants in the IR experiments. The words in the queries and source documents were replaced by the corresponding morpheme analyses provided by the participants, and the search was then based on morphemes instead of words. Any word that did not have a morpheme analysis was left un-replaced.

The evaluation was performed using a state-of-the-art retrieval method (the latest version of the freely available LEMUR toolkit[2]). We utilized two standard retrieval methods: Tfidf and Okapi term weighting. The Tfidf implementation in LEMUR applies BM25 based term frequency weights for both query and document and the Euclidean dot-product as similarity measure. Okapi in LEMUR is an implementation of the BM25 retrieval function as described in [12].

The evaluation criterion was Uninterpolated Average Precision. There were several different categories and the winner with the highest Average Precision was selected separately for each language and each category:

1. All morpheme analyses from the training data are used as index terms *"withoutnew"* vs. additionally using also the morpheme analyses for new words that existed in the IR data but not in the training data *"withnew"*.
2. Tfidf was utilized for all index terms vs. Okapi for all index terms excluding an automatic stop-list. The stop-list was constructed for each run separately and consisted of the most common terms (frequency threshold was 75,000 for Finnish and 150,000 for German and English). The stop-list was necessary for the Okapi weighting, because otherwise the retrieval favoured documents that had many very common terms.

5 Results and Discussions

The IR evaluation results are shown in Table 1. Here we have selected only the best runs from each participant (in bold) and reference method. For the full results see [13]. Indexing was performed using Tfidf weighting for all morphemes (left)

[2] http://www.lemurproject.org/

Table 1. The obtained average precision (AP%) in the IR task for the best method of each participant and the reference

Tfidf weighting for all morphemes			Okapi weighting and a stop-list		
Finnish:	WORDLIST	AP%	**Finnish:**	WORDLIST	AP%
Morfessor baseline	withnew	0.4105	**Bernhard** 2	withnew	0.4915
Bernhard 1	withoutnew	0.4016	Morfessor baseline	withnew	0.4412
grammatical first	withoutnew	0.3995	**Bordag** 5a	withnew	0.4309
Bordag 5	withnew	0.3831	grammatical all	withoutnew	0.4307
McNamee 5	withoutnew	0.3646	**McNamee** 5	withnew	0.3684
Porter	withnew	0.3566	Porter	withnew	0.3517
dummy	withnew	0.3559	dummy	withnew	0.3274
Zeman	withoutnew	0.2494	**Zeman**	withoutnew	0.2813
German:	WORDLIST	AP%	**German:**	WORDLIST	AP%
Morfessor baseline	withnew	0.3874	**Bernhard** 1	withnew	0.4729
Bernhard 1	withoutnew	0.3777	**Monson** Morfessor	withnew	0.4602
Porter	withnew	0.3725	Morfessor MAP	withnew	0.4571
Monson Morfessor	withnew	0.3520	**Bordag** 5	withnew	0.4308
dummy	withnew	0.3496	Porter	withnew	0.3866
Bordag 5a	withnew	0.3496	**McNamee** 5	withoutnew	0.3617
McNamee 5	withoutnew	0.3442	grammatical first	withoutnew	0.3467
grammatical first	withoutnew	0.3223	dummy	withnew	0.3228
Zeman	withoutnew	0.2828	**Zeman**	withoutnew	0.2568
English:	WORDLIST	AP%	**English:**	WORDLIST	AP%
Porter	withnew	0.3052	Porter	withnew	0.4083
McNamee 5	withoutnew	0.2888	**Bernhard** 2	withnew	0.3943
Morfessor baseline	withnew	0.2863	Morfessor baseline	withnew	0.3882
Tepper	withoutnew	0.2784	grammatical first	withoutnew	0.3774
dummy	withnew	0.2783	Tepper	withoutnew	0.3728
Bernhard 1	withoutnew	0.2781	**Monson** Morfessor	withoutnew	0.3721
Monson Morfessor	withoutnew	0.2676	**Pitler**	withoutnew	0.3652
Pitler	withoutnew	0.2666	**McNamee** 4	withoutnew	0.3577
grammatical all	withoutnew	0.2619	**Bordag** 5	withoutnew	0.3427
Zeman	withoutnew	0.2297	dummy	withnew	0.3123
Bordag 5	withoutnew	0.2210	**Zeman**	withoutnew	0.2674

and Okapi weighting for all morphemes except the most common ones (stop-list) (right).

In the Finnish task, the highest average precision was obtained by the "Bernhard 2" algorithm, which was also the best in [4]. The highest average precision 0.49 was obtained using the Okapi weighting and stop-list for both the originally submitted morpheme analysis [4] and the morpheme analysis for the added new words. The "Bernhard 1" algorithm obtained the highest average precision 0.47 for the German task using the new words, Okapi and stop-list. For English, the highest average precision was obtained by the "Bernhard 2" algorithm, which was also won the evaluation in [4]. As in Finnish and German, the highest average precision 0.39 was obtained with the new words and using the Okapi weighting and stop-list.

As expected, the "grammatical" reference method based on linguistic gold standard morpheme analysis [4] did not perform very well. However, with stop-list and Okapi term weighting it did achieve better results than the "dummy" method in all languages. In Finnish and English the performance was better than average, but quite poor in German. The "grammatical first" that utilized only the first of the alternative analyses in indexing was at least as good or better than the "grammatical all", which seems to indicate that the alternative analyses are not very useful here.

For the "Morfessor" references it is interesting to note that they always performed better than the "grammatical", which seems to suggest that the coverage of the analysis ("Morfessor" has no out-of-vocabulary words) is more important for IR than the grammatical correctness. In general, the old "Morfessor Baseline" seems to provide a very good baseline in all tested languages also for the IR tasks as it did for the language modeling and speech recognition in [11].

The comparison of the results in the Tfidf and Okapi categories show that the Okapi with stop-list performed significantly better for all languages. We also run Tfidf with stop-list (the results were not included here) which achieved results that were better than the plain Tfidf and only slightly inferior to Okapi with stop-list. However, we report the original Tfidf to show the performance and the relative ranking of the methods without stop-list.

When comparing the results in the "withnew" and "withoutnew" categories, we see that with stop-list (and Okapi) the addition of the analysis of the new words helps in Finnish, but in German and in English it does not seem to affect the results. Probably this just indicates that in Finnish the vocabulary explosion is more severe and the new corpus introduced a significant amount of important new words. In general, the new words can be analyzed in two different ways: either use the trained analyzer method as such, or train it first with the new words. In this evaluation both ways were actually possible for the participants, and many of them probably already applied the second one.

The Porter stemmer that is a standard word preprocessing tool in IR, remained unbeaten (by a narrow margin) in our evaluations in English, but in German and especially in Finnish, the unsupervised morpheme analysis methods clearly dominated the evaluation. There might exist better stemming algorithms for those languages, but because of the more complex morphology, their development might not be an easy task.

As future work in this field it should be relatively straight-forward to evaluate the unsupervised morpheme analysis in several other interesting languages, because it is not bounded to only those languages where rule-based grammatical analysis can be performed. It would also be interesting to try to combine the rival analyses to produce something better.

6 Conclusions and Acknowledgments

The objective of Morpho Challenge is to design a statistical machine learning algorithm that discovers which morphemes (smallest individually meaningful units of language) words consist of. Ideally, these are basic vocabulary units suitable for different tasks, such as text understanding, machine translation, IR, and statistical language modeling. The recent challenges are successful follow-ups to our first Morpho Challenge 2005 (Unsupervised Segmentation of Words into Morphemes). The task has become more general in that instead of looking for an explicit segmentation of words, the focus is in the morpheme analysis of the word forms in the data.

The scientific goals of the challenge are to learn of the phenomena underlying word construction in natural languages, to discover approaches suitable for a wide range of languages and to advance machine learning methodology. The analysis and evaluation of the submitted machine learning algorithm for unsupervised morpheme analysis have shown that these goals are quite nicely met. There have been several novel unsupervised methods that have achieved good results in several test languages, both with respect to finding meaningful morphemes and useful units for IR. The results show that the morpheme analysis has a significant effect in IR performance in all languages, and that the performance of the best unsupervised methods can be superior to the supervised reference methods.

The algorithms and results were presented in Morpho Challenge Workshops, arranged in connection with CLEF 2007 and CLEF 2008 Workshops. Morpho Challenge is part of the EU Network of Excellence PASCAL Challenge Program and organized in collaboration with CLEF. Our work has been supported by the Academy of Finland in the projects *Adaptive Informatics* and *New Adaptive and Learning Methods in Speech Recognition* and in part by the IST Programme of the European Community, under the PASCAL Network of Excellence, IST-2002-506778.

References

1. Bilmes, J.A., Kirchhoff, K.: Factored language models and generalized parallel backoff. In: Proceedings of the Human Language Technology, Conference of the North American Chapter of the Association for Computational Linguistics (HLT-NAACL), Edmonton, Canada, pp. 4–6 (2003)
2. Lee, Y.S.: Morphological analysis for statistical machine translation. In: Proceedings of the Human Language Technology, Conference of the North American Chapter of the Association for Computational Linguistics (HLT-NAACL), Boston, MA, USA (2004)
3. Zieman, Y., Bleich, H.: Conceptual mapping of user's queries to medical subject headings. In: Proceedings of the 1997 American Medical Informatics Association (AMIA) Annual Fall Symposium (October 1997)
4. Kurimo, M., Creutz, M., Varjokallio, M.: Morpho Challenge evaluation using a linguistic Gold Standard. In: Peters, C., Jijkoun, V., Mandl, T., Müller, H., Oard, D.W., Peñas, A., Petras, V., Santos, D. (eds.) CLEF 2007. LNCS, vol. 5152, pp. 864–872. Springer, Heidelberg (2008)
5. Creutz, M., Lagus, K.: Inducing the morphological lexicon of a natural language from unannotated text. In: Proceedings of the International and Interdisciplinary Conference on Adaptive Knowledge Representation and Reasoning (AKRR 2005), Espoo, Finland, pp. 106–113 (2005)

6. Creutz, M., Lagus, K.: Unsupervised discovery of morphemes. In: Proceedings of the Workshop on Morphological and Phonological Learning of ACL 2002, pp. 21–30 (2002)
7. Creutz, M., Lagus, K.: Unsupervised morpheme segmentation and morphology induction from text corpora using Morfessor. Technical Report A81, Publications in Computer and Information Science, Helsinki University of Technology (2005), `http://www.cis.hut.fi/projects/morpho/`
8. Tepper, M.: A Hybrid Approach to the Induction of Underlying Morphology. PhD thesis, University of Washington (2007)
9. Creutz, M., Linden, K.: Morpheme segmentation gold standards for finnish and english. Technical Report A77, Publications in Computer and Information Science, Helsinki University of Technology (2004), `http://www.cis.hut.fi/projects/morpho/`
10. Porter, M.: An algorithm for suffix stripping. Program 14(3), 130–137 (1980)
11. Kurimo, M., Creutz, M., Varjokallio, M., Arisoy, E., Saraclar, M.: Unsupervised segmentation of words into morphemes - Challenge 2005, an introduction and evaluation report. In: PASCAL Challenge Workshop on Unsupervised segmentation of words into morphemes, Venice, Italy (2006)
12. Robertson, S., Walker, S., Jones, S., Hancock-Beaulieu, M., Gatford, M.: Okapi at TREC-3. In: Proceedings of the Third Text Retrieval Conference (TREC-3), pp. 109–126 (1994)
13. Kurimo, M., Creutz, M., Turunen, V.: Unsupervised morpheme analysis evaluation by IR experiments – Morpho Challenge 2007. In: Working Notes for the CLEF 2007 Workshop, Budapest, Hungary (2007)

Author Index

Kolibri 360 GmbH
Ferdinand-Jühlke-Straße 7
99095 Erfurt, Deutschland
produktsicherheit@kolibri360.de